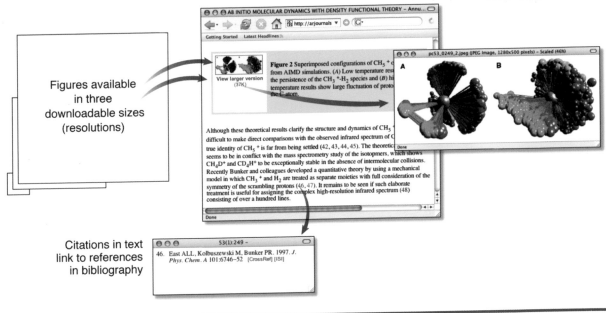

Figures available in three downloadable sizes (resolutions)

Citations in text link to references in bibliography

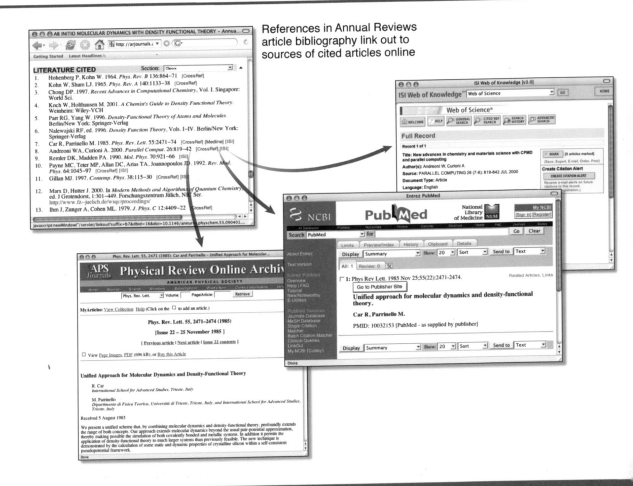

References in Annual Reviews article bibliography link out to sources of cited articles online

Annual Review of
Analytical Chemistry

Annual Review of Analytical Chemistry

Volume 1, 2008

Edward S. Yeung, *Co-Editor*
Iowa State University

Richard N. Zare, *Co-Editor*
Stanford University

www.annualreviews.org • science@annualreviews.org • 650-493-4400

Annual Reviews
4139 El Camino Way • P.O. Box 10139 • Palo Alto, California 94303-0139

Annual Reviews
Palo Alto, California, USA

International Standard Serial Number: 0732-0582
International Standard Book Number: 978-0-8243-4401-6

TYPESET BY APTARA, INC.
PRINTED AND BOUND BY FRIESENS CORPORATION, ALTONA, MANITOBA, CANADA

Preface

Experimental measurements are the foundation of all of the natural sciences. In launching the *Annual Review of Analytical Chemistry*, we aim to provide in-depth and timely perspectives on the fundamental, technological, and applied aspects of measurement science. We seek to stimulate interaction and collaboration among scientists in fields as diverse as physics, chemistry, biology, mathematics, materials, medicine, and environment. In doing so, we hope to facilitate the acceleration of applications of novel measurement concepts, as well as to address the challenges and limitations of current measurement protocols. Such interdisciplinary discourse is essential to understanding and engaging with global challenges such as climate change, energy, and health.

For critical, thorough reviews of the rapidly developing fields that constitute modern analytical chemistry, and for thoughtful perspectives on the future directions of these fields, we contend that there is no better resource than the *Annual Review of Analytical Chemistry*.

We look forward to receiving feedback and suggestions from our readers, students, and colleagues. Finally—and most importantly—we extend our gratitude to each of the authors who contributed to the inaugural volume of this exciting new series.

The Editors

Annual Review of
Analytical
Chemistry

Volume 1, 2008

Contents

Related Articles

From the *Annual Review of Materials Research*, Volume 37 (2007)

From the *Annual Review of Physical Chemistry*, Volume 58 (2007)

Leroy Hood

A Personal Journey of Discovery: Developing Technology and Changing Biology

Lee Hood

Institute for Systems Biology, Seattle, Washington 98103;
email: lhood@systemsbiology.org

Annu. Rev. Anal. Chem. 2008. 1:1–43

First published online as a Review in Advance on February 27, 2008

The *Annual Review of Analytical Chemistry* is online at anchem.annualreviews.org

This article's doi:
10.1146/annurev.anchem.1.031207.113113

1936-1327/08/0719-0001$20.00

Key Words

biological instrumentation and strategies, automated DNA, protein sequencers and synthesizers, cross-disciplinary biology, systems biology, P4 medicine

Abstract

This autobiographical article describes my experiences in developing chemically based, biological technologies for deciphering biological information: DNA, RNA, proteins, interactions, and networks. The instruments developed include protein and DNA sequencers and synthesizers, as well as ink-jet technology for synthesizing DNA chips. Diverse new strategies for doing biology also arose from novel applications of these instruments. The functioning of these instruments can be integrated to generate powerful new approaches to cloning and characterizing genes from a small amount of protein sequence or to using gene sequences to synthesize peptide fragments so as to characterize various properties of the proteins. I also discuss the five paradigm changes in which I have participated: the development and integration of biological instrumentation; the human genome project; cross-disciplinary biology; systems biology; and predictive, personalized, preventive, and participatory (P4) medicine. Finally, I discuss the origins, the philosophy, some accomplishments, and the future trajectories of the Institute for Systems Biology.

New directions in science are launched by new tools much more often than by new concepts.
The effect of a concept-driven revolution is to explain old things in new ways.
The effect of a tool-driven revolution is to discover new things that have to be explained.
Freeman J. Dyson, Imagined Worlds

1. INTRODUCTION

I began to appreciate the beauty of chemistry in high school when, as a senior, I helped teach a sophomore biology class. I remember teaching the class using a 1956 *Scientific American* article on the structure of DNA. That article catalyzed the realization that the core of life was the DNA molecule with its fascinating chemistry of complementarity. Although I doubt I fully understood what that meant at the time, it was clear DNA was a beautiful molecule. This awakening pushed me toward the study of biology, but biology was always embedded in the context of chemistry. With this frame of reference grounded in chemisty, I was able to pioneer technology developments in biology and to grow up with the biotechnology industry. What follows is a personal overview of my career trajectories as they relate to biology, technology, paradigm changes, the creation of new companies, and the founding of organizations to change how biology is done. I discuss these topics chronologically as they emerged to provide the context for their origins.

I subscribe completely to Dyson's comment (see quotations above) (1) that new directions in science are launched by new tools much more often than by new concepts. Much of my career has been focused on developing instruments to decipher biological information, then applying them to fascinating biological problems—an endeavor made possible by a series of wonderful colleagues who were largely responsible for our success in pioneering new instruments and strategies for doing biology. Many of my colleagues (indicated in **Table 1**) went on to become leaders in technology development and application in both academia and industry.

Before continuing, let me say that in looking back it is quite easy to revise history in the context of our current understanding of the issues. Although I have tried to avoid this, I suspect it is impossible not to occasionally provide a compelling rationale for what were often intuitive decisions or decisions made on different grounds.

2. THE BEGINNINGS

I grew up in small towns in Montana where my parents always encouraged me to do well in school and gave me the freedom to explore many different dimensions of life. My father was an electrical engineer with the Mountain States Telephone Company and taught courses in electrical circuitry that I took while in high school. I was not interested in engineering, but these early courses probably provided me with a conceptual framework for my later thinking about systems biology. In high school I was encouraged to explore my potential by three outstanding teachers (in math, chemistry, and social studies and history), one of whom was instrumental in persuading me to attend the California Institute of Technology (Caltech).

Table 1 Colleagues in instrument and strategy development

Ruedi Aebersold	Protein blotting, protein microsequencing
Bruce Birren	Pulse-field gel electrophoresis
Alan Blanchard	Ink-jet DNA synthesizer
Richard Bonneau	Computational tools for protein folding
Ian Clark-Lewis	Long peptides
Cecilie Boysen	BAC shotgun sequencing
Nat Goodman	Database development and computational biology
Pat Griffin	Mass spectrometry and proteins
Mike Harrington	Two-dimensional gels and proteins
Suzanne Horvath	DNA synthesizer
Henry Huang	DNA sequencer (Maxam–Gilbert and Sanger approaches)
Mike Hunkapiller	Gas-liquid-phase protein sequencer, DNA synthesizer, DNA sequencer
Tim Hunkapiller	Computational tools, DNA sequencing
Daehee Hwang	Computational tools for systems biology
Trey Ideker	Computational tools for systems biology
Karen Jonscher	Mass spectrometry and proteins
Rob Kaiser	Labeling DNA
Steven Kent	Protein synthesizer, long peptides
Joan Kobori	Primer-directed sequencing
Eric Lai	Pulse-field gel electrophoresis
Ulf Landegren	OLA/SNP analyses
Steve Lasky	Ink-jet synthesizer
Chris Lausted	Ink-jet technology and surface plasmon resonance
Greg Mahairas	BAC-end sequence mapping
Debbie Nickerson	DNA polymorphism analyses
Jared Roach	Strategies for genomic analyses
Lee Rowen	Shotgun sequencing
Jack Silver	Radiolabeled microsequencing
Lloyd Smith	DNA sequencer (Sanger)
Mark Stolowitz	Protein attachment chemistry
Bingyun Sun	Prefractionation for mass spectrometry
Paul Tempst	Protein microsequencing
David Teplow	Protein chemistry
Mike Waterfield	Solid-phase protein sequencing
John Yates	Spectrometry and proteins
Hyuntae Yoo	Mass spectrometry and blood diagnostic

Abbreviations: BAC, bacterial artificial chromosome; OLA, oligonucleotide ligase assay; SNP, single nucleotide polymorphism.

The move from Shelby, Montana, to Pasadena, California, provided a striking culture shock, but after a year at Caltech I began to appreciate my exceptional classmates and the easy access to outstanding faculty. I had Richard Feynman for physics, Linus Pauling occasionally for chemistry, and George Beadle for biology, and from the beginning I appreciated the power of conceptually oriented teaching. Ray Owen, an immunologist, and James Bonner, a plant physiologist, also helped me to appreciate the marvelous beauty and complexity of biology. At Caltech my career expectations were raised, and I was provided with an excellent background in math, physics, and chemistry. By my senior year I had decided that I was primarily interested in human biology, so I went to Johns Hopkins Medical School in an accelerated three-year program and was immediately immersed in science (as well as medicine). I also found immunology, cancer biology, and diseases of the nervous system fascinating. Although all these areas became central to my later career, I found immunology particularly intriguing because of my readings, specifically a detailed topic paper associated with a fantastic microbiology course taught by Barry Wood, one of the early pioneers in infectious disease.

As my studies progressed I became convinced that the mechanisms of antibody diversity could be explored by characterizing the homogeneous blood immunoglobulin proteins derived from tumors of the antibody-producing cells (plasma cells) present in both mice and humans. In my search for a graduate school that would allow me to follow this direction, I ran into Bill Dreyer, who had recently moved to Caltech. Bill proposed a simple and supposedly noncompetitive "Saturday afternoon project" ideal for a new graduate student, namely that I sequence the homogeneous immunoglobulins that could be purified from the blood of mice with plasma cell tumors induced by interperitoneal injections of mineral oil. I accepted the proposal and became Bill's first graduate student. My first task was to learn the many aspects of protein chemistry, including protein purification, protein peptide mapping, and protein or Edman sequencing. The Edman chemistry that I learned in exhaustive detail would become the mainstay of our efforts to automate and increase the sensitivity of protein sequencing.

My conviction in conducting this research was that by determining the antibody sequences of the light and heavy immunoglobulin chains, one could reverse-translate back to gene sequences and begin to decipher the mysteries of the origins of antibody diversity. I immediately started to characterize immunoglobulin diversity, which catapulted me into the middle of one of the most exciting and rapidly moving periods of molecular immunology. With these amino acid sequences I was able to begin making fundamental hypotheses about antibody diversity (e.g., two genes encoded one antibody chain, diversity was encoded in many germ line genes, and recombinational and possibly somatic mutational events contributed significantly to antibody diversity) (2–5). I had now sensed the excitement of discovery and the satisfaction of personally formulating new conceptual ideas in biology—a heady opportunity for a second-year graduate student—and I was hooked forever on science as a career. At that time, Bill Dreyer also gave me two dicta that have guided my career: (*a*) Always practice biology at the leading edge; it is more fun there. (*b*) If you really want to transform a biological discipline, invent a new technology that permits you to explore new dimensions of data space.

3. THE NATIONAL INSTITUTES OF HEALTH YEARS: 1967–1970

After Caltech I wanted to do a postdoctoral fellowship in Europe, but I had a medical degree and Vietnam War–era policies dictated that all young doctors go either to Vietnam or into the Public Health Service. I chose the latter course and went to the Cancer Branch of the National Cancer Institute at the National Institutes of Health (NIH). I was given an independent position, established a first-rate protein chemistry lab, and continued working on molecular immunology. My tenure at the NIH resulted in two additional, unexpected benefits: First, I had the pleasure of meeting many young physicians who would become leaders in U.S. medicine. Second, I had time to think about what I would like to do with the rest of my career. I decided that I would like to create a laboratory where I spent half my time on molecular immunology (or other biologies) and half my time developing new technologies. After looking at several schools I realized that the Division of Biology at Caltech provided an ideal environment for this dual-research approach.

4. THE CALTECH RESEARCH YEARS: 1970–1992

At the outset of my Caltech career, I told the Chairman of Biology who hired me, Bob Sinsheimer, that I wanted to devote half my time to biology and half to technology. Based on my deliberations at NIH and my graduate experience with Bill Dreyer, my thinking about technology development was driven by two convictions: (*a*) The frontier needs of biology should determine which new technologies should be developed (and once developed, those technologies can revolutionize biology). (*b*) New technologies should focus on deciphering one of the several different types of molecular (or chemical) biological information (e.g., DNA, RNA, proteins). Initially, because of my expertise in protein chemistry, I thought about how to improve the Edman sequencing chemistry and its automation. By the mid-1970s I began to include protein synthesis, DNA synthesis, and DNA sequencing in my grand vision of biological instrumentation. "Decipher," in this case, meant determining the linear order of subunits in nucleic acid or protein digital strings so as to formulate hypotheses about the nature of gene structure, function, or evolution. It could also mean synthesizing smaller fragments of these DNA or protein strings so as to use clever new biological strategies (later in conjunction with recombinant DNA techniques) to further enable digital string characterizations.

Later I became interested in instrumentation for the quantification of mRNAs (ink-jet technology) and proteins (mass spectrometry). I wanted to develop instruments that would automate the chemistry of these processes. In doing so, their analytic throughput as well as the repetitive efficiencies of these synthetic or sequencing chemistries would be increased. This would in turn enable researchers either to synthesize or sequence longer strings (e.g., proteins and DNA), or to sequence regular strings with less starting material (e.g., proteins). For example, molecular immunologists in the 1970s and 1980s needed to be able to sequence small amounts of protein and, from such a protein sequence, clone the corresponding genes (see the following section). Likewise, immunologists needed to characterize the large and complex gene families that encoded the antibody, major histocompatibility complex, and T cell receptors. These were the major biological drivers in my decision to develop the DNA and protein sequencers and synthesizers. Below, I describe each of the four instruments we developed during the Caltech years.

4.1. Protein Sequencer

At the time I moved to Caltech, my chemical expertise resided in the field of protein chemistry, so I logically began with the development of an automated protein sequencer with greater sensitivity and reliability. The challenge was twofold: (*a*) better efficiency of the cyclic Edman chemistry that cleaved amino acid residues one at a time from the N terminus of the protein and (*b*) the development of valves for the automated sequencing instrument that were leak proof and resistant to the corrosive Edman reagents and that prevented the reagent mixing that decreased the repetitive yield of the cyclic sequencing process and, hence, the length of chains that could be sequenced. The ultimate objective was to develop an instrument that was 100-fold (or higher) more sensitive than existing approaches for analyzing proteins. We played with a variety of approaches to improve the Edman sequencing process, including unsuccessful attempts involving solid-phase protein sequencing and radioactive protein sequencing (6). When Mike Hunkapiller joined the lab, progress in automated protein sequencing accelerated. Mike was an organic chemist who quickly gained engineering expertise, solving the valve challenge and getting a handle on the Edman chemistry. When we combined these efforts with Bill Dreyer's concept of a gas-liquid-phase protein sequencer (rather than the conventional spinning cup), we created an instrument that was approximately 100-fold more sensitive than its predecessors (7, 8).

The gas-liquid-phase protein sequencer, because of its increased sensitivity, allowed us to explore a series of new fields in biology because of the availability in low levels of many fascinating proteins that previously could not be sequenced. With the advent of recombinant DNA techniques, the sequencing of low-abundant proteins could lead to the cloning and characterization of the entire corresponding genes. We sequenced a human blood hormone, platelet-derived growth factor, and showed that its N-terminal sequence was nearly identical to that of an avian oncogene, *v-cis*. This observation generated for the first time the hypothesis that oncogenes are normal genes of human growth and development subject to control by a cancer virus (9). It was also the first time that a string search of a new protein sequence was carried out against a database of preexisting protein strings to learn interesting new biology—the beginning of bioinformatics.

With Stanley Prusiner, we sequenced the prion protein (10), which enabled the cloning of the gene (11). This eventually led to the hypothesis of an infectious protein—infectious because its misfolded structure encoded a catalytic ability to convert normal prion proteins to the infectious form, thus generating an autocatalytic disease that led to neural degeneration. Stanley Prusiner won a Nobel Prize in 1996 for this work. Prion disease later became our flagship initial study of a systems approach to disease (see below).

We also sequenced erythropoietin, providing useful information for Amgen in its eventual cloning of the gene and the creation of biotechnology's first billion-dollar drug. Our sequencing of the four chains of the torpedo acetylcholine receptor (12) led to the now-classic work of Professor Nomura at Kyoto University, who cloned the genes encoding this receptor and many other neuroreceptors, thereby transforming our understanding of important aspects of how the brain functions.

We were the first to sequence the α and β interferon proteins (13, 14), facilitating the cloning of their genes as well as their eventual deployment as useful drugs for cancer and multiple sclerosis. We were also the first to sequence a hematopoietic colony-stimulating factor, which led to cloning and its application as a useful drug.

In each case these sequences initiated new approaches to biology, and in many cases they generated interesting biotech drugs. These are but a few examples of the many other fascinating proteins that we microsequenced during the late 1970s and 1980s (15–18).

Below I describe the steps we took to commercialize the protein sequencer, which resulted in the creation of an extraordinarily successful company, Applied Biosystems, Inc. (ABI). The productive partnership between the Hood lab at Caltech and ABI led to the development of robust instruments for each of the three technologies described below.

4.2. DNA Synthesizer

In the late 1970s, Marvin Caruthers at the University of Colorado had just developed the phosphoramidite chemistry for DNA synthesis. I proposed that he teach my technician, Suzanne Horvath, how to do manual DNA synthesis so that we, in turn, could automate this process. Marvin argued that he could easily teach the procedure in a week to anyone who needed to synthesize DNA, and because he felt the demand for oligonucleotides was never going to be significant, why bother with automation? I convinced him otherwise. Suzanne learned the procedure and then, together with Mike Hunkapiller, designed an instrument for the repetitive DNA synthetic procedure (19). The challenge, once again, was designing valves that could withstand corrosive chemicals and maintain the separation of reagents from various cycle steps to improve the repetitive yield. Mike's earlier experience with the protein sequencer valves was useful in this regard. Caltech put together a prototype, and ABI moved quickly to develop robust DNA synthesis instrumentation.

The DNA synthesis platform provided critical oligonucleotides for many aspects of the emerging recombinant DNA technologies. We quickly suggested a new sequencing strategy that employed oligonucleotide primers for primer-directed DNA sequencing (20). We saw that genes could be synthesized by joining overlapping oligonucleotides in a sequential and hierarchical manner. We cloned interesting genes by synthesizing degenerate oligonucleotides, reverse-translated directly from the protein sequence analysis of low-abundance proteins (see, e.g., Reference 11). It was obvious that the ability to synthesize DNA primers enabled the conceptualization and development of the DNA amplification procedure—the polymerase chain reaction—because this procedure required pairs of oligonucleotides. So once again a new instrument provided many new possibilities for biology.

4.3. Peptide Synthesizer

Steven Kent came to my laboratory in 1983 with a wealth of protein synthesis experience from the Merrifield lab at Rockefeller University. Kent was attracted to the idea that the repetitive chemistry of protein synthesis could be improved through superior instrumentation to the point that modest-sized proteins could be synthesized with high yields, thus both improving the chemistry and developing another instrument utilizing leak-proof and corrosion-resistant valves. Steve worked in the Hood lab together with ABI to develop this instrumentation (21). Once again, many of his challenges were similar to those arising from the automation of protein sequencing and DNA synthesis. The chemistry in each case was, of course, unique, and each improvement presented special chemical challenges. I remember a time when an ABI executive came to me with a field representative's assessment that there was a market for only a few peptide-synthesis machines per year (fewer than ten). Fortunately, my contrary convictions overrode this erroneous prediction (ABI sold approximately 70 in the first year), and development of the peptide synthesizer proceeded.

Steve and ABI developed a superb peptide synthesizer with high repetitive yields and hence the ability to synthesize long peptides or even small proteins. Perhaps the most spectacular result arose from Steve's collaboration with Merck to synthesize the HIV protease (99 residues). They did this so effectively that after purification, it was possible to crystallize the chemically synthesized protein and solve the crystal structure to a resolution of a few angstroms (22). From these data, Merck developed its antiprotease drug, which came to be one of the most successful AIDS drugs. Steve went on to synthesize several interleukins, as well as a variety of other proteins, and carried out a fascinating series of structure-function studies (23, 24), and we did some interesting peptide/DNA binding studies with zinc fingers (25).

4.4. DNA Sequencer

Certainly the most difficult Caltech instrument to develop was the automated DNA sequencer. In 1975, Maxam & Gilbert (26) and Sanger (27) developed the chemical and enzymatic (di-deoxy) manual approaches to DNA sequencing, respectively. In the late 1970s, Henry Huang, a postdoctoral fellow in my lab, began working to automate first the chemical and later the enzymatic approach. Henry was a biologist with some interest in engineering, but it gradually became clear that better knowledge of engineering and a far more sophisticated chemical expertise were required to complete this project. In 1982, I assembled a team including a chemist/laser expert (Lloyd Smith), an engineer/chemist (Mike Hunkapiller), a biologist with knowledge of computer science (Tim Hunkapiller), and myself, a molecular biologist—and one spring day we had a transforming conversation. Four central ideas emerged about a new proposed approach to automated DNA sequencing: (*a*) The DNA fragments of the Sanger reactions could probably be separated nicely by capillary gel electrophoresis. (*b*) The fragments could be labeled with one of four different fluorescent dyes, according to which base terminated the fragment. (*c*) All four colored bases could be detected together in a single capillary channel (the manual sequencing approaches used radioactive reporter groups that required four separate lanes, one for each DNA fragment ending in a distinct base), thus standardizing the DNA fragment comparisons and increasing the efficiency of the sequencing process. (d) The four distinct classes of fluorescence-labeled DNA fragments could be distinguished by laser detection of the dyes, and this four-parameter dye space signal could be converted into DNA sequence by computational algorithms. Although these simple ideas emerged in a single afternoon, it took another three years before the practical details were solved and a prototype, capillary-based automated DNA sequencing instrument was developed (28). We had to develop a chemistry for coupling the dyes to DNA (29), identify four good dyes, design laser instrumentation for reading dye space, generate algorithms for converting dye space into DNA sequence, optimize the enzymology of polymerases for the extension sequencing reactions, and solve a host of other chemical and engineering challenges. Lloyd Smith played a central role in solving many of these problems. This team effort illustrated the power of a cross-disciplinary approach and the need for team science in solving a challenging technical problem (two points I return to below). During the latter stages of this effort, Mike Hunkapiller moved to ABI, which was very much a partner in developing the automated DNA sequencer. ABI took the lead in pioneering the robust instrument necessary for its commercialization.

The DNA sequencer truly changed biology by making the human genome project possible. The DNA sequencer also pointed the way toward the development of even higher throughput DNA sequencing instrumentation for the generation of the massive amounts of data essential for eventually finishing the human genome project. The genome project, through its genetic parts list, eventually paved the way for systems biology. It also made possible the current analyses of thousands of genome sequences from microbes, plants, animals, and even multiple humans; these, in turn, have transformed and are transforming many different fields of biology and medicine (30–32).

4.5. Integrated Microchemical Facility

In the early 1980s, my lab suggested that developing and coordinating or integrating the respective functions of these four automated instruments (DNA and protein sequencers and synthesizers) would lead to the creation of an integrated microchemical facility with a powerful capacity for moving from genes to proteins and vice versa (33). For example, if the protein sequencer was used to determine the amino acid sequence of an interesting protein, this protein sequence could then be translated via the genetic code dictionary into a degenerate oligonucleotide sequence; degenerate oligonucleotides could then be generated by the DNA synthesizer, and these DNA probes could then be employed with recombinant DNA techniques to clone the corresponding genomic or cDNA clones that could then be sequenced by the automated DNA sequencer. In a similar vein, genes could be translated into an amino acid sequence that could then be synthesized by the protein synthesizer, and the resulting peptide fragments could be used as antigens to generate specific antibodies for protein localization and characterization. Likewise, multiple degenerate oligonucleotide probes could be used to clone large gene families. Thus, the integrated microchemical facility could be used to coordinate the chemistries of these four instruments in conjunction with appropriate recombinant DNA and biological techniques.

Interestingly, publication of the *Nature* paper that described this integrated microchemical facility (33) was delayed for almost two years because of reviewers' skepticism about several of the predicted integrated strategies. In fact, microchemical facilities became common throughout the academic and industrial worlds. The automation and chemical optimization of these sequencing and synthesis procedures also led naturally to the emergence of other high-throughput technologies, which arose either by parallelization (from 1 sequencing capillary tube to 96) or by more rapid serial procedures (often from miniaturization of reaction procedures or more efficient chemistries), thus providing a push for the eventual development of both genomic and proteomic high-throughput platforms capable of generating the large data sets required by systems biology. Below, I discuss our development of one of the most powerful high-throughput technologies in biology today: the ink-jet DNA synthesis technology that enables the synthesis of DNA chips. The microchemical facility became an early embodiment of the Caltech vision for transforming biology through automated instrumentation.

4.6. Commercialization of the Four Instruments

By the late 1970s we had developed the protein sequencer, were working on automated DNA sequencing and synthesis, and were thinking about peptide synthesis. A friend suggested that we think about commercializing these instruments to make them available to others. Caltech's president, Murph Goldberger, was skeptical: He argued that the role of an academic institution was scholarship and education, not commercialization. "But," he said, "you can try to commercialize if you wish." I went to 19 different instrument companies over the next year and all said no. I visited Beckman, Inc., three times before they told me not to come back. Then Bill Bowes, a venture capitalist from San Francisco, called me and said he knew of my failed attempts with the instrumentation companies, and he offered to provide $2 million to start a venture-backed company. Murph, once again, was cautious; he was reluctant to accept venture money because Harvard was then going through a messy debate about the venture-backed creation of a company called Genetics Institute. In the end, however, Murph relented, and just as we were about to sign an agreement with Bill Bowes, I gave a talk to the Caltech trustees on the vision of how these four instruments would change biology. Arnold Beckman, founder of Beckman, Inc., and a Caltech trustee, immediately approached me afterward and said that this new instrumentation was just what Beckman, Inc., needed. An awkward moment followed, and I said that his company had not appeared interested. A complex and difficult series of misunderstandings ensued between Arnold and Caltech (Beckman was a large donor to Caltech), but in the end, Bill Bowes and I started ABI—and it grew to be the most successful biotech instrumentation company worldwide. Indeed, the protein sequencer had been so well engineered that it was successfully produced in the second quarter of the first year of ABI's existence, and the company was in the black by the end of that year. All turned out well for Caltech as well, because Arnold Beckman later donated more than $100 million to create the now well-known Beckman Institute at Caltech.

This experience taught me three important lessons. First, always discuss visions with the highest-level administrator or leader (e.g., CEOs, founders)—for they, if anybody, will be able to comprehend the vision's future potential and relevance to the company or organization. Second, it was fortunate that none of the extant companies accepted my offer of the instruments, as none would have been able to acquire the scientific talent, provide the sufficient resources, or focus entirely on the problem of developing these four instruments. ABI did a superb job in all these regards and represented the new organizational structure needed to realize this new instrumentation vision or paradigm change of developing and commercializing automated and integrated instrumentation—a vision that later encompassed high-throughput data-production instruments. This was the first of several paradigm changes in which I participated, creating new organizational structures when they were needed to catalyze the paradigm change. I went on to play a founding or cofounding role in creating more than 13 additional biotech companies (including Amgen, Systemix, Darwin Molecular, and Rosetta Inpharmatics) and thus effectively transferred academic knowledge to society. Third, I realized that there are five stages to the development of new instruments: (*a*) conceptualization; (*b*) the development of a prototype instrument to provide proof of principle; (*c*) the development of robust instruments that any biologist could use; (*d*) the conversion of the robust instrument into a high-throughput platform; and (*e*) the development of a completely novel and better approach for the relevant chemical instrument (where better could mean faster, cheaper, miniaturized, more highly parallelized, more effectively automated, and/or integrated with other chemistries or procedures), followed by repetition of the first four stages.

I came to realize that academia is superb at conceptualization and prototype development, but generally has neither the resources nor the systems engineering skills for robust commercial development and high-throughput conversion. Novel approaches to the instruments in development will likely arise once again in academia. I can illustrate this point by noting that conceptualization of the automated DNA sequencer was very inexpensive; prototype development probably cost approximately $500,000 (and I failed in two attempts to get this funded by NIH); commercial development cost approximately $75 million; high-throughput conversion probably cost several hundred million dollars altogether; and several new approaches to very-high-throughput DNA sequencing instrumentation are now under way. Conceptually, these new approaches came initially from academic labs. Thus, the creation of ABI was the critical step in realizing the potential of our instruments, and their widespread use throughout the scientific community resulted from our commercialization.

4.7. New Strategies Employing the Four Instruments

Not only did the four commercialized instruments open up new areas of biology when used directly as they were designed, they also enabled the pioneering of new strategies for carrying out biology, such as primer-directed DNA sequencing, the synthesis and assembly of long DNA strings including genes, the use of degenerate oligonucleotides reverse-translated from protein sequences to clone genes, and the production of peptide antibodies (all discussed above). In addition, we developed a series of additional genomic and recombinant DNA strategies (34–42). At the protein level, Mike Hunkapiller, Ruedi Aebersold, Steven Kent, Mike Harrington, and Paul Tempst developed new approaches for the purification of proteins and their microsequence analyses (43–53). John Yates and Pat Griffin pioneered the use of the mass spectrometer as a tool for protein and peptide identification and characterization. Yeats also developed some of the earliest computational proteomics techniques (54–57). Also, we developed a pulse-field instrument for the size measurement of very large DNA fragments (58, 59). The significance of these developments is that our new instrumentation made it possible to create many new strategies for generating data at both the DNA and protein levels. Two of these strategies deserve special mention because they initially faced considerable skepticism from the relevant scientific communities as to their utility, but eventually played important roles in genomic analyses: the oligonucleotide ligation assay and the use of mate pairs (e.g., sequence data from both ends of large insert clones) to facilitate the mapping and sequencing of entire genomes.

When Ulf Landegren was a postdoctoral fellow in my lab, he proposed that enzymatic oligonucleotide ligation could be used as a means for identifying single nucleotide polymorphisms (SNPs) (60). His idea was that if two adjacent oligonucleotides were complementary to a target DNA, then they could be covalently joined at their abutting ends by the enzyme DNA ligase. If, however, a polymorphism were present at the 5′ end of the 3′ oligonucleotide, a mismatch in complementarity would prevent the ligase from joining the two oligonucleotides. Hence, one could synthesize two 3′ oligonucleotides, each with a distinct 5′ base complementary to one of the two SNP variants. Accordingly, if each of the two 3′ oligonucleotides were labeled at its 3′ end with a different fluorescent dye, then one could identify the SNP variants by the color of the 3′ probe ligated to its 5′ counterpart (a heterozygous individual with both SNP variants would ligate both colored 3′ ends, and a homozygous individual with just one variant would ligate the complementary 3′ end and exhibit just a single color). Debbie Nickerson, a visiting professor in my lab, later applied the oligonucleotide ligation assay in a variety of ways for effective and revealing SNP analyses (61). This procedure has been the basis for a variety of very-high-throughput SNP typing procedures, including one first developed by the company Illumina, which used it to make major contributions to the large-scale SNP mapping required to delineate the human haplotype SNP map (ascertaining the SNP linkage relationships on the maternal and paternal chromosome sets from individual humans).

The use of paired ends, or mate-pair information, to facilitate the assembly of genomic regions from collections of shotgun sequencing reads was initially proposed by Jared Roach, a graduate student in my lab (62). In shotgun sequencing, a genome, or a large insert clone derived from a genome, is randomly fragmented into short pieces, which are then subcloned into a suitable cloning vector and propagated as DNA templates for sequencing; from these, one obtains sequence reads of several hundred bases in length. These reads are then assembled by aligning and conjoining overlapping sequence strings, with the goal of reconstructing the original genome or large insert clone sequence. This task is facilitated by obtaining redundant coverage of the genome from the shotgun reads, and is hindered by the presence of duplicated regions (repeats) or by regions with no sequence read coverage. My colleague Lee Rowen employed this approach in a classic sequence analysis of the 650-kb-long human T cell receptor locus (63)—by far the longest segment of DNA sequenced at the time—and these efforts were followed in our lab by the sequence analysis of the >1-Mb human (64) and mouse α T cell receptor loci and the major histocompatibility complex locus (65). We also did a fascinating species comparison of the T cell receptor loci (66). In the early days of genome sequencing, a single-stranded virus (M13) was used as the cloning vector because of its ease of propagation and purification. Roach rightly pointed out that if a double-stranded vector, such as a plasmid, were used for the propagation of DNA templates, then a sequence could be obtained from both ends of the genomic fragment, rather than obtaining only the single read provided with the M13 vector. The advantage to using a double-stranded vector lay in the ability of these mate pairs to increase the accuracy of the assembly process by localizing repeats (if one half of the pair aligned to a unique region) and gaps (if the mate pairs spanned a gap in the assembled sequence, then one could infer that these stretches of sequence were in proximity in the genome). Thus, the mate pairs allowed for the ordering and orienting of reconstructed blocks of genomic sequence assembled from shotgun reads.

The next leap for using mate-pair information came about with the development in Mel Simon's lab at Caltech of a cloning vector called bacterial artificial chromosomes (BACs), which can propagate large inserts up to ~250 kb (67). Cecilie Boysen, a graduate student in my lab, was the first to demonstrate that inserts propagated in BACs could be successfully sequenced and assembled (68). This success led Craig Venter and me to propose using BACs and BAC-end mate-pair sequences as reagents for mapping genomes in a way that would facilitate high-throughput sequencing of large chromosomes (69). The basic idea was the following: We can begin with the assumption that 30,000 randomly generated BAC insert clones, each approximately 200 kb in length, could be sequenced for 500 base pairs at either end. If the inserts of the BAC clones were randomly generated, then there would be a 500-base-pair sequence tag, on average, every 50,000 base pairs across the genome. Once a 200-kb BAC was completely sequenced, on average, it would overlap with the end sequences of four other BAC clones—and one could simultaneously sequence out in both directions from tens or hundreds of nucleating BAC clones to generate dual growing points of overlapping sequences, sequentially allowing individual chromosomes to be sequenced. Although this approach was initially unpopular, we were able to obtain funding to sequence the ends of hundreds of thousands of human BACs (reduced to practice by Greg Mahairas in my lab), and thereby create what proved to be an invaluable resource for completing the human genome (70). In spite of some technical limitations to this strategy, the BAC-end sequencing approach has been and is widely used in whole-genome sequencing.

4.8. Systems Biology: Beginning Thoughts and Conceptualization

My labs at Caltech were located next to those of Max Delbruck, a Nobel Prize–winning physicist who pioneered the field of quantitative biology. Max was never very impressed with immunology. He was skeptical about whether the deep problems of immunology—the immune response, tolerance, and autoimmunity—could ever be solved by a one-gene- or one-protein-at-a-time approach. I argued that we had done very well in coming to understand the mechanisms of antibody diversity, and he replied that diversity was an easy problem and not a deep problem of immunology. I gradually concluded that Max was correct at least in part, and I started thinking about what it would take to successfully attack these big problems in immunology (which remain largely unsolved today). I decided that one needed to study the systems of genes or proteins that mediate immune phenotypes, and not just individual genes or their protein counterparts. But we did not have the tools for identifying the components of biological systems and their interactions in the mid-1980s, nor did we have a complete parts list of all genes nor, by inference, all proteins. We needed to be able to do far more comprehensive analyses of the behaviors (changes in structure, expression levels, interactions, and cellular localizations) of mRNAs and proteins (and this also led in part to the conceptualization of the ink-jet DNA array technology, discussed below).

This systems thinking also made the idea of the genome project attractive because one could more or less completely define a parts list of genes and, by inference, mRNAs and proteins—and thus hope to carry out comprehensive or global analyses (see below). I submitted a couple of systems-like biology grants to the National Science Foundation (NSF) that did not do very well because the reviewers were unclear as to what I was proposing (it was not clear that I completely understood what I was proposing either). In 1991 I wrote a chapter in the book *The Code of Codes* (71) (which was about the human genome project and was coedited by Dan Kevles and me) that contained a description of systems biology that could be used today (although I did not start using the term systems biology until a few years later). Our current view of systems biology emerged slowly over the late 1980s and early 1990s, fueled by the creation of an NSF Science and Technology Center (STC) at Caltech in 1989 and then further matured by the establishment of the cross-disciplinary Department of Molecular Biotechnology (MBT) at the University of Washington (see below).

4.9. The Human Genome Project

The development of the DNA sequencer presented me with two interesting new opportunities to change how biology is done: (*a*) the genome project and (*b*) cross-disciplinary biology. The first opportunity arose when, because of our development of automated DNA sequencing, Bob Sinsheimer, now the Chancellor of the University of California at Santa Cruz, invited me to the first-ever meeting on the human genome project in the spring of 1985. Bob was considering setting up an institute at Santa Cruz to sequence the human genome, and he had invited 12 experts to discuss this possibility. Over several days the group came to two conclusions: first, that the human genome sequence was technically feasible, although difficult (the prototype DNA sequencer had just been developed), and second, that the group was evenly split as to the advisability of carrying out this project. I came away from this meeting with several impressions: (*a*) The genome project would certainly drive the development of DNA sequencing and other technologies; (*b*) it would require the development of an array of new computational tools; and (*c*) it would provide a complete list of all human genes, a necessity for carrying out some of the systems approaches to biology I was just starting to consider (discussed in detail below).

In 1985, Wally Gilbert, Charlie Cantor, and I, along with others, began talking to the community of biologists about the genome project. I was surprised to find that perhaps 90% of the biologists bitterly opposed it on several grounds, namely that it was big science and that big science was inherently bad and would take the funding from hypothesis-driven small science. Moreover, the opponents argued, nothing interesting would come from the genome project; therefore, it would be impossible to recruit talented scientists to work on it. In my view, neither argument had any merit. NIH was initially firmly opposed to the project, and, indeed, it was the Department of Energy that really championed this proposal in its early days. A National Academy of Sciences committee, comprising both opponents and champions of the project, unanimously endorsed it after a year of deliberation. This endorsement brought NIH to the genome table, and an essential component of NIH's subsequent success with the genome project was the establishment of a brand-new institute at NIH to oversee this process. The proposed 15-year project began in 1990 and was more or less complete in 2004.

The striking lesson I learned from this experience was how conservative most scientists are, and how difficult it can be to reason with those who operate primarily from the biases of their past experiences, rather than thinking about the potential of future possibilities. In retrospect, another interesting point emerged. A friend told me in 1985 that it would take 100 years to sequence the human genome; with the technology we had at that time, he was probably correct. The real driver of change in biology (or science in general) is technology (see the quotation by Dyson at the beginning of this article), and what most people fail to understand is that some technologies can change exponentially at certain periods of their development. Thus, our ability to predict the future of a field depends very much on understanding how rapidly the driver technologies for a field are changing. For DNA sequencing technology, the period from 1986 to approximately 1998 was one of those periods of exponential change. Hence, if one could gauge the nature of this change, one could make predictions about future possibilities that would appear excessive to those who do not understand the dynamics of exponentially changing relevant technologies. One of my strengths was that I often saw which were the important technologies (and important ideas in biology) and how these technologies were changing or were going to change. Interestingly, today we are going through a similar period of exponential change in the increasing throughput of DNA sequencing, which will transform predictive medicine (as I discuss below).

We played an important role in learning how to sequence large DNA fragments (62–66): We were the first to do an evolutionary analysis of a 100-kb fragment from an important region in the human and mouse α/δT cell receptor loci and indeed defined most of the parameters used today in cross-species comparisons (72). We then established one of the 16 centers to sequence the human genome and sequenced significant portions of chromosomes 14 (73) and 15 (74), which contributed to the complete human genome sequence (75). Thus we developed the instrumentation and pioneered its applications to biology.

4.10. Cross-Disciplinary Biology

The second insight I had from developing the DNA sequencer was the power of bringing together scientists from different fields to attack a technically challenging problem. I applied this insight in the context of my lab and as a result it grew to be very large (as we were doing both biology of several different flavors and broad-spectrum technology development). In 1987, I decided to compete for an NSF STC—an idea pioneered by Dick Zare, a first-class chemist at Stanford. The idea was to integrate science (biology in my case) with the development of the appropriate technology. In addition, the STC strategy emphasized industrial strategic partnerships and facilitating K–12 science education. This STC grant was the most effective grant I ever received because of its flexibility and the breadth and relevance of its requirements. This grant, together with my experience in developing instrumentation, continued to transform my thinking about how to do science. I realized the importance of bringing to biology a cross-disciplinary environment wherein biologists, chemists, computer scientists, engineers, mathematicians, and physicists could focus on using biological challenges to drive the development of relevant technologies and computational or mathematical tools. In such a cross-disciplinary environment, it is imperative for each scientist to learn to speak the languages of the other scientists and engineers, and work together in teams to attack hard biological problems in an integrated manner. I approached a different Caltech president, Tom Everhart, to argue that Caltech should start a new cross-disciplinary biology department with a cross-disciplinary faculty focused on biology. Tom said that it sounded like a good idea, but that I would have to persuade my colleagues. The chemists and engineers thought it was a good idea, but the biologists opposed it for reasons that were never completely clear to me.

Let me stress that not all Caltech biologists were opposed or indifferent to my interest in technology. Eric Davidson, a long-time loyal friend and wonderful collaborator on the genomics and systems biology of the sea urchin (76, 77), was a strong supporter of my various efforts throughout my 22 years at Caltech. Eric always appreciated and readily adapted and employed new technologies, and he continues to do so today. Just last year, he led a magnificent effort to sequence and analyze in enormous detail the sea urchin genome (see the Nov. 10, 2006, issue of *Science*). Over the past 35 years, Eric has fashioned the sea urchin into a superb model for development in metazoan organisms by dint of brilliant experimentation, always applying cutting-edge technologies. Sea urchin development today, as pioneered by Eric, is one of the outstanding examples of a systems approach to biological complexity.

However, the opposition of the leadership in biology to a possible cross-disciplinary department at Caltech presented me with the most difficult professional decision I ever had to make. Should I stay at Caltech with great students and colleagues, or should I move elsewhere to create a new cross-disciplinary department? I decided to follow my cross-disciplinary convictions. After looking at several schools, the University of Washington (UW) became a possibility, partly because a former student—Roger Perlmutter, then Chairman of Immunology at the medical school—pushed the idea of my recruitment. I interviewed with the dean of the school of medicine, Phil Fialkow, and at the end of our interview he said that a cross-disciplinary biology department would be far too complex for a medical school. I went home disappointed, but Phil called me a few days later and said that he had made a mistake and that he wanted to fly to Pasadena and talk to me about rectifying it (this was a remarkable event—a dean admitting he had made a mistake—I was really impressed).

We met and came to an agreement that I would consider a move to the UW. I would first build the cross-disciplinary department (MBT) and then, after four or five years, I would receive additional space and could expand to build up systems biology (this latter agreement was informal and never documented). Phil arranged for me to give three John Dantz lectures at the UW on the future of biology—all were attended by Bill Gates. After the last lecture, Bill and I had a fascinating dinner that lasted about three hours; we talked at length about our views of science and engineering. Bill then offered $12 million to support this new department. So in 1992, after a wonderful 22 years at Caltech (in addition to eight more as an undergraduate and graduate student), I moved to the UW to found MBT.

5. THE MOLECULAR BIOTECHNOLOGY YEARS: 1992–2000

5.1. Getting Started

Moving to the UW was difficult: Not only was the cultural and scientific environment quite different from Caltech, but our space was not going to be ready for several years. The department was crammed into a modest space in the Washington Technology Center, staffed by a diffuse collection of many different types of scientists and engineers—some of whom focused more on commercialization than on academic science. Fortunately, many of my key colleagues moved with me from Caltech, some to be faculty members. We recruited wonderful additional faculty from a variety of institutions with many different areas of expertise, including genomics, computational biology, protein chemistry, and later proteomics, biology, computer science, and mathematics. At MBT my lab developed its fifth instrument: the ink-jet DNA arrayer.

5.2. Ink-Jet DNA Array Technology

Alan Blanchard was a computational biologist who came with me to MBT from Caltech. At Caltech he had started thinking about using common ink-jet printer technology for synthesizing arrays of oligonucleotides on glass or silicon chips (78, 79). The biological impetus was the desire to quantitatively analyze the concentrations of potentially all the mRNAs of humans or other organisms—one of the global technologies that is vital to systems biology. Prior to our efforts, Steve Fodor began developing a photolithography approach to DNA array synthesis for the company Affymetrix. The ink-jet project called for sophisticated surface chemistry: It required struggling with DNA synthesis chemistry in the context of the ink-jet printer environment (oxygen was lethal to the synthesis reactions) and the ability to computationally drive and engineer the ink-jet printer. Alan persevered, learning chemistry and employing his knowledge of computer science, and ultimately produced a functioning prototype instrument. However, we were a long way from an effective and robust instrument, and commercialization to achieve these goals was essential. Ultimately we started a company, Rosetta, that continued developing this technology for commercial application. Rosetta eventually licensed it to Agilent, which made further significant improvements. One advantage of the ink-jet DNA arrays over their photolithography counterparts is that the efficiency of the synthesis process permits long oligonucleotides to be generated (60–70 mers versus 25 mers), and these have significant advantages for accurate hybridization and special applications. For example, the creation of ink-jet DNA arrays allowed for striking new opportunities in synthetic biology for the assembly of extremely long DNA fragments—the ability to synthesize everything from genes to gene families to genomes. The ink-jet approach is the core technology Agilent uses today for DNA chip synthesis. In my lab, Chris Lausted and Steve Lasky developed a less sophisticated but quite functional ink-jet synthesizer instrument (80) whose design was copied by several other institutions.

I stress another important point about high-throughput instrumentation: If one can sequence, synthesize, measure, or localize in a high-throughput mode, one will transform the nature of the biology that can be done. We can see this in the ways automated DNA sequencing and DNA arrays have changed evolutionary biology, developmental biology, genetics, and physiology over the past 15 years. Indeed, as I discuss below, in perhaps ten years we will be able to sequence individual human genomes rapidly and inexpensively; this will be an important element in transforming medicine.

5.3. The Struggle to Realize the Systems Biology Vision

In 1996—at about the time I was beginning to think of developing systems biology— Phil Fialkow tragically died in a blizzard in the Himalayas. Phil was a wonderful human geneticist who fostered a marvelous environment for the development of basic science as dean of the University of Washington Medical School. Thanks to Phil, MBT really thrived during our first four years at the UW (1992–1996).

I was beginning to realize that although a cross-disciplinary environment represented an essential foundation for systems biology, systems biology required many additional cultural changes quite at odds with the traditional practice of biology and its governance by traditional academic bureaucracies. In addition to a superb cross-disciplinary environment, one had to create a culture of teamwork for hard problems, high-throughput facilities for biological measurements, a strong computational infrastructure and the ability to create strategic partnerships with academia, research institutes and industry to fill in missing biological and technical skills. The cross-disciplinary culture required that the scientists learn one another's languages, that the nonbiologists learn relevant biology deeply, and that the scientists effectively work together in teams with continual feedback and interactions. One had to develop expensive high-throughput genomic and proteomic facilities (requiring space, capital investment, and ongoing support) and to possess the ability to manage these rapidly changing technologies and keep them up to date. It was necessary to have extensive computational facilities. As noted above, one had to carry out team science, which led to potential difficulties with tenure for younger faculty. One needed to be able to readily make outside strategic partnerships to bring in needed engineering or scientific expertise for hard problems. Clearly, systems approaches were going to spin off an enormous amount of intellectual property, and managing that intellectual property in a manner that could successfully create additional support for the science was essential. It was also clear that extensive fundraising was necessary to enable this vision (but this was impossible in the context of a large state university where presidents and deans jealously guard the right to raise funds from private individuals). Each of these issues posed serious problems for a traditional academic bureaucracy.

Here I present two examples of the challenges faced in building systems biology at the UW. First, at Caltech I had helped raise the money to build two biology buildings; at the UW I was never once asked to do significant fundraising, apart from my initial interactions with Bill Gates, whose support created MBT. Yet, systems biology clearly was going to require significant resources. Second, the MBT cross-disciplinary environment needed to be significantly expanded. This proved to be challenging: For example, the new dean refused to let me hire a wonderful surface chemist from Penn State who could have contributed enormously to nanotechnology measurement approaches for systems biology, approaches we have subsequently collaborated on with Jim Heath at Caltech (discussed below). The dean claimed that surface chemistry was irrelevant to the medical school; this was, once again, a perfectly reasonable position for someone who did not understand how changing technologies would transform biology and medicine.

After trying to compromise on many of these issues, it became obvious that the administrative structure of the university could not accommodate most of our requirements (the workings of bureaucracies are honed by past experience and can rarely accommodate future change with ease). Hence, I decided in late 1999 to resign from the UW and start the independent Institute for Systems Biology (ISB), one of the first, if not the first, systems-biology centers in the world. After raising some funds, I persuaded Alan Aderem (an immunologist) and Ruedi Aebersold (a protein chemist and pioneer in proteomics) to join me as cofounders of ISB, which started in early 2000. These scientists, together with ISB faculty member John Aitchison, have played fundamental roles in the emergence of systems biology and/or technology development at the institute. But before discussing ISB, I here summarize the record of MBT.

5.4. Molecular Biotechnology's Remarkable Record of Accomplishment

MBT, arguably the first truly cross-disciplinary biology department in the United States, was strikingly successful. Ruedi Aebersold and John Yates advanced the field of proteomics by developing, respectively, an isotope-labeling technique known as isotope-coded affinity tags that permitted relative or absolute quantification of proteins by mass spectrometry (81) and an important computational technique called Sequest that allowed proteins to be identified from a database search of tryptic peptides present in all proteins (as determined computationally from complete genome sequence) against the actual mass spectrometry measurements (82). Phil Green developed two computational tools that proved essential for the human genome project: (*a*) a tool that allowed short sequenced DNA fragments to be assembled into larger fragments, or contigs, and (*b*) software that assessed the quality of DNA sequence data (83). Ger van den Engh pioneered the development of a multiparameter, very high speed cell sorter. Maynard Olson and I each directed 2 of the 16 international genome centers sequencing the human genome—and in this regard, Lee Rowen did a wonderful job managing the Hood center's efforts to sequence significant portions of human chromosomes 14 and 15 (as noted above) together with Anup Madan and Shizen Qin. As mentioned above, Alan Blanchard developed the ink-jet array technology. Debbie Nickerson began her pioneering work on SNP analyses. Barbara Trask pushed the development of technologies for chromosomal in situ hybridization (gene localization) so essential to human genetics.

After I resigned, MBT subsequently merged with the Department of Genetics to create the new Department of Genome Sciences. Some MBT faculty stayed at the university, and others (generally those who were more more technically oriented) moved on.

6. THE INSTITUTE FOR SYSTEMS BIOLOGY YEARS: 2000–PRESENT

As I discuss in this section, ISB embodies much of my philosophy on doing science and really represents the summation of much of what I have learned in my career. Hence, I take the liberty of discussing in more detail its historical context, its rationale, its culture, and how systems biology drove the development of new technologies and computational tools in my lab.

6.1. General Comments about Systems Biology in the Twenty-First Century

Let me set the general context for how I think about systems biology and its central role in twenty-first-century biology. Biology will be a dominant science in the twenty-first century—just as chemistry was in the nineteenth century and physics was in the twentieth century—and for a fascinating reason. The dominant challenge for all the scientific and engineering disciplines in the twenty-first century will be complexity, and biology is now in a unique position to solve the deep problems arising from its complexity and to begin to apply this knowledge to the most challenging issues of humankind. Biology will use systems approaches (holistic, as opposed to atomistic) and powerful new measurement and visualization technologies, as well as the new computational and mathematical tools that are emerging in the aftermath of the human genome project and the emergence of systems biology. Biology will make use of the fact that our models of biological complexity can be tested by experimentation.

Biology will also use the emerging insight that it can be viewed as an informational science (an idea articulated in References 71 and 84–88).

Solving the complexities of biology will enable scientists to achieve two important objectives: (*a*) Biology will begin to solve some of humankind's most challenging problems, including health care for all, agriculture, nutrition, and bioenergy. (*b*) It will bring to the other scientific and engineering disciplines solutions to many of their most vexing problems, such as integrative computing strategies to computer science, striking new chemistries to chemistry, molecular-level machines for manipulating matter and measurements to engineering, new materials to material sciences, new ways of thinking about complexity to physics, and new ways of deducing relevant historical pasts to geology and archeology. Thus, twenty-first-century biology will enrich most of the other scientific and engineering disciplines. This biological treasure trove of knowledge exists because biology has had three billion years of evolutionary trial and error to create, test, and perfect these scientific strategies and engineering solutions.

6.2. Biology as an Informational Science and the Institute for Systems Biology

ISB, created in 2000, articulated the vision of transforming modern biology by pioneering the systems approaches to decipher biological complexity and by viewing biology as an informational science. ISB argued that through the convergence of comprehensive systems approaches to biology (combining both holistic and reductionist approaches), the development of new technologies, and the creation of powerful new computational/mathematical tools, the complexity of biology could be penetrated and understood. ISB sought to pioneer and integrate approaches and tools for each of these areas.

The view of biology as an informational science provides a powerful conceptual framework for dealing with complexity. First, there are two fundamental types of biological information, the digital information of the genome and the outside or environmental information that impinges on and modifies the digital information. Indeed, this digital core of knowable information distinguishes biology from all the other scientific disciplines: None of the others have this digital and hence readily decipherable core of information. Systems biology attempts to understand the integration of the digital and environmental information that mediates the three fundamental processes of life: evolution, development (e.g., in humans growing from one cell in the fertilized egg to 10^{14} cells of many different types in the adult), and physiology (e.g., the immune response to an infection). Hence, the purpose of systems biology is to integrate the information from the digital genome and the environment to understand how life unfolds.

Second, biological information is captured, processed, integrated, and transferred by biological networks—interacting sets of RNAs, proteins, the control regions of genes, and small molecules—to the simple and complex molecular machines that actually execute the functions of life. Thus, a central focus of ISB is an understanding of the dynamical operation of biological networks in the context of evolution, development, physiology, or even disease, as is the understanding of the construction and function of molecular machines. Third, biological information is encoded by a multiscale information hierarchy: DNA, RNA, proteins, interactions, biological networks, cells, tissues and organs, individuals, and, finally, ecologies. Importantly, the environment impinges upon each level of the hierarchy and modulates the digital informational output from the genome. Hence, to understand how systems operate at a particular level—say understanding the 50 or so proteins that mediate the cell cycle—one should capture in a global manner each level of information that lies between the phenotypic measurements (features of the cell cycle) and the core digital genome. The information at each level should then be integrated in such a manner that the environmental modifications are identified so as to understand how they impact the functioning of the systems.

Note that each of these levels of information poses chemical and technical challenges for their global analysis; this allows us the possibility to bring in emerging technologies, such as microfluidics, nanotechnology, in vivo and in vitro molecular imaging, and new chemistries for creating protein capture agents. For example, we need to develop many new global chemistries to study proteins as they dynamically execute their functions; these measurements include their structure, expression patterns, chemical processing and modification, half-lives, interactions with other informational molecules and small molecules, locations in the cell, and dynamically changing three-dimensional structures. Obviously, many different aspects of chemistry will play a vital role in creating these new tools of systems biology. The capture, validation, storage, analysis, integration, visualization, and graphical or mathematical modeling of data sets—dynamically captured and global in nature—pose a host of computational and mathematical challenges. This intellectual context, and the formation of the modern version of what we call systems biology, is at the heart of the creation of ISB and its work over the past seven years. The implementation of the scientific program that has emerged from this context is at the heart of the future of ISB, and requires an organizational and cross-disciplinary cultural context to assist in its development.

We have had the opportunity to pioneer systems approaches, genomic (75), proteomic (94–96), and single-cell technologies and a wide variety of computational tools essential for systems approaches (77, 84, 89–93, 98–100). The Hood lab has been involved in many aspects of systems biology, such as systems approaches to biology (76, 77, 84, 87, 103), disease (104–109), proteomics (110, 111), technology development (77–109), and the development of relevant computational and mathematical tools (84, 99, 112–124). Trey Ideker, one of my former graduate students and now a professor at the University of California at San Diego, played a special role in catalyzing the experimental beginnings of a systems approach to galactose utilization in yeast and in pioneering the development of the systems-biology standard for graphical network visualization, the software program Cytoscape (86; in my view one of the pioneering papers of systems biology).

ISB has established effective facilities for high-throughput data generation (genomic, proteomic, and single cell). In general, the frontier challenges of the biology that the ISB faculty are studying have driven all that ISB does.

6.3. The Culture and Philosophy of the Institute for Systems Biology

Some powerful new ideas that have begun to effectively integrate biology, medicine, technology, and computation/mathematics are key to the development of ISB as an institution. The first three of these ideas are fundamentally scientific; the rest are strategic and guide the implementation of the institution's development. Some of these points partially repeat earlier discussion, but I include them here for the sake of completeness.

1. The frontier problems of biology should dictate what technologies should be developed (making it possible to view new dimensions of biological data space). Likewise, the creation of new data sets from these technologies should drive the development of new computational and mathematical techniques for analyzing them. New technologies and computational tools in turn enable the understanding of new levels of biological complexity. Thus, the dynamics of emerging new technologies frame the rate at which new biological insights are generated.

2. This integration of biology, technology, and computation necessitates the creation of a cross-disciplinary environment that brings biologists, chemists, computer scientists, engineers, mathematicians, physicians, and physicists together to learn one another's languages and work together in teams and allows the nonbiologists to learn relevant biology in a deep manner. ISB is still struggling to achieve these objectives.

3. Biology spans a spectrum of complexity, ranging from simpler model organisms such as single-celled bacteria or yeast to more complex model organisms such as mice and, ultimately, humans. Most importantly, biological experimentation is easier in simple species. Hence, ISB develops new tools and approaches using simpler model organisms and then must learn how to apply these tools to higher model organisms—and ultimately to human complexity.

4. ISB is deeply committed to an open-source philosophy, that is, making our data and tools readily available to the scientific community and taking advantage of the collective input of this community to improve those tools.

5. ISB is committed to remaining small, focused, and highly interactive. To compensate for the limitations this philosophy imposes upon faculty size, ISB requires an immediately available reservoir of expertise. This critical mass of knowledge in relevant biology, chemistry, computer science, engineering, mathematics, and medicine resides with our 25 senior scientists as well as our 13 faculty members.

6. One of ISB's approaches to attacking big scientific problems while remaining relatively small is to create strategic partners that bring us the scientific, technological, computational, and medical expertise that we lack. We focus on partnering with the best. ISB searches for these partners among academics, industry, and research institutions, as well as in relevant institutions in foreign countries.

7. ISB is determined to transfer its relevant knowledge to society, be it through K–12 science education, spin-off companies employing ISB's novel biological or technology insights, or educating society in science and technology.

8. ISB believes in providing the scientific leadership necessary for catalyzing paradigm changes in how biology and medicine are carried out and in how biology and medicine are organized—and we have a long history of doing so (see Section 6.11 below).

The following examples provide a glimpse into the ways the Hood lab at ISB plays a leading role in catalyzing the emergence of new organizational structures and creating powerful strategic partnerships.

6.4. Prostate Cancer

At the UW, my work turned to more medically focused problems, and I continued and extended those efforts at ISB. I took a systems approach to cancer biology, with a particular focus on prostate cancer—both from the viewpoint of genome-wide genetic mapping (125–133), which we carried out with our wonderful collaborators Elaine Ostrander and Janet Stanford at the Fred Hutchinson Cancer Research Center, and from the viewpoint of using DNA array technology to understand the dynamics of gene expression in the cancer disease process (134–144). This latter effort included productive collaborations with a prostate cancer clinician, Paul Lange at the UW; an MD postdoctoral fellow in my lab, Pete Nelson (now on the faculty at the Fred Hutchinson Cancer Research Center); and Biaoyang Lin, a senior scientist and long-time colleague at ISB. This project really drove the application of DNA array technologies (see, e.g., Reference 109): It pioneered applications using the then-new digital transcript counting technology known as multiple parallel signature sequencing (144), and we (in collaboration with the biotech company Helicos) have begun to explore the use of the powerful new single-strand DNA–sequencing technology to study interesting tissue transcriptomes.

In the context of the genetic mapping prostate project, I learned an important lesson about thinking outside the box. I received support from Michael Milken's Prostate Cancer Foundation and got to know Michael reasonably well. I remember in the mid-1990s talking with him about the challenge of obtaining families with a history of prostate cancer—collecting sufficient numbers of such families to provide adequate statistical power for the genetic analyses often took clinicians 10–20 years. Michael's solution was typically simple, rapid, and effective. He proposed that he and I, together with General Norman Schwarzkopf (who had prostate cancer, as had Michael), go on Larry King Live and make an hour-long pitch for these prostate-cancer families. I initially thought this idea was ridiculous. However, we did the program with Larry King, which was one of his most popular that year (1995). In six weeks we had recruited about 250 prostate cancer families from around the world (most from the United States), and together with wonderful collaborators—Elaine Ostrander and Janet Stanford at the Fred Hutchinson Cancer Research Center—we collected appropriate DNA samples from the families, then carried out a long-term series of genetic mapping studies.

The King show was a brilliant and imaginative solution to the challenging problem of disease-family collection. Frustration with the limited success of these family studies led us at ISB to begin thinking about a new genetic approach to understanding disease; we term this approach systems genetics (see Section 6.5). More recently, we have begun analyzing the biology and therapeutic responses of human glioblastomas and ovarian cancer using these same powerful genomic and proteomic technologies.

These cancer biology studies are driving tool development in several respects: (*a*) software algorithms to statistically assess and integrate different data sets both of the same data type and of different data types (112); (*b*) computational techniques for visualizing the dynamics of disease-perturbed networks, and ultimately their graphical or mathematical modeling; (*c*) new computational approaches to reducing the enormous data dimensionality of DNA array studies to simple hypotheses about health and disease (see Reference 115 for a fascinating successful example); and (*d*) high-throughput (digital) DNA-sequencing methods (e.g., single-stranded DNA sequencing) to quantitatively delineate dynamically changing transcriptomes. Ultimately, rapid DNA sequencing will completely replace DNA arrays in this task and will eventually determine the sequences of individual genomes (see below).

6.5. Systems Genetics

The genome-wide genetic mapping results on prostate cancer were frustrating in that, as is generally the case in the genetic studies of complex diseases, signal-to-noise problems posed significant challenges. I assume that for most complex genetic diseases there may be 30 or more potential variant genes involved in a combinatorial manner in smaller subsets, for example, where the appropriate six or so genetic variants can work together in various combinations to generate the distinct disease phenotypes. With most genome-wide association studies now being carried out on hundreds of thousands to millions of SNPs and thousands of patients, in conjunction with the recently completed human haplotype map, the signal-to-noise ratio is so poor that experimenters are fortunate if they can identify even one dominant gene out of the many involved in the complex disease process.

Of course, this problem can be somewhat mitigated by combining the association studies with other types of genomic data (transcriptome quantitation, indels, amplifications, or deletions of genomic regions, etc.). However, we feel that an entirely new approach is required for high-resolution disease-gene finding in complex genetic diseases. Indeed, one of the efforts that is just now getting under way at ISB is the creation of a new field we call systems genetics, wherein we attempt to connect genotype and phenotype together through an understanding of biological network behaviors of relevant variant genes. We are beginning these systems-genetics efforts using yeast as a simple model system (pioneered by ISB faculty members Aimee Dudley, Tim Galitski, John Aitchison, and David Galas). These efforts are driving us to develop (or to collaborate with those developing) two emerging technologies: (*a*) very-high-throughput DNA sequencing, which should allow the determination of an individual's genome sequence quickly and for less than one thousand dollars within the next ten years with the use of one or more of the emerging next-generation sequencing strategies, and (*b*) the detailed dynamic analysis of thousands of single cells obtained from an individual and perturbed by hundreds of appropriate environmental agents to interrogate differing subsets of the function networks of individual cells to reveal the activity of the underlying biological cellular networks in the context of an individual's known genome sequence. This analysis uses the microfluidic approaches that Steve Quake has pioneered with soft polydimethylsiloxane (PDMF) materials, allowing one to create miniaturized valves, pumps, and mixing chambers, which are needed to analyze fluids and cells. Perhaps we can do systems genetics by looking at what single cells can tell us if appropriately perturbed. We are working together with strategic partners in each of these important areas.

6.6. Neural Degenerative Diseases

I have also become interested in applying a systems approach to the study of neural degenerative diseases. We have been studying prion disease in mice with George Carlson (a wonderfully interactive Montana colleague) and Stanley Prusiner for the past 25 years, and more recently with Inyoul Lee, a senior scientist in my lab.

Recently we looked at mice infected intercranially with infectious prion proteins. We studied the dynamic brain transcriptomes (populations of mRNAs in the brain) in five different mouse inbred strains, two congenic strains, and one knockout strain at ten different time points across the progression of their prion disease. We studied the large number of different strains because each serves as a fascinating biological filter to deal with the signal-to-noise challenges of transcriptome data; this strategy allowed us to identify and assess the core set of genes encoding the prion disease process. We compared and integrated the dynamically changing disease and normal control brain transcriptomes to identify the key genes that have changed activity as a consequence of the disease. We then integrated the transcriptome data and mapped them onto known biological networks together with phenotypic data (the histopathology and the clinical signs). These studies led to several interesting conclusions. For instance, the dynamically changing brain networks can explain much of the pathophysiology of prion disease, but these networks also provide a new systems approach to thinking about blood diagnostics. Systems medicine is predicated on the simple idea that disease arises from one or more disease-perturbed networks in the affected organ (perturbed genetically and/or environmentally) (108). This alters the patterns of dynamically expressed information from these networks, and these altered dynamically changing patterns of transcription encode the corresponding dynamically changing pathophysiology of the disease. Systems medicine is driving the development of new measurement technologies [DNA sequencing to measure genomes and transcriptomes; protein-measurement technologies (see below); single-cell analyses of DNA, RNA, proteins and protein-protein, protein-DNA and protein–small molecule interactions; and in vivo and in vitro molecular-imaging technologies] to determine how networks are changing in individual cells or individual organisms. Once again, ISB is itself developing several of these technologies and is collaborating with strategically chosen partners on others.

6.7. Organ-Specific Blood Protein Fingerprints: A Systems Approach to Disease Diagnostics

I describe the systems approach to disease diagnostics in detail because I believe it is going to be one of the most transformational approaches in the new medicine and because it beautifully embodies the essence of systems medicine. The dynamically changing prion networks revealed that levels of some transcripts were elevated (or decreased) many weeks before the detection of clinical signs. If some of these transcripts encode proteins that are secreted into the blood, then their elevated levels in blood might be an early preclinical sign of incipient disease. With proteomics discovery approaches, it is relatively easy to identify quantitative changes in levels of many proteins that distinguish, for example, individuals with ovarian cancer from their healthy counterparts. However, if these same markers are examined in the blood of individuals with, say, ten other diseases, they behave in unpredictable ways because these markers are generally synthesized in multiple organs and hence are responsive to different environmental stimuli.

Our idea has been to use transcriptome analyses to identify the organ-specific transcripts in, for example, the brain, by comparing the brain transcriptome against the transcriptomes of 40 other organs and determining which transcripts are primarily expressed in the brain. If this is done with every organ and tissue, we anticipate that most will contain 100–200 organ-specific transcripts. If the protein products of some of these transcripts are secreted into the blood, they constitute an organ-specific blood protein fingerprint wherein the levels of the individual proteins reflect the operations of the corresponding organ networks that encode them. Accordingly, a normal individual will have one set of levels of brain-specific proteins in his or her blood fingerprint, whereas the levels of some of these proteins will change in ways that are specific to each different brain disease (i.e., brain cancer or brain infection) because each is encoded by different combinations of disease-perturbed networks. Because we would like to deduce the nature of the disease-perturbed networks from the dynamically changing concentrations of proteins in the organ-specific fingerprint, we need to measure at least 50 proteins per fingerprint. The technical challenge, then, is twofold: (*a*) One must measure 50 proteins for the organ fingerprints for each of the 50 or so different human organs (2500 measursements), and (*b*) the assay should be carried out from a single droplet of blood. Thus, this measurement technology must be miniaturized and highly parallelized: Technology using microfluidic and/or nanotechnology approaches appears to be required.

The analysis of biomarkers in the blood represents an enormous technical challenge to proteomics. Blood is a mixture of millions of proteins whose concentrations span a dynamic range of perhaps 10^{12}, with 21 proteins constituting approximately 99% of the blood protein mass. [Keep in mind we are talking only about the identification and quantification of individual biomarkers (proteins). There remain enormous chemical challenges with regard to developing techniques that can characterize the additional types of protein diversity that arise posttranscriptionally, delineate different forms from alternative RNA splicing, characterize the processing of proteins by enzymes, detect the 400 or more different potential chemical modifications of proteins, and measure the different half-lives of proteins. In addition, proteins change their structures dynamically during the execution of their functions and migrate to different regions of the cell to carry out specific functions; global technologies to measure these features also need to be developed.] We need to quantify proteins secreted from large organs (e.g., the liver) and small organs (e.g., the prostate); hence, we need measurement techniques for proteins that span a very large dynamic range in the blood.

The challenge to the identification of blood biomarkers is twofold. First, one must discover appropriate biomarkers (in our case, the organ-specific proteins), which requires the analysis of tens to hundreds of samples. This can be done by a variety of techniques including mass spectrometry and antibody-based assays (enzyme-linked immunosorbent assay, Western blot, surface plasmon resonance, etc.). My lab is now collaborating with the biotech company Plexera to develop a surface plasmon resonance instrument that can measure 800 different antibody interactions in just five minutes and can repeat the cycle without signal degradation every ten minutes for up to 40 cycles—another example of a key high-throughput measurement technology, in this case for proteomics. (Ruedi Aebersold has been a cutting-edge pioneer in developing a wide variety of proteomics techniques and strategies for analyzing biomarkers; see References 92–95.) Also, there remain striking chemical challenges for the development of these discovery approaches, which need to be more sensitive, more global, and more specific. Second, one must eventually be capable of large-scale typing—perhaps the quantification of 2500 blood proteins from one droplet of blood in hundreds of millions of patients per year. Typing mandates the use of microfluidic and nanotechnology measurement strategies.

Four years ago, I started a collaboration with Jim Heath, a young chemist at Caltech, using microfluidic and nanotechnology approaches to quantify organ-specific (and other) blood proteins. Jim's lab has recently developed a new type of protein chip, known as a DNA-encoded antibody library (145). This protein chip appears to have a dynamic range of 10^8 and a sensitivity in the low femtomole range, and it can potentially be manufactured inexpensively in large quantities. We currently have protein chips with approximately 20 protein-capture features—a scale that can eventually be expanded to thousands of features. This feature scale is limited only by the availability of protein-capture agents with high specificity and affinity (currently antibodies, but we are exploring alternative chemical possibilities for synthesizing highly specific reagents). This collaboration has progressed rapidly over the past four years and illustrates the power of carefully chosen strategic partnerships wherein all parties bring together complementary scientific skills. The collaboration with Jim has been one of the best I have ever had.

6.8. Predictive, Personalized, Preventive, and Participatory (P4) Medicine

The convergence of systems approaches to disease, new measurement and visualization technologies, and new computational and mathematical tools suggests that our current largely reactive mode of medicine (i.e., wait until one is sick before responding) will over the next 10–20 years be transformed to predictive, personalized, preventive, and participatory (P4) medicine (104–106, 108). Two components of predictive medicine will emerge over the next ten years: (*a*) Individual genome sequences will be available, and (*b*) increasingly, we will be able to determine the likelihood of one's future health (e.g., 50% probability of ovarian cancer by age 50). Hand-held devices to prick the finger and quantify 2500 organ-specific proteins from all human organs will send this information via wireless communication to a server, which in turn will analyze the information and email a report to the patient and their physician. This rapid communication, done perhaps twice a year, will thus permit an instantaneous assessment of current health status. These measurements will themselves personalize medicine. And we must remember that each of us differs on average by approximately six million nucleotides from our neighbors; hence, we are susceptible to differing combinations of diseases and, once again, must be treated as individuals.

From the assessment of genomes and environmental exposures will emerge initially a predictive and personalized medicine. Later, physicians will learn how to identify drugs to re-engineer disease-perturbed networks, causing them to behave in a more normal fashion, or at least abrogating the most deleterious of their effects. In the future, we will be able to design drugs to prevent networks from becoming disease perturbed. For example, if there is an 80% change of prostate cancer at age 50, taking these preventive drugs beginning at age 35 may reduce disease probability to 2%. Finally, if we can educate patients and their physicians as to the nature of P4 medicine, then patients will be in a position to take more responsibility for charting and participating in their own future health choices. The realization of P4 medicine is a major strategic goal of ISB.

The vision of P4 medicine has emerged from ISB, but it has also emerged from a strategic partnership (described above) that brings together three critical skills: (*a*) systems biology and medicine (Hood and Galas, ISB); (*b*) microfluidics and nanotechnology (Heath, Caltech); and (*c*) molecular imaging (Mike Phelps, inventor of positron emission tomography scanning, University of California at Los Angeles). This partnership, termed the NanoSystems Biology Alliance, has facilitated the creation of a series of NIH centers (e.g., the Systems Biology Center at ISB and the NanoSystems Biology Cancer Center at Caltech). Our P4 vision has been delineated in a series of papers (104, 106–108) and has been a stimulating and broadening opportunity for all involved. ISB recently has elected to make P4 medicine one of its central strategic projects, and all of its faculty are now in discussion about the convergence and focus of our collective talents on this challenging problem.

6.9. P4 Medicine and Its Implications for Society

P4 medicine has several fascinating implications. It will, over the next ten years, transform the business plans of virtually every sector of the health care industry—pharmaceuticals, biotech, medical instrumentation, diagnostics, health care information technology, payers, providers, medical centers, medical schools, and so forth. For example, pharmaceutical companies are generally acknowledged as failing in their quest to produce effective and reasonably priced drugs. Systems medicine will bring presymptomatic diagnostics and the ability to stratify disease so that effective therapies can be successfully matched against specific diseases. It will provide powerful new approaches to assessing drug toxicity early in clinical trials. It will also provide new approaches to assessing drug doses for individual patients and evaluating drug toxicities at a very early stage. A fascinating question is whether the pharmaceutical companies will be able to effectively employ these strategies of systems medicine. Another challenging issue for P4 medicine concerns medical schools, which are currently teaching physicians that will be practicing P4 medicine in 10–20 years. However, these students are not learning the background and concepts that they will need for P4 medicine. Will medical schools be able to transform their teaching, research, and eventually their patient care to encompass the P4 concepts? Similar issues apply to every sector of the health care industry.

P4 medicine will lead to the digitalization of medicine, that is, the ability to extract disease- or health-relevant information from single molecules, single cells, or single individuals. The digitization of medicine will have a far greater impact on society than will the digitization of communications or information technology. The reason for this is because at some time in the future (depending upon the rate at which technologies emerge and the extent of federal and private resources that can be focused on P4 medicine), there will be a sharp turnaround in the ever-escalating costs of health care to the point that we will be able to export P4 medicine to the developing world. Indeed, P4 medicine will, in the near future, form the foundation of global medicine.

P4 medicine poses significant technical and social challenges that are amenable to powerful cross-disciplinary scientific attack and societal education and debate. The societal challenges must be dealt with at the same time the technical challenges are being solved if P4 medicine is to successfully emerge in the next 10–20 years.

ISB is in the process of generating strategic partnerships to attack the technical and societal challenges to P4 medicine. My collaborations with David Galas and Diane Isonoka at ISB have led to a series of fascinating possibilities for strategic partnerships—with individuals, academic centers, companies, and even with countries—that will not only bring critical scientific, engineering, and clinical expertise, but will also provide striking new funding opportunities, some of which are global in nature. Indeed, I envision a globalization of science emerging over the next ten years or so—just as Tom Friedman described the globalization of the economy. And in a similar manner, there are striking opportunities for those who will be at the leading edge of scientific globalization.

6.10. Using Intellectual Property to Support Science: The Accelerator

About three years ago I decided to create a for-profit company called the Accelerator, whose mission was to create successful new biotech companies with the help of ISB. Those companies, in turn, could generate the resources for a long-term ISB endowment. Venture capital companies at that time were focusing primarily on companies generating late-phase drugs, and I wanted to facilitate the emergence of new companies with large-scale discovery platforms or new strategies for extracting relevant biological information for medicine. I went to about 20 venture capital companies with a proposal for a group of five or so venture capitalists to join with ISB to create the Accelerator. Each venture group would contribute $3–5 million. A board would be established with representative membership from each venture company, the CEO of the Accelerator (Carl Weissman, who has done a superb job), and myself. This board would select suitable companies to support for two to three years to prove the principles of their scientific approaches. Then the new company would graduate by raising series B (second-round) money and would go out on their own—or, if the company could not raise money after a few years, it would be terminated. The Accelerator would handle all the management for each company.

We also recruited Alexandria, a real estate company specializing in laboratory construction. They built beautiful facilities for us at a reasonable cost with the expectation of providing space for some of the successful companies as they graduated from the Accelerator. ISB played a special role in creating the Accelerator, as well as in providing faculty support for due diligence and the scientific advisory boards, ISB's excess capacity for high-throughput genomic and proteomic measurements, access to our outstanding computational facilities, and my expertise of almost 30 years in starting companies. For this, ISB receives an equity position in each company that will be used to build our endowment in the next 10–15 years; some of these companies will hopefully mature to become as successful as some of the past companies we have founded. To date we have screened more than 400 business plans—those coming from the venture companies as well as those identified by Carl Weissman and myself—and we have selected six. The results have been outstanding. Three of the companies have successfully raised series B money—from $30 million to $55 million. We believe that two of the remaining three companies will also do very well—a remarkable record when compared with most venture efforts.

Recently we raised $22.5 million for the second Accelerator round, with the expectation that we will bring five or six new companies to the Accelerator over the next few years. Indeed, in the last year and a half, the venture money raised in the state of Washington for biotech has been approximately $160 million—and the Accelerator has raised more than 70% of this money. This testifies to what a small, focused, and knowledgeable effort can do. This will probably be a far more successful approach to converting scientific knowledge into support for science than the vast majority of the efforts directed at licensing intellectual property to pre-existing companies. Indeed, in my experience, only the licensing of the automated DNA sequencer to ABI generated significant income (on the order of $100 million for Caltech), and it was the most successful licensing in Caltech's history. But significant financial licensing success is rare indeed.

6.11. Paradigm Changes, or New Ideas Often Need New Organizational Structures for Their Realization

Whether instrument integration and high data throughput, the human genome project, cross-disciplinary biology, systems biology, and P4 medicine are really paradigm changes or just interesting new ideas is partly in the eye of the beholder. What I can say for certain is that each of these concepts was initially greeted with considerable skepticism from the scientific or medical communities. I believe each of these innovations has impacted biology in major ways, and the first four required new organizations to begin to realize their potential. We do not yet know what types of new organizations P4 medicine will require for its realization. I stress now, as I did at the beginning of this article, that I am but one player of many who participated in each of these five paradigm changes. I summarize my experiences as follows:

1. To realize the full development, commercialization, and widespread application of the first four chemical technologies that we developed, we needed to set up a new organization: ABI. We came to this realization with striking clarity when Bob Sinsheimer, then Chairman of Biology at Caltech, stopped by my office in mid-1973 to suggest in the strongest possible terms that I give up technology development and focus only on biology. This did not have to do with tenure, as I was awarded tenure just a few months later (quite early for Caltech). Rather, it appeared to have to do with taste. I refused, of course, and 20 years later finally asked Bob why he had paid me that visit. He said the senior biologists at Caltech felt it was inappropriate to mix engineering (and probably commercialization) with biology. I later created ABI to take on much of this engineering and commercialization.

2. When I (along with several others) first went into the biological community in 1985 to push the human genome project, most biologists and NIH were firmly opposed. Rationality eventually held sway, but only after a new NIH institute, the National Human Genome Research Institute, was created to manage the genome project.

3. When I decided to build a cross-disciplinary biology department, I had to move from Caltech to the UW to create MBT. As illustrated above, this new department—the first of its kind—flourished.

4. In building an environment for systems biology it once again became clear that a new and independent organization, ISB, was necessary to realize systems biology, and it has done so with considerable success.

5. Finally, it will be interesting to see how P4 medicine emerges. Will the established companies in the health care sector be capable of the transformational change required to take advantage of the new opportunities emerging from systems medicine and a systems approach to disease? Or will many new organizations emerge focused precisely and totally on the objectives of P4 medicine with appropriate technologies? The jury is out, but I expect that many new health care companies will form over the next 10–20 years in response to the disruptive technologies and strategies emerging from P4 medicine. There is also the question of how the academic medical environment will embrace these new opportunities; doing so in an incremental fashion, as is usually the case, generally leads to modest evolutionary change and not the desired revolutionary changes. The future will be very exciting.

7. A FEW CLOSING COMMENTS FOR STUDENTS

I leave students (and even some of my colleagues) with several pieces of advice. First, I stress the importance of a good cross-disciplinary education. Ideally, I suggest a double major with the two fields being orthogonal—say, biology with computer science or applied physics. Some argue that there is insufficient time to learn two fields deeply at the undergraduate level. I argue that this is not true. If we realize that many undergraduate courses now taught are filled with details that are immediately forgotten after the course is finished, we must then learn to teach in an efficiently conceptual manner. As I noted above, as an undergraduate at Caltech I had Feynman for physics and Pauling for chemistry, and both provided striking examples of the power of conceptual teaching. Second, I argue that students should grow accustomed to working together in teams: In the future, there will be many hard problems (like P4 medicine) that will require the focused integration of many different types of expertise. Third, I suggest that students acquire an excellent background in mathematics and statistics and develop the ability to use various computational tools. Fourth, I argue that a scholar, academic, scientist, or engineer should have four major professional objectives: (*a*) scholarship, (*b*) education (teaching), (*c*) transferring knowledge to society, and (*d*) playing a leadership role in the local community to help it become the place in which one would like one's children and grandchildren to live. Fifth, with regard to the scientific careers of many scientists—they can be described as bell-shaped curves of success—they rise gradually to a career maximum and then slowly fall back toward the base line. To circumvent this fate, I propose a simple solution: a major change in career focus every 10 or so years. By learning a new field and overcoming the attendant insecurities that come from learning new areas, one can reset the career clock. Moreover, with a different point of view and prior experience, one can make fundamental new contributions to the new field by thinking outside the box. Then the new career curve can be a joined series of the upsides of the bell-shaped curve, each reinvigorated by the ten-year changes. Finally, science is all about being surrounded by wonderful colleagues and having fun with them, so I recommend choosing one's science, environment, and colleagues carefully. I end this discussion with what I stressed at the beginning—I am so fortunate to have been surrounded by outstanding colleagues who loved science and engineering (**Table 1**). Science for each of us is a journey with no fixed end goal. Rather, our goals are continually being redefined.

8. PERSONAL THOUGHTS ABOUT MY CAREER EXPERIENCES

In retrospect I see the 40 years of my career as embodying and leading toward the same principles that we brought to ISB. But at the very foundation of all the science and technology I have enjoyed are the wonderful colleagues I have associated with throughout my career—they are the true pioneers of virtually all that has been accomplished (**Table 1**). Biology is the central focus and driver of the technologies and strategies to be developed. Indeed, I see my career as having pointed toward the creation of the cross-disciplinary environments that enabled this virtuous cycle of biology driving technology, and technology in turn generating data that drive the development of computational and mathematical tools. Thus biology drives both technology and computation.

Also, I see myself moving throughout much of my career toward a systems approach, first to biology and then, more recently, to medicine. Each of the technologies that we developed focused upon solving chemical problems—for the essence of life is the chemistry of how living organisms deal with biological information. Another major aspect of my career has been to participate in the creation of new visions for how biology should be carried out—and I was often able to create new organizational structures that could allow these visions to manifest themselves. I wonder what new organizational structures for practicing both biology and P4 medicine will emerge over the next ten years. What a fascinating time to be in science and technology!

DISCLOSURE STATEMENT

The author is not aware of any biases that might be perceived as affecting the objectivity of this review.

ACKNOWLEDGMENTS

I would like to thank Lee Rowen for her critical review of this manuscript and always-insightful feedback and Shirley Meinecke for assisting in the manuscript's production.

LITERATURE CITED

1. Dyson FJ. 1998. *Imagined Worlds*. Cambridge, MA: Harvard Univ. Press
2. Hood LE, Gray WR, Dreyer WJ. 1966. On the mechanism of antibody synthesis: a species comparison of l-chains. *Proc. Natl. Acad. Sci. USA* 55:826–32
3. Hood L, Gray WR, Dreyer WJ. 1966. On the evolution of antibody light chains. *J. Mol. Biol.* 22:179–82
4. Gray WR, Dreyer WJ, Hood L. 1967. Mechanism of antibody synthesis: size differences between mouse kappa chains. *Science* 155:465–67
5. Hood L, Gray WR, Sanders BG, Dreyer WJ. 1967. Light chain evolution. *Cold Spring Harbor Symp. Quant. Biol.* 32:133–46
6. Silver J, Hood L. 1975. Automated microsequence analysis by use of radioactive phenylisothiocyanate. *Anal. Biochem.* 67:392–96
7. Hunkapiller MW, Hood L. 1980. New protein sequenator with increased sensitivity. *Science* 207:523–25
8. Hewick RM, Hunkapiller MW, Hood LE, Dreyer WJ. 1981. A gas-liquid-solid-phase peptide and protein sequenator. *J. Biol. Chem.* 256:7990–97
9. Doolittle RF, Hunkapiller MW, Hood LE, Devare SG, Robbins KC, et al. 1983. Simian sarcoma virus one gene, v-sis, is derived from the gene (or genes) encoding a platelet-derived growth factor. *Science* 221:275–77
10. Prusiner SB, Groth DF, Bolton DC, Kent S, Hood LE. 1984. Purification and structural studies of a major scrapie prion protein. *Cell* 38:127–34

11. Desch B, Westaway D, Walchli M, McKinley MP, Kent SBH, et al. 1985. A cellular gene encodes scrapie PrP 27–30 protein. *Cell* 40:735–46

12. Raftery MA, Hunkapiller MW, Strader CD, Hood L. 1980. Acetylcholine receptor: complex of homologous subunits. *Science* 208:1454–57

13. Hunkapiller MW, Hood LE. 1980. Human fibroblast interferon: amino acid analysis and amino terminal amino acid sequence. *Science* 207:525–26

14. Zoon KC, Smith ME, Bridgen PJ, Anfinsen CB, Hunkapiller MW, Hood LE. 1980. Amino terminal sequence of the major component of human lymphoblastoid interferon. *Science* 207:527–28

15. Goldstein AS, Tachibana L, Lowney I, Hunkapiller MW, Hood LE. 1979. Dynorphin (1–13), an extraordinarily potent opioid peptide. *Proc. Natl. Acad. Sci. USA* 76:6666–70

16. Schally AV, Huang WY, Chang RCC, Arimura A, Redding TW, et al. 1980. Isolation and structure of pro-somatostatin: a putative somatostatin precursor from pig hypothalamus. *Proc. Natl. Acad. Sci. USA* 77:4489–93

17. Heller EL, Kaczmarek K, Hunkapiller MW, Hood LE, Strumwasser F. 1980. Purification and primary structure of two neuroactive peptides that cause bag cell after discharge and egg-laying in *Aplysia*. *Proc. Natl. Acad. Sci. USA* 77:2328–32

18. Goldstein A, Fischli W, Lowney LI, Hunkapiller M, Hood L. 1981. Porcine pituitary dynorphin: complete amino acid sequence of the biologically active heptadecapeptide. *Proc. Natl. Acad. Sci. USA* 78:7219–23

19. Horvath SJ, Firca JR, Hunkapiller T, Hunkapiller MW, Hood L. 1987. An automated DNA synthesizer employing deoxynucleoside 3′ phosphoramidites. *Methods Enzymol.* 154:314–26

20. Strauss EC, Kobori JA, Siu G, Hood LE. 1986. Specific primer-directed DNA sequencing. *Anal. Biochem.* 154:353–60

21. Kent SB, Hood LE, Beilan H, Marriot M, Meister S, Geiser T. 1984. A novel approach to automated peptide synthesis based on new insights into solid phase chemistry. In *Proceedings of the Japanese Peptide Symposium*, ed. N. Isymiya, pp. 217–22. Osaka: Protein Res. Found.

22. Miller M, Schneider J, Sathyanarayana BK, Toth MV, Marshall GR, et al. 1989. Structure of a complex of synthetic HIV-1 protease with a substrate-based inhibitor at 2.3 A resolution. *Science* 246:1149–52

23. Clark-Lewis I, Hood LE, Kent SB. 1988. Role of disulfide bridges in determining the biological activity of interleukin 3. *Proc. Natl. Acad. Sci. USA* 85:7897–901

24. Clark-Lewis I, Lopez A, Luen B, Vadas M, Schrader JW, et al. 1989. Structure-function studies of human granulocyte-macrophage colony-stimulating factor: identification of amino acids required for activity, and an 84-residue active fragment. *J. Immunol.* 141:881–89

25. Pärraga G, Horvath SJ, Eisen A, Taylor WE, Hood L, et al. 1988. Zinc-dependent structure of a single-finger domain of yeast ADM. *Science* 241:1489–92

26. Maxam AM, Gilbert W. 1977. A new method for sequencing DNA. *Proc. Natl. Acad. Sci. USA* 74:560–64

27. Sanger F, Coulson AR. 1975. A rapid method for determining sequences in DNA by primed synthesis with DNA polymerase. *J. Mol. Biol.* 94:441–48

28. Smith LM, Sanders JZ, Kaiser RJ, Hughes P, Dodd C, et al. 1986. Fluorescence detection in automated DNA sequence analysis. *Nature* 321:674–79

29. Kaiser RJ, MacKellar SL, Vinayak RS, Sanders JZ, Saavedra RA, Hood LE. 1989. Specific-primer-directed DNA sequencing using automated fluorescence detection. *Nucleic Acids Res.* 17:6087–102

30. Int. Hum. Genome Seq. Consort. 2001. Initial sequencing and analysis of the human genome. *Nature* 409:860–921

31. Hood L, Rowen L. 1997. The impact of genomics on 21st century medicine. *Contemp. Urol.* 9:86–98

32. Hood L. 2002. A personal view of molecular technology and how it has changed biology. *J. Proteome Res.* 1:399–409

33. Hunkapiller M, Kent S, Caruthers M, Dreyer W, Firca J, et al. 1984. A microchemical facility for the analysis and synthesis of genes and proteins. *Nature* 310:105–11

34. Olson M, Hood L, Cantor C, Botstein D. 1989. A common language for physical mapping of the human genome. *Science* 245:1434–35

35. Wang K, Koop BF, Hood L. 1994. A simple method using T4 DNA polymerase to clone polymerase chain reaction products. *Benchmarks* 17(2):236–38

36. Wilson R, Chen C, Hood L. 1990. Optimization of asymmetric polymerase chain reaction for rapid fluorescent DNA sequencing. *BioTechniques* 8:184–89

37. Lai E, Wang K, Avdalovic N, Hood L. 1991. Rapid restriction map constructions using a modified pWE15 cosmid vector and a robotic workstation. *BioTechniques* 11:212–17

38. Huang GM, Wang K, Kuo C, Paeper B, Hood L. 1994. A high-throughput plasmid DNA preparation method. *Anal. Biochem.* 223:35–38

39. Wang K, Gan L, Boysen C, Hood L. 1995. A microtiter plate–based high-throughput DNA purification method. *Anal. Biochem.* 226:85–90

40. Suzuki M, Baskin D, Hood L, Loeb LA. 1996. Random mutagenesis of *Thermus aquaticus* DNA polymerase I: concordance of immutable sites in vivo with the crystal structure. *Proc. Natl. Acad. Sci. USA* 93:9670–75

41. Wang K, Boysen C, Shizuya H, Simon MI, Hood L. 1997. Complete nucleotide sequence of two generations of a bacterial artificial chromosome cloning vector. *Benchmarks* 23:992–93

42. Guo Z, Gatterman MS, Hood L, Hansen JA, Petersdorf EW. 2002. Oligonucleotide arrays for high-throughput SNPs detection in the MHC class I genes: HLA-B as a model system. *Genome Res.* 12:447–57

43. Hunkapiller MW, Hood LE. 1978. Direct microsequence analysis of polypeptides using an improved sequenator, a nonprotein carrier (polybrene), and high-pressure liquid chromatography. *Biochemistry* 17:2124–33

44. Hunkapiller MW, Lujan F, Ostrander F, Hood L. 1983. Isolation of microgram quantities of proteins from polyacrylamide gels for amino acid sequence analysis. *Methods Enzymol.* 91:227–36

45. Aebersold RH, Teplow DB, Hood LE, Kent SBH. 1986. Electroblotting onto activated glass: high efficiency preparation of proteins from analytical sodium dodecyl sulfate–polyacrylamide gels for direct sequence analysis. *J. Biol. Chem.* 261:4229–38

46. Aebersold R, Teplow DB, Hood LE, Kent SBH. 1986. Electroblotting from immobiline isoelectric focusing gels for direct protein sequence determination. *Peptides Biol. Fluids* 34:715–18

47. Kent S, Hood L, Aebersold R, Teplow D, Smith L, et al. 1987. Approaches to sub-picomole protein sequencing. *BioTechniques* 5:314–21

48. Aebersold R, Pipes GD, Wettenhall REH, Mika H, Hood LE. 1990. Covalent attachment of peptides for high-sensitivity solid-phase sequence analysis. *Anal. Biochem.* 187:56–65

49. Wettenhall REH, Aebersold RH, Hood LE. 1991. Solid-phase sequencing of ^{32}P-labeled phosphopeptides at picomole and subpicomole levels. *Methods Enzymol.* 201:186–99

50. Tempst P, Woo DL, Teplow DB, Aebersold R, Hood L, Kent SBH. 1986. Microscale structure analysis of a high-molecular-weight, hydrophobic membrane glycoprotein fraction with platelet-derived growth factor–dependent kinase activity. *J. Chromatogr.* 359:403–12

51. Aebersold RH, Pipes G, Hood LE, Kent SBH. 1988. N-terminal and internal sequence determination of microgram amounts of proteins separated by isoelectric focusing in immobilized pH gradients. *Electrophoresis* 9:520–30

52. Harrington MG, Gudeman D, Zewert T, Hood L. 1991. Analytical and micropreparative two-dimensional electrophoresis of proteins. *METHODS: Companion Methods Enzymol.* 3:98–108

53. Harrington MG, Hood L, Puckett C. 1991. Simultaneous analysis of phosphoproteins and total cellular proteins from PC12 cells. *METHODS: Companion Methods Enzymol.* 3:135–41

54. Mononen I, Heisterkamp N, Kaartinen V, Williams JC, Yates JR III, et al. 1991. Aspartylglycosaminuria in the Finnish population: identification of two point mutations in the heavy chain of glycoasparaginase. *Proc. Natl. Acad. Sci. USA* 88:2941–45

55. Kaartinen V, Williams J, Tomich J, Yates JR III, Hood L, Mononen I. 1991. Glycosaparaginase from human leukocytes. Inactivation and covalent modification with diazo-oxoriorvaline. *J. Biol. Chem.* 266:5860–69

56. Griffin PR, Coffman JA, Hood LE, Yates JR III. 1991. Structural analysis of proteins by capillary HPLC electrospray tandem mass spectrometry. *Int. J. Mass Spectrom. Ion Process.* 111:131–49

57. Yates JR III, Zhou J, Griffin PR, Hood LE. 1991. Computer aided interpretation of low energy MS/MS mass spectra of peptides. *Tech. Protein Chem. II* 46:477–85

58. Lai E, Davi NA, Hood LE. 1989. Effect of electric field switching on the electrophoretic mobility of single-stranded DNA molecules in polyacrylamide gels. *Electrophoresis* 10:65–67

59. Birren BW, Hood L, Lai E. 1989. Pulsed field gel electrophoresis: studies of DNA migration made with the programmable, autonomously controlled electrode (PACE) apparatus. *Electrophoresis* 10:302–9

60. Landegren U, Kaiser R, Sanders J, Hood L. 1988. A ligase-mediated gene detection technique. *Science* 241:1077–80

61. Nickerson DA, Kaiser R, Lappin S, Stewart J, Hood L, Landegren U. 1990. Automated DNA diagnostics using an ELISA-based oligonucleotide ligation assay. *Proc. Natl. Acad. Sci. USA* 87:8923–27

62. Roach JC, Boysen C, Wang K, Hood L. 1995. Pairwise end sequencing: a unified approach to genomic mapping and sequencing. *Genomics* 26:345–53

63. Rowen L, Koop BF, Hood L. 1996. The complete 685-kb DNA sequence of the human β T cell receptor locus. *Science* 272:1755–62

64. Boysen C, Simon MI, Hood L. 1997. Analysis of the 1.1-Mb human α/δ T-cell receptor locus with bacterial artificial chromosome clones. *Genome Res.* 7:330–38

65. MHC Consortium. 1999. Complete sequence and gene map of a human major histocompatibility complex: the MHC sequencing consortium. *Nature* 401:921–23

66. Glusman G, Rowen L, Lee I, Boysen C, Roach JC, et al. 2001. Review: comparative genomics of the human and mouse T-cell receptor loci. *Immunity* 15:337–49

67. Shizuya H, Birren B, Kim UJ, Mancino V, Slepak T, et al. 1992. Cloning and stable maintenance of 300-kb-pair fragments of human DNA in *Escherichia coli* using an F-factor-based vector. *Proc. Natl. Acad. Sci. USA* 89:8794–97

68. Boysen C, Simon MI, Hood L. 1997. Fluorescence-based sequencing directly from bacterial and P1-derived artificial chromosomes. *Benchmarks* 23:978–82

69. Venter JC, Smith HO, Hood L. 1996. A new strategy for genome sequencing. *Nature* 381:364–66

70. Mahairas GG, Wallace JC, Smith K, Swartzell S, Holzman T, et al. 1999. Sequence-tagged connectors: a sequence approach to mapping and scanning the human genome. *Proc. Natl. Acad. Sci. USA* 96:9739–44

71. Hood L. 1992. Biology and medicine in the twenty-first century. In *The Code of Codes: Scientific and Social Issues in the Human Genome Project*, ed. DJ Kevles, L Hood, pp. 136–63. Cambridge, MA: Harvard Univ. Press

72. Koop BF, Hood L. 1994. Striking sequence similarity over almost 100 kilobases of human and mouse T-cell receptor DNA. *Nat. Genet.* 7:48–53

73. Heilig R, Eckenberg R, Petit JL, Fonknechten N, Da Silva C, et al. 2003. The DNA sequence and analysis of human chromosome 14. *Nature* 421:601–7

74. Zody MC, Garber M, Sharpe T, Young SK, Rowen L, et al. 1006. Analysis of the DNA sequence and duplication history of human chromosome 15. *Nature* 440:671–75

75. Int. Hum. Genome Seq. Consort. 2004. Finishing the euchromatic sequence of the human genome. *Nature* 431:931–45

76. Davidson EH, Rost JP, Oliveri P, Ransick A, Calestani C, et al. 2002. A genomic regulatory network for development. *Science* 295:1669–78

77. Davidson EH, McClay DR, Hood L. 2003. Regulatory gene networks and the properties of the developmental process. *Proc. Natl. Acad. Sci. USA* 100:1475–80

78. Blanchard AP, Kaiser RJ, Hood LE. 1996. High-density oligonucleotide arrays. *Biosens. Bioelectron.* 11:687–90

79. Blanchard AP, Hood L. 1996. Sequence to array: probing the genome's secrets. *Nat. BioTechnol.* 14:1649

80. Lausted C, Dahl T, Warren C, King K, Smith K, et al. 2004. POSaM: a fast, flexible, open source, ink-jet oligonucleotide synthesizer and microarrayer. *Genome Biol.* 5:R58

81. Gygi SP, Rist B, Gerber SA, Turecek F, Gelb MH, Aebersold R. 1999. Quantitative analysis of complex protein mixtures using isotope-coded affinity tags. *Nat Biotechnol.* 17:994–99

82. Eng JK, McCormack AL, Yates JR III. 1994. An approach to correlate tandem mass spectral data of peptides with amino acid sequences in a protein database. *J. Am. Soc. Mass Spectrom.* 5:976–89

83. Ewing B, Hillier L, Wendl MC, Green P. 1998. Base-calling of automated sequencer traces using phred. I. Accuracy assessment. *Genome Res.* 8:175–85

84. Ideker T, Galitski T, Hood L. 2001. A new approach to decoding life: systems biology. *Annu. Rev. Genomics Hum. Genet.* 2:343–72

85. Hood L, Galas DJ. 2003. The digital code of DNA. *Nature* 421:444–48

86. Ideker T, Thorsson V, Ranish JA, Christmas R, Buhler J, et al. 2001. Integrated genomic and proteomic analyses of a systematically perturbed metabolic network. *Science* 292:929–34

87. Weston AD, Baliga NS, Bonneau R, Hood L. 2003. Systems approaches applied to the study of *Saccharomyces cerevisiae* and *Halobacterium* sp. *Cold Spring Harbor Symp. Quant. Biol.* 68:345–57

88. Hood L, Galas DJ, Dewey G, Wilson J, Veras R. 2008. *Biology as an Informational Science and the Emergence of Systems Biology.* Greenwood Village, CO: Roberts & Co. In preparation

89. Ramsey SA, Smith JJ, Orrell D, Marelli M, Petersen TW, et al. 2006. Dual feedback loops in the GAL regulon suppress cellular heterogeneity in yeast. *Nat. Genet.* 38:1082–87

90. Gilchrist M, Thorsson V, Li B, Rust AG, Korb M, et al. 2006. Systems biology approaches identify ATF3 as a negative regulator of Toll-like receptor 4. *Nature* 441:173–78

91. Taylor RJ, Siegel AF, Galitski T. 2007. Network motif analysis of a multi-mode genetic-interaction network. *Genome Biol.* 8:R160

92. Bonneau R, Facciotti MT, Reiss DJ, Schmid AK, Pan M, et al. 2007. A predictive model for transcriptional control of physiology in a free living cell. *Cell* 131:1354–65

93. Facciotti MT, Reiss DJ, Pan M, Kaur A, Vuthoori M, et al. 2007. General transcription factor specified global gene regulation in archaea. *Proc. Natl. Acad. Sci. USA* 104:4630–35

94. Nesvizhskii AI, Vitek O, Aebersold R. 2007. Analysis and validation of proteomic data generated by tandem mass spectrometry. *Nat. Methods* 4:787–97

95. Klimek J, Eddes JS, Hohmann L, Jackson J, Peterson A, et al. 2008. The standard protein mix database: a diverse data set to assist in the production of improved peptide and protein identification software tools. *J. Proteome Res.* 7(1):96–103

96. Zhou Y, Aebersold R, Zhang H. 2007. Isolation of N-linked glycopeptides from plasma. *Anal. Chem.* 79:5826–37

97. Mallick P, Schirle M, Chen SS, Flory MR, Lee H, et al. 2007. Computational prediction of proteotypic peptides for quantitative proteomics. *Nat. Biotechnol.* 25:125–31

98. Bonneau R, Reiss DJ, Shannon P, Facciotti M, Hood L, et al. 2006. The inferelator: an algorithm for learning parsimonious regulatory networks from systems biology data sets de novo. *Genome Biol.* 7:R36

99. Shannon PT, Reiss DJ, Bonneau R, Baliga NS. 2006. The Gaggle: an open-source software system for integrating bioinformatics software and data sources. *BMC Bioinformatics* 7:176

100. Shannon P, Markiel A, Ozier O, Baliga NS, Wang JT, et al. 2003. Cytoscape: a software environment for integrated models of biomolecular interaction networks. *Genome Res.* 13:2498–504

101. Reiss DJ, Baliga NS, Bonneau R. 2006. Integrated biclustering of heterogeneous genome-wide datasets for the inference of global regulatory networks. *BMC Bioinformatics* 7:280

102. Longabaugh WJ, Davidson EH, Bolouri H. 2005. Computational representation of developmental genetic regulatory networks. *Dev. Biol.* 283:1–16

103. Ng WV, Ciufo SA, Smith TM, Bumgarner RE, Baskin D, et al. 1998. Snapshot of a large dynamic replicon in a halophilic archaeon: megaplasmid or minichromosome? *Genome Res.* 8:1131–41

104. Hood L, Heath JR, Phelps ME, Lin B. 2004. Systems biology and new technologies enable predictive and preventative medicine. *Science* 306:640–43

105. Heath JR, Phelps ME, Hood L. 2003. Nanosystems biology. *Mol. Imaging Biol. Official Publ. Acad. Mol. Imaging* 5:312–25

106. Weston AD, Hood L. 2004. Systems biology, proteomics, and the future of health care: toward predictive, preventative, and personalized medicine. *J. Proteome Res.* 3:179–96

107. Lin B, White JT, Lu W, Xie T, Utleg AG, et al. 2005. Evidence for the presence of disease-perturbed networks in prostate cancer cells by genomic and proteomic analyses: a systems approach to disease. *Cancer Res.* 65:3081–91

108. Price ND, Edelman LB, Lee I, Yoo H, Hwang D, et al. 2007. Systems biology and the emergence of systems medicine. In *Genomic and Personalized Medicine: From Principles to Practice*, ed. G Ginsburg, H Willard. San Diego: Academic. In press

109. Nelson PR, Goulter AB, Davis RJ. 2005. Effective analysis of genomic data. *Methods Mol. Med.* 104:285–312

110. Sun B, Ranish JA, Utleg AG, White JT, Yan X, et al. 2006. Shotgun glycopeptide capture approach coupled with mass spectrometry for comprehensive glycoproteomics. *Mol. Cell. Proteomics* 6:141–49

111. Griffin TJ, Gygi SP, Ideker T, Rist B, Eng J, et al. 2002. Complementary profiling of gene expression at the transcriptome and proteome levels in *Saccharomyces cerevisiae*. *Mol. Cell. Proteomics* 4:323–33

112. Hwang D, Smith JJ, Leslie DM, Weston AD, Rust AG, et al. 2005. A data integration methodology for systems biology: experimental verification. *Proc. Natl. Acad. Sci. USA* 102:17302–7

113. Hwang D, Rust AG, Ramsey S, Smith JJ, Leslie DM, et al. 2005. A data integration methodology for systems biology. *Proc. Natl. Acad. Sci. USA* 102:17296–301

114. Glusman G, Qin S, Raafat El-Gewely M, Siegel AF, Roach JC, et al. 2006. A third approach to gene prediction suggests thousands of additional human transcribed regions. *PLoS Comput. Biol.* 2:e18

115. Price ND, Trent J, El-Naggar AK, Cogdell D, Taylor E, et al. 2007. Highly accurate two-gene classifier for differentiating gastrointestinal stromal tumors and leiomyosarcomas. *Proc. Natl. Acad. Sci. USA* 104:3414–19

116. Huang GM, Farkas J, Hood L. 1996. High-throughput cDNA screening utilizing a low order neural network filter. *BioTechniques* 21:1110–14

117. Smith TM, Abajian C, Hood L. 1997. Hopper: software for automating data tracking and flow in DNA sequencing. *CABIOS* 13:175–82

118. Smith TM, Hood L. 1999. What are biologists going to do with all these data? *Math. Model. Sci. Comput.* 9:155–62

119. Siegel AF, van den Engh G, Hood L, Trask B, Roach JC. 2000. Modeling the feasibility of whole genome shotgun sequencing using a pairwise end strategy. *Genomics* 68:237–46

120. Bonneau R, Baliga NS, Deutsch EW, Shannon P, Hood L. 2004. Comprehensive de novo structure prediction in a systems-biology context for the archaea *Halobacterium* sp. NRC-1. *Genome Biol.* 5:R52

121. Facciotti MT, Bonneau R, Hood L, Baliga NS. 2004. Systems biology experimental design—considerations for building predictive gene regulatory network models for prokaryotic systems. *Curr. Genomics* 5:1389–2029

122. Ideker T, Thorsson V, Seigel AF, Hood L. 2000. Testing for differentially expressed genes by maximum-likelihood analysis of microarray data. *J. Comput. Biol.* 7:805–17

123. Thorsson V, Hörnquist M, Siegel AF, Hood L. 2005. Reverse engineering galactose regulation in yeast through model selection. *Stat. Appl. Genet. Mol. Biol.* 4:article 28

124. Davidson EH, McClay DR, Hood L. 2003. Regulatory gene networks and the properties of the developmental process. *Proc. Natl. Acad. Sci. USA* 100:1475–80

125. McIndoe RA, Stanford JL, Gibbs M, Jarvik GP, Brandzel S, et al. 1997. Linkage analysis of 49 high-risk families does not support a common familial prostate cancer–susceptibility gene at 1q24–25. *Am. J. Hum. Genet.* 61:347–353

126. Jarvik G, Stanford JL, Goode EL, Hood L, Ostrander EA. 1999. Confirmation of prostate cancer susceptibility genes using high-risk families. *J. Natl. Cancer Inst.* 26:81–87

127. Gibbs M, Stanford JL, McIndoe RA, Jarvik GP, Kolb S, et al. 1999. Evidence for a rare prostate cancer–susceptibility locus at chromosome 1p36. *Am. J. Hum. Genet.* 64:776–87

128. Goode E, Stanford JL, Chakrabarti L, Gibbs M, Kolb S, et al. 2000. Linkage analysis of 150 high-risk prostate cancer families at 1q24–25. *Genet. Epidemiol.* 18:251–75

129. Gibbs M, Stanford JL, Jarvik GP, Janer M, Badzioch M, et al. 2000. A genomic scan of families with prostate cancer identifies multiple regions of interest. *Am. J. Hum. Genet.* 67:100–9

130. Janer M, Friedrichsen DM, Stanford JL, Badzioch MD, Kolb S, et al. 2003. Genomic scan of 254 hereditary prostate cancer families. *Prostate* 57:309–19

131. Friedrichsen DM, Stanford JL, Isaacs SD, Janer M, Chang BL, et al. 2004. Identification of a prostate cancer susceptibility locus on chromosome 7q11–21 in Jewish families. *Proc. Natl. Acad. Sci. USA* 101:1939–44

132. Pierce BL, Friedrichsen-Karyadi DM, McIntosh L, Deutsch K, Hood L, et al. 2007. Genomic scan of 12 hereditary prostate cancer families having an occurrence of pancreas cancer. *Prostate* 67:410–15

133. Johanneson B, Deutsch K, McIntosh L, Friedrichsen-Karyadi DM, Janer M, et al. 2007. Suggestive genetic linkage to chromosome 11p11.2-q12.2 in hereditary prostate cancer families with primary kidney cancer. *Prostate* 67:732–42

134. Hood L, Lange PH. 1997. The coming revolution in urology. *Contemp. Urol.* 9:33–50

135. Hood L, Lange PH. 1997. Preparing for the urologic revolution. *Contemp. Urol.* 9:39–58

136. Liu AY, True LD, LaTray L, Nelson PS, Ellis WJ, et al. 1997. Cell-cell interaction in prostate gene regulation and cytodifferentiation. *Proc. Natl. Acad. Sci. USA* 94:10705–10

137. Liu AY, Corey E, Vessella RL, Lange PH, True LD, et al. 1997. Identification of differentially expressed prostate genes: increased expression of transcription factor ETS-2 in prostate cancer. *Prostate* 30:145–53

138. Nelson P, Ng W-L, Schummer M, True LD, Liu AY, et al. 1998. An expressed-sequence-tag database of the human prostate: sequence analysis of 1,168 cDNA clones. *Genomics* 47:12–25

139. Hawkins V, Doll D, Bumgarner R, Smith T, Abajian C, et al. 1999. PEDB: the prostate expression database. *Nucleic Acids Res.* 27:204–8

140. Nelson PS, Hawkins V, Schummer M, Bumgarner R, Ng W, et al. 1999. Negative selection: a method for obtaining low-abundance cDNAs using high-density cDNA clone arrays. *Genet. Anal. Biomol. Eng.* 15:209–15

141. Lin B, White JT, Ferguson C, Bumgarner R, Friedman C, et al. 2000. PART-1: a novel human prostate–specific, androgen-regulated gene that maps to chromosome 5q121. *Cancer Res.* 60:858–63

142. Liu AY, Nelson PS, van den Engh G, Hood L. 2002. Human prostate epithelial cell-type cDNA libraries and prostate expression patterns. *Prostate* 50:92–103

143. Nelson PS, Clegg N, Arnold H, Ferguson C, Bonham M, et al. 2002. The program of androgen-responsive genes in neoplastic prostate epithelium. *Proc. Natl. Acad. Sci. USA* 99:11890–95

144. Lin B, White JT, Lu W, Xie T, Utleg AG, et al. 2005. Evidence for the presence of disease-perturbed networks in prostate cancer cells by genomic and proteomic analyses: a systems approach to disease. *Cancer Res.* 65:3081–91
145. Bailey RC, Kwong GA, Radu CG, Witte ON, Heath JR. 2007. DNA-encoded antibody libraries: a unified platform for multiplexed cell sorting and detection of genes and proteins. *J. Am. Chem. Soc.* 129:1959–67

Spectroscopic and Statistical Techniques for Information Recovery in Metabonomics and Metabolomics

John C. Lindon and Jeremy K. Nicholson

Department of Biomolecular Medicine, Faculty of Medicine, Imperial College London, London SW7 2AZ, United Kingdom; email: j.nicholson@imperial.ac.uk

Annu. Rev. Anal. Chem. 2008. 1:45–69

First published online as a Review in Advance on December 20, 2007

The *Annual Review of Analytical Chemistry* is online at anchem.annualreviews.org

This article's doi:
10.1146/annurev.anchem.1.031207.113026

Key Words

nuclear magnetic resonance, mass spectrometry, chromatography, chemometrics, multivariate statistics, systems biology

Abstract

Methods for generating and interpreting metabolic profiles based on nuclear magnetic resonance (NMR) spectroscopy, mass spectrometry (MS), and chemometric analysis methods are summarized and the relative strengths and weaknesses of NMR and chromatography-coupled MS approaches are discussed. Given that all data sets measured to date only probe subsets of complex metabolic profiles, we describe recent developments for enhanced information recovery from the resulting complex data sets, including integration of NMR- and MS-based metabonomic results and combination of metabonomic data with data from proteomics, transcriptomics, and genomics. We summarize the breadth of applications, highlight some current activities, discuss the issues relating to metabonomics, and identify future trends.

1. INTRODUCTION

NMR: nuclear magnetic resonance

MS: mass spectrometry

Metabonomics: the quantitative measurement of the dynamic multiparametric metabolic response of living systems to pathophysiological stimuli or genetic modification

Metabolomics: a comprehensive analysis in which all the metabolites of a biological system are identified and quantified

Molecular biology has mainly centered on the determination of multiple gene expression changes, either between subjects or following drug treatment or other interventions (termed transcriptomics). Much effort has also been put into determination of multiple protein expression changes in a cell or tissue (termed proteomics). However, the main problem with interpreting transcriptomic and proteomic data is the difficulty of relating observed gene expression fold changes or protein level (not activity) changes to conventional disease and pharmaceutically relevant endpoints.

At the metabolic (small-molecule) level, similar developments have been taking place. Many years before the development of the various "-omics" approaches, simultaneous analysis of the plethora of metabolites seen in biological fluids had been carried out largely using nuclear magnetic resonance (NMR) spectroscopy (1) and mass spectrometry (MS) (2), and these complex data sets were interpreted in detail using multivariate statistics (3, 4). This approach to understanding the metabolic responses of complex systems to some sort of stimulus was christened metabonomics and was carefully defined with respect to the measurement of biological effects (5). Metabonomics encompasses the comprehensive and simultaneous systematic profiling of metabolite levels and their systematic and temporal changes in whole organisms through effects such as diet, lifestyle, environment, genetic effects, and pharmaceutical effects both beneficial and adverse, and is achieved through the study of biofluids and tissues (5, 6). A parallel approach arose in the late 1990s. Mainly from plant and microbial sciences and originally from the study of in vitro cellular systems, this approach has led to the coining of the similar term metabolomics (7). This term has a broadly analytical definition, but the methods and approaches used for cells, plants, and animals are now highly convergent. The main techniques and applications of both approaches have been reviewed in a recent book (8).

As discussed below, there is a need to integrate information at the transcriptomic, proteomic, and metabonomic levels to provide a full systems biology understanding; this goal has been, at best, only partially fulfilled. For example, environmental and lifestyle effects have a large effect upon all levels of molecular biology. Animals, including humans, can be considered "superorganisms" with an internal ecosystem of diverse symbiotic gut microflora (often with unknown genomes and functional ecologies) whose metabolic processes interact with the host. The complexity of mammalian biological systems and the diverse features that need to be measured to allow -omics data to be fully interpreted have been reviewed recently (9), and novel approaches are required to measure and model such cometabolic processes (10).

Metabonomic studies generally use biofluids or cell or tissue extracts, which are usually readily available. Urine and plasma are obtained essentially noninvasively, and hence can be obtained more easily for use in disease diagnosis and in clinical trials for monitoring drug therapy. However, many other fluids have been studied, including seminal fluids, amniotic fluid, cerebrospinal fluid, synovial fluid, digestive fluids, blister and cyst fluids, lung aspirates, and dialysis fluids (8). In addition, numerous metabonomics studies have analyzed tissue biopsy samples and their lipid and aqueous extracts, as well as in vitro cell systems such as Caco-2 cells (11), model systems such as yeast (12), tumor cells, and spheroids (13).

UPLC: ultraperformance liquid chromatography

2. ANALYTICAL TECHNOLOGIES FOR METABONOMICS AND METABOLOMICS

2.1. Introduction

The main analytical techniques employed for metabonomic studies are based on NMR spectroscopy and MS. This is for the good reason that both technologies can deliver "high-density" spectroscopic/structural information on a wide range of compound classes and chemistries simultaneously with high analytical precision. The use of MS requires a preseparation of the metabolic components using either gas chromatography (GC) after chemical derivatization or liquid chromatography (LC), with the newer method of ultraperformance LC (UPLC) also being used increasingly. Additionally, the use of capillary electrophoresis (CE) coupled to MS has also shown some promise. Other more specialized techniques, such as Fourier transform infrared spectroscopy and arrayed electrochemical detection, have been used in some cases. The main limitation of the use of these latter techniques is the low level of detailed molecular identification that can be achieved, and, as a result, MS is also employed for metabolite identification.

All metabonomic studies yield complex multivariate data sets; these usually require visualization software and chemometric and bioinformatic methods for interpretation and production of biochemical fingerprints that are of diagnostic or other classification value. The next step is to identify the substances causing the diagnosis or classification, as these are the combination of biomarkers that define the biological or clinical situation.

2.2. Nuclear Magnetic Resonance Spectroscopy

NMR spectroscopy provides detailed information on molecular structure, both for pure compounds and in complex mixtures, but it can also be used to probe metabolite molecular dynamics and mobility through the interpretation of NMR spin relaxation times and by the determination of molecular diffusion coefficients (14). Commercially available instruments up to an observation frequency for ^{1}H of 950 MHz are available (with higher frequencies on the way). For small-molecule studies, the increased sensitivity and dispersion that result from the higher magnetic fields are of great value and there are no disadvantages as there could be in macromolecule studies.

Automatic sample preparation involves simply buffering and adding D_2O as a magnetic field lock signal for the spectrometer; standard NMR spectra typically take only a few minutes to acquire using robotic flow-injection methods. For large-scale studies, bar-coded vials containing the biofluid can be used, and the contents can be transferred into 96-well plates under LIMS system control and prepared for analysis using robotic liquid-handling technology. Currently, the use of these approaches allows well over 100 samples per day to be be measured on one spectrometer.

Alternatively, for more precious samples or for those of limited volume, conventional 5-mm or capillary NMR tubes are usually used, either individually or with a variety of commercially available sample tube changers with automatic data acquisition. The large NMR signal that arises from water in all biofluids is easily eliminated with the use of appropriate standard NMR solvent suppression methods. Absolute concentrations can be obtained if the sample contains an added internal standard of known concentration, if a standard addition of the analyte of interest is added to the sample, or if the concentration of a substance is determined by independent means (e.g., glucose in plasma can be quantified by a conventional biochemical assay).

The ^{1}H NMR spectra of urine show thousands of sharp peaks from predominantly small-molecule metabolites, whereas spectra of blood plasma and serum show broad bands from protein and lipoprotein signals, with sharp peaks from small molecules superimposed thereon (15). Standard NMR pulse sequences, where the observed peak intensities are edited on the basis of molecular diffusion coefficients or NMR relaxation times (such as the Carr-Purcell-Meiboom-Gill spin-echo sequence), can be used to select only the contributions from macromolecules, or alternatively to select only the signals from the small molecule metabolites. A typical 950-MHz ^{1}H NMR spectrum of urine showing the degree of spectral complexity is given in **Figure 1**. Most of the major peaks shown in this image have now been assigned (16).

The development of cryogenic probes wherein the detector coil and preamplifier (but not the samples) are cooled to around 20° K has improved spectral signal-to-noise ratios by up to a factor of five by reducing the thermal noise from the electronics of the spectrometer. Conversely, because the NMR signal-to-noise ratio is proportional to the square root of the number of co-added scans, shorter (by up to a factor of 25) data acquisition times become possible for the same amount of sample. Using NMR spectroscopy of biofluids to detect the much less sensitive ^{13}C nuclei, which have a natural abundance of 1.1%, also becomes possible because of the increase in signal-to-noise ratio (17). This technology also makes the use of tissue-specific microdialysis samples more feasible (18).

Two-dimensional (2D) NMR spectroscopy is useful for increasing signal dispersion and for elucidating the connectivities between signals, thus helping to identify metabolites. These include the ^{1}H-^{1}H 2D J-resolved experiment, which attenuates the peaks from macromolecules and yields information on the multiplicity and coupling patterns of resonances. A projection of such a spectrum onto the chemical shift axis yields a fingerprint of peaks from only the most highly mobile small molecules, with all spin-coupling peak multiplicities removed. Other 2D experiments [e.g., correlation spectroscopy (COSY) and total correlation spectroscopy (TOCSY)] provide ^{1}H-^{1}H spin-spin coupling connectivities, providing information as to which hydrogens in a molecule are close in chemical bond terms. Use of other types of nuclei, such as naturally abundant ^{13}C or ^{15}N, or, where present, ^{31}P, can be important in helping to assign NMR peaks through inverse-detected heteronuclear correlation NMR experiments. In these experiments, the lower-sensitivity or less-abundant nucleus NMR spectrum (such as ^{13}C) is detected indirectly using the more sensitive or abundant nucleus (^{1}H) by utilizing spin-spin interactions such as the one-bond ^{13}C-^{1}H spin-spin coupling. These yield both ^{1}H and ^{13}C NMR chemical shifts of

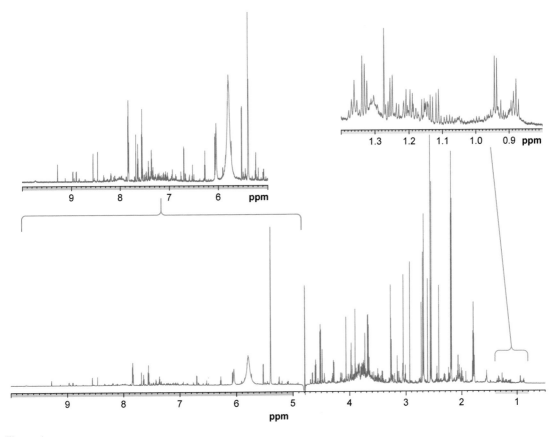

Figure 1

A 950-MHz ^{1}H nuclear magnetic resonance spectrum of human urine with expansions showing the degree of spectral complexity.

CH, CH$_2$, and CH$_3$ groups, which are useful for identification purposes (19). There is also a sequence that allows correlation of protons to quaternary carbons based on long-range ^{13}C-^{1}H spin-spin coupling.

Using a technique called high-resolution ^{1}H magic angle spinning (MAS) NMR spectroscopy, it is possible to acquire high-resolution NMR data on small pieces of intact tissues with no pretreatment (8, 20, 21). Rapid spinning of the sample (typically at ~4–6 kHz) at an angle of 54.7° relative to the applied magnetic field serves to reduce the loss of information caused by the line-broadening effects seen in nonliquid samples such as tissues (such effects are caused by sample heterogeneity and residual anisotropic NMR parameters that are normally averaged out in free solution where molecules can tumble isotropically and rapidly). MAS NMR spectroscopy requires straightforward, but manual, sample preparation. NMR spectroscopy on a tissue sample in an MAS experiment is the same as solution-state NMR, and all common-pulse techniques can be employed in order to study metabolic changes and to perform molecular structure elucidation and molecular dynamics studies.

2.3. Mass Spectrometry

MS, including tandem MS methods, has also been widely used in metabolic finger-printing and metabolite identification. Although most studies to date have been on plant extracts and model cell system extracts, the application of MS to mammalian studies is becoming more common. In general, a prior separation of the complex mixture sample using chromatography is required. MS is inherently much more sensitive than NMR spectroscopy, but it is generally necessary to employ different separation techniques (e.g., different LC column packings) for different classes of substances. Analyte quantitation by MS in complex mixtures of highly variable composition can be impaired by variable ionization and ion suppression effects. For plant metabolic studies, most investigations have used chemical derivatization to ensure volatility and analytical reproducibility, followed by GC-MS analysis. Some approaches using MS rely on more targeted studies, for example a detailed analysis of lipids (22).

For the application of metabonomics to biofluids such as urine, a high-performance liquid chromatography (HPLC) chromatogram is generated with MS detection, usually utilizing electrospray ionization, and both positive and negative ion chromatograms. Typically a time-of-flight (TOF) instrument is used, but other studies have used ion cyclotron resonance (also known as Fourier transform) MS (FT-MS). At each sampling point in the chromatogram, there is a full mass spectrum and so the data are three dimensional in nature (i.e., retention time, mass, and intensity).

The recently introduced technique known as UPLC is a combination of a 1.7-μm reversed-phase packing material and a chromatographic system operating at around 12,000 psi. UPLC provides approximately a ten-fold increase in speed and a three- to five-fold increase in sensitivity compared to a conventional stationary phase. Because of the much-improved chromatographic resolution of UPLC, the problem of ion suppression from co-eluting peaks is greatly reduced. A typical UPLC-MS chromatogram from a rat serum extract is shown in **Figure 2**; also shown for comparison is a chromatogram from a conventional HPLC-MS trace.

More recently, capillary electrophoresis coupled to mass spectrometry has also been explored as a suitable technology for metabonomics studies (23). Using this technique, metabolites are first separated by CE based on their charge and size and then selectively detected using MS monitoring. This method has been used to measure 352 metabolic standards and has been employed for the analysis of 1692 metabolites from *Bacillus subtilis* extracts, revealing changes in metabolite levels during the bacterial growth.

2.4. Hyphenated Systems

For biomarker identification, it is also possible to separate out substances of interest on a larger scale from a complex biofluid sample using techniques such as solid-phase-extraction or HPLC. For metabolite identification, directly coupled chromatography–NMR spectroscopy methods can also be used. The most general of these "hyphenated" approaches is HPLC-NMR-MS (24), in which the eluting HPLC peak is split, with parallel analysis by directly coupled NMR and MS techniques.

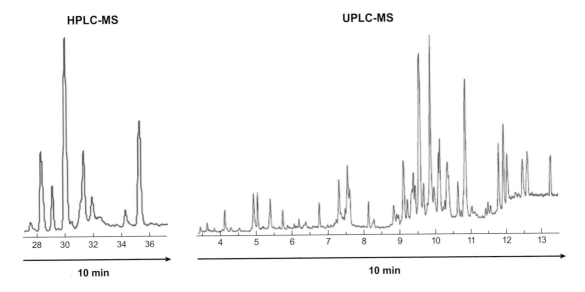

HPLC-MS

UPLC-MS

28 30 32 34 36

4 5 6 7 8 9 10 11 12 13

10 min

10 min

Figure 2

A comparison of high-performance liquid chromatography (HPLC)–mass spectrometry (MS) (*left*) and ultraperformance liquid chromatography (UPLC)–MS (*right*) chromatograms showing a 10-m segment derived from a methanolic extract of rat serum. Reproduced courtesy of E. Want, Imperial College London.

HPLC-NMR-MS can be operated in on-flow, stopped-flow, and loop-storage modes and thus can provide the full array of NMR- and MS-based molecular identification tools. These include 2D NMR spectroscopy as well as MS-MS for identification of fragment ions and FT-MS or TOF-MS for accurate mass measurement and hence derivation of molecular empirical formulae.

2.5. Chemometric Methods

To achieve optimal characterization of the samples and enable efficient biomarker detection, GC-MS and HPLC-MS data sets particularly, but also NMR data sets generally, require several steps of data preprocessing prior to any multivariate analysis. Considerable effort has gone into the development of algorithms to align signals from the same compound in data acquired from different samples (25, 26). This problem has been partially solved for NMR spectroscopic data (which can be subject to peak position variation problems) by dividing the spectrum into frequency windows and integrating signal intensity within these segments (27, 28), or by means of advanced principal components analysis (PCA) techniques (29). Analogous approaches have been applied to MS data. In one study, a series of prominent marker peaks (30) was used to divide chromatograms into time windows in which the retention time scale was then adjusted piecewise by an interpolation scheme. Another approach (31) divided chromatograms into time windows separated by regions showing a baseline response in all samples, constructed the equivalent direct-injection mass spectra, and then, using alternating regression, extracted the significantly varying m/z values. For NMR spectroscopy, the need to segment the data in this way has largely been eliminated by the use of correlation techniques (see Section 4 below).

The data point intensities in an NMR spectrum or in an LC-MS data set can be considered as a multidimensional graph of metabolic coordinates; the spectrum is therefore a point in a multidimensional metabolic hyperspace. The initial objective in metabonomics is primarily to classify a sample based on identification of its detailed spectral patterns and then to identify those metabolic features responsible for the classification. The approach can also be used for reducing the dimensionality of complex data sets, for example by 2D or 3D mapping procedures, to enable easy visualization of any clustering of the various samples. Alternatively, in what are known as supervised methods, multiparametric data sets can be modeled so that the class of separate samples (known as a validation set) can be predicted based on a series of mathematical models derived from the original data (the training set).

One simple technique used extensively in metabonomics is PCA (32). Conversion of the data matrix to principal components results in two matrices known as scores and loadings. Scores, the linear combinations of the original variables, represent the coordinates for the samples in the established model and may be regarded as new latent variables. In a scores plot, each point represents a single sample spectrum. The PC loadings define the way in which the old variables are linearly combined to form the new variables and indicate those variables carrying the greatest weight in transforming the position of the original samples from the data matrix into their new position in the scores matrix. In the loadings plot, each point represents a different spectral intensity. Thus, the cause of any spectral clustering observed in a PC scores plot is interpreted by examination of the loadings that cause any cluster separation. In addition, there are many other visualization (i.e., unsupervised) methods, such as nonlinear mapping and hierarchical cluster analysis.

One widely used supervised method (i.e., using a training set of data with known endpoints) is partial least squares (PLS) (32). This method relates a data matrix containing independent variables from samples, such as spectral intensity values (an X matrix) to a matrix containing dependent variables, such as measurements of response, for those samples (a Y matrix). PLS can also be combined with discriminant analysis (DA) to establish the optimal position in which to place the discriminant surface that best separates classes. It is possible to use such supervised models to provide classification probabilities and quantitative response factors for a wide range of sample types, but given the strong possibility of chance correlations when the number of descriptors is large, it is important to build and test such chemometric models using independent training data and validation data sets. Orthogonal signal correction (OSC) can be used to remove irrelevant parts of the data that are uncorrelated with the endpoints; OSC has been integrated into the PLS algorithm for optimum use (33).

PCA and PLS use linear combinations of parameters for dimension reduction or classification, but other methods do exist. For example, in neural network analysis a training set of data is used to develop algorithms, which "learn" the structure of the data and can cope with complex functions. Recently, probabilistic neural networks, which represent an extension to the approach, have shown promise for metabonomics applications in toxicity (34). Other approaches currently being tested include genetic algorithms (35), machine learning (36), and Bayesian modeling (37).

STOCSY: statistical total correlation spectroscopy

3. THE STRENGTHS AND WEAKNESSES OF NUCLEAR MAGNETIC RESONANCE AND MASS SPECTROMETRY FOR METABOLIC PROFILING

Although both NMR spectroscopy and mass spectrometry have been widely used in metabolic profiling studies, each has its champions and often the real benefits of each are not wholly appreciated. **Table 1** presents a comparison of these two highly complementary approaches.

4. THE CONCEPT OF STATISTICAL SPECTROSCOPY AND ITS IMPLEMENTATION

A recent development has been the implementation of the statistical total correlation spectroscopy (STOCSY) analysis method for aiding the identification of potential biomarker molecules in metabonomic studies based on NMR spectroscopic data (38). STOCSY takes advantage of the multicollinearity of the intensity variables in a set of spectra (e.g., ^{1}H NMR spectra) to generate a pseudo-two-dimensional NMR spectrum that displays the correlation among the intensities of the various peaks across the whole sample. This method is not limited to the usual connectivities that are deducible from more standard 2D NMR spectroscopic methods, such as TOCSY. Moreover, two or more molecules involved in the same pathway can also present high intermolecular correlations because of biological covariance due to common pathway control or can even be anticorrelated for the same reason.

A combination of STOCSY with supervised pattern recognition, particularly orthogonal projection on latent structure–discriminant analysis (OPLS-DA), offers a powerful new framework for the analysis of metabonomic data. First, OPLS-DA extracts the parts of NMR spectra related to discrimination. This information is then cross-combined with the STOCSY results to help identify the molecules responsible for the metabolic variation. This method has been applied to ^{1}H NMR spectra of urine from a metabonomic study of a model of insulin resistance based on the administration of a carbohydrate diet to three different mice strains, in which a series of metabolites of biological importance was conclusively assigned and identified (38).

The background to this method was introduced by Sasic et al. (39) and is based on another method, proposed by Noda, for generalized 2D correlation spectroscopy (40, 41, 42, 43). Successful previous applications of this correlative approach include infrared, Raman, near-infrared, and fluorescence spectroscopies (41, 42), with which correlations between different spectral features were identified.

STOCSY is based on the properties of the correlation matrix **C**, computed from a set of sample spectra according to where **X**1 and **X**2 denote the autoscaled experimental matrices of $n \times v1$ and $n \times v2$, respectively; n is the number of spectra (one for each sample) and $v1$ and $v2$ are the number of variables in the spectra for each matrix. **C** is therefore a matrix of $v1 \times v2$, where each value is a correlation

Table 1 The relative strengths and weaknesses of nuclear magnetic resonance and mass spectrometry for metabolic profiling[a]

	NMR	MS
Detection limits	Low-micromolar at typical observation frequencies (600 MHz), but nanomolar using cryoprobes	Picomolar with standard techniques, but can be much lower with special techniques
Universality of metabolite detection	If metabolite contains hydrogens it will be detected, assuming the concentration is sufficient or protein binding does not cause marked line broadening	Usually needs a more targeted approach. There can be problems with poor chromatographic separation; with the loss of metabolites in void volumes; with ion suppression (but this is reduced when using UPLC); lack of ionization; ability to run both +ve and −ve ion detection gives extra information
Sample handling	Whole sample analyzed in one measurement	Different LC packings and conditions for different classes of metabolite; usually samples have to be extracted into a suitable solvent; samples have to be aliquoted but some recent studies have avoided the need for chromatography
Amount of sample used	Typically 200–400 μL, but much less for microcoil probes, down to 5–10 μL	Low μL range
Sample recovery	Technique is nondestructive	Technique is destructive but only small amounts used
Analytical reproducibility	Very high	Fair
Sample prepreparation	Minimal: addition of buffer, D_2O and chemical shift reference (not always required)	Can be substantial; often needs different LC columns and protein precipitation
Ease of molecular identification	High, both from databases of authentic material and by self-consistent analysis of 1D and 2D spectra	Difficult, often only the molecular ion is available; this needs extra experiments, such as routine tandem MS; GC-MS is generally better with accurate retention times and comprehensive databases of spectra
Time to collect basic data	5 min for 1D ^1H NMR	10 min for UPLC-MS run
Quantitation	1–5%	5% intraday and interday is now common with or without prior chromatography
Robustness of instruments	High	Low
Molecular dynamics information	Yes, from T1, T2 relaxation time and diffusion coefficient measurements	No
Analysis of tissue samples	Yes, using MAS NMR	No
Availability of databases	Not yet comprehensive but increasing; several are available freely on the web; some commercial products also exist	Comprehensive databases for electron impact MS allow spectral comparisons; For electrospray ionization, as is usual in LC-MS, only mass values can be compared

[a]Abbreviations: GC, gas chromatography; LC, liquid chromatography; MAS, magic angle spinning; MS, mass spectrometry; NMR, nuclear magnetic resonance; UPLC; ultraperformance liquid chromatography.

coefficient between two variables of the matrices $\mathbf{X}1$ and $\mathbf{X}2$. The simplest case is the autocorrelation analysis where $\mathbf{X}1 = \mathbf{X}2$. Because the different resonance intensities from a single molecule will always have the same ratio, if the spectrometer conditions are kept identical between samples the relative intensities will theoretically be totally correlated (correlation coefficient $r = 1$). In real samples of biofluids, r will always be <1 because of spectral noise or peak overlaps from other molecules. In practice, however, the correlation matrix from a set of spectra containing different amounts of the same molecule shows very high correlations between the variables corresponding to the resonances of the same molecule. Plotting the correlation matrix provides a graphic representation of the multi-sample spectroscopic data set comparable to that of a 2D correlation NMR experiment conducted on one sample containing all the molecules of all the samples.

The method is not restricted to ^{1}H-^{1}H NMR correlation, but can be applied to different nuclei. If these include different NMR-active nuclei (^{13}C-^{13}C, ^{1}H-^{13}C, ^{13}C-^{31}P, etc.), then heteronuclear correlation is also possible and will yield novel molecular connectivity information using both types of nuclear spin properties. We have recently applied this approach to correlation of ^{1}H and ^{31}P MAS NMR spectra of tissues (44). An illustration of the correlation between ^{1}H and ^{31}P MAS NMR spectra of rat liver tissues, following administration of a model liver toxin galactosamine, is shown in **Figure 3** (44). The one-dimensional ^{1}H and ^{31}P correlation plots constructed from the two-dimensional STOCSY cross-peak at heteronuclear chemical shifts of δ_{31P} 12.772 and δ_{1H} 5.9822 are given in **Figure 3a,b,c**. The two high-field ^{31}P resonances correlate to a host of ^{1}H resonances and the ^{31}P spectral metabolites have been assigned to UDP-galNAc and UDP-glcNAc. These STOCSY-led assignments have been confirmed via spiking of standard compounds into an aqueous extract of liver.

Finally, it should be noted that STOCSY can be used to derive NMR spectral splittings and J couplings with the same theoretical precision as the 1D spectral properties from which the 2D data set was derived; this derivation is not limited by the generally lower resolution in the $F1$ domain of most correlation 2D experiments, provided, of course, that any physicochemical environment variation between samples does not induce variation of the peak positions.

An approach to enhancing information recovery from cryogenic probe on-flow LC-NMR spectroscopic analyses of complex biological mixtures has been demonstrated using a variation on the STOCSY method (45). Cryoflow probe technology enables more sensitive and hence faster NMR detection of metabolites using on-flow HPLC-NMR or UPLC-NMR, and the rapid spectral scanning allows multiple spectra to be collected over unresolved chromatographic peaks that can contain several species with similar, but nonidentical, retention times. This enables the identification of ^{1}H NMR signal connectivities between close-eluting metabolites, resulting in a "virtual" chromatographic resolution enhancement visualized directly in the NMR spectral projection. This approach is of wide general applicability to any complex mixture analysis problem involving chromatographic peak overlap, with particular application in metabolomics and metabonomics for identifying both endogenous and xenobiotic metabolites.

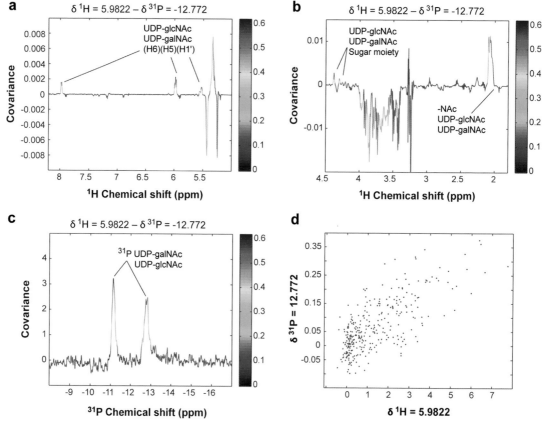

Figure 3

The use of statistical total correlation spectroscopy (STOCSY) to aid identification of linked nuclear magnetic resonance (NMR) peaks from ^{1}H and ^{31}P magic-angle-spinning NMR spectra of rat liver tissue. The ^{1}H and ^{31}P NMR chemical shifts for the anomeric proton and one of the two phosphorus atoms, respectively, are around 6 ppm and -12 ppm for both UDP-GlcNac and UDP-GalNac. The correlations from the peak pair at 5.982 and -12.77 ppm to aromatic protons are shown in panel a, and the correlation to aliphatic protons is shown in panel b. Panel c indicates that the intensities of the two high-field ^{31}P NMR peaks are highly correlated as expected, as both UDP-GlcNac and UDP-GalNac have two phosphorus atoms and the ^{31}P chemical shifts of these are at the same positions in both substances. The correlation of the intensities of the peaks at 5.98 and -12 ppm is given in panel d.

The complementarity of NMR and MS methods as structural tools has encouraged their parallel use in structure elucidation studies for natural product research, drug metabolite analysis, and other complex mixture analysis problems for many years. Typically, NMR and MS spectra of various types are examined together, and structural parameters such as chemical shifts and coupling constants (NMR) and exact molecular mass and fragmentation patterns (MS) are compared, often for a single sample, to generate structural assignment information that is consistent with the outputs of both technologies. However, in many metabonomic studies, multiple samples with a wide range of biochemical variation are available for both NMR and MS analysis, creating the opportunity for statistical analysis of signal amplitude covariation between technologies and direct cross-correlation of data for assignment purposes.

SHY: statistical heterospectroscopy

Statistical heterospectroscopy (SHY) is an extension of STOCSY for the coanalysis of multispectroscopic data sets acquired on multiple samples (46). This method also operates through the analysis of the intrinsic covariance between signal intensities in the same and related molecules measured by different techniques across cohorts of samples. The potential of SHY has been illustrated using both 600 MHz ^{1}H NMR and UPLC-TOF-MS data obtained from control rat urine samples and from a corresponding group treated with hydrazine, a model liver toxin (46). We have shown that direct cross-correlation of spectral parameters, e.g., chemical shifts from NMR and m/z data from MS, is readily achievable for a variety of metabolites, leading to improved efficiency of molecular biomarker identification. In addition to structural information, higher-level biological information can be obtained on metabolic pathway activity and connectivities by examination of different levels of the NMR-to-MS correlation and anticorrelation matrixes. The SHY approach can be used if two or more independent spectroscopic data sets are available for any sample cohort. This approach is of wide applicability and can be extended beyond NMR and LC-MS to include any spectroscopic, electrochemical, or other multivariate analytical measurements where multiple samples are measured by more than one technology. An example of the use of SHY to correlate NMR and MS data is given in **Figure 4** (46).

5. A DATA SPACE REDUCTION APPROACH FOR IMPROVING THE SEARCH FOR BIOMARKERS

One of the major problems with interpreting metabonomic UPLC-MS data sets is the sheer number of variables to consider. A strategy for biomarker recovery from such data sets from biofluids has been presented and exemplified using a study on hydrazine-induced liver toxicity (47). A key step in this strategy involves a novel procedure for reducing the spectroscopic search space by differential analysis of cohorts of normal and pathological samples using orthogonal projection-to-latent structures discriminant analysis (O-PLS-DA). This efficiently sorts principal discriminators of toxicity from the background of thousands of metabolic features commonly observed in the data sets generated by UPLC-MS analysis of biological fluids; it is thus a powerful tool for biomarker discovery.

The advent of UPLC has led to the detection of many thousands of metabolic features by mass spectrometry. Paradoxically, this creates a new statistical problem related to biomarker discovery: it becomes difficult to recover key discriminating biomarker information in the background of the thousands of detectable metabolic variables, the majority of which do not actually contribute to class separation or diagnosis. We have approached the processing of UPLC-MS data acquired on urine samples from an exemplar toxicology study by applying O-PLS-DA to matrices formed by binning the raw data over windows in both retention time and m/z. The sample

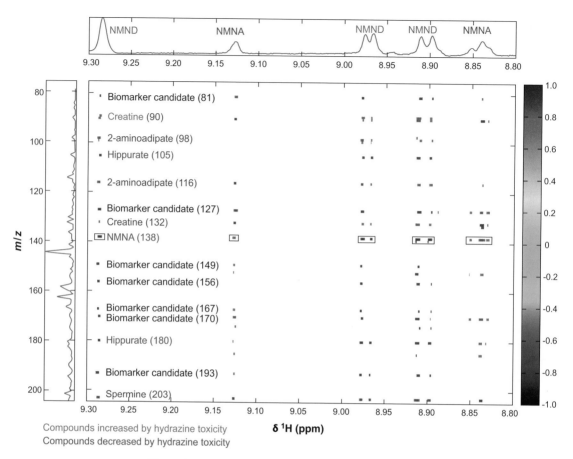

Compounds increased by hydrazine toxicity
Compounds decreased by hydrazine toxicity

δ ¹H (ppm)

Figure 4

The use of statistical heterospectroscopy to correlate ¹H nuclear magnetic resonance (NMR) spectral data with mass spectrometry (MS) *m/z* values. NMR-MS correlation for urine data from a group of rats that had been treated with the liver toxin hydrazine, expanded to show an N-methylnicotinic acid (NMNA)/N-methylnicotinamide (NMND) region of the ¹H NMR spectrum. The insets show mean NMR and mass spectra. All identified ions (with their *m/z* values in paretheses) show directions of correlation consistent with known effects of hydrazine. For example, NMNA and NMND are negatively correlated, as they are transaminase related (inhibited by hydrazine). Also, NMNA correlates positively with itself, as must be the case. These NMNA/NMND correlations are shown boxed. The newly identified spermine ion is shown to correlate positively with toxicity. All unidentified ions (also with their *m/z* values in paretheses) are also candidate biomarkers.

set analyzed contained urine specimens from both control animals and from those exhibiting a strong response to administration of the model toxin hydrazine, which elicits a steatotic effect in the liver (48). The aim of the study was to demonstrate the stratification of the many UPLC-MS metabolite peaks according to their relevance to toxicity (and, in principle, to any other pathological or disease state). In this way, subsequent metabolite identification and quantitation efforts were focused on the most important peaks. Use of this approach efficiently avoids the vexed problems of deconvolution or peak alignment between samples and the calculation of relative concentrations before applying the multivariate analysis, and allows significant peaks that are close to the detection limit to be recovered statistically without time-consuming inspection of individual spectra. We have shown that, given a suitable choice of window size, useful peak selection can be obtained rapidly and the peaks can be treated as biomarker candidates for further analysis and structure characterization.

The use of a cross-validated method like O-PLS-DA provides greater confidence than simple correlation analysis of variable against class. In addition, the method detects and models separately any variation that is orthogonal to class, so this can be separately analyzed and does not affect the class dependency results. The analytical strategy employed is summarized in **Figure 5**. Following the binning of data comes the selection step, which consists of filtration of variables by weight in an O-PLS-DA model; this yields a measure of correlation between peak intensity and dosage. The weights, color coded by value, are displayed in 2D diagrams labeled by retention time bin and *m/z* bin, which are expanded as required. Once the discriminating peaks are found, a library search is conducted to identify the possible compounds involved. Given that a single bin may contain more than one peak and that accurate masses and retention times are required, it is still necessary to interrogate the

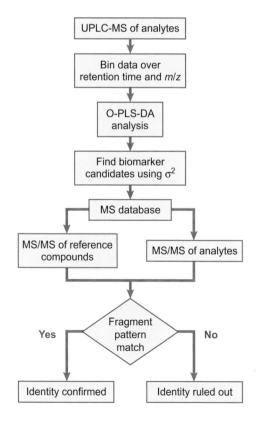

Figure 5

The procedure for data space reduction in the search for biomarkers separating two classes of sample.

original UPLC-MS spectra to find the exact values for the biomarker candidates. In this study (47), the library search against accurate m/z was done manually for the list of peaks of interest; however, this search may be automated. The final step compares MS/MS fragment patterns of pure compounds with urine samples from the study to either confirm or exclude candidate compounds highlighted by the search procedure.

6. RECENT MAMMALIAN APPLICATIONS AND TRENDS

Numerous studies have used metabonomics to characterize normal metabolic variation, caused by a range of inherent and external factors, in experimental animals such as mice and rats (49 and references therein). Such differences may help explain differential toxicity of drugs between strains and interanimal variation within a study. Many effects can be distinguished using NMR- and MS-based metabonomics, including male/female differences, wild-type and genetically modified animal models, age-related changes, estrus cycle effects in females, diet, diurnal effects, and interspecies differences and similarities (8, 49). Analogous studies have also been undertaken in humans (50, 51, 52). The importance of the symbiotic relationship between mammals and their gut microfloral populations has been recognized (9) and studied extensively using axenic animals (53), different animal colonies (54), parasitic infections in animals (55), and probiotics (56).

Minimizing the occurrence of drug adverse effects is one of the most important aims of pharmaceutical research and development, and the pharmaceutical industry is now embracing metabonomics for evaluating the adverse effects of candidate drugs (e.g., through the COMET project; see below). Metabonomics classification of the target organ or region of toxicity, the biochemical mechanism of that toxin, the identification of combination biomarkers of toxic effect, and evaluation of the time-course of the effect (e.g., the onset, evolution, and regression of toxicity) can all be determined (57, 58). The usefulness of metabonomics for the evaluation of xenobiotic toxicity effects has recently been comprehensively explored by the Consortium for Metabonomic Toxicology (COMET). COMET was formed among five pharmaceutical companies and Imperial College London (59) with the aim of developing methodologies for the acquisition and evaluation of metabonomic data generated using ^{1}H NMR spectroscopy of rat and mouse urine and blood serum for preclinical toxicological screening of candidate drugs. New methodologies for analyzing and classifying the complex data sets were developed. For example, because the predictive expert system takes into account the metabolic trajectory over time, a new way of comparing and scaling these multivariate trajectories was developed (60). Additionally, a novel classification method, termed Classification Of Unknowns by Density Superposition (CLOUDS), has been generated in order to identify the class of toxicity based on all of the NMR data for a given study (61).

COMET showed that it is possible to construct predictive and informative models of toxicity using NMR-based metabonomic data, delineating the whole time course of toxicity. This successful outcome is evidenced by the generated databases of spectral and conventional results for a wide range of model toxins (147 in total) that served as the basis for computer-based expert systems for toxicity prediction (62). The project's goals, namely the generation of comprehensive metabonomic databases (now containing approximately 35,000 × 600 MHz ^{1}H NMR spectra) and the creation of successful, robust multivariate statistical models (expert systems) for prediction of toxicity—initially for liver and kidney toxicity in the rat and mouse—have now been achieved, and the predictive systems and databases have been transferred to the sponsoring companies (63).

Many examples exist in the literature (reviewed in 8) of the use of NMR-based metabolic profiling to aid human disease diagnosis. A promising use of NMR spectroscopy of urine and plasma, as evidenced by the number of publications on the subject (8, chapter 14), is in the diagnosis of inborn errors of metabolism in children. In addition, tissues themselves can be studied by metabonomics through the MAS technique and there are many published examples (8, chapter 4, and references therein). In this field, MS is also a well-established technique and many diseases are routinely diagnosed using MS-based methods, but investigation of rare and unexpected diseases often benefits from an exploratory NMR spectroscopic approach.

The value of obtaining multiple NMR spectroscopic (or indeed other types of analysis) data sets from various biofluid samples and tissues of the same animals collected at different time-points has been demonstrated. This procedure, termed integrated metabonomics (6), can be used to describe the changes in metabolism in different body compartments affected by exposure to, for example, toxic drugs (64, 65). Such timed profiles in multiple compartments are themselves characteristic of particular types and mechanisms of pathology and can be used to give a more complete description of the biochemical consequences than can be obtained from one fluid or tissue alone.

Integration of metabonomic data with data from multivariate techniques in molecular biology, such as gene array experiments or proteomics, is also feasible. Thus, it is also possible to integrate data from transcriptomics and metabonomics in order to find, for example after acetaminophen administration to mice, common metabolic pathways implicated by both gene expression changes and changes in metabolism (66). It has also been found that evaluation of both transcriptomic and metabolic changes following administration of the toxin bromobenzene provides a more sensitive approach for detecting the effects of the toxin (67). Similarly, changes in gene expression detected in microarray experiments can lead to the identification of changed enzyme activity; this can also be achieved by analysis of metabolic perturbations (68).

A novel method of integrating proteomic and metabonomic data has been developed and applied to a human tumor xenograft mouse model of prostate cancer (69). Parallel 2D–difference gel electrophoresis proteomic data and ^{1}H NMR metabolic profile data were collected on blood plasma from mice implanted with a prostate cancer xenograft and from matched control animals. To interpret the xenograft-induced differences in proteomic and metabonomic plasma profiles, multivariate statistical algorithms, including OPLS, were applied to generate models characterizing the disease profile and to elucidate metabolite concentrations and protein abundances that could be directly related to the disease model.

Clearly, characterizing the relationships between genomic and phenotypic variation is an essential step in understanding disease processes. To this end, the first real transcriptomic-metabonomic cross-correlations have been achieved (70). Data sets derived from pathophysiological, proteomic, and transcriptomic profiling were used to map so-called quantitative trait loci (QTLs). Metabolic traits, as used in plant QTL studies, can be used to define phenotypes in mammalian genetics to characterize disease biomarkers. Untargeted plasma metabolic fingerprints derived from NMR spectroscopic analysis were mapped to chromosomes found in a rat strain derived from crossing diabetic and control animals. Identifying such metabotypic QTLs appears to be a practical approach to understanding genome-phenotype relationships in mammals and might uncover deeper biological complexity, such as extended genome (microbiome) perturbations that could affect disease processes through transgenomic effects.

7. THE FUTURE

Developments in NMR spectroscopy will continue to provide improvements in sensitivity from the use of cryogenic probes and increased magnetic field strengths. Improved identification of metabolites will come from inspection of databases of NMR spectra of standard substances. An increasing use of MS-based analyses in mammalian systems is expected and improvements in detection, sensitivity, and reliability will undoubtedly be achieved. Given the need for data sets to provide information on molecular identity, it is unlikely that techniques other than NMR and MS will find widespread use.

The main pharmaceutical areas in which metabonomics is already having an impact include validation of animal models of disease (as in genetically modified animals); preclinical evaluation of candidate drugs in safety studies; assessment of safety in humans in clinical trials; and, after product launch, quantitation or ranking of the beneficial effects of pharmaceuticals; improved understanding of the causes of highly sporadic idiosyncratic toxicity of marketed drugs; and patient stratification for clinical trials and drug treatment. In addition, in terms of disease studies, metabonomics is playing a role in improved, differential diagnosis and prognosis of human diseases, particularly for chronic and degenerative diseases and for diseases caused by genetic effects. A better understanding of large-scale human population differences through epidemiological studies is also being achieved. Other applications where major expansion is expected are nutritional studies; sports medicine and lifestyle studies, including the effects of diet, exercise, and stress; and evaluation of the effects of interactions among drugs and between drugs and diet.

If personalized healthcare is to become a reality, an individual's drug treatments must be tailored so as to achieve maximal efficacy and avoid adverse drug reactions. One of the long-term goals of pharmacogenomics is to understand the genetic makeup of different individuals (their genetic polymorphisms) and how well they are able to handle pharmaceuticals—this is important for identifying both beneficial and adverse effects. An alternative approach to understanding intersubject variability in response to drug treatment involves using a combination of multivariate metabolic profiling and chemometrics to predict the metabolism and toxicity of a dosed substance, based solely on the analysis and modeling of a predose metabolic profile (71). Unlike pharmacogenomics, this approach, termed pharmacometabonomics, is sensitive to both the genetic and environmental influences that determine the basal metabolic fingerprint of an individual, as these will also influence the outcome of a chemical intervention. This new approach has been illustrated with studies of the toxicity and metabolism of compounds, administered to rats, with very different modes of action. The next challenge is to adapt pharmacometabonomic approaches to humans, and although we are more metabolically variable than rats, so are our range of responses, meaning that pharmacometabonomics is likely to be successful in predictive drug metabolism and toxicity for at least some compound classes. Analogous to this application to pharmaceuticals, it has very recently been shown to be possible to distinguish individuals with different food preferences from their basal metabolic profiles of urine and blood plasma (72).

A major initiative is under way to make consensual recommendations for standardizing reporting arrangements for metabonomics studies. This research was initiated by the Standard Metabolic Reporting Structures group (**http://www.smrsgroup.org**), which produced a draft policy document covering all of the aspects of a metabolic study that are recommended for recording, including the origin of a biological sample, the analysis of material from that sample and chemometric and statistical approaches, and retrieval of information from the sample data; a summary publication has appeared (73). The various levels of (and consequent detail for) reporting needs, including journal submissions, public databases, and regulatory submissions, have also been addressed. In parallel, a scheme called ArMet for capturing data and metadata from metabolic studies has been proposed and developed (74). This has been followed up with a workshop and discussion meeting sponsored by the U.S. National Institutes of Health (75), from which the Metabolomics Standards Initiative was born (see **http://msi-workgroups.sourceforge.net**). This organization has now produced draft reports on many aspects of standardization and reporting in the subject and a number of papers have been published covering characterization of sample-related metadata, technical standards and related data, metadata and QC matters for the analytical instrumentation, data transfer methodologies, schema for the implementation of such activities, and development of standard vocabularies to enable transparent exchange of data (76).

8. CHALLENGES AND OPPORTUNITIES

In summary, it is clear that metabonomics will have an increasing and major impact. The analytical procedures used are stable and robust and have a high degree of reproducibility. Although advances will obviously be made in the future, current data will always be readable and interpretable. In contrast to other -omics, metabonomics enjoys a good level of biological reproducibility and the cost per sample and per analyte is low. It has the advantage of not having to preselect analytes, and through use of biofluids it is minimally invasive, creating the possibility of hypothesis-generation studies. Metabolic biomarkers are identifiable with real biological endpoints and provide a global systems interpretation of biological effects, including interactions between multiple genomes (such as those of humans and their gut microflora). One major potential strength of metabonomics is the possibility that metabolic biomarkers will be more easily used across species than transcriptomic or proteomic biomarkers; this flexibility should be important for pharmaceutical studies.

On the other hand, metabonomics suffers from the application of multiple analytical technologies, there are questions of the sensitivity and dynamic range of the technologies used, and the data sets are complex. Through the use of chemometrics, it is possible to overinterpret the data, but this is easily avoided by applying correct statistical rigor. Currently, the groups using metabonomics are moving towards defining standards for data and operations; a good start has been made, but there remains a need for the regulatory agencies to be trained in the data interpretation and for better-trained practitioners.

The reality of the complexity of disease and drug effects means that the use of biomarker combinations will increase; thus, there are many opportunities for metabonomics that have yet to be explored, such as its use in environmental toxicity studies, in directing the timing of transcriptomic and proteomic experiments, and for deriving theranostic biomarkers. Metabonomics will surely be an integral part of any multi-omics study where all the data sets are combined in order to derive an optimum set of biomarkers.

The ultimate goal of systems biology is the integration of data acquired from living organisms at the gene, protein, and metabolite levels. Hopefully, through the combination of transcriptomics, proteomics, and metabonomics, an improved understanding of an organism's total biology will result; subsequently, better understanding of the causes and progression of human diseases will ensue.

DISCLOSURE STATEMENT

The authors are not aware of any biases that might be perceived as affecting the objectivity of this review.

LITERATURE CITED

1. Nicholson JK, Wilson ID. 1989. High-resolution proton magnetic resonance spectroscopy of biological fluids. *Prog. NMR Spectrosc.* 21:449–501

2. van der Greef J, Tas AC, Bouwman J, Ten Noever de Brauw MC, Schreurs WHP. 1983. Evaluation of field-desorption and fast atom–bombardment mass-spectrometric profiles by pattern-recognition techniques. *Anal. Chim. Acta* 150:45–52

3. Gartland KPR, Sanins SM, Nicholson JK, Sweatman BC, Beddell CR, Lindon JC. 1990. Pattern recognition analysis of high resolution ^{1}H NMR spectra of urine: a nonlinear mapping approach to the classification of toxicological data. *NMR Biomed.* 3:166–72

4. Gartland KPR, Beddell CR, Lindon JC, Nicholson JK. 1991. The application of pattern recognition methods to the analysis and classification of toxicological data derived from proton NMR spectroscopy of urine. *Mol. Pharmacol.* 39:629–42

5. Nicholson JK, Lindon JC, Holmes E. 1999. "Metabonomics": understanding the metabolic responses of living systems to pathophysiological stimuli via multivariate statistical analysis of biological NMR spectroscopic data. *Xenobiotica* 29:1181–89

6. Nicholson JK, Connelly J, Lindon JC, Holmes E. 2002. Metabonomics: a platform for studying drug toxicity and gene function. *Nat. Rev. Drug Disc.* 1:153–62

7. Fiehn O. 2002. Metabolomics: the link between genotypes and phenotypes. *Plant Mol. Biol.* 48:155–71

8. Lindon JC, Nicholson JK, Holmes E, eds. 2007. *The Handbook of Metabonomics and Metabolomics*. Amsterdam: Elsevier

9. Nicholson JK, Wilson ID. 2003. Understanding 'global' systems biology: metabonomics and the continuum of metabolism. *Nat. Rev. Drug Disc.* 2:668–76

10. Nicholson JK, Holmes E, Lindon JC, Wilson ID. 2004. The challenges of modeling mammalian biocomplexity. *Nat. Biotech.* 22:1268–74

11. Lamers RJAN, Wessels ECHH, van der Sandt JJM, Venema K, Schaafsma G, et al. 2003. A pilot study to investigate effects of inulin on Caco-2 cells through in vitro metabolic fingerprinting. *J. Nutr.* 133:3080–84

12. Villas-Boas SG, Hojer-Pedersen J, Akesson M, Smedsgaard J, Nielsen J. 2005. Global metabolite analysis of yeast: evaluation of sample preparation methods. *Yeast* 22:1155–69

13. Bollard ME, Xu JS, Purcell W, Griffin JL, Quirk C, et al. 2002. Metabolic profiling of the effects of D-galactosamine in liver spheroids using ^{1}H NMR and MAS-NMR spectroscopy. *Chem. Res. Toxicol.* 15:1351–59

14. Liu M, Nicholson JK, Lindon JC. 1996. High resolution diffusion- and relaxation-edited one- and two-dimensional ^{1}H NMR spectroscopy of biological fluids. *Anal. Chem.* 68:3370–76

15. Nicholson JK, Foxall PJD, Spraul M, Farrant RD, Lindon JC. 1995. 750 MHz ^{1}H and ^{1}H-^{13}C NMR spectroscopy of human blood plasma. *Anal. Chem.* 67:793–811

16. Lindon JC, Nicholson JK, Everett JR. 1999. NMR spectroscopy of biofluids. In *Annual Reports on NMR Spectroscopy*, volume 38, ed. GA Webb, pp. 1–88

17. Keun HC, Beckonert O, Griffin JL, Richter C, Moskau D, et al. 2002. Cryogenic probe ^{13}C NMR spectroscopy of urine for metabonomic studies. *Anal. Chem.* 74:4588–93

18. Price KE, Vandaveer SS, Lunte CE, Larive CK. 2005. Tissue targeted metabonomics: metabolic profiling by microdialysis sampling and microcoil NMR. *J. Pharmaceut. Biomed. Anal.* 38:904–9

19. Claridge TDW. 2004. *High Resolution NMR Techniques in Organic Chemistry.* Amsterdam, Netherlands: Elsevier

20. Garrod SL, Humpfer E, Spraul M, Connor SC, Polley S, et al. 1999. High resolution magic angle spinning ^{1}H NMR spectroscopic studies on intact rat renal cortex and medulla. *Magn. Reson. Med.* 41:1108–18

21. Cheng LL, Chang IW, Louis DN, Gonzalez RG. 1998. Correlation of high-resolution magic angle spinning proton magnetic resonance spectroscopy with histopathology of intact human brain tumor specimens. *Cancer Res.* 58:1825–32
22. Morris M, Watkins SM. 2005. Focused metabolomic profiling in the drug development process: advances from lipid profiling. *Curr. Opin. Chem. Biol.* 9:407–12
23. Soga T, Ohashi Y, Ueno Y, Nasaoka H, Tomita M, Nishioka T. 2003. Quantitative metabolome analysis using capillary electrophoresis mass spectrometry. *J. Proteome Res.* 2:488–94
24. Lindon JC, Nicholson JK, Wilson ID. 2000. Directly-coupled HPLC-NMR and HPLC-NMR-MS in pharmaceutical research and development. *J. Chromatog. B* 748:233–58
25. Jonsson P, Johansson AI, Gullberg J, Trygg J, AJ, Grung B, et al. 2005. High-throughput data analysis for detecting and identifying differences between samples in GC/MS-based metabolomic analyses. *Anal. Chem.* 77:5635–42
26. Idborg-Bjorkman H, Edlund P-O, Kvalheim OM, Schuppe-Koistinen I, Jacobsson SP. 2003. Screening of biomarkers in rat urine using LC/electrospray ionization-MS and two-way data analysis. *Anal. Chem.* 75:4784–92
27. Farrant RD, Lindon JC, Rahr E, Sweatman BC. 1992. An automatic data reduction and transfer method to aid pattern-recognition analysis and classification of NMR spectra. *J. Pharm. Biomed. Anal.* 10:141–44
28. Holmes E, Foxall PJ, Nicholson JK, Neild GH, Brown SM, et al. 1994. Automatic data reduction and pattern-recognition methods for analysis of H-1 nuclear magnetic resonance spectra of human urine from normal and pathological states. *Anal. Biochem.* 220:284–96
29. Stoyanova R, Nicholls AW, Nicholson JK, Lindon JC, Brown TR. 2004. Automatic alignment of individual peaks in large high-resolution spectral data sets. *J. Magn. Reson.* 170:329–35
30. Yang J, Xu G, Zheng Y, Kong H, Wang C, et al. 2005. Strategy for metabonomics research based on high-performance liquid chromatography and liquid chromatography coupled with tandem mass spectrometry. *J. Chromatogr. A* 1084:214–21
31. Jonsson P, Bruce SJ, Moritz T, Trygg J, Sjostrom M, et al. 2005. Extraction, interpretation and validation of information for comparing samples in metabolic LC/MS data sets. *Analyst* 130:701–7
32. Lindon JC, Holmes E, Nicholson JK. 2001. Pattern recognition methods and applications in biomedical magnetic resonance. *Prog. NMR Spectrosc.* 39:1–40
33. Trygg J, Wold S. 2002. Orthogonal projections to latent structures (O-PLS). *J. Chemomet.* 16:119–28
34. Holmes E, Nicholson JK, Tranter G. 2001. Metabonomic characterization of genetic variations in toxicological and metabolic responses using probabilistic neural networks. *Chem. Res. Toxicol.* 14:182–91
35. Kell DB. 2002. Metabolomics and machine learning: explanatory analysis of complex metabolome data using genetic programming to produce simple, robust rules. *Mol. Biol. Rep.* 29:237–41
36. Kell DB. 2006. Metabolomics, modelling and machine learning in systems biology: towards an understanding of the languages of cells. *FEBS J.* 273:873–94

37. Vehtari A, Makinen V-P, Soininen P, Ingman P, Makela S, et al. 2007. A novel Bayesian approach to quantify clinical variables and to determine their spectroscopic counterparts in H-1 NMR metabonomic data. *BMC Bioinformatics* 8(Suppl. 2):S8

38. Cloarec O, Dumas ME, Craig A, Barton RH, Trygg J, et al. 2005. Statistical total correlation spectroscopy (STOCSY): a new approach for individual biomarker identification from metabonomic NMR datasets. *Anal. Chem.* 77:1282–89

39. Sasic S, Muzynski A, Ozaki Y. 2000. A new possibility of the generalized two-dimensional correlation spectroscopy. 1. Sample-sample correlation spectroscopy. *J. Phys. Chem. A* 104:6380–87

40. Noda I, Dowrey AE, Marcott C, Story GM, Ozaki Y. 2000. Generalized two-dimensional correlation spectroscopy. *Appl. Spectrosc.* 54:236A–248A

41. Noda I. 1989. Two-dimensional infrared-spectroscopy. *J. Am. Chem. Soc.* 111:8116–18

42. Noda I. 1993. Generalized 2-dimensional correlation method applicable to infrared, raman, and other types of spectroscopy. *Appl. Spectrosc.* 47:1329–36

43. Noda I. 1990. 2-dimensional infrared (2D IR) spectroscopy: theory and applications. *Appl. Spectrosc.* 44:550–61

44. Coen M, Hong Y-S, Cloarec O, Rhode C, Reily MD, et al. 2007. Magic angle spinning ^{31}P-^{1}H hetero-nuclear statistical total correlation spectroscopy (STOCSY). *J. Proteome Res.* 6:2711–19

45. Cloarec O, Campbell A, Tseng L-H, Braumann U, Spraul M, et al. 2007. Virtual chromatographic resolution enhancement in cryoflow LC-NMR experiments via statistical total correlation spectroscopy. *Anal. Chem.* 79:3304–11

46. Crockford DJ, Holmes E, Lindon JC, Plumb RS, Zirah S, et al. 2006. Statistical HeterospectroscopY (SHY), a new approach to the integrated analysis of NMR and UPLC-MS datasets: application in metabonomic toxicology studies. *Anal. Chem.* 78:363–71

47. Crockford DJ, Lindon JC, Cloarec O, Plumb RS, Bruce SJ, et al. 2006. Statistical search space reduction and 2-dimensional data display approaches for UPLC-MS in biomarker discovery and pathway analysis. *Anal. Chem.* 78:4398–408

48. Jenner AM, Timbrell JA. 1994. Effect of acute and repeated exposure to low-doses of hydrazine on hepatic-microsomal enzymes and biochemical parameters in-vivo. *Arch. Toxicol.* 68:240–45

49. Bollard ME, Stanley EG, Lindon JC, Nicholson JK, Holmes E. 2005. NMR-based metabonomics approaches for evaluating physiological influences on biofluid composition. *NMR Biomed.* 18:143–62

50. Dumas M-E, Maibaum EC, Teague C, Ueshima H, Zhou B, et al. 2006. Assessment of analytical reproducibility of ^{1}H NMR spectroscopy based metabonomics for large-scale epidemiological research: the INTERMAP study. *Anal. Chem.* 78:2199–208

51. Lenz EM, Bright J, Wilson ID, Hughes A, Morrison J, et al. 2004. Metabonomics, dietary influences and cultural differences: a H-1 NMR-based study of urine samples obtained from healthy British and Swedish subjects. *J. Pharmaceut. Biomed. Anal.* 36:841–49

52. Lenz EM, Bright J, Wilson ID, Morgan SR, Nash AFP. 2003. A H-1 NMR-based metabonomic study of urine and plasma samples obtained from healthy human subjects. *J. Pharmaceut. Biomed. Anal.* 33:1103–15

53. Nicholls AW, Mortishire-Smith RJ, Nicholson JK. 2003. NMR spectroscopic-based metabonomic studies of urinary metabolite variation in acclimatizing germ-free rats. *Chem. Res. Toxicol.* 16:1395–404

54. Robosky LC, Wells DF, Egnash LA, Manning ML, Reily MD, Robertson DG. 2005. Metabonomic identification of two distinct phenotypes in Sprague-Dawley (Crl: CD(SD)) rats. *Toxicol. Sci.* 87:277–84

55. Wang Y, Holmes E, Nicholson JK, Cloarec O, Chollet J, et al. 2004. Metabonomic investigations in mice infected with *Schistosoma* mansoni: an approach for biomarker identification. *Proc. Nat. Acad. Sci. USA* 101:12676–81

56. Martin FPJ, Wang Y, Sprenger N, Holmes E, Lindon JC, et al. 2007. Effects of probiotic *Lactobacillus paracasei* treatment on the host gut tissue metabolic profiles probed via magic-angle-spinning NMR spectroscopy. *J. Proteome Res.* 4:1471–81

57. Lindon JC, Holmes E, Bollard ME, Stanley EG, Nicholson JK. 2004. Metabonomics technologies and their applications in physiological monitoring, drug safety evaluation and disease diagnosis. *Biomarkers* 9:1–31

58. Lindon JC, Holmes E, Nicholson JK. 2004. Toxicological applications of magnetic resonance. *Prog. NMR Spectrosc.* 45:109–43

59. Lindon JC, Nicholson JK, Holmes E, Antti H, Bollard ME, et al. 2003. Contemporary issues in toxicology—the role of metabonomics in toxicology and its evaluation by the COMET project. *Toxicol. Appl. Pharmacol.* 187:137–46

60. Keun HC, Ebbels TMD, Bollard ME, Beckonert O, Antti H, et al. 2004. Geometric trajectory analysis of metabolic responses to toxicity can define treatment specific profiles. *Chem. Res. Toxicol.* 17:579–87

61. Ebbels T, Keun H, Beckonert O, Antti H, Bollard M, et al. 2003. Toxicity classification from metabonomic data using a density superposition approach: 'CLOUDS.' *Anal. Chim. Acta* 490:109–22

62. Ebbels TMD, Keun HC, Beckonert OP, Bollard ME, Lindon JC, et al 2003. Prediction and classification of drug toxicity using probabilistic modeling of temporal metabolic data: The Consortium on Metabonomic Toxicology screening approach. *J. Proteome Res.* 6:3944–51

63. Lindon JC, Keun HC, Ebbels TMD, Pearce JMT, Holmes E, Nicholson JK. 2005. The Consortium for Metabonomic Toxicology (COMET): aims, activities and achievements. *Pharmacogenomics* 6:691–99

64. Waters NJ, Holmes E, Williams A, Waterfield CJ, Farrant RD, Nicholson JK. 2001. NMR and pattern recognition studies on the time-related metabolic effects of α-naphthylisothiocyanate on liver, urine, and plasma in the rat: an integrative metabonomic approach. *Chem. Res. Toxicol.* 14:1401–12

65. Coen M, Lenz EM, Nicholson JK, Wilson ID, Pognan F, Lindon JC. 2003. An integrated metabonomic investigation of acetaminophen toxicity in the mouse using NMR spectroscopy. *Chem. Res. Toxicol.* 16:295–303

66. Coen M, Ruepp SU, Lindon JC, Nicholson JK, Pognan F, et al. 2004. Integrated application of transcriptomics and metabonomics yields new insight into the

toxicity due to paracetamol in the mouse. *J. Pharmaceut. Biomed. Anal.* 35:93–105

67. Heijne WHM, Lamers RJAN, van Bladeren PJ, Groten JP, van Nesselrooij JHJ, Van Ommen B. 2005. Profiles of metabolites and gene expression in rats with chemically induced hepatic necrosis. *Toxicol. Pathol.* 33:425–33

68. Griffin JL, Bonney SA, Mann C, Hebbachi AM, Gibbons GF, et al. 2004. An integrated reverse functional genomic and metabolic approach to understanding orotic acid–induced fatty liver. *Physiol. Genomics* 17:140–49

69. Rantaleinen M, Cloarec O, Beckonert O, Wilson ID, Jackson D, et al. 2006. Statistically integrated metabonomic-proteomic studies on a human prostate cancer xenograft model in mice. *J. Proteome Res.* 5:2642–55

70. Dumas ME, Wilder SP, Bihoreau MT, Barton RH, Fearnside JF, et al. 2007. Direct quantitative trait locus mapping of mammalian metabolic phenotypes in diabetic and normoglycemic rat models. *Nat. Genetics* 39:666–72

71. Clayton TA, Lindon JC, Antti H, Charuel C, Hanton G, et al. 2006. Pharmaco-metabonomic phenotyping and personalised drug treatment. *Nature* 440:1073–77

72. Rezzi S, Ramadan Z, Martin F-PJ, Fay LB, van Bladeren P, et al. 2007. Human metabolic phenotypes link directly to specific dietary preferences in healthy individuals. *J. Proteome Res.* 6:4469–77

73. Lindon JC, Nicholson JK, Holmes E, Keun HC, Craig A, et al. 2005. Summary recommendations for standardization and reporting of metabolic analyses. *Nat. Biotech.* 23:833–38

74. Jenkins H, Hardy N, Beckmann M, Draper J, Smith AR, et al. 2004. A proposed framework for the description of plant metabolomics experiments and their results. *Nat. Biotech.* 22:1601–6

75. Castle AL, Fiehn O, Kaddurah-Daouk R, Lindon JC. 2006. Metabolomics standards workshop and the development of international standards for reporting metabolomics experimental results. *Briefing Bioinformatics* 7:159–65

76. Sumner LW, Amberg A, Barrett D, Beale MH, Daykin C, et al. 2007. Proposed minimum reporting standards for chemical analysis. *Metabolomics* 3:211–21

Mass Spectrometry for Rapid Characterization of Microorganisms

Plamen A. Demirev[1] and Catherine Fenselau[2]

[1]Johns Hopkins University, Applied Physics Laboratory, Laurel, Maryland 20723; email: plamen.demirev@jhuapl.edu

[2]University of Maryland, Department of Chemistry and Biochemistry, College Park, Maryland 20742; email: fenselau@umd.edu

Annu. Rev. Anal. Chem. 2008. 1:71–93

First published online as a Review in Advance on December 20, 2007

The *Annual Review of Analytical Chemistry* is online at anchem.annualreviews.org

This article's doi: 10.1146/annurev.anchem.1.031207.112838

Key Words

MALDI, ESI, tandem MS, bacteria, viruses, spores, proteomics, bioinformatics, proteolysis

Abstract

Advances in instrumentation, proteomics, and bioinformatics have contributed to the successful applications of mass spectrometry (MS) for detection, identification, and classification of microorganisms. These MS applications are based on the detection of organism-specific biomarker molecules, which allow differentiation between organisms to be made. Intact proteins, their proteolytic peptides, and nonribosomal peptides have been successfully utilized as biomarkers. Sequence-specific fragments for biomarkers are generated by tandem MS of intact proteins or proteolytic peptides, obtained after, for instance, microwave-assisted acid hydrolysis. In combination with proteome database searching, individual biomarker proteins are unambiguously identified from their tandem mass spectra, and from there the source microorganism is also identified. Such top-down or bottom-up proteomics approaches permit rapid, sensitive, and confident characterization of individual microorganisms in mixtures and are reviewed here. Examples of MS-based functional assays for detection of targeted microorganisms, e.g., *Bacillus anthracis*, in environmental or clinically relevant backgrounds are also reviewed.

1. INTRODUCTION

The worldwide threat to human health from existing as well as emerging infectious pathogens has not diminished despite spectacular advances in medicine. Thus, the need for novel molecular-level technologies for rapid and confident microorganism detection, identification, and characterization is constantly growing. Current technologies based on antibody recognition or DNA detection after polymerase chain reaction (PCR) amplification typically require hours to yield results (1). However, mass spectrometry (MS) possesses several unique features that make it an attractive complementary technology for microorganism detection, identification, and characterization.

First and foremost, MS is broad band: Its capability to detect organisms is not restricted to prespecified targets, whereas most other molecular detection technologies must rely on molecular recognition events by either antibodies or DNA primers and probes to selectively bind predetermined and specified targets. Second, MS is rapid: A typical experiment from sample collection and preparation to final result requires as little as a few minutes. In contrast, days are needed for (often retrospective) detection and verification via the classical microbiology methods that rely on microorganism culture. Third, MS is sensitive: A signal with a sufficient signal-to-noise ratio can be generated from a sample containing less than 10^4 microorganisms. A variety of sample-collection and sample-processing modules for sampling of microorganisms and their constituents from different environments—from aerosols to biofluids—can be interfaced to MS instruments. Finally, tandem MS in conjunction with chromatography and bioinformatics is the most rapidly growing analytical technology in the postgenomic era, fueling rapid global advances in proteomics. Most proteomics tools, developed initially for protein characterization, can be successfully adapted for rapid characterization of microorganisms.

For more than 30 years, the capabilities of MS have been utilized to characterize bacteria, bacterial spores, and other kinds of microorganisms with minimal sample preparation and without fractionation, centrifugation, or chromatography (2–9). The development of rapid MS methods for microorganism detection has been driven, in large part, by government interest in the identification of biowarfare agents in the "detect to protect" time frame of under several minutes and the "detect to treat" time frame of several hours (10, 11). The capability of MS to rapidly characterize microorganisms has potential applications in a number of areas beside medical diagnostics, biodefense, and homeland security, such as environmental monitoring, agricultural stewardship, food quality control, occupational safety, and culture typing.

Every molecule has a mass, an immutable intrinsic property reflecting its elemental composition. The masses of intact molecular ions and their fragments can be facilely determined by various kinds of MS. In addition, in tandem MS the intact molecular ions can be internally excited to produce fragment ions, which provide information about molecular structures and sequences. The current paradigm for rapid MS identifications of intact microorganisms relies on determination of the masses of unique biomarker molecules from experimental mass spectra of intact organisms or their extracts. This paradigm can be traced back to Anhalt and Fenselau (2) who demonstrated (*a*) that biomolecules from different pathogenic bacteria, introduced intact in a mass spectrometer, could be vaporized and directly ionized; (*b*) that these molecules could be structurally identified; and (*c*) that taxonomic distinctions can be made based on the characteristic mass spectral "fingerprint" signatures for individual organisms. The experimentally observed signatures can be related to either intact biomarker molecules or their fragments, and each organism-specific signature can be derived experimentally by acquiring mass spectra for a particular microorganism under a variety of conditions. Alternatively, these signatures can be deduced/derived by means of bioinformatics. **Figure 1** illustrates the confluence of experimental and theoretical analyses in these strategies.

In the following sections, we discuss biomarkers that are predominantly peptides or proteins. Peptides and proteins are more abundant in microbial cells compared to other classes of molecules. In addition, proteins are ionized efficiently and resolve some of the ambiguities of DNA sequences (6, 12). At the same time, protein sequences are linked to gene sequences and are therefore more characteristic biomarkers than lipids and metabolites. We point out that there are MS-based applications in microbiology utilizing biomarkers other than proteins, such as DNA (13) and 16s

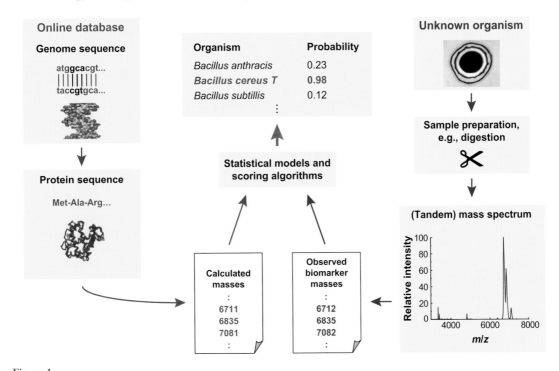

Figure 1

Strategy for rapid characterization of microorganisms by mass spectrometry (MS) and bioinformatics. An experimental mass spectrum of an unknown is matched against the in silico–predicted masses of organisms with sequenced genomes. The "best" match is determined by a statistical algorithm.

RNA (14, 15). Low-mass molecules and molecular fragments have also been reported as MS-based biomarkers for microorganism detection and identification—e.g., heme as a biomarker for malaria (16)—or in bacterial spore forensics (17), phospholipids (18) or fatty acid methyl esters (19).

TOF: time of flight

Electrospray ionization (ESI): multiply charged biomolecular ions are generated by transporting solution containing analyte through a capillary needle biased at high voltage

Matrix-assisted laser desorption/ionization (MALDI): ionization method for large nonvolatile biomolecules, wherein the analyte is mixed with a matrix, then dried and irradiated with laser pulses

AP: atmospheric pressure

1.1. Instrumentation

A number of custom-built and commercially available ion analyzers have been demonstrated for rapid analysis of proteins from unfractionated bacteria, notably time of flight (TOF), Fourier transform, and ion trap analyzers. Small TOF- and ion trap–based instruments have been designed for field portability (e.g., 19–24), whereas larger instruments with higher mass-resolving power have been proposed for work in reference laboratories (25–29). Providers of fieldable systems have incorporated small and rugged pumps and detectors, as well as devices for sample collection and processing. The major advances that have enabled reliable rapid analysis of microorganisms are two techniques particularly suited for ionization and transfer into the gas phase of large nonvolatile biomolecules such as intact proteins: laser desorption and electrospray (30). Successful characterization of an unfractionated nonenveloped virus using electrospray ionization (ESI) was reported in 1996 (31).

Challenges to broadening the application of ESI in this area have included mechanical clogging and obstruction of charge state deconvolution by the presence of multiple proteins, each with a distribution of charge states. Matrix-assisted laser desorption/ionization (MALDI) generates mostly singly charged ions, and thus MALDI spectra accommodate larger numbers of proteins than those of ESI. Both MALDI and ESI are selective: Depending on their intrinsic properties, some proteins in a mixture are more easily ionized than others. MALDI mass spectra from intact microorganisms are not readily reproducible, being sensitive to sample:matrix ratios, the presence of contaminants, and the irregular surfaces of bacterial samples (32).

Newer techniques for direct sample ionization from ambient sources have also been interfaced with different types of mass spectrometers for analysis of intact bacteria or rapidly digested bacterial extracts. These ionization techniques are capable of interrogating a sample directly in the atmosphere before transferring the ions into the instrument for mass analysis. They include atmospheric pressure (AP) MALDI (23, 33), desorption electrospray ionization (DESI) (34), and direct analysis in real time (DART) (35).

1.2. Sample Preparation

Characterization of environmental or clinical microorganisms usually requires enriching or purifying the bacteria and viruses obtained from air, water, foodstuffs, urine, blood, et cetera. The use of antibodies offers a universal method for separating and enriching targeted species (36, 37). The use of lectin/carbohydrate affinity binding has also been demonstrated (38, 39); this technique offers a broad-range, less selective recovery. Microorganisms have been recovered from air (the cleanest medium) on filters, in aqueous solutions, and directly onto MALDI sample holders via particle impactors. In the latter case, the sample is cleaned up (for instance, separated from fine sand) by transmitting particles only in a preselected size range. Several laboratories have undertaken analysis of single particles such as bacterial rafts introduced directly into mass spectrometers (40–43). Coating particles in flight is proposed to permit MALDI analysis of protein biomarkers (40, 41, 43).

Intact bacteria have been separated by capillary electrophoresis (44) and field flow fractionation (45), which are compatible with subsequent mass spectrometric analysis, albeit in a longer time frame compared to direct sample deposition. A small virus (MS 2) as well as individual *Escherichia coli* cells have been weighed intact (and non-routinely) by MS (46, 47). Exposure to MALDI matrix solutions lyses many bacteria types in situ (i.e., on the sample holder) and the released proteins cocrystallize with the matrix. MALDI matrix solutions and most electrospray solutions contain both aqueous acid and organic solvents. Rapid extraction of protein biomarkers from bacterial spores and some viruses may be achieved by the use of strong acid or organic solvents (12, 48–54). An automated system would be expected to provide adequate conditions for effective cell lysis and also for biomarker ionization.

Below we describe several ways in which proteins released from or accessed in intact cells are hydrolyzed in situ to peptides for proteomic analysis. Immobilized trypsin can be used in high concentrations for rapid proteolysis (53, 55, 56), and microwave-assisted residue-specific acid cleavage has recently been demonstrated (57–59).

2. RAPID ANALYSES BASED ON INTACT PROTEINS FROM UNFRACTIONATED BACTERIA

2.1. Exploiting Molecular Masses of Intact Biomarkers

Most MALDI spectra of proteins from intact microorganisms contain intense peaks below *m/z* 25,000. These spectra are sufficiently distinctive to establish species-characteristic fingerprint libraries (60). Both the mass range and species differences are illustrated in **Figure 2**. To identify the microorganism, experimental MALDI mass spectra from unknowns are compared with such reference libraries. However, MALDI MS of intact microorganisms is influenced by many experimental factors, such as individual protein biomarker solubility, variation of expression levels due to culture time and media, ionization efficiencies for different biomarker molecules (e.g., as a function of matrix:biomarker protein ratios), variation in laser pulse energy, and instrument-dependent detection efficiency (32, 61, 62). A round robin study of MALDI mass spectral reproducibility from intact microorganisms has been performed in three independent laboratories with three different commercial instruments using identical aliquots of *E. coli* culture, matrix, and calibration standards as well as automated data processing and analysis algorithms (63). Only 25% of the biomarker ions attributed to this bacterium were found in common by all three laboratories, whereas more than 50% were observed in spectra from only one of the three laboratories.

Several fingerprint libraries are available commercially, as are directions on how to culture and prepare samples to ensure reproducible, searchable spectra. Numerous

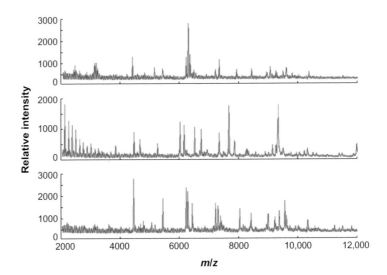

reports advocate the use of this approach for environmental monitoring, microbial forensics, and potential clinical applications (12, 64–72). A standardized method, which includes culture time and medium, number of passages, and matrix type and concentration, has been proposed for MALDI MS–based strain typing of clinical isolates of methicillin-resistant *Staphylococcus aureus* (66). MALDI MS for cultured environmental samples has been compared with PCR for rapid discrimination between closely related strains of *E. coli* and for tracking bacterial contamination in surface water (71). Although MALDI MS reproducibility appears to be lower than that of PCR, the MS approach was superior in correctly assigning *E. coli* isolates to a specific contamination source.

Difficulties in reproducing the MALDI mass spectra of intact microorganisms, without the possibility of culturing or controlling other experimental parameters, have prompted the development of a bioinformatics approach based on proteome database queries (4, 73–78). In this approach, peaks in the spectra need not be reproducible if the protein masses can be related to the relevant genome (73). The experimentally observed biomarker masses are matched to predicted suites of protein molecular masses from a proteome database. A bioinformatics-based tool has been developed to introduce additional constraints on candidate protein biomarkers by determining the number of methionine residues in a sequence from the mass shifts before and after rapid in situ sample oxidation (79). The method has been predominantly applied to microorganism identification at the genus and species levels. Depending on database availability, even strain specificity has been reported for *Campylobacter* (80) and *Bacillus* (81).

The bioinformatics method has been strengthened with the introduction of hypothesis testing to quantify the significance of microorganism identifications (74). For that purpose, the probability (p-value) for a random match between an experimentally observed biomarker mass and the calculated mass of an unrelated database protein is introduced. The lower probability reflects a lower likelihood for misidentification due to accidental matches. Therefore, lower p-values provide more confident overall microorganism identifications. Both theoretical analysis of the bioinformatics approach and statistical modeling (in silico MS) demonstrate the dependence of p-values on experimental parameters, e.g., mass accuracy, number of experimentally observed independent biomarkers, and database properties (74). Increasing the mass accuracy of the experimentally observed biomarkers by external/internal calibration (82), higher resolution mass measurements (27), and/or isotope depletion improves confidence in microorganism identification.

We identify several important features and specific requirements for a proteome database to be used successfully in microorganism identification by mass spectrometry. The first important feature of the database is its completeness, characterized by the number of organisms with sequenced genomes. Obviously, for an organism to be identified by this approach its genome has to be sequenced (at least partially). Currently there are more than 550 bacteria and archaea and more than 1800 reference virus genome sequences publicly available (83).

Another important feature of the database is its fidelity, namely the correlation (or lack thereof) between observed and predicted biomarker masses from the amino acid sequences. The observed and predicted masses of intact proteins are different if posttranslational modifications (PTMs) are present. The occurrence of the most common PTM in prokaryotes—loss of the N-terminal Met (start codon)—can be predicted with a large degree of certainty from the penultimate N-terminal amino acid (75). Microorganism identification (evaluated from the p-values) is improved by at least an order of magnitude if this PTM is accounted for (75).

The third important database feature is biomarker density: the number of database proteins per unit mass interval for a given microorganism. Both statistical modeling (74, 77) and experimental results (76) predict that reducing the database density by rationally constraining the entire proteome of a sequenced organism allows its successful identification from its experimental mass spectrum at the mass accuracy typical for linear TOF instruments. For instance, including only ribosomal proteins (a class of highly expressed proteins) in a microbial biomarker database reduces the number of biomarkers in the range of 4 to 20 kDa by more than two orders of magnitude. In a blind study, microorganisms represented in this database were correctly identified from their experimental MALDI spectra 100% of the time at the 95% confidence level, with no incorrect identifications (76). In addition, a truncated protein biomarker database has been created using only bioinformatics tools and microbial genome sequence information (77). That approach relies on the correlation between the statistics of synonymous codon usage in a gene and the expression level of the protein coded by that gene (84, 85). Truncated databases have also been constructed using mass spectrometry to identify highly expressed biomarkers in an organism (86, 87). Data compression from a two-dimensional array of high-performance liquid chromatography (HPLC) retention times and electrospray mass spectra into a one-dimensional mass/intensity spectral format allows the creation of biomarker profiles for bacteria from electrosprayed extracts of intact proteins (87). Microorganism-specific search algorithms based on experimental mass spectra, proteomics, and bioinformatics are available on a publicly accessible website (88).

PTM: posttranslational modification

HPLC: high-performance liquid chromatography

2.2. Protein Fragmentation by Tandem Mass Spectrometry

SASP: small acid-soluble spore protein

A complement to the analysis of proteins ionized from unfractionated bacteria is offered by instruments that can fragment intact proteins to provide sequence-specific fragment ions. In this "top-down" proteomics approach (89), an intact protein is identified by deducing its partial amino acid sequence after fragmentation in a tandem MS experiment and subsequent homology search. MALDI and top-down proteomics have recently been demonstrated on a tandem TOF instrument for the rapid and high-confidence identification of intact *Bacillus* spore species (29). Unlike ESI/FT-ICR (Fourier transform ion cyclotron resonance), MALDI TOF top-down proteomics does not require protein biomarker enrichment and separation prior to analysis. It can directly interrogate intact microorganisms, both pure and in mixtures, and switching from MS to tandem MS mode takes just a few seconds. Dissociation of precursor protein ions results in sequence-specific backbone cleavages, with spectra dominated by ions formed by cleavages on the C-terminal side of aspartic or glutamic acid residues.

To identify the precursor ion, the spectra are compared with tandem spectra generated in silico from all the protein sequences in a proteome database, whose masses correspond to the intact precursor ion mass observed. By inference, the source microorganism is then identified, based on the identification of one or more individual protein biomarker(s). For example, the major biomarker ions between *m/z* 6000 and 9000 observed in MALDI spectra of mixtures of intact *Bacillus* spores can all be unambiguously identified (29). From here, the presence of, for instance, intact *B. anthracis* spores in the mixture can be confirmed. **Figure 3** illustrates the fragmentation of the protein with MH+ at *m/z* 6680 in the MALDI spectrum of a mixture of *B. anthracis* Sterne spores and *B. cereus* T spores. Following the procedure outlined in Reference 29, a small acid-soluble spore protein (SASP) from *B. anthracis* is unambiguously identified from the tandem spectrum (p-value 7.4×10^{-10}), thus indicating the presence of intact *B. anthracis* spores in the sample. The MALDI tandem MS spectra of intact biomarkers are fairly reproducible, and library fingerprint matching of such tandem mass spectra can be implemented for intact microorganism identification. This may be particularly advantageous when the sequence of the microorganism genome is not available, or when only a few protein biomarkers, such as virus coat proteins, are present in a regular mass spectrum.

Top-down proteomics, using ESI on an ion trap instrument to study intact SASPs extracted from *Bacillus* spores, has recently been applied to grouping ten different *B. cereus* strains into two clusters, one closely and one distantly related to *B. anthracis* (90). The strains from the closely related cluster contained a major SASP with only a single amino acid substitution (close to either the C-terminus or the N-terminus). The more distantly related cluster displayed both amino acid substitutions, as previously observed by MALDI TOF MS (51, 77). Interestingly, a *B. cereus* isolate from a patient with severe pneumonia (an anthrax-like disease) has been grouped into the more distantly related cluster based on masses of observed SASPs alone.

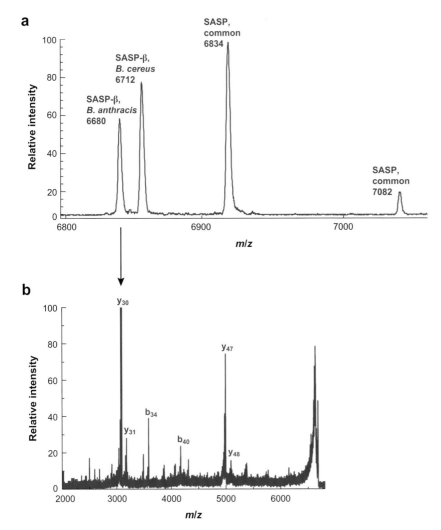

Figure 3

(*a*) Matrix-assisted laser desorption/ionization time-of-flight (MALDI TOF) mass spectrum of a 1:1 mixture of intact *Bacillus cereus* T spores and *Bacillus anthracis* Sterne spores. (*b*) Tandem mass spectrum (MALDI TOF/TOF) of the precursor ion at *m/z* 6680. SASP, small acid-soluble spore protein.

3. RAPID CHARACTERIZATION OF BACTERIA BASED ON PEPTIDES FROM BIOMARKER PROTEINS

3.1. Sequence Tags

The identification of one or more proteins from partial sequences readily allows the identification of a bacterium or closely related family at the species level. Parallel to the top-down approach, a bottom-up strategy (91) has also been adopted from proteomics for the rapid identification of bacteria and spores. Proteins fractionated from lysed cells, or accessed in situ in automated rapid strategies, are cleaved to peptides, which are then fragmented by collisionally induced dissociation (CID) or laser-induced dissociation. These tandem mass spectra provide the basis for identifying the peptides. Identified peptides provide identification of proteins and, in turn, proteins provide identification of microorganisms. This strategy has been implemented using both MALDI (92–94) and electrospray ionization (6, 28, 95–99). **Figure 4***a* presents the MALDI TOF spectrum of a 1:1 mixture of *Bacillus anthracis* Ames spores and *Bacillus subtilis* 168 spores, which had been treated for 90 s with hot formic acid. Both molecular ions and peptide ions were detected. **Figure 4***b* presents a portion of the spectrum with an expanded abscissa. In **Figure 4***c* we see the tandem mass spectrum obtained from the MH+ peak at *m/z* 2462 using laser-induced dissociation (58). From this spectrum, the peptide was identified as originating from extracellular antigen 1 surface-layer protein in *B. anthracis*. All spore components of the mixture were identified by considering the molecular masses and tandem spectra of the other peptides and small proteins observed in **Figure 4***a,b*.

One benefit of interrogating peptides is the increased mass accuracy and sensitivity of ion analyzers and detectors for smaller ions such as peptides compared to protein molecular ions. This strategy provides access to heavier proteins (via their peptides) and has been suggested as one solution to the challenges of identifying components of mixtures of microorganisms (53, 92, 93, 96) and of recognizing plasmid insertion in genetically engineered bacteria (94). When this bottom-up approach is implemented in a rapid automatable MALDI strategy, residue-specific cleavage of biomarker proteins is required as an extra step and is carried out enzymatically or chemically in situ. The strategy works best when only a limited set of proteins is preferentially solubilized.

Traditional bottom-up proteomics approaches, which involve extensive sample preparation and peptide analysis by chromatography/ESI/tandem MS, have also been successfully applied to the characterization of individual microorganisms in often complex mixtures (6, 95–99). Although these methods are slower than the rapid MALDI approach, one- or two-dimensional chromatography improves considerably the MS capability to identify many individual peptides. Shotgun proteomics and dedicated software have been recently applied to *B. anthracis* strain differentiation (99). Selective interrogation of subsets of targeted characteristic peptides has been proposed in both MALDI (94) and electrospray (96) workflows in order to reduce the overall analysis time.

3.2. Small Acid-Soluble Spore Proteins

The SASP family of proteins constitutes up to 15% of the mass of *Bacillus* spores. These basic proteins can be selectively solubilized in acid, and they are readily protonated to provide strong signals in MALDI and electrospray spectra. They have been studied by many researchers interested in protein-based identification of *B. anthracis* spores, and numerous publications have considered the question of how specific the biomarkers provided by the SASP family are (e.g., 11, 12, 29, 77, 92, 94, 100). A survey of

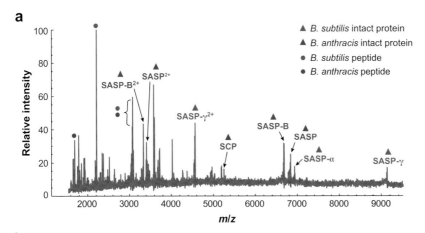

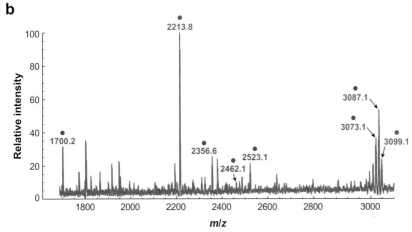

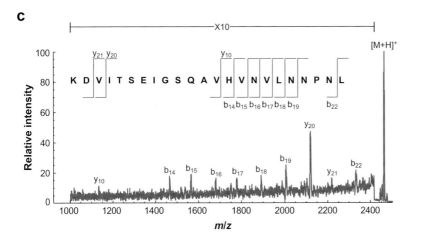

Figure 4

(*a*) Matrix-assisted laser desorption/ionization mass spectrum of proteins and peptides obtained by 90-s microwave-assisted acid digestion of a 1:1 mixture of *Bacillus anthracis* Sterne spores and *Bacillus subtilis* 168 spores. (*b*) Spectrum from panel *a* expanded between *m/z* 1700 and 3100. (*c*) Tandem mass spectrum obtained by laser-induced dissociation of a *B. anthracis* Sterne peptide with MH+ at *m/z* 2462. SASP, small acid-soluble spore protein. Reproduced with permission from Reference 58. Copyright 2006, American Chemical Society.

publicly available genomes of ten *B. anthracis* strains has evaluated the specificity of the SASPs and their peptides obtained by either immobilized trypsin or aspartate-specific chemical digestion to characterize each strain (94). Five of the eight *B. anthracis* SASPs have exactly the same sequences as SASPs from different, albeit closely related species, such as *B. thuringiensis*, *B. cereus*, and others. Two of the remaining SASPs are distinct in mass from the respective SASPs in the other two species. However, each individual tryptic peptide from these SASPs can be found in at least one other *B. cereus* or *B. thuringiensis* strain (94). The sequence of only one SASP (γ) indicates that both the molecular ions and peptide products allow the source of this protein to be distinguished. As has been shown (18, 58, 66, 100), use of the SASP family for species identification is optimized when measurements can be made of both the molecular masses and critical peptides or microsequences.

In general, the probability for positive matching of a protein but false matching of a microorganism is lower in top-down proteomics than in bottom-up proteomics. In both cases, correct microorganism identification relies on the identification of unique versus degenerate biomarker ions. In this context, degenerate means either intact proteins (i.e., orthologs) or peptides found in two or more microorganisms.

3.3. Peptide Mass Maps

By analogy with intact protein mass searching for microorganism identification, the idea of utilizing the masses of proteolytic peptides generated in situ has been evaluated for identification of unfractionated viruses (101) and *Bacillus* spores (56). Databases were established against which experimentally generated suites of peptide masses were mapped. This approach provides successful identifications; however, the experimental step required for rapid residue-specific cleavage can be more productively linked with microsequencing by tandem MS, which allows more reliable protein and thus microorganism identification.

4. RAPID ANALYSIS OF NONGENOMIC CYCLIC LIPOPEPTIDES

Cyclic lipopeptides, e.g., polymixins, surfactins, fengycins, and kurstakins, are often abundant in bacteria and constitute a class of potential biomarker molecules that has been proposed to differentiate some microorganisms at the species and even at the subspecies levels (102–107). The structures of these secondary metabolites cannot be predicted from genome sequences, and it is necessary to employ library fingerprinting methods for identification. Both AP MALDI ion trap (23) and MALDI TOF/TOF (29) tandem spectra of these biomarkers are reproducible; therefore, it may be possible to incorporate tandem mass spectra for automated microorganism identification. For lipopeptides in which the ring structure is still intact, CID tandem MS results in fewer fragments, which reflect the structure of the fatty acid moiety (105).

5. AMPLIFYING PROTEIN SIGNALS FOR MASS SPECTRAL ANALYSIS

It has been recognized that MS, employing protein biomarkers for microorganism detection and identification, can never rival the sensitivity of PCR-based methods, which amplify the target DNA fragment many thousands of times. This extra sensitivity is especially important for the analysis of the complex mixtures of microorganisms typical in clinical infections. Recently, however, several MS-based functional assays have been developed and demonstrated for targeted bacteria, resulting in significant amplification of the signal.

5.1. Phage-Based Assay for Bacteria Identification

Targeted microorganisms have been identified by MALDI TOF MS through bacteriophage amplification (108). In this approach, samples containing pathogenic bacteria are infected with bacteria-specific bacteriophages (e.g., MS2 and MPSS-1 phages specific for *E. coli* and *Salmonella*, respectively). Proteins indicative of the progeny phages are then detected and utilized as secondary biomarkers for the target pathogen. For instance, *E. coli* mixed with both MS2 and MPSS-1 produces only an MS2 biomarker protein, as detected by MALDI TOF MS. Amplification of both phages in a mixture of the two bacteria leads to detection of biomarkers characteristic for both MS2 and MPSS-1.

5.2. Endopeptidase Assay for Anthrax

Anthrax lethal factor is a protein complex of a protective antigen and a zinc-dependent endoproteinase, which is known to cleave five protein kinases. The complex forms when the two proteins are secreted by the bacterium in an infected host. The protease activity of anthrax lethal factor has been exploited by Barr and colleagues for detection of anthrax in serum (109). Lethal factor is recovered from serum with monoclonal antibodies immobilized on magnetic protein G beads, and exposed to an optimized synthetic peptide substrate. The protease reiteratively hydrolyses peptide substrate molecules into two smaller peptides; this produces biomarkers amplified through time, which are detected by MALDI TOF mass spectrometry. Thus, three layers of selectivity contribute to the reliability of the analysis. The method can be carried out in four hours, with a detection limit of 0.05 ng/ml of the lethal factor complex. The cleavage reaction is shown in **Figure 5a**, along with the masses of the protonated peptide products. **Figure 5b** shows the spectrum obtained from a control sample of peptide, processed in the absence of lethal factor. The ions detected represent the intact peptide with one and two charges. **Figure 5c** shows the spectrum obtained when 2 nmol of peptide is incubated with 1 ng lethal factor for two hours. The addition of known amounts of the peptide carrying stable isotope labels on the alanine residues has allowed quantitative analysis, and this assay is being used to study the progression of infection and disease in animals (109).

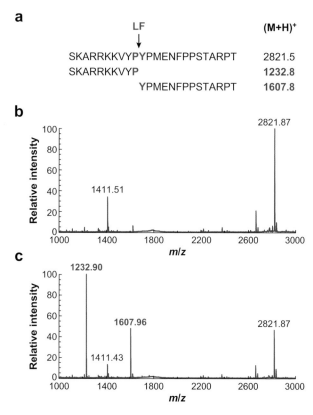

Figure 5

Anthrax lethal factor (LF) detection using an amplifying protease reaction. (*a*) The LF cleavage reaction of a target peptide with predicted masses of cleaved products denoted in the column. (*b*) Matrix-assisted laser desorption/ionization time-of-flight (MALDI TOF) spectrum of a peptide sample without addition of LF. (*c*) MALDI TOF spectrum of a peptide sample incubated with LF for 2 h. Reproduced with permission from Reference 109. Copyright 2007, American Chemical Society.

6. CONCLUSIONS

A number of modern MS instrumental systems—from MALDI TOF to tandem HPLC/ESI ion traps—have been applied for successful characterization of microorganisms at the species and, occasionally, the subspecies levels. Bioinformatics and proteomics approaches, initially developed for large-scale characterization of proteins, have been adapted to provide rapid and confident microorganism identification. Such approaches have a number of practical advantages; most notably, they do not require rigorous control of all culturing and experimental variables. Although the number of available prokaryotic genome sequences is still limited, the genomics community is moving aggressively to extend that inventory. In the meantime, it is clear that environmental, regulatory, and diagnostic analyses can be effectively carried out by MS using more limited or bounded sets of reference microbial genomes.

SUMMARY POINTS

1. MS determines with unprecedented accuracy an intrinsic molecular property: the masses of a set of biomarker molecules, which can uniquely characterize a microorganism.

2. Success has been demonstrated in the rapid characterization of microorganisms isolated/captured from various environments (e.g., air, water, culture medium, bodily fluids, and food) by mass spectrometry.

3. MS provides speed, sensitivity, and specificity for microorganism detection and identification. It is also broad band, and can be automated and multiplexed by, for instance, rapid screening of samples in 96- or 384-well plates. It is, however, expensive.

4. The applications of MS for microorganism characterization are significantly enhanced when it is combined with proteomic and bioinformatic strategies.

FUTURE ISSUES

1. Improved identification of individual microorganisms in mixtures (finding the "needle in the haystack") will be critical for future clinical applications.

2. Rapid characterization of bioengineered and mutated organisms is needed.

3. Contributions by MS will allow studies of microbial community metaproteomics to progress rapidly.

4. Rugged protein arrays (e.g., antibody or lectin), which can be read by MS, will be valuable for targeted applications.

5. Developments in microfluidics are expected to provide improved interfaces for sample preparation and transfer to ESI and MALDI.

6. Single microorganism trapping (e.g., in ion traps) and accurate mass determination will provide insights into the variability among individual cells and into the changes in total mass during, for instance, sporulation.

DISCLOSURE STATEMENT

A US patent (no. 7020559; listed as Reference 78 in this review) was issued in 2006. The patent rights holder is the University of Maryland. So far the patent has not been licensed.

ACKNOWLEDGMENTS

We extend special thanks to our collaborators in this area through the last 30 years, and to colleagues in other institutions and countries who have made outstanding contributions to this important field. We thank Dr. John Barr from the Centers for Disease Control for sharing a prepublication manuscript. Work in our laboratories has been funded at various times by a number of US Government agencies. PAD has been supported in part by Applied Physics Laboratory funds during the preparation of this review.

LITERATURE CITED

1. Tang YW, Stratton CW, eds. 2006. *Advanced Techniques in Diagnostic Microbiology*. New York, NY: Springer-Verlag
2. Anhalt JP, Fenselau C. 1975. Identification of bacteria using mass spectrometry. *Anal. Chem.* 47:219–25
3. Fenselau C, ed. 1994. *Mass Spectrometry for the Characterization of Microorganisms*. Washington, DC: Am. Chem. Soc.
4. Fenselau C, Demirev P. 2001. Characterization of intact microorganisms by MALDI mass spectrometry. *Mass Spectrom. Rev.* 20:157–71
5. Lay J. 2001. MALDI-TOF mass spectrometry of bacteria. *Mass Spectrom. Rev.* 20:172–94
6. Dworsanski JP, Snyder PA. 2005. Classification and identification of bacteria using mass spectrometry-based proteomics. *Expert Rev. Proteomics* 2:863–78
7. Wilkins C, Lay J, eds. 2006. *Identification of Microorganisms by Mass Spectrometry*. Hoboken, NJ: Wiley
8. Jungblut PR, Hecker M, eds. 2007. *Proteomics of Microbial Pathogens*. Weinheim, Ger.: Wiley
9. Demirev PA. 2008. Mass spectrometry of infectious pathogens. In *Medical Applications of Mass Spectrometry*, ed. K Vekey, A Telekes, A Vertes, pp. 283–300. Amsterdam: Elsevier
10. National Research Council. 2005. *Sensor Systems for Biological Agent Attacks*. Washington, DC: Natl. Acad. Press
11. Demirev PA, Feldman AB, Lin JS. 2005. Chemical and biological weapons: current concepts for future defenses. *Johns Hopkins APL Tech. Digest* 26:321–33
12. Dickinson DN, LaDuc MT, Haskins WE, Gorhushkin I, Winefordner JD, et al. 2004. Species differentiation of a diverse suite of *Bacillus* spores by mass spectrometry–based protein profiling. *Appl. Environ. Microbiol.* 70:475–82
13. Ecker DF, Sampath R, Blyn LB, Eshoo MW, Ivy C, et al. 2005. Rapid identification and strain-typing of respiratory pathogens for epidemic surveillance. *Proc. Natl. Acad. Sci. USA* 102:8012–17
14. Von Wintzingerode F, Bocker S, Schlotelburg C, Chiu NH, Storm N, et al. 2002. Base-specific fragmentation of amplified 16S rRNA genes analyzed by mass spectrometry: a tool for rapid bacterial identification. *Proc. Natl. Acad. Sci. USA* 99:7039–44
15. Jackson GW, McNichols RJ, Fox GE, Willson RC. 2007. Universal bacterial identification by mass spectrometry of 16S ribosomal RNA cleavage products. *Intern. J. Mass Spectrom.* 261:218–26

16. Demirev PA, Feldman AB, Kongkasuriyachai D, Scholl P, Sullivan D Jr, Kumar N. 2002. Detection of malaria parasites by laser desorption mass spectrometry. *Anal. Chem.* 74:3262–66

17. Whiteaker JR, Fenselau C, Fetterolf D, Steele D, Wilson D. 2004. Quantitative determination of heme for forensic characterization of *Bacillus* spores using matrix-assisted laser desorption/ionization time-of-flight mass spectrometry. *Anal. Chem.* 76:2836–41

18. Heller DN, Murphy CM, Cotter RJ, Fenselau C, Uy OM. 1988. Constant neutral loss scanning for the characterization of bacterial phospholipids desorbed by fast atom bombardment. *Anal. Chem.* 60:2787–91

19. Griest WH, Lammert SA. 2006. The development of the block II chemical biological mass spectrometer. In *Identification of Microorganisms by Mass Spectrometry*, ed. C Wilkins, J Lay, pp. 61–90. Hoboken, NJ: Wiley

20. Bryden WA, Benson RC, Ecelberger SA, Phillips TE, Cotter RJ, Fenselau C. 1995. The tiny-TOF mass-spectrometer for chemical and biological sensing. *Johns Hopkins APL Tech. Digest* 16:296–310

21. Cotter RJ, Fancher C, Cornish TJ. 1999. Miniaturized time-of-flight mass spectrometer for peptide and oligonucleotide analysis. *J. Mass Spectrom.* 34:1368–72

22. Cooks RG, Badman ER. 2000. Miniature mass analyzers. *J. Mass Spectrom.* 35:659–71

23. Madonna AJ, Voorhees KJ, Taranenko NI, Laiko VV, Doroshenko VM. 2003. Detection of cyclic lipopeptide biomarkers from *Bacillus* species using atmospheric pressure matrix-assisted laser desorption/ionization mass spectrometry. *Anal. Chem.* 75:1628–37

24. Cornish TJ, Antoine MD, Ecelberger SA, Demirev PA. 2005. Arrayed time-of-flight mass spectrometry for time-critical detection of hazardous agents. *Anal. Chem.* 77:3954–59

25. Demirev P, Ramirez J, Fenselau C. 2001. Tandem mass spectrometry of intact proteins for characterization of biomarkers from *Bacillus cereus* T spores. *Anal. Chem.* 73:5725–31

26. Meng F, Cargile BJ, Miller LM, Forbes AJ, Johnson JR, Kelleher NL. 2001. Informatics and multiplexing of intact protein identification in bacteria and the archaea. *Nat. Biotechnol.* 19:952–57

27. Jones JJ, Stump MJ, Fleming RC, Lay JO, Wilkins CL. 2003. Investigation of MALDI-TOF and FT-MS techniques for analysis of *Escherichia coli* whole cells. *Anal. Chem.* 75:1340–47

28. VerBerkmoes NC, Hervey WJ, Shah M, Land M, Hauser L, et al. 2005. Evaluation of "shotgun" proteomics for identification of biological threat agents in complex environmental matrixes: experimental simulations. *Anal. Chem.* 77:923–32

29. Demirev P, Feldman A, Kowalski P, Lin J. 2005. Top-down proteomics for rapid identification of intact microorganisms. *Anal. Chem.* 77:7455–61

30. Nobel Laureates. 2002. *Chemistry*. **http://www.Nobelprize.org**

31. Despeyroux D, Phillpotts R, Watts P. 1996. Electrospray mass spectrometry for detection and characterization of purified cricket paralysis virus (CrPV). *Rapid Commun. Mass Spectrom.* 10:937–41

32. Ramirez J, Fenselau C. 2001. Factors contributing to peak broadening and mass accuracy in the characterization of intact spores using matrix-assisted laser desorption/ionization coupled with time of flight mass spectrometry. *J. Mass Spectrom.* 36:929–36

33. Pribil PA, Patton E, Black G, Doroshenko V, Fenselau C. 2005. Rapid characterization of *Bacillus* spores targeting species-unique peptides produced with an atmospheric pressure matrix-assisted laser desorption/ionization source. *J. Mass Spectrom.* 40:464–74

34. Song YS, Talaty N, Tao WA, Pan ZZ, Cooks RG. 2007. Rapid ambient mass spectrometric profiling of intact, untreated bacteria using desorption electrospray ionization. *Chem. Commun.* 61–63

35. Pierce CY, Barr JR, Cody RB, Massung RF, Woolfitt AR, et al. 2007. Ambient generation of fatty acid methyl ester ions from bacterial whole cells by direct analysis in real time (DART) mass spectrometry. *Chem. Commun.* 807–9

36. Madonna AJ, Basile F, Furlong E, Voorhees KJ. 2001. Detection of bacteria from biological mixtures using immunomagnetic separation combined with matrix-assisted laser desorption/ionization time-of-flight mass spectrometry. *Rapid Commun. Mass Spectrom.* 15:1068–74

37. Whiteaker J, Karns J, Fenselau C, Perdue M. 2004. Analysis of *Bacillus anthracis* spores in milk using mass spectrometry. *Foodborne Pathogens Dis.* 1:185–94

38. Bundy J, Fenselau C. 1999. Lectin-based affinity capture for MALDI-MS analysis of bacteria. *Anal. Chem.* 71:1460–63

39. Bundy JL, Fenselau C. 2001. Lectin and carbohydrate affinity capture surfaces for mass spectrometric analysis of microorganisms. *Anal. Chem.* 73:751–57

40. Stowers MA, van Wuijckhuijse AL, Marijnissen JC, Scarlett B, van Baar BL, Kientz CE. 2000. Application of matrix-assisted laser desorption/ionization to on-line aerosol time-of-flight mass spectrometry. *Rapid Commun. Mass Spectrom.* 14:829–33

41. Kim JK, Jackson SN, Murray KK. 2005. Matrix-assisted laser desorption/ionization mass spectrometry of collected bioaerosol particles. *Rapid Commun. Mass Spectrom.* 19:1725–29

42. Tobias HJ, Schafer MP, Pitesky M, Fergenson DP, Horn J, et al. 2005. Bioaerosol mass spectrometry for rapid detection of individual airborne *Mycobacterium tuberculosis* H37Ra particles. *Appl. Environ. Microbiol.* 71:6086–95

43. van Wuijckhuijse AL, Stowers MA, Kleefsman WA, van Baar BL, Kientz CE, Marijnissen JC. 2005. Matrix-assisted laser desorption/ionization aerosol time-of-flight mass spectrometry for the analysis of bioaerosols: development of a fast detector for airborne biological pathogens. *J. Aerosol Sci.* 36:677–87.

44. Desai MJ, Armstrong DW. 2003. Separation, identification and characterization of microorganisms by capillary electrophoresis. *Microbiol. Mol. Biol. Rev.* 67:38–51

45. Lee H, Williams SK, Wahl KL, Valentine NB. 2003. Analysis of whole bacterial cells by field-flow fractionation and matrix-assisted laser desorption/ionization time-of-flight mass spectrometry. *Anal. Chem.* 75:2746–52

46. Fuerstenau SD, Benner WH, Thomas JJ, Brugidou C, Bothner B, Siuzdak G. 2001. Mass spectrometry of an intact virus. *Angew. Chem. Int. Ed.* 40:542–44

47. Peng WP, Yang YC, Kang MW, Lee YT, Chang HC. 2004. Measuring masses of single bacterial whole cells with a quadrupole ion trap. *J. Am. Chem. Soc.* 126:11766–67

48. Thomas JJ, Falk B, Fenselau C, Jackman J, Ezzell J. 1998. Viral characterization by direct analysis of capsid proteins. *Anal. Chem.* 70:3863–67

49. Scholl PF, Leonardo MA, Rule AM, Carlson MA, Antoine MD, Buckley TJ. 1999. The development of matrix-assisted laser desorption/ionization time-of-flight mass spectrometry for the detection of biological warfare agent aerosols. *Johns Hopkins APL Tech. Digest* 20:343–51

50. Kim YJ, Freas A, Fenselau C. 2001. Analysis of viral glycoproteins by MALDI-TOF mass spectrometry. *Anal. Chem.* 73:1544–48

51. Hathout Y, Setlow B, Cabrera-Marrinez RM, Fenselau C, Setlow P. 2003. Small acid-soluble proteins as biomarkers in mass spectrometry analysis of *Bacillus* spores. *Appl. Environ. Microbiol.* 69:1100–7

52. Antoine MD, Carlson MA, Drummond WR, Doss OW, Hayek CS, et al. 2004. Mass spectral analysis of biological agents using the BioTOF mass spectrometer. *Johns Hopkins APL Tech. Digest* 25:20–26

53. Warscheid B, Fenselau C. 2004. A targeted proteomics approach to the rapid identification of bacterial cell mixtures by MALDI mass spectrometry. *Proteomics* 4:2877–92

54. Wunschel DS, Hill EA, LcLean JS, Jarman K, Gorby YA, et al. 2005. Effects of varied pH, growth rate and temperature using controlled fermentation and batch culture on matrix assisted laser desorption/ionization whole cell protein fingerprints. *J. Microbiol. Methods* 62:259–71

55. Warscheid B, Jackson K, Sutton C, Fenselau C. 2003. MALDI analysis of *Bacilli* in spore mixtures by applying a quadrupole ion trap time of flight tandem mass spectrometer. *Anal. Chem.* 75:5608–17

56. English R, Warscheid B, Fenselau C, Cotter R. 2003. *Bacillus* spore identification via proteolytic peptide mapping with a miniaturized MALDI TOF mass spectrometer. *Anal. Chem.* 75:6886–93

57. Hauser N, Zhang S, Smetana K, Basile F. 2004. Non-enzymatic cell digestion procedure for proteomic-based microorganism identification. *Proc. Am. Soc. Mass Spectrom., 52nd, Nashville, TN* . Santa Fe: Am. Soc. Mass Spectrom.

58. Swatkoski S, Russell S, Edwards N, Fenselau C. 2006. Rapid chemical digestion of small acid-soluble spore proteins for analysis of *Bacillus* spores. *Anal. Chem.* 78:181–88

59. Swatkoski S, Russell S, Edwards N, Fenselau C. 2007. Analysis of a model virus using residue-specific chemical cleavage and MALDI TOF mass spectrometry. *Anal. Chem.* 79:654–58

60. Wahl KL, Wunschel SC, Jarman KH, Valentine NB, Petersen CE, et al. 2002. Analysis of microbial mixtures by matrix-assisted laser desorption/ionization time-of-flight mass spectrometry. *Anal. Chem.* 74:6191–99

61. Williams TL, Andrzejewski D, Lay JO, Musser SM. 2003. Experimental factors affecting the quality and reproducibility of MALDI TOF mass spectra obtained from whole bacteria cells. *J. Am. Soc. Mass Spectrom.* 14:342–51

62. Valentine N, Wunschel S, Wunschel D, Petersen C, Wahl K. 2005. Effect of culture conditions on microorganism identification by matrix-assisted laser desorption ionization mass spectrometry. *Appl. Environ. Microbio.* 71:58–64

63. Wunschel SC, Jarman KH, Petersen CE, Valentine NB, Wahl KL, et al. 2005. Bacterial analysis by MALDI-TOF mass spectrometry: an interlaboratory comparison. *J. Am. Soc. Mass Spectrom.* 16:456–62

64. Marvin LF, Roberts MA, Fay LB. 2003. Matrix-assisted laser desorption/ionization time-of-flight mass spectrometry in clinical chemistry. *Clin. Chim. Acta* 337:11–21

65. Dieckmann R, Graeber I, Kaesler I, Szewzyk U, von Dohren H. 2005. Rapid screening and dereplication of bacterial isolates from marine sponges of the Sula Ridge by Intact-Cell-MALDI-TOF mass spectrometry (ICM-MS). *Appl. Microbio. Biotech.* 67:539–48

66. Jackson KA, Edwards-Jones V, Sutton CW, Fox AJ. 2005. Optimization of intact cell MALDI method for fingerprinting of methicillin-resistant *Staphylococcus aureus*. *J. Microbio. Methods* 62:273–84

67. Mazzeo MF, Sorrentino A, Gaita M, Cacace G, Di Stasio M, et al. 2006. Matrix-assisted laser desorption ionization-time of flight mass spectrometry for the discrimination of food-borne microorganisms. *Appl. Environ. Microbio.* 72:1180–89

68. Wilkes JG, Buzatu DA, Dare DJ, Dragan YP, Chiarelli MP, et al. 2006. Improved cell typing by charge-state deconvolution of matrix-assisted laser desorption/ionization mass spectra. *Rapid Commun. Mass Spectrom.* 20:1595–603

69. Hettick JM, Kashon ML, Slaven JE, Ma Y, Simpson JP, et al. 2006. Discrimination of intact *mycobacteria* at the strain level: a combined MALDI-TOF MS and biostatistical analysis. *Proteomics* 6:6416–25

70. Pierce CY, Barr JR, Woolfitt AR, Moura H, Shaw EI, et al. 2007. Strain and phase identification of the US category B agent *Coxiella burnetii* by matrix assisted laser desorption/ionization time-of-flight mass spectrometry and multivariate pattern recognition. *Anal. Chim. Acta* 583:23–31

71. Siegrist TJ, Anderson PD, Huen WH, Kleinheinz GT, McDermott CM, Sandrin TR. 2007. Discrimination and characterization of environmental strains of *Escherichia coli* by matrix-assisted laser desorption/ionization time-of-flight mass spectrometry (MALDI-TOF-MS). *J. Microbio. Methods* 68:554–62

72. Carbonnelle E, Beretti JL, Cottyn S, Quesne G, Berche P, et al. 2007. Rapid identification of *Staphylococci* isolated in clinical microbiology laboratories by matrix-assisted laser desorption ionization-time of flight mass spectrometry. *J. Clin. Microbio.* 45:2156–61

73. Demirev PA, Ho YP, Ryzhov V, Fenselau C. 1999. Microorganism identification by mass spectrometry and protein database searches. *Anal. Chem.* 71:2732–38

74. Pineda F, Lin JS, Fenselau C, Demirev P. 2000. Testing the significance of microorganism identification by mass spectrometry and proteome database search. *Anal. Chem.* 72:3739–44

75. Demirev P, Lin J, Pineda F, Fenselau C. 2001. Bioinformatics and mass spectrometry for microorganism identification: proteome-wide post-translational modifications and database search algorithms for characterization of intact *H. Pylori. Anal. Chem.* 73:4566–73

76. Pineda FJ, Antoine MD, Demirev PA, Feldman AB, Jackman J, et al. 2003. Microorganism identification by matrix-assisted laser/desorption ionization mass spectrometry and model-derived ribosomal protein biomarkers. *Anal. Chem.* 75:3817–22

77. Demirev P, Feldman A, Lin J. 2004. Bioinformatics-based strategies for rapid microorganisms identification by mass spectrometry. *Johns Hopkins APL Tech. Digest* 25:27–37

78. Demirev P, Fenselau C. 2006. *U.S. Patent No. 7020559*

79. Demirev P. 2004. Enhanced specificity of bacterial spore identification by oxidation and mass spectrometry. *Rapid Commun. Mass Spectrom.* 18:2719–22

80. Fagerquist CK, Bates AH, Heath S, King BC, Garbus BR, et al. 2006. Subspeciating *Campylobacter jejuni* by proteomic analysis of its protein biomarkers and their post-translational modifications. *J. Prot. Res.* 5:2527–38

81. Sun LW, Teramoto K, Sato H, Torimura M, Tao H, Shintani T. 2006. Characterization of ribosomal proteins as biomarkers for matrix-assisted laser desorption/ionization mass spectral identification of *Lactobacillus plantarum. Rapid Commun. Mass Spectrom.* 20:3789–98

82. Gantt SL, Valentine NB, Saenz AJ, Kingsley MT, Wahl KL. 1999. Use of an internal control for matrix-assisted laser desorption/ionization time-of-flight mass spectrometry analysis of bacteria. *J. Am. Soc. Mass Spectrom.* 10:1131–1137

83. Natl. Cent. Biotech. Info. *Genome.* **http://www.ncbi.nlm.nih.gov/sites/entrez?db=genome**

84. Karlin S, Campbell AM, Mrazek J. 1998. Comparative DNA analysis across diverse genomes. *Annu. Rev. Genetics* 32:185–225

85. Smith RD, Anderson GA, Lipton MS, Pasa-Tolic L, Shen Y, et al. 2002. An accurate mass tag strategy for quantitative and high-throughput proteome measurements. *Proteomics* 2:513–23

86. Wang ZP, Dunlop K, Long SR, Li L. 2002. Mass spectrometric methods for generation of protein mass database used for bacterial identification. *Anal. Chem.* 74:3174–82

87. Williams TL, Leopold P, Musser S. 2002. Automated postprocessing of electrospray LC/MS data for profiling protein expression in bacteria. *Anal. Chem.* 74:5807–13

88. Rapid Microorganism Identification Database. **http://www.RMIDb.org**

89. Mortz E, O'Connor P, Roepstorff P, Kelleher N, Wood T, McLafferty F. 1996. Sequence tag identification of intact proteins by matching tandem mass spectral data against sequence data bases. *Proc. Nat. Acad. Sci. USA* 93:8264–67

90. Castanha ER, Vestal M, Hattan S, Fox A, Fox KF, Dickinson D. 2007. *Bacillus cereus* strains fall into two clusters (one closely and one more distantly related to *Bacillus anthracis*) according to amino acid substitutions in small acid-soluble proteins as determined by tandem mass spectrometry. *Mol. Cell. Probes* 21:190–201

91. Eng JK, McCormack AL, Yates JR. 1994. An approach to correlate tandem mass spectral data of peptides with amino acid sequences in a protein database. *J. Am. Soc. Mass Spectrom.* 5:976–89

92. Warscheid B, Fenselau C. 2003. Characterization of *Bacillus* spore species and their mixtures using postsource decay with a curved-field reflectron. *Anal. Chem.* 75:5618–27

93. Russell SC, Edwards N, Fenselau C. 2007. Detection of plasmid insertion in *Escherichia coli* by MALDI-TOF mass spectrometry. *Anal. Chem.* 79:5399–406

94. Fenselau C, Russell S, Swatkoski S, Edwards N. 2007. Proteomic strategies for rapid analysis of microorganisms. *Eur. J. Mass Spectrom.* 13:35–39

95. Dworzanski JP, Snyder AP, Chen R, Zhang HY, Wishart D, Li L. 2004. Identification of bacteria using tandem mass spectrometry combined with a proteome database and statistical scoring. *Anal. Chem.* 76:2355–66

96. Hu A, Tsai PJ, Ho YP. 2005. Identification of microbial mixtures by capillary electrophoresis/selective tandem mass spectrometry. *Anal. Chem.* 77:1488–95

97. Lo AL, Hu A, Ho YP. 2006. Identification of microbial mixtures by LC-selective proteotypic-peptide analysis (SPA). *J. Mass Spectrom.* 41:1049–60

98. Norbeck AD, Callister SJ, Monroe ME, Jaitly N, Elias DA, et al. 2006. Proteomic approaches to bacterial differentiation. *J. Microbio. Methods* 67:473–86

99. Krishnamurthy T, Deshpande S, Hewel J, Liu HB, Wick CH, Yates JR. 2007. Specific identification of *Bacillus anthracis* strains. *Inter. J. Mass Spectrom.* 259:140–46

100. Castanha ER, Fox A, Fox KF. 2006. Rapid discrimination of *Bacillus anthracis* from other members of the *B. cereus* group by mass and sequence of "intact" small acid soluble proteins (SASPs) using mass spectrometry. *J. Microbio. Methods* 67:230–40

101. Yao Z, Demirev P, Fenselau C. 2002. Mass spectrometry-based proteolytic mapping for rapid virus identification. *Anal. Chem.* 744:2529–34

102. Leenders F, Stein TH, Kablitz B, Franke P, Vater J. 1999. Rapid typing of *Bacillus subtilis* strains by their secondary metabolites using matrix-assisted laser desorption ionization mass spectrometry. *Rapid Commun. Mass Spectrom.* 14:2393–400

103. Hathout Y, Ho YP, Ryzhov V, Demirev P, Fenselau C. 2000. Kurstakins: a new class of lipopeptides isolated from *Bacillus thuringiensis*. *J. Nat. Prod.* 63:1492–96

104. Elhanany E, Barak R, Fisher M, Kobiler D, Altboum Z. 2001. Detection of specific *Bacillus anthracis* spore biomarkers by matrix-assisted laser desorption/ionization time-of-flight mass spectrometry. *Rapid Commun. Mass Spectrom.* 15:2110–16

105. Williams BH, Hathout Y, Fenselau C. 2002. Structural characterization of lipopeptide biomarkers isolated from *Bacillus globigii*. *J. Mass Spectrom.* 37:259–64

106. Jegorov A, Haiduch M, Sulc M, Havlicek V. 2006. Nonribosomal cyclic peptides: specific markers of fungal infections. *J. Mass Spectrom.* 41:563–76

107. Price NP, Rooney AP, Swezey JL, Perry E, Cohan FM. 2007. Mass spectrometric analysis of lipopeptides from *Bacillus* strains isolated from diverse geographical locations. *FEMS Microbiol. Lett.* 271:83–89

108. Rees JC, Voorhees KJ. 2005. Simultaneous detection of two bacterial pathogens using bacteriophage amplification coupled with matrix-assisted laser desorption/ionization time-of-flight mass spectrometry. *Rapid Commun. Mass Spectrom.* 19:2757–61

109. Boyer AD, Quinn CP, Woolfitt AR, Pirkle JL, McWilliams LG, et al. 2007. Detection and quantification of anthrax lethal factor in serum by mass spectrometry. *Anal. Chem.* 79:8463–70

Scanning Electrochemical Microscopy

Shigeru Amemiya,[1] Allen J. Bard,[2]
Fu-Ren F. Fan,[2] Michael V. Mirkin,[3]
and Patrick R. Unwin[4]

[1] University of Pittsburgh, Department of Chemistry, Pittsburgh, Pennsylvania 15260
[2] University of Texas at Austin, Department of Chemistry and Biochemistry, Austin, Texas 78712; email: ajbard@mail.utexas.edu
[3] Queens College, Department of Chemistry and Biochemistry, Flushing, New York 11367
[4] University of Warwick, Department of Chemistry, Coventry CV4 7AL, United Kingdom

Annu. Rev. Anal. Chem. 2008. 1:95–131

First published online as a Review in Advance on December 20, 2007

The *Annual Review of Analytical Chemistry* is online at anchem.annualreviews.org

This article's doi:
10.1146/annurev.anchem.1.031207.112938

1936-1327/08/0719-0095$20.00

Key Words

ultramicroelectrode, charge transport, interface, scanning probe microscopy, electroanalytical chemistry

Abstract

This review describes work done in scanning electrochemical microscopy (SECM) since 2000 with an emphasis on new applications and important trends, such as nanometer-sized tips. SECM has been adapted to investigate charge transport across liquid/liquid interfaces and to probe charge transport in thin films and membranes. It has been used in biological systems like single cells to study ion transport in channels, as well as cellular and enzyme activity. It is also a powerful and useful tool for the evaluation of the electrocatalytic activities of different materials for useful reactions, such as oxygen reduction and hydrogen oxidation. SECM has also been used as an electrochemical tool for studies of the local properties and reactivity of a wide variety of materials, including metals, insulators, and semiconductors. Finally, SECM has been combined with several other nonelectrochemical techniques, such as atomic force microscopy, to enhance and complement the information available from SECM alone.

1. INTRODUCTION

Scanning electrochemical microscopy (SECM) employs an ultramicroelectrode (UME), also known as a tip, to scan in close proximity to a surface of interest. The electrochemical response of the tip (or of the substrate in response to the tip) provides quantitative information about the interfacial region. Since the technique was first reported in 1989, approximately 1000 papers have been published on its methodology and applications, and numerous reviews have appeared. The basics of the technique, the instrumentation, and many applications through about 2000 have been thoroughly covered in a monograph (1). This review mainly covers work that has been done since 2000 and emphasizes the new and important trends in SECM.

2. NANOMETER-SIZED TIPS AND THEIR APPLICATIONS

2.1. Overview

Nanometer-sized tips offer important advantages for various SECM applications. In kinetic experiments, nanoelectrodes offer high mass-transfer rates under steady-state conditions in combination with practically negligible effects of the resistive potential drop in solution, double layer charging current, and low levels of reactant adsorption. The high spatial resolution achievable with such tips makes nanoscale electrochemical imaging possible. The small size of the scanning probe is essential for subcellular-level studies of biological systems and single-molecule experiments.

2.2. Preparation and Characterization

A simple way to produce nm-sized conical tips is to etch a metal wire and then coat it with an insulator—e.g., molten Apiezon wax (2)—while leaving the apex exposed. Investigators have tried other types of coating, including varnish, molten paraffin, silica coating, poly(α-methylstyrene), polyimide (3), electropolymerized phenol, and electrophoretic paint (4). Sub-μm-sized conical carbon tips have been prepared using flame etching (5).

Disk-type tips can be prepared using a micropipette puller (6–8). Recently, it was shown that a tip as small as $\sim$10-nm-radius can be polished on a lapping tape under video-microscopic control (8). The polished flat nanotips yield more reliable and reproducible data and can be used for fast kinetic measurements. Another type of nm-sized SECM probe—the slightly recessed nanotip—has also been prepared by etching polished Pt nanoelectrodes (9).

The shape and size of a nanotip can be characterized by a combination of voltammetry and SECM (6, 8–10). In **Figure 1**, an SECM approach curve (*a*) and a steady-state voltammogram (*b*) were obtained with the same polished Pt electrode. The experimental current versus distance curve in **Figure 1a** fits the theoretical curve calculated with $a = 46$ nm that was obtained from the limiting current in **Figure 1b**. This radius value is reliable because of the high positive feedback current (up to a normalized current of 8) obtained in **Figure 1a**. The distance of the closest

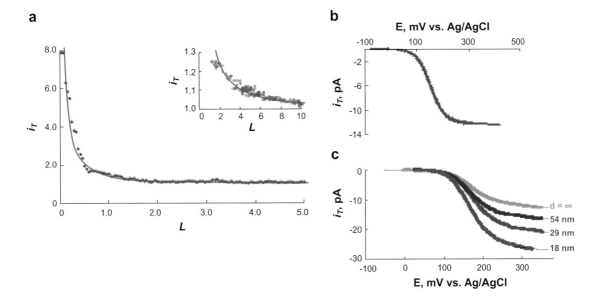

Figure 1

(*a*) Scanning electrochemical microscopy current versus distance curve; (*b*) steady-state voltammogram obtained in the bulk solution; and (*c*) four voltammograms for different tip/substrate separation distances. Aqueous solution contained 1 mM ferrocenemethanol and 0.2 M NaCl. (*a*) Solid line indicates theoretical approach curve for diffusion-controlled positive feedback; symbols represent experimental data. The tip approached the unbiased Au film substrate with a speed of 5 nm/s. The tip was either a 46-nm (panels *a* and *b*) or a 36-nm (panel c) polished Pt electrode. The inset in panel a shows an approach curve for a 13-nm tip. (*c*) Experimental (*symbols*) and theoretical (*solid lines*) voltammograms obtained at the tip-substrate distance, $d = \infty$ (*1*), 54 nm (*2*), 29 nm (*3*), and 18 nm (*4*). Adapted with permission from Reference 8. Copyright 2007, American Chemical Society.

approach in **Figure 1a** (~5 nm) could be achieved only with the essentially flat and well-polished tip surface.

2.3. Penetration Experiments

Nm-sized tips can be used to penetrate a microstructure (e.g., a sub-μm-thick polymer film containing fixed redox centers or loaded with a redox mediator) and can extract spatially resolved information about concentrations and kinetic and mass-transport parameters (11, 12). With a tip inside the film, relatively far from the underlying conductor or insulator, solid-state voltammetry at the tip can be carried out similarly to conventional voltammetric experiments in solution. At smaller distances, the tip current either increases or decreases depending on the rate of the mediator regeneration at the substrate. If the film is homogeneous and not very resistive, the current-distance curves are similar to those obtained in solution. In this way, kinetic parameters, diffusion coefficients, and formal potential for $Os(bpy)_3^{2+/3+}$ in Nafion were extracted from steady-state voltammograms at a nm-sized conical tip (11). For poly(vinylferrocene), the depth profile of film resistance was obtained by scanning the tip across a 300-nm-thick layer (12).

2.4. Imaging

The attainable resolution in SECM depends upon the tip radius and separation distance between tip and sample. The use of nm-sized tips allows one to increase the lateral resolution from micrometers to nanometers. However, feedback-mode constant-height SECM imaging on the nanoscale is challenging because of the possibility of a tip crash. Few examples of such images have been published to date (13). Constant-distance imaging (14) can be used to overcome this problem.

The generation/collection mode imaging with nm-sized probes is easier because the tip/substrate separation distance does not have to be very small. It is useful for mapping fluxes and concentration profiles of electroactive species and studying diffusion and convection processes in electrochemical systems (15) (**Figure 2**).

The highest spatial SECM resolution reported to date was achieved in imaging insulators in humid air (16). In this mode, a conical tip as used for scanning tunneling microscopy partially penetrates an Å-thick aqueous layer that is present on the surface. The tip is scanned over the surface in a constant-current mode to obtain nanoscale topographic images. This arrangement was used to image molecular steps on the mica surface as well as DNA cast on mica. Another approach to nanoscale electrochemical imaging—SECM-AFM imaging (17, 18)—is discussed in Section 8 below.

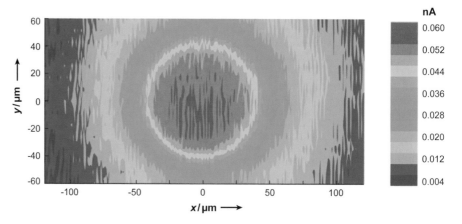

Figure 2

Two-dimensional map of currents (concentrations) measured by an 80-nm tip (collector) in the vicinity of a 40-μm-radius disk electrode (generator). The nanoprobe was scanned parallel to the microelectrode plane at a constant vertical distance $d = 5$ μm with a speed of 8 μm/s. Aqueous solution contained 5 mM $[Ru(NH_3)_6]Cl_3$. The generator potential of the microelectrode was placed on the plateau of the reduction wave of $[Ru(NH_3)_6]^{3+}$, and the nanoelectrode probe potential was set on the plateau of the $[Ru(NH_3)_6]^{2+}$ oxidation wave. Reprinted with permission from Reference 15. Copyright 2004, Wiley-VCH Verlag GmbH & Co. KGaA, Weinheim.

2.5. Feedback-Mode Kinetic Measurements and Single-Molecule Experiments

Feedback mode of the SECM was employed to measure the kinetics of several fast electron transfer (ET) reactions at nm-sized electrodes (8). Polished Pt nanoelectrodes were characterized by a combination of voltammetry, scanning electron microscopy (SEM), and SECM, as discussed above. The substrate was kept at a constant potential, such that the ET reaction at its surface was diffusion-controlled, while the tip potential was swept slowly to record a steady-state voltammogram. A number of voltammograms were obtained at the same nanoelectrode at different tip/substrate separation distances (**Figure 1c**). In this way, the self-consistent kinetic parameter values for several rapid ET reactions, of which the fastest was $Ru(NH_3)_6^{3+}$–reduction in KCl ($k° = 17.0 \pm 0.9$ cm/s), were obtained at electrodes of different radii (3 to 400 nm) with an uncertainty margin of ~10%.

In single molecule experiments, positive feedback was used to amplify low signals, which were produced by oxidation/reduction of single molecules (19, 20). The active area of a ~15-nm-diameter Apiezon wax-sealed Pt/Ir tip was slightly recessed into the insulating sheath so that, at small tip/substrate distances, solution was trapped in a nm-sized wax pocket between two electrodes. When this tip was held at ~10 nm from the indium-tin oxide (ITO) substrate, the pA-range tip current exhibited slow (on the order of tens of s) fluctuations. Although the signal was noisy, the tip current reproducibly showed peaks of 0.7 and 1.4 pA corresponding to the trapping of one and two electroactive molecules, respectively, and periods of essentially zero average current between them. When the tip potential was swept linearly, the stochastic process produced well-defined steady-state voltammograms with the half-wave potential expected for the employed redox mediator.

3. LIQUID/LIQUID INTERFACE

3.1. Principles

The SECM approaches to probing heterogeneous charge-transfer reactions at the interface between two immiscible electrolyte solutions (ITIES) are similar to those used to investigate solid/liquid electrochemistry (21, 22). To study ET at the ITIES, a UME tip is usually placed in the upper liquid phase (e.g., organic solvent) containing one form of the redox species (e.g., the reduced form, R_1). With the tip held at a sufficiently positive potential, R_1 reacts at its surface to produce the oxidized form of the species, O_1. When the tip approaches the ITIES, the mediator can be regenerated at the interface via the bimolecular redox reaction between O_1 in the organic phase (o) and R_2 in the aqueous phase (w):

$$O_1(o) + R_2(w) \rightarrow R_1(o) + O_2(w), \tag{1}$$

and i_T increases with the decrease in the tip/interface distance (positive feedback). The kinetics of such a reaction can be evaluated from approach curves (23).

In the ion-transfer feedback mode, a μm- or nm-sized pipette is filled with a solvent immiscible with the outer solution and used as a tip to approach a macroscopic ITIES.

The interfacial charge-transfer reactions producing the tip current and the feedback are either facilitated (24) or simple (i.e., unassisted) (25) ion transfers rather than ET. Selzer and Mandler (26) used micropipette tips to probe the coupling of ion transfer and ET at the ITIES.

In most SECM experiments, a nonpolarizable ITIES is poised by the concentrations of the potential-determining ion providing a constant driving force for either ET or ion-transfer process. Alternatively, a polarizable ITIES can be externally biased to provide a wider range of interfacial voltages for the study of ET (27) or facilitated ion transfer (28).

3.2. Studies of Electron Transfer

The kinetics of ET between zinc porphyrin dissolved in the organic phase and different aqueous redox species (e.g., $Ru(CN)_6^{4-}$, $Fe(CN)_6^{4-}$, or V^{2+}) were studied at the water/benzene interface (29, 30). The results were in agreement with the main predictions of existing ET theory including the linear dependence of $log(k_f)$ on the interfacial voltage (Tafel plot) with a transfer coefficient, $\alpha = 0.5$. Similar results were obtained for other ET reactions (21, 31, 32). However, the rate of the reverse reaction between ZnPor and $Ru(CN)_6^{3-}$ was found to be essentially potential-independent at the interfaces between water and three organic solvents of different polarities (33). Barker et al. (34) developed a theoretical treatment of the ET at a nonpolarizable ITIES, in which the finite diffusion rate of the reactant in the bottom phase was taken into account. Using this approach, one can measure much faster ET rate constants at the ITIES (31).

Two predictions of the Marcus theory at the liquid/liquid interface were tested, namely the exponential distance dependence of the ET rate and the inverted region behavior (30). The ET rate decreased with the number of C atoms in the hydrocarbon chain of lipid adsorbed at the ITIES, as predicted by ET theory, and the inverted Marcus behavior was observed. Sun et al. (35) used an externally biased polarizable ITIES to control the ET driving force and also observed the inverted Marcus behavior for different combinations of aqueous and organic redox mediators.

The kinetics of interfacial ET between ferrocene dissolved in ionic liquid and aqueous ferrocyanide was probed and found to follow the Butler-Volmer equation (36, 37). The measured bimolecular rate constant was much larger than that obtained at the water/1,2-dichloroethane interface.

Figure 3 shows two possible experimental setups for studying ET reactions between monolayer-protected gold clusters (MPCs) and redox molecules confined to two immiscible liquid phases by SECM (38, 39). Either MPCs (**Figure 3a**) (38) or conventional redox molecules (**Figure 3b**) (39) can be employed as a mediator shuttling the charge between the tip and the ITIES. The latter approach was used to measure the heterogeneous rate constant of the ET between organic soluble Au_{38} clusters and an aqueous $IrCl_6^{2-}$ oxidant.

3.3. Ion Transfer Reactions

In an SECM study of facilitated ion transfer (24), the transfer of K^+ from the nanopipette filled with aqueous solution to the external 1,2-dichloroethane (DCE)

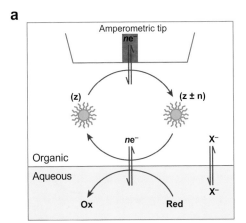
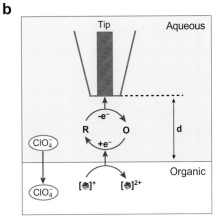

Figure 3

Schematic diagram of two scanning electrochemical microscopy (SECM) approaches to the measurement of the electron transfer rate between an organic-soluble monolayer-protected gold cluster (MPC) and an aqueous redox species. (*a*) The MPC is used as an SECM mediator species. (*b*) Redox molecules shuttle electrons between the tip and the interface between two immiscible electrolyte solutions. Adapted with permission from References 38 and 39. Copyright 2003 and 2004, American Chemical Society.

phase was assisted by dibenzo-18-crown-6 (DB18C6):

$$K^+_{(w)} + DB18C6_{(DCE)} \rightarrow [KDB18C6]^+_{(DCE)} \qquad \text{(at the pipette tip).} \qquad (2)$$

When the tip approached the ITIES, the regeneration of DB18C6 occurred:

$$[KDB18C6]^+_{(DCE)} \rightarrow K^+_{(w)} + DB18C6_{(DCE)} \qquad \text{(at the ITIES),} \qquad (3)$$

and positive feedback current was observed. SECM at an externally biased non-polarizable ITIES was used to measure the kinetics of Reaction 2 (28). The results were in good agreement with those obtained previously by nanopipette voltammetry at a polarizable interface.

Similar SECM experiments can be performed using a simple IT process (25). In this case, both the top and the bottom phases contain the same ion at equilibrium. The micropipette tip is used to decrease the concentration of this common ion in the top solvent near the ITIES.

Some applications of SECM at the ITIES, e.g., works on molecular partitioning (37, 40), and studies of phase-transfer processes by combination of the SECM with a Langmuir trough (41), could not be covered here because of space limitations.

4. MEMBRANES, THIN FILMS AND SELF-ASSEMBLED MONOLAYERS

4.1. Electron Transfer on Self-Assembled Monolayers

SECM has been used for the studies of electron transfer through monolayers supported at solid/liquid (42, 43), liquid/liquid (44, 45) and air/water interfaces (46). In

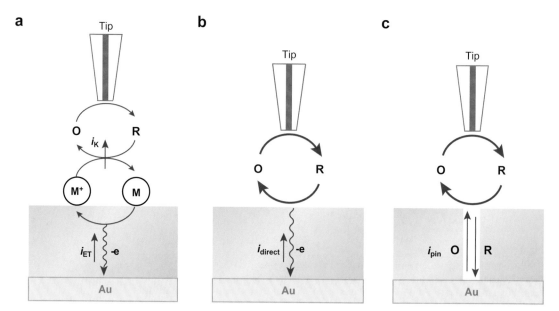

Figure 4

Schematic view of the processes involved in the scanning electrochemical microscopy measurements of electron transfer (ET) across an electroactive self-assembled monolayer: (*a*) mediated ET; (*b*) direct electron tunneling through monolayer; and (*c*) ET through pinholes. Adapted with permission from Reference 42. Copyright 2004, American Chemical Society.

the case of a solid/liquid interface, the rates of ET mediated by monolayer-attached redox moieties and direct ET through the film as well as the rate of a bimolecular ET reaction between the attached and dissolved redox species have been measured using SECM (42). The monolayer may contain redox centers (**Figure 4*a***) or simply act as a blocking layer (**Figure 4*b***).

ET rate kinetics through n-alkanethiol SAMs on Au and Hg have been studied using cyclic voltammetry (47) and SECM and compared with impedance spectroscopy (48, 49) and potentiometry (50). A theoretical model has been developed in an attempt to independently measure the rates of all of these processes (42). According to the model, provided an appropriate mediator is used and the mediator concentration is high such that the bimolecular ET is much faster than the ET through the monolayer, the rate constant of electron tunneling through the monolayer, k_b, is determined by Equation 4:

$$k_b \cong k_{eff}C^*/\Gamma^*, \tag{4}$$

where C^* is the bulk concentration of the redox mediator in solution, Γ^* is the surface coverage of the monolayer-bound redox centers, and k_{eff} is the effective rate constant obtained by fitting an experimental SECM approach curve to SECM theory.

When the tip is very close to the substrate, i.e., $L \ll 1$, L is the normalized tip/substrate distance ($L = d/a$, where d is the distance between the tip and the substrate and a is the tip radius), the mass transfer coefficient is $m_o \sim D/d$ (where D is the diffusion coefficient), and the upper limit for measurable k_{eff} is $\sim 5 m_o = 5\, D/d$. The tip can be brought down to an $L \approx 0.1$. Thus, for a 1-μm-radius tip, $d = 10^{-5}$ cm. Assuming a typical diffusion coefficient $D = 10^{-5}$ cm^2/s, $\Gamma^* \sim 10^{-11}$ mol/cm^2, and $C^* = 2 \times 10^{-4}$ mol/cm^3, the upper limit for measurable k_b is $\sim 10^8$ s^{-1}, calculated using Equation 4. If an nm-sized tip is used, a faster ET can be measured.

On the other hand, if the bimolecular electron transfer is slow in comparison to electron tunneling, and/or the concentration of the redox mediator in solution is low, the following equation holds:

$$k_{Ox} \cong k_{eff}(k_b + k_f)/k_b \Gamma^*, \tag{5}$$

where k_{Ox} is the bimolecular ET rate constant (mol^{-1}cm^{3}s^{-1}), and k_b and k_f are backward and forward electron tunneling rate constants (s^{-1}), respectively. Thus, the bimolecular rate constant for ET between the monolayer-bound and dissolved redox species can be determined based on Equation 5. For a 1-μm-radius tip, the upper limit for measurable k_{Ox} is 10^{11} mol^{-1}cm^{3}s^{-1} (or 10^8 M^{-1}s^{-1}).

4.2. Mass Transfer Across and Lateral Diffusion in Monolayers

SECM (51) allows interfacial dynamics and diffusion to be studied on small length and time scales approaching those relevant to cellular membranes. In recent applications to assemblies at the water/air (W/A) interface, SECM has been used to study lateral proton diffusion along a stearic acid monolayer (46), the effect of a 1-octadecanol monolayer on oxygen transfer across the W/A interface (52), and lateral conductivity in assemblies of metal nanoparticles (53). In these studies, the response of the probe either translated toward or held close to a spot at a target interface was used to obtain quantitative data on a local scale.

For studies involving W/A interface monolayers or Langmuir films, a submarine UME is required. This electrode is immersed in the solution and approaches the layer from below (46, 54). The submarine electrode consists simply of a conventional UME of the desired metal and size fixed to a glass J tube by Teflon tape or epoxy (**Figure 5**); this electrode has also been used for ion channel SECM experiments (55).

Lateral proton diffusion was investigated by steady-state approach curves (51), as surface diffusion contributes primarily to the long-time SECM current response (56). Examples are monolayers comprising either acidic DL-R-phosphatidyl-L-serine, di-palmitoyl (DPPS) or zwitterionic L-R-phosphatidylcholine, dipalmitoyl (DPPC) at a range of surface pressures (57). Typical results, as shown in **Figure 6a**, indicated an enhanced reduction current for the probe approaching the DPPS monolayer. These results were attributed to lateral proton diffusion providing an additional proton source, which is detected by SECM proton feedback mediated through $H_2PO_4^-/HPO_4^{2-}$. In contrast, approach curves for both a native W/A interface (**Figure 6a**) and a DPPC monolayer (**Figure 6b**) showed a current response due only to the diffusion of $H_2PO_4^-$ through solution. For comparison, **Figure 6** also shows the simulated behavior for the DPPS system when the lateral proton diffusion coefficient is comparable

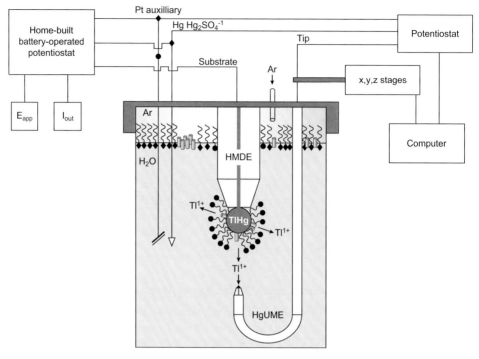

Figure 5

Schematic diagram of an ion channel scanning electrochemical microscopy experiment where a submarine ultramicroelectrode is used. Adapted with permission from Reference 55. Copyright 2002, American Chemical Society.

to the bulk proton diffusion coefficient. In this case, a significant increase in current is predicted as the probe approaches the monolayer that was not observed experimentally. Lateral proton diffusion coefficients at DPPS derived from approach curve measurements are summarized in **Figure 6c**. As the surface pressure increased from 1 to 20 mN/m, the lateral proton diffusion coefficient, D_{lat}, increased from approximately 1.5×10^{-6} cm^2 s^{-1} to 6×10^{-6} cm^2 s^{-1}. At the higher pressures, where A changed only slightly, the lateral diffusion coefficient was reasonably uniform.

4.3. Mass Transport through Phospholipid Monolayer, Bilayer Lipid Membrane, or Liposome

The self-assembling ability of amphiphilic molecules, such as phospholipids, to form bilayers has been widely studied because of the similar physical properties of such artificial membranes and membranes found in biological cells (58). Tsionsky et al. (59) first used SECM to study the kinetics of charge-transfer processes at bilayer lipid membranes (BLMs). Analysis of the SECM tip response demonstrates that an unmodified BLM behaves as an insulator, whereas a BLM doped with iodine shows some positive feedback. Later, using a similar setup, a voltammetric K$^+$-selective

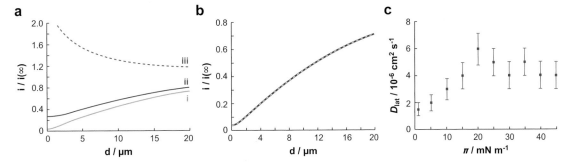

Figure 6

Typical approach curves for the measurement of lateral proton diffusion. (*a*) The solid experimental curves are for the reduction of 0.5 mM $H_2PO_4^-$ at a UME approaching (*i*) a native W/A interface and (*ii*) a DPPS monolayer (p = 20 mN m^{-1}). The dashed curves represent simulations for (*i*) an inert interface; (*ii*) the DPPS system with $D_{lat} = 6 \times 10^{-6}$ cm^2 s^{-1} ($A = 50$ Å^2); and (*iii*) $D_{lat} = 1 \times 10^{-4}$ cm^2 s^{-1} ($A = 50$ Å^2). (*b*) The experimental curve for a DPPC monolayer (*solid*) was similar for all pressures (p = 1, 5, 10, 20, 30, and 40 mN m^{-1}) and the dashed curve (coincident) represents the simulation for an inert interface. (*c*) D_{lat} as a function of surface pressure, *p*, derived from approach curve measurements for the DL-R-phosphatidyl-L-serine, dipalmitoyl system. Adapted with permission from Reference 57. Copyright 2002, American Chemical Society.

micropipette electrode based on valinomycin and ETH 500 in dichloroethane was used as the probe to study K^+ transfer through gramicidin channels in a horizontal BLM in the SECM feedback and generation-collection modes (60).

The difficulty with this measurement is controlling the time at which the ion transport occurs. To have a controlled release of the ion of interest, an ampero-metric approach can also be used but requires the use of Tl(I) as a surrogate for K^+. Ion transport across gramicidin channels embedded in a dioleoylphosphatidyl-choline monolayer adsorbed onto a Tl/Hg amalgam can be used as a membrane system. This arrangement allows one to control the release of Tl(I) into gramicidin channels by preconcentration of the Tl as an amalgam on a hanging mercury drop electrode with controlled oxidation of the amalgam using a potential step. An Hg/Pt submarine UME positioned close to the membrane, as shown in **Figure 5**, collects the generated Tl(I) following its diffusion from the gramicidin.

Giant liposomes encapsulating redox molecules have also been prepared and probed by SECM using sub-µm conical carbon fiber tips, which were employed to approach, image, and puncture individual liposomes (5). Characteristic breakthrough (puncture) current transients corresponded to liposomes of different compartmental configurations, and voltammetric curves with the tip inside the liposome were obtained. The leakage of the encapsulated redox molecules from liposomes was also studied by positioning electrodes close to (<200 nm) the surface of a single liposome.

As shown in **Figure 7** (*right*), different characteristic breakthrough curves were observed, consistent with the different cases shown schematically in **Figure 7** (*left*) (5). In all three cases shown in **Figure 7**, a decrease of current was first observed as the tip traveled down to the surface. After the apex of the tip touched the membrane,

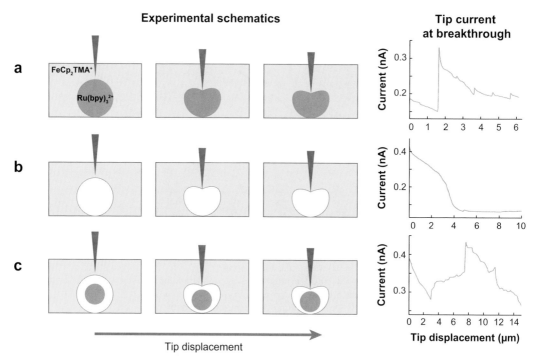

Figure 7

(*Left*) Schematic presentation of different liposomes that undergo shape deformation during the course of tip breakthrough experiments (from left to right). Liposomes are divided into three kinds based on their $Ru(bpy)_3^{2+}$ content and distribution. (*Right*) Scanning electrochemical microscopy monitoring of tip current when it breaks through the bilayer(s) of different liposomes, corresponding to each of the schematics. In this series of experiments, the tip potential was held at 1.2 V (versus AgQRE) where $Ru(bpy)_3^{2+}$ is oxidized to $Ru(bpy)_3^{3+}$. The regular, 1-μm-spaced current spikes in the approach curves are due to the mechanical clicks of the inchworm motor used. A 1-mM ferrocenylmethyltrimethylammonium perchlorate ($FeCp_2TMA^+$) in Tris-HCl buffer (10 mM Tris and 0.1 M NaCl, pH 7.0) was used as the redox mediator. Adapted with permission from Reference 5. Copyright 2006, American Chemical Society.

a further lowering of the tip deformed the lipid bilayers downward, accompanied by a further drop of the current. This effect is reminiscent of the SECM behavior noted when a tip approaches the interface between an aqueous solution and an immiscible liquid and a thin layer of solution is trapped. The breakthrough of the liposome normally took place a few micrometers below the point where the tip first touched the liposome surface, which varied from one experiment to another due to differences in the tip sharpness and liposome size.

At the breakthrough point, a transient with a sharply rising current was obtained for a liposome [**Figure 7** (***right***, **panel *a***)]. This event was attributed to the sudden immersion of the tip into a solution of $Ru(bpy)_3^{2+}$ and a Cottrell current transient along with a contribution from charging current. For a liposome that did not contain electroactive species, Ru(bpy)$_3^{2+}$, a decrease of current was observed on breakthrough [**Figure 7** (*right*, **panel b**)] (5). A rise of current was only observed following a decrease in current [(**Figure 7** (*right*, **panel c**)] when the tip broke into the core of a multilamellar liposome whose core was filled with Ru(bpy)$_3^{2+}$.

4.4. Transport Activity in Thin Films or Membranes

In one type of SECM study on thin films or membranes, the tip is held at a potential where it can detect an electroactive ion released from the film during a redox process (61–63). For example, SECM was used to detect the release of Br$^-$ during the reduction of oxidized polypyrrole (PP), in the form of PP$^+$Br$^-$ (63). During a reductive cyclic voltammetric scan, Br$^-$ was found to be released after an appreciable amount of cathodic charge had passed, suggesting that during the early stage of the reduction the uptake of cations, rather than the release of anions, maintained charge balance in the film.

The SECM studies of polymer films described above rely on using the tip to probe the solution environment directly above the polymer to investigate electron transfer and ion ejection and incorporation into the film. SECM can also be used for direct electrochemical measurements by recording the tip current and CVs as the tip is moved from the solution into the polymer phase and ultimately contacts the substrate. (See Section 2.3 for more detailed discussion.)

Early SECM work on the studies of charge and mass transport activity of membranes has been extensively reviewed by Bath et al. (64). We mainly focus here on work carried out since then. Recently, SECM utilizing an amperometric microbiosensor based on coimmobilization of the enzymes glucose oxidase and hexokinase has been applied for imaging ATP transport through a porous polycarbonate membrane under physiologically relevant conditions (65). Integration of the amperometric ATP microbiosensor into a dual-microdisk electrode configuration has been achieved by immobilizing the enzymes at one of the microdisk electrodes while the second disk serves as an unmodified amperometric probe for controlled positioning of the microbiosensor in close proximity to the sample surface, enabling quantification of the obtained current signal. SECM operated in feedback mode has also been used in combination with silver nanoparticles–tagging to achieve visualization of proteins adsorbed on a polyvinylidene difluoride (PVDF) membrane, which was detected with osmium tris-bipyridine as a redox mediator in aqueous solution (66).

Another application of SECM is to investigate counter-ion transport through four different proton-conducting membranes with poly(styrene sulfonic acid) side chains (67). These membranes, intended for the polymer electrolyte fuel cell, are based on PVDF and PVDF-co-HFP matrix materials and have been prepared by an irradiation grafting method. SECM is found to be suitable for mapping variations in proton diffusion coefficient and concentration in these inhomogeneous membranes.

In their investigations of membrane transport, White et al. (68, 69) developed two imaging modes. In the forward imaging mode (FIM), the SECM tip is placed in the receptor compartment and poised at a potential such that the electroactive molecule is oxidized or reduced as it emerges from pores in the membrane. In the reverse imaging mode (RIM), the diffusion of molecules into the pore depletes solute in the donor solution immediately adjacent to the pore entrance, resulting in a decrease in the tip current as the tip is rastered across the area above the pore. In principle, RIM has an inherently lower sensitivity than FIM because the signal is measured relative to a large background. However, RIM provides a means of imaging molecular transport into biological membranes, which is not possible using FIM (68). This is a key advantage for monitoring the uptake of molecular species into single cells and biological tissues.

SECM imaging of a membrane containing conical-shaped pores (60-nm and 2.5-μm diameter openings) by alternating current (ac) impedance mode has been demonstrated (70). Impedance images of the pore openings were obtained by rastering a glass-sealed conical-shaped Pt tip (~1-μm radius) above the membrane surface, while measuring the total impedance between the tip and a large-area Pt electrode located on the opposite side of the membrane. Image contrast in ac impedance SECM of membranes increases dramatically as the membrane resistance decreases. Experiments and theory (71) demonstrate that, in instances where the SECM tip resistance is small relative to the internal pore resistance, the total impedance changes by a negligible amount during imaging, resulting in insufficient image contrast to obtain images. As reported (71), a simple and highly effective solution to this problem is to shunt the ion current around the membrane using a low-impedance bypass channel.

Finally, conductance measurements at solid/liquid interfaces with scanning probes are interesting, especially under conditions in which the ionic species under investigation are not electroactive. Besides those ac impedance techniques described above (72–74), a bipolar conductance (BICON) technique for the measurement of solution resistance by SECM was used (75), based on the application of μs-current pulses originally described by Enke et al. (76) for measurements with conventional electrodes. This technique was applied to measuring conductance changes during irradiation of photoresist film in 18-MΩ-cm water, an important process in immersion lithography (77). BICON/SECM approach curves over insulating substrates followed SECM negative feedback theory. Approach curves to a conducting substrate at open circuit potential are influenced by the solution time constant (i.e., solution resistance at the tip multiplied by electric double-layer capacitance), which is a function of tip/substrate distance as well as the substrate size (75). With BICON/SECM, it is possible to obtain approach curves in solution with no intentionally added electrolyte; however, its sensitivity limit is only in the micromolar region (77).

5. BIOLOGICAL CELLS

In recent years, SECM has been extensively applied to studies of a variety of biological cells (78–81). In these new biological applications of SECM, a chemical substance that participates in a cellular process is monitored selectively and sensitively to obtain quantitative information of biological significance. A single cell, under either a closely packed or an isolated condition, can be identified utilizing SECM imaging with μm/subμm spatial resolution. Moreover, without any cellular fixation or invasion, chemical reactivity of a living cell can be investigated in real time by SECM.

These unique advantages, illustrated by the following examples, make SECM a powerful tool for biological and biomedical research.

5.1. Single-Cell Viability

The viability of single cells can be assessed by SECM on the basis of their respiration activity. Employing x-y SECM scans, a tip current based on oxygen reduction decreased near living human colon cells because of oxygen consumption in cellular respiration (82). Spatial resolution and sensitivity of SECM were high enough to image an oxygen concentration profile for individual cells. Recently, the noninvasive SECM measurement of respiration activity was applied to the monitoring of oxygen consumption by identical bovine embryos during their development under a cultured condition (83). The oxygen consumption rate at the initial developing stage was found to be a measure of embryo quality.

Effects of toxic substances on the viability of a HeLa cell were monitored by SECM (84). The cell was loaded with a Calcein dye for a simultaneous fluorescence viability assay. Upon exposure to CN^-, which functions as a respiration inhibitor, the oxygen concentration around the cell increased immediately and reached a bulk concentration after 30 min. After the complete loss of respiration activity, the fluorescence intensities at each cell declined, thereby indicating that the dye had leaked through collapsed cell membranes. On the other hand, an SECM measurement was necessary to detect cytotoxicity of antibiotic Antimycin A, addition of which to a HeLa cell resulted in immediate respiration breakdown without membrane deformation for at least 2 h.

SECM was also applied to the investigation of interactions of fibroblast cells with micromolar amounts of Ag^+ known as antibiotics (85). An Ag^+ concentration profile around a single cell was monitored using an amperometric ion-selective probe based on a calixarene-type Ag^+ ionophore. A comparison between SECM images and SEM images revealed that Ag^+ had accumulated in the cell nucleus.

5.2. Cellular Redox Activity

SECM was utilized to distinguish between metastatic and nontransformed human breast cells by their redox activity (86). In the SECM feedback measurements, a hydrophobic redox mediator was electrolyzed at the electrode tip so that a tip-generated species entered a cell and was oxidized or reduced at intracellular redox centers. Using menadione as a mediator, a smaller feedback effect was observed when comparing a metastatic cell with a nontransformed cell. A detailed mechanistic study demonstrated that the smaller redox response for a metastatic cell was due to a lower intracellular concentration of oxidative moieties available for menadione regeneration (87). The suppressed redox activity was related to high endogenous levels of protein kinase Cα in a metastatic cell; this lower activity was also observed with a human breast cell engineered to overproduce the protein kinase. In fact, expression of cellular reductants such as NAD(P)H:quinone oxidoreductase was induced by the protein kinase, thereby resulting in the diminished oxidative activity (88).

SECM imaging of a mixed field of metastatic and nontransformed cells allows for identification of metastatic cells on the basis of their lower reactivity to quinone mediators (89). Such a redox activity map was obtained using SECM imaging and confirmed using a fluorescence microscopic assay, where cancer cells were tagged with fluorescent nanospheres without alteration of their redox activity. Interestingly, a cellular redox response to a mediator, N,N,N',N'-tetramethyl-1,4-p-phenylenediamine, was found to depend on the cell density. Whereas densely packed cancer cells became slightly more active, the redox activity of nontransformed cells decreased in a confluent culture, suggesting contact inhibition of cellular metabolism (88).

5.3. Channel-Mediated Membrane Transport

Membrane transport mediated by protein channels/pumps is crucial for cellular functions. SECM was applied to the investigation of menadione metabolism by monitoring the export of a menadione-glutathione conjugate, i.e., thiodione, by an ATP-dependent pump. Exposure of yeast cells to menadione was immediately followed by the active export of thiodione (90). A thiodione efflux was measured amperometrically and analyzed using a constant flux model, yielding an average single cell flux of 3×10^4 molecules/s. The active thiodione export was also investigated for hepatoblastoma (91). Successive SECM images of a single Hep G2 cell were obtained (**Figure 8**) to demonstrate that a thiodione efflux decreased gradually to fall below detectable limits after 1 h. This result agrees well with a cytotoxicity assay.

Passive molecular transport through the nuclear pore complex (NPC) was investigated by SECM (92). The NPC is a large protein complex that exclusively mediates molecular transport across the nuclear envelope between the nucleus and the cytoplasm. SECM approach curves at intact nuclei isolated from *Xenopus laevis* oocytes

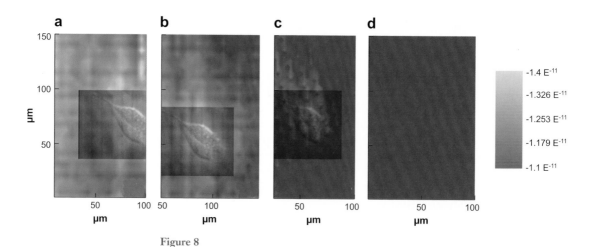

Figure 8

Superimposed optical micrographs on the corresponding scanning electrochemical microscopy (SECM) images based on thiodione export from the Hep G2 cells (91). The SECM images were obtained successively (panels *a–d*). It took ~7 min to acquire each image. Adapted from Reference 91 with permission. Copyright 2004, National Academy of Sciences, USA.

demonstrated that the permeability of the nuclear envelope to small redox molecules such as ferrocene derivatives is at least two orders of magnitude larger than the permeability of BLMs and cell membranes to the same molecules. The large permeability, which was noninvasively determined by SECM, is consistent with the high density and large size of open NPC pores.

5.4. Imaging Model Neurons

Sophisticated imaging techniques were developed to map the exocytosis of neuro-transmitters from individual vesicles at cultured model neurons (93). High-resolution imaging of model neurons with high relief requires dynamic tip positioning at a short and constant distance from the cell surface. A shear force between the tip and the cell was utilized as a feedback signal for constant-distance control (93). A topographic image of a PC12 cell was obtained by employing the constant-force mode, where a resonant frequency of a vibrating carbon fiber probe was monitored to maintain a constant shear force. Neurotransmitters released from individual vesicles were detected amperometrically by positioning the tip above the cell surface in the constant-force mode. A fluorescence viability assay confirmed that PC12 cells were not damaged during shear force–based topographic imaging (94).

An electrochemical tip response can also be used for distance control (95). In constant-current mode, the tip is moved vertically during an x-y scan so that a constant tip current generated from the electrolysis of redox molecules is maintained. A detailed topographic image of a differentiated PC12 cell was acquired using the constant-current mode (**Figure 9**), clearly demonstrating the height differences at the cell body and neutrites. Alternatively, a constant distance can be maintained by monitoring the tip impedance as a feedback signal. In contrast to the constant-current

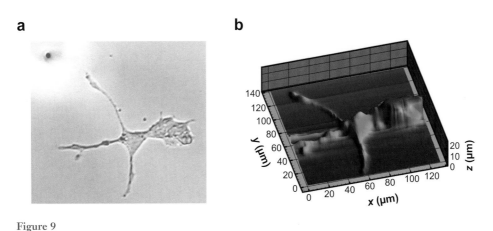

a   **b**

Figure 9

(*a*) Optical and (*b*) constant-current topographic images of a differentiated PC12 cell (95). The scanning electrochemical microscopy image was acquired using a ~1-μm-diameter carbon ring electrode and 1.0-mM $Ru(NH_3)_6^{3+}$ mediator. Adapted from Reference 95 with permission. Copyright 2005, American Chemical Society.

mode, the constant-impedance mode allows for simultaneous amperometric recording of neurotransmitter releases.

5.5. Advanced Electrochemical and Optical Imaging

The combination of SECM and optical microscopy is a powerful approach to obtaining multidimensional information about complicated cellular processes. An SECM stage mounted on an inverted optical microscope is a standard setup for simultaneous electrochemical and optical imaging. Recently, such a setup permitted a single-molecule fluorescence study of F_0F_1-ATP syntheses in liposomes (96). ATP synthesis was driven by a local pH gradient created by H^+ reduction at an SECM tip. The resulting intersubunit rotation within a single enzyme molecule was able to be monitored by intramolecular fluorescence resonance energy transfer with a confocal laser microscope.

A ring UME based on an optical fiber was developed as a probe of a combined scanning electrochemical/optical microscope (97). An optical image of freshwater diatoms, *Navicula minima*, was obtained using the optical fiber probe as a light source. At the same time, a topographic image of the soft biological samples was obtained in the constant-current mode, yielding better image quality than in the constant-force mode. Moreover, a recent study demonstrated single-cell imaging by SECM integrated into a commercial near-field optical microscope, which also has the capabilities of atomic force microscopy (AFM) and confocal microscopic measurements (98). Construction of a smaller optical fiber probe will enable nm-resolution electrochemical and optical imaging.

6. ELECTROCATALYSTS AND SCREENING

6.1. Basic Concepts of High-Throughput Screening of Electrocatalysts

The search for new materials for electrochemical applications (e.g., electrocatalysts for complex electrode reactions like the oxygen reduction reaction (ORR), hydrogen or methanol oxidation reactions for fuel cells, or the water oxidation reaction for photoelectrochemical systems) continues to be an important activity. SECM has found growing application in this area, as it is excellent for rapid screening of arrays of materials and directly produces electrochemical data that can be used as quantitative estimates of rate and activity. This approach is somewhat different than the traditional one where one carefully synthesizes and characterizes a material and then tests it on a macroscale, for example with the rotating disk electrode, but it allows one to screen a large number of different compositions rapidly. Of equal importance to the screening is the rapid preparation of the arrays to yield small samples (spots or lines) that are of reasonably uniform, but varying, composition on a substrate that has very low electrocatalytic properties. This section reviews briefly the principles involved in such studies and examples of the work that has been carried out in this area.

6.2. Fabrication of Arrays

To be a useful technique for high-throughput screening, arrays of samples should ideally be capable of being prepared in an automated way, rapidly and inexpensively, while covering a wide range of compositions, with good reproducibility of the test spot composition. Numerous different array substrates and methods of preparation for use in SECM have been reported. These are summarized in **Table 1**.

A substrate that uses individually addressable spots would, in principle, allow the closest control of each location potential and thus the smallest background current, as the substrate itself and all of the spots except the one addressed by the tip would not contribute to the substrate current. For example, an array of 25 Pt spots, 10 μm in diameter and spaced 100 μm apart, has been imaged by SECM using ferrocene methanol as a mediator with all of the spots at the same potential or at open circuit (99). However in a rapid screening mode, one would have to synchronize the application of potential at the spots with the tip position over a given spot by some sort of potentiostat multiplexing arrangement. Moreover such arrays are relatively expensive. To date, such arrays have not been used for SECM screening purposes. This is similarly true for an array of eight individually addressable lines, 25 mm long, 1 mm wide, and with a 1.5-mm separation formed on an indium-tin oxide electrode by photolithography and etching (100). In this case the individual elements could be plated with different compositions Pt_xRu_y by controlling the potential of one element at a time and changing the solution for electrodeposition. However, in SECM screening all of the elements were maintained at the same potential.

The usual approach is to use a single substrate material, often carbon or ITO, that shows poor catalytic properties (so the background current density at an interspot region is small) and prepare an array of spots or lines thereon. This approach was used to prepare bi- and trimetallic spots on glassy carbon (GC) substrates (105–107). This made use of a piezo dispenser that could deliver pL-drops of appropriate

Table 1   Representative approaches to forming sample arrays[a]

Substrate	Composition and deposition method	Reaction studied	Reference(s)
Addressable array, Pt	Pt, proof of concept (5×5 array)	FcMeOH, feedback, SG/TC	99
Patterned ITO	$Pt_xRu_yMo_z$, electrodeposition (8 lines)	H_2 oxidation, feedback	100
Au	Pt, electrodeposition proof of concept	H_2 oxidation, feedback	101
ITO	Pt_xRu_y, gradient, agar, electrodeposition	H_2 oxidation, feedback	102
ITO	Pt coverage, gradient, electrodeposition	H_2 oxidation, feedback	103
HOPG	Pt, random, drop coating	H^+ reduction, feedback, TG/SC	104
GC	Pd_xM_y (M=Co, Ti), PZT dispenser	O_2 reduction, TG/SC	105–107
GC	Cu(II)-polyhis films, PZT dispenser	O_2 reduction, TG/SC	108
GC	Hydrogels, wired enzymes (laccase)	O_2 reduction, TG/SC	109
Si, TiN layer	Pt_xRu_y, sputtering	H_2 oxidation, feedback	110

[a]Abbreviations: ITO, indium-tin oxide; HOPG, highly ordered pyrolytic graphite; GC, glassy carbon; PZT, piezoelectric; TG/SC, tip generation/substrate collection; SG/TC, substrate generation/tip collection.

solutions of the metal salts that were computer-controlled to move and dispense at precise locations. After one metal salt was deposited, the solution was drained from the dispenser and a second metal salt solution (and perhaps a third) was used over the same array. This resulted in an array of drops containing different programmed amounts of the metals. These were vortex mixed and then treated with H_2 at an elevated temperature to produce alloy spots. Alternatively, one can use sodium borohydride solution for the reduction. The exact solution composition (choice of metal salt and solvent) and the array-processing procedure are important for obtaining reproducible and well-formed electrocatalyst spots (106).

A similar approach utilizing a single inert substrate (GC) and piezo dispensing was used with enzymes and biomimetic-type spots (108, 109). Instead of the piezo dispenser, one can also employ an ink jet printing device such as those used in combinatorial array preparations with other methods of screening. A disadvantage of approaches using a common conductive substrate is that the background current from this material can be significant. However, one can prepare a covering layer of an insulator such as Teflon with holes for the spots, which is much simpler than preparing an individually addressable array (J. Rodriguez and A.J. Bard, unpublished experiments).

An alternative approach to preparing different electrocatalysts is to prepare a continuous gradient of compositions. This has been accomplished with electrodeposition across a resistive element (101, 102), by changing the composition of a thin-layer film (e.g., agar) by controlled solution treatment, or by sputtering (110).

6.3. Screening by Scanning Electrochemical Microscopy

As illustrated in **Figure 10**, there are four different SECM approaches to examining electrocatalysis on a substrate. Determining which technique to employ depends upon the reaction to be studied at the substrate and whether the reactant of interest can be generated at the tip. Each approach is briefly discussed and some examples of their use are given.

6.3.1. Feedback approach. If the species to be studied at the substrate can be generated at the tip so that the redox pair of interest represents a typical SECM mediator, the effect of substrate potential, E_S, on the tip current, i_T, can be used to determine the activity of the substrate and the heterogeneous rate constant. Consider the examination of electrocatalysts for H_2 oxidation (110–112): The tip is held at a potential where H^+ is reduced at the diffusion controlled rate in a solution where its concentration is ~0.01 M. This concentration is consistent with typical SECM measurements and is at a value where the H_2 generated does not form bubbles, but rather dissolves in the solution and diffuses to the substrate. As with other kinetic studies by SECM, approach curves as a function of substrate potential allow one to calculate the heterogeneous rate constant, $k(E)$. Similarly, a substrate can be scanned at a constant d to determine relative activities. A disadvantage of this approach is that one is limited to a rather small range of H^+ concentration (or pH). At much lower concentrations the currents become too small to measure above background, and at higher concentrations hydrogen

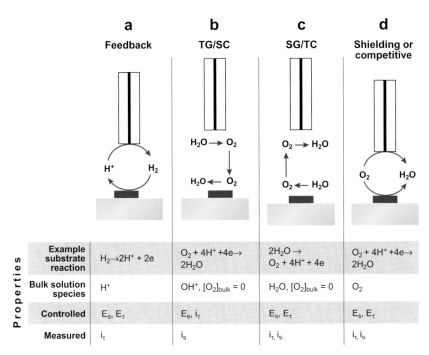

	a Feedback	**b** TG/SC	**c** SG/TC	**d** Shielding or competitive
Example substrate reaction	$H_2 \rightarrow 2H^+ + 2e$	$O_2 + 4H^+ + 4e \rightarrow 2H_2O$	$2H_2O \rightarrow O_2 + 4H^+ + 4e$	$O_2 + 4H^+ + 4e \rightarrow 2H_2O$
Bulk solution species	H^+	OH^+, $[O_2]_{bulk} = 0$	H_2O, $[O_2]_{bulk} = 0$	O_2
Controlled	E_s, E_τ	E_s, i_τ	E_s, E_τ	E_s, E_τ
Measured	i_τ	i_s	i_τ, i_s	i_τ, i_s

(left label: **Properties**)

Figure 10

Schematic representation of different modes of substrate screening in scanning electrochemical microscopy. TG, tip generation; SC, substrate collection.

bubbles form. The effect of other species on the reaction, such as CO, can also be studied (112). A similar approach can be used to study the O_2 reduction reaction at the substrate by oxidation of 0.01 M OH^-, which is oxidized before water oxidation, at the tip (113). Again, one is limited to a rather small pH range and therefore would be unable to study the ORR in 0.1 M acid, which would be a more appropriate medium for fuel cell studies.

6.3.2. Tip generation/substrate collection. An alternative approach wherein the tip generates the substrate reactant with considerably more flexibility is tip generation/substrate collection (TG/SC) (**Figure 10b**) (106, 114, 115). In this approach i_T is controlled, usually as a constant current that generates a constant flux of reactant. When the tip is close to the substrate and the array spots are sufficiently far apart, only the spot immediately below the tip is addressed. The substrate current at that spot at a given potential is a measure of the activity of the spot. For example, for a study of the ORR at the substrate, the solution is de-aerated so that the bulk concentration of O_2 is low. Then O_2 is generated at the tip and reduced at the substrate and the x-y scan shows i_S as a function of position and potential (105–107, 116). A typical result for Pd-Co electrocatalysts for the ORR is shown in **Figure 11**.

6.3.3. Substrate generation/tip collection. The above approaches are not available under some circumstances, for instance screening an array with electrocatalysts for water oxidation (**Figure 10c**). One cannot generate the reactant, water, in an aqueous

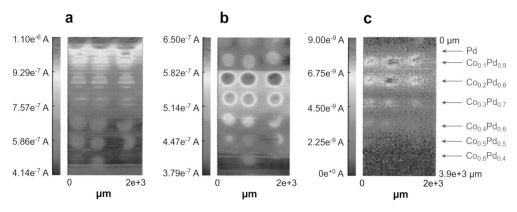

Figure 11

Scanning electrochemical microscopy tip generation/substrate collection images for the oxygen reduction reaction for an array of Pd_xCo_y spots on glassy carbon in 0.5 M H_2SO_4. Oxygen generated at the tip (25-μm diameter W) at a constant current of 183 nA and reduced at the spots with substrate potentials of (*a*) 0.2; (*b*) 0.45; and (*c*) 0.7 V versus standard hydrogen electrode (SHE). Distance between tip and substrate, *d*, is 50 μm. Tip scanned in *x-y* plane at 600 μm/s. Adapted from Reference 106 with permission. Copyright 2007, American Chemical Society.

solution at the tip nor involve it in a feedback arrangement. In this case, substrate generation/tip collection (SG/TC) can be used, where the substrate potential is held at different potentials and O_2 is detected at the tip (A. Minguzzi, S. Rondinini, and A.J. Bard, unpublished experiments). In this approach, one cannot have O_2 in the bulk solution and must provide for its removal or another method of preventing interference from O_2 generated at neighboring spots.

6.3.4. Shielding (competitive). If the species to be tested cannot be conveniently generated at an electrode and no product of the electrode reaction can be detected, e.g., with methanol, it is still possible to test an array using SECM shielding. In shielding, the same reaction occurs at the tip and the substrate, so that the current at the tip is decreased compared to what would be observed with an insulating substrate (117, 118), i.e., the tip is shielded by and competes with the substrate for the same reactant (**Figure 10***d*). Thus, a reactant in bulk solution can be studied by noting how well the substrate competes with the tip for consumption (e.g., oxidation of methanol) of the reactant. This approach was used to study the ORR as an alternative to TG/SC (119, 120).

7. PROPERTIES AND REACTIVITY OF MATERIALS

The ability of SECM to map physicochemical processes at interfaces has proved particularly powerful in elucidating the local properties and reactivity of a wide variety of materials. Applications in this area continue to expand and embrace metals, semiconductors, electrically insulating inorganic and organic materials, soft matter, immobilized nanoparticles, and nanotubes.

7.1. Redox Activity of Complex Electrode Surfaces

The application of SECM to the investigation of the heterogeneous activity of electrodes is well illustrated by studies of conducting (boron-doped) polycrystalline diamond (BDD) (121–123). This material is attracting attention as a novel electrode due to its wide potential window in many solvents and resistance to fouling in complex media, among other attributes. BDD has often been treated as uniformly active, but SECM studies of several sources of synthetic diamond (with varying dopant levels and surface termination) have demonstrated that different grains in polycrystalline BDD show characteristics that depend on the dopant level of individual grains (121–123). SECM data (approach curves, feedback imaging, and SG/TC measurements) have been linked to results from complementary techniques, such as conducting probe (CP)-AFM, cathodoluminescence imaging, and Raman microscopy on the same areas of samples to provide a wealth of quantitative information on the nature of active sites. These studies have demonstrated how the local composition of a grain and its properties influence the electrochemical response (122, 123).

SECM has proved especially powerful in identifying precursor sites for corrosion in otherwise passive oxide films that protect certain metals and alloys (124–126). In these studies, the substrate of interest is generally biased to carry out a redox reaction and the imaging SECM tip detects the product of the reaction, thereby allowing visualization of redox activity. The reactivity of grain structures in titanium alloys has also been explored using SECM feedback measurements as a function of substrate potential (127), whereas, at higher resolution, CP-AFM has provided insight into the conductivity of passive films that complement SECM measurements (128). The complementarity of SECM and CP-AFM is also well illustrated in studies of other materials, such as dimensionally stable $Ti/TiO_2/Pt$ electrodes (129), wherein two types of active Pt sites were identified and shown to be correlated with the nature of the contact to the underlying substrate.

The SECM tip can be used to initiate local reactions, such as corrosion, and can monitor the consequences via the tip current response and/or topographical analysis of the resulting surface. This approach is evident in studies in which Cl^- was generated by electrolysis of a precursor compound at an SECM tip in order to induce the corrosion of iron (130). SECM tips can also detect the species released or taken up during corrosion via amperometry or ion-selective electrode probes (131).

7.2. Heterogeneous Photochemical Reactions at Semiconductors

In early work, SECM was used to probe electron transfer kinetics for a range of outer-sphere redox couples at n- and $p-WSe_2$ electrodes in aqueous electrolytes and at n-Si in acetonitrile and methanol (132). These studies surmounted the difficulties and problems inherent in alternative conventional techniques, for example ohmic effects, charging currents, and parallel processes such as corrosion. In subsequent studies, SECM provided insights into processes at illuminated semiconductor-electrolyte interfaces, exemplified by feedback measurements to probe the photoelectron transfer kinetics for the methyl viologen redox couple ($MV^{2+/+}$) at CdS films (133).

Nanostructured semiconductor films, relevant to applications in photocatalysis and for photovoltaics, have recently been investigated with SECM. In one study, Cl^- was detected with an ion-selective electrode during the photomineralization of chlorophenols (134), and the depletion of O_2 at TiO_2 during photomineralization has also been monitored (135). In both cases, the well-defined mass transport between the UME probe and illuminated surfaces allowed quantitative interpretation of measured fluxes in terms of the underlying kinetics. A time-of-flight SECM technique has been used to measure the diffusion coefficient of redox species in a nanostructured film operated under solar cell conditions (136), whereas the feedback mode has been used to probe the reduction kinetics of photo-oxidized immobilized dye molecules (eosin-Y) by iodide (137). The latter examples illustrate the wide range of processes that can be addressed with SECM.

7.3. Nanostructured Interfaces

The properties of two-dimensional nanoparticle assemblies and nanotube networks have been explored using SECM. The kinetics of hydrogen evolution at gold nanoparticles immobilized on an insulating glass surface were investigated using the SECM feedback mode, in which MV^{2+} was reduced to $MV^{+\cdot}$ at an Hg-Au amalgam UME tip (138). $MV^{+\cdot}$ served as an electron donor in the reduction of protons to hydrogen at the nanoparticle assembly, generating MV^{2+} at the substrate surface that was then fed back to the SECM tip. These feedback current measurements provided quantitative information on H_2 evolution kinetics, which could be interpreted at the single-nanoparticle level.

An important matter in the study of nanoparticle assemblies is the degree of connectivity and the associated lateral conductivity. SECM feedback measurements have proved powerful in monitoring the connectivity of a monolayer of monodisperse thiol-capped silver nanoparticles at the A/W interface via approach curve measurements as a function of surface pressure (53). These experiments involved the deployment of SECM in a Langmuir trough, in an arrangement similar to that discussed in Section 4.2 above for studies of monolayer molecular films. Using the oxidation of ferrocene methanol as the tip redox process, a transition from negative to positive feedback (connected substrate) was observed as the surface pressure was increased. This type of approach has also been used to examine the effect of surface pressure on the conductivity of a polyaniline monolayer (139), in which an insulating-to-conducting transition was observed at a threshold surface pressure.

The conductivity of films on solid supports has also been examined using the SECM feedback mode. Example applications include measuring the conductivity of a monolayer of hexanethiol stabilized gold nanoparticles (140) and polymer-nanoparticle composite films (141, 142). Transient SECM techniques, which involve measuring the current-time response of the SECM tip as a function of a defined potential-step time program, have been used to provide complementary information on charge injection and lateral charge propagation in ultrathin polymer films (143).

Single-walled carbon nanotubes (SWNTs) are attracting considerable attention for a variety of applications in electroanalysis and electrocatalysis, although there are questions concerning the nature of the active sites and intrinsic electroactivity. SECM feedback measurements have allowed an assessment of electrochemical processes at pristine SWNTs grown on an insulating support without any further processing of the SWNTs (144). One interpretation of these measurements is that SWNT networks may essentially behave as metallic films on typical voltammetric time scales even though the fractional surface coverage of SWNTs is only ~1%.

7.4. Crystal Dissolution, Adsorption, and Desorption

Transient SECM techniques are versatile as a means of introducing a short perturbation to an interfacial system, initially at equilibrium, and measuring the kinetic consequences of the perturbation. In these applications, a UME tip is used to drive desorption processes (56), adsorption processes (145), and dissolution reactions (146) as consequences of an electrochemical process at the tip which changes the local solution composition and thus the equilibrium. The kinetics of the process, initiated electrochemically, is reflected in the SECM current-time response, which can be analyzed using appropriate mass transport models to provide kinetic information. In the case of dissolution, the process naturally leads to a change in surface topography, which can be investigated using complementary techniques. The combination of SECM with AFM, discussed in Section 8.2 below, has provided a means of rapidly assessing dissolution processes in situ at high spatial resolution.

8. HYBRID TECHNIQUES

SECM has been combined with several other techniques to enhance and complement the information available from SECM alone. Considerable effort has gone into developments that provide nonelectrochemical means of monitoring topography, as highlighted in Sections 8.1 and 8.2, and combinations with other techniques (see Section 8.3) have produced multifunctional instruments.

8.1. Shear Force Positioning Methods

The use of shear force positioning opens up the possibility of imaging a sample at constant separation between the tip and surface so that the electrochemical signal is not convoluted by variations in the tip/substrate separation, which may complicate simple SECM experiments in certain applications. However, the implementation of shear force positioning remains nonroutine and rather complex. The UME can be excited to undergo a lateral motion by either a piezoelectric actuator (147) or by attaching the UME to one arm of a tuning fork (97). Both the amplitude and the phase of the tip motion are sensitive to the tip/substrate separation, but it is important to note that the shear force signal arises from the entire probe end, which usually leads to lower spatial resolution compared to that of the electrochemical signal.

8.2. Scanning Electrochemical–Atomic Force Microscopy

Recent work has seen the integration of UMEs into AFM tips for use in a technique termed SECM-AFM (148). The benefits of SECM-AFM are that one can use the AFM component of the tip for high-resolution topographical imaging or for maintaining a desired tip/substrate separation, whereas the SECM component can be used to electrochemically measure and/or perturb a system. These probes can be used in conjunction with any commercially available AFM instruments. The origins of SECM-AFM can be traced to work in which conventional AFM tips were sputter-coated with a metal so that the entire tip and cantilever could function as an electrode (149). The latter probes have proved beneficial for inducing and monitoring surface reactions such as dissolution (150) and for imaging electroactive surfaces in air (151).

The original high-resolution SECM-AFM probe was produced by coating an etched and flattened microwire with an electrophoretic paint (17, 148). The flattened section provided a flexible cantilever (force sensor), and the coated etched tip served as an electrode. SECM-AFM probes with conical-shaped electrodes of size 10–1000 nm have been fabricated in this way and have found application in reactivity mapping (149) and in studies of transport through membranes (152). This style of probe has also been used to image the conductivity of heterogeneous surfaces, particularly for hard surfaces where thin-film metal-coated probes may be subject to wear (122).

A second method involves coating a silicon nitride tip in a metal, as for integrated electrochemical AFM probes (149), and then insulating the entire probe. The electrode is exposed using a focused ion beam (FIB) (153) to leave a small electrode area. This type of probe has been used to map immobilized enzyme activity and substrate topography (154, 155). Ultrahigh-resolution imaging probes have been produced by attaching bundles of single-walled carbon nanotubes to the tip of a metal-coated AFM tip and using these as a template for the deposition of a metal nanowire (156). The probe is insulated and the end of the nanowire is cut by a FIB to expose a disk-shaped electrode, with a diameter typically in the range of 50–100 nm.

There have been recent advances toward the development of batch fabrication procedures. Using electron beam lithography, it has been possible to produce probes of a reproducible nature from a silicon wafer (157). Ring electrodes integrated into AFM tips have also been fabricated in a batch process (158). This type of methodology offers great promise for the parallel fabrication of many probes with similar characteristics.

8.3. Coupling Scanning Electrochemical Microscopy and Other Methods

In Section 5.5 above we highlight advances in combining SECM with optical techniques. SECM has also been integrated with laser scanning confocal microscopy, which allows direct visualization of three-dimensional concentration profiles on a rapid time scale and similar length scale to many SECM investigations (96, 159). Studies in this area are at an early stage, but the ability to follow concentration profiles optically, and to follow simultaneously the local electrochemical signal from SECM, is anticipated to provide additional insights into a myriad of interfacial physicochemical processes.

SECM has also been combined with surface-sensitive techniques such as the quartz-crystal microbalance (QCM) (160) and surface plasmon resonance (SPR) (161). These techniques are employed on macroscopic samples, and SECM can be used simultaneously with the QCM or SPR signal to provide local information on associated chemical processes, such as uptake and release from a surface of interest.

DISCLOSURE STATEMENT

The authors are not aware of any biases that might be perceived as affecting the objectivity of this review.

LITERATURE CITED

1. Bard A, Mirkin M, eds. 2001. *Scanning Electrochemical Microscopy*. New York: Marcel Dekker
2. Nagahara L, Tundat T, Lindsay S. 1989. Preparation and characterization of STM tips for electrochemical studies. *Rev. Sci. Instrum.* 60:3128–30
3. Sun P, Zhang Z, Guo J, Shao Y. 2001. Fabrication of nm-sized electrodes and tips for scanning electrochemical microscopy. *Anal. Chem.* 73:5346–51
4. Slevin C, Gray N, Macpherson J, Webb M, Unwin P. 1999. Fabrication and characterization of nm-sized platinum-electrodes for voltammetric analysis and imaging. *Electrochem. Commun.* 1:282–88
5. Zhan W, Bard A. 2006. Scanning electrochemical microscopy. 56. Probing outside and inside single giant liposomes containing Ru(bpy)$_3^{2+}$. *Anal. Chem.* 78:726–33
6. Shao Y, Mirkin M, Fish G, Kokotov S, Palanker D, Lewis A. 1997. Nm-sized electrochemical sensors. *Anal. Chem.* 69:1627–34
7. Katemann B, Schuhmann W. 2002. Fabrication and characterization of needle-type Pt-disk nanoelectrodes. *Electroanalysis* 14:22–28
8. Sun P, Mirkin M. 2006. Kinetics of electron-transfer reactions at nanoelectrodes. *Anal. Chem.* 78:6526–34
9. Sun P, Mirkin M. 2007. Scanning electrochemical microscopy with slightly recessed nanotips. *Anal. Chem.* 79:5809–16
10. Mirkin M, Fan F-R, Bard A. 1992. Scanning electrochemical microscopy. 13. Evaluation of the tip shapes of nm size microelectrodes. *J. Electroanal. Chem.* 328:47–62
11. Mirkin M, Fan F-R, Bard A. 1992. Direct electrochemical measurements inside a 2000-angstrom thick polymer film by scanning electrochemical microscopy. *Science* 257:364–66

12. Fan F-R, Mirkin M, Bard A. 1994. Polymer-films on electrodes. 25. Effect of polymer resistance on the electrochemistry of poly(vinylferrocene): scanning electrochemical microscopic, chronoamperometric, and cyclic voltammetric studies. *J. Phys. Chem.* 98:1475–81

13. Bard A, Fan F-R. 1992. Introductory lecture studies of the liquid-solid interface by scanning-tunneling-microscopy and scanning electrochemical microscopy. *Faraday Discuss. Chem. Soc.* 94:1–22

14. Katemann B, Schulte A, Schuhmann W. 2004. Constant-distance mode scanning electrochemical microscopy. Part II: High-resolution SECM imaging employing Pt nanoelectrodes as miniaturized scanning probes. *Electroanalysis* 16:60–65

15. Baltes N, Thouin L, Amatore C, Heinze J. 2004. Imaging concentration profiles of redox-active species with nanometric amperometric probes: effect of natural convection on transport at microdisk electrodes. *Angew. Chem. Int. Ed.* 43:1431–35

16. Fan F-R, Bard A. 1995. STM on wet insulators: electrochemistry or tunneling? *Science* 270:1849–52

17. Macpherson J, Unwin P. 2001. Noncontact electrochemical imaging with combined scanning electrochemical atomic force microscopy. *Anal. Chem.* 73:550–57

18. Kueng A, Kranz C, Lugstein A, Bertagnolli E, Mizaikoff B. 2003. Integrated AFM-SECM in tapping mode: simultaneous topographical and electrochemical imaging of enzyme activity. *Angew. Chem. Int. Ed.* 42:3238–40

19. Fan F-R, Bard A. 1995. Electrochemical detection of single molecules. *Science* 267:871–74

20. Fan F-R, Kwak J, Bard A. 1996. Single-molecule electrochemistry. *J. Am. Chem. Soc.* 118:9669–75

21. Barker A, Gonsalves M, Macpherson J, Slevin C, Unwin P. 1999. Scanning electrochemical microscopy: beyond the solid/liquid interface. *Anal. Chim. Acta* 385:223–40

22. Mirkin M, Tsionsky M, eds. 2001. Charge transfer at the liquid/liquid interface. In *Scanning Electrochemical Microscopy*, ed. A Bard, M Mirkin, pp. 299–342. New York: Marcel Dekker

23. Wei C, Bard A, Mirkin M. 1995. Scanning electrochemical microscopy. 31. Application of SECM to the study of charge-transfer processes at the liquid-liquid interface. *J. Phys. Chem.* 99:16033–42

24. Shao Y, Mirkin M. 1997. Scanning electrochemical microscopy (SECM) of facilitated ion transfer at the liquid/liquid interface. *J. Electroanal. Chem.* 439:137–43

25. Shao Y, Mirkin M. 1998. Probing ion transfer at the liquid/liquid interface by scanning electrochemical microscopy (SECM). *J. Phys. Chem. B* 102:9915–21

26. Selzer Y, Mandler D. 2000. Probing the coupling of charge-transfer processes across liquid/liquid interfaces by the scanning electrochemical microscope. *J. Phys. Chem. B* 104:4903–10

27. Zhang Z, Yuan Y, Sun P, Su B, Guo J, et al. 2002. Study of electron-transfer reactions across an externally polarized water/1,2-dichloroethane interface by scanning electrochemical microscopy. *J. Phys. Chem. B* 106:6713–17

28. Sun P, Zhang Z, Gao Z, Shao Y. 2002. Probing fast facilitated ion transfer across an externally polarized liquid-liquid interface by scanning electrochemical microscopy. *Angew. Chem. Int. Ed.* 41:3445–48

29. Tsionsky M, Bard A, Mirkin M. 1996. Scanning electrochemical microscopy. 34. Potential dependence of the electron-transfer rate and film formation at the liquid/liquid interface. *J. Phys. Chem.* 100:17881–88

30. Tsionsky M, Bard A, Mirkin M. 1997. Long-range electron-transfer through a lipid monolayer at the liquid/liquid interface. *J. Am. Chem. Soc.* 119:10785–92

31. Ding Z, Quinn B, Bard A. 2001. Kinetics of heterogeneous electron-transfer at liquid/liquid interfaces as studied by SECM. *J. Phys. Chem. B* 105:6367–74

32. Zhang J, Unwin P. 2002. Microelectrochemical measurements of electron transfer rates at the interface between two immiscible electrolyte solutions: Potential dependence of the ferro/ferricyanide-7,7,8,8-tetracyanoquinodimethane (TCNQ)/TCNQ system. *Phys. Chem. Chem. Phys.* 4:3820–27

33. Liu B, Mirkin M. 1999. Potential-independent electron transfer rate at the liquid/liquid interface. *J. Am. Chem. Soc.* 121:8352–55

34. Barker A, Unwin P, Amemiya S, Zhou J, Bard A. 1999. Scanning electrochemistry microscopy (SECM) in the study of electron transfer kinetics at liquid/liquid interfaces: Beyond the constant composition approximation. *J. Phys. Chem. B* 103:7260–69

35. Sun P, Li F, Chen Y, Zhang M, Gao Z, Shao Y. 2003. Observation of the Marcus inverted region of electron transfer reactions at a liquid/liquid interface. *J. Am. Chem. Soc.* 125:9600–1

36. Laforge F, Kakiuchi T, Shigematsu F, Mirkin M. 2004. Comparative study of electron transfer reactions at the ionic liquid/water and organic/water interfaces. *J. Am. Chem. Soc.* 126:15380–81

37. Laforge F, Kakiuchi T, Shigematsu F, Mirkin M. 2006. SECM study of solute partitioning and electron transfer at the ionic liquid/water interface. *Langmuir* 22:10705–10

38. Quinn B, Liljeroth P, Kontturi K. 2002. Interfacial reactivity of monolayer-protected clusters studied by scanning electrochemical microscopy. *J. Am. Chem. Soc.* 124:12915–21

39. Georganopoulou D, Mirkin M, Murray RW. 2004. SECM measurement of the fast electron transfer dynamics between Au-38(1+) nanoparticles and aqueous redox species at a liquid/liquid interface. *Nano Lett.* 4:1763–67

40. Barker A, Unwin P. 2001. Measurement of solute partitioning across liquid/liquid interfaces using scanning electrochemical microscopy–double potential step chronoamperometry (SECM-DPSC): principles, theory, and application to ferrocenium ion transfer across the 1,2-dichloroethane/aqueous interface. *J. Phys. Chem. B* 105:12019–31

41. Zhang J, Strutwolf J, Cannan S, Unwin P. 2003. Combined scanning electrochemical microscopy–Langmuir trough technique for investigating phase

transfer kinetics across liquid/liquid interfaces modified by a molecular monolayer. *Electrochem. Commun.* 5:105–10

42. Liu B, Bard A, Mirkin M, Creager S. 2004. Electron transfer at self-assembled monolayers measured by scanning electrochemical microscopy. *J. Am. Chem. Soc.* 126:1485–92

43. Cannes C, Kanoufi F, Bard A. 2003. Cyclic voltammetry and scanning electrochemical microscopy of ferrocenemethanol at monolayer and bilayer-modified gold electrodes. *J. Electroanal. Chem.* 547:83–91

44. Delville M, Tsionsky M, Bard A. 1998. Scanning electrochemical microscopy studies of electron transfer through monolayers containing conjugated species at the liquid-liquid interface. *Langmuir* 14:2774–79

45. Liu B, Mirkin M. 2002. Electron transfer at liquid/liquid interfaces. The effects of ionic adsorption, electrolyte concentration, and spacer length on the reaction rate. *J. Phys. Chem. B* 106:3933–40

46. Zhang J, Slevin C, Morton C, Scott P, Walton D, Unwin P. 2001. New approach for measuring lateral diffusion in Langmuir monolayers by scanning electrochemical microscopy (SECM): theory and application. *J. Phys. Chem.* 105:11120–30

47. Finklea H. 1996. Electrochemistry of organized monolayers of thiols and related molecules on electrodes. In *Electroanalytical Chemistry*, vol. 19, ed. AJ Bard, H Lund, pp. 109–335. New York: Marcel Dekker

48. Lesia V, Protsailo W, Fawcett R. 2000. Studies of electron transfer through self-assembled monolayers using impedance spectroscopy. *Electrochim. Acta* 45:3497–505

49. Cohen-Atiya M, Nelson A, Mandler D. 2006. Characterization of n-alkanethiol self-assembled monolayers on mercury by impedance and potentiometric measurements. *J. Electroanal. Chem.* 593:227–40

50. Cohen-Atiya M, Mandler D. 2006. Studying electron transfer through alkanethiol self-assembled monolayers on a hanging mercury drop electrode using potentiometric measurements. *Phys. Chem. Chem. Phys.* 8:4405–409

51. Slevin C, Unwin P. 2000. Lateral proton diffusion rates along stearic acid monolayers. *J. Am. Chem. Soc.* 122:2597–602

52. Slevin C, Ryley S, Walton D, Unwin P. 1998. A new approach for measuring the effect of a monolayer on molecular transfer across an air/water interface using scanning electrochemical microscopy. *Langmuir* 14:5331–34

53. Quinn B, Prieto I, Haram S, Bard A. 2001. Electrochemical observation of a metal/insulator transition by scanning electrochemical microscopy. *J. Phys. Chem. B* 105:7474–76

54. Slevin C, Umbers J, Atherton J, Unwin P. 1996. A new approach to the measurement of transfer rates across immiscible liquid/liquid interfaces. *J. Chem. Soc., Faraday Trans.* 92:5177–80

55. Mauzeroll J, Buda M, Bard A, Prieto F, Rueda M. 2002. Detection of Tl(I) transport through a gramicidin-dioleoylphosphatidylcholine monolayer using the substrate generation-tip collection mode of scanning electrochemical microscopy. *Langmuir* 18:9453–61

56. Unwin P, Bard A. 1992. Scanning electrochemical microscopy. 14. Scanning electrochemical microscope induced desorption: a new technique for the measurement of adsorption/desorption kinetics at the solid/liquid interface. *J. Phys. Chem.* 96:5035–45

57. Zhang J, Unwin P. 2002. Proton diffusion at phospholipid assemblies. *J. Am. Chem. Soc.* 124:2379–83

58. Tien H. 1978. *Bilayer Lipid Membranes (BLM): Theory and Practice*. New York: Marcel Dekker

59. Tsionsky M, Zhou J, Amemiya S, Fan F-R, Bard A, Dryfe R. 1999. Scanning electrochemical microscopy. 38. Application of SECM to the study of charge transfer through bilayer lipid membranes. *Anal. Chem.* 71:4300–5

60. Amemiya S, Bard A. 2000. Scanning electrochemical microscopy. 40. Voltammetric ion-selective micropipet electrodes for probing ion transfer at bilayer lipid membranes. *Anal. Chem.* 72:4940–48

61. Lee C, Bard A. 1990. Scanning electrochemical microscopy: application to polymer and thin metal oxide films. *Anal. Chem.* 62:1906–13

62. Lee C, Anson F. 1992. Use of electrochemical microscopy to examine counter ion ejection from Nafion coatings on electrodes. *Anal. Chem.* 64:528–33

63. Arca M, Mirkin M, Bard A. 1995. Polymer films on electrodes. 26. Study of ion transport and electron transfer at polypyrrole films by scanning electrochemical microscopy. *J. Phys. Chem.* 99:5040–50

64. Bath B, White H, Scott E. 2001. Imaging molecular transport across membranes. In *Scanning Electrochemical Microscopy*, ed. A Bard, M Mirkin, pp. 343–96. New York: Marcel Dekker

65. Kueng A, Kranz C, Mizaikoff B. 2005. Imaging of ATP membrane transport with dual micro-disk electrodes and scanning electrochemical microscopy. *Biosens. Bioelectron.* 21:346–53

66. Carano M, Lion N, Abid J, Girault H. 2004. Detection of proteins on poly(vinylidene difluoride) membranes by scanning electrochemical microscopy. *Electrochem. Commun.* 6:1217–21

67. Kallio T, Slevin C, Sundholm G, Holmlund P, Kontturi K. 2003. Proton transport in radiation-grafted membranes for fuel cells as detected by SECM. *Electrochem. Commun.* 5:561–65

68. Uitto O, White H. 2001. Scanning electrochemical microscopy of membrane transport in the reverse imaging mode. *Anal. Chem.* 73:533–39

69. Uitto O, White H, Aoki K. 2002. Diffusive-convective transport into a porous membrane. A comparison of theory and experiment using scanning electrochemical microscopy operated in reverse imaging mode. *Anal. Chem.* 74:4577–82

70. Ervin E, White H, Baker L. 2005. Alternating current impedance imaging of membrane pores using scanning electrochemical microscopy. *Anal. Chem.* 77:5564–69

71. Ervin E, White H, Baker L, Martin C. 2006. Alternating current impedance imaging of high-resistance membrane pores using a scanning electrochemical microscope. Application of membrane electrical shunts to increase measurement sensitivity and image contrast. *Anal. Chem.* 78:6535–41

72. Horrocks B, Schmidtke D, Heller A, Bard A. 1993. Scanning electrochemical microscopy. 24. Enzyme ultramicroelectrodes for the measurement of hydrogen peroxide at surfaces. *Anal. Chem.* 65:3605–14

73. Alpuche-Aviles M, Wipf D. 2001. Impedance feedback control for scanning electrochemical microscopy. *Anal. Chem.* 73:4873–81

74. Katemann B, Inchaispe C, Castro P, Schulte A, Calvo E, Schuhmann W. 2003. Precursor sites for the localized corrosion on lacquered tinplates visualized by means of alternating current scanning electrochemical microscopy. *Electrochim. Acta* 48:1115–21

75. LeSuer R, Fan F-R, Bard A. 2004. Scanning electrochemical microscopy. 52. Bipolar conductance technique at ultramicroelectrodes for resistance measurements. *Anal. Chem.* 76:6894–901

76. Johnson D, Enke C. 1970. Bipolar pulse technique for fast conductance measurements. *Anal. Chem.* 42:329–35

77. Taylor J, LeSuer R, Chambers C, Fan F-R, Bard A, et al. 2005. Experimental techniques for detection of components extracted from model 193 nm immersion lithography photoresists. *Chem. Mater.* 17:4194–203

78. Yasukawa T, Kaya T, Matsue T. 2000. Characterization and imaging of single cells with scanning electrochemical microscopy. *Electroanalysis* 12:653–59

79. Bard A, Li X, Zhan W. 2006. Chemical imaging of living cells by scanning electrochemical microscopy. *Biosens. Bioelectron.* 22:461–72

80. Amemiya S, Guo J, Xiong H, Gross D. 2006. Biological applications of scanning electrochemical microscopy: Chemical imaging of single living cell and beyond. *Anal. Bioanal. Chem.* 386:458–71

81. Edwards M, Martin S, Whitworth A, Macpherson J, Unwin P. 2006. Scanning electrochemical microscopy: principles and applications to biophysical systems. *Physiol. Meas.* 27:R63–R108

82. Yasukawa T, Kondo Y, Uchida I, Matsue T. 1998. Imaging of cellular activity of single cultured cells by scanning electrochemical microscopy. *Chem. Lett.* 8:767–68

83. Shiku H, Shiraishi T, Ohya H, Matsue T, Abe H, et al. 2001. Oxygen consumption of single bovine embryos probed by scanning electrochemical microscopy. *Anal. Chem.* 73:3751–58

84. Kaya T, Torisawa Y, Oyamatsu D, Nishizawa M, Matsue T. 2003. Monitoring the cellular activity of a cultured single cell by scanning electrochemical microscopy (SECM). A comparison with fluorescence viability monitoring. *Biosens. Bioelectron.* 18:1379–83

85. Zhan D, Li X, Zhan W, Fan F-R, Bard A. 2007. Scanning electrochemical microscopy. 58. Application of a micropipet-supported ITIES tip to detect Ag^+ and study its effect on fibroblast cells. *Anal. Chem.* 79:5225–31

86. Liu B, Rotenberg S, Mirkin M. 2000. Scanning electrochemical microscopy of living cells: Different redox activities of nonmetastatic and metastatic human breast cells. *Proc. Natl. Acad. Sci. USA* 97:9855–60

87. Liu B, Rotenberg S, Mirkin M. 2002. Scanning electrochemical microscopy of living cells. 4. Mechanistic study of charge transfer reactions in human breast cells. *Anal. Chem.* 74:6340–48

88. Rotenberg S, Mirkin M. 2004. Scanning electrochemical microscopy: detection of human breast cancer cells by redox environment. *J. Mammary Gland Biol.* 9:375–82

89. Feng W, Rotenberg S, Mirkin M. 2003. Scanning electrochemical microscopy of living cells. 5. Imaging of fields of normal and metastatic human breast cells. *Anal. Chem.* 75:4148–54

90. Mauzeroll J, Bard A. 2004. Scanning electrochemical microscopy of menadione-glutathione conjugate export from yeast cells. *Proc. Natl. Acad. Sci. USA* 101:7862–67

91. Mauzeroll J, Bard A, Owhadian O, Monks T. 2004. Menadione metabolism to thiodione in hepatoblastoma by scanning electrochemical microscopy. *Proc. Natl. Acad. Sci. USA* 101:17582–87

92. Guo J, Amemiya S. 2005. Permeability of the nuclear envelope at isolated *Xenopus* oocyte nuclei studied by scanning electrochemical microscopy. *Anal. Chem.* 77:2147–56

93. Hengstenberg A, Blöchl A, Dietzel I, Schuhmann W. 2001. Spatially resolved detection of neurotransmitter secretion from individual cells by means of scanning electrochemical microscopy. *Angew. Chem. Int. Ed.* 40:905–908

94. Takahashi Y, Hirano Y, Yasukawa T, Shiku H, Yamada H, Matsue T. 2006. Topographic, electrochemical, and optical images captured using standing approach mode scanning electrochemical/optical microscopy. *Langmuir* 22:10299–306

95. Kurulugama R, Wipf D, Takacs S, Pongmayteegul S, Garris P, Baur J. 2005. Scanning electrochemical microscopy of model neurons: constant distance imaging. *Anal. Chem.* 77:1111–17

96. Boldt F-M, Heinze J, Diez M, Petersen J, Börsch M. 2004. Real-time pH microscopy down to the molecular level by combined scanning electrochemical microscopy/single-molecule fluorescence spectroscopy. *Anal. Chem.* 76:3473–81

97. Lee Y, Ding Z, Bard A. 2002. Combined scanning electrochemical/optical microscopy with shear force and current feedback. *Anal. Chem.* 74:3634–43

98. Zhao X, Petersen N, Ding Z. 2007. Comparison study of live cells by atomic force microscopy, confocal microscopy, and scanning electrochemical microscopy. *Can. J. Chem.* 85:175–83

99. Zoski C, Simjee N, Guenat O, Koudelka-Hep M. 2004. Addressable microelectrode arrays: Characterization by imaging with scanning electrochemical microscopy. *Anal. Chem.* 76:62–72

100. Jayaraman S, Hillier A. 2003. Screening the reactivity of Pt_xRu_y and $Pt_xRu_yMo_z$ catalysts toward the hydrogen oxidation reaction with the scanning electrochemical microscope. *J. Phys. Chem. B* 107:5221–30

101. Shah B, Hillier A. 2000. Imaging the reactivity of electro-oxidation catalysts with the scanning electrochemical microscope. *J. Electrochem. Soc.* 147:3043–48

102. Jayaraman S, Hillier A. 2004. Construction and reactivity screening of a surface composition gradient for combinatorial discovery of electro-oxidation catalysts. *J. Comb. Chem.* 6:27–31

103. Jayaraman S, Hillier A. 2001. Construction and reactivity mapping of a platinum catalyst gradient using the scanning electrochemical microscope. *Langmuir* 17:7857–64

104. Kucernak A, Chowdhury P, Wilde C, Kelsall G, Zhu Y, Williams D. 2000. Scanning electrochemical microscopy of a fuel-cell electrocatalyst deposited onto highly oriented pyrolytic-graphite. *Electrochim. Acta* 45:4483–91

105. Fernández J, Raghuveer V, Manthiram A, Bard A. 2005. Pd-Ti and Pd-Co-Au electrocatalysts as a replacement for platinum for oxygen reduction in proton exchange membrane fuel cells. *J. Am. Chem. Soc.* 127:13100–1

106. Fernández J, Walsh D, Bard A. 2004. Thermodynamic guidelines for the design of bimetallic catalysts for oxygen electroreduction and rapid screening by scanning electrochemical microscopy. M-Co (M: Pd, Ag, Au). *J. Am. Chem. Soc.* 127:357–65

107. Walsh D, Fernández J, Bard A. 2006. Rapid screening of bimetallic electrocatalysts for oxygen reduction in acidic media by scanning electrochemical microscopy. *J. Electrochem. Soc.* 153:E99–E103

108. Weng Y, Fan F-R, Bard A. 2005. Combinatorial biomimetics. Optimization of a composition of copper(II) poly-L-histidine complex as an electrocatalyst for O_2 reduction by scanning electrochemical microscopy. *J. Am. Chem. Soc.* 127:17576–77

109. Fernández J, Mano N, Heller A, Bard A. 2004. Optimization of "wired" enzyme O_2-electroreduction catalyst compositions by scanning electrochemical microscopy. *Angew. Chem. Int. Ed.* 43:6355–57

110. Black M, Cooper J, McGinn P. 2005. Scanning electrochemical microscope characterization of thin film combinatorial libraries for fuel cell electrode applications. *Meas. Sci. Technol.* 16:174–82

111. Zhou J, Zu Y, Bard A. 2000. Scanning electrochemical microscopy. 39. The proton/hydrogen mediator system and its application to the study of the electrocatalysis of hydrogen oxidation. *J. Electroanal. Chem.* 491:22–29

112. Jambunathan K, Shah B, Hudson J, Hillier A. 2000. Scanning electrochemical microscopy of hydrogen electro-oxidation. Rate constant measurements and carbon monoxide poisoning on platinum. *J. Electroanal. Chem.* 500:279–89

113. Liu B, Bard A. 2002. Scanning electrochemical microscopy. 45. Study of the kinetics of oxygen reduction on platinum with potential programming of the tip. *J. Phys. Chem. B* 106:12801–806

114. Zoski C. 2003. Investigation of hydrogen electrocatalysis at polycrystalline noble metal electrodes by scanning electrochemical microscopy. *J. Phys. Chem. B* 107:6401–5

115. Fernández J, Bard A. 2004. Scanning electrochemical microscopy. 50. Kinetic study of electrode reactions by the tip generation-substrate collection mode. *Anal. Chem.* 76:2281–89

116. Lu G, Cooper J, McGinn P. 2007. SECM imaging of electrocatalytic activity for oxygen reduction reaction on thin film materials. *Electrochim. Acta* 52:5172–81

117. Zoski C, Aguilar J, Bard A. 2003. Scanning electrochemical microscopy. 46. Shielding effects on reversible and quasireversible reactions. *Anal. Chem.* 75:2959–66

118. Zoski C, Luman C, Fernández J, Bard A. 2007. Scanning electrochemical microscopy. 57. SECM tip voltammetry at different substrate potentials under quasi-steady-state and steady-state conditions. *Anal. Chem.* 79:4957–66

119. Eckhard K, Chen X, Turcu F, Schuhmann W. 2006. Redox-competition mode of scanning electrochemical microscopy (SECM) for visualisation of local catalytic activity. *Phys. Chem. Chem. Phys.* 8:5359–65

120. Karnicka K, Eckhard K, Guschin D, Stoica L, Kulesza P, Schuhmann W. 2007. Visualisation of the local bio-electrocatalytic activity in biofuel cell cathodes by means of redox competition scanning electrochemical microscopy (RC-SECM). *Electrochem. Commun.* 9:1998–2002

121. Holt K, Bard A, Show Y, Swain G. 2004. Scanning electrochemical microscopy and conductive probe atomic force microscopy studies of hydrogen-terminated boron-doped diamond electrodes with different doping levels. *J. Phys. Chem. B* 108:15117–27

122. Wilson N, Clewes S, Newton M, Unwin P, Macpherson J. 2006. Impact of grain-dependent boron uptake on the electrochemical and electrical properties of boron doped polycrystalline diamond electrodes. *J. Phys. Chem. B* 110:5639–46

123. Szunerits S, Mermoux M, Crisci A, Marcus B, Bouvier P, et al. 2006. Raman imaging and Kelvin probe microscopy for the examination of the heterogeneity of doping in polycrystalline boron-doped diamond electrodes. *J. Phys. Chem. B* 110:23888–97

124. Basame S, White H. 1998. Scanning electrochemical microscopy: measurement of the current density at microscopic redox-active sites on titanium. *J. Phys. Chem. B* 102:9812–19

125. Basame S, White H. 1999. Chemically-selective and spatially-localized redox activity at Ta/Ta$_2$O$_5$ electrodes. *Langmuir* 15:819–25

126. Serebrennikova I, White H. 2001. Scanning electrochemical microscopy of electroactive defect sites in the native oxide film on aluminium. *Electrochem. Solid State Lett.* 4:B4–B6

127. Zhu R, Nowierski C, Ding Z, Noel J, Shoesmith D. 2007. Insights into grain structures and their reactivity on grade-2 Ti alloy surfaces by scanning electrochemical microscopy. *Chem. Mater.* 19:2533–43

128. Boxley C, White H, Gardner C, Macpherson J. 2003. Nanoscale imaging of the electronic conductivity of the native oxide film on titanium using conducting atomic force microscopy. *J. Phys. Chem. B* 107:9677–80

129. Macpherson J, de Mussy J, Delplancke J. 2002. High-resolution electrochemical, electrical, and structural characterization of a dimensionally stable Ti/TiO$_2$/Pt electrode. *J. Electrochem. Soc.* 149:B306–B313

130. Still J, Wipf D. 1997. Breakdown of the iron passive layer by use of the scanning electrochemical microscope. *J. Electrochem. Soc.* 144:2657–65

131. Gray N, Unwin P. 2000. Simple procedure for the fabrication of silver/silver chloride potentiometric electrodes with micrometre and smaller dimensions: application to scanning electrochemical microscopy. *Analyst* 125:889–93

132. Horrocks B, Mirkin M, Bard A. 1994. Scanning electrochemical microscopy. 25. Application to the kinetics of heterogeneous electron transfer at semiconductor (WSe$_2$ and Si) electrodes. *J. Phys. Chem.* 98:9106–14

133. Haram S, Bard A. 2001. Scanning electrochemical microscopy. 42. Studies of the kinetics and photoelectrochemistry of thin film CdS/electrolyte interfaces. *J. Phys. Chem. B* 105:8192–95

134. Fonseca S, Barker A, Ahmed S, Kemp T, Unwin P. 2004. Scanning electrochemical microscopy investigation of the photodegradation kinetics of 4-chlorophenol sensitised by TiO$_2$ films. *Phys. Chem. Chem. Phys.* 6:5218–24

135. Fonseca S, Barker A, Ahmed S, Kemp T, Unwin P. 2003. Direct observation of oxygen depletion and product formation during photocatalysis at a TiO$_2$ surface using scanning electrochemical microscopy. *Chem. Comm.* 1002–1003

136. Bozic B, Figgemeier E. 2006. Scanning electrochemical microscopy under illumination: an elegant tool to directly determine the mobility of charge carriers within dye-sensitized nanostructured semiconductors. *Chem. Comm.* 2268–70

137. Shen Y, Nonomura K, Schlettwein D, Zhao C, Wittstock G. 2006. Photo-electrochemical kinetics of eosin Y-sensitized zinc oxide films investigated by scanning electrochemical microscopy. *Chem. Eur. J.* 12:5832–39

138. Zhang J, Lahtinen R, Kontturi K, Unwin P, Schiffrin D. 2001. Electron transfer reactions at gold nanoparticles. *Chem. Comm.* 1818–19

139. Zhang J, Barker A, Mandler D, Unwin P. 2003. Effect of surface pressure on the insulator to metal transition of a Langmuir polyaniline monolayer. *J. Am. Chem. Soc.* 125:9312–13

140. Liljeroth P, Quinn B, Ruiz V, Kontturi K. 2003. Charge injection and lateral conductivity in monolayers of metallic nanoparticles. *Chem. Comm.* 1570–71

141. Ruiz V, Liljeroth P, Quinn B, Kontturi K. 2003. Probing conductivity of polyelectrolyte/nanoparticle composite films by scanning electrochemical microscopy. *Nano Lett.* 3:1459–62

142. Ruiz V, Nicholson P, Jollands S, Thomas P, Macpherson J, Unwin P. 2005. Molecular ordering and 2D conductivity in ultrathin poly(3-hexylthiophene)/gold nanoparticle composite films. *J. Phys. Chem. B* 109:19335–44

143. O'Mullane A, Macpherson J, Unwin P, Cervera-Montesinos J, Manzanares J, et al. 2004. Measurement of lateral charge propagation in [Os(bpy)(2)(PVP)(n)Cl]Cl thin films: a scanning electrochemical microscopy approach. *J. Phys. Chem. B* 108:7219–27

144. Wilson N, Guille M, Dumitrescu I, Fernandez V, Rudd N, et al. 2006. Assessment of the electrochemical behavior of two-dimensional networks of single-walled carbon nanotubes. *Anal. Chem.* 78:7006–15

145. Burt D, Cervera J, Mandler D, Macpherson J, Manzanares J, Unwin P. 2005. Scanning electrochemical microscopy as a probe of Ag$^+$ binding kinetics at Langmuir phospholipid monolayers. *Phys. Chem. Chem. Phys.* 7:2955–64

146. Macpherson J, Unwin P. 1994. A novel approach to the study of dissolution kinetics using the scanning electrochemical microscope: theory and application to copper sulfate pentahydrate dissolution in aqueous sulfuric acid solutions. *J. Phys. Chem.* 98:1704–13

147. Ludwig M, Kranz C, Schuhmann W, Gaub H. 1995. Topography feedback mechanism for the scanning electrochemical microscope based on hydrodynamic forces between the tip and sample. *Rev. Sci. Instrum.* 66:2857–60

148. Macpherson J, Unwin P. 2000. Combined scanning electrochemical-atomic force microscopy. *Anal. Chem.* 72:276–85

149. Macpherson J, Unwin P, Hillier A, Bard A. 1996. In-situ imaging of ionic crystal dissolution using an integrated electrochemical/AFM probe. *J. Am. Chem. Soc.* 118:6445–52

150. Jones C, Unwin P, Macpherson J. 2003. In situ observation of the surface processes involved in dissolution from the cleavage surface of calcite in aqueous solution using combined scanning electrochemical-atomic force microscopy (SECM-AFM). *Chem. Phys. Chem.* 4:139–46

151. Macpherson J, Jones C, Barker A, Unwin P. 2002. Electrochemical imaging of diffusion through single nanoscale pores. *Anal. Chem.* 74:1841–48

152. Gardner C, Unwin P, Macpherson J. 2005. Correlation of membrane structure and transport activity using combined scanning electrochemical-atomic force microscopy. *Electrochem. Commun.* 7:612–18

153. Kranz C, Friedbacher G, Mizaikoff B. 2001. Integrating an ultramicroelectrode in an AFM cantilever: combined technology for enhanced information. *Anal. Chem.* 73:2491–500

154. Kranz C, Kueng A, Lugstein A, Bertagnolli E, Mizaikoff B. 2004. Mapping of enzyme activity by detection of enzymatic products during AFM imaging with integrated SECM-AFM probes. *Ultramicroscopy* 100:127–34

155. Kueng A, Kranz C, Lugstein A, Bertagnolli E, Mizaikoff B. 2005. AFM-tip-integrated amperometric microbiosensors: high-resolution imaging of membrane transport. *Angew. Chem. Int. Ed.* 44:3419–22

156. Burt D, Wilson N, Weaver J, Dobson P, Macpherson J. 2005. Nanowire probes for high resolution combined scanning electrochemical microscopy: atomic force microscopy. *Nano Lett.* 5:639–43

157. Dobson P, Weaver J, Holder M, Unwin P, Macpherson J. 2005. Characterization of batch-microfabricated scanning electrochemical–atomic force microscopy probes. *Anal. Chem.* 77:424–34

158. Shin H, Hesketh P, Mizaikoff B, Kranz C. 2007. Batch fabrication of atomic force microscopy probes with recessed integrated ring microelectrodes at a wafer level. *Anal. Chem.* 79:4769–77

159. Rudd N, Cannan S, Bitziou E, Ciani L, Whitworth A, Unwin P. 2005. Fluorescence confocal laser scanning microscopy as a probe of pH gradients in electrode reactions and surface activity. *Anal. Chem.* 77:6205–17

160. Gollas B, Bartlett P, Denuault G. 2000. An instrument for simultaneous EQCM impedance and SECM measurements. *Anal. Chem.* 72:349–56

161. Xiang J, Guo J, Zhou F. 2006. Scanning electrochemical microscopy combined with surface plasmon resonance: studies of localized film thickness variations and molecular conformation changes. *Anal. Chem.* 78:1418–24

Novel Detection Schemes of Nuclear Magnetic Resonance and Magnetic Resonance Imaging: Applications from Analytical Chemistry to Molecular Sensors

Elad Harel,[1] Leif Schröder,[1] and Shoujun Xu[2]

[1] Materials Sciences Division, Lawrence Berkeley National Laboratory, and Department of Chemistry, University of California, Berkeley, California 94720; email: elharel@berkeley.edu; schroeder@waugh.cchem.berkeley.edu

[2] Department of Chemistry, University of Houston, Houston, Texas 77204; email: sxu7@uh.edu

Annu. Rev. Anal. Chem. 2008. 1:133–163

First published online as a Review in Advance on January 15, 2008

The *Annual Review of Analytical Chemistry* is online at anchem.annualreviews.org

This article's doi:
10.1146/annurev.anchem.1.031207.113018

Key Words

microfluidics, flow imaging, low field, atomic magnetometer, molecular imaging, chemical exchange saturation transfer (CEST)

Abstract

Nuclear magnetic resonance (NMR) is a well-established analytical technique in chemistry. The ability to precisely control the nuclear spin interactions that give rise to the NMR phenomenon has led to revolutionary advances in fields as diverse as protein structure determination and medical diagnosis. Here, we discuss methods for increasing the sensitivity of magnetic resonance experiments, moving away from the paradigm of traditional NMR by separating the encoding and detection steps of the experiment. This added flexibility allows for diverse applications ranging from lab-on-a-chip flow imaging and biological sensors to optical detection of magnetic resonance imaging at low magnetic fields. We aim to compare and discuss various approaches for a host of problems in material science, biology, and physics that differ from the high-field methods routinely used in analytical chemistry and medical imaging.

1. INTRODUCTION

NMR: nuclear magnetic
resonance

MRI: magnetic resonance
imaging

Remote detection: general
methodology in which
encoding and detection
occur in different physical
or molecular environments

Although nuclear magnetic resonance (NMR) is perhaps the most powerful analytical method in chemistry, its usefulness is limited in many applications by poor sensitivity. As a general rule, approximately 10^{15}–10^{18} spins are needed for an inductively detectable signal at high magnetic field strengths. This is mainly due to the very small nuclear magnetic moment relative to $k_B T$ at room temperature. Consequently, NMR is performed on relatively large samples a few cubic centimeters in volume; for magnetic resonance imaging (MRI), the sample is typically several hundred cubic centimeters. Much of the work on magnetic resonance over the past 60 years has gone into combating this very problem, as the rewards for doing so are far-reaching.

High-resolution one-dimensional NMR is an everyday tool of the organic chemist (1). Multidimensional NMR allows for the structural determination of proteins in solution (2). MRI is one of the most valuable diagnostic tools available in medicine (3) and materials characterization (4).

Here, we describe a very general method, which we call remote detection, for increasing the sensitivity in NMR or MRI experiments; this method is applicable when excitation and detection can be separated in some manner. This process may involve actual physical separation of these stages of the experiment, or it may involve a chemical exchange in which information about a species in one environment is detected by examination of it in another.

First, we give a general overview of the concept. We then provide examples of the methodology's utility in various applications, including materials characterization, flow imaging, lab-on-a-chip devices, analytical chemistry, solid-state NMR, low-field MRI, optical detection of NMR and MRI, and biological sensors.

2. TRADITIONAL NUCLEAR MAGNETIC RESONANCE

The phenomenon of NMR is based on the interaction between an external magnetic field and the nonzero nuclear spin of certain atomic isotopes (5); this is known as the Zeeman interaction. This interaction splits otherwise degenerate energy levels and is responsible for the NMR spectrum. For molecules, the orbiting electrons of nearby atoms slightly alter the magnetic field felt by the nuclear spin. The resulting chemical shift can only be observed at very high magnetic fields and under conditions of extremely high field homogeneity. These opposing experimental conditions are responsible for the very high costs and maintenance associated with high-field NMR. Other interactions between nuclear spins that are independent of the magnetic field strength, the so-called indirect interactions, give rise to J-couplings and raise the possibility of multidimensional NMR, which would allow unambiguous assignment of peaks through a network of coupled spins. The power of NMR is in its ability to precisely control these spin interactions and to observe them by applying sequences of accurately timed radio frequency (rf) pulses and magnetic field gradients.

Such rf pulse sequences have been the focus of a tremendous amount of research since pulsed NMR was first proposed nearly 30 years ago (6). In pulsed NMR, the spins are encoded into the desired quantum mechanical state. Detection is accomplished by the reverse process, whereas a changing magnetization resulting from the ensemble spin state is converted into a small but detectable electronic signal. Typically, encoding and detection are performed with a single rf coil, which surrounds the sample of interest. The coil may be ideal for encoding but will suffer drastically for detection if the region of interest is small in comparison to the coil (7). This is often the case for heterogeneous samples, including organs of the body, porous materials, and microfluidic devices. This restriction may also apply in the case of a very dilute chemical species surrounded by a large solvent bath, in which the former is difficult to detect.

3. REMOTE DETECTION BASICS

The first modality of remote detection that we discuss challenges this paradigm by physically separating the encoding and detection steps of an NMR pulse sequence so as to optimize each step individually. In remote detection NMR, a large rf coil encompasses the sample while a smaller coil concentrates the signal for more efficient and sensitive detection (8, 9). This separation allows for the possibility of moving away from inductive detection altogether and implementing optical detection and other noninductive schemes (discussed below). Regardless of the detection modality, the principle of separate encoding and decoding remains the same.

Figure 1 illustrates a generic remote detection scheme. At equilibrium the nuclear spins align in the direction of the static field. A selective pulse excites the spins according to their location in space or chemical shift or both. As in any other NMR or MRI pulse sequence, the spins experience either free evolution or evolution according to a well-defined sequence of rf and gradient pulses, which places the magnetization into a certain desired state. In traditional magnetic resonance experiments, readout is performed at this junction. In remote detection, the phase of the spins in the transverse plane must be stored along the longitudinal direction so as to be affected only by spin-lattice relaxation, T_1. Because T_1 is typically much longer than T_2^*, the effective transverse relaxation time, the spins can effectively flow out to the detector before an irreversible loss in signal occurs. At the detection stage, readout of the stored magnetization provides information about the phase, which in turn provides a complete description of the desired spin state (as in traditional NMR). Because the detector typically has a smaller volume than the encoding volume, several pulses are needed to completely detect the encoded magnetization. This results in a travel curve, which measures the time of flight (TOF) of encoded spins to the detector.

Because the first spins to arrive are typically unencoded, the signal is at a maximum value. As encoded spins arrive they mix with unencoded spins, causing a decrease in the overall signal that can be observed by a dip in the TOF curve. As the encoded spins leave the detector, the signal is restored. The conventional time dimension is no longer available as the direct dimension because the encoding coil is used only to manipulate the magnetization, so encoding must proceed point by point. For example, a free induction decay (FID) is usually measured by exciting the spins into the transverse plane and then acquiring the signal over time. In remote, points along the FID are recorded by delaying the time between the storage and detection pulses. Each point in the travel curve thus contributes one point to the indirect interferogram

Effective transverse relaxation: spin dephasing in the transverse plane due to the combined effects of spin-spin interactions and magnetic field inhomogeneities

Time of flight (TOF): time needed for an encoded spin to reach the detection region

FID: free induction decay

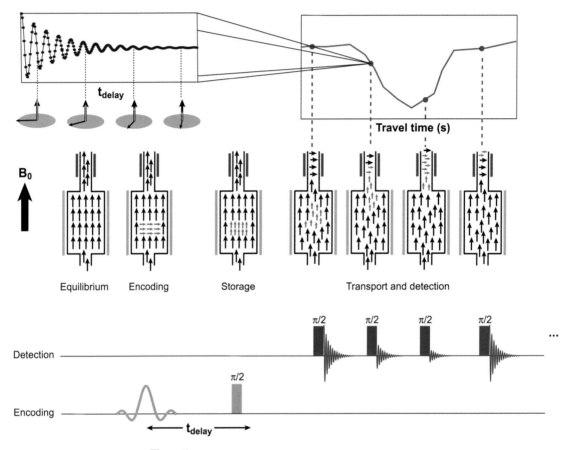

Figure 1

General remote detection scheme. A selective pulse excites the spins of interest. The magnetization precesses about the static field, acquiring phase which gives either spectroscopic (free evolution) or imaging (evolution in the presence of gradients) information about the sample. A storage pulse converts the phase information into longitudinal magnetization, which is subject only to T_1 relaxation. The spins then flow to the detector, which reads out the amplitudes by a train of hard pulses. Each detection pulse contributes one point to the time-of-flight (TOF) curve. Repeating the experiment point by point creates an interferogram (*upper left*) that decays, i.e. the free induction decay (FID). In the case of spectroscopy, this corresponds to incrementing the delay between the encoding and storage pulses, creating an indirect FID. In the case of imaging, the delay is fixed and the gradient strength is incremented which covers k-space (the reciprocal space in which the signal is acquired prior to Fourier transformation). Each detection pulse then creates a partial image corresponding to the spin density in the sample, which arrives at the same TOF to the detector. The sum of these partial images creates the full image of the sample.

(e.g., FID) built up by repeating the remote experiment, each time with a different indirect point (e.g., time delay). In the following sections we illustrate these ideas more concretely with several examples.

Hyperpolarization: polarization of spin states that significantly exceeds the state populations given by the Boltzmann distribution

EPI: echo planar imaging

3.1. Gas Flow Imaging in Porous Materials

Figure 2a,b illustrates the results of a remote experiment using hyperpolarized xenon gas flowing through a porous rock (10). Optical pumping methods allow thermally polarized xenon to be enhanced from a thermal polarization of about 10^{-5} to 2–5% (11, 12), which corresponds to an increase in sensitivity of over three orders of magnitude. The increased sensitivity allows gas mixtures containing xenon to be visualized with MRI even though the density of the gas is much lower than that of a liquid. For applications related to crude oil recovery, pore-level structural determination, and materials characterization, gas flow imaging is critical (13). The images in **Figure 2** were obtained using a simple phase-encoding scheme applied to xenon gas flowing through a Bentheimer sandstone rock with a porosity of about 22.5% and pore size of approximately 100 μm.

The entire data set consists of the spatial dimensions and the additional TOF dimension. Projecting along the TOF dimension provides the full three-dimensional image as shown in **Figure 2b**. Cutting a wedge out of the data set reveals flow inhomogeneities, such as regions of spins that rapidly dephase either because of incoherent flow or T_2^* relaxation or because of static spins that never reach the detector within the T_1 relaxation time of the gas. A direct image would show such inhomogeneities in the former case, but not the latter. Thus, by comparison with direct imaging, insights into the flow can be elucidated.

Notably, the partial images corresponding to different points along the TOF dimension are not real-time images of the flow as in fast, direct imaging sequences such as echo planar imaging (EPI) (14). Rather, they are a spatial representation of spins that take an equal amount of time to reach the detection region. Remote detection, therefore, is much more similar to chromatography used in chemical analysis. As a result, the first images represent spins that typically initiate from the outlet region and that take the least amount of time to reach the detection region. Later TOF values correspond to spins that take longer to reach the detector. Therefore, the geometry of the sample plays an important role in determining the spatial profiles of each partial image, as it determines to a large degree the necessary flow path for a given packet of encoded spins.

In this example, investigators utilized hyperpolarized xenon to trace gas flow through a porous material. However, xenon has another remarkable property in addition to its ability to become highly polarized: It displays a wide range of chemical shifts that are very sensitive to the chemical or physical environment (15). Proton NMR exhibits a chemical shift range of about 8 ppm. Xenon, on the other hand, can exhibit a range of more than 7000 ppm arising from its highly polarizable electron cloud. In nanoporous materials such as silica aerogel, xenon shows a chemical shift range of up to ~120 ppm depending on the size of the pore within which it is enclosed (16). In the large macropores of the aerogel, the xenon chemical shift approximates

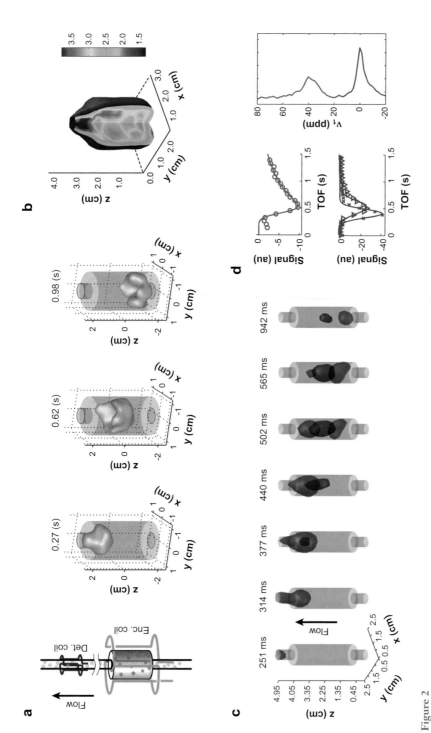

Figure 2

Xenon flow through a porous Bentheimer rock. (*a*) Images acquired by applying a hard pulse and then incrementing the gradient strength prior to the storage pulse, resulting in the creation of a full three-dimensional image. The first partial image corresponds to the spin density that took 0.27 s to reach the detector. Similarly, the last partial image corresponds to spins that took 0.98 s to reach the detector. (*b*) Full image, reconstructed by projecting the data along the time-of-flight (TOF) dimension. Regions of low intensity arise either from low spin density, fast T_2^* relaxation, stagnant flow in which spins never reach the detector within T_1, or incoherent flow (e.g., turbulence). (*c*) Three-dimensional isosurfaces of xenon flow through aerogel. Spin density from free xenon gas (*red*) and occluded gas (*blue*) are represented. (*d*) TOF curves for regions inside the aerogel (*blue circles*) and inlet (*red triangles*) and outlet (*red squares*). Dispersion causes a broadening of the Gaussian profiles from outlet to inlet as the latter must travel a longer distance to the detector. The width of the aerogel TOF curve is broader than that of the inlet although the distance to the detector is shorter, indicating the presence of two distinct flow paths. The long exponential tail is the result of dead-end pores that are poorly connected to the main flow field, which provides a path to the detector.

that of the free gas, which is 0 ppm. In the mesoporous region, at about 20–30 nm, the xenon chemical shift measures ∼38 ppm at room temperature (17). The correlation between chemical shift and pore size allows MRI to indirectly probe the nanoporous environment of the aerogel even when the actual MRI spatial resolution is several orders of magnitude lower than that needed to resolve the pore structure directly.

Radio frequency pulses can easily tag one spin species over another at large frequency separations by employing short bandwidth pulses. **Figure 2c,d** illustrates the results of two experiments of xenon gas flow through aerogel, each selecting a different resonance peak (18). The flow of the free gas (0 ppm) can thus be distinguished from the flow of the occluded gas (38 ppm).

Dispersion can be described in either the spatial domain or the time domain (19). In the former, the spreading of a fluid in a given time interval is measured, whereas in the latter, the arrival time distribution of a tagged fluid is examined. In remote detection, this time distribution is represented by the TOF curve, the width of which is a measure of dispersion. If we assume that the flow in an unrestricted geometry is described by a normal distribution, then we can assign a width and mean time for each voxel in the image. The fit of a normal distribution to the TOF curve is fairly good, validating this approximation. Mathematically, we have

$$s(t, \boldsymbol{r}) = s_0(\boldsymbol{r}) \Delta t \, \exp(-(t - t_0(\boldsymbol{r}))^2 / 2\sigma(\boldsymbol{r})^2) / \sqrt{2\pi} \sigma(\boldsymbol{r}). \qquad (1)$$

Therefore, $\sigma(\mathbf{r})$ and $t_0(\mathbf{r})$ can be assigned to each position $\boldsymbol{r}$. For voxels inside the aerogel itself, however, this model no longer applies. Instead, we find that a Gaussian convoluted with an exponential decay accurately models the TOF curves. Once again, we can assign $\sigma(\boldsymbol{r})$ and $t_0(\boldsymbol{r})$ to each position $\boldsymbol{r}$. Additionally, we can assign a time constant $\tau(\boldsymbol{r})$ to yield a measure of how well the occluded gas is connected to the main flow field, which provides a coherent path to the detection region. Therefore, by selecting each peak separately we are able to identify two completely separate flow paths and image each individually with a high time resolution (**Figure 2c**). The physical explanation for this observation is that the xenon inside the small mesopores of the aerogel is not well connected to the main flow field, most likely due to the presence of dead-end pores inside the sample.

Unlike other methods that identify porosity (20), pore-size distribution (21, 22), and other bulk values, remote detection flow imaging provides material heterogeneities. That is, material properties can be measured for each voxel in the image. It should be noted that, even in principle, optical techniques could never perform such an experiment because the occluded and free gas have identical physical properties. Furthermore, many samples of interest are opaque and thus inaccessible to optical wavelengths. Numerous porous materials show a large chemical shift range for xenon (23) and are amenable to these methods.

3.2. Lab-on-a-Chip: Imaging and Spectroscopy

As a noninvasive spectroscopic technique, NMR presents an attractive complementary approach to optics-based detection methods for lab-on-a-chip devices (24). Microfluidics shows great promise in the fields of biology, chemistry, and physics due to its unique ability to precisely control fluids at very small volumes (25). Its applications range from DNA analysis (26), proteomics (27), and synthetic chemistry (28) to fundamental studies of physical processes (29). Optics-based detection methods generally suffer from a lack of chemical specificity and limit the use of microfluidic devices to only those made from optically transparent materials (30). Although some spectroscopic techniques (such as Raman and IR) have been applied to microfluidic devices (31), they are inhibited to a large extent by the very short optical path length through the microchannels of the devices (32). Furthermore, they are generally limited to examining only a single point on the chip at a time. NMR also suffers from a significant loss in sensitivity due to the very small volume of analyte on the chip and the necessarily large coil (which is several orders of magnitude larger than the individual channels) surrounding the entire microfluidic device. Remote detection allows this sensitivity limitation to be overcome and has proved ideal for the analysis of microfluidic devices. For microfluidic chips it offers the ability of full-chip imaging, spectroscopy, and flow analysis—all at very high sensitivity.

The experiment described below (also see **Figure 3**) involves the mixing of two fluids inside a simple mixing microfluidic chip. In one channel flows ethanol, in the

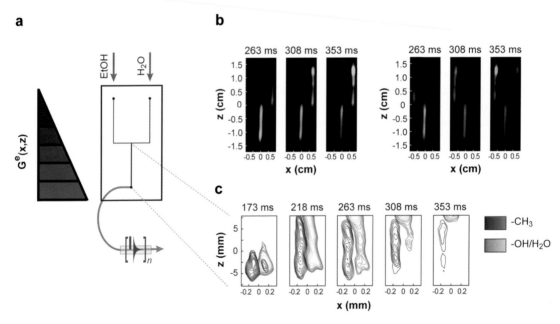

Figure 3

Liquid flow imaging through a complex microfluidic chip. (*a*) Ethanol (*blue arrow*) and water (*red arrow*) flow through a T-mixer chip. The detection volume is less than 50 nL. Also shown is the symbol for the magnetic field gradient necessary for imaging (*red triangle*). (*b*) Spectra. Even though the spectra are not resolvable inside the chip, they are highly resolved outside the chip. The flow of each species can be distinguished and subsequently imaged by selecting the appropriate resonance peak in the detector. (*c*) A zoomed-in view of the mixing region shows that the fluids in fact do not mix at this flow rate.

other flows water. As mentioned above, the chemical shift range of protons is rather small, thus magnetic field homogeneity is critical. Unfortunately, the interface between the glass and fluid causes severe local magnetic field gradients because of the difference in magnetic susceptibility between the different materials and their geometry (33). This effectively makes resolving the chemical shifts of the different species impossible directly on the chip. Although there may be methods of overcoming this limitation, such as using susceptibility-matched materials, they are not applicable to existing microfluidic chip designs.

Remote detection can circumvent this inhomogeneity problem. Once the fluids exit the chip and pass through the small solenoid detection coil, the spectrum is highly resolvable (34). For this particular chip, the resolution decreases from 5 kHz to less than 50 Hz upon exiting the chip—an improvement of two orders of magnitude. This allows the flow of each species to be resolved even when the chemical shift is not directly resolvable. In this way, any number of species can be resolved and imaged in a single experiment as long as the residence time in the detection coil is long enough to resolve the peaks.

Furthermore, unlike other methods that perform NMR on a chip, remote detection is capable of imaging the entire chip or a section thereof with very high sensitivity. Direct NMR-on-a-chip methods fabricate the detection coil directly onto the chip, greatly enhancing sensitivity (35). However, this means that only a few coils can be placed on the chip at once as their fingerprint is rather large (>1 mm in diameter); furthermore, the chip fabrication must be significantly altered. Again, magnetic susceptibility becomes a major problem as the addition of a new element, the copper surface coil, creates severe magnetic field distortions. Additionally, this method is not compatible with most of the chip fabrication techniques already in place. Nonetheless, NMR on a chip, whether remote or direct, promises to become a growing area of research.

3.3. Solid-State Nuclear Magnetic Resonance

So far we have discussed enhancing the detection sensitivity of liquid samples. It is also possible to employ a type of remote detection for very high sensitivity detection of solid samples. Due to the dipolar interaction between spins, which are averaged away in isotropic samples, NMR spectra of solids are very broad unless spun at the "magic" angle, $\theta \sim 54.7°$. This is because of the dominant term in the dipolar Hamiltonian, which has the form $(3 \cos^2 \theta - 1)$, where θ is the angle between the internuclear vector and the external field. Magic angle spinning, which averages away these broadening terms, does have a drawback: The spinning rotor needed to hold the sample is large compared to mass-limited samples. The solution to this problem has been demonstrated in an experiment by Sakellariou et al. (36) in which two coils were used, a static coil for sample spin manipulation and detection that was inductively coupled to a very sensitive microcoil wrapped directly around a glass capillary containing the sample. Susceptibility problems were eliminated due to the presence of coil material and sample interface in liquid microcoil designs (37). This method has important applications to the analysis of biological tissue, organic powders, and radioactive material.

3.4. Perspectives and In Vivo Applications

Remote detection, which was initially utilized solely to enhance sensitivity, is a very powerful method for elucidating the fluid flow properties of porous materials. Adding spectroscopy in the encoding and detection regions allows the applicability of this method to a large class of systems, from porous rocks and nanoporous materials to microfluidics. In vivo applications, where spins are excited in the brain and detected in the jugular where the blood collects, may also be possible (38). Such applications would allow much higher sensitivity imaging in regions of the brain far below the surface of the skull where the head coil is most insensitive. Furthermore, such a modality would allow for optical detection of the NMR signal and, possibly, low-field encoding (see **Figure 4** for an illustration of this potential medical application). Because flow is ubiquitous in nature, NMR remote detection may also be applicable to many other systems of biological significance, including diffusion across biological membranes and the study of protein/ligand binding events.

In the following section, we provide an overview of another remote detection method in which chemical exchange replaces the physical transport of a fluid, as in the applications described earlier. This method has very important implications for molecular and biological sensors.

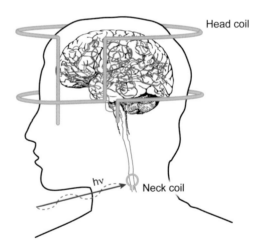

Figure 4

Potential in vivo applications of remote detection. Although magnetic resonance imaging (MRI) offers very high resolution images of the brain, it suffers from a severe lack of sensitivity of regions buried deep inside the brain. Such regions, however, are critically important in stroke and other diseases. Coil sensitivity falls steeply with the distance of the coil to the region of interest. Blood flow in the brain, however, transfers blood from most regions of the brain and collects at the jugular vein near the surface of the neck where a sensitive neck coil could potentially be positioned for an increase in sensitivity. This may provide a modality for very high resolution imaging in regions not currently accessible by current MRI methods. Detection could also proceed by optical detection which would measure the magnetic flux of the encoded blood.

4. INDIRECT NUCLEAR MAGNETIC RESONANCE DETECTION BASED ON CHEMICAL EXCHANGE

Biomedical NMR applications require enhanced sensitivity for diagnostic purposes in the rapidly expanding field of molecular imaging. In this context, a special version of remote detection comes into play. The inherently low sensitivity of NMR limits the detection of molecules (other than water) for imaging in order to reveal biochemical abnormalities at very early stages of diseases. Hence, proton MRI currently is not the first choice for molecular imaging (39). To expand the capabilities of biomedical NMR, the problem of sensitivity can be tackled by indirect detection methods that rely on chemical exchange of selectively saturated magnetization. This is, in principle, an in situ version of remote detection that separates the molecular environment of the detection from the encoded molecular information. There are two ways to implement this approach in order to significantly amplify the signal from a low-concentration target molecule: either via exchange with a huge reservoir of thermally polarized spins, or by using a reservoir of only medium size that is, however, hyperpolarized.

Historically, the thermal polarization method was first realized by using chemical exchange saturation transfer (CEST) (40) to detect exchangeable protons of the OH, NH_2, and NH groups of several sugars, amino acids, nucleosides, and other compounds after selective saturation (i.e., depletion of the magnetization) at 1–6.5 ppm downfield of the water resonance and subsequently observing changes in the abundant water signal. An expansion of the concept involving xenon gas that was hyperpolarized through laser polarization was presented recently (41) and differs from the former method in various aspects (discussed below).

Both concepts have the following principle in common: The chemical exchange in combination with adjustable, selective rf pulses allows detection of information from the molecular species at low concentration by encoding it in a way that enables remote amplification, resulting in a high-intensity signal (**Figure 5a**). Whereas conventional detection acts like a snapshot technique, reading only signal from the few nuclei that are actually bound to the target molecule, the CEST approach involves many hundreds to thousands of participating nuclei per second at each exchange site in the labeling process and thus "stores" the preamplified information.

4.1. Theory of the Chemical Exchange Saturation Transfer Enhancement Method

The basic idea of the CEST technique was first demonstrated in a double resonance experiment by Forsén & Hoffman (42) in a study of the chemical exchange between two compounds of similar concentration. The authors analyzed their data based on differential equations describing the dynamics of a two-site exchange system as proposed by McConnell in terms of modified Bloch equations (43). The detection pool is denoted by index d and parameters of the saturated pool are indexed s; the observed longitudinal magnetization, M_z^d, decreases upon irradiation on the frequency of M_z^s. The parameters to quantify the chemical exchange are the lifetimes τ^d and τ^s of the exchangeable nuclei in the two environments. Under the assumption of achieving

Molecular imaging: illustrates the spatial distribution of specific molecules of biochemical relevance

CEST: chemical exchange saturation transfer

Figure 5

(*a*) Schematics of the chemical exchange saturation transfer (CEST) method using molecular separation of encoding and detection for significant signal amplification. The resonance of the detection molecule at high concentration (in this case, water) is observed after off-resonance saturation (*blue spectrum*) and after on-resonance saturation (*red spectrum*) of a highly diluted CEST agent. (*b*) Decrease in the observed signal assigned to M_0^d, assuming complete saturation of M_z^s. The time constant τ_{sat} and the amplitude of the new steady-state magnetization are determined by the longitudinal relaxation rate $R_1^d = 1/T_1^d$ and the exchange rate from the detection pool into the saturation pool, $k^{ds} = 1/\tau^d$. (*c*) Proton transfer ratio (PTR) from the data in (*b*) assuming an incomplete saturation with $\alpha = 0.8$.

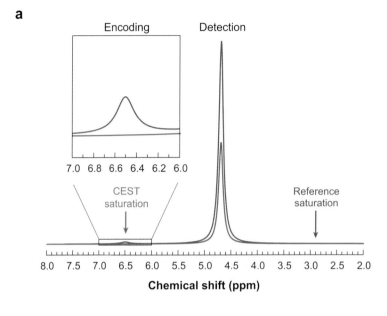

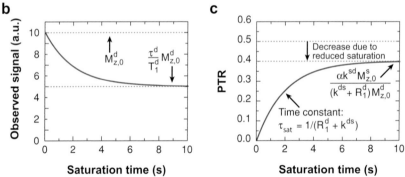

$M_z^s = 0$ instantaneously, the new equilibrium of M_z^d is reached in a monoexponential decay (**Figure 5*b***).

An important precondition for CEST is that the saturation field be selective enough to neglect direct saturation of M_z^d. The first papers on CEST illustrated the so-called spillover effect of the detected water resonance at 4.7 ppm in so-called z-spectra (42, 44), which allow a comparison of the desired saturation effect at 4.7 ppm + δ with a control signal after saturation at 4.7 ppm − δ. Although proton CEST experiments deal with chemical shift differences of at least 300 Hz, pulsed rf signals have a non-negligible minimum bandwidth. Therefore, their amplitude is adjusted to minimize spillover effects. Realistic conditions are given by the weak saturation pulse approximation that allows for incomplete (but still instantaneous) saturation of M_z^s. For this purpose, the concept of the proton transfer ratio (PTR) (45)

Spillover effect: unwanted direct saturation of a spin ensemble while saturating another spin ensemble at different resonance frequency

PTR: proton transfer ratio

was introduced, which includes the saturation efficiency α of the saturated resonance to modify the above-mentioned signal change (**Figure 5c**). Theoretical analysis of the underlying dynamics of CEST agents (45) yields:

$$\text{PTR} \equiv \frac{M_{z,0}^d - M_z^d(t)}{M_{z,0}^d} = \frac{\alpha k^{sd} M_{z,0}^s}{(k^{ds} + R_1^d) M_{z,0}^d} \left(1 - e^{-(k^{ds} + R_1^d)t}\right). \tag{2}$$

The time constant for the exponential behavior of both parameters is still $1/\tau_{\text{sat}}$. The magnetizations $M_z^d(t)$ and $M_{z,0}^d$ are determined from the signal intensities with on- and off-resonant saturation and subsequent correction of so-called magnetic transfer (MT) effects that possibly overlap with the CEST effect (46, 47) (see **Figure 6a**). Equation 2 shows that the PTR is directly proportional to the concentration of the

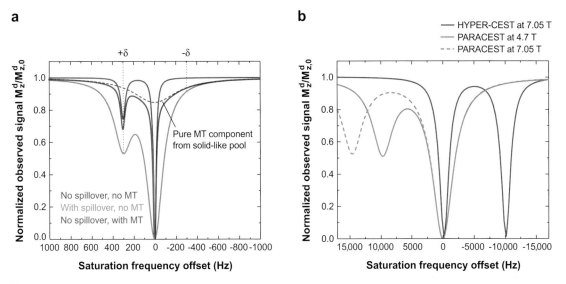

Figure 6

(*a*) Schematic representation of z-spectra for different scenarios of chemical exchange saturation transfer (CEST) with thermally polarized protons and no chemical shift reagents. Low saturation powers (*black curve*) yield profiles with one sharp peak at the resonance of the CEST agent ($+\delta$) and one peak for direct water saturation with no spillover effect observed at $-\delta$. Higher saturation powers (*red curve*) increase the saturation bandwidths and lead to unwanted saturation at $-\delta$. Magnetic transfer (MT) effects can yield an additional signal decrease that must be separated from the CEST effect. Such MT effects are only relevant in the presence of a solid-like spin ensemble, such as immobilized water in vivo (*dashed resonance profile*), which causes an extremely broad dip in z-spectra. For further details, see References 46 and 47. (*b*) z-spectra comparing a paramagnetic CEST (PARACEST) agent at 4.7 T (schematic representation according to results in Reference 55) with hyperpolarized xenon CEST (HYPER-CEST) of a functionalized xenon biosensor at 7.05 T. Both show frequency separations from the detection signal on the order of 10 kHz, but PARACEST is related to fast exchange, thus showing broad dips in the saturation profiles. The spillover effect, around -10 kHz, is negligible for both agents. Simulation of the PARACEST spectrum for 7.05 T shows a better separation of the two saturation dips. However, the HYPER-CEST profile is still sharper.

APT: amide proton transfer

saturated pool. α is defined via the power $\omega_1 = \gamma B_1$ of the saturation pulse in Hz and two parameters p and q that take transverse and longitudinal relaxation effects into account (for details, see Reference 48):

$$\alpha = \frac{\omega_1^2}{\omega_1^2 + pq}. \tag{3}$$

Hence, powers must be adjusted to minimize the relevance of pq and to achieve $\alpha \rightarrow 1$ (full saturation). One important contribution to incomplete saturation is exchange-line broadening that reduces T_2 and increases p.

4.2. Diamagnetic Chemical Exchange Saturation Transfer Agents and Applications

CEST experiments using thermally polarized protons can be divided into two groups: diamagnetic CEST (DIACEST) and paramagnetic CEST (PARACEST). The first group represents the "classic" CEST agents that contain no paramagnetic shift reagent. Several kinds of biochemical compounds were studied and yielded different signal enhancements due to various exchange characteristics. As seen in Equation 3, a high exchange rate k^{ds} and a slow (long) T_1 relaxation are favorable to achieving a high PTR. Hence, such experiments are predestined for high magnetic fields to assure both small $R_1^d = 1/T_1^d$ and large chemical shift separation. The first CEST candidates investigated by Ward and colleagues were studied at concentrations between 62.5 mM and 250 mM to achieve saturations between 10–67% of the 111-M water proton signal (40). The enhancement of, for instance, barbituric acid under physiological conditions (pH = 7.4, $T = 37°C$) was quantified as 264-fold and used for demonstration of pH-sensitive imaging. Because the contrast depends on both concentration and the pH-adjustable exchange rate, Balaban and colleagues later introduced a method using agents with more than one type of exchangeable proton to calibrate for pure pH–dependence while eliminating concentration effects (49).

The second generation of DIACEST agents included compounds with high numbers of exchangeable protons. Cationic polymers at concentrations of $\sim 10^{-4}$ M were reported to cause water signal changes of 40–50% (50). Poly-L-lysine (PLL), for example, involves more than 7000 exchangeable sites, thus yielding an enhancement factor of more than 486,000. Polyuridilic acid, a polymer of 2000 uridine units, was observed at 10 μM with an enhancement factor of $\sim 10.8 \times 10^6$ via its imino protons (51).

The amide protons of some endogenous compounds were used in a method called amide proton transfer (APT) to detect pH changes via the exchange rate shown in Equation 3. This approach is of special biomedical interest because the rates reflect tissue pH and local biochemical parameters such as salt or metal content. It was used to study ischemic brain related to acute stroke (46) and to reveal brain tumors in mice (52). Two CEST-related methods have been proposed (53) to determine the base catalyzed, the acid catalyzed, and the spontaneous exchange rate constants for very low concentrations of amide protons in polyamide-based APT agents. In such applications, conventional methods such as the quantification of exchange line broadening would be very time consuming because of the low signal-to-noise ratio related to direct observation of such dilute protons.

4.3. Limitations of Conventional ^{1}H Chemical Exchange Saturation Transfer Detection

One limiting factor of DIACEST compounds is the small chemical shift range of protons. This has two consequences: (*a*) the slow exchange condition needed for re-solving signals of two sites in exchange is relatively strict, thus limiting the maximum observable rate; and (*b*) the spillover effect due to direct saturation of the relatively close detection signal requires a trade-off between saturation power and unwanted signal reduction. Agents with a larger chemical shift separation would therefore sig-nificantly improve the capabilities of the CEST enhancement and could be detected at lower concentrations.

Spillover effects are of minor importance as long as the amplitude of the satura-tion pulse (given in Hz) is negligible compared to the chemical shift difference (in Hz) between the saturated and the observed resonance (54). At 4.7 T, the frequency separation of the saturated protons of PLL from the detected water signal is ap-proximately +700 Hz and a saturation pulse amplitude of 0.5 μT (corresponding to ~21 Hz) causes no spillover. In contrast, a 3-μT pulse (~128 Hz) yields a CEST decrease of ~30%, but the off-resonance saturation at –700 Hz is also related to a signal change of –25%. Performing such studies at 11.7 T is one loophole (albeit an expensive one) exploited to increase the chemical shift difference to approximately 1.8 kHz. Then, saturation pulses up to 250 Hz (i.e., ~6 μT) can be used, as the spillover-based signal decrease is only 3% (53).

4.4. Improvement through Paramagnetic Chemical Exchange Saturation Transfer

Incorporation of paramagnetic shift reagents close to the exchangeable protons im-proves proton CEST agents by increasing the frequency difference between water and the saturated pool. Introduced by Zhang et al. (55), the first agent was based on saturation of the bound water signal of a Eu^{3+} complex formed by a DOTA-tetra(amide) derivative. With a chemical shift separation of 49.7 ppm (9.8 kHz at 4.7 T) relative to free water, a saturation of 61% was achieved, whereas the reference experiment with saturation at –9.8 kHz showed no spillover effect. Some PARACEST agents are now used with saturation powers up to 250 μT (~10.6 kHz) at 7 T (56).

Reviews by Sherry et al. (57, 58) summarize the factors that allow for construct-ing such agents by slowing down the usually very fast exchange of water at the inner-sphere coordination sites of lanthanide(III) derivatives of DOTA. However, the short lifetimes of the bound water cause broad saturation dips in the z-spectra (see **Figure 6***b*) and decrease the selectivity of the PARACEST agents. A comparison of different complexes showed that only those with Eu^{3+}, Tb^{3+}, Dy^{3+}, and Ho^{3+} have the right combination of water-exchange time and induced chemical shift to be potential candidates for $B_0 < \sim 10$ T (57). Nevertheless, these agents hold great potential in terms of detection threshold, as agents with highly shifted bound water (~500 ppm) can be detected at concentrations of only 10 µM.

Xenon biosensor: a molecular construct using encapsulated xenon and a targeting unit to sense biochemical targets

Exchangeable protons of NH groups in addition to the coordinated water molecules in the same complex also turn PARACEST agents into sensitive pH probes (59, 60). Moreover, the use of different lanthanides makes PARACEST agents potential candidates for multiplexing (i.e., detecting different targets in the same setup). By using the different saturation resonances of bound water in an Eu^{3+} (+50 ppm induced shift) and a Tb^{3+} (–600 ppm induced shift) dotamGly complex, Aime and colleagues demonstrated selective in vitro cell imaging of two marked cell populations incubated with either of the PARACEST probes (40 mM concentration) (56).

4.5. Combining Chemical Exchange Saturation Transfer with Hyperpolarized Nuclei

Exchangeable hyperpolarized nuclei also allow for transferring NMR information from a molecule into a different environment in terms of in situ remote detection (61, 62). Hence, the concept of combining the CEST amplification scheme with the advantages of hyperpolarized nuclei such as ^{129}Xe can resolve some of the remaining limitations of PARACEST. As has been demonstrated (41), this combination significantly increases the capabilities of NMR for sensing biochemical targets, especially for molecular imaging (**Figure 7a,b**). The main improvements compared to thermally polarized nuclei are as follows:

1. Signals of xenon associated with other molecules show an intrinsically huge chemical shift range compared to free dissolved xenon in solution (63). Hence, no paramagnetic agents are required to reduce spillover effects to negligible, even if experiments would be be performed at field strengths of ~1 T. Unlike the protons of PARACEST agents, however, the exchange rates of atomic xenon are relatively slow and yield more selective saturation dips in z-spectra (**Figure 6b**) and can be used for multiplexing with different sensors at the same time (64).

2. At the same time, the slow exchange causes no problems in terms of transfer efficiency because T_1 relaxation for hyperpolarized xenon in solution is extremely slow compared to the saturation time. Therefore, the depolarization is stored much more effectively in the bulk pool than in the case of thermally polarized protons that underlie competing relaxation effects. Complete saturation can be easily achieved, and CEST with hyperpolarized xenon (HYPER-CEST) benefits from the full dynamic range of the bulk signal.

Implementation of HYPER-CEST was made possible by using molecular cages such as cryptophanes (65) as hosts to trap the noble gas for some milliseconds and make it sensitive to selective saturation. To obtain biochemical specificity, such cages can be incorporated in so-called xenon biosensors (66), in which the cages are linked to targeting units for reporting a specific binding event (**Figure 7c**). This modular setup allows, in principle, functionalization of the CEST-active site for any target that can bind a ligand or antibody. The prototype was a biotinylized cage constructed for sensing avidin. Increasing temperature is known to accelerate chemical exchange

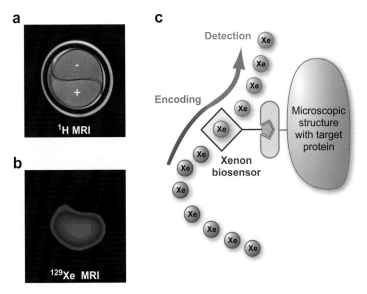

Figure 7

(*a*) Proton magnetic resonance image (MRI) showing a two-compartment phantom with avidin-labeled agarose beads. The lower compartment also contains a xenon biosensor that targets the protein. (*b*) ^{129}Xe MRI obtained after subtracting the hyperpolarized xenon chemical exchange saturation transfer (HYPER-CEST) data set with on-resonant sensor signal saturation from control data with off-resonant saturation. Only the area containing the sensor shows the CEST contrast; other areas remain dark in the difference image. (*c*) Functionalized biosensors act as host for exchangeable xenon that can be labeled using the CEST method for in situ remote detection of biochemical binding events. Data in (*b*) is from a biotinylized sensor that interacts with the protein avidin, which is attached to the surface of microscopic agarose beads.

significantly for $T > 25°C$ (67) and can thus be used to enhance the contrast. Changes in pH are, however, negligible, but competing guest molecules can alter the exchange dynamics of xenon into the cage (65).

The signal transfer is very efficient and can reveal target molecules even at fairly low concentrations. With molecular imaging, a high signal contrast of 50% was achieved with only ~1.3-µM concentration of NMR-active sensor and ~2.1 µM of dissolved, detectable xenon. Typical CEST experiments with signal changes of a few percent from thermally polarized protons require contrast agent concentrations of at least ~10 µM and are based on significantly higher concentrations of detectable nuclei (factor 2600 or 4400 for protons at 7 or 11 T, respectively, compared to xenon in Reference 41). The concentration of the detected target protein in the molecular imaging application was as low as 325 nM, and HYPER-CEST reduced acquisition time by 99.97% compared to conventional biosensor detection (41, 68).

For further optimization, the number of exchange sites can be significantly increased by dendrimeric amplification (69). Alternative detection techniques that are not based on the weak NMR signal of Faraday induction will also help to improve sensitivity for biochemical/biomedical applications. For example, low-field detection (discussed in the following section) is a powerful tool in this context, especially because of the advantages of the xenon chemical shift range over the small-proton chemical shift range. Thus, only moderate prepolarization fields are required to encode molecular information in the xenon signal, and such techniques can be combined with optical magnetometer detection.

5. DETECTING NUCLEAR MAGNETIC RESONANCE IN LOW FIELD

As discussed above, NMR and MRI are usually carried out in a high magnetic field (>1 T). A strong magnetic field leads both to high polarization of the sample and to good detection sensitivity for conventional inductive coils. Furthermore, chemical shifts of most nuclei are only resolvable in a sufficiently strong magnetic field, as chemical shifts scale with field strength (5).

There are several limitations for high-field NMR/MRI, however. First, samples possessing large magnetic-susceptibility gradients disturb the field homogeneity, broadening spectral lines and distorting images (70). Examples include materials with metal components, such as metallic implants. Second, when the sample is enclosed by conductors, the small skin depth of the rf excitation pulses prohibit NMR measurements (71). For example, at a proton Larmor frequency of 100 MHz (2.4 T), the penetration depth of copper is only 10 μm. A copper sheet thicker than 10 μm will effectively prevent excitation of the nuclear spins and detection of the NMR signal.

In addition to these fundamental physical problems, there are also engineering and cost issues, such as:

1. The superconducting coils that generate the strong magnetic fields for high-field NMR and MRI are immobile, making measurements achievable only in a laboratory or hospital environment.
2. The availability of the cryogenics associated with superconducting magnets may also be an issue in many circumstances.
3. The bore of a high-field magnet is often not feasible for large objects.
4. The cost of the magnets and cryogenics is considerably high, limiting the use of high-field NMR/MRI to major facilities.

To overcome these limitations and therefore expand the applicability of NMR/MRI, researchers have extensively explored performing NMR/MRI at low fields (72–74). Here, we define low fields as magnetic fields lower than a few mT. (The Earth's magnetic field is ~0.05 mT.) Intermediate-field NMR/MRI, also a booming research area, is not discussed in this review.

In general, low-field NMR/MRI can be divided into three stages: prepolarization, encoding, and detection. Because the nuclear polarization of the sample in a low field is insignificant, a prepolarization stage is usually required to provide sufficient initial polarization. Several methods are available for prepolarization, the most common of which involves strong (~1-T) permanent magnets. High homogeneity of the magnetic field is not required in this case, as it is not used for spectral or spatial encoding.

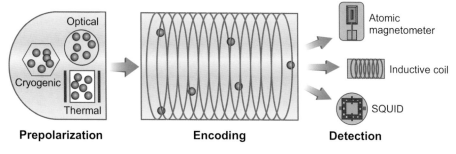

Prepolarization   **Encoding**   **Detection**

Figure 8

A schematic of low-field nuclear magnetic resonance/magnetic resonance imaging showing three stages: prepolarization, encoding, and detection. Prepolarization methods include thermal polarization by a strong magnetic field, optical hyperpolarization for xenon or helium, and polarization transfer from para-H_2 (generated by cryogenic cooling), among others. Detection can be conducted with an inductive coil, a superconducting quantum interference device (SQUID), or an atomic magnetometer.

Furthermore, cryogenics are not needed. Other prepolarization methods include optical pumping processes, such as those that produce hyperpolarized ^{129}Xe and ^{3}He gases (11), polarization transfer from para-H_2 (75), and dynamic nuclear polarization (76).

The encoding process, which is conducted in a weak and homogeneous field, is analogous to high-field encoding except for two differences. One is the Larmor frequency, which is in audio frequency range. Because of this, the penetration of the excitation pulse and the signal through conductive materials is significantly improved. The other difference is that only very weak gradients are required ($\sim$0.1 mT/m) for imaging, as they are much easier to generate than strong gradients.

Detection can be achieved via several possible techniques: conventional inductive detection, detection with superconducting quantum interference devices (SQUIDs), and laser detection using optical atomic magnetometers. A schematic of low-field NMR/MRI is illustrated in **Figure 8**. In the following sections, we discuss and compare these three detection techniques.

5.1. Faraday Inductive Detection

Using an inductive coil for low-field NMR/MRI detection is conceptually straightforward. By eliminating the superconducting magnets and cryogenics, the spectrometers become portable and less expensive. Early work on inductive detection of low-field NMR was attempted by Packard & Varian (77). However, accurate measurements had not been achieved until recent work by Appelt and colleagues (78), who reported NMR spectra detected by a tuned resonant circuit in the Earth's field in a remote area. The detection coil was composed of thousands of turns for improved efficiency. Because of the excellent homogeneity of the Earth's field, a linewidth of several millihertz was obtained, compared to 0.1 Hz in modern high-field NMR spectrometers. Heteronuclear J-coupling for a range of compounds was resolved with extraordinary

SQUID: superconducting quantum interference device

precision, i.e., ± 5 mHz, which is more than one order of magnitude better than analogous high-field measurements. The J-coupling values are thus capable of serving as key parameters to distinguish chemicals containing heteronuclei, as chemical shift information is not resolvable in such low fields.

Multidimensional NMR is a powerful tool for further clarification of chemical structures. Robinson et al. reported the first two-dimensional correlation spectroscopy spectra of 1,4-difluorobenzene and 2,2,2-trifluoroethanol in the Earth's field (79). Both proton and fluorine nuclei appeared in the spectra: Higher frequency resonances arose from the protons and lower ones from fluorine. The fact that both types of nuclei appeared in the same spectrum is a unique feature of low-field NMR; this occurs because their Larmor frequencies are only slightly different compared to the excitation pulse width and detection bandwidth. This technique thus allows measurements of the relative abundance of these different nuclei in one chemical by a single NMR spectrum. The diagonal peaks reveal molecular orbital connectedness and atomic proximity, providing decisive and unambiguous information for structural identification.

Conventional inductive detection is also used for low-field MRI. Mohorič et al. reported a nicely designed MRI system for the Earth's field (80). A reference signal from a separate spectrometer was used to monitor the fluctuation of the Earth's field, creating a stable rotating frame for the measured signal. Sensitivity was improved by a rational design of the receiving coil and an audio frequency shield which resulted in an image with the smallest voxel on the order of 50 mm^3. On the application front, Halse and colleagues showed MR images of a pepper in the Earth's field (81). Because the B_0 is only ~0.05 mT, the imaging gradient is on the order of ~0.1 mT/m, which is 2–3 orders of magnitudes weaker than conventional MRI. Thus, no high-power gradient amplifiers are required, meaning that low-field MRI apparatus consume less power and are more portable. The exceptional size and excellent homogeneity of the Earth's field enable imaging of samples of very large sizes. Practical applications include plant analysis and food inspection.

The major drawback of inductive detection in low field is its limited sensitivity, which is proportional to the strength of the magnetic field B_0. In the examples mentioned above, milliliters of samples are usually required, making inductive detection in low field not feasible for analyzing trace amounts of chemicals. Trace chemical analysis is important for applications such as monitoring chemical reactions or biochemical processes occurring on lab-on-a-chip devices. In order for low-field NMR and MRI techniques to be applicable for efficient chemical analysis, alternative detection techniques need to be developed. For this purpose, among others, NMR detection using SQUIDs and atomic magnetometers has been explored.

5.2. Superconducting Quantum Interference Devices

SQUIDs are based on the voltage oscillation of a superconducting loop interrupted by Josephson junctions (82). This oscillation is a function of the magnetic flux through the loop. Therefore, a SQUID acts as a flux-to-voltage transformer. Unlike a conventional inductive coil, SQUID can detect magnetic signal in an arbitrarily low magnetic field without losing sensitivity; in other words, the sensitivity is independent of the magnetic field strength. The current state-of-the-art SQUID has reached sensitivities below 1 fT/(Hz)$^{1/2}$ for ac signal (83), sufficient for NMR measurements.

Initially, SQUID was used for detecting NMR signals of samples at liquid helium temperature, which was of limited practical potential. In the last decade, researchers have significantly expanded the applications of SQUID in NMR/MRI measurements for samples at room temperature. For example, McDermott et al. showed a SQUID-detected NMR spectrum of trimethyl phosphate (84). A linewidth of ~1 Hz is obtained in a 1.8-μT B_0 without shimming. The signature J-coupling between phosphorous and protons in trimethyl phosphate is obtained.

To demonstrate the advantage of low-field MRI in imaging objects with large magnetic-susceptibility gradients and objects enclosed in metal, Mößle and colleagues reported SQUID-detected MR images of a grid with a titanium bar and images of a pepper enclosed in an aluminum can, and compared them with corresponding high-field results (85). The first set of images revealed that the susceptibility distortion to the images due to the titanium bar was not pronounced in low field, in contrast to the images taken in high field. The images of the canned pepper showed no shielding factor of the metal for audio frequency excitation and signal, whereas images of the pepper were not obtainable in high field because of the effective shielding by the aluminum can.

Recent technical advances enable SQUID to be implemented for in vivo MRI. Mößle and coworkers employed a second-order SQUID gradiometer to obtain images of human forearm in ~100 μT (86). The subject was surrounded by a 3-mm-thin aluminum sheet to eliminate environmental noise. A 100-mT magnetic field was used for prepolarization. The resulting images clearly showed different tissues, with three-dimensional spatial resolution of $2 \times 2 \times 8$ mm^3. Thus, SQUID MRI in low field offers a viable alternative for medical diagnosis.

One unique characteristic of MRI is its capacity to distinguish between different sections of the imaging object via contrast. In low field, contrast based on T_1 relaxation is greatly enhanced (87). SQUID detection of low-field MRI is well-suited to realize this advantage because of its high sensitivity. Lee et al. (88) were able to demonstrate the outstanding T_1 contrast of agarose solution (compared to solution without agarose) with a concentration as low as 0.25% at 0.01 mT, whereas no observable contrast was seen at 300 mT. Even though inductive detection is also applicable to measuring the enhanced low-field T_1 contrast via field cycling, direct detection with SQUID drastically simplifies the measuring procedure and reduces the requirement of field homogeneity.

The ultrahigh sensitivity and maturity with regard to engineering make SQUID the current state-of-the-art technique for low-field NMR/MRI detection. It is applicable to numerous fields, ranging from chemical analysis to medical imaging. Like any other technique, however, SQUID also has several drawbacks. One is that the cryogens needed for the superconducting loop impose constraints for the portability and feasibility of this technique. In addition, the electronic components, which improve sensitivity and detection efficiency, become very complicated for arrayed SQUIDs. These issues will likely be addressed in the next generation of SQUIDs.

5.3. Optical Atomic Magnetometer

Another high-sensitivity magnetometer is the optical atomic magnetometer, which has long been used for precise measurement of magnetic fields and fundamental physics parameters (89, 90). An atomic magnetometer utilizes the magneto-optical interactions between a laser beam and an alkali vapor. First, the incident laser beam, which has a near-resonance frequency to one of the electronic transitions of the alkali, generates a coherent ground state in the alkali vapor enclosed in a glass cell. The coherent alkali atoms interact with the probe beam (which can come from the same laser for pumping), resulting in an optical rotation of the polarization of the laser, or a dispersive absorption spectrum. The magneto-optical effects depend on the external magnetic field from the sample to be measured. Therefore, detection of the magnetic field is dependent upon the detection of photons, which can be measured with very high sensitivity.

For a given alkali element, the sensitivity of atomic magnetometers depends on the coherence time of the ground electronic state and the vapor density of the alkali (91). Two means of optimizing the sensitivity have been demonstrated. The Budker group used paraffin-coated alkali cells to minimize spin-destructive wall collisions and improve the coherence time of the ground state. The vapor cell was filled with pure vapor of an alkali metal, so the atomic density was in the high-vacuum regime. The ground-state coherence was preserved during thousands of collisions between the alkali atoms and the cell wall and lasted for seconds. The sensitivity of the magnetometer was thereby significantly improved. In this method, a single laser beam is often used for both pumping and probing, as the low-density alkali vapor only requires a low-power laser. A near-dc sensitivity of ~ 50 fT/(Hz)$^{1/2}$ has been achieved with a 1-cm^3 ^{87}Rb cell (92). For ac signal detection, the sensitivity is several fT/(Hz)$^{1/2}$ (93).

The other approach, developed by Romalis and colleagues, uses buffer gases to eliminate the coherence-destroying collisions (94). The alkali atoms reach a spin-exchange relaxation-free regime in near-zero magnetic fields. Here the atomic density is much higher than that of the other approach. Two laser beams are usually used, one with relatively high power for pumping the high-density alkali vapor, one with low power for detection. Recently, Savukov et al. demonstrated that subfemtotesla sensitivity at ~ 20 Hz can be achieved (95), a rate comparable to that of a SQUID magnetometer.

Application of atomic magnetometers in NMR was initially demonstrated by Yashchuk et al. (96), who measured the longitudinal magnetization of hyperpolarized xenon and obtained the T_1 relaxation time of 14 min. Shortly after, a proton NMR spectrum at 20 Hz was obtained by Savukov et al. in ~ 0.5 μT (97). These pioneering works demonstrated that atomic magnetometry would become an alternative technique for the detection of faint NMR signals.

Since those experiments, other groups have employed novel NMR and MRI techniques with atomic magnetometry. Coupled with remote detection (described above), atomic magnetometers have been used in obtaining both NMR spectra and images. In a study of flow, Xu et al. (98) first demonstrated MR images detected by an atomic magnetometer with 1.6-mm spatial resolution and 0.1-s temporal resolution (**Figure 9**).

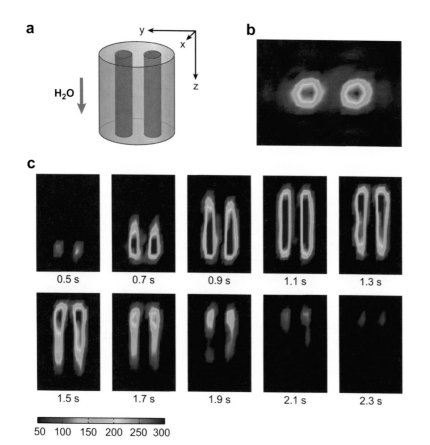

a

y ← → x

z

H$_2$O

b

c

| 0.5 s | 0.7 s | 0.9 s | 1.1 s | 1.3 s |

| 1.5 s | 1.7 s | 1.9 s | 2.1 s | 2.3 s |

50 100 150 200 250 300

Figure 9

Magnetic resonance images detected with an atomic magnetometer. (*a*) The encoding volume. The two channels are 3.2 mm in diameter and 25 mm long, with a center-to-center spacing of 5.1 mm. (*b*) Image of the cross section perpendicular to the flow (*xy* plane) at t = 1.1 s. (*c*) Time-resolved images in the *yz* plane. Measurements were obtained with a temporal interval of 0.1 s. Reproduced with permission from Reference 98. Copyright PNAS.

The spatial resolution is considerably better than that by inductive detection at low fields. It demonstrated the capability of atomic magnetometry as a viable and inexpensive alternative to high-field MRI. Extensive applications of this detection scheme can be expected, such as slow dynamics in microchannels or porous materials. Recent developments have shown that submillimeter resolution is also achievable in the Earth's field (S. Xu, et al., unpublished data).

For the in situ detection of NMR and MRI, atomic magnetometers need to be operated in rf mode to measure the Larmor precession of the nuclei. Romalis and colleagues pioneered rf atomic magnetometers and reported a sensitivity of ~1 fT/Hz in a frequency range up to 99 kHz (99). A static water sample was prepolarized in situ and NMR spectra were obtained directly (100). The Romalis group offered a detailed discussion on implementing atomic magnetometry for NMR detection. Based on their simplified model, they concluded that atomic magnetometers have the advantage in sensitivity over inductive coils when B_0 is less than ~1 T.

A major motivation for developing alternative NMR/MRI detection techniques involves searching for a miniaturizable method for monitoring biological processes and chemical reactions on lab-on-a-chip devices. Atomic magnetometry has been shown to possess such characteristics. Schwindt et al. used a magnetometer, with a total height of less than 3 mm, with an outstanding sensitivity of 5 pT (101). Because the distance of this detector to the sample is greatly reduced, the overall detection limit is improved. Applications in microanalysis are currently under way.

5.4. Perspectives for Low-Field Magnetic Resonance

Among the three detection devices described above, SQUIDs and atomic magnetometers offer much better sensitivity than inductive coils in low fields. These devices, therefore, have greater potential in low-field NMR and MRI. Developments in SQUID techniques have advanced much further than have those in atomic magnetometry. Not surprisingly, the applications of SQUID in this field far surpass those of atomic magnetometers. However, compared to SQUID modality, atomic magnetometry requires no cryogenics, a significant advantage in both cost and portability. In terms of electronics, atomic magnetometers require much less complicated instrumentation as the signal is measured via photon detection, which allows for easier multichannel detection and miniaturization.

In addition to the techniques mentioned herein, other novel tools have been investigated for NMR detection without superconducting magnets. Rugar et al. (102) developed magnetic resonance force microscopy that is capable of measuring a single electron spin. Direct magneto-optical interaction between polarized nuclei and a laser beam has also been studied (103). All such innovative work will lead to more possibilities for low-field NMR and MRI.

SUMMARY POINTS

1. Remote detection NMR and MRI allows significant signal amplification for low-porosity systems due to the optimal filling factor of the detector.

2. TOF information allows material heterogeneities to be elucidated, making TOF an ideal tool for studying porous systems and microfluidic devices.

3. Chemical exchange of saturated spin systems represents an in situ version of remote detection with significant signal amplification.

4. Hyperpolarized, exchangeable xenon can be functionalized to encode specific molecular information and is an ideal candidate for saturation transfer experiments with further sensitivity enhancement in molecular imaging.

5. SQUIDs and atomic magnetometers are viable for alternative detection of low-field NMR and MRI.

FUTURE ISSUES

1. Remote detection will be applied for in vivo applications utilizing the natural blood flow of the body's circulatory system.

2. High-throughput chemical analysis and imaging for use with lab-on-a-chip devices will be necessary.

3. Which of the many exogenous CEST agents will be suitable for in vivo applications?

4. Will optimized CEST agents be able to compete with radionuclide tracers?

5. HYPER-CEST sensors for cell tracking will be demonstrated.

6. Complex pulse sequences in SQUID-detected and laser-detected NMR and MRI will be implemented.

7. SQUIDs and atomic magnetometers will be miniaturized.

DISCLOSURE STATEMENT

The authors are not aware of any biases that might be perceived as affecting the objectivity of this review.

ACKNOWLEDGMENTS

We remain, as always, grateful to Alex Pines for his guidance, advice, and support which reflect decades of innovation and the impact of magnetic resonance in chemistry. Much of the data presented here is representative of the work of many of the so-called Pinenuts collected over the past few years. We are grateful for all their countless hours of hard work, which has led to these exciting developments in magnetic resonance.

This work is supported by the Director, Office of Science, Office of Basic Energy Sciences and the Materials Sciences Division of the U.S. Department of Energy under contract DE-AC03–76SF0098. E.H. is supported by a fellowship from the U.S. Department of Homeland Security under DOE contract number DE-AC05-00OR22750. L.S. is supported by the Deutsche Forschungsgemeinschaft (SCHR 995/1-1) through an Emmy Noether Fellowship.

LITERATURE CITED

1. Breitmaier E. 1993. *Structure Elucidation by NMR in Organic Chemistry*. New York: Wiley
2. Wüthrich K. 1986. *NMR of Proteins and Nucleic Acids*. New York: Wiley
3. Edelman R, Hesselink J, Zlatkin M. 1996. *Clinical Magnetic Resonance Imaging*. Philadelphia: Saunders
4. Callaghan PT. 1991. *Principles of Nuclear Magnetic Resonance Microscopy*. Oxford: Oxford Univ. Press
5. Abragham A. 1961. *Principles of Nuclear Magnetism*. Oxford: Oxford Univ. Press
6. Farrar TC, Becker ED. 1971. *Pulse and Fourier Transform NMR*. New York: Academic

7. Hoult DI, Richards RE. 1976. The signal-to-noise ratio of the nuclear magnetic resonance experiment. *J. Magn. Reson.* 24:71–85

8. Moule AJ, Spence MM, Han SI, Seeley JA, Pierce KL, et al. 2003. Amplification of xenon NMR and MRI by remote detection. *Proc. Natl. Acad. Sci. USA* 100:9122–27

9. Seeley JA, Han SI, Pines A. 2004. Remotely detected high-field MRI of porous samples. *J. Magn. Reson.* 167:282–90

10. Granwehr J, Harel E, Han SI, Garcia S, Pines A, et al. 2005. Time-of-flight flow imaging using NMR remote detection. *Phys. Rev. Lett.* 95:075503

11. Walker TG, Happer W. 1997. Spin-exchange optical pumping of noble gas nuclei. *Rev. Mod. Phys.* 69:629

12. Goodson BM. 2002. Nuclear magnetic resonance of laser-polarized noble gases in molecules, materials, and organisms. *J. Magn. Reson.* 155:157–216

13. Cohen MH, Mendelson KS. 1982. Nuclear magnetic resonance and the internal geometry of sedimentary rocks. *J. Appl. Phys.* 53:1127–35

14. Mansfield P, Turner R, Stehling MK. 1991. Echo-planar imaging: magnetic resonance imaging in a fraction of a second. *Science* 254:43–50

15. Gregory DM, Gerald RE, Botto RE. 1998. Pore-structure determinations of silica aerogels by ^{129}Xe NMR spectroscopy and imaging. *J. Magn. Reson.* 131:327–35

16. Terskikh VV, Moudrakovski IL, Mastikhin VM. 1993. ^{129}Xe nuclear magnetic resonance studies of the porous structure of silica gels. *J. Chem. Soc. Faraday Trans.* 89:4239–43

17. Terskikh VV, Moudrakovski IL, Breeze SR, Lang S, Ratcliffe CI, et al. 2002. A general correlation for the ^{129}Xe NMR chemical shift-pore size relationship in porous silica-based materials. *Langmuir* 18:5653–56

18. Harel E, Granwehr J, Seeley JA. 2006. Multiphase imaging of gas flow in a nanoporous material using remote-detection NMR. *Nat. Mater.* 5:321–27

19. Taylor G. 1953. Dispersion of soluble matter in solvent flowing slowly through a tube. *Proc. R. Soc. London Ser. A* 219:186–203

20. Reichenauer G, Stumpf C, Fricke J. 1995. Characterization of SiO$_2$, rf and carbon aerogels by dynamic cas expansion. *J. Non-Cryst. Solids* 186:334–41

21. Emmerling A, Fricke J. 1992. Small angle scattering and the structure of aerogels. *J. Non-Cryst. Solids* 145:113–20

22. Schaefer DW, Keefer KD. 1986. Structure of random porous materials: silica aerogel. *Phys. Rev. Lett.* 56:2199–202

23. Bonardet JL, Fraissard J, Gédéon A, Springuel-Huet MA. 1999. Nuclear magnetic resonance of physisorbed Xe-129 used as a probe to investigate porous solids. *Catal. Rev.-Sci. Eng.* 41:115–225

24. Hilty C, McDonnell EE, Granwehr J, Pierce KL, Han SI, Pines A. 2005. Microfluidic gas-flow profiling using remote-detection NMR. *Proc. Natl. Acad. Sci. USA* 102:14960–63

25. Stone HA, Stroock AD, Ajdari A. 2004. Engineering flows in small devices: microfluidics toward a lab-on-a-chip. *Annu. Rev. Fluid Mech.* 36:381–411

26. Koop MU, de Mello AJ, Manz A. 1998. Chemical amplifications: continuous-flow PCR on a chip. *Science* 280:1046–48

27. Figeys D, Pinto D. 2001. Proteomics on a chip: promising developments. *Electrophoresis* 22:208–16

28. Haswell SJ, Middleton RJ, O'Sullivan B, Skelton V, Watts P, Styring P. 2001. The application of micro reactors to synthetic chemistry. *Chem. Commun.* 2001:391–98

29. Hong JW, Studer V, Hang G, Anderson WF, Quake SR. 2004. A nanoliter-scale nucleic acid processor with parallel architecture. *Nat. Biotechnol.* 22:435–39

30. Stroock AD, Dertinger SK, Ajdari A, Mezic I, Stone HA, Whitesides GM. 2002. Chaotic mixer for microchannels. *Science* 295:647–51

31. Cristobal G, Arbouet L, Sarrazin F, Talaga D, Bruneel J-L, et al. 2006. On-line laser Raman spectroscopic probing of droplets engineered in microfluidic devices. *Lab Chip* 6:1140–46

32. Zhu L, Lee CS, DeVoe DL. 2006. Integrated microfluidic UV absorbance detector with attomol-level sensitivity for BSA. *Lab Chip* 6:115–20

33. Wensink H, Benito-Lopez F, Hermes DC, Verboom W, Gardeniers HJGE, et al. 2005. Measuring reaction kinetics in a lab-on-a chip by microcoil NMR. *Lab Chip* 5:280–84

34. Harel E, Hilty C, Koen K, McDonnell EE, Pines A. 2007. Time-of-flight flow imaging of two-component flow inside a microfluidic chip. *Phys. Rev. Lett.* 98:017601

35. Ehrmann K, Gersbach M, Pascoal P, Vincent F, Massin C, et al. 2006. Sample patterning on NMR surface microcoils. *J. Magn. Reson.* 178:96–105

36. Sakellariou D, Le Goff G, Jacquinot J-F. 2007. High-resolution, high-sensitivity NMR of nanolitre anisotropic samples by coil spinning. *Nature* 447:694–97

37. Olson DL, Peck TL, Webb AG, Magin RL, Sweedler JV. 1995. High-resolution microcoil ^{1}H-NMR for mass-limited, nanoliter-volume samples. *Science* 270:1967–70

38. Bouchard LS, Budker D, Harel E, Ledbetter M, Lowery T, et al. 2007. U.S. patent pending

39. Massoud TF, Gambhir SS. 2003. Molecular imaging in living subjects: seeing fundamental biological processes in a new light. *Genes Dev.* 17:545–80

40. Ward KM, Aletras AH, Balaban RS. 2000. A new class of contrast agents for MRI based on proton chemical exchange dependent saturation transfer (CEST). *J. Magn. Res.* 143:79–87

41. Schröder L, Lowery TJ, Hilty C, Wemmer DE, Pines A. 2006. Molecular imaging using a targeted magnetic resonance hyperpolarized biosensor. *Science* 314:432–33

42. Forsén S, Hoffman R. 1963. Study of moderately rapid chemical exchange reactions by means of nuclear magnetic double resonance. *J. Chem. Phys.* 39(11):2892–901

43. McConnell HM. 1958. Reaction rates by nuclear magnetic resonance. *J. Chem. Phys.* 28:430–31

44. Bryant RG. 1996. The dynamics of water-protein interactions. *Annu. Rev. Biophys. Biomol. Struct.* 25:29–53

45. Zhou J, Wilson DA, Zhe Sun P, Klaus JA, van Zijl PCM. 2004. Quantitative description of proton exchange processes between water and endogenous and exogenous agents for WEX, CEST, and APT experiments. *Magn. Reson. Med.* 51:945–52

46. Zhou J, Payen J-F, Wilson DA, Traystman RJ, van Zijl PCM. 2003. Using the amide proton signals of intracellular proteins and peptides to detect pH effects in MRI. *Nat. Med.* 9:1085–90

47. van Zijl PCM, Zhou J, Mori N, Payen J-F, Wilson D, Mori S. 2003. Mechanism of magnetization transfer during on-resonance water saturation: a new approach to detect mobile proteins, peptides, and lipids. *Magn. Reson. Med.* 49:440–49

48. Zhou J, van Zijl PCM. 2006. Chemical exchange saturation transfer imaging and spectroscopy. *Progr. NMR Spectrosc.* 48:109–36

49. Ward KM, Balaban RS. 2000. Determination of pH using water protons and chemical exchange dependent saturation transfer (CEST). *Magn. Reson. Med.* 44:799–802

50. Goffeney N, Bulte JW, Duyn J, Bryant LH Jr, van Zijl PC. 2001. Sensitive NMR detection of cationic-polymer-based gene delivery systems using saturation transfer via proton exchange. *J. Am. Chem. Soc.* 123:8628–29

51. Snoussi K, Bulte JW, Guéron M, van Zijl PCM. 2003. Sensitive CEST agents based on nucleic acid imino proton exchange: detection of poly(rU) and of a dendrimer-poly(rU) model for nucleic acid delivery and pharmacology. *Magn. Reson. Med.* 49:998–1005

52. Zhou J, Lal B, Wilson DA, Laterra J, van Zijl PCM. 2003. Amide proton transfer (APT) contrast for imaging of brain tumors. *Magn. Reson. Med.* 50:1120–26

53. McMahon MT, Gilad AA, Zhou J, Sun PZ, Bulte JW, van Zijl PC. 2006. Quantifying exchange rates in chemical exchange saturation transfer agents using the saturation time and saturation power dependencies of the magnetization transfer effect on the magnetic resonance imaging signal (QUEST and QUESP): Ph calibration for poly-L-lysine and a starburst dendrimer. *Magn. Reson. Med.* 55:836–47

54. Sun PZ, van Zijl PCM, Zhou J. 2005. Optimization of the irradiation power in chemical exchange dependent saturation transfer experiments. *J. Magn. Reson.* 175:193–200

55. Zhang S, Winter P, Wu K, Sherry DK. 2001. A novel europium(III)-based MRI contrast agent. *J. Am. Chem. Soc.* 123:1517–18

56. Aime S, Carrera C, Delli Castelli D, Geniatti Crich S, Terreno E. 2005. Tunable imaging of cells labeled with MRI-PARACEST agents. *Angew. Chem. Int. Ed.* 44:1813–15

57. Zhang S, Merritt M, Woessner DE, Lenkinski RE, Sherry DA. 2003. PARACEST agents: modulating MRI contrast via water proton exchange. *Acc. Chem. Res.* 36:783–90

58. Woods M, Woessner DE, Sherry DA. 2006. Paramagnetic lanthanide complexes as PARACEST agents for medical imaging. *Chem. Soc. Rev.* 35:500–11

59. Aime S, Delli Castelli D, Terreno E. 2002. Novel pH-reporter MRI contrast agents. *Angew. Chem. Int. Ed.* 41(22):4334–36

60. Terreno E, Delli Castelli D, Cravotto G, Milone L, Aime S. 2004. Ln(III)-DOTAMGly complexes: a versatile series to assess the determinants of the efficacy of paramagnetic chemical exchange saturation transfer agents for magnetic resonance imaging applications. *Invest. Radiol.* 39:235–43

61. Spence MM, Ruiz EJ, Rubin SM, Lowery TJ, Winssinger N, et al. 2004. Development of a functionalized xenon biosensor. *J. Am. Chem. Soc.* 126:15287–94

62. Garcia S, Chavez L, Lowery TJ, Han S-I, Wemmer DE, Pines A. 2007. Sensitivity enhancement by exchange mediated magnetization transfer of the xenon biosensor signal. *J. Magn. Reson.* 184(1):72–77

63. Goodson BM. 1999. Using injectable carriers of laser-polarized noble gases for enhancing NMR and MRI. *Concepts Magn. Reson.* 11(4):203–23

64. Lowery TJ, Rubin SM, Ruiz EJ, Spence MM, Winssinger N, et al. 2003. Applications of laser-polarized ^{129}Xe to biomolecular assays. *Magn. Reson. Imag.* 21(10):1235–39

65. Bartik K, Luhmer L, Dutasta J-P, Collet A, Reisse J. 1998. ^{129}Xe and ^{1}H NMR study of the reversible trapping of xenon by cryptophane-A in organic solution. *J. Am. Chem. Soc.* 120(4):784–91

66. Spence MM, Rubin SM, Dimitrov IE, Ruiz EJ, Wemmer DE, et al. 2001. Functionalized xenon as a biosensor. *Proc. Natl. Acad. Sci. USA* 98:10654–57

67. Lowery TJ, Garcia S, Chavez L, Ruiz EJ, Wu T, et al. 2006. Optimization of xenon biosensors for detection of protein interactions. *ChemBioChem* 7:65–73

68. Hilty C, Lowery TJ, Wemmer DE, Pines A. 2006. Spectrally resolved magnetic resonance imaging of a xenon biosensor. *Angew. Chem. Int. Ed.* 45(1):70–73

69. Mynar JL, Lowery TJ, Wemmer DE, Pines A, Frechet JM. 2006. Xenon biosensor amplification via dendrimer-cage supramolecular constructs. *J. Am. Chem. Soc.* 128(19):6334–35

70. Ludecke KM, Roschmann P, Tischler R. 1993. Susceptibility artefacts in NMR imaging. *Magn. Reson. Imaging* 3:329–43

71. Bennett CR, Wang PS, Donahue MJ. 1996. Artifacts in magnetic resonance imaging from metals. *J. Appl. Physiol.* 79:4712–14

72. Béné GJ. 1980. Nuclear magnetism of liquid system in the Earth field range. *Phys. Rep.* 58:213–67

73. Stepišnik J, Kos M, Planinšič G, Eržen V. 1994. Strong non-uniform magnetic field for self-diffusion measurement by NMR in the earth's magnetic field. *J. Magn. Reson. A* 107:167–72

74. Appelt S, Häsing FW, Kühn H, Perlo J, Blümich B. 2005. Mobile high resolution xenon nuclear magnetic resonance spectroscopy in the earth's magnetic field. *Phys. Rev. Lett.* 94:197602

75. Bowers CR, Weitekamp DP. 1986. Transformation of symmetrization order to nuclear-spin magnetization by chemical reaction and nuclear magnetic resonance. *Phys. Rev. Lett.* 57:2645–48

76. Slichter CP. 1990. *Principles of Nuclear Magnetic Resonance.* New York: Springer-Verlag

77. Packard M, Varian R. 1954. Free nuclear induction in the Earth's magnetic field. *Phys. Rev.* 93:941

78. Appelt S, Kühn H, Häsing FW, Blümich B. 2006. Chemical analysis by ultrahigh-resolution nuclear magnetic resonance in the Earth's magnetic field. *Nat. Phys.* 2:105–9

79. Robinson JN, Coy A, Dykstra R, Eccles CD, Hunter MW, et al. 2006. Two-dimensional NMR spectroscopy in Earth's magnetic field. *Magn. Reson.* 182:343–47

80. Mohorič A, Planinšič G, Kos M, Duh A, Stepišnik J. 2004. Magnetic resonance imaging system based on Earth's magnetic field. *Instrum. Sci. Technol.* 32:655–67

81. Halse ME, Coy A, Dykstra R, Eccles C, Hunter M, et al. 2006. A practical and flexible implementation of 3D MRI in the Earth's magnetic field. *J. Magn. Reson.* 182:75–83

82. Clarke J. 1996. SQUID fundamentals. In *SQUID Sensors: Fundamentals, Fabrication, and Applications*, ed. H Weinstock, pp. 1–62. Dordrecht, Netherlands: Kluwer Acad.

83. Kleiner R, Koelle F, Ludwig F, Clarke J. 2004. Superconducting quantum interference devices: state of the art and applications. *Proc. IEEE* 92:1534–48

84. McDermott R, Trabesinger AH, Mück M, Hahn EL, Pines A, et al. 2002. Liquid-state NMR and scalar couplings in microtesla magnetic fields. *Science* 295:2247–49

85. Mößle M, Han S-I, Myers WR, Lee S-K, Kelso N, et al. 2006. SQUID-detected microtesla MRI in the presence of metal. *J. Magn. Reson.* 179:146–51

86. Mößle M, Myers WR, Lee S-K, Kelso N, Hatridge M, et al. 2005. SQUID-detected in vivo MRI at microtesla magnetic fields. *IEEE Trans. Appl. Supercond.* 15:757–60

87. Macovski A, Conolly S. 1993. Novel approaches to low-cost MRI. *Magn. Reson. Med.* 30:221–30

88. Lee SK, Mößle M, Myers WR, Kelso N, Trabesinger AH, et al. 2005. SQUID-detected MRI at 132 microT with T1-weighted contrast established at 10 microT–300 mT. *Magn. Reson. Med.* 53:9–14

89. Dehmelt HG. 1957. Modulation of a light beam by precessing absorbing atoms. *Phys. Rev.* 105:1924–25

90. Dupont-Roc J, Haroche S, Cohen-Tannoudji C. 1969. Detection of very weak magnetic fields (10–9 gauss) by Rb 87 zero-field level crossing resonances. *Phys. Lett. A* 28:638–39

91. Budker D, Romalis MV. 2007. Optical magnetometry. *Nat. Phys.* 3:227–34

92. Xu S, Rochester SM, Yashchuk VV, Donaldson MH, Budker D. 2006. Construction and applications of an atomic magnetic gradiometer based on nonlinear magneto-optical rotation. *Rev. Sci. Instrum.* 77:083106

93. Ledbetter MP, Acosta VM, Rochester SM, Budker D, Pustelny S, et al. 2007. Detection of radio-frequency magnetic fields using nonlinear magneto-optical rotation. *Phys. Rev. A* 75:023405

94. Allred JC, Lyman RN, Komack TW, Romalis MV. 2002. High-sensitivity atomic magnetometer unaffected by spin-exchange relaxation. *Phys. Rev. Lett.* 89:130801

95. Kominis IK, Kornack TW, Allred JC, Romalis MV. 2003. A subfemtotesla multichannel atomic magnetometer. *Nature* 422:596–99

96. Yashchuk VV, Granwehr J, Kimball DF, Rochester SM, Trabesinger AH, et al. 2004. Hyperpolarized xenon nuclear spins detected by optical atomic magnetometry. *Phys. Rev. Lett.* 93:160801

97. Savukov IM, Romalis MV. 2005. NMR detection with an atomic magnetometer. *Phys. Rev. Lett.* 94:123001

98. Xu S, Yashchuk VV, Donaldson MH, Rochester SM, Budker D, Pines A. 2006. Magnetic resonance imaging with an optical atomic magnetometer. *Proc. Natl. Acad. Sci. USA* 103:12668–71

99. Savukov IM, Seltzer SJ, Romalis MV, Sauer KL. 2005. Tunable atomic magnetometer for detection of radio-frequency magnetic fields. *Phys. Rev. Lett.* 95:063004

100. Savukov IM, Selter SJ, Romalis MV. 2007. Detection of NMR signals with a radio-frequency atomic magnetometer. *J. Magn. Reson.* 185:227–33

101. Schwindt PDD, Lindseth B, Knappe S, Shah V, Kitching J, et al. 2007. A chip-scale atomic magnetometer with improved sensitivity using the Mx technique. *Appl. Phys. Lett.* 90:081102

102. Rugar D, Budakian R, Mamin HJ, Chui BW. 2004. Single spin detection by magnetic resonance force microscopy. *Nature* 430:329–32

103. Savukov IM, Lee SK, Romalis MV. 2006. Optical detection of liquid-state NMR. *Nature* 442:1021–24

Chemical Cytometry: Fluorescence-Based Single-Cell Analysis

Daniella Cohen,[1] Jane A. Dickerson,[1]
Colin D. Whitmore,[1] Emily H. Turner,[1]
Monica M. Palcic,[2] Ole Hindsgaul,[2]
and Norman J. Dovichi[1]

[1]Department of Chemistry, University of Washington, Seattle, Washington 98195;
email: dovichi@chem.washington.edu

[2]Carlsberg Laboratory, Valby Copenhagen, DK-2500, Denmark

Annu. Rev. Anal. Chem. 2008. 1:165–90

First published online as a Review in Advance on
January 15, 2008

The *Annual Review of Analytical Chemistry* is online
at anchem.annualreviews.org

This article's doi:
10.1146/annurev.anchem.1.031207.113104

Key Words

fluorescence, capillary electrophoresis, single-cell analysis,
cytometry

Abstract

Cytometry deals with the analysis of the composition of single cells.
Flow and image cytometry employ antibody-based stains to characterize a handful of components in single cells. Chemical cytometry,
in contrast, employs a suite of powerful analytical tools to characterize a large number of components. Tools have been developed
to characterize nucleic acids, proteins, and metabolites in single
cells. Whereas nucleic acid analysis employs powerful polymerase
chain reaction–based amplification techniques, protein and metabolite analysis tends to employ capillary electrophoresis separation and
ultrasensitive laser-induced fluorescence detection. It is now possible
to detect yoctomole amounts of many analytes in single cells.

1. INTRODUCTION

1.1. Cellular Heterogeneity

The cell is the fundamental unit of life, and individual cells often differ significantly from their neighbors. As an example, **Figure 1** presents a photomicrograph of a population of the bacterium *Deinococcus radiodurans*; the fusion protein shown plays a key role in the organism's response to genetic damage. Only a small fraction of the population shows green fluorescence, and these cells have superior survival when challenged with a genotoxin. Classic analytical methods would report the average composition of cells, which would conceal the cell-to-cell differences that determine each cell's survival under stress. Other methods are required to characterize the composition of individual cells in order to understand the behavior of a cellular population.

Although the study of microbes is important, more exciting opportunities may be found in the characterization of cells from complex organisms, which can provide insights into the function of the cell and its responses to changes in its environment. We provide three examples that motivate this work. First, a zygote undergoes astonishing changes as it develops from a single cell into a fully functional organism; cataloging the changes in cellular composition at each stage of an embryo's development provides powerful insights into ontogeny. Second, the brain is the most complex organ in a vertebrate, and characterization of differences in neuronal composition not only yields an opportunity to systematically classify cells but also provides some guidance as to the function of each cell. Third, many tumors are highly heterogeneous, and pathologists grade tumors based on the subjective interpretation of the heterogeneous appearance of the tissue. We hypothesize that heterogeneity in cellular composition increases as the disease progresses, which may serve as a prognostic indicator to guide individualized treatment. In essence, we are supplementing the subjective prognosis provided by the pathologist with an objective catalog of the components that define that heterogeneity.

Analysis of the composition of a cell is important, but cells are usually small, and their analysis presents technological challenges. **Table 1** summarizes the approximate composition of a single bacterium, yeast cell, mammalian cell, giant neuron, and

Figure 1

Combined phase contrast and fluorescence image of *Deinococcus radiodurans* that expresses the fusion between green fluorescent protein and RecA protein. Only a subset of cells expresses high levels of the repair protein, and these cells are primed to survive exposure to DNA-damaging agents.

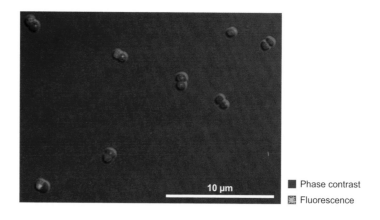

10 μm

■ Phase contrast
▨ Fluorescence

Table 1 Approximate composition of a single cell

Component	Bacterium or mitochondrion	Yeast cell	Human somatic cell	Giant neuron	*Drosophila* egg
Diameter	1 μm	5 μm	10 μm	100 μm	500 μm
Mass	500 fg	50 pg	0.5 ng	0.5 μg	50 μg
Protein (~15%)	100 fg	10 pg	100 pg	100 ng	10 μg
Nucleic acids (~7%)	40 fg	4 pg	40 pg	40 ng	4 μg
Carbohydrates (~3%)	15 fg	1.5 pg	15 pg	15 ng	1.5 μg
Lipids (~2%)	10 fg	1 pg	10 pg	10 ng	1 μg
Small ions and molecules (~3%)	15 fg	1.5 pg	15 pg	15 ng	1.5 μg

Drosophila egg. Even in large cells, such as giant neurons, only a few nanograms of material are available for analysis. Microbes and organelles contain mere femtograms of material, presenting a formidable analytical challenge to the investigator.

1.2. Classic Cytometry

Cytometry employs instrumental methods to determine the amount of selected components in single cells; Shapiro's classic text provides outstanding background on the technology (1). Early work employed microspectrophotometers to obtain rough estimates of the nucleic acid and protein content of single cells based on absorbance at 260 nm and 280 nm, respectively (2). These initial studies laboriously measured absorbance from cells one at a time.

Two important technological advances arose from this early work. The first was the incorporation of a flow system to pass single cells sequentially through the microspectrophotometer and to record signals electronically; this system is the basis of flow cytometry. This technology benefited greatly from electronics developed by the atomic physics community for pulse height analysis of events generated by radioactive decay; these electronics were easily modified for use in recording the signals generated by single cells. The second advance was to incorporate a scanning stage to the microscope so that an entire field of view could be automatically passed through the microspectrophotometer, thereby allowing analysis of large numbers of cells; this technology is the basis of image cytometry. The incorporation of a confocal aperture improved resolution significantly, and the incorporation of digital cameras allowed rapid capture across the field of view. Image cytometry developed more slowly than flow cytometry because appropriate optical and data manipulation tools were not widely available until late in the twentieth century (1).

Flow and image cytometry are used to measure both the physical properties and the chemical composition of single cells, and are very important in both clinical and fundamental research. Physical measurements are usually related to cell size and include forward- and side-light scatter and the electrical resistance associated with a cell obscuring a narrow aperture in a flowing system, which is the basis of the venerable Coulter counter.

The chemical composition of a cell is characterized by use of affinity reagents in both image and flow cytometry. For example, intercalating dyes, such as Hoechst 33342 and propidium iodine, are used to quantitate the DNA content of a cell. These dyes are weakly fluorescent in an aqueous environment, but are highly fluorescent when intercalated between base pairs in double-stranded DNA. Recently divided cells are in the G1 phase of the cell cycle and have one pair of each chromosome. Cells that are about to divide are in the G2 or M phase of the cell cycle and have four copies of each chromosome. The fluorescence from intercalating dyes is twice as large for tetraploid cells compared to diploid cells. Cancer cells often have abnormal chromosomal composition due to the presence of additional chromosomal copies. These aneuploid cells generate a fluorescence signal that is larger than that generated by normal cells progressing through the cell cycle; aneuploid fluorescence often does not correspond to an integral number of chromosomes due to deletions and duplications of a portion of the genome. An increase in the heterogeneity of the chromosomal content and an increase in the fraction of aneuploid cells are often taken as negative prognostic indicators for cancers.

Fluorescently labeled antibodies are commonly used to determine the distribution of the expression of a particular protein in a cellular population. For instance, CD4-expressing lymphocytes are the target for HIV, and the determination of CD4+ lymphocytes is a common tool for determining the efficacy of treatment of AIDS patients. Cells are treated with the labeled antibody, washed, and then analyzed by flow cytometry.

Multiparameter methods can be employed to simultaneously measure two or more signals from a single cell. For example, blood can be treated with both an intercalating dye and a fluorescently labeled anti-CD4 antibody. Only nucleated cells that express the CD4 antigen will generate fluorescence from both reagents, thereby allowing discrimination against non-nucleated cells that happen to express the CD4 antigen (3).

Conventional cytometry methods are common tools in biomedical research and clinical practice. They are able to measure five or more properties from a single cell, and they can process thousands of cells per second (4). Cytometers can be modified to select cells with a specific fluorescence signature and deliver them to wells on a microtiter plate; these cell sorters are powerful, nondestructive tools that are used to isolate rare cells for further study. However, conventional cytometry has two important limitations. First, only a limited number of components can be analyzed from a single cell; it is difficult to resolve the fluorescence signal from more than five components. Second, conventional cytometry employs affinity probes against known targets; the unexpected is therefore invisible.

2. CHEMICAL CYTOMETRY

2.1. Overview

Chemical cytometry employs powerful analytical methods to characterize the composition of a lysed cell. The analysis is destructive and slow, but provides the opportunity to characterize hundreds of components from a single cell. Examples of chemical cytometry considered in this review employ a capillary-based separation technique, such as microbore liquid chromatography or capillary electrophoresis, to separate components from a single lysed cell. Ultrasensitive detection methods, typically laser-induced fluorescence, are used to measure the amount of the separated components.

2.2. History of Chemical Cytometry before Ultrasensitive Detection

Microscale electrophoresis methods were developed in the 1950s and 1960s to separate components from small amounts of sample, including single cells. The earliest study considered rRNA analysis of single cells based on electrophoresis on a silk fiber and detection by UV absorbance (5). A second study employed an acrylamide fiber to study hemoglobin from single erythrocytes (6).

A series of papers published in the 1960s and early 1970s reported using more reproducible separation methods for the study of single cells. For example, Hyden and colleagues performed sodium dodecyl sulfate polyacrylamide gel electrophoresis (SDS-PAGE) in 200-µm-diameter capillaries to separate nanogram amounts of proteins from single cells (7), Marchalonis & Nossal separated antibodies from single cells (8), and Wilson and Ruchel separated proteins from single giant neurons of *Aplaysia* (9–10). Repin and colleagues monitored lactate dehydrogenase from single oocytes (11). As a particularly interesting example, Neukirchen and colleagues utilized a miniaturized two-dimensional gel electrophoresis separation method and silver staining to characterize proteins separated from a single *Drosophila* egg (12). Several hundred proteins were resolved from the micrograms of protein contained in the egg.

2.3. History of Chemical Cytometry with Ultrasensitive Detection

Jorgenson inaugurated the modern era of single-cell analysis by combining capillary chromatography and electrophoresis with ultrasensitive detection methods for analysis of single giant neurons (13). A single giant neuron was isolated, homogenized, and centrifuged. The supernatant was injected directly onto the chromatography column and detected with either amperometry or fluorescence of naphthalene-2,3-dicarboxyaldehyde (NDA)–labeled compounds.

Ewing also performed experiments on single giant neurons, using a 10-µm-inner-diameter capillary to sample the cytoplasm of a single neuron. The sample was separated electrophoretically in the capillary and detected by a carbon fiber microelectrode (14).

In Jorgenson's early experiments, an entire cell was homogenized in a relatively large volume of buffer (13). Only a portion of this homogenate was then used for subsequent analysis. Although it is acceptable for the relatively large amount of protein present in a giant neuron, dilution must occur for this approach to be useful for analysis of typical somatic cells. Rather, it is more useful to inject a cell directly into the capillary, where it can be lysed before subsequent analysis.

In an early example, Hogan & Yeung employed monobromobimane to derivatize glutathione in individual erythrocytes; the cell was lysed and its labeled contents were separated by capillary electrophoresis and detected by laser-induced fluorescence (15). Ewing's group reported an on-column lysis and labeling scheme for chemical cytometry. In this experiment, the cell was injected into the capillary, lysed, and its contents labeled with a fluorogenic reagent. By performing the labeling chemistry within the capillary, excessive dilution was avoided and the entire reaction product was available for subsequent electrophoretic separation (16).

Rather than relying on chemical derivatization, Yeung used native fluorescence to monitor the separation of hemoglobin variants from a single erythrocyte, catecholamines from single adrenal chromaffin cells, serotonin in neurons, and proteins from neurons (17–21). Sweedler has employed native fluorescence to study neurotransmitters in single neurons (22–23). Native fluorescence eliminates the need for derivatization, but usually requires excitation in the UV portion of the spectrum using relatively expensive and temperamental lasers. Krylov and colleagues have also used native fluorescence to monitor expression of green fluorescent protein (GFP) from single eukaryotes (24), and our group has studied GFP expression in single prokaryote cells (25). GFP is excited in the blue portion of the spectrum and can be detected with very high sensitivity. GFP-fusion proteins can be prepared by genetic engineering, which represents a highly precise means of fluorescently labeling a specific protein.

Most work has used surfactant or other reagents to lyse cells. Allbritton and colleagues used a pulsed laser to lyse a cell before analysis of kinase activities by capillary electrophoresis. The pulsed laser allows very fast lysis of the cell, eliminating potential artifacts associated with cellular response to the stress associated with a slower lysis process (26–30). Alternatively, an electric field can be used to induce rapid cell lysis, which has the potential for high-throughput analysis (28). Use of hydrodynamic, rather than electrokinetic, injection of reagents leads to improved mixing and efficient lysis (31). For an overview of methods used in single-cell analysis, see the article by the Allbritton group in this volume (32).

Zare and colleagues reported a microfabricated device to automate cell lysis, labeling, and separation. This system has the potential to automate cell isolation with the subsequent processing necessary to characterize cell composition (33).

3. CHEMICAL CYTOMETRY OF PROTEINS BY ONE- AND TWO-DIMENSIONAL CAPILLARY ELECTROPHORESIS

3.1. Instrumentation

Our group has performed chemical cytometry based on capillary electrophoresis with ultrasensitive laser-induced fluorescence detection (**Figure 2a**). In our experiments, a cell is aspirated into the capillary, lysed, and labeled if necessary (34–35). Components are separated electrophoretically and detected by fluorescence.

Cells are chosen and injected into a capillary with the aid of an inverted microscope (**Figure 2b**) (35). The operator views a drop of cell suspension through the microscope. The capillary is attached to a three-axis micromanipulator, which allows the capillary to be centered over a cell of interest. To inject the cell, a 1-s-long pulse of vacuum is applied to the sheath-flow cuvette (35); the vacuum pulls the cell into

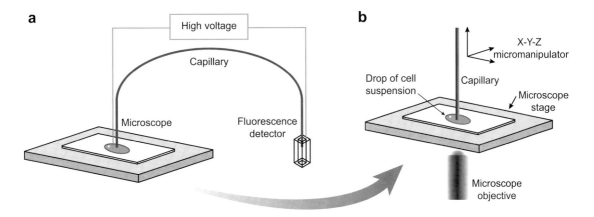

a

High voltage

Capillary

Microscope

Fluorescence detector

b

X-Y-Z micromanipulator

Drop of cell suspension

Capillary

Microscope stage

Microscope objective

Figure 2

(*a*) Schematic of instrumentation for chemical cytometry. (*b*) Instrument for injection of a cell into the separation capillary.

the capillary for lysis, labeling if necessary, and electrophoretic analysis. The material used to coat the microscope slide influences the injection of the cell. If the cell adheres too strongly to the surface, it will not be possible to inject the cell, and if the cell does not adhere to the surface, the cell will float free in suspension, making capture of the cell difficult (36).

This group has focused on the development of extremely high sensitivity fluorescence detection systems for chemical cytometry, and we routinely obtain detection limits of a few hundred copies of a wide range of analytes. A schematic of our detector is shown in **Figure 3** (37–39). The detector design is based on a flow cytometer, where the analyte migrates from the separation capillary into a flowing stream in a sheath-flow cuvette. Fluorescence is excited with a low-power laser beam ~100 μm downstream from the capillary in the postcolumn detector (we used a 10-mW diode laser pumped solid-state laser operating at 473 nm for most experiments). Fluorescence is collected with a 60 × 0.7 NA microscope objective, spectrally filtered to reduce scattered laser light, and imaged onto a gradient-index lens, which acts as a limiting aperture and which couples fluorescence to a fiber optic. We employed a high-quantum-yield single-photon-counting avalanche photodiode to detect fluorescence (40).

3.2. Fluorescent Labeling of Proteins and Biogenic Amines

We have used chemical cytometry based on capillary electrophoresis to study proteins and biogenic amines in a wide range of cell types, including the HT29 colon cancer cell line; the MCF7 breast cancer cell line; the AtT20 adrenal gland cancer cell line; the MC3T3 osteoprecursor cell line; the hTERT and the CP18821 telomerase–expressing Barrett's esophagus cell lines; the A549 lung cancer cell line; the SupT1 T-cell line; MC3T3 osteoprecursor cells; RAW 264.7 macrophage cells; *D. radiodurans* cells; primary neurons, neural stem-cells, macrophages, and T-cells; and single-cell

Figure 3

Sheath-flow cuvette as a postcolumn laser-induced fluorescence detector.

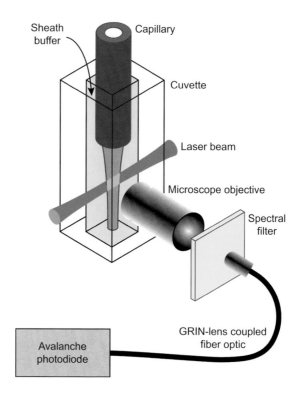

C. *elegans* and mouse embryos (41–51). We rely on covalent labeling of primary amines to fluorescently label biogenic amines and the ε-amine of lysine residues on proteins. Lysine is a common amino acid in most organisms, and the vast majority of proteins contains one or more lysine residues, at least in yeast (52).

Fluorescent labeling allows the use of low power and rugged solid-state lasers as excitation sources, which is preferable to the expensive and temperamental UV lasers required for excitation of most native fluorescence. However, labeling is not without its challenges. The presence of several labeling sites on a protein can produce an experimental artifact. Most derivatizing reagents convert the cationic lysine residue to a neutral or anionic product. For example, fluorescein isothiocyanate converts lysine residues to the fluorescein thiocarbamyl product, which corresponds to a change in the residue's charge state from +1 to –2. If every lysine residue on the protein is labeled, or if the same residues are labeled on each copy of the protein, then the derivatized products will all carry the same charge and have the same electrophoretic mobility. However, exhaustive labeling of proteins is challenging (53), and the labeling reaction inevitably produces a complex mixture of products. If there are n possible labeling sites, then there are 2^n-1 possible fluorescent products (54). This mixture forms a charge ladder for the protein, where each charge state has its own mobility, leading to complex electrophoretic profiles (55–59). Each charge state can be formed by reaction with different combinations of residues, which can lead to additional band broadening.

For example, we labeled enhanced GFP (eGFP) with 3-(2-furoyl)quinoline-2-carboxaldehyde (FQ), which converts the cationic lysine residue to a neutral product, and we performed capillary electrophoresis on the reaction products while monitoring GFP fluorescence (57). There are 20 lysine and 6 arginine residues for a total of 26 positive charges. We observed a set of peaks appear with lower mobility than the unlabeled protein; additional peaks appeared with lower mobility as the reaction progressed, eventually forming a very broad, featureless peak. We interpreted these peaks as the reactions product for increasing numbers of labeled lysine residues. The mobility decreased linearly ($r < -0.999$) by $2.8 \pm 0.2\%$ for addition of a label, which is close to the mobility change expected simply because of charge neutralization.

Multiple labeling can severely degrade separation efficiency. Native GFP generated over 100,000 theoretical plates in our capillary electrophoresis experiment; the labeled product barely generated 400 plates after extensive labeling. Fortunately, we discovered that peak broadening due to multiple labeling can be reduced by using FQ as the labeling reagent and by adding an anionic surfactant, such as sodium dodecyl sulfate (SDS), to the separation buffer (60). The surfactant presumably ion-pairs with unreacted lysine residues, neutralizing the charge just as reaction with FQ does.

FQ has another fortuitous property. It is a fluorogenic reagent that is nonfluorescent when excited at 473 nm and generates a very low reagent blank. In contrast, conventional labeling reagents, such as fluorescein, the rhodamines, BODIPYs, and Cy dyes, generate unacceptable reagent blanks due to unreacted dye, hydrolysis products, and impurities. FQ allows detection of picomolar concentrations of proteins with a very low background signal, which is required for chemical cytometry of proteins from somatic cells.

3.3. One-Dimensional Capillary Electrophoresis for Chemical Cytometry

Figure 4a presents a one-dimensional analysis of the proteins from single AtT20 cells. In this separation, a cell is injected into the capillary and lysed, and primary amines are labeled with FQ. Components are separated by capillary sieving electrophoresis (CSE). Data from two cells are shown. With the exception of a peak at 7 min, the electropherograms are quite similar to each other, each generating perhaps 20 features differing only in intensity. The integrated intensity for the top cell is twice the integrated intensity of the bottom cell. The phase of these cells in the cell cycle was not synchronized, and it is likely that the former cell is in the G2/M phase of the cell cycle while the latter cell is in the G1 phase. We have observed that cells taken from most cell cultures tend to generate single-cell electropherograms that strongly resemble one another.

In contrast, primary cells tend to generate much more heterogeneous electropherograms. **Figure 4b** presents three electropherograms generated from single mouse neural stem cells. Although some components appear in each electropherogram, there are dramatic differences in the data, particularly in the 10–12 min region. These differences presumably reflect the different ways in which these cells would develop.

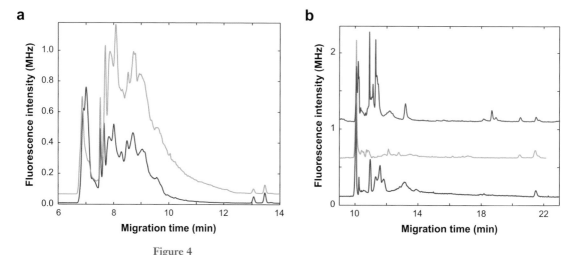

Figure 4

(*a*) Capillary sieving electrophoresis (CSE) single-cell electropherograms generated from two AtT20 cells. Electropherograms were aligned using a two-point method (61).

(*b*) One-dimensional CSE analysis of three mouse neural stem cells. Traces offset for clarity. Electropherograms were aligned using a two-point method (61).

3.4. Two-Dimensional Capillary Electrophoresis for Chemical Cytometry

One-dimensional electrophoresis provides a relatively rapid measure of the composition of a single cell. However, one-dimensional separation is insufficient to resolve the complex composition of a single cell. Two-dimensional electrophoresis provides a powerful tool for increasing the number of components resolved in a separation.

Two-dimensional electrophoresis is traditionally performed first by separating proteins by isoelectric focusing in a thin tube or strip. The strip is then placed at the top of a large gel, where the components are further separated by SDS-PAGE. Silver staining, or other visualization chemistry, is used to observe the separated proteins as a set of spots distributed across the gel, where the position of each protein corresponds to its isoelectric point and molecular weight. As noted above, 25 years ago Neukirchen used a miniaturized two-dimensional electrophoresis system to analyze ~1 μg of protein from a single *Drosophila* egg (12). Unfortunately, the amount of protein in a typical human somatic cell is five orders of magnitude smaller than that contained in a *Drosophila* egg. The proteins from a single somatic cell are diluted to invisibility when separated in a classic two-dimensional gel.

We have coupled two capillary electrophoresis capillaries to generate a separation method that is compatible with the minute amount of sample contained in a single cell (47, 62–63). In this method, labeled components are subjected sequentially to two stages of separation. This method is related to a method developed by Jorgenson to couple capillary liquid chromatography with capillary electrophoresis for the comprehensive separation of complex mixtures (64). In our system (**Figure 5***a*), both dimensions of the separation are based on capillary electrophoresis, which is much simpler than the use of chromatography for chemical cytometry. A cell is aspirated into the first capillary, the cell is lysed, components are labeled, and the labeled components are separated by CSE. Once components approach the end of the first capillary, a voltage program is used to sequentially transfer components across an interface into a second capillary. Once a fraction is transferred, the two power supplies are adjusted so that there is no voltage drop across the first capillary, where components remain stationary. High voltage is then applied across the second capillary, where components undergo separation by micellar electrokinetic capillary chromatography (MEKC), a form of capillary electrophoresis where separation is based on interaction with SDS-micelles. This series of fraction transfers and second dimension separation is typically performed 200 times to comprehensively separate the cell's components (**Figure 5b**).

The data are continually recorded, producing a long one-dimensional data vector (**Figure 6a**). The data consist of a set of ~200 separate MEKC separations, concatenated as a long data stream. Each MEKC separation lasted 14 s, and the entire experiment required just over 1 h to complete.

Figure 6b presents a close-up of the data near 24 min corresponding to the MEKC analysis of three successive fractions. The data consist of a set of peaks generated during the MEKC separation. The peaks from each MEKC separation are similar because components are sampled several times as they are transferred from the CSE capillary.

There is a dead time in the MEKC separation during the period from injection to the appearance of the first component at the detector. To speed the overall separation, we transfer fractions to the MEKC capillary before all components from the previous fraction have reached the detector. Analyte from two transfers are present in the MEKC capillary simultaneously; the fastest components from the second fraction do not quite catch up to the slowest components from the first fraction before detection.

The data can be shown as a spiral wrapped around a cylinder (**Figure 6c**) where the axial distance corresponds to the first dimension fraction and the angle corresponds to the second dimension separation time. The image is presented as a false color related to the fluorescence intensity.

To generate an image, the cylinder is cut and flattened, creating a gel-image of the separation (**Figure 6c**). Any component that happens to be cut will appear as a portion of a spot located on opposite sides of the image. The cut location is arbitrary and is chosen to minimize the number of components that are intersected. **Figure 6d** presents such an image, which resembles a silver-stained two-dimensional gel, presented here in false color. The image consists of a set of spots distributed across the surface. The positions of these spots are highly reproducible from cell to cell, with typical precision in spot position being on the order of the size of the spot itself (49–50).

The gel image is convenient for comparing spot position. However, the image, particularly when overexposed, is less useful in comparing the amplitude of spots or characterizing the dynamic range of the experiment. Instead, the data can be plotted as a surface in the form of a landscape, where height is proportional to the fluorescence intensity (**Figure 6e**).

a

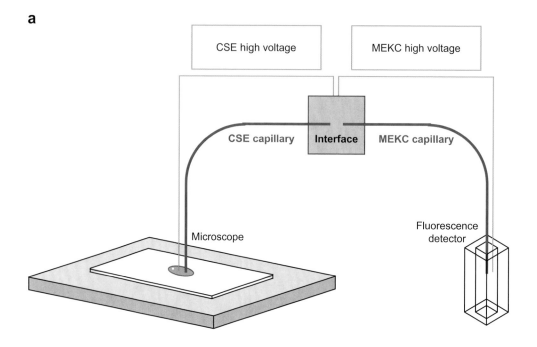

b

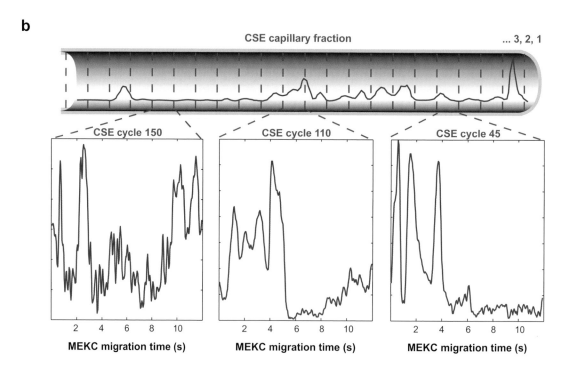

CSE capillary fraction   ... 3, 2, 1

CSE cycle 150

CSE cycle 110

CSE cycle 45

MEKC migration time (s)

MEKC migration time (s)

MEKC migration time (s)

This two-dimensional separation is reminiscent of classic two-dimensional gel electrophoresis, where proteins are separated first by isoelectric focusing and second by sodium dodecyl sulfate–polyacrylamide gel electrophoresis (SDS-PAGE). In our case, we employ CSE in the first dimension of the separation; this form of separation is the capillary version of SDS-PAGE, which separates proteins based on their molecular weight. Unfortunately, isoelectric focusing is not compatible with our labeling chemistry (58). We employ MEKC as the second dimension in this separation. MEKC separates components based on their interaction with surfactant, and is based, in part, on the hydrophobicity of the protein.

The spot capacity of a two-dimensional separation is analogous to the peak capacity of a one-dimensional separation. If the separation mechanisms are uncorrelated and if the separation is not degraded by coupling the two separation mechanisms, then the spot capacity of the two-dimensional separation is given by the product of the peak capacities of the one-dimensional separations. In our case, we are able to achieve reasonable peak capacity in the CSE dimension. However, our MEKC separation tends to produce mediocre peak capacity, and improvements are certainly desirable. Our overall spot capacity typically approaches 500 (49–50), which is roughly one order of magnitude poorer than classic two-dimensional gels, but with six orders of magnitude higher sensitivity.

Our published work has primarily focused on cultured cells. These cells generate remarkably reproducible electropherograms, which are useful for validating the precision of the experimental protocol. The next step in the development of chemical cytometry will focus on primary cells taken from living organisms. These cells are expected to show great heterogeneity. We are particularly excited about applying this technology for analysis of cells isolated from human biopsies for early detection of cancer and for generation of prognoses with improved accuracy to guide individualized therapy.

4. CHEMICAL CYTOMETRY OF METABOLITES: METABOLIC CYTOMETRY

Characterization of a cell's protein content provides insight into the function and behavior of a cell. Greater insight can be provided by characterization of cellular metabolism, which ultimately reflects the action of those proteins. Classic cytometry can be used to monitor a single metabolic transformation; fluorogenic reagents can be used to monitor the activity of a specific hydrolytic enzyme in a single cell. Hydrolysis of a quenching group releases a fluorescent product that can be monitored by flow

Figure 5

(*a*) Two-dimensional capillary electrophoresis instrument. (*b*) Schematic of two-dimensional electrophoresis. The cell's contents undergo a preliminary separation in the first capillary by capillary sieving electrophoresis (CSE). Fractions are sequentially transferred to the second capillary, where components undergo further separation by micellar electrokinetic capillary chromatography (MEKC).

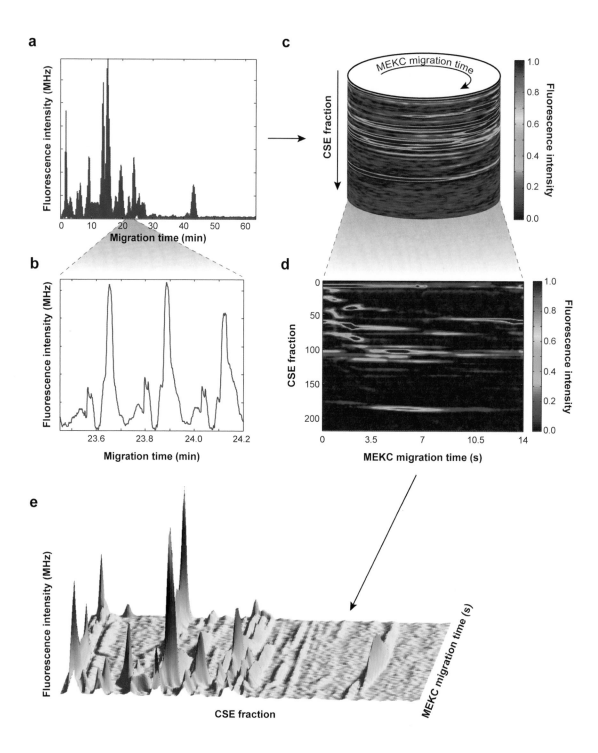

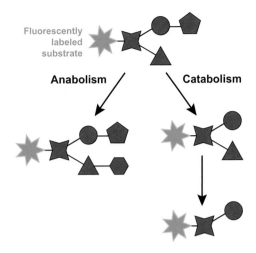

Fluorescently
labeled
substrate

Anabolism   **Catabolism**

Figure 7

Metabolic cytometry. A
fluorescent substrate is
shown with its label denoted
by the red star. This
substrate can undergo
anabolism to create larger
structures (*left*) or
catabolism to degrade the
substrate to smaller
metabolites (*right*). As long
as the fluorescent tag
remains intact, the
metabolic products can be
monitored by chemical
cytometry.

cytometry. However, conventional enzymatic probes are unable to characterize more
than one step in an enzymatic pathway and classic cytometry methods are limited in
the insight that they can provide on complex metabolic cascades.

Metabolic cytometry refers to the study of a selected set of metabolic pathways in
single cells (26–30, 34, 65–70). In metabolic cytometry, cells are treated with a fluorescent substrate, which is taken up and enzymatically processed to create products
that, in turn, act as substrates for other enzymes (**Figure 7**). Cells are then aspirated
into a capillary and lysed. Fluorescent metabolites are separated by electrophoresis
and detected by laser-induced fluorescence. As long as the metabolic products retain
their fluorescent tag, then the reaction products can be determined with exquisite
sensitivity.

Metabolic cytometry differs from chemical cytometry of proteins and biogenic
amines in one important respect. In metabolic cytometry, a fluorescent substrate is
prepared at high concentration in the laboratory. This substrate can be subjected to
exhaustive purification, which allows the use of highly fluorescent dyes as labels. In
particular, we find tetramethylrhodamine to be a very useful dye in these applications.
This molecule is highly fluorescent, and compounds labeled with it appear to be taken

---

Figure 6

(*a*) Time trace of a two-dimensional capillary electrophoresis analysis of a single CP18821 cell.
The sample underwent a preliminary separation of 5 min in the capillary sieving
electrophoresis (CSE) capillary before fractions transfers began, during which time signal was
not recorded. (*b*) Close-up of the three successive micellar electrokinetic capillary
chromatography (MEKC) separations shown in (*a*). (*c*) The raw electrophoresis data presented
in (*a*) are wrapped as around a cylinder, where the optical density is related to the fluorescence
intensity. (*d*) The raw electrophoresis data presented in (*a*) as wrapped around a cylinder that
is cut and flattened, creating a gel-image where the color is mapped to fluorescence intensity.
The image is overexposed to highlight low-amplitude components. (*e*) Landscape view, where
height is proportional to fluorescence intensity.

up efficiently by cells. In contrast, proteins and biogenic amines must be labeled after the cell is lysed, which places constraints (discussed above) on the labeling chemistry.

4.1. Metabolic Cytometry of Sphingolipids

We have been studying glycosphingolipid metabolism in single AtT20 cells. Glycosphingolipids are among the most common molecules on neural cell surfaces and consist of several components. Ceramide is the nonpolar tail that comprises a long-chain amino alcohol linked to a long-chain fatty acid. A chain of saccharide groups attached to the ceramide forms the constituent glycosphingolipids, and the presence of one to four sialic acids turns these glycosphingolipids into gangliosides. In the common naming of gangliosides, the subscript letter indicates the quantity of sialic acid groups: M for monosialo, D for disialo, et cetera. A stands for asialo, which refers to glycosphingolipids without sialic acids. The number refers to the order of peak appearance in thin-layer chromatography and thus the length of the sugar chain; the sequential removal of zero, one, and two terminal saccharides decreases retention to yield G_{M1}, G_{M2}, and G_{M3}, respectively.

These glycolipids form a large fraction of neuronal cell membranes. Defects in their metabolism lead to a series of devastating diseases, the best known of which is Tay-Sachs. The compounds are synthesized from the lipid ceramide by successive additions of monosaccharides (**Figure 8a**). A defect in the degradation of G_{M2} to G_{M3} leads to accumulation of the former, and is the underlying cause of Tay-Sachs disease.

We have synthesized the fluorescently labeled G_{M1} (**Figure 8b**) (70). We modified the ceramide group by incorporating the fluorescent label tetramethylrhodamine. We have enzymatically transformed the tetramethylrhodamine-labeled G_{M1} to labeled G_{M2}, G_{A1}, and G_{A2}, and have produced labeled G_{M3}, Lac-Cer, Glc-Cer, and Cer by additional chemical and enzymatic methods. These compounds are used as standards to tentatively identify metabolic cytometry products based on comigration.

Figure 9 presents metabolic cytometry data generated from single AtT20 cells. In these experiments, a cell was incubated with the substrate. The substrate was taken up, which was confirmed by fluorescence microscopy. Enzymatic transformations converted the substrate to products. To analyze the products, the cell was aspirated into a capillary and lysed, its components were separated by capillary electrophoresis, and products were detected by laser-induced fluorescence.

AtT20 cells were incubated with the substrate shown in **Figure 8b**, aspirated into a capillary, lysed, and analyzed by capillary electrophoresis. Components were identified by comigration with standard compounds. Three unknown components were observed that did not migrate with standards and likely are more complex structures (see **Figure 8a**).

4.2. Metabolic Cytometry of Kinase Activity

Allbritton's group has also performed a form of metabolic cytometry to assay kinase activity in single cells (26–29). In those experiments, a cell was treated with a

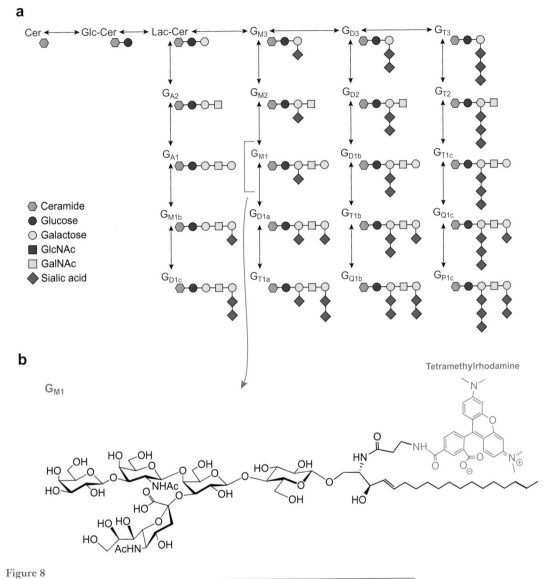

a

Cer ⟷ Glc-Cer ⟷ Lac-Cer ⟷ G_{M3} ⟷ G_{D3} ⟷ G_{T3}

G_{A2}   G_{M2}   G_{D2}   G_{T2}

G_{A1}   G_{M1}   G_{D1b}   G_{T1c}

G_{M1b}   G_{D1a}   G_{T1b}   G_{Q1c}

G_{D1c}   G_{T1a}   G_{Q1b}   G_{P1c}

⬡ Ceramide
● Glucose
○ Galactose
■ GlcNAc
□ GalNAc
◆ Sialic acid

b

G_{M1}

Tetramethylrhodamine

Figure 8

(*a*) Partial metabolic pathway associated with sphingolipids. Arrows denote known metabolic transformations in primates. (*b*) Structure of tetramethylrhodamine-labeled G_{M1}. The tetramethylrhodamine structure is shown.

fluorescent peptide that is a substrate for a specific kinase. Phosphorylation of the peptide led to a mobility shift for the peptide, which was easily detected by capillary electrophoresis and laser-induced fluorescence. By employing a suite of peptides, a large number of kinases can be monitored simultaneously, provided that the substrates and products generate nonoverlapping electrophoresis peaks.

Figure 9

Metabolic cytometry
analysis of four cells and
the blank (*bottom trace*).

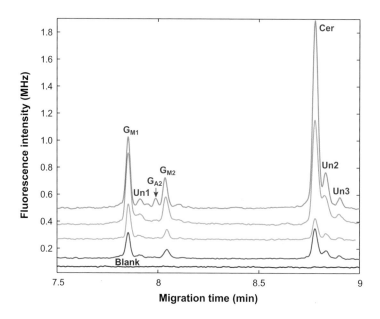

5. CHEMICAL CYTOMETRY OF DNA AND mRNA

The polymerase chain reaction (PCR) has made possible the study of DNA and
mRNA expression levels in single cells. In genomic sequencing experiments, a single
cell was selected and lysed in a nanoliter microfabricated well (71). The lysed cell
was subjected to whole genome amplification and the products were sequenced using
Pyrosequencing. For mRNA studies, the cell was placed in a cocktail containing
lysis buffer and poly-dT-coated magnetic beads (72–76). The captured mRNA was
converted to complemetary DNA. Hot-start PCR was performed for 40 cycles with
random primers; the PCR products were studied either by Western blotting, by
hybridization to an oligonucleotide array, or by direct sequencing.

This technology has been fairly widely disseminated. In perhaps its most famous
application, Dulac and Axel screened single neurons from the rat vomeronasal organ
and identified a family of putative pheromone receptor genes (73). Other examples
include the characterization of neuronal progenitors, of cancer micrometastatic cells,
and of difficult-to-culture microbes (71, 77, 78).

6. SUMMARY: CHEMICAL CYTOMETRY PROVIDES HIGH-RESOLUTION CHARACTERIZATION OF SINGLE CELLS

Although chemical cytometry has a 50-year history, the technology is nevertheless in
its infancy. As might be expected, single-cell nucleic acid analysis is the most mature
form of chemical cytometry; powerful nucleic acid amplification technology facilitates
the study of the minute amount of material present in a single cell. Analysis of proteins,
biogenic amines, and metabolites in single cells represents a much more formidable

challenge. Most work to date has dealt with cultured cells, which is necessary for method validation and for system characterization. Large-scale analyses of primary cells will be required to demonstrate the value of the technology.

Several issues remain to be addressed. First, it is becoming obvious that living cells cannot be stored for long periods before analysis; cellular response to storage likely leads to profound changes in composition. Instead, cells must be fixed with a preservative after removal from culture or from an experimental animal to eliminate artifacts associated with storage. Formalin crosslinks proteins and nucleic acids but not glycolipids, which makes it an ideal fixative for characterization of most metabolites. However, formalin is not compatible with protein analysis; we employ ethanol to fix cells for protein analysis. Although ethanol precipitates proteins, our cell lysis solution contains SDS, which efficiently solubilizes proteins.

Second, throughput of chemical cytometry is much lower than in classical cytometry. Flow cytometry can process many thousands of cells per second. We are developing multiple capillary instruments for high-throughput chemical cytometry (79). These instruments are based on the multiple-capillary electrophoresis systems that have become ubiquitous in DNA sequencing (80–83), and should allow the study of hundreds to thousands of cells per day.

Third, separation efficiency must be improved for single-cell protein analysis. Current two-dimensional electrophoresis systems are generating a spot capacity of a few hundred components; an improvement of an order of magnitude is desirable. MEKC is not an ideal separation mode for proteins, and should be replaced with a technique such as isoelectric focusing (IEF), which can provide exquisite resolution. A major hurdle deals with protein labeling. FQ, and presumably other dyes that produce an anionic or neutral reaction product, generates very complex IEF electropherograms due to multiple labeling (58). Anionic surfactants, which eliminate the peak broadening that results from multiple labeling in CSE and MEKC, are not compatible with IEF. It will be necessary to employ a labeling reagent that produces a cationic product. The recently developed chameleon dyes are interesting examples of fluorogenic reagents that produce cationic products (84–85).

Fourth, component identification must be improved. Current methods rely on comigration with standards (34, 45). Although these methods are practical for metabolic cytometry and biogenic amines due to the relatively small number of target compounds, spiking is extremely tedious for protein analysis. Ideally, an on-line mass spectrometer can be used to monitor compounds as they migrate from the capillary (86). Although the current generation of mass spectrometers does not have sufficient sensitivity to monitor the minute amount of protein present in a single cells, mass spectrometry could be used to study concentrated homogenates prepared from large numbers of cells. Based on the excellent run-to-run reproducibility of the two-dimensional electrophoresis system, component identification should be practical in fluorescence-based detection of single cells. As a complication, fluorescently labeled proteins are required for single-cell analysis. To use mass spectrometry to identify components with the same mobility, software will need to be modified to reflect the mass change induced by the fluorescent tag. Peptides are much easier to analyze than proteins by mass spectrometry, and Sweedler's group has done impressive work on the identification of novel neuropeptides in single neurons by matrix-assisted laser desorption/ionization mass spectrometry (87–92).

Finally, improved sensitivity is required to perform chemical cytometry on a single bacterium or subcellular organelle, which contain a million-fold smaller amount of material than does a typical eukaryote cell (25, 93–101).

DISCLOSURE STATEMENT

The authors are not aware of any biases that might be perceived as affecting the objectivity of this review.

ACKNOWLEDGMENTS

This work was funded by National Institutes of Health grant numbers R01NS061767-01, R01GM071666, R33CA122900, and P50HG002360.

LITERATURE CITED

1. Shapiro HM. 2003. *Practical Flow Cytometry*. 4th ed. Hoboken: Wiley
2. Caspersson TO. 1987. History of the development of cytophotometry from 1935 to the present. *Anal. Quant. Cytol. Histol.* 9:2–6
3. Baumgarth N, Roederer M. 2000. A practical approach to multicolor flow cytometry for immunophenotyping. *J. Immunol. Methods* 243:77–97
4. Ibrahim SF, van den Engh G. 2003. High-speed cell sorting: fundamentals and recent advances. *Curr. Opin. Biotech.* 14:5–12
5. Edstrom JE. 1953. Nucleotide analysis on the cyto-scale. *Nature* 172:809
6. Matioli GT, Niewisch HB. 1965. Electrophoresis of hemoglobin in single erythrocytes. *Science* 150:1824–26
7. Hyden H, Bjurstam K, McEwen B. 1966. Protein separation at the cellular level by microdisk electrophoresis. *Anal. Biochem.* 17:1–15
8. Marchalonis JJ, Nossal GJ. 1968. Electrophoretic analysis of antibody produced by single cells. *Proc. Natl. Acad. Sci. USA* 61:860–67
9. Wilson DL. 1971. Molecular weight distribution of proteins synthesized in single, identified neurons of *Aplysia*. *J. Gen. Physiol.* 57:26–40
10. Ruchel R. 1976. Sequential protein analysis from single identified neurons of *Aplysia californica*. A microelectrophoretic technique involving polyacrylamide gradient gels and isoelectric focusing. *J. Histochem. Cytochem.* 24:773–91
11. Repin VS, Akimova IM, Terovskii VB. 1975. Detection of lactate dehydrogenase isoenzymes in single mammalian oocytes during cleavage by a micromodification of disc electrophoresis. *Bull. Exp. Biol. Med.* 77:767–69
12. Neukirchen RO, Schlosshauer B, Baars S, Jackle H, Schwarz U. 1982. Two-dimensional protein analysis at high resolution on a microscale. *J. Biol. Chem.* 257:15229–34
13. Kennedy RT, Oates MD, Cooper BR, Nickerson B, Jorgenson JW. 1989. Microcolumn separations and the analysis of single cells. *Science* 246:57–63

14. Wallingford RA, Ewing AG. 1988. Capillary zone electrophoresis with electrochemical detection in 12.7 μm diameter columns. *Anal. Chem.* 60:1972–75

15. Hogan BL, Yeung ES. 1992. Determination of intracellular species at the level of a single erythrocyte via capillary electrophoresis with direct and indirect fluorescence detection. *Anal. Chem.* 64:2841–45

16. Gilman SD, Ewing AG. 1995. Analysis of single cells by capillary electrophoresis with on-column derivatization and laser-induced fluorescence detection. *Anal. Chem.* 67:58–64

17. Lillard SJ, Yeung ES. 1996. Analysis of single erythrocytes by injection-based capillary isoelectric focusing with laser-induced native fluorescence detection. *J. Chromatogr. B* 687:363–69

18. Tong W, Yeung ES. 1997. On-column monitoring of secretion of catecholamines from single bovine adrenal chromaffin cells by capillary electrophoresis. *J. Neurosci. Methods* 76:193–201

19. Ho AM, Yeung ES. 1998. Capillary electrophoretic study of individual exocytotic events in single mast cells. *J. Chromatogr. A* 817:377–82

20. Parpura V, Tong W, Yeung ES, Haydon PG. 1998. Laser-induced native fluorescence (LINF) imaging of serotonin depletion in depolarized neurons. *J. Neurosci. Methods* 82:151–58

21. Yeung ES. 1999. Study of single cells by using capillary electrophoresis and native fluorescence detection. *J. Chromatogr. A* 830:243–62

22. Lapainis T, Scanlan C, Rubakhin SS, Sweedler JV. 2007. A multichannel native fluorescence detection system for capillary electrophoretic analysis of neurotransmitters in single neurons. *Anal. Bioanal. Chem.* 387:97–105

23. Zhang X, Sweedler JV. 2001. Ultraviolet native fluorescence detection in capillary electrophoresis using a metal vapor NeCu laser. *Anal. Chem.* 73:5620–24

24. Hu K, Ahmadzadeh H, Krylov SN. 2004. Asymmetry between sister cells in a cancer cell line revealed by chemical cytometry. *Anal. Chem.* 76:3864–66

25. Turner EH, Lauterbach K, Pugsley HR, Palmer VR, Dovichi NJ. 2007. Detection of green fluorescent protein in a single bacterium by capillary electrophoresis with laser-induced fluorescence. *Anal. Chem.* 79:778–81

26. Lee CL, Linton J, Soughayer JS, Sims CE, Allbritton NL. 1999. Localized measurement of kinase activation in oocytes of *Xenopus laevis*. *Nat. Biotechnol.* 17:759–62

27. Meredith GD, Sims CE, Soughayer JS, Allbritton NL. 2000. Measurement of kinase activation in single mammalian cells. *Nat. Biotechnol.* 18:309–12

28. Nelson AR, Allbritton NL, Sims CE. 2007. Rapid sampling for single-cell analysis by capillary electrophoresis. *Methods Cell Biol.* 82:709–22

29. Sims CE, Allbritton NL. 2003. Single-cell kinase assays: opening a window onto cell behavior. *Curr. Opin. Biotechnol.* 14:23–28

30. Han F, Wang Y, Sims CE, Bachman M, Chang R, et al. 2003. Fast electrical lysis of cells for capillary electrophoresis. *Anal. Chem.* 75:3688–96

31. Berezovski MV, Mak TW, Krylov SN. 2007. Cell lysis inside the capillary facilitated by transverse diffusion of laminar flow profiles (TDLFP). *Anal. Bioanal. Chem.* 387:91–96

32. Borland LM, Kottegoda S, Phillips KS, Allbritton NL. 2008. Chemical analysis of single cells. *Annu. Rev. Anal. Chem.* 1:191–227

33. Wu H, Wheeler A, Zare RN. 2004. Chemical cytometry on a picoliter-scale integrated microfluidic chip. *Proc. Natl. Acad. Sci. USA* 101:12809–13

34. Krylov SN, Zhang Z, Chan NWC, Arriaga E, Palcic MM, Dovichi NJ. 1999. Correlating cell cycle with metabolism in single cells: the combination of image and metabolic cytometry. *Cytometry.* 37:15–20

35. Krylov SN, Starke DA, Arriaga EA, Zhang Z, Chan NW, et al. 2000. Instrumentation for chemical cytometry. *Anal. Chem.* 72:872–77

36. Krylov SN, Dovichi NJ. 2000. Single-cell analysis using capillary electrophoresis: influence of surface support properties on cell injection into the capillary. *Electrophoresis* 21:767–73

37. Cheng YF, Dovichi NJ. 1988. Subattomole amino acid analysis by capillary zone electrophoresis and laser-induced fluorescence. *Science* 242:562–64

38. Wu S, Dovichi NJ. 1989. High-sensitivity fluorescence detector fluorescein isothiocyanate derivatives of amino acids separated by capillary zone electrophoresis. *J. Chromatogr.* 480:141–55

39. Chen DY, Dovichi NJ. 1996. Single-molecule detection in capillary electrophoresis: molecular shot noise as a fundamental limit to chemical analysis. *Anal. Chem.* 68:690–96

40. Kraly JR, Jones MR, Gomez DG, Dickerson JA, Harwood MM, et al. 2006. Reproducible two-dimensional capillary electrophoresis analysis of Barrett's esophagus tissues. *Anal. Chem.* 78:5977–86

41. Dovichi NJ, Hu S. 2003. Chemical cytometry. *Curr. Opin. Chem. Biol.* 7:603–8

42. Zhang Z, Krylov S, Arriaga EA, Polakowski R, Dovichi NJ. 2000. One-dimensional protein analysis of an HT29 human colon adenocarcinoma cell. *Anal. Chem.* 72:318–22

43. Hu S, Lee R, Zhang Z, Krylov SN, Dovichi NJ. 2001. Protein analysis of an individual *Caenorhabditis elegans* single-cell embryo by capillary electrophoresis. *J. Chromatogr. B* 752:307–10

44. Hu S, Zhang L, Cook LM, Dovichi NJ. 2001. Capillary sodium dodecyl sulfate-DALT electrophoresis of proteins in a single human cancer cell. *Electrophoresis* 22:3677–82

45. Hu S, Le Z, Newitt R, Aebersold R, Kraly JR, et al. 2003. Identification of proteins in single-cell capillary electrophoresis fingerprints based on comigration with standard proteins. *Anal. Chem.* 75:3502–5

46. Hu S, Le Z, Krylov S, Dovichi NJ. 2003. Cell cycle-dependent protein fingerprint from a single cancer cell: image cytometry coupled with single-cell capillary sieving electrophoresis. *Anal. Chem.* 75:3495–501

47. Hu S, Michels DA, Fazal MA, Ratisoontorn C, Cunningham ML, Dovichi NJ. 2004. Capillary sieving electrophoresis/micellar electrokinetic capillary chromatography for two-dimensional protein fingerprinting of single mammalian cells. *Anal. Chem.* 76:4044–49

48. Harwood MM, Christians ES, Fazal MA, Dovichi NJ. 2006. Single-cell protein analysis of a single mouse embryo by two-dimensional capillary electrophoresis. *J. Chromatogr. A* 1130:190–94

49. Harwood MM, Bleecker JV, Rabinovitch PS, Dovichi NJ. 2007. Cell cycle-dependent characterization of single MCF-7 breast cancer cells by 2-D CE. *Electrophoresis* 28:932–37

50. Sobhani K, Fink SL, Cookson BT, Dovichi NJ. 2007. Repeatability of chemical cytometry: 2-DE analysis of single RAW 264.7 macrophage cells. *Electrophoresis* 28:2308–13

51. Chen X, Fazal MA, Dovichi NJ. 2007. CSE-MECC two-dimensional capillary electrophoresis analysis of proteins in the mouse tumor cell (AtT-20) homogenate. *Talanta* 71:1981–85

52. Pinto D, Arriaga EA, Schoenherr RM, Chou SS, Dovichi NJ. 2003. Kinetics and apparent activation energy of the reaction of the fluorogenic reagent 5-furoylquinoline-3-carboxaldehyde with ovalbumin. *J. Chromatogr. B* 793:107–14

53. Liu HJ, Cho BY, Krull IS, Cohen SA. 2001. Homogeneous fluorescent derivatization of large proteins. *J. Chromatogr. A* 927:77–89

54. Zhao JY, Waldron KC, Miller J, Zhang JZ, Harke H, Dovichi NJ. 1992. Attachment of a single fluorescent label to peptides for determination by capillary zone electrophoresis. *J. Chromatogr.* 608:239–42

55. Gao J, Gomez FA, Harter R, Whitesides GM. 1994. Determination of the effective charge of a protein in solution by capillary electrophoresis. *Proc. Natl. Acad. Sci. USA* 91:12027–30

56. Gao J, Whitesides GM. 1997. Using protein charge ladders to estimate the effective charges and molecular weights of proteins in solution. *Anal. Chem.* 69:575–80

57. Craig DB, Dovichi NJ. 1998. Multiple labeling of proteins. *Anal. Chem.* 70:2493–94

58. Richards D, Stathakis C, Polakowski R, Ahmadzadeh H, Dovichi NJ. 1999. Labeling effects on the isoelectric point of green fluorescent protein. *J. Chromatogr. A* 853:21–25

59. Anderson JR, Chemiavskaya O, Gitlin I, Engel GS, Yuditsky L, Whitesides GM. 2002. Analysis by capillary electrophoresis of the kinetics of charge ladder formation for bovine carbonic anhydrase. *Anal. Chem.* 74:1870–78

60. Pinto DM, Arriaga EA, Craig D, Angelova J, Sharma N, et al. 1997. Picomolar assay of native proteins by capillary electrophoresis: precolumn labeling, submicellar separation and laser induced fluorescence detection. *Anal. Chem.* 69:3015–21

61. Li XF, Ren H, Le X, Qi M, Ireland ID, Dovichi NJ. 2000. Migration time correction for the analysis of derivatized amino acids and oligosaccharides by micellar capillary electrochromatography. *J. Chromatogr. A* 869:375–84

62. Michels DA, Hu S, Schoenherr RM, Eggertson MJ, Dovichi NJ. 2002. Fully automated two-dimensional capillary electrophoresis for high sensitivity protein analysis. *Mol. Cell. Proteomics.* 1:69–74

63. Michels DA, Hu S, Dambrowitz KA, Eggertson MJ, Lauterbach K, Dovichi NJ. 2004. Capillary sieving electrophoresis–micellar electrokinetic chromatography fully automated two-dimensional capillary electrophoresis analysis of *Deinococcus radiodurans* protein homogenate. *Electrophoresis* 25:3098–105

64. Opiteck GJ, Lewis KC, Jorgenson JW, Anderegg RJ. 1997. Comprehensive on-line LC/LC/MS of proteins. *Anal. Chem.* 69:1518–24

65. Le XC, Tan W, Scaman CH, Szpacenko A, Arriaga EA, et al. 1999. Single cell studies of enzymatic hydrolysis of a tetramethylrhodamine labeled trisaccharide in yeast. *Glycobiology* 9:219–25

66. Krylov SN, Arriaga EA, Zhang Z, Chan NW, Palcic MM, Dovichi NJ. 2000. Single-cell analysis avoids sample processing bias. *J. Chromatogr. B* 741:31–35

67. Krylov SN, Arriaga EA, Chan NW, Dovichi NJ, Palcic MM. 2000. Metabolic cytometry: monitoring oligosaccharide biosynthesis in single cells by capillary electrophoresis. *Anal. Biochem.* 283:133–35

68. Arkhipov SN, Berezovski M, Jitkova J, Krylov SN. 2005. Chemical cytometry for monitoring metabolism of a Ras-mimicking substrate in single cells. *Cytometry A* 63:41–47

69. Whitmore CD, Hindsgaul O, Palcic MM, Schnaar RL, Dovichi NJ. 2007. Metabolic cytometry: glycosphingolipid metabolism in single cells. *Anal. Chem.* 79:5139–42

70. Larsson EA, Olsson U, Whitmore CD, Martins R, Tettamanti G, et al. 2007. Synthesis of reference standards to enable single cell metabolomic studies of tetramethylrhodamine-labeled ganglioside GM1. *Carbohydr. Res.* 342:482–89

71. Marcy Y, Ouverney C, Bik EM, Losekann T, Ivanova N, et al. 2007. Dissecting biological "dark matter" with single-cell genetic analysis of rare and uncultivated TM7 microbes from the human mouth. *Proc. Natl. Acad. Sci. USA* 104:11889–94

72. Brady G, Barbara M, Iscove NN. 1990. Representative in vitro cDNA amplification from individual hemopeoitic cells and colonies. *Meth. Mol. Cell. Biol.* 2:17–25

73. Dulac C, Axel R. 1995. A novel family of genes encoding putative pheromone receptors in mammals. *Cell* 83:195–206

74. Salk JJ, Sanchez JA, Pierce KE, Rice JE, Soares KC, Wangh LJ. 2006. Direct amplification of single-stranded DNA for pyrosequencing using linear-after-the-exponential (LATE)-PCR. *Anal. Biochem.* 353:124–32

75. Tietjen I, Rihel JM, Cao YX, Koentges G, Zakhary L, Dulac C. 2003. Single-cell transcriptional analysis of neuronal progenitors. *Neuron* 38:161–75

76. Klein CA, Seidl S, Petat-Dutter K, Offner S, Geigl JB, et al. 2002. Combined transcriptome and genome analysis of single micrometastatic cells. *Nat. Biotechnol.* 20:387–92

77. Wells D. 2007. Use of real-time polymerase chain reaction to measure gene expression in single cells. *Methods Mol. Med.* 132:125–33

78. Pantel K, Otte M. 2001. Occult micrometastasis: enrichment, identification and characterization of single disseminated tumour cells. *Semin. Cancer Biol.* 11:327–37

79. Zhu C, He X, Kraly JR, Jones MR, Whitmore CD, et al. 2007. Instrumentation for medium-throughput two-dimensional capillary electrophoresis with laser-induced fluorescence detection. *Anal. Chem.* 79:765–68

80. Zhang J, Voss KO, Shaw DF, Roos KP, Lewis DF, et al. 1999. A multiple-capillary electrophoresis system for small-scale DNA sequencing and analysis. *Nucleic Acids Res.* 27:e36

81. Crabtree HJ, Bay SJ, Lewis DF, Zhang J, Coulson LD, et al. 2000. Construction and evaluation of a capillary array DNA sequencer based on a micromachined sheath-flow cuvette. *Electrophoresis* 21:1329–35

82. Zhang J, Yang M, Puyang X, Fang Y, Cook LM, Dovichi NJ. 2001. Two-dimensional direct-reading fluorescence spectrograph for DNA sequencing by capillary array electrophoresis. *Anal. Chem.* 73:1234–39

83. Dovichi NJ. 1997. DNA sequencing by capillary electrophoresis. *Electrophoresis* 18:2393–99

84. Wetzl BK, Yarmoluk SM, Craig DB, Wolfbeis OS. 2004. Chameleon labels for staining and quantifying proteins. *Angew. Chem. Int. Ed. Engl.* 43:5400–2

85. Craig DB, Wetzl BK, Duerkop A, Wolfbeis OS. 2005. Determination of pi-comolar concentrations of proteins using novel amino reactive chameleon labels and capillary electrophoresis laser-induced fluorescence detection. *Electrophoresis* 26:2208–13

86. Schoenherr RM, Ye M, Vannatta M, Dovichi NJ. 2007. CE-microreactor-CE-MS/MS for protein analysis. *Anal. Chem.* 79:2230–38

87. Li L, Garden RW, Romanova EV, Sweedler JV. 1999. In situ sequencing of peptides from biological tissues and single cells using MALDI-PSD/CID analysis. *Anal. Chem.* 71:5451–58

88. Li L, Romanova EV, Rubakhin SS, Alexeeva V, Weiss KR, et al. 2000. Peptide profiling of cells with multiple gene products: combining immunochemistry and MALDI mass spectrometry with on-plate microextraction. *Anal. Chem.* 72:3867–74

89. Li L, Garden RW, Sweedler JV. 2000. Single-cell MALDI: a new tool for direct peptide profiling. *Trends Biotechnol.* 18:151–60

90. Rubakhin SS, Garden RW, Fuller RR, Sweedler JV. 2000. Measuring the peptides in individual organelles with mass spectrometry. *Nat. Biotechnol.* 18:172–75

91. Rubakhin SS, Churchill JD, Greenough WT, Sweedler JV. 2006. Profiling signaling peptides in single mammalian cells using mass spectrometry. *Anal. Chem.* 78:7267–72

92. Rubakhin SS, Sweedler JV. 2007. Characterizing peptides in individual mammalian cells using mass spectrometry. *Nat. Protoc.* 2:1987–89

93. Strack A, Duffy CF, Malvey M, Arriaga EA. 2001. Individual mitochondrion characterization: a comparison of classical assays to capillary electrophoresis with laser-induced fluorescence detection. *Anal. Biochem.* 294:141–47

94. Duffy CF, Gafoor S, Richards DP, Admadzadeh H, O'Kennedy R, Arriaga EA. 2001. Determination of properties of individual liposomes by capillary electrophoresis with postcolumn laser-induced fluorescence detection. *Anal. Chem.* 73:1855–61

95. Fuller KM, Duffy CF, Arriaga EA. 2002. Determination of the cardiolipin content of individual mitochondria by capillary electrophoresis with laser-induced fluorescence detection. *Electrophoresis* 23:1571–76

96. Duffy CF, Fuller KM, Malvey MW, O'Kennedy R, Arriaga EA. 2002. Determination of electrophoretic mobility distributions through the analysis of individual mitochondrial events by capillary electrophoresis with laser-induced fluorescence detection. *Anal. Chem.* 74:171–76

97. Fuller KM, Arriaga EA. 2003. Analysis of individual acidic organelles by capillary electrophoresis with laser-induced fluorescence detection facilitated by the endocytosis of fluorescently labeled microspheres. *Anal. Chem.* 75:2123–30

98. Fuller KM, Arriaga EA. 2003. Advances in the analysis of single mitochondria. *Curr. Opin. Biotechnol.* 14:35–41

99. Chen Y, Arriaga EA. 2006. Individual acidic organelle pH measurements by capillary electrophoresis. *Anal. Chem.* 78:820–26

100. Johnson RD, Navratil M, Poe BG, Xiong G, Olson KJ, et al. 2007. Analysis of mitochondria isolated from single cells. *Anal. Bioanal. Chem.* 387:107–18

101. Meany DL, Thompson L, Arriaga EA. 2007. Simultaneously monitoring the superoxide in the mitochondrial matrix and extramitochondrial space by micellar electrokinetic chromatography with laser-induced fluorescence. *Anal. Chem.* 79:4588–94

Chemical Analysis
of Single Cells

Laura M. Borland, Sumith Kottegoda,
K. Scott Phillips, and Nancy L. Allbritton

Department of Chemistry, University of North Carolina at Chapel Hill, Chapel Hill,
North Carolina 27599; email: nlallbri@unc.edu

Annu. Rev. Anal. Chem. 2008. 1:191–227

First published online as a Review in Advance on
February 7, 2008

The *Annual Review of Analytical Chemistry* is online
at anchem.annualreviews.org

This article's doi:
10.1146/annurev.anchem.1.031207.113100

Key Words

capillary electrophoresis, laser-induced fluorescence, chemical
cytometry, cellular analytes

Abstract

Chemical analysis of single cells requires methods for quickly and
quantitatively detecting a diverse array of analytes from extremely
small volumes (femtoliters to nanoliters) with very high sensitivity
and selectivity. Microelectrophoretic separations, using both traditional capillary electrophoresis and emerging microfluidic methods,
are well suited for handling the unique size of single cells and limited numbers of intracellular molecules. Numerous analytes, ranging
from small molecules such as amino acids and neurotransmitters to
large proteins and subcellular organelles, have been quantified in
single cells using microelectrophoretic separation techniques. Mi-
croseparation techniques, coupled to varying detection schemes including absorbance and fluorescence detection, electrochemical detection, and mass spectrometry, have allowed researchers to examine
a number of processes inside single cells. This review also touches
on a promising direction in single cell cytometry: the development
of microfluidics for integrated cellular manipulation, chemical processing, and separation of cellular contents.

1. INTRODUCTION

Capillary electrophoresis
(CE): typically utilizes high
electric fields for separation
of an analyte, which may be
based on its charge, size, or
hydrophobicity

A plethora of technologies now exist for analyzing the chemical constituents in single cells. Among the tools used for these analyses are capillary electrophoresis (CE), mass spectrometry, electrochemistry, flow cytometry, and fluorescence microscopy. Functional reagents are frequently key components of these technologies and, in combination with the instrumentation, enable the sensitivity and specificity for single-cell measurements. Examples include the development of ion indicators for the measurement of intracellular Ca^{2+} by fluorescence microscopy and the high-specificity, antiphosphoprotein antibodies for the quantification of intracellular phosphoproteins by flow cytometry (1, 2). These single-cell technologies have also had a substantial impact on the conduct of biomedical research, for example in the use of electrochemistry and mass spectrometry for the measurement of neurotransmitters and other secreted molecules such as peptides.

Full coverage of all technologies utilized in the chemical analysis of single cells is beyond the scope of this review. Rather, this work focuses on the analysis of single cells using microelectrophoretic separations. Both microfluidic and capillary-based electrophoresis are well suited for handling the femtoliter-to-nanoliter volumes of cells and their organelles. The methods are fast and quantitative, provide multicomponent analysis, and are compatible with high-sensitivity detection. Separation of the cellular components also enables the quantification of large numbers of analytes simultaneously without the interference of the background cellular matrix. All of these attributes are required for measurements on single cells and their components.

The separation of single cells using microelectrophoretic methods was pioneered by the labs of Jorgenson, Ewing, and Yeung (3–5). The early contributions of these labs are summarized in an excellent review by Dovichi and Hu (6). Since this time, much of the instrumentation has changed, although the central goal of introducing a cell's contents into a small-diameter tube and inducing separation in an electric field remains unchanged. Innovations in cell sampling have expanded microelectrophoretic assays to a wide range of cell types as well as to a variety of subcellular fractions (7). Most commonly, the entirety of a cell is introduced by hydrodynamic or electroosmotic fluid flow into a channel. The cell may be lysed either before or after entry into the microchannel (7, 15) by one of several methods: chemical lysis, typically by a detergent (8, 10); hypotonic lysis by a dilute aqueous solution yielding cell swelling and rupture (9, 11); mechanical lysis by a shock wave and cavitation bubble of a focused, pulsed laser beam (14, 16); or electrical lysis due to membrane breakdown by an applied voltage (12, 13). Various methods for subcellular sampling have also been developed. These include direct insertion of a capillary into a cell followed by hydrodynamic or electrokinetic withdrawal of cytoplasm into a tube (17, 18), and use of a laser to detach a cellular process for analysis (19). Permeabilization of cellular membranes by electroporation or chemical reagents followed by limited withdrawal of cellular contents has been mated with microelectrophoresis (20–22). Organelles are also appropriate for analysis when injected into a capillary either by optical trapping or by direct loading of the purified fraction (23–27). These sample loading methods have greatly broadened the range of analytes accessible for quantitation and have also increased the diversity of cell types suitable for analysis by microelectrophoretic methods (**Table 1**).

Table 1 Cell types used in single-cell analysis[a]

Cell Type	Analyte	Reference(s)
Synechococcus sp. (cyanobacteria) (PCC 7942)	phycobiliprotein	138
Deinococcus radiodurans (bacteria)	protein-GFP	86
Saccharomyces cerevisiae (yeast, baker's)	trisaccharide metabolism	48
	α-glucosidase I activity	48
Arabidopsis embryo (plant, rockcress)	DNA (LFY and AP2)	89
	α-glucosidase I	51, 52
	α-glucosidase II	52
	GTA	52
Spodoptera frugiperda (Fall armyworm)	diglucoside metabolism	53
	α-glucosidase I	52
	α-glucosidase II	52, 53
	native proteins	135
	GFP	134
	YFP	133
	β-adrenergic receptors	138
	GTA	52
Helix aspersa neuron (common garden snail)	amino acids	4
Planorbis corneus neuron (red ramshorn snail)	amino acids	39
	catechols	18, 37
	dopamine	18, 37–39
	serotonin	18, 37
Lymnaea stagnalis (great pond snail)	NOS metabolites	106
	NO_2^- and NO_3^-	106
	NO	107
Pleurobranchaea californica neuron (sea slug)	serotonin	44
	NO_2^- and NO_3^-	104
	NOS metabolites	105
Aplysia californica (California sea slug)		
neuron	NO	107
	NOS metabolites	105
	ascorbic acid	141
	D-amino acid peptide	56
	D-/L-glutamate	30
	serotonin	44
atrial gland	secretory vesicles	23
Xenopus laevis oocyte (African clawed frog)	IP$_3$	95, 96
	β-galactosidase activity	46
	kinases: phosphorylation of peptide substrate	17

(Continued)

Table 1 (*Continued*)

Cell Type	Analyte	Reference(s)
Mus musculus (mouse)		
Peritoneal macrophages	amino acids	34
	GSH	62
β-TC3 (mouse pancreatic β cell)	insulin	57, 58
3T3 (mouse fibroblast)	kinases: phosphorylation of peptide substrate	100
	farnseyltransferase: farnesylation of substrate	102
4T1 (mouse mammary tumor)	farnseyltransferase: farnesylation of substrate	102
NG 108–15 (mouse neuroblastoma)	GSH	61
NS1 (mouse myeloma)	mitochondria	24, 116
	cardiolipin from mitochondria	118
Embryo (mouse)	proteins	80
Cricetulus griseus (Chinese hamster)		
CHO (oocytes)	calcein-AM	139
	RNA	51, 87, 91
	DNA	87
	mitochondria	116
Mesocricetus Auratus (Syrian hamster)		
BHK-21 (baby kidney)	Ca^{2+}	95, 96
Rattus norvegicus (rat)		
PC12 (pheochromocytoma)	Ca^{2+}	95, 96
	amino acids	8, 15, 20, 39, 142
	dopamine	8, 15, 20, 142
Hippocampi cells	glutamate	36
Cardiomyocyte	DNA fragmentation	140
Islet of Langerhans	insulin	59, 60
C2C12 (myoblast)	kinases: phosphorylation of endogenous GFP substrate	101
peritoneal mast cells (RPMCs)	histamine	33
	serotonin	42, 43
R2C (testicle)	steroids: progesterone	94
RBL (basophilic leukemia)	kinases: phosphorylation of peptide substrate	100
Hepatocytes	Tryptophan	35
	GSH	35
Canis familiaris (dog)		
MDCK (Madin Darby canine kidney)	protein-GFP	143
Bos taurus (cow)		
Chromaffin cells (adrenal gland)	norepinephrine and epinephrine	40, 41

(*Continued*)

Table 1 *(Continued)*

Cell Type	Analyte	Reference(s)
Homo sapiens (human)		
Erythrocyte	reactive oxygen species	144
	glutamate	36
	reduced GSH	3, 63, 131
	hemoglobin	11, 66–69
	carbonic anhydrase	29
	glucose 6-phosphate dehydrogenase	97
K562 (erythroleukemia)	glycoprotein	83
	proteins	145
HEK (embryonic kidney)	protein-GFP	84
Jurkat (T cell leukemia)	fluorescein	12
	Oregon green	12
	NDA-derivatized amino acids	137
AML-5 (acute myloid leukemia)	calcein-AM	136
CEM-C2 (leukemia)	single nuclei: doxorubicin content	123
	acid organelles	112
CCRF-CEM (lymphoblast leukemia)	single nuclei: doxorubicin content	123
	acid organelles	112
Lymphocyte	amino acids	32
CSF lymphocytes	dopamine	38
	DOPAC	38
	uric acid	38
lymphoblasts	DNA (β-actin)	90
Neutrophils	ascorbic acid	146
AtT20 (pituitary adenoma)	gangliosides	54, 55
LNCaP (prostate cancer)	DNA (β-actin)	87
MCF-7 (breast cancer)	DNA (β-actin and ERα)	88
	protein	78
HT 29 (colon adenocarcinoma)	protein	64, 65, 72, 73, 75
	α-glucosidase I	52
	α-glucosidase II	52
	GTA	52
NK	IFN-γ	98
ΔH2–1 (osteosarcoma)	nuclei: via eGFP conjugated nuclear-tracking proteins	25, 122

[a]Abbreviations: DOPAC, dihydroxyphenylacetic acid; eGFP, enhanced green fluorescent protein; GFP, green fluorescent protein; GSH, glutathione; GTA, α-1,3-*N*-acetylgalactosaminyltransferase; NDA, naphthalene-2,3-dicarboxaldehyde; NK, natural killer; YFP, yellow fluorescent protein.

A key need in the analysis of single cells by microelectrophoresis is increased throughput. Most electrophoretic separations on single cells have been low in throughput, with fewer than 35 cells analyzed per day. Many biological studies, however, require data from large numbers (≥ 1000) of cells. Some key biological states, for example those of stem cells, are occupied by low numbers of cells, and many cells must be examined to identify these rare cell types. In other instances, understanding the diversity present within a large population of cells, for example cancer cells, is critical to understanding the dynamics of the disease and requires assessing a multitude of cells. Lastly, because cells are highly heterogeneous, analyzing only a few cells may yield misleading results, especially if the cells are extreme outliers in the population.

Both parallel and serial measurements have been explored as means of increasing the throughput of cellular analyses by microelectrophoretic methods. Dovichi and colleagues have recently constructed an array of five capillaries for parallel cellular analyses (28). Impressively, the Dovichi group also developed a juncture between two five-capillary arrays so that this system could be utilized for two-dimensional capillary electrophoresis of cellular proteins. Although this work analyzed cell homogenates, taking the next step to array-based analyses of single cells appears readily attainable. Rapid, serial analysis of single cells has been performed in both capillaries and microfluidic devices. Chen and Lillard analyzed hemoglobin and carbonic anhydrase in red blood cells at a rate of ~ 0.3 cells/min using one capillary to supply a stream of red blood cells to a second separation capillary (29). However, the highest-throughput measurements on single cells (10 cells/min) in which a separation of analytes was demonstrated are those achieved by Ramsey and colleagues in a microfluidic device (12). In this study, leukemic cells were lysed at a fluidic junction by a combination of an electric field and a detergent, followed by the electrophoretic separation of two intracellular fluorescent dyes.

Despite these advances, there is a continued need to develop strategies for high-throughput analysis of cells using microelectrophoresis. Microfluidic devices will likely be the best option for high-throughput measurements on cells not adherent to a surface, as the cells are easily moved to the separation channel; however, capillary-based measurements may be optimal for cells that are attached to a surface because the capillary can be readily transported to the immobilized cell.

2. ANALYTE DETECTION AND MEASUREMENT IN SINGLE CELLS

2.1. Amino Acids

Amino acids play important roles in cellular signaling and act as building blocks or intermediates in the synthesis of bioamines, nucleic acids, and proteins. Thus there is widespread interest in the quantitation of intracellular and intraorganelle amino acids. Currently, laser-induced fluorescence (LIF) and electrochemical detection (EC) are the most sensitive detection methods associated with CE for single-cell analysis. However, most amino acids are neither electroactive nor fluorescent. Therefore, covalent labeling with fluorescent or electroactive tags is necessary to achieve sufficient sensitivity for single-cell analyses. Naphthalene-2,3-dicarboxaldehyde (NDA) is the most common derivatizing agent used in the analysis of amino acids. In the presence of cyanide, NDA reacts with primary amine groups to produce cyano[*f*]benzoisoidole products that are both fluorescent and electroactive. Work by the Jorgenson group (4) heralded the modern era of amino acid analysis in single cells by combining NDA-derivatization with CE and open tubular liquid chromatography (OTLC). Individual neurons of *Helix aspersa* were derivatized with NDA, then loaded into a capillary and analyzed either by chromatography with EC or by CE-LIF. A total of 17 amino acids were separated by the OTLC-EC method and concentrations were estimated at 40 pmol to 40 fmol. The separation of six amino acids, including tryptophan, serine, alanine, glycine, glutamine, and aspartic acid, was achieved using CE-LIF (4).

Most recent studies utilized on-column cell lysis and derivatization with NDA prior to chemical separation. An intact single cell was loaded into the capillary by electroosmotic migration, then injected with a plug of NDA/CN⁻ for derivatization and lysis buffer for cell lysis. This method was utilized by the Liu group (30), which used micellar electrokinetic chromatography (MEKC) with LIF detection to separate glutamate enantiomers, as both L- and D-amino acids play a role in neurotransmission. Individual neurons from the abdominal ganglion of *Aplysia californica* exhibited higher levels of D-glutamate than of L-glutamate (**Figure 1a**) (30). In a separate

Micellar electrokinetic chromatography (MEKC): separation of an analyte based on the differential partitioning between a micellar phase and an aqueous phase

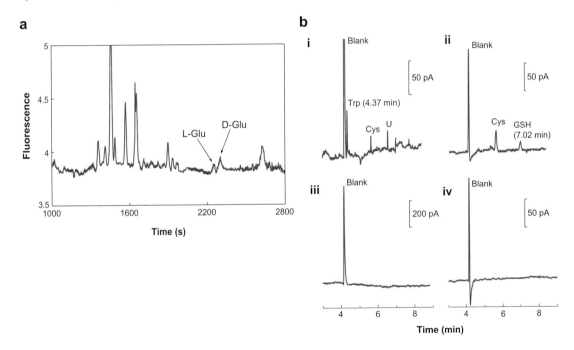

Figure 1

(*a*) Separation of D/L-Glu in a single abdominal cell from an *Aplysia* ganglion. Reprinted with permission from Reference 30. (*b*) Electropherograms of tryptophan and glutathione (GSH) from single hepatocytes. Detection of tryptophan and an unknown analyte (U) detected at the carbon fiber bundle electrode (panel *i*). Detection of GSH at the Au/Hg electrode (panel *ii*). Shown for comparison: (*iii*) Electropherograms of blank runs (no cells) at the carbon fiber bundle and (*iv*) the Au/Hg electrode. Reprinted with permission from Reference 35.

GSH: glutathione

experiment, glutamate, aspartic acid, and glycine were identified in giant dopamine neurons of *Planorbis corneus* by CE-EC (31).

Pheochromocytoma (PC12) cells, isolated from the adrenal glands of rats, have been used extensively as a model system for the study of neuronal differentiation. CE-LIF was used to measure four amino acids—alanine, glycine, glutamate, and aspartic acid—and the cysteine derivative taurine in PC12 cells. The average quantities of amino acids in single PC12 cells ranged from 200 amol to 5 fmol (8). Swanek and colleagues (31) conducted a similar study of amino acid levels in PC12 cells by CE-EC. Because background signals from the surrounding cell media contributed to the overall electrochemically detected signal, internal standards were used to aid in the quantification of the amino acids originating inside the cell.

The Jin group (32–34) analyzed the amino acid levels in different cells of the immune system. Custom-made cell injectors were used to load cells into the capillary after which the cells were lysed and then reacted with NDA. Carbon fiber microdisk array electrodes were used for EC detection of five amino acids—alanine, aspartic acid, glycine, glutamate, and serine—plus the derivative taurine in mouse peritoneal macrophages. The amino acid levels in each cell ranged from 0.3 to 6.5 fmol (34). In related work, serine, alanine, glycine, and taurine were identified in human lymphocytes, with the amounts per cell ranging from femtomoles to attomoles (32). Histamine, a chemical mediator implicated in allergic reactions derived from the amino acid histadine, was monitored in single rat peritoneal mast cells by CE. The average histamine level was 96 fmol/cell (33).

Tryptophan and glutathione (GSH) are important indicators of liver activity and can reflect liver function when measured in single liver cells. Although CE-EC is usually a very sensitive method for the detection of electroactive species such as tryptophan and GSH, these two species cannot be detected simultaneously. Because tryptophan and GSH are uniquely sensitive to different electrode materials and thus cannot be detected simultaneously using traditional methods, Jin and colleagues constructed a carbon fiber bundle–Au/Hg dual-electrode detection system. Tryptophan and GSH were identified and quantified from individual rat hepatocytes by the dual electrode (**Figure 1***b*) (35). Other compounds derivatized with NDA (e.g., cysteine, tyrosine, histamine, L-dopa, dopamine, serotonin, epinephrine, and norepinephrine) were measured and monitored simultaneously with tryptophan and GSH using this dual-electrode system.

Wang and Yeung (36) measured glutamate in human erythrocytes and single neurons without the need for labeling with NDA. A dual-enzyme on-column reaction method combined with CE-LIF was used to detect the formation of reduced nicotinamide adenine dinucleotide (NADH), the product of the reaction of glutamate dehydrogenase and glutamic pyruvic transaminase with cellular glutamate. Sub-femtomole levels of glutamate were detected in both cell types (36).

2.2. Neurotransmitters

Dopamine is an important neurotransmitter implicated in drug abuse, schizophrenia and Parkinson's disease. Determining how dopamine is synthesized, packaged, and released from single cells is essential to understanding how the brain as a whole works. Neuroscientists are particularly interested in the concentrations of dopamine in different compartments of the cell.

Early studies by Ewing's group used tapered capillaries, etched to 2–5 μm in diameter, to sample neurons for quantification of their intracellular dopamine content. These tapered capillaries were inserted into single *Planorbis corneus* neurons in order to sample ultrasmall intracellular volumes (~270 fL). Neurotransmitters, including dopamine and serotonin, were electrophoretically injected and separated by CE-EC (18, 37). Even smaller tapered capillaries, with 770-nm diameter tips, have been used to sample even smaller cell lines, such as PC12 cells. An ultrasmall capillary can sample 8% of the total volume of a PC12 cell, which can then be separated and detected by CE-EC. This sampling and detection method was used to determine the dopamine content of a typical PC12 cell (240 ± 60 μM) (15). Subsequently, electroporation was coupled to the use of the 770-nm-diameter sampling capillaries as a nondestructive method of sampling single cells (20).

Ewing and colleagues identified dopamine in human lymphocytes by performing CE-EC on whole, single-cell human lymphocytes (38). Dopamine synthesis inhibitors decreased dopamine levels in the lymphocytes, whereas the lymphocytes also showed evidence of cellular uptake as extracellular dopamine increased intracellular dopamine (38). Whole *Planorbis corneus* neurons sampled by CE-EC led to the discovery of two compartments of dopamine: readily available cytoplasmic dopamine and dopamine in storage vesicles (9). *Planorbis corneus* cells were also used with a new CE detection technique, scanning electrochemical detection. This technique allows for secondary verification of the identity of the analyte (39). Although electrochemical detection is the preferred method for dopamine detection, dopamine has also been derivatized with NDA for use in CE-LIF (8).

Norepinephrine and epinephrine from single bovine chromaffin cells have been separated by Yeung and colleagues using CE. When illuminated in UV, these analytes fluoresced in the 320–350 nm wavelength range (40). In an interesting experiment, CE was used to monitor the native fluorescence of norepinephrine and epinephrine released from single chromaffin cells following application of acetylcholine. Immediately following the sampling of extracellular norepinephrine and epinephrine, the cells were lysed and the intracellular neurotransmitter content was analyzed, revealing the relative amounts of secreted and nonsecreted neurotransmitter (41).

Another neurotransmitter, serotonin, has been examined in single cells using CE by both Yeung's and Sweedler's groups. CE with native fluorescence detection was used to quantify serotonin from rat peritoneal mast cells (RPMCs). On-capillary monitoring of the release of serotonin following the application of polymyxin B sulfate showed a 28% release of serotonin from RPMCs (42, 43). CE with native fluorescence detection was also used to separate and detect serotonin in *Aplysia californica* and *Pleurobranchaea californica* neurons following homogenization of single neurons in small vials. In these experiments, wavelength-resolved fluorescence detection, which compares fluorescence emission spectra, was used to better identify the analyte being separated (44). Recently, the Sweedler group designed a novel CE-LIF system that used a hollow cathode metal vapor laser emitting at 224 nm for fluorescence excitation. The emitted fluorescence was spectrally distributed into three wavelength channels, 250–310 nm, 310–400 nm, and >400 nm, and was collected by separate photomultiplier tubes. This new system more accurately detects, quantifies, and identifies serotonin, dopamine, and related neurotransmitters in single *Lymnaea stagnalis* neurons based on the neurons' multichannel spectral signatures (45).

2.3. Oligosaccharides

Oligosaccharide metabolism has been extensively examined in single cells by capillary electrophoresis. Precedence for oligosaccharide work in single cells was established by the development of in vitro CE-based assays for examining the enzymes responsible for sugar metabolism, such as β-galactosidase and the family of α-glucosidases (46, 48). The commercially available fluorescent probe, fluorescein di-β-D-galactopyranoside, was used to examine β-galactosidase activity in *Xenopus* oocytes microinjected with RNA encoding for β-galactosidase. CE-LIF was used to monitor fluorescein, released as a result of β-galactosidase-induced hydrolysis, from 100-pL samples microaspirated from a single oocyte (46).

Palcic and colleagues synthesized oligosaccharide substrates labeled with 5-carboxytetramethyl rhodamine (TMR) for use in examining α-glucosidase activity in vitro and in vivo (47). TMR has been extensively used to label oligosaccharides as it is pH-insensitive, easy to attach, photostable, inexpensive, easily matches with the He-Ne laser line, and has superior detection limits over other fluorophores. Yeast cells, loaded with a TMR-labeled trigluocoside, were isolated and sampled whole by CE-LIF to investigate α-glucosidase I activity (48). Confocal microscopy confirmed loading of the TMR-labeled oligosaccharide probe over several hours. Pharmacological inhibition of hydrolysis by castanospermine was used to confirm α-glucosidase I activity towards the oligosaccharide probe (48).

The lactose-derived disaccharide probe LacNAc-TMR was used by Dovichi's group to monitor α-glucosidase activity in HT-29 (human colon adenocarcinoma) cell lines. The group used CE-LIF to examine the metabolism of the LacNAc-TMR disaccharide in single HT-29 cells. Metabolism of the fluorescent oligosaccharide probe was correlated with the cell cycle by confocally imaging the cells loaded with a DNA-intercalating dye prior to lysis (49). Work with LacNAc-TMR in HT-29 cells revealed not only hydrolyzed monosaccharide products, but also the formation of monofucosylated trisaccharides, Le^x-TMR and Le^y-TMR. This discovery demonstrated the activity of Golgi-located fucosyltransferases within this cell line (47). Studies of LacNAc-TMR in HT-29 cells also showed the importance of single-cell assays as means of avoiding or eliminating sample bias. LacNAc-TMR metabolism in cellular extracts yielded an array of enzymatic reactions not seen in the single cell, suggesting that the time (1.5 h) and procedures required to prepare cellular extracts upregulated a cascade of enzymatic activity not produced or observed in whole, single cells, which are quickly lysed before CE-LIF examination (50).

More recent work by Palcic and colleagues employing micromanipulators and nanoscale reaction vessels has advanced the study of single cell oligosaccharide metabolism to new levels. The activity of three enzymes, α-glucosidase I, α-glucosidase II, and α-1,3-*N*-acetylgalactosaminyl-transferase (GTA), was probed using a TMR-labeled diglucoside substrate (DG-TMR) in Sf9 cells, HT-29 cells, and *Arabidopsis* embryos (51–53). A single cell loaded with DG-TMR substrate was placed into a nanovessel and lysed with 1 nL of buffer. Low-nanoliter aliquots of the cell lysate were examined for the hydrolytic conversion of DG-TMR to monogluco-side. An advantage of this type of assay is that multiple measurements can be made from a single cell (51–53).

Gangliosides are glycosphingolipids comprising sialic acid–containing saccharide groups attached to a fatty acid–containing sphingosine tail, known as a ceramide. Ganglioside metabolites are used as disease markers in Tay-Sachs disease and related genetic disorders. Following synthesis and characterization of TMR-labeled GM1 and its metabolites, Dovichi and colleagues used CE-LIF to measure galactosidase activity in single AtT20 cells following formalin fixing, with 11 metabolic products detected at low zeptomole levels (54, 55).

2.4. Peptides

Peptides are an important class of cellular molecule that act as neuropeptides or hormones, form during the metabolism of proteins, and play a role in immune surveillance. Although most peptides and proteins present in animals are made of L-amino acids, CE-based systems have been used to identify D-amino acid–containing peptides in single cells. The D-amino acid–containing peptide, NdWFa (NH_2-Asn-D-Trp-Phe-$CONH_2$), was identified in individual peptidergic neurons in the abdominal ganglion of *Aplysia californica* mollusks by Sweedler and colleagues (56). Matrix-assisted laser desorption/ionization imaging mass spectrometry (MALDI-MS) studies confirmed the mass and sequence of NWFa/NdWFa whereas CE-LIF-based chiral separation of fluorescamine-labeled individual neurons identified NdWFa as the dominant form of the peptide.

Insulin is a naturally occurring hormone secreted by the pancreas. An increase in blood glucose concentration stimulates secretion of insulin from pancreatic β cells; this process signals the peripheral tissues to consume the excess glucose. Because the insulin:glucose regulatory system is disrupted in diabetics, there is substantial interest in understanding the control of insulin secretion. Tong and Yeung (57) used CE-LIF to detect the native fluorescence of insulin in individual rat (RINm5F) and mouse (β-TC3) cells. The average insulin content of a single β cell was 1.6 ± 0.7 fmol. In another study, insulin release from single β-TC3 cells was quantified after on-capillary permeabilization with digitonin (58). The amount of insulin released and the amount remaining in the cell were measured simultaneously by CE-LIF. Similarly, the Kennedy group (15, 59) developed a competitive immunoassay based on the CE-LIF method to monitor the glucose-induced secretion of insulin from single rat islets of Langerhans. This method was later automated to measure insulin release with a temporal resolution of 3 s (60).

The tripeptide glutathione, γ-Glu-Cys-Gly, is the most abundant low-molecular-weight thiol in mammalian cells. Due to its biological and clinical significance in limiting cellular oxidative damage, numerous single-cell assays have been developed to measure and monitor GSH. Hogan and Yeung (3) demonstrated the on-capillary lysis and detection of GSH from human erythrocytes by CE-LIF. When erythrocytes were prelabeled with monobromobimane, a sulfhydryl-reactive fluorophore, detection limits for GSH were in the attomole-to-femtomole range. Orwar and colleagues (61) used NDA as a derivatizing agent to quantitate GSH in single mouse neuroblastoma (NG 108–15) cells. Under physiological conditions, NDA reacts rapidly with GSH without the addition of CN^-, permitting the detection of low femtomole levels of GSH in single cells by CE-LIF.

The Jin group introduced a derivatization-free CE-EC system for the direct analysis of GSH in single cells. They utilized a gold-mercury amalgam microelectrode for end-column amperometric detection following CE separation. Average cellular GSH levels were 5.8 fmol for single mouse peritoneal macrophages (62), 11 fmol for single rat hepatocytes (35), and 0.1 fmol for single human erythrocytes (63). As described above, a carbon fiber bundle–Au/Hg dual electrode was used to simultaneously quantify both tryptophan and GSH in single rat hepatocytes (**Figure 1***b*) (35).

2.5. Proteins

Recently, proteins have become one of the most investigated analyte groups in single-cell CE-based analysis. The increased focus on proteins is due to the biological and clinical significance of many classes of proteins, as well as to the complexity and scope of their action in cells. A typical eukaryotic somatic cell expresses about 10,000 different proteins averaging 600 copies each (6); therefore, extremely sensitive detection techniques and advanced separation methods are necessary to analyze the minute amount of protein present in a single cell. Capillary-sieving electrophoresis and MEKC are the most successful separation methods for use with proteins and are most often used in combination with capillaries possessing covalent or dynamic surface coatings. Although this chapter cannot cover the entire scope of single-cell proteomics, excellent reviews on this topic are available in this volume and elsewhere (64, 65). This section focuses predominantly on the different fluorescence-labeling methods for protein detection.

2.5.1. Detection of proteins by their native fluorescence. Hemoglobin and carbonic anhydrase were the first proteins analyzed from individual erythrocytes (11). Because they are at extremely high concentrations in erythrocytes, they can be detected using their native fluorescence. A single erythrocyte was injected into a capillary with the aid of negative pressure. Cells were then lysed by exposure to the hypotonic electrophoretic buffer and their contents were separated and detected by CE-LIF without a sieving matrix. Lillard and colleagues (66) employed CE-LIF to study hemoglobin variants in human erythrocytes. In this study, hemoglobin from fetal, normal adult, and adult A_1 red blood cells was analyzed. Fluorocarbon-coated capillaries and fluorocarbon surfactants were used to separate the α, β, and glycated polypeptide chains from denatured hemoglobin molecules without the use of a sieving matrix. Erythrocytes from diabetics showed higher glycated hemoglobin levels than did erythrocytes from nondiabetics. In a subsequent study, Lillard and Yeung (67) coupled capillary isoelectric focusing with LIF to study hemoglobin variants in human erythrocytes. Hofstadler and colleagues (68) used CE coupled to mass spectrometry to analyze hemoglobin from human erythrocytes. They detected both the α and β chains of hemoglobin by ESI-FTICR-MS. The Ren group (69) employed CE with chemiluminescence detection to study hemoglobin from human erythrocytes. The chemiluminescence detection was based on the catalytic effects of the iron group in hemoglobin on the luminol-hydrogen peroxide reaction.

GFP: green fluorescent protein

2.5.2. Detection of proteins by covalent labeling with fluorophores. Most proteins cannot be detected by high-sensitivity LIF because of the low abundance of tryptophan and tyrosine in cellular proteins (70). Therefore, amine-reactive fluorescent tags are commonly used to covalently attach fluorophores to both the ε-amine of lysine and the amino terminus of proteins. Lysine is one of the most abundant amino acids and at least one lysine is present in a majority of proteins. Because the covalently attached fluorophores possess substantially better absorption coefficients and quantum efficiencies than the native protein fluorophores, detection limits are improved by several orders of magnitude, permitting the detection of low copy number proteins.

The Dovichi group has extensively studied the proteins of HT-29 cells. In most of their studies, cell injection was performed with a multipurpose single-cell injector (71) followed by cell lysis and fluorescent labeling of cellular proteins. Under those conditions, the fluorophore 3-(2-furoyl)quinoline-2-carboxaldehyde reacted with all available free amino groups (72). Proteins were separated by CE with LIF detection in a postcolumn sheath flow cuvette. This one-dimensional separation with a surfactant-containing buffer yielded separation efficiencies of greater than 400,000 theoretical plates with resolution of 30 components from a single cell (73). The same single-cell approach was used to study protein expression in a *Caenorhabditis elegans* zygote (74). The protein profile of individual HT-29 cells was also generated using a similar one-dimensional CE separation with a sieving matrix (64, 65).

CE-based assays of single cells were also used to examine variations in the protein profile of individual cells resulting from changes in cell physiology. The Dovichi group observed that cell-to-cell variation in the protein profile correlated with the phase of the cell cycle. On average, 60% of the cell-to-cell variation in protein expression was due to differences in the cell cycle state (**Figure 2a**). For single cells in the same cell cycle state, the variability in protein expression was less than 30% (75). Similar patterns were observed for green fluorescent protein (GFP) expression in 4T1 sister cells (**Figure 2b**) (76).

Because the separation power of one-dimensional electrophoresis is insufficient to resolve the large number of proteins present in a single cell, a 2D-CE method was developed to study the protein content of single cells. The method developed by Dovichi and colleagues utilized capillary sieving electrophoresis in the first dimension and capillary-based MEKC separation in the second dimension (77, 78). This method provides rapid and reproducible separations up to a 600-spot capacity (79). In one study, different protein expression patterns were observed for normal and drug-treated MCF-7 and MC3T3-E1 cells (77). Cell cycle–dependent characterization of single MCF-7 cells in G1 and G2/M cell cycles were also studied using 2D-CE. Cells of different phases showed a 2.5x increase in variability over cells of the same phase. Over 100 components were resolved by 2D-CE with a 260-spot capacity (80).

a

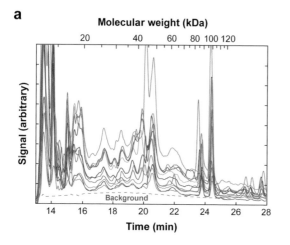

Molecular weight (kDa)

Signal (arbitrary)

Background

Time (min)

b

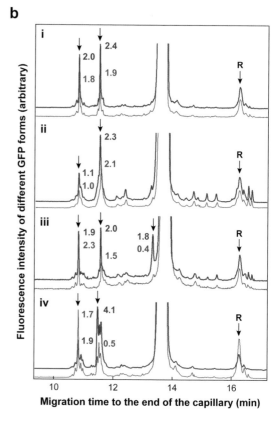

Fluorescence intensity of different GFP forms (arbitrary)

Migration time to the end of the capillary (min)

Figure 2

(*a*) Protein profile of 11 different HT-29 human cancer cells analyzed by capillary-sieving electrophoresis with laser-induced fluorescence. The dashed curve represents the background signal of the cellular supernatant. Each trace represents a different single cell. Reproduced with permission from Reference 75. (*b*) Differences in GFP (green fluorescent protein) expression patterns between 4T1 sister cells stably transfected with GFP. Each panel shows an electropherogram of a different pair of sister cells (*red, blue traces*). Numbers indicate relative intensities of adjacent peaks. Reproduced with permission from Reference 76.

Additionally, cell-to-cell heterogeneity was observed in a cell cycle phase study of mouse monocytic-macrophage cells (RAW 264.7) (81). See also the Dovichi group's article on chemical cytometry in this volume (82).

2.5.3. Detection of single proteins in a single cell. Identification of a single specific protein from among the many thousands of proteins within a single cell requires either that the protein be at a very high concentration relative to other proteins or that the protein be tagged to distinguish it from other proteins. Proteins can be selectively tagged with an antibody or with a molecularly engineered fluorophore, i.e., GFP. These two strategies have been pursued by several different investigators. Xiao and colleagues (83) created a CE-based immunoassay to measure P-glycoprotein (PGP) levels in normal and multi-drug-resistant K562 cells. Single cells were loaded with a goat IgG antibody against PGP and a fluorescein-tagged antibody directed against goat IgG. The cells were then were analyzed by CE-LIF to detect the PGP-antibody complex.

The genetic fusion of GFP to a protein of interest followed by expression of the GFP-protein construct in cells is used throughout biology to examine gene expression, protein mobility and localization, and other aspects of protein function. Malek and Khaledi (84) utilized this technology to examine GFP expression in HEK293 cells by CE-LIF. GFP was detected at very high sensitivity with detection limits of 50 zmol in single cells. Brown and Audet employed a novel approach to estimating the cell-sampling efficiency of the single-cell laser lysis CE-LIF system. They investigated the effects of laser focus placement and laser pulse energies on the sampling of GFP expressed in HEX 293T cells (16). Hu and colleagues combined fluorescence microscopy and chemical cytometry to measure total GFP fluorescence from single cells. In this study, total fluorescence of GFP expressed by 4T1 cells was first measured by an inverted fluorescence microscope, then the cells were analyzed by CE-LIF. This investigation demonstrated that the total fluorescence intensity depended on the number of intracellular fluorophores and was independent of the cell size, cell shape, and intracellular GFP distribution (85). The expression pattern of GFP was observed in dividing cells by two-channel CE-LIF to show that GFP was not equally distributed to the daughter cells (76). Turner and colleagues performed CE with ultrasensitive fluorescence detection of GFP from a single bacterium (*Deinococcus radiodurans*) by CE-LIF. The system had detection limits of 100 ymol (60 GFP copies) (86).

2.6. RNA

Measurement of RNA from single cells is generally performed using an amplification step known as reverse transcription–polymerase chain reaction (RT-PCR), which also converts the RNA into complementary DNA (cDNA). Single-cell (SC) RT-PCR was performed by Lillard and Zabzdyr in vials containing a single cell with low μL volumes of PCR–nucleotide mixture. The first reported use of CE-LIF and SC-RT-PCR investigated β-actin expression in human prostate carcinoma (LNCaP) cells (87). A gene-specific primer pair for β-actin was used to selectively amplify β-actin mRNA from the cell, while hydroxypropylmethylcellulose and ethidium bromide were used as a sieving matrix and a fluorescent label for DNA, respectively (87). Lillard and Zabzdyr (88) also measured the expression of two genes, β-actin and αER, in single human breast cancer (MCF-7) cells using CE-LIF and SC-RT-PCR. In similar experiments, Lu and collaborators measured APETELA2 (AP2) and LEAFY (LFY) gene expression in the shoot apical meristem, leaf, root, and stem of single *Arabidopsis* cells in response to the growth regulator gibberellic acid (89). In this instance, YO-PRO1 replaced ethidium bromide as the DNA label of choice.

The Yeung group (90) took SC-RT-PCR to a new level by developing an on-line capillary PCR method with CE-LIF, all on a single capillary. A 100-cm-long capillary,

Chemical cytometry: uses high-sensitivity analytical tools, including mass spectrometry, electrochemistry, and capillary separation methods, to chemically characterize single cells

RT-PCR: reverse transcription–polymerase chain reaction

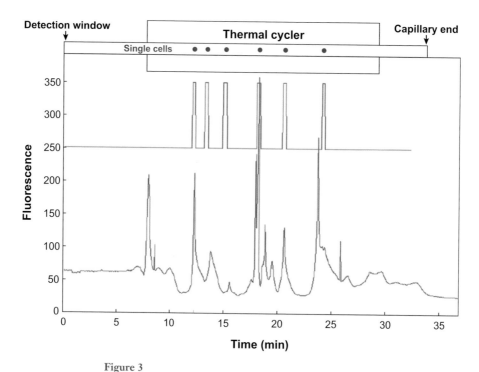

Figure 3

On-line capillary polymerase chain reaction with capillary electrophoresis–laser-induced fluorescence. An electropherogram (*bottom*) shows amplified β-actin DNA from single cells. The spacing and position of the DNA peaks in the electropherogram correspond to the position of single cells (*circles*) within the capillary, as calculated by migration velocity and starting position of the cell prior to reaction. Reproduced with permission from Reference 90.

placed into a PCR thermocycler, was loaded with human lymphoblast cells mixed with β-actin primer, SYBR Green I, and the PCR mixture. The cells were monitored by fluorescence microscopy to ensure sufficient spacing between individual cells in the capillary, after which RT-PCR was initiated. Following RT-PCR, the sieving buffer was added, an electric field was applied, and the amplified DNA encoding β-actin was measured. Discrete fluorescent peaks in the capillary corresponded to each cell originally loaded into the capillary, as shown in **Figure 3** (90).

Lillard & Han (91) performed direct measurement of RNA from single cells without conversion to DNA and the consequent amplification. Ethidium bromide–labeled RNA fragments were first detected using CE-LIF in single Chinese hamster ovary (CHO-K1) cells. Separation of an RNA ladder was used to characterize sieving performance while RNase I was used to verify that only RNA was detected. RNA degradation, following application of hydrogen peroxide, was observed in single cells. In subsequent work, Lillard's group tested alternative fluorophores (SYBR Green I and SYBR Gold) in labeling of RNA in CHO-K1 cells (92). CE-LIF was also used to examine RNA expression in CHO-K1 cells in different stages of the cell cycle (M, G1, S, G2). The total amount of RNA increased over each phase; however, the rate of increase of individual RNA sequences varied with the cell cycle state (93).

2.7. Miscellaneous

2.7.1. Steroids. Malek and Khaledi measured steroidal compounds in single R2C cells (Leydig tumor cells from rat testicle) (94). DMSO was used to permeabilize the cell membrane, allowing entry of dansylhydrazine into the cell. Within the cell, dansylhydrazine reacted with progesterone, converting it to a fluorescent compound detectable by CE-LIF. Progesterone in the R2C cell was calculated at 0.62 μM.

2.7.2. Inositol-1,4,5-triphosphate. Inositol-1,4,5-triphosphate (IP_3) is a second messenger that is generated in response to the binding of many hormones, growth factors, and neurotransmitters to their cognate receptors on the plasma membrane of cells. IP_3 binds to receptor/channels to release Ca^{2+} from the endoplasmic reticulum (ER) into the cytosol. Allbritton and colleagues developed a CE-based system to measure IP_3 in single cells (95). A permeabilized PC12 or BHK-21 cell loaded with the calcium-sensitive dye mag-fura-2 was used at the outlet of the capillary as a sensor for detection of IP_3. Cytoplasm sampled from a *Xenopus* oocyte was separated by CE and the cell components were delivered from the capillary onto the "detector" cell. When IP_3 eluted onto the detector cell, Ca^{2+} was released from the ER, altering the fluorescence of the mag-fura-2 contained in the ER. Thus, the fluorescence of the detector cell was used to quantitate the concentration of IP_3 in the cytoplasmic sample. **Figure 4** shows a schematic of this technique. IP_3 was measured in *Xenopus* oocyte both with and without the addition of the agonist lysophosphatidic acid to the oocyte. The basal concentration of IP_3 was 40 nM, rising to 650 nM after agonist addition. IP_3 production was also measured at varying times following sperm infusion with a *Xenopus laevis* egg. These measurements demonstrated that a wave of increased IP_3 accompanied the Ca^{2+} fertilization wave as it traversed the cell (96).

2.7.3. Antigen-antibody complexes. In an interesting experiment combining nanomaterials with single cells, antibody-coated latex particles were used to bind glucose 6–phosphate dehydrogenase proteins and act as intracellular protein counters. After microinjection of the antibody-coated particles into erythrocytes, CE with light scattering as a detection scheme was used to monitor the resulting antigen-antibody particle complexes. The performance of this CE-based assay was comparable to more traditional fluorescence assays, with a detection limit of 620 molecules of glucose 6–phosphate dehydrogenase (97). In a separate experiment, CE with LIF was used to detect IFN-γ in natural killer (NK) cells. Electroporation was used to permeabilize the NK cells, allowing fluorescently-labeled anti-IFN-γ antibodies entry into the cells. Zeptomole detection limits for the antigen-antibody complex were attained and new alloantigens of IFN-γ were discovered (98).

3. MONITORING ENZYMATIC ACTIVITY IN SINGLE CELLS

3.1. Kinases

Kinases play an important role in numerous cell signaling pathways and have been implicated in numerous diseases, including cancer. Examination of kinase activity is best

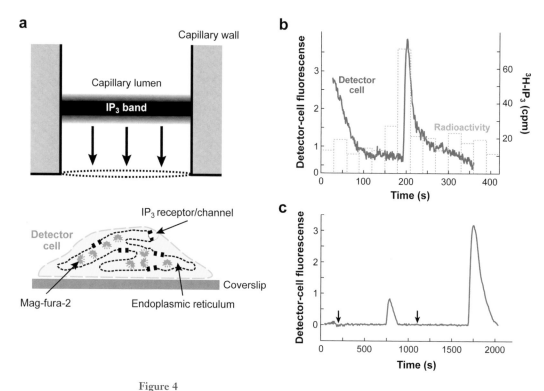

Figure 4

Inositol-1,4,5-triphosphate (IP$_3$) detection using capillary electrophoresis (CE) and a detector cell. (*a*) Schematic of detector cell–CE system. IP$_3$ eluted onto a detector cell causes Ca^{2+} release from the endoplasmic reticulum of the detector cell, triggering an increase in mag-fura-2 fluorescence. (*b*) Measurement of IP$_3$ elution time by a detector cell (mag-fura-2 fluorescence) (*line graph*) and radioactivity (*dotted bar graph*). (*c*) Detection of nanomolar IP$_3$ by a detector cell. At the first arrow, 10 nL of 100 nM IP$_3$ is eluted onto a detector cell. At the second arrow, 10 nL of 2.5 μM IP$_3$ is eluted onto the same detector cell. Reproduced with permission from Reference 95.

performed when these enzymes are in their native state inside cells; hence a capillary electrophoresis–based assay relying on designer substrate peptides was developed to monitor kinase activity in single cells (17). Fluorescein-labeled peptides were selected and synthesized as substrates for a number of different kinases, including protein kinase C, protein kinase A, CamKII, and cdc2 kinase. After microinjection and incubation of these substrates into *Xenopus* oocytes, CE-LIF was used to separate and measure the phosphorylation of the peptide substrates. Using this method, kinase activity was assessed based on the rate and percentage of phosphorylation in response to different stimuli (17). Protein kinase C activity was assessed using CE-LIF following stimulation with ATP or IgE-antigen in both rat basophilic leukemia (RBL) and mouse embryonic fibroblast (3T3) cells. The effect of the kinase inhibitor staurosporin on overall kinase activity was evaluated using CE-LIF with numerous fluorescently labeled peptide substrates loaded simultaneously into single RBL cells (99, 100).

The Krylov group used similar CE-LIF methodology to examine the phosphorylation of a molecularly engineered substrate in C2C12 rat myoblast cells (101). The substrate was composed of GFP coupled to a series of protein kinase A phosphorylation sites. When the cells were stimulated with 8 bromo–cAMP, which activates protein kinase A, only the fully phosphorylated form of the substrate was detected. Although this method eliminated the substrate-loading step, it was complicated somewhat by the nearly complete phosphorylation of the substrate in unstimulated cells (101).

3.2. Farnesyl Transferases

Farnesylation, the transfer of a 15-carbon farnesyl moiety to the sulfhydryl group of a cysteine, is a posttranslational modification that increases the hydrophobicity of proteins. Protein farnesyltransferase (PFTase) catalyzes this irreversible modification. Ras protein, which is implicated in 30% of cancers, is farnesylated by PFTase. Drugs targeting the farnesylation of ras are currently being screened for their effectiveness in treating cancer, necessitating the development of new PFTase assays. CE-LIF was used to investigate the PFTase activity in mouse mammary tumor (4T1) cells and 3T3 cells loaded with a ras-derived fluorescently labeled pentapeptide substrate. Confocal microscopy was used to confirm substrate loading, while CE-LIF was used to monitor fluorescent metabolic products of the peptide. The pentapeptide substrate was converted to products; however, none of the products matched the products expected after farnesylation, proteolysis, or methylation (102, 103).

3.3. Nitric Oxide Synthase via Metabolite Detection

Nitric oxide synthase (NOS) is responsible for producing nitric oxide (NO), a highly reactive biological molecule responsible for vasodilation, neurotransmission, and the inhibition of platelet aggregation. Because NO is quickly consumed and metabolized due to its high reactivity, NOS activity is monitored by measuring the metabolites NO_2^- and NO_3^-. Capillary electrophoresis with direct UV absorbance was used to detect NO_2^- and NO_3^- in *Pleurobranchaea californica* neurons. CE studies confirmed that high levels (2 and 12 mM, respectively) of both NO_2^- and NO_3^- existed in neurons histochemically positive for NADH-diaphorase, a signature of NOS activity. There were no detectable levels of NO_2^- or NO_3^- in neurons lacking NADH-diaphorase activity (104). CE-LIF was also used to monitor the metabolites from the arginine-citrate cycle, which is modulated by NO. L-citrulline, L-arginine, L-argininosuccinate, L-ornithine, and L-arginine phosphate were measured in *Pleurobranchaea californica* and *Aplysia californica* neurons, using fluorescamine labeling (105). *Lymnaea stagnalis* neurons were used to correlate NOS expression with metabolite presence (L-citrulline, L-arginine, L-argininosuccinate, NO_2^- and NO_3^-), using CE with LIF (106).

In a novel experiment, the fluorescent NO indicator DAF-2 was used to monitor NO in single *Aplysia californica* and *Lymnaea stagnalis* neurons (107). Because DAF-2–based measurements can be altered by ascorbic acid, which occurs at high levels in neurons, ascorbic acid oxidase was added to remove endogenous ascorbic acid.

CE-LIF was then used to separate the DAF-2-NO adjunct from DAF-2 bound to dihydroxyascorbic acid (product of the reaction of ascorbic acid oxidase with ascorbic acid) (107).

4. ANALYSIS OF SUBCELLULAR MACROCOMPLEXES

Capillary electrophoresis is suitable for the analysis of samples ranging from attoliter to nanoliter volumes. Most eukaryotic cells measure several picoliters in volume, with larger cell types such as eggs and oocytes in the nanoliter range. Within cells, however, the volume of many organelles can be measured in attoliters or femtoliters; these compartments are therefore suitable for analysis by CE. The use of CE in examining acidic organelles, mitochondria, nuclei, and secretory vesicles is described in this section.

4.1. Acidic Organelles

The effect of subcellular drug distribution on cytotoxicity and drug efficacy is the driving force behind the Arriaga group's pioneering methods for the separation and detection of acidic organelles from single cells (108–113). The group's early work perfected the detection of the antitumor drug doxorubicin (DOX) and its metabolites in single NS-1 cells by MEKC with LIF (108). The use of MEKC-LIF and on-capillary lysis of human leukemic cells (CEM-C2 and CCRF-CEM) treated with DOX revealed a unique metabolic profile for each cell line (109). The ultimate goal, to interrogate individual organelles, was achieved when acidic organelles, identified by their accumulation of fluorescent nanospheres, were separated from crude subcellular fractions. Thereafter, their DOX content was quantified by CE-LIF (110). CE-LIF was also used to detect DOX and a fluorescent reporter of reactive oxygen species within acidic organelles (111). Arriaga and colleagues have also utilized the ratiometric dye fluorescein tetramethylrhodium dextran to monitor the pH of acidic organelles from individual CEM-C2 and CCRF-CEM cells (112). Recently, this group measured the maturation and DOX content of acidic organelles from individual CEM-C2 and CCRF-CEM cells using Alexa Fluor 488 Dextran (113).

4.2. Mitochondria

Arriaga and colleagues have also used CE-LIF to monitor mitochondria from single cells. CE-LIF was used to measure DOX in mitochondria labeled with MitoTracker Green (114, 115) and to study the electrophoretic mobility of individual mitochondria from mouse hybridoma (NS1) and CHO cells (116, 117). Interestingly, these studies showed that there are many different types, shapes, and sizes of mitochondria, which may have different mobilities within a single cell (116). A schematic of cell lysis and release of mitochondria from single cells is shown in **Figure 5**. In an extension of this experiment, the Arriaga group examined the chemical composition of mitochondria, particularly the concentration of cardiolipin (diphosphatidylglycerol) therein. CE-LIF was used to quantify the amount of cardiolipin in mitochondria

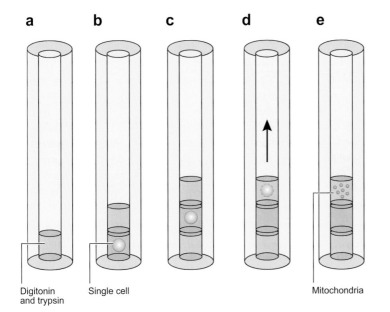

a b c d e

Digitonin
and trypsin

Single cell

Mitochondria

Figure 5

Cell lysis and release of mitochondria from single cells. (*a*) Digitonin and trypsin are hydrodynamically injected into the capillary. (*b*) A plug containing a single cell is loaded into the capillary, followed by (*c*) another plug of digitonin and trypsin. (*d*) Electrophoretic flow is used to move the cell into the first plug, resulting in permeation and proteolysis of the cell. (*e*) The mitochondria are released from the cell. Reproduced with permission from Reference 120.

from NS1 cells, using 10-nonyl acridine orange (NAO), a fluorophore which binds cardiolipin (24, 118, 119). Recently, a mitochondria-targeted form of DsRed2, a fluorescent protein, was used to label mitochondria in single osteosarcoma 143B cells. This targeted DsRed2 localized to the mitochondrial matrix; however, the protein leaked from the mitochondria in the presence of microchondrial disruption. When used in conjunction with NAO, intact DsRed2-labeled mitochondria were separated from damaged NAO-labeled mitochondria and mitochondrial fragments by CE-LIF (120). In a separate experiment, MEKC-LIF was used to detect the superoxide probe hydroethidine. Superoxide generation was measured in mitochondria isolated from rat skeletal tissue (121).

4.3. Nuclei

The nuclear-targeted protein nuDsRed2 and the plasma membrane–bound farnesylated enhanced green fluorescent protein (EGFP) were used as fluorescent probes to monitor the interaction of the nucleus and the plasma membrane when ΔH2–1 cells were treated with digitonin. CE-LIF of the nuclei demonstrated the absence of EGFP, demonstrating that the plasma membrane was separated from the nuclei (25). Subsequent work used two different EGFP-labeled human aminoacyl–tRNA synthetases (aaRSs), expressed in human-derived ΔH2–1 osteosarcoma cells, to examine subcellular localization of proteins within the cell, particularly the nucleus. CE-LIF of the cytosolic and nuclear fractions revealed that only one EGFP-aaRS localized in the nucleus, along with the nuclear marker nuDsRed (122).

In a separate experiment, CE-LIF was used to quantify DOX in single nuclei sampled from human leukemia CEM-C2 cells and CCRF-CEM cells. Cells were lysed with SDS and the isolated nuclei were loaded into the capillary, eliminating the problem of detecting DOX in other cell compartments (123).

4.4. Secretory Vesicles

Secretory vesicles are used by cells to store biological molecules, such as neurotransmitters, for immediate delivery after a particular signal or trigger. Capillary electrophoresis was used to examine the contents of secretory vesicles obtained from the atrial gland of *Aplysia californica*. Single attoliter-sized secretory vesicles were optically trapped and loaded into a capillary. The vesicles were lysed and fluorescently labeled in the capillary with NDA prior to separation. CE-LIF revealed the presence of numerous low-mass compounds, including the dominant component, taurine, inside single secretory vesicles (23).

5. MICROCHIP CAPILLARY ELECTROPHORESIS

Implementation of chemical cytometry in the microfluidic format confers several advantages associated with lab-on-a-chip devices. The small footprint of microdevices and their ability to control cell movement enables high-throughput designs with scores of separation channels. With further refinement of these technologies, it should be possible to analyze thousands of cells per hour using multiplexed continuous flow methods like those described below. Because the size and shape of microchannels can be engineered for a specific task, cells can be lysed with minimal dispersion of pL content volumes, producing high-efficiency separations. Automation of the sample-handling steps inside a microdevice has great potential for eliminating many of the labor-intensive procedures associated with analyzing the contents of single cells. The successful realization of single-cell CE on-chip is tantalizing, and a number of reviews have already touched on the subject of single cells in microchips (124–129), with two devoted specifically to progress in chemical cytometry (13, 130). This work is highly interdisciplinary, requiring expertise in engineering, optics, chemistry, and cell biology; and there is a need for significant improvement in the manufacture of working devices. One of the most challenging aspects of performing single-cell cytometry in a microfluidic format is the strategy used for positioning a cell at the time of lysis. The positioning method varies greatly among different groups and impacts the analytes that can be measured. In the following sections we discuss five different methods of cell positioning for lysis, summarizing the setup, method, cells and analytes, and results.

5.1. Docking on Channel Walls

Allowing a cell to settle on a solid wall, then lysing it with SDS or an electric field can serve as a quick proof of concept without sophisticated fabrication techniques. In one study by Gao and colleagues, a low electric potential was used to position a cell in a 12×48 μm glass separation channel. NDA-derivatized GSH and reactive oxygen species were separated from erythrocytes after lysis in 40 ms using a field strength of 280 V/cm (131). The separation efficiency for GSH was five times better than that obtained by pinched injection. A similar strategy of immobilizing cells on a channel wall was also used by Shi and collaborators to measure neurotransmitters in a (PC12) pheochromocytoma cell line (132).

5.2. Positioning Cells in Traps

An engineered weir or obstruction at the start of a separation channel can be used to successfully integrate trapping, lysis, and separation in a single microchip design. With the cell aligned for optimal injection of its contents, the operator is free to switch buffer solutions and/or apply potentials to a separation channel to initiate the analysis process. Ros and colleagues (133) used optical tweezers to move *Spodoptera frugiperda* (Sf9) cells to an injection cross in a chip with 95×6 µm channels and $\sim 7.4 \times \sim 2.2$ µm containment posts on three sides. After trapping the cell, SDS or an electrical field was used for lysis. When Sf9 cells expressing GFP and YFP proteins were lysed with 1% SDS and an electric field was applied, two broad unresolved peaks were observed using LIF. When cells expressing a single GFP protein were lysed by application of 1250 V/cm for more than 50 ms, then separated at 830 V/cm, a single sharp peak with tailing was observed, followed by an increase in the baseline fluorescence. The same group also used UV-LIF with Sf9 cells for label-free detection. Black carbon was incorporated into the PDMS microchips to increase the sensitivity (134, 135).

Munce and colleagues (136) used optical tweezers for selection and transportation of acute myloid leukemic (AML) cells on chip. Cells were incubated with calcein AM and Hoechst dyes in a chamber and docked in tapered 10-µm microfluidic channels. Application of a 300-V/cm electric field resulted in cell lysis in <0.3 s, followed by electrophoresis of the cell contents into a 10×12 µm channel. Four unidentified peaks were seen in the electropherogram, possibly due to hydrolysis of the calcein dye by the cell. With four channels, a throughput of 24 cells/h was achieved.

5.3. Immobilizing Cells with Valves

A technically innovative area of cell manipulation is the use of valves to control cell and reagent flows. Some of the best separations have been obtained with this technique. In 2004, the Zare group (137) revealed a three-state PDMS valve design (**Figure 6a**) to trap an individual Jurkat T cell in a 70-pL chamber. Because the volume of a typical Jurkat T cell is 1 pL, the chamber results in a 70-fold dilution of cell contents. Lysis reagents, followed by derivatization reagents, were metered into the cell chamber through a partially closed valve acting as a "picopipette" and were allowed to react with the cell. After filling the channels with electrophoretic buffer, the valve was opened and MEKC was initiated. The separation quality for the detected amino acids was improved relative to that of the analytes detected when cell traps and docking were employed.

With a similar valve design, Huang and colleagues (138) implemented sophisti-cated single-molecule detection optics to increase the sensitivity for target analytes. They examined the number of β-adrenergic receptors expressed in Sf9 cells. The N-terminus of the receptors was tagged with FLAG and the Cy5-labeled monoclonal

a

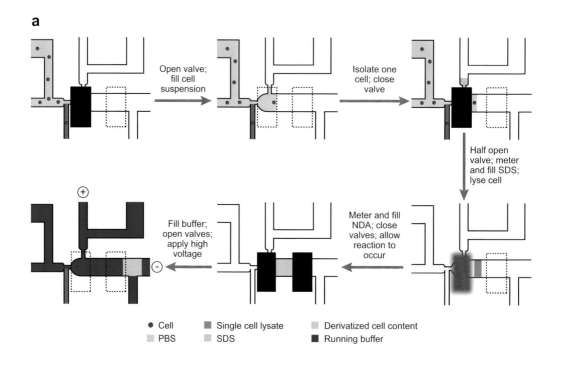

Open valve; fill cell suspension

Isolate one cell; close valve

Half open valve; meter and fill SDS; lyse cell

Fill buffer; open valves; apply high voltage

Meter and fill NDA; close valves; allow reaction to occur

⊕

⊖

- ● Cell
- ▨ PBS
- ▨ Single cell lysate
- ▨ SDS
- ▨ Derivatized cell content
- ■ Running buffer

b

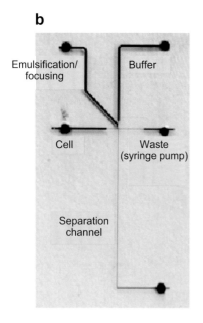

Emulsification/
focusing

Buffer

Cell

Waste
(syringe pump)

Separation
channel

c

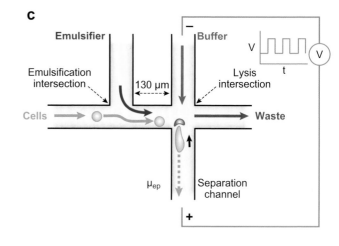

Emulsifier

Buffer

Emulsification
intersection

130 μm

Lysis
intersection

Cells

Waste

μ_{ep}

Separation
channel

−

+

V

t

V

antibody was used as a fluorescent label. In a study of five cells, they found between 2000 and 60,000 receptors per cell. The variation is likely due to the nature of virus infection in the transfection process. In another experiment, the authors studied phycobiliprotein complexes in cyanobacterial cells grown in different nitrogen environments. Up to three cells at a time were lysed and the contents separated. Cells grown in a nitrogen-rich environment had much higher levels of phycobiliprotein.

A drawback to some of the above-mentioned methods is the use of mannitol or a similar reagent as a replacement for salt in the cell buffer. Because separations must be performed partially in the cell buffer, this permits the separation buffer to be low in ionic strength and avoid excessive joule heating. However, while the isotonic conditions prevent the cells from bursting, they also induce cell death or severe damage, resulting in alteration of many analytes within the cells.

5.4. Positioning Cells in a Continuous Flow Stream

The continuous single cell lysis and separation in a modified cross channel was an exciting development by McClain and coworkers (12) in 2003. The glass chip design (shown in **Figure 6b,c**) introduced cells by suction into a tapered channel perpendicular to the separation channel. At 130 μm before the cross area, a focusing channel confined the cells into a single file near the separation side of the channel and added an emulsification reagent. At the 30×30 μm cross, the cells were subjected to an AC field that served to rapidly lyse (<33 ms) and inject the contents downstream towards a detector. Analysis of Oregon Green and carboxyfluorescein loaded into cells demonstrated that efficient injection of the cellular contents into the separation channel was obtained when the cell velocity, driven by hydrodynamic flow, was less than the electrophoretic velocity of the injected analyte. The throughput was 7–12 cells/min for up to 45 min. The separation performance was excellent when compared with other separations of single-cell contents on-chip. In addition, a physiologic buffer was used to keep the cells in an unstressed state before the rapid lysis event.

In another twist on this strategy, Wang and coworkers (139) recently showed lysis of CHO cells in a PDMS cross channel with perpendicular hydrodynamic and electric field driven flows. Detection of calcein AM products was reported at rates up to

---

Figure 6

(*a*) Stepwise diagram showing the process for handling single cells with a valve system. Valves are operated to trap a single cell in the chamber. Sodium dodecyl sulfate (SDS) and a dye are metered using the "picopipette" and three-state valve system. Valves are opened and an electrical potential is applied to separate the derivatized cell contents. Reproduced with permission from Reference 137. (*b*) Image of continuous-flow separation microchip and (*c*) schematic showing its operation. The cells are carried by hydrodynamic flow (*solid arrows*) toward the waste. At the focusing junction, emulsifier is introduced. Cells are lysed by square waveform potential at the lysis intersection and the contents with high mass-to-charge ratio travel down the separation channel while larger organelles continue toward the waste. Abbreviation: NDA, naphthalene-2,3-dicarboxaldehyde. Reproduced with permission from Reference 12.

85 cells/min. Although the throughput is about 10× higher than the study mentioned above, it was accomplished using a buffer containing 250 mM sucrose and 10 mM phosphate to minimize joule heating, which may be unsuitable for the analysis of many cellular analytes.

5.5. Retarding Cell Movement with a Viscous Medium

Kleparnik and colleagues used a viscous medium to retard cell movement prior to lysis (140). Microchannels were fabricated on a CD-like plastic disc and then filled with polyacrylamide before injection of cardiomyocytes. Cells were moved into the separation cross by vacuum and movement was retarded by the polymer sieve. In each experiment, a cell was lysed in an alkaline environment for 3 min. The released DNA was separated with 60 V/cm in a 2% polyacrylamide sieving solution and the ethidium bromide–labeled contents were detected with LIF. The effect of incubation of the cells with 30 μm doxorubicin was investigated. Increased doxorubicin incubation times resulted in poorly resolved DNA fragment peaks, thought to be caused by the onset of cellular necrosis.

SUMMARY POINTS

1. Many strategies for cell lysis and loading into microchannels have been developed; these methods are suitable for all almost all conceivable cellular analytes.

2. Microelectrophoretic methods have been successfully combined with a variety of detection methods (e.g., LIF, absorbance, EC, light scatter, mass spectrometry) to quantitate analytes in single cells.

3. A large number of analytes can be measured in single cells, including neurotransmitters, amino acids, oligosaccharides, nucleotide polymers, proteins, and a variety of metabolites.

4. Macromolecular entities, such as antigen-antibody complexes, as well as organelles can be sampled from single cells and electrophoretically separated.

5. Chemical labeling of endogenous cellular analytes such as proteins, steroids, peptides, and metabolites can be performed efficiently either before or after cell lysis and yields excellent detection sensitivity.

6. Introduction of nonnative molecules such as enzyme substrates or antibodies into cells permits the measurement of many cellular attributes such as enzymatic activity and protein concentration.

7. Hundreds of components from a single cell can be resolved by 2D-CE with LIF.

8. The highest throughput for single cell analysis (10 × 80 cells/min) has been achieved on microfluidic devices.

FUTURE ISSUES

1. The vast majority of small molecule analytes within cells cannot be detected by current methods. There is tremendous opportunity for microelectrophoretic techniques to close this gap through the development of new chemical labeling and identification methods.

2. Current labeling and separation techniques make monitoring and quantifying specific proteins in cells difficult. New methods are needed to distinguish specific proteins from the thousands of other proteins within a cell.

3. Microfluidics demonstrates great promise in single-cell analysis due to the potential to integrate multiple cell- and lysate-handling steps on a single miniaturized device; however, substantial challenges must be overcome to reach these goals.

4. Strategies for high-throughput analysis of single cells need to be developed for nonadherent as well as adherent cell lines. Microfluidic devices will likely be the best option for high-throughput analysis of cells not adherent to a surface, as the cells are easily moved to the separation channel. However, capillary-based measurements may be optimal for cells that are attached to a surface, as the capillary can be readily transported to the immobilized cell.

5. Methods of quantifying large numbers of analytes from the same single cell would be of great value to cell biologists and would be complementary to many other established single-cell techniques, such as microscopy methods.

6. Although microelectrophoretic methods can reliably quantitate a large number of analytes in single cells, these techniques have not yet moved into mainstream biology. Understanding the reasons for the slow adoption by biologists will inspire chemists to design more biologically friendly applications for use in the laboratory and in related biomedical settings.

DISCLOSURE STATEMENT

The authors are not aware of any biases that might be perceived as affecting the objectivity of this review.

ACKNOWLEDGMENTS

We apologize in advance to all the investigators whose research could not be cited due to space limitations. We would like to acknowledge the NIH for support (EB004436 and EB004597 to NLA and F32GM078768 to LMB).

LITERATURE CITED

1. Giepmans BNG, Adams SR, Ellisman MH, Tsien RY. 2006. Review: The fluorescent toolbox for assessing protein location and function. *Science* 312:217–24

2. Hale MB, Nolan GP. 2006. Phospho-specific flow cytometry: intersection of immunology and biochemistry at the single-cell level. *Curr. Opin. Mol. Ther.* 8:215–24

3. Hogan BL, Yeung ES. 1992. Determination of intracellular species at the level of a single erythrocyte via capillary electrophoresis with direct and indirect fluorescence detection. *Anal. Chem.* 64:2841–45

4. Kennedy RT, Oates MD, Cooper BR, Nickerson B, Jorgenson JW. 1989. Microcolumn separations and the analysis of single cells. *Science* 246:57–63

5. Wallingford RA, Ewing AG. 1988. Capillary zone electrophoresis with electrochemical detection in 12.7-mu-m diameter columns. *Anal. Chem.* 60:1972–75

6. Dovichi NJ, Hu S. 2003. Chemical cytometry. *Curr. Opin. Chem. Biol.* 7:603–8

7. Arcibal I, Santillo M, Ewing A. 2007. Recent advances in capillary electrophoretic analysis of individual cells. *Anal. Bioanal. Chem.* 387:51–57

8. Gilman SD, Ewing AG. 1995. Analysis of single cells by capillary electrophoresis with on column derivatization and laser-induced fluorescence detection. *Anal. Chem.* 67:58–64

9. Kristensen HK, Lau YY, Ewing AG. 1994. Capillary electrophoresis of single cells: observation of two compartments of neurotransmitter vesicles. *J. Neurosci. Methods* 51:183–88

10. Krylov SN, Dovichi NJ. 2000. Single-cell analysis using capillary electrophoresis: Influence of surface support properties on cell injection into the capillary. *Electrophoresis* 21:767–73

11. Lee TT, Yeung ES. 1992. Quantitative determination of native proteins in individual human erythrocytes by capillary zone electrophoresis with laser-induced fluorescence detection. *Anal. Chem.* 64:3045–51

12. McClain MA, Culbertson CT, Jacobson SC, Allbritton NL, Sims CE, Ramsey JM. 2003. Microfluidic devices for the high-throughput chemical analysis of cells. *Anal. Chem.* 75:5646–55

13. Sims CE, Allbritton NL. 2007. Analysis of single mammalian cells on-chip. *Lab Chip* 7:423–40

14. Sims CE, Meredith GD, Krasieva TB, Berns MW, Tromberg BJ, Allbritton NL. 1998. Laser-micropipet combination for single cell analysis. *Anal. Chem.* 70:4570–77

15. Woods LA, Roddy TP, Ewing AG. 2004. Capillary electrophoresis of single mammalian cells. *Electrophoresis* 25:1181–87

16. Brown RB, Audet JA. 2007. Sampling efficiency of a single-cell capillary electrophoresis system. *Cytometry Part A* 71(A):882–88

17. Lee CL, Linton J, Soughayer JS, Sims CE, Allbritton NL. 1999. Localized measurement of kinase activation in oocytes of *Xenopus laevis*. *Nat. Biotechnol.* 17:759–62

18. Olefirowicz TM, Ewing AG. 1990. Capillary electrophoresis in 2-mu-m and 5-mu-m diameter capillaries: application to cytoplasmic analysis. *Anal. Chem.* 62:1872–76

19. Li HN, Sims CE, Wu HY, Allbritton NL. 2001. Spatial control of cellular measurements with the laser micropipet. *Anal. Chem.* 73:4625–31

20. Woods LA, Gandhi PU, Ewing AG. 2005. Electrically assisted sampling across membranes with electrophoresis in nanometer inner diameter capillaries. *Anal. Chem.* 77:1819–23

21. Gao N, Wang WL, Zhang XL, Jin WR, Yin XF, Fang ZL. 2006. High-throughput single-cell analysis for enzyme activity without cytolysis. *Anal. Chem.* 78:3213–20

22. Ocvirk G, Salimi-Moosavi H, Szarka RJ, Arriaga EA, Andersson PE, et al. 2004. Beta-galactosidase assays of single-cell lysates on a microchip: a complementary method for enzymatic analysis of single cells. *Proc. IEEE* 92:115–25

23. Chiu DT, Lillard SJ, Scheller RH, Zare RN, Rodriguez-Cruz SE, et al. 1998. Probing single secretory vesicles with capillary electrophoresis. *Science* 279:1190–93

24. Duffy CF, MacCraith B, Diamond D, O'Kennedy R, Arriaga EA. 2006. Fast electrophoretic analysis of individual mitochondria using microchip capillary electrophoresis with laser induced fluorescence detection. *Lab Chip* 6:1007–11

25. Gunasekera N, Olson KJ, Musier-Forsyth K, Arriaga EA. 2004. Capillary electrophoretic separation of nuclei released from single cells. *Anal. Chem.* 76:655–62

26. Fuller KM, Arriaga EA. 2003. Analysis of individual acidic organelles by capillary electrophoresis with laser-induced fluorescence detection facilitated by the endocytosis of fluorescently labeled microspheres. *Anal. Chem.* 75:2123–30

27. Anderson AB, Ciriacks CM, Fuller KM, Arriaga EA. 2003. Distribution of zeptornole-abundant doxorubicin metabolites in subcellular fractions by capillary electrophoresis with laser-induced fluorescence detection. *Anal. Chem.* 75:8–15

28. Zhu CR, He XY, Kraly JR, Jones MR, Whitmore CD, et al. 2007. Instrumentation for medium-throughput two-dimensional capillary electrophoresis with laser-induced fluorescence detection. *Anal. Chem.* 79:765–68

29. Chen S, Lillard SJ. 2001. Continuous cell introduction for the analysis of individual cells by capillary electrophoresis. *Anal. Chem.* 73:111–18

30. Quan Z, Liu YM. 2003. Capillary electrophoretic separation of glutamate enantiomers in neural samples. *Electrophoresis* 24:1092–96

31. Swanek FD, Anderson BB, Ewing AG. 1998. Capillary electrophoresis with NDA derivatization and electrochemical detection for the analysis of cellular amino acids. *J. Microcolumn Sep.* 10:185–92

32. Weng QF, Jin WR. 2003. Assay of amino acids in individual human lymphocytes by capillary zone electrophoresis with electrochemical detection. *Anal. Chim. Acta* 478:199–207

33. Weng QF, Xia FQ, Jin WR. 2002. Measurement of histamine in individual rat peritoneal mast cells by capillary zone electrophoresis with electrochemical detection. *J. Chromatogr. B: Anal. Technol. Biomed. Life Sci.* 779:347–52

34. Weng QF, Jin WR. 2001. Determination of free intracellular amino acids in single mouse peritoneal macrophages after naphthalene-2,3-dicarboxaldehyde derivatization by capillary zone electrophoresis with electrochemical detection. *Electrophoresis* 22:2797–803

35. Jin WR, Li XJ, Gao N. 2003. Simultaneous determination of tryptophan and glutathione in individual rat hepatocytes by capillary zone electrophoresis with electrochemical detection at a carbon fiber bundle–Au/Hg dual electrode. *Anal. Chem.* 75:3859–64

36. Wang ZQ, Yeung ES. 1997. Dual-enzyme assay of glutamate in single cells based on capillary electrophoresis. *J. Chromatogr. B* 695:59–65

37. Olefirowicz TM, Ewing AG. 1990. Dopamine concentration in the cytoplasmic compartment of single neurons determined by capillary electrophoresis. *J. Neurosci. Methods* 34:11–15

38. Bergquist J, Tarkowski A, Ekman R, Ewing A. 1994. Discovery of endogenous catecholamines in lymphocytes and evidence for catecholamine regulation of lymphocyte function via an autocrine loop. *Proc. Natl. Acad. Sci. USA* 91:12912–16

39. Swanek FD, Chen GY, Ewing AG. 1996. Identification of multiple compartments of dopamine in a single cell by CE with scanning electrochemical detection. *Anal. Chem.* 68:3912–16

40. Chang HT, Yeung ES. 1995. Determination of catecholamines in single adrenal-medullary cells by capillary electrophoresis and laser-induced native fluorescence. *Anal. Chem.* 67:1079–83

41. Tong W, Yeung ES. 1997. On-column monitoring of secretion of catecholamines from single bovine adrenal chromaffin cells by capillary electrophoresis. *J. Neurosci. Methods* 76:193–201

42. Lillard SJ, Yeung ES, McCloskey MA. 1996. Monitoring exocytosis and release from individual mast cells by capillary electrophoresis with laser-induced native fluorescence detection. *Anal. Chem.* 68:2897–904

43. Ho AM, Yeung ES. 1998. Capillary electrophoretic study of individual exocytotic events in single mast cells. *J. Chromatogr. A* 817:377–82

44. Fuller RR, Moroz LL, Gillette R, Sweedler JV. 1998. Single neuron analysis by capillary electrophoresis with fluorescence spectroscopy. *Neuron* 20:173–81

45. Lapainis T, Scanlan C, Rubakhin SS, Sweedler JV. 2007. A multichannel native fluorescence detection system for capillary electrophoretic analysis of neurotransmitters in single neurons. *Anal. Bioanal. Chem.* 387:97–105

46. Luzzi V, Lee CL, Allbritton NL. 1997. Localized sampling of cytoplasm from *Xenopus* oocytes for capillary electrophoresis. *Anal. Chem.* 69:4761–67

47. Krylov SN, Arriaga EA, Chan NWC, Dovichi NJ, Palcic MM. 2000. Metabolic cytometry: monitoring oligosaccharide biosynthesis in single cells by capillary electrophoresis. *Anal. Biochem.* 283:133–35

48. Le XC, Tan W, Scaman CH, Szpacenko A, Arriaga E, et al. 1999. Single cell studies of enzymatic hydrolysis of a tetramethylrhodamine labeled triglucoside in yeast. *Glycobiology* 9:219–25

49. Krylov SN, Zhang ZR, Chan NWC, Arriaga E, Palcic MM, Dovichi NJ. 1999. Correlating cell cycle with metabolism in single cells: combination of image and metabolic cytometry. *Cytometry* 37:14–20

50. Krylov SN, Arriaga E, Zhang Z, Chan NWC, Palcic MM, Dovichi NJ. 2000. Single-cell analysis avoids sample processing bias. *J. Chromatogr. B: Biomed. Sci. Appl.* 741:31–35

51. Gillmor CS, Poindexter P, Lorieau J, Palcic MM, Somerville C. 2002. α-Glucosidase I is required for cellulose biosynthesis and morphogenesis in *Arabidopsis*. *J. Cell Biol.* 156:1003–13

52. Shoemaker GK, Lorieau J, Lau LH, Gillmor CS, Palcic MM. 2005. Multiple sampling in single-cell enzyme assays using CE-laser-induced fluorescence to monitor reaction progress. *Anal. Chem.* 77:3132–37

53. Shoemaker GK, Palcic MM. 2007. Multiple sampling in single-cell enzyme assays using capillary electrophoresis with laser-induced fluorescence detection. *Anal. Bioanal. Chem.* 387:13–15

54. Larsson EA, Olsson U, Whitmore CD, Martins R, Tettamanti G, et al. 2007. Synthesis of reference standards to enable single cell metabolomic studies of tetramethylrhodamine-labeled ganglioside GM1. *Carbohydrate Res.* 342:482–89

55. Whitmore CD, Hindsgaul O, Palcic MM, Schnaar RL, Dovichi NJ. 2007. Metabolic cytometry: glycosphingolipid metabolism in single cells. *Anal. Chem.* 79:5139–42

56. Sheeley SA, Miao H, Ewing MA, Rubakhin SS, Sweedler JV. 2005. Measuring d-amino acid-containing neuropeptides with capillary electrophoresis. *The Analyst* 130:1198–203

57. Tong W, Yeung ES. 1996. Determination of insulin in single pancreatic cells by capillary electrophoresis and laser-induced native fluorescence. *J. Chromatogr. B: Biomed. Sci. Appl.* 685:35–40

58. Tong W, Yeung ES. 1997. Monitoring single-cell pharmacokinetics by capillary electrophoresis and laser-induced native fluorescence. *J. Chromatogr. B* 689:321–25

59. Schultz NM, Huang L, Kennedy RT. 1995. Capillary electrophoresis–based immunoassay to determine insulin content and insulin-secretion from single islets of Langerhans. *Anal. Chem.* 67:924–29

60. Tao L, Aspinwall CA, Kennedy RT. 1998. On-line competitive immunoassay based on capillary electrophoresis applied to monitoring insulin secretion from single islets of Langerhans. *Electrophoresis* 19:403–8

61. Orwar O, Fishman HA, Ziv NE, Scheller RH, Zare RN. 1995. Use of 2,3-naphthalenedicarboxaldehyde derivatization for single-cell analysis of glutathione by capillary electrophoresis and histochemical-localization ion by fluorescence microscopy. *Anal. Chem.* 67:4261–68

62. Jin WR, Dong Q, Ye XY, Yu DQ. 2000. Assay of glutathione in individual mouse peritoneal macrophages by capillary zone electrophoresis with electrochemical detection. *Anal. Biochem.* 285:255–59

63. Jin WR, Li W, Xu Q. 2000. Quantitative determination of glutathione in single human erythrocytes by capillary zone electrophoresis with electrochemical detection. *Electrophoresis* 21:774–79

64. Hu S, Jiang J, Cook LM, Richards DP, Horlick L, et al. 2002. Capillary sodium dodecyl sulfate-DALT electrophoresis with laser-induced fluorescence detection for size-based analysis of proteins in human colon cancer cells. *Electrophoresis* 23:3136–42

65. Hu S, Zhang L, Cook LM, Dovichi NJ. 2001. Capillary sodium dodecyl sulfate-DALT electrophoresis of proteins in a single human cancer cell. *Electrophoresis* 22:3677–82

66. Lillard SJ, Yeung ES, Lautamo RMA, Mao DT. 1995. Separation of hemoglobin variants in single human erythrocytes by capillary electrophoresis with laser-induced native fluorescence detection. *J. Chromatogr. A* 718:397–404

67. Lillard SJ, Yeung ES. 1996. Analysis of single erythrocytes by injection-based capillary isoelectric focusing with laser-induced native fluorescence detection. *J. Chromatogr. B: Biomed. Sci. Appl.* 687:363–69

68. Hofstadler SA, Severs JC, Smith RD, Swanek FD, Ewing AG. 1996. Analysis of single cells with capillary electrophoresis electrospray ionization Fourier transform ion cyclotron resonance mass spectrometry. *Rapid Commun. Mass Spectrometry* 10:919–22

69. Zhi Q, Xie C, Huang X, Ren J. 2007. Coupling chemiluminescence with capillary electrophoresis to analyze single human red blood cells. *Anal. Chim. Acta* 583:217–22

70. Pinto D, Arriaga EA, Schoenherr RM, Chou SSH, Dovichi NJ. 2003. Kinetics and apparent activation energy of the reaction of the fluorogenic reagent 5-furoylquinoline-3-carboxaldehyde with ovalbumin. *J. Chromatogr. B: Anal. Technol. Biomed. Life Sci.* 793:107–14

71. Krylov SN, Starke DA, Arriaga EA, Zhang Z, Chan NWC, et al. 2000. Instrumentation for chemical cytometry. *Anal. Chem.* 72:872–77

72. Lee IH, Pinto D, Arriaga EA, Zhang ZR, Dovichi NJ. 1998. Picomolar analysis of proteins using electrophoretically mediated microanalysis and capillary electrophoresis with laser-induced fluorescence detection. *Anal. Chem.* 70:4546–48

73. Zhang Z, Krylov S, Arriaga EA, Polakowski R, Dovichi NJ. 2000. One-dimensional protein analysis of an HT29 human colon adenocarcinoma cell. *Anal. Chem.* 72:318–22

74. Hu S, Lee R, Zhang Z, Krylov SN, Dovichi NJ. 2001. Protein analysis of an individual *Caenorhabditis elegans* single-cell embryo by capillary electrophoresis. *J. Chromatogr. B: Biomed. Sci. Appl.* 752:307–10

75. Hu S, Zhang L, Krylov S, Dovichi NJ. 2003. Cell cycle–dependent protein fingerprint from a single cancer cell: image cytometry coupled with single-cell capillary sieving electrophoresis. *Anal. Chem.* 75:3495–501

76. Hu K, Ahmadzadeh H, Krylov SN. 2004. Asymmetry between sister cells in a cancer cell line revealed by chemical cytometry. *Anal. Chem.* 76:3864–66

77. Hu S, Michels DA, Fazal MA, Ratisoontorn C, Cunningham ML, Dovichi NJ. 2004. Capillary sieving electrophoresis/micellar electrokinetic capillary chromatography for two-dimensional protein fingerprinting of single mammalian cells. *Anal. Chem.* 76:4044–49

78. Harwood MM, Bleecker JV, Rabinovitch PS, Dovichi NJ. 2007. Cell cycle–dependent characterization of single MCF-7 breast cancer cells by 2-D CE. *Electrophoresis* 28:932–37

79. Kraly JR, Jones MR, Gomez DG, Dickerson JA, Harwood MM, et al. 2006. Reproducible two-dimensional capillary electrophoresis analysis of Barrett's esophagus tissues. *Anal. Chem.* 78:5977–86

80. Harwood MM, Christians ES, Fazal MA, Dovichi NJ. 2006. Single-cell protein analysis of a single mouse embryo by two-dimensional capillary electrophoresis. *J. Chromatogr. A* 1130:190–94

81. Sobhani KFS, Cookson BT, Dovichi NJ. 2007. Repeatability of chemical cytometry: 2-DE analysis of single RAW 264.7 macrophage cells. *Electrophoresis* 28:2308–13

82. Cohen D, Dickerson JA, Whitmore CD, Turner EH, Palcic MM, et al. 2008. Chemical cytometry: fluorescence-based single-cell analysis. *Annu. Rev. Anal. Chem.* 1:165–190

83. Xiao H, Li X, Zou HF, Yang L, Yang YQ, et al. 2006. Immunoassay of P-glycoprotein on single cell by capillary electrophoresis with laser induced fluorescence detection. *Anal. Chim. Acta* 556:340–46

84. Malek A, Khaledi MG. 1999. Expression and analysis of green fluorescent proteins in human embryonic kidney cells by capillary electrophoresis. *Anal. Biochem.* 268:262–69

85. Hu K, Zarrine-Afsar A, Ahmadzadeh H, Krylov SN. 2004. Single-cell analysis by chemical cytometry combined with fluorescence microscopy. *Instrum. Sci. Technol.* 32:31–41

86. Turner EH, Lauterbach K, Pugsley HR, Palmer VR, Dovichi NJ. 2007. Detection of green fluorescent protein in a single bacterium by capillary electrophoresis with laser-induced fluorescence. *Anal. Chem.* 79:778–81

87. Zabzdyr JL, Lillard SJ. 2001. Measurement of single-cell gene expression using capillary electrophoresis. *Anal. Chem.* 73:5771–75

88. Zabzdyr JL, Lillard SJ. 2005. A qualitative look at multiplex gene expression of single cells using capillary electrophoresis. *Electrophoresis* 26:137–45

89. Liu X, Ma L, Zhang J-F, Lu Y-T. 2004. Determination of single-cell gene expression in *Arabidopsis* by capillary electrophoresis with laser induced fluorescence detection. *J. Chromatogr. B* 808:241–47

90. Li HL, Yeung ES. 2002. Selective genotyping of individual cells by capillary polymerase chain reaction. *Electrophoresis* 23:3372–80

91. Han FT, Lillard SJ. 2000. In-situ sampling and separation of RNA from individual mammalian cells. *Anal. Chem.* 72:4073–79

92. Zabzdyr JL, Lillard SJ. 2001. UV- and visible-excited fluorescence of nucleic acids separated by capillary electrophoresis. *J. Chromatogr. A* 911:269–76

93. Han FT, Lillard SJ. 2002. Monitoring differential synthesis of RNA in individual cells by capillary electrophoresis. *Anal. Biochem.* 302:136–43

94. Malek A, Khaledi MG. 1999. Steroid analysis in single cells by capillary electrophoresis with collinear laser-induced fluorescence detection. *Anal. Biochem.* 270:50–58

95. Luzzi V, Sims CE, Soughayer JS, Allbritton NL. 1998. The physiologic concentration of inositol 1,4,5-trisphosphate in the oocytes of *Xenopus laevis*. *J. Biol. Chem.* 273:28657–62

96. Wagner J, Fall CP, Hong F, Sims CE, Allbritton NL, et al. 2004. A wave of IP3 production accompanies the fertilization Ca^{2+} wave in the egg of the frog, *Xenopus laevis*: theoretical and experimental support. *Cell Calcium* 35:433–47

97. Rosenzweig Z, Yeung ES. 1994. Laser-based particle-counting microimmunoassay for the analysis of single human erythrocytes. *Anal. Chem.* 66:1771–76

98. Zhang H, Jin W. 2006. Single-cell analysis by intracellular immuno-reaction and capillary electrophoresis with laser-induced fluorescence detection. *J. Chromatogr. A* 1104:346–51

99. Li H, Wu HY, Wang Y, Sims CE, Allbritton NL. 2001. Improved capillary electrophoresis conditions for the separation of kinase substrates by the laser micropipet system. *J. Chromatogr. B: Biomed. Sci. Appl.* 757:79–88

100. Meredith GD, Sims CE, Soughayer JS, Allbritton NL. 2000. Measurement of kinase activation in single mammalian cells. *Nat. Biotechnol.* 18:309–12

101. Zarrine-Afsar A, Krylov SN. 2003. Use of capillary electrophoresis and endogenous fluorescent substrate to monitor intracellular activation of protein kinase A. *Anal. Chem.* 75:3720–24

102. Arkhipov SN, Berezovski M, Jitkova J, Krylov SN. 2005. Chemical cytometry for monitoring metabolism of a Ras-mimicking substrate in single cells. *Cytometry Part A* 63(A):41–47

103. Jitkova J, Carrigan CN, Poulter CD, Krylov SN. 2004. Monitoring the three enzymatic activities involved in posttranslational modifications of Ras proteins. *Anal. Chim. Acta* 521:1–7

104. Cruz L, Moroz LL, Gillette R, Sweedler JV. 1997. Nitrite and nitrate levels in individual molluscan neurons: single-cell capillary electrophoresis analysis. *J. Neurochem.* 69:110–15

105. Floyd PD, Moroz LL, Gillette R, Sweedler JV. 1998. Capillary electrophoresis analysis of nitric oxide synthase related metabolites in single identified neurons. *Anal. Chem.* 70:2243–47

106. Moroz LL, Dahlgren RL, Boudko D, Sweedler JV, Lovell P. 2005. Direct single cell determination of nitric oxide synthase related metabolites in identified nitrergic neurons. *J. Inorg. Biochem.* 99:929–39

107. Kim WS, Ye XY, Rubakhin SS, Sweedler JV. 2006. Measuring nitric oxide in single neurons by capillary electrophoresis with laser-induced fluorescence: use of ascorbate oxidase in diaminofluorescein measurements. *Anal. Chem.* 78:1859–65

108. Anderson AB, Gergen J, Arriaga EA. 2002. Detection of doxorubicin and metabolites in cell extracts and in single cells by capillary electrophoresis with laser-induced fluorescence detection. *J. Chromatogr. B: Anal. Technol. Biomed. Life Sci.* 769:97–106

109. Anderson AB, Arriaga EA. 2004. Subcellular metabolite profiles of the parent CCRF-CEM and the derived CEM/C2 cell lines after treatment with doxorubicin. *J. Chromatogr. B: Anal. Technol. Biomed. Life Sci.* 808:295–302

110. Chen Y, Walsh RJ, Arriaga EA. 2005. Selective determination of the doxorubicin content of individual acidic organelles in impure subcellular fractions. *Anal. Chem.* 77:2281–87

111. Eder AR, Arriaga EA. 2006. Capillary electrophoresis monitors enhancement in subcellular reactive oxygen species production upon treatment with doxorubicin. *Chem. Res. Toxicol.* 19:1151–59

112. Chen Y, Arriaga EA. 2006. Individual acidic organelle pH measurements by capillary electrophoresis. *Anal. Chem.* 78:820–26

113. Chen Y, Xiong G, Arriaga EA. 2007. CE analysis of the acidic organelles of a single cell. *Electrophoresis* 28:2406–15

114. Anderson AB, Xiong GH, Arriaga EA. 2004. Doxorubicin accumulation in individually electrophoresed organelles. *J. Am. Chem. Soc.* 126:9168–69

115. Presley AD, Fuller KM, Arriaga EA. 2003. MitoTracker Green labeling of mitochondrial proteins and their subsequent analysis by capillary electrophoresis with laser-induced fluorescence detection. *J. Chromatogr. B: Anal. Technol. Biomed. Life Sci.* 793:141–50

116. Duffy CF, Fuller KM, Malvey MW, O'Kennedy R, Arriaga EA. 2002. Determination of electrophoretic mobility distributions through the analysis of individual mitochondrial events by capillary electrophoresis with laser-induced fluorescence detection. *Anal. Chem.* 74:171–76

117. Fuller KM, Arriaga EA. 2004. Capillary electrophoresis monitors changes in the electrophoretic behavior of mitochondrial preparations. *J. Chromatogr. B: Anal. Technol. Biomed. Life Sci.* 806:151–59

118. Fuller KM, Duffy CF, Arriaga EA. 2002. Determination of the cardiolipin content of individual mitochondria by capillary electrophoresis with laser-induced fluorescence detection. *Electrophoresis* 23:1571–76

119. Ahmadzadeh H, Andreyev D, Arriaga EA, Thompson LV. 2006. Capillary electrophoresis reveals changes in individual mitochondrial particles associated with skeletal muscle fiber type and age. *J. Gerontol. Ser. A-Biol. Sci. Med. Sci.* 61:1211–18

120. Johnson RD, Navratil M, Poe BG, Xiong GH, Olson KJ, et al. 2007. Analysis of mitochondria isolated from single cells. *Anal. Bioanal. Chem.* 387:107–18

121. Meany DL, Thompson L, Arriaga EA. 2007. Simultaneously monitoring the superoxide in the mitochondrial matrix and extramitochondrial space by micellar electrokinetic chromatography with laser-induced fluorescence. *Anal. Chem.* 79:4588–94

122. Gunasekera N, Lee SW, Kim S, Musier-Forsyth K, Arriaga E. 2004. Nuclear localization of aminoacyl-tRNA synthetases using single-cell capillary electrophoresis laser-induced fluorescence analysis. *Anal. Chem.* 76:4741–46

123. Xiong GH, Chen Y, Arriaga EA. 2005. Measuring the doxorubicin content of single nuclei by micellar electrokinetic capillary chromatography with laser-induced fluorescence detection. *Anal. Chem.* 77:3488–93

124. Auroux PA, Iossifidis D, Reyes D, Manz A. 2002. Micro total analysis systems. 2. Analytical standard operations and applications. *Anal. Chem.* 74:2637–52

125. Dittrich PS, Tachikawa K, Manz A. 2006. Micro total analysis systems: latest advancements and trends. *Anal. Chem.* 78:3887–908

126. El-Ali J, Sorger PK, Jensen KF. 2006. Cells on chips. *Nature* 442:403–11

127. Price A. 2007. Chemical analysis of single mammalian cells with microfluidics. *Anal. Chem.* 79:2614–21

128. Roman GT, Chen Y, Viberg P, Culbertson AH, Culbertson CT. 2007. Single-cell manipulation and analysis using microfluidic devices. *Anal. Bioanal. Chem.* 387:9–12

129. Yi C, Li C-W, Ji S, Yang M. 2006. Microfluidics technology for manipulation and analysis of biological cells. *Anal. Chim. Acta* 560:1–23

130. Culbertson CT. 2006. Single cell analysis on microfluidic devices. *Methods Mol. Biol.* 339:203–16

131. Gao J, Yin XF, Fang ZL. 2004. Integration of single cell injection, cell lysis, separation and detection of intracellular constituents on a microfluidic chip. *Lab Chip* 4:47–52

132. Shi BX, Huang WH, Cheng JK. 2007. Determination of neurotransmitters in PC 12 cells by microchip electrophoresis with fluorescence detection. *Electrophoresis* 28:1595–600

133. Ros A, Hellmich W, Regtmeier J, Duong TT, Anselmetti D. 2006. Bioanalysis in structured microfluidic systems. *Electrophoresis* 27:2651–58

134. Hellmich W, Greif D, Pelargus C, Anselmetti D, Ros A. 2006. Improved native UV laser induced fluorescence detection for single cell analysis in poly(dimethylsiloxane) microfluidic devices. *J. Chromatogr. A* 1130:195–200

135. Hellmich W, Pelargus C, Leffhalm K, Ros A, Anselmetti D. 2005. Single cell manipulation, analytics, and label-free protein detection in microfluidic devices for systems nanobiology. *Electrophoresis* 26:3689–96

136. Munce NR, Li J, Herman PR, Lilge L. 2004. Microfabricated system for parallel single-cell capillary electrophoresis. *Anal. Chem.* 76:4983–89

137. Wu H, Wheeler A, Zare RN. 2004. Chemical cytometry on a picoliter-scale integrated microfluidic chip. *Proc. Natl. Acad. Sci. USA* 101:12809–13

138. Huang B, Wu H, Bhaya D, Grossman A, Granier S, et al. 2007. Counting low-copy number proteins in a single cell. *Science* 315:81–84

139. Wang HY, Lu C. 2006. Microfluidic chemical cytometry based on modulation of local field strength. *Chem. Commun.(Camb.)* 2006(33):3528–30

140. Kleparnik K, Horky M. 2003. Detection of DNA fragmentation in a single apoptotic cardiomyocyte by electrophoresis on a microfluidic device. *Electrophoresis* 24:3778–83

141. Kim WS, Dahlgren RL, Moroz LL, Sweedler JV. 2002. Ascorbic acid assays of individual neurons and neuronal tissues using capillary electrophoresis with laser-induced fluorescence detection. *Anal. Chem.* 74:5614–20

142. Woods LA, Powell PR, Paxon TL, Ewing AG. 2005. Analysis of mammalian cell cytoplasm with electrophoresis in nanometer inner diameter capillaries. *Electroanalysis* 17:1192–97

143. Pang ZL, Al-Mahrouki A, Berezovski M, Krylov SN. 2006. Selection of surfactants for cell lysis in chemical cytometry to study protein-DNA interactions. *Electrophoresis* 27:1489–94

144. Ling YY, Yin XF, Fang ZL. 2005. Simultaneous determination of glutathione and reactive oxygen species in individual cells by microchip electrophoresis. *Electrophoresis* 26:4759–66

145. Xiao H, Li X, Zou HF, Yang L, Wang YL, et al. 2006. CE-LIF coupled with flow cytometry for high-throughput quantitation of fluorophores in single intact cells. *Electrophoresis* 27:3452–59

146. Jin WR, Jiang L. 2002. Measurement of ascorbic acid in single human neutrophils by capillary zone electrophoresis with electrochemical detection. *Electrophoresis* 23:2471–76

Ion Chemistry in the Interstellar Medium

Theodore P. Snow and Veronica M. Bierbaum

Center for Astrophysics and Space Astronomy, Department of Astrophysical and Planetary Sciences and Department of Chemistry and Biochemistry, University of Colorado, Boulder, Colorado 80309; email: Theodore.Snow@colorado.edu, Veronica.Bierbaum@colorado.edu

Annu. Rev. Anal. Chem. 2008. 1:229–59

The *Annual Review of Analytical Chemistry* is online at anchem.annualreviews.org

This article's doi:
10.1146/annurev.anchem.1.031207.112907

1936-1327/08/0719-0229$20.00

Key Words

ion-atom reactions, interstellar clouds, interstellar species

Abstract

We present an overview of the interstellar medium, including physical and chemical conditions, spectroscopic observations, and current challenges in characterizing interstellar chemistry. Laboratory studies of ion-atom reactions, including experimental approaches and instrumentation, are described. We also tabulate and discuss comprehensive summaries of ion-neutral reactions involving hydrogen, nitrogen, and oxygen atoms that have been studied since Sablier and Rolando's 1993 review.

1. INTRODUCTION TO THE INTERSTELLAR MEDIUM

The physical conditions of the interstellar medium (ISM) are unlike any that occur naturally on Earth, and they are extremely difficult to reproduce in laboratory experiments. The ISM particle densities in our galaxy range from 10^{-4} cm^{-3} in diffuse regions to 10^5 cm^{-3} in dense clouds, the latter comparable to the best laboratory vacuums. Temperatures in these regions range from 10 K to 150 K. But, strange as it may seem, an active chemistry converts these atoms to molecules, some large and complex.

Despite the harsh, seemingly unproductive conditions in the ISM, two aspects of astrochemistry emerge as critical. One is that ions, formed by particle collisions, intense UV starlight, and cosmic rays, have high ion-neutral reaction rates. The other pertains to the timescales over which reactions can proceed. In a typical diffuse interstellar cloud, the average time between particle collisions is measured in decades—but cloud lifetimes can be millions of years. In denser interstellar clouds, collision times are reduced to a few hours.

Laboratory experiments are a valid means of exploring the chemistry of the ISM because steady-state conditions can be achieved and pressure and temperature effects are generally understood. The relevant reactions are two-body processes with no activation barrier; the neutral partner has a dipole moment that either occurs naturally or is induced by the ionic collision partner, so there is an attractive force between the particles. The frequency of reactive collisions can be determined by cloud densities and reaction rate constants. Unfortunately, most rate constants have been measured at room temperature, not at the very cold temperatures of interstellar space. However, temperature-variable experiments have often demonstrated either a simple temperature dependence or no dependence at all.

Evidence that a rich chemistry does operate in the ISM (1, 2) has been found in the observations of ~140 molecules to date (**Table 1**). There is evidence for the existence of far larger molecular species, the identities of which currently elude investigators. Laboratory results enable theorists to construct accurate chemical models of these clouds. These models refine the calculated abundances of observed species and make predictions about other species not yet observed.

Because this review is intended for a general scientific audience, we first give an overview of the physical and chemical conditions in the interstellar medium, then address several major challenges in astrochemistry that can be explored in the laboratory.

1.1. Physical and Chemical Conditions in the Interstellar Medium

The ISM is extremely heterogeneous. Instead of a rarefied, uniform environment, astronomers find a patchy, clumpy medium with extremes of temperature and a broad range of densities. **Table 2** presents physical data characteristic of the various regimes in interstellar clouds, derived from observations made throughout the electromagnetic spectrum using ground- and space-based telescopes.

1.1.1. Observations. The ISM was first detected by astronomers who observed that the Milky Way is punctuated by dark clouds (**Figure 1**). The interstellar dust is visible, and dark clouds obscure the stars within and behind these regions. The interstellar gas

Table 1 Detected interstellar molecules[a,b]

H$_2$[c,d]	CF$^+$	SiCN	C$_4$H$^-$	CH$_3$NH$_2$
AlF	C$_3$[e,f]	AlNC	HC$_2$NC	c-C$_2$H$_4$O
AlCl	C$_2$H	SiNC	HCOOH	H$_2$CCHOH
C$_2$[d,e]	C$_2$O	HCP	H$_2$CNH	CH$_2$CHCN
CH[e]	C$_2$S	c-C$_3$H	H$_2$C$_2$O	CH$_3$C$_3$N
CH$^+$[e]	CH$_2$[d,f]	l-C$_3$H	H$_2$NCN	HC(O)OCH$_3$
CN[e]	HCN[f,g]	C$_3$N	HNC$_3$	CH$_3$COOH
CO[c,d,f]	HCO	C$_3$O	SiH$_4$	C$_7$H
CO$^+$	HCO$^+$	C$_3$S	H$_2$COH$^+$	H$_2$C$_6$
CP	HCS$^+$	C$_2$H$_2$[c]	HC$_3$N	CH$_2$OHCHO
SiC	HOC$^+$	NH$_3$	C$_5$H	CH$_2$CCHCN
HCl[d]	H$_2$O	HCCN	l-H$_2$C$_4$	CH$_3$C$_4$H
KCl	H$_2$S	HCNH$^+$	C$_2$H$_4$	CH$_3$CH$_2$CN
NH[e]	HNC	HNCO	CH$_3$CN	(CH$_3$)$_2$O
NO	HNO	HNCS	CH$_3$NC	CH$_3$CH$_2$OH
NS	MgCN	HOCO$^+$	CH$_3$OH	HC$_7$N
NaCl	MgNC	H$_2$CO	CH$_3$SH	CH$_3$C(O)NH$_2$
OH[d,f]	N$_2$H$^+$	H$_2$CN	HC$_3$NH$^+$	C$_8$H
PN	N$_2$O	H$_2$CS	HC$_2$CHO	C$_8$H$^-$
SO	NaCN	H$_3$O$^+$	NH$_2$CHO	CH$_3$C$_5$N
SO$^+$	OCS	c-SiC$_3$	C$_5$N	(CH$_3$)$_2$CO
SiN	SO$_2$	CH$_3$[c]	l-HC$_4$N	(CH$_2$OH)$_2$
SiO	c-SiC$_2$	C$_5$	c-H$_2$C$_3$O	CH$_3$CH$_2$CHO
SiS	CO$_2$[c]	C$_4$H	C$_6$H	HC$_9$N
CS	NH$_2$[c]	l-C$_3$H$_2$	C$_6$H$^-$	CH$_3$C$_6$H
HF[c,f]	H$_3$$^+$[c]	c-C$_3$H$_2$	CH$_3$C$_2$H	HC$_{11}$N
SH	H$_2$D$^+$	H$_2$CCN	HC$_5$N	
O$_2$[g]	HD$_2$$^+$	CH$_4$	CH$_3$CHO	

Annotations:

[a] Most of this information was gathered from Reference 13. Many of these species have detected isotopologues, which are not listed. Additional detections that are questionable or probable include N$_2$, FeO, SiH, l-HC$_4$H, H$_2$CCNH, l-HC$_6$H, CH$_2$CHCHO, C$_6$H$_6$, C$_2$H$_5$OCH$_3$, H$_2$NCH$_2$COOH (glycine), and 1,3-dihydroxypropanone.

[b] Molecules without annotations have been detected only by rotational transitions in the radiofrequency spectrum.

[c] Detected by vibrational transitions in the infrared region.

[d] Detected by electronic transitions in the UV region.

[e] Detected by electronic transitions in the visible region.

[f] Detected by rotational transitions in the radiofrequency spectrum, as well as another method or methods, as specified.

[g] Detected by rotational transitions at submillimeter wavelengths.

is visually evident only when it is hot and glowing (the reddish regions of **Figure 1**); spectroscopic observations are usually needed to explore the properties of the gas.

On its journey to us, starlight is selectively absorbed by atoms, ions, and molecules in interstellar space. There are few absorption lines at visible wavelengths because only transitions arising from the ground electronic state are seen, and these occur

Table 2 Physical conditions in molecule-bearing interstellar clouds[a,b]

Property	Cloud type			
	Diffuse atomic	**Diffuse molecular**	**Translucent**	**Dense molecular**
Defining characteristic	$f^n_{H_2} < 0.1$	$f^n_{H_2} > 0.1$ $f^n_{C+} > 0.5$	$f^n_{CO} < 0.9$ $f^n_{C+} < 0.5$	$f^n_{CO} > 0.9$
A_V (minimum)	0–~0.2	~0.2–~1	~1–~5	~5–~10
Typical n_H (cm^{-3})	1–100	100–500	500–5000	10^4–10^6
Typical T (K)	30–150	30–100	15–50	10–50
Observational techniques	UV/visible absorption, H I radio emission	UV/visible/IR absorption, radio absorption	Visible/UV, IR absorption, radio absorption/emission	IR absorption, radio emission

[a]Abbreviations: $f^n_{H_2}$, the fraction of hydrogen nuclei contained in the form of H_2; $f^n_{H_2} = 2N(H_2)/[N(H) + 2N(H_2)]$, where the N symbolizes column density, which is the product of the volume density and the path length between Earth and the observed star; f^n_{C+} and f^n_{CO}, the ionized carbon or CO molecular fraction, as compared to the total hydrogen, using observed column densities; A_V, the dust extinction parameter, a logarithmic value expressed in units of stellar magnitudes, refers to the extinction at visual wavelengths; n_H, the volume density of hydrogen nuclei, which is usually inferred by indirect means; H I, the astronomical term for H-atom; IR, infrared.
[b]Based on table 1 in Reference 26.

mostly in the UV region. The density, and therefore the collision rate, is too low to maintain populations in excited electronic states. Typically, in the visible region, interstellar atomic lines are observed for Na (the D lines), K, Fe, and Ca, and for some ions, such as Ca^+ (the H and K lines). About 70 years ago, three molecules were identified in visible-wavelength spectra: CH, CH^+, and CN (3–6). These observations led to the first discussions concerning how molecules could form in the low-density environment of interstellar space; ion-neutral reactions were proposed as efficient mechanisms (7–9). For a review of this topic, see Dalgarno (10).

Because most electronic transitions arising from the ground state of atoms, ions, and molecules occur at UV wavelengths, most interstellar absorption lines can only be observed from telescopes above the Earth's atmosphere. There have been many orbiting space-based observatories that obtain UV spectra, culminating with the Hubble Space Telescope. With these telescopes, most common elements (except helium) can be observed in either neutral atomic or ionic form. Moreover, the most abundant molecules in space, such as H_2 and CO, have absorption lines in the UV and are easily observed, at least in diffuse clouds where UV radiation can penetrate. A typical UV spectrum, including both atomic and H_2 lines, is shown in **Figure 2**.

A new means of observing interstellar molecules was developed in the 1960s, when the first rotational emission lines were detected at millimeter wavelengths (11, 12). Despite the low gas densities, occasional collisions—especially in denser clouds—have enough energy and frequency to maintain a population of rotationally excited molecules. Only molecules that have a nonzero dipole moment can be detected through millimeter-wave emission lines when the rotationally excited molecules cascade down to the ground state. Homonuclear molecules, such as H_2 and C_2, have no dipole moment and no rotational (or vibrational) transitions, and can be observed only through their electronic dipole transitions or through forbidden (quadrupole) vibrational transitions. About 140 molecules have now been detected through their

Figure 1

The interstellar medium. This CCD image, taken by Adam Block at the Caelum Observatory, shows many aspects of the interstellar medium, including dark clouds, ionized hydrogen regions (known as H II regions due to H-α emission) (*red*), a newly-formed star cluster (*lower center*), nebulae caused by light scattered from dust grains (*blue*), and stars (*reddish color*) seen through the interstellar dust. Most molecules identified by their radio emission lines are found in the dark clouds; however, some are found in more diffuse regions through optical and UV absorption along the lines of sight toward reddened stars. Copyright Adam Block and Tim Puckett. Reproduced with permission.

radio emission lines, from diatomics to polyynes with 13 atoms (**Table 1**) (13). The vast majority of the detected species contains carbon. Although no biological molecules have been conclusively detected, the search for simple amino acids is ongoing and shows some promise (14); however, the one reported detection is probably not valid (15).

Until recently, only neutral and singly charged positive ions had been found, but there have now been exciting discoveries of several negative ions (16–18). These anions may be important in forming some interstellar neutrals through ion chemistry; however, studies of such mechanisms are in their infancy. To date, the largest unambiguously identified interstellar molecules contain carbon chains, but there are

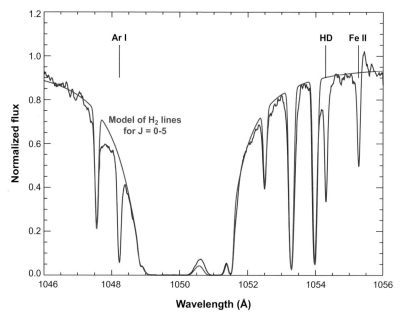

Figure 2

UV spectrum showing interstellar absorption lines obtained by NASA's Far Ultraviolet Spectroscopic Explorer. In this image, most of the line absorption features are due to interstellar H_2 molecules, two lines are the result of interstellar atoms and ions, and one line represents interstellar HD molecules. The smooth profile is a fit to the H_2 band, with rotational levels from $J = 0$ to $J = 5$. Both the H_2 and HD features are due to transitions from the ground electronic and vibrational states, but some excitation to low-lying rotational states is evident. Abbreviations: Ar I, neutral form of argon; Fe II, singly ionized form of iron. Copyright NASA and the Center for Astrophysics and Space Astronomy, University of Colorado. Reproduced with permission.

hints that much larger species exist. Interstellar polycyclic aromatic hydrocarbons (PAHs) are almost certainly present, as shown by their infrared (IR) emission bands (19–21).

IR emission and absorption constitute another spectral window for observing interstellar molecules; the observed lines are predominantly the result of transitions between vibrational levels. Only in irradiated, relatively dense regions (known as photon-dominated regions or PDRs) are vibrational lines seen in emission. Interestingly, H_2, which has no allowed vibrational transitions, is seen in emission in PDRs through low-probability quadrupole transitions (22, 23); this detection is possible because of the high abundance of H_2.

A few IR interstellar absorption lines from gas-phase species are observed, usually from the ground- and vibrationally excited levels of the ground electronic state. Despite its high abundance, H_2 is rarely seen in IR absorption (24); however, CO, which has allowed vibrational transitions and is second in abundance among interstellar molecules, is observed in absorption (25).

PAH: polycyclic aromatic hydrocarbon

PDR: photon-dominated region

1.1.2. Physical conditions. As already noted, the ISM is heterogeneous, with relatively dense clouds or nebulae that account for most of its mass. At the other extreme, the intercloud medium has extremely low densities and occupies most of the volume of the galaxy. The dense molecular clouds are often components of large aggregates containing many hundreds of solar masses. These complexes, such as those seen in **Figure 1**, are the birthplaces of new stars. **Table 2** summarizes the conditions in these clouds, where the temperatures can be as low as 15–20 K and the densities as high as 10^5–10^6 cm^{-3}.

The dense clouds are largely molecular in nature and are composed primarily of H_2 followed by CO, which constitutes about 10^{-4} of the abundance of H_2 by number. Other simple species such as OH, CH, CN, and H_2O represent only 10^{-7} of the abundance of H_2, at most. The density of PAHs in the general ISM is unknown, but some astronomers estimate the abundance as comparable to the simple species mentioned above, thus accounting for 15 to 20% of the carbon in the ISM (19–21).

Both translucent and diffuse molecular clouds (**Table 2**) have molecular populations. **Figure 3** illustrates several types of clouds. Although the sequence is not necessarily evolutionary, one can imagine a set of conditions smoothly changing from

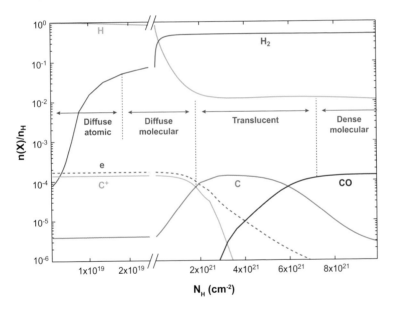

Figure 3

Physical and chemical conditions in interstellar clouds. The ordinate specifies the density of the given species relative to total hydrogen; the abscissa indicates the total hydrogen column density (the product of the hydrogen volume density and the path length). Moving from the diffuse atomic clouds (*left*) to the dense molecular clouds (*right*), there are major transitions of the dominant atoms, ions, and molecules. Molecules in the dense molecular and translucent clouds can be observed by radio-wave emission lines, and some molecules (such as CO) can be detected by the same method in diffuse molecular clouds. In diffuse atomic and molecular clouds, the interstellar species are primarily observed through absorption lines at visible or UV wavelengths. Reproduced with permission from Reference 26.

diffuse atomic to dense molecular clouds. **Figure 3** and **Table 2** indicate that diffuse atomic clouds contain mostly atomic hydrogen and ionic species, with a molecular fraction (i.e., H_2 as compared by number to all hydrogen nuclei) of 0.0 to 0.1. For diffuse molecular clouds, the molecular hydrogen fraction lies between 0.1 and ~0.5, and increases to nearly 1.0 in translucent and dense molecular clouds. The distinction between these latter clouds relates to the form of carbon: In translucent clouds, carbon exists as the ionized or neutral atomic form with the CO fraction beginning to rise. In dense molecular clouds, all of the hydrogen and most of the carbon are in molecular form.

The main factors that determine the physical and chemical conditions in the various cloud types are density, temperature, and the radiation field. The density is controlled by gas motions caused by random perturbations, stellar winds, and stellar explosions such as supernovae (the ISM is a violent environment!). The temperatures in the various regimes are determined by heat input, which depends upon radiation and kinetics of the gas, and heat losses, which primarily depend upon radiative losses from excited states of atoms, ions, and molecules.

The intensity of the radiation field is driven by the field's proximity to hot stars and the attenuation (i.e., extinction) of starlight by dust. When very close to a hot star, molecules can still form despite the high intensity of the radiation field as long as the density is sufficiently high (high density may result from shocks compressing the gas). Interstellar dust further away from hot stars causes wavelength-dependent extinction (**Figure 4**), which greatly affects the ambient radiation field and hence the molecular

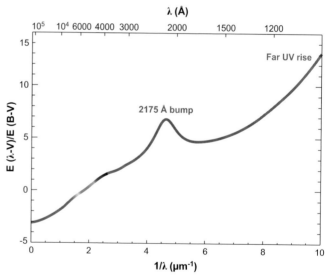

Figure 4

The UV interstellar dust extinction curve. The effect of interstellar dust in blocking starlight rises toward short wavelengths; therefore, in the UV portion of the spectrum—where the majority of interstellar atoms, ions, and molecules have electronic transitions—it is difficult to observe the interstellar absorption lines. The prominent "bump" centered at 2175 Å is thought to be caused by carbonaceous dust or perhaps by polycyclic aromatic hydrocarbons.

population. As **Figure 4** shows, the extinction rises toward UV wavelengths, so the effect of dust is greatest in the spectral window where molecules are mostly likely to be either dissociated or prevented from forming. As shown in **Table 2**, dust extinction (denoted by A_V, a logarithmic value) is a critical factor that varies with cloud type.

1.1.3. Chemical modeling of interstellar clouds.

A complete chemical model of an interstellar cloud should include the physical and chemical conditions described above, namely (*a*) the gas kinetics; (*b*) the depth-dependence of density and temperature from the exterior to the interior of the cloud; (*c*) the time variability as the cloud evolves; and (*d*) the correct reaction rates. These are demanding criteria, and no complete model has been developed, although progress is being made. See Snow & McCall (26) for a recent summary of chemical modeling.

1.2. Current Challenges

We identify several problems in constructing better models and gaining more insight into the chemistry of the ISM. In this section, we outline some of those challenges.

1.2.1. The interdependence of related reactions.

A stable population of a given molecular species is maintained through the combination of formation and destruction processes. As many abundant species are formed through a sequence of steps, the rate constants of many intermediate processes can be important; however, many of these rate constants are poorly understood, if at all.

Recently, Markwick-Kemper compiled a list of the most important reactions, ranked by potential impact due to uncertainties in the rate constants (27). Some of the most abundant species have been shown to have uncertainties in their theoretical abundances of 50% or greater; these uncertainties can be propagated through the network of sequential reactions. Wakelam et al. (28) have recently considered the effect of uncertainties on the chemical models of dark clouds.

There are at least two major databases of reaction rate constants that are relevant to ISM chemistry models: One database is maintained by Woodall et al. (29) and the other by Herbst (30). There is occasional disagreement between these databases, which highlights uncertainties and identifies the rate constants that should be re-examined.

1.2.2. The formation and abundance of molecular anions.

The first molecular anions (C_4H^-, C_6H^-, and C_8H^-) have recently been detected (16–18) in interstellar space and circumstellar envelopes (i.e., shells or bubbles of gas ejected by stellar winds). Observations of other carbon chain anions are anticipated. Millar et al. (31), using the code for PDRs proposed by Le Petit et al. (32), have calculated that $C_{10}H^-$ should be comparable in abundance to C_6H^- and C_8H^- ($\sim 10^{-9}$ of the total particle number density). As the length of the carbon chain increases, the abundance is expected to diminish only slowly; thus, even larger anions could potentially be observed.

Negative molecular ions, in addition to being sufficiently abundant for detection in space, can play a crucial role in ion chemistry. Our work has shown that some observed interstellar neutrals can be formed by reactions involving anions as reagents. These reaction sequences can form, for example, cyanoacetylene and glycine, the simplest amino acid (33). Positive ion gas-phase syntheses for interstellar carboxylic and amino acids have been proposed by Bohme and colleagues (34).

1.2.3. The search for biological molecules.
A new field of astrochemistry, known variously as bioastronomy or astrobiology, has been born. Astrobiology has generated new courses, conferences, faculty lines, and institutes (e.g., the Search for Extraterrestrial Intelligence Institute), as well as funding from the National Aeronautics and Space Administration. Although no sign of extraterrestrial life has been found to date, many scientists remain optimistic.

One step toward the discovery of extraterrestrial life would be the positive detection of biological molecules in interstellar space. As noted above, mechanisms for producing simple amino acids in interstellar clouds have been found to be feasible, and there has been a reported (14) but contested (15) detection of interstellar glycine. However, models based on experimental work predict that glycine, and perhaps other biogenic molecules, should be sufficiently abundant for detection. Such discoveries could yield insights into the origin of life.

1.2.4. The diffuse interstellar bands.
Diffuse interstellar bands (DIBs) have a long history, having first been reported in 1922 (35). DIBs consist of a set of absorption features (**Figure 5**), are definitely interstellar, and are almost certainly caused by

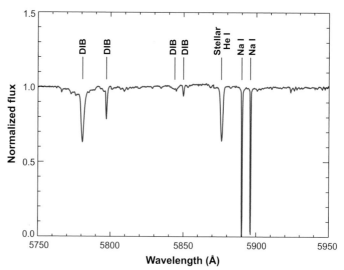

Figure 5

Diffuse interstellar bands (DIBs). This figure shows four broad, weak features that are part of the 300+ DIBs. At right are the interstellar sodium D-lines and one stellar line (He I represents helium in its neutral state). The two DIBs toward the left end of the spectrum are those first noticed more than 85 years ago (35). Data from the 3.5-m telescope at Apache Point, New Mexico, operated and owned by Astronomical Research Corporation. Used with permission.

large organic molecules (36, 37). The identity of the DIB carriers is the oldest astronomical spectroscopic mystery by far. After decades of being ignored or viewed as a fringe problem, beginning in the 1980s DIBs have been recognized by astronomers as holding the key to numerous issues in ISM chemistry.

Based on observational properties and the probable presence of PAHs in the ISM, along with experimental data and physical models of the ionization state of interstellar PAHs (38, 39), PAH cations have emerged as very strong candidates for DIB carriers. Carbon chains are also considered candidates as they comprise the majority of identified interstellar molecules (see **Table 1**). However, carbon chains have not yet been detected in diffuse atomic clouds in which DIBs are strongest.

The models and experiments point to PAH cations having 30 or more carbon atoms as the most likely DIB carriers (39). However, such large PAHs present difficulties to kineticists and spectroscopists as they are difficult to vaporize and retain in the gas phase as cations. From solid-state spectroscopy, we know that PAH cations having 30 or more carbon atoms have strong spectra at visible wavelengths and that neutral PAHs tend to have their lines and bands in the UV; however, only gas-phase spectroscopy can identify the specific species responsible for the DIBs.

Estimates of the PAH abundance in the ISM suggest that 15–20% of the entire galactic ISM carbon budget could exist in this form (19–21); this abundance would be sufficient to account for the observed absorption of starlight through DIBs. Whatever molecules are responsible for DIBs, their identification will yield new insights into the chemistry of the ISM. It will reveal their role in the energy budget of the ISM and in the formation of other species, possibly including biogenic molecules. Great acclaim surely awaits the scientists who finally identify the DIB carriers, and lab work will play a major role in that eventual success.

1.2.5. Identifying polycyclic aromatic hydrocarbons in the interstellar medium.
In addition to the DIB problem, the PAHs themselves represent a challenge to ISM chemists. The infrared emission bands associated with PAHs arise from vibrational transitions (such as C-C or C-H stretching or bending modes), which are relatively similar for most PAH molecules (19–21). However, the electronic spectra are unique; therefore, if the electronic spectra of PAHs or their cations were known, astronomers could search for specific molecules, perhaps even those responsible for DIBs. More relevant to this review, laboratory studies of reaction rates with appropriate modeling can identify PAHs for spectroscopic study, which in turn can enable identification of PAHs in the ISM.

Another branch of the PAH family tree is represented by the heterocyclic PAHs and their cations. Nitrogen, in particular, can be exchanged with CH groups to form polycyclic aromatic nitrogen heterocycles (PANHs) (40). As much as half of the nitrogen content of the galactic ISM is unaccounted for (41); thus a substantial quantity of nitrogen may be hidden away in PANHs, which are currently estimated to incorporate more than 1% of the available nitrogen.

Obtaining experimental reaction rates or spectra of PANH cations is challenging as these species—especially those with interior nitrogen atoms, which are the most likely sources of the observed IR emission spectra (40)—are difficult to synthesize.

Some neutral PANHs are unstable in the lab because they have unpaired electrons; however, these radicals can survive in the ISM because of the low collision rates. The performance of experimental studies of PANHs and PANH⁺s remains an important challenge to the laboratory astrophysics community.

SIFT: selected ion flow tube

1.2.6. Reactions forming the first molecules in the universe. Cosmological observations and theory both implicate H_2 as the trigger for the formation of the first stars and galaxies, in short, for the creation of all the complex matter in the universe. In the absence of H_2, the universe would contain only atomic hydrogen, a small amount of ^3He, and traces of other elements (such as Li and Be) formed in the initial big bang. All of the other elements and isotopes that make up the matter in the universe were created by nucleosynthesis in stellar interiors. And without H_2, no stars would ever have formed.

H_2 serves as a coolant in star formation through its quadrupole transitions, allowing atomic clouds to condense (42). However, gas-phase reactions forming H_2 are very slow, so most models assume H_2 formation on solid-grain surfaces (43, 44). Although this mechanism is adequate for interstellar clouds containing dust, there was no dust in the beginning. Thus, scientists have invoked associative detachment of H with H^- (45): $H + H^- \rightarrow H_2 + e^-$.

The rate constant for this crucial reaction has been measured in only one laboratory and has large error bars. Recent work (46) highlights the importance of this reaction rate and explores the consequences of values at either end of the uncertainty range. Clearly, this reaction rate constant must be refined.

2. REVIEW OF LABORATORY STUDIES OF ION-ATOM REACTIONS

2.1. Methods

The development of sophisticated analytical instrumentation has enabled the detailed characterization of the interstellar medium described above. Both ground-based and space-based telescopes with spectroscopic detection capabilities throughout the electromagnetic spectrum have allowed the determination of the chemical and physical properties of the ISM.

Similarly, laboratory studies of relevant chemical reactions have been dependent on the development of analytical techniques with increasingly powerful capabilities. For studies of charged particles, mass spectrometry has provided an exceptionally versatile approach. Studies during the last five decades have utilized ion cyclotron resonance (ICR), flowing afterglow, flowing afterglow–Langmuir probe, Fourier transform mass spectrometry, ion beams, ion traps, ion storage rings, selected ion flow tube (SIFT) instruments, and a low-temperature supersonic flow technique, which has allowed studies of ion chemistry at 10 K (47). Moreover, an array of ionization methods has allowed the generation of both common and exotic ions (48). These approaches have explored a wide variety of processes such as bimolecular ion chemistry ($A^\pm + B \rightarrow C^\pm + D$), associative detachment ($A^- + B \rightarrow AB + e^-$), radiative association ($A^+ + B \rightarrow AB^+ + h\nu$), and dissociative recombination ($AB^+ + e^- \rightarrow A + B$), among others, all of which are important in the ISM. Several reviews and compilations have described the analytical approaches (49, 50) and summarized the available data (51–53).

One of the most challenging areas of laboratory research involves the gas-phase reactions between ions and neutral atoms, as both partners are often highly reactive species. A review of gas-phase ion-atom reactions was published in 1993 by Sablier & Rolando (54). Our paper provides a comprehensive review of laboratory studies done since 1993 of positive ion and negative ion reactions with three atomic species that are important in the ISM: H, N, and O. These studies have been carried out almost exclusively with the SIFT technique; however, several reactions of positive ions with H-atoms have recently been reported utilizing a newly developed multi-electrode ion trap. These instruments, as well as the methods of atom production and measurement, are described below.

2.1.1. The selected ion flow tube.
Flow tube techniques for studying gas-phase ion chemistry were first introduced 40 years ago by Ferguson and colleagues (55) at the National Oceanic and Atmospheric Administration laboratories in Boulder, CO. Since that time, the technique has been extended and refined (56, 57) to incorporate temperature (58, 59) and kinetic energy variability (60), a variety of ionic and neutral sources (61), Langmuir probe detection (62), laser interrogation methods (63), triple quadrupole detection (64), and mass selection of the reactant ion (SIFT) (65).

Figure 6 presents a schematic diagram of the SIFT instrument used in our laboratory for the study of ion-atom reactions of interstellar relevance. The apparatus consists of regions for (*a*) ion production, (*b*) mass selection and injection of ions, (*c*) reaction of the ions with neutral reagents, and (*d*) detection of the ionic reactant and products. Positive and negative ions are generated in a flow of helium by a variety of methods. For example, electron ionization of N_2O generates O^-, which reacts with CH_4 to form HO^-. Subsequent addition of acetylene produces $HC{\equiv}C^-$ by proton abstraction. Carbon chain anions, such as C_7^-, are readily formed by a dc discharge between a graphite rod and the stainless steel flow tube. Molecular cations of PAHs are formed by the Penning ionization reaction of metastable argon (formed in a cold cathode discharge) with the parent PAH. For compounds of low volatility, such as coronene, a resistively heated oven can be utilized to increase the vapor pressure of the neutral in the source flow tube.

The ions are extracted through a nose cone and focused by lenses; the desired ion is mass-selected by the SIFT quadrupole mass filter, refocused by lenses, and injected into the reaction flow tube through a Venturi inlet. The ions are entrained in a flow of helium buffer gas (P = 0.5 torr, F = 200 atm cm^3 s^{-1}) that provides thermalization of the ions and transport along the flow tube. The neutral reactants (H, N, or O) are added downstream of the ion injection, and the ion-neutral reaction occurs throughout the remainder of the flow tube. Most of the gas mixture is exhausted by a large Roots blower system, and the ions are sampled through a nose cone, mass-analyzed, and detected with an electron multiplier. Rate constants

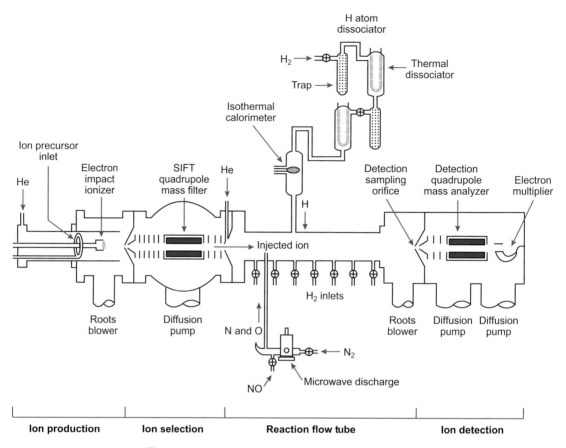

Figure 6

The flowing afterglow–selected ion flow tube (SIFT) instrument. The tandem flow tube instrument allows the production of ions by a variety of ionization techniques, followed by selection of the desired reactant ion with the SIFT quadrupole mass filter, interaction of the ion with atomic or molecular reagents in the reaction flow tube, and detection of reactant and product ions with a quadrupole mass analyzer coupled to an electron multiplier detector.

are determined from the decrease of reactant ion signal with increasing flow rate of the atomic reactant, and other experimental parameters including the pressure and flow of helium, the temperature, and the reaction time. Alternatively, for stable neutral reactants such as molecular hydrogen, the reagent can be introduced into the flow tube through a manifold of inlets, and the reactant ion signal can be monitored as a function of increasing reaction distance. Product distributions are determined by the ratio of the product ions under conditions where mass discrimination is minimized.

One of the most powerful features of flow systems is their capability for studying atomic neutral reactants, including H, N, and O atoms. Hydrogen atoms can be formed by passing molecular hydrogen through a microwave discharge (66).

However, as shown in **Figure 6** our SIFT experiments usually utilize thermal dissociation (67), which produces $H(^2S)$ atoms in the absence of metastable species. High-purity H_2 flows through a molecular sieve trap immersed in liquid nitrogen, over a heated tungsten filament, and through a second cooled sieve trap. This process efficiently removes trace impurities. The molecular hydrogen is dissociated by passage over a second heated filament, and dissociation ratios of about 30% can be achieved. To introduce the hydrogen atoms into the reaction flow tube, the "isothermal calorimeter" is withdrawn into a side arm. From the exit of the second thermal dissociator to the entrance of the reaction flow tube, the H/H_2 mixture flows through Teflon tubing in order to minimize recombination. To measure the H-atom flow rate, the Teflon tubing is positioned above the isothermal calorimeter, which consists of a platinum wire mesh that efficiently recombines the H-atoms. A self-balancing Kelvin bridge accurately measures the heat released in this process, thereby also measuring the flow rate of the hydrogen atoms. Alternatively, a calibration reaction with known rate constant can be examined to determine the atom flow rate.

$N(^4S)$ atoms are formed by flowing pure molecular nitrogen through a microwave discharge operating between 10 and 50 W. Phosphoric acid coating of the discharge tube can be utilized to minimize wall recombination, and dissociation ratios of about 2% can be achieved. $O(^3P)$ atoms are formed by adding NO (5% in helium) immediately downstream of the nitrogen discharge (**Figure 6**) to initiate the quantitative reaction $N + NO \rightarrow O + N_2$ (68). This method is preferred over direct microwave dissociation of molecular oxygen, which is known to also generate reactive O_2 $(^1\Delta_g)$ metastables. The flow rates of the nitrogen and oxygen atoms are determined by the endpoint of the $N + NO$ titration.

2.1.2. The multi-electrode ion trap. Gerlich and colleagues (69, 70) have recently described an innovative instrument for the study of ion-neutral reactions, including atomic neutral reagents, at temperatures as low as 10 K. A schematic diagram of the Atomic Beam 22-Pole Trap apparatus is shown in **Figure 7**. The hydrogen atoms are generated in a radiofrequency (rf) plasma source, which has been optimized to maximize dissociation and minimize recombination. The atoms are cooled by passage through a glass tube at 100 K; they then pass through a copper accommodator with a temperature range of 10–300 K, as determined by a silicon temperature sensor. The effusive beam of H-atoms is skimmed, differentially pumped twice, and focused by two hexapole magnets into the 22-pole trap. The number density of atomic and molecular hydrogen in the reaction region is determined using a calibrated universal detector based on ionization via electron bombardment.

Primary ions are generated in a standard storage ion source, mass-selected with an rf quadrupole, and injected into the trap via an electrostatic quadrupole bender. The trap is maintained at a temperature between 10–300 K. A variable storage time (milliseconds to seconds) allows the extent of the ion-neutral reaction to be varied and monitored. The ions are then extracted, mass-analyzed, and detected. Due to the high efficiency of trapping, reaction rate constants as low as 10^{-13} cm^3 s^{-1} can be determined by this approach.

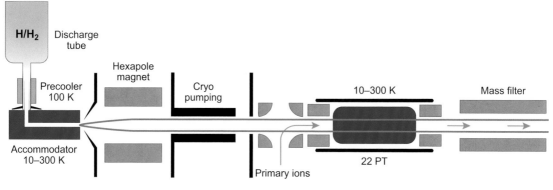

Figure 7

The Atomic Beam 22-Pole Ion Trap. Hydrogen atoms are generated in a radiofrequency plasma source, cooled to temperatures as low as 10 K, and focused by two hexapole magnets into a temperature-variable 22-pole trap (see text for details). Reproduced with permission from Dieter Gerlich.

2.2. Reactions of Positive Ions with Hydrogen, Nitrogen, and Oxygen

Tables 3, **4**, and **5** summarize the laboratory studies of the reactions of positive ions with H, N, and O atoms, respectively; these studies have been carried out since the comprehensive review of Sablier & Rolando (54). The reaction products, branching ratios, reaction rate constants, and literature references are tabulated. Ions for which no reactions were observed are also identified.

2.2.1. Positive ions with H-atoms. Although this review focuses on the reactivity of ions with H-atoms, some discussion of reactivity with molecular hydrogen is included as these species inevitably coexist in laboratory experiments. The reactions of aromatic cations with hydrogen atoms are particularly important because PAH cations have been implicated as the carriers of the DIBs. The parent molecular cations of benzene ($C_6H_6^+$), naphthalene ($C_{10}H_8^+$), pyrene ($C_{16}H_{10}^+$), chrysene ($C_{18}H_{12}^+$), and coronene ($C_{24}H_{12}^+$) are radical cations, and they react readily with H-atoms by addition (71–73); only in the case of benzene is there a competing H-atom abstraction channel. Similarly, the radical cations $C_6H_4^+$ and $C_{10}H_6^+$ react efficiently with H-atoms. In contrast, the singlet cations $C_6H_5^+$ and $C_{10}H_7^+$ are essentially unreactive with hydrogen atoms, but react with molecular hydrogen by addition. The observation of this latter process at the low pressures of an ICR indicates that radiative association is occurring, even for these relatively small systems (74); thus these reactions are likely to be important in the diffuse ISM. In analogy to these smaller systems, $C_{16}H_9^+$ might be expected to be a singlet cation and therefore unreactive with H-atoms; however, both theoretical calculations and its high reactivity with hydrogen atoms confirm that $C_{16}H_9^+$ is a triplet cation in its ground state. The reactions of the singlet protonated naphthalene cation ($C_{10}H_9^+$) and protonated pyrene cation ($C_{16}H_{11}^+$) with H-atoms or molecular hydrogen are very slow, suggesting that these protonated species will be terminal ionic species in interstellar environments where

Table 3 Experimental results for the reactions of positive ions[a] with H-atoms at 298 (± 5) K, unless otherwise specified

Ionic reactant	Reaction products	Branching ratio	Rate constant (cm^3 s^{-1})	Reference
$C_6H_6^+$ (benzene)	$C_6H_7^+$	~ 0.65	2.1×10^{-10}	76
	$C_6H_5^+ + H_2$	~ 0.35		
$C_{10}H_8^+$ (naphthalene)	$C_{10}H_9^+$	1.0	1.9×10^{-10}	71
$C_{16}H_{10}^+$ (pyrene)	$C_{16}H_{11}^+$	1.0	1.4×10^{-10}	72
$C_{18}H_{12}^+$ (chrysene)	$C_{18}H_{13}^+$	1.0	1.8×10^{-10}	73
$C_{24}H_{12}^+$ (coronene)	$C_{24}H_{13}^+$	1.0	1.4×10^{-10}	73
$C_{10}H_6^+$	$C_{10}H_7^+$	1.0	$\sim 2 \times 10^{-10}$	71
$C_{10}H_7^+$	$C_{10}H_8^+$	1.0	$\leq 5 \times 10^{-11}$	71
$C_{10}H_9^+$	$C_{10}H_{10}^+$	1.0	$\sim 4 \times 10^{-12}$	71
$C_{16}H_9^+$	$C_{16}H_{10}^+$	1.0	$\sim 1.6 \times 10^{-10}$	72
$C_{16}H_{11}^+$	$C_{16}H_{12}^+$	1.0	$\sim 3 \times 10^{-12}$	72
$C_2H_3^+$	$C_2H_2^+ + H_2$	1.0	6.8×10^{-11}	76
$H_2C_3H_2^+$	$C_3H_3^+ + H_2$	1.0	1.7×10^{-10}	76
$HC_3H_3^+$	$C_3H_3^+ + H_2$	1.0	3.0×10^{-10}	76
$C_3H_5^+$	$C_3H_6^+$	1.0	1.6×10^{-10}	76
$C_3H_7^+$	$C_3H_6^+ + H_2$	1.0	3.2×10^{-11}	76
C_4H^+	$C_4H_2^+$	1.0	$\sim 5.8 \times 10^{-10}$	76
$C_4H_2^+$	$C_4H_3^+$	1.0	2.6×10^{-10}	76
$C_4H_3^+$	$C_4H_4^+$	1.0	$\sim 5 \times 10^{-11}$	76
$C_4H_6^+$	$C_2H_3^+ + C_2H_4$	~ 0.15	1.9×10^{-10}	76
	$C_2H_5^+ + C_2H_2$	~ 0.65		
	$C_4H_5^+ + H_2$	~ 0.20		
$C_4H_8^+$	$C_4H_7^+ + H_2$	1.0	1.1×10^{-10}	76
c-$C_6H_4^+$	$C_6H_5^+$	1.0	3.3×10^{-11}	76
CO^+	$H^+ + CO$	1.0	4.0×10^{-10}	78
CO_2^+	$HCO^+ + O$	>0.95	4.7×10^{-10}	78
	$H^+ + CO_2$	<0.05		
SO_2^+	$SO^+ + OH$	1.0	4.2×10^{-10}	78
CS_2^+	$HCS^+ + S$	1.0	2.8×10^{-10}	78
CN^+	$H^+ + CN$	1.0	6.4×10^{-10}	78
$C_2N_2^+$	$HNC^+ + CN$	0.8	6.2×10^{-10}	78
	$C_2H^+ + N_2$	0.2		
$C_2H_3^+$	$C_2H_2^+ + H_2$	1.0	6.8×10^{-11}	78
CH^+	$C^+ + H_2$	1.0	1.3×10^{-9} at $T_H = T_{ion} = 50$ K	70
			8.7×10^{-10} at $T_H = 100$ K, $T_{ion} = 80$ K	
CH_4^+	$CH_3^+ + H_2$	1.0	6.0×10^{-10} at $T_H = T_{ion} = 50$ K	69, 70
			5.1×10^{-10} at $T_H = 100$ K, $T_{ion} = 80$ K	
CH_5^+	$CH_4^+ + H_2$	1.0	9×10^{-12} at $T_H = 12$ K, $T_{ion} = 10$ K	70
			1.3×10^{-11} at $T_H = T_{ion} = 50$ K	
			2.3×10^{-11} at $T_H = T_{ion} = 100$ K	

[a]No reactions were observed for C_2^+, C_2H^+, $HC_3H_2^+$, c-$C_3H_3^+$, $C_4H_5^+$, $C_4H_9^+$, ac-$C_6H_4^+$, ac-$C_6H_5^+$, c-$C_6H_5^+$ (76), or NO_2^+ (78).

Table 4 Experimental results for the reactions of positive ions[a] with N-atoms at 298 (± 5) K

Ionic reactant	Reaction products	Branching ratio	Rate constant (cm^3 s^{-1})	Reference
$C_6H_6^+$ (benzene)	$C_5H_5^+ + HCN$	1.0	1.2×10^{-10}	71
$C_6H_6^+$ (benzene)	$C_5H_5^+ + HCN$	>0.95	1.4×10^{-10}	79
	$C_3H_3^+ + C_3H_3N$	<0.05		
$C_{10}H_8^+$ (naphthalene)	$C_9H_7^+ + HCN$	0.3	2.3×10^{-11}	71
	$C_{10}H_8N^+$	0.7		
$C_{16}H_{10}^+$ (pyrene)	$C_{15}H_9^+ + HCN$	<0.05	1.5×10^{-12}	72
	$C_{16}H_{10}N^+$	>0.95		
$C_{24}H_{12}^+$ (coronene)	None observed		$<1 \times 10^{-12}$	73
$C_{16}H_9^+$	$C_{16}H_9N^+$	>0.8	$\sim 3 \times 10^{-11}$	72
CH_3^+	$HCNH^+ + H$	0.65	9.4×10^{-11}	79
	$HCN^+ + H_2$	0.35		
$C_2H_2^+$	$HC_2N^+ + H$	0.60	2.4×10^{-10}	79
	$HCNH^+ + C$	0.25		
	$C_2N^+ + H_2$	0.10		
	$CH^+ + HCN$	0.05		
$C_2H_3^+$	$HCCN^+ + H_2$	>0.90	2.2×10^{-11}	79
	$H_2CCN^+ + H$	<0.10		
$C_2H_4^+$	$CH_2CNH^+ + H$	1.0	3.0×10^{-10}	79
C_3^+	C_3N^+	1.0	2.3×10^{-10}	79
C_3H^+	$C_3N^+ + H$	0.9	2.7×10^{-10}	79
	HC_3N^+	0.1		
$C_3H_2^+$	$C_2H_2^+ + CN$	0.85	4.4×10^{-11}	79
	$HCNH^+ + C_2$	0.15		
ac-$C_3H_3^+$	$HC_3N^+ + H_2$	1.0	5.8×10^{-11}	79
$C_4H_2^+$	$C_3H^+ + HCN$	0.90	1.9×10^{-10}	79
	$C_4HN^+ + H$	0.05		
	$HCNH^+ + C_3$	0.05		
$C_6H_2^+$	$C_5H^+ + HCN$	1.0	1.9×10^{-10}	79
c-$C_6H_5^+$	$C_5H_4^+ + HCN$	1.0	3.7×10^{-11}	79
CN^+	$N_2^+ + C$	1.0	6.1×10^{-10}	80
HCN^+	$CH^+ + N_2$	1.0	2.2×10^{-10}	80
HC_3N^+	$C_2N^+ + HCN$	0.6	2.4×10^{-10}	80
	$C_3H^+ + N_2$	0.4		
H_2O^+	$NOH^+ + H$	0.8	1.4×10^{-10}	80
	$NO^+ + H_2$	0.2		
N_2^+	N_3^+	1.0	1.4×10^{-11}	80
CO^+	$NO^+ + C$	1.0	8.2×10^{-11}	80
O_2^+	$NO^+ + O$	1.0	1.0×10^{-10}	80
CO_2^+	$CO^+ + NO$	1.0	3.4×10^{-10}	80
H_3^+	$NH_2^+ + H$		4.5×10^{-10}	94
			$<5 \times 10^{-11}$	81

[a]No reactions were observed for $C_{10}H_7^+$ (71); $C_{16}H_{11}^+$ (72); $C_2H_5^+$, c-$C_3H_3^+$, $C_3H_5^+$, $C_4H_3^+$, ac-$C_6H_5^+$ (79); $HCNH^+$, HC_3NH^+, H_3O^+, HCO^+, HCO_2^+ (80).

Table 5 Experimental results for the reactions of positive ions[a] with O-atoms at 298 ($\pm$ 5) K

Ionic reactant	Reaction products	Branching ratio	Rate constant (cm^3 s^{-1})	Reference
$C_6H_6^+$ (benzene)	$C_5H_6^+ + CO$	1.0	9.5×10^{-11}	71
$C_6H_6^+$ (benzene)	$C_5H_6^+ + CO$	0.9	1.4×10^{-10}	82
	$C_4H_4O^+ + C_2H_2$	0.1		
$C_{10}H_8^+$ (naphthalene)	$C_9H_8^+ + CO$	0.55	1.0×10^{-10}	71
	$C_{10}H_8O^+$	0.45		
$C_{16}H_{10}^+$ (pyrene)	$C_{15}H_{10}^+ + CO$	<0.05	9.5×10^{-11}	72
	$C_{16}H_{10}O^+$	>0.95		
$C_{24}H_{12}^+$ (coronene)	$C_{24}H_{12}O^+$	1.0	1.3×10^{-10}	73
$C_{16}H_9^+$	$C_{15}H_9^+ + CO$	~0.5	~2×10^{-10}	72
	$C_{16}H_9O^+$	~0.5		
CH_3^+	$HCO^{+b} + H_2$	1.0	4.1×10^{-10}	82
$C_2H_2^+$	$HCO^{+b} + CH$	0.5	2.0×10^{-10}	82
	$HC_2O^+ + H$	0.5		
$C_2H_3^+$	$H_2CCO^+ + H$	0.85	1.0×10^{-10}	82
	$C_2H_3O^+$	0.10		
	$CH_3^+ + CO$	0.05		
$C_2H_4^+$	$CH_3^+ + HCO$	0.45	2.4×10^{-10}	82
	$HCO^+ + CH_3$	0.35		
	$HC_2O^+ + H_2 + H$	0.10		
	$CH_3CO^+ + H$	~0.05		
	$H_2CCO^+ + H_2$	~0.05		
ac-$C_3H_3^+$	$C_3H_2O^+ + H$	0.30	1.5×10^{-10}	82
	$C_2H_3^+ + CO$	0.30		
	$C_2H_2^+ + HCO$	0.25		
	$HC_3O^+ + H_2$	0.15		
$C_4H_2^+$	$C_4HO^+ + H$	0.50	2.7×10^{-10}	82
	$C_3H_2^+ + CO$	0.40		
	$C_3HO^+ + CH$	~0.05		
	$C_4H_2O^+$	~0.05		
c-$C_6H_5^+$	$C_5H_5^+ + CO$	0.6	1.0×10^{-10}	82
	$C_3H_3^+ + C_3H_2O$	0.4		
HC_3N^+	$C_3NO^+ + H$	0.50	4.1×10^{-10}	83
	$HC_2N^+ + CO$	0.40		
	HC_3NO^+	0.10		
N_2^+	$NO^+ + N$	0.95	1.4×10^{-10}	83
	$O^+ + NO$	0.05		
H_3^+	$OH^+ + H_2$	0.70	1.2×10^{-9}	84
	$H_2O^+ + H$	0.30		

[a]No reactions were observed for $C_{10}H_7^+$ (71); $C_{16}H_{11}^+$ (72); $C_2H_5^+$, c-$C_3H_3^+$, ac-$C_3H_5^+$, $C_4H_3^+$, ac-$C_6H_5^+$ (82); $HCNH^+$, $H_2C_3N^+$, H_2O^+, HCO^+, HCO_2^+ (83).
[b]and/or HOC^+.

they and their derivatives are able to survive. For a comprehensive review of earlier studies of ion-neutral chemistry of PAHs and fullerenes, see Bohme (75).

Based on these and other results, we have modeled (38, 39) the hydrogenation and charge states of PAHs in the ISM for molecules ranging from benzene to those containing 200 carbon atoms. In diffuse clouds, neutral and positively charged species dominate, and the degree of hydrogenation depends strongly on molecular size. Small PAHs (<15–20 carbon atoms) are destroyed in most environments while intermediate-size PAHs (20–30 carbon atoms) are stripped of most peripheral hydrogen atoms. Larger PAHs (>30 carbon atoms) primarily have normal hydrogen coverage (each peripheral carbon atom bears a single hydrogen atom) with competition from the protonated form; these PAHs are good candidates for DIB carriers. Finally, very large PAHs may be fully hydrogenated to give products with two hydrogens on every peripheral carbon atom.

The reactions of a variety of small hydrocarbon cations $C_mH_n^+$ have been reported (76) (**Table 3**). Two types of reaction occur with hydrogen atoms: (1) H-atom transfer to produce $C_mH_{n-1}^+ + H_2$, if this process is exothermic; and (2) association to form $C_mH_{n+1}^+$. With molecular hydrogen as a reactant, highly unsaturated cations undergo primarily hydrogen atom abstraction to form $C_mH_{n+1}^+ + H$; more highly saturated ions are unreactive. McEwan et al. (77) have utilized these data and other experimental and computational results in their model of dense cloud interstellar chemistry. They found that reaction with hydrogen atoms is an important mechanism for increasing the saturation of $C_mH_n^+$ ions with m > 3 in dense interstellar clouds, even though the H/H$_2$ abundance ratio is 10^{-4}. They also concluded that a significant abundance of benzene could be produced in dense clouds through ion-neutral chemistry followed by ion-electron recombination.

The reactions of several oxygen-, sulfur-, and nitrogen-containing cations with hydrogen atoms (78) are also shown in **Table 3**. Reactions proceed by charge transfer, atom transfer, and rearrangement. The rate constants are consistently below the Langevin capture rate, and spin statistics has been identified as the cause in several cases. Notably, the reactant ion CO^+ and three of the product ions (SO^+, HCO^+, and HCS^+) have been detected in dense clouds (**Table 1**).

Recent major advances are the development of the 22-pole ion trap instrument and the measurement of ion-hydrogen atom reactions at low temperatures by Gerlich et al. (69, 70). These researchers have reported the reactions of CH^+, CH_4^+, and CH_5^+ with hydrogen atoms, which proceed by hydrogen abstraction to form molecular hydrogen and C^+, CH_3^+, and CH_4^+, respectively. The reactions of both CH^+ and CH_4^+ are rapid at low temperatures and decrease slightly with increasing temperature. In contrast, the reaction of CH_5^+ is extremely slow at 10 K and increases slightly as the temperature increases. A more complete understanding of these experimental results remains a significant challenge for theorists.

2.2.2. Positive ions with N-atoms. Although molecular nitrogen is relatively unreactive with most ions due to its strong triple bond and high ionization energy, nitrogen atoms exhibit moderate reactivity and rich chemistry. Reaction of N-atom with $C_6H_6^+$ proceeds by abstraction of CH to generate $C_5H_5^+$ and the interstellar neutral hydrogen cyanide (HCN). This pathway rapidly decreases in importance as the aromatic cation increases in size, and simple addition dominates (71–73). For hydrocarbon cations (79) and other small positive ions (80), various product channels are formed in which nitrogen is incorporated into the ionic or neutral product. These reactions provide potential mechanisms for generating the nitriles and other nitrogen-containing species detected in the ISM. The rate constant for the important reaction of $H_3^+ + N$ has been recently refined to be less than 5×10^{-11} cm^3 s^{-1} (81).

2.2.3. Positive ions with O-atoms. Atomic oxygen is remarkably reactive with small aromatic and PAH cations. For $C_6H_6^+$ the major pathway is extrusion of a carbon atom from the ring to form $C_5H_6^+$ and CO. Although the rate constant remains moderately large, simple addition dominates the reactivity for the pyrene and coronene cations (71–73). The reactions of hydrocarbon cations with O-atoms generally proceed at a substantial fraction of the collision rate; the processes occur by multiple channels and show C-O bond formation in the ionic or neutral product (82). The rate constants and product branching ratios for the important reaction of O-atoms with N_2^+ (83) and H_3^+ (84) have been remeasured using the SIFT technique.

2.3. Reactions of Negative Ions with Hydrogen, Nitrogen, and Oxygen

Tables 6, 7, and **8** summarize the laboratory studies of the reactions of negative ions with H, N, and O atoms, respectively, which have been carried out since 1993. The reaction products, branching ratios, reaction rate constants, and literature references are tabulated. Ions for which no reactions were observed are listed in the footnotes.

2.3.1. Negative ions with H-atoms. Our studies (85) of the reactions of carbon chain anions and hydrogenated carbon chain anions were prompted by the proposal (86) that C_7^- might form some of the DIBs, and by the fact that unsaturated hydrocarbon species have been detected in the ISM. The recent discovery of C_4H^-, C_6H^-, and C_8H^- in the ISM confirms the relevance of these anions. We found that C_n^- anions react rapidly with H-atoms by associative detachment, with larger ions (n $\geq$ 7) also reacting by association. The hydrogenated carbon chain anions react exclusively by rapid associative detachment. None of these ions reacts with molecular hydrogen. Similarly, Viggiano and colleagues (87) have found that the small fluorinated hydrocarbon anions CF_3^-, $C_2F_5^-$, and $C_3F_3^-$ react by associative detachment with H and are unreactive with H_2. This research group has also explored the reactivity of fluorinated anions containing sulfur (88) and chlorinated anions containing phosphorus (89) with H-atoms as a function of temperature. Halogen atom abstraction dominates, and the reactions show only small changes in reactivity between 300 and 500 K.

2.3.2. Negative ions with N-atoms. The carbon chain anions and hydrogenated carbon chain anions are unreactive with N_2 but react at moderates rates with N-atoms (90). Associative detachment occurs for all ions, and several atom transfer and

Table 6 Experimental results for the reactions of negative ions[a] with H-atoms at 298 ($\pm$ 5) K, unless otherwise specified

Ionic reactant	Reaction products	Branching ratio	Rate constant (cm^3 s^{-1})	Reference
C_2^-	$C_2H + e^-$	1.0	7.7×10^{-10}	85
C_4^-	$C_4H + e^-$	1.0	6.2×10^{-10}	85
C_5^-	$C_5H + e^-$	1.0	6.2×10^{-10}	85
C_6^-	$C_6H + e^-$	1.0	6.1×10^{-10}	85
C_7^-	$C_7H + e^-$	0.41	6.9×10^{-10}	85
	C_7H^-	0.59		
C_8^-	$C_8H + e^-$	0.33	7.3×10^{-10}	85
	C_8H^-	0.67		
C_9^-	$C_9H + e^-$	0.17	7.2×10^{-10}	85
	C_9H^-	0.83		
C_{10}^-	$C_{10}H + e^-$	0.24	7.5×10^{-10}	85
	$C_{10}H^-$	0.76		
HC_2^-	$C_2H_2 + e^-$	1.0	1.6×10^{-9}	85
HC_4^-	$C_4H_2 + e^-$	1.0	8.3×10^{-10}	85
HC_6^-	$C_6H_2 + e^-$	1.0	5.0×10^{-10}	85
HC_7^-	$C_7H_2 + e^-$	1.0	7.4×10^{-10}	85
CF_3^-	Product $+ e^-$	1.0	5.3×10^{-10}	87
$C_2F_5^-$	Product $+ e^-$	1.0	6.0×10^{-10}	87
$C_3F_3^-$	Product $+ e^-$	1.0	9.0×10^{-10}	87
F^-	$HF + e^-$	1.0	1.4×10^{-9} at 500 K[b]	88
SF_6^-	$SF_5^- + HF$	1.0	3.1×10^{-10} at 298 K	88
			2.5×10^{-10} at 500 K	
SOF_4^-	$SOF_3^- + HF$	1.0	1.8×10^{-10} at 298 K	88
			1.2×10^{-10} at 500 K	
SO_2^-	cis-HOSO $+ e^-$	1.0	2.3×10^{-10} at 298 K	88
			2.6×10^{-10} at 500 K	
$SO_2F_2^-$	$SO_2F^- + HF$	1.0	3.2×10^{-10} at 298 K	88
			2.6×10^{-10} at 500 K	
PO_2Cl^-	$PO_2^- + HCl$	1.0	3.6×10^{-10} at 300 K	89
			3.7×10^{-10} at 500 K	
$POCl_2^-$	$POCl^- + HCl$	0.94	1.6×10^{-10} at 300 K	89
	$Cl^- + HPOCl$	0.06	1.6×10^{-10} at 500 K	
$POCl_3^-$	$POCl_2^- + HCl$	1.0	3.8×10^{-10} at 300 K	89
			3.7×10^{-10} at 500 K	

[a]No reactions were observed for SF_5^-, SOF_3^-, SO_2F^- (88); $PO_2Cl_2^-$ (89).
[b]Measured relative to the rate constant (1.6×10^{-9}) at 298 K (95).

fragmentation channels appear, with formation of C-N bonds. Several of these nitrile species, including CN, C_3N, C_5N, HC_2N, and HC_4N, have been detected in the ISM. Unfortunately, it was not possible to determine the branching ratios for these processes due to the occurrence of rapid secondary reactions.

Recent measurements (91) have found that the rate constant for $O_2^- + N$ is a factor of two slower than previously determined, and that a second product channel ($O^- + NO$, 35%) occurs in addition to the associative detachment reported earlier.

Table 7 Experimental results for the reactions of negative ions[a] with N-atoms at 298 (± 5) K

Ionic reactant	Reaction products	Branching ratio	Rate constant (cm^3 s^{-1})	Reference
C_2^-	$CN^- + C$		2.3×10^{-10}	90
	$C_2N + e^-$			
C_4^-	$CN^- + C_3$		2.0×10^{-10}	90
	$C_3^- + CN$			
	$C_4N + e^-$			
C_5^-	$CN^- + C_4$		2.7×10^{-10}	90
	$C_4^- + CN$			
	$C_3N^- + C_2$			
	$C_5N + e^-$			
C_6^-	$CN^- + C_5$		1.5×10^{-10}	90
	$C_5^- + CN$			
	$C_3N^- + C_3$			
	$C_6N + e^-$			
C_7^-	$CN^- + C_6$		2.2×10^{-10}	90
	$C_6^- + CN$			
	$C_3N^- + C_4$			
	$C_5N^- + C_2$			
	$C_7N + e^-$			
HC_2^-	$HCCN + e^-$	1.0	5×10^{-11}	90
HC_4^-	$CN^- + HC_3$		$\sim 6 \times 10^{-12}$	90
	$CN + HC_3^-$			
	$HC_4N + e^-$			
HC_6^-	$CN^- + HC_5$		$\sim 1 \times 10^{-11}$	90
	$C_3N^- + HC_3$			
	$HC_6N + e^-$			
O_2^-	$NO_2 + e^-$	0.65	2.3×10^{-10}	91
	$O^- + NO$	0.35		
SO_2^-	$SO^- + NO$	>0.90	1.8×10^{-10}	88
	$S^- + NO_2$	<0.10		
PO_2Cl^-	$PO_2^- + NCl$	0.34	8.0×10^{-11}	92
	$PO_2N^- + Cl$	0.66		

[a]No reactions were observed for SF_5^-, SF_6^-, SO_2F^-, $SO_2F_2^-$, SOF_3^-, SOF_4^- (88); $POCl_3^-$, $POCl_2^-$ (92).

Rate constants and product branching ratios were recently determined for the reactions of SO_2^- (88) and PO_2Cl^- (92) with N-atoms; various other ions (**Table 7**) were found to be unreactive.

2.3.3. Negative ions with O-atoms. Oxygen atoms react rapidly with carbon chain anions and hydrogenated carbon chain anions by carbon atom abstraction to generate carbon monoxide; associative detachment also occurs (90). Moreover, for the hydrogenated anions, oxygen/hydrogen exchange is observed. The rate constants for several of these reactions exceed the Langevin collision rate; a new approach that includes the polarizability of the anion has been developed and provides good agreement with experiments (93).

Table 8 Experimental results for the reactions of negative ions with O-atoms at 298 ($\pm$5) K

Ionic reactant	Reaction products	Branching ratio	Rate constant (cm^3 s^{-1})	Reference
C_2^-	$C^- + CO$		5.8×10^{-10}	90
	$C_2O + e^-$			
C_4^-	$C_3^- + CO$		5.6×10^{-10}	90
	$C_4O + e^-$			
C_5^-	$C_4^- + CO$		6.4×10^{-10}	90
	C_5O^-			
	$C_5O + e^-$			
C_6^-	$C_5^- + CO$		4.7×10^{-10}	90
	$C_6O + e^-$			
C_7^-	$C_6^- + CO$		5.3×10^{-10}	90
	$C_7O + e^-$			
HC_2^-	$HC^- + CO$		6.2×10^{-10}	90
	$H + C_2O^-$			
	$HC_2O + e^-$			
HC_4^-	$HC_3^- + CO$		5.3×10^{-10}	90
	$H + C_4O^-$			
	$HC_4O + e^-$			
HC_6^-	$HC_5^- + CO$		5.4×10^{-10}	90
	$H + C_6O^-$			
	$HC_6O + e^-$			
O_2^-	$O_3 + e^-$	<0.55	3.9×10^{-10}	91
	$O^- + O_2$	>0.45		
SF_5^-	$F^- + SOF_4$	1.0	5.4×10^{-11}	88
SF_6^-	$O^- + SF_6$	1.0	1.1×10^{-10}	88
SO_2F^-	$F^- + SO_3$	0.60	1.5×10^{-10}	88
	$SO_3^- + F$	0.40		
$SO_2F_2^-$	$SO_3F^- + F$	1.0	9.1×10^{-11}	88
SO_2^-	$O^- + SO_2$	1.0	4.0×10^{-10}	88
SOF_3^-	$SO_2F_2^- + F$	1.0	7.7×10^{-11}	88
SOF_4^-	$F^- + (SO_2F_2 + F)$	1.0	7.9×10^{-11}	88
$POCl_3^-$	$POCl_2^- + ClO$	1.0	3.9×10^{-10}	92
$POCl_2^-$	$PO_2Cl^- + Cl$	$\geq$0.84	3.7×10^{-10}	92
	$Cl^- + PO_2Cl$	$\leq$0.09		
	$PO_2^- + Cl_2$	$\leq$0.04		
	$Cl_2^- + PO_2$	$\leq$0.03		
PO_2Cl^-	$PO_2^- + ClO$	0.34	2.6×10^{-10}	92
	$PO_3^- + Cl$	0.66		

The newly determined rate constant (91) for reaction of O_2^- with oxygen atom is slightly larger than the previously reported value. Reactions of fluorinated sulfur anions show charge transfer and F-atom or F^- exchange in the reactant ion (88). Reactions of $PO_xCl_y^-$ are rapid and proceed primarily by loss of chlorine from the reactant ion (92).

SUMMARY POINTS

1. The ISM is extremely heterogeneous with a broad range of temperatures, densities, and extinction parameters.

2. Spectroscopic observations throughout the electromagnetic spectrum have detected more than 140 molecular species in the ISM, despite the relatively hostile conditions.

3. Chemical modeling of interstellar clouds relies on accurate laboratory data, including rates and products of ion-neutral reactions.

4. Studies of ion-atom reactions pose unique experimental challenges. Two innovative techniques, the SIFT and the multi-electrode ion trap, have proven to be powerful approaches for these studies.

5. The past fifteen years have seen active research in experimental studies of both positive and negative ions reacting with hydrogen, nitrogen, and oxygen atoms; many of these systems are relevant to interstellar chemistry.

FUTURE ISSUES

1. Laboratory studies of ion-atom reactions at extremely low temperatures with newly developed instrumentation will provide unprecedented data.

2. The detection of negative ions in the ISM highlights the need for a comprehensive understanding of negative ion chemistry.

3. Experiments with other atoms, especially carbon, and with molecular radicals detected in interstellar clouds pose challenges for future studies of ion reactions.

4. Large PAHs, as well as PAHs containing nitrogen and oxygen, are possible carriers of the diffuse interstellar bands; experimental studies of their ion chemistry are essential.

5. Additional studies of other gas-phase reactions, including radiative association and dissociative recombination, as well as full characterization of their products, are critical for developing accurate chemical models of interstellar clouds. Synthetic routes to biological models remain an intriguing challenge.

6. The future of astrochemistry is bright: The synergy of astronomical observations and laboratory studies will continue to advance our understanding of both the ISM and its fundamental chemical processes.

DISCLOSURE STATEMENT

The authors are not aware of any biases that might be perceived as affecting the objectivity of this review.

ACKNOWLEDGMENTS

We gratefully acknowledge support of our research by the National Aeronautics and Space Administration. We thank Dieter Gerlich for generously supplying **Figure 7** as well as unpublished results, and we acknowledge many colleagues for responding to our request for information: Nigel Adams, Peter Barnes, Edwin Bergin, Stephen Blanksby, Diethard Bohme, Andre Canosa, Eric Herbst, Jan M. Hollis, Eva Kovacevic, Harvey Liszt, Murray McEwan, Tom Millar, Holger Mueller, Mark Smith, Albert Viggiano, Valentine Wakelam, Adam Walters, Mark Wolfire, Daniel Wolf-Savin, Paul Woods, and David Woon. We thank Joshua Destree, Kimberly Johnson, Oscar Martinez, and Teresa Ross for assistance with figures and tables.

LITERATURE CITED

1. Herbst E. 2001. The chemistry of interstellar space. *Chem. Soc. Rev.* 30:168–76
2. Smith D. 1992. The ion chemistry of interstellar clouds. *Chem. Rev.* 92:1473–85
3. Dunham T. 1937. Interstellar neutral potassium and neutral calcium. *Pub. Astron. Soc. Pacific* 49:26–28
4. McKellar A. 1940. Evidence for the molecular origin of some hitherto unidentified interstellar lines. *Pub. Astron. Soc. Pacific* 52:187–92
5. Adams WS. 1941. Some results with the coudé spectrograph of the Mount Wilson Observatory. *Astrophys. J.* 93:11
6. Douglas AE, Herzberg G. 1941. CH^+ in interstellar space and in the laboratory [note]. *Astrophys. J.* 94:381
7. Bates DR, Spitzer L. 1951. The density of molecules in interstellar space. *Astrophys. J.* 113:441–63
8. Solomon PM, Klemperer W. 1972. The formation of diatomic molecules in interstellar clouds. *Astrophys. J.* 178:389–422
9. Watson WD. 1973. Interstellar molecular reactions. *Rev. Modern Phys.* 48:13–52
10. Dalgarno A. 2000. In *Astrochemistry: Historical Perspective and Future Challenges*. Presented at IAU Symp. 197, p. 1, Aug. 23–27, 1999, Sogwipo, Cheju, Korea
11. Weinreb S, Barrett AH, Meeks ML, Henry JC. 1963. Radiospectroscopic observations of OH in interstellar space. *Nature* 200:829–31
12. Rydbeck OEH, Hjalmarson A. 1985. Radio observations of interstellar molecules, of their behavior, and of their physics. In *Molecular Astrophysics: State of the Art and Future Directions*, pp. 45–174. Bad Windsheim, Germany: Dordrecht: Reidel
13. Thorwirth S. 2007. et seq. The Cologne Database for Molecular Spectroscopy. **http://www.ph1.uni-koeln.de/vorhersagen/**
14. Kuan Y-J, Charnley SB, Huang H-C, Tseng W-L, Kisiel Z. 2003. Interstellar glycine. *Astrophys. J.* 593:848–67
15. Snyder LE, Lovas FJ, Hollis JM, Friedel DN, Jewell PR, et al. 2005. A rigorous attempt to verify interstellar glycine. *Astrophys. J.* 619:914–30
16. Cernicharo J, Guélin M, Agúndez M, Kawaguchi K, McCarthy M, Thaddeus P. 2007. Astronomical detection of C_4H^-, the second interstellar anion. *Astron. Astrophys. J.* 467:L37–L40

17. McCarthy MC, Gottlieb CA, Gupta H, Thaddeus P. 2006. Laboratory and astronomical identification of the negative molecular ion C_6H^-. *Astrophys. J. Lett.* 652:L141–44

18. Remijan AJ, Hollis JM, Lovas FJ, Cordiner MA, Millar TJ. 2007. Detection of C_8H^- and comparison with C_8H toward IRC +10 216. *Astrophys. J. Lett.* 664:L47–L50

19. Allamandola LJ, Tielens AGGM, Barker JR. 1989. Interstellar polycyclic aromatic hydrocarbons: the infrared emission bands, the excitation/emission mechanism, and the astrophysical implications. *Astrophys. J. Supp.* 71:733–75

20. Puget JL, Léger A. 1989. A new component of the interstellar matter: small grains and large aromatic molecules. *Ann. Rev. Astron. Astrophys.* 27:161–98

21. Léger A, D'Hendecourt L, Boccara N, eds. 1987. *Polycyclic Aromatic Hydrocarbons and Astrophysics*, vol. 191. Dordrecht, Neth.: Reidel

22. Gautier TN, Fink U, Larson HP, Treffers RR. 1976. Detection of quadrupole emission in the Orion Nebula. *Astrophys. J. Lett.* 207:L129–L33

23. Shull JM, Beckwith S. 1982. Interstellar molecular hydrogen. *Ann. Rev. Astron. Astrophys.* 20:163–90

24. Lacy JH, Knacke R, Geballe TR, Tokunaga AT. 1994. Detection of absorption by H_2 in molecular clouds: a direct measurement of the H_2:CO ratio. *Astrophys. J. Lett.* 428:L69–L72

25. Black JH, Willner SP. 1984. Interstellar absorption lines in the infrared spectrum of NGC 2024 IRS 2. *Astrophys. J.* 279:673–78

26. Snow TP, McCall BJ. 2006. Diffuse atomic and molecular clouds. *Ann. Rev. Astron. Astrophys.* 44:367–414

27. Markwick-Kemper AJ. 2005. The most important reactions in gas-phase astrochemical models. In *Astrochemistry: Recent Successes and Current Challenges*. Presented at IAU Symp. 231, Aug. 29-Sept. 2, 2004, Pacific Grove, CA

28. Wakelam V, Herbst E, Selsis F. 2006. The effect of uncertainties on chemical models of dark clouds. *Astron. Astrophys.* 451:551–62

29. Woodall J, Agúndez M, Markwick-Kemper AJ, Millar TJ. 2007. The UMIST database for astrochemistry 2007. *Astron. Astrophys. J.* 466:1197–204

30. Herbst E. 2007. **www.physics.ohio-state.edu/~eric/**

31. Millar TJ, Walsh C, Cordiner MA, Chuimín RN, Herbst E. 2007. Hydrocarbon anions in interstellar clouds and circumstellar envelopes. *Astrophys. J. Lett.* 662:L87–L90

32. Le Petit F, Nehmé C, Le Bourlot J, Roueff E. 2006. A model for atomic and molecular interstellar gas: The Meudon PDR Code. *Astrophys. J. Suppl.* 164:506–29

33. Stepanovic M, Betts NB, Eichelberger BR, Snow TP, Bierbaum VM. 2008. Gas phase reactions of organic anions with atoms. Manuscript in preparation

34. Blagojevic V, Petrie S, Bohme DK. 2003. Gas-phase syntheses for interstellar carboxylic and amino acids. *Mon. Not. R. Astron. Soc.* 339:L7–L11

35. Heger ML. 1922. Further study of the sodium lines in class B stars; The spectra of certain class B stars in the regions 5630Å-6680Å and 3280Å-3380Å; Note on the spectrum of γ Cassiopeiae between 5860Å and 6600Å. *Lick Observatory Bull. No.* 337:141–48

36. Herbig GH. 1995. The diffuse interstellar bands. *Ann. Rev. Astron. Astrophys.* 33:19–73

37. Snow TP. 2001. The unidentified diffuse interstellar bands as evidence for large organic molecules in the interstellar medium. *Spectrchem. Acta, Part A* 75:615–26

38. Le Page V, Snow TP, Bierbaum VM. 2001. Hydrogenation and charge states of PAHs in diffuse clouds. I. Development of a model. *Astrophys. J. Suppl.* 132:233–51

39. Le Page V, Snow TP, Bierbaum VM. 2003. Hydrogenation and charge states of PAHs in diffuse clouds. II. Results. *Astrophys. J.* 584:316–30

40. Hudgins DM, C. W. Bauschlicher J, Allamandola LJ. 2005. Variations in the peak position of the 6.2 μm interstellar emission feature: a tracer of N in the interstellar polycyclic aromatic hydrocarbon population. *Astrophys. J.* 632:316–32

41. Jensen AG, Snow TP. 2007. Is there enhanced depletion of gas-phase nitrogen in moderately reddened lines of sight? *Astrophys. J.* 654:955–70

42. Lepp S, Stancil PC, Dalgarno A. 2002. Atomic and molecular processes in the early universe. *J. Phys. B: At. Mol. Opt. Phys.* 35:R57–R80

43. Gould RJ, Salpeter EE. 1963. The interstellar abundance of the hydrogen molecule. I. Basic processes. *Astrophys. J.* 138:393–407

44. Hollenbach D, Salpeter EE. 1971. Surface recombination of hydrogen molecules. *Astrophys. J.* 163:155–64

45. McDowell MRC. 1961. On the formation of H_2 in H I regions. *Observatory* 81:240–43

46. Glover SC, Savin DW, Jappsen A-K. 2005. Cosmological implications of the uncertainty in H^- destruction rate coefficients. *Astrophys. J.* 640:553–68

47. Smith IWM, Rowe BR. 2000. Reaction kinetics at very low temperatures: laboratory studies and interstellar chemistry. *Acc. Chem. Res.* 33:261–68

48. Vestal ML. 2001. Methods of ion generation. *Chem. Rev.* 101:361–75

49. McLuckey SA, Wells JM. 2001. Mass analysis at the advent of the 21st century. *Chem. Rev.* 101:571–606

50. Petrie S, Bohme DK. 2003. Mass spectrometric approaches to interstellar chemistry. *Top. Curr. Chem.* 225:37–75

51. Ikezoe Y, Matsuoka S, Takebe M, Viggiano A. 1987. *Gas Phase Ion-Molecule Reaction Rate Constants Through 1986.* Tokyo: Maruzen

52. Anicich VG. 1993. A survey of bimolecular ion-molecule reactions for use in modeling the chemistry of planetary atmospheres, cometary comae, and interstellar clouds: 1993 supplement. *Astrophys. J. Supp.* 84:215–315

53. Anicich VG. 1993. Evaluated bimolecular ion-molecule gas phase kinetics of positive ions for use in modeling planetary atmospheres, cometary comae, and interstellar clouds. *J. Phys. Chem. Ref. Data* 22:1469–569

54. Sablier M, Rolando C. 1993. Gas-phase ion-atom reactions. *Mass Spectrom. Rev.* 12:285–312

55. Ferguson EE, Fehsenfeld FC, Schmeltekopf AL. 1969. Flowing afterglow measurements of ion-neutral reactions. *Adv. At. Mol. Phys.* 5:1–56

56. Graul ST, Squires RR. 1988. Advances in flow reactor techniques for the study of gas-phase ion chemistry. *Mass Spectrom. Rev.* 7:263–358

57. Bierbaum VM. 2003. Instrumentation: flow tubes. In *The Encyclopedia of Mass Spectrometry*, ed. PB Armentrout, pp. 98–109. Amsterdam, Neth.: Elsevier

58. Dunkin DB, Fehsenfeld FC, Schmeltekopf AL, Ferguson EE. 1968. Ion-molecule reaction studies from 300 to 600 K in a temperature-controlled flowing afterglow system. *J. Chem. Phys.* 49:1365–71

59. Hierl PM, Friedman JF, Miller TM, Dotan I, Mendendez-Barreto M, et al. 1996. Flowing afterglow apparatus for the study of ion-molecule reactions at high temperatures. *Rev. Sci. Instrum.* 67:2142–48

60. McFarland M, Albritton DL, Fehsenfeld FC, Ferguson EE, Schmeltekopf AL. 1973. Flow-drift technique for ion mobility and ion-molecule reaction rate constant measurements. I. Apparatus and mobility measurements. *J. Chem. Phys.* 59:6610–19

61. Poutsma JC, Seburg RA, Chyall LJ, Sunderlin LS, Hill BT, et al. 1997. Combining electrospray ionization and the flowing afterglow method. *Rapid Commun. Mass Spectrom.* 11:489–93

62. Smith D, Adams NG. 1984. Studies of plasma reaction processes using a flowing-afterglow/Langmuir probe apparatus. In *Swarms of Ions and Electrons in Gases*, ed. W Lindinger, TD Märk, F Howorka, pp. 194–217. Vienna, Austria: Springer

63. Bierbaum VM, Ellison GB, Leone SR. 1984. Flowing afterglow studies of ion reaction dynamics. In *Gas Phase Ion Chemistry*, ed. MT Bowers, pp. 1–39. New York: Academic

64. Squires RR, Lane KR, Lee RE, Wright LG, Wood KV, Cooks RG. 1985. A tandem flowing afterglow-triple quadrupole instrument. *Int. J. Mass Spectrom. Ion Proc.* 64:185–91

65. Adams NG, Smith D. 1976. The selected ion flow tube: a technique for studying ion-neutral reactions. *Int. J. Mass Spectrom. Ion Phys.* 21:349–59

66. Setser DW, Kolts JH. 1979. Electronically excited long-lived states of atoms and diatomic molecules in flow systems. In *Reactive Intermediates in the Gas Phase*, ed. DW Setser, pp. 152. New York: Academic

67. Trainor DW, Ham DO, Kaufman F. 1973. Gas phase recombination of hydrogen and deuterium atoms. *J. Chem. Phys.* 58:4599–609

68. Goldan PD, Schmeltekopf AL, Fehsenfeld FC, Schiff HL, Ferguson EE. 1966. Thermal energy ion-neutral reaction rates. II. Some reactions of ionospheric interest. *J. Chem. Phys.* 44:4095–103

69. Gerlich D, Smith M. 2006. Laboratory astrochemistry: studying molecules under inter- and circumstellar conditions. *Phys. Scr.* 73:C25–C31

70. Luca A, Borodi G, Gerlich D. 2006. Interactions of ions with hydrogen atoms. In *Progress Rep. XXIV ICPEAC 2005*, ed. FD Colavecchia, PD Fainstein, J Fiol, MAP Lima, JE Miraglia, et al., p. 20. Rosario, Argent.

71. Le Page V, Keheyan Y, Snow TP, Bierbaum VM. 1999. Reactions of cations derived from naphthalene with molecules and atoms of interstellar interest. *J. Am. Chem. Soc.* 121:9435–46

72. Le Page V, Keheyan Y, Snow TP, Bierbaum VM. 1999. Gas phase chemistry of pyrene and related cations with molecules and atoms of interstellar interest. *Int. J. Mass Spectrom.* 185/186/187:949–59

73. Betts NB, Stepanovic M, Snow TP, Bierbaum VM. 2006. Gas-phase study of coronene cation reactivity of interstellar relevance. *Ap. J.* 651:L129–L131

74. Snow TP, Le Page V, Keheyan Y, Bierbaum VM. 1998. The interstellar chemistry of PAH cations. *Nature* 391:259–60

75. Bohme DK. 1992. PAH and fullerene ions and ion/molecule reactions in interstellar and circumstellar chemistry. *Chem. Rev.* 92:1487–508

76. Scott GBI, Fairley DA, Freeman CG, McEwan MJ, Adams NG, Babcock LM. 1997. $C_mH_n^+$ reactions with H and H_2: an experimental study. *J. Phys. Chem. A* 101:4973–78

77. McEwan MJ, Scott GBI, Adams NG, Babcock LM, Terzieva R, Herbst E. 1999. New H and H_2 reactions with small hydrocarbon ions and their roles in benzene synthesis in dense interstellar clouds. *Ap. J.* 513:287–93

78. Scott GB, Fairley DA, Freeman CG, McEwan MJ, Spanel P, Smith D. 1997. Gas phase reactions of some positive ions with atomic and molecular hydrogen at 300 K. *J. Chem. Phys.* 106:3982–87

79. Scott GBI, Fairley DA, Freeman CG, McEwan MJ, Anicich VG. 1999. $C_mH_n^+$ reactions with atomic and molecular nitrogen: an experimental study. *J. Phys. Chem. A* 103:1073–77

80. Scott GBI, Fairley DA, Freeman CG, McEwan MJ, Anicich VG. 1998. Gas-phase reactions of some positive ions with atomic and molecular nitrogen. *J. Chem. Phys.* 109:9010–14

81. Milligan DB, Fairley DA, Freeman CG, McEwan MJ. 2000. A flowing afterglow selected ion flow tube (FA/SIFT) comparison of SIFT injector flanges and H_3^+ + N revisited. *Int. J. Mass Spectrom.* 202:351–61

82. Scott GBI, Milligan DB, Fairley DA, Freeman CG, McEwan MJ. 2000. A selected ion flow tube study of the reactions of small $C_mH_n^+$ ions with O atoms. *J. Chem. Phys.* 112:4959–65

83. Scott GBI, Fairley DA, Milligan DB, Freeman CG, McEwan MJ. 1999. Gas phase reactions of some positive ions with atomic and molecular oxygen and nitric oxide at 300 K. *J. Phys. Chem. A* 103:7470–73

84. Milligan DB, McEwan MJ. 2000. H_3^+ + O: an experimental study. *Chem. Phys. Lett.* 319:482–85

85. Barckholtz C, Snow TP, Bierbaum VM. 2001. Reactions of C_n^- and C_nH^- with atomic and molecular hydrogen. *Ap. J.* 547:L171–74

86. Tulej M, Kirkwood DA, Pachkov M, Maier JP. 1998. Gas-phase electronic transitions of carbon chain anions coinciding with diffuse interstellar bands. *Ap. J.* 506:L69–L73

87. Morris RA, Viggiano AA, Paulson JF. 1994. Electron detachment reactions of fluorinated carbanions with atomic hydrogen. *J. Chem. Phys.* 100:1767–68

88. Midey AJ, Viggiano AA. 2007. Kinetics of sulfur oxide, sulfur fluoride, and sulfur oxyfluoride anions with atomic species at 298 and 500 K. *J. Phys. Chem. A* 111:1852–59

89. Midey AJ, Miller TM, Morris RA, Viggiano AA. 2005. Reactions of $PO_xCl_y^-$ ions with H and H_2 from 298 to 500 K. *J. Phys. Chem. A* 109:2559–63

90. Eichelberger BR, Snow TP, Barckholtz C, Bierbaum VM. 2007. Reactions of H, N, and O atoms with carbon chain anions of interstellar interest: an experimental study. *Ap. J.* 667:1283–89

91. Poutsma JC, Midey AJ, Viggiano AA. 2006. Absolute rate coefficients for the reactions of $O_2^- + N(^4S_{3/2})$ and $O_2^- + O(^3P)$ at 298 K in a selected-ion flow tube instrument. *J. Chem. Phys.* 124:074301

92. Poutsma JC, Midey AJ, Thompson TH, Viggiano AA. 2006. Absolute rate coefficients and branching percentages for the reactions of $PO_xCl_y^- + N\ (^4S_{3/2})$ and $PO_xCl_y^- + O\ (^3P)$ at 298 K in a selected-ion flow tube instrument. *J. Phys. Chem. A* 110:11315–19

93. Eichelberger BR, Snow TP, Bierbaum VM. 2003. Collision rate constants for polarizable ions. *J. Am. Soc. Mass Spectrom.* 14:501–5

94. Scott GBI, Fairley DA, Freeman CG, McEwan MJ. 1997. The reaction $H_3^+ + N$: a laboratory measurement. *Chem. Phys. Lett.* 269:88–92

95. Fehsenfeld FC, Howard CJ, Ferguson EE. 1973. Thermal energy reactions of negative ions with H atoms in the gas phase. *J. Chem. Phys.* 58:5841–42

Plasma Diagnostics for Unraveling Process Chemistry

Joshua M. Stillahn, Kristina J. Trevino, and Ellen R. Fisher

Department of Chemistry, Colorado State University, Fort Collins, Colorado 80523-1872; email: jstill@lamar.colostate.edu, ktrevino@lamar.colostate.edu, erfisher@lamar.colostate.edu

Annu. Rev. Anal. Chem. 2008. 1:261–91

First published online as a Review in Advance on February 26, 2008

The *Annual Review of Analytical Chemistry* is online at anchem.annualreviews.org

This article's doi: 10.1146/annurev.anchem.1.031207.112953

1936-1327/08/0719-0261$20.00

Key Words

deposition, etching, surface modification, gas-surface interactions

Abstract

This review focuses on the use of diagnostic tools to examine plasma processing chemistry, primarily plasma species energetics, dynamics, and molecule-surface reactions. We describe the use of optical diagnostic tools, mass spectrometry, and Langmuir probes in measuring species densities, rotational and kinetic energies, and plasma-surface reactions. Molecule-surface interactions for MX_n species (M = C, Si, N; X = H, F, Cl) are presented and interpreted with respect to the molecule's electronic configuration and dipole moments.

1. INTRODUCTION

T_s: substrate temperature

Plasmas, or partially ionized gases, are complex systems containing a range of reactive species including radicals, metastables, ions, electrons, and photons. Although plasmas can be generated in several ways, they are most commonly created in the laboratory using radio frequency (rf), microwave, or direct current (dc) applied electrical power. Because of the reactive nature of plasma species, a multitude of reactions can occur in the gas phase or at gas-surface interfaces. The result of the interactions of gas-phase plasma species with surfaces (either reactor walls or substrates) generally falls into three categories: (1) etching, or removal of material, often in a selective manner; (2) deposition, wherein a distinctly different chemical material is formed on a substrate; and (3) surface modification, which refers to implantation of chemical functional groups in the outermost surface layer. These processes all rely on a complex set of intertwined chemical reactions that are difficult to understand on a molecular level.

An additional factor contributing to plasma chemistry complexity is the number of system variables used for processes optimization, including equipment variables (reactor size and configuration, materials of construction, method and amount of power applied); gas variables (pressure, flow, and gas ratios); and substrate variables [substrate temperature (T_S), material, and location in reactor]. Often, small changes in a single parameter can result in large changes in the overall process chemistry. In some systems, the balance between etching and deposition is so sensitive to these parameters that controlling the outcome is challenging and miscalculations can result in significant waste. Thus, it is imperative to develop analytical diagnostic tools capable of providing straightforward and reliable process chemistry data.

Understanding fundamental plasma chemistry has been an elusive goal for plasma scientists, primarily because of the systems' complexity. Consequently, relatively little is known about mechanisms for plasma processing. Studies focusing on correlating surface properties with plasma parameters can reveal the nature of the relationships between gas-phase species and processed surface composition. Although these simplistic relationships often offer support for assumptions about deposition and etching mechanisms, they do not directly address gas-surface interface chemistry and thus provide an incomplete picture. A more global representation of molecular-level chemistry is critical, but requires diagnostic tools that (1) allow for the identification and quantification of all plasma species (charged and neutral), preferably temporally and spatially resolved; (2) can be performed in a nonintrusive manner; (3) provide data on the gas phase, surface, and gas-surface interface; and (4) characterize the internal and kinetic energies of plasma species to estimate rate constants and provide energy partitioning information.

Clearly, no one diagnostic tool can fulfill all of these needs. Consequently, many studies that focus on unraveling process chemistry rely on a combination of techniques, affording broader descriptions of plasma processes. As a result of the need for diagnostic tools in both industrial processes and fundamental studies, plasma diagnostics have been the subject of numerous review articles and books (1–4). Here, we focus on a few complex plasma systems that have benefited from the application of highly sophisticated diagnostic tools to unraveling the underlying molecular-level chemistry.

OES: optical emission spectroscopy

2. DIAGNOSTIC TECHNIQUES

An array of techniques can be applied to the examination of plasma chemistry. Here, we focus primarily on nonintrusive, in situ optical gas-phase diagnostics, although some nonoptical and surface techniques are included. This is by no means an exhaustive list, and detailed descriptions of the instruments can be found elsewhere (1, 5–7).

2.1. Optical Emission Spectroscopy

Optical emission spectroscopy (OES) analyzes light emitted from a given medium in the absence of external excitation via collection, dispersion, and detection of the light (1). In a plasma, gas-phase species are promoted to excited electronic states by collisions with energetic electrons and relaxation is accompanied by emission of a photon. In OES, emitted radiation is spectrally dispersed and detected. In its simplest configuration, OES requires only a means of collecting the light emitted (e.g., an optical fiber), a dispersing element (a grating), and a detector [a photomultiplier tube (PMT) or charge-coupled device (CCD)]. Thus, OES is an inexpensive, real-time monitoring system that can identify emitting plasma species.

OES can be employed quantitatively or qualitatively for plasma species identification and determination of absolute or relative species densities. Identification requires knowledge of the emission lines of a given plasma species (**Table 1**). OES has proven useful in understanding gas-phase kinetics and reaction mechanisms, etching endpoint detection, and performing spatial and temporal measurements of species densities (3, 8, 9). Although quantitative OES is possible, it must be used cautiously because signal intensity is not always directly related to concentration. This is often addressed with actinometry, wherein emission intensities are compared to the relatively constant emission of an actinometer, generally an inert gas (e.g., Ar) or a combination of actinometers added in small quantities. Time-resolved OES (TR-OES) has also been employed (10).

OES is limited to excited-state species that radiatively decay and whose detectable wavelength range is hindered by the collection window, generally precluding vacuum UV (VUV) emitters. Instrument resolution (often only ~0.5–1 nm) and signal-to-noise ratios also limit the utility of OES.

2.2. Optical Absorption Spectroscopy

Optical absorption spectroscopy (OAS) is an alternate probe for excited-state species that measures the light absorbed by a sample at a particular wavelength. OAS is used to probe highly excited molecules because these states are long lived, yet do not decay via emission of a visible photon. Thus, they are not easily probed by OES (1). OAS spectrometers consist of a light source, typically a tungsten-filament or gas-discharge

Table 1 Emission lines observed in optical emission spectroscopy studies of selected plasma species

Species	Wavelength (nm)	Transition	Reference(s)
Ar	696.5	$1s_5 - 2p_2$	136
	706.7	$1s_5 - 2p_3$	
	738.4	$1s_4 - 2p_3$	
	750.4	$1s_2 - 2p_1$	
	751.5	$1s_4 - 2p_5$	
	763.5	$1s_5 - 2p_3$	
	772.4	$1s_5 - 2p_7$	
	794.8	$1s_3 - 2p_4$	
	800.6	$1s_4 - 2p_6$	
	801.5	$1s_5 - 2p_8$	
	810.4	$1s_4 - 2p_7$	
	811.5	$1s_5 - 2p_9$	
	826.5	$1s_2 - 2p_2$	
	840.8	$1s_2 - 2p_3$	
	842.5	$1s_4 - 2p_8$	
	852.1	$1s_2 - 2p_3$	
C_2	469.8, 471.6, 473.7	$^3\Pi \rightarrow {}^3\Pi$	137
	512.9, 516.5, 558.6	$^3\Pi \rightarrow {}^3\Pi$	
	563.6	$^3\Pi \rightarrow {}^3\Pi$	
C_3	405.1	$A^1\Pi_u \rightarrow X^1\Sigma^+_g$	138
CF_2	251.9	$A^1B_1 \rightarrow X^1A_1$	65, 84
CH	389	$B^2\Sigma \rightarrow X^2\Pi$	116
	430, 431.4	$A^2\Delta \rightarrow X^2\Pi$	117, 139
CN	304.2, 387	$B^2\Sigma^+ \rightarrow X^2\Sigma^+$	116, 140
CO	283, 292.2, 297	$b^3\Sigma \rightarrow a^3\Pi$	137
	302.8, 313.8, 325.3	$b^3\Sigma \rightarrow a^3\Pi$	
	451.1, 483.5, 518.6, 561	$B^1\Sigma^+ \rightarrow A^1\Pi$	
F	685.4	$3p^4D_{7/2} \rightarrow 3s^4P_{5/2}$	141
	703.7, 712.8	$2p^43p \rightarrow 2p^43s$	142
H	434, 486.1, 656.5	$^2P^\circ \rightarrow {}^2D$	137
N	674	$4d^4P \rightarrow 3p^4P^\circ$	116
O	777.2, 844.7	$^3S^\circ \rightarrow {}^3P$	137
NO	247.9, 288.5, 289.3, 303.5, 304.3	$A^2\Sigma^+ \rightarrow {}^2\Pi$	137
	319.8, 320.7, 337.7, 338.6	$A^2\Sigma^+ \rightarrow {}^2\Pi$	
N_2	315.9, 337.1	$C^3\Pi \rightarrow B^3\Pi$	137
OH	281.1, 306.4, 307.8, 308.9	$^2\Sigma \rightarrow {}^2\Pi$	137

lamp, which is directed into a sample chamber, and a detector placed on the opposite side of the sample to analyze transmitted light (11). Thus, OAS is comparable to OES in equipment simplicity.

For gas-phase OAS analyses, the light source can be tuned to a specific optical transition and time-dependent information about a particular species can be obtained. Alternatively, gas-phase absorption can be measured using self absorption, which has been described in detail elsewhere (12). This allows for quantitative analysis of absorbing species, provided that line shapes and spatial distributions are known.

2.3. Laser-Induced Fluorescence

LIF: laser-induced fluorescence

A common optical plasma diagnostic is laser-induced fluorescence (LIF), which probes ground-state species with sensitivities on the order of 10^8 cm^{-3}. An LIF apparatus generally consists of a tunable laser (e.g., excimer or Nd:YAG-pumped dye laser) and a detector situated orthogonal to the source beam to collect fluorescence. LIF occurs when molecules in the sample volume undergo resonant absorption upon interaction with laser light of the correct wavelength (1, 13). Relaxation via spontaneous emission generates photons that are collected by the detector. The relationship between LIF intensity and the number density of a species depends on intensity of the laser, the transition's quantum efficiency, and the detector's spectral response (7, 13). **Table 2** lists spectral details for specific LIF-probable transitions of plasma species.

Table 2 Spectroscopic properties, dipole moments, relative surface reactivites, and selected laser-induced fluorescence studies of plasma species

Species	Plasma sources	Excited transition	λ(nm)[a]	Radiative lifetime (ns)	Dipole moment (D)	Relative surface reactivity[b]	Reference(s)
C_2	C_xH_y	$A^1\Pi \leftarrow X^1\Sigma^+$	691	1.85×10^4	—	—	143
C_3	C_xH_y	$A^1\Pi \leftarrow X^1\Sigma^+$	410	200	0.44	low/moderate	126, 139
CH	C_xH_y, CH_3OH	$A^2\Delta \leftarrow X^2\Pi$	430	537	0.55	high	117, 139
CHF	CH_xF_{4-x}	$A^1A'' \leftarrow X^1A'$	571	2.45×10^3	1.30	low/moderate	144
CF	C_xF_y	$A^2\Sigma^+ \leftarrow X^2\Pi$	224	26.7	0.64	low/moderate	65
CF_2	C_xF_y	$A^1B_1 \leftarrow X^1A_1$	226	61	0.44	low	65, 84
CCl	CCl_4, CH_4/Cl_2	$A^2\Delta \leftarrow X^2\Pi$	279	105	—	—	145
CN	CH_3CN, CH_4/N_2	$B^2\Sigma^+ \leftarrow X^2\Sigma^+$	387	65	0.50	high	140, 146
NH	NH_3, N_2/H_2	$A^3\Pi \leftarrow X^3\Sigma^-$	336	440	1.39	low/moderate	130
NH_2	NH_3, N_2/H_2	$A^2A_1 \leftarrow X^2B_1$	598	10×10^3	1.82	moderate	130
NO	NO, N_2/O_2	$A^2\Delta \leftarrow X^2\Pi$	226	205	0.16	—	147, 148
OH	H_2O, H_2/O_2	$A^2\Delta \leftarrow X^2\Pi$	308	686	1.80	moderate	28, 149
SiCl	$SiCl_4$, Cl_2[c]	$B^2\Sigma^+ \leftarrow X^2\Pi$	297	10	—	—	150
$SiCl_2$	$SiCl_4$, Cl_2[c]	$A^1B_1 \leftarrow X^1A_1$	320	4.5×10^3	1.46	low	135
SiF	SiF_4, CF_4[c], SF_6[c]	$A^2\Sigma \leftarrow X^2\Pi$	437	230	1.07	moderate	24, 101
SiF_2	SiF_4, CF_4[c], SF_6[c]	$A^1B_1 \leftarrow X^1A_1$	225	6.2	1.23	low	24, 98
SiH	SiH_4, Si_2H_6	$A^2\Delta \leftarrow X^2\Pi$	413	534	0.14	high	20, 107
SiH_2	SiH_4, Si_2H_6	$A^1B_1 \leftarrow X^1A_1$	580	111	0.16	moderate	109
SO	SO_2, SF_6/O_2	$B^3\Sigma \leftarrow X^3\Sigma$	235	16.2	1.55	—	151
SO_2	SO_2, SF_6/O_2	$A^1B_1 \leftarrow X^1A_1$	300	10×10^3	1.63	—	151

[a]Excitation wavelength for listed transition.
[b]Relative reactivity scale: low = < 0.1; low/moderate = ~0.1–0.3; moderate = ~0.3–0.7; high = ~0.7–1.0.
[c]Species of interest is produced during Si processing.

T_g: gas temperature in a plasma

IRIS: imaging of radicals interacting with surfaces, an LIF-based technique

LIF plasma experiments can provide relative and absolute number densities (7, 13–15), gas temperature (T_g), and surface reactivities. LIF techniques that make use of optical evanescent waves, which allow characterization of plasma processes in the near-surface region, can provide relative densities and surface reactivities (16). Other LIF techniques provide data on kinetics (17) and electric fields in plasmas (18). Spatial resolution of LIF signals allows measurement of velocity distributions (19–21), yielding convection and diffusion data (18) and mechanisms for energy partitioning (19, 20). LIF only probes ground-state species, and only those that possess a fluorescing excited state. LIF is not well suited to process control, primarily because of the required laser-system maintenance, and is also hindered by nonradiative relaxation processes.

2.4. Imaging of Radicals Interacting with Surfaces

One special adaptation of LIF as a plasma diagnostic is the imaging of radicals interacting with surfaces (IRIS) technique, which combines molecular beams with spatially resolved LIF to explore radical-surface interactions during plasma processing (22). Gas-phase density and surface reactivity measurements are made as a function of parameters such as applied power (P), substrate material, and T_S. The effects of ion bombardment can be explored using a grounded mesh in the molecular beam path to remove charged species (23), biasing the substrate (19, 24, 25), or employing an alternate, ion-free molecular beam source such as hot-filament CVD (26). IRIS can determine velocities by exploiting the time resolution of the CCD (20, 21, 27), and uses mass spectrometry (MS) to study plasma ions (23, 25).

Figure 1 shows a schematic of the IRIS apparatus (22). In a typical IRIS experiment, feed gases enter a tubular reactor and rf power is applied to produce a plasma. Expansion into a differentially pumped vacuum chamber generates an effusive molecular beam containing virtually all plasma species. A tunable laser intersects the molecular beam and spatially resolved LIF signals are collected by a CCD located perpendicular to the interaction region. For reactivity measurements, a substrate is rotated into the molecular beam path (**Figure 1a**), and LIF signals are again collected. Differences between spatial distributions with and without the surface (**Figure 1b**) provide radical-surface interaction data.

Spatially resolved LIF images are interpreted using a quantitative model that reproduces the scattering data in one dimension (22, 28). The model calculates the scattering coefficient (S), which represents the fraction of incident radicals scattered from the surface and is adjusted to best fit the experimental data. For molecules produced at the surface, $S > 1$, whereas for molecules lost at the surface, $S < 1$. Surface reactivity, R, is defined as $1-S$ and is equivalent to surface loss probability, β, the fraction of gas-phase molecules lost upon interaction with a surface, regardless of loss mechanism.

2.5. Cavity Ringdown Absorption Spectroscopy

Cavity ringdown absorption spectroscopy (CRDS) uses laser pulses to measure absorption of a sample placed directly in the optical cavity of a laser. This is accomplished

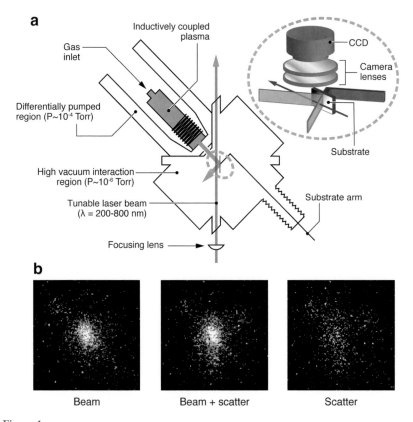

a

Inductively coupled plasma

Gas inlet

CCD

Camera lenses

Differentially pumped region (P~10⁻⁴ Torr)

Substrate

High vacuum interaction region (P~10⁻⁶ Torr)

Substrate arm

Tunable laser beam (λ = 200-800 nm)

Focusing lens

b

Beam   Beam + scatter   Scatter

Figure 1

(*a*) Schematic of imaging of radicals interacting with surfaces (IRIS) apparatus. Detail of the interaction region shows the spatial orientation of the optics and detector relative to the molecular and laser beams. Specular scattering of the molecular beam is illustrated. (*b*) Two-dimensional charge-coupled device (CCD) images of CN laser-induced fluorescence (LIF) signals in a CH_3CN plasma molecular beam; CN with a Si substrate rotated into the path of the molecular beam (beam + scatter); and the difference between the other two images, representing only CN molecules scattered from the surface.

by measuring the temporal decay in the intensity of light leaving the laser's output coupler. The time required for the output signal to decay to a fraction, 1/e, of the initial value is the ringdown time, which provides a single-pass transmission coefficient. At nonresonant wavelengths, the ringdown time can be used to correct for effects of mirror reflectivities and cavity dimensions. At resonant wavelengths, sample absorption causes additional signal decay, which can be converted to an absolute absorption (6, 29). Because of its time dependence, CRDS offers increased sensitivity over conventional OAS. CRDS is well suited to plasma systems because the sample is in the laser cavity, providing long effective path lengths and permitting analysis of strongly absorbing species in trace amounts or weakly absorbing species in larger concentrations (30). Although CRDS is less sensitive than "background-free" techniques such

CRDS: cavity ringdown absorption spectroscopy

as LIF and resonant enhanced multiphoton ionization (REMPI), it can be applied when fluorescence and ionization are not practical (29). CRDS can provide absolute densities (31), gas-phase loss rates, and β values (32–34).

2.6. Langmuir Probes

The most widely used electrical plasma probe is the Langmuir probe (11, 35), which measures the plasma current potential (*I-V*) relationship and functions similar to electrodes in electrochemical cells. Probes can be run effectively in potential sweep mode, similar to cyclic voltammetry. From the resulting data, electron and ion distribution functions are derived, providing electron temperature (T_e), electron density (n_e), and plasma potential (V_p). Langmuir probes often have reference electrodes or double probes, wherein a second electrode forms part of the electrical circuit (11, 36). Double probes are preferred as they do not induce as large a perturbation to the plasma, which is critical for probe theory validation. A distinct disadvantage arises from probe contamination from sputtered or deposited material, which can affect the probe's ability to accurately describe V_p (35).

2.7. Fourier Transform Infrared Spectroscopy

Fourier transform infrared spectroscopy (FTIR) is a vibrational spectroscopy applicable across a range of chemistries. FTIR plasma diagnostics focus on both the gas phase and processed surfaces (5, 37). The gas phase can be analyzed as a nonintrusive in situ probe in the exhaust region to monitor plasma effluent (38), which also provides information on plasma-generated species. Determination of parent molecule breakdown in the reactor helps ensure that the desired chemistry is occurring. FTIR analysis of a plasma-processed sample can occur either in situ, by monitoring the gain or loss of a particular species on the substrate, or ex situ, by removing the substrate from the system.

2.8. Mass Spectrometry

Quadrupole mass spectrometers (QMSs) (39) are the most common MS instruments used in plasma diagnostics (40). QMSs provide several analytical advantages, including fast analysis with good sensitivity, compactness, and robustness. In general, a mass spectrometer consists of an ion source, an analyzer, and a detector. A common ionization source is electron impact, which allows adjustment of the electron energy. Threshold ionization mass spectrometry [TIMS, or appearance potential mass spectrometry (APMS)] uses this source for detection of neutral radicals by controlling the ionization energy to selectively ionize plasma radicals over the parent species (39, 41–43).

In addition to providing information about the identity and concentration of plasma species, retarding field analyzers or ion optic elements can be incorporated to characterize ion energy distributions (IEDs). Analysis of IEDs can be used to study the angular distribution of plasma ions (44). This is especially relevant in etching, where angles of incidence influence etch rates and profiles. Other MS adaptations include cryotrapping-assisted MS, in which successive outgassing of cryotrapped plasma species provides mass spectra in complex deposition plasmas (45), and temperature-programmed desorption, in which a QMS is used to detect desorbing species (46).

2.9. Spectroscopic Ellipsometry and Second Harmonic Generation

Spectroscopic ellipsometry (SE) involves the measurement of changes in the polarization of light when reflected from a surface (1). It is noninvasive and can provide in situ characterization of plasma-deposited films. The experimental ratio of reflection coefficients can be compared to a model of the material structure to determine thickness and dielectric function. Refractive index, microstructure, and density data can also be determined. SE can achieve submonolayer resolution ($\sim$0.01 nm), but data quality depends largely on how well the model approximates the experimental system. Although SE is useful in characterizing buried interfaces, analysis becomes increasingly difficult with thicker films. Phase-modulated SE is used as an in situ probe of surface bonds during etching (47, 48).

Second harmonic generation (SHG) is related to SE in that incident light causes variations in the polarization of affected molecules, thereby generating a second electromagnetic field. When this occurs via a linear process, the polarization is linearly proportional to the incoming field, and the generated and incident fields have the same frequency. With more intense light (i.e., from a laser), additional terms must be included; in SHG, these are associated with generation of a field whose frequency is twice that of the incident light (1, 49). SHG signals from an interface are reflected and the polarization components are analyzed, the intensity of which depends on the sample's susceptibility (50). With centrosymmetric media, bulk molecules possess zero susceptibility and do not create a second harmonic response. This symmetry breaks down at a surface (49), and the sensitivity of SHG to interfaces (exposed and buried) is a particular strength. SHG can be applied under many experimental conditions, and does not require sample isolation in vacuum. Thus, it can be performed in situ to obtain time-resolved data on adsorption dynamics, orientation, surface charge, and number density of interfacial species with submonolayer sensitivity (51–53).

3. CHEMISTRY OF SPECIFIC SYSTEMS

Many plasma systems have benefited from the application of a combination of diagnostic tools, which aids the understanding of the synergy between gas-phase chemistry and the resulting process (etching, deposition, or surface modification). Here we review a few selected major systems, with an emphasis on understanding the plasma-surface interface.

3.1. Fluorocarbon Plasmas

Early spectroscopic studies in fluorocarbon (FC) plasmas focused on CF and CF_2, examining spectral characteristics using an FC electrodeless discharge as the CF_x

source (54, 55); these studies represent the beginning of an understanding of CF_x plasma chemistry and properties. Mathias and Miller examined polytetrafluoroethylene (PTFE) decomposition in microwave discharges, identifying C_xF_y products along with SiF_4, CO_2, CO and COF (56). Proposed mechanisms for plasma-surface interactions included both thermal and radiation-induced reactions. This early work provided a foundation for subsequent studies of FC plasma-surface interactions.

Due to the exceptional etching capabilities of FC plasmas, interest in these plasmas rose dramatically with the development of the microelectronics industry. This is reflected in comprehensive reviews of the recombination chemistry and parameters affecting etched surfaces (57–59). Numerical modeling has also offered insight into FC plasma etching (60, 61). Key findings clearly demonstrate that both ions and neutrals are central to etching and deposition mechanisms. Moreover, it is the synergistic activity of FC species in the gas phase and at the processed substrate surface that allows for the development of FC plasma applications (62).

3.1.1. CF_x density measurements.

Identification of gas-phase species and changes in density as a function of plasma parameters can help elucidate plasma chemistry. Thus, many FC plasma studies have measured CF_x and CF_x^+ densities using multiple techniques and by exploring parameter-based trends. Kiss et al. used LIF and OES to measure the relative densities of CF and CF_2 in CF_4/Ar plasmas (63). The two techniques yielded comparable results and showed that LIF results were linearly correlated with actinometric OES results. Booth et al. measured relative CF_x densities and internal temperatures with UV-OAS (64, 65), and employed LIF and CRDS in CF_4 capacitively coupled plasmas (CCPs) to measure absolute CF and CF_2 concentrations (30, 66). CRDS resulted in $[CF] = 6 \times 10^{11}$ cm^{-3} in the reactor center, considerably lower than the result obtained with LIF. The LIF studies demonstrated, however, that [CF] decreased with distance from the electrode. $[CF_2]$ could not be determined with CRDS, but LIF yielded $[CF_2] \sim 6.8 \times 10^{12}$ cm^{-3} near the powered electrode, decreasing with distance from the electrode, similar to CF.

In related studies, [F] in a CF_4 CCP was measured by combined LIF and OES measurements (15). Using $[CF_2]$ measured by LIF and the actinometric OES ratio of I_F/I_{Ar}, where I_F and I_{Ar} are emission intensities of the F and Ar signals, respectively, Cunge et al. (15) determined the absolute [F] with and without a Si substrate. [F] without a Si substrate increased as a function of P to a maximum of $\sim 6.5 \times 10^{14}$ cm^{-3}. With a Si substrate, [F] was substantially lower because etching reactions allow for recombination with etch products. Graves and colleagues measured neutral and ionic number densities in CF_4 inductively coupled plasmas (ICPs) using APMS, QMS, and a Langmuir probe (41, 67). A key result was that $[CF_x]$ varied with wall conditions, suggesting that wall interactions control FC gas–phase chemistry.

FC plasmas are known to selectively etch Si over SiO_2, primarily as a result of different plasma-surface interactions. Miyoshi and colleagues used computerized tomography OES (CT-OES) to measure the spatiotemporal structure of etchants, etch products, and their daughter products during Si and SiO_2 etching in CF_4/Ar mixtures (68). They measured the number densities of excited Ar, Si, SiF, and F and found that $[F^*]$ decreased as the Si etching rate increased, whereas $[SiF^*]$ increased during SiO_2 etching. This was attributed to the deposition regime in SiO_2 etching, wherein FC film deposition prevents reactive etchants such as F atoms and CF_x^+ ions from reaching the underlying SiO_2 layer. Studies such as these provide insight into selective etching mechanisms in FC plasmas.

$\Theta_R(MX_n)$: rotational temperature of a given species

3.1.2. Fluorocarbon plasma dynamics.
Understanding the dynamic processes that influence plasma species behavior is critical to refining overall etch or deposition processes in FC plasmas. The kinetic behavior of plasmas relies heavily on the energies of its constituents; thus, measurement of T_e and T_g provides one method for examining energy distribution in a plasma. Donnelly and colleagues applied trace rare gases OES (TRG-OES) to measure T_e and T_g in C_2F_6/C_4F_8 plasmas (69). The feed gas also included a carrier gas and multiple inert probe gases (He, Ne, Ar, Kr, and Xe). The emission behavior of probe gases is predictable from excitation cross sections and relaxation processes. Each gas is sensitive to a different part of the electron energy distribution function; collectively, the gases' emission intensities provide a description of T_e. Emission spectra with N_2 as the probe were used to characterize T_g from analysis of N_2 rotational spectra. Both T_e and T_g were strongly dependent on carrier gas, suggesting that careful selection of carrier gas could provide another degree of control over the energies of plasma species.

Measurements of internal and translational energies of plasma species also provide estimates of T_g. Nagai and Hori measured the rotational temperature (Θ_R) of CF in CF_4 and CF_4/Ar plasmas using OES and infrared laser absorption spectroscopy (IRLAS) (70). Θ_R increased from 300 to 380 K as P increased from 375 to 1500 W, and was accompanied by a ~threefold increase in [CF]. With CF_4/Ar, Θ_R and [CF] were substantially lower, with Θ_R only increasing by ~20 K over the same P range. This likely resulted from increased elastic collisions leading to CF rotational cooling, and demonstrates that rotational heating is not appreciable in these systems. Similar values were obtained using planar LIF to create two-dimensional maps of $\Theta_R(CF)$ in CF_4 CCPs (71). Θ_R displayed strong gradients, increasing with distance from electrodes. These results have implications for density and kinetics studies that examine only a single rotational state.

Both negative and positive ions significantly influence FC plasma etching processes. Negative ion decay occurs only through ion-ion recombination in the plasma bulk. Hebner and colleagues used photodetachment spectroscopy to measure the F^- absolute density in CF_4, C_2F_6, and CHF_3 plasma afterglows (72). The [F^-] time-dependence yielded ion-ion recombination rates of 0.88×10^{-6}, 1.5×10^{-6}, and 3.9×10^{-6} cm^3/s for CF_4, C_2F_6 and CHF_3, respectively. Positive ions can strongly influence the net rate of ion-ion recombination. The dominant positive ion is CF_3^+ for CF_4 and C_2F_6 plasmas, but CF_2^+ dominates CHF_3 plasmas, suggesting that CF_2^+ enhances ion-ion recombination rates in FC plasmas.

Hancock et al. measured F-atom emission at different delay times after the power-off in FC plasmas using TR-OES (10). Decay was attributed to gas-phase recombination processes and losses at the reactor walls. Addition of a Si substrate increased etching reactions, thereby dramatically increasing the decay, whereas addition of O_2 significantly increased [F]. UV-OAS results from studies by Sasaki et al. (73)

β(MX$_n$): surface loss probability; the fraction of molecules lost from the gas phase upon interaction with a substrate

contained rapid decay in [F] in the initial afterglow of their CF_4 plasma, followed by an exponential decay in [F] at longer times. This was attributed to the initial reaction of CF_x on reactor walls and simple diffusion to, and loss at, wall surfaces in the afterglow. These mechanisms are supported by the QMS, Langmuir probe, and OES studies by Sugai et al. who found that heated reactor walls (100–200°C) resulted in large increases in [CF_x] (74). Using in situ FTIR, OES, and ex situ attenuated total reflectance (ATR)–FTIR, Goeckner and colleagues (75) found that CF_x was lost on low-temperature walls in CF_4 plasmas, but preferentially desorbed at high temperatures. They also concluded that both film deposition and etch rate were determined by two major competing processes, direct ion incorporation and ion-assisted surface desorption.

3.1.3. Surface interactions. In FC plasmas, etching and deposition are competitive processes and often occur simultaneously (57). Species density studies as a function of plasma parameters reveal that film formation and etching are controlled by the interactions of plasma species with a substrate. The balance between highly energetic ions and neutral radicals bombarding the substrate being processed dictate the dominant process.

3.1.3.1. Radical-surface interactions. Radical-surface interactions are key steps leading to FC film deposition and to etching of Si substrates. In general, it is assumed that when radicals impinge on a surface, they react with unit probability, essentially the first step in all three radical-surface processes depicted in **Figure 2a**. Indeed, this assumption is regularly made in computer simulations where no experimental data are available. Surface loss, however, can occur via several different processes, including dissociative adsorption (**Figure 2a**, *process 2*), or surface atom abstraction (**Figure 2a**, *process 3*). In both cases, the newly formed species can subsequently desorb (as shown), but whether desorption occurs or not, these processes contribute to β. Although the first step of process 1 depicts simple adsorption, which contributes to β, subsequent desorption would result in higher S. CF_x (x = 1–3) species are proposed critical components in FC polymerization (76–78). Radicals contribute to polymer growth by reacting with "activated" sites on the polymeric surface or by forming addition compounds through gas-phase reactions. Despite these predicted behaviors, growth mechanisms remain unclear, and β values measured *during* plasma-surface interactions are critical to full understanding (79).

β(CF_x) appears to vary dramatically with plasma gas chemistry, surface conditions, and wall history. For example, Booth and colleagues used LIF to measure β(CF) = 0.06–0.24, depending on [F(g)] (66, 80). APMS measurements of β(F) and β(CF_x) on the walls of an ICP reactor were strongly dependent on wall conditions (81). Hikosaka et al. found that β(CF_2) and β(CF_3) measured by TIMS were nearly identical in CF_4 plasmas (~0.13), but that β decreased by ~10^2 for both species when H_2 was added (82). In general, CF_x (x = 1–3) species have low β, and all appear to be strongly coupled to the conditions under which measurements are made.

IRIS studies demonstrated that CF_2 is generated at surfaces (i.e., $S > 1$; β < 0) under etching conditions (e.g., 100% C_2F_6 plasmas); however, β ~ 0.2 when a-C:F,H films are produced from 50/50 C_2F_6/H_2 plasmas (76). IRIS studies of CF_2

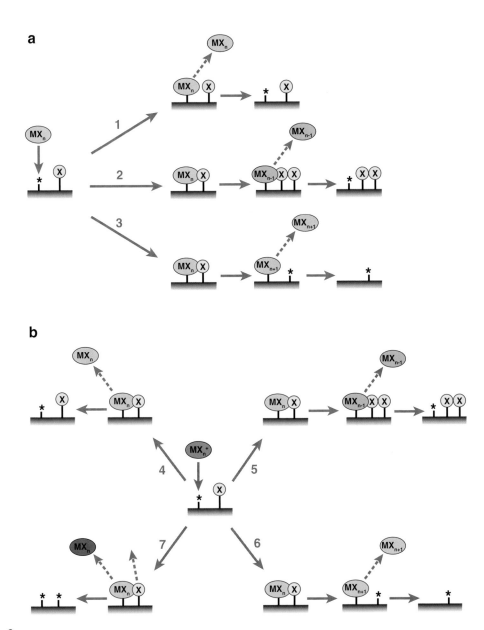

Figure 2

Generalized schematic representation of (*a*) neutral plasma species and (*b*) ionic plasma species interacting with a substrate, where M = C, Si, or N; X = H, F, or Cl; and asterisk represents an active surface site. Process 1 represents simple adsorption-desorption from a surface, usually with thermal equilibration. Processes 2 and 3 represent dissociative adsorption with subsequent desorption of MX_{n-1} and adsorption followed by surface recombination and desorption of MX_{n+1}, respectively. Processes 4–6 are similar to processes 1–3, but include surface neutralization of the incident ion. Process 7 represents ion-induced sputtering, wherein multiple volatile reaction products can be formed.

$\langle E_i \rangle$: mean ion energy

$S(MX_n)$: surface scattering coefficient; represents the fraction of incident molecules scattered from a substrate

using hexafluoropropylene oxide (HFPO), CHF_3, C_3F_8, and C_4F_8 plasmas (76, 77, 83, 84) have yielded three key observations: (1) $S(CF_2)$ is nearly always greater than unity, indicating surface generation of CF_2 during FC processing; (2) $S(CF_2)$ increases with P; and (3) $S(CF_2)$ decreases under ion-free conditions. Time-resolved UV-OAS data for C_2F_4 and HFPO CCPs also found significant CF_2 surface production (85). Booth and colleagues determined that both CF and CF_2 are produced at surfaces in CF_4 plasmas (15, 66, 80). Their proposed mechanism relies on energetic ion-surface interactions producing CF_x, either through CF_x^+ neutralization, (**Figure 2b**, *process 4*) or via film sputtering (**Figure 2b**, *process 7*). With no ion bombardment, many surfaces act as CF_x sinks. These observations indicate CF_2 surface production is strongly correlated to plasma ions. Indeed, Martin et al. found a positive linear correlation between mean ion energy ($\langle E_i \rangle$) in C_3F_8 and C_4F_8 plasmas and $S(CF_2)$ (23). An additional correlation was observed between $S(CF_2)$ and FC film crosslinking for several FC systems (86). Conditions leading to highly crosslinked films result in higher $S(CF_2)$. Overall, when the surface is not bombarded by energetic species (e.g., ions), CF_2 is contributing to film formation, resulting in a more ordered (high CF_2 content) material (26).

3.1.3.2. Ion-surface interactions. As the above evidence demonstrates, ions clearly play a significant role in both FC plasma etching and deposition; possible ion-induced mechanisms are shown in **Figure 2b**. To clarify this role, the behavior of ions or groups of ions in an isolated environment has been investigated. Sawin and colleagues used individual beams of CF_2, F atoms (created from HFPO pyrolysis and F_2/Xe discharges, respectively), and Ar^+ to simulate the CF plasma etching environment (87, 88). Butterbaugh et al. found that the etch yield saturates at lower radical flux values when the surface is bombarded with higher-energy ions (87). This was attributed to ion-induced surface roughness producing more active sites and higher β.

Hanley et al. examined ion-surface interactions using mass-selected beams of $C_3F_5^+$ or $C_2F_4^+$ ions (89, 90). Ions were accelerated or decelerated into a surface to examine the effects of incident ion energy on the properties of the resulting films, as measured by X-ray photoelectron spectroscopy. Ion identity and energy strongly affected film thickness and morphology, but not film composition. Goyette et al. used an electrostatic energy analyzer to examine IEDs and ion flux of CF^+ and CF_3^+ in C_2F_6 and C_4F_8 plasmas (91). At higher pressures, ion flux increased, as did rf modulations, and IEDs were highly dependent on ion mass. Martin et al. measured IEDs and determined $<E_i>$ in C_3F_8 and c-C_4F_8 plasmas (23). IEDs did not exhibit a strong dependence on ion mass, but clearly showed evidence of sheath effects, as $<E_i>$ was relatively high. Increases in $<E_i>$ were linearly correlated to $S(CF_2)$.

3.2. SiF_x Species

In plasma etching (e.g., FC plasmas), SiF_x species can be volatile etch byproducts, but they can also act as deposition precursors. Thus, there are many similarities between CF_x and SiF_x plasma species, and optical diagnostics can elucidate both parent FC gas dissociation and etch-product behavior (92, 93). Early SiF_x studies originated with spectroscopic and thermodynamic measurements using plasma sources (94–96). Subsequent studies focused on parameter dependence in processing systems. Hebner measured $[CF_x]$ and $[SiF_x]$ (x = 1, 2) with LIF, and found that $[CF_x]$ decreased with P and increased with pressure (97). In contrast, $[SiF_x]$ increased with P and pressure in C_2F_6 and CHF_3 plasmas, but was independent of pressure for C_4F_8 plasmas. These results suggest that C_2F_6 and CHF_3 are more efficient Si etchants than C_4F_8. CT-OES studies of Si and SiF demonstrated that production of SiF occurs via gas-phase dissociation of etch products. Other LIF studies distinguished between SiF production in the presence of ion bombardment and by chemical etching of F atoms (98).

3.2.1. SiF_x dynamics. Giapis and colleagues obtained gas-phase and surface kinetics for SiF_x species using MS (99, 100). For SiF_x^+ (x = 1–3), the time-of-flight (TOF) distributions contained two components, which were attributed to thermally equilibrated species and hyperthermal products of surface reactions. Increases in T_S caused a sharp drop in the SiF_3^+ high-energy component, suggesting that the mechanism favors direct surface reactions for SiF_3 formation.

IRIS studies measured Θ_R for SiF and SiF_2 by comparing surface reactivities for different rotational transitions over a range of T_S (101). This method accounts for changes in rotational state populations for molecules that thermally equilibrate at a substrate, which can result in measured β values being larger or smaller than the true value. When $T_S = \Theta_R$ for the beam species, the state population distribution will not change upon equilibration at the surface, and β for all transitions should coincide. This analysis yielded $\Theta_R(SiF) = 450 \pm 50$ K and $\Theta_R(SiF_2) = 752 \pm 100$ K for a 170-W SiF_4 plasma. Assuming that the SiF_4 parent gas measured ~300 K, these values suggest that thermal equilibration did not take place during surface interactions of SiF_x (27).

3.2.2. Surface interactions. Cunge et al. measured $[SiF_2]$ above Si and SiO_2 substrates in a CF_4 CCP using spatially resolved LIF (98). SiF_2 surface generation should produce a linear decrease in $[SiF_2]$ away from the substrate, and gas-phase loss or production processes would predictably alter concentration profiles. For Si and SiO_2, no net production of SiF_2 was observed, suggesting that SiF_2 is produced primarily at the substrate as an etch product. Data from the plasma afterglow characterized concentration profiles under ion-free conditions. At 50 mTorr, profiles collected 2 ms after rf power interruption revealed that SiF_2 continued to be produced at the Si substrate, but was lost on SiO_2. This suggests that SiF_2 production from Si results from chemical processes, but production from SiO_2 relies on other species such as ions. At 50 mTorr, $\beta(SiF_2)$ estimates were 0.8–1.0 and 0.5–0.6 for Si and SiO_2, respectively. At 200 mTorr, β decreased to 0.04 and 0.11 on Si and SiO_2 substrates, respectively.

Williams & Fisher used LIF, QMS, OES, and ex situ FTIR to study SiF_x surface interactions during etching and deposition from SiF_4 plasmas (24, 101). They found that T_S significantly affected mechanisms for SiF_x surface desorption and surface adlayer composition (101). Higher T_S resulted in increased SiF and SiF_2 scatter, presumably from increases in process 1 (**Figure 2a**), although processes 2 and 3 could also contribute. Thus, elevated T_S increases etching efficiency by enhancing etch product formation. Another IRIS study examined ion influence on SiF_2 surface interactions (24) using a grounded mesh and application of ± 200 V substrate bias. Changes in $S(SiF_2)$ demonstrated that SiF_2 surface generation is ion induced (**Figure 2b**) and that the amount of SiF_2 generated depends more on energy transfer to the surface than on the availability of chemical etching species (e.g., F).

Interestingly, ion-induced processes did not strongly influence SiF surface interactions, with $\beta(SiF) \leq \sim 0.8$, depending on the overall plasma chemistry (25, 101). The combination of SiF_2 and SiF IRIS data suggests that SiF is primarily a deposition precursor, whereas SiF_2 is a major etch product. IRIS studies at elevated T_S found that $\beta(SiF_2)$ decreased with T_S (300–575 K), a change attributed to changing surface adlayer composition (101). For SiF and SiF_2, β increased for $T_S >$ 575 K as a result of exceeding the desorption energy threshold. TOF-MS studies of SiF_x etch products suggest that ion-surface interactions have two components (99, 100), one that corresponds to ion equilibration at the surface (**Figure 2b**, *process 4*), and another that produces energetic surface reaction products. Thus, by controlling processes that generate high-energy species, anisotropy and undercutting issues in reactive ion etching processes may be resolved.

In another study, SiF and SiF_2 velocities derived from changes in time-delayed LIF images were converted to translational temperatures (Θ_T) (27). As P increased from 80 to 200 W, $\Theta_T(SiF)$ and $\Theta_T(SiF_2)$ increased from ~ 570 K to 870 K and from 430 K to 560 K, respectively. Differences in Θ_T for the two species were attributed to differences in mass, as lighter species gain more translational energy. Increases with P were ascribed to increases in ion density, n_e, and collision frequency that accompany higher P. These studies established that plasmas are not always thermally equilibrated and that T_g estimates from a single species in any plasma should be applied cautiously.

3.3. SiH$_x$ Systems

Amorphous, hydrogenated silicon (a-Si:H) has been extensively studied because of its widespread use as an inexpensive solar cell material (102). Although a-Si:H can be produced by a variety of methods, it is most often grown using SiH_4 or Si_2H_6 plasmas, which have afforded a plethora of fundamental experimental and theoretical studies of gas-phase kinetics, thermodynamics, film characterization, and SiH_x measurements (103–105).

3.3.1. SiH$_x$ energetics. Stamou et al. measured $\Theta_R(SiH)$ in SiH_4 CCPs using spatially resolved OES (106). Θ_R linearly decreased with increasing distance from the rf electrode from ~ 2500 K to ~ 1975 K at 5–16 mm. This was attributed to the sensitivity of Θ_R to high-energy electrons, which are more abundant near the electrode. $\Theta_R(SiH)$ peaked 5 mm from the electrode, whereas $[SiH^*]$ was highest 7 mm from the electrode. This suggests that the region closer to the electrode has more high-energy electrons, even though n_e is lower. Further from the electrode, emission intensity decreased because less SiH_4 dissociation occurs.

IRIS studies determined $\Theta_R(SiH)$ and $\Theta_T(SiH)$ as a function of P and Ar dilution (20, 107). $\Theta_T(SiH) > \Theta_R(SiH)$ at all P and both are relatively constant at $P > 20$ W

(20). $\Theta_R \sim 500–600$ K in all feed gas mixtures, indicating that thermal equilibration of rotational energy was established (107). Differences in $\Theta_R(SiH)$ in different studies likely result from differences in plasma configurations. $\Theta_T(SiH) > \Theta_R(SiH)$ under most Ar dilution conditions, averaging ~ 1000 K, although in the Si_2H_6/Ar system, $\Theta_T(SiH) \sim \Theta_R(SiH)$ at the highest Ar dilutions. This suggested equilibration occurs more slowly for translational energy than for rotational energy, and translational energy is equilibrated at different rates in the two silane systems.

3.3.2. Surface interactions. van de Sanden and colleagues utilized an aperture-well assembly to measure a global β value for SiH_x in $Ar/H_2/SiH_4$ plasmas (108). This apparatus is formed by two substrates 0.5 mm apart, with a 0.1-mm slit in the upper substrate. Radicals adsorb at the lower substrate or are reflected to the underside of the top substrate. Deposition profiles as a function of H_2 flow were related to a global β of < 0.5, which decreased with H_2 flow. This may indicate a role for SiH_3, which decreases in density with the addition of H_2.

Individual contributions of Si and SiH to film growth were derived from the comparison of CRDS-measured densities with simulated data (31). Experimental and simulation results suggest that at low SiH_4 flows, SiH radicals are lost to charge transfer with Ar^+. SiH production reached a maximum at high SiH_4 flows, where the molecules are predominantly lost in reactions with SiH_4. $\beta(Si)$ and $\beta(SiH)$ were used to correlate gas-phase densities to film growth contributions, which revealed that Si and SiH contributions are only weakly dependent on H_2 flow. Time-resolved CRDS measurements also provided estimates for β (33). Si and SiH_3 loss rates yielded $\beta(Si)$ ~ 1 and $\beta(SiH_3) \sim 0.3$ in $Ar/H_2/SiH_4$ plasmas. The latter value was independent of T_S (104), indicating that surface reactions occur during film growth. IRIS results confirm that $\beta(SiH) \sim 1$, regardless of plasma parameters, feed gas, and T_S (20).

Few studies have focused on SiH_2 surface interactions, primarily because of its high reactivity and therefore low density in silane plasmas. Decay in $[SiH_2]$ measured by LIF in the afterglow of $Ar/H_2/SiH_4$ plasmas found that $\beta(SiH_2) = 0.6 \pm 0.15$ (109). TIMS measurements of SiH_3 and Si_2H_5 decay rates in the same apparatus were related to β values using assumptions about surface processes, yielding $\beta(SiH_3) = 0.28 \pm 0.03$ and $\beta(Si_2H_5) \sim 0.1–0.3$ (110). In general, molecular dynamics simulations compare favorably with experimental values for $\beta(SiH_x)$ (111–113).

3.4. CH_x Plasma Chemistry

Although methane plasmas are widely used to deposit a-C:H films, most studies have focused on film properties, providing scant data on gas-phase processes (114). The predominant species in methane plasmas are CH_3 and H atoms (115). MS studies show, however, that larger C_xH_y (x > 1) molecules are formed in hydrocarbon plasmas (114, 116).

3.4.1. Energetics. Energy distribution characterization in CH_4 plasmas focused on measuring $\Theta_R(CH)$ and on measuring $\Theta_T(CH)$ in CH_4/Ar plasmas with LIF (117). Θ_R was ~ 1450 K, independent of P and [Ar]. However, $\Theta_T(CH)$ was significantly higher than Θ_R, decreasing from 9000 K to 2500 K as P increased from 20 to 100 W, plateauing at $P > 100$ W. A similar trend was observed in [CH], and collectively, these results were attributed to plasma coupling modes. [Ar] did not significantly affect Θ_T, but lower Θ_T occurred at higher pressures, mimicking T_e. Thus, CH is rotationally very hot and is not thermally equilibrated, with the disparity becoming more pronounced at lower P and lower pressures. In situ FTIR spectra were used to derive $\Theta_R(C_2H_2)$ in an expanding thermal arc plasma (118). Although these measurements were unreliable where C_2H_2 consumption was high (low absorbances), simulated spectra reproduced the experimental data, yielding $\Theta_R \sim 300$–450 K, with no dependence on gas flow.

3.4.2. Surface interactions. van de Sanden and colleagues used TIMS, Rutherford backscattering, and ellipsometry to study film properties and C_xH_y species as a function of the C_2H_2/Ar flux ratio, F (119, 120). Radical densities' ($\sim 10^{16}$–10^{17} m^{-3}) dependence on F was modeled and ultimately related to β. For $F < 1$, C and C_2 dominate the gas phase and are strong contributors to film growth, whereas for $F > 1$, radicals with an odd number of carbons contribute the most. Ellipsometric refractive indices were used to correlate C_3 radical density to a-C:H film quality.

CH and H radical beams in conjunction with isotope labeling and in situ ellipsometry studies reveal a-C:H deposition mechanisms (115, 121, 122). Under CH_3 radical flux, the surface coverage of methyl groups reaches a steady state. With H-beam flux, generation of active sites occurs via surface hydrogen abstraction. This leads to film erosion as neighboring active sites relax to form C-C bonds. Alternatively, active sites promote chemisorption of CH_3 and subsequent film growth. Ion-induced processes were studied by exposing surfaces to CH_3 and He$^+$ beams, wherein film growth was promoted via ion-induced creation of reactive sites. In plasmas, the presence of both ions and H influences film formation. The importance of H flux is supported by $\beta(CH_3)$, which increases from 10^{-4} to 10^{-2} with H atom flux, as well as by isotope-labeling studies. Controlled exposure to D-labeled beams results in surface composition changes that suggest that active sites can be created several monolayers deep. Deeper active sites form C-C bonds, whereas shallow sites promote CH_3 chemisorption.

Time-resolved TIMS in CH_4 plasma afterglows provide $\beta(CH_3)$ values from [CH_3] decay (110). Results show that $\beta(CH_3)$ decreases from $\sim 10^{-2}$ in the plasma to $\sim 10^{-3}$ ms after discharge termination, indicating that active site quenching occurs rapidly. The spatially resolved TIMS results published by Sugai et al. yielded $\beta(CH_3) = 0.001$ and $\beta(CH_2) = 0.028$ (123, 124). Measurements by Loh & Capelli using UV-OAS resulted in $\beta(CH_3) = 10^{-2}$ in CH_4/H_2 plasmas (125), and recent IRIS studies yielded $\beta(CH)$ near unity in CH_4/Ar plasmas (117) and $\beta(C_3) = 0.1$–0.4 in CH_2F_2/C_3F_8 plasmas, depending on substrate bias and gas ratios (126).

3.5. NH$_x$ Species

N_2 and NH_3 are used for nitride film deposition and polymer surface modification. Tahara et al. used spatially resolved OES to investigate NH_3 and N_2/H_2 plasmas (127) and found that $T_g \sim 10^3$ K. In N_2/H_2 plasmas, T_g gradually decreases downstream from the plasma arc. For NH_3 plasmas, $\Theta_R(NH)$ drops precipitously. Electrostatic probe measurements suggest that n_e is lower in the NH_3 system; this disparity is likely responsible for changes in $\Theta_R(NH)$. CRDS was applied to NH_x species in Ar/NH_3 plasmas to extract $\Theta_R(NH) = 1920 \pm 100$ K (128). This is comparable to $\Theta_T(NH) = 1750 \pm 100$ K, suggesting energy equilibration of NH. Temperatures and trends found in OES and CRDS studies were comparable and suggest that thermal equilibration of NH_x species occurs readily.

3.5.1. Energetics. Several IRIS studies have focused on NH_x energetics in NH_3 plasmas. Changes in rotational spectra upon interaction with a heated substrate yielded $\Theta_R(NH_2) \sim 340$ K (129), indicating that formation of NH_2 does not involve significant rotational heating. $\Theta_T(NH_2)$ in the molecular beam averaged 500–650 K and was positively correlated with P (19, 21). Θ_T of NH_2 scattered from a substrate, $\Theta_{Tsc}(NH_2)$, was lower than in the beam (300 K $< \Theta_{Tsc} <$ 550 K), but higher than T_S (300 K). Using an ion-free molecular beam, Θ_{Tsc} was much closer to T_S, indicating that ion-induced processes are responsible for elevated Θ_{Tsc} (19). In addition, Θ_{Tsc} was dependent on substrate material (metals or polymers), with polymeric substrates having a stronger P dependence. This suggests that substrate material influences energy transfer during plasma-surface interactions.

3.5.2. Surface interactions. $\beta(NH)$ and $\beta(NH_2)$ measured in NH_3 plasmas show that the behavior of the two species are affected in very different ways by P, substrate material, and the presence of ions (19, 21, 130). Increases in $S(NH_2)$ with P are accompanied by decreased $S(NH)$, suggesting that NH surface generation comes at the expense of NH_2, potentially via process 2 shown in **Figure 2a**. As $\Theta_{Tsc}(NH_2) > T_S$, NH_2 surface generation must occur via processes that do not allow full thermal equilibration at the surface (19, 21).

Processing of polymers such as polyethylene (PE) in NH_3/H_2 plasmas is controlled by gas-phase species (131, 132). OES data indicate that N_2, NH, and H are dominant NH_3 decomposition products. Increases in $[H_2]$ in the feed were accompanied by decreases in PE N/C ratios because hydrogen reduces surface NH_x, ultimately promoting formation of volatile NH_x. P dependence data revealed that although low-energy ion bombardment activates the surface for grafting, high-energy bombardment increases sputtering (**Figure 2a**, *process* 7) and limits grafting efficiency (133). Ishikawa et al. characterized film etching in N_2/H_2 plasmas using in situ ATR-FTIR spectroscopy and electron spin resonance (134). With no substrate bias, treatment resulted in formation of CN and NH moieties in the film. As bias voltage was increased, these disappeared from the spectra, and at bias >200 V, etching was observed. Although surface nitriding may produce etch resistance, N-containing etch products were detected, suggesting that nitrogen both facilitates and inhibits etching. [NH] and [NH_2] CRDS measurements in NH_3/Ar plasmas indicate that Ar^+ undergoes charge transfer reactions with NH_3; subsequent NH_3 decomposition occurs via dissociative recombination. Thus, Ar^+ is largely responsible for the NH_3 breakdown.

4. GLOBAL REMARKS ON PLASMA-SURFACE INTERACTIONS

The studies reviewed above clearly demonstrate that diagnostic tools reveal details of plasma process chemistry. Currently, however, only a limited number of techniques are available to determine β values for plasma species. Given the complexity of plasma-surface interactions, it would be useful to establish some global trends. One characteristic that could be critical is the electronic configuration of unsaturated species (135).

In addition to spectral properties, **Table 2** presents dipole moments and relative surface reactivities (also known as "stickiness" factors) for plasma species. Trends that can be gleaned from these data suggest that several doublet species (SiH, CH, CN) are highly reactive, with $\beta \sim 1$ under all conditions. Other doublets, such as NH_2, OH, and SiF, display moderate β during film deposition (**Table 2**). The observed differences in β of these doublet species could result from relative electronegativities. Molecules with stronger dipole moments (SiF, OH, NH_2) appear to be less reactive than those with smaller dipole moments (CH, SiH, CN). Reactivities may also be related to the availability of surface reaction partners, for example H atoms (28). The sole triplet species listed in **Table 2**, NH, has very low β despite having two unpaired electrons, suggesting that NH may not be an active film precursor. Finally, the isoelectronic singlet species CF_2, SiF_2, and $SiCl_2$ are clearly generated during plasma-surface interactions ($S \gg 1$). This may be related to these species' inherent stability, such that they are extremely probable reaction products. The low β values strongly suggest that MX_2 species (M = Si, C; X = F, Cl) may not be film precursors, although ion-induced production of MX_2 is a significant contributor. Notably, singlet species without strongly electron-withdrawing substituents (SiH_2, C_3) exhibit moderate reactivity, indicating that these species likely contribute to film growth.

Although these statements are generalizations, they represent a beginning for the difficult task of characterizing plasma-surface interactions. Areas requiring continued study include:

1. Measurement of internal and kinetic temperatures for multiple species within a plasma system to determine energy partitioning and kinetics. As evident from the Θ_R and Θ_T measurements discussed above, knowing only one of these does not representatively assess T_g. Moreover, data for only a single plasma species can result in a limited picture of energy partitioning.
2. Continued development of diagnostics applicable to a wider range of plasma species. CRDS is perhaps the closest to meeting this need as it has the widest applicability.
3. Increased focus on measurement of β values during processing to provide broader views of plasma chemistry.

Thus, in situ measurement methods and creation of radical-surface interaction databases to describe contributions of **Figure 2** processes to the overall plasma chemistry are critical. Future efforts will significantly enhance our understanding of process chemistry, with the goal of controlling plasma-surface interactions to create tailored materials.

SUMMARY POINTS

1. Application of multiple diagnostic tools is key to improving and controlling plasma process chemistry.

2. Continued development of diagnostics applicable to a wider range of plasma species is critical to enhancing our understanding of plasma processing.

3. Measurement of internal and kinetic temperatures for multiple species within a plasma system to determine energy partitioning and kinetics will significantly aid numerical modeling efforts.

4. Surface interaction data suggest that molecule-surface interactions are strongly influenced by the molecule's electronic configuration and dipole moment, suggesting possible generalizations for these reactions.

FUTURE ISSUES

1. An increased focus on the measurement of β values during processing to provide broader views of plasma chemistry is needed. Thus, in situ measurement methods and creation of radical-surface interaction databases are critical.

2. Additional progress will be made through development of numerical models utilizing recent experimental data derived from plasma diagnostic tools, which provide more accurate descriptions of the fate of individual species during plasma processing.

DISCLOSURE STATEMENT

The authors are not aware of any biases that might be perceived as affecting the objectivity of this review.

ACKNOWLEDGMENTS

Much of this work was supported by the National Science Foundation. We also acknowledge the assistance of past and current coworkers, Dr. Dongping Liu, Dr. Ina T. Martin, Dr. Patrick R. McCurdy, Ms. Michelle Morgan, Dr. Keri L. Williams, Dr. Jianming Zhang, and Dr. Jie Zhou.

LITERATURE CITED

1. Herman IP. 1996. *Optical Diagnostics for Thin Film Processing*. San Diego: Academic. 783 pp.
2. Selwyn GS. 1993. *Optical Diagnostic Techniques for Plasma Processing*. New York: AVS. 161 pp.

3. Donnelly VM. 1989. Optical diagnostic techniques for low pressure plasmas and plasma processing. In *Plasma Diagnostics*, ed. O Auciello, DF Flamm. Boston: Academic

4. Gottscho RA, Miller TA. 1984. Optical techniques in plasma diagnostics. *Pure Appl. Chem.* 36:189–208

5. Hershkowitz N, Breun RA. 1997. Diagnostics for plasma processing (etching plasmas). *Rev. Sci. Instrum.* 68:880–85

6. Schere JJ, Paul JB, O'Keefe A, Saykally RJ. 1997. Cavity ringdown laser absorption spectroscopy: history, development, and application to pulsed molecular beams. *Chem. Rev.* 97:25–51

7. Freegarde TGM, Hancock G. 1997. A guide to laser-induced fluorescence diagnostics in plasmas. *J. Phys. IV* 7:15–29

8. Donnelly VM, Malyshev MV, Schabel M, Kornblit A, Tai W, et al. 2002. Optical plasma emission spectroscopy of etching plasmas used in Si-based semiconductor processing. *Plasma Sources Sci. Technol.* 11:A26–A30

9. Malyshev MV, Donnelly VM. 1999. Trace rare gases optical emission spectroscopy: nonintrusive method for measuring electron temperatures in low-pressure, low-temperature plasmas. *Phys. Rev. E* 60:6016–29

10. Hancock G, Sucksmith JP, Toogood MJ. 1990. Plasma kinetic measurements using time-resolved actinometry: comparisons with laser-induced fluorescence. *J. Phys. Chem.* 94:3269–72

11. Hebner GA, Miller PA, Woodworth JR. 2000. Overview of Plasma Diagnostic Techniques. In *Handbook of Advanced Plasma Processing Techniques*, ed. RJ Shul, SJ Pearton. Berlin: Springer-Verlag

12. Miller PA, Hebner GA, Jarecki RL. 1998. Optical self-absorption technique for qualitative measurement of excited-state densities in plasma reactors. *J. Vac. Sci. Technol. A* 16:3240–46

13. Amorim J, Baravian G, Jolly J. 2000. Laser-induced resonance fluorescence as a diagnostic technique in nonthermal equilibrium plasmas. *J. Phys. D: Appl. Phys.* 33:R51–65

14. Cunge G, Booth JP, Derouard J. 1996. Absolute concentration measurements by pulsed laser-induced fluorescence in low-pressure gases: allowing for saturation effects. *Chem. Phys. Lett.* 263:645–50

15. Cunge G, Chabert P, Booth JP. 2001. Absolute fluorine atom concentrations in fluorocarbon plasmas determined from NH_3 loss kinetics. *J. Appl. Phys.* 89:7750–55

16. Sakurai T. 2007. Laser-based plasma particle analysis of the surface in a discharge. *Plasma Sources Sci. Technol.* 16:S101–6

17. Amorim J, Baravian G, Sultan G. 1998. Absolute density measurements of ammonia synthesized in N_2-H_2 mixture discharges. *Appl. Phys. Lett.* 68:1915–17

18. Bowles J, McWilliams R, Rynn N. 1994. Direct measurement of velocity space transport in a plasma. *Phys. Plasmas* 1:3814–25

19. Butoi CI, Steen ML, Peers JRD, Fisher ER. 2001. Mechanisms and energy transfer for surface generation of NH_2 during NH_3 plasma processing of metal and polymer substrates. *J. Phys. Chem. B* 105:5957–67

20. Kessels WMM, McCurdy PR, Williams KL, Barker GR, Venturo VA, Fisher ER. 2002. Surface seactivity and plasma energetics of SiH radicals during plasma deposition of silicon-based materials. *J. Phys. Chem. B* 106:2680–89

21. McCurdy PR, Venturo VA, Fisher ER. 1997. Velocity distributions of NH_2 radicals in an NH_3 plasma molecular beam. *Chem. Phys. Lett.* 274:120–26

22. McCurdy PR, Bogart KHA, Dalleska NF, Fisher ER. 1997. A modified molecular beam instrument for the imaging of radicals interacting with surfaces during plasma processing. *Rev. Sci. Instrum.* 68:1684–93

23. Martin IT, Zhou J, Fisher ER. 2006. Correlating ion energies and CF_2 surface production during fluorocarbon plasma processing of silicon. *J. Appl. Phys.* 100:013301

24. Williams KL, Fisher ER. 2003. Mechanisms for deposition and etching in fluorosilane plasma processing. *J. Vac. Sci. Technol. A* 21:1688–701

25. Williams KL, Martin IT, Fisher ER. 2002. On the importance of ions and ion-molecule reactions to plasma-surface interface reactions. *J. Am. Soc. Mass. Spectrom.* 13:518–29

26. Liu D, Martin IT, Fisher ER. 2006. CF_2 surface reactivity during hot filament and plasma-enhanced chemical vapor deposition of fluorocarbon films. *Chem. Phys. Lett.* 430:113–16

27. Zhang J, Williams KL, Fisher ER. 2003. Velocity distributions of SiF and SiF_2 radicals in an SiF_4 plasma molecular beam. *J. Phys. Chem. A* 107:593–97

28. Bogart KHA, Cushing JP, Fisher ER. 1997. Effects of plasma processing parameters on the surface reactivity of $OH(X^2P)$ in tetraethoxysilane/O_2 plasmas during deposition of SiO_2. *J. Phys. Chem. B* 101:10016–23

29. Berden G, Peeters R, Meijer G. 2000. Cavity ring-down spectroscopy: experimental schemes and applications. *Internat. Rev. Phys. Chem.* 19:565–607

30. Booth JP, Cunge G, Biennier L, Romanini D, Kachanov A. 2000. Ultraviolet cavity ring-down spectroscopy of free radicals in etching plasmas. *Chem. Phys. Lett.* 317:631–36

31. Kessels WMM, Hoefnagels JPM, Boogaarts MGH, Schram DC, van de Sanden MCM. 2001. Cavity ring down study of the densities and kinetics of Si and SiH in a remote Ar-H_2-SiH_4 plasma. *J. Appl. Phys.* 89:2065–73

32. Hoefnagels JPM, Barrell Y, Kessels WMM, van de Sanden MCM. 2004. Time-resolved cavity ringdown study of the Si and SiH_3 surface reaction probability during plasma deposition of a-Si:H at different substrate temperatures. *J. Appl. Phys.* 96:4094–106

33. Hoefnagels JPM, Stevens AAE, Boogaarts MGH, Kessels WMM, van de Sanden MCM. 2002. Time-resolved cavity ring-down spectroscopic study of the gas phase and surface loss rates of Si and SiH_3 plasma radicals. *Chem. Phys. Lett.* 360:189–93

34. Yalin AP, Zare RN, Laux CO, Kruger CH. 2002. Temporally resolved cavity ring-down spectroscopy in a pulsed nitrogen plasma. *Appl. Phys. Lett.* 81:1408–10

35. Piejak R, Godyak V, Alexandrovich B. 2001. Validation of current density measurements with a B-dot probe. *Rev. Sci. Instrum.* 72:4002–4

36. Tuszewski M, Tobin JA. 1996. The accuracy of Langmuir probe ion density measurements in low-frequency RF discharges. *Plasma Sources Sci. Technol.* 5:640–47
37. van Hest MFAM, de Graaf A, van de Sanden MCM, Schram DC. 2000. Use of in situ FTIR spectroscopy and mass spectrometry in an expanding hydrocarbon plasma. *Plasma Sources Sci. Technol.* 9:615–24
38. Karecki S, Chatterjee R, Pruette L, Reif R, Vartanian V, et al. 2001. Characterization of iodoheptafluoropropane as a dielectric etchant. III. Effluent analysis. *J. Vac. Sci. Technol. B* 19:1306–18
39. Schmidt M, Foest R, Basner R. 2001. Mass spectrometric diagnostics. *Low Temp. Plasma Phys.* 199:199–227
40. Eddy CR Jr. 2000. Mass spectrometric characterization of plasma etching processes. In *Handbook of Advanced Plasma Processing Techniques*, ed. RJ Shul, SJ Pearton. Berlin: Springer-Verlag
41. Singh H, Coburn JW, Graves DB. 1999. Mass spectrometric detection of reactive neutral species: Beam-to-background ratio. *J. Vac. Sci. Technol. A* 17:2447–55
42. Agarwal S, Quax GWW, van de Sanden MCM, Maroudas D, Aydil ES. 2004. Measurement of absolute radical densities in a plasma using modulated-beam line-of-sight threshold ionization mass spectrometry. *J. Vac. Sci. Technol. A* 22:71–81
43. Benedikt J, Agarwal S, Eijkman D, Vandamme W, Creatore M, van de Sanden MCM. 2005. Threshold ionization mass spectrometry of reactive species in remote Ar/C_2H_2 expanding thermal plasma. *J. Vac. Sci. Technol. A* 23:1400–12
44. Janes J, Banzinger U, Huth C, Hoffmann P, Neumann G, Scheer H-C. 1992. Analysis of large-area beam attacks on surfaces and testing of etching reactions. *Rev. Sci. Instrum.* 63:48–55
45. Ferreira JA, Tabares FL. 2007. Cryotrapping assisted mass spectrometry for the analysis of complex gas mixtures. *J. Vac. Sci. Technol. A* 25:246–51
46. Lopez-Garzon FJ, Domingo-Garcia M, Perez-Mendoza M, Alvarez PM, Gomez-Serrano V. 2003. Textural and chemical surface modifications produced by some oxidation treatments of glassy carbon. *Langmuir* 19:2838–44
47. Motomura H, Imai S, Tachibana K. 2001. Surface reaction processes in C_4F_8 and C_5F_8 plasmas for selective etching of SiO over photo-resist. *Thin Solid Films* 390:134–38
48. Shirafuji T, Motomura H, Tachibana K. 2004. Fourier transform infrared phase-modulated ellipsometry for in situ diagnostics of plasma-surface interactions. *J. Phys. D: Appl. Phys.* 37:R49–R73
49. Schneider L, Peukert W. 2007. Second harmonic generation spectroscopy as a method for in situ and online characterization of particle surface properties. *Part. Part. Sys. Charact.* 23:351–59
50. Simpson GJ. 2001. New tools for surface second-harmonic generation. *Appl. Spectrosc.* 55:16A–32A
51. Kessels WMM, Gielis JJH, Aarts IMP, Leewis CM, van de Sanden MCM. 2004. Spectroscopic second harmonic generation measured on plasma-deposited hydrogenated amorphous silicon thin films. *Appl. Phys. Lett.* 85:4049–51

52. Aarts IMP, Gielis JJH, van de Sanden MCM, Kessels WMM. 2006. Probing hydrogenated amorphous silicon surface states by spectroscopic and real-time second-harmonic generation. *Phys. Rev. B: Condens. Matter* 73:045327

53. Gielis JJH, Gevers PM, Stevens AAE, Beijerinck HCW, van de Sanden MCM, Kessels WMM. 2006. Spectroscopic second-harmonic generation during Ar$^+$-ion bombardment of Si(100). *Phys. Rev. B: Condens. Matter* 74:1665311

54. Laird RK, Andrews EB, Barrow RF. 1950. Absorption spectrum of CF$_2$. *Trans. Faraday Soc.* 46:803–5

55. Venkateswarlu P. 1959. On the emission bands of CF$_2$. *Phys. Rev.* 77:676–80

56. Mathias E, Miller GH. 1965. The decomposition of polytetrafluoroethylene in a glow discharge. *J. Phys. Chem.* 71:2671–75

57. Winters HF, Coburn JW. 1992. Surface science aspects of etching reactions. *Surf. Sci. Rep.* 14:161–269

58. Coburn JW, Winters HF. 1983. Plasma-assisted etching in microfabrication. *Ann. Rev. Mater. Sci.* 13:91–116

59. Coburn JW, Winters HF. 1981. Plasma-etching: a discussion of mechanisms. *CRC Crit. Rev. Solid State Mater. Sci.* 10:119–41

60. Graves DB, Humbird D. 2002. Surface chemistry associated with plasma etching processes. *Appl. Surf. Sci.* 192:72–87

61. Zhang D, Kushner MJ. 2000. Surface kinetics and plasma equipment model for Si etching by fluorocarbon plasmas. *J. Appl. Phys.* 87:1060–69

62. Martin IT, Dressen B, Boggs M, Liu Y, Henry CS, Fisher ER. 2007. Plasma modification of PDMS microfluidic devices for control of electroosmotic flow. *Plasma Process. Polym.* 4:414–24

63. Kiss LDB, Nicolai J-P, Conner WT, Sawin HH. 1992. CF and CF$_2$ actinometry in a CF$_4$/Ar plasma. *J. Appl. Phys.* 71:3186–92

64. Booth JP, Cunge G, Neuilly F, Sadeghi N. 1998. Absolute radical densities in etching plasmas determined by broad-band UV absorption spectroscopy. *Plasma Sources Sci. Technol.* 7:423–30

65. Booth JP, Abada H, Chabert P, Graves DB. 2005. CF and CF$_2$ radical kinetics and transport in a pulsed CF$_4$ ICP. *Plasma Sources Sci. Technol.* 14:273–82

66. Booth JP, Cunge G, Chabert P, Sadeghi N. 1999. CF$_x$ radical production and loss in a CF$_4$ reactive ion etching plasma: fluorine rich conditions. *J. Appl. Phys.* 85:3097–107

67. Singh H, Coburn JW, Graves DB. 2001. Measurements of neutral and ion composition, neutral temperature, and electron energy distribution function in a CF$_4$ inductively coupled plasma. *J. Vac. Sci. Technol. A* 19:718–29

68. Miyoshi Y, Miyauchi M, Oguni A, Makabe T. 2006. Optical diagnostics for plasma-surface interaction in CF$_4$/Ar radio-frequency inductively coupled plasma during Si and SiO$_2$ etching. *J. Vac. Sci. Technol. A* 24:1718–24

69. Schabel MJ, Donnelly VM, Kornblit A, Tai WW. 2002. Determination of electron temperature, atomic fluorine concentration, and gas temperature in inductively coupled fluorocarbon/rare gas plasmas using optical emission spectroscopy. *J. Vac. Sci. Technol. A* 20:555–63

70. Nagai M, Hori M. 2007. Temperature and density of CF radicals in 60 MHz capacitively coupled fluorocarbon gas plasma. *Jpn. J. Appl. Phys., Part 1* 46:1176–80

71. Steffens KL, Sobolewski MA. 2004. A technique for temperature mapping in fluorocarbon plasmas using planar laser-induced fluorescence of CF. *J. Appl. Phys.* 96:71–81

72. Hebner GA, Miller PA. 2000. Electron and negative ion densities in C_2F_6 and CHF_3 containing inductively coupled discharges. *J. Appl. Phys.* 87:7660–66

73. Sasaki K, Kawai Y, Suzuki C, Kadota K. 1997. Kinetics of fluorine atoms in high-density carbon tetrafluoride plasmas. *J. Appl. Phys.* 82:5938–43

74. Sugai H, Nakamura K, Hikosaka Y, Nakamura M. 1995. Diagnostics and control of radicals in an inductively coupled etching reactor. *J. Vac. Sci. Technol. A* 13:887–93

75. Zhou B, Joseph EA, Sant SP, Liu Y, Radhakrishnan A, et al. 2005. Effect of surface temperature on plasma-surface interactions in an inductively coupled modified gaseous electronics conference reactor. *J. Vac. Sci. Technol. A* 23:1657–67

76. Butoi CI, Mackie NM, Williams KL, Capps NE, Fisher ER. 2000. Ion and substrate effects on surface reactions of CF_2 using C_2F_6, C_2F_6/H_2, and hexafluoropropylene oxide plasmas. *J. Vac. Sci. Technol. A* 18:2685–98

77. Mackie NM, Venturo VA, Fisher ER. 1997. Surface reactivity of CF_2 radicals measured using laser-induced fluorescence and C_2F_6 plasma molecular beams. *J. Phys. Chem. B* 101:9425–28

78. Senesi GS, Aloia ED, Gristina R, Favia P, d'Agostino R. 2007. Surface characterization of plasma deposited nano-structured fluorocarbon coatings for promoting in vitro cell growth. *Surf. Sci.* 601:1019–25

79. Hori M, Goto T. 2007. Insights into sticking of radicals on surface for smart plasma nano-processing. *Appl. Surf. Sci.* 253:6657–71

80. Cunge G, Booth JP. 1999. CF_2 production and loss mechanisms in fluorocarbon discharges: fluorine-poor conditions and polymerization. *J. Appl. Phys.* 85:3952–59

81. Singh H, Coburn JW, Graves DB. 2000. Surface loss coefficients of CF_x and F radicals on stainless steel. *J. Vac. Sci. Technol. A* 18:2680–84

82. Hikosaka Y, Toyoda H, Sugai H. 1993. Drastic change in CF_2 and CF_3 kinetics induced by hydrogen addition into a CF_4 etching plasma. *Jpn. J. Appl. Phys., Part 1* 32:L690–93

83. Capps NE, Mackie NM, Fisher ER. 1998. Surface interactions of CF_2 radicals during deposition of amorphous fluorocarbon films from CHF_3 plasmas. *J. Appl. Phys.* 84:4736–43

84. Martin IT, Fisher ER. 2004. Ion effects on CF_2 surface interactions during C_3F_8 and C_4F_8 plasma processing of Si. *J. Vac. Sci. Technol. A* 22:2168–76

85. Cruden BA, Gleason KK, Sawin HH. 2001. Time resolved UV absorption spectroscopy of pulsed fluorocarbon plasmas. *J. Appl. Phys.* 89:915–22

86. Fisher ER. 2004. A review of plasma-surface interactions during processing of polymeric materials measured using the IRIS technique. *Plasma Process. Polym.* 1:13–27

87. Butterbaugh JW, Grey DC, Sawin HH. 1991. Plasma-surface interactions in fluorocarbon etching of silicon dioxide. *J. Vac. Sci. Technol. B* 9:1461–70

88. Gray DC, Sawin HH, Butterbaugh JW. 1991. Quantification of surface film formation effects in fluorocarbon plasma etching of polysilicon. *J. Vac. Sci. Technol. A* 9:779–85

89. Fuoco ER, Hanley L. 2002. Large fluorocarbon ions can contribute to film growth during plasma etching of silicon. *J. Appl. Phys.* 92:37–44

90. Wijesundara MBJ, Zajac G, Fuoco ER, Hanley L. 2001. Aging of fluorocarbon thin films deposited on polystyrene from hyperthermal $C_3F_5^+$ and CF_3^+ ion beams. *J. Adhes. Sci. Technol.* 15:599–612

91. Goyette AN, Wang Y, Misakian M, Olthoff JK. 2000. Ion fluxes and energies in inductively coupled radio-frequency discharges containing C_2F_6 and c-C_4F_8. *J. Vac. Sci. Technol. A* 18:2785–90

92. Lee HU, Deneufville JP, Ovshinsky SR. 1983. Laser-induced fluorescence detection of reactive intermediates in diffusion flames and in glow-discharge deposition reactors. *J. Non-Crystal. Solids* 59–6:671–74

93. Mutsukura N, Ohuchi M, Satoh S, Machi Y. 1983. The analysis of an SiF_4 plasma in an R.F. glow discharge for preparing fluorinated amorphous silicon thin films. *Thin Solid Films* 109:47–57

94. Martin RW, Merer AJ. 1973. Electronic transition of SiF. *Can. J. Phys.* 51:634–43

95. Johns JWC, Barrow RF. 1958. The band spectrum of silicon monofluoride, SiF. *Proc. Phys. Soc. Lond.* 71:476–84

96. Lee S, Tien Y-C, Hsu C-F. 1999. Direct spectroscopic evidence of the influence of chamber wall condition on oxide etch rate. *Plasma Chem. Plasma Process.* 19:285–98

97. Hebner GA. 2002. Spatially resolved CF, CF_2, SiF and SiF_2 densities in fluorocarbon containing inductively driven discharges. *Appl. Surf. Sci.* 192:161–75

98. Cunge G, Chabert P, Booth JP. 1997. Laser-induced fluorescence detection of SiF_2 as a primary product of Si and SiO_2 reactive ion etching with CF_4 gas. *Plasma Sources Sci. Technol.* 6:349–60

99. Giapis KP, Minton TK. 1996. Monitoring of direct reactions during etching of silicon. *Mater. Res. Soc. Symp. Proc.* 406:33–38

100. Giapis KP, Moore TA, Minton TK. 1995. Hyperthermal neutral beam etching. *J. Vac. Sci. Technol. A* 13:959–65

101. Williams KL, Fisher ER. 2003. Substrate temperature effects on surface reactivity of SiF_x (x = 1, 2) radicals in fluorosilane plasmas. *J. Vac. Sci. Technol. A* 21:1024–32

102. Matsuda A. 2004. Thin-film silicon growth process and solar cell application. *Jpn. J. Appl. Phys., Part 1* 43:7909–20

103. Sriraman S, Aydil ES, Maroudas D. 2004. Growth and characterization of hydrogenated amorphous silicon thin films from SiH_2 radical precursor: atomic-scale analysis. *J. Appl. Phys.* 95:1792–804

104. Kessels WMM, Hoefnagels JPM, van den Oever PJ, Barrell Y, van de Sanden MCM. 2003. Temperature dependence of the surface reactivity of SiH_3 radicals and the surface silicon hydride composition during amorphous silicon growth. *Surf. Sci.* 547:L865–70

105. Kessels WMM, Marra DC, van de Sanden MCM, Aydil ES. 2002. In situ probing of surface hydrides on hydrogenated amorphous silicon using attenuated total reflection infrared spectroscopy. *J. Vac. Sci. Technol. A* 20:781–89

106. Stamou S, Mataras D, Rapakoulias D. 1998. Spatial rotational temperature and emission intensity profiles in silane plasmas. *J. Phys. D: Appl. Phys.* 31:2513–20

107. Zhou J, Zhang J, Fisher ER. 2005. Effects of argon dilution on the translational and rotational temperatures of SiH in silane and disilane plasmas. *J. Phys. Chem. A* 109:10521–26

108. Kessels WMM, van de Sanden MCM, Severens RJ, Schram DC. 2000. Surface reaction probability during fast deposition of hydrogenated amorphous silicon with a remote silane plasma. *J. Appl. Phys.* 87:3313–20

109. Hertl M, Jolly J. 2000. Laser-induced fluorescence detection and kinetics of SiH_2 radicals in $Ar/H_2/SiH_4$ RF discharges. *J. Phys. D: Appl. Phys.* 33:381–88

110. Perrin J, Shiratani M, Kae-Nune P, Videlot H, Jolly J, Guillon J. 1998. Surface reaction probabilities and kinetics of H, SiH_3, Si_2H_5, CH_3, and C_2H_5 during deposition of a-Si:H and a-C:H from H_2, SiH_4, and CH_2 discharges. *J. Vac. Sci. Technol. A* 16:278–89

111. Bakos T, Valipa MS, Maroudas D. 2007. Interactions between radical growth precursors on plasma-deposited silicon thin-film surfaces. *J. Chem. Phys.* 126:114704

112. Singh T, Valipa MS, Mountziaris TJ, Maroudas D. 2007. First-principles theoretical analysis of sequential hydride dissociation on surfaces of silicon thin films. *Appl. Phys. Lett.* 90:251915

113. Sriraman S, Ramalingam S, Aydil ES, Maroudas D. 2000. Abstraction of hydrogen by Si radicals from hydrogenated amorphous silicon wafers. *Surf. Sci.* 459:L475–81

114. Zhou J, Martin IT, Adams E, Liu D, Fisher ER. 2006. Investigation of inductively coupled Ar and CH_4/Ar plasmas and the effect of ion energy on DLC film properties. *Plasma Sources Sci. Technol.* 15:714–26

115. von Keudell A, Kim I, Consoli A, Schulze M, Yangua-Gil A, Bendikt J. 2007. The search for growth precursors in reactive plasmas: from nanoparticles to microplasmas. *Plasma Sources Sci. Technol.* 16:S94–100

116. Liu D, Zhou J, Fisher ER. 2007. Correlation of gas-phase composition with film properties in the plasma-enhanced chemical vapor deposition of hydrogenated amorphous carbon nitride films. *J. Appl. Phys.* 101:023304

117. Zhou J, Fisher ER. 2006. Surface reactivity and energetics of CH radicals during plasma deposition of hydrogenated diamond-like carbon films. *J. Phys. Chem. B* 110:21911–19

118. van Hest MFAM, Haartsen JR, van Weert MHM, Schram DC, van de Sanden MCM. 2003. Analysis of the expanding thermal argon–oxygen plasma gas phase. *Plasma Sources Sci. Technol.* 12:539–53

119. Benedikt J, Schram DC, van de Sanden MCM. 2005. Detailed TIMS study of Ar/C_2H_2 expanding thermal plasma: identification of a-C:H film growth precursors. *J. Phys. Chem. A* 109:10153–67

120. Benedikt J, Woen RV, van Mensfoort SLM, Perina V, Hong J, van de Sanden MCM. 2003. Plasma chemistry during the deposition of a-C:H films and its influence on film properties. *Diamond Rel. Mater.* 12:90–97

121. von Keudell A, Schwarz-Selinger T, Jacob W. 2001. Simultaneous interaction of methyl radicals and atomic hydrogen with amorphous hydrogenated carbon films. *J. Appl. Phys.* 89:2979–86

122. von Keudell A, Meier M, Schwarz-Selinger T. 2001. Simultaneous interaction of methyl radicals and atomic hydrogen with amorphous hydrogenated carbon films, as investigated with optical in situ diagnostics. *Appl. Phys. A* 72:551–56

123. Sugai H, Toyoda H. 1992. Appearance mass spectrometry of neutral radicals in radio frequency plasmas. *J. Vac. Sci. Technol. A* 10:1193–200

124. Kojima H, Toyoda H, Sugai H. 1989. Observation of CH_2 radical and comparison with CH_3 radical in an rf methane discharge. *Appl. Phys. Lett.* 55:1292–94

125. Loh MH, Capelli MA. 1997. CH_3 detection in a low-density supersonic arcjet plasma during diamond synthesis. *Appl. Phys. Lett.* 70:1052–54

126. Liu D, Fisher ER. 2007. Surface reactivities of C_3 radicals during the deposition of fluorocarbon and hydrocarbon films. *J. Vac. Sci. Technol. A* 25:1519–23

127. Tahara H, Ando Y, Onoe K, Yoshikawa T. 2002. Plasma plume characteristics of supersonic ammonia and nitrogen/hydrogen-mixture DC plasma jets for nitriding under low-pressure environment. *Vacuum* 65:311–18

128. van den Oever PJ, van Helden JH, Lamers CCH, Engeln R, Schram DC, et al. 2005. Density and production of NH and NH_2 in an Ar-NH_3 expanding plasma jet. *J. Appl. Phys.* 98:093301

129. McCurdy PR, Butoi CI, Williams KL, Fisher ER. 1999. Surface interactions of NH_2 radicals in NH_3 plasmas. *J. Phys. Chem. B* 103:6919–29

130. Steen ML, Kull KR, Fisher ER. 2002. Comparison of surface interactions for NH and NH_2 radicals on polymer and metal substrates during NH_3 plasma processing. *J. Appl. Phys.* 92:55–63

131. Cicala G, Bruno G, Capezzuto P, Favia P. 1996. Photoelectron spectroscopy study of amorphous silicon-carbon alloys deposited by plasma-enhanced chemical vapor deposition. *J. Mater. Res.* 11:3017–23

132. Creatore M, Cicala G, Favia P, Lamendola R, d'Agostino R. 1999. Selective grafting of amine groups on polyethylene by means of modulated RF NH_3 plasmas. *Mater. Res. Soc. Symp. Proc.* 544:115–20

133. Cicala G, Creatore M, Favia P, Lamendola R, d'Agostino R. 1999. Modulated rf discharges as an effective tool for selecting exciting species. *Appl. Phys. Lett.* 75:37–39

134. Ishikawa K, Yamaoka Y, Nakamura M, Yamazaki Y, Yamasaki S, et al. 2006. Surface reactions during etching of organic low-*k* films by plasmas of N_2 and H_2. *J. Appl. Phys.* 99:083305

135. Liu D, Martin IT, Zhou J, Fisher ER. 2006. Radical surface interactions during film deposition: a sticky situation? *Pure Appl. Chem.* 78:1187–202

136. Iordanova S, Koleva I. 2007. Optical emission spectroscopy diagnostics of inductively-driven plasmas in argon gas at low pressures. *Spectrochim. Acta B* 62:344–56

137. Hueso JL, Gonzalez-Flipe AR, Cotrino J, Caballero A. 2005. Plasma chemistry of NO in complex gas mixtures excited with a surfatron launcher. *J. Phys. Chem. A* 109:4930–38

138. Balfour WJ, Cao JY, Prasad CVV, Qian CXW. 1994. Laser-induced fluorescence spectroscopy of the a^1P_u-$X^1S^+_g$ transition in jet-cooled C_3. *J. Chem. Phys.* 101:10343–49

139. Luque J, Juchmann W, Jeffries JB. 1997. Spatial density distributions of C_2, C_3, and CH radicals by laser-induced fluorescence in a diamond depositing dc-arcjet. *J. Appl. Phys.* 82:2072–81

140. Ito H, Ichimura SY, Namiki KC, Saitoh H. 2003. Absolute density and sticking probability of the $CN(X^2S^+)$ radicals produced by the dissociative excitation reaction of BrCN with the microwave discharge flow of Ar. *Jpn. J. Appl. Phys., Part 1* 42:7116–21

141. Ushirozawa Y, Matsuda H, Wagatsuma K. 2004. Determination of fluorine in copper by radio-frequency-powered glow-discharge plasma source emission spectroscopy associated with laser ablation sampling. *Jpn. Soc. Anal. Chem.* 53:699–703

142. Foest R, Olthoff JK, VanBrunt RJ, Benck EC, Roberts JR. 1996. Optical and mass spectrometric investigations of ions and neutral species in SF_6 radio-frequency discharges. *Phys. Rev. E* 54:1876–87

143. Suzuki C, Sasaki K, Kadota K. 1999. Formation of C_2 radicals in high-denisty C_4F_8 plasmas studied by laser-induced fluorescence. *Jpn. J. Appl. Phys., Part 1* 38:6896–901

144. Liu D, Martin IT, Fisher ER. 2008. *J. Appl. Phys.* Comparison of CH, C_3, CHF, and CF_2 surface reactivities during plasma enhanced chemical vapor deposition of fluorocarbon films. Work. Pap., Dep. Chem., Colo. State Univ.

145. Gottscho RA, Burton RH, Davis GP. 1982. Radiative lifetime and collisional quenching of carbon monochloride (A^2D) in an alternating current glow discharge. *J. Chem. Phys.* 77:5298–301

146. Skromme BJ, Liu W, Jensen KF, Giapis KP. 1994. Effects of C incorporation on the luminescence properties of the ZnSe grown by metalorganic chemical vapor deposition. *J. Cryst. Growth* 138:338–45

147. Hu X, Zhao G-B, Janardhan Garikipati SVB, Nicholas K, Legowski SF, Radosz S. 2005. Laser-induced fluorescence (LIF) probe for in-situ nitric oxide concentration measurement in a nonthermal pulsed corona discharge plasma reactor. *Plasma Chem. Plasma Process.* 25:351–70

148. Fresnet F, Baravian G, Pasquiers S, Postel C, Puech V, et al. 2000. Time-resolved laser-induced fluorescence study of NO removal plasma technology in N_2/NO mixtures. *J. Phys. D: Appl. Phys.* 33:1315–22

149. Fisher ER, Ho P, Breiland WG, Buss RJ. 1993. Temperature dependence of the reactivity of OH($X^2\Pi$) with oxidized silicon nitride and PMMA film surfaces. *J. Phys. Chem.* 97:10287–94

150. Donnelly VM, Herman IP, Cheng CC, Guinn KV, Donnelly VM, et al. 1996. Surface chemistry during plasma etching of silicon. *Pure Appl. Chem.* 68:1071–74

151. Greenberg KE, Hargis PJ Jr. 1990. Laser-induced-fluorescence detection of sulfur monoxide and sulfur dioxide in sulfur hexafluoride/oxygen plasma-etching discharges. *J. Appl. Phys.* 68:505–11

Biomolecule Analysis by Ion Mobility Spectrometry

Brian C. Bohrer, Samuel I. Merenbloom,
Stormy L. Koeniger,[*] Amy E. Hilderbrand,[**]
and David E. Clemmer

Department of Chemistry, Indiana University, Bloomington, Indiana 47405;
email: bbohrer@indiana.edu, smerenbl@indiana.edu, stormy.koeniger@abbott.com,
ahilderb@email.arizona.edu, clemmer@indiana.edu

Annu. Rev. Anal. Chem. 2008. 1:293–327

First published online as a Review in Advance on
February 27, 2008

The *Annual Review of Analytical Chemistry* is online
at anchem.annualreviews.org

This article's doi:
10.1146/annurev.anchem.1.031207.113001

*Present address: Abbott Laboratories, Abbott
Park, Illinois 60064
**Present address: University of Arizona, Tucson,
Arizona 85721

Key Words

mass spectrometry, electrospray ionization, protein conformation, multidimensional analysis

Abstract

Although nonnative protein conformations, including intermediates along the folding pathway and kinetically trapped misfolded species that disfavor the native state, are rarely isolated in the solution phase, they are often stable in the gas phase, where macromolecular ions from electrospray ionization can exist in varying charge states. Differences in the structures of nonnative conformations in the gas phase are often large enough to allow different shapes and charge states to be separated because of differences in their mobilities through a gas. Moreover, gentle collisional activation can be used to induce structural transformations. These new structures often have different mobilities. Thus, there is the possibility of developing a multidimensional separation that takes advantage of structural differences of multiple stable states. This review discusses how nonnative states differ in the gas phase compared with solution and presents an overview of early attempts to utilize and manipulate structures in order to develop ion mobility spectrometry as a rapid and sensitive technique for separating complex mixtures of biomolecules prior to mass spectrometry.

1. INTRODUCTION

1.1. Scope of This Review

IMS: ion mobility
spectrometry

**Plasma/ion
chromatography:**
gas-phase separation of ions
based on size and charge;
synonym for ion mobility
spectrometry

MS: mass spectrometry

LC: liquid chromatography

Ion mobility spectrometry (IMS), sometimes called plasma (1, 2) or ion chromatography (3, 4), has been utilized for many analytical applications, ranging from the detection of chemical warfare agents (5, 6) to particle sizing (7, 8). In the 1990s, several important advances made it possible to use IMS for analyzing biomolecules. Specifically, researchers coupled soft macromolecular sources (9–11) with IMS (12–16), developed theoretical methods for elucidating ion structure from comparisons of mobility measurements with calculated mobilities for computer-generated structures[1] (17–21), and implemented a range of mass spectrometry (MS) techniques (22–24). In the past five years, combinations of IMS with MS and liquid chromatography (LC) have emerged as powerful, hyphenated platforms for examining complex biomolecular mixtures (25–30), and mobility-based MS instruments have recently become commercially available [e.g., the Sionex Corporation microDMx differential mobility sensor (31), the Ionalytics Selectra FAIMS system (32), and the Synapt HDMS system (33)]. Applications ranging from the high-throughput detection of tens of thousands of peptide ions (30) to the obtaining of structural insight about large protein complexes (34, 35) have stimulated significant excitement.

Whereas many important applications of IMS to biomolecular analysis are now becoming routine, the early development of this field was driven by curiosity about the structures, stabilities, and reactivities of protein ions in the absence of solvent (36–47). In the gas phase, proteins display many nonnative conformations that are stable during the millisecond timescales of ion mobility experiments. Resolving different conformations is readily achieved and requires that some structures are not in equilibrium with one another. This behavior is different from the solution-phase equilibrium of states that is responsible for the cooperative transitions normally observed in solution (48, 49). The ability to examine stable populations raises an opportunity to select specific ion shapes for activation and then examine the new conformations that are formed in nondissociative collisions. The vacuum environment not only allows one to examine intramolecular interactions in the absence of solvent effects, it arguably may be the only place where it is possible to select many different types of precursors and intermediates in a way that allows the step-by-step motions of folding and unfolding transitions to be delineated. These curiosity-driven studies are intimately tied to the development of the next generation of multidimensional [IMS-MS, IMS-IMS and IMS-IMS-IMS (50–55)] techniques that are likely to lead to new forms of biomolecule analysis.

---

[1]A number of software packages are available for calculating molecular structure, including the molecular mechanics package available through the Insight II suite of programs [*Insight II 2000* (Accelerys Software, San Diego, California, 2001)] or the AMBER suite of programs [*AMBER* 7 (University of California, San Francisco, California, 2002)] and quantum chemical calculations from Gaussian [*GAUSSIAN 03* (Gaussian, Inc., Wallingford, Connecticut, 2004)] or Jaguar [*Jaguar* 5.5 (Schrödinger, LLC, Portland, Oregon, 1991–2003)].

This review begins with a short background discussion of protein structure intended as a framework for understanding IMS separations of biomolecules. We present general theoretical considerations associated with IMS and describe a modern instrumental design as an operational example. We have tried to provide references that are representative of the many advances within the past decade that have led to the flurry of activity in this field. Instrumental operation is illustrated with several example data sets, including applications involving what is perhaps one of the most complex biomolecular mixtures: proteins from plasma. Anderson & Anderson (56) have argued that because plasma is in contact with all cells, it may contain proteins from the entire human proteome. It presents an extraordinary challenge for analysis, and although IMS analyses are at an early stage, including such work in a review is an important milestone. Although we mention these experiments, the focus of this review is intended to illuminate the next steps of IMS evolution that will allow new generations of multidimensional IMS experiments to be conducted. The analysis of biomolecules by IMS is an area that builds on many diverse fields and important studies by others. We refer the interested reader to several other valuable reviews about ion thermochemistry (41, 57, 58), macromolecular conformations (59, 60), MS instrumentation (61, 62), condensed-phase separations (63, 64), and proteomics (65).

1.2. General Features of Macromolecular Structures in Solution and in the Gas Phase

The structures of large biological molecules such as proteins are often described as native or denatured. The native state implies a conformation that is capable of biological function; investigators have now obtained thousands of detailed geometries (66, 67) of what are usually assumed to be native structures from nuclear magnetic resonance (68–70) and crystallographic techniques (71, and references therein; 72–74). Far less is conveyed about structure from the term denatured. Lumry & Eyring (75) began their 1954 paper "Conformation Changes of Proteins" with "[t]he term protein denaturation even in its original meaning included all those reactions destroying the solubility of native proteins and has since acquired so many other meanings as to become virtually useless." In the half-century since this statement, little has changed. The nature of macromolecules in solution is such that transitions between denatured and native states occur rapidly and presumably through a plurality of intermediates; however, intermediates along folding pathways are rarely stabilized to a degree that allows them to be isolated in the quantities and lifetimes necessary for structural characterization by conventional methods. Rather, macromolecular systems appear to rapidly establish an equilibrium in distributions of structures to optimize appropriate interactions with the environment (76–78).

As an example of such behavior, let us consider the acid denaturation curve for cytochrome c in **Figure 1** that is measurable by a number of techniques, in this case Soret absorption (79). As the pH of the solution is lowered to a value of $\sim$4, the fraction of the native state begins to decrease; this continues as pH is dropped to $\sim$2 until essentially no signal associated with the native conformation is detectable. The sigmoidal

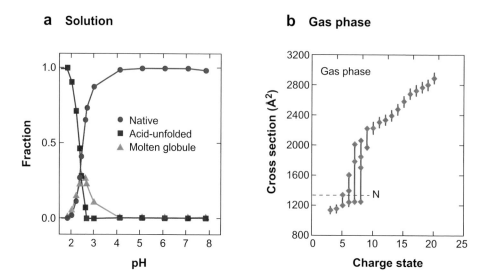

a Solution

b Gas phase

Figure 1

(*a*) The relative abundances of three ensembles of structures (native, acid-unfolded, and molten globule) observed by Soret absorption analysis of the acid denaturation of cytochrome *c* in water. (*b*) The collision cross sections for the most intense features observed as a function of charge state for protonated cytochrome *c* in the gas phase. The dotted line denotes the cross section of the native state at 1334 Å^2. **Figure 1*a*** reprinted from Reference 79. Copyright American Chemical Society 1993. The value for the dotted line in **Figure 1*b*** comes from Reference 88.

shape associated with this transition is characteristic of many macromolecules, regardless of the approach used to induce denaturation (e.g., temperature or solvent denaturation) (80–83), and is considered a signature of a cooperative transition (48, 49). In this case, the transition involves two other types of states: (*a*) a molten globule state (48, 80, 84) observed from pH $\sim$2 to 4 and believed to correspond to a set of relatively compact, although not native, distributions and (*b*) another corresponding to a distribution of unfolded states, which dominates the pH scale below a pH of 2.

Such transitions for macromolecules in solution are well known, and usually only few states coexist. In the absence of solvent, large molecules display many properties that resemble those of their solution-phase counterparts, but also many that are different. As an example, we consider the gas-phase collision cross sections of cytochrome *c* as a function of the protonation state produced by electrospray ionization (ESI) (42, 60). The overall appearance of these data shows a roughly sigmoidal shape, as observed for decreasing pH in solution. In the absence of solvent, low-charge states of cytochrome *c* (e.g., the $[M + 3H]^{3+}$ to $[M + 7H]^{7+}$ species) show features in the ion mobility distributions corresponding to ions with cross sections ranging from $\sim$1000 to 1200 Å^2, values that are near the cross section expected for compact states that are similar in conformation to the native solution structure. As the number of protons

Molten globule: compact form of a protein that does not retain its biological function

ESI: electrospray ionization

added during electrospray increases, the ions adopt geometries with larger cross sections. For example, cross sections for highly charged ions $[M + 12H]^{12+}$ indicate that extended states (with cross sections that are more than twice the value anticipated for the native conformations) are favored. The $[M + 6H]^{6+}$ to $[M + 9H]^{9+}$ species exist as structures that range in cross sections from ~1200 to 2000 Å^2.

The transition from compact to extended states observed with an increasing protonation state has been explained by considering the forces involved in stabilizing conformations. The folding free energy of cytochrome c in solution is -37.1 kJ mol^{-1} (85), whereas in vacuo values range from -2182 to -3497 kJ mol^{-1} (85–87). In the gas phase, the energetics of these structural differences are not mitigated by solvation effects. The structure of a gas-phase protein having a net charge of zero is established only by intramolecular interactions, such as zwitterion formation, hydrogen bonding, and van der Waals contacts (88). Excess protons presumably disrupt solution-phase structure; that is, a protonated basic site that would normally be solvated in solution must be accommodated by intramolecular interactions, primarily involving polar side chains and backbone N-H or C-O groups. This internal solvation of charged residues in the gas phase causes the conformations of low-charge-state ions in the gas phase to contract and become more compact than the native solution structures (85, 89, 90). As the number of excess protons increases, the structure becomes sensitive to differences in the dielectric of the surrounding media (~80 for water and 1.0 for a vacuum). In the low dielectric of the vacuum, high-charge states adopt highly extended conformations to minimize repulsive Coulombic interactions that are induced upon desolvation, giving rise to the sigmoidal shaped curve in **Figure 1**.

Having pointed out the similarities in the shapes of these curves associated with changes in structure upon acid denaturation in solution and the increased state of protonation in the gas phase, we need to stress a key difference. Although the sigmoidal shape associated with the solution-phase denaturation implies an equilibrium and cooperativity, the similar shape of the curve as a function of protonation state for ions in the gas phase does not. Ion shapes in the gas phase are often stable for extended time periods (substantially longer than the millisecond time periods necessary for analysis). This difference makes it possible to utilize the gas-phase conformations for a number of different applications.

2. EXPERIMENTAL CONSIDERATIONS

2.1. Mobility Measurements

When a packet of ions in a buffer gas is exposed to a weak electric field (E), it drifts with a velocity $v_D = K \cdot E$, where K corresponds to the mobility constant of a specific ion in the buffer gas. Because K is specific to interactions between the ion and the gas, individual components within packets that contain a mixture of species may be separated owing to differences in the mobilities of the components. One determines the value of K by measuring the time (t_D) required for ions to drift through a specified distance. Measurements between laboratories can be compared by normalizing values to standard conditions that produce the reduced mobility (K_0) using the relation

$$K_0 = \frac{L^2}{t_D V} \times \frac{273.2}{T} \times \frac{P}{760},\qquad(1)$$

where the variables L and V correspond to the length of the drift region and the voltage applied across it, respectively; and P and T correspond to buffer gas pressure and temperature, respectively (91).

It is also possible to report values as experimental cross sections by the relation

$$\Omega = \frac{(18\pi)^{1/2}}{16} \frac{ze}{(k_B T)^{1/2}} \left[\frac{1}{m_I} + \frac{1}{m_B} \right]^{1/2} \frac{t_D E}{L} \frac{760}{P} \frac{T}{273.2} \frac{1}{N},\qquad(2)$$

where ze corresponds to the charge on the ion; k_B is Boltzmann's constant; N is the number density of the buffer gas; and m_I and m_B correspond to the mass of the ion and buffer gas, respectively (91). One can easily rearrange this equation to solve for t_D. For a macromolecular ion drifting in He buffer gas, the expression associated with the reduced mass shows that this separation is dominated by the cross section (rather than the ion mass). We point this out because when coupling IMS separations with MS techniques, the strong correlation of the cross section and mass arises because the mass and size increase are intrinsically coupled (92). It is often useful to convert the time axis of an IMS directly to a cross-section axis via Equation 2. We note that this can be done only for ions having the same net charge.

2.2. Weak and Strong Fields

An important definition is associated with the drift field. The applied field is considered to be weak if the buffer gas density is high enough to collisionally dampen ions such that the internal ion temperature is that of the bulk buffer gas (60, 91, 93). Under these conditions, v_D is small compared with the thermal velocity of the gas, and ions are not expected to align with the field in the drift region. In this case, the cross-section measurement corresponds to the average of all orientations of the ion as it passes through the drift region and can be used to infer information about the average shape of the ion.

Although a wide range of applied fields satisfies the low-field definition, many interesting phenomena occur as the applied field reaches and then exceeds the low-field limit and ions move in the high-field regime. For example, the field associated with the transition region is different for each ion and depends on the ion charge state, as well as the stability and dynamics of different structures. Under high-field conditions, reduced mobilities may increase or decrease relative to those measured at low fields owing to processes that are not completely understood and are currently under investigation (94–98). It is possible to induce structural transitions as well as dissociation in this region. Differences in mobilities at low and high fields are analytically valuable and have led to the development of the field asymmetric (FA)IMS technique (31, 32, 94–96), as well as hybrid FAIMS-IMS approaches (24).

2.3. Peak Shape and Resolving Power Considerations

As a complement to MS methods, we are especially interested in utilizing IMS for separating isobaric ions. To this end, highly folded (compact) conformations have smaller cross sections (or larger mobilities) than unfolded (extended) states. Often ions with similar structures exist over several charge states. In this case, more highly charged ions have higher mobilities because they experience a different drift force (qeE).

For a single isomer, the theoretical shape of a packet of ions exiting the drift region is determined by the flux

$$\Phi(t) = \int \frac{C}{(Dt)^{1/2}} (v_D + L/t) \left[1 - \exp\left(\frac{-r_0^2}{4Dt}\right)\right] \exp\left[\frac{-(L - v_Dt)^2}{4Dt}\right] P(t_p) \, dt_p,$$

(3)

where r_0 is the radius of the drift tube entrance aperture; $P(t_p)dt_p$ is the time-dependent shape of the packet as it enters the drift region; C is a constant; and D is the diffusion constant, given by $K_0 k_b T/ze$ (99). If more than one structure is present, the experimental peak shape may be broader than that calculated for a single structure. Although good agreement with the experimental and calculated peak shapes is suggestive of a single structure, this is not required because two geometries could have identical mobilities; additionally, if multiple structures interconvert on timescales that are much shorter than the millisecond timescale associated with the experiment, then they appear as a single sharp peak.

Revercomb & Mason (93) showed that the theoretical resolving power ($R = t/\Delta t$, where Δt corresponds to the full width at half-maximum of a peak) of a drift tube can be approximated from

$$\frac{t}{\Delta t} \approx \left(\frac{LEze}{16 k_B T \ln 2}\right)^{\frac{1}{2}},$$

(4)

which shows that increasing the drift field or length, or decreasing the temperature, leads to an increase in resolution. It is unfortunate that this increase scales as the square root of the experimental parameters. However, it provides an understanding of the impetus to build longer, high-field drift regions.

Although not shown in Equation 4, a parameter that ultimately plays a key role in defining instrument design and resolving power is buffer gas pressure. At high fields the gas must remain stable, so it is necessary to operate at a pressure where the gas does not discharge. Above a few torr, the breakdown potential of most buffer gasses increases with increasing pressure. Thus, it is possible to obtain very high resolving powers (~80 to 300) by utilizing relatively high drift voltages (10 to 30 kV) over ~1-m drift regions at high buffer gas pressures (~100 to 760 torr). One drawback of the high-pressure approach is that it is difficult to store ions. Thus, for a continuous ion source, the introduction of a short pulse of ions upon initiation of experiments often limits duty cycles. Researchers have used a number of approaches (including utilizing multiple injections, as well as Fourier and Hadamard transform approaches)

TOF: time of flight

to improve duty cycles (100–102). Alternatively, at low pressures, it is possible to store ions for extended times to accumulate continuous signals into concentrated packets. However, at low pressures, drift fields are typically limited to ~10 to 30 V cm^{-1}.

2.4. Utilizing Nonuniform Fields

In the past decade, a number of separation devices that utilize nonuniform fields to separate ions based on differences in mobilities have emerged. These include FAIMS (31, 32, 94–96), as well as an approach in which ions are exposed to a sequence of accelerating voltages and dampening collisions, referred to as a traveling-wave IMS (33, 103). These instruments often yield results for biomolecules that are similar in appearance to those that utilize uniform fields. Some theoretical treatments of the separation mechanisms have been presented. However, in general, this treatment is at a relatively early stage, and calibrations to uniform field instruments are often employed in data interpretation.

2.5. Combining Ion Mobility Spectrometry with Mass Spectrometry Technologies

The analysis of biomolecules by IMS has been accelerated by advances in MS-based technologies. Specifically, macromolecular ions are created by gentle ionization sources, such as ESI [and more recently desorption (D)ESI (104)] and matrix-assisted laser desorption/ionization. Interestingly, Dole and colleagues' (105) early work to develop an ESI source for biomolecules investigated lysozyme and utilized IMS rather than MS detection. This pioneering analysis did not resolve charge-state distributions; a later interpretation suggests that although protein ions were observed, the assignment of peaks as low-charge-state protein clusters is probably incorrect (92). Hill and coworkers (12) were the first to resolve ESI charge states with IMS. Clemmer et al. (13) were the first to resolve different protein ion conformations. Bowers's (14, 15) and Russell's (16) groups developed early matrix-assisted laser desorption/ionization sources with IMS. Recently, Clemmer's (106) group coupled desorption ESI with IMS. Once the ion source is separated from the tube, it is possible to couple essentially any ionization source or MS technique with IMS. To this end, IMS has been coupled with Fourier transform ion cyclotron resonance (107), linear quadrupoles (108), and trapping devices (109–111), as well as time-of-flight (TOF) (112) mass spectrometers.

2.6. Methods for Determining Ion Structure

Arguably one of the most important advances achieved with IMS technologies is associated with the ability to determine information about ion shape. The first description of such a comparison was given in 1925 by Mack (17), who projected the shadows of models at different orientations onto a screen to determine an orientationally averaged cross section, which could then be used to back out a calculated diffusion constant. In the mid-1990s, Jarrold's (19) and Bowers's (20) groups developed computer algorithms for calculating cross sections. These calculations were initially used to investigate the structures of a number of atomic clusters and attracted considerable attention with the elucidation of a series of structures associated with carbon clusters as a function of cluster size (including a family of fullerenes). At about this time, advances in biological ion sources and computations of molecular structure that made it possible to rapidly generate molecular coordinates of biomolecules became available. The combination of technologies led to rapid advances in understanding biomolecular ion structure. Most of what is now understood about the shapes of macromolecules in the absence of solvent was determined by comparing calculated cross sections for trial geometries with experimental values. In many cases, it is only possible to estimate a general conformation type; however, in favorable cases, the comparison helps to guide theory in such a way that a low-energy structure that fits the experiment can be obtained. There is now substantial evidence for sequences that form compact globular, helical, and helical-coil conformations. Recently, Robinson and coworkers (34) used a similar approach to determine that the overall geometry of the trp RNA binding protein complex favored a ring-like structure.

2.7. Combining Ion Mobility Spectrometry with Liquid Chromatography Separations

Because of the complexity of biological systems, essentially all analyses involve some form of chromatography. A number of studies have utilized combinations of LC with IMS-MS. As discussed in more detail below, a primary advantage of these techniques for the analysis of complex samples is that multiple dimensions provide enhanced analytical peak capacity. With more analytical space available for peaks upon inclusion of IMS, there are advantages of reduced spectral congestion and new information content. Recently, attempts to modulate conditions between precursor and fragment ion formation using two-dimensional LC, combined with an additional IMS separation, have resulted in a highly parallel approach for identifying large mixtures of peptides. Early examples include analyses of the complex mixtures of proteins from human plasma (as tryptic peptides) (29, 30), as well as attempts to characterize the proteome of the model *Drosophila* organism (26). Two-dimensional LC (strong cation exchange coupled offline with reversed-phase LC) has been combined with IMS-MS to produce a comprehensive list of peptides detected from human plasma. The list includes more than 9000 entries, 2928 of which are believed to be high-confidence assignments (30).

3. INSTRUMENTATION, EXAMPLES OF ION MOBILITY SPECTROMETRY DATA, AND APPLICATIONS

3.1. Description of an Ion Mobility Spectrometry–Mass Spectrometry Instrument

The constraints associated with buffer gas discharge upon the application of high fields have led to two general instrumental designs: high-pressure, high-field instruments

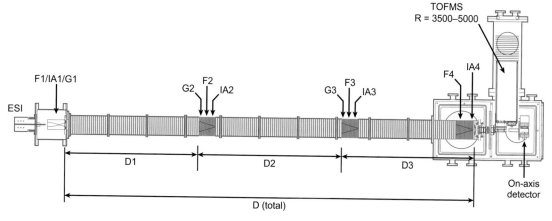

Figure 2

Schematic of the IMS-IMS-IMS-TOF instrument built at Indiana University. Ions are accumulated in the ion funnel F1 and pulsed into the drift tube for experiments by G1. Ions can be mobility selected in two regions (G2, G3); the mobility-selected structures can then be collisionally activated (IA2, IA3). Fragmentation of ions can be performed at IA4, prior to mass analysis in the time-of-flight (TOF) mass analyzer.

that are capable of generating high resolving powers (113, 114) and low-pressure, low-field configurations with lower resolving powers (115). There are advantages to each type of instrument; many of the desirable features as well as limitations have been discussed previously (60). A significant attraction to low-pressure instruments is that it is straightforward to couple them with existing MS ion sources, trapping devices, and analyzers.

Figure 2 shows an example of an instrument we constructed at Indiana University in 2005 (50, 51), which builds on a design that incorporated an ion funnel at the exit of the drift tube (23). This instrument utilizes a 3-m drift tube with several ion funnels and activation regions along the drift axis. The drift tube can be operated as a single long instrument or broken into three independent drift regions. We begin by considering operation as a single IMS separation in combination with the TOFMS analysis. This is intended to capture the advances in IMS-TOFMS techniques developed from ~1997 to 2005. The utilization of independent drift regions is described below.

Because flight times in the evacuated flight tube of the mass spectrometer (on the order of microseconds) are substantially shorter than drift times through the He-filled drift tube (milliseconds), it is possible to record mass spectra within individual drift windows across the IMS spectrum. This is referred to as a nested measurement because m/z information is nested within individual time windows of the IMS distribution (112). The collection of nested IMS-MS data commences when ions are gated into the entrance of the drift region (using a standard electrostatic gating strategy). The pulse that initiates this gate is synchronized with a pulse in the source region of the TOF mass spectrometer. The repetition rate of the MS pulser is fixed to allow

Nested measurement: recording a longer timescale experiment as increments of a shorter timescale experiment

the mass range of interest to be analyzed. This range (usually 5 to 50 μs) becomes the width of the individual time elements that make up the IMS distribution.

3.2. Nested Ion Mobility Spectrometry–Mass Spectrometry Data Set Recorded for a Mixture of Tryptic Peptides

A nested IMS-MS measurement is best illustrated by an example. **Figure 3** shows a nested IMS-MS data set for a mixture of tryptic peptides (in this case obtained upon digestion of a standard mixture of commercially available proteins). The drift times of ions (associated with the IMS separation) range from 14 to 22.2 ms, whereas flight times in the mass spectrometer range from 19.5 to 30.5 μs. As individual components associated with the mixture of ions exit the drift region, they enter the source region of the MS instrument where they are orthogonally accelerated into the TOF instrument and the flight times are recorded. For example, ions that exit the drift tube at a single drift time (e.g., 18 ms as in **Figure 3**) split into multiple peaks over the ~10.5-μs range of flight times shown.

There are a number of apparent advantages associated with this approach. The nested measurement allows mobilities and m/z values for all the ions present in the mixture to be recorded in a single experiment. This has led to large databases of cross sections for different peptide sequences and charge states (116). The availability of cross sections for many sequences makes it possible to extract information about how the amino acid composition, as well as the position of a specific amino acid in a sequence, influences cross section (117). The ability to predict mobilities from sequences is an exciting advance as it provides a constraint for assignment that is not available from MS or MS/MS studies.

Additionally, the IMS separation reduces spectral congestion. Also plotted vertically in **Figure 3** is the mass spectrum that would be obtained if no IMS separation was used. Although many peaks in the mass spectrum are apparent, the ability to pull the distribution of peaks apart prior to MS analysis allows many features (especially small peaks) that would otherwise overlap to be resolved (118). In total, 98 resolved peaks were observed in the two-dimensional data set, corresponding to 60 of the 187 peptides expected upon complete digestion of all five proteins.

Finally, in many cases, ions appear to fall into families. **Figure 3** illustrates families of ions having the same charge states (in this case either $[M + H]^+$, $[M + 2H]^{2+}$, or $[M + 3H]^{3+}$ produced during ESI). In other cases, it is possible to resolve families of molecules that have different chemical properties (119). For example, mixtures of lipids (120), glycans (121), peptides with specific post-translational modifications (122), and other types of polymers often fall into families. Thus, there is significant new information associated with the combined measurement.

3.3. Instruments That Incorporate Additional Dimensionality

The relatively long times associated with IMS compared with MS offer other advantages for resolving complex mixtures. As long as dissociation techniques are carried out rapidly (on submillisecond timescales and with no disruption of the time resolution

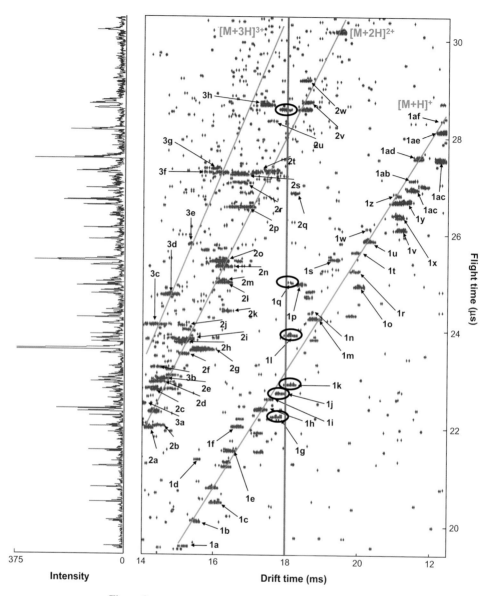

Figure 3

Nested $t_d(t_f)$ distribution for the direct infusion of a mixture of ions produced by electrospray of a tryptic peptide mixture from the digestion of dog and pig hemoglobin, bovine and pig albumin, and horse cytochrome c. The solid lines in the two-dimensional plot indicate the positions of the [M + H]$^+$, [M + 2H]$^{2+}$, and [M + 3H]$^{3+}$ charge-state families. The vertical line at 18 ms highlights the overlap of different m/z species within the mobility distribution; several of the ions having drift times similar to this value are circled. The numbers and letters label peaks and correspond to the assignments given in table 1 of Reference 25. On the left is a mass spectrum obtained by summing the intensities at a given flight time across all drift time windows. These data were acquired using a drift field of 171.7 V cm^{-1} with a 300-K helium pressure of 150.3 torr. The data were acquired in ~100 min. Figure adapted from Reference 25.

after the fragments are formed), it is possible to gain a parallel advantage of generating fragmentation spectra. In this case, fragment ions appear along the vertical dimension of IMS-MS data sets. By modulating voltages at the exit of the drift tube, it is possible to use conditions that favor precursors and those that favor collision-induced dissociation (CID). The coincidence of drift times is used to unite peaks formed during dissociation with their antecedent precursors (from back-to-back, modulated condition experiments). This is called a parallel fragmentation approach (123, 124) to emphasize that all ions are examined as precursors and fragments after ionization (unlike MS selection of precursor ions used for traditional MS/MS studies). However, the approach is really a very fast (submillisecond timescale) serial approach for those ions having different mobilities. Similar to MS/MS, the fidelity of the information is limited by the ability to resolve components prior to CID. There is clearly a need to improve IMS resolving power as mixture complexity increases.

CID: collision-induced dissociation

A range of dissociation approaches has been demonstrated, including traditional collisions in an octopole collision cell (123, 124), surface-induced dissociation (125, 126), and fragmentation induced by an orifice skimmer cone mounted at the exit of a drift tube (127), as well as fragmentation induced in the high-pressure exit region of the drift tube (128, 129). It should be straightforward to include photodissociation approaches (130) and, to the extent that time resolution is not lost, new techniques such as electron capture (131) or transfer (132) dissociation as well. Parallel dissociation is now obtainable using the traveling-wave IMS approach that is available commercially, in which it is referred to as time-aligned-parallel fragmentation (133).

Although 1 to 100 ms is a long time to manipulate ions in a mass spectrometer, the combined IMS-MS timescale allows it to still function as a very-high-speed detector for condensed-phase analyses. Thus, one can utilize the nested advantage for higher-dimensionality experiments by coupling IMS with slower techniques. We discuss the data set in **Figure 3** above in terms of an IMS-MS analysis; however, it was actually the first IMS-MS data set recorded as a single frame of a higher-dimensionality separation involving LC. In this case, as peptides eluted from the LC column, they were ionized by ESI and analyzed by IMS-MS. The combined LC-IMS-MS approach can be modulated to include CID between the IMS and MS instruments [an LC-IMS(CID)MS analysis]. Thus, although no MS is used for precursor ion selection, the combined resolution of the LC and IMS dimensions often provides an opportunity to correlate precursor and fragment ion spectra for ion identification. The approach has been automated to an extent that database-searching methods (134, 135) can now be employed.

The combination of IMS with LC, parallel CID, and MS is emerging as an especially powerful approach for examining complex mixtures of proteins from biological sources. One of the most formidable challenges in proteomics experiments is to develop reproducible, information-rich, tremendously high-peak-capacity analytical platforms that operate quickly. **Figure 4a** shows a plot of the LC and IMS dimensions for an LC-IMS(CID)MS analysis of peptides obtained upon tryptic digestion of proteins isolated from plasma. In this case, we show only the precursor data set (the modulated data associated with the CID spectra appear similar when plotted in these dimensions). We have used a high-resolution LC separation that over several hours

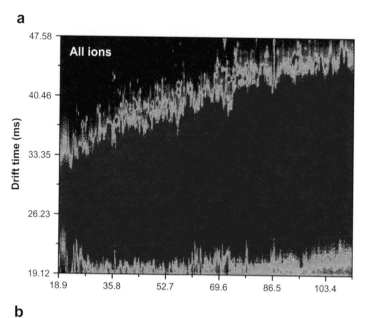

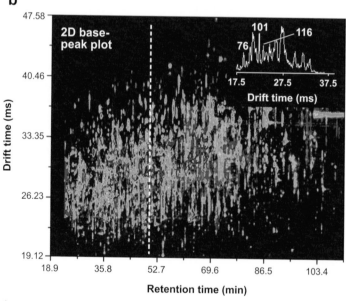

Figure 4

Two-dimensional retention time (drift time) contour plots of the high-resolution ion mobility spectrometry–mass spectrometry analysis of a digest of proteins from plasma. (*a*) The summed intensity of all mass spectral features at each retention time and drift time value. (*b*) A two-dimensional base-peak plot in which only the most intense mass spectral feature is plotted at each retention time and drift time value. An intensity threshold of four was used to construct both plots. The inset in panel *b* shows the drift time distribution for species observed at a retention time of 48 min, along with the resolution for several species. Figure reprinted from S.J. Valentine, S.L. Koeniger & D.E. Clemmer, unpublished data.

shows a peak capacity of >300. Clearly this sample is complex, and the number of components far exceeds this capacity. Even with the additional high-resolution IMS separation (having a resolving power in excess of 100 for most peaks and providing a two-dimensional peak capacity of >15,000), the main portion of the spectrum appears as a large unresolved feature. We note that a few species on the leading and trailing edges of the IMS distribution are resolved. A feeling for the shapes of peaks associated with individual components within the data set can be obtained by extracting and plotting the most intense features (a two-dimensional base-peak plot) shown in **Figure 4b**.

A primary consideration in developing this analysis is that no additional time is required for the experiment. This is not the case for the interpretation of results. It is possible to pick large features (precursors and CID data sets) quickly and carry out database-search analyses for assignments relatively quickly (and, in doing so, several hundred proteins can be confidently identified); however, a more detailed analysis is time intensive owing to the large size of the data sets.

3.4. New Developments Involving IMS-IMS and IMS-IMS-IMS Instrumentation

In analogy with condensed-phase chromatography, in which it is possible to change the mobile phase during a separation (or use different stationary phases to influence the ability to separate different components), a number of studies have attempted to change the IMS separation of ions in the gas phase by changing the buffer gas (136). There are many interesting factors that arise as the buffer gas is varied, and it is possible to substantially shift the mobilities of ions by using different gasses. Asbury & Hill (136) have shown that it is possible to change the order of the drift times of chloroaniline and iodoaniline by varying the buffer gas from helium to carbon dioxide. However, in general, mobility separations are dominated by momentum transfer processes during the ion-neutral collisions; because this depends largely on the shapes of ions, the ability to substantially change the resolution of two ions that have similar components is limited (137).

As described above, an important difference between the behavior of macromolecules in the gas phase compared with solution is that in the absence of solvent, many different conformations appear to be stable. Early ESI-IMS work on proteins showed that upon high-energy injection of ions into low-pressure drift tubes, it was possible to induce unfolding and refolding transitions (115, 138). Other studies of ions in Paul-type and ion-cyclotron-resonance traps also indicated that different types of conformations were stable over long time periods (139–142).

This curious property of biomolecules in the absence of solvent provided the rationale to develop a multidimensional IMS-IMS instrument that would utilize differences in ion cross sections (before and after activation) for separation. The schematic shown for the 3-m long drift tube in **Figure 2** is designed specifically for this type of measurement. Although the instrument can be operated as a single separation device to produce IMS-MS data, it comprises three separate drift regions (D1–D3) and ion gates (G1–G3), as well as four ion activation regions (IA1–IA4). These regions are designed so that following an initial IMS separation (in the D1 region), ions of a specified mobility can be selected at G2, activated at IA2, and separated again in D2 and D3. If desired, one can carry out an additional selection and activation process in the G3/IA3 region and follow this with a final separation in D3. Finally, after the final IMS separation, it is possible to induce fragmentation at IA4 (and carry this out in a modulated fashion to generate precursor and CID spectra).

A drift tube of this length and capability is possible because of the development of the ion funnel by Smith's group (22, 23). The instrument shown in **Figure 2** uses four of these funnels, F1 to F4. A combination of radio frequency and direct current fields in the funnel allows ions to be focused in the off-axis dimensions so that ion transmission is high. Tang et al. (23) demonstrated that inclusion of a funnel inside a drift region does not significantly change the resolving power of the measurements.

The ability to select and activate ions inside of the drift tube has led to a number of interesting findings. Koeniger et al. (52) examined the broad IMS peaks observed for specific charge states of ubiquitin (a small 8.6-kDa protein that has been studied extensively). **Figure 5** shows that the IMS spectrum for $[M + 7H]^{7+}$ (obtained by transmitting ions through the entire 3-m drift region) is essentially identical to $[M + 7H]^{7+}$ distributions recorded using earlier low-pressure, low-resolution as well as high-pressure, high-resolution instruments. Interestingly in the prior work, although the instrumental resolving powers differed by an order of magnitude, the spectra looked similar in that the $[M + 7H]^{7+}$ ion exhibited at most three fairly broad features. Comparisons of the cross sections with values calculated for the native solution coordinates (as well as other structures generated by molecular modeling) suggested that these peaks corresponded to compact (tightly folded) conformers and partially folded states; the position of a peak corresponding to an elongated state (approximately 1600 to 1700 Å^2) is also indicated. This state can be observed by injecting ions into a low-pressure drift tube at high energies or heating the capillary ion source inlet, but it is not observed in significant abundance in the IMS distribution shown.

The question that emerged from these and early data was, what limits the widths of these features? Are the peaks broad because multiple conformations interconvert during the experiment on timescales that are similar to the separation? Or are the peaks broad because there are many unresolved conformations within the broader conformation type? Selection of a narrow ion pulse at G2 reveals that the peak remains sharp (**Figure 5b**) as selected ions drift the remaining distances through the drift tube. **Figure 5c** shows that this is the case at essentially any point across the distribution, requiring that many unresolved structures within the broader conformation type are stable (thus peaks are broad because these structures are not resolved).

Once it was known that it is possible to select subsets of conformations, it is possible to investigate how activation influences IMS distributions. **Figure 6** shows a study demonstrating the ability to select and vary the conformation of $[M + 7H]^{7+}$ across a broad range of conformations (53). As in **Figure 5**, it is possible to select any region of ions (here a subset of compact states); upon activation (in this case an 80-V bias applied across a short 0.3-cm activation region), the compact states produce broad peaks associated with a range of partially folded and elongated states. Which states are produced depend on what part of the distribution is selected, as well as

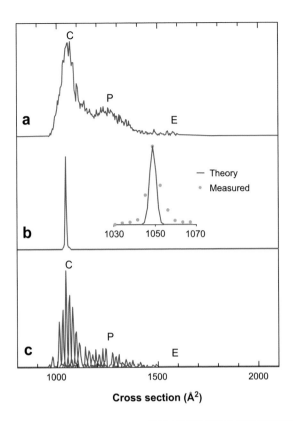

Figure 5

Drift time distributions of the $[M + 7H]^{7+}$ ion of ubiquitin. (*a*) The total mobility distribution shows that the +7 charge state exists mostly as compact structures (C), some partially folded structures (P), and minimal elongated structures (E). (*b*) A single peak is observed when a narrow distribution (50 μs) of the compact structure is isolated with mobility selection at 7.8 ms. The inset compares the theoretical to the experimentally measured peak shape. (*c*) The total mobility distribution for the +7 charge state is reconstructed with 28 mobility-selected distributions acquired every ~0.125 ms. This demonstrates that this distribution arises from many overlapping structures that are stable over the course of the two-dimensional acquisition. Figure reprinted from Reference 52.

the solution conditions used to produce the ions. From the distribution shown in **Figure 6c**, it is possible to further isolate ions (**Figure 6d**) by selection at G3. For example, activation of the partially folded conformers can produce a range of other structures for separation. Under low-energy activation conditions, partially folded intermediates (formed from compact precursors) primarily lead to products that are partially folded. At higher energies along this pathway, elongated states can be favored (53).

These pathways are clearly complex, and we are at an early stage in understanding these data. However, it is apparent that IMS-IMS and IMS-IMS-IMS studies can offer information about how ions with specified cross sections change through specified intermediates to reach product states. Such step-by-step structural transitions are rare,

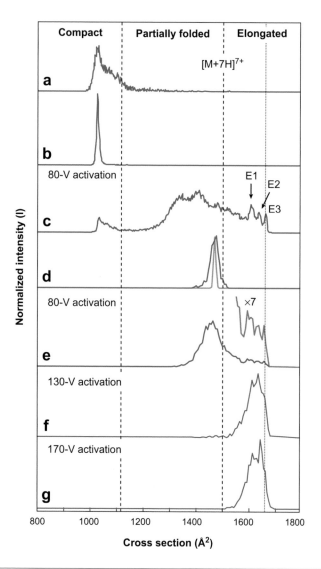

Figure 6

Ion mobility distributions of ubiquitin $[M + 7H]^{7+}$ ions obtained in IMS-IMS-IMS/MS experiments. The distribution in panel *a* is the initial distribution consisting of primarily compact conformers. Upon selection of a narrow distribution of compact ions at G2 (100 μs), the distribution in panel *b* is obtained. Activation of the selected compact ions at IA2 produces the distribution in panel *c*. A second selection of an intermediate within the partially folded structures performed at G3 (150 μs) is shown in panel *d*, with the diffusion-limited peak width of the selected ion (*gray line*). Upon activation of the partially folded structures at IA3, the distribution in panel *e* is produced, consisting of a broader distribution of partially folded structures and a smaller distribution of elongated states (*gray line*). The distributions in panels *f* and *g* are obtained upon higher-energy activation of the partially folded structures shown in panel *d*. Figure reprinted from Reference 53.

and this type of study is impossible in solution because of difficulties in trapping the intermediate states. One intriguing finding of the early work is that the distribution of elongated states appears to retain information relevant to the compact precursor ions that produced it. Presumably, this memory exists as retained elements of secondary structure (53, 54).

3.5. Extension of IMS-IMS Techniques for Complex Mixture Analysis

The ability to separate, select, and activate different conformations may also have merit for the analysis of complex mixtures. **Figure 7** shows an example of such an experiment for a mixture of tryptic peptides from human hemoglobin. Here, we

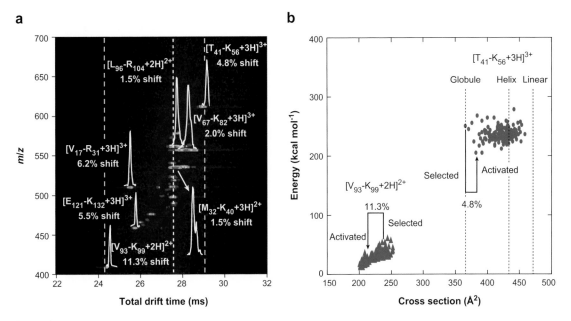

Figure 7

(*a*) An expanded view of an activated selection of human hemoglobin tryptic peptides. The white dotted line denotes the time at which mobility-selected ions with no activation are observed, whereas the dashed yellow lines show the new effective separation space of the second ion mobility spectrometry experiment. Drift distributions for several peptides are shown, along with shifts from original (inactivated) drift times for all ions. (*b*) Energy versus conformer cross section of the 150 lowest-energy structures post–simulated annealing for two peptides from panel *a* ([V$_{93}$ − K$_{99}$ + 2H]$^{2+}$ and [T$_{41}$ − K$_{56}$ + 3H]$^{3+}$) as well as cross sections of the energy-minimized structures of the [T$_{41}$ − K$_{56}$ + 3H]$^{3+}$ ion modeled as a helix and linear structure, denoting the range of structures available to each sequence. Cross sections of the selected and activated structures for each sequence are highlighted, along with percent shifts from the mobility-selected structure. Note that the nomenclature used refers to the position of the peptide by providing the location (with respect to the intact protein sequence) and single letter abbreviation of the N- and C-terminal residues, respectively. Figure reprinted from Reference 55.

represent the final IMS-IMS separation for a subset of separated ions that were selected and activated at the G2/IA2 region. With no activation, this set of ions would arrive along the line at 27.5 ms. With activation, some ions shift to slightly shorter or longer times. These shifts require that the conformations have changed such that ions have higher or lower mobilities through the remaining D2 and D3 regions of the instrument.

In the distribution shown in **Figure 7**, peaks have been assigned based on their masses and drift times, and are consistent with previous work in our laboratory. The origin shifts for peaks associated with two peptides, $[V_{93} - K_{99} + 2H]^{2+}$ and $[T_{41} - K_{56} + 3H]^{3+}$, have been studied using molecular modeling and collision cross-section calculations (and the results are also shown in **Figure 7**). The molecular modeling protocol is intended to start with a range of states and simulate an annealing process. In the case of $[V_{93} - K_{99} + 2H]^{2+}$, an array of relatively high-energy structures generated by molecular modeling anneals into lower-energy structures upon application of the heating/cooling protocol used in the calculation. The data are consistent with the experimental findings. It appears that upon exiting the ion source, $[V_{93} - K_{99} + 2H]^{2+}$ favors a relatively open conformation having a cross section of 240 $Å^2$; as these ions are accelerated through the IA2 region (at 80 V $\times$ 0.3 cm^{-1}), they are rapidly heated, and as they exit the region, they cool down. This process allows the ions to anneal to a lower-energy state. In this case, $[V_{93} - K_{99} + 2H]^{2+}$ adopts a conformation that has a cross section that is 11.3% smaller than the state favored from the source. In contrast, the $[T_{41} - K_{56} + 3H]^{3+}$ species shifts to lower mobilities because it can form more extended states upon activation.

3.6. Peak Capacity Estimates for IMS-IMS Separations

For multidimensional separations, it is relevant to describe the analytical peak capacity. Specifically, it is important to understand what can be gained by the two-dimensional IMS-IMS compared with the single dimension of IMS separation. Giddings and others (25, 143, 144) have discussed these ideas for multidimensional chromatography. **Figure 7** shows insets and shifts associated with the peaks for specific peptides after initial IMS separation, selection, and activation, followed by additional IMS separation. A careful analysis of the peak shapes and ranges over which peaks are observed (before and after activation) indicates that two-dimensional IMS-IMS peak capacities can be surprisingly high. This increase occurs because the ion's mobility is influenced by the change in structure. In general, separations are carried out under equilibrium conditions such that adding a second identical separation would lead to little, if any, improvement. Values of ~480 to 1360 have been reported for mixtures of tryptic peptides. We note that this is for IMS-IMS alone. The inclusion of MS and LC will greatly increase these values.

3.7. Developing Shift Reagents

In the case of peptide ions, structural differences between states are much smaller than for larger protein ions. As we look to the future of these techniques, it would be advantageous to develop approaches that maximize changes in structures that can be used for multidimensional IMS separations. Another approach to modifying the ion structure is to produce noncovalent molecular adducts and then remove them at specific locations within the IMS instrument. The nature of the ion-neutral interactions is relatively long range; thus, the choice of system allows a means to tune dissociation energies and may also make it possible to access different predissociative states having different structures. A number of studies have investigated peptide-ion interactions with neutrals that may yield binding to specific structural motifs (145–147). One set of molecules to emerge as candidates for shift reagents for IMS-IMS (and higher-order IMS) separations is the crown ethers.

Figure 8 shows the first study specifically designed to incorporate molecular adducts as shift reagents for IMS. Hilderbrand et al. (148) examined three dipeptides ions: $[RA + H]^+$, $[KV + H]^+$, and $[LN + H]^+$. These ions are isobaric and also have the same mobilities; thus, even with a two-dimensional high-resolution IMS-MS analysis, only a single peak is observed. Upon the addition of 18-crown-6

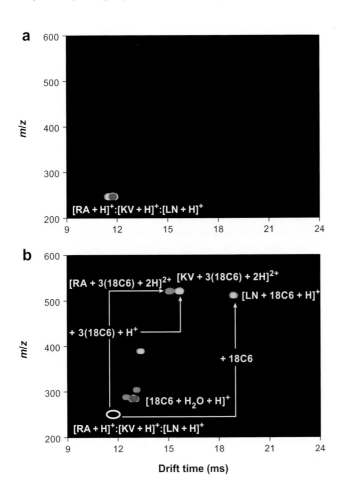

Figure 8

(*a*) Nested plot showing the protonated forms of the dipeptides RA, KV, and LN, which are irresolvable in the drift dimension. (*b*) The addition of 18-crown-6 ether (18C6) to the electrospray solution creates noncovalent complexes that shift the mobilities of the three peptides from that of the bare species (*open white oval*) enough that they are now mobility resolved under the same instrumental conditions. Figure reprinted from Reference 148.

ether (18C6) to the mixture of analytes, we find substantially different properties in the ESI-generated species. Specifically RA and KV accommodate three crowns to produce $[M + 3(18C6) + 2H]^{2+}$, whereas LN incorporates only a single 18C6. The result is that all three peptides can be resolved as adducts. In the same paper, Hilderbrand et al. also showed that it was possible to remove adducts (by low-energy CID) from tripeptides designed by combinatorial synthesis to selectively bind different numbers of crowns. By incorporating CID following the IMS separation (but prior to MS analysis), the mobilities of the peptide-adduct ions are shifted relative to the peptides (because of adduct formation), but the measured m/z values are those of peptides without adducts.

Bohrer and coworkers (B.C. Bohrer, S.J. Valentine, S. Naylor & D.E. Clemmer, unpublished data) have extended the use of crown ethers as shift regents for the analysis of peptides from human plasma. **Figure 9** shows one result for an analysis of peptides obtained by digesting a mixture of proteins from human plasma (following depletion of six abundant proteins) that was carried out using the instrument in **Figure 2**. IMS-MS analysis of ions produced by direct-infusion ESI of the peptide-crown mixture leads to a complex pattern of peaks (**Figure 9a**). Examination of the mass spectrum shows that the overlap of many features leads to a large baseline signal. This baseline is reduced substantially when the G2 gate is used to introduce a small fraction of ions into the remaining portion of the drift tube (**Figure 9b**). Selection at G2 and activation at IA2 lead to large changes in mobilities (associated with losses of different numbers of 18C6 units from the adducts). The simplification of the spectrum is remarkable. It is possible to extract regions that have only a few abundant ions (**Figure 9c**). When the selected and collisionally activated ions are exposed to the parallel CID analysis (at the exit of the drift tube), there is clear evidence for many different peptides; however, some ranges still show fragmentation data that can be interpreted. For example, a database search of the peaks found in the extracted regions of **Figure 9c,d** allows us to assign the $[RHPDYSVVLLLR + 3H]^{3+}$ ion, a sequence from albumin. The potential to selectively move across this spectrum with a controlled approach holds promise for developing analyses aimed at determining the presence of a component in a complex sample.

Figure 9

Ion mobility spectrometry–mass spectrometry nested plot of human plasma peptides, electrosprayed from solution containing 18-crown-6 ether (18C6) (*a*) with integrated mass spectrum (*right column*). The white box indicates the region focused on in the other panels of this figure. Mobility-selected complex ions gated through G2 (*b*) are activated with 160 V at IA2 to dissociate the complex ions into bare peptide ions, which resolve further in mobility (*c*). Integration of the data between the dashed lines shows that $[RHPDYSVVLLLR + 3H]^{3+}$ is practically isolated in the drift dimension (*c, right column*). Parallel dissociation at the end of the drift tube renders mobility-labeled fragments (*d*). Integration about the dashed lines now provides fragmentation spectrum for $[RHPDYSVVLLLR + 3H]^{3+}$ (*d, right column*). Abbreviation: CID, collision-induced dissociation. Figure taken from B.C. Bohrer, S.J. Valentine, S. Naylor & D.E. Clemmer, unpublished manuscript.

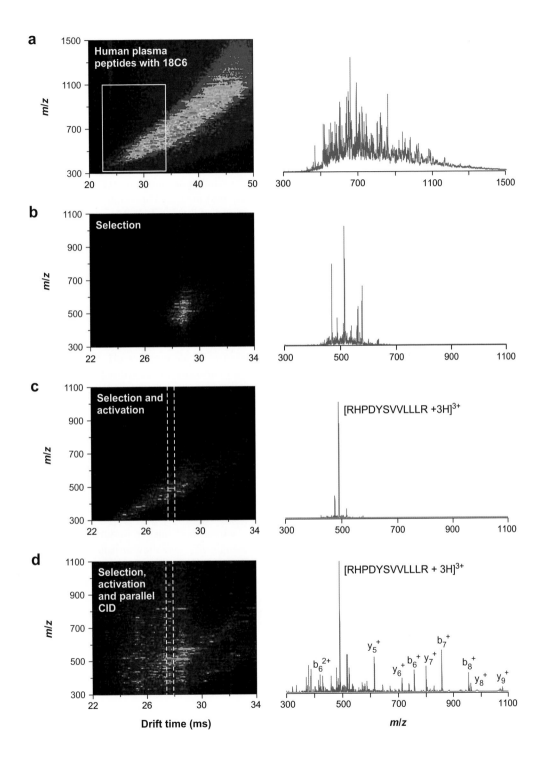

4. CONCLUSIONS

The above discussion investigates the analytical utility of exploring the curious nature of macromolecules in the gas phase. Primarily, the importance of these studies arose because the shapes of large ions in the absence of solvent were relatively unknown and the new information obtained appears to complement MS analyses. That the ion shapes do not appear to readily reach equilibrium is also interesting as this makes the IMS separation highly complementary to condensed-phase-separation approaches. Another interesting parameter that has been stressed is the ability to place the IMS separation between an LC separation and MS detection. In general IMS separations require longer acquisition times than the flight times required for MS analysis, but are still much faster than those required for LC. This results from the differences in number density for the environment of each analysis. Moreover, the experimental variables associated with the IMS measurement can be controlled in a manner that makes measurements highly reproducible. Alternatively, the coupling of FAIMS to IMS, as well as the ability to change the structures of mobility-selected ions between IMS drift regions, introduces the concept of high-speed, high-peak capacity multidimensional analyses solely in the gas phase.

One imagines that the availability of commercial instruments will lead to rapid advances and many new applications for IMS as it pertains to analyzing biomolecules. This also leads to opportunities to take advantage of the coupling of IMS to other orthogonal approaches, including ion-molecule reactions (41, 59) and fluorescence (149). This type of analysis provides additional dimensionality for biomolecules with several stable structures to the extent that these structures react differently. An early example of this combination of experiments involved hydrogen-deuterium exchange reactions on different conformations of protein ions (109). Another area gaining interest is the examination of large proteins and macromolecular complexes with IMS techniques, as demonstrated by work in the Robinson (34) and Loo (35) groups.

With growing interest in higher-dimensionality IMS, we (S.I. Merenbloom & D.E. Clemmer, unpublished results) have recently constructed the circular instrument shown in **Figure 10**. The idea behind this design is to bring ions in from the source and allow them to separate around the circle multiple times before changing the focusing such that ions exit to the mass spectrometer. Results from a previous instrument configuration featuring a single 90° curve show little loss in drift resolution (**Figure 10b**). We also have demonstrated that it is possible to get ions around multiple turns with little loss of signal. We anticipate that once functional, this instrument will be of great importance because it allows ions to be introduced and analyzed for extended periods. Initial studies of the curved regions suggest that it will be possible to move ions through curves and, by doing so, take advantage of longer drift lengths to enhance IMS resolution. One can see that the limit of resolution of ions moving in this system occurs when the leading edge of the diffusing ion cloud catches up with its own trailing edge. When this occurs, one can eject the ion packet into the remaining regions of the instrument for subsequent mass analysis. The ability to cycle ions should allow higher-order IMS experiments to be carried out. It should also allow

a

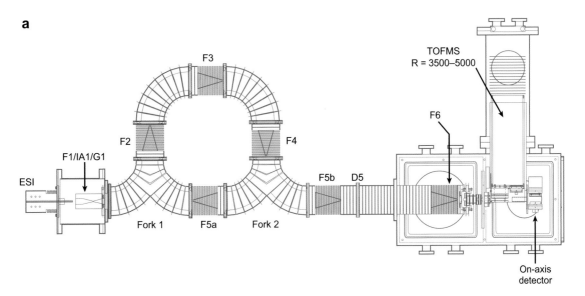

b

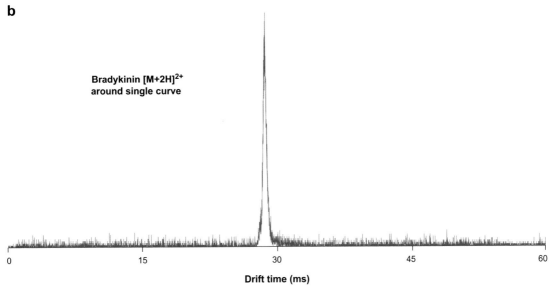

Figure 10

(*a*) Schematic of a circular ion mobility spectrometry time-of-flight instrument. (*b*) The
mobility distribution for mobility-selected bradykinin [M + 2H]$^{2+}$ ions that have traversed a
drift tube containing a single 90° turn. Figure reprinted from S.I. Merenbloom & D.E.
Clemmer, unpublished data.

ions to be stored at well-defined temperatures in the presence of reactant gasses for much longer times.

SUMMARY POINTS

1. Large macromolecules in the gas phase exhibit a myriad of structures that can be monitored by IMS and are related to solution-phase structures, including nonnative states that may be intermediates along the folding pathway or trapped species.

2. IMS coupled to MS provides a two-dimensional separation space in which ions can be resolved on the basis of size, charge state, and chemical properties.

3. Mobility selection of a macromolecular ion between drift regions that results in a sharp peak is indicative of a stable structure, collisional activation of which can probe folding or unfolding pathways.

4. The coupling of IMS with additional separations such as LC or additional IMS leads to very high peak capacities without significant increases in acquisition time. Incorporating shift reagents improves mobility-based separation performance.

FUTURE ISSUES

1. IMS may be coupled to other orthogonal techniques including ion-molecule reactions, such as electron-transfer dissociation and hydrogen-deuterium exchange, and fluorescence analyses.

2. Increasingly higher IMS resolutions and dimensionalities will drive the further development of emerging technologies and instrumentation.

3. Analysis of large protein complexes by IMS will aid in purifying and understanding structures.

DISCLOSURE STATEMENT

D.E.C. is a scientific cofounder of BG Medicine and founder of Predictive Physiology and Medicine, two early-stage biotechnology companies. He is also currently a consultant for Waters Corporation. The authors have filed a number of patents related to ion mobility.

ACKNOWLEDGMENTS

The authors are grateful for many valuable contributions of our colleagues and reviewers of our papers, specifically, the early work involving protein conformations and instrumentation carried out by Stephen J. Valentine, Anne E. Counterman, Cherokee S. Hoaglund-Hyzer, Catherine Srebalus-Barnes, John Taraszka, and Sunnie Myung. We also thank Emily Vincent for her meticulous editing and proofreading work. New instrumentation has been supported by grants from the NIH (P41 RR018942) and the Indiana 21st Century fund.

LITERATURE CITED

1. Cohen MJ, Karasek FW. 1970. Plasma chromatography: a new dimension for gas chromatography and mass spectrometry. *J. Chromatogr. Sci.* 8:330–37

2. Karasek FW. 1974. Plasma chromatography. *Anal. Chem.* 46:A710–20

3. Kemper PR, Bowers MT. 1991. Electronic state chromatography: application to 1st-row transition-metal ions. *J. Phys. Chem.* 95:5134–46

4. Bowers MT, Kemper PR, von Helden G, van Koppen PAM. 1993. Gas-phase ion chromatography: transition metal state selection and carbon cluster formation. *Science* 260:1446–51

5. Hill HH, Siems WF, St. Louis RH, McMinn DG. 1990. Ion mobility spectrometry. *Anal. Chem.* 62:A1201–9

6. Chen YH, Hill HH, Wittmer DP. 1996. Thermal effects on electrospray ionization ion mobility spectrometry. *Int. J. Mass Spectrom. Ion Process.* 154:1–2

7. Sanders TM, Forrest SR. 1989. Small particle-size distributions from mobility measurements. *J. Appl. Phys.* 66:3317–23

8. Whitby KT, Clark WE. 1966. Electronic aerosol particle counting and size distribution measuring system for 0.015 to 1 mu size range. *Tellus* 18:573

9. Karas M, Hillenkamp F. 1988. Laser desorption ionization of proteins with molecular masses exceeding 10000 daltons. *Anal. Chem.* 60:2299–301

10. Tanaka K, Waki H, Ido Y, Akita S, Yoshida Y, Yoshida T. 1988. Protein and polymer analyses up to *m/z* 100000 by laser ionization time-of-flight mass spectrometry. *Rapid Commun. Mass Spectrom.* 2:151–53

11. Whitehouse CM, Dreyer RN, Yamashuta M, Fenn JB. 1985. Electrospray interface for liquid chromatographs and mass spectrometers. *Anal. Chem.* 57:675–79

12. Wittmer D, Luckenbill BK, Hill HH, Chen YH. 1994. Electrospray-ionization ion mobility spectrometry. *Anal. Chem.* 66:2348–55

13. Clemmer DE, Hudgins RR, Jarrold MF. 1995. Naked protein conformations: cytochrome *c* in the gas phase. *J. Am. Chem. Soc.* 117:10141–42

14. von Helden G, Wyttenbach T, Bowers MT. 1995. Conformation of macromolecules in the gas phase: use of matrix-assisted laser desorption methods in ion chromatography. *Science* 267:1483–85

15. von Helden G, Wyttenbach T, Bowers MT. 1995. Inclusion of a MALDI ion-source in the ion chromatography technique: conformational information on polymer and biomolecular ions. *Int. J. Mass Spectrom. Ion Process.* 146:349–64

16. Gillig KJ, Ruotolo B, Stone EG, Russell DH, Fuhrer K, et al. 2000. Coupling high-pressure MALDI with ion mobility/orthogonal time-of-flight mass spectrometry. *Anal. Chem.* 72:3965–71

17. Mack E. 1925. Average cross-sectional areas of molecules by gaseous diffusion methods. *J. Am. Chem. Soc.* 47:2468–82

18. Shvartsburg AA, Jarrold MF. 1996. An exact hard-spheres scattering model for the mobilities of polyatomic ions. *Chem. Phys. Lett.* 261:86–91

19. Mesleh MF, Hunter JM, Shvartsburg AA, Schatz GC, Jarrold MF. 1996. Structural information from ion mobility measurements: effects of the long-range potential. *J. Phys. Chem.* 100:16082–86

20. Wyttenbach T, von Helden G, Batka JJ, Carlat D, Bowers MT. 1997. Effect of the long-range potential on ion mobility measurements. *J. Am. Chem. Soc.* 8:275–82

21. Shvartsburg AA, Hudgins RR, Dugourd P, Jarrold MF. 2001. Structural information from ion mobility measurements: applications to semiconductor clusters. *Chem. Soc. Rev.* 30:26–35

22. Shaffer SA, Tang KQ, Anderson GA, Prior DC, Udseth HR, Smith RD. 1997. A novel ion funnel for focusing ions at elevated pressure using electrospray ionization mass spectrometry. *Rapid Commun. Mass Spectrom.* 11:1813–17

23. Tang K, Shvartsburg AA, Lee HN, Prior DC, Buschbach MA, Li FM, et al. 2005. High-sensitivity ion mobility spectrometry/mass spectrometry using electrodynamic ion funnel interfaces. *Anal. Chem.* 77:3330–39

24. Tang KQ, Li FM, Shvartsburg AA, Strittmatter EF, Smith RD. 2005. Two-dimensional gas-phase separations coupled to mass spectrometry for analysis of complex mixtures. *Anal. Chem.* 77:6381–88

25. Valentine SJ, Kulchania M, Barnes CA, Clemmer DE. 2001. Multidimensional separations of complex peptide mixtures: a combined high-performance liquid chromatography/ion mobility/time-of-flight mass spectrometry approach. *Int. J. Mass Spectrom.* 212:97–109

26. Taraszka JA, Kurulagama R, Sowell RA, Valentine SJ, Koeniger SL, et al. 2005. Mapping the proteome of *Drosphila melanogaster*: analysis of embryos and adult heads by LC-ion mobility-MS methods. *J. Proteome Res.* 4:1223–37

27. Valentine SJ, Liu XY, Plasencia MD, Hilderbrand AE, Kurulugama RT, et al. 2005. Developing liquid chromatography ion mobility mass spectrometry techniques. *Expert Rev. Proteomics* 2:553–65

28. Venne K, Bonneil E, Eng K, Thibault P. 2005. Improvement in peptide detection for proteomics analyses using nanoLC-MS and high-field asymmetry waveform ion mobility mass spectrometry. *Anal. Chem.* 77:2176–86

29. Valentine SJ, Plasencia MD, Liu X, Krishnan M, Naylor S, et al. 2006. Toward plasma proteome profiling with ion mobility-mass spectrometry. *J. Proteome Res.* 5:2977–84

30. Liu X, Valentine SJ, Plasencia MD, Trimpin S, Naylor S, Clemmer DE. 2007. Mapping the human plasma proteome by SCX-LC-IMS-MS. *J. Am. Soc. Mass Spectrom.* 18:1249–64

31. Miller RA, Eiceman GA, Nazarov EG, King AT. 2000. A novel micromachined high-field asymmetric waveform-ion mobility spectrometer. *Sens. Actuators B* 67:300–6

32. Guevremont R, Barnett DA, Purves RW, Vandermey J. 2000. Analysis of a tryptic digest of pig hemoglobin using ESI-FAIMS-MS. *Anal. Chem.* 72:4577–84

33. Pringle SD, Giles K, Wildgoose JL, Williams JP, Slade SE, et al. 2007. An investigation of the mobility separation of some peptide and protein ions using a new hybrid quadrupole/travelling wave IMS/oa-ToF instrument. *Int. J. Mass Spectrom.* 261:1–12

34. Ruotolo BT, Giles K, Campuzano I, Sandercock AM, Bateman RH, Robinson CV. 2005. Evidence for macromolecular protein rings in the absence of bulk water. *Science* 310:1658–61

35. Loo JA, Berhane B, Kaddis CS, Wooding KM, Xie YM, et al. 2005. Electrospray ionization mass spectrometry and ion mobility analysis of the 20S proteasome complex. *J. Am. Soc. Mass Spectrom.* 16:998–1008

36. Chowdhury SK, Katta V, Chait BT. 1990. Probing conformational changes in proteins by mass spectrometry. *J. Am. Chem. Soc.* 112:9012–13

37. Suckau D, Shi Y, Beu SC, Senko MW, Quinn JP, et al. 1993. Coexisting stable conformations of gaseous protein ions. *Proc. Natl. Acad. Sci. USA* 90:790–93

38. Loo RRO, Smith RD. 1994. Investigation of the gas-phase structure of electrosprayed proteins using ion-molecule reactions. *J. Am. Soc. Mass Spectrom.* 5:207–20

39. Wolynes PG. 1995. Biomolecular folding in vacuo. *Proc. Natl. Acad. Sci. USA* 92:2426–27

40. Gross DS, Schnier PD, Rodriguez-Cruz SE, Fagerquist CK, Williams ER. 1996. Conformation and folding of lysozyme ions in vacuo. *Proc. Natl. Acad. Sci. USA* 93:3143–48

41. Williams ER. 1996. Proton transfer reactivity of large multiply charged ions. *J. Mass Spectrom.* 31:831–42

42. Shelimov KB, Clemmer DE, Hudgins RR, Jarrold MF. 1997. Protein structure in vacuo: gas-phase conformations of BPTI and cytochrome *c*. *J. Am. Chem. Soc.* 119:2240–48

43. Hudgins RR, Woenckhaus J, Jarrold MF. 1997. High resolution ion mobility measurements for gas phase proteins: correlation between solution phase and gas phase conformations. *Int. J. Mass Spectrom. Ion Process.* 165:497–507

44. Gross DS, Zhao YX, Williams ER. 1997. Dissociation of heme-globin complexes by blackbody infrared radiative dissociation: molecular specificity in the gas phase? *J. Am. Soc. Mass Spectrom.* 8:519–24

45. Mirza UA, Chait BT. 1997. Do proteins denature during droplet evolution in electrospray ionization? *Int. J. Mass Spectrom. Ion Process.* 162:173–81

46. Hoaglund-Hyzer CS, Counterman AE, Clemmer DE. 1999. Anhydrous protein ions. *Chem. Rev.* 99:3037–79

47. Li J, Taraszka JA, Counterman AE, Clemmer DE. 1999. Influence of solvent composition and capillary temperature on the conformations of electrosprayed ions: unfolding of compact ubiquitin conformers from pseudonative and denatured solutions. *Int. J. Mass Spectrom. Ion Process.* 185–187:37–47

48. Kuwajima K. 1989. The molten globule state as a clue for understanding the folding and cooperativity of globular protein structure. *Proteins* 6:87–103

49. Dill KA, Fiebig KM, Chan HS. 1993. Cooperativity in protein folding kinetics. *Proc. Natl. Acad. Sci. USA* 90:1942–46

50. Koeniger SL, Merenbloom SI, Valentine SJ, Jarrold MF, Udseth H, et al. 2006. An IMS-IMS analogue of MS-MS. *Anal. Chem.* 78:4161–74

51. Merenbloom SI, Koeniger SL, Valentine SJ, Plasencia MD, Clemmer DE. 2006. IMS-IMS and IMS-IMS-IMS/MS for separating peptide and protein fragment ions. *Anal. Chem.* 78:2802–9

52. Koeniger SL, Merenbloom SI, Clemmer DE. 2006. Evidence for many resolvable structures within conformation types of electrosprayed ubiquitin ions. *J. Phys. Chem. B* 110:7017–21

53. Koeniger SL, Merenbloom SI, Sevugarajan S, Clemmer DE. 2006. Transfer of structural elements from compact to extended states in unsolvated ubiquitin. *J. Am. Chem. Soc.* 128:11713–19

54. Koeniger SL, Clemmer DE. 2007. Resolution and structural transitions of elongated states of ubiquitin. *J. Am. Soc. Mass Spectrom.* 18:322–31

55. Merenbloom SI, Bohrer BC, Koeniger SL, Clemmer DE. 2007. Assessing the peak capacity of IMS-IMS separations of tryptic peptide ions in 300 K He. *Anal. Chem.* 79:515–22

56. Anderson NL, Anderson NG. 2002. The human plasma proteome: history, character, and diagnostic prospects. *Mol. Cell. Proteomics* 1:845–67

57. Harrison AG. 1997. The gas-phase basicities and proton affinities of amino acids and peptides. *Mass Spectrom. Rev.* 16:201–17

58. McLuckey SA, Goeringer DE. 1997. Slow heating methods in tandem mass spectrometry. *J. Mass Spectrom.* 32:461–74

59. Green MK, Lebrilla CB. 1997. Ion-molecule reactions as probes of gas-phase structures of peptides and proteins. *Mass Spectrom. Rev.* 16:53–71

60. Clemmer DE, Jarrold MF. 1997. Ion mobility measurements and their applications to clusters and biomolecules. *J. Mass Spectrom.* 32:577–92

61. Marshall AG, Grosshans PB. 1991. Fourier transform ion cyclotron resonance mass spectrometry: the teenage years. *Anal. Chem.* 63:A215–59

62. March RE. 1997. An introduction to quadrupole ion trap mass spectrometry. *J. Mass Spectrom.* 32:351–69

63. Gorg A, Obermaier C, Boguth G, Harder A, Scheibe B, et al. 2000. The current state of two-dimensional electrophoresis with immobilized pH gradients. *Electrophoresis* 21:1037–53

64. Wang H, Hanash S. 2003. Multi-dimensional liquid phase based separations in proteomics. *J. Chromatogr. B* 787:11–18

65. Aebersold R, Mann M. 2003. Mass spectrometry–based proteomics. *Nature* 422:198–207

66. Berman HM, Westbrook J, Feng Z, Gilliland G, Bhat TN, et al. 2000. The Protein Data Bank. *Nucleic Acids Res.* 28:535–42

67. Berman H, Henrick K, Nakamura H. 2003. Announcing the worldwide Protein Data Bank. *Nat. Struct. Biol.* 10:980

68. Wüthrich K. 1986. *NMR of Proteins and Nucleic Acids*. New York: Wiley. 320 pp.

69. Bax A. 1989. Two-dimensional NMR and protein structure. *Annu. Rev. Biochem.* 58:223–56

70. Wüthrich K. 1989. The development of nuclear magnetic resonance spectroscopy as a technique for protein structure determination. *Acc. Chem. Res.* 22:36–44

71. Lipscomb WN. 1953. Experimental crystallography. *Annu. Rev. Phys. Chem.* 4:253–66

72. Perutz MF, Rossmann MG, Cullis AF, Muirhead H, Will G, North ACT. 1960. Structure of haemoglobin: a three-dimensional Fourier synthesis at 5.5 Å resolution, obtained by X-ray analysis. *Nature* 185:416–22

73. Kendrew JC, Dickerson RE, Strandberg BE, Hart RG, Davies DR, et al. 1960. Structure of myoglobin: a three-dimensional Fourier synthesis at 2 Å resolution. *Nature* 185:422–27

74. Blundell TL, Johnson LN. 1976. *Protein Crystallography*. New York: Academic. 565 pp.

75. Lumry R, Eyring H. 1954. Conformation changes of proteins. *J. Phys. Chem.* 58:110–20

76. Levinthal C. 1968. Are there pathways for protein folding? *J. Chim. Phys.* 65:44–45

77. Anfinsen CB. 1973. Principles that govern folding of protein chains. *Science* 181:223–30

78. Roder H, Elöve GA, Englander SW. 1988. Structural characterization of folding intermediates in cytochrome c by H-exchange labelling and proton NMR. *Nature* 335:700–4

79. Goto Y, Hagihara Y, Hamada D, Hoshino M, Nishii I. 1993. Acid-induced unfolding and refolding transitions of cytochrome c: a three-state mechanism in H_2O and D_2O. *Biochemistry* 32:11878–85

80. Semisotnov GV, Rodionova NA, Razgulyaev OI, Uversky VN, Gripas AF, Gilmanshin RI. 1991. Study of the molten globule intermediate state in protein folding by a hydrophobic fluorescent probe. *Biopolymers* 31:119–28

81. Mirza UA, Cohen SL, Chait BT. 1993. Heat-induced conformational changes in proteins studied by electrospray ionization mass spectrometry. *Anal. Chem.* 65:1–6

82. Bai YW, Milne JS, Mayne L, Englander SW. 1994. Protein stability parameter measured by hydrogen exchange. *Proteins* 20:4–14

83. Dobson CM, Sali A, Karplus M. 1998. Protein folding: a perspective from theory and experiment. *Angew. Chem. Int. Ed. Engl.* 37:868–93

84. Ohgushi M, Wada A. 1983. 'Molten-globule' state: a compact form of globular proteins with mobile side-chains. *FEBS Lett.* 164:21–24

85. Makhatadze GI, Privalov PL. 1995. Energetics of protein structure. *Adv. Protein Chem.* 47:307–425

86. Lazaridis T, Archontis G, Karplus M. 1995. Enthalpic contribution to protein stability: insights from atom-based calculations and statistical mechanics. *Adv. Protein Chem.* 47:231–306

87. Woenckhaus J, Mao Y, Jarrold MF. 1997. Hydration of gas phase proteins: folded +5 and unfolded +7 charge states of cytochrome *c*. *J. Phys. Chem. B* 101:847–51

88. Mao Y, Woenkhaus J, Kolafa J, Ratner MA, Jarrold MF. 1999. Thermal unfolding of unsolvated cytochrome *c*: experiment and molecular dynamics simulations. *J. Am. Chem. Soc.* 121:2712–21

89. van Gunsteren WF, Karplus M. 1982. Protein dynamics in solution and in a crystalline environment: a molecular dynamics study. *Biochemistry* 21:2259–74

90. Levitt M, Sharon R. 1988. Accurate simulation of protein dynamics in solution. *Proc. Natl. Acad. Sci. USA* 85:7557–61

91. Mason EA, McDaniel EW. 1988. *Transport Properties of Ions in Gases*. New York: Wiley & Sons. 560 pp.

92. Counterman AE, Hilderbrand AE, Srebalus Barnes CA, Clemmer DE. 2001. Formation of peptide aggregates during ESI: size, charge, composition, and contributions to noise. *J. Am. Soc. Mass Spectrom.* 12:1020–35

93. Revercomb HW, Mason EA. 1975. Theory of plasma chromatography/gaseous electrophoresis: a review. *Anal. Chem.* 47:970–83

94. Buryakov IA, Krylov EV, Nazarov EG, Rasulev UK. 1993. A new method of separation of multi-atomic ions by mobility at atmospheric pressure using a high-frequency amplitude-asymmetric strong electric field. *Int. J. Mass Spectrom. Ion Process.* 128:143–48

95. Purves RW, Guevremont R, Day S, Pipich CW, Matyjaszczyk MS. 1998. Mass spectrometric characterization of a high-field asymmetric waveform ion mobility spectrometer. *Rev. Sci. Instrum.* 69:4094–105

96. Guevremont R, Purves RW. 1999. Atmospheric pressure ion focusing in a high-field asymmetric waveform ion mobility spectrometer. *Rev. Sci. Instrum.* 70:1370–83

97. Shvartsburg AA, Mashkevich SV, Smith RD. 2006. Feasibility of higher-order differential ion mobility separations using new asymmetric waveforms. *J. Phys. Chem. A* 110:2663–73

98. Shvartsburg AA, Bryskiewicz T, Purves RW, Tang KQ, Guevremont R, Smith RD. 2006. Field asymmetric waveform ion mobility spectrometry studies of proteins: dipole alignment in ion mobility spectrometry? *J. Phys. Chem. B* 110:21966–80

99. Moseley JT, Gatland IR, Martin DW, McDaniel EW. 1969. Measurement of transport properties of ions in gases: results for K^+ ions in N_2. *Phys. Rev.* 178:234–39

100. Knorr FJ, Eatherton RL, Siems WF, Hill HH. 1985. Fourier transform ion mobility spectrometry. *Anal. Chem.* 57:402–6

101. Szulmas AW, Ray SJ, Hieftje GM. 2006. Hadamard transform ion mobility spectrometry. *Anal. Chem.* 78:4474–81

102. Belov ME, Buschbach MA, Prior DC, Tang KQ, Smith RD. 2007. Multiplexed ion mobility spectrometry-orthogonal time-of-flight mass spectrometry. *Anal. Chem.* 79:2451–62

103. Giles K, Pringle SD, Worthington KR, Little D, Wildgoose JL, Bateman RH. 2004. Applications of a traveling wave-based radio frequency-only stacked ring ion guide. *Rapid Commun. Mass Spectrom.* 18:2401–14

104. Cooks RG, Ouyang Z, Takats Z, Wiseman JM. 2006. Ambient mass spectrometry. *Science* 311:1566–70

105. Gieniec J, Mack LL, Nakamae K, Gupta C, Kumar V, Dole M. 1984. Electrospray mass spectroscopy of macromolecules: application of an ion-drift spectrometer. *Biomed. Mass Spectrom.* 11:259–68

106. Myung S, Wiseman JM, Valentine SJ, Zoltán T, Cooks RG, Clemmer DE. 2006. Coupling desorption electrospray ionization (DESI) with ion mobility/mass spectrometry for analysis of protein structure: evidence for desorption of folded and denatured states. *J. Phys. Chem. B* 110:5045–51

107. Bluhm BK, Gillig KJ, Russell DH. 2000. Development of a Fourier-transform ion cyclotron mass spectrometer-ion mobility spectrometer. *Rev. Sci. Instrum.* 71:4078–86

108. Lawrence AH, Barbour RJ, Sutcliffe R. 1991. Identification of wood species by ion mobility spectrometry. *Anal. Chem.* 63:1217–21

109. Valentine SJ, Clemmer DE. 1997. H/D exchange levels of shape-resolved cytochrome *c* conformers in the gas phase. *J. Am. Chem. Soc.* 119:3558–66

110. Creaser CS, Benyezzar M, Griffiths JR, Stygall JW. 2000. A tandem ion trap/ion mobility spectrometer. *Anal. Chem.* 72:2724–29

111. Clowers BH, Hill HH. 2005. Mass analysis of mobility-selected ion populations using dual gate, ion mobility, quadrupole ion trap mass spectrometry. *Anal. Chem.* 77:5877–85

112. Hoaglund CS, Valentine SJ, Sporleder CR, Reilly JP, Clemmer DE. 1998. Three-dimensional ion mobility TOFMS analysis of electrosprayed biomolecules. *Anal. Chem.* 70:2236–42

113. Howorka F, Fehsenfeld FC, Albritton DL. 1979. H$^+$ and D$^+$ ions in He: observations of a runaway mobility. *J. Phys. B* 12:4189–97

114. Dugourd Ph, Hudgins RR, Clemmer DE, Jarrold MF. 1997. High-resolution ion mobility measurements. *Rev. Sci. Instrum.* 68:1122–29

115. Valentine SJ, Clemmer DE. 1997. H/D exchange levels of shape-resolved cytochrome *c* conformers in the gas phase. *J. Am. Chem. Soc.* 119:3558–66

116. Valentine SJ, Counterman AE, Clemmer DE. 1999. A database of 660 peptide ion cross sections: use of intrinsic size parameters for bona fide predictions of cross sections. *J. Am. Soc. Mass Spectrom.* 10:1188–211

117. Counterman AE, Clemmer DE. 1999. Volumes of individual amino acid residues in gas-phase peptide ions. *J. Am. Chem. Soc.* 121:4031–39

118. Valentine SJ, Counterman AE, Hoaglund CS, Reilly JP, Clemmer DE. 1998. Gas-phase separations of protease digests. *J. Am. Soc. Mass Spectrom.* 9:1213–16

119. Woods AS, Ugarov M, Egan T, Koomen J, Gillig KJ, et al. 2004. Lipid/peptide/nucleotide separation with MALDI-ion mobility-TOF MS. *Anal. Chem.* 76:2187–95

120. Jackson SN, Wang HYJ, Woods AS. 2005. Direct tissue analysis of phospholipids in rat brain using MALDI-TOFMS and MALDI-ion mobility-TOFMS. *J. Am. Soc. Mass Spectrom.* 16:133–38

121. Dwivedi P, Bendiak B, Clowers BH, Hill HH. 2007. Rapid resolution of carbohydrate isomers by electrospray ionization ambient pressure ion mobility spectrometry-time-of-flight mass spectrometry (ESI-APIMS-TOF). *J. Am. Soc. Mass Spectrom.* 18:1163–75

122. Ruotolo BT, Verbeck GF, Thomson LM, Woods AS, Gillig KJ, Russell DH. 2002. Distinguishing between phosphorylated and nonphosphorylated peptides with ion mobility mass spectrometry. *J. Proteome Res.* 1:303–6

123. Hoaglund-Hyzer CS, Li JW, Clemmer DE. 2000. Mobility labeling for parallel CID of ion mixtures. *Anal. Chem.* 72:2737–40

124. Hoaglund-Hyzer CS, Clemmer DE. 2001. Ion trap/ion mobility/quadrupole/time of flight mass spectrometry for peptide mixture analysis. *Anal. Chem.* 73:177–84

125. Stone E, Gillig KJ, Ruotolo B, Fuhrer K, Gonin M, et al. 2001. Surface-induced dissociation on a MALDI-ion mobility-orthogonal time-of-flight mass spectrometer: sequencing peptides from an "in-solution" protein digest. *Anal. Chem.* 73:2233–38

126. Sun WJ, May JC, Russell DH. 2007. A novel surface-induced dissociation instrument for ion mobility-time-of-flight mass spectrometry. *Int. J. Mass Spectrom.* 259:79–86

127. Lee YJ, Hoaglund-Hyzer CS, Taraszka JA, Zientara GA, Counterman AE, Clemmer DE. 2001. Collision-induced dissociation of mobility-separated ions using an orifice-skimmer cone at the back of a drift tube. *Anal. Chem.* 73:3549–55

128. Steiner WE, Clowers BH, Fuhrer K, Gonin M, Matz LM, et al. 2001. Electrospray ionization with ambient pressure ion mobility separation and mass analysis by orthogonal time-of-flight mass spectrometry. *Rapid Commun. Mass Spectrom.* 15:2221–26

129. Valentine SJ, Koeniger SL, Clemmer DE. 2003. A split-field drift tube for separation and efficient fragmentation of biomolecular ions. *Anal. Chem.* 75:6202–8

130. Little DP, Speir JP, Senko MW, O'Connor PB, McLafferty FW. 1994. Infrared multiphoton dissociation of large multiply charged ions for biomolecule sequencing. *Anal. Chem.* 66:2809–15

131. Zubarev RA, Kelleher NL, McLafferty FW. 1998. Electron capture dissociation of multiply charged protein cations: a nonergodic process. *J. Am. Chem. Soc.* 120:3265–66

132. Syka JEP, Coon JJ, Schroeder MJ, Shabanowitz J, Hunt DF. 2004. Peptide and protein sequence analysis by electron transfer dissociation mass spectrometry. *Proc. Natl. Acad. Sci. USA* 101:9528–33

133. Chen W, Olivova P, Doneanu CE, Gebler JC. 2007. *A novel approach for identification and characterization of glycoproteins using a quadrupole ion-mobility time-of-flight mass spectrometer.* Presented at Annu. Meet. Am. Soc. Mass Spectrom. 55th, Indianapolis

134. Eng JK, McCormack AL, Yates JR. 1994. An approach to correlate tandem mass spectral data of peptides with amino acid sequences in a protein database. *J. Am. Soc. Mass Spectrom.* 5:976–89

135. Perkins DN, Pappin DJC, Creasy DM, Cottrell JS. 1999. Probability-based protein identification by searching sequence databases using mass spectrometry data. *Electrophoresis* 20:3551–67

136. Asbury GR, Hill HH. 2000. Using different drift gases to change separation factors (α) in ion mobility spectrometry. *Anal. Chem.* 72:580–84

137. Ruotolo BT, McLean JA, Gillig KJ, Russell DH. 2004. Peak capacity of ion mobility mass spectrometry: the utility of varying drift gas polarizability for the separation of tryptic peptides. *J. Mass Spectrom.* 39:361–67

138. Shelimov KB, Jarrold MF. 1997. Conformations, unfolding, and refolding of apomyoglobin in vacuum: an activation barrier for gas-phase protein folding. *J. Am. Chem. Soc.* 119:2987–94

139. Freitas MA, Hendrickson CL, Emmett MR, Marshall AG. 1999. Gas-phase bovine ubiquitin cation conformations resolved by gas-phase hydrogen/deuterium exchange rate and extent. *Int. J. Mass Spectrom.* 187:565–75

140. Badman ER, Hoaglund-Hyzer CS, Clemmer DE. 2001. Monitoring structural changes of proteins in an ion trap over ~10–200 ms: unfolding transitions in cytochrome *c* ions. *Anal. Chem.* 73:6000–7

141. Myung S, Badman ER, Young JL, Clemmer DE. 2002. Structural transitions of electrosprayed ubiquitin ions stored in an ion trap over ~10 ms to 30 s. *J. Phys. Chem. A* 106:9976–82

142. Badman E, Myung S, Clemmer DE. 2005. Evidence for unfolding and refolding of gas phase cytochrome *c* ions in a Paul trap. *J. Am. Soc. Mass Spectrom.* 16:1493–97

143. Giddings JC. 1987. Concepts and comparisons in multidimensional separation. *J. High Resolut. Chromatogr. Chromatogr. Commun.* 10:319–23

144. Giddings JC. 1991. Steady-state, two-dimensional, and overlapping zones. In *Unified Separation Science*, pp. 112–40. New York: Wiley. 352 pp.

145. Julian RR, Beauchamp JL. 2001. Site specific sequestering and stabilization of charge in peptides by supramolecular adduct formation with 18-crown-6 ether by way of electrospray ionization. *Int. J. Mass Spectrom.* 210:613–23

146. Colgrave ML, Bramwell CJ, Creaser CS. 2003. Nanoelectrospray ion mobility spectrometry and ion trap mass spectrometry studies of the noncovalent complexes of amino acids and peptides with polyethers. *Int. J. Mass Spectrom.* 229:209–16

147. Ly T, Julian RR. 2006. Using ESI-MS to probe protein structure by site-specific noncovalent attachment of 18-crown-6. *J. Am. Soc. Mass Spectrom.* 17:1209–15

148. Hilderbrand AE, Myung S, Clemmer DE. 2006. Exploring crown ethers as shift reagents for ion mobility spectrometry. *Anal. Chem.* 78:6792–800

149. Iavarone AT, Patriksson A, van der Spoel D, Parks JH. 2007. Fluorescence probe of Trp-cage protein conformation in solution and in gas phase. *J. Am. Chem. Soc.* 129:6726–35

In Vitro Electrochemistry of Biological Systems

Kelly L. Adams,[1,2] Maja Puchades,[2] and Andrew G. Ewing[1,2]

[1] Pennsylvania State University, Department of Chemistry, University Park, Pennsylvania 16802; email: kellya@chem.gu.se, andrewe@chem.gu.se

[2] Göteborg University, Department of Chemistry, SE-412 96 Göteborg, Sweden; email: maja.puchades@chem.gu.se

Annu. Rev. Anal. Chem. 2008. 1:329–55

First published online as a Review in Advance on February 27, 2008

The *Annual Review of Analytical Chemistry* is online at anchem.annualreviews.org

This article's doi:
10.1146/annurev.anchem.1.031207.113038

1936-1327/08/0719-0329$20.00

Key Words

amperometry, chronoamperometry, electroporation, microfluidics, microelectrodes

Abstract

This article reviews recent work involving electrochemical methods for in vitro analysis of biomolecules, with an emphasis on detection and manipulation at and of single cells and cultures of cells. The techniques discussed include constant potential amperometry, chronoamperometry, cellular electroporation, scanning electrochemical microscopy, and microfluidic platforms integrated with electrochemical detection. The principles of these methods are briefly described, followed in most cases with a short description of an analytical or biological application and its significance. The use of electrochemical methods to examine specific mechanistic issues in exocytosis is highlighted, as a great deal of recent work has been devoted to this application.

1. INTRODUCTION

Electrochemistry in ultrasmall environments has emerged as an increasingly important technique for fundamental studies of single-cell neuronal communication and release and reuptake of chemical messenger molecules as well as cellular imaging and small-scale electroporation applications. The development of electrochemical methods for detection of neurotransmitters began with the groundbreaking work of Adams (1) and has progressed to the point where it is now possible to detect the release of a neurotransmitter from a single vesicle, as first demonstrated in the seminal work by Wightman et al. (2, 3). In these pioneering experiments, a carbon fiber electrode measuring 5 μm in diameter was placed adjacent to a bovine adrenal chromaffin cell isolated in a culture dish. The cell was then stimulated to release by either chemical or mechanical means.

Understanding the chemistry and structure at the single-cell level is of great interest in the biological and medical sciences; indeed, books have been written on this broad topic (4). In neuroscience, knowledge of the chemical composition and dynamics of single nerve cells leads to better models of the cellular neurotransmission process. The key dynamic event in neuronal communication is exocytosis, a process that has been extensively investigated for several decades (5, 6). The process of exocytosis can be summarized as the docking of vesicles (storage compartments) to the cell membrane and subsequently releasing the contents to the extracellular space by fusion of the vesicular and cellular membranes. This process allows the conversion of an electrical signal (action potential) to a chemical signal (messenger release and receptor recognition), which is necessary for exocytotic communication between cells.

Methods to observe and quantify individual exocytotic events have traditionally revolved around electron microscopy and patch-clamp capacitance measurements (7). In 1990, Wightman and colleagues showed that they could directly monitor individual exocytotic events involving easily oxidized messengers occurring on the millisecond time scale by use of amperometric measurements at microelectrodes (3). This method was applied to adrenal chromaffin cells first by Wightman's group (2), and later by Neher's group (8).

This review focuses on the use of electrochemical methodologies at single cells or isolated cultures of cells in a "biologically simplified" in vitro setting. These methods, emphasizing constant potential amperometry at microelectrodes and the electrical phenomenon of electroporation, are here reviewed based on their in vitro applicability and are discussed in terms of what new information has been obtained for cellular environment studies that could not have been as eloquently delivered by any analytical method other than electrochemistry.

2. DEVELOPMENTS IN MODIFICATIONS OF MICROELECTRODES

Carbon fiber microelectrodes were developed in several laboratories in the late 1970s for work in vivo. Leaders among these researchers included the Wightman (9, 10) and Gonon (11, 12) groups, who applied this tool to neuroscience. The method was a major breakthrough for several reasons. First, the carbon fiber electrodes were biocompatible and could therefore carry a current while maintaining sensitivity to reductants, thus increasing the working lifetime of an electrode. Second, carbon fibers as small as 5 μm became available, enabling the development of very small probes that minimize tissue damage. For later in vitro work, the carbon fiber electrodes were advantageous in that they were highly resistant to strain and could be placed firmly against cell surfaces without physically breaking, thus providing greater sensitivity and reproducible measurements. For a more in-depth discussion of the factors affecting electrode sensitivity, selectivity, and temporal response, refer to the paper by Cahill et al. (13).

Electrodes used for studying single cells are usually constructed by first aspirating a single carbon fiber through a glass capillary. The capillary is then pulled using a commercial pipette puller, producing two long-tapered fiber-containing pipettes. Each is then cut at a cross-sectional diameter of 8–10 μm on a microscope using a scalpel blade. Cut electrodes are immersed in freshly prepared high-quality epoxy to create a tight seal between the glass tip and carbon fiber. Finally, no more than a few hours before the experiment, electrodes are beveled to 45° on a rotary microgrinder to produce a fine-polished, angled tip (14, 15). Working electrodes are generally prepared from 5- to 10-μm-diameter carbon fibers. The signal-to-noise ratio improves as the electrode size approaches the size of the detection area of interest, but larger electrodes can detect a greater number of electrochemical events (13, 16).

Electrode sensitivity requires that the electrode surface be free of adsorbed molecules, such as proteins and oxidized products (13, 17). Several groups, however, have modified the working electrode surface in hopes of enhancing electrode sensitivity and/or selectivity (18–20). A common modification to enhance cation detection is through the application of a thin coating of Nafion®, a perfluorinated cation-exchange polymer. Anions are primarily excluded by this thin barrier, whereas cations are able to pass and accumulate at the electrode surface, thus improving their detection, particularly in central nervous system studies (17, 21–23). Due to this unique exclusionary property, Nafion®-coated carbon fiber microelectrodes can be utilized to differentiate between neurochemicals possessing similar oxidation potentials, for example dopamine and ascorbic acid. Electrode response time, however, can be compromised when coating with a thin film of Nafion® and a significant loss in temporal resolution may occur. Wightman's group recently modified 10-μm carbon fiber microelectrodes via the reduction of 4-sulfobenzenediazonium tetrafluoroborate to create an electroactive surface with enhanced sensitivity to cationic species, specifically dopamine, without a severe loss of electrode response time (24). These modified electrodes, unlike those coated with Nafion®, also allowed anionic species to diffuse to the electrode surface, thus not limiting detection to cationic species. Moreover, the sensitivity for dopamine and other positively charged analytes was increased by a factor of 5 compared to an unmodified electrode, a significant increase compared to the enhancement of 1.5 at best attainable with Nafion® coating (24).

Single-walled carbon nanotubes have also been used to modify carbon fiber electrodes in an effort to improve detection of molecules at the single-cell and subcellular levels. Cheng and colleagues fabricated ultrasmall (100–300 nm in diameter) carbon fiber nanoelectrodes via flame etching, then immersed the tips into a solution of suspended single-walled carbon nanotubes. The electrodes were subsequently dried, yielding tips covered with sheets of single-walled carbon nanotubes that effectively increased the working surface area of the electrode (25). Combined with the fine intrinsic electrical properties of single-walled carbon nanotubes, these modified electrodes improved the detection limits of select neurotransmitters when characterized with cyclic voltammetry. For example, unmodified ultrasmall carbon fiber nanoelectrodes exhibited detection limits on the order of 76 nM, whereas the same electrode geometry modified with single-walled carbon fiber nanotubes demonstrated a detection limit of 7.7 nM (25). Because of the improved sensitivity and the small size of these particular electrodes, they may be potentially useful in ultrasmall environments as high-quality detectors with good temporal and spatial resolution.

Fabrication of electrode surfaces in conjunction with a photoresist has also been explored. Orwar, Weber, and colleagues controlled the electroactive area of a carbon fiber working electrode by electrodepositing a negative photoresist onto an exposed carbon fiber microelectrode surface (26). The face of the tip was exposed using a xenon arc lamp, then the resist layer was subsequently removed upon immersing the tip in the developing solution. Experimental parameters, such as the temperature of the photoresist solution and the length of time the photoresist was exposed to light, strongly influenced the photoresist thickness and the area of exposed electroactive surface, respectively. Thus, fine adjustments to the photoresist removal procedure would allow one to reproducibly fabricate an electrode of a specified size, as well as to effectively increase the electroactive surface area of the same electrode without making a fresh electrode (26). Likewise, Wightman's group used a photoresist/pyrolysis method to fabricate carbon ultramicroelectrodes from tungsten microelectrode substrates (27). Electrodes made in this way still exhibited electrochemical properties comparable to their glass-enclosed fiber counterparts, but possessed the additional properties of flexibility and rigidity inherent in tungsten microelectrodes. Furthermore, because these electrodes lack bulky glass capillary encasements, a tightly configured array format of these individual electrodes may be possible, thus allowing spatial and temporal information to be conserved within an ultrasmall cellular environment (27).

3. SINGLE-CELL AMPEROMETRY: A METHOD TO EXPLORE EXOCYTOTIC RELEASE

Amperometric experiments involve holding the electrode at potentials sufficient to oxidize molecules near the electrode at a diffusion-limited rate. In constant potential amperometric mode, a change in the concentration of an easily oxidized species results in a concomitant change in the oxidation current. Integrating the current of such a transient provides the charge passed which, according to Faraday's law ($N = Q/nF$), is directly related to the number of molecules oxidized. Amperometry has been used to study exocytosis in primary cultures (3, 10, 28, 29), in immortalized cell lines (30, 31), in brain slices (32), and at intact neurons in vivo (33, 34). As previously stated, this review focuses on single cells and cell cultures.

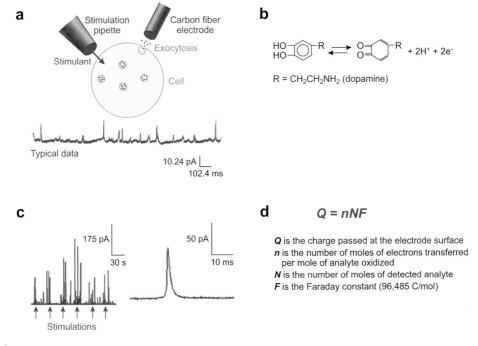

Figure 1

Basic introduction to amperometric detection of exocytosis at single cells. (*a*) The top image is the typical setup for amperometry of a single cell. Exocytosis is stimulated by a pipette containing a stimulant, and the release is monitored by a carbon fiber electrode. The bottom image shows typical amperometric data. (*b*) The oxidation reaction for catecholamines. The catecholamine is oxidized to the orthoquinone form, losing two electrons. (*c*) Left-hand trace shows a series of stimulations, represented by arrows, and the electrochemical responses detected after each stimulation. Right-hand trace shows a single amperometric current transient. (*d*) Faraday's equation, which is used to determine the amount of material released during exocytosis.

In general, candidate cell systems have been limited to those that release an oxidizable substance, usually a catecholamine-, serotonin-, or tyrosine/tryptophan-containing peptide. Immortalized cultures permit single, isolated cells to be studied, whereas primary cultures offer the advantage of modeling cells that can be studied in the context of belonging to a network (i.e., continuing to receive information from adjacent cells). Amperometry is well suited to measuring secretion from cells because of its ability to quantify release from vesicles on the millisecond time scale (**Figure 1**) (16). To carry out amperometry at single cells, a small electrode is placed near the cell and held at a potential where oxidation is diffusion limited. Stimulant is applied with a small pipette (**Figure 1*a***), resulting in current transients when oxidizable species are released. The first experiment to measure exocytosis used adrenal cells and the catechols norepinephrine and epinephrine were detected (2, 3). The general oxidation reaction for catechols is shown in **Figure 1*b*** and a typical current-time trace (in this case at a pheochromocytoma cell) is shown in **Figure 1c**. The number of molecules detected can be calculated by Faraday's law (**Figure 1d**). Estimates of the distance between an electrode placed flush against the cell and the cell membrane surface suggest that a small solution-filled gap of about 300 nm exists (35, 36).

The shape of amperometric peaks is determined by the various aspects of the release event. Specifically, the half-width of the peak (full width at half maximum) represents the duration of the release event. The rise time, typically the time it takes to change from 10% to 90% of the peak height, relates to the time it takes the fusion pore to open. Thus, the basic amperometric measurement is suitable for providing data about the amount of transmitter released, the duration of each event, and the opening of the fusion pore.

4. RECENT ACHIEVEMENTS IN EXPLORING DOPAMINE AND CATECHOLAMINE RELEASE

4.1. Lipid Incubation Appears to Affect the Biophysics of Exocytosis

Recent work employing amperometric detection in vitro aims to capitalize on the ability of amperometry to detect small oxidative current changes of molecules, such as dopamine (corresponding to quantities as minute as a few zeptomoles), to obtain submillisecond kinetic information about the secretory event release process, or both (28). Recently, this methodology has been used to study the biophysics of the exocytosis process with a focus on membrane composition. Amatore and colleagues have used short (2–3 min) incubations of adrenal cells with different lipids to demonstrate that the shape of the lipid might affect the rate of release (37). Our laboratory has used the amperometric method to demonstrate that dopamine release from PC12 (pheochromocytoma) cells is altered upon incubation for 3 d with 100 μM phospholipid (38). This neuron-like, immortalized cell line is an excellent model system to work with due to its large, circular size, ease in culturing and manipulation, and robust nature (39). Although these cells do not form functional synapses to neighboring cells, they provide an excellent model for the examination of presynaptic cellular machinery. In these experiments, phosphotidylserine incubation increased the number of events elicited by a high potassium stimulus, whereas phosphotidylcholine reduced the quantal size (total amount of transmitter released) per event. Furthermore, phosphotidylethanolamine accelerated the rate of the release process, resulting in a shortened average half-width and decay time and increasing the average peak amplitude of amperometric measurements. In contrast, phosphotidylcholine incubation decreased the rate of the release process by lengthening the half-width and decay time while decreasing the average peak amplitude. These data are exciting and support a phospholipid-based mechanism that regulates cell-to-cell communication by altering local membrane composition, thereby impacting exocytosis machinery.

Enhancements of secretion were also observed for a similar incubation protocol. PC12 cells differentiated with nerve growth factor and subsequently treated with 1 mM lithium for 2 d exhibited a greater frequency of stimulant-evoked release without noticeable changes to the quantity and rate of release (40). Electron microscopy, however, revealed no increase in the number of secretory vesicles per unit area in these treated cells, but an overall increase in vesicular diameter ($\sim$15%). Alongside an immunoblotting assay, electrochemical measurements in this work complemented the observation of lithium, a common therapeutic for psychological disorders, changing dense core vesicle protein expression by altering secretion machinery and generating a therapeutic response (40).

4.2. 3,4-dihydroxy-L-phenylalanine Loading of Cells Fills the Halo, Not the Dense Core

Release of dopamine from PC12 cells was augmented by loading them with 3,4-dihydroxy-L-phenylalanine (L-DOPA), and exocytosis under both physiological and hypertonic conditions has been monitored with amperometry (41). A majority of the loaded dopamine was hypothesized to be preferentially stored in the halo region of the vesicle rather than the dense protein core of the vesicle. Under high-osmolarity (hypertonic) conditions, the dense core matrix does not fully dissociate during exocytosis (42). Therefore, a relatively larger amount of the release observed under hypertonic conditions is attributed to expulsion of material held in the clear halo region of the vesicle, an area not commonly known as an important storage location for transmitter when compared to the dense core matrix.

Amperometry data reflected statistically similar peak area increases (26–36%) for both isotonic and hypertonic conditions after loading with L-DOPA, implying that supplemental transmitter preferentially stays in the halo regardless of osmolaric conditions. However, Amatore and colleagues have also investigated exocytosis under varying osmolaric conditions using carbon fiber amperometry on bovine chromaffin cells that release catecholamines (43) and found that the release frequency and the average amount of catecholamine release notably increased under hypotonic conditions (low osmolarity). These findings support the swelling of the dense core more favorably under such conditions. Moreover, this finding may be explained by the presence of two different populations of vesicles: the first of small content and which favorably release under isotonic conditions (83% of events versus 35% at hypotonic), the second of larger content and which favorably releases at hypotonic conditions (65% of events versus 17% at isotonic conditions) (43).

4.3. Multiple Populations of Vesicles in Catecholamine Cells

Multiple population types of catecholamine vesicles have been suggested, based on results obtained with carbon fiber amperometry in several cell lines, including the large dopamine cell of *Planorbis corneus* (33) and PC12 cells (44). Tse and colleagues have also hypothesized the existence of multiple vesicle populations in rat chromaffin cells (45). In these cells, each population [small, medium, and large quantal release (Q) granules] can be manipulated separately by lengthening culture duration or simultaneously via coculturing with cyclic AMP. When culture duration was extended from 1 d to 3 d, the overall detected average quantal size decreased, with both small and large Q granule populations changing their distribution proportionalities (each to a different degree). Additionally, introduction of cAMP to the culture media resulted in an increase in the overall average Q for the cell as well as an increased release amount for each population without significant changes to the distribution proportions for each Q subpopulation.

Probable mechanisms for the increased release from all granule sizes, which is apparently triggered by addition of cAMP, include: (*a*) provoked increase in catecholamine synthesis, (*b*) improved reuptake of free catecholamine to all granules sizes, (*c*) fused multiple granules, and (*d*) dissolving of the dense core matrix, among others (45). Similarly, enhanced catecholamine release from bovine chromaffin cells has been attributed to molecular interaction with D1 dopaminergic receptors (46). The time required to reestablish initial cellular calcium levels after stimulation was optimized by decreasing the length of high-potassium stimulus application (from 2 s to ~0.5 s). This decrease resulted in the observation of more release events after the second potassium stimulation; this effect could be blocked by a D1 antagonist (SCH-23390). Additional release events following the second potassium stimulation could be evoked by a D1 agonist (SKF-38393), yet they remained unaffected by a D2 antagonist (raclopride). These findings imply that a D1-like receptor on bovine chromaffin cells plays a key role in catecholamines promoting release of themselves within the same cell (46).

4.4. Is Exocytotic Release the Same at the Top and Bottom of a Cell?

All of the amperometric data discussed up to this point have been acquired from the top of the cell surface. A more detailed discussion on amperometric data collected from the bottom of adhered cells onto microfluidic devices integrated with electrodes is presented later in this review. It is worth mentioning here a recent study by Amatore and colleagues in which they discovered that events detected from the top of the cell differed in terms of release dynamics and frequency from those monitored at the bottom of the cell (47). This disparity is extremely important when results are compared for techniques that examine similar phenomena from different geometric setups (i.e., amperometry versus total internal reflectance fluorescence microscopy) (47). When monitoring events with amperometry above and below the cell, those events originating from the top of the cell demonstrated a faster rise time and half-width, but the average flux of electroactive material was 2.5 times less compared to that of events monitored below the cell. Several explanations for these findings were suggested, the most convincing of which centered on membrane dynamics. The cell surface is firmly affixed to the electrode surface when monitoring release from the bottom, whereas for the top of the cell the electrode is lightly contacted with the cell membrane. Thus, the cell membrane under the cell is not as efficient at incorporating new membrane from fusing vesicles and might not flow as easily. It may also be possible that there are different vesicle populations at each pole of chromaffin cells. These data, collected in different geometric configurations, must be carefully interpreted both analytically and biologically to avoid erroneous conclusions (47).

5. REPRODUCIBLE AMPEROMETRIC MONITORING OF SEROTONIN IN VITRO

A great deal of work to date has been carried out on dopamine as it is relatively easy to detect and is present in a discrete and large part of the mammalian brain. Some recent work has aimed to detect more challenging molecules, particularly serotonin, which is notorious for rapidly fouling carbon fiber electrodes in constant potential mode. Fast-scan cyclic voltammetry (FSCV) and chronoamperometry methods (see Section 6, below) can be used to minimize electrode fouling from serotonin oxidation products.

More recently, however, boron-doped diamond microelectrodes have been used to provide more stable, sensitive, and reproducible measurements in the constant potential mode (48–52). Serotonin overflow was monitored from mechanically or electrically stimulated enterochromaffin cells (found in the mucosa of the intestine and known to be rich in gastrointestinal serotonin), and these data were directly compared to data obtained with carbon fiber microelectrodes (53). Several advantages of the former method were found. First, diamond electrodes exhibited a slower rate of fouling compared to carbon fiber electrodes (50% versus 85% loss of initial signal for diamond and carbon fiber electrodes, respectively, after 10 injections of serotonin); however, diamond electrodes can be cleaned with alcohol and, essentially, returned to their initial sensitivity level. Second, diamond electrodes outperformed carbon fiber electrodes when measuring serotonin overflow from different regions of the intestinal mucosa (see **Figure 2a**). The electrode was positioned away from the mucosa to establish a stable background current. Then the electrode was placed within 1 mm of the tissue surface to determine the "over tissue current" (OT_c) resulting from fluid flow over the tissue. Finally, neighboring mucosa was stimulated either mechanically or electrically to determine the "touching tissue current" (TT_c). To validate that the observed change in oxidative current could be attributed to changes in serotonin, the experiments were repeated in the presence of a serotonin transporter (SERT) antagonist, fluoxetine. By blocking the SERT, the over tissue current was significantly higher as to be expected with a decreased ability to clear excess serotonin (see **Figure 2b**). In a recent follow-up study, Swain and colleagues determined that SERT expression was lower in neonatal guinea pig ileum than in adult guinea pigs when fluoxetine was used as the SERT antagonist. The antagonist did not substantially increase the over tissue current observed, implying that initial SERT function was significantly low in guinea pig ileum in the neonatal development stage (see **Figure 2c**) (54).

6. SYNAPTOSOME MODELS FOR REUPTAKE OF DOPAMINE AND SEROTONIN USING CHRONOAMPEROMETRY

Chronoamperometric experiments differ from those using constant potential amperometry in that repetitive potential pulses are applied to the working electrode in a square wave fashion. A typical 1-Hz square wave profile begins at a resting potential (commonly 0 V) and is stepped to a potential where the oxidation of the molecule of interest is diffusion limited (similar to the overpotential selected for steady-state

a

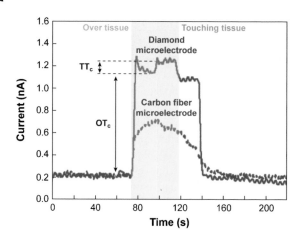

b

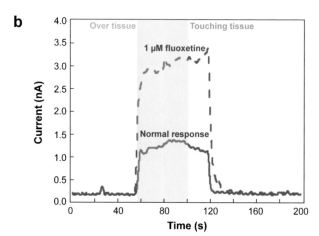

c

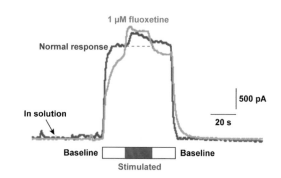

amperometry recordings). This overpotential is maintained briefly (for 100 ms) before the potential is returned to 0 V, essentially reducing any oxidized species still near the electrode back to their original form. The potential is held here for 100 ms plus an additional 800-ms resting period before the cycle is repeated. This essentially prevents fouling of the electrode surface by freshly oxidized species. The changes in both oxidative and reductive currents can be monitored over time, providing the temporal resolution necessary to study biological uptake systems, particularly of monoamine transporters. Synaptosomes, or neuronal liposomes containing one or more mitochondria and fully functioning respiratory capabilities, have also been shown to be suitable in vitro models for determining uptake of both dopamine and serotonin using chronoamperometric methods (55–57). For further reading on synaptosomes and their properties and preparation, please see References 58 and 59.

Two experimental considerations worth noting when working with a synaptosomal preparation are (*a*) the oxygenation level of the assay buffer containing the experimental synaptosomes, and (*b*) the rate of convection (stirring) in the solution throughout the experiment. Andrews and colleagues determined that an oxygen-rich assay buffer was necessary for optimal SERT function as interpreted through the more reliable clearance rates of serotonin (126 $\pm$ 8 pmol/mg protein-min, n = 6) compared to those obtained in oxygen-depleted buffer (21 $\pm$ 8 pmol/mg protein-min, n = 6 (55). Oxygen is necessary to maintain cellular ion gradients through the breakdown of glucose to ATP; therefore, depriving synaptosomes of oxygen hinders their ability to sustain their inherent ion gradient, thus impairing transport activity (58, 60, 61). The effects of stirring to minimize mass transport issues that have also been circumvented by use of rotating disk electrodes (62) have also been addressed; surprisingly, gentle stirring eliminated serotonin clearance (even in oxygen-enriched assay buffer), suggesting that constant agitation forces any serotonin that may have been reuptaken into the synaptosomes to leak out, thereby producing a steady, unchanging level of residual serotonin (55).

The submillisecond time resolution provided with chronoamperometry and the synaptosomal preparations derived from brain tissue have provided more precise

---

Figure 2

Measurements of serotonin overflow from enterochromaffin cells at adult or neonatal guinea pig ileum tissue. (*a*) Current versus time trace of serotonin overflow from adult guinea pig ileum tissue recorded with a diamond electrode (*solid line*) and carbon fiber electrode (*dotted line*). (*b*) Current versus time trace of serotonin overflow recorded with a diamond electrode for normal response (*solid line*) and in the presence of serotonin transporter (SERT) antagonist, fluoxetine (*dotted line*) for adult guinea pig ileum tissue. The current increased dramatically in the presence of fluoxetine, signifying that the measured current was indeed serotonin. (*c*) Current versus time trace of serotonin overflow recorded from neonatal guinea pig ileum tissue (*dark blue*) and the same tissue in the presence of fluoxetine (*light blue*). No difference in current profile was noted, suggesting that SERT activity in neonates is significantly low or underdeveloped as compared with adult. Abbreviations: OT$_c$, over tissue current; TT$_c$, touching tissue current. Panels *a* and *b* reproduced from Reference 53 with permission. Panel *c* reproduced from Reference 54 with permission.

determination of serotonin uptake rates. Uptake rates in mice lacking one or both alleles of the SERT gene have been compared in three different brain regions; the resulting data were then compared to results obtained using traditional radiochemical methods (56). Briefly, synaptosomes were prepared from three mice genotypes for SERT function: SERT +/+, +/−, and −/−. The amount of time needed for each preparation to clear 1 μM of introduced serotonin was monitored chronoamperometrically and an estimated uptake rate was extrapolated from the slope of the line constructed between the point of 20% and 60% decrease in maximum current signal. **Figure 3** illustrates representative serotonin uptake traces for brain stem synaptosomes prepared from three genotypes of SERT-deficient mice (56). It was determined that the clearance rate for SERT +/− mice was reduced by 60% compared to the uptake rate obtained chronoamperometrically for SERT +/+ mice (**Figure 3**, top and center panels). This is a minute difference from the radiochemical method determination of a 50% reduction rate for SERT +/− mice, but it is a key difference presumed to be based on the potential loss of sensitivity of uptake inherent in the filtration step used in the tissue preparation of the radiochemical method (56). The sensitivity provided with chronoamperometry allows this level of difference to be determined. As expected, no detectable uptake of serotonin was observed for SERT −/− mice (**Figure 3**, bottom panel).

Dopamine uptake has also been investigated with chronoamperometry, specifically with regard to an abnormality of the α-synuclein protein, which is potentially linked to Parkinson's disease (63). Mice have been genetically modified to overexpress this protein, and striatum synaptosomes have been prepared and tested for dopamine transport rates using the same approach described above for serotonin. These genetically modified mice exhibited hyperlocomotive activity, and the same mutant mouse synaptosomes showed a significant decrease (40%) in the clearance rate of 1 μM dopamine introduced into the medium. This observation possibly links the loss of transporter function (versus expression) to observed behavior (57).

7. SINGLE-CELL ELECTROPORATION FOR CELL MANIPULATION AND ANALYSIS

The examination of cell composition and the functioning chemical processes inside intact cellular environments remains an exciting area of study in which there is still much to be learned. To probe the internal environment of a cell while maintaining cell viability, electroporation methods have been developed. Using these methods, it is possible to passively introduce into a cell a numerous collection of antisense agents,

Figure 3

Representative chronoamperometric recordings for serotonin uptake in brain stem synaptosomes prepared from serotonin transporter (SERT) +/+ (*a*), SERT +/− (*b*), and SERT −/− (*c*). The average uptake rates for SERT +/+ and SERT +/− were 162 ± 13 and 67.7 ± 4.1 pmol/mg protein-min, respectively, whereas SERT −/− showed no detectable uptake. Reproduced from Reference 56 with permission.

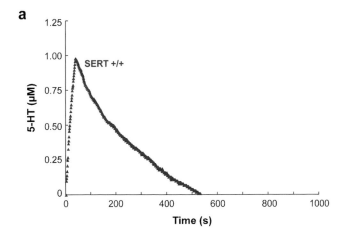

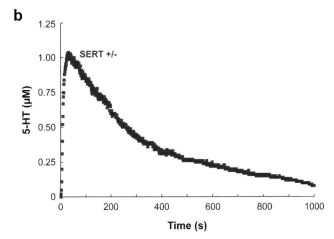

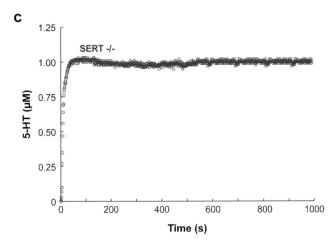

fluorescent dyes, pharmacological stimulants, and other relatively small molecules unable to self-permeate the membrane surface (64–67). Electroporation is not exactly electroanalysis, but it does involve application of a potential across electrodes and so falls under the scope of this review.

In traditional electroporation, a strong electric field capable of exceeding the transmembrane potential (and thus creating small pores in the plasma membrane) is applied in pulses to a bulk population of suspended cells and the agent to be introduced is administered via the bathing solution (68–70). Typically, the field is generated by passing large voltage pulses between two large, fixed millimeter-sized electrodes positioned on either side of the cells to be electroporated. This creates a broad, unfocused electric field useful for bulk transfection applications. Lundqvist and colleagues developed the first single-cell electroporation scheme, exemplifying the small dimensions of carbon fiber electrodes to generate more localized, high electric fields (71). With a concentrated electric field produced at the electrode tip, individual or smaller groupings of cells could then be addressed.

Since then, other single-cell electroporation methods have been developed (for a more detailed discussion of the various single-cell electroporation methodologies available, please see Reference 72). Here, we highlight a method that generates a highly concentrated electric field at the outlet of a fused silica capillary that also transports the agent to be loaded into the cells (73, 74). These electrolyte-filled capillaries (EFCs) had previously been shown to project an intense electric field from the capillary outlet when used in capillary electrophoresis work (75). By use of this finely focused electric field near single cells or small groups of cells, materials in small defined quantities can be introduced into single cells via electroporation. Electroporation via an EFC was successfully demonstrated with the introduction of YOYO-1, an RNA-/DNA-intercalating dye, into NG108–15 cells (glial-neuronal hybrids) (73). In that experiment, single cells and/or single-cell processes were selectively loaded with YOYO-1 using various potential ranges at a constant cell–EFC tip distance. Different success rates were obtained for each applied potential, most notably when focusing poration efforts to small processes. For an applied potential of 2 kV, only 20% of poration attempts were successful, probably due to insufficient pore formation with such a low potential. However, a potential of 6 kV increased success to 92%. Still higher potentials did not yield greater success, with 10 kV only producing a 70% success rate, revealing a potential-dependence on both poration success rate and cell viability.

In addition to the magnitude of the applied potential, other key parameters have been carefully considered, including pulse duration, distance between the EFC tip and the cell surface, cell size, and cell morphology (76, 77). In these experiments, the effectiveness of the method to introduce fluorescent dyes to the cells or process was examined. Longer pulses (120–150 ms) produced greater decreases in internal fluorescence intensity of cells loaded with fluorescent molecules. This suggests that more extensive pore formation is occurring allowing labeled molecules to escape the cell. This decreases the total concentration of labeled molecule inside the cell, thus decreasing its fluorescence intensity. EFC tip distances on the order of multiple (3 or 4) micrometers also yielded favorable poration results. With these two parameters simultaneously optimized, more than 90% of the cells were electroporated and more than 80% remained viable (76). Moreover, big semicircular cells were more likely to be successfully permeated and remain viable for subsequent studies (77).

In light of the above parameters significant to poration success, a computer-controlled scanning electroporation configuration aiming to more controllably and selectively electroporate adhered cell cultures has recently been presented (78). In brief, the EFC was initially positioned in a dish of adhered cells using a micromanipulator. Then, a motorized stage was used to position the cell of interest underneath the EFC and subsequently porate the cell as described in the previous paragraph. With the addition of the motorized stage, this configuration can be performed in two modes: stationary and scanning. The stationary mode has been used to examine the shape of the electric field when fluorescein diphosphate–loaded WSS cells are electroporated (see **Figure 4a**). In addition, simple repositioning of the stage has been used to show that repetitive ring-shaped fluorescent areas can be patterned within the same culture. In the scanning mode, however, the potential is maintained while the stage is moved, resulting in a "drawing" effect (**Figure 4b,c**) for another selection

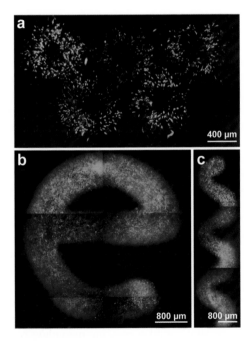

Figure 4

Examples of confluent WSS cells electroporated with fluorescein diphosphate in stationary mode (*a*) and scanning mode (*b, c*). Five rings can be seen in image *a*, demonstrating the electrolyte-filled capillary's ability to independently electroporate different places within one cell culture in the stationary mode. Fluorescent "e" and "snake" patterns in images *b* and *c*, respectively, were drawn using the scanning mode. Reproduced from Reference 78 with permission.

of WSS cells electroporated with fluorescein diphosphate. This approach allows the electroporation agent to be changed many times, and several different experiments can be performed simultaneously within the same cell culture (78).

8. SCANNING ELECTROCHEMICAL MICROSCOPY

Scanning electrochemical microscopy (SECM) is a scanning probe technique that has been used at the cellular surface. Briefly, an ultramicroelectrode (UME) is used to detect local electrochemical activity. When the UME is moved over the cell, electrochemical data are recorded at multiple positions and an image is constructed based on the local electrochemical properties of the area of interest. SECM provides topographic information and chemical analysis of cellular functions such as respiration, photosynthesis, and membrane transport. The SECM technique and its applications have been thoroughly reviewed (79–84). In this section, we discuss the latest work showing applications of SECM to cells [for an in-depth review on this subject, see the article by Amemiya et al. in this volume (85)].

SECM has been successfully used to study exocytosis from neuronal cells. Use of carbon fiber microelectrodes yields excellent time resolution and allows detection of zeptomole quantities of dopamine released by a single neuron (86). Recently, Baur and colleagues successfully developed a technique for imaging the topography of single PC12 cells using cell-impermeable $Ru(NH_3)_6^{3+}$ as a mediator (87). Use of a constant height mode based on the negative feedback effect clearly revealed distinct regions of decreased current on neurites, suggesting the presence of raised structures not visible in optical images. However, the constant height mode could be disadvantageous for imaging differentiated PC12 cells because the neurites are too low to be imaged when the electrode is placed at the apex of the cell body. The authors of the study, however, achieved an improved image resolution of PC12 cells and neurites by using a constant distance mode and two types of feedback signal for distance control. They argued that the highest resolution for these images was achieved with constant current mode using a 1-μm-diameter carbon ring electrode, although specific numbers were not provided (88).

Several groups have combined electrochemical and optical imaging to obtain more information-rich experiments at single cells. For example, Takahashi et al. developed an optical fiber electrode via Ti/Pt sputtering, allowing simultaneous topographic, electrochemical, and optical imaging (89). Optical fibers sputtered with gold were used by Suzuki's group to simultaneously record electrochemical and optical images with nanometer-scale resolution. Neurites of PC12 cells have also been successfully imaged with high resolution. In that work, a pencil-shaped microelectrode was used for combined electrochemical/optical microscopy; the authors reported a spatial resolution of 300 nm (90).

Recently, Wipf, Baur and colleagues combined FSCV with SECM, which provided important advantages over SECM used alone (91–93). In summary, cyclic voltammetry was carried out at the SECM tip at rapid scan rates (10–1000 V s^{-1}); 3D data sets were produced for line scans and 4D data sets were produced for imaging experiments. By combining those two techniques, the researchers were able to

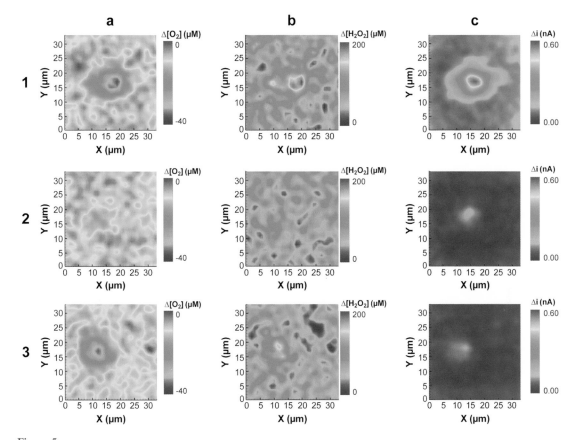

a   b   c

Figure 5

Fast-scan cyclic voltammetry–scanning electrochemical microscopy images of a single RAW 264.7 macrophage cell following stimulation with zymosan particles (*row 1*), followed by the addition of 0.3 IU of catalase (*row 2*), and after washing with fresh Hank's buffer (*row 3*). Column *a* shows the potential for reduction of O_2 average of 14 points between -1.2 and -1.3 V on cathodic sweep). Column *b* shows the potential for H_2O_2 oxidation (average of 12 points between $+1.2$ and $+1.3$ V on the first anodic sweep). Column *c* represents the average of -0.45 to -0.55 V on the cathodic sweep. The images in each row were recorded simultaneously. The potential applied to the tip consisted of a three-segment waveform (0.0 to $+1.2$ to -1.4 to 0.0 V) with a scan rate of 450 V·s^{-1}. Reproduced from Reference 92 with permission.

detect multiple electroactive species simultaneously and obtain micrometer spatial resolution of cells (**Figure 5**). Both oxygen consumption by macrophages and their release of hydrogen peroxide upon stimulation were monitored (92). In conventional SECM, the tip must be held at a constant potential, resulting in oxygen depletion. With the FSCV approach, oxygen depletion is minimized by use of low-duty cycle; therefore, this approach is well suited for imaging oxygen consumption at specific regions across the cell surface. In the future, FSCV combined with SECM will likely be used for specific single-cell applications.

9. THE INTERFACE OF CELL-BASED MICROFLUIDIC SYSTEMS WITH ELECTROCHEMICAL DETECTION

Using microfluidic devices to investigate cellular systems has become very popular, especially with the recent development of elastomeric materials such as polydimethylsiloxane (PDMS) for the fabrication of devices (94, 95). Although the fabrication of many 2D or 3D microfluidic platforms preferably requires access to a cleanroom in a nanofabrication laboratory, PDMS has the advantages of being inexpensive, easy to handle, and able to connect to plastic tubing. This material is perfectly suited for on-chip cell culturing because it is permeable to gas, allowing cells to grow on it; it also has an excellent optical transparency that allows, for instance, detection with fluorescence microscopes. Several excellent recent reviews on microfluidics (96–99) are recommended for in-depth coverage of this topic.

Within bioanalytical studies, microfluidic devices are used for many different purposes including cell sorting, DNA analysis, immunoassays, and drug discovery (97, 99–102). The diversity of design in the various microfluidic platforms reflects the flourishing imagination of their creators. Indeed, the cells being studied are grown either at the surface, inside microchannels, or in different compartments as illustrated by Taylor et al. (103), who separated nerve cell bodies and axons and thereby stimulated only the nerve terminals.

An important aspect of cell-based microfluidics is, of course, the detection method used to characterize analytes of interest. The main detection techniques are optical (fluorescence being the most popular), electrochemical, and mass spectrometric. In this section we discuss the recent work illustrating the use of microfluidics with electrochemical detection.

Typically, an amperometric electrode is created using metal deposition techniques during the fabrication process of the microfluidic device; the most common electrode materials are carbon, platinum, and gold. Microfluidic devices with integrated electrochemical detection possess several advantages over external electrodes used to directly measure substances at single cells in stagnant solution. In the microfluidic device, the electrode does not need to be manually positioned close to the cell, and experiments using this device are not limited to one cell at a time. The Amatore group has developed a microfluidic device for measuring the oxidative stress generated by macrophages, which are key cells in the immune system. A platinized band microelectrode was used to detect oxidative bursts of reactive oxygen species or reactive nitrogen species with great sensitivity and reproducibility (104). In these experiments, the electrode surface is regenerated in situ by reducing hydrogen hexachloroplatinate in the presence of lead acetate at -60 mV versus a sodium-saturated calomel electrode (105) between experiments, thus circumventing electrode fouling effects and the subsequent decrease in current. In another set of experiments, an elegant microfluidic chip with a built-in amperometric detector array was recently described by the Matsue group (106). They used this system to measure oxygen consumption of single bovine embryos. The embryo was placed in the chip and immobilized near four platinum working electrodes with a constant flow stream. The oxygen consumption was measured with chronoamperometry (see **Figure 6**) (106). They compared

a

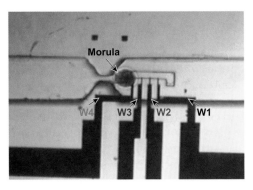

b

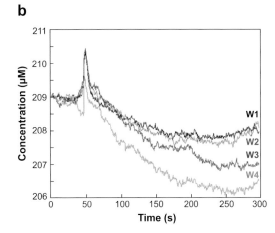

Figure 6

(*a*) An optical image and (*b*) amperometric recordings of oxygen consumption obtained from a bovine embryo adhered to a microfluidic device integrated with four working electrodes (labeled W1, W2, W3, and W4 in the optical image). Panel *a* shows the immobilized embryo at the gate position; the plot of the oxygen concentration profile consumed by the day-6 embryo at the morula stage was obtained with the amperometric detector array at room temperature against time. The monitored concentration profiles for each electrode are shown in panel *b*. Reproduced from Reference 106 with permission.

this method to SECM and argued that experiments with the microfluidic chip are as precise as those with SECM, and are more cost effective and easier to use.

Recently, microfluidic devices have been developed to study nerve cells. Owing to their fragility, culturing and studying nerve and nerve-like cells in chip-based devices present a significant challenge. PC12 cells, nerve-like cells that have been used to study dopamine release by electrochemical detection, contain dopamine-filled vesicles and release their contents upon stimulation via exocytosis. They have active dopamine uptake transporters as well and so may be considered a good model system of the release and removal process for dopamine.

Traditionally, dopamine release by PC12 cells has been monitored using carbon fiber amperometry (described above); however, with microfluidic chip technology, more complex systems can be studied. Cells can be grown in conditions that are more similar to in vivo. Huang et al. constructed a device capable of on-chip transport, location and monitoring of Q for a single PC12 cell (107). Martin's group used integrated Nafion®-coated microelectrodes in their microchip. PC12 cells were reversibly sealed inside a collagen-coated microchannel and dopamine levels were measured by amperometry after calcium stimulation (108). Cui et al. (2006) presented a microelectrode array biochip to investigate the effect of several drugs such as reserpine, L-DOPA, and nomifensine on the level of dopamine release upon multiple potassium stimulations in PC12 cells (109). On-chip determination of dopamine release by PC12 cells with a gold electrode was achieved by Spégel et al. (110). Normally, gold and platinum electrodes are not suitable for dopamine measurements, as dopamine oxidation products form a nonconducting polymer at the surface of the electrode. This problem was solved by modifying the electrode surface by mercaptopropionic acid to decrease the rate of dopamine polymerization, thereby promoting the stability of the electrodes while providing good reversibility and sensitivity of the measurement.

Microfluidic devices allow more precise control of both flow and concentration of molecules than can be obtained with a micropipette and a pressure-based (often called "puffed") injection. In addition, several detection methods can be used simultaneously. Sun and Gillis recently developed a microchip that uses transparent indium tin oxide (ITO) electrodes to measure quantal exocytosis of catecholamines by bovine adrenal chromaffin cells. The cells adhered to the ITO electrode within the microchannel. Amperometric spikes were recorded with a signal-to-noise ratio comparable to that of carbon fiber electrodes. Also, because the electrodes are transparent simultaneous fluorescence-based measures could be implemented (111). It appears certain that in the future several detection techniques will be combined and integrated within microfluidic chips used for cellular measurements, allowing better imaging quality of the cells and more precise quantitative measurement of the substances released by the cells.

10. SUMMARY AND FUTURE DIRECTIONS

In vitro electrochemistry at cells clearly produces results pertinent to a variety of biologically relevant questions. It has been used to study the exocytosis of zeptomole-to-attomole levels of transmitters from single cells in isolated cultures, and to reveal fine details about the rate of transmitter uptake using synaptosomal models. It has been manipulated to electroporate single cells and to introduce multiple biomolecules unable to permeate the cell membrane. Finally, it has been coupled to a microfluidic platform as a detection scheme. The numerous fields that have been impacted by this technique testify to the versatility of electrochemistry in bioanalysis.

DISCLOSURE STATEMENT

The authors are not aware of any biases that might be perceived as affecting the objectivity of this review.

ACKNOWLEDGMENTS

The authors gratefully acknowledge support from the National Institutes of Health and the National Science Foundation for work reviewed here. Additionally, the hard work of all our former coworkers referenced herein is gratefully acknowledged. A.G.E. gratefully acknowledges the support of the European Union in the form of a Marie Curie Chair.

LITERATURE CITED

1. Adams RN. 1976. Probing brain chemistry with electroanalytical techniques. *Anal. Chem.* 48:1126A–38A

2. Leszczyszyn DJ, Jankowski JA, Viveros OH, Diliberto EJ Jr, Near JA, Wightman RM. 1990. Nicotinic receptor-mediated catecholamine secretion from individual chromaffin cells: chemical evidence for exocytosis. *J. Biol. Chem.* 265:14736–37

3. Wightman RM, Jankowski JA, Kennedy RT, Kawagoe KT, Schroeder TJ, et al. 1991. Temporally resolved catecholamine spikes correspond to single vesicle release from individual chromaffin cells. *Proc. Natl. Acad. Sci. USA* 88:10754–58

4. Durack GR, Robinson JP, eds. 2000. *Emerging Tools for Single-Cell Analysis: Advances in Optical Measurement Technologies Cytometric Cellular Analysis.* New York: Wiley

5. Stamford JA, Justice JBJ. 1996. Probing brain chemistry. *Anal. Chem.* 68:359A–63A

6. Valtorta F, Fesce R, Grohovaz F, Haimann C, Hurlbut WP, et al. 1990. Neurotransmitter release and synaptic vesicle recycling. *Neuroscience* 35:477–89

7. Neher E, Marty A. 1982. Discrete changes of cell membrane capacitance observed under conditions of enhanced secretion in bovine adrenal chromaffin cells. *Proc. Natl. Acad. Sci. USA* 79:6712–16

8. Chow RH, von Ruden L, Neher E. 1992. Delay in vesicle fusion revealed by electrochemical monitoring of single secretory events in adrenal chromaffin cells. *Nature* 356:60–63

9. Dayton MA, Ewing AG, Wightman RM. 1981. Evaluation of amphetamine-induced in vivo electrochemical response. *Eur. J. Pharmacol.* 75:141–44

10. Ewing AG, Wightman RM, Dayton MA. 1982. In vivo voltammetry with electrodes that discriminate between dopamine and ascorbate. *Brain Res.* 249:361–70

11. Cespuglio R, Faradji H, Ponchon JL, Riou F, Buda M, et al. 1981. In vivo measurements by differential pulse voltammetry of extracellular 5-hydroxyindoleacetic acid in the rat brain. *J. Physiol.* 77:327–32

12. Gonon F, Buda M, Cespuglio R, Jouvet M, Pujol JF. 1980. In vivo electrochemical detection of catechols in the neostriatum of anaesthetized rats: dopamine or DOPAC? *Nature* 286:902–4

13. Cahill PS, Walker QD, Finnegan JM, Mickelson GE, Travis ER, Wightman RM. 1996. Microelectrodes for the measurement of catecholamines in biological systems. *Anal. Chem.* 68:3180–86

14. Kawagoe KT, Zimmerman JB, Wightman RM. 1993. Principles of voltammetry and microelectrode surface states. *J. Neurosci. Methods* 48:225–40

15. Pothos EN, Davila V, Sulzer D. 1998. Presynaptic recording of quanta from midbrain dopamine neurons and modulation of the quantal size. *J. Neurosci.* 18:4106–18

16. Travis ER, Wightman RM. 1998. Spatio-temporal resolution of exocytosis from individual cells. *Annu. Rev. Biophys. Biomol. Struct.* 27:77–103

17. Gerhardt GA, Oke AF, Nagy G, Moghaddam B, Adams RN. 1984. Nafion-coated electrodes with high selectivity for CNS electrochemistry. *Brain Res.* 290:390–95

18. Friedemann MN, Robinson SW, Gerhardt GA. 1996. O-phenylenediamine-modified carbon fiber electrodes for the detection of nitric oxide. *Anal. Chem.* 68:2621–28

19. Gmucova K, Weis M, Barancok D, Cirak J, Tomcik P, Pavlasek J. 2007. Ion selectivity of a poly(3-pentylmethoxythiophene) LB-layer modified carbon-fiber microelectrode as a consequence of the second order filtering in voltcoulometry. *J. Biochem. Biophys. Methods* 70:385–90

20. Wang Y, Joshi PP, Hobbs KL, Johnson MB, Schmidtke DW. 2006. Nanostructured biosensors built by layer-by-layer electrostatic assembly of enzyme-coated single-walled carbon nanotubes and redox polymers. *Langmuir* 22:9776–83

21. Kristensen EW, Kuhr WG, Wightman RM. 1987. Temporal characterization of perfluorinated ion exchange coated microvoltammetric electrodes for in vivo use. *Anal. Chem.* 59:1752–57

22. Nagy G, Gerhardt GA, Oke AF, Rice ME, Adams RN, et al. 1985. Ion exchange and transport of neurotransmitters in Nafion films on conventional and microelectrode surfaces. *J. Electroanal. Chem.* 188:85–94

23. Rice ME, Oke AF, Bradberry CW, Adams RN. 1985. Simultaneous voltammetric and chemical monitoring of dopamine release in situ. *Brain Res.* 340:151–55

24. Hermans A, Seipel AT, Miller CE, Wightman RM. 2006. Carbon-fiber microelectrodes modified with 4-sulfobenzene have increased sensitivity and selectivity for catecholamines. *Langmuir* 22:1964–69

25. Chen RS, Huang WH, Tong H, Wang ZL, Cheng JK. 2003. Carbon fiber nanoelectrodes modified by single-walled carbon nanotubes. *Anal. Chem.* 75:6341–45

26. Lambie BA, Orwar O, Weber SG. 2006. Controlling the electrochemically active area of carbon fiber microelectrodes by the electrodeposition and selective removal of an insulating photoresist. *Anal. Chem.* 78:5165–71

27. Hermans A, Wightman RM. 2006. Conical tungsten tips as substrates for the preparation of ultramicroelectrodes. *Langmuir* 22:10348–53

28. Hochstetler SE, Puopolo M, Gustincich S, Raviola E, Wightman RM. 2000. Real-time amperometric measurements of zeptomole quantities of dopamine released from neurons. *Anal. Chem.* 72:489–96

29. Pothos EN, Mosharov E, Liu K-P, Setlik W, Haburcak M, et al. 2002. Stimulation-dependent regulation of the pH, volume and quantal size of bovine and rodent secretory vesicles. *J. Physiol.* 542:453–76

30. Chen TK, Luo G, Ewing AG. 1994. Amperometric monitoring of stimulated catecholamine release from rat pheochromocytoma (PC12) cells at the zeptomole level. *Anal. Chem.* 66:3031–35

31. Colliver TL, Pyott SJ, Achalabun M, Ewing AG. 2000. VMAT-mediated changes in quantal size and vesicular volume. *J. Neurosci.* 20:5276–82

32. Schmitz Y, Lee CJ, Schmauss C, Gonon F, Sulzer D. 2001. Amphetamine distorts stimulation-dependent dopamine overflow: effects on D2 autoreceptors, transporters, and synaptic vesicle stores. *J. Neurosci.* 21:5916–24

33. Chen G, Gavin PF, Luo G, Ewing AG. 1995. Observation and quantitation of exocytosis from the cell body of a fully developed neuron in *Planorbis corneus*. *J. Neurosci.* 15:7747–55

34. Schonfu D, Reum T, Olshausen P, Fischer T, Morgenstern R. 2001. Modelling constant potential amperometry for investigations of dopaminergic neurotransmission kinetics in vivo. *J. Neurosci. Methods* 112:163–72

35. Anderson BB, Chen G, Gutman DA, Ewing AG. 1999. Demonstration of two distributions of vesicle radius in the dopamine neuron of *Planorbis corneu*s from electrochemical data. *J. Neurosci. Methods* 88:153–61

36. Cans AS, Wittenberg N, Eves D, Karlsson R, Karlsson A, et al. 2003. Amperometric detection of exocytosis in an artificial synapse. *Anal. Chem.* 75:4168–75

37. Amatore C, Arbault S, Bouret Y, Guille M, Lemaître F, Verchier Y. 2006. Regulation of exocytosis in chromaffin cells by *trans*-insertion of lysophosphatidylcholine and arachidonic acid into the outer leaflet of the cell membrane. *Chem. Bio. Chem.* 7:1998–2003

38. Uchiyama Y, Maxson MM, Sawada T, Nakano A, Ewing AG. 2007. Phospholipid mediated plasticity in exocytosis observed in PC12 cells. *Brain Res.* 1151:46–54

39. Greene LA, Tischler AS. 1976. Establishment of a noradrenergic clonal line of rat adrenal pheochromocytoma cells which respond to nerve growth factor. *Proc. Natl. Acad. Sci. USA* 73:2424–28

40. Umbach JA, Zhao Y, Gundersen CB. 2005. Lithium enhances secretion from large dense-core vesicles in nerve growth factor–differentiated PC12 cells. *J. Neurochem.* 94:1306–14

41. Sombers LA, Maxson MM, Ewing AG. 2005. Loaded dopamine is preferentially stored in the halo portion of PC12 cell dense core vesicles. *J. Neurochem.* 93:1122–31

42. Troyer KP, Wightman RM. 2002. Temporal separation of vesicle release from vesicle fusion during exocytosis. *J. Biol. Chem.* 277:29101–7

43. Amatore C, Arbault S, Bonifas I, Lemaître F, Verchier Y. 2007. Vesicular exocytosis under hypotonic conditions shows two distinct populations of dense core vesicles in bovine chromaffin cells. *Chem. Phys. Chem.* 8:578–85

44. Westerink RH, de Groot A, Vijverberg HP. 2000. Heterogeneity of catecholamine-containing vesicles in PC12 cells. *Biochem. Biophys. Res. Commun.* 270:625–30

45. Tang KS, Tse A, Tse FW. 2005. Differential regulation of multiple populations of granules in rat adrenal chromaffin cells by culture duration and cyclic AMP. *J. Neurochem.* 92:1126–39

46. Villanueva M, Wightman RM. 2007. Facilitation of quantal release induced by a D1-like receptor on bovine chromaffin cells. *Biochem. J.* 46:3881–87

47. Amatore C, Arbault S, Lemaître F, Verchier Y. 2007. Comparison of apex and bottom secretion efficiency at chromaffin cells as measured by amperometry. *Biophys. Chem.* 127:165–71

48. Cvacka J, Quaiserova V, Park J, Show Y, Muck A, Swain GM. 2003. Borondoped diamond microelectrodes for use in capillary electrophoresis with electrochemical detection. *Anal. Chem.* 75:2678–87

49. Muna GW, Quaiserova-Mocko V, Swain GM. 2005. Chlorinated phenol analysis using off-line solid-phase extraction and capillary electrophoresis coupled with amperometric detection and a boron-doped diamond microelectrode. *Anal. Chem.* 77:6542–48

50. Park J, Galligan JJ, Fink GD, Swain GM. 2006. In vitro continuous amperometry with a diamond microelectrode coupled with video microscopy for simultaneously monitoring endogenous norepinephrine and its effect on the contractile response of a rat mesenteric artery. *Anal. Chem.* 78:6756–64

51. Park J, Quaiserova-Mocko V, Peckova K, Galligan JJ, Fink GD, Swain GM. 2006. Fabrication, characterization, and application of a diamond microelectrode for electrochemical measurement of norepinephrine release from the sympathetic nervous system. *Diam. Relat. Mater.* 15:761–72

52. Park J, Show Y, Quaiserova V, Galligan JJ, Fink GD, Swain GM. 2005. Diamond microelectrodes for use in biological environments. *J. Electroanal. Chem.* 583:56–68

53. Patel BA, Bian X, Quaiserova-Mocko V, Galligan JJ, Swain GM. 2007. In vitro continuous amperometric monitoring of 5-hydroxytryptamine release from enterochromaffin cells of the guinea pig ileum. *Analyst* 132:41–47

54. Bian X, Patel B, Dai X, Galligan JJ, Swain G. 2007. High mucosal serotonin availability in neonatal guinea pig ileum is associated with low serotonin transporter expression. *Gastroenterology* 132:2438–47

55. Perez XA, Andrews AM. 2005. Chronoamperometry to determine differential reductions in uptake in brain synaptosomes from serotonin transporter knockout mice. *Anal. Chem.* 77:818–26

56. Perez XA, Bianco LE, Andrews AM. 2006. Filtration disrupts synaptosomes during radiochemical analysis of serotonin uptake: comparison with chronoamperometry in SERT knockout mice. *J. Neurosci. Methods* 154:245–55

57. Unger EL, Eve DJ, Perez XA, Reichenbach DK, Xu Y, et al. 2006. Locomotor hyperactivity and alterations in dopamine neurotransmission are associated with overexpression of A53T mutant human α-synuclein in mice. *Neurobio. Dis.* 21:431–43

58. Erecinska M, Nelson D, Silver IA. 1996. Metabolic and energetic properties of isolated nerve ending particles (synaptosomes). *Biochim. Biophys. Acta. Bioenerg.* 1277:13–34

59. Hyde CE, Bennett BA. 1994. Similar properties of fetal and adult amine transporters in the rat brain. *Brain Res.* 646:118–23

60. Scott ID, Nicholls DG. 1980. Energy transduction in intact synaptosomes: influence of plasma-membrane depolarization on the respiration and membrane potential of internal mitochondria determined in situ. *Biochem. J.* 186:21–33

61. Whittaker VP. 1993. Thirty years of synaptosome research. *J. Neurocytol.* 22:735–42

62. Schenk JO, Wright C, Bjorklund N. 2005. Unraveling neuronal dopamine transporter mechanisms with rotating disk electrode voltammetry. *J. Neurosci. Methods* 143:41–47

63. Kruger R, Muller T, Riess O. 2000. Involvement of α-synuclein in Parkinson's disease and other neurodegenerative disorders. *J. Neural. Transm.* 107:31–40

64. Templeton NS, Roberts DD, Safer B. 1997. Efficient gene targeting in mouse embryonic stem cells. *Gene Ther.* 4:700–9

65. Tsong TY. 1991. Electroporation of cell membranes. *Biophys. J.* 60:297–306

66. Weaver JC. 1993. Electroporation: a general phenomenon for manipulating cells and tissues. *J. Cell. Biochem.* 51:426–35

67. Zimmermann U. 1982. Electric field-mediated fusion and related electrical phenomena. *Biochim. Biophys. Acta. Rev. Biomem.* 694:227–77

68. Gabriel B, Teissie J. 1999. Time courses of mammalian cell electropermeabilization observed by millisecond imaging of membrane property changes during the pulse. *Biophys. J.* 76:2158–65

69. Golzio M, Teissie J, Rols M-P. 2002. Direct visualization at the single-cell level of electrically mediated gene delivery. *Proc. Natl. Acad. Sci. USA* 99:1292–97

70. Hibino M, Itoh H, Kinosita K. 1993. Time courses of cell electroporation as revealed by submicrosecond imaging of transmembrane potential. *Biophys. J.* 64:1789–800

71. Lundqvist JA, Sahlin F, Aberg MAI, Stromberg A, Eriksson PS, Orwar O. 1998. Altering the biochemical state of individual cultured cells and organelles with ultramicroelectrodes. *Proc. Natl. Acad. Sci. USA* 95:10356–60

72. Olofsson J, Nolkrantz K, Ryttsen F, Lambie BA, Weber SG, Orwar O. 2003. Single-cell electroporation. *Curr. Opin. Biotechnol.* 14:29–34

73. Nolkrantz K, Farre C, Brederlau A, Karlsson RID, Brennan C, et al. 2001. Electroporation of single cells and tissues with an electrolyte-filled capillary. *Anal. Chem.* 73:4469–77

74. Nolkrantz K, Farre C, Hurtig KJ, Rylander P, Orwar O. 2002. Functional screening of intracellular proteins in single cells and in patterned cell arrays using electroporation. *Anal. Chem.* 74:4300–5

75. Lu W, Cassidy RM. 1994. Background noise in capillary electrophoretic amperometric detection. *Anal. Chem.* 66:200–4

76. Agarwal A, Zudans I, Orwar O, Weber SG. 2007. Simultaneous maximization of cell permeabilization and viability in single-cell electroporation using an electrolyte-filled capillary. *Anal. Chem.* 79:161–67

77. Agarwal A, Zudans I, Weber EA, Olofsson J, Orwar O, Weber SG. 2007. Effect of cell size and shape on single-cell electroporation. *Anal. Chem.* 79:3589–96

78. Olofsson J, Levin M, Stromberg A, Weber SG, Ryttsen F, Orwar O. 2007. Scanning electroporation of selected areas of adherent cell cultures. *Anal. Chem.* 79:4410–18

79. Amemiya S, Guo J, Xiong H, Gross DA. 2006. Biological applications of scanning electrochemical microscopy: chemical imaging of single living cells and beyond. *Anal. Bioanal. Chem.* 386:458–71

80. Wightman RM. 2006. Detection technologies: probing cellular chemistry in biological systems with microelectrodes. *Science* 311:1570–74

81. Edwards MA, Martin S, Whitworth AL, Macpherson JV, Unwin PR. 2006. Scanning electrochemical microscopy: principles and applications to biophysical systems. *Physiol. Meas.* 27:R63–R108

82. Bard AJ, Li X, Zhan W. 2006. Chemically imaging living cells by scanning electrochemical microscopy. *Biosens. Bioelectron.* 22:461–72

83. Zhan D, Li X, Zhan W, Fan FR, Bard AJ. 2007. Scanning electrochemical microscopy. 58. Application of a micropipet-supported ITIES tip to detect Ag$^+$ and study its effect on fibroblast cells. *Anal. Chem.* 79:5225–31

84. Sun P, Laforge FO, Mirkin MV. 2007. Scanning electrochemical microscopy in the 21st century. *Phys. Chem. Chem. Phys.* 9:802–23

85. Amemiya S, Bard AJ, Fan FRF, Mirkin MV, Unwin PR. Scanning electrochemical microscopy. *Annu. Rev. Anal. Chem.* 1:95–131

86. Hochstetler SE, Puopolo M, Gustincich S, Raviola E, Wightman RM. 2000. Real-time amperometric measurements of zeptomole quantities of dopamine released from neurons. *Anal. Chem.* 72:489–96

87. Liebetrau JM, Miller HM, Baur JE, Takacs SA, Anupunpisit V, et al. 2003. Scanning electrochemical microscopy of model neurons: imaging and real-time detection of morphological changes. *Anal. Chem.* 75:563–71

88. Kurulugama RT, Wipf DO, Takacs SA, Pongmayteegul S, Garris PA, Baur JE. 2005. Scanning electrochemical microscopy of model neurons: constant distance imaging. *Anal. Chem.* 77:1111–17

89. Takahashi Y, Hirano Y, Yasukawa T, Shiku H, Yamada H, Matsue T. 2006. Topographic, electrochemical, and optical images captured using standing approach mode scanning electrochemical/optical microscopy. *Langmuir* 22:10299–306

90. Maruyama K, Ohkawa H, Ogawa S, Ueda A, Niwa O, Suzuki K. 2006. Fabrication and characterization of a nanometer-sized optical fiber electrode based on selective chemical etching for scanning electrochemical/optical microscopy. *Anal. Chem.* 78:1904–12

91. Diaz-Ballote L, Alpuche-Aviles M, Wipf DO. 2007. Fast-scan cyclic voltammetry-scanning electrochemical microscopy. *J. Electroanal. Chem.* 604:17–25

92. Schrock DS, Baur JE. 2007. Chemical imaging with combined fast-scan cyclic voltammetry–scanning electrochemical microscopy. *Anal. Chem.* 79:7053–61

93. Schrock DS, Wipf DO, Baur JE. 2007. Feedback effects in combined fast-scan cyclic voltammetry-scanning electrochemical microscopy. *Anal. Chem.* 79:4931–41

94. McDonald JC, Duffy DC, Anderson JR, Chiu DT, Wu H, et al. 2000. Fabrication of microfluidic systems in poly(dimethylsiloxane). *Electrophoresis* 21:27–40

95. Whitesides GM, Ostuni E, Takayama S, Jiang X, Ingber DE. 2001. Soft lithography in biology and biochemistry. *Annu. Rev. Biomed. Eng.* 3:335–73

96. Dittrich PS, Tachikawa K, Manz A. 2006. Micro total analysis systems: latest advancements and trends. *Anal. Chem.* 78:3887–908

97. Fiorini GS, Chiu DT. 2005. Disposable microfluidic devices: fabrication, function, and application. *Biotechniques* 38:429–46

98. Martin RS, Root PD, Spence DM. 2006. Microfluidic technologies as platforms for performing quantitative cellular analyses in an in vitro environment. *Analyst* 131:1197–206

99. Yi C, Zhang Q, Li CW, Yang J, Zhao J, Yang M. 2006. Optical and electrochemical detection techniques for cell-based microfluidic systems. *Anal. Bioanal. Chem.* 384:1259–68

100. El-Ali J, Sorger PK, Jensen KF. 2006. Cells on chips. *Nature* 442:403–11

101. Dittrich PS, Manz A. 2006. Lab-on-a-chip: microfluidics in drug discovery. *Nat. Rev. Drug Discov.* 5:210–18

102. Sims CE, Allbritton NL. 2007. Analysis of single mammalian cells on-chip. *Lab Chip* 7:423–40

103. Taylor AM, Blurton-Jones M, Rhee SW, Cribbs DH, Cotman CW, Jeon NL. 2005. A microfluidic culture platform for CNS axonal injury, regeneration and transport. *Nat. Methods* 2:599–605

104. Amatore C, Arbault S, Chen Y, Crozatier C, Tapsoba I. 2007. Electrochemical detection in a microfluidic device of oxidative stress generated by macrophage cells. *Lab Chip* 7:233–38

105. Amatore C, Arbault S, Bouton C, Coffi K, Drapier J, et al. 2006. Monitoring in real time with a microelectrode the release of reactive oxygen and nitrogen species by a single macrophage stimulated by its membrane mechanical depolarization. *Chem. Bio. Chem.* 7:653–61

106. Wu C-C, Saito T, Yasukawa T, Shiku H, Abe H, et al. 2007. Microfluidic chip integrated with amperometric detector array for in situ estimating oxygen consumption characteristics of single bovine embryos. *Sensors Actuators B: Chem.* 125:680–87

107. Huang WH, Cheng W, Zhang Z, Pang DW, Wang ZL, et al. 2004. Transport, location, and quantal release monitoring of single cells on a microfluidic device. *Anal. Chem.* 76:483–88

108. Li MW, Spence DM, Martin RS. 2005. A microchip-based system for immobilizing PC 12 cells and amperometrically detecting catecholamines released after stimulation with calcium. *Electroanalysis* 17:1171–80

109. Cui HF, Ye JS, Chen Y, Chong SC, Sheu FS. 2006. Microelectrode array biochip: tool for in vitro drug screening based on the detection of a drug effect on dopamine release from PC12 cells. *Anal. Chem.* 78:6347–55

110. Spégel C, Heiskanen A, Acklid J, Wolff A, Taboryski R, et al. 2007. On-chip determination of dopamine exocytosis using mercaptopropionic acid modified microelectrodes. *Electroanalysis* 19:263–71

111. Sun X, Gillis KD. 2006. On-chip amperometric measurement of quantal catecholamine release using transparent indium tin oxide electrodes. *Anal. Chem.* 78:2521–25

Current Applications of Liquid Chromatography/ Mass Spectrometry in Pharmaceutical Discovery After a Decade of Innovation

Bradley L. Ackermann,[1] Michael J. Berna,[1] James A. Eckstein,[1] Lee W. Ott,[1] and Ajai K. Chaudhary[2]

[1]Drug Disposition, Eli Lilly and Company, Greenfield Laboratories, Greenfield, Indiana 46140; email: brad.ackermann@lilly.com

[2]Drug Disposition, Eli Lilly and Company, Lilly Corporate Center, Indianapolis, Indiana 46285

Annu. Rev. Anal. Chem. 2008. 1:357–396

First published online as a Review in Advance on March 4, 2008

The *Annual Review of Analytical Chemistry* is online at anchem.annualreviews.org

This article's doi: 10.1146/annurev.anchem.1.031207.112855

Key Words

proteomics, metabonomics, biomarker, medicinal chemistry, bioanalytical, biotransformation

Abstract

Current drug discovery involves a highly iterative process pertaining to three core disciplines: biology, chemistry, and drug disposition. For most pharmaceutical companies the path to a drug candidate comprises similar stages: target identification, biological screening, lead generation, lead optimization, and candidate selection. Over the past decade, the overall efficiency of drug discovery has been greatly improved by a single instrumental technique, liquid chromatography/mass spectrometry (LC/MS). Transformed by the commercial introduction of the atmospheric pressure ionization interface in the mid-1990s, LC/MS has expanded into almost every area of drug discovery. In many cases, drug discovery workflow has been changed owing to vastly improved efficiency. This review examines recent trends for these three core disciplines and presents seminal examples where LC/MS has altered the current approach to drug discovery.

1. INTRODUCTION

LC/MS: liquid chromatography/mass spectrometry

Drug disposition: the study of the fate of drugs in living systems; typically involves the combined use of pharmacokinetics and drug metabolism

LG: lead generation

Proteomics (PTX): global analysis of protein mixtures used in target validation and biomarker discovery

Metabonomics (MTX): comprehensive analysis of endogenous small molecules, or specific molecular classes, used in the study of pharmacology and toxicology (synonymous with the term metabolomics)

Much has been written about the role of mass spectrometry (MS) in the pharmaceutical industry (1, 2). At present, there is hardly a discipline or laboratory engaged in drug discovery that has not benefited in some way from MS technology. The "hyphenated" technique liquid chromatography with MS (LC/MS) represents the most widely used tool in the MS arsenal. In many instances, it has dramatically altered the workflow of modern drug discovery owing to improved efficiency or the ability to provide essential data at earlier stages in discovery with rapid turnaround.

This article reviews advances in the three core disciplines of drug discovery: biology, chemistry, and drug disposition. In addition to recent trends, it cites key events over the past decade to provide an understanding of the impact of LC/MS on drug discovery. Prior to the aforementioned review, we provide background information in two areas. The following section provides an overview of the stages of drug discovery. This is followed by a brief discussion of LC/MS instrumentation currently used in drug discovery.

Drug discovery proceeds through a series of fairly well-defined stages: target identification, biochemical screening, lead generation (LG) (hit-to-lead), lead optimization, and candidate selection. **Table 1** lists representative activities associated with each stage, which are categorized according to the three core disciplines.

Drug discovery begins with the selection of a target. Most targets are proteins (e.g., enzymes, receptors) and are chosen based on available knowledge associating a target with a certain disease state. Drugs are agents designed to selectively regulate a biological pathway or process believed to be associated with the disease state in question. The goal of drug discovery is to identify a safe chemical entity that can be predictably delivered to the drug target to allow the hypothesized association between a target and a disease to be tested in the clinic.

Drug target validation often uses biological tools such as gene knockout animals or RNA interference to implicate the role of the target in the disease state. Investigators analyze specific biochemicals and study their perturbation in specific pharmacology or disease models. LC/MS has played a vital role in the analysis of the biomolecular classes listed in **Table 1**. Of particular significance is the field of proteomics (PTX), in which MS is the primary tool used. Metabonomics (MTX), the small-molecule counterpart to PTX, has also benefited from advances in MS technology. Both PTX and MTX have been used throughout drug discovery to assist in the identification of biomarkers for efficacy and/or toxicity. Owing to the cost and complexity of these methods, their deployment is reserved for specific applications (i.e., those not found on the critical path of project flow schemes).

In the next stage of drug discovery, high-throughout screening (HTS) identifies core structural motifs that bind to the target. Increasingly, LC/MS is being used as an alternative to ligand-binding methods for HTS, although it is not likely to supplant existing methods [e.g. enzyme-linked immunosorbent assay (ELISA), scintillation proximity assay, etc.], that are less expensive and have higher throughput. LC/MS is typically applied in cases in which selectivity and/or expedited method development is important.

Table 1 Activities in the major stages of drug discovery supported by liquid chromatography/mass spectrometry

Stage	Target ID	Biochemical screening		Lead generation (hit-to-lead)			Lead optimization		
Function	**Biology**	**Biology**	**Chemistry**	**Biology**	**Chemistry**	**ADME**	**Biology**	**Chemistry**	**ADME**
Activities	PTX	HTS	Hit ID	In vitro pharmacology	Library QC	Physical properties	In vitro pharmacology	HTOS support	Physical properties
	MTX	Target validation	Library QC	In vivo model development	HTOS support	ADME screening	In vivo pharmacology	Med chem support	ADME screening
			HTOS support	BM discovery	COS	Exposure screening	BM discovery	Scale-up synthesis	In vitro assays
						PK	BM qualification	COS	Exposure screening
							PK/PD		PK
									TK

Abbreviations used: ADME, absorption, distribution, metabolism, and excretion; BM, biomarker; COS, confirmation of structure; HTOS, high-throughput organic synthesis; HTS, high-throughput screening; Med chem, medicinal chemistry; MTX, metabonomics; PD, pharmacodynamics; PK, pharmacokinetics; PTX, proteomics; QC, quality control; TK, toxicokinetics.

HTOS: high-throughput organic synthesis

Structure-activity relationship (SAR): a correlation established between molecular structure and activity toward the target of interest

LO: lead optimization

ADME: absorption, distribution, metabolism, and excretion

Biotransformation: the official term used to describe the metabolism of drugs or other xenobiotics

The involvement of chemistry begins during the biochemical screening stage because of the need to deliver compounds for HTS. LC/MS is heavily involved in the characterization of the chemical libraries generated from high-throughput organic synthesis (HTOS) or from other sources, including natural products and corporate compound repositories. As discussed below, LC/MS can support HTOS in several ways, including library quality control, hit identification, and compound purity assessment.

As hits are identified, an expanded effort known as LG interrogates the chemical space around the identified hits and the influence of structural changes on biological activity. This activity leads to a correlation referred to as a structure-activity relationship (SAR). LC/MS is used to support chemical synthesis and to purify and register the compounds tested. Typically, a fraction of each compound is stored in a corporate repository. During LG, biological models follow a progression from in vitro assays to cell culture screens and ultimately to live-phase animal models. Much effort is expended to understand these models prior to lead optimization (LO). Because activity in more complex biological models relies on the delivery of the drug to the target, investigators direct additional effort in LG at understanding the ADME (absorption, distribution, metabolism, and excretion) characteristics of the molecules in the SAR. LC/MS has been extensively applied to create ADME-related screens and to assess relevant physical properties. Assessing drug exposure in the in vivo pharmacology models also begins in LG because of the throughput established by LC/MS methods. Initial investigation of the biotransformation of the lead molecules also occurs in LG to identify metabolically labile positions on molecules. In addition, chemical scaffolds found to generate reactive (potentially toxic) metabolites are also removed from active consideration.

The final stage leading to a selected candidate is LO. In many ways LO is an expanded version of LG that follows a specific testing cascade appropriate for the project. Although the previously developed in vitro methods are still used to screen compounds, LO places more emphasis on live-phase models for efficacy, as well as establishing pharmacokinetic (PK), pharmacodynamic, and toxicokinetic information for the lead compounds. ADME screening still occurs in LO and is used to more fully understand the factors that control drug clearance. In addition, researchers often conduct more definitive experiments on the compounds that advance to understand their potential for drug-drug interactions. LC/MS is also used in LO to study the major metabolic routes for leads that have the potential to become drug candidates. In addition, interspecies metabolism comparisons occur to guide toxicology assessment and to understand the disposition in live-phase pharmacology models.

Chemical supply is profoundly important in LO. Although HTOS still occurs in the early stages of LO, it is quickly supplanted by traditional medicinal chemistry so that more precise synthesis can occur and because larger amounts of highly purified material are required. In later-stage LO, process chemistry is used to produce sufficient material to perform toxicokinetic and other compound-intensive studies.

Investigators give increased attention in LO to identifying biomarkers that can be used in the clinic. Therefore, extensive effort occurs in parallel during the advancement of specific compounds to qualify biomarkers that can predict useful biological outcomes. This work coupled with the combined effort cited from the three core disciplines leads to the selection of a candidate for clinical development.

2. LIQUID CHROMATOGRAPHY/MASS SPECTROMETRY INSTRUMENTATION

UHPLC: ultrahigh-pressure liquid chromatography

ESI: electrospray ionization

The integration of LC/MS into the pharmaceutical industry has benefited from continuous innovation in instrumentation. The section below provides an overview of the major forms of LC/MS instrumentation. Further information on this topic is available for interested readers (3).

2.1. Liquid Chromatography

A decade ago, pharmaceutical applications of LC/MS primarily employed reversed-phase (RP) columns with 5-μm particles. As documented by a recent survey comparing high-pressure LC trends in 1997 and 2007 (4), RP-LC is still the dominant technique; however, there is a trend toward smaller particle sizes, driven by the current interest in sub-2-μm particles and ultrahigh-pressure LC (UHPLC). The survey also indicated that today's chromatographer has access to more tools, including monolithic phases, hydrophilic interaction liquid chromatography (HILIC), and supercritical fluid chromatography (SFC). In addition, stationary phases have improved significantly over the past decade (5).

For qualitative applications of LC/MS involving the analysis of complex mixtures, resolution and peak capacity are critical attributes. PTX (6), MTX (7), and drug biotransformation (8) have all benefited from UHPLC through improved resolution, shortened analysis times, and improved electrospray ionization (ESI) sensitivity. For example, using pellicular phases and UHPLC, Wang et al. (6) observed a 50% increase in peak capacity compared to a conventional porous phase for PTX analysis. Similarly, newer 1.7-μm RP materials were combined with UHPLC to improve resolution, speed, and sensitivity for metabolic profiling in urine (7). Superficially porous stationary phases offer another way to improve peak capacity using normal operating pressures and are used primarily for protein separations (9).

HILIC, which is an aqueous-containing format for normal phase separation, provides an excellent complement to RP-LC for highly polar molecules. It is also operationally simpler than conventional normal-phase LC for MS applications. Because HILIC uses high organic mobile phases, improved LC/MS sensitivity is obtained owing to improved desolvation. HILIC has been successfully applied to qualitative applications, such as MTX (10), and has provided improved quantification of polar analytes (11).

Quantitative LC/MS applications focus on sensitivity and throughput. Hence, LC formats such as monolithic columns (12) and ballistic gradients with short columns (13) have been widely used for screening applications and drug bioanalysis. Gradient elution has become the default for high-throughput applications because it combines fast run times with the flexibility to rapidly accommodate high chemical diversity. Investigators have also extended the cited benefits of UHPLC to quantitative applications (14). In addition, multidimensional LC applications linked via column-switching methods have been extensively deployed throughout drug discovery for automation and improved efficiency. Column switching is commonly used with PTX to permit large injections onto capillary columns and to carry out on-line two-dimensional (2D) separations coupling cation exchange and RP-LC (15). As discussed in Section 5, column switching has been applied extensively for on-line sample cleanup in bioanalytical applications. Information on this topic is available in a current review on bioanalysis (16).

2.2. Mass Spectrometric Ionization Methods

A vast majority of all LC/MS analyses use two ionization methods: ESI and atmospheric pressure chemical ionization (APCI). We can trace the robustness of these methods to the common use of atmospheric pressure ionization (API), meaning that spray initiation and desolvation occur at atmospheric pressure, separated from the high-vacuum internal region of the mass spectrometer by a pinhole or capillary orifice. Under ESI, droplets containing a charge excess are produced by the application of high electric fields to a needle-like sprayer. The resultant droplets contain a net charge having the same polarity as the voltage placed on the needle. As the droplets shrink from evaporation, they become unstable owing to excess charge, and they dissociate into droplets of smaller radius (Rayleigh dissociation). For small molecules, including peptides, gas-phase analyte ions are ultimately produced by ion desorption from solution driven by the intense electric fields (ion evaporation). Alternatively, ion formation for large molecules (e.g. proteins) is believed to occur by a competing mechanism (charged residue model) in which gas phase ions are produced by complete desolvation of the charged droplets. The popularity of ESI is attributed to its sensitivity, the wide range of molecules ionized, and the ease of interfacing to LC.

APCI involves ionization of gas-phase analyte molecules produced by heated nebulization of the LC effluent. Chemical ionization, typically by proton transfer from mobile phase ions, occurs in a plasma created by corona discharge. Because of the requirement for analyte volatilization before ionization, APCI is limited to smaller molecules than ESI. Nonetheless, APCI provides an excellent complement to ESI and has the advantage of being less influenced by sample matrix (17). More recently, a variant of APCI called atmospheric pressure photoionization was introduced that induces ionization by a xenon arc lamp instead of a corona discharge. A review by Hsieh (18) discusses further details about atmospheric pressure photoionization and other LC/MS ionization methods.

2.3. Mass Analyzers

One can differentiate mass analyzers by several attributes: scan speed, duty cycle, mass resolution, mass range, and cost. Quadrupole mass filters are the most common and cost-effective mass analyzer. Because of their durability and low cost, they are used for several routine LC/MS applications in chemistry and drug disposition. In comparison with other mass analyzers, quadrupoles have limited mass resolution and are less sensitive for full-scan work because they do not store ions that are not being detected. A tandem quadrupole instrument, known as the triple quadrupole MS, is the instrument of choice for many bioanalytical applications (2, 16). This instrument consists of two mass-filtering quads coaligned at opposite ends of a center, non-mass-filtering quad serving as a collision cell. Extremely high selectivity is derived by monitoring a characteristic fragment transition for a target analyte using the first and third quadrupoles. This technique, referred to as selected reaction monitoring (SRM), makes the triple quadrupole MS a powerful option for trace analysis in complex biological matrices (3).

Quadrupole ion-trapping devices provide an alternative to conventional quads and offer improved full-scan sensitivity with the ability to perform tandem mass spectrometry (MS/MS). Because ionization, collision-induced dissociation, and mass analysis occur within the filter, these devices utilize staged pulse sequences and have the ability to store ions. This latter feature gives ion traps improved sensitivity over quadrupoles for full-scan experiments. The recent introduction of linear 2D ion traps has led to improved overall performance (19, 20). These instruments have found extensive use in biotransformation studies and are among the most widely used tools for PTX experiments.

Time of flight (TOF) is another popular mass analyzer that, as its name implies, sorts ions according to their arrival time to the detector after receiving a common kinetic energy. TOF offers the advantage of higher mass resolution, allowing for exact mass determination. In addition, TOF is capable of extremely high acquisition rates (50 μs per mass spectrum). A hybrid known as the Q-TOF combines a quadrupole mass filter for mass selection, an intermediate collision cell, and TOF mass analysis (21). Fast acquisition and the ability to perform exact mass determination on both precursor and product ions make this instrument extremely effective for structural elucidation in complex mixtures. As a result, it has been extensively deployed for PTX, MTX, and drug biotransformation.

Recently, much interest has been generated by high-end ion-trapping devices capable of extremely high mass resolution, to go along with high duty cycle. Although Fourier transform MS (FT-MS) based on ion cyclotron resonance was commercially introduced in the 1980s, the unparalleled resolving power of this instrument was not compatible with the timescale of LC. This changed with the recent introduction of a hybrid instrument that employs dedicated ion-cyclotron-resonance detection at the back end of a linear ion-trap mass spectrometer that handles the bulk of the LC/MS duties (22). More recently, Markarov and colleagues (23) introduced a variation of high-resolution ion trapping called the orbitrap. Because this instrument traps ions by electrostatic fields, it is less expensive than FT-MS, which requires a superconducting magnet. Although the orbitrap has slightly lower performance specifications than FT-MS, it is finding increased utility in application space shared by Q-TOF and FT-MS instruments.

A current trend in LC/MS instrumentation involves the incorporation of gas-phase ion mobility as an orthogonal means for resolving ions. Clemmer and coworkers (24) pioneered the use of ion mobility for PTX, demonstrating its use for enhanced separation of protein digests. Additional applications of ion mobility are on the rise

SRM: selected reaction monitoring (also known as multiple reaction monitoring)

Tandem mass spectrometry (MS/MS): the use of multiple, consecutive stages of mass analysis to derive structural information or to improve analytical selectivity

owing to the commercial introduction of two instrumental formats. High-field asymmetric waveform ion-mobility spectrometry uses ion mobility in the API ion source region to improve the signal-to-noise ratio of target analyte ions (25). In this method, ions drift through a field produced by a pair of electrodes to which a radio frequency voltage (alternating current) is applied. Application of a fixed supplemental voltage (direct current) selects for ions of a given mobility by allowing a stable trajectory through the electrodes. The second method, based on traveling-wave technology, sorts ions according to their mobility as they pass through a grid formed by a series of planar ring electrodes of alternating polarity (radio frequency). A voltage gradient (direct current) applied across the lens stack enables the ions to traverse through a series of potential wells created by the electrodes at a rate determined by their mobility. The incorporation of traveling-wave technology into the design of a Q-TOF mass spectrometer has been reported (26).

3. BIOLOGY

Of the three core disciplines, biology plays the most central role in the discovery of new pharmaceutical agents. Biology drives the initial stages of drug discovery through the identification and validation of drug targets (**Table 1**). LC/MS is used throughout discovery to assist biology, including the pivotal transition from in vitro to in vivo models. It is also used to support the increased effort now given to the discovery and qualification of biomarkers.

In contrast to chemistry and ADME applications (which analyze drugs or drug-related products), biological applications measure endogenous molecules. The applications discussed below are divided into two categories: profiling and targeted analysis. An example of the first category is the use of LC/MS for PTX and MTX analysis. These methods integrate large-scale profiling of endogenous molecules with bioinformatics to derive a more comprehensive understanding of the biology associated with the target and are applied later in drug discovery in the search for biomarkers.

The second category, targeted analysis, analyzes fewer molecules with higher throughput and is generally more quantitative. Targeted analysis by LC/MS is used to support in vitro screening, in vivo pharmacology, and ultimately biomarker qualification. Both categories of LC/MS for biological applications are addressed in the sections that follow.

3.1. Proteomics

PTX involves the large-scale analysis of protein mixtures usually by MS following multiple stages of off-line and/or on-line separation. Most PTX methods attempt a comprehensive analysis referred to as nonbiased. As discussed below, targeted approaches are also used and are becoming more common. To achieve greater sensitivity, investigators typically analyze proteins following enzymatic digestion to yield peptides. This approach is commonly referred to as bottom-up PTX because proteins are identified by the detection of one or more tryptic fragments using sophisticated algorithms to carry out searches of protein databases (**Figure 1**). Yates has pioneered

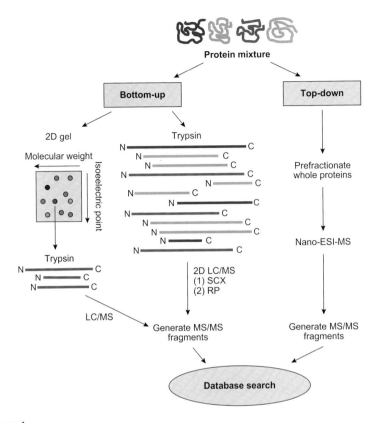

Figure 1

Schematic diagram comparing bottom-up and top-down proteomics. In bottom-up methods, protein identification occurs through the detection of predicted tryptic fragments. In contrast, top-down methods measure intact proteins. Although the top-down approach adds a useful complement to the more common bottom-up approach, it has comparatively lower sensitivity. Two bottom-up methods are compared in this figure. One method is based on two-dimensional gel electrophoresis, whereas the second method uses two-dimensional liquid chromatography (LC) for separation. The latter method, often referred to as shotgun proteomics, has become popular owing to its higher automation and throughput. Abbreviations: ESI, electrospray ionization; MS, mass spectrometry; RP, reversed phase; SCX, strong cation exchange.

many developments in the bottom-up approach, including a commonly used 2D LC approach called MudPIT (multidimensional protein identification technology), which fractionates complex peptide mixtures on-line using cation exchange prior to RP-LC. Yates and coworkers (27) recently published the latest version of this approach. Once sorted, the peptides are sprayed directly into the mass spectrometer (Q-TOF or quadrupole ion trap) to generate MS/MS spectra. These spectra are then searched against a database generated in silico using search algorithms (e.g., SEQUEST, MASCOT, X-Tandem). More recently, FT-MS and orbitrap mass analyzers have provided powerful interrogation of complex peptide mixtures under high mass resolution (22, 28).

We can trace the widespread use of bottom-up PTX directly to the automation provided by on-line LC/MS/MS, representing a great savings in labor relative to older methodologies involving 2D gel electrophoresis followed by MS analysis of excised spots. A disadvantage to the bottom-up approach is that information about post-translational modifications (PTMs) is lost. As a complement to the bottom-up approach, Kelleher (29) has advocated the use of top-down PTX (**Figure 1**), whereby intact proteins are analyzed directly by MS and MS/MS under high mass resolution following chromatographic fractionation. Although top-down methods permit PTM detection, sensitivity currently limits the use of this technique.

One chief concern about PTX is the ability to derive quantitative information. Reported approaches to quantification are divided into label and label-free methods. Label strategies draw on the well-established precedent of using stable isotope-labeled (SIL) molecules to improve MS precision. Label-free methods, conversely, also give acceptable precision and have the advantages of lower complexity and cost. Higgs and coworkers (22) demonstrated that high reproducibility (<10% relative standard deviation) can be obtained using label-free quantitation employing a linear ion trap and bottom-up methodology. More recently, Weiner et al. (30) reported a different label-free approach called differential MS. This method identifies ions that have differential intensity in samples compared by LC-FT/MS under high resolution. Ions that show statistically significant differences are further interrogated by MS/MS for sequence identification. Differential MS offers streamlined computation because only the statistically different products are targeted for identification.

Quantitative strategies incorporating SIL peptides are largely represented by three main approaches: stable isotope labeling by amino acids in cell culture (SILAC), isotope-coded affinity tags (ICAT), and isobaric tagging for relative and absolute quantitation (iTRAQ). The SILAC approach introduces isotope labels into proteins by growing mammalian cells in media containing an SIL version of an essential amino acid (31). Differential protein expression is quantified by comparing normal cells with cells grown using SILAC. Because of the m/z shift introduced by the SIL, the samples can be combined and analyzed in a single run. A limitation of SILAC is that it is restricted to tissue culture applications.

ICAT, introduced in 1999 by Gygi et al. (32), was the first application of chemical labeling for quantitative PTX. ICAT reagents insert stable isotopes into proteins through the derivatization of specific amino acids (e.g., Cys) in protein mixtures with a biotin-containing structure to facilitate affinity cleanup. Differential protein expression is observed by derivatizing samples with alternate versions of the ICAT reagent (with and without SIL). As with SILAC, the samples are combined prior to analysis. The use of ICAT with MS/MS allows sequence identification and accurate quantification of proteins in complex mixtures, and it has been applied to the analysis of global protein expression changes, protein changes in subcellular fractions, components of protein complexes, and protein secretion in body fluids.

iTRAQ is a newer labeling approach that allows the simultaneous comparison of up to eight samples. Under iTRAQ, the N-terminus and lysine residues are derivatized with a reagent that introduces a common mass shift to each free amino group.

Differential analysis occurs by the appearance of different product ions that range from m/z 114 to 121 according to the reagent used (33). Samples to be compared are each derivatized with a different iTRAQ reagent, allowing differential expression to be assessed by the relative intensities of the reporter ions.

One of the most important applications of PTX technology is the study of PTMs. Although several known PTMs exist, the most widely investigated are phosphorylation (34) and glycosylation (35). Phosphorylation is the more relevant PTM to drug discovery because of its importance to cell signaling and the high interest in kinase targets. Carr et al. (34) provide a recent thorough review of phosphoproteomics.

Because the ability to detect PTMs can be compromised by the liability of the phosphopeptide (or sugar-peptide) linkage under collision-induced dissociation, extensive interest has been generated by two related techniques that selectively fragment the peptide backbone leaving the PTM intact. Electron capture dissociation (36) and electron transfer dissociation (37) use electrons and negatively charged reagent ions, respectively, to induce c- and z-type cleavage. Although these techniques were originally introduced using FT-MS, capabilities are now being implemented on lower-cost mass spectrometers.

3.2 Metabonomics

In the simplest sense, one can think of MTX as the small-molecule counterpart to PTX. This fast-growing area has been the subject of previous reviews (38, 39) and has advantages over PTX owing to the comparative simplicity of the metabolome and the fact that metabolites are highly conserved between species. A limitation to MTX is that the changes observed, particularly in homeostatic fluids, are often displaced from the biological target or stimulus. Hence, researchers commonly employ sophisticated bioinformatic tools to map observed changes in metabolite flux to specific pathways.

The original applications of MTX in the pharmaceutical industry involved the use of nuclear magnetic resonance (39), although LC/MS, along with other MS techniques, has rapidly increased in popularity owing to the need to analyze complex mixtures and perform structural analysis. LC/MS offers several advantages, including sensitivity, selectivity, and dynamic range. Several LC formats, including HILIC (10) and UHPLC (7), in addition to RP-LC, can maximize the information obtained by MTX. Kind et al. (40) provide an excellent recent example. This group used each of these tools in conjunction with gas chromatography TOF-MS to detect signals of renal cell carcinoma in urine from affected patients.

Along with multiple LC formats, researchers have employed several mass analyzers for MTX, including quadrupole (41), quadrupole ion trap (42), Q-TOF (7), FT-MS (43), and orbitrap (44). Owing to the complexity of the mixtures analyzed, high-resolution mass analyzers are generally favored. The ability to identify unknown metabolites using MS/MS is also an important consideration.

MTX, similar to PTX, is divided into nonbiased and targeted methods. Nonbiased methods attempt to analyze as many molecules as possible in a single injection. Targeted methods, discussed below, profile a specific molecular class. Multivariate

NCE: new chemical entity

statistical tools, such as principal components analysis (PCA), are typically used to analyze the complex data sets acquired by nonbiased methods. Although such methods can differentiate experimental groups and identify the spectral features responsible for clustering, they do not directly result in identified metabolites. Hence molecular libraries, based on retention time and mass spectral information, are often used to facilitate metabolite identification.

Plumb and coworkers (45) reported the first pharmaceutical application of MS-based MTX, using negative-ion RP-LC-ESI/MS with TOF mass analysis to compare the urinary metabolome of rats following a single oral dose of one of three new chemical entities (NCEs). Using PCA, they showed clear differences between each drug and a vehicle control. Five unique m/z values were identified from the PCA loadings plot as strong contributors to the observed clustering, although specific metabolites were not assigned.

Another representative illustration of nonbiased MTX is the investigation of drug-induced phospholipidosis by the antidepressant citalopram (41). In this study, rats were given 125 mg kg^{-1} day^{-1} citalopram by oral gavage for 14 days. **Figure 2a** displays the PCA scores plot constructed from the LC-ESI/MS data acquired from

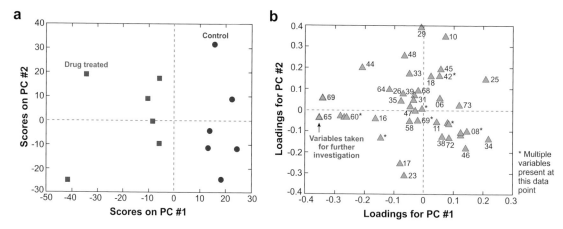

Figure 2

Principal component analysis (PCA) plots from a nonbiased metabonomics investigation of drug-induced phospholipidosis. In this study 12 male Wistar rats were dosed daily for 2 weeks by oral gavage with either tap water or citalopram (125 mg kg^{-1} day^{-1}). Urine collected prior to dosing (day –5) and on days 1, 3, 7, 10, and 14 was analyzed by single quadrupole liquid chromatography–electrospray ionization/mass spectrometry with alternating positive- and negative-ion detection. (*a*) The PCA scores plot obtained at day 7 clearly shows separation of the six drug-treated rats (*squares*) from the six control animals (*circles*) when the data are plotted along axes defined by principal components 1 and 2. (*b*) The loadings plot for PCA of day 7 displays the spectral features most responsible for the observed differentiation, as indicated by displacement from the origin. The two highlighted variables (65 and 69) were taken for further structural investigation using tandem mass spectrometry. Figure reproduced from Reference 42 with permission from the American Chemical Society.

urine collected at day 7. This plot clearly shows that the treated rats were shifted to the left of the control animals. The PCA loadings plot (**Figure 2b**) identifies spectral features having the largest contribution to the observed separation. Two variables were selected for further analysis with MS/MS; however, structural assignments for the metabolites were not provided.

Additional examples of nonbiased MTX have been reported for the study of toxicity (42, 46), including a novel approach by Gamache et al. (47) that provided structural insight for redox-active metabolites in rat urine using parallel electrochemical array detection on-line with LC-ESI-TOF/MS. Nonbiased MTX has also been used to study pharmacology models and disease states with examples involving cancer (48), clinical depression (49), and diabetes (50).

The most active area for targeted MTX is lipid profiling, often referred to as lipidomics. Generally, nonbiased approaches are not viable for lipidomics owing to the prevalence of isomers and the lack of universal analysis conditions for structurally divergent lipid classes. Information about the field of lipidomics has been summarized in recent reviews (51, 52).

LC/MS is one of the main tools used to profile lipids. One can find a representative pharmaceutical application in the work of Mortuza et al. (53) who profiled phospholipids in rat urine after exposure to the phospholipidosis-inducing compound amiodarone. Investigators have also used LC/MS to profile other lipid classes including steroids, eicosanoids, and fatty acids. Because of the limited resolution provided by RP-LC for lipid mixtures, UHPLC has been applied to improve peak capacity for lipid profiling (54). Ion mobility offers another option, as demonstrated by Kapron and coworkers (25), who used high-field asymmetric waveform ion-mobility spectrometry with RP-LC-ESI/MS to improve the sensitivity and selectivity of eicosanoid profiling. Lee et al. (55) introduced the most selective and sensitive LC/MS method for eicosanoid profiling with a novel methodology based on negative-ion APCI and normal-phase chiral LC. In addition to offering superior chromatographic resolution, low picogram-per-milliliter detection was obtained by enhancing electron capture through chemical derivatization.

3.3. Pharmacology Screening

MS is increasingly being incorporated into strategies for pharmacology screening. Despite the inherent throughput limitation compared to traditional ligand-binding methods, which are parallel in nature, LC/MS is being considered for select applications. Typically LC/MS warrants investigation for cases in which reagents are rate limiting (e.g., antibodies), fast method development is needed, and/or medium throughput is required. In addition, LC/MS offers a label-free method that accommodates several molecular classes and allows for the simultaneous detection of substrate and product. The selectivity of LC/MS is another important consideration and is vital when screening involves the detection of multiple chemical structures, a topic addressed in greater detail in Section 5.1.

LC/MS has been used in several applications of pharmacology screening, beginning with hit identification and extending through the support of cellular-based screens. A comprehensive review by Siegel (56) serves as an excellent resource on this subject. Additional information is available in a review by Geoghegan & Kelly (57).

A variety of LC/MS methods assess ligand binding by capture and release. Although direct information about ligand binding can be obtained by the analysis of gas-phase noncovalent interactions observed by ESI-MS (58), this topic is beyond the scope of this review. A common method for capture and release involves on-line affinity capture. After washing to remove weak binders, the affinity interaction is disrupted and captured ligands are measured by LC/MS. Numerous formats have embraced this general concept, including affinity chromatography, frontal chromatography, affinity ultrafiltration, pulsed ultrafiltration, and gel-permeation spin columns. Although MS can directly identify compounds that bind in such formats, the more common approach involves repetitive detection of a single molecular entity to maximize throughput.

Enzyme targets often lend themselves to LC/MS-based screening because it is possible to monitor the appearance of a small-molecule product in the presence of chemicals being screened as potential inhibitors. A representative example is the cellular assay reported by Xu and coworkers (59), in which they screened inhibitors of the metabolic syndrome target 11-β hydroxysteroid dehydrogenase, measuring the ratio of cortisol (product) to cortisone (substrate). In this screen, they achieved a throughput of 1 min per sample. Other examples in the literature have been produced using an LC/MS/MS screening technology introduced by a commercial vendor. In one such application, LC/MS/MS compared favorably with the scintillation proximity assay for the cancer target AKT1/PKB alpha (60).

One of the most novel applications of LC/MS for pharmacology is the assessment of receptor occupancy. Recently, Chernet and coworkers (61) demonstrated the ability to measure the receptor occupancy of neurochemical drugs by displacing a selective, nonlabeled tracer from the receptor of interest. Tracer measurements in brain tissue offer a viable alternative to radiolabeling and can assist in the development of positron-emission-tomography ligands for clinical use.

3.4. Biomarker Quantification

LC/MS/MS is frequently used to quantify endogenous molecules to assess their ability to serve as useful predictors of biological outcomes. We recently reviewed this subject, covering both large- and small-molecule applications (62). The use of quantitative LC/MS for small molecules pulls from a rich tradition of using MS to screen for inborn errors in metabolism (63). To date, investigators have used LC/MS to profile or quantify almost every conceivable small-molecule class, including amino acids, organic acids, sugars, nucleotides, neurotransmitters, and lipids. Unfortunately, the breadth of this topic exceeds the capacity of this review.

Because of the need to qualify protein biomarkers, intense interest has been focused on the application of LC/MS/MS for protein quantification. To facilitate analysis, a peptide surrogate generated by enzymatic cleavage is quantified using SRM detection. In addition, SIL peptide standards are used to improve assay precision and to provide an absolute rather than a relative measurement. Triple quadrupoles are commonly used because they provide the best balance of selectivity and sensitivity; however, quadrupole ion traps (64) and FT-MS instruments (65) have also been used owing to their prevalence in PTX laboratories. Although applications of peptide quantification using stable isotope dilution data date back to the late 1980s, the resurgence has been traced to a 2003 paper by Gerber et al. (66), in which the authors used SIL peptides to quantify target proteins with a method referred to as AQUA, standing for absolute quantification. They used LC/MS/MS with SRM detection to quantify the native and phosphorylated forms of a tryptic fragment, derived from a protein isolated from cell lysate by sodium dodecyl sulfate polyacrylamide gel electrophoresis. An overview of AQUA appears in **Figure 3**.

A common application of AQUA is the interrogation of biomarker leads identified using nonbiased PTX. Because of the number of proteins involved, methods with multiple SRM transitions are employed, hence the frequently used term multiple reaction monitoring. In one example, Anderson & Hunter (67) used 137 multiple reaction monitoring channels to monitor 53 medium-to-high-abundance proteins in human plasma. In this example, they prepared a concatenated protein made from all the individual SIL peptide standards using recombinant methods.

One fundamental limitation of nonbiased PTX is that the identified proteins are of extremely high abundance (microgram per milliliter). Although examples of AQUA-type analysis of serological biomarkers have been published (68, 69), abundant proteins are unlikely candidates for disease biomarkers (70). In a recent review, Rifai et al. (70) covered this issue and other challenges involved in using MS to discover, qualify, and ultimately verify protein biomarkers. One solution to address the need to access lower-abundance proteins is to incorporate immunoaffinity isolation. Anderson and colleagues were the first to apply immunoaffinity enrichment by a technique named SISCAPA (stable isotope standards and capture by antipeptide antibodies). They successfully applied this method, which allowed for an enrichment of greater than 100-fold, to lower-abundance proteins in plasma (71). We recently published a complementary strategy that uses antiprotein antibodies or antibodies against a target peptide (72). This strategy is outlined in the flow scheme displayed in **Figure 4**. By simple analogy, one can think of this scheme as a sandwich ELISA in which MS replaces the detection antibody. We have used this methodology to expedite the qualification of protein biomarkers with lower quantification limits in the range of 100 pg ml^{-1} (73–75). Whiteaker et al. (76) recently published a similar approach, using immunoprecipitation on magnetic beads to assay for TNF-α.

4. CHEMISTRY

In many ways, one can think of medicinal chemistry as the engine that drives small-molecule drug discovery. Although the goal of pharmaceutical discovery is indeed biological in nature, the flow of drug discovery is built around the need to explore diverse chemical space coupled with the need to synthesize increasingly larger amounts of material, with requisite purity, to allow stage-specific experiments to be conducted enroute to a drug candidate. Over the past decade, MS has greatly impacted all facets of chemical synthesis involved in drug discovery. This section reviews key

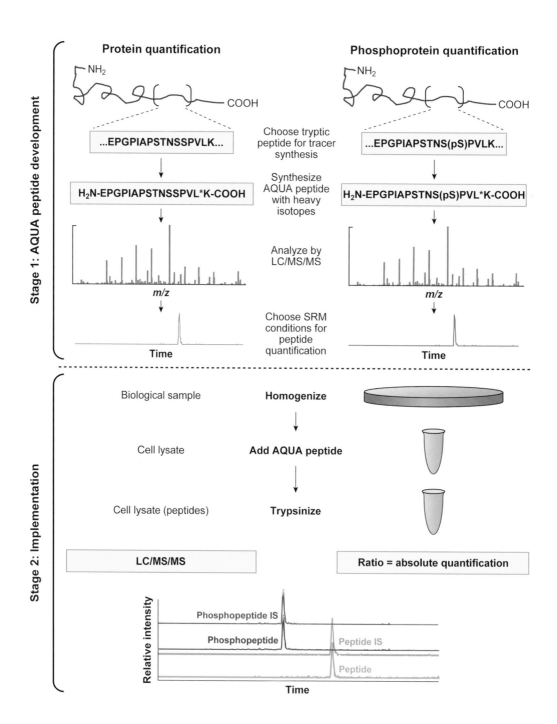

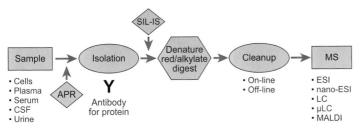

Figure 4

Flow scheme for liquid chromatography/tandem mass spectrometry (LC/MS/MS)-based protein quantification using immunoaffinity capture via an antiprotein antibody. As with the AQUA method, quantification occurs using a peptide surrogate with a stable isotope-labeled peptide internal standard. This strategy allows for expedited qualification of low-abundance protein biomarkers due to a 1000-fold enrichment that occurs through the use of immunoaffinity isolation. This strategy also is useful in guiding the development of sandwich ELISA methods and verifying their selectivity. Figure adapted from Reference 78. Abbreviations: APR, abundant protein removal; CSF, cerebrospinal fluid; ESI, electrospray ionization; MALDI, matrix-assisted laser desorption/ionization; SIL-IS, stable isotope-labeled internal standard.

contributions, organized according to the primary ways LC/MS can support organic synthesis.

For effective small-molecule drug discovery, it is essential to have access to large numbers of structurally diverse compounds for both the generation and optimization of leads. Pharmaceutical companies rely on two primary sources for molecular libraries: corporate compound repositories and HTOS. Although natural products have long been a source for molecular diversity (77), we do not cover this topic in this review.

All major pharmaceutical companies maintain corporate collections of registered compounds and use these collections to screen new targets for hits. Because corporate compound collections are stored in dimethyl sulfoxide (DMSO) over extended periods, significant effort has gone into understanding the storage stability of

Figure 3

Absolute quantification of proteins and phosphoproteins using the AQUA strategy. In this strategy protein quantification occurs through the analysis of surrogate peptides produced from enzymatic digestion. Selective analysis occurs via liquid chromatography/tandem mass spectrometry (LC/MS/MS) in the selected reaction monitoring (SRM) mode, typically performed using triple quadrupole MS. The AQUA strategy is illustrated for the quantification of a native protein and its phosphorylated counterpart. In the first stage of the strategy, surrogate peptides are identified, standards are synthesized, and LC/MS/MS conditions are obtained. In addition, stable isotope-labeled forms of the peptides are also prepared for use as internal standards. The second stage of the process involves the actual isolation of the target protein followed by digestion to liberate the peptides of interest. Quantitative estimates of protein concentration are obtained from peak area ratios of the target peptides to their corresponding internal standards spiked during the digestion step. Figure reproduced from Reference 71 with permission from the National Academy of Sciences.

compounds in corporate compound collections (78, 79). For more detail on this and other related topics, we refer the reader to a review by Cheng & Hochlowski (80).

HTOS is used both to generate libraries for initial screening and to explore chemical space around distinct chemical scaffolds identified from HTS. HTOS consists of two basic formats: combinatorial chemistry and parallel synthesis, both using solid-phase and solution techniques. Combinatorial chemistry, popular in the mid-1990s, has largely given way to the more focused parallel synthesis. Combinatorial chemistry uses variations on a split-couple-recombine strategy to allow almost unlimited permutations of known building blocks to be coupled (81). Although such approaches can yield unparalleled diversity, combinatorial chemistry suffers from a number of logistical problems, including the need to identify hits to obtain structure [it often requires a chemical decoding strategy (82)] and difficulty in controlling reaction yield and purity. Despite these drawbacks, the fundamental limitation of combinatorial chemistry is that it does not yield sufficient material to permit follow-up on hits, making resynthesis necessary. Today, most HTOS involves parallel synthesis whereby discrete chemical reactions occur simultaneously in an array format. Parallel synthesis has become the default approach because it allows for more controlled synthesis and generates larger quantities for testing. As leads progress through discovery, HTOS gives way to traditional organic synthesis to enable targeted synthesis and the production of larger quantities. More in-depth information about the role of MS in HTOS can be obtained from several reviews (80, 83, 84).

4.1. Strategies for Increased Throughput

The ruggedness of the API interface combined with the encompassing scope of ESI transformed organic synthesis in many ways. The first significant impact occurred in the form of open-access (OA) analysis. Beginning in the mid-1990s, chemists started analyzing their own samples, either using flow injection analysis (FIA) with ESI (85) or APCI (86). LC/MS was soon added to the repertoire (87), and most organizations today still have a blend of FIA and LC/MS for their OA needs. High-resolution mass analysis has also been offered for OA analysis (88). Current trends in OA along with other areas pharmaceutical synthesis support can be found in a recent survey (89).

With the uptake of HTS fueled by HTOS methods, the serial nature of MS made analytical characterization rate limiting in the generation of leads. Over the past decade, significant innovation has occurred to rectify this problem. As one might expect, the highest attainable throughput is obtained by FIA. Rapid methods for FIA characterization of compound libraries were introduced using modified autosamplers and multiport injectors. The best example is work by Morand et al. (90), who used a modified eight-port Gilson 215 autosampler to permit a throughput of <4 s per sample, allowing an entire 96-well plate to be analyzed in approximately 5 min. Although FIA is often used to analyze LC fractions or the products of a discrete chemical reaction, Yates and coworkers (91) demonstrated that FIA can also be used for expedient quality control of entire compound libraries.

Despite the speed of FIA, LC/MS is still the workhorse for synthesis support because it can provide a more detailed assessment of complex mixtures, particularly when combined with other forms of detection. Because of the need for generic methods that accommodate wide structural diversity, gradient elution has become the default approach. Several methods can increase the throughput of LC/MS. For rapid LC/MS analysis, most laboratories use short columns (2–3 cm), high flow rates (2–5 ml min^{-1}), small particles (<4 μm), and generic gradients (10%–90% acetonitrile) to achieve full analysis in 3–5 min per sample. As early as 1997, Weller et al. (92) demonstrated the ability to perform 300 LC/MS analyses per day on a single instrument. More recently, Kyranos and coworkers (93) demonstrated the ability to do full-gradient characterization in 1 min using LC/MS interfaced to ultraviolet (UV) and evaporative light scattering detection (ELSD) for the rapid characterization of high-throughput parallel synthesis. A discussion of the theory and application of fast gradient elution LC/MS has been published (94).

SFC offers a viable alternative to LC/MS and is capable of even higher throughput (approximately threefold) owing to lower viscosity (reduced back pressure). The most frequently used mobile phase composition incorporates carbon dioxide and methanol, which exhibits behavior similar to normal-phase LC. Wang et al. (95) first reported the use of packed SFC/MS for HTOS. Since this time, there have been a number of improvements, particularly with regard to mobile phase additives to permit the analysis of more polar compounds. Recently, Pinkston and coworkers (96) performed a detailed comparison of 2266 compounds using LC/MS and SFC/MS and found both methods to be very comparable with respect to compound coverage. These results were consistent with a previous comparison by Searle et al. (97).

To achieve the next level in throughput, investigators needed to find ways to interface parallel LC methodologies to MS. Given the relative expense of a mass spectrometer, multiplexed methods were introduced to enable the effluent from multiple LC columns to be sequentially sampled by a single mass spectrometer. An approach taken by several laboratories is to employ multi-injector autosamplers, permitting the operation of as many as eight columns in parallel. This mode uses a single LC system to provide a common gradient that is split prior to the injectors. The effluent from each column is sent to a single mass spectrometer in which rapid sampling occurs in succession via a multiport switching valve. Because this approach introduces significant intersample carryover, two groups simultaneously introduced indexed ESI ion sources that use a dedicated ESI sprayer for each column (98, 99). In these designs, discrete sampling of individual effluent streams occurs by using a notched rotating plate assembly that at any instant occludes all but a single sprayer. Micromass introduced a commercial form of this design known as MUX^{TM}. **Figure 5a** displays a schematic representation of the MUX^{TM} interface. **Figure 5b** shows a diagram of the four-column parallel LC/MS system with MUX^{TM} detection used by Xu et al. (100) for compound analysis and purification. This system also incorporated multiple UV and ELSD detectors for purity estimation and automated fraction collection (we review these topics below).

a

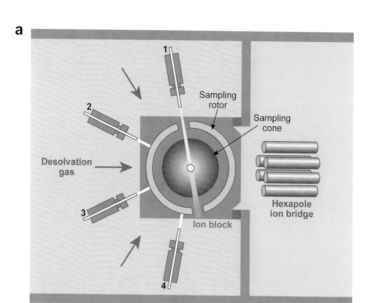

1

2

Sampling
rotor

Sampling
cone

Desolvation
gas

3

**Hexapole
ion bridge**

Ion block

4

b

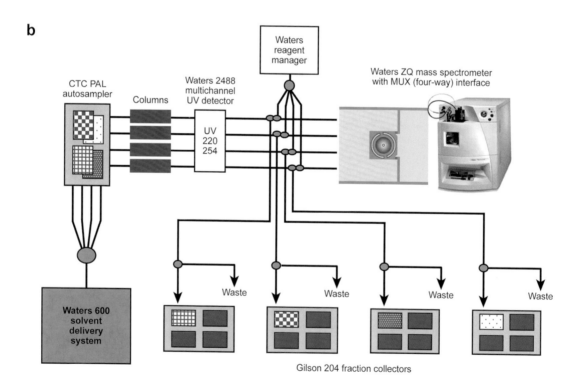

Waters
reagent
manager

CTC PAL
autosampler

Columns

Waters 2488
multichannel
UV detector

Waters ZQ mass spectrometer
with MUX (four-way) interface

UV
220
254

Waste

Waste

Waste

Waste

Waters 600
solvent
delivery
system

Gilson 204 fraction collectors

Although MUX[TM] can provide up to an eightfold increase in throughput, drawbacks include slight intersprayer cross-talk and the intersprayer dwell time (50 ms), which makes the system incompatible with ultrafast LC methods. Some practitioners have employed multiple LC systems to avoid the deleterious effect from having a single column plug.

4.2. Purity Assessment via Hyphenated Techniques

The issue of compound purity is paramount to successful LG and optimization. During LG, a detailed assessment of the influence of chemical structure on biological activity occurs, known as an SAR. From this perspective, it is easy to see how an SAR could be confounded by the introduction of impure compounds. This issue becomes even more important in LO in which a higher utilization of animal models occurs, including preliminary toxicological assessment. As a result, most pharmaceutical companies have adopted guidelines to govern the level of compound purity required as a function of stage in drug discovery. However, because of the nonuniform nature of MS response for different compounds, other forms of detection are required to provide purity assessment. Several detectors have been used on-line with LC/MS for this purpose, including UV (83), ELSD (101), and chemiluminescence nitrogen detection (CLND) (102). Yurek et al. (103) provide a powerful illustration of the utilization of alternative detectors by coupling LC-ESI-MS with TOF detection with UV (diode array detection), ELSD, and CLND. The authors used this combination of detectors to simultaneously provide compound identification, purity assessment, and an estimate of abundance for compounds prepared by parallel synthesis. In addition, they used the exact mass capability of the system with information obtained from UV and CLND to determine the structure of side products.

In addition to purity, solubility is another important factor that influences screening results derived from compounds synthesized by HTOS. Issues can arise both from insolubility in DMSO upon freeze-thaw cycles and from incomplete dissolution during dilution for HTS. To address this concern, one group routinely uses UV and CLND data, acquired as part of an overall LC/MS-based strategy for library analysis and purification, to provide direct estimates of compound abundance presented for HTS (104).

←──

Figure 5

(*a*) Schematic diagram of the MUX[TM] electrospray ionization (ESI) source (Waters Corp., Milford, MA) showing four ESI sprayers positioned around an indexed sample rotor. At any given instant, the rotor occludes all but a single sprayer, allowing for sequential sampling of each effluent stream. Figure 5*a* used with permission from Waters Corp. (*b*) Diagram of the mass-triggered purification system designed by Xu and coworkers, incorporating the MUX[TM] ESI source. Figure 5*b* reproduced from Reference 107 with permission from the American Chemical Society.

4.3. Compound Purification

As molecules progress through the testing cascades used in drug discovery, increasing amounts of material are needed for testing. In the early stages of LG, semipreparative columns [10-mm internal diameter (i.d.)] are used to purify low-milligram quantities. Eventually, during LO, preparative LC (50-mm i.d.) is used to purify hundreds of milligrams to gram quantities to fuel compound-intensive in vivo experiments. Over the years, sophisticated systems have been built and are now commercially available for automated purification (105). Weller and coworkers (92) are credited with the first published system that used UV triggering to perform automated fraction collection to purify products from parallel synthesis. Unfortunately, systems using UV triggering collect an inordinate number of fractions that must be analyzed by FIA-MS. Later, Kibbey (106) interfaced LC/MS in parallel with UV-guided fraction collection to avoid this problem and also introduced the idea of using an analytical LC/MS reconnaissance run prior to prep-LC to optimize collection. Cheng & Hochlowski (80) published a variation of this theme, using a single LC/MS equipped with ELSD and UV to perform the reconnaissance to feed four prep-LC systems using either UV or ELSD triggering. The system described had the capacity to purify 200 compounds per day.

In subsequent years, researchers introduced several methods based on MS-triggered fraction collection. Zeng et al. (107) described the first such system, referred to as parallel analyt/prepLC/MS. This system used a valve to switch between dual parallel analytical columns with UV detection and dual parallel preparative columns connected to fraction collectors. In either mode of operation, the effluent streams were coupled to a dual-sprayer ESI interface for MS detection. Initial analysis by analytical LC/MS was used for structural confirmation with on-line UV detection (220 and 254 nm) to estimate purity. Compounds less than 85% pure were scheduled for preparative LC and MS-triggered fraction collection.

Increasingly, several laboratories are using SFC for mass-guided fraction collection (97, 108, 109). In addition to improved separation speed, SFC significantly reduces mobile phase cost and expedites solvent stripping from collected fractions because of the use of carbon dioxide. SFC has also become the mainstay for chiral purification because most chiral columns operate under normal phase conditions (110). Ultimately, because there are advantages to both LC and SFC, many laboratories utilize a blend of these techniques. For example, Ventura and colleagues (109) reported a highly automated strategy for optimal use of the two methods.

In an effort to maximize throughput for compound purification, Xu and coworkers (100) applied parallel LC with MUX™ detection (**Figure 5b**). In this example, they used mass triggering with semipreparative LC to purify up to 10 mg of material. Interestingly, although this application used mass triggering, there is still no consensus on the preferred mode for triggering fraction collection. Mass triggering has been suggested as being preferred for early discovery applications to minimize the number of collected fractions given the need to purify larger numbers of compounds (83). Regardless of the method used, the current utilization of LC/MS for compound purification represents a blend of walk-up use by organic chemists and dedicated core laboratories.

5. DRUG DISPOSITION

According to Kola & Landis's (111) 2004 report on pharmaceutical attrition rates, clinical attrition due to poor PK/bioavailability dropped from just over 40% to under 10% in the period from 1991 to 2000. The reason for this dramatic decline is straightforward: ADME properties are investigated far earlier in the drug-discovery process. In the past, ADME properties were not fully investigated until after candidate selection largely owing to an analytical bottleneck. More than any other single factor, LC/MS is responsible for this profound improvement. Using LC/MS, routine methods can be rapidly developed and executed for diverse structural sets within the cycle time mandated by LG and LO. This section is organized according to quantitative and qualitative applications, the latter represented by drug metabolite identification and profiling.

5.1. Quantitative In Vitro Applications

Because ADME applications trace the fate of NCEs in biological systems, tools are needed that can rapidly dial in analytical conditions specific for each compound studied. This situation is further complicated by the frequent need for low-level detection in a variety of biological matrices. LC/MS relieves this bottleneck in two ways. First, the implementation of fast gradient elution methods with on-line sample cleanup by one of several column-switching formats (16, 112) provides a means for rapid method development along with high injection-to-injection throughput. Secondly, the versatility of ESI combined with the analytical selectivity of MS/MS delivers nearly universal detection when judged by historical standards. Amazingly, in what has become a landmark paper, Janiszewski and coworkers (113) demonstrated the ability to analyze 2000 samples per day per mass spectrometer for routine hepatic metabolic stability assessment. This same group also developed methods for automated instrument tuning to optimize the conditions for each NCE tested, a process that eventually became rate limiting (114). Since this time, a variety of instrumental variations have been introduced to improve throughput dealing with methods for sample preparation, injection, chromatography, and detection. Two reviews on ADME screening contain accounts of these advances (115, 116).

The current mandate across the pharmaceutical industry is to factor drug-like properties into the development of chemical leads. Medicinal chemists have become increasingly educated about the need to incorporate favorable ADME characteristics into molecules (117) and routinely utilize a variety of medium-throughput ADME screens in the process of refining an SAR. Although such screens may be run at any time during drug discovery, the limiting factor with respect to starting early is the need for low-milligram quantities of purified compounds. To address this constraint and to reduce costs, researchers have found increased use for in silico models built from empirical data sets in guiding early SAR development (118).

Typically, ADME screens are run by highly automated centralized laboratories. A number of integrated strategies have appeared in the literature (119–121) using a variety of mass analyzers, including single (119) and triple quadrupole (120) and quadrupole ion traps (121). Because the bioavailability of small molecules largely depends on absorption and hepatic metabolism, most primary screens are aimed at understanding these two attributes. Tools used to examine absorption range from an assessment of physical properties [e.g., solubility, log D (122)] to transport studies across intestinal cell monolayers (i.e., Caco-2). The technique known as PAMPA (parallel artificial membrane permeability assay) has become popular because it provides an estimate of passive diffusion across cell membranes without incurring the overhead of tissue culture (123).

Active transport of drugs across membranes (e.g., intestinal, hepatocyte, blood-brain barrier) often affects drug absorption, metabolism, and distribution and has become an active area of research (124). Although several important transporters have been identified, the efflux protein p-glycoprotein is the most commonly studied. Smalley and colleagues (125) developed an LC/MS/MS assay for p-glycoprotein inhibition by measuring the influence of NCEs on the bidirectional transport of the known p-glycoprotein substrate digoxin in Caco-2 cells.

The most widely used ADME screen is metabolic stability, which measures the extent or rate of disappearance for NCEs incubated in hepatic media (e.g., microsomes, hepatocytes). Since the introduction of LC/MS-based metabolic stability screening in the late 1990s (126), several variations and refinements have been reported (119, 121). An important secondary screen, useful for assessing the potential for drug-drug interactions, is cytochrome P450 (CYP) inhibition. During these studies, NCEs are coincubated with chemical probes known to be selectively metabolized by a single CYP isoform. Inhibition is indicated by the decreased production of a specific metabolite for the probe in the presence of the NCE. Dierks et al. (127) demonstrated that increased throughput can be obtained by simultaneous incubation of seven probe substrates, each selective for a different CYP isoform. They monitored metabolite formation for each probe in a single LC/MS/MS run. More recently, Kerns and colleagues (128) reviewed common methods used for CYP inhibition and advocated the use of a double cocktail system.

Distribution refers to a drug's ability to spread beyond the vasculature into various tissues and organs. Because distribution often depends on protein binding, several methods in discovery can estimate the binding of NCEs to plasma proteins. The most popular screening methods use either ultrafiltration or equilibrium dialysis with detection by LC/MS. An interesting alternative is to estimate protein binding from retention on columns prepared using immobilized human serum albumin. Cheng et al. (129) described this approach, including favorable comparisons with the other methods mentioned above. Estimating penetration across the blood-brain barrier is another important issue related to distribution for which LC/MS has had profound impact. We refer readers interested in this topic to a comprehensive review of in vitro and in vivo models to assess brain penetration (130).

5.2. Quantitative In Vivo Applications

The monitoring of drugs in biological fluids and tissues is important during both preclinical and clinical development and is typically referred to as bioanalysis. The selectivity and sensitivity of LC/MS/MS have significantly transformed the practice of bioanalysis by allowing methods to be quickly developed and executed in a time frame consistent with drug discovery. By far the most common fluid analyzed is plasma as it is the accepted surrogate for modeling drug exposure. Today, exposure determination starts in LG and continues through candidate selection. This topic has been the subject of prior reviews (16, 112).

As mentioned above, bioanalysis has benefited from several advances in LC, including monolithic columns (12), UHPLC (14), and HILIC (11). Moreover, these methodologies are used in conjunction with fast gradient elution methods to achieve run times often under 1 min (13). Whereas LC-ESI-MS/MS using triple quadrupole MS is the default tool for bioanalysis, APCI and atmospheric pressure photoionization are also commonly used and are less prone to matrix effects (17, 18, 131). Various instrumental techniques have also been applied to increase the selectivity of LC/MS/MS, including a high mass resolution triple quadrupole (132) and high-field asymmetric waveform ion-mobility spectrometry (133).

Investigators usually extract analytes from a biological matrix using traditional off-line techniques such as protein precipitation (PP), liquid-liquid extraction, or solid-phase extraction. PP using organic solvents is the most common approach in drug discovery because it is the least expensive and the most universal. Although PP yields dirtier samples than the other methods, low nanogram-per-milliliter levels can be achieved routinely for most drugs.

With shortened LC/MS/MS run times, sample preparation has become more of a bottleneck for high-throughput bioanalysis. Most laboratories employ semiautomated approaches to sample preparation using robotic liquid-handling systems in conjunction with a 96-well plate format to increase the speed of bioanalysis. Automation using on-line extraction techniques is also widely applied. A common theme to all on-line methods is the application of column switching to couple the extraction and analysis steps. Column switching can provide additional cleanup for samples prepared by PP and can be viewed as on-line solid-phase extraction. Indeed, commercial versions of on-line solid-phase extraction have been available for several years (134). Other on-line methods permit direct plasma injection, allowing sample preparation to be even further minimized. The most common methods are turbulent flow chromatography (135) and the use of restricted access media (136). Recently, monolithic columns have been shown to be another choice for direct plasma injection (137).

One inefficiency of LC/MS is that actual MS detection occupies only a small fraction of the total LC duty cycle. Numerous instrumental variations take advantage of this dead time. One of the simplest variations is to use parallel extraction columns with a single analytical column in alternate-regenerate mode (**Figure 6**). Other variations include dual extraction and two analytical columns (138), four columns with staggered injections (139, 140), and the four-channel MUX™ system introduced above (**Figure 5a**) (141).

Noninstrumental strategies have also increased bioanalytical capacity, including cassette dosing (142) and sample pooling (143). [We refer interested readers to White & Manitpisitkul's (144) discussion of PK considerations for cassette dosing.]

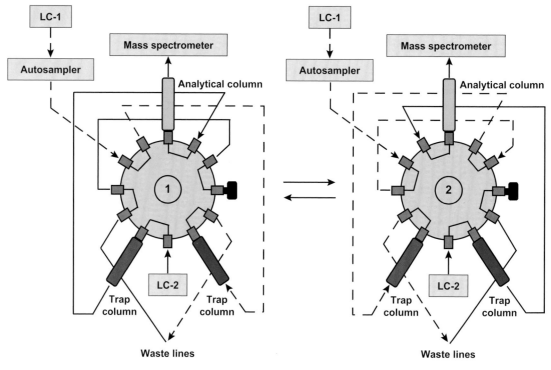

Figure 6

Schematic illustration of a liquid chromatography/mass spectrometry (LC/MS) system configured to perform alternate-regenerate column switching using a 12-port valve. This system uses two LC systems. LC-1 is connected to the autosampler and is responsible for sample loading (*dotted line*). The second system (LC-2) performs gradient elution for sample analysis (*solid line*). In valve position 1, the blue trap column is being loaded and washed, while the red trap column is back-flushed and its contents sent to the analytical column (*yellow*) for gradient elution LC/MS/MS analysis. As the valve switches to position 2, the red column is loaded from LC-1, while the blue column is eluted for analysis. When short columns and fast gradient elution are used with the configuration shown, an injection-to-injection duty cycle of less than 1 min is easily obtained. Alternate-regenerate column switching has been employed with several column formats for on-line sample cleanup and offers a simple means of doubling LC/MS throughput.

One key limitation of quantitative LC/MS is the issue of matrix effects caused by the competition for ionization that occurs when coeluting molecules are simultaneously ionized. Beginning with the publication of landmark papers in the late 1990s that identified and characterized the issue of ionization suppression (145, 146), extensive research has occurred and is the subject of a recent review by Mei (147). To date, several sources for ion suppression have been identified, including plasma phospholipids, sample containers, anticoagulants, and the dosing vehicle (147).

Using current-generation instruments, quantification in the low picogram-per-milliliter range in plasma is not uncommon. Owing to this exquisite sensitivity, investigators have now demonstrated the use of LC/MS/MS to support PK assessment for microdosing studies (148). A microdose has been defined as 1/100th of the proposed pharmacological dose or less than 100 μg, whichever is lower (148). Microdosing has been proposed as a means for accelerated clinical assessment of PK behavior. Another study achieved a lower limit of quantitation of 0.1 nM for fluconazole, tolbutamide, and an investigational drug (MLNX) using 100-μL plasma (149). This allowed PK parameters to be calculated in rats for doses as low as 0.001 mg kg^{-1} using current-generation LC/MS/MS instrumentation.

5.3. Drug Metabolite Profiling and Identification

A number of comprehensive reviews dealing with the use of MS to study drug metabolism have appeared in the literature (150–152). LC/MS is the primary analytical tool used to study drug biotransformation, which is frequently categorized into phase I– and phase II–type reactions. In phase I, metabolizing enzymes modify the NCE through hydroxylation, epoxidation, dealkylation, deamination, N- or S-oxidation, reduction, and hydrolysis. In phase II, polar groups are added to the parent drug itself or to one of the phase I metabolites (152). Some examples of phase II reactions include glucuronidation, sulfation, methylation, acetylation, and conjugation with glutathione (GSH) or amino acids such as glycine, glutamic acid, and taurine (152). These reactions result in either the addition or subtraction of known mass increments that can be easily monitored by MS.

A common approach to study drug metabolism involves the use of a radiolabeled drug (e.g., ^{14}C, ^{3}H) to facilitate metabolite detection in biomatrices using LC/MS coupled with radioactive detection. Because radiolabeled drugs are typically not available during discovery, LC/MS/MS techniques coupled with additional forms of detection (e.g., UV) are used to study the metabolic fate of NCEs in drug discovery. In LG and early LO, metabolic hot spots are identified to guide the SAR toward more metabolically stable molecules. This occurs by incubating NCEs with in vitro metabolizing systems (e.g., recombinant enzymes, liver microsomes, hepatocytes, liver slices, or liver S9 fractions) followed by metabolite profiling and identification by LC/MS/MS. MS detection is frequently carried out in both the positive and negative mode to cover the range of basic and acidic metabolites. Relevant metabolite peaks identified by full-scan MS are subsequently interrogated by MS/MS experiments (often conducted via automated data-dependent analysis) to obtain structural information. Common MS/MS scans include product ion, MS^n product ion (quadrupole ion trap), precursor ion, and constant neutral loss scans (triple quadrupole). A representative example is the use of neutral loss scanning to detect glucuronide conjugates, in which a loss of 176 mass units is monitored.

A prerequisite for successful LC/MS/MS metabolite identification is good chromatographic separation. Although significant advances have occurred in LC modes and stationary phases, the separation of isobaric metabolites, such as different hydroxylated forms, can require long run times. UHPLC has enabled these analyses to be carried out at higher speed and with better sensitivity (8).

One limitation of LC/MS for metabolite profiling is that different ESI response factors are observed for metabolites, making LC/MS at best a semiquantitative

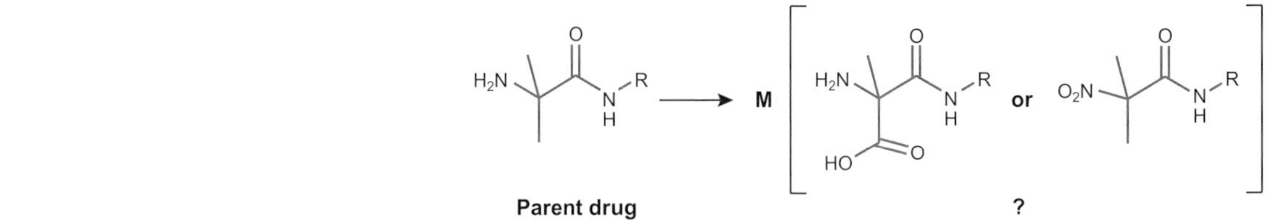

Figure 7

Isobaric chemical structures showing two possibilities for metabolite M, a biotransformation product observed in a recent metabolism study. These two structures that differ by 0.0125 mass units could not be distinguished using nominal mass analyzers such as a quadrupole or an ion trap. Accurate mass–product ion data acquired using a Q-TOF mass analyzer confirmed that the nitro metabolite was the correct structure.

tool. This problem is further influenced by gradient elution because the desolvation efficiency increases during the run due to increased organic solvent content. In 2005, Hop et al. (153) demonstrated that ESI response factors could be normalized using nanoelectrospray (ESI at 200–1000 nL min^{-1} using 10–20-μm i.d. emitters). Building on this work, Ramanathan and coworkers (154) recently published a method that efficiently normalizes ESI response for metabolite profiling. Their method utilizes nanoelectrospray in combination with a postcolumn makeup solution having the inverse composition to the analytical gradient. Their examples showed excellent agreement between radioactivity and LC/MS data by this method (154).

Accurate mass measurements have significantly impacted the study of biotransformation. Today, instruments such as Q-TOF, FT-MS, and the orbitrap are used routinely to solve structural problems that are refractory to LC/MS instruments with unit mass resolution. **Figure 7** illustrates a representative example of the power of accurate mass. In this example, a product ion mass spectrum for a metabolite exhibited as loss of 46 mass units. Because this loss could correspond to the removal of H_2CO_2 (accurate mass = 46.0054) or NO_2 (accurate mass = 45.9929), accurate mass–product ion data were acquired using a Q-TOF mass analyzer. Accurate mass MS/MS indicated a neutral loss of 45.9908 mass units confirming the structure of the metabolite as a nitro compound instead of a carboxylic acid. This example illustrates the power of accurate mass to quickly solve structural problems without resorting to metabolite isolation and nuclear magnetic resonance analysis.

Drugs can undergo metabolic activation to form reactive metabolites that bind to endogenous nucleophiles (e.g., protein, DNA), potentially causing adverse drug reactions. Molecules in drug discovery are routinely screened for their propensity to form reactive metabolites, by the detection of drug-related GSH adducts. LC/MS/MS analysis of reactive metabolites trapped by GSH is now a common practice (155–156). Typically, the signature neutral loss of 129 Da (pyroglutamic acid) is used to detect GSH adducts. GSH derivatives can be used to enhance the detection of trapped adducts. Examples include dansyl-GSH (157), as well as trapping with stable isotope-labeled GSH (158). In the latter method, isotopic doublets differing by 3 Da signal the presence of GSH adducts. In addition to these methods, a linear quadrupole ion trap MS was used to screen GSH metabolites in a recent publication that emphasized the selectivity and sensitivity of multiple reaction monitoring detection (159).

6. CONCLUSIONS

This review discusses both recent and historic advances regarding the use of LC/MS for drug discovery for three core functional areas involved in drug discovery: biology, chemistry, and drug disposition. LC/MS clearly has transformed the way drug discovery is conducted in each of these components.

Now that commercial LC/MS instruments based on the API interface have a history spanning more than a decade, it is not surprising that innovation has slowed for applications that have become fully reduced to practice. Several medicinal chemistry applications now fall into this category along with selected applications in drug disposition and biology. At the same time, other areas are still finding fertile ground for research. For instance, the use of LC/MS to discover and qualify novel biomarkers is rapidly expanding. Other applications that push the limits of MS technology for complex mixture analysis (e.g. biotransformation), sensitivity (e.g. microdosing), or throughput (e.g. pharmacology screening) are also receiving attention. Irrespective of the applications that lie ahead, the path forward will continue to rely on sustained improvements in LC/MS instrumentation. For example, a reduction in cost (and perhaps size) is clearly needed for LC/MS to be more broadly introduced into pharmaceutical applications. Unfortunately, despite the introduction of microfluidics and chip-based MS (160, 161), the impact of the nanorevolution has yet to fully materialize in the modern pharmaceutical laboratory. This represents an obvious opportunity. The need for multiplexed MS detection can be cited as another gap, although there has been some pioneering work (162). Lastly, there is always a need for improved ionization techniques. In this sense, the recent development of in situ ionization methods warrants further attention (163).

DISCLOSURE STATEMENT

The authors are not aware of any biases that might be perceived as affecting the objectivity of this review.

LITERATURE CITED

1. Lee MS, Kerns EH. 1999. LC/MS applications in drug development. *Mass Spectrom. Rev.* 18:187–279
2. Korfmacher WA. 2005. Principles and applications of LC-MS in new drug discovery. *Drug Discov. Today* 10:1357–67
3. Watson JT, Sparkman OD. 2007. *Introduction to Mass Spectrometry: Instrumentation, Applications, and Strategies for Data Interpretation.* New York: Wiley. 832 pp. 4th ed.

4. Majors RE. 2007. Current trends in HPLC column usage. *LG-GC North Am.* **http://www.chromatographyonline.com/lcgc/article/articleDetail.jsp?id=434997**

5. Kazakevich Y, LoBrutto R. 2007. Stationary phases. In *HPLC for Pharmaceutical Scientists*, ed. Y Kazakevich, R LoBrutto, 3:75–138. New York: Wiley

6. Wang X, Barber WE, Carr PW. 2006. A practical approach to maximizing peak capacity by using long columns packed with pellicular stationary phases for proteomic research. *J. Chromatogr. A* 1107:139–51

7. Wilson ID, Plumb R, Granger J, Major H, Williams R, Lenz EM. 2005. HPLC-MS-based methods for the study of metabonomics. *J. Chromatogr. B Anal. Technol. Biomed. Life Sci.* 817:67–76

8. Castro-Perez J, Plumb R, Granger JH, Beattie I, Joncour K, Wright A. 2005. Increasing throughput and information content for in vitro drug metabolism experiments using ultraperformance liquid chromatography coupled to a quadrupole time-of-flight mass spectrometer. *Rapid Commun. Mass Spectrom.* 19:843–48

9. Kirkland JJ, Truszkowski FA, Dilks CH Jr, Engel GS. 2000. Superficially porous silica microspheres for fast high-performance liquid chromatography of macromolecules. *J. Chromatogr. A* 890:3–13

10. Idborg H, Zamani L, Edlund PO, Schuppe-Koistinen I, Jacobsson SP. 2005. Metabolic fingerprinting of rat urine by LC/MS. Part 1. Analysis by hydrophilic interaction liquid chromatography–electrospray ionization mass spectrometry. *J. Chromatogr. B Anal. Technol. Biomed. Life Sci.* 828:9–13

11. Song Q, Naidong W. 2006. Analysis of omeprazole and 5-OH omeprazole in human plasma using hydrophilic interaction chromatography with tandem mass spectrometry (HILIC-MS/MS): eliminating evaporation and reconstitution steps in 96-well liquid/liquid extraction. *J. Chromatogr. B. Anal. Technol. Biomed. Life Sci.* 830:135–42

12. Wu JT, Zeng H, Deng Y, Unger SE. 2001. High-speed liquid chromatography/tandem mass spectrometry using a monolithic column for high-throughput bioanalysis. *Rapid Commun. Mass Spectrom.* 15:1113–19

13. Romanyshyn L, Tiller PR, Alvaro R, Pereira A, Hop CE. 2001. Ultra-fast gradient vs fast isocratic chromatography in bioanalytical quantification by liquid chromatography/tandem mass spectrometry. *Rapid Commun. Mass Spectrom.* 15:313–19

14. Shen JX, Wang H, Tadros S, Hayes RN. 2006. Orthogonal extraction/chromatography and UPLC, two powerful new techniques for bioanalytical quantitation of desloratadine and 3-hydroxydesloratadine at 25 pg/ml. *J. Pharm. Biomed. Anal.* 40:689–706

15. Wolters DA, Washburn MP, Yates JR 3rd. 2001. An automated multidimensional protein identification technology for shotgun proteomics. *Anal. Chem.* 73:5683–90

16. Xu RN, Fan L, Rieser MJ, El-Shourbagy TA. 2007. Recent advances in high-throughput quantitative bioanalysis by LC-MS/MS. *J. Pharm. Biomed. Anal.* 44:342–55

17. King R, Bonfiglio R, Fernandez-Metzler C, Miller-Stein C, Olah T. 2000. Mechanistic investigation of ionization suppression in electrospray ionization. *J. Am. Soc. Mass Spectrom.* 11:942–50

18. Hsieh Y. 2004. APPI: a new ionization source for LC-MS/MS assays. See Ref. 164, pp. 253–76

19. Biringer RG, Amato H, Harrington MG, Fonteh AN, Riggins JN, Hühmer AF. 2006. Enhanced sequence coverage of proteins in human cerebrospinal fluid using multiple enzymatic digestion and linear ion trap LC-MS/MS. *Brief. Funct. Genomics Proteomics* 5:144–53

20. Hopfgartner G, Varesio E, Tschäppät V, Grivet C, Bourgogne E, Leuthold LA. 2004. Triple quadrupole linear ion trap mass spectrometer for the analysis of small molecules and macromolecules. *J. Mass Spectrom.* 39:845–55

21. van den Heuvel RH, van Duijn E, Mazon H, Synowsky SA, Lorenzen K, et al. 2006. Improving the performance of a quadrupole time-of-flight instrument for macromolecular mass spectrometry. *Anal. Chem.* 78:7473–83

22. Higgs RE, Knierman MD, Freeman AB, Gelbert LM, Patil ST, Hale JE. 2007. Estimating the statistical significance of peptide identifications from shotgun proteomics experiments. *J. Proteome Res.* 6:1758–67

23. Markarov A, Denisov E, Kholomeev A, Balschun W, Lange O, et al. 2006. Performance evaluation of a hybrid linear ion trap/orbitrap mass spectrometer. *Anal. Chem.* 78:2113–20

24. Liu X, Valentine SJ, Plasencia MD, Trimpin S, Naylor S, Clemmer DE. 2007. Mapping the human plasma proteome by SCX-LC-IMS-MS. *J. Am. Soc. Mass Spectrom.* 18:1249–64

25. Kapron J, Wu J, Mauriala T, Clark P, Purves RW, Bateman KP. 2006. Simultaneous analysis of prostanoids using liquid chromatography/high-field asymmetric waveform ion mobility spectrometry/tandem mass spectrometry. *Rapid Commun. Mass Spectrom.* 20:1504–10

26. Giles K, Pringle SD, Worthington KR, Little D, Wildgoose JL, Bateman RH. 2004. Applications of a traveling wave-based radio-frequency-only stacked ring ion guide. *Rapid Commun. Mass Spectrom.* 18:2401–14

27. Bailey AO, Miller TM, Dong MQ, Velde CV, Cleveland DW, Yates JR. 2007. RCADiA: simple automation platform for comparative multidimensional protein identification technology. *Anal. Chem.* 79:6410–18

28. Venable JD, Wohlschlegel J, McClatchy DB, Park SK, Yates JR 3rd. 2007. Relative quantification of stable isotope labeled peptides using a linear ion trap-orbitrap hybrid mass spectrometer. *Anal. Chem.* 79:3056–64

29. Kelleher NL. 2004. Top-down proteomics. *Anal. Chem.* 76:A197A–203

30. Wiener MC, Sachs JR, Deyanova EG, Yates NA. 2004. Differential mass spectrometry: a label-free LC-MS method for finding significant differences in complex peptide and protein mixtures. *Anal. Chem.* 76:6085–96

31. Ong SE, Mann M. 2007. Stable isotope labeling by amino acids in cell culture for quantitative proteomics. *Methods Mol. Biol.* 359:37–52

32. Gygi SP, Rist B, Gerber SA, Turecek F, Gelb MH, Aebersold R. 1999. Quantitative analysis of complex protein mixtures using isotope-coded affinity tags. *Nat. Biotechnol.* 17:994–99

33. Ross PL, Huang YN, Marchese JN, Williamson B, Parker K, et al. 2004. Multiplexed protein quantitation in *Saccharomyces cerevisiae* using amine-reactive isobaric tagging reagents. *Mol. Cell. Proteomics* 3:1154–69

34. Carr SA, Annan RS, Huddleston MJ. 2005. Mapping posttranslational modifications of proteins by MS-based selective detection: application to phosphoproteomics. *Methods Enzymol.* 405:82–115

35. Durham M, Regnier FE. 2006. Targeted glycoproteomics: serial lectin affinity chromatography in the selection of O-glycosylation sites on proteins from the human blood proteome. *J. Chromatogr. A* 1132:165–73

36. Zubarev RA. 2004. Electron-capture dissociation tandem mass spectrometry. *Curr. Opin. Biotechnol.* 15:12–16

37. Syka JE, Coon JJ, Schroeder MJ, Shabanowitz J, Hunt DF. 2004. Peptide and protein sequence analysis by electron transfer dissociation mass spectrometry. *Proc. Natl. Acad. Sci. USA* 101:9528–33

38. Robertson DG, Reily MD, Baker JD. 2007. Metabonomics in pharmaceutical discovery and development. *J. Proteome Res.* 6:526–39

39. Lindon JC, Holmes E, Nicholson JK. 2007. Metabonomics in pharmaceutical R&D. *FEBS J.* 274:1140–51

40. Kind T, Tolstikov V, Fiehn O, Weiss RH. 2007. A comprehensive urinary metabolomic approach for identifying kidney cancer. *Anal. Biochem.* 363:185–95

41. Idborg-Björkman H, Edlund PO, Kvalheim OM, Schuppe-Koistinen I, Jacobsson SP. 2003. Screening of biomarkers in rat urine using LC/electrospray ionization-MS and two-way data analysis. *Anal. Chem.* 75:4784–92

42. Lafaye A, Junot C, Ramounet-Le Gall B, Fritsch P, Tabet JC, Ezan E. 2003. Metabolite profiling in rat urine by liquid chromatography/electrospray ion trap mass spectrometry: application to the study of heavy metal toxicity. *Rapid Commun. Mass Spectrom.* 17:2541–49

43. Sanders M, Shipkova PA, Zhang H, Warrack BM. 2006. Utility of the hybrid LTQ-FTMS for drug metabolism applications. *Curr. Drug Metab.* 7:547–55

44. Ding J, Sorensen CM, Zhang Q, Jiang H, Jaitly N, et al. 2007. Capillary LC coupled with high-mass measurement accuracy mass spectrometry for metabolic profiling. *Anal. Chem.* 79:6081–93

45. Plumb RS, Stumpf CL, Gorenstein MV, Castro-Perez JM, Dear GJ, et al. 2002. Metabonomics: The use of electrospray mass spectrometry coupled to reversed-phase liquid chromatography shows potential for the screening of rat urine in drug development. *Rapid Commun. Mass Spectrom.* 16:1991–96

46. Williams RE, Major H, Lock EA, Lenz EM, Wilson ID. 2005. D-serine-induced nephrotoxicity: a HPLC-TOF/MS-based metabonomics approach. *Toxicology* 207:179–90

47. Gamache PH, Meyer DF, Granger MC, Acworth IN. 2004. Metabolomic applications of electrochemistry/mass spectrometry. *J. Am. Soc. Mass Spectrom.* 15:1717–26

48. Yang J, Xu G, Zheng Y, Kong H, Pang T, et al. 2004. Diagnosis of liver cancer using HPLC-based metabonomics avoiding false-positive result from hepatitis

and hepatocirrhosis diseases. *J. Chromatogr. B Anal. Technol. Biomed. Life Sci.* 813:59–65

49. Paige LA, Mitchell MW, Krishnan KR, Kaddurah-Daouk R, Steffens DC. 2007. A preliminary metabolomic analysis of older adults with and without depression. *Int. J. Geriatr. Psychiatry* 22:418–23

50. Wang C, Kong H, Guan Y, Yang J, Gu J, et al. 2005. Plasma phospholipid metabolic profiling and biomarkers of type 2 diabetes mellitus based on high-performance liquid chromatography/electrospray mass spectrometry and multivariate statistical analysis. *Anal. Chem.* 77:4108–16

51. Wenk MR. 2005. The emerging field of lipidomics. *Nat. Rev. Drug Discov.* 4:594–610

52. Fahy E, Subramaniam S, Brown HA, Glass CK, Merrill AH Jr, et al. 2005. A comprehensive classification system for lipids. *J. Lipid Res.* 46:839–61

53. Mortuza GB, Neville WA, Delaney J, Waterfield CJ, Camilleri P. 2003. Characterisation of a potential biomarker of phospholipidosis from amiodarone-treated rats. *Biochim. Biophys. Acta* 1631:136–46

54. Rainville PD, Stumpf CL, Shockcor JP, Plumb RS, Nicholson JK. 2007. Novel application of reversed-phase UPLC-oaTOF-MS for lipid analysis in complex biological mixtures: a new tool for lipidomics. *J. Proteome Res.* 6:552–58

55. Lee SH, Williams MV, DuBois RN, Blair IA. 2003. Targeted lipidomics using electron capture atmospheric pressure chemical ionization mass spectrometry. *Rapid Commun. Mass Spectrom.* 17:2168–76

56. Siegel MM. 2005. Mass-spectrometry-based drug screening assays for early phases in drug discovery. See Ref. 165, pp. 27–70

57. Geoghegan KF, Kelly MA. 2005. Biochemical applications of mass spectrometry in pharmaceutical drug discovery. *Mass Spectrom. Rev.* 24:347–66

58. Hofstadler SA, Sannes-Lowery KA. 2006. Applications of ESI-MS in drug discovery: interrogation of noncovalent complexes. *Nat. Rev. Drug Discov.* 5:585–95

59. Xu R, Sang BC, Navre M, Kassel DB. 2006. Cell-based assay for screening 11β-hydroxysteroid dehydrogenase inhibitors using liquid chromatography/tandem mass spectrometry detection. *Rapid Commun. Mass Spectrom.* 20:1643–47

60. Quercia AK, LaMarr WA, Myung J, Ozbal CC, Landro JA, Lumb KJ. 2007. High-throughput screening by mass spectrometry: comparison with the scintillation proximity assay with a focused-file screen of AKT1/PKBα. *J. Biomol. Screen.* 12:473–80

61. Chernet E, Martin LJ, Li D, Need AB, Barth VN, et al. 2005. Use of LC/MS to assess brain tracer distribution in preclinical, in vivo receptor occupancy studies: dopamine D2, serotonin 2A and NK-1 receptors as examples. *Life Sci.* 78:340–46

62. Ackermann BL, Hale JE, Duffin KL. 2006. The role of mass spectrometry in biomarker discovery and measurement. *Curr. Drug Metab.* 7:525–39

63. Koeberl DD, Young SP, Gregersen NS, Vockley J, Smith WE, et al. 2003. Rare disorders of metabolism with elevated butyryl- and isobutyryl-carnitine detected by tandem mass spectrometry newborn screening. *Pediatr. Res.* 54:219–23

64. Lin S, Shaler TA, Becker CH. 2006. Quantification of intermediate-abundance proteins in serum by multiple reaction monitoring mass spectrometry in a single-quadrupole ion trap. *Anal. Chem.* 78:5762–77

65. Hawkridge AM, Heublein DM, Bergen HR 3rd, Cataliotti A, Burnett JC Jr, Muddiman DC. 2005. Quantitative mass spectral evidence for the absence of circulating brain natriuretic peptide (BNP-32) in severe human heart failure. *Proc. Natl. Acad. Sci. USA* 102:17442–47

66. Gerber SA, Rush J, Stemman O, Kirschner MW, Gygi SP. 2003. Absolute quantification of proteins and phosphoproteins from cell lysates by tandem MS. *Proc. Natl. Acad. Sci. USA* 100:6940–45

67. Anderson L, Hunter CL. 2006. Quantitative mass spectrometric multiple reaction monitoring assays for major plasma proteins. *Mol. Cell. Proteomics* 5:573–88

68. Barnidge DR, Goodmanson MK, Klee GG, Muddiman DC. 2004. Absolute quantification of the model biomarker prostate-specific antigen in serum by LC-MS/MS using protein cleavage and isotope dilution mass spectrometry. *J. Proteome Res.* 3:644–52

69. Kuhn E, Wu J, Karl J, Liao H, Zolg W, Guild B. 2004. Quantification of C-reactive protein in the serum of patients with rheumatoid arthritis using multiple reaction monitoring mass spectrometry and ^{13}C-labeled peptide standards. *Proteomics* 4:1175–86

70. Rifai N, Gillette MA, Carr SA. 2006. Protein biomarker discovery and validation: the long and uncertain path to clinical utility. *Nat. Biotechnol.* 24:971–83

71. Anderson NL, Anderson NG, Haines LR, Hardie DB, Olafson RW, Pearson TW. 2004. Mass spectrometric quantitation of peptides and proteins using Stable Isotope Standards and Capture by Anti-Peptide Antibodies (SISCAPA). *J. Proteome Res.* 3:235–44

72. Ackermann BL, Berna MJ. 2007. Coupling immunoaffinity techniques with MS for quantitative analysis of low-abundance protein biomarkers. *Expert Rev. Proteomics* 4:175–86

73. Berna MJ, Zhen Y, Watson DE, Hale JE, Ackermann BL. 2007. Strategic use of immunoprecipitation and LC/MS/MS for trace-level protein quantification: myosin light chain 1, a biomarker of cardiac necrosis. *Anal. Chem.* 79:4199–205

74. Oe T, Ackermann BL, Inoue K, Berna MJ, Garner CO, et al. 2006. Quantitative analysis of amyloid β peptides in cerebrospinal fluid of Alzheimer's disease patients by immunoaffinity purification and stable isotope dilution liquid chromatography/negative electrospray ionization tandem mass spectrometry. *Rapid Commun. Mass Spectrom.* 20:3723–35

75. Berna M, Schmalz C, Duffin K, Mitchell P, Chambers M, Ackermann B. 2006. Online immunoaffinity liquid chromatography/tandem mass spectrometry determination of a type II collagen peptide biomarker in rat urine: investigation of the impact of collision-induced dissociation fluctuation on peptide quantitation. *Anal. Biochem.* 356:235–43

76. Whiteaker JR, Zhao L, Zhang HY, Feng LC, Piening BD, et al. 2007. Antibody-based enrichment of peptides on magnetic beads for mass-spectrometry-based quantification of serum biomarkers. *Anal. Biochem.* 62:44–54

77. Gilbert JR, Lewer P, Duebelbeis DO, Carr AW. 2005. A central role of mass spectrometry in natural products discovery. See Ref. 165, pp. 149–88

78. Kozikowski BA, Burt TM, Tirey DA, Williams LE, Kuzmak BR, et al. 2003. The effect of freeze/thaw cycles on the stability of compounds in DMSO. *J. Biomol. Screen.* 8:210–15

79. Cheng X, Hochlowski J, Tang H, Hepp D, Beckner C, et al. 2003. Studies on repository compound stability in DMSO under various conditions. *J. Biomol. Screen.* 8:292–304

80. Cheng X, Hochlowski J. 2005. Application of mass spectrometry to compound library generation, analysis, and management. See Ref. 165, pp. 189–230

81. Lebl M. 1999. Parallel personal comments on "classical" papers in combinatorial chemistry. *J. Comb. Chem.* 1:3–24

82. Wagner DS, Wagner RW, Schonen F, Geysen HM. 2005. A combinatorial process for drug discovery. See Ref. 165, pp. 231–59

83. Kassel DB. 2001. Combinatorial chemistry and mass spectrometry in the 21st century drug discovery laboratory. *Chem. Rev.* 101:255–67

84. Zhao Y, Semin DJ. 2005. New approaches for method development and purification in lead optimization. See Ref. 165, pp. 403–32

85. Pullen FS, Perkins GL, Burton KJ, Ware RS, Teague MS, Kiplinger JP. 1995. Putting mass spectrometry in the hands of the end user. *J. Am. Soc. Mass Spectrom.* 6:394–99

86. Taylor LCE, Johnson RL, Raso L. 1995. Open-access atmospheric pressure chemical ionization mass spectrometry for routine sample analysis. *J. Am. Soc. Mass Spectrom.* 6:387–93

87. Mallis LM, Sarkahian AB, Kulishoff JM Jr, Watts WL Jr. 2002. Open-access liquid chromatography/mass spectrometry in a drug discovery environment. *J. Mass Spectrom.* 37:889–96

88. Thomas SR, Gerhard U. 2004. Open-access high-resolution mass spectrometry in early drug discovery. *J. Mass Spectrom.* 39:942–48

89. Peake DA, Ackermann BL. 2005. Results from a bench marking survey on supporting chemical synthesis and structural elucidation in the pharmaceutical industry. *J. Am. Soc. Mass Spectrom.* 16:599–605

90. Morand KL, Burt TM, Regg BT, Chester TL. 2001. Techniques for increasing the throughput of flow injection mass spectrometry. *Anal. Chem.* 73:247–52

91. Yates N, Wislocki D, Roberts A, Berk S, Klatt T, et al. 2001. Mass spectrometry screening of combinatorial mixtures, correlation of measured and predicted electrospray ionization spectra. *Anal. Chem.* 73:2941–51

92. Weller HN, Young MG, Michalcyzk SJ, Reitnauer GH, Cooley RS, et al. 1997. High throughput analysis and purification in support of parallel synthesis. *Mol. Divers.* 3:61–70

93. Kyranos JN, Lee H, Goetzinger WK, Li LY. 2004. One-minute full-gradient HPLC/UV/ELSD/MS analysis to support high-throughput parallel synthesis. *J. Comb. Chem.* 6:796–804

94. Pereira L, Ross P, Woodruff M. 2000. Chromatographic aspects in high throughput liquid chromatography/mass spectrometry. *Rapid Commun. Mass Spectrom.* 14:357–60

95. Wang T, Barber M, Hardt I, Kassel DB. 2001. Mass-directed fractionation and isolation of pharmaceutical compounds by packed-column supercritical fluid chromatography/mass spectrometry. *Rapid Commun. Mass Spectrom.* 15:2067–75

96. Pinkston JD, Wen D, Morand KL, Tirey DA, Stanton DT. 2006. Comparison of LC/MS and SFC/MS for screening of a large and diverse library of pharmaceutically relevant compounds. *Anal. Chem.* 78:7467–72

97. Searle PA, Glass KA, Hochlowski JE. 2004. Comparison of preparative HPLC/MS and preparative SFC techniques for the high-throughput purification of compound libraries. *J. Comb. Chem.* 6:175–80

98. Wang T, Cohen, Kassel DB, Zeng L. 1999. A multiple electrospray interface for parallel mass spectrometric analyses of compound libraries. *Comb. Chem. High Throughput Screen.* 2:327–34

99. De Biasi V, Haskins N, Organ A, Bateman R, Giles K, Jarvis S. 1999. High throughput liquid chromatography/mass spectrometric analyses using a novel multiplexed electrospray interface. *Rapid Commun. Mass Spectrom.* 13:1165–68

100. Xu R, Wang T, Isbell J, Cai Z, Sykes C, et al. 2002. High-throughput mass-directed parallel purification incorporating a multiplexed single quadrupole mass spectrometer. *Anal. Chem.* 74:3055–62

101. Kibbey CE. 1996. Quantitation of combinatorial libraries of small organic molecules by normal-phase HPLC with evaporative light scattering detection. *Mol. Divers.* 1:247–58

102. Taylor EW, Qian MG, Dollinger GD. 1998. Simultaneous online characterization of small organic molecules derived from combinatorial libraries for identity, quantity, and purity by reversed-phase HPLC with chemiluminescent nitrogen, UV, and mass spectrometric detection. *Anal. Chem.* 70:3339–47

103. Yurek DA, Branch DL, Kuo M-S. 2002. Development of a system to evaluate compound identity, purity, and concentration in a single experiment and its application in quality assessment of combinatorial libraries and screening hits. *J. Comb. Chem.* 4:138–48

104. Popa-Burke IG, Issakova O, Arroway JD, Bernasconi P, Chen M, et al. 2004. Streamlined system for purifying and quantifying a diverse library of compounds and the effect of compound concentration measurements on the accurate interpretation of biological assay results. *Anal. Chem.* 76:7278–87

105. Edwards C, Liu J, Smith TJ, Brooke D, Hunter DJ, et al. 2003. Parallel preparative high-performance liquid chromatography with on-line molecular mass characterization. *Rapid Commun. Mass Spectrom.* 17:2027–33

106. Kibby CE. 1997. An automated system for purification of combinatorial libraries by preparative LC/MS. *Lab Robot. Autom.* 9:309–21

107. Zeng L, Kassel DB. 1998. Development of a fully automated parallel HPLC/mass spectrometry system for the analytical characterization and preparative purification of combinatorial libraries. *Anal. Chem.* 70:4380–88

108. Kyranos JN, Cai H, Zhang B, Goetzinger WK. 2001. High-throughput techniques for compound characterization and purification. *Curr. Opin. Drug Discov. Dev.* 4:719–28

109. Ventura M, Farrell W, Aurigemma C, Tivel K, Greig M, et al. 2004. High-throughput preparative process utilizing three complementary chromatographic purification technologies. *J. Chromatogr. A* 1036:7–13

110. Maftouh M, Granier-Loyaux C, Chavana E, Marini J, Pradines A, et al. 2005. Screening approach for chiral separation of pharmaceuticals. Part III. Supercritical fluid chromatography for analysis and purification in drug discovery. *J. Chromatogr. A* 1088:67–81

111. Kola I, Landis J. 2004. Can the pharmaceutical industry reduce attrition rates? *Nat. Rev. Drug Discov.* 3:711–15

112. Ackermann BL, Berna MJ, Murphy AT. 2005. Advances in high throughput quantitative drug discovery bioanalysis. See Ref. 165, pp. 315–58

113. Janiszewski JS, Rogers KJ, Whalen KM, Cole MJ, Liston TE, et al. 2001. A high-capacity LC/MS system for the bioanalysis of samples generated from plate-based metabolic screening. *Anal. Chem.* 73:1495–501

114. Whalen KM, Rogers KJ, Cole MJ, Janiszewski JS. 2000. AutoScan: an automated workstation for rapid determination of mass and tandem mass spectrometry conditions for quantitative bioanalytical mass spectrometry. *Rapid Commun. Mass Spectrom.* 14:2074–79

115. Kerns EH, Di L. 2006. Utility of mass spectrometry for pharmaceutical profiling applications. *Curr. Drug Metab.* 7:457–66

116. Kassel DB. 2004. High throughput strategies for in vitro ADME assays: How fast can we go? See Ref. 164, pp. 35–82

117. Lipinski CA, Lombardo F, Dominy BW, Feeney PJ. 2001. Experimental and computational approaches to estimate solubility and permeability in drug discovery and development settings. *Adv. Drug Deliv. Rev.* 46:3–26

118. van de Waterbeemd H, Gifford E. 2003. ADMET in silico modeling: towards prediction paradise? *Nat. Rev. Drug Discov.* 2:192–204

119. Jenkins KM, Angeles R, Quintos MT, Xu R, Kassel DB, Rourick RA. 2004. Automated high throughput ADME assays for metabolic stability and cytochrome P450 inhibition profiling of combinatorial libraries. *J. Pharm. Biomed. Anal.* 34:989–1004

120. Whalen K, Gobey J, Janiszewski J. 2006. A centralized approach to tandem mass spectrometry method development for high-throughput ADME screening. *Rapid Commun. Mass Spectrom.* 20:1497–503

121. Drexler DM, Belcastro JV, Dickinson KE, Edinger KJ, Hnatyshyn SY, et al. 2007. An automated high throughput liquid chromatography–mass spectrometry process to assess the metabolic stability of drug candidates. *Assay Drug Dev. Technol.* 5:247–64

122. Wilson DM, Wang X, Walsh E, Rourick RA. 2001. High throughput log D determination using liquid chromatography–mass spectrometry. *Comb. Chem. High Throughput Screen.* 4:511–19

123. Kerns EH, Di L, Petusky S, Farris M, Ley R, Jupp P. 2004. Combined application of parallel artificial membrane permeability assay and Caco-2 permeability assays in drug discovery. *J. Pharm. Sci.* 93:1440–53

124. Xia CQ, Milton MN, Gan LS. 2007. Evaluation of drug-transporter interactions using in vitro and in vivo models. *Curr. Drug Metab.* 8:341–63

125. Smalley J, Marino AM, Xin B, Olah T, Balimane PV. 2007. Development of a quantitative LC-MS/MS analytical method coupled with turbulent flow chromatography for digoxin for the in vitro P-gp inhibition assay. *J. Chromatogr. B Anal. Technol. Biomed. Life Sci.* 854:260–67

126. Korfmacher WA, Palmer CA, Nardo C, Dunn-Meynell K, Grotz D, et al. 1999. Development of an automated mass spectrometry system for the quantitative analysis of liver microsomal incubation samples: a tool for rapid screening of new compounds for metabolic stability. *Rapid Commun. Mass Spectrom.* 13:901–7

127. Dierks EA, Stams KR, Lim HK, Cornelius G, Zhang H, Ball SE. 2001. A method for the simultaneous evaluation of the activities of seven major human drug-metabolizing cytochrome P450s using an in vitro cocktail of probe substrates and fast gradient liquid chromatography tandem mass spectrometry. *Drug Metab. Dispos.* 29:23–29

128. Di L, Kerns EH, Li SQ, Carter GT. 2007. Comparison of cytochrome P450 inhibition assays for drug discovery using human liver microsomes with LC-MS, rhCYP450 isozymes with fluorescence, and double cocktail with LC-MS. *Int. J. Pharm.* 335:1–11

129. Cheng Y, Ho E, Subramanyam B, Tseng JL. 2004. Measurements of drug-protein binding by using immobilized human serum albumin liquid chromatography–mass spectrometry. *J. Chromatogr. B Anal. Technol. Biomed. Life Sci.* 809:67–73

130. Dash AK, Elmquist WF. 2003. Separation methods that are capable of revealing blood-brain barrier permeability. *J. Chromatogr. B Anal. Technol. Biomed. Life Sci.* 797:241–54

131. Matuszewski BK. 2006. Standard line slopes as a measure of a relative matrix effect in quantitative HPLC-MS bioanalysis. *J. Chromatogr. B Anal. Technol. Biomed. Life Sci.* 830:293–300

132. Jemal M, Ouyang Z. 2003. Enhanced resolution triple-quadrupole mass spectrometry for fast quantitative bioanalysis using liquid chromatography/tandem mass spectrometry: investigations of parameters that affect ruggedness. *Rapid Commun. Mass Spectrom.* 17:24–38

133. Hatsis P, Brockman AH, Wu JT. 2007. Evaluation of high-field asymmetric waveform ion mobility spectrometry coupled to nanoelectrospray ionization for bioanalysis in drug discovery. *Rapid Commun. Mass Spectrom.* 21:2295–300

134. Alnouti Y, Srinivasan K, Waddell D, Bi H, Kavetskaia O, Gusev AI. 2005. Development and application of a new on-line SPE system combined with LC-MS/MS detection for high throughput direct analysis of pharmaceutical compounds in plasma. *J. Chromatogr. A* 1080:99–106

135. Grant RP, Cameron C, Mackenzie-McMurter S. 2002. Generic serial and parallel on-line direct injection using turbulent flow liquid chromatography/tandem mass spectrometry. *Rapid Commun. Mass Spectrom.* 16:1785–92

136. Needham SR, Cole MJ, Fouda HG. 1998. Direct plasma injection for high-performance liquid chromatographic–mass spectrometric quantitation of the anxiolytic agent CP-93 393. *J. Chromatogr. B Biomed. Sci. Appl.* 718:87–94

137. Naxing Xu R, Fan L, Kim GE, El-Shourbagy TA. 2006. A monolithic-phase based on-line extraction approach for determination of pharmaceutical components in human plasma by HPLC-MS/MS and a comparison with liquid-liquid extraction. *J. Pharm. Biomed. Anal.* 40:728–36

138. Xia YQ, Hop CE, Liu DQ, Vincent SH, Chiu SH. 2001. Parallel extraction columns and parallel analytical columns coupled with liquid chromatography/tandem mass spectrometry for on-line simultaneous quantification of a drug candidate and its six metabolites in dog plasma. *Rapid Commun. Mass Spectrom.* 15:2135–44

139. Van Pelt CK, Corso TN, Schultz GA, Lowes S, Henion J. 2001. A four-column parallel chromatography system for isocratic or gradient LC/MS analyses. *Anal. Chem.* 73:582–88

140. King RC, Miller-Stein C, Magiera DJ, Brann J. 2002. Description and validation of a staggered parallel high performance liquid chromatography system for good laboratory practice level quantitative analysis by liquid chromatography/tandem mass spectrometry. *Rapid Commun. Mass Spectrom.* 16:43–52

141. Yang L, Mann TD, Little D, Wu N, Clement RP, Rudewicz PJ. 2001. Evaluation of a four-channel multiplexed electrospray triple quadrupole mass spectrometer for the simultaneous validation of LC/MS/MS methods in four different preclinical matrixes. *Anal. Chem.* 73:1740–47

142. Sadagopan N, Pabst B, Cohen L. 2005. Evaluation of online extraction/mass spectrometry for in vivo cassette analysis. *J. Chromatogr. B Anal. Technol. Biomed. Life Sci.* 820:59–67

143. Korfmacher WA, Cox KA, Ng KJ, Veals J, Hsieh Y, et al. 2001. Cassette-accelerated rapid rat screen: a systematic procedure for the dosing and liquid chromatography/atmospheric pressure ionization tandem mass spectrometric analysis of new chemical entities as part of new drug discovery. *Rapid Commun. Mass Spectrom.* 15:335–40

144. White RE, Manitpisitkul P. 2001. Pharmacokinetic theory of cassette dosing in drug discovery screening. *Drug Metab. Dispos.* 29:957–66

145. Buhrman DL, Price PI, Rudewicz PJ. 1996. Quantitation of SR 27417 in human plasma using electrospray liquid chromatography–tandem mass spectrometry: a study of ion suppression. *J. Am. Soc. Mass Spectrom.* 7:1099–105

146. Bonfiglio R, King RC, Olah TV, Merkle K. 1999. The effects of sample preparation methods on the variability of the electrospray ionization response for model drug compounds. *Rapid Commun. Mass Spectrom.* 13:1175–85

147. Mei H. 2004. Matrix effects: causes and solutions. See Ref. 164, pp. 103–50

148. McLean MA, Tam CJ, Barrata MT, Holliman CL, Ings RM, Galluppi GR. 2007. Accelerating drug development: methodology to support first-in-man pharmacokinetic studies by the use of drug candidate microdosing. *Drug Dev. Res.* 68:14–22

149. Balani SK, Nagaraja NV, Qian MG, Costa AO, Daniels JS, et al. 2006. Evaluation of microdosing to assess pharmacokinetic linearity in rats using liquid chromatography–tandem mass spectrometry. *Drug Metab. Dispos.* 34:384–88

150. Cox K. 2004. Special requirements for metabolites characterization. See Ref. 164, pp. 229–52

151. Prakash C, Shaffer CL, Nedderman A. 2007. Analytical strategies for identifying drug metabolites. *Mass Spectrom. Rev.* 26:340–69

152. Kamel A, Prakash C. 2006. High performance liquid chromatography/atmospheric pressure ionization/tandem mass spectrometry (HPLC/API/MS/MS) in drug metabolism and toxicology. *Curr. Drug Metab.* 7:837–52

153. Hop CE, Chen Y, Yu LJ. 2005. Uniformity of ionization response of structurally diverse analytes using a chip-based nanoelectrospray ionization source. *Rapid Commun. Mass Spectrom.* 19:3139–42

154. Ramanathan R, Zhong R, Blumenkrantz N, Chowdhury SK, Alton KB. 2007. Response normalized liquid chromatography nanospray ionization mass spectrometry. *J. Am. Soc. Mass Spectrom.* 18:1891–99

155. Soglia JR, Contillo LG, Kalgutkar AS, Zhao S, Hop CECA, et al. 2006. A semiquantitative method for the determination of reactive metabolite conjugate levels in vitro utilizing liquid chromatography–tandem mass spectrometry and novel quaternary ammonium glutathione analogues. *Chem. Res. Toxicol.* 19:480–90

156. Castro-Perez J, Plumb R, Liang L, Yang E. 2005. A high-throughput liquid chromatography/tandem mass spectrometry method for screening glutathione conjugates using exact mass neutral loss acquisition. *Rapid Commun. Mass Spectrom.* 19:798–804

157. Gan J, Harper TW, Hsueh MM, Qu Q, Humphreys WG. 2005. Dansyl glutathione as a trapping agent for the quantitative estimation and identification of reactive metabolites. *Chem. Res. Toxicol.* 18:896–903

158. Yan Z, Caldwell GW. 2004. Stable-isotope trapping and high-throughput screenings of reactive metabolites using the isotope MS signature. *Anal. Chem.* 76:6835–47

159. Zheng J, Ma L, Xin B, Olah T, Humphreys WG, Zhu M. 2007. Screening and identification of GSH-trapped reactive metabolites using hybrid triple quadruple linear ion trap mass spectrometry. *Chem. Res. Toxicol.* 20:757–66

160. Srbek J, Eickhoff J, Effelsberg U, Kraiczek K, van de Goor T, Coufal P. 2007. Chip-based nano-LC-MS/MS identification of proteins in complex biological samples using a novel polymer microfluidic device. *J. Sep. Sci.* 30:2046–52

161. Wickremsinhe ER, Singh G, Ackermann BL, Gillespie TA, Chaudhary AK. 2006. A review of nanoelectrospray ionization applications for drug metabolism and pharmacokinetics. *Curr. Drug Metab.* 7:913–28

162. Tabert AM, Goodwin MP, Duncan JS, Fico CD, Cooks RG. 2006. Multiplexed rectilinear ion trap mass spectrometer for high-throughput analysis. *Anal. Chem.* 78:4830–88

163. Cooks RG, Ouyang Z, Takats Z, Wiseman JM. 2006. Detection technologies: ambient mass spectrometry. *Science* 311:1566–70

164. Korfmacher WA, ed. 2004. *Using Mass Spectrometry for Drug Metabolism Studies.* Boca Raton, FL: CRC. 370 pp.

165. Lee M, ed. 2005. *Integrated Strategies for Drug Discovery Using Mass Spectrometry.* New York: Wiley. 568 pp.

Optical Probes for Molecular Processes in Live Cells

Yoshio Umezawa

Department of Chemistry, School of Science, University of Tokyo, Tokyo 113-0033, Japan; email: umezawa@chem.s.u-tokyo.ac.jp

Annu. Rev. Anal. Chem. 2008. 1:397–421

First published online as a Review in Advance on March 4, 2008

The *Annual Review of Analytical Chemistry* is online at anchem.annualreviews.org

This article's doi:
10.1146/annurev.anchem.1.031207.112757

1936-1327/08/0719-0397$20.00

Key Words

fluorescent/bioluminescent indicators, single living cells, cellular signaling

Abstract

In this review, I summarize the development over the past several years of fluorescent and/or bioluminescent indicators to pinpoint cellular processes in living cells. These processes involve second messengers, protein phosphorylations, protein-protein interactions, protein-ligand interactions, nuclear receptor–coregulator interactions, nucleocytoplasmic trafficking of functional proteins, and protein localization.

1. PROBING CELLULAR SIGNALING PATHWAYS IN LIVING CELLS

Organic fluorescent probe molecules have been developed for nondestructive analysis of chemical processes in living cells, including ions and small molecules such as Ca^{2+} (1), NO (2), Mg^{2+} (3), and Zn^{2+} (4). In addition, green fluorescent protein (GFP) and its analogs have been used to probe proteins to determine their structural and locational changes after genetically labeling them to proteins of interest (5).

Many intracellular chemical processes and cellular signaling processes are still studied essentially by destructive analysis. Such methods disrupt hundreds of thousands of cells prior to separation, purification, and detection of intracellular components. It is necessary, therefore, to develop methods for direct nondestructive analysis of cellular signaling steps in live cells.

Intercellular signaling substances include neurotransmitters, which are cytokines and hormones functioning in the nerve, immune, and endocrine systems. These substances bind either to ion-channel-, kinase-, or G protein–coupled membrane receptor proteins and trigger the respective downstream intracellular signaling processes. Intracellular signaling can be monitored in vivo in living cells (6) by genetically encoded intracellular fluorescent and bioluminescent probes or indicators. Scientists have reported a number of these probes for visualizing cellular signaling. The probes include second messengers such as Ca^{2+} (7), camp (8), nitric acid (NO) (9), inositol 1,4,5-trisphosphate (IP_3) (10, 11), cyclic guanosine $3',5'$-monophosphate (cGMP) (12), and phosphatidylinositol-3,4,5-trisphosphate (13), protein phosphorylation (14–16), protein-protein interactions (17–21), and protein localizations in organelles (22–26). These probes are useful not only for fundamental biological studies, but also for the assay and screening of possible pharmaceutical or toxic chemicals that inhibit or facilitate cellular signaling pathways.

2. SECOND MESSENGERS

2.1. Nitric Oxide

Nitric oxide (NO) is a small uncharged free radical that is involved in diverse physiological and pathophysiological mechanisms. NO is generated by three isoforms (endothelial, neuronal, and inducible) of NO synthase (NOS). When generated in vascular endothelial cells, NO plays a key role in vascular tone regulation. An amplifier-coupled fluorescent indicator for NO was developed (9) to visualize physiological nanomolar dynamics of NO in living cells (detection limit of 0.1 nM). Earlier, a cGMP fluorescent indicator, cyan fluorescent protein/protein kinase G/yellow fluorescent protein (known as CGY), was developed, which was combined with soluble guanylate cyclase (sGC) for the amplified detection of NO in living cells (9). This amplifier-coupled fluorescent indicator was named NOA-1. NOA-1 binds with single NO molecules and generates a large number of cGMPs in single living cells. The increased amount of cGMP in situ is detected by the cGMP fluorescence resonance energy transfer (FRET) sensor built into NOA-1. NOA-1 unbound to cGMP does not emit the FRET signal. The vascular endothelial cell stably generates 1 nM of the basal NO. This genetically encoded high-sensitivity indicator revealed that approximately 1 nM of NO, which is enough to relax blood vessels, is generated in vascular endothelial cells even in the absence of shear stress. The nanomolar range of basal endothelial NO thus revealed appears to be fundamental to vascular homeostasis (9) (**Figure 1**).

We report a novel cell-based indicator that is able to visualize picomolar dynamics of NO release from living cells. Cells from a pig kidney–derived cell line (PK15) endogenously express sGC, which is a receptor protein for the selective recognition of NO. Binding of NO by sGC causes the amplified generation of cGMP. To make the PK15 cells into NO indicators, the cells are transfected with a plasmid vector encoding a fluorescent indicator for cGMP, and FRET is recorded at 480 ± 15 and 535 ± 12.5 nm upon excitation of the cells at 440 ± 10 nm. The cell-based indicator exhibits exceptional sensitivity (detection limit of 20 pM), selectivity, reversibility, and reproducibility. The outstanding sensitivity of the present indicator has led us to uncover an oscillatory release of picomolar concentrations of NO from hippocampal neurons. We present evidence that Ca^{2+} oscillations in hippocampal neurons underlie the oscillatory NO release from the neurons during neurotransmission. We have also succeeded in visualizing the extent of diffusing NO from single vascular endothelial cells. The present cell-based indicator provides a powerful tool for uncovering picomolar dynamics of NO that regulate a wide range of cell functions in biological systems (27).

2.2. Phosphatidylinositol-3,4,5-Trisphosphate

Phosphatidylinositol-3,4,5-trisphosphate (PIP$_3$) regulates diverse cellular functions, including cell proliferation and apoptosis, and has roles in the progression of diabetes and cancer. However, little is known about its production. Fluorescent indicators for PIP$_3$ have been developed based on FRET (13). These novel PIP$_3$ indicators are composed of two distinctly colored mutants of GFP and a PIP$_3$-binding domain. The PIP$_3$ level was observed by dual-emission ratio imaging, thereby allowing stable observation without the problem of artifacts. Furthermore, these indicators were fused with localization sequences to direct them to the plasma membrane or endomembranes, allowing localized analysis of PIP$_3$ concentrations. Using these fluorescent indicators, we analyzed the spatiotemporal regulation of the PIP$_3$ production in single living cells. To examine PIP$_3$ dynamics, a pleckstrin homology (PH) domain from GRP1 was used, which selectively binds PIP$_3$, fused between cyan and yellow fluorescent protein (CFP and YFP, respectively) variants through rigid α-helical linkers, 12 of which consist of repeated EAAAR sequences. Within one of the rigid linkers, a single diglycine motif was introduced as a hinge. We then tethered the chimeric indicator protein to the membrane by fusing it with a membrane localization sequence through the rigid α-helical linker. Thus, after PI(3)K activation, the PH domain binds to PIP$_3$ and a significant conformational change of the indicator protein occurs through the flexible diglycine motif introduced into the rigid α-helical linker. This "flip-flop-type" conformational change of the indicator protein changes

a

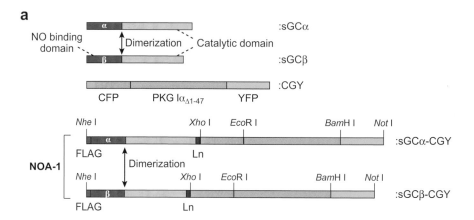

FLAG: MDYKDDDDK
Ln: GGEQKLISEEDLLESR

b

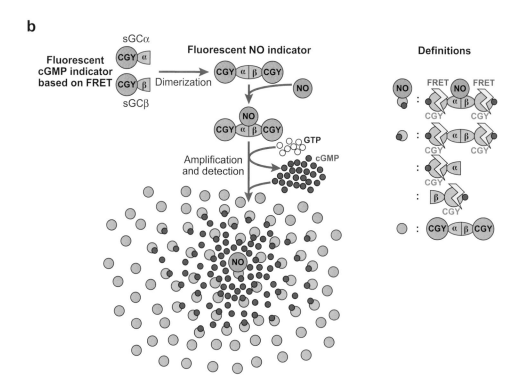

intramolecular FRET from CFP to YFP, allowing detection of PIP$_3$ dynamics at the membrane. We named this indicator fllip (fluorescent indicator for a lipid second messenger that can be tailor made).

The developed fllip allows a spatiotemporal examination of PIP$_3$ production in single living cells. After ligand stimulation, PIP$_3$ levels increased to a larger extent at the endomembranes (i.e., the endoplasmic reticulum and the Golgi) than at the plasma membrane. This increase was found to originate from in situ production at the endomembranes, a process stimulated directly by receptor tyrosine kinases endocytosed from the plasma membrane to the endomembranes. The demonstration of PIP$_3$ production through receptor endocytosis addresses a long-standing question about how signaling pathways downstream of PIP$_3$ are activated at intracellular compartments remote from the plasma membrane (13) (**Figure 2**).

We described fluorescent indicators for a lipid second messenger, diacylglycerol (DAG), which allow the localized analysis of DAG dynamics at subcellular membranes. We have shown that DAG concentrations increase and/or decrease at not only the plasma membrane, but also at organelle membranes such as endomembranes and mitochondrial outer membranes (13).

3. PROTEIN PHOSPHORYLATIONS

Protein phosphorylation by intracellular kinases plays one of the most pivotal roles in signaling pathways within cells. The kinase proteins catalyze transfer of the phosphate of ATP and phosphorylation of the hydroxy groups of serines, threonines, and/or tyrosines on the substrate proteins. Upon phosphorylation, the substrate proteins undergo conformational changes caused by the negative charges of the phosphates, which subsequently trigger their enzymatic activation and interaction with their respective target proteins. To reveal the biological issues related to the kinase proteins, electrophoresis, immunocytochemistry, and in vitro kinase assay have been used. However, these conventional methods do not provide sufficient information about the spatial and temporal dynamics of signal transduction based on protein phosphorylation and dephosphorylation in living cells. To overcome the limitations

---

Figure 1

An amplifier-coupled fluorescent indicator for visualizing nitric oxide (NO) in single living cells. (*a*) Schematic representations of domain structures of soluble guanylate cyclase (sGC), CGY, sGC-CGY, and sGC-CGY. The amino acid sequence of FLAG tag and linker (Ln) is shown at the bottom. The heterodimer of sGC-CGY and sGC-CGY has been named NOA-1. (*b*) Principle of the NO indicator NOA-1. sGC-CGY and sGC-CGY are spontaneously associated to form a matured heterodimer, NOA-1. NOA-1 binds with NO and generates cyclic guanosine 3′,5′-monophosphate (cGMP) at the rate of 3000–6000 molecules/min. Thus, generated cGMP binds to the CGY domain in NOA-1 and causes NOA-1 to emit a fluorescence resonance energy transfer (FRET) signal. About 99.9% of cGMP molecules is thus generated diffusely and is bound to NO-free NOA-1. As a result, even a single NO molecule can trigger a large amount of NOA-1 to emit FRET signals. Even if sGC-CGY and sGC-CGY exist as monomers, the monomers also emit FRET signals upon binding with generated cGMP. Abbreviations: GFP, green fluorescent protein; PKG, protein kinase G.

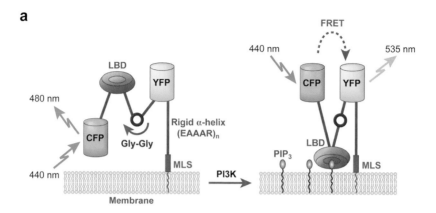

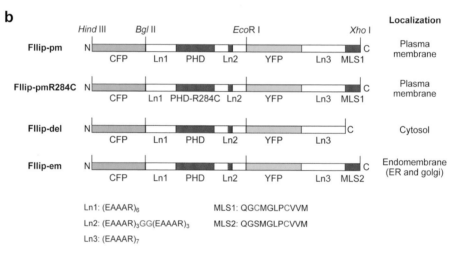

Ln1: (EAAAR)₆   MLS1: QGCMGLPCVVM

Ln2: (EAAAR)₃GG(EAAAR)₃   MLS2: QGSMGLPCVVM

Ln3: (EAAAR)₇

Figure 2

Fluorescent indicators for PtdInsP3 in single living cells. (*a*) Principle of fluorescent indicator for a lipid second messenger that can be tailor made (fllip) for visualizing PtdInsP3. Cyan fluorescent protein (CFP) and yellow fluorescent protein (YFP) are different-colored mutants of green fluorescent protein with mammalian codons and additional mutations. After binding of PtdInsP3 to the pleckstrin homology (PH) domain within fllip, a flip-flop-type conformational change occurs, changing the efficiency of fluorescence resonance energy transfer from CFP to YFP. (*b*) Schematic representations of domain structures within the present fllips. The PH domain is derived from human GRP1 (amino acids 261–382) and selectively binds to PtdInsP3. PH domain R284C is a mutant PH domain, in which Arg 284 is replaced with cysteine, abolishing binding to PtdInsP3. Abbreviations: ER, endoplasmic reticulum; FRET, fluorescence resonance energy transfer; Ln1, Ln2, and Ln3, linkers (the amino-acid sequences of which are shown at bottom); MLS1 and MLS2, membrane localization sequences for the plasma membrane and endomembranes, respectively, the amino-acid sequences of which are shown at bottom.

of investigating kinase signaling, genetically encoded fluorescent indicators have been developed for visualizing protein phosphorylation in living cells (14).

Besides traditional genomic pathways of sex steroid receptors in the nucleus, the extranuclear nongenomic pathways of these receptors have also been shown to strongly relate to many biological consequences, including vascular protection and cell proliferation (28–30). These nongenomic pathways are rapidly mediated through several critical protein kinases. A nonreceptor protein tyrosine kinase, Src, is known to be activated immediately after a steroid stimulation (31, 32). The activated Src phosphorylates various substrate proteins, such as Shc, that finally induce ERK-dependent transcription (33, 34). To determine how the Src activity is nongenomically regulated by steroid receptors in single living cells, we developed a fluorescent indicator for Src kinase activity and named it Srcus. This indicator can monitor substrate phosphorylation by activated endogenous Src as a FRET response in single living cells.

Based on the fluorescence imaging with the present fluorescent indicators, we demonstrated that E2-induced Src activation takes place in not only plasma- but also endomembranes. This was ascribed to the existence of epidermal growth factor (EGF) and occurrence of EGF receptor– (EGFR-)involved endocytosis of estrogen receptor (ER) together with Src. EGFR, ER, and Src were found to form a ternary complex upon E2 stimulation. The cell growth of breast cancer–derived MCF-7 cells increased markedly through the above EGF-involved estrogen-signaling process. In contrast to estrogen-activated Src signaling, the male steroid hormone, 5α-dihydroxytestosterone (DHT), was found to activate Src only in the plasma membrane free from the interaction of EGFR with androgen receptor (AR). The cell growth occurred only moderately as a result. The spatial difference in Src activation between E2 and DHT may be responsible for the different rates of MCF-7 cell growth between E2 and DHT (39) (**Figure 3**).

Extracellular signal–regulated kinase (ERK) is a serine/threonine protein kinase that regulates a wide variety of cell functions such as cell growth and differentiation. To study the spatiotemporal dynamics of protein phosphorylation by activated ERK in living cells, we have developed genetically encoded fluorescent indicators for ERK (39). The present indicators change their conformation upon protein phosphorylation by activated ERK and then emit fluorescence signals based on FRET. We visualized the cytosolic and nuclear activity of ERK using the present indicators. We thus found that the activation duration of ERK is considerably different between the cytosol and nucleus in living cells. The subcellular differences in ERK activity may be fundamental to the regulation of cell functions by ERK (**Figure 4**).

Cholesterol-enriched nanodomains called lipid rafts are thought to act as a platform for protein signaling in cells, but the physiologic significance of lipid rafts in cells and tissues is still unknown (35, 36). The main point of the present work is to show the physiologic significance of kinase activity in lipid rafts. Src family kinases (SFKs) are known to be distributed throughout the cell membranes and to regulate many biological processes (37, 38). To locate with high resolution where SFK activation occurs in the cell membranes, we developed a genetically encoded transmembrane fluorescent indicator for detecting the SFK activation at the cell membranes, and named it TM-Srcus (14, 39). TM-Srcus can monitor the substrate

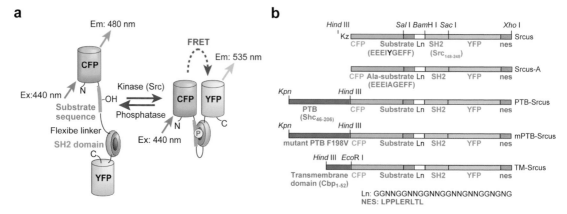

Figure 3

Fluorescent indicators visualize rapid Src signaling stimulated by sex steroids. (*a*) Principle of Srcus for monitoring Src activation. Cyan fluorescent protein (CFP) and yellow fluorescent protein (YFP) are different-colored mutants of green fluorescent protein (GFP). Upon phosphorylation of the substrate sequence within Srcus by Src, the adjacent tyrosine phosphorylation recognition (SH2) domain binds with the phosphorylated substrate sequence, which increases the efficiency of fluorescence resonance energy transfer between the GFP mutants within Srcus. (*b*) Schematic representations of domain structures of Srcuses. Restriction sites are also shown with the constructs. A Kozak sequence (*Kz*) allows optimal translation initiation in mammalian cells. The amino acid sequence of the substrate sequence is EEEIYGEFF, which is preferentially phosphorylated by Src. The SH2 domain is derived from Src-(148–248). The phosphotyrosine binding (PTB) domain is derived from Shc-(46–206). The transmembrane domain is from Cbp-(1–52). In the mutant PTB F198V domain, an amino acid residue at position 198 of Shc-(46–206) is mutated from phenylalanine to valine. Abbreviations: FRET, fluorescence resonance energy transfer; Ln, flexible linker sequence (GGNNGGNNGGNNGGNNGGNNGGNGNG); NES, nuclear-export-signal sequence derived from the human immunodeficiency virus protein Rev.

phosphorylation by activated SFKs as the decrease in the CFP/YFP emission ratio through FRET. The total internal reflection fluorescence imaging of the SFK activation on the plasma membrane by TM-Srcus showed that SFK activation takes place in lipid rafts.

Based on this finding, we developed a lipid raft–targeted SFK inhibitory fusion protein (LRT-SIFP) to inhibit SFK activity in lipid rafts. The LRT-SIFP contains the peptide inhibitor of SFK and the targeting sequence for localizing the SIFP to lipid rafts. The significance of the subcellular locations of kinase activity has hardly been studied by conventional inhibition methods such as small interfering RNA and chemical or peptide inhibitors, which inhibit kinase activity in cells. The present LRT-SIFP highlighted the importance of SFK activation in lipid rafts in the function of breast cancer cells. Although it is highly potent, the previously developed peptide inhibitor of SFK does not affect cell functions of MCF-7 cells derived from human breast cancer. In contrast to this conventional peptide inhibitor, the present LRT-SIFP inhibits cell adhesion and cell cycle progression of MCF-7 and MDA-MB231 cells. In addition, these inhibitory effects of LRT-SIFP on cell functions are

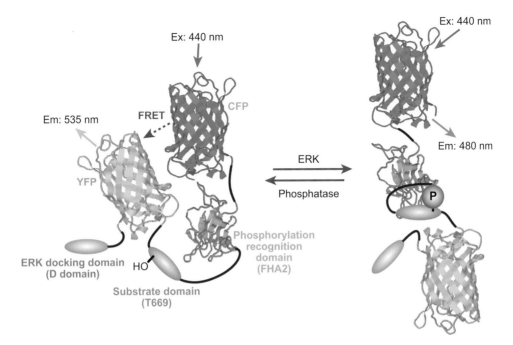

Figure 4

Design of the present fluorescent indicator to visualize protein phosphorylation by extracellular signal–related kinase (ERK), Erkus. Erkus consists of CFP, FHA2 domain, substrate domain, yellow fluorescent protein (YFP), and D domain. Cyan fluorescent protein (CFP) and YFP are different-colored mutants of *Aequorea victoria* fluorescent proteins. Upon phosphorylation, the substrate domain binds with the FHA2 domain, resulting in the decrease of fluorescence resonance energy transfer (FRET) efficiency from CFP to YFP.

specific for tumor cell lines. The lipid raft–specific knockdown of SFK activity would potentially be useful for selective cancer therapy to prevent tumorigenesis and metastasis of breast cancer cells (**Figure 5**).

4. PROTEIN-PROTEIN INTERACTIONS

Protein-protein interactions play pivotal roles in many chemical processes in living cells, yet they have been among the most difficult aspects of molecular and cellular biology to be studied. Monitoring protein-protein interactions in living cells is important for screening and assaying chemicals that increase or inhibit cellular signaling processes. To promote a greater understanding of the chemical processes, several methods have been developed for detecting protein-protein interactions. Available information about protein-protein interactions was obtained mostly via biochemical methods, but these methods required destructive analysis, which did not provide us with live-cell dynamics. The yeast two-hybrid system (40, 41) and the mammalian two-hybrid system (42, 43) use a "bait" protein fused to a DNA binding domain with a nuclear localization signal (NLS) in order to find their "prey" protein connected

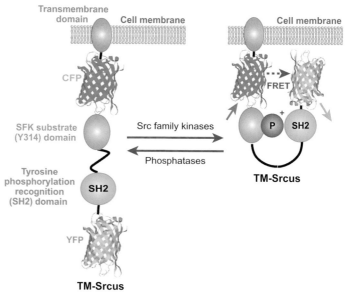

Figure 5

TM-Srcus, a fluorescent indicator for detecting Src family kinase (SFK) activation in cell membranes. The principle of TM-Srcus for visualizing SFK activation in cell membranes. On SFK activation, a conformational change in TM-Srcus occurs as a result of the binding of a tyrosine phosphorylation recognition (SH2) domain to a phosphorylated SFK substrate (Y314) domain, which results in an intramolecular fluorescence resonance energy transfer response. Abbreviation: FRET, fluorescence resonance energy transfer.

to a transcription activation domain with an NLS. The interaction between bait and prey accumulates the transcription activation domain on a specific sequence of DNA located upstream of a reporter gene; thus, the reporter gene expression is transactivated. Although significant signals for detection are obtained with the two-hybrid systems, they are limited in that detectable protein-protein interactions occur only in the nucleus to transactivate the reporter gene (44–46).

The split ubiquitin system (47–49) for detecting an interaction between a membrane protein and a cytosolic protein also limits the detection of interactions between cytoplasm proteins or nuclear localizing proteins. Several methods have been reported as useful in overcoming these limitations, including the protein complementation system such as using firefly luciferase (50), and *Renilla* luciferase (20, 51). We previously proposed a novel concept, a protein reconstitution system based on protein splicing, for detecting protein-protein interactions (18, 19, 52) and protein localization in organelles (22, 24). Although the protein complementation system and the protein reconstitution system allow us to monitor interactions between cytoplasm and membrane-proximal proteins, they suffer from low-endpoint signals for the interactions due to the much lower activity of complemented or reconstituted split reporters compared with that of intact reporter proteins. Here we describe a protein splicing–based reporter gene assay to monitor protein-protein interactions

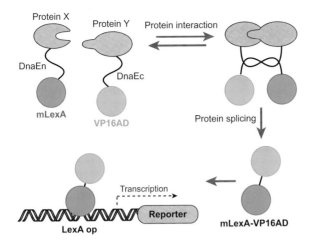

Figure 6

Principle for the intein-mediated reporter gene assay. DnaEn (amino acids 1–123) and DnaEc (amino acids 1–36) are connected with modified LexA (mLexA) (amino acids 1–229) and the transcription activation domain of a herpes simplex virus protein (VP16AD) (amino acids 411–456), respectively. Interested proteins X and Y are linked to the ends of DnaEn and DnaEc, respectively. Interaction between X and Y accelerates the folding of DnaEn and DnaEc, and protein splicing results. mLexA and VP16AD are linked together by a peptide bond to obtain a transcriptional activity.

in mammalian cells. Protein splicing is a posttranslational autocatalytic process in which an intein is excised with the concomitant ligation of the flanking exteins (53–56). An important property of the protein splicing is that the substitution of exteins for different peptides does not interfere with the splicing process (57, 58). We chose N- and C-terminal halves of an Ssp-DnaE intein as the protein-splicing elements, and modified LexA (mLexA) and a transcription activation domain of a herpes simplex virus protein (VP16AD) as the transcription factors. The present reporter gene assay allowed us to detect EGF-induced membrane-proximal Ras-Raf-1 interactions that could not be detected with the previous reporter gene assay, the two-hybrids method. In addition, the present reporter gene assay enabled us to obtain sufficient signals for the interactions that were not identified with the firefly luciferase complementation system (**Figure 6**).

We developed an approach for discriminating agonist and antagonist in a nongenomic steroid-signaling pathway using an association of AR with Src. We constructed a pair of genetically encoded indicators, where N- and C-terminal fragments of split firefly luciferase (FLuc) were fused to AR and Src, respectively. The proteins fused with AR and Src are localized in the cytoplasm and on the plasma membrane, respectively. Upon being activated with androgen, AR undergoes an intramolecular conformational change and binds with Src. The association causes the complementation of the split FLuc and recovery of FLuc activity. The resulting luminescence intensities were taken as a measure of the rapid hormonal activity of steroids in the nongenomic AR signaling (**Figure 7**) (24, 25, 59, 60).

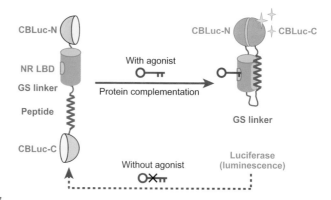

Figure 7

Schematic diagram showing the detection scheme of the single-molecule-format bioluminescent indicator based on an intramolecular complementation strategy of split click beetle luciferase (CBLuc) for monitoring bioactive small molecules. An agonist induces conformational change in the ligand binding domain of a nuclear receptor (NR LBD). It subsequently activates the association of NR LBD with the specific recognition peptide sequence. The association triggers the recovery of the CBLuc activities by an intramolecular complementation of split CBLuc. The recovered luciferase activities are taken as a measure of the androgenicity of ligands. On the other hand, removal of the agonist dissociates the complementation between NR LBD and the motif and cancels the developed CBLuc activities. The agonist was animated with a key, whereas the split CBLuc was drawn as a half-segmented ball. Abbreviations: CBLuc-N, N-terminal fragment of click beetle luciferase; CBLuc-C, C-terminal fragment of click beetle luciferase; GS linker, a flexible amino acid sequence consisting of glycines and serines.

5. PROTEIN-LIGAND INTERACTIONS

Click beetle luciferase (CBLuc) is insensitive to pH, temperature, and heavy metals, and emits a stable, highly tissue-transparent red light with luciferin in physiological circumstances. Thus, the luminescence signal is optimal for a bioanalytical index reporting the magnitude of a signal transduction of interest. We validated a single-molecule-format complementation system of split CBLuc to study signal-controlled protein-protein (peptide) interactions (61). First, we generated 10 pairs of N- and C-terminal fragments of CBLuc to examine whether a significant recovery of the activity occurs through the intramolecular complementation. The ligand binding domain of androgen receptor (AR LBD) was connected to a functional peptide sequence through a flexible linker. The fusion protein was then sandwiched between the dissected N- and C-terminal fragments of CBLuc. Androgen induces the association between AR LBD and a functional peptide and the subsequent complementation of N- and C-terminal fragments of split CBLuc inside the single-molecule-format probe, which restores the activities of CBLuc. Examination of the dissection sites of CBLuc revealed that the dissection positions next to the amino acids D412 and I439 admit a stable recovery of CBLuc activity through an intramolecular complementation.

Ligand-induced conformational changes of nuclear hormone receptors (NRs) are important initiators of various kinds of hormone signaling. However, little is known of the bioanalytical use of the hormone-induced conformational changes of NRs. Here, we describe a generally applicable bioluminescence assay with a genetically encoded bioluminescent indicator to determine androgenicity of ligands based on the intramolecular association of the ligand-binding domain of androgen receptor (AR LBD) with the "FQNLF" motif in the N-terminal domain of AR (AR NTD). FLuc was dissected into N-terminal (1–415 AA) and C-terminal (416–550 AA) fragments. The AR LBD and FQNLF motif of AR NTD were sandwiched between the dissected fragments of FLuc to construct a single-molecule-format bioluminescent probe. Androgens induce the association of AR LBD with the FQNLF motif in the NTD, and the subsequent complementation of N- and C-terminal fragments of FLuc partially restores the activities of FLuc. A 10^{-5} M solution of DHT induced a quick increase in the luminescence intensities from cervical carcinoma–derived HeLa cells carrying the genetic indicator, which reached a plateau in 9 min, whereas DHT withdrawal from the cells by a medium change decreased the luminescence more slowly (i.e., 2 h elapsed until luminescence returned to the background level) (61). The present luminescent indicator was found to exhibit high agonist selectivity and reproducible recovery of the luminescence to a repeated androgen addition and withdrawal. This is the first contribution that cellular signaling steps can be imaged with bioluminescence using a single-molecule-format bioluminescence probe (Simbi), in which all the components required for signal sensing and visualization are integrated. Simbi is applicable to developing biotherapeutic agents effective to the AR signaling, and for screening adverse chemicals that possibly influence the signal transduction of AR (17, 24, 26, 62–64).

Firefly luciferase connected with a substrate sequence for caspase-3 (DEVD) is cyclized by a DnaE intein. When the cyclic luciferase is expressed in living cells, its activity is greatly decreased because of a steric effect. Activated caspase-3 cleaves the substrate sequence in the cyclic luciferase and the luciferase activity is restored. Quantitative sensing of time-dependent caspase-3 activity in living cells and in mice upon the application of extracellular stimuli has been demonstrated (**Figure 8**) (65–71).

6. NUCLEAR RECEPTOR–COREGULATOR INTERACTIONS

A sensitive fluorescent indicator was designed to visualize, in real time, the activities of the AR ligands in the physiological environment of single living cells (17). An androgen promotes interaction between the androgen receptor ligand binding domain (AR LBD) and coactivator protein. This results in an increase in FRET from CFP to YFP. The indicator is capable of distinguishing ligands of different potencies for the AR. The present assay is intended to indicate not the binding affinity of a drug, but rather the efficacy of a drug as either an antagonist or partial agonist in vivo. The permeability of a drug into cells and the conformational changes induced in the AR all determine its efficacy, much more so than a simple binding assay. Progesterone, glucocorticoid, and peroxisome proliferator-activated receptors (PR, GR, and PPAR, respectively) also belong to the NR family and play important roles in mediating the

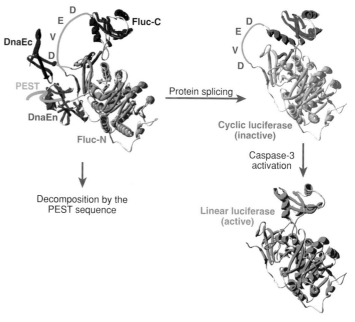

Figure 8

Strategy for the detection of caspase-3 activity. Principle for monitoring the activity of caspase-3 by using split firefly luciferase (Fluc). Abbreviation: PEST, prolinc, glutamic acid, serine, and threonine-rich.

actions of drugs for contraception (by PR), inflammation (by GR), and type-2 diabetes (by PPARγ). Using the present strategy, indicators for PR, GR, and PPARγ can be developed for screening and characterization of their ligands. The indicators would be helpful in the development of NR-based pharmaceutical drugs for the treatment of different diseases (**Figure 9**).

Selective nuclear receptor modulators (SNRMs), which are used clinically for the treatment of NR-related diseases, display mixed agonistic/antagonistic activity in a tissue-selective manner depending on the cellular concentrations of coregulator proteins, (e.g., coactivators and corepressors). The molecular details of the SNRM function provided us with an idea for a rational method for the high-throughput screening of SNRMs in real time in intact living cells. We have developed genetically encoded fluorescent indicators based on the principle of ligand-induced co-activator and/or corepressor recruitment to NR ligand-binding domain in single living cells. We demonstrated that an SNRM induces a distinct conformational change in the NR LBD, which differ from that induced by a full agonist or antagonist, favorable for the recruitment of a coactivator or corepressor protein to the NR. The molecular details of an SNRM-binding NR and the subsequently induced conformational changes are important to the understanding of SNRM action in the living body. Our fluorescent indicators are capable of distinguishing among agonists, antagonists, and SNRMs, and can therefore serve as versatile molecular sensors that predict

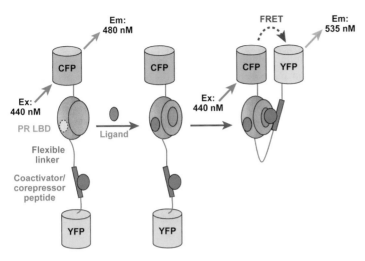

Figure 9

Fluorescent indicator for the ligand-induced coactivator/corepressor recruitment to the progesterone ligand binding domain (PR LBD) in living cells. Principle of the coactivator/corepressor-based ligand-induced/fluorescent indicator based on intramolecular fluorescence resonance energy transfer (FRET) to visualize the ligand-dependent interaction between the PR LBD and the steroid receptor coactivator 1 peptide/silencing mediator for retinoid and thyroid hormone receptor. Upon ligand binding, the PR LBD and coactivator/corepressor interact with each other. Consequently, yellow fluorescent protein (YFP) is oriented in close proximity to cyan fluorescent protein (CFP); this results in an increase in the FRET response. The magnitude of the FRET increase strongly depends on the relative orientation and distance between the donor (CFP) and acceptor (YFP) fluorophore.

the pharmacological character ligands, which is important for accurate treatment of disease (9, 72–75).

One member of the NR superfamily, PPAR (described above) plays an important role in modulation of insulin sensitivity in type 2 diabetes. Ligand-dependent protein-protein interactions between NRs and NR coactivators are critical in regulation of transcription. To visualize the ligand-induced coactivator recruitment to PPAR in live cells, we developed a genetically encoded fluorescent indicator in which PPAR ligand binding domain (PPAR LBD) was connected to a steroid receptor coactivator peptide that contains LXXLL motif (where L = leucine and X = any amino acid) through a flexible linker. This fusion protein was inserted between CFP and YFP, the donor and acceptor fluorophores, respectively. Monitoring real-time ligand-induced conformational change in the PPAR LBD to interact with the coactivator allowed screening of natural and synthetic ligands (drugs for the treatment of type 2 diabetes) in single living cells using intramolecular FRET microscopy. The high sensitivity of the present indicator made it possible to distinguish between strong and weak affinity ligands for PPAR in a dose-dependent fashion immediately after adding a ligand to live cells. The indicator can discriminate agonist from antagonist compounds efficiently within a few minutes. The present system may be promising for the development of PPAR-targeted drugs for type 2 diabetes and inflammation.

7. NUCLEOCYTOPLASMIC TRAFFICKING OF FUNCTIONAL PROTEINS

Nucleocytoplasmic trafficking of functional proteins plays a key role in regulating gene expressions in response to extracellular signals. A genetically encoded bioluminescent indicator was developed for monitoring the nuclear trafficking of target proteins in vitro and in vivo (24). The principle is based on reconstitution of split fragments of *Renilla reniformis* (Rluc) by protein splicing with a DnaE intein. A target cytosolic protein fused to the amino-terminal half of Rluc is expressed in mammalian cells. If the protein translocates into the nucleus, the Rluc moiety meets the C-terminal half of Rluc, which is localized in the nucleus with a fused NLS, and full-length Rluc is reconstituted by protein splicing. The bioluminescence is thereby emitted with coelenterazine as the substrate. The principle of the approach is an extension of the method developed earlier for identifying mitochondrial proteins (22) (**Figure 10**).

The method of cell-based screening with the genetically encoded indicator provided a quantitative measure of the extent of nuclear translocation of AR upon stimulation with various chemicals. Currently, high-throughput screening tools for protein translocation into the nucleus have mostly depended upon the GFP- (or its variant) tagged approach in combination with the fluorescence microscopy and computer-driven imaging system. The system offers only semiquantitative information, as it is difficult to accurately distinguish the fluorescence of GFP-tagged proteins localized only in the nucleus from that left in the cytosol. In addition, the precision of the observed fluorescence intensities from the nucleus obtained with the statistical analysis is not high because the number of cells examined under a

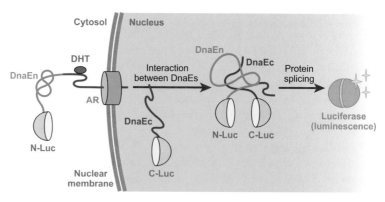

Figure 10

When androgen receptor (AR) is bound to 5α-dihydrotestosterone (DHT), it translocates into the nucleus and brings the N- and C-terminal halves of DnaEs close enough to fold correctly, thereby initiating protein splicing to link the concomitant Rluc halves with a peptide bond. The C-terminal half of split Rluc was located beforehand in the nucleus by a fused nuclear localization signal. The cells containing this reconstituted Rluc allow monitoring of nuclear translocation of AR with its luminescence using coelenterazine as the substrate.

fluorescence microscope is limited. However, the present method enabled determination of the subcellular localization of AR by the luminescence signals generated only when the AR localized in the nucleus. AR remaining in the cytosol did not induce reconstitution of split Rluc and therefore no background luminescence was observed (24).

Polychlorinated biphenyl (PCB) and procymidone have been suspected of having neurotoxic and antiandrogenic effects, respectively, and they possibly adversely influence hormonal activities in living animals' brains. We demonstrated the usefulness of the split rLuc reporter for monitoring AR translocation into the nucleus in living mice by implanting COS-7 cells in the mouse brain at a depth of 3 mm and measuring emitted bioluminescence with a cooled CCD camera; we thereby investigated the distribution of these chemicals in the brains of living mice. As expected, 2 h after intraperitoneally injecting PCB or procymidone, both chemicals were found to completely inhibit the DHT-stimulated translocation of AR, where coelenterazine was injected intracerebrally (24).

Similar genetically encoded bioluminescent probes were developed for illuminating protein nuclear transport induced by phosphorylation or proteolysis (25). A genetically encoded stress indicator was also reported as noninvasively imaging endogenous corticosterone in living mice (26) (**Figure 11**).

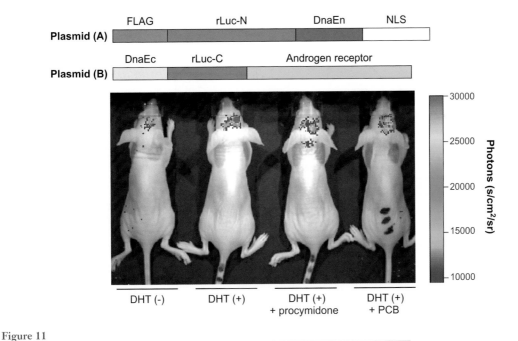

Figure 11

Effects of inhibitors on androgen receptor (AR) translocation into the nucleus in the mouse brain. Polychlorinated biphenyl (PCB) and procymidone were found to have an ability to pass through the blood-brain barrier, to reach the brain, and to inhibit the AR signal transduction in the organ.

8. PROTEIN LOCALIZATIONS

Smac/DIABLO is an intermembrane space– (IMS-)localized protein that is conserved in mammals (76, 77). The Smac/DIABLO protein derived from *Mus musculus* is synthesized as a precursor molecule of 237 amino acids; the N-terminal 53 residue serves as the mitochondrial targeting sequence, which is removed by the inner membrane peptidase complex after import (78). Thus, the mature Smac/DIABLO protein has 184 amino acids, and the four residues of the N terminus are Ala-Val-Pro-Ile. These four N-terminal residues play an indispensable role in Smac/DIABLO function: They promote apoptosis by eliminating the inhibitory effect of the inhibitor of apoptosis protein (IAP) through physical interaction (79–81). A point mutation in the four residues leads to a loss of interaction with IAP and a concomitant loss of the Smac/DIABLO function. The structural and functional aspects of Smac/DIABLO have been extensively investigated, but how the protein targets into the mitochondrial IMS remains unknown. To identify the amino acids or domains of Smac/DIABLO that are important for targeting into the IMS, we have developed a high-throughput screening system that enables discrimination between the proteins in the IMS and

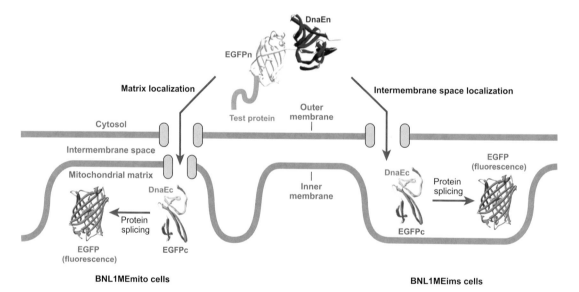

Figure 12

Enhanced green fluorescent protein (EGFP) reconstitution by protein splicing in submitochondrial compartments. When a test protein is localized in the mitochondrial matrix (*left*) or in the intermembrane space (IMS) (*right*), N- and C-terminal DnaE are brought close together, and EGFP is formed by protein splicing in the respective compartments. BNL1MEmito cells permanently express a fusion composed of C-terminal fragments of DnaE and EGFP in the mitochondrial matrix (*left*), and BNL1MEims cells express the same fusion in the IMS (*right*). The colored three-dimensional structures represent DnaEn and DnaEc, and the gray structures represent N- and C-terminal fragments of EGFP, respectively. The attached strand represents a test protein.

those in the mitochondrial matrix or cytosol. Using this system, we identified from Smac/DIABLO mutant libraries the amino acids necessary for the localization of the IMS. Further, we showed that amino acid residues 54–57 (Ala-Val-Pro-Ile), which are crucial for the apoptotic function, are also important for the IMS localization. We found that N-terminal amino acid residues 10–57 (RSVCSLFRYRQRFPVLAN-SKKRCFSELIKPWHKTVLTGFGMTLCAVPI) are the minimal sequence that functions as the IMS-targeting signal. We demonstrated that this IMS-targeting signal is able to deliver different probe proteins and intrabodies into the IMS (**Figure 12**).

We developed genetically encoded RNA probes for characterizing localization and dynamics of mitochondrial RNA (mtRNA) in single living cells. The probes consist of two RNA-binding domains of PUMILIO1, each connected with split fragments of a fluorescent protein capable of reconstituting upon binding to a target RNA. We

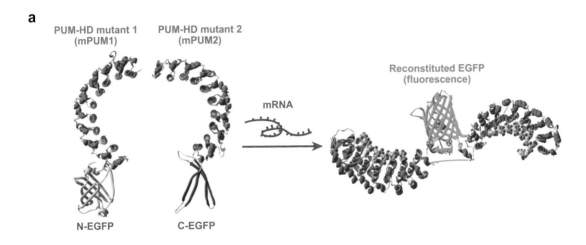

a

PUM-HD mutant 1 (mPUM1) PUM-HD mutant 2 (mPUM2)

mRNA

Reconstituted EGFP (fluorescence)

N-EGFP C-EGFP

b

mtRNA sequences for binding of PUM-HD mutants

ND6 mRNA (186-210): 5'- AA UGAUGGUU GUCUU UGGAUAUA CU-3'
                              mPUM1   mPUM2

ND1 mRNA (540-551): 5'-CU UGGCCAUA AU-3'
                             mPUM2

Figure 13

Detection of a target mRNA based on complementation of split enhanced green fluorescent protein (EGFP) fragments. (*a*) Schematic of the basic strategy for detecting target mRNAs. Two RNA-binding domains of pumilio (PUM)-HD are engineered to recognize specific sequences on a target mRNA (mPUM1-RNA and mPUM2-RNA). In the presence of the target mRNA, mPUM1 and mPUM2 bind to their target sequences, bringing together the N- and C-terminal fragments of EGFP. This results in functional reconstitution of the fluorescent protein. (*b*) Sequences of mitochondrial RNA (mtRNA) that are recognized by PUM-HD mutants.

designed the probes to specifically recognize a 16-base sequence of mtRNA-encoding nicotinamide adenine dinucleotide (NADH) dehydrogenase subunit 6 (ND6) and to be targeted into the mitochondrial matrix, which allowed real-time imaging of ND6 mtRNA localization in living cells. We showed that ND6 mtRNA is localized within mitochondria and is concentrated particularly on mitochondrial DNA. Movement of the ND6 mtRNA is restricted, but oxidative stress induces the mtRNA to disperse in the mitochondria and gradually decompose. These probes provide a means of studying spatial and temporal mtRNA dynamics in intracellular compartments in living mammalian cells (82) (**Figure 13**).

DISCLOSURE STATEMENT

The author is not aware of any biases that might be perceived as affecting the objectivity of this review.

ACKNOWLEDGMENTS

This work was supported by the Japan Science and Technology Agency (JST), and the Japan Society for the Promotion of Science (JSPS).

LITERATURE CITED

1. Grynkiewicz G, Poenie M, Tsien RY. 1985. A new generation of Ca^{2+} indicators with greatly improved fluorescence properties. *J. Biol. Chem.* 260:3440–50

2. Kojima H, Nakatsubo N, Kikuchi K, Kawahara S, Kirin Y, et al. 1998. Detection and imaging of nitric oxide with novel fluorescent indicators: diaminofluoresceins. *Anal. Chem.* 70:2446–53

3. Komatsu H, Iwasawa N, Citterio D, Suzuki Y, Kubota T, et al. 2004. Design and synthesis of highly sensitive and selective fluorescein-derived magnesium fluorescent probes and application to intracellular 3D Mg^{2+} imaging. *J. Am. Chem. Soc.* 126:16353–60

4. Hirano T, Kikuchi K, Urano Y, Higuchi T, Nagano T. 2000. Novel zinc fluorescent probes excitable with visible light for biological applications. *Angew. Chem. Int. Ed.* 39:1052–54

5. Chalfie M, Kain S, eds. 1998. *Green Fluorescent Protein: Properties, Applications, and Protocols.* New York: Wiley-Liss

6. Zacharias DA, Baird GS, Tsien RY. 2000. Recent advances in technology for measuring and manipulating call signals. *Curr. Opin. Neurobiol.* 10:416–21

7. Miyawaki A, Llopis J, Heim R, McCaffery JM, Adams JA, et al. 1997. Fluorescent indicators for Ca^{2+} based on green fluorescent proteins and calmodulin. *Nature* 388:882–87

8. Zaccolo M, Giorgi FD, Cho CY, Feng L, Knapp T, et al. 2000. A genetically encoded, fluorescent indicator for cyclic AMP in living cells. *Naure Cell Biol.* 2:25–29

9. Sato M, Hida N, Umezawa Y. 2005. Imaging the nanomolar range of nitric oxide with an amplifier-coupled fluorescent indicator in living cells. *Proc. Natl. Acad. Sci. USA* 102:14515–20

10. Hirose K, Kadowaki S, Tanabe M, Takeshima H, Iino M. 1999. Spatiotemporal dynamics of inositol 1,4,5-trisphosphate that underlies complex Ca^{2+} mobilization patterns. *Science* 284:1527–30

11. Sato M, Ueda Y, Shibuya M, Umezawa Y. 2005. Locating inositol 1,4,5-trisphosphate in the nucleus and neuronal dendrites with genetically encoded fluorescent indicators. *Anal. Chem.* 77:4751–58

12. Sato M, Hida N, Ozawa T, Umezawa Y. 2000. Fluorescent indicators for cyclic GMP based on cyclic GMP-dependent protein kinase Iα and green fluorescent proteins. *Anal. Chem.* 72:5918–24

13. Sato M, Ueda Y, Takagi T, Umezawa Y. 2003. Production of PtdInsP3 at endomembranes is triggered by receptor endocytosis. *Nat. Cell Biol.* 5:1016–22

14. Sato M, Ozawa T, Inukai K, Asano T, Umezawa Y. 2002. Fluorescent indicators for imaging protein phosphorylation in single living cells. *Nat. Biotechnol.* 20:287–94

15. Sasaki K, Sato M, Umezawa Y. 2003. Fluorescent indicators for Akt/protein kinase B and dynamics of Akt activity visualized in living cells. *J. Biol. Chem.* 247:30945–51

16. Sato M, Umezawa Y. 2004. Imaging protein phosphorylation by fluorescence in single living cells. *Methods* 32:451–55

17. Awais M, Sato M, Lee X, Umezawa Y. 2006. A fluorescent indicator to visualize activities of the androgen receptor ligands in single living cells. *Angew. Chem. Int. Ed.* 45:2707–12

18. Ozawa T, Nogami S, Sato M, Ohya Y, Umezawa Y. 2000. A fluorescent indicator for detecting protein-protein interactions in vivo based on protein splicing. *Anal. Chem.* 72:5151–57

19. Ozawa T, Kaihara A, Sato M, Tachihara K, Umezawa Y. 2001. Split luciferase as an optical probe for detecting protein-protein interactions in mammalian cells based on protein splicing. *Anal. Chem.* 73:2516–21

20. Kaihara A, Kawai Y, Sato M, Ozawa T, Umezawa Y. 2003. Locating a protein-protein interaction in living cells via split *Renilla* luciferase complementation. *Anal. Chem.* 75:4176–81

21. Kanno A, Ozawa T, Umezawa Y. 2006. Intein-mediated reporter gene assay for detecting protein-protein interactions in living mammalian cells. *Anal. Chem.* 78:556–60

22. Ozawa T, Sako Y, Sato M, Kitamura T, Umezawa Y. 2003. Intein-mediated reporter gene assay for detecting protein-protein interactions in living mammalian cells. *Nat. Biotechnol.* 21:287–93

23. Ozawa T, Nishitani K, Sako Y, Umezawa Y. 2005. A high-throughput screening of genes that encode proteins transported into the endoplasmic reticulum in mammalian cells. *Nucleic Acids Res.* 33:e34

24. Kim SB, Ozawa T, Watanabe S, Umezawa Y. 2004. High-throughput sensing and noninvasive imaging of protein nuclear transport by using reconstitution of split *Renilla* luciferase. *Proc. Natl. Acad. Sci. USA* 101:11542–47

25. Kim SB, Takao R, Ozawa T, Umezawa Y. 2005. Quantitative determination of protein nuclear transport induced by phosphorylation or by proteolysis. *Anal. Chem.* 77:6928–34

26. Kim SB, Ozawa T, Umezawa Y. 2005. Genetically encoded stress indicator for noninvasively imaging endogenous corticosterone in living mice. *Anal. Chem.* 77:6588–93

27. Sato M, Nakajima T, Goto M, Umezawa Y. 2006. A cell-based indicator visualizes picomolar dynamics of nitric oxide release from living cells. *Anal. Chem.* 78:8175–82

28. Losel R, Wehling M. 2003. Nongenomic actions of steroid hormones. *Nat. Rev. Mol. Cell Biol.* 4:46–56

29. Marquez DC, Pietras RJ. 2001. Membrane-associated binding sites for estrogen contribute to growth regulation of human breast cancer cells. *Oncogene* 20:5420–30

30. Kousteni S, Chen JR, Bellido T, Han L, Ali AA, et al. 2002. Reversal of bone loss in mice by nongenotropic signaling of sex steroids. *Science* 298:843–46

31. Migliaccio A, Castoria G, Di Domenico M, de Falco A, Bilancio A, et al. 2000. Steroid-induced androgen receptor-oestradiol receptor β-Src complex triggers prostate cancer cell proliferation. *EMBO J.* 19:5406–17

32. Castoria G, Migliaccio A, Bilancio A, Di Domenico M, de Falco A, et al. 2001. PI3-kinase in concert with Src promotes the S-phase entry of oestradiol-stimulated MCF-7 cells. *EMBO J.* 20:6050–59

33. Migliaccio A, Di Domenico M, Castoria G, de Falco A, Bontempo P, et al. 1996. Tyrosine kinase/p21(ras)/MAP-kinase pathway activation by estradiol-receptor complex in MCF-7 cells. *EMBO J.* 15:1292–300

34. Kousteni S, Bellido T, Plotkin LI, O'Brien CA, Bodenner DL, et al. 2001. Nongenotropic, sex-nonspecific signaling through the estrogen or androgen receptors: dissociation from transcriptional activity. *Cell* 104:719–30

35. Simons K, Toomre D. 2000. Lipid rafts and signal transduction. *Nat. Rev. Mol. Cell Biol.* 1:31–39

36. Cary LA, Cooper JA. 2000. Molecular switches in lipid rafts. *Nature* 404:945–47

37. Sargiacomo M, Sudol M, Tang Z, Lisanti MP. 1993. Signal transducing molecules and glycosyl-phosphatidylinositol-linked proteins form a caveolin-rich insoluble complex in MDCK cells. *J. Cell Biol.* 122:789–807

38. Liang X, Nazarian A, Erdjument-Bromage H, Bornmann W, Tempst P, Resh MD. 2001. Heterogeneous fatty acylation of Src family kinases with polyunsaturated fatty acids regulates raft localization and signal transduction. *J. Biol. Chem.* 276:30987–94

39. Hitosugi T, Sasaki K, Sato M, Umezawa Y. 2007. Epidermal growth factor directs sex-specific steroid signaling through Src activation. *J. Biol. Chem.* 282:10697–706

40. Fields S, Song OK. 1989. A novel genetic system to detect protein protein interactions. *Nature* 340:245–46

41. Chien CT, Bartel PL, Sternglanz R, Fields S. 1991. The 2-hybrid system: a method to identify and clone genes for proteins that interact with a protein of interest. *Proc. Natl. Acad. Sci. USA* 88:9578–82

42. Dang CV, Barrett J, Villagarcia M, Resar LMS, Kato GJ, Fearon ER. 1991. Intracellular leucine zipper interactions suggest c-Myc hetero-oligomerization. *Mol. Cell. Biol.* 11:954–62

43. Fearon ER, Finkel T, Gillison ML, Kennedy SP, Casella JF, et al. 1992. Karyoplasmic interaction selection strategy: a general strategy to detect protein-protein interactions in mammalian cells. *Proc. Natl. Acad. Sci. USA* 89:7958–62

44. Flores A, Briand JF, Gadal O, Andrau JC, Rubbi L, et al. 1999. A protein-protein interaction map of yeast RNA polymerase III. *Proc. Natl. Acad. Sci. USA* 96:7815–20

45. Ito T, Tashiro K, Muta S, Ozawa R, Chiba T, et al. 2000. Toward a protein-protein interaction map of the budding yeast: a comprehensive system to examine two-hybrid interactions in all possible combinations between the yeast proteins. *Proc. Natl. Acad. Sci. USA* 97:1143–47

46. Walhout AJM, Sordella R, Lu XW, Hartley JL, Temple GF, et al. 2000. Protein interaction mapping in *C. elegans* using proteins involved in vulval development. *Science* 287:116–22

47. Johnsson N, Varshavsky A. 1994. Split ubiquitin as a sensor of protein interactions in-vivo. *Proc. Natl. Acad. Sci. USA* 91:10340–44

48. Stagljar I, Korostensky C, Johnsson N, te Heesen S. 1998. A genetic system based on split-ubiquitin for the analysis of interactions between membrane proteins in vivo. *Proc. Natl. Acad. Sci. USA* 95:5187–92

49. Dunnwald M, Varshavsky A, Johnsson N. 1999. Detection of transient in vivo interactions between substrate and transporter during protein translocation into the endoplasmic reticulum. *Mol. Biol. Cell* 10:329–44

50. Luker KE, Smith MCP, Luker GD, Gammon ST, Piwnica-Worms H, et al. 2004. Kinetics of regulated protein-protein interactions revealed with firefly luciferase complementation imaging in cells and living animals. *Proc. Natl. Acad. Sci. USA* 101:12288–93

51. Paulmurugan R, Gambhir SS. 2003. Monitoring protein-protein interactions using split synthetic *Renilla* luciferase protein-fragment-assisted complementation. *Anal. Chem.* 75:1584–89

52. Ozawa T, Takeuchi M, Kaihara A, Sato M, Umezawa Y. 2001. Protein splicing–based reconstitution of split green fluorescent protein for monitoring protein-protein interactions in bacteria: improved sensitivity and reduced screening time. *Anal. Chem.* 73:5866–74

53. Hirata R, Ohsumi Y, Anraku Y. 1989. Functional molecular masses of vacuolar membrane H^+-ATPase from *Saccharomyces cerevisiae* as studied by radiation inactivation analysis. *FEBS Lett.* 244:397–401

54. Kane PM, Yamashiro CT, Wolczyk DF, Neff N, Goebl M, et al. 1990. Protein splicing converts the yeast TFP1 gene-product to the 69-kD subunit of the vacuolar $H^{(+)}$-adenosine triphosphatase. *Science* 250:651–57

55. Noren CJ, Wang JM, Perler FB. 2000. Dissecting the chemistry of protein splicing and its applications. *Angew. Chem. Int. Ed.* 39:450–66

56. Paulus H. 2000. Protein splicing and related forms of protein autoprocessing. *Annu. Rev. Biochem.* 69:447–96

57. Cooper AA, Chen YJ, Lindorfer MA, Stevens TH. 1993. Protein splicing of the yeast TFP1 intervening protein-sequence: a model for self-excision. *EMBO J.* 12:2575–83

58. Chong SR, Xu MQ. 1997. Protein splicing of the *Saccharomyces cerevisiae* VMA intein without the endonuclease motifs. *J. Biol. Chem.* 272:15587–90

59. Migliaccio A, Castoria G, Di Domenico M, de Falco A, Bilancio A, et al. 2000. Steroid-induced androgen receptor–oestradiol receptor-Src complex triggers prostate cancer cell proliferation. *EMBO J.* 19:5406–17

60. Luker KE, Smith MC, Luker GD, Gammon ST, Piwnica-Worms H, Piwnica-Worms D. 2004. Kinetics of regulated protein-protein interactions revealed with firefly luciferase complementation imaging in cells and living animals. *Proc. Natl. Acad. Sci. USA* 101:12288–93

61. Kim SB, Otani Y, Umezawa Y, Tao H. 2007. Bioluminescent indicator for determining protein/protein interactions using intramolecular complementation of split click beetle luciferase. *Anal Chem.* 79:4820–26

62. Schaufele F, Carbonell X, Guerbadot M, Borngraeber S, Chapman MS, et al. 2005. The structural basis of androgen receptor activation: intramolecular and intermolecular amino-carboxy interactions. *Proc. Natl. Acad. Sci. USA* 102:9802–7

63. Langley E, Zhou ZX, Wilson EM. 1995. Evidence for an antiparallel orientation of the ligand-activated human androgen receptor dimmer. *J. Biol. Chem.* 270:29983–90

64. He B, Kemppainen JA, Wilson EM. 2000. FXXLF and WXXLF sequences mediate the NH2-terminal interaction with the ligand binding domain of the androgen receptor. *J. Biol. Chem.* 275:22986–94

65. Reits E, Griekspoor A, Neijssen J, Groothuis T, Jalink K, et al. 2003. Peptide diffusion, protection, and degradation in nuclear and cytoplasmic compartments before antigen presentation by MHC class I. *Immunity* 18:97–108

66. Nagai T, Miyawaki A. 2004. A high-throughput method for development of FRET-based indicators for proteolysis. *Biochem. Biophys. Res. Commun.* 319:72–77

67. Takemoto K, Nagai T, Miyawaki A, Miura M. 2003. Spatio-temporal activation of caspase revealed by indicator that is insensitive to environmental effects. *J. Cell Biol.* 160:235–43

68. Xu X, Gerard ALV, Huang BCB, Anderson DC, Payan DG, et al. 1998. Detection of programmed cell death using fluorescence energy transfer. *Nucleic Acids Res.* 26:2034–35

69. Mahajan NP, Harrison-Shostak DC, Michaux J, Herman B. 1999. Novel mutant green fluorescent protein protease substrates reveal the activation of specific caspases during apoptosis. *Chem. Biol.* 6:401–9

70. Weissleder R, Ntziachristos V. 2003. Shedding light onto live molecular targets. *Nat. Med.* 9:123–28

71. Laxman B, Hall DE, Bhojani MS, Hamstra DA, Chenevert TL, et al. 2002. Noninvasive real-time imaging of apoptosis. *Proc. Natl. Acad. Sci. USA* 99:16551–55

72. Xu HE, Stanley TB, Montana VG, Lambert MH, Shearer BG, et al. 2002. Structural basis for antagonist-mediated recruitment of nuclear corepressors by PPAR. *Nature* 415:813–17

73. Williams SP, Sigler PB. 1998. Transcriptional activation independent of TFIIH kinase and the RNA polymerase II mediator in vivo. *Nature* 393:392–96

74. Heery DM, Kalkhoven E, Hoare S, Parker MG. 1997. A signature motif in transcriptional coactivators mediates binding to nuclear receptor. *Nature* 387:733–36

75. Onate SAO, Tsai SY, Tsai MU, O'Malley BW. 1995. Sequence and characterization of a coactivator for the steroid hormone receptor superfamily. *Science* 270:1354–57

76. Du C, Fang M, Li Y, Li L, Wang X. 2000. Smac, a mitochondrial protein that promotes cytochrome c-dependent caspase activation by eliminating IAP inhibition. *Cell* 102:33–42

77. Verhagen AM, Ekert PG, Pakusch M, Silke J, Connolly LM, et al. 2000. Identification of DIABLO, a mammalian protein that promotes apoptosis by binding to and antagonizing IAP proteins. *Cell* 102:43–53

78. Burri L, Strahm Y, Hawkins CJ, Gentle IE, Puryer MA, et al. 2005. Mature DIABLO/Smac is produced by the IMP protease complex on the mitochondrial inner membrane. *Mol. Biol. Cell* 16:2926–33

79. Liu Z, Sun C, Olejniczak ET, Meadows RP, Betz SF, et al. 2000. Structural basis for binding of Smac/DIABLO to the XIAP BIR3 domain. *Nature* 408:1004–8

80. Chai J, Du C, Wu JW, Kyin S, Wang X, Shi Y. 2000. Structural and biochemical basis of apoptotic activation by Smac/DIABLO. *Nature* 406:855–62

81. Wu G, Chai J, Suber TL, Wu JW, Du C, et al. 2000. Structural basis of IAP recognition by Smac/DIABLO. *Nature* 408:1008–12

82. Ozawa T, Natori Y, Sato M, Umezawa Y. 2007. Imaging dynamics of endogenous mitochondrial RNA in single living cells. *Nature Methods* 4:413–19

Cell Culture Models in Microfluidic Systems

Ivar Meyvantsson[1] and David J. Beebe[2]

[1]Bellbrook Labs, LLC, Madison, Wisconsin 53711;
email: ivar.meyvantsson@bellbrooklabs.com

[2]Department of Biomedical Engineering, University of Wisconsin at Madison,
Madison, Wisconsin 53706; email: djbeebe@wisc.edu

Annu. Rev. Anal. Chem. 2008. 1:423–49

First published online as a Review in Advance on
March 4, 2008

The *Annual Review of Analytical Chemistry* is online
at anchem.annualreviews.org

This article's doi:
10.1146/annurev.anchem.1.031207.113042

1936-1327/08/0719-0423$20.00

Key Words

microenvironment, extracellular matrix, microfabrication

Abstract

Microfluidic technology holds great promise for the creation of advanced cell culture models. In this review, we discuss the characterization of cell culture in microfluidic systems, describe important biochemical and physical features of the cell microenvironment, and review studies of microfluidic cell manipulation in the context of these features. Finally, we consider the integration of analytical elements, ways to achieve high throughput, and the design constraints imposed by cell biology applications.

1. INTRODUCTION

We continuously strive toward a better understanding of human biology and disease. Because direct observation of humans is only possible in epidemiological studies, experimental work must rely on biological models. Models are, to varying degrees, able to manipulate and analyze biological systems. They can be categorized in order of biological relevance as follows: (*a*) biochemical models using purified biomolecules, (*b*) cell lines, (*c*) cultured primary cells, (*d*) model organisms such as yeast and fruit flies expressing endogenous or human genes, (*e*) human tissue explants, and (*f*) animals (often rodents or nonhuman primates). However, increasing biological relevance goes hand in hand with increases in cost, labor, and experiment duration. Consequently, the simplest model that sufficiently represents the system of interest is usually chosen so that the largest possible parameter space can be explored using available resources.

Mammalian cell culture can in many cases provide both the desired biological relevance and throughput, and it represents a large fraction of human biology research. The year 2007 marked the one-hundredth anniversary of in vitro cell culture. Although our understanding of molecular and cell biology has increased tremendously over the past 100 years, the methods used today are surprisingly similar to those employed by Harrison in his 1907 study of frog neurons cultured in hanging drops of clotted frog lymph on a depression slide (1). We still rely on undefined biological material (e.g., fetal bovine serum) and simple containers such as dishes, bottles, and flasks. Currently, studies of model organisms and animal models are more biologically relevant than those using cultured human cells, but this may change in the future with the introduction of improved cell culture systems.

Various influences determine the phenotype of cells in vivo, including interactions with neighboring cells, interactions with the extracellular matrix (ECM), and systemic factors. Ease of use and low price notwithstanding, dishes and flasks allow no control over the spatial distribution of the cells and biomolecules needed to model many chemical and physical influences cells experience in vivo. Approaches that increase the biological relevance of cell culture models while maintaining or increasing the throughput of current methods are of great interest to the life sciences community.

Microfluidic systems represent a new kind of cell culture vessel that expands our ability to control the local cellular microenvironment (2, 3). Microfluidic systems enable patterning of molecules and cells (4) as well as both passive (5) and active (6) cell handling and environmental control. Temporal and spatial control on the micrometer scale (0.1–100 μm) have been used in fundamental studies from the subcellular (7) to the organismal (8) level, for instance in studies of cell division axis orientation (9) and geometric influence on cell survival (10).

Applications of soft lithography in cell biology have been reviewed (11), as have methods of engineering cellular interactions via microtechnology (4). Microfabricated cell cultures were reviewed by Voldman et al. (2), Park and Shuler (12), and more recently by El-Ali et al. (3). This review explicitly focuses upon cell biology and the manner in which microfluidic structures have been or may be employed to manipulate the cellular microenvironment in order to better understand cell biology or to build cellular models and assays. We also discuss analytical methods used with microfluidic cell culture, ways in which throughput can be increased, and the constraints imposed upon microfluidic system design by cell biology applications.

2. MICROFLUIDIC CELL CULTURE

The cell culture methods in use today have been founded on over a century of work. It is important to put microfluidic cell culture in context with this work to determine which assumptions hold true and which do not when methods are scaled down to microchannels. Although this characterization is just beginning, several research groups have already contributed to a better understanding of the multiple aspects of microfluidic environments.

The physical design of microfluidic devices affects the cell microenvironment of cultured cells. Design considerations for useful application of microfluidic devices in cell biology were described by Walker et al. (13), who introduced the concept of effective culture volume as an indicator of cellular control over the microenvironment in the culture device. The complex but predictable patterns formed by growth factors and other solutes have been described as secondary interfaces (14), the shapes of which are affected by perfusion of the cell culture medium. A comprehensive review of perfusion culture system design, material choices, and operation was recently presented by Kim et al. (15), whose discussion of both the engineering aspects and biological application considerations of these factors provided a practical overview of design, fabrication, sterilization, culture, and analysis. In this section, we describe relevant microfluidic cell culture work to date both in two dimensions (i.e., suspension and monolayer) and in three dimensions (i.e., cells in a polymer matrix).

2.1. Cell Culture in Suspension and Monolayers

Fluid suspension is the natural environment of several cell types, including yeast cells and mammalian blood cells. Other cell types retain important phenotypic characteristics in monolayer culture. This is true for cell-cell attachments that are generally conserved in monolayer culture of epithelial cells. Suspension cell culture is traditionally performed in roller bottles or spinner flasks, whereas monolayer cell culture is done in Petri dishes, multiwell plates, or culture flasks.

One of the first studies of adherent cell culture in microfluidic channels was performed by Tilles et al. (16). They constructed a microchannel cell culture system from polycarbonate and glass with channels 85–500 μm in height and a cell culture area 25 mm wide × 75 mm long. Primary rat hepatocytes were seeded in coculture with 3T3-J2 fibroblasts in channels that had either a gas exchange membrane top or a polycarbonate top. A significant difference in cell viability and hepatocyte function was seen after only 8 h when the devices with and without gas exchange membrane were compared. Monitoring albumin and urea production (markers of hepatocyte function) in devices with a gas exchange membrane over time for different flow rates, the authors found that increasing flow rate led to diminished viability and function. At low flow rates, hepatocyte function remained stable for 10 d.

As the liver is a highly perfused organ, perfusion culture is an appropriate way to model the mass transport aspects of liver biology in vitro. The culture of a human hepatocarcinoma cell line (Hep G2) in a perfusion culture device was reported by Leclerc et al. (17). They built a network of 270-μm-high channels constructed from two layers of poly(dimethylsiloxane) (PDMS). Leclerc et al. used 4-(2-hydroxyethyl)-1-piperazineethanesulfonic acid (HEPES)–buffered media (i.e., independent of carbon dioxide) and reported successful attachment, spreading, and growth on the PDMS surface; they also reported that gas transport through PDMS was sufficient to maintain a viable culture over days. They further compared perfusion and static conditions. Glucose consumption was similar in both cases for 3 d, after which the cells in static culture started to die. In perfusion culture, the cells continued to grow and consume glucose at an increasing rate until confluence was reached around day 7. The authors also monitored albumin expression, which is an indicator of normal phenotype of the Hep G2 cell line. Albumin production was comparable for the first 3–4 d in perfusion and static culture, after which it dropped sharply in static culture, consistent with the observed cell death. Although nutrient requirements and rate of waste production were shown to vary widely for different cell types, this study demonstrates that there is a time window wherein these needs are met in static culture.

Yu et al. studied the effects of microchannel dimensions on the proliferation of suspended insect cells (Sf9) (18) as well as the effects of cell density, exogenous growth factors, and media change frequency on the growth rate of normal murine mammary gland cells (NMuMG) (19). They found that the growth rate of Sf9 cells decreased with increasing cell density and decreasing channel height, whereas channel width and length did not affect cell proliferation. Interestingly, the authors showed that NMuMG cells proliferated more rapidly in microchannels than in 96-well plates, all else being equal. The difference between microfluidic channels and wells was decreased when the cell culture medium was changed frequently (either every 1 h or every 4 h). The frequent media change also reduced the growth advantage provided by fetal bovine serum supplementation over epidermal growth factor (EGF) supplementation only. Consistent with the effective culture volume concept discussed above (13), the authors hypothesized that the secreted growth factor accumulation facilitated by the diffusion-dominated environment of microfluidic channels was responsible for this effect.

Other indications of the environmental differences between microfluidic systems and conventional culture come from embryo studies. Raty et al. cultured murine embryos in microfluidic devices and observed a greater proliferation rate with daily media changes in the microfluidic device compared to conventional methods (20).

2.2. Three-Dimensional Cell Culture

In vivo tissue organization and three-dimensional cell culture play an important role in cells' behavior. Three-dimensional cell culture can increase the biological relevance of cell-based models beyond that achievable in monolayer culture. Dolberg's and Bissell's study of Rous sarcoma virus (RSV) infection is a striking example of the powerful influence exerted by complex three-dimensional environments in vivo (21). RSV is known to cause neoplastic transformation of chick tissues and cells derived from chick embryos, but when Dolberg and Bissell infected cells of the early chick embryo, the activity of the highly potent oncogene possessed by the virus was completely inhibited, even in the fully grown organism. When removed from the animal the cells expressed a transformed phenotype after only 24 h in culture, demonstrating that the ability of in vivo three-dimensional architecture to control structure and function is lost when the cells are removed from the animal. These aspects of cell behavior can only be modeled in three-dimensional culture (22).

Several approaches to seeding cells in gels inside microchannels have been reported. Toh et al. produced a device with a middle channel for cell seeding in a gel flanked on each side by aqueous channels for perfusion. The perfusion channels were separated from the culture channel by rows of micropillars (23). The authors cultured several cell types in the device and, over several days, analyzed cell functions including albumin secretion of primary hepatocytes and differentiation competence of mesenchymal stem cells. Laminar flow was used to pattern a microchannel with a hydrophobic coating, which led to the formation of virtual walls along the channel (J.P. Puccinelli & D.J. Beebe, manuscript in preparation). Subsequently, laminar flow was again used to pattern two separate cell types in collagen gel along the hydrophilic center line of the channel, leaving the hydrophobic edges of the channel unwetted. The unwetted area was subsequently filled with culture media, thereby providing an aqueous interface to the cell culture.

Kim et al. created another device that employed laminar flow to create perfusion channels on either side of a gel compartment (24). They cultured hepatocellular carcinoma cells in the device and compared their albumin secretion when cultured in collagen, PuraMatrix[TM] (a synthetic peptide provisional matrix), or polylactic acid (PLA). They found that albumin secretion was initially lower in PuraMatrix[TM] than either collagen or PLA, but became higher by day three and remained high throughout the eight-day duration of the study. Cells cultured in collagen and PLA showed similar rates of albumin secretion.

Paguirigan and Beebe molded channels from gelatin by crosslinking with the enzyme transglutaminase (25). This enzyme occurs naturally; thus, the device contained no synthetic crosslinking or photoinitiating agents. The authors reported that the morphology of cells grown on the gelatin surface of these channels was significantly different from that of cells in monolayer culture on tissue culture–treated polystyrene. The former showed nuclear staining results, indicating cell entry into the three-dimensional crosslinked collagen matrix.

3. CELL MICROENVIRONMENT

The cell microenvironment comprises a complex of biochemical and physical influences (which may be synergistic or antagonistic), the frequency (26) and time sequence (27) of which can be remembered by cells. However, the cell microenvironment and the part it plays in homeostasis are still poorly understood (28). With regard to the soil nematode *Caenorhabditis elegans*, for example, every single cell division leading to the formation of the hatched larvae has been mapped (29), the entire genome of the organism has been sequenced (29), and numerous published reports have described various details of *C. elegans* cell biology. Still, we cannot take a cell or tissue from the nematode and maintain its natural phenotype in culture because the biochemical and physical influences necessary to reproduce the cell microenvironment that maintains that phenotype remain unknown. Microfluidics provides a set of tools that may enable us to define specific features of in vitro cell culture environments.

3.1. Biochemical Environment

The healing of wounded skin illustrates several biochemical aspects of cell microenvironments including gradients, soluble factor signaling, gas transport, and interactions with the ECM (**Figure 1**). When a blood vessel breaks, a cascade of events leads to the formation of a clot that plugs the vessel to prevent blood loss. Immediately, platelets

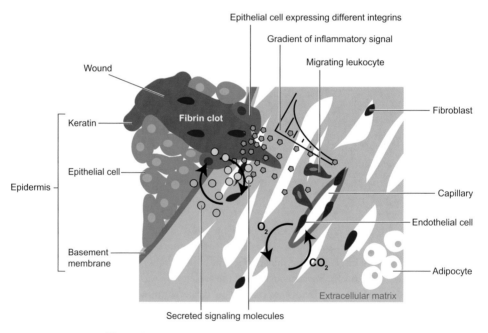

Figure 1

The biochemical environment of wound healing. A fibrin clot has formed to plug the severed blood vessel. Macrophages and platelets in the clot release cytokines that form a gradient and mediate recruitment of leukocytes from the blood stream via chemotaxis. As the epithelium regenerates the skin in the wounded area, epithelial cells interact with the extracellular matrix via integrin receptors and with stromal cells (e.g., fibroblasts and adipocytes) via soluble factors.

and resident macrophages begin to release soluble factors, known as cytokines, that mediate inflammation. As cytokine molecules diffuse away from the wound, they form a gradient. When the gradient reaches the closest intact blood vessel, the cytokine signal is sensed by local endothelial cells and soon by leukocytes in the blood stream. This signal prompts endothelial cells to mediate leukocyte entry into tissue and leukocytes to enter the tissue (30). Leukocytes move up the cytokine gradient, a behavior known as chemotaxis.

Later in the healing process, skin epithelial cells proceed to regenerate the epidermis (31). The cells of the dermal connective tissue, including fibroblasts and adipocytes (collectively known as stromal cells), regulate epithelial proliferation and differentiation via soluble factors (32). Integrin receptors allow epithelial cells to sense and bind to the ECM and are specific for a given set of ECM proteins. In normal skin, epithelial cells lie on the basement membrane, a layer of ECM rich in laminin. During regeneration, the epithelial cells express new integrin receptors in order to interact with the fibrin clot and the dermis and re-epithelialize the wounded area (31). Another important biochemical influence is gas concentration: As the wound continues to heal, new blood vessels are formed to provide gas exchange to the regenerated tissue. Oxygen plays important roles in metabolism, in the regulation of angiogenesis (discussed in detail below in the context of physical environmental influences), and in the zonal differentiation in the liver (33). These biochemical aspects of the cell microenvironment and the ways in which they have been explored with microfluidic systems are described in this section.

3.1.1. Soluble factors. Various biomolecules exist dissolved intracellularly and in the interstitial space. Many actions in the body are regulated, at least partially, by soluble factors including angiogenesis (34) and embryonic morphogenesis (35). A molecule secreted from a cell is transported away from the source via diffusion, the convective flow of the blood stream, and the continuous flow from capillaries into lymphatic vessels. The spatial and temporal distribution of soluble factors is further affected by the lifetime of the factor dictated by the molecule's inherent stability or enzymatic degradation, sequestration by the ECM, and binding endocytosis by other cells. Microfluidic systems have been employed to perform spatially defined treatment. For instance, Sawano et al. used laminar streams to expose only a small part of a cell to EGF. They subsequently observed the intracellular propagation of cell signaling (7). Similarly, Blake et al. demonstrated the use of laminar flow for treatment of specific regions of perfused rat brain slices (36). They measured electrical activity related to respiratory motor function and showed that this activity could be suppressed in a specific area of the slice by applying the appropriate solution.

Spatially defined treatment is often impossible to achieve with conventional means. Microfluidic systems provide new opportunities in achieving spatially defined treatment by facilitating readout (see, e.g., Reference 7) and in studies of spatial heterogeneity of responses and spatially defined signals (see, e.g., Reference 36).

3.1.2. Gradients. Many soluble factor signals, in particular those associated with chemotaxis, exist as gradients in vivo. Using microfluidic networks with two inlets, Jeon et al. generated gradients via successive flow splitting and diffusive mixing (37). Under continuous flow the concentration distribution can be maintained stably over time. The authors studied the migration of primary neutrophils in gradients of interleukin-8 (IL-8) and observed the differential sensitivity of neutrophils to a sharp versus gradual spatial drop in concentration. Wang et al. demonstrated that although MDA-MB-231 cells did not show a chemotactic response to linear gradients of EGF, they did show significant directional migration when exposed to polynomial gradients (38).

In another exploration of the temporal aspects of neutrophil migration, Irimia et al. employed a gradient-switching device to study the response of neutrophils to time-dependent step-up, step-down, and reversal of IL-8 gradients (39). Interestingly, the authors observed a period of depolarization followed by repolarization when stepping down the gradient to half of the initial concentration, thereby revealing dynamics that had not been reported before. Employing an extremely low flow rate (314 nl/min), Mao et al. quantified the migration of *Escherichia coli* bacteria (40). In contrast to previous work with mammalian cells, the *E. coli* bacteria were suspended during the assay and were collected in different channels downstream depending on their position within the laminar flow. In the absence of gradients the bacteria were distributed symmetrically around the center channel. The authors successfully characterized the responses of wild-type and chemotactic mutant strains to several gradients, yielding results consistent with the known behavior of the strains. Flow-based gradients have also been used to explore the concentration dependence of proliferation and differentiation of human neural stem cells (41).

All of the above-mentioned approaches to gradient generation rely on continuous flow to maintain the gradient. Walker et al. found that although total migration distance was not influenced by flow rate, the direction of migration was biased in the direction of flow in a flow rate–dependent manner (42). An additional concern for flow-based gradients is the dissipation of endogenous signaling molecules. Cells generally do not act alone, but rather orchestrate their concerted action through soluble factor signals. Under flow conditions, even relatively slow flow, these signaling molecules are transported away from the cells that would otherwise respond (43). Thus, flow-free gradient devices facilitate the inclusion of cell-cell signaling in microfluidic cell culture models.

Several methods have been reported to produce gradients in microfluidic devices without flow. Abhyankar et al. characterized a device employing a 0.2-μm pore-sized membrane to limit flow in a narrow channel between a large volume source and a sink (44). Using a hydrogel sandwiched between two layers of PDMS, one of which contained a channel network, Wu et al. demonstrated the formation of a gradient (45). The gradient varied linearly in one dimension across the hydrogel and different concentration profiles were achieved along the length of the microchannels by producing channels that traced a variety of curves across the plane of the hydrogel.

Generation of three-dimensional gradients in gels has also been shown. Rosoff et al. demonstrated the formation of arbitrary gradients via micropump printing on the top face of a collagen gel (46). Using three parallel channels molded in an agarose gel, Cheng et al. formed a gradient in the center channel by using the two outermost channels as chemoattractant source and sink (47). They studied the behavior of *E. coli* as well as HL-60 cells exposed to appropriate gradient stimuli.

A variety of gradient generation tools have been reported, providing researchers with a choice of several approaches, including continuous-flow devices for fast switching and no-flow devices for inclusion of cell-cell signaling. Microfluidic gradient devices are a great improvement over pipette-based methods, Dunn and Boyden chambers, and transwell plates in terms of precision and variety of gradient shapes. However, the ease of use and throughput of microfluidic gradient devices can still be improved.

3.1.3. The extracellular matrix. The coupling of secreted factor distribution with flow in gel culture is considered in a recent review by Griffith and Swartz (48). Aided by computational models, the authors demonstrate the complex patterns formed by convective transport of secreted factors that interact with the ECM to enzymatically release secondary sequestered factors. These are some of the many functions of ECM in cell biology. As mentioned above, cell attachments to the ECM are not merely mechanical anchors, but receptors that sense the biochemistry and mechanics of the microenvironment. The use of synthetic biomaterials containing bioactive ligands to guide tissue morphogenesis in vitro was recently reviewed (49).

A surface-bound gradient of the ECM protein laminin was produced using a flow-based gradient (50) in the manner described above [see, e.g., Jeon et al. (37)]. Rat hippocampal neurons were grown on this gradient and the orientation of axon growth was observed. The orientation was found to be in the direction of increasing laminin concentration. Also, Tan and Desai produced multilayer cocultures with the goal of incorporating the cellular heterogeneity of blood vessels (51). They seeded each cell type in a gel and relied on the inherent contraction of the gel to form a channel on top of the gel for the next layer. Frisk et al. also took advantage of gel shrinkage to form an aqueous channel for perfusion of three-dimensional culture and relied on an array of micropillars to hold the gel in place (52). They also demonstrated rapid switching between perfusion solutions. The duration from the point at which 10% relative concentration of the new solution was reached to the point at which 90% relative concentration was reached (the rise time) was only 2 min. The authors further demonstrated viability of fetal monkey kidney (COS 7) cells after 72 h in culture. Evans et al. employed micropatterned three-dimensional ECM to study the direction of spiral ganglion neurite outgrowth by laminin and fibronectin (53). They observed outgrowth patterns consistent with the hypothesized role of ECM patterns in neurite guidance.

Tsang et al. produced three-dimensional hepatic tissues by photopatterning of poly(ethylene glycol) (PEG) hydrogels containing cells (54). The researchers incorporated cell-adhesive peptides, representing specific ECM proteins, in the hydrogels to support hepatocyte survival. From the same research group, Underhill et al. studied the function of liver cells embedded in PEG hydrogels (55) and found that albumin secretion of mouse embryonic liver cells was significantly influenced by the peptide sequence incorporated in the PEG hydrogel.

As described above, several methods have been reported to seed cells in three-dimensional gels. The results discussed in this section clearly show the exciting opportunities that exist for combining microfluidic patterning of biological matrices and rapid solution-switching with three-dimensional cell culture. These methods are important in the effort to establish appropriate spatial and temporal patterns of biochemical influences.

3.1.4. Gas concentration. Mammalian cell metabolism depends upon a regulated oxygen supply and the removal of carbon dioxide. To perform these processes in microfluidic cell culture systems, we must design the systems such that supply and use are well balanced. Supply depends on the device geometry, material, and perfusion conditions, whereas use depends on cell type and cell density. Oxygen concentration in microfluidic bioreactors has been measured during cell culture by means of an oxygen-sensitive ruthenium dye (56) and was found to decrease with increasing cell density.

A perfusion bioreactor system was demonstrated to form steady-state oxygen gradients in cell culture (57). The bioreactor was fabricated by machining from polycarbonate, which is relatively impermeable to oxygen. The reactor was perfused with an oxygenated cell culture medium and the flow rate was adjusted to achieve the desired gradient between the input and output due to cell consumption. It was shown that the pattern of oxygen tension–related enzyme expression (cytochrome P450 2B) in primary rat hepatocyte cultures is consistent with that shown in vivo. This model was further extended by adding fibroblasts in coculture with hepatocytes (58), and was recently applied to study the expression profile of hypoxic primary hepatocytes (59).

Multicompartment devices (cell culture analogs) have been constructed for toxicology studies (12). In some cases these devices have a gas-permeable area intended to model gas exchange in the lungs. An alternate way to regulate oxygen concentration was presented by Park et al., who employed water electrolysis to control the oxygen tension in a separate microfluidic chamber (60). Arbitrary spatial concentration profiles can be created by varying electrode geometry. Hyperoxic apoptosis of C2C12 myoblasts was also demonstrated.

A range of oxygen levels can be established in conventional cell culture, but this requires external gas regulation or culture media level adjustment. Microfluidic systems offer faster response times than external gas regulation; also, they do not depend upon cell culture media volume. Furthermore, perfusion of oxygenated media into gas-impermeable microfluidic devices with active culture has been shown to allow the formation of gradients that promote zonation in liver cell culture (57).

3.2. Physical Environment

Physical influences such as force and temperature can be sensed directly by cells, and other physical factors—including geometry—affect cells indirectly. Many physical influences exist in the vascular system. As discussed in the previous section, cells depend on constant transport and exchange of oxygen and carbon dioxide. When the geometry of the capillary network is structured such that transport in a certain

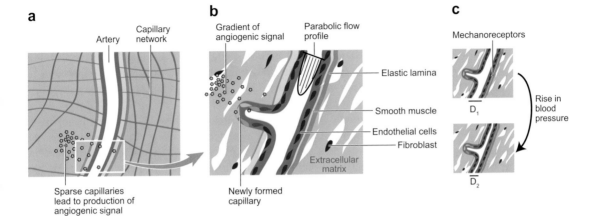

a
Artery
Capillary network
Sparse capillaries lead to production of angiogenic signal

b
Gradient of angiogenic signal
Parabolic flow profile
Elastic lamina
Smooth muscle
Endothelial cells
Fibroblast
Extracellular matrix
Newly formed capillary

c
Mechanoreceptors
D_1
D_2
Rise in blood pressure

Figure 2

Physical environment of blood vessels. (*a*) An insufficiently oxygenated area of an artery and its associated capillary network. Oxygen-sensitive transcription factors cause the cells in that area to express angiogenic factors. (*b*) Local blood vessels respond to the angiogenic factors by expanding the capillary network in a "sprouting" fashion. The flow velocity profile in the artery is shown, as is the construction of arterial walls with endothelial cells lining the vessels and with an external covering of smooth muscle and elastic lamina. (*c*) The smooth muscle cells have mechanoreceptors that contribute locally to the regulation of blood pressure. When blood pressure rises, the receptors allow the smooth muscle cells to sense the change in vessel wall tension and respond by contracting the vessel to bring flow back to the previous level.

area is insufficient, cells sense the condition via oxygen-sensitive transcription factors such as hypoxia-inducible factors (HIFs) (61). In areas that are insufficiently perfused, HIFs lead to the expression of factors that promote angiogenesis, such as vascular endothelial growth factor (**Figure 2*a,b***).

The velocity profile of blood flow is shown in **Figure 2*b***. Blood flow causes shear stress on the endothelial cells lining blood vessels as well as on the leukocytes attached to vessel walls (62). The smooth muscle cells that surround arteries and arterioles are responsible for regulating blood flow; when blood pressure changes suddenly, smooth muscle cells detect the change in vessel expansion via mechanoreceptors and act to maintain the same flow rate to the downstream capillary network (63) (**Figure 2*c***). In this section we discuss the ability of microfluidic systems to control these physical parameters of the cell microenvironment.

3.2.1. Fluid flow. Fluid flow in cell biological systems leads to shear stresses as well as local transport and sometimes systemic distribution of soluble factors. A wide range of fluid velocities exist in the human body. The velocity in large blood vessels may measure up to 0.3 m/s (300,000 μm/s) in the aorta (64), but then slows down immensely as the vascular network expands over the entire volume of the body at the capillary level. Interstitial flow from the vascular system to the lymphatic system is typically in the range of 0.1–1 μm/s (48).

Bone tissue responds to mechanical signals, including shear stress in culture. The activity of the enzyme alkaline phosphatase (ALP) is a marker of bone cell function. Leclerc et al. studied the ALP activity of mouse calvarial osteoblastic (MC3T3-E1) cells in a PDMS microfluidic network (65). The authors found that although high flow rates led to loss of activity and eventually cell death, lower flow rates increased ALP activity up to threefold after 13 d in culture compared to static culture with media changes each day.

Using an array of 12 microfluidic bioreactors, Figallo et al. explored the differentiation of human embryonic stem cells to endothelial cells as determined by expression of α–smooth muscle actin (66). By comparing two different culture chamber designs, the authors showed that cultures in chambers producing higher shear stress developed a greater fraction of differentiated cells. For both chamber designs, lower seeding density also resulted in a greater fraction of differentiated cells.

Schaff et al. constructed a device to monitor leukocyte interaction with a biological substrate under flow conditions (67). The authors validated the device using primary human neutrophils and mouse fibroblasts transfected with human E-selectin. The capture, rolling, and deceleration to arrest were monitored under flow conditions. The cells were found to exhibit the expected response to activation by IL-8.

Microfluidic systems provide a variety of methods to control flow rate and modulate shear stress. These methods have been used to model physical influences on cells including endothelial cells, leukocytes, and osteoblasts.

3.2.2. Tissue mechanics.

The mechanical properties of the cell microenvironment are an important influence on cell behavior (68). The cell-surface molecules that are responsible for anchoring cells to two-dimensional substrates, three-dimensional matrices, and neighboring cells act as receptors as well as anchors. For example, each member of the integrin family binds a specific set of ECM proteins (69). Integrin receptor complexes are coupled to the cytoskeleton. Ingber has described the processes of mechanical and chemical signal integration on the whole-cell level in the context of tensegrity models (70, 71). Such models present a view of sensing on the whole-cell level rather than on the individual mechanoreceptor level. In three-dimensional cell culture, for instance, the density of the ECM has been shown to influence cell differentiation (72). Similarly, McBeath et al. found that cell shape as dictated by the adhesive area available to cells in two-dimensional culture influenced lineage commitment of human mesenchymal stem cells (73). The authors showed that commitment to adipose or osteoblastic lineages was signaled through the cytoskeletal regulation protein RhoA. Micropatterning has also been used to elucidate various aspects of cell mechanical and spatial sensing (9, 10, 74, 75).

3.2.3. Geometry.

The formation of branched duct structures (branching morphogenesis) in the breast may be guided by a geometrical influence analogous to the vascular remodeling discussed above. Nelson et al. used three-dimensional micropatterning to define mammary epithelial cell colonies in a collagen gel (76) and found that the synthetic tubules branched out at positions determined by the geometry of the tubule. The observed branching pattern was consistent with the pattern of an unknown diffusible inhibitor.

Embryoid bodies (EBs) are spherical aggregates of embryonic stem cells (ES cells) and are commonly formed as a first step in ES cell experiments. Torisawa et al. demonstrated synchronized formation of uniformly sized EBs using a microfluidic device (77). They showed that the size of EBs could be controlled by channel cross-sectional geometry. Similarly, Karp et al. employed PEG microwells to control EB shape (78).

In an effort to reproduce aspects of liver tissue structure, Powers et al. constructed three-dimensional culture scaffolds via deep reactive ion etching in silicon (79). They selectively deposited a cell-adhesive coating on the inside walls of the culture chambers, as the top surface of the device and the bottom surface of each chamber did not support cell attachment. Interestingly, the authors found that cells precultured as spheroids fared better in the devices than those seeded from a single-cell suspension. This suggests the importance of the ECM in three-dimensional culture: During spheroid formation cells will synthesize ECM, which then becomes an integral part of the spheroid. Similarly, Gottwald et al. produced polymer microcontainer arrays for three-dimensional cell culture, and analyzed gene expression of hepatocellular carcinoma cells cultured in the device (80). They identified several genes that were upregulated in cells cultured in the device rather than in cells in conventional mono-layer culture.

Using differently sized patches of clot-inducing tissue factor, Kastrup et al. studied the threshold response of blood clotting (81). They compared the threshold responses of platelet-rich and platelet-poor plasma as well as the importance of specific coagulation cascade factors.

Recently, Hui and Bhatia described a micromechanical device that allowed temporal control of cell-cell interaction (82). Their system was based on a structure made of microfabricated silicon combs, two of which could be separated and brought into close contact or brought into proximity without contact. Using this system, Hui and Bhatia analyzed the communication of hepatocytes and supporting stromal cells and found that a short period of contact followed by soluble factor communication is sufficient to maintain hepatocellular phenotype.

Tao et al. compared the fate of retinal progenitor cells (RPCs) cultured on porous and nonporous poly(methylmethacrylate) scaffolds after implantation into the sub-retinal space of mice (83). Although RPCs grew equally well on both substrates in vitro, cell survival, differentiation, and integration into host tissue in vivo was significantly greater for porous substrates than nonporous. Upon integration with the microenvironment provided by the mouse eye, the RPCs differentiated and expressed neuronal, glial, and retina-specific markers.

Precise microfluidic geometries have been utilized to model important influences in the body, ranging from mammary morphogenesis to blood-clotting threshold response. As evident from the studies discussed above, geometric effects are often the result of biochemical influences. Microfluidics enables the study of unknown biochemical influences [see, e.g., Nelson et al. (76)], as well as influences with complex kinetics [see, e.g., Kastrup et al. (81)].

3.2.4. Temperature. Cell behavior is influenced to a large extent by the temperature of the cells' environment. Given the small thermal mass and the large surface area:volume ratio of microfluidic systems, temperature is a particularly important consideration for microfluidic cell culture devices. Along with gene expression (84), diffusion and biochemical reaction kinetics change as a function of temperature. Microfluidic devices have been used to provide temperature-controlled environments for biological studies.

By placing a *Drosophila* embryo in a microfluidic device straddling the interface between two streams at different temperatures, Lucchetta et al. (8) observed the dynamics of embryonic patterning arising from the bicoid morphogen. The authors detected lower cell density and slower patterning in the cooler half of the embryo's body, followed by compensation that evened out the pattern across the entire body.

Differential flow control from two inputs at different temperatures was used by Pearce et al. to control the temperature in a cell culture chamber (85). The chamber had an electrode array on the bottom surface to monitor the electrical activity of neurons. The firing rates of dorsal root ganglion neurons were measured to demonstrate the operation of the system.

Microfluidic systems have proven useful for high-resolution control of temperature in biological studies. These would be difficult to do in conventional culture.

3.3. Compartmentalization

Two of the advantages of PDMS are its elasticity and its ability to conform to surfaces. These properties have been exploited for valving (86) and loading of bacteria into miniature culture chambers (87). For example, valves were employed by Balagadde et al. to construct small-volume bioreactors (88). The authors used the bioreactors to study quorum sensing and reported that population density via broadcasting and sensing of bacteria population was autonomously regulated, resulting in population oscillation. The small isolated volume of the reactor enabled monitoring at single-cell resolution. Compartmentalization has also been employed to specifically manipulate or analyze different parts of a single cell or individual cells that are biochemically connected to other cells via gap junctions.

Neural cell bodies and axons have distinct properties and functions. Taylor et al. employed a two-compartment microfluidic system to culture polarized neurons with cell bodies and axons in separate compartments (89). The authors studied axon-specific gene expression and regeneration after physical axotomy.

One symptom common to several diseases of the nervous system is distal-to-proximal axonal degeneration. Recently, Ravula et al. demonstrated the division of the environment surrounding a neuron into two distinct fluid compartments, one containing the cell body and the other containing the axon (90). Using the compartmentalized culture device, Ravula et al. settled a long-standing debate regarding the source of this degeneration and showed that a chemical insult to the axon, but not the cell body, led to the death of the axons.

In a similar experiment, Klauke et al. placed cardiomyocytes in microfluidic devices with two individually addressable compartments (91). The authors harvested pairs of rabbit cardiomyocytes maintaining end-to-end connections, used micromanipulators to place the pair such that the cell-cell junction was located between two dam structures, and surrounded the junction region with mineral oil to separate the two compartments. Pulled-glass pipette-tip perfusion and integrated electrodes were used to explore the mechanical and electrical coupling of the cells as well as the transmission of Ca^{+2} waves under a variety of physiological and nonphysiological conditions.

The ability to contain cell populations or specific parts of a cell provides opportunities to focus biochemical influences on specific biological elements and to study the causal relationships among elements. Although a certain level of compartmentalization can be achieved using pipettes, and although pipettes remain a superior method for electrical isolation of different compartments (i.e., patch clamping), microfluidic systems have greatly increased the number of available compartmentalization methods.

4. SYSTEMS FOR CELL-BASED MODELS

4.1. Integrated Analysis

Every cell-based experiment requires a reliable method of extracting information. Thus far, we have considered the ways in which microfluidic systems serve to construct environments that yield improved cellular models. Analytical methods may also be integrated with microfluidic systems to produce information from cell models. Analyses of single mammalian cells in microfluidic systems were recently reviewed by Sims and Allbritton (92), and analyses of cell cultures were described in reviews by Anderson and van den Berg (6) and El-Ali et al. (3).

Many microfluidic cell culture studies have relied on traditional microscopy techniques including phase contrast imaging, histological stains, fluorescent dyes, and immunocytochemistry. Yu et al. employed a plate reader to quantify cell number (93). Thompson et al. studied dynamic gene expression profiles using green fluorescent protein reporters, and King et al. (94) studied the same for multiple genes and stimuli in a multiplexed microfluidic system.

Additionally, several methods rely on PCR. Wilding et al. reported an application of PCR to analyze cells captured in a microfabricated filter (95). Easley et al. presented a microfluidic system integrating all the components necessary to go from a complex biological sample to a final PCR result (96). Marcy et al. demonstrated genomic analysis of complex populations of microbes at the organism level using a microfluidic genome amplification system (97). This integrated microfluidic system contained nine amplification units and allowed selection of individual cells, lysis, sample preparation, and amplification via PCR. The results from this study demonstrate the heterogeneity of biological samples and the future utility of highly parallel microfluidic systems for analysis.

Capillary electrophoresis was one of the first systems to employ microfluidic channels (98), and it has continued to develop high-performance tools for use in various applications, including sequencing (99). Electrophoresis has also been combined with cell culture systems to identify and quantify the activities of intracellular enzymes (100).

Biosensors have also been employed to study cells in culture. Bratten et al. measured purine release from cardiomyocytes under conditions mimicking cardiac ischemia (101). The authors used a cascade of three enzymes to convert the analyte into hydrogen peroxide that could be detected by the sensor. Chen et al. employed amperometric sensors to detect quantal secretion of catecholamines from individual cells in microfabricated wells (102).

Other demonstrations of integrated analytical tools include cell culture on cantilevers coupled with an atomic force microscopy detection apparatus used to study myotube function (103), application of cylindrical optics to count proteins in single-cell lysates (104), and integrated spectrometry for fluorescence analysis (105).

4.2. Throughput

Microfluidic systems have the potential to increase the biological relevance of cell culture models and to contribute to improved assay content. For large-scale biology research applications (106), as well as for industry applications such as in drug discovery, throughput is an important consideration. Various methods have been used to increase microfluidic system throughput, some of which were discussed in a recent review by Dittrich and Manz (107). The approaches that have been applied to cell culture can be divided into four general categories: (*a*) microarrays, (*b*) gradient devices, (*c*) valved arrays, and (*d*) individually addressable channel arrays.

Several tools exist for spotting arrays that can be employed to spot different matrices for cell culture or possibly the cells themselves. Microarrays provide a high spatial density of different surface coatings and/or different cell types (78, 108, 109), which represents a great advantage for screening growth surfaces, et cetera. In microarrays, however, all of the cells will be exposed to the same fluid environment. Therefore, they are not suitable for screening soluble factors, including drug compounds.

Gradient devices can provide different concentrations of a single compound for a single cell type. Hung et al. reported the use of a microfluidic gradient generator to address an array of 10×10 cell culture chambers (110). All 10 columns were seeded with the same cell type, and a gradient generator was used to produce a linearly varying concentration across the rows. A similar design was reported by Thompson et al. (111), who used the same inputs for cell seeding and perfusion. A scalable linear gradient generator for cell assays was presented by Walker et al. (112).

Valved arrays enable simultaneous testing of multiple cell types and multiple compounds. With the use of valves, Wang et al. demonstrated multiplexed cytotoxicity testing in a high-density microfluidic array with six distinct cell inputs and 12 distinct compound inputs (113). After 24 h in culture, three different cell lines were exposed to a panel of five toxins. Cell morphology and viability were comparable to those of 96-well control cultures. Using a similar approach for selecting between seeding and perfusion mode, King et al. studied dynamic responses to soluble factor stimuli in an 8×8 array (94).

Meyvantsson and Beebe demonstrated cell culture in arrays of 192 individually addressable microchannels (114, 115). Droplet-based passive pumping (116) was employed to seed cells, change media, and treat and stain cells. This approach requires no physical connections between external instruments and the microfluidic device, and is compatible with hand pipettes as well as generic liquid-handling robotics. Warrick et al. analyzed the fluid replacement in these arrays (117) and Berthier et al. presented an analytical model (118) that indicates that perfusion culture will be feasible using passive pumping.

A comparison of different approaches to high-throughput microfluidic cell assays is presented in **Figure 3**. A wide range of flow velocities can be applied in perfusion culture using microarrays, gradient devices, and valved arrays. In channel arrays operated by passive pumping, however, the range is smaller. As mentioned above, the natural perfusion velocity for most cells is very slow and falls within the range available via passive pumping. Whereas microarray cell culture allows screening of a large number of surface bound elements, soluble factors, if any, are uniform across the entire array. The opposite is true for gradient devices, where several different concentrations are possible but where all the cells and other surface-bound elements are identical. Valved arrays represent a compromise between these two methods, and allow several cell types and several culture environments to be assayed simultaneously. However, the constraints imposed by the physical connections required for valved arrays severely limit throughput. In contrast, individually addressable channels can have a large number of access ports (384 ports are shown in **Figure 3d**, and we have demonstrated up to 1536 access ports). When using passive pumping, the density of assay chambers is limited only by the precision of the liquid-handling equipment in its ability to control evaporation in small volumes. Thus, a number of different approaches are being pursued, each of which has inherent advantages and disadvantages that must be taken into account with regard to the intended application.

5. FABRICATION OF MICROFLUIDIC CELL CULTURE DEVICES

Currently, there are several fabrication challenges facing microfluidics researchers. A wide variety of macro-to-micro interfaces exist (119) that need to be standardized (120). Two important aspects of microfluidic interfacing are automation (3) and ease of handling (107), both of which must be considered when designing microfluidic systems.

Also related to interfacing is choice of materials. Many research devices have been made from PDMS (121), and some interfacing strategies have relied on its elastic properties. However, the material properties of PDMS may be problematic for cell culture models (122, 123). Because several molecules involved in cell microenvironments are small and hydrophobic, it is necessary to characterize the effect of PDMS absorption of such molecules upon cell culture in microfluidic devices. Although glass is very common for chemical analysis devices, the cost is too high for disposable

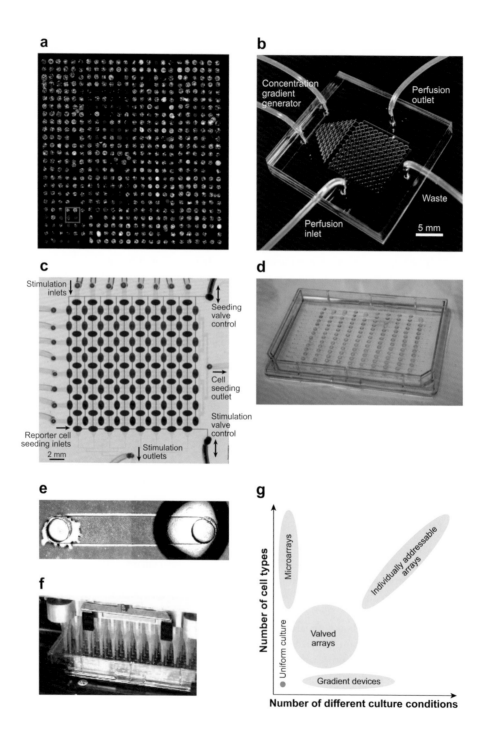

devices. Therefore, thermoplastic materials are the most desirable for microfluidic cell culture devices.

To that end, fabrication of microfluidic structures using thermoplastic materials has been demonstrated (124). However, application in cell culture may impose unique constraints upon materials and fabrication methods. For instance, materials to be used for construction should not leach monomers or additives into the cell culture medium. Similarly, adhesives and solvents should be avoided when choosing a chip-bonding method. In every case, all materials employed in device fabrication must be tested with the cells of intended use.

Other factors to consider when selecting materials include the materials' background fluorescence (125), their ability to produce appropriate surface chemistry for cell attachment (referred to as tissue culture treatment), and their compatibility with standard sterilization methods such as autoclaving, gamma irradiation, ethylene oxide exposure, or low-pressure plasma treatment.

6. SUMMARY AND FUTURE OUTLOOK

Microfluidic cell culture devices enable both basic cell biology research and development of engineered tissues. Careful characterization of cell culture in microfluidic devices is critical in order to understand which macroscale assumptions hold true when culture is scaled down. The number of reports describing microfluidic methods to control cell microenvironments is ever increasing, and these methods are being used to make novel discoveries in cell biology. Despite this trend, we have a long way to go to before we can reproduce in vivo microenvironments.

If microfluidic systems can be employed to establish a sufficiently complete microenvironment for human cells in culture, this may lead to in vitro models that would outperform both conventional cell culture models and animal models in predicting tissue-specific responses in humans. Such a cell culture platform would greatly enhance basic research in cell biology and would improve our ability to study disease.

An important challenge for the microfluidics community is to find ways of simplifying the use of microfluidic devices. Although the utility of microfluidic devices is becoming increasingly evident, such devices will not be adopted by researchers in

---

Figure 3

Throughput in microfluidic cell culture. (*a*) Microarray cell culture. Reprinted with permission from Reference 108. Copyright 2004, Macmillan. (*b*) Gradient device. Reprinted with permission from Reference 110. Copyright 2005, Wiley-Liss. (*c*) Valved array. Reprinted with permission from Reference 94. Copyright 2007, Royal Society of Chemistry. (*d*) Microconduit array using passive pumping to interface an array of individually addressable microchannels. (*e*) 3T3-L1 cells on day 12 in culture in a microconduit array. (*f*) A microconduit array interfaced with a generic high-throughput liquid-handling system. (*g*) A comparison of parallel microfluidic cell culture approaches in terms of number of different cell types and culture conditions achievable.

medicine or the life sciences unless they can be used off the shelf without the aid of engineers.

DISCLOSURE STATEMENT

The authors have an ownership interest in Bellbrook Labs, LLC, which has licensed technology presented in this review.

ACKNOWLEDGMENTS

The authors would like to thank members of the University of Wisconsin, Madison MMB laboratory for their support. **Figure 3a** is reprinted with permission from Macmillan Publishers Ltd. **Figure 3b** is reprinted with permission from Wiley-Liss, Inc., a subsidiary of John Wiley & Sons, Inc. **Figure 3c** is reprinted with permission from the Royal Society of Chemistry.

LITERATURE CITED

1. Harrison R. 1907. Observations on the living developing nerve fiber. *Anat. Rec.* 1:116–28
2. Voldman J, Gray ML, Schmidt MA. 1999. Microfabrication in biology and medicine. *Annu. Rev. Biomed. Eng.* 1:401–25
3. El-Ali J, Sorger PK, Jensen KF. 2006. Cells on chips. *Nature* 442:403–11
4. Folch A, Toner M. 2000. Microengineering of cellular interactions. *Annu. Rev. Biomed. Eng.* 2:227–56
5. Beebe DJ, Mensing GA, Walker GM. 2002. Physics and applications of microfluidics in biology. *Annu. Rev. Biomed. Eng.* 4:261–86
6. Anderson H, van den Berg A. 2003. Microfluidic devices for cellomics: a review. *Sens. Actuators B* 92:315–25
7. Sawano A, Takayama S, Matsuda M, Miyawaki A. 2002. Lateral propagation of EGF signaling after local stimulation is dependent on receptor density. *Dev. Cell* 3:245–57
8. Lucchetta EM, Lee JH, Fu LA, Patel NH, Ismagilov RF. 2005. Dynamics of *Drosophila* embryonic patterning network perturbed in space and time using micofluidics. *Nature* 434:1134–38
9. Thery M, Racine V, Pepin A, Piel M, Chen Y, et al. 2005. The extracellular matrix guides the orientation of the cell division axis. *Nat. Cell Biol.* 7:947–57
10. Chen CS, Mrksich M, Huang S, Whitesides GM, Ingber DE. 1997. Geometric control of cell life and death. *Science* 276:1425–28
11. Whitesides GM, Ostuni E, Takayama S, Jiang X, Ingber DE. 2001. Soft lithography in biology and biochemistry. *Annu. Rev. Biomed. Eng.* 3:335–73
12. Park TH, Shuler ML. 2003. Integration of cell culture and microfabrication technology. *Biotechnol. Prog.* 19:243–53
13. Walker GM, Zeringue HC, Beebe DJ. 2004. Microenvironment design considerations for cellular scale studies. *Lab Chip* 4:91–97

14. Atencia J, Beebe DJ. 2005. Controlled microfluidic interfaces. *Nature* 437:648–55

15. Kim L, Toh Y, Voldman J, Yu H. 2007. A practical guide to microfluidic perfusion culture of adherent mammalian cells. *Lab Chip* 7:681–94

16. Tilles AW, Baskaran H, Roy P, Yarmush ML, Toner M. 2001. Effects of oxygenation and flow on the viability and function of rat hepatocytes cocultured in a microchannel flat-plate bioreactor. *Biotechnol. Bioeng.* 73:379–89

17. Leclerc E, Sakai Y, Fujii T. 2003. Cell culture in a three-dimensional network of PDMS (polydimethylsiloxane) microchannels. *Biomed. Microdev.* 5:109–14

18. Yu H, Meyvantsson I, Shkel IA, Beebe DJ. 2005. Diffusion dependent cell behavior in microenvironments. *Lab Chip* 5:1089–95

19. Yu H, Alexander CM, Beebe DJ. 2007. Understanding microchannel culture: parameters involved in soluble factor signaling. *Lab Chip* 7:726–30

20. Raty S, Walters EM, Davis J, Zeringue H, Beebe DJ, et al. 2004. Embryonic development in the mouse is enhanced via microchannel culture. *Lab Chip* 4:186–90

21. Dolberg DS, Bissell MJ. 1984. Inability of Rous sarcoma virus to cause sarcomas in the avian embryo. *Nature* 309:552–56

22. Abbott A. 2003. Cell culture: biology's new dimension. *Nature* 424:870–72

23. Toh YC, Zhang C, Zhang J, Khong YM, Chang S, et al. 2007. A novel 3D mammalian cell perfusion–culture system in microfluidic channels. *Lab Chip* 7:302–9

24. Kim MS, Yeon JH, Park JK. 2007. A microfluidic platform for 3-dimensional cell culture and cell-based assays. *Biomed. Microdev.* 9:25–34

25. Paguirigan A, Beebe DJ. 2006. Gelatin-based microfluidic devices for cell culture. *Lab Chip* 6:407–13

26. Udy GB, Towers RP, Snell RG, Wilkins RJ, Park SH, et al. 1997. Requirement of STAT5b for sexual dimorphism of body growth rates and liver gene expression. *Proc. Natl. Acad. Sci. USA* 94:7239–44

27. Yan Y, Yang D, Zarnowska ED, Du Z, Werbel B, et al. 2005. Directed differentiation of dopaminergic neuronal subtypes from human embryonic stem cells. *Stem Cells* 23:781–90

28. Davenport RJ. 2005. What controls organ regeneration? *Science* 309:84

29. Blaxter M. 1998. *Caenorhabditis elegans* is a nematode. *Science* 282:2041–46

30. Wu D. 2005. Signaling mechanisms for regulation of chemotaxis. *Cell Res.* 15:52–56

31. Martin P. 1997. Wound healing: aiming for perfect skin regeneration. *Science* 276:75–81

32. Mackenzie IC, Fusenig NE. 1983. Regeneration of organized epithelial structure. *J. Invest. Dermatol.* 81:189–94

33. Jungermann K, Kietzmann T. 2000. Oxygen: modulator of metabolic zonation and disease of the liver. *Hepatology* 31:255–60

34. Montesano R, Pepper MS, Orci L. 1993. Paracrine induction of angiogenesis in vitro by Swiss 3T3 fibroblasts. *J. Cell Sci.* 105:1013–24

35. Tabata T, Takei Y. 2004. Morphogens, their identification and regulation. *Development* 131:703–12

36. Blake AJ, Pearce TM, Rao NS, Johnson SM, Williams JC. 2007. Multilayer PDMS microfluidic chamber for controlling brain slice microenvironment. *Lab Chip* 7:842–49

37. Jeon NL, Baskaran H, Dertinger SKW, Whitesides GM, Van de Water L, Toner M. 2002. Neutrophil chemotaxis in linear and complex gradients of interleukin-8 formed in a microfabricated device. *Nat. Biotechnol.* 20:826–30

38. Wang S-J, Saadi W, Lin F, Nguyen CM-C, Jeon NL. 2004. Differential effect of EGF gradient profiles on breast cancer cell chemotaxis. *Exp. Cell Res.* 300:180–89

39. Irimia D, Liu SY, Tharp WG, Samadani A, Toner M, Poznansky MC. 2006. Microfluidic system for measuring neutrophil migratory responses to fast switches of chemical gradients. *Lab Chip* 6:191–98

40. Mao H, Cremer PS, Manson MD. 2003. A sensitive, versatile microfluidic assay for bacterial chemotaxis. *Proc. Natl. Acad. Sci. USA* 100:5449–54

41. Chung BG, Flanagan LA, Rhee SW, Schwartz PH, Lee AP, et al. 2005. Human neural stem cell growth and differentiation in a gradient-generating microfluidic device. *Lab Chip* 5:401–6

42. Walker GM, Sai J, Richmond A, Stremler M, Chung CY, Wikswo JP. 2005. Effects of flow and diffusion on chemotaxis studies in a microfabricated gradient generator. *Lab Chip* 5:611–18

43. Berthier ES, Warrick JW, Yu H, Beebe DJ. 2008. Managing evaporation for more robust microscale assays. Part 2: Characterization of convection and diffusion for cell biology. *Lab Chip.* doi: 10.1039/b717423c

44. Abhyankar VV, Lokuta MA, Huttenlocher A, Beebe DJ. 2006. Characterization of a membrane-based gradient generator for use in cell-signaling studies. *Lab Chip* 6:389–93

45. Wu HK, Huang B, Zare RN. 2006. Generation of complex, static solution gradients in microfluidic channels. *J. Am. Chem. Soc.* 128:4194–95

46. Rosoff WJ, McAllister R, Esrick MA, Goodhill GJ, Urbach JS. 2005. Generating controlled molecular gradients in 3D gels. *Biotechnol. Bioeng.* 91:754–59

47. Cheng S, Heilman S, Wasserman M, Archer S, Shuler ML, Wu M. 2007. A hydrogel-based microfluidic device for the studies of directed cell migration. *Lab Chip* 7:763–69

48. Griffith LG, Swartz MA. 2006. Capturing complex 3D tissue physiology in vitro. *Nat. Rev. Mol. Cell Biol.* 7:211–24

49. Lutolf MP, Hubbell JA. 2005. Synthetic biomaterials as instructive extracellular microenvironments for morphogenesis in tissue engineering. *Nat. Biotechnol.* 23:47–55

50. Dertinger SKW, Jiang X, Li Z, Murthy VN, Whitesides GM. 2002. Gradients of substrate-bound laminin orient axonal specification of neurons. *Proc. Natl. Acad. Sci. USA* 99:12542–47

51. Tan W, Desai TA. 2004. Microscale multilayer cocultures for biomimetic blood vessels. *J. Biomed. Mater. Res.* 72A:146–60

52. Frisk T, Rydholm S, Andersson H, Stemme G, Brismar H. 2005. A concept for miniaturized 3-D cell culture using an extracellular matrix gel. *Electrophoresis* 26:4751–58

53. Evans AR, Euteneuer S, Chavez E, Mullen LM, Hui EE, et al. 2007. Laminin and fibronectin modulate inner ear spiral ganglion neurite outgrowth in an in vitro alternate choice assay. *Dev. Neurobiol.* 67:1721–30

54. Tsang VL, Chen AA, Cho LM, Jadin KD, Sah RL, et al. 2007. Fabrication of 3D hepatic tissues by additive photopatterning of cellular hydrogels. *FASEB J.* 21:790–801

55. Underhill GH, Chen AA, Albrecht DR, Bhatia SN. 2007. Assessment of hepatocellular function within PEG hydrogels. *Biomaterials* 28:256–70

56. Sud D, Mehta G, Mehta K, Linderman J, Takayama S, Mycek MA. 2006. Optical imaging in microfluidic bioreactors enables oxygen monitoring for continuous cell culture. *J. Biomed. Opt.* 11:050504

57. Allen JW, Bhatia SN. 2003. Formation of steady-state oxygen gradients in vitro: application to liver zonation. *Biotechnol. Bioeng.* 82:253–62

58. Allen JW, Khetani SR, Bhatia SN. 2005. In vitro zonation and toxicity in a hepatocyte bioreactor. *Toxicol. Sci.* 84:110–19

59. Allen JW, Khetani SR, Johnson RS, Bhatia SN. 2006. In vitro liver tissue model established from transgenic mice: role of HIF-1α on hypoxic gene expression. *Tissue Eng.* 12:3135–47

60. Park J, Bansal T, Pinelis M, Maharbiz MM. 2006. A microsystem for sensing and patterning oxidative microgradients during cell culture. *Lab Chip* 6:611–22

61. Pugh CW, Radcliffe PJ. 2003. Regulation of angiogenesis by hypoxia: role of the HIF system. *Nat. Med.* 9:677–84

62. Li YS, Haga JH, Chien S. 2005. Molecular basis of the effects of shear stress on vascular endothelial cells. *J. Biomech.* 38:1949–71

63. Davis MJ, Hill MA. 1999. Signaling mechanisms underlying the vascular myogenic response. *Physiol. Rev.* 79:387–423

64. Widmaier EP, Raff H, Strang KT. 2004. *Human Physiology*, ed. SI Fox, pp. 375–466. New York: McGraw-Hill. 9th ed.

65. Leclerc E, David B, Griscom L, Lepioufle B, Fujii T, et al. 2006. Study of osteoblastic cells in a microfluidic environment. *Biomaterials* 27:586–95

66. Figallo E, Cannizzaro C, Gerecht S, Burdick JA, Langer R, et al. 2007. Microbioreactor array for controlling cellular microenvironments. *Lab Chip* 7:710–19

67. Schaff UY, Xing MM, Lin KK, Pan N, Jeon NL, Simon SI. 2007. Vascular mimetics based on microfluidics for imaging the leukocyte–endothelial inflammatory response. *Lab Chip* 7:448–56

68. Discher DE, Janmey P, Wang YL. 2005. Tissue cells feel and respond to the stiffness of their substrate. *Science* 310:1139–43

69. Li S, Guan JL, Chien S. 2005. Biochemistry and biomechanics of cell motility. *Annu. Rev. Biomed. Eng.* 7:105–50

70. Ingber DE. 2003. Tensegrity I. Cell structure and hierarchical systems biology. *J. Cell Sci.* 116:1157–73

71. Ingber DE. 2003. Tensegrity II. How structural networks influence cellular information processing networks. *J. Cell Sci.* 116:1397–408

72. Wozniak MA, Desai R, Solski PA, Der CJ, Keely PJ. 2003. ROCK-generated contractility regulates breast epithelial cell differentiation in response to the

physical properties of a three-dimensional collagen matrix. *J. Cell Biol.* 163:583–95

73. McBeath R, Pirone DM, Nelson CM, Bhadriraju K, Chen CS. 2004. Cell shape, cytoskeletal tension, and RhoA regulate stem cell lineage commitment. *Dev. Cell* 6:483–95

74. Balaban NQ, Schwarz US, Riveline D, Goichberg P, Tzur G, et al. 2001. Force and focal adhesion assembly: a close relationship studied using elastic micropatterned substrates. *Nat. Cell Biol.* 3:466–72

75. Tan JL, Tien J, Pirone DM, Gray DS, Bhadriraju K, Chen CS. 2003. Cells lying on a bed of microneedles: an approach to isolate mechanical force. *Proc. Natl. Acad. Sci. USA* 100:1484–89

76. Nelson CM, Van Duijn MM, Inman JL, Fletcher DA, Bissell MJ. 2006. Tissue geometry determines sites of mammary branching morphogenesis in organotypic cultures. *Science* 314:298–300

77. Torisawa Y, Chueh BH, Huh D, Ramamurthy P, Roth TM, et al. 2007. Efficient formation of uniform-sized embryoid bodies using a compartmentalized microchannel device. *Lab Chip* 7:770–76

78. Karp JM, Yeh J, Eng G, Fukuda J, Blumling J, et al. 2007. Controlling size, shape and homogeneity of embryoid bodies using poly(ethylene glycol) microwells. *Lab Chip* 7:786–94

79. Powers MJ, Domansky K, Kaazempur-Mofrad MR, Kalezi A, Capitano A, et al. 2002. A microfabricated array bioreactor for perfused 3D liver culture. *Biotechnol. Bioeng.* 78:257–69

80. Gottwald E, Giselbrecht S, Augspurger C, Lahni B, Dambrowsky N, et al. 2007. A chip-based platform for the in vitro generation of tissues in three-dimensional organization. *Lab Chip* 7:777–85

81. Kastrup CJ, Shen F, Runyon MK, Ismagilov RF. 2007. Characterization of the threshold response of initiation of blood clotting to stimulus patch size. *Biophys. J.* 93:2969–77

82. Hui EE, Bhatia SN. 2007. Micromechanical control of cell-cell interactions. *Proc. Natl. Acad. Sci. USA* 104:5722–26

83. Tao S, Young C, Redenti S, Zhang Y, Klassen H, et al. 2007. Survival, migration and differentiation of retinal progenitor cells transplanted on micro-machined poly(methyl methacrylate) scaffolds to the subretinal space. *Lab Chip* 7:695–701

84. Sonna LA, Fujita J, Gaffin SL, Lilly CM. 2002. Invited review: Effects of heat and cold stress on mammalian gene expression. *J. Appl. Physiol.* 92:1725–42

85. Pearce TM, Wilson JA, Oakes SG, Chiu SY, Williams JC. 2005. Integrated microelectrode array and microfluidics for temperature clamp of sensory neurons in culture. *Lab Chip* 5:97–101

86. Thorsen T, Maerkl SJ, Quake SR. 2002. Microfluidic large-scale integration. *Science* 298:580–84

87. Groisman A, Lobo C, Cho H, Campbell JK, Dufour YS, et al. 2005. A microfluidic chemostat for experiments with bacterial and yeast cells. *Nat. Methods* 2:685–49

88. Balagaddé FK, You L, Hansen CL, Arnold FH, Quake SR. 2005. Long-term monitoring of bacteria undergoing programmed population control in a microchemostat. *Science* 309:137–40

89. Taylor AM, Blurton-Jones M, Rhee SW, Cribbs DH, Cotman CW, Jeon NL. 2005. A microfluidic culture platform for CNS axonal injury, regeneration and transport. *Nat. Methods* 2:599–605

90. Ravula SK, Wang MS, McClain MA, Asress SA, Frazier B, Glass JD. 2007. Spatiotemporal localization of injury potentials in DRG neurons during vincristine-induced axonal degeneration. *Neurosci. Lett.* 415:34–39

91. Klauke N, Smith G, Cooper JM. 2007. Microfluidic systems to examine intercellular coupling of pairs of cardiac myocytes. *Lab Chip* 7:731–39

92. Sims CE, Allbritton NL. 2007. Analysis of single mammalian cells on-chip. *Lab Chip* 7:423–40

93. Yu H, Alexander CM, Beebe DJ. 2007. A platereader-compatible microchannel array for cell biology assays. *Lab Chip* 7:388–91

94. King KR, Wang S, Irimia D, Jayaraman A, Toner M, Yarmush ML. 2007. A high-throughput microfluidic real-time gene expression living cell array. *Lab Chip* 7:77–85

95. Wilding P, Kricka LJ, Cheng J, Hvichia G, Shoffner MA, Fortina P. 1998. Integrated cell isolation and polymerase chain reaction analysis using silicon microfilter chambers. *Anal. Biochem.* 257:95–100

96. Easley CJ, Karlinsey JM, Bienvenue JM, Legendre LA, Roper MG, et al. 2006. A fully integrated microfluidic genetic analysis system with sample-in-answer-out capability. *Proc. Natl. Acad. Sci. USA* 103:19272–77

97. Marcy Y, Ouverney C, Bik EM, Lösekann T, Ivanova N, et al. 2007. Dissecting biological "dark matter" with single-cell genetic analysis of rare and uncultivated TM7 microbes from the human mouth. *Proc. Natl. Acad. Sci. USA* 104:11889–94

98. Harrison DJ, Fluri K, Seiler K, Fan Z, Effenhauser CS, Manz A. 1993. Micromachining a miniaturized capillary electrophoresis–based chemical analysis system on a chip. *Science* 261:895–97

99. Aborn JH, El-Difrawy SA, Novotny M, Gismondi EA, Lam R, et al. 2005. A 768-lane microfabricated system for high-throughput DNA sequencing. *Lab Chip* 5:669–74

100. Meredith GD, Sims CE, Soughayer JS, Allbritton NL. 2000. Measurement of kinase activation in single mammalian cells. *Nat. Biotechnol.* 18:309–12

101. Bratten CDT, Cobbold PH, Cooper JM. 1998. Single-cell measurements of purine release using a micromachined electroanalytical sensor. *Anal. Chem.* 70:1164–70

102. Chen P, Xu B, Tokranova N, Feng X, Castracane J, Gillis KD. 2003. Amperometric detection of quantal catecholamine secretion from individual cells on micromachined silicon chips. *Anal. Chem.* 75:518–24

103. Wilson K, Molnar P, Hickman J. 2007. Integration of functional myotubes with a Bio-MEMS device for non-invasive interrogation. Integration of functional myotubes with a Bio-MEMS device for non-invasive interrogation. *Lab Chip* 7:920–22

104. Huang B, Wu H, Bhaya D, Grossman A, Granier S, et al. 2007. Counting low-copy number proteins in a single cell. *Science* 315:81–84

105. Schmidt O, Bassler M, Kiesel P, Knollenberg C, Johnson N. 2007. Fluorescence spectrometer-on-a-fluidic-chip. *Lab Chip* 7:626–29

106. Abraham VC, Taylor DL, Haskins JR. 2004. High content screening applied to large-scale cell biology. *Trends Biotechnol.* 22:15–22

107. Dittrich PS, Manz A. 2006. Lab-on-a-chip: microfluidics in drug discovery. *Nat. Rev. Drug Discov.* 5:210–18

108. Anderson DG, Levenberg S, Langer R. 2004. Nanoliter-scale synthesis of arrayed biomaterials and application to human embryonic stem cells. *Nat. Biotechnol.* 22:863–66

109. Flaim CJ, Chien S, Bhatia SN. 2005. An extracellular matrix microarray for probing cellular differentiation. *Nat. Methods* 2:119–25

110. Hung PJ, Lee PJ, Sabounchi P, Lin R, Lee LP. 2005. Continuous perfusion microfluidic cell culture array for high-throughput cell-based assays. *Biotechnol. Bioeng.* 89:1–8

111. Thompson DM, King KR, Wieder KJ, Toner M, Yarmush ML, Jayaraman A. 2004. Dynamic gene expression profiling using a microfabricated living cell array. *Anal. Chem.* 76:4098–103

112. Walker GM, Monteiro-Riviere N, Rouse J, O'Neill AT. 2007. A linear dilution microfluidic device for cytotoxicity assays. *Lab Chip* 7:226–32

113. Wang Z, Kin M, Marquez M, Thorsen T. 2007. High-density microfluidic arrays for cell cytotoxicity analysis. *Lab Chip* 7:740–45

114. Meyvantsson I, Beebe DJ. 2005. High throughput microfluidic system for heterogeneous assays. *Proc. Annu. Int. IEEE/EMBS Conf. Microtech. Med. Biol., 3rd, Oahu*, pp. 42–44. Piscataway: IEEE

115. Meyvantsson I, Warrick JW, Hayes S, Skoien A, Beebe DJ. 2008. Automated cell culture in high density tubeless microfluidic device arrays. *Lab Chip*. doi: 10.1039/b71537a

116. Walker GM, Beebe DJ. 2002. A passive pumping method for microfluidic devices. *Lab Chip* 2:131–34

117. Warrick JW, Meyvantsson I, Ju J, Beebe DJ. 2007. High-throughput microfluidics: improved sample treatment and washing over standard wells. *Lab Chip* 7:316–21

118. Berthier ES, Berthier JC, Beebe DJ. 2007. Flow rate analysis of a surface tension driven micropump. *Lab Chip* 7:1475–78

119. Fredrickson CK, Fan ZH. 2004. Macro-to-micro interfaces for microfluidic devices. *Lab Chip* 4:526–33

120. Whitesides GM. 2006. The origins and the future of microfluidics. *Nature* 442:368–73

121. Duffy D, McDonald J, Schueller O, Whitesides GM. 1998. Rapid prototyping of microfluidic systems in poly(dimethylsiloxane). *Anal. Chem.* 70:4974–84

122. Toepke MW, Beebe DJ. 2006. PDMS absorption of small molecules and consequences in microfluidic applications. *Lab Chip* 6:1484–86

123. Mukhopadhyay R. 2007. When PDMS isn't the best: What are its weaknesses, and which other polymers can researchers add to their toolboxes? *Anal. Chem.* 79:3248–53

124. Becker H, Gärtner C. 2000. Polymer microfabrication methods for microfluidic analytical applications. *Electrophoresis* 21:12–26

125. Hawkins KR, Yager P. 2003. Nonlinear decrease of background fluorescence in polymer thin-films: a survey of materials and how they can complicate fluorescence detection in microTAS. *Lab Chip* 3:248–52

Peptides in the Brain: Mass Spectrometry–Based Measurement Approaches and Challenges

Lingjun Li[1] and Jonathan V. Sweedler[2]

[1]School of Pharmacy and Department of Chemistry, University of Wisconsin, Madison, Wisconsin 53705-2222; email: lli@pharmacy.wisc.edu

[2]Department of Chemistry, University of Illinois at Urbana-Champaign, Urbana, Illinois 61801; email: jsweedle@uiuc.edu

Annu. Rev. Anal. Chem. 2008. 1:451–83

First published online as a Review in Advance on April 10, 2008

The *Annual Review of Analytical Chemistry* is online at anchem.annualreviews.org

This article's doi: 10.1146/annurev.anchem.1.031207.113053

Key Words

neuropeptides, hormones, cytokines, mass spectrometric imaging, single-cell measurements, quantitation

Abstract

The function and activity of almost every circuit in the human brain are modified by the signaling peptides (SPs) surrounding the neurons. As the complement of peptides can vary even in adjacent neurons and their physiological actions can occur over a broad range of concentrations, the required figures of merit for techniques to characterize SPs are surprisingly stringent. In this review, we describe the formation and catabolism of SPs and highlight a range of mass spectrometric techniques used to characterize SPs. Approaches that supply high chemical information content, direct tissue profiling, spatially resolved data, and temporal information on peptide release are also described. Because of advances in measurement technologies, our knowledge of SPs has greatly increased over the last decade, and SP discoveries will continue as the capabilities of modern measurement approaches improve.

1. INTRODUCTION

Animal behaviors are controlled by the actions of the networks of neurons making up their brains. The activity and response of a particular neuron in a network depends not only on the actions of the other cells interacting with that neuron, but also on the neuromodulators surrounding it. The neurons in our brain are bathed in a complex suite of signaling peptides (SPs), which affects the activities of the networks in many profound and subtle ways.

SPs have a variety of functions in the body. They are involved in neuromodulation, neurotransmission, cell outgrowth, cell survival, and hormonal signaling between organs (1–3). SPs contribute to multiple aspects of behavior. For example, SPs have been reported to play important roles in the physiological mechanisms of feeding (neuropeptide Y and galanin) (4, 5), thirst (angiotensin) (6), and pain (enkephalins) (7–9). Improving our understanding of these various physiological processes requires knowledge of the SPs present in the appropriate anatomical regions. The simultaneous detection, identification, and quantitation of the peptide complement (the peptidome) of complex structures such as the brain present a significant challenge that is exacerbated by the inherently small quantities of SPs, their broad dynamic range, and their diverse chemical and physical properties.

Here we explore the analytical methods used to characterize neuropeptides, neurohormones, and other brain peptides (i.e., the SPs), with an emphasis on mass spectrometry (MS)–based approaches. Although our discussion of several of the techniques overlaps with other reviews in this volume, we focus upon what makes the measurement of SPs unique and the analytical requirements for such studies. We also provide an overview of the approaches used to characterize SPs in terms of their structure, function, and dynamics within the brain. Interested readers are encouraged to examine the articles in this volume on high-resolution MS (10), imaging MS (11), and the chemical analysis of single cells (12).

For most animal models studied in neuroscience, dozens to hundreds of distinct brain peptides have been reported. The diversity of peptides characterized is a relatively recent phenomenon; the original peptide experiments involved valiant efforts on sample preparation and characterization. For example, Roger Guillemin required 5 million hypothalamic fragments from sheep, and Andrew Schally used a similar amount of material from pigs; the output of their heroic work was the structure of a doubly modified tripeptide (pyroGlu-His-Pro-NH$_2$), the thyroid releasing factor (13) (see **http://nobelprize.org/nobel_prizes/medicine/laureates/1977/press.html**). Because of advances in instrumentation, peptides can now be characterized in samples consisting merely of individual neurons, and are routinely characterized in samples nearly 15 orders of magnitude smaller than those used by the original neuropeptide pioneers.

Often, studies are directed toward several previously defined peptides in a sample. Other studies seek to measure the brain peptidome. The recently coined term peptidomics arose simultaneously from three separate groups in 2001 (14–16). Since then, numerous studies characterizing the subset of bioactive peptides related to the brain and other organs have appeared (17–21).

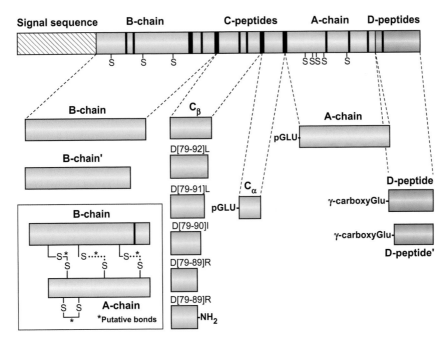

Figure 1

Representation of a single preprohormone for *Aplysia californica* insulin (*long rectangle*); the basic sites that can be cleaved by processing enzymes are shown as vertical black lines. The colored bars represent the partially or fully processed peptides arising from this prohormone; the final products include peptides modified with cleavages, amidation, disulfide bond formation, pyroglutamylation, and gamma-carboxylation (25, 26).

2. WHAT IS A SIGNALING PEPTIDE?

As with all gene products, the creation of a neuropeptide occurs when its RNA is translated into a large protein. In the case of SPs, this protein is a preprohormone that contains a signal sequence. The signal sequence is cleaved from the preprohormone and targets the remaining protein, a prohormone, to the secretory pathway (22, 23). The prohormone is packaged into dense core secretory vesicles, and several processing steps then occur to convert the prohormone into a suite of peptides within the secretory vesicles (24). The prohormone often encodes multiple bioactive neuropeptides within a single vesicle, including replicates of an individual peptide and single copies of other peptides (**Figure 1**) (25, 26). These prohormone cleavages occur under the control of a suite of intracellular prohormone convertases. Because the convertases responsible for the cleavage of the peptides are endogenous, no additional enzymes, such as trypsin, need to be added in order to produce the peptides. As a result, tissues can be dissected from an animal and directly analyzed by MS to observe the final neuropeptides.

The processing of neuropeptide prohormones into biologically active neuropeptides requires a surprisingly large number of enzymatic processing steps, with the

Prohormone: gene product packed in dense core vesicles and processed using a suite of intracellular processing enzymes, resulting in a series of peptides final products depending on the particular enzymes and the order in which they act. As these steps can occur in a brain region–specific manner, one cannot normally predict the final products without measuring them via MS or another information-rich approach. Whereas such measurements of neuropeptides are critical, there are no experimentally confirmed lists of peptides for most animals—even for animals with genomic information—nor for most brain regions. Genomic information coupled with knowledge of enzymatic processing enables an educated guess as to the resulting SPs from a novel prohormone. For example, cleavage of a prohormone at basic amino acid sites is common; although most of the prohormone cleavages occur at basic amino acid residues (lysine and arginine) (27, 28), only a small percentage of basic sites are actually cleaved (29). Approximately one-third of the ten basic sites found in a survey of mammalian prohormones have been observed to be cleaved during prohormone processing (30). But which of the ten are cleaved? Several research groups have formulated guidelines, based on the frequency of amino acids appearing near cleaved and noncleaved sites, to help determine the location of these basic processing sites (29, 31–33). We describe a statistics-based approach to predicting prohormone processing in Section 8, below.

2.1. Brain Peptides Versus Signaling Peptides

SPs can function as neurotransmitters acting across a synapse, act as hormones with non-neuronal targets, or act at intermediate distant targets by passive diffusion via neuroendocrine pathways. Often, the same peptide can function as both a hormone and as a neurotransmitter at different sites. Peptides can inhibit, excite, or modulate small-molecule transmission. Other functions include cytokine and trophic actions (24). Although neuropeptides can have diverse functions, peptides must have a function associated with cell-cell signaling to be considered neuropeptides—their presence in neural tissue is not enough. Cell-cell signaling is most commonly associated with the site of classical transmitter chemical communication, the synapse. However, neuropeptides are normally packaged into dense core vesicles (DCVs), which lack many of the specialized proteins that cause localized release in classical transmitter-active zones. Thus, neuropeptides can be released anywhere along a nerve terminal membrane; in fact, some peptides are more often released in the soma or dendrites than in axons/synaptic regions (34, 35).

An interesting observation arises when examining most prohormone processing schemes such as the one shown for *Aplysia* insulin (**Figure 1**): A prohormone often encodes multiple peptides. An important question is: Which of the peptides derived from a prohormone are involved in cell-cell signaling? Often, prohormones contain SPs separated by "linker" peptides with no reported function. If these peptides play no role in cell-cell signaling, what is their function? Additional suggested roles for specific peptides or prohormone domains highlight the other functions of these peptide sequences within prohormones. For example, several prohormones have linker peptides that contain a preponderance of acidic residues (36–38); at times these peptides provide charge neutrality during the acidification of the DCVs. In this (nonsignaling) role, the atrial natriuretic peptide's acidic domain affects the size and shape of the resulting DCVs (39). Also, specific peptides in a prohormone may be required for correct folding and disulfide bond formation, as in the case of the insulin C-peptide, which links the A- and B-chains to promote their effective folding and assembly into insulin and may aid in the solubilization of insulin after vesicular release (40, 41). Thus, the functions of the peptides derived from a single prohormone are diverse, and only a subset of these functions pertain to cell-cell signaling. Although nonsignaling functions are certainly important, it is necessary to develop a more precise nomenclature for brain peptides, as many such peptides are not involved in cell-cell signaling.

Catabolism: extracellular processing and degradation of signaling peptides as a mechanism for peptide deactivation

In addition to the final peptides formed from a prohormone, one can also detect processing intermediates. When preparing a tissue sample, one normally freezes, heats, or chemically denatures the tissue to stop enzymatic processing. Thus, one would expect the sample to contain both final fully processed peptides and prohormone processing intermediates, with relative amounts of each dictated by the enzymatic formation rates and the sampling protocols. Often, these partially processed forms may be difficult to distinguish from final peptides; however, they may contain an intermediate noncleaved basic residue. If these peptides represent processing intermediates, can they be considered bioactive? They may well be: Several SPs interact with their cognate receptor with only a portion of their length; that portion often includes the C-terminal end of the peptide. In such a case, N-terminally extended forms may affect receptor binding, but do not abolish it. For example, FMRFamide peptides can exist in many extended forms in some prohormones; many overlap with regard to receptor binding, and they may have overlapping or distinct functions (42, 43). Peptides can be detected and tested, but without appropriate controls (such as sampling the peptides released after appropriate physiological stimulations), the data obtained might be misleading. Developing and employing appropriate sampling protocols are often key to data collection and interpretation, and thus are highlighted throughout this review.

Another complication in sampling is becoming more important because of the increased performance of modern MS. Many processing enzymes, like other enzymes, may have a high (e.g., 95%) specificity for the correct cleavage but may cleave some sites in error. This may not be an issue if only a small fraction of a prohormone is cleaved incorrectly. However, as the dynamic range and limits of detection for MS techniques improve, more of these peptides can be detected. Using the lower-sensitivity techniques of the past, most of the chemically sequenced peptides were observed to be bioactive. Currently hundreds of peptides can be detected, many of which do not perform easily determined functions.

2.2. Catabolism

So far, our discussion has focused on SP formation (anabolism). After an SP is released, it can be removed or degraded via catabolic processes, such as cellular (re)uptake by target neurons or glia. Degradation of peptides also has profound effects on the ability to measure them. Much of the extracellular degradation is caused by endogenous C-terminal and N-terminal peptidases. Researchers often optimize a technique by maximizing the number of compounds characterized; however, in the case of SPs,

PTM: posttranslational modification

MALDI: matrix-assisted laser desorption/ionization

this may create a protocol that allows the characterization of extended extracellular degradation products. Many SPs contain posttranslational modifications (PTMs), such as C-terminal amidation, N-terminal acetylation, or pyroglutamylation, that slow extracellular degradation; such PTMs are often used as hallmarks of bioactivity.

Whereas extracellular processing of a peptide is normally considered a process of catabolism and peptide deactivation, it has been shown that some of these extracellularly processed peptides become important SPs. For instance, the intracellular processing of the angiotensinogen prohormone produces angiotensinogen. After the angiotensinogen is released from the cell, the enzyme rennin interacts with it to produce angiotensin I. One of the more bioactive forms of angiotensin, angiotensin II, is then produced by the action of angiotensin-converting enzyme (2). These processes can occur hormonally in many locations throughout the body. The brain may use this pathway, but also appears to use other angiontensin processing systems such as alternative processing enzymes and an intracellular rennin/angiotensin-converting enzyme system (44). Thus, the same peptide can follow different routes to formation depending on its specific location, and what may be an extracellular process in one region may be intracellular in others. The complex interplay of SP anabolism and catabolism allows the fine-tuning of the cellular response to the cocktail of SPs that surround a neuron and makes SP measurements critical in determining neuronal function. These overlapping needs for measuring the peptides in the brain require approaches with high chemical, spatial, and dynamic information content, approaches that are not yet available within the context of a single measurement platform.

3. SAMPLE PREPARATION STRATEGIES

Although MS is a useful technique due to its high sensitivity, speed, and chemical specificity, mass spectral characterization of SPs from tissue samples can be complicated by extracellular proteases, high salt content, lipids, and high sample complexity, which may cause ionization suppression and/or problems with dynamic range. Therefore, proper sample handling from animal dissection through analysis is key to obtaining relevant results. Depending on the type of chemical information sought from a given nervous system or neuronal structure, different sample preparation strategies can be employed. **Figure 2** presents three major sample preparation methodologies for the detection of neuropeptides from neural tissues.

Many experiments are performed by direct tissue profiling or by molecular ion imaging from tissue sections using matrix-assisted laser desorption/ionization (MALDI)–based MS. Direct tissue investigation involves less sample preparation and allows comparison among individual samples or animals. The sample preparation can be described as a simple sequence of steps that includes placing the tissue of interest on the MALDI plate, applying a droplet of matrix, and irradiating the cocrystallized tissue to cause desorption/ionization of the analyte for subsequent detection. To obtain quality mass spectra, tissues are commonly rinsed with water (45, 46) or matrix solution (47–49) to prevent interference from the high concentration of salt and lipids in the tissue. This technique is sensitive enough for SP analysis from single organs (50, 51) and even single cells (48, 52–54). In many cases, sufficient signal is obtained for

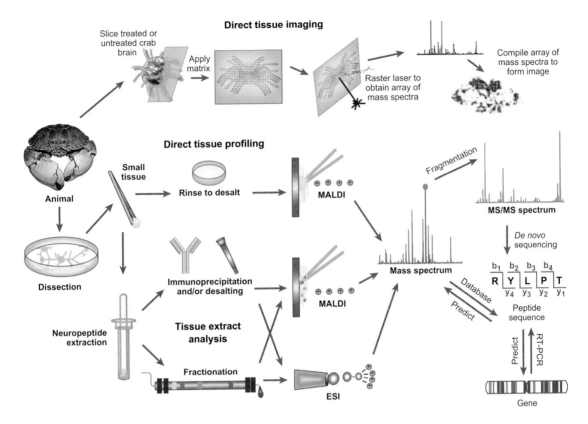

Figure 2

Overview of sample preparation strategies for mass spectrometry (MS)–based neuropeptide analysis. Three major sample preparation approaches are outlined: (1) molecular ion imaging from tissue sections, (2) direct tissue profiling of isolated neuronal tissues and/or clusters of cells, and (3) neuropeptide extraction from the tissue sample. This extract can then be further processed by desalting with C_{18} microcolumns, enriched for a particular peptide family through immunoprecipitation and/or fractionated prior to matrix-assisted laser desorption/ionization (MALDI) or electrospray ionization (ESI) MS analysis. When novel neuropeptide sequences are determined for species with unknown genomic information, the use of reverse transcriptase polymerase chain reaction (RT-PCR) can enable the discovery of novel genes. Adapted and modified from Reference 52 with permission.

peptide fragmentation information to be collected (46, 50, 55–57). Rinsing the tissue with dilute matrix solution and acidified methanol allows the extraction of peptides to the tissue surface and the removal of salts, resulting in improved detection. Direct tissue analysis of neuropeptides has been applied extensively in invertebrate systems (49), and to a lesser extent in more complex vertebrate systems (51, 57). An alternative direct tissue peptide-mapping experiment involves obtaining a collection of molecular ion images across a tissue section by rastering the laser beam at an ordered array of locations from the tissue.

Direct tissue analysis: MALDI MS analysis of freshly dissected and mounted tissue samples with minimal sample preparation

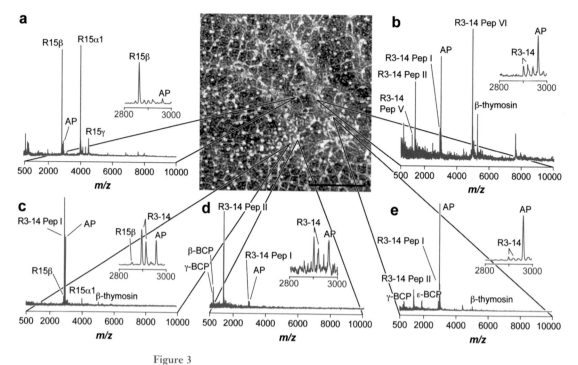

Figure 3

A high-throughput, small-volume sampling approach for direct tissue analysis of neuropeptides. *Center*: Schematic of the massively parallel sample preparation using the stretching sample method. Panels *a–e*: Representative mass spectra obtained from a tissue section showing characteristic peptide profiles at discrete locations. Adapted from Reference 60 with permission.

To improve the sensitivity of in situ peptide analysis, matrix volume must be minimized to avoid dilution of the sample. Although the dried-droplet method of matrix application is sufficient for many applications (49, 58), application of nano-liter volumes of matrix is often desirable when working with smaller samples so as to limit peptide dilution (53, 59). For example, more than 20 FMRFamide-related peptides were detected at nanomolar concentration in single cockroach cells by applying nanoliter volumes of matrix solution (53). Another approach to minimizing peptide dilution and migration is the combined use of an array of glass beads and a stretchable hydrophobic membrane such as Parafilm M®. As shown in **Figure 3**, by adhering a thin tissue section to the glass bead array and then stretching the membrane, one can rapidly create a set of single-cell-sized tissue samples that maximize peptide extraction while limiting analyte diffusion (60). With a precisely controlled micromanipulator and an optimized sampling protocol involving the use of a glycerol cell stabilization technique, single mammalian cells can be investigated for their peptide content by MALDI MS (61, 62).

Although direct tissue experiments allow investigators a quick glimpse at peptide profiles of individual organs or a small cluster of neurons with minimal sample preparation, a homogenization- and extraction-based strategy offers several advantages for obtaining neuropeptide-rich samples through the pooling of many organs/cells. However, increased sample handling and processing may lead to sample dilution, loss, and/or chemically induced artifacts.

Due to the wide variation in model systems and in the chemical properties of neuropeptides, numerous extraction schemes have been developed. Neuropeptide extraction procedures require the reduction of protease activity. Postmortem protease activity yields protein degradation products that interfere with the identification of SPs. This is especially true for mammalian tissues, which require rapid postmortem protease deactivation to maintain sample integrity. Protease heat deactivation employing microwave irradiation or boiling extraction buffers are common techniques employed to reduce protease activity (20, 63, 64). Microwave irradiation is generally performed on the entire brain and dissection is performed by hand after irradiation. Alternatively, the whole brain can be snap-frozen immediately following sacrifice and subsequently dissected by a combination of cryostat sectioning and tissue punching. In this manner, the tissue is kept frozen during the dissection process. The proteases are deactivated immediately prior to extraction by placing the tissue punches in a boiling extraction buffer (65).

As an alternative to heat deactivation, proteases can also be deactivated using acidified organic extraction buffers such as acidified acetone, ethanol, and methanol. These acidic buffers deactivate proteases via denaturation and precipitation. This technique is more commonly used for invertebrate species but is also effective for mammalian tissues (66–69).

4. QUALITATIVE ANALYSIS

Characterizing the SPs in a sample requires simplifying the chemical complexity of the sample. As described in the previous section, the end result of sample preparation for SPs normally consists of either a small tissue sample (as small as a single neuron) or an extract that is well suited to performing a separation prior to MS. Just as the sampling protocols are distinct for these approaches, so too are their analyses. In this section, we discuss the benefits and drawbacks of each approach.

4.1. Mass Spectrometric Analysis of Tissue Extracts and Extracellular Fluids

After SPs are extracted from tissue samples, the chemical complexity and wide dynamic range of SPs present in the crude extracts require a separation prior to MS detection. Given the low abundance of most endogenous SPs, a capillary-based chromatographic approach for peptide fractionation is often employed (70–72). In this fractionation technique, a nanoflow liquid chromatography (LC) column (e.g., 75-μm internal diameter) is typically connected to a wider-bore precolumn for sample loading and desalting via a column-switching device. This arrangement enables fast sample loading and sample cleanup/concentration prior to eluting on the analytical column that is interfaced to the mass spectrometer. A wide range of peptidomic applications have been reported to utilize this scheme for effective identification of numerous peptides in a single chromatographic run (66, 67, 73, 74).

ESI: electrospray ionization

To improve the sensitivity for monitoring in vivo peptide secretion, Kennedy and colleagues (75, 76) employed capillary LC columns (25-μm inner diameter) interfaced to a quadrupole ion trap mass spectrometer. The miniaturization of the LC column, coupled with a smaller electrospray emitter and automated two-pump system for high flow-rate sample loading and low flow-rate peptide elution, enabled the detection of attomole-level peptides and the identification of close to 30 SPs in brain extracellular fluid collected in vivo from live rats.

Multidimensional chromatography has been used to reduce sample complexity by improving the resolution of complex mixtures of proteins or peptides (70). A popular two-dimensional LC scheme involves the on-line coupling of strong-cation exchange followed by reverse-phase separation; this technique has been successfully applied to the peptidomic analysis of both *Drosophila melanogaster* and *Caenorhabditis elegans*, which resulted in enhanced detection of low-abundance, novel peptides (69, 77). More recently, an alternative two-dimensional LC scheme employing differential pH selectivity in the first- and second-dimensional reverse-phase separations was used to enhance rat brain neuropeptidome coverage (65).

Although on-line coupling is often preferred to other schemes, LC fractionation off-line coupled to MALDI MS detection offers several unique advantages. These include the flexibility to independently optimize separation and MS detection and the availability of the majority of individual LC fractions for subsequent in-depth chemical and biochemical characterization. Furthermore, compared to electrospray ionization (ESI)–MS detection, which is often coupled on-line to LC separation, MALDI analysis often provides additional information on the peptides in the sample (65, 78). In addition, an array of sample deposition methods ranging from electrospray deposition to a pulsed electric field deposition, in conjunction with modification of the MALDI target surface, has been developed to enable trace-level detection of SPs from brain tissues (21, 79, 80).

Besides coupling to LC separation, MALDI MS has also been coupled with capillary electrophoresis (CE) for SP detection. Due to its characteristic small injection volume, CE is well suited for peptide analysis in single cells. For example, a post-capillary deposition technique involving a dual capillary configuration that mixes the MALDI matrix with CE eluent from the separation capillary outlet enabled single-cell analysis from nanovial preparations (81, 82). To enhance sample concentration detection limits for in vivo analysis, a combined solid-phase preconcentration CE and matrix-precoated membrane target was used to study intracerebral metabolic processing of neuropeptide in rat, with an overall concentration limit of 10 pM reported (83, 84).

4.2. Direct Tissue Profiling

Compared with methods that involve pooling and extracting tissue samples for neuropeptide analysis, direct tissue analysis offers several distinct benefits. This methodology simplifies sample preparation by eliminating the separation step, thereby

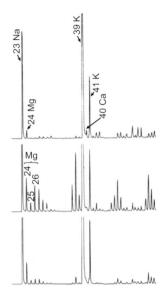

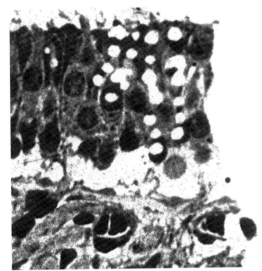

Figure 4

Laser desorption/ionization using a focused laser allows the probing of areas from a tissue section only a few microns in diameter. *Right*: A section of the inner ear used to investigate the homeostasis of the endolymph, with the individual cell outlines and sampling locations clearly visible. *Left*: Several of the associated mass spectra. Adapted with permission from Reference 86 and courtesy of F. Hillenkamp.

minimizing sample preparation artifacts and reducing sample loss and contamination. Perhaps more importantly, direct tissue analysis offers higher sensitivity and can be used with smaller samples such as single brain nuclei and even single cells. Finally, direct tissue analysis provides unique information on the spatial distribution of peptides in a given tissue or cell sample that is often lost in the pooled tissue homogenates.

For peptides that are only made in a rare cell, pooling samples adds chemical complexity but may not result in a greater amount of specific peptides; basically, pooling the sample dilutes the peptides of interest in a complex matrix of the peptides and proteins common to all the cells in the pooled samples. Perhaps one of the more successful advances in neuropeptide research has been the development of single-cell MS. Historically, single-cell MS measurements predate MALDI. Hillenkamp (85, 86) developed the laser microprobe mass analysis technique (LAMMA) that allowed a range of lower-molecular-weight compounds to be detected directly from thin tissue sections. **Figure 4** shows a tissue section with several < 5-μm holes ablated from the tissue by the laser, with each spot producing a mass spectrum (87).

One of the first neuropeptide characterizations using single-cell MALDI MS was described in 1994 and used large neurons from the freshwater invertebrate *Lymnaea stagnalis* (88, 89). Several neurons were examined for peptides encoded by known prohormones. Since then, single-cell MALDI has enabled the characterization of hundreds of new peptides from a range of animals spanning a number of model organisms (48, 49).

CID: collision-induced dissociation

PSD: postsource decay

4.3. Structurally Characterizing the Signaling Peptide

4.3.1. Mass spectrometry–based neuropeptide identification and characterization.
Whether using direct tissue characterization or analyzing separated fractions, characterizing the resulting peptides via MS is of paramount importance. For an organism whose genome has been sequenced and knowledge of neuropeptide prohormone sequence is available, an accurate mass measurement may be employed in conjunction with database searching to identify the amino acid composition of a given neuropeptide (49, 52). However, even with knowledge of the genome sequence, prediction of the final bioactive peptide sequences is often difficult due to extensive PTMs, tissue-specific prohormone processing, and unusual processing sites. Therefore, fragmentation techniques such as collision-induced dissociation (CID) and postsource decay (PSD) are often required to obtain a peptide fingerprint or to discover novel prohormone processing products (52). The combined use of accurate mass measurement and gas-phase fragmentation analysis in conjunction with web-based searching tools has enabled several large-scale peptidomic studies in organisms with sequenced genomes (20, 69, 77, 90, 91).

How does one confirm identifications with direct tissue measurements? One way involves the detection of the multiple peptides that result from processing of a single prohormone (see **Figure 1**). For example, the identification of pro-opiomelanocortin (POMC)-derived peptides from a single pituitary cell was achieved because a dozen peptides from the same prohormone were characterized in every cell measured (61, 62). However, when studying samples with unknown prohormones, additional information is required for structural characterization; unfortunately, the tandem MS (MS/MS) approaches that can be adapted to direct tissue measurements are limited.

The first fragmentation approach applied to single neurons used PSD on identified molluscan neurons (48, 55, 92). More recently, direct single-organ neuropeptide fragmentation was performed by MALDI-time-of-flight (TOF) PSD on cockroach (45) and flesh fly (93). MALDI CID of neuropeptides taken directly from single organs has been performed on lobster and crab by Fourier transform (FT)–MS (50, 94). Furthermore, Neupert et al. (46) recently used a combination of PSD and CID fragmentation to de novo sequence a new insect periviscerokinin peptide in a single insect cell.

To improve the mass measurement accuracy for peptide identification, studies utilizing direct tissue analysis have employed high-resolution, high-accuracy MALDI-FT-MS (50, 66, 94). Using an in-cell accumulation technique to improve sensitivity, mass measurement accuracy can also be improved to sub-parts-per-million levels by incorporating calibrants on a separate spot from the sample of interest without premixing calibration standards with the sample. This technique is highly beneficial for tissue analysis (50).

4.3.2. De novo sequencing assisted by microscale chemical derivatization. For neuropeptide identification and discovery from organisms lacking genomic information, de novo sequencing is often required; however, it represents a significant analytical challenge. Several chemical derivatization schemes have been developed to facilitate de novo sequencing by introducing a mass shift at either the N terminus (56, 95) or the C terminus (96) of a peptide, thus enabling differentiation between b- and y-type fragment ion series in a complex MS/MS spectrum. Most of these derivatization methods work well for tryptic peptides, but their utility for native neuropeptide sequencing is somewhat problematic and requires further evaluation. For example, acetylation of primary amine groups often changes the ionic states of peptides. This may lead to substantial loss of sensitivity in peptide analysis due to reduced ionization and fragmentation efficiencies for lower charge state ions. Likewise, a C-terminal methyl esterification method would not work for peptides lacking free C termini. Because C-terminal amidation is the most common PTM in SPs, the utility of this labeling approach for SP de novo sequencing is limited. However, a recent MALDI-FT MS/MS study using methyl esterification to characterize acidic amino acid–containing orcokinin peptides in crustacean showed significant improvement in the fragmentation efficiency of derivatized orcokinins due to blockage of the aspartate-selective cleavage pathway of the native orcokinin peptides (97).

Recently, a reductive methylation-labeling method using formaldehyde has shown great promise for de novo sequencing due to its simplicity and speed (98, 99). This derivatization method labels the N terminus and ε-amino group of lysine (Lys) in peptides through reductive amination and produces peaks differing by 28 Da for each derivatized site. The intensity of the N-terminal fragment a_1 ion is also substantially enhanced upon labeling, which is beneficial for de novo peptide sequencing. Furthermore, the ionic state of the modified peptides is not changed. By incorporating isotopic formaldehyde labeling, several unique features for SP sequencing applications are revealed. For example, reductive methylation enables differentiation of isobaric amino acid residues, such as Lys versus glutamine, due to selective dimethylation of the Lys side chain. Additionally, N-terminal blockage such as pyroglutamate modification can be readily assessed. Perhaps a more significant feature is the simplification of fragmentation pattern for singly charged peptides after dimethylation (**Figure 5**), which is beneficial for SP de novo sequencing because many SPs are singly charged due to the lack of a basic amino acid residue at their C termini. This N-terminal isotopic methylation strategy enabled de novo sequencing of 55 peptides, including 25 novel peptides from a neurohemal organ extract from the crab *Cancer borealis*, an organism whose genomic information remains unknown (99, 100).

Several on-target approaches have been used for tissue measurements. For example, on-target acetylation reacts with primary amine groups and results in a 42-Da mass shift for each added acetyl group. Similarly, treatment of worm head ring ganglion with hydrogen peroxide caused the oxidation of methionine groups to methionine sulfoxide, assayed as a 16-Da mass shift for each incorporation of an oxygen atom. These in situ chemical derivatizations provide a rapid method for confirming the presence of a free amine at the N terminus of a peptide and for counting the

a **Reductive methylation reaction scheme**

b **GGAYSFGLamide** (770.29⁺)

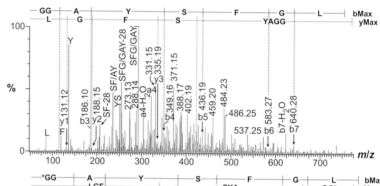

c **(CH₃)₂GGAYSFGLamide** (798.45⁺)

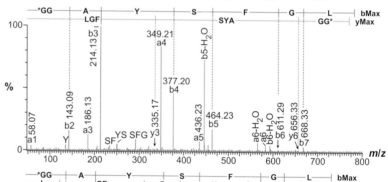

d **(CHD₂)₂GGAYSFGLamide** (802.51⁺)

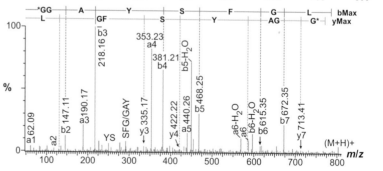

number of Lys or methionine residues in a sequence, thereby serving as an additional constraint for peptide identification with higher confidence (101).

4.3.3. Posttranslational modification characterization. Neuropeptide prohormones undergo extensive posttranslational cleavages and modifications to generate final bioactive peptides. The discovery and characterization of PTMs often rely on the detection of characteristic mass shifts during mass analysis or sequence-specific fragmentation analysis. Sometimes enzymatic treatment can be combined with mass analysis to probe a specific PTM. For example, by comparing the MALDI mass spectra of an *Aplysia californica* nervous tissue homogenate before and after incubation of pyroglutamate aminopeptidase, a peptide with an N-terminal pyroglutamate modification was readily characterized (102). Alternatively, direct multistage MS/MS analysis enabled the discovery of the first carboxylated neuropeptide (26). Furthermore, for several labile PTMs, such as nitrotyrosine-containing SPs, it appears that the choice of MALDI matrix plays a critical role in detection (103). Finally, a global reduction/alkylation procedure followed by mass profiling and sequencing allowed high-throughput characterization of heavily modified conotoxins containing multiple disulfide bonds (104, 105). For a more detailed examination of PTM characterization in SPs, please refer to a previous review (49).

5. IMAGING PEPTIDE DISTRIBUTIONS

The previous sections describe the process of characterizing the peptides in a tissue sample. In addition to knowledge of the chemical form of an SP, an understanding of its specific localization can be important. Several characterization methods, including (perhaps the most obvious method) isolating a particular structure, yield localization information for measurement. Another common approach involves specific stains and microscopy.

For the past century, the ability to image the distribution of a compound within a tissue or single cell has been a powerful tool for scientists, and has often necessitated the development of antibodies or other selective probes. Several MS-based approaches, including secondary ion mass spectrometry (SIMS), have been used for

---

Figure 5

Reductive methylation via isotopic formaldehyde labeling assists de novo sequencing of neuropeptides by simplifying fragmentation patterns upon derivatization. (*a*) Reductive methylation reaction scheme. (*b–d*) Tandem mass spectrometry (MS/MS) de novo sequencing of GGAYSFGLamide (770.29$^+$). The native peptide displays a highly complex MS/MS spectrum that is difficult to interpret even with the assistance of de novo sequencing software (*b*). However, the formaldehyde labeling results in much cleaner fragmentation spectra (*c* and *d*). By comparing the similar MS/MS fragmentation patterns of the H$_2$- and D$_2$-formaldehyde-labeled peptides, the almost-complete a- and b-ion series allowed complete sequencing of the full-length peptide, thereby revealing a new peptide sequence belonging to the A-type allatostatin peptide family. Adapted and modified from Reference 99 with permission.

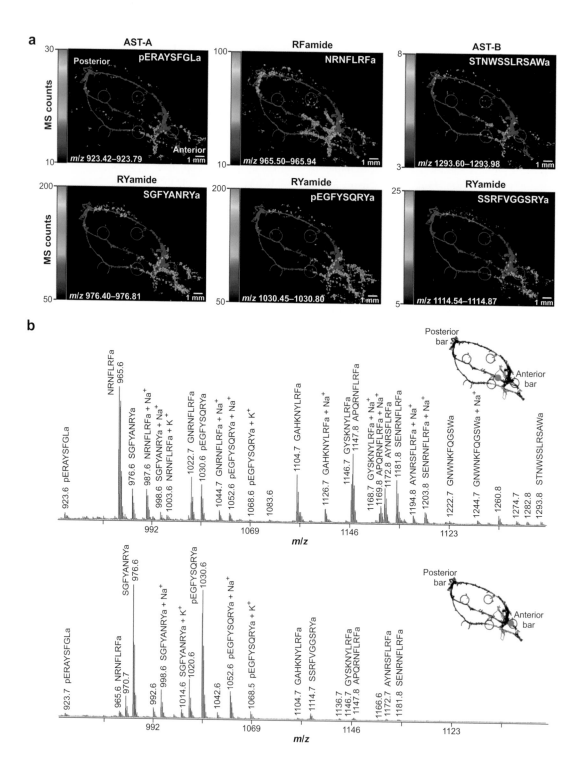

cellular-level profiling and imaging for decades (106). However, it was MALDI MS imaging (MSI), first demonstrated by Caprioli and advanced by many groups (107–110), that allowed the study of larger peptides and proteins in biological materials. As described by Caprioli et al. (see Reference 11, in this volume), one creates images with MALDI MS by coating the sample with a thin, homogenous layer of matrix prior to the collection of a series of mass spectra from an ordered array of locations across the sample. Following data collection, the intensities of selected signals are plotted for each mass spectrum to produce a series of two-dimensional distribution maps/ion images for each individual selected *m/z*. Because a complete mass spectrum is collected for each location in the sample, literally hundreds of ion images, each of a different compound, are created from each experiment. A recent special issue of the *International Journal of Mass Spectrometry* devoted to MSI highlights the capabilities of this approach (111), and several recent reviews describe the state of the art in greater detail for a range of applications including characterization of peptides (109, 112, 113).

MSI is useful for localizing SPs in brain slices and has produced impressive results from a wide range of samples. MSI has been used to characterize a range of peptides, including endogenous peptides and tryptic fragments of proteins (114, 115). Several earlier MSI studies described the characterization of SPs (116, 117). Because of the cell-cell heterogeneity of peptides, efforts to improve the spatial resolution have also focused on SPs. These include methods to achieve cellular-length scales using stigmatic imaging (112, 118), a spatial resolution at a cellular scale smaller than the laser beam diameter using oversampling (119), and massively parallel cell-sized sampling via the stretched-sample method (60). Besides MALDI MS, SIMS has been used to characterize the distribution of APGWamide, a smaller neuropeptide, in the central nervous system of *L. stagnalis* at high spatial resolution (120). Using *A. californica* as an example, both profiling and imaging of individual neurons have been achieved (121).

The current state of the art for peptide MSI is exemplified by three recent studies that demonstrated MS/MS on selected signals to verify peptide assignments, optimized sample preparation to minimize analyte migration during tissue preparation, and used well-characterized neuronal samples so that the performance could be evaluated. Specifically, Altelaar et al. (112) examined a series of pituitary peptides using stigmatic imaging and obtained an effective resolving power of 4 μm. Combining high-resolution MS with MSI, DeKeyser et al. (122) characterized more than 30 SPs belonging to 10 neuropeptide families, several with distinct spatial patterns (see **Figure 6**). Finally, Verhaert (123) characterized a complex suite of peptides in the brain–corpus cardiacum axis of insects. Future research will lead to improvements in throughput, sample preparation strategies, and spatial resolution of such

Mass spectrometric imaging (MSI): creates molecular ion images by rastering laser beam across a tissue sample and collecting the resultant mass spectra from an ordered array of locations across the sample

Figure 6

Images from the pericardial organ of a crab and the resultant mass spectra showing a differential distribution of peptides from the RYamide and RFamide peptide families. The colors in the images relate to the intensity of the specific mass spectral peak intensities. Adapted from Reference 122 with permission.

neuropeptide MSI experiments, and will also introduce a range of additional applications. We expect a great increase in the knowledge of regional differences in peptide processing as the use of this technology becomes more widespread.

6. QUANTITATION

Quantitative analysis by MS is complicated by the heterogeneity of ionization and the unpredictable discrimination and suppression effects of complex mixtures. Several strategies have been developed to address these issues. An accurate approach employs a corresponding stable isotope–incorporated synthetic peptide for each peptide of interest as the internal standard (124, 125). The drawback of this approach, however, is the need to synthesize isotope-incorporated peptide for each native peptide of interest. For a global-scale survey of relative quantitative changes, numerous isotope-labeling methods have been reported (95, 96, 126, 127). These methodologies produce chemical counterparts of the analytes of interest; these counterparts possess an identical chemical structure but a distinguishable mass difference.

Although the isotope-coded affinity tag (ICAT) approach is widely applicable for protein quantitation, in its original form it is limited to cysteine-containing proteins/peptides. Few SPs contain cysteine residues. Other labeling reagents have been designed for more generic labeling, such as acetylation of the N terminus and the ε-amino group of Lys residues by acetic anhydride (-H_6 and -D_6) (128). Acetylation of basic amino groups, however, may cause a substantial loss of sensitivity for peptides without a C-terminal charge due to the reduction of charge states. An alternative general labeling approach relies on methylation using d0- or d3-methanol (96). This method suffers from the limitation that peptides with C-terminal amidation and without aspartic or glutamic acids will not be labeled. Recently, Fricker and colleagues (129) reported the use of active esters of trimethylammoniumbutyrate containing either nine deuteriums (heavy) or nine hydrogen atoms (light) for quantitative peptidomic applications. The N-hydroxysuccinimide ester of the trimethylammoniumbutyrate reagent reacted with primary amines at the N terminus and Lys side chain. Successful applications of this isotopically encoded reagent enabled the assessment of relative peptidomic changes in response to food intake or drug treatment in mouse models (17, 130, 131).

Another labeling technique is reductive methylation, which targets the N-terminal amine and Lys side chain using isotopically encoded formaldehyde reagents. This dimethyl labeling strategy labels the N terminus and ε-amino group of Lys through reductive amination, producing peaks that differ by 28 Da for each derivatized site and 4 Da (H_2/D_2) for each derivatized isotopic pair (132). As outlined in the de novo sequencing section, isotopic formaldehyde labeling provides both enhanced quantitation and improved fragmentation capability. Of course, this labeling approach would not work for SPs with blocked N termini (pyroglutamate or acetylated amino acid).

Recently, a novel analyte-ion combination scheme that has the potential to enable direct tissue mass spectral quantitation was reported (133). The potential application of this method for quantitative comparison of neuropeptides directly from individual neurons or tissues isolated from different physiological states will have a great impact in SP research.

7. PEPTIDE FUNCTION: DYNAMIC MEASUREMENTS

Microdialysis: in vivo sampling technique based on size-selective diffusion of the analyte through a semipermeable membrane that collects peptides and/or smaller molecules for further characterization

An SP is not active in cell-cell signaling while inside the cell; it must be released in order to function. Under what conditions and in what chemical form is a specific SP released? Such information is not easy to obtain by directly sampling a brain region, and so a number of approaches have been developed to sample SPs released from neurons. Stated differently, whereas a microliter of brain may contain 100,000 cells and literally tens of thousands of proteins and peptides, only a small fraction of those peptides are released as SPs. How can one selectively measure the release of peptides from a tissue?

Investigations of SP secretion from biological tissues have employed a number of approaches, each with its own advantages and disadvantages (see, e.g., References 134 and 135). Direct detection of SPs can be accomplished by sampling extracellular media such as the blood or the extracellular fluids surrounding neurohemal sites to study hormones.

Preselected peptides in these releasates are often detected via radioimmunoassays. Although it is an exquisitely sensitive technique, radiochemical detection of peptidergic release requires preselection of the peptide of interest, and has a limited ability to monitor multiple peptides during the same experiment. Other approaches, such as amperometric detection (136), offer improved temporal and spatial resolution. The development of approaches that combine release sampling with MS has become particularly important for characterizing unknown SPs.

Both microdialysis and push-pull perfusion are exemplary methods for collecting SP release from localized regions (137–139). These approaches rely on the insertion of a sampling probe into a specific brain region that collects the extracellular peptides. Greater detail on the technology, protocols, and capabilities of microdialysis sampling of SPs can be found in several recent reviews (140–142). Because the peptides are released from a cell/neurohemal release area and are rapidly diluted by diffusion from localized release sites, the concentrations are often in the picomolar-to-nanomolar range, with sample volumes in the nanoliter-to-microliter range. Thus, the collected material is often introduced to capillary-scale separations and requires high-sensitivity MS (75, 76, 140, 143, 144).

What other methods can be used? For brain slices and semi-intact preparations, a number of additional strategies are available for sampling SP release. For example, detection of activity-dependent release of peptides from brain regions as well as from single cultured neurons has been reported using a combination of electrophoretic separation and solid-phase extraction (SPE) (47, 145). Such approaches have been improved by using individual SPE beads placed near the cell or region of interest (146). Because the brain slice is accessible and the bead can be precisely located even over a specific release zone, localized release can be followed, and precise chemical stimulations can be applied while collecting peptide releasate. Using this approach, the release of unexpected peptides has been confirmed for several systems (26, 147).

We expect continued improvement in sampling methods adapted to collecting SP release from the brain. Dynamic information is crucial to understanding peptide function and is among the most difficult types of data to obtain from SP measurements.

8. BIOINFORMATICS

Identification and characterization of neuropeptides can be achieved in several ways, depending on the type of MS data and database availability. The simplest way is by comparing experimental masses against a list of known neuropeptide masses via manual comparison or by automated searching using a database such as SwePep (**http://www.swepep.org/**) (148).

As mentioned previously, a neuropeptide prohormone is processed into multiple gene products by the actions of a number of enzymes. During the past several years, we have created bioinformatics tools to aid in predicting the final peptide forms from a prohormone; these tools use binary logistic regression models, trained on the neuropeptides that we and others have characterized, to identify cleaved basic sites among the many possible cleavage sites in a prohormone (30, 149). Thus, one can make well-informed decisions about the expected neuropeptides from a novel prohormone. These prediction tools are accessible via our NeuroPred suite, available at **http://neuroproteomics.scs.uiuc.edu/neuropred.html** (150). The site also contains lists of prohormones and their associated peptides for a variety of common neuronal models.

As described above, there are two common methods of performing MS analysis of brain tissues: direct profiling via MALDI and ESI-MS/MS of extracts. Direct profiling of a cell or tissue has an advantage in that many peptides from a specific prohormone are detected at the same time. When multiple peptides are detected using accurate mass measurements even without MS/MS, the ability to confidently assign the peaks is enhanced. Single-cell MS is an extreme example where often a significant fraction of a prohormone sequence can be covered with accurate mass MALDI MS measurements.

For extracts or larger brain regions, one gains the benefits of working with a larger amount of material but loses the advantage of prohormone coverage due to greater sample complexity. In such cases, the ability to sequence novel neuropeptides can be critical. Neuropeptide sequencing requires an instrument capable of producing fragment ions (via MS/MS). Depending on instrument availability and the species of interest, a number of different data analysis strategies can be employed for MS/MS neuropeptide identification. If the species of interest has a sequenced genome, a standard protein database search can be performed using proteomic-based search engines such as Mascot (Matrix Science, **http://www.matrixscience.com/**) or SEQUEST (Thermo Corp., **http://www.thermo.com/**) (151). Although these search engines were designed for proteomics applications, they are easily adapted to neuropeptidomics.

Another way to identify neuropeptides using a database is by sequence tag searching. In contrast to Mascot and SEQUEST, which base identifications on parent masses and MS/MS fragmentation patterns, sequence tag searches are performed using small contiguous strings of identified amino acids (sequence tags) compared against sequences in the database (152). Because sequence tag searching does not rely on parent mass identification, it is a powerful technique for identifying SPs containing PTMs. For example, MS-Seq (Protein Prospector, **http://prospector.ucsf.edu/**) and Mascot offer sequence tag searching for single tag queries. Both interfaces are flexible and powerful but can be time consuming if analyzing a large volume of data.

If the species of interest does not have a sequenced genome, de novo sequencing of the neuropeptide is required. In contrast to database searching, de novo peptide sequencing represents the direct "reading" of the amino acid sequence from the MS/MS spectrum. With a great deal of experience, de novo sequencing can be performed manually. However, this is extremely labor intensive and time consuming. Fortunately, there exist several software packages that can perform de novo sequencing directly on MS/MS data, including LuteFisk (**http://www.hairyfatguy.com/lutefisk/**), PepSeq (packaged with Waters's MassLynx software, **http://www.waters.com/**), Mascot Distiller (Matrix Science), and PEAKS (Bioinformatics Solutions, **http://www.bioinformaticssolutions.com/**). Many of these programs, such as PEAKS and Mascot Distiller, include automated de novo processing to enhance sequencing speed and flexibility.

De novo sequencing does not always produce an accurate sequence for the entire SP. In this case, a BLAST homology search (**http://www.ncbi.nlm. nih.gov/BLAST/**) can be performed. A BLAST search compares a partial SP sequence against the database of a closely related species. Although a homology search rarely yields a complete sequence, it can provide useful information about the evolutionary origins and potential function of the partially sequenced SP. More recently, the use of pattern finding software such as SPIDER (Bioinformatics Solutions) and MEME (**http://meme.sdsc.edu/meme/**) has improved partial sequence homology searches (153). These programs are optimized for MS/MS–derived data and are more tolerant of sequencing errors than BLAST searches. We certainly expect that bioinformatics tools tailored to the unique features of SPs will continue to expand.

9. FUTURE DIRECTIONS

It is evident that the increasing number of SPs described for every animal model reflects significant recent advances in MS. We expect that the number of identified SPs, as well as information on their localization and processing, will continue to increase for all animals. Although peptide characterization by MS is generally a well-developed area, the field of neuropeptidomics requires additional developments in instrumentation and protocol, especially for dynamic and spatially resolved information on very small samples. As the methods of localized sampling and even activity-dependent sampling are further refined, information on dynamics will be combined with information on chemical form and spatial localization to enable a more comprehensive understanding of peptidergic signaling.

It is interesting to speculate whether the vast diversity of peptides found in the brain all contribute to and influence behavior, or whether many of these shortened and processed forms are simply peptides on their way to being catabolized. In other words, even though many peptides are present, we do not yet know how many of the peptides are actually used by each brain circuit, nor how the circuit activity depends on dynamic changes in the extracellular peptide cocktail. Only with improved analytical measurement tools such as MSI, single-cell MS, and bioinformatics tools can these questions be addressed. Thirty years after the beginning of neuropeptide research, the vast scope of SP complexity is still being investigated.

SUMMARY POINTS

1. MS has become the method of choice for neuropeptide analysis due to its high speed, low detection limits, and the impressive chemical information it provides. The ever-increasing number of new peptides detected in a wide range of animal model systems reflects the advancements of mass spectrometric measurements and improved sampling approaches.

2. No single measurement platform can simultaneously provide chemical, spatial, and temporal information content for probing SPs in the brain. Therefore, a multifaceted approach that combines direct tissue profiling/imaging, peptide sequencing, and dynamic release measurements using a wide array of separation- and MS-based instrumentation is needed to provide a more complete picture of peptide signaling.

3. Given the sensitivity of MS-based approaches and their capability to measure a large number of peptides simultaneously, it is important to develop specialized sample preparation strategies for different types of neural samples in order to obtain biologically relevant results.

4. The development of direct tissue profiling and single-cell MS over the past decade has contributed significantly to neuropeptide research. Further improvements in sampling protocols, mass measurement accuracy, and microscale chemical derivatization will further improve their application to single-cell peptidomics.

FUTURE ISSUES

1. Despite continued development of microscale sampling approaches, dynamic measurement of SP release remains an analytical challenge. Future developments should address issues regarding sensitivity and temporal resolution.

2. MSI has become a powerful tool for exploring the spatial distribution of SPs throughout tissue sections. Future work should focus on the development of improved spatial resolution to enable imaging of peptide distribution at the subcellular level. Instrumentation advancement and data handling capability will be equally important for higher-throughput imaging analysis.

3. Although numerous chemical reagents have been developed for improved de novo sequence analysis of SPs, only a limited range of SPs can be sequenced. Several electron-based MS/MS fragmentation methods have shown great promise for sequencing large peptides and small proteins while preserving labile PTMs. These newer MS/MS techniques, coupled with improvements in instrumentation, will prompt a new generation of studies in peptidomics.

4. Various bioinformatic tools and algorithms have been developed to derive peptide/protein matches via either identity or homology searching. Currently, it is not efficient to query protein databases constructed largely from translated DNA sequences with neuropeptide MS/MS data. New and improved bioinformatics tools tailored to the unique features of SPs are needed and have the potential to accelerate neuropeptide discovery.

DISCLOSURE STATEMENT

The authors are not aware of any biases that might be perceived as affecting the objectivity of this review.

ACKNOWLEDGMENTS

The support of the National Science Foundation through grant no. CHE-05–26692 to JVS and the National Institute on Drug Abuse under award no. P30 DA 018310 to the UIUC Neuroproteomics Center on Cell to Cell Signaling is gratefully acknowledged. Preparation of this manuscript was supported in part by the National Science Foundation through a CAREER Award (CHE-0449991), the National Institutes of Health through grant no. 1R01DK071801, and an Alfred P. Sloan Research Fellowship to LL.

LITERATURE CITED

1. Sandman CA, Strand FL, Beckwith B, Chronwall BM, Flynn FW, Nachman RJ, eds. 1999. *Neuropeptides: Structure and Function in Biology and Behavior*, Vol. 897. New York: Ann. N.Y. Acad. Sci. 443 pp.

2. Strand FL. 1999. *Neuropeptides: Regulators of Physiological Processes*. Cambridge, MA: MIT Press. 658 pp.

3. Kastin AJ. 2006. *Handbook of Biologically Active Peptides*. Amsterdam/Boston: Academic. 1595 pp.

4. Jensen J. 2001. Regulatory peptides and control of food intake in nonmammalian vertebrates. *Comp. Biochem. Physiol. A Mol. Integr. Physiol.* 128:471–79

5. Gundlach AL, Burazin TC, Larm JA. 2001. Distribution, regulation and role of hypothalamic galanin systems: renewed interest in a pleiotropic peptide family. *Clin. Exp. Pharmacol. Physiol.* 28:100–5

6. Levin ER, Hu RM, Rossi M, Pickart M. 1992. Arginine vasopressin stimulates atrial natriuretic peptide gene expression and secretion from rat diencephalic neurons. *Endocrinology* 131:1417–23

7. Okada Y, Tsuda Y, Bryant SD, Lazarus LH. 2002. Endomorphins and related opioid peptides. *Vitam. Horm.* 65:257–79

8. Kieffer BL, Gaveriaux-Ruff C. 2002. Exploring the opioid system by gene knockout. *Prog. Neurobiol.* 66:285–306

9. Stefano GB, Fricchione G, Goumon Y, Esch T. 2005. Pain, immunity, opiate and opioid compounds and health. *Med. Sci. Monit.* 11:MS47–53

10. Marshall AG, Hendrickson CL. 2008. High-resolution mass spectrometers. *Annu. Rev. Anal. Chem.* 1:579–99

11. Burnum KE, Frappier SL, Caprioli R. 2008. Matrix-assisted laser desorption/ionization imaging mass spectrometry for the investigation of proteins and peptides. *Annu. Rev. Anal. Chem.* 1:689–705

12. Borland LM, Kottegoda S, Phillips KS, Allbritton NL. 2008. Chemical analysis of single cells. *Annu. Rev. Anal. Chem.* 1:191–227

13. Lindsten J, ed. 1992. *Nobel Lectures, Physiology or Medicine 1971–1980.* Singapore: World Sci. Publ. Co.

14. Clynen E, Baggerman G, Veelaert D, Cerstiaens A, Van der Horst D, et al. 2001. Peptidomics of the pars intercerebralis–corpus cardiacum complex of the migratory locust, *Locusta migratoria. Eur. J. Biochem.* 268:1929–39

15. Schulz-Knappe P, Zucht HD, Heine G, Jurgens M, Hess R, Schrader M. 2001. Peptidomics: the comprehensive analysis of peptides in complex biological mixtures. *Comb. Chem. High Throughput Screen.* 4:207–17

16. Verhaert P, Uttenweiler-Joseph S, de Vries M, Loboda A, Ens W, Standing KG. 2001. Matrix-assisted laser desorption/ionization quadrupole time-of-flight mass spectrometry: an elegant tool for peptidomics. *Proteomics* 1:118–31

17. Che FY, Vathy I, Fricker LD. 2006. Quantitative peptidomics in mice: effect of cocaine treatment. *J. Mol. Neurosci.* 28:265–75

18. Jost MM, Budde P, Tammen H, Hess R, Kellmann M, et al. 2005. The concept of functional peptidomics for the discovery of bioactive peptides in cell culture models. *Comb. Chem. High Throughput Screen.* 8:767–73

19. Soloviev M, Finch P. 2006. Peptidomics: bridging the gap between proteome and metabolome. *Proteomics* 6:744–47

20. Svensson M, Skold K, Svenningsson P, Andren PE. 2003. Peptidomics-based discovery of novel neuropeptides. *J. Proteome Res.* 2:213–19

21. Wei H, Nolkrantz K, Parkin MC, Chisolm CN, O'Callaghan JP, Kennedy RT. 2006. Identification and quantification of neuropeptides in brain tissue by capillary liquid chromatography coupled off-line to MALDI-TOF and MALDI-TOF/TOF-MS. *Anal. Chem.* 78:4342–51

22. Emanuelsson O, Brunak S, von Heijne G, Nielsen H. 2007. Locating proteins in the cell using TargetP, SignalP and related tools. *Nat. Protoc.* 2:953–71

23. Emanuelsson O, Nielsen H, Brunak S, von Heijne G. 2000. Predicting subcellular localization of proteins based on their N-terminal amino acid sequence. *J. Mol. Biol.* 300:1005–16

24. Kandel E, Schwartz JJ, Jessell TM. 2000. *Principles of Neural Science*. New York: McGraw-Hill

25. Floyd PD, Li L, Rubakhin SS, Sweedler JV, Horn CC, et al. 1999. Insulin prohormone processing, distribution, and relation to metabolism in *Aplysia californica*. *J. Neurosci.* 19:7732–41

26. Jakubowski JA, Hatcher NG, Xie F, Sweedler JV. 2006. The first gamma-carboxyglutamate-containing neuropeptide. *Neurochem. Int.* 49:223–29

27. Seidah NG. 2001. Cellular limited proteolysis of precursor proteins and peptides. In *The Enzymes*, ed. RE Dalbey, DS Sigman, pp. 237–58. San Diego, CA: Academic

28. Steiner D. 2001. The prohormone convertases and precursor processing in protein biosynthesis. In *The Enzymes*, ed. RE Dalby, DS Sigman, pp. 163–98. San Diego, CA: Academic

29. Devi L. 1991. Consensus sequence for processing of peptide precursors at monobasic sites. *FEBS Lett.* 280:189–94

30. Amare A, Hummon AB, Southey BR, Zimmerman TA, Rodriguez-Zas SL, Sweedler JV. 2006. Bridging neuropeptidomics and genomics with bioinformatics: prediction of mammalian neuropeptide prohormone processing. *J. Proteome Res.* 5:1162–67

31. Lindberg I, Hutton JC. 1991. Peptide processing proteinases with selectivity for paired basic residues. In *Peptide Biosynthesis and Processing*, ed. LD Fricker, pp. 141–74. Boca Raton, FL: CRC

32. Rholam M, Brakch N, Germain D, Thomas DY, Fahy C, et al. 1995. Role of amino acid sequences flanking dibasic cleavage sites in precursor proteolytic processing: the importance of the first residue C-terminal of the cleavage site. *Eur. J. Biochem.* 227:707–14

33. Cameron A, Apletalina EV, Lindberg I. 2001. The enzymology of PC1 and PC2. In *The Enzymes*, ed. RE Dalbey, DS Sigman, pp. 291–332. San Diego, CA: Academic

34. Pow DV, Morris JF. 1989. Dendrites of the hypothalamic magnocellular neurons release neurohypophysal peptides by exocytosis. *Neuroscience* 32:435–39

35. Sabatier N, Caquineau C, Douglas AJ, Leng G. 2003. Oxytocin released from magnocellular dendrites. *Ann. N.Y. Acad. Sci.* 994:218–24

36. Fujisawa Y, Furukawa Y, Ohta S, Ellis TA, Dembrow NC, et al. 1999. The *Aplysia mytilus* inhibitory peptide-related peptides: identification, cloning, processing, distribution, and action. *J. Neurosci.* 19:9618–34

37. Furukawa Y, Nakamaru K, Wakayama H, Fujisawa Y, Minakata H, et al. 2001. The enterins: a novel family of neuropeptides isolated from the enteric nervous system and CNS of *Aplysia*. *J. Neurosci.* 21:8247–61

38. Sweedler JV, Li L, Rubakhin SS, Alexeeva V, Dembrow NC, et al. 2002. Identification and characterization of the feeding circuit-activating peptides, a novel neuropeptide family of *Aplysia*. *J. Neurosci.* 22:7797–808

39. Baertschi AJ, Monnier D, Schmidt U, Levitan ES, Fakan S, Roatti A. 2001. Acid prohormone sequence determines size, shape, and docking of secretory vesicles in atrial myocytes. *Circ. Res.* 89:E23–29

40. Shafqat J, Melles E, Sigmundsson K, Johansson BL, Ekberg K, et al. 2006. Proinsulin C-peptide elicits disaggregation of insulin resulting in enhanced physiological insulin effects. *Cell Mol. Life. Sci.* 63:1805–11

41. Steiner DF. 2004. The proinsulin C-peptide: a multirole model. *Exp. Diabesity Res.* 5:7–14

42. Edison AS, Espinoza E, Zachariah C. 1999. Conformational ensembles: the role of neuropeptide structures in receptor binding. *J. Neurosci.* 19:6318–26

43. Santama N, Benjamin PR. 2000. Gene expression and function of FMRFamide-related neuropeptides in the snail *Lymnaea. Microsc. Res. Tech.* 49:547–56

44. von Bohlen und Halbach O. 2005. The renin-angiotensin system in the mammalian central nervous system. *Curr. Protein Pept. Sci.* 6:355–71

45. Predel R. 2001. Peptidergic neurohemal system of an insect: mass spectrometric morphology. *J. Comp. Neurol.* 436:363–75

46. Neupert S, Predel R, Russell WK, Davies R, Pietrantonio PV, Nachman RJ. 2005. Identification of tick periviscerokinin, the first neurohormone of *Ixodidae*: single cell analysis by means of MALDI-TOF/TOF mass spectrometry. *Biochem. Biophys. Res. Commun.* 338:1860–64

47. Garden RW, Moroz LL, Moroz TP, Shippy SA, Sweedler JV. 1996. Excess salt removal with matrix rinsing: direct peptide profiling of neurons from marine invertebrates using matrix-assisted laser desorption/ionization time-of-flight mass spectrometry. *J. Mass Spectrom.* 31:1126–30

48. Li L, Garden RW, Sweedler JV. 2000. Single-cell MALDI: a new tool for direct peptide profiling. *Trends Biotechnol.* 18:151–60

49. Hummon AB, Amare A, Sweedler JV. 2006. Discovering new invertebrate neuropeptides using mass spectrometry. *Mass Spectrom. Rev.* 25:77–98

50. Kutz KK, Schmidt JJ, Li L. 2004. In situ tissue analysis of neuropeptides by MALDI FTMS in-cell accumulation. *Anal. Chem.* 76:5630–40

51. Fournier I, Day R, Salzet M. 2003. Direct analysis of neuropeptides by in situ MALDI-TOF mass spectrometry in the rat brain. *Neuro. Endocrinol. Lett.* 24:9–14

52. DeKeyser SS, Li L. 2007. Mass spectrometric charting of neuropeptides in arthropod neurons. *Anal. Bioanal. Chem.* 387:29–35

53. Neupert S, Predel R. 2005. Mass spectrometric analysis of single identified neurons of an insect. *Biochem. Biophys. Res. Commun.* 327:640–45

54. Li L, Romanova EV, Rubakhin SS, Alexeeva V, Weiss KR, et al. 2000. Peptide profiling of cells with multiple gene products: combining immunochemistry and MALDI mass spectrometry with on-plate microextraction. *Anal. Chem.* 72:3867–74

55. Li L, Garden RW, Romanova EV, Sweedler JV. 1999. In situ sequencing of peptides from biological tissues and single cells using MALDI-PSD/CID analysis. *Anal. Chem.* 71:5451–58

56. Yew JY, Dikler S, Stretton AO. 2003. De novo sequencing of novel neuropeptides directly from *Ascaris suum* tissue using matrix-assisted laser desorption/ionization time-of-flight/time-of-flight. *Rapid Commun. Mass Spectrom.* 17:2693–98

57. Jespersen S, Chaurand P, van Strien FJ, Spengler B, van der Greef J. 1999. Direct sequencing of neuropeptides in biological tissue by MALDI-PSD mass spectrometry. *Anal. Chem.* 71:660–66

58. Redeker V, Toullec JY, Vinh J, Rossier J, Soyez D. 1998. Combination of peptide profiling by matrix-assisted laser desorption/ionization time-of-flight mass spectrometry and immunodetection on single glands or cells. *Anal. Chem.* 70:1805–11

59. Rubakhin SS, Garden RW, Fuller RR, Sweedler JV. 2000. Measuring the peptides in individual organelles with mass spectrometry. *Nat. Biotechnol.* 18:172–75

60. Monroe EB, Jurchen JC, Koszczuk BA, Losh JL, Rubakhin SS, Sweedler JV. 2006. Massively parallel sample preparation for the MALDI MS analyses of tissues. *Anal. Chem.* 78:6826–32

61. Rubakhin SS, Churchill JD, Greenough WT, Sweedler JV. 2006. Profiling signaling peptides in single mammalian cells using mass spectrometry. *Anal. Chem.* 78:7267–72

62. Rubakhin SS, Sweedler JV. 2007. Characterizing peptides in individual mammalian cells using mass spectrometry. *Nat. Protoc.* 2:1987–97

63. Che FY, Lim J, Pan H, Biswas R, Fricker LD. 2005. Quantitative neuropeptidomics of microwave-irradiated mouse brain and pituitary. *Mol. Cell Proteomics* 4:1391–405

64. Dockray GJ. 1976. Immunochemical evidence of cholecystokinin-like peptides in brain. *Nature* 264:568–70

65. Dowell JA, Vander Heyden W, Li L. 2006. Rat neuropeptidomics by LC-MS/MS and MALDI-FTMS: enhanced dissection and extraction techniques coupled with 2D RP-RP HPLC. *J. Proteome Res.* 5:3368–75

66. Fu Q, Kutz KK, Schmidt JJ, Hsu YW, Messinger DI, et al. 2005. Hormone complement of the *Cancer productus* sinus gland and pericardial organ: an anatomical and mass spectrometric investigation. *J. Comp. Neurol.* 493:607–26

67. Fu Q, Goy MF, Li L. 2005. Identification of neuropeptides from the decapod crustacean sinus glands using nanoscale liquid chromatography tandem mass spectrometry. *Biochem. Biophys. Res. Commun.* 337:765–78

68. Schoofs L, Holman GM, Hayes TK, Nachman RJ, De Loof A. 1991. Isolation, primary structure, and synthesis of locustapyrokinin: a myotropic peptide of *Locusta migratoria*. *Gen. Comp. Endocrinol.* 81:97–104

69. Baggerman G, Boonen K, Verleyen P, De Loof A, Schoofs L. 2005. Peptidomic analysis of the larval *Drosophila melanogaster* central nervous system by two-dimensional capillary liquid chromatography quadrupole time-of-flight mass spectrometry. *J. Mass Spectrom.* 40:250–60

70. Washburn MP, Wolters D, Yates JR 3rd. 2001. Large-scale analysis of the yeast proteome by multidimensional protein identification technology. *Nat. Biotechnol.* 19:242–47

71. Wolters DA, Washburn MP, Yates JR 3rd. 2001. An automated multidimensional protein identification technology for shotgun proteomics. *Anal. Chem.* 73:5683–90

72. Peng J, Gygi SP. 2001. Proteomics: the move to mixtures. *J. Mass Spectrom.* 36:1083–91

73. Baggerman G, Cerstiaens A, De Loof A, Schoofs L. 2002. Peptidomics of the larval *Drosophila melanogaster* central nervous system. *J. Biol. Chem.* 277:40368–74

74. Skold K, Svensson M, Kaplan A, Bjorkesten L, Astrom J, Andren PE. 2002. A neuroproteomic approach to targeting neuropeptides in the brain. *Proteomics* 2:447–54

75. Haskins WE, Wang Z, Watson CJ, Rostand RR, Witowski SR, et al. 2001. Capillary LC-MS2 at the attomole level for monitoring and discovering endogenous peptides in microdialysis samples collected in vivo. *Anal. Chem.* 73:5005–14

76. Haskins WE, Watson CJ, Cellar NA, Powell DH, Kennedy RT. 2004. Discovery and neurochemical screening of peptides in brain extracellular fluid by chemical analysis of in vivo microdialysis samples. *Anal. Chem.* 76:5523–33

77. Husson SJ, Clynen E, Baggerman G, De Loof A, Schoofs L. 2005. Discovering neuropeptides in *Caenorhabditis elegans* by two dimensional liquid chromatography and mass spectrometry. *Biochem. Biophys. Res. Commun.* 335:76–86

78. Floyd PD, Li L, Moroz TP, Sweedler JV. 1999. Characterization of peptides from *Aplysia* using microbore liquid chromatography with matrix-assisted laser desorption/ionization time-of-flight mass spectrometry guided purification. *J. Chromatogr. A* 830:105–13

79. Wei H, Nolkrantz K, Powell DH, Woods JH, Ko MC, Kennedy RT. 2004. Electrospray sample deposition for matrix-assisted laser desorption/ionization (MALDI) and atmospheric pressure MALDI mass spectrometry with attomole detection limits. *Rapid Commun. Mass Spectrom.* 18:1193–200

80. Wei H, Dean SL, Parkin MC, Nolkrantz K, O'Callaghan JP, Kennedy RT. 2005. Microscale sample deposition onto hydrophobic target plates for trace level detection of neuropeptides in brain tissue by MALDI-MS. *J. Mass Spectrom.* 40:1338–46

81. Page JS, Rubakhin SS, Sweedler JV. 2002. Single-neuron analysis using CE combined with MALDI MS and radionuclide detection. *Anal. Chem.* 74:497–503

82. Page JS, Rubakhin SS, Sweedler JV. 2000. Direct cellular assays using off-line capillary electrophoresis with matrix-assisted laser desorption/ionization time-of-flight mass spectrometry. *Analyst* 125:555–62

83. Zhang H, Caprioli RM. 1996. Capillary electrophoresis combined with matrix-assisted laser desorption/ionization mass spectrometry: continuous sample deposition on a matrix-precoated membrane target. *J. Mass Spectrom.* 31:1039–46

84. Zhang H, Stoeckli M, Andren PE, Caprioli RM. 1999. Combining solid-phase preconcentration, capillary electrophoresis and off-line matrix-assisted laser desorption/ionization mass spectrometry: intracerebral metabolic processing of peptide E in vivo. *J. Mass Spectrom.* 34:377–83

85. Kupka KD, Schropp WW, Schiller C, Hillenkamp F. 1980. Laser-micro-mass analysis (LAMMA) of metallic and organic ions in medical samples. *Scan. Electron Microsc.* II: 635–40

86. Meyer zum Gottesberge-Orsulakova A, Kaufmann R. 1985. Recent advances in laser microprobe mass analysis (LAMMA) of inner ear tissue. *Scan. Electron Microsc.* VII:393–405

87. Kaufmann R, Hillenkamp F, Nitsche R, Schurmann M, Wechsung R. 1978. The laser microprobe mass analyser (LAMMA): biomedical applications. *Microsc. Acta* Suppl.(2):297–306

88. Jimenez CR, van Veelen PA, Li KW, Wildering WC, Geraerts WP, et al. 1994. Neuropeptide expression and processing as revealed by direct matrix-assisted laser desorption ionization mass spectrometry of single neurons. *J. Neurochem.* 62:404–7

89. Li KW, Hoek RM, Smith F, Jimenez CR, van der Schors RC, et al. 1994. Direct peptide profiling by mass spectrometry of single identified neurons reveals complex neuropeptide-processing pattern. *J. Biol. Chem.* 269:30288–92

90. Clynen E, De Loof A, Schoofs L. 2003. The use of peptidomics in endocrine research. *Gen. Comp. Endocrinol.* 132:1–9

91. Hummon AB, Richmond TA, Verleyen P, Baggerman G, Huybrechts J, et al. 2006. From the genome to the proteome: uncovering peptides in the *Apis* brain. *Science* 314:647–49

92. Perry SJ, Dobbins AC, Schofield MG, Piper MR, Benjamin PR. 1999. Small cardioactive peptide gene: structure, expression and mass spectrometric analysis reveals a complex pattern of cotransmitters in a snail feeding neuron. *Eur. J. Neurosci.* 11:655–62

93. Predel R, Russell WK, Tichy SE, Russell DH, Nachman RJ. 2003. Mass spectrometric analysis of putative capa-gene products in *Musca domestica* and *Neobellieria bullata*. *Peptides* 24:1487–91

94. Stemmler EA, Provencher HL, Guiney ME, Gardner NP, Dickinson PS. 2005. Matrix-assisted laser desorption/ionization Fourier transform mass spectrometry for the identification of orcokinin neuropeptides in crustaceans using metastable decay and sustained off-resonance irradiation. *Anal. Chem.* 77:3594–606

95. Munchbach M, Quadroni M, Miotto G, James P. 2000. Quantitation and facilitated de novo sequencing of proteins by isotopic N-terminal labeling of peptides with a fragmentation-directing moiety. *Anal. Chem.* 72:4047–57

96. Goodlett DR, Keller A, Watts JD, Newitt R, Yi EC, et al. 2001. Differential stable isotope labeling of peptides for quantitation and de novo sequence derivation. *Rapid Commun. Mass Spectrom.* 15:1214–21

97. Ma M, Kutz-Naber KK, Li L. 2007. Methyl esterification assisted MALDI FTMS characterization of the orcokinin neuropeptide family. *Anal. Chem.* 79:673–81

98. Hsu JL, Huang SY, Shiea JT, Huang WY, Chen SH. 2005. Beyond quantitative proteomics: signal enhancement of the a1 ion as a mass tag for peptide sequencing using dimethyl labeling. *J. Proteome Res.* 4:101–8

99. Fu Q, Li L. 2005. De novo sequencing of neuropeptides using reductive isotopic methylation and investigation of ESI QTOF MS/MS fragmentation pattern of neuropeptides with N-terminal dimethylation. *Anal. Chem.* 77:7783–95

100. Cruz-Bermudez ND, Fu Q, Kutz-Naber KK, Christie AE, Li L, Marder E. 2006. Mass spectrometric characterization and physiological actions of GAHKNYLRFamide, a novel FMRFamide-like peptide from crabs of the genus *Cancer*. *J. Neurochem.* 97:784–99

101. Yew JY, Kutz KK, Dikler S, Messinger L, Li L, Stretton AO. 2005. Mass spectrometric map of neuropeptide expression in *Ascaris suum*. *J. Comp. Neurol.* 488:396–413

102. Garden RW, Moroz TP, Gleeson JM, Floyd PD, Li L, et al. 1999. Formation of N-pyroglutamyl peptides from N-Glu and N-Gln precursors in *Aplysia* neurons. *J. Neurochem.* 72:676–81

103. Sheeley SA, Rubakhin SS, Sweedler JV. 2005. The detection of nitrated tyrosine in neuropeptides: a MALDI matrix-dependent response. *Anal. Bioanal. Chem.* 382:22–27

104. Jakubowski JA, Sweedler JV. 2004. Sequencing and mass profiling highly modified conotoxins using global reduction/alkylation followed by mass spectrometry. *Anal. Chem.* 76:6541–47

105. Jakubowski JA, Keays DA, Kelley WP, Sandall DW, Bingham JP, et al. 2004. Determining sequences and post-translational modifications of novel conotoxins in *Conus victoriae* using cDNA sequencing and mass spectrometry. *J. Mass Spectrom.* 39:548–57

106. Bellhorn MB, Lewis RK. 1976. Localization of ions in retina by secondary ion mass spectrometry. *Exp. Eye Res.* 22:505–18

107. Caprioli RM, Farmer TB, Gile J. 1997. Molecular imaging of biological samples: localization of peptides and proteins using MALDI-TOF MS. *Anal. Chem.* 69:4751–60

108. Chaurand P, Fouchecourt S, DaGue BB, Xu BJ, Reyzer ML, et al. 2003. Profiling and imaging proteins in the mouse epididymis by imaging mass spectrometry. *Proteomics* 3:2221–39

109. Rubakhin SS, Jurchen JC, Monroe EB, Sweedler JV. 2005. Imaging mass spectrometry: fundamentals and applications to drug discovery. *Drug Discov. Today* 10:823–37

110. Stoeckli M, Chaurand P, Hallahan DE, Caprioli RM. 2001. Imaging mass spectrometry: a new technology for the analysis of protein expression in mammalian tissues. *Nat. Med.* 7:493–96

111. Heeren RM, Sweedler JV. 2007. Imaging mass spectrometric imaging. *Int. J. Mass Spectrom.* 260:89

112. Altelaar AF, Luxembourg SL, McDonnell LA, Piersma SR, Heeren RM. 2007. Imaging mass spectrometry at cellular length scales. *Nat. Protoc.* 2:1185–96

113. McDonnell LA, Heeren RM. 2007. Imaging mass spectrometry. *Mass Spectrom. Rev.* 26:606–43

114. Chaurand P, Schwartz SA, Caprioli RM. 2002. Imaging mass spectrometry: a new tool to investigate the spatial organization of peptides and proteins in mammalian tissue sections. *Curr. Opin. Chem. Biol.* 6:676–81

115. Groseclose MR, Andersson M, Hardesty WM, Caprioli RM. 2007. Identification of proteins directly from tissue: in situ tryptic digestions coupled with imaging mass spectrometry. *J. Mass Spectrom.* 42:254–62

116. Garden RW, Sweedler JV. 2000. Heterogeneity within MALDI samples as revealed by mass spectrometric imaging. *Anal. Chem.* 72:30–36

117. Kruse R, Sweedler JV. 2003. Spatial profiling invertebrate ganglia using MALDI MS. *J. Am. Soc. Mass Spectrom.* 14:752–59

118. Altelaar AFM, Taban IM, McDonnell LA, Verhaert PDEM, Lange RPJD, et al. 2007. High-resolution MALDI imaging mass spectrometry allows localization of peptide distributions at cellular length scales in pituitary tissue sections. *Int. J. Mass Spectrom.* 260:203–21

119. Jurchen JC, Rubakhin SS, Sweedler JV. 2005. MALDI-MS imaging of features smaller than the size of the laser beam. *J. Am. Soc. Mass Spectrom.* 16:1654–59

120. Altelaar AF, van Minnen J, Jimenez CR, Heeren RM, Piersma SR. 2005. Direct molecular imaging of *Lymnaea stagnalis* nervous tissue at subcellular spatial resolution by mass spectrometry. *Anal. Chem.* 77:735–41

121. Rubakhin SS, Greenough WT, Sweedler JV. 2003. Spatial profiling with MALDI MS: distribution of neuropeptides within single neurons. *Anal. Chem.* 75:5374–80

122. DeKeyser SS, Kutz-Naber KK, Schmidt JJ, Barrett-Wilt GA, Li L. 2007. Imaging mass spectrometry of neuropeptides in decapod crustacean neuronal tissues. *J. Proteome Res.* 6:1782–91

123. Verhaert PD, Conaway MCP, Pekar TM, Miller K. 2007. Neuropeptide imaging on an LTQ with vMALDI source: the complete "all-in-one" peptidome analysis. *Int. J. Mass Spectrom.* 260:177–84

124. Desiderio DM, Zhu X. 1998. Quantitative analysis of methionine enkephalin and beta-endorphin in the pituitary by liquid secondary ion mass spectrometry and tandem mass spectrometry. *J. Chromatogr. A* 794:85–96

125. Gobom J, Kraeuter KO, Persson R, Steen H, Roepstorff P, Ekman R. 2000. Detection and quantification of neurotensin in human brain tissue by matrix-assisted laser desorption/ionization time-of-flight mass spectrometry. *Anal. Chem.* 72:3320–26

126. Oda Y, Huang K, Cross FR, Cowburn D, Chait BT. 1999. Accurate quantitation of protein expression and site-specific phosphorylation. *Proc. Natl. Acad. Sci. USA* 96:6591–96

127. Gygi SP, Rist B, Gerber SA, Turecek F, Gelb MH, Aebersold R. 1999. Quantitative analysis of complex protein mixtures using isotope-coded affinity tags. *Nat. Biotechnol.* 17:994–99

128. Che F-Y, Fricker LD. 2002. Quantitation of neuropeptides in Cpe[fat]/Cpe[fat] mice using differential isotopic tags and mass spectrometry. *Anal. Chem.* 74:3190–98

129. Fricker LD, Lim J, Pan H, Che FY. 2006. Peptidomics: identification and quantification of endogenous peptides in neuroendocrine tissues. *Mass Spectrom. Rev.* 25:327–44

130. Che F-Y, Yuan Q, Kalinina E, Fricker LD. 2005. Peptidomics of Cpe[fat/fat] mouse hypothalamus: effect of food deprivation and exercise on peptide levels. *J. Biol. Chem.* 280:4451–61

131. Decaillot FM, Che FY, Fricker LD, Devi LA. 2006. Peptidomics of Cpe[fat/fat] mouse hypothalamus and striatum: effect of chronic morphine administration. *J. Mol. Neurosci.* 28:277–84

132. Hsu JL, Huang SY, Chow NH, Chen SH. 2003. Stable-isotope dimethyl labeling for quantitative proteomics. *Anal. Chem.* 75:6843–52

133. DeKeyser SS, Li L. 2006. Matrix-assisted laser desorption/ionization Fourier transform mass spectrometry quantitation via in-cell combination. *Analyst* 131:281–90

134. Anderson L. 1996. Intracellular mechanisms triggering gonadotrophin secretion. *Rev. Reprod.* 1:193–202

135. Lessmann V, Gottmann K, Malcangio M. 2003. Neurotrophin secretion: current facts and future prospects. *Prog. Neurobiol.* 69:341–74

136. Paras CD, Kennedy RT. 1995. Electrochemical detection of exocytosis at single rat melanotrophs. *Anal. Chem.* 67:3633–37

137. Maidment NT, Brumbaugh DR, Rudolph VD, Erdelyi E, Evans CJ. 1989. Microdialysis of extracellular endogenous opioid peptides from rat brain in vivo. *Neuroscience* 33:549–57

138. Ramirez VD, Chen JC, Nduka E, Lin W, Ramirez AD. 1986. Push-pull perfusion of the hypothalamus and the caudate nucleus in conscious, unrestrained animals. *Ann. N.Y. Acad. Sci.* 473:434–48

139. Ungerstedt U, Hallstrom A. 1987. In vivo microdialysis: a new approach to the analysis of neurotransmitters in the brain. *Life Sci.* 41:861–64

140. Andrén PE, Farmer TB, Klintenberg R. 2002. Endogenous release and metabolism of neuropeptides utilizing in vivo microdialysis microelectrospray mass spectrometry. In *Mass Spectrometry and Hyphenated Techniques in Neuropeptide Research*, ed. J Silberring, R Ekman, pp. 193–213. New York: Wiley

141. Horn TF, Engelmann M. 2001. In vivo microdialysis for nonapeptides in rat brain—a practical guide. *Methods* 23:41–53

142. Kennedy RT, Watson CJ, Haskins WE, Powell DH, Strecker RE. 2002. In vivo neurochemical monitoring by microdialysis and capillary separations. *Curr. Opin. Chem. Biol.* 6:659–65

143. Baseski HM, Watson CJ, Cellar NA, Shackman JG, Kennedy RT. 2005. Capillary liquid chromatography with MS3 for the determination of enkephalins in microdialysis samples from the striatum of anesthetized and freely-moving rats. *J. Mass Spectrom.* 40:146–53

144. Jakubowski JA, Hatcher NG, Sweedler JV. 2005. Online microdialysis–dynamic nanoelectrospray ionization–mass spectrometry for monitoring neuropeptide secretion. *J. Mass Spectrom.* 40:924–31

145. Rubakhin SS, Page JS, Monroe BR, Sweedler JV. 2001. Analysis of cellular release using capillary electrophoresis and matrix-assisted laser desorption/ionization–time of flight mass spectrometry. *Electrophoresis* 22:3752–58

146. Hatcher NG, Richmond TA, Rubakhin SS, Sweedler JV. 2005. Monitoring activity-dependent peptide release from the CNS using single-bead solid-phase extraction and MALDI TOF MS detection. *Anal. Chem.* 77:1580–87

147. Jing J, Vilim FS, Horn CC, Alexeeva V, Hatcher NG, et al. 2007. From hunger to satiety: reconfiguration of a feeding network by *Aplysia* neuropeptide Y. *J. Neurosci.* 27:3490–502

148. Falth M, Skold K, Norrman M, Svensson M, Fenyo D, Andren PE. 2006. SwePep, a database designed for endogenous peptides and mass spectrometry. *Mol. Cell Proteomics* 5:998–1005

149. Hummon AB, Hummon NP, Corbin RW, Li L, Vilim FS, et al. 2003. From precursor to final peptides: a statistical sequence–based approach to predicting prohormone processing. *J. Proteome Res.* 2:650–56

150. Southey BR, Amare A, Zimmerman TA, Rodriguez-Zas SL, Sweedler JV. 2006. NeuroPred: a tool to predict cleavage sites in neuropeptide precursors and provide the masses of the resulting peptides. *Nucleic Acids Res.* 34:W267–72

151. Ducret A, Van Oostveen I, Eng JK, Yates JR 3rd, Aebersold R. 1998. High throughput protein characterization by automated reverse-phase chromatography/electrospray tandem mass spectrometry. *Protein Sci.* 7:706–19

152. Mann M, Wilm M. 1994. Error-tolerant identification of peptides in sequence databases by peptide sequence tags. *Anal. Chem.* 66:4390–99

153. Baggerman G, Liu F, Wets G, Schoofs L. 2005. Bioinformatic analysis of peptide precursor proteins. *Ann. N.Y. Acad. Sci.* 1040:59–65

Analysis of Atmospheric Aerosols

Kimberly A. Prather,[1] Courtney D. Hatch,[2]
and Vicki H. Grassian[2]

[1] Department of Chemistry and Biochemistry, Scripps Institution of Oceanography,
University of California, San Diego, California 92093-0314; email: kprather@ucsd.edu

[2] Departments of Chemistry and Chemical and Biochemical Engineering, Center for
Global and Regional Environmental Research, University of Iowa, Iowa City,
Iowa 52242; email: Courtney-Hatch@uiowa.edu, Vicki-Grassian@uiowa.edu

Annu. Rev. Anal. Chem. 2008. 1:485–514

First published online as a Review in Advance on
March 4, 2008

The *Annual Review of Analytical Chemistry* is online
at anchem.annualreviews.org

This article's doi:
10.1146/annurev.anchem.1.031207.113030

1936-1327/08/0719-0485$20.00

Key Words

mass spectrometry, sources, mixing state, climate, heterogeneous
chemistry, dust, organic aerosol, carbonaceous aerosol

Abstract

Aerosols represent an important component of the Earth's atmosphere. Because aerosols are composed of solid and liquid particles
of varying chemical complexity, size, and phase, large challenges
exist in understanding how they impact climate, health, and the
chemistry of the atmosphere. Only through the integration of field,
laboratory, and modeling analysis can we begin to unravel the roles
atmospheric aerosols play in these global processes. In this article,
we provide a brief review of the current state of the science in the
analysis of atmospheric aerosols and some important challenges
that need to be overcome before they can become fully integrated.
It is clear that only when these areas are effectively bridged can we
fully understand the impact that atmospheric aerosols have on our
environment and the Earth's system at the level of scientific certainty
necessary to design and implement sound environmental policies.

1. INTRODUCTION

Atmospheric aerosols in the troposphere are composed of solid and liquid particles of varying composition and phase (1). Examples include smoke, fog, clouds, and smog (2). Particles play major, yet poorly understood, roles in affecting human health, visibility, air quality, and our overall climate (3). **Figure 1** shows a schematic of the major processes contributing to atmospheric aerosols. Primary particles can be directly emitted into the atmosphere by combustion sources (e.g., coal combustion, biomass burning, and vehicle emissions) or other wind-driven processes such as the resuspension of dust and sea salt. Secondary particles are formed from photochemical reactions of gas-phase species that are emitted directly into the atmosphere, producing more highly oxidized, less volatile species, which can then form new particles or condense on existing particle surfaces. The most common example of this occurs when gas-phase organics undergo oxidation producing less volatile organic species that condense on particle surfaces; these particles are termed secondary organic aerosols

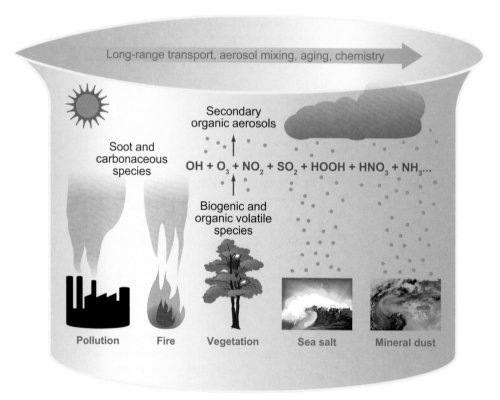

Figure 1

Illustration of the Earth's atmosphere as a chemical reactor. This figure shows examples of the natural and anthropogenic processes producing gases and particles in the atmosphere. Subsequent processes can transform and age these particles as they are transported through the atmosphere.

(SOA). Temperature reduction, as well as reactive uptake via heterogeneous processes and adsorption of chemical species, also shift species from the gas to particle phase.

Details on the variability of chemical species within individual particles remain limited. Additionally, our understanding of chemical speciation within particles is limited by the availability of analytical techniques used to measure them. We understand far more about gas-phase atmospheric components, including ozone, NO_x (NO + NO_2), and volatile organic species (4). Analytical methods are typically designed to target a specific molecule or class of molecules in the gas phase. In contrast, each particle is composed of many chemical species, and it is difficult to find a single analytical method that can measure all of them. As discussed below, mass spectrometry (MS) has become the most commonly used method for analyzing the composition of individual particles in recent years as it serves as a universal on-line detector of a broad range of chemical components.

Recent atmospheric chemistry studies have focused much effort on developing new techniques for measuring aerosols. A strong size dependence exists with submicrometer particles comprising primarily carbonaceous species, including elemental carbon (i.e., soot) and organic carbon, as well as inorganic species such as metals, sulfate, nitrate, and ammonium. Supermicrometer particles are made up of inorganic species such as sea salt and dust and thus are composed mostly of transition metals, silicates, and alkali metals (5). Heterogeneous processing adds secondary species to primary particles such as dust and sea salt, resulting in nitrate and sulfate in the particles (6, 7).

Aerosols represent the largest uncertainty in understanding how humans are changing our climate. Thus, more studies are aimed at developing analytical techniques that can be used in laboratory and field studies at the necessary level to better understand and quantify their role in affecting climate. Particles affect climate through several mechanisms: (*a*) scattering and absorbing solar radiation; (*b*) scattering, absorbing, and emitting thermal radiation; and (*c*) acting as cloud condensation nuclei. The first two mechanisms are called the direct climate effect and relate to the optical properties of particles based on their size, shape, and chemistry. The third mechanism is called the indirect climate effect.

Currently, the study of atmospheric aerosols includes three scientific approaches: (*a*) field measurements of aerosol particles; (*b*) laboratory studies of the chemistry and physical properties of model systems representative of atmospheric aerosols; and (*c*) modeling analysis that includes gas-phase chemistry, photochemistry, heterogeneous chemistry, and optical properties so that models can be effectively used to predict important global climate impacts of aerosols. Integration of these three approaches is essential if we are to fully understand the roles of atmospheric aerosols in climate and air pollution.

One complicating factor in the integration of field, laboratory, and modeling studies is the different treatment of aerosol complexity. Field measurements show that nearly every particle differs in size, physical properties, and chemical characteristics. Laboratory studies typically use model systems to decrease the complexity of different particle types. For example, these studies often use NaCl as a surrogate for sea salt, metal oxides to mimic mineral dust, and single-component mixtures as

MS: mass spectrometry

Secondary particles: gases or particles formed by chemical reactions in the atmosphere

Direct climate effect: the manner in which atmospheric particles directly impact climate through the absorption and scattering of solar radiation

Indirect climate effect: the manner in which atmospheric particles indirectly impact climate through their influence on cloud formation and cloud properties

Heterogeneous chemistry: reactions between the gas and particle phases of the aerosol

Mixing state: associations of chemical species within individual particles; in internal mixtures, all particles of a given size have the same chemical composition, whereas in external mixtures, they have distinct combinations of chemical species

surrogates to study properties of organic aerosols. Many atmospheric chemistry models treat aerosols as a single entity. Even studies that distinguish between aerosol types (e.g., mineral aerosol) treat particles as a single component without considering that different compositions have different reactivities (8, 9). Because the complexity of aerosols is treated so differently, a key question is how to achieve integration of these activities when such different and seemingly disparate approaches to complexity are being taken.

This review provides some examples of the state of the science in the analysis of atmospheric aerosols. We describe some important results that have emerged when the integration problem has been addressed. Due to space limitations, this is not an exhaustive review of previous studies on aerosols, but instead it provides examples of how analytical chemistry performed in the field and laboratory has shed new light on our understanding of the role aerosols play in air pollution and climate change. The final half of this review provides some key issues and needs in the field of analytical chemistry of atmospheric aerosols.

2. SELECT EXAMPLES OF FIELD MEASUREMENTS AND LABORATORY STUDIES OF ATMOSPHERIC AEROSOLS

Field measurements provide an opportunity to determine the atmospheric species and processes that are indeed driving the chemistry of the atmosphere. Investigators then use field observations to develop laboratory studies that can be conducted under well-controlled conditions in an effort to quantify these atmospheric chemical processes. Additionally, laboratory studies can provide insightful data that can aid the interpretation of field observations. Furthermore, laboratory data can also be expressed in a form that is useful for modeling analysis so that models can predict atmospheric composition. Synergy is thus required between field and laboratory studies to perform the studies required to better understand the atmosphere. In this give-and-take approach, one can more fully understand the important processes and the parameters affecting different measurements. The ultimate goal is to incorporate detailed laboratory mechanistic and kinetic data, acquired in studies motivated by field study findings, into models that can be used to accurately predict and reproduce field measurements. Also, models allow the results to be extrapolated over much larger global scales, giving broad spatial data that would be impossible to obtain in field studies. This closure will allow us to quantify the impact of gases and particles on atmospheric chemistry and climate on global and regional scales.

2.1. Field and Laboratory Measurement Approaches and Questions

Individual field studies are designed to address specific scientific objectives; thus, researchers select the region in which the study is to be conducted in order to best suit these objectives. Some of the questions regarding aerosol chemistry addressed in coupled field-modeling studies are listed below.

1. How does the single-particle mixing state of particles impact the physical, optical, and chemical properties of the particles, as well as the regional climate of specific regions (e.g., marine versus urban, Indo-Asian haze) (10, 11)?

2. How do ocean-atmosphere interactions play a role in cloud formation in marine environments (12)?
3. How does the presence of mineral dust affect the overall resulting chemistry and clouds in the atmosphere?
4. What are the major carbonaceous components in atmospheric aerosols, and are most produced from primary sources, or are they formed through secondary reactions?
5. What is the spatial variability of the chemistry and sources of atmospheric aerosols, and how much does long-range transport of pollutants and aerosols play a role in affecting air pollution and climate on a global scale?

As many laboratory studies typically use simplified model systems to begin to ask fundamental questions about the chemistry and physics of atmospheric aerosols, some key issues and questions in laboratory studies include the following:

1. Which model systems best describe a particular class of atmospheric aerosols?
2. How close can these model systems come to mimicking atmospheric aerosols, and how close do they need to be to provide answers to important scientific issues, such as the impact of aerosol on the chemical balance of the atmosphere, Earth's climate, and the health of humans and ecosystems?
3. As laboratory studies probing aerosol chemistry are often performed using powders, thin films, and isolated particles deposited on substrates as well as suspended particles, questions arise, such as do results from these different model systems agree, and are they applicable to aerosol chemistry?
4. Which combination of analytical techniques is best suited for bridging the results from laboratory studies with field studies of atmospheric aerosols?

Early field studies focused on deriving a better understanding of gas-phase reaction processes, typically measuring ozone and its precursors. More recently, field studies have begun focusing on the particle component of air pollution via platforms such as airplanes, balloons, ships, ground-based sites, unmanned aerial vehicles, and even space shuttles. Advancing an instrument from the laboratory to a field version can be quite challenging. Smaller instruments with lower power requirements and robust operation that can handle extreme temperatures are required. Interestingly, many of the state-of-the-art instruments used in the laboratory are now being developed for the field (13). Field studies serve as an opportunity to intercompare multiple instruments using different approaches for measuring the same, or slightly different, components of the aerosol (14, 15). However, field studies are limited to studying a relatively small spatial scale, and thus models and satellite measurements are important for extrapolating findings to a larger scale.

The ultimate goal of any analytical technique developed for atmospheric studies is to quantitatively identify all species within each individual particle as rapidly as possible to monitor real-time changes in aerosol chemistry. This represents an enormous challenge as particles come in many shapes and sizes and can be quite small and highly complex. Their size range spans from 1 nm up to 10 μm, corresponding to a mass range of 10^{-21} g to 10^{-9} g, respectively. Different instruments are used to measure particle size in various ranges: a scanning mobility particle sizer for particles in the

3- to 600-nm size range and an aerodynamic particle sizer or optical particle counter for larger particles (>500 nm). Size metrics include optical, aerodynamic, geometric, and mobility diameters. [We refer the reader to a review that details the methods used to obtain particle size (16).] In the atmosphere, particle chemistry undergoes dynamic changes owing to heterogeneous reactions, the partitioning of species to and from the gas phase, and aqueous phase processes. Capturing these changes within a minute volume and for such a small mass represents an enormous challenge, requiring high time-resolution measurements, especially for airborne platforms.

The sections below provide a brief overview of traditional aerosol chemistry measurements with a focus on the types of questions and techniques used to address them in two rapidly growing areas: (*a*) the characterization of organic species in aerosols and (*b*) single-particle measurements of mixing state. Finally, we describe results from a case study, ACE-Asia, to demonstrate how a combination of laboratory studies, field measurements, and modeling analysis can provide unique insights into atmospheric aerosols.

2.2. Off-Line Aerosol Analysis

Traditionally, field studies use filter-based techniques to collect particles for subsequent analysis. To obtain enough material for quantification and identification by standard analytical methods, researchers collect samples for periods of hours to weeks. Once in the laboratory, multiple traditional analytical techniques, such as gas chromatography/MS and liquid chromatography/MS, are used to analyze major components.

One downside to the off-line approach is sampling artifacts, including the gain or loss of species and reactions that can alter the sample. Thus, questions exist as to how representative the filter samples are of the actual atmospheric aerosol. Also, one goal of field studies is to capture real-time changes in particle chemistry and relate this to other parameters. For these reasons, there has been a major push over the past decade to shift to on-line measurements of aerosol chemistry coupled with other physical and optical properties with high time resolution.

2.3. Characterization of Organic Aerosols

On-line MS analysis has become the method of choice for many environmental studies, including the study of atmospheric aerosols (17). Large research efforts focus on characterizing the organic fraction because it is the most poorly understood component of atmospheric aerosols (18, 19). The organic fraction can account for 50% of the fine particle mass (PM$_{2.5}$) on an annual average, yet only 10%–15% of the organic species have been identified. As a result of photochemical processing, many organic aerosol species are highly oxidized (20). The low-volatility properties that cause substances to exist predominately in the particle phase make it challenging to desorb them into the gas phase for analysis and identification. Additionally, heating the sample to induce vaporization can cause organic species to break down or even polymerize in the presence of an acidic aerosol (21). Thus, electrospray ionization and atmospheric pressure chemical ionization coupled with MS have emerged to identify new organic species in aerosols via off-line analysis (22–24). Water-soluble organic carbon and other inorganic ions can now be analyzed in real time using an ion chromatograph instrument adapted for the field (25, 26). Until recently, the majority of organic species identified in aerosols had molecular weights of less than 300 Da. However, recent smog chamber photochemical studies, as well as field studies, have shown the production of high-mass oligomeric species, which have recently been analyzed using off-line and on-line laser-desorption MS (27–29). Additionally, humic substances similar to those found in aqueous environments have been shown to exist in atmospheric aerosols (30). These complex high-mass species are quite difficult to analyze by conventional analytical methods.

Carbonaceous aerosols: elemental and organic carbon species

2.3.1. Field studies of organic aerosols. High-sensitivity, universal detection of virtually any chemical species and the recent demonstration of smaller, portable, and rugged instruments make MS an ideal choice for field measurements of aerosols. The basic design can be subdivided into those mass spectrometers that measure the chemistry of particle ensembles and those that measure the chemistry of one particle at a time. Different methods of ionization are inherently linked to each class of mass spectrometer (31). One type measures the chemistry of particle ensembles by using a filament to flash vaporize species from the particles and electron impact ionization to form ions. More recently, thermal desorption was coupled with chemical ionization to ionize species evaporated from the particles (32, 33). A commercial version of the electron impact ionization–based mass spectrometer is an aerosol mass spectrometer developed by Aerodyne, Inc. This instrument uses a quadrupole mass spectrometer, and most recently a time-of-flight mass spectrometer, to provide mass concentrations of submicrometer nonrefractory species including ammonium, nitrate, sulfate, and organic carbon with high time resolution (34, 35). The aerosol mass spectrometer has been used in aircraft- and ground-based measurements in a number of field studies around the world as well as in the laboratory. The instrument has been used to distinguish between different forms of nonrefractory carbonaceous species, namely oxidized organic aerosol and hydrocarbon organic aerosol. By examining these basic organic types, investigators have used the aerosol mass spectrometer for extensive studies focusing on the importance of oxidized organic species and SOA formation (36, 37). Recent high-resolution measurements using a newly added time-of-flight mass spectrometer have the potential to delve more deeply into the complexity of the organic fraction (34). Additionally, thermal desorption has been coupled with gas chromatography/MS to separate and identify organic species in atmospheric aerosols (38).

2.3.2. Laboratory studies of organic aerosols. Laboratory studies of the properties of organic aerosols including carbonaceous aerosols (which consist of soot, black and brown carbon, and biomass-burning aerosols) are of particular interest from a global climate perspective as their impact on climate forcing depends on their physiochemical properties, which vary significantly depending on the source. In the case of carbonaceous aerosols, recent laboratory studies have attempted to determine the most appropriate reference materials for these species (30, 39–41).

Recent studies using a state-of-the-art single-particle analysis technique, scanning transmission X-ray microscopy/near-edge X-ray absorption fine structure spectroscopy, of various laboratory surrogates concluded that carbonaceous aerosols are extremely diverse, and atmospheric aging processes alter their composition (40). Thus, no single standard is available that accurately represents atmospheric carbonaceous aerosols. However, efforts are promising as the International Steering Committee for Black Carbon Reference Materials believes that n-hexane soot is the most appropriate standard available (42). Additionally, an inverted methane/air diffusion flame produces soot that appears to be representative of atmospheric black carbon and thus may also be a suitable standard reference material, particularly for light-absorption studies (39, 41). As the physiochemical properties depend strongly on the composition and atmospheric processing, these types of studies are important and should be continued so that appropriate surrogates are identified.

As biomass-burning particles collected in the field show highly complex and heterogeneous particles, it is difficult to study the physiochemical properties of these aerosols because there is currently no appropriate surrogate for laboratory use. Some studies utilize single components observed in biomass plumes (43, 44). Tivanski et al. (40) have shown that biomass-burning particles resemble atmospheric humic-like substances (HULIS) in terms of composition and structural ordering and that they are dissimilar to black carbon standard reference materials. Therefore, atmospheric HULIS may be more representative of biomass-burning aerosol than black carbon standards that have been previously used as surrogates. However, as representative HULIS atmospheric samples are not easily attainable, many laboratory studies substitute atmospheric HULIS with humic and fulvic acids from terrestrial and aquatic sources. Although these surrogates are similar to HULIS as they comprise large-molecular-weight, multifunctional organic molecules and appear to have similar infrared spectral characteristics, there are many differences between terrestrial and atmospheric humic substances (30, 45). For example, water-soluble organic carbon–derived HULIS from fine atmospheric aerosol tend to have higher surface activity, less aromatic functionalities, smaller molecular sizes, and weaker acidity than aquatic fulvic acids. Thus, there remains a need for appropriate surrogates for atmospheric HULIS and biomass-burning aerosols so that we can better understand the properties of carbonaceous aerosols.

Organic aerosols are complex, and many factors play a role in affecting their cloud condensation nuclei (CCN) potential, including surface tension, impurities, contact angle, deliquescence, morphology, age, and volatility (46). Researchers use all these factors when theoretically predicting and interpreting field results for CCN activation. Obviously, deconvoluting the relative impact of each property on CCN activation is more straightforward in well-controlled laboratory experiments. In an effort to better integrate laboratory and modeling studies, Roberts et al. (47) suggested reporting the hygroscopic growth and CCN activity of aerosols studied in the laboratory relative to pure ammonium sulfate, a well-known CCN active aerosol. Additionally, Petters & Kreidenweis (48) theoretically developed a single parameter, κ, related to hygroscopicity that can integrate both laboratory and field CCN properties. As more such systems are developed and utilized, our understanding of atmospheric processing and climate effects of various types of aerosols will strengthen. Thus, laboratory studies are essential for providing necessary information for input into more detailed atmospheric models. However, as many laboratory studies focus on the CCN properties of single-component aerosol, future studies of more complex, multicomponent systems are needed to understand the effects of mixing state on cloud formation.

The poorly understood factors affecting SOA formation are being investigated through careful smog chamber studies (49). However, most early studies used orders of magnitude higher concentrations of reactants/oxidants than are present in the atmosphere to simulate atmospheric aging processes that occur over days in the relatively short time period of a smog chamber experiment (hours). These unrealistic reactant concentrations influence the actual aerosol products that form (50). Smog chamber studies, made feasible by the advent of on-line instruments that provide higher sensitivity, are now being conducted at lower, more realistic oxidant concentrations (51). A question still exists as to how to replicate atmospheric concentrations and timescales, reduce wall losses, and extend the results of smog chamber studies using relatively simplistic aerosols to the real atmosphere. Recently, several research groups have moved in a direction that will better bridge this gap by making aging measurements of aerosols formed under more realistic conditions and emitted from specific sources (29, 52). Although these are still relatively complex chemical mixtures, they are more representative of processes occurring in the atmosphere, and the results are easier to interpret relative to field studies. Smog chambers have been used to study the CCN activation of SOAs, providing insight into how CCN activation changes for slightly more complex organic mixtures (53). These studies represent important steps in bridging field and laboratory studies, and more studies such as these are needed.

2.4. Single-Particle Measurements of Mixing State

An alternative approach for measuring the chemistry of aerosols involves measurements of single particles rather than collections of particles. Such measurements are being used to directly address the diversity of different compositions within the aerosol mix of the atmosphere. A key requirement to answer many climate-related models involves insight into the properties of atmospheric aerosols as a function of chemical mixing state (54, 55). Mixing state refers to the distribution of chemical species within individual particles. The combination of species in a single particle is important in determining the reactivity, water uptake, and optical properties of the particles. Perhaps the most important example of how mixing state can influence the optical properties and radiative forcing of aerosols involves the soot-sulfate mixture (56, 57). Soot is the strongest absorber of tropospheric solar radiation and leads to warming (58, 59). In contrast, sulfate scatters light back to space and leads to cooling. Internally mixed soot-sulfate mixtures are predicted to absorb up to three times more strongly than soot and sulfate in separate particles (57, 60). Soot and sulfate concentrations vary in different regions of the world; thus a better understanding of the spatial variability of the soot-sulfate mixing state is critically needed. If more accurate input on mixing state is provided for models, field observations of aerosols and their radiative properties may be in better agreement with model predictions (61, 62).

To obtain single-particle chemical information, investigators use electron microscopy, MS, and laser-induced breakdown spectroscopy (63). Electron microscopy has been used extensively to characterize the size, morphology, water uptake properties, and chemistry of individual particles (64, 65). Techniques such as micro-proton-induced X-ray emission can provide insights on species distribution within individual particles (66). High-resolution scanning electron microscopy/energy dispersive X-ray spectroscopy has been used to estimate the complex refractive index of individual atmospheric particles. One issue with electron microscopy is that most studies examine very few particles; thus statistical questions arise as to whether an atmospherically representative sample was obtained. In addition, other semivolatile components evaporate under vacuum and the intense energy of the electron beam. To overcome losses in vacuum and to study the hygroscopic properties of single particles, a number of researchers use environmental scanning electron microscopy (67).

To establish an understanding of real-time changes in particle size and mixing state, investigators developed single-particle MS. Statistics are not an issue as the size and chemistry of millions of particles can be analyzed. Two recent reviews summarize the different designs of single-particle mass spectrometers and how each provides different types of data (31, 68). **Figure 2a** shows a typical mass spectrum obtained via bulk (ensemble) averaging, which analyzes, by necessity, multiple particles. A reasonable interpretation of the spectrum is that all particles are composed of the same average chemical composition. In contrast, mass spectra of single particles (**Figure 2b**) clearly show that this interpretation is incorrect and that there are distinct particles with different compositions. Thus, the bulk composition analysis predicts incorrect particle properties in this case. The ion fingerprints in the deconvoluted mass spectra show unique combinations of OC, EC, K, and metals (i.e., Zn, Pb, Cr, Ag) that provide unique insight into the specific source of each particle type (**Figure 2b**). The presence of nitrate and sulfate demonstrates which particle types have undergone atmospheric aging processes. These single-particle mass spectral signatures provide critical insights into the partitioning of species to different particle types, the reactions different particle types undergo in the atmosphere, and the original source that produced each particle.

Friedlander and coworkers (69) developed the original design of these single-particle instruments in the early 1980s. Early mass spectrometers used a filament to desorb the particle and electron impact ionization coupled with a quadrupole mass spectrometer. Today, the majority of single-particle mass spectrometers use time-of-flight technology to obtain the entire mass spectrum of each particle. In fact, many systems now use a dual polarity time-of-flight mass spectrometer, so both positive and negative ions can be obtained from each particle (70). The time-of-flight mass spectrometer is ideally suited for pulsed laser desorption/ionization (LDI), offering the advantage of high-throughput analysis. The particles are under vacuum for less than 1 ms before analysis, and thus minimal artifacts are encountered. The advantage to using a pulsed laser is that all chemical species in each particle can be analyzed.

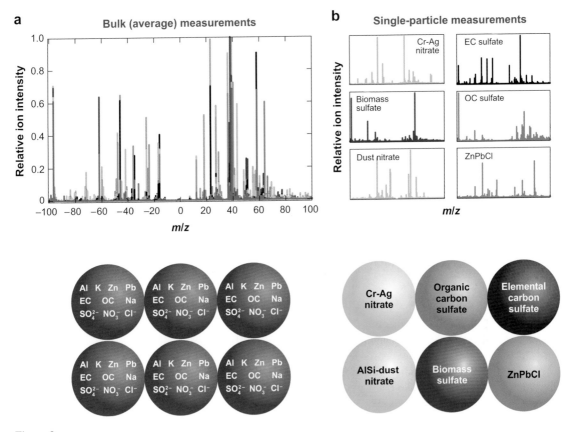

Figure 2

Comparison of mass spectra obtained using (*a*) bulk ensemble analysis versus (*b*) single-particle analysis. The interpretations of the bulk versus single-particle analyses are shown below.

Quantification is more challenging, but various groups are working on this aspect (71–76). One key to improving the quantification step involves homogenizing the LDI laser beam to reduce the shot-to-shot variability of the ion intensity between particles (77).

LDI at the proper wavelengths can produce molecular ions, but the quantification of individual organic species has not been accomplished (78). Two-step LDI shows promise; a pulsed infrared laser heats the particle and rapidly desorbs intact neutral molecules from the particles, and a second, lower-power, ultraviolet laser can ionize the desorbed gas-phase species (79). Investigators have also used extremely high-powered lasers to obtain quantitative results on total elemental composition in individual particles by ablating species from the particles and fragmenting molecules down to elemental forms (80). One recent approach involves desorption with a laser followed by the introduction of low-energy photoelectrons that attach to the desorbed neutrals (81). The Johnston group (82) has shown that vacuum UV photoionization can be used for soft ionization and the characterization of organic species in aerosols.

Details on the surface chemistry and structure of single particles are critical to understanding their reactivity, optical properties, and water uptake potential (83–85). Off-line analysis by time-of-flight secondary ion mass spectrometry has been used to study hydrophobic compounds on the surfaces of atmospheric particles (86). Baer and coworkers (87) demonstrated in laboratory studies that one can study the surface versus core chemistry of individual particles using two-step LDI.

Single-particle mass spectrometers are a relatively new approach for determining the major sources of atmospheric pollution (88, 89). One can use the mass spectral fingerprints (**Figure 2b**) to identify and determine the relative fractions of particles from different sources (89, 90). Using a combination of unique gas-phase tracers coupled with unique single-particle MS signatures, Guazzotti et al. (91) showed that biomass/biofuel emissions were the most abundant aerosol particle type in the region in shipboard measurements during the Indian Ocean Experiment.

Some promising recent advances in single-particle MS studies couple the instrument with other measurements. Researchers are now using information on particle volatility, hygroscopicity, density, and optical properties to obtain linked chemical-optical or chemical-physical information on individual particles (92–96).

2.5. Case Study: ACE-Asia

Advances in analytical techniques for aerosol analysis pave the way for a greater understanding of atmospheric aerosols through both field and laboratory studies. However, a complete understanding will only be achieved upon integration of these studies with modeling efforts. Below, we discuss important results and remaining issues that have emerged via integrated efforts during ACE-Asia.

2.5.1. Field studies. Heard (97) reviewed measurements, techniques, and locations for a number of major field campaigns. ACE-Asia represents an excellent example of an intensive field study aimed at quantifying the spatial and vertical distributions of aerosol concentrations, processes controlling aerosol formation and evolution, and the radiative impacts of aerosols in spring 2001. Over 250 publications describe results from this campaign, reflecting how a combination of laboratory, field, and modeling efforts can help unravel the contributions of aerosols to climate change. Three aircraft, two research ships, a lidar network, and many surface sites measured Asian aerosols during the spring dust storm season. A wide range of aerosol models (including microphysical, radiative transfer, chemical transport models, and global climate models) was used to assess how Asian aerosols are influencing pollution and climate in this region. Satellite data provided a larger scale view of the radiative impacts of aerosols over the region under different air mass conditions (98). One major facet of ACE-Asia involved studying the impact of large Asian dust plumes lofted into the free troposphere by dust storms emanating from the desert regions. Mineral dust aerosol represents an important component of the Earth's system that links land, air, and oceans in a unique way, and although it is a naturally occurring aerosol, anthropogenic activities clearly influence the effects of dust in the atmosphere (99). For example, dust serves as a major sink for trace gases in the atmosphere by providing a large surface area upon which heterogeneous chemistry can occur (100).

Mineral dust aerosol: soil particles primarily from desert or semiarid regions mobilized by wind currents and entrained in the atmosphere

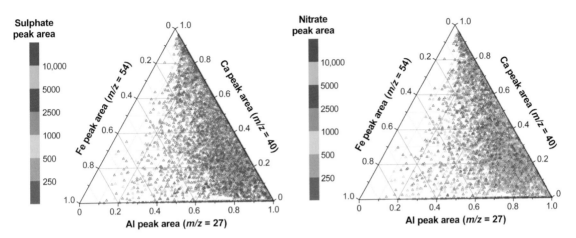

Figure 3

The relative distribution of peak areas for three major mineral dust components sampled during ACE-Asia (Fe, Ca, and Al) mixed with secondary acids (6). Copyright 2007. Reproduced with permission. The same dust particles are displayed in both panels with the symbol color corresponding to (*a*) the sulfate absolute peak area and (*b*) the nitrate absolute peak area. This figure shows that AlSi dust particles have a higher abundance of sulfate, whereas Ca dust particles contain more nitrate.

Arimoto and coworkers' (101) review presents an overview of the findings on the impacts of dust on tropospheric chemistry during ACE-Asia. A unique finding from aerosol time-of-flight mass spectrometry (ATOFMS) measurements made onboard the R/V Ronald H. Brown during ACE-Asia was that, when large amounts of sulfate and nitrate were present, they tended to segregate and partition to dust particles with differing mineralogy (6). The ternary plots in **Figure 3** demonstrate this, displaying the relative ATOFMS single-particle mass spectral ion signals from three major dust mineral components: Al, Ca, and Fe. **Figure 3** shows dramatically different mixing behavior for nitrate versus sulfate and reflects the segregation of nitrate from sulfate in dust particles. The sulfate-rich dust particles lie predominantly near the Al vertex, indicating an association with AlSi dust particles. The nitrate dust particles are mostly located toward the Ca vertex and thus are most likely associated with calcite-rich dust. Notably, this field observation confirmed predictions based on carefully controlled laboratory studies (102). In light of this example, we note that current models treat dust as a single entity that reacts with species such as sulfate and nitrate with one rate constant. Including more kinetic rate data for heterogeneous reactions on dust of differing mineralogy is one specific area in which models can be significantly improved.

Another interesting observation during ACE-Asia occurred upon the arrival of the dust front: Prior to the dust front, high amounts of nitrate were observed on sea salt particles. When the dust front descended into the boundary layer and mixed with the ground-level pollution, nitrate suddenly shifted from being associated with sea

ATOFMS: aerosol time-of-flight mass spectrometer

salt to the dust particles (103, 104). This rapid change could result from a variety of factors, including available surface area, the arrival of a new air mass, or the fact that dust reacts faster with NO_y precursors than sea salt. To sort out the possible factors leading to these significant changes, we have begun to design and perform laboratory experiments focusing on heterogeneous chemistry processes observed during ACE-Asia. Ultimately the findings from these laboratory studies will be used as inputs for chemical transport models.

2.5.2. Laboratory studies of dust. Laboratory studies of mineral dust typically focus on the common components, including clay, carbonate, and oxide minerals, and therefore can address important issues with respect to dust mineralogy. Recent reviews on heterogeneous reactions of mineral dust describe how the mechanisms for the reaction chemistry of mineral aerosol need to be better understood (105, 106). One important consideration is the need for laboratory studies measuring heterogeneous reaction kinetics of mineral dust as a function of relative humidity (RH). For example, several studies have shown that adsorbed water, which increases as a function of RH, can play an important role in the heterogeneous reactions of nitric acid on Arizona test dust (107), a complex, well-characterized dust mixture; montmorillonite (108), a swellable clay mineral; and calcite (109), a reactive component of mineral dust. Furthermore, laboratory studies analyzing the uptake of other trace gases (such as HCl, CO_2, and SO_2) have also shown that adsorbed water can significantly increase the uptake of these gases on major mineral components (110).

Recently, in addition to the role adsorbed water plays in the heterogeneous chemistry of mineral aerosols, laboratory studies have begun to explore water adsorption over the full dynamic range, including water uptake and hygroscopicity under subsaturated conditions and CCN activity under supersaturated conditions. **Figure 4** shows a schematic of a multianalysis aerosol reaction system developed at the University of Iowa for studying CCN activation and water uptake (111, 112). This instrument has the capability of measuring CCN activity and hygroscopic growth of size-selected aerosol, as well as heterogeneous processing and aerosol infrared extinction over a range of RH values. For example, $CaCO_3$ is highly reactive and can be converted to $Ca(NO_3)_2$ upon heterogeneous reaction with gas-phase HNO_3. The $Ca(NO_3)_2$ reaction product is more soluble and thus takes up significantly more water than the unprocessed $CaCO_3$ aerosol. **Figure 4c** shows the hygroscopic growth curves for 100-nm size-selected $CaCO_3$ and the reaction products with HNO_3, $Ca(NO_3)_2$. The CCN activity of $Ca(NO_3)_2$ is significantly greater than that of $CaCO_3$ and is comparable with common CCN such as ammonium sulfate (**Figure 4d**).

Many recent surface characterization studies have begun to investigate the speciation of adsorbed products on the surfaces of processed mineral aerosol using techniques such as X-ray photoelectron spectroscopy (113, 114). However, more studies are needed to work out details of surface speciation and the role of adsorbed water in the heterogeneous mechanism of atmospheric gas uptake on mineral aerosol.

2.5.3. Model-laboratory-field comparisons. Model analysis was also an important component of the ACE-Asia campaign, both from a forecasting perspective during the

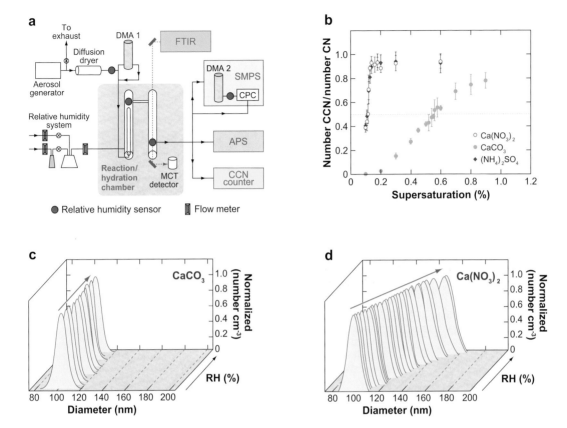

Figure 4

(*a*) A schematic of a laboratory apparatus that measures hygroscopic growth and cloud condensation nuclei (CCN) activity of aerosols representative of mineral dust (CaCO₃) and reacted mineral dust [Ca(NO₃)₂], as well as the infrared extinction spectra and size distributions of aerosols. Ca(NO₃)₂ forms as a result of heterogeneous reactions of CaCO₃ with nitrogen oxides in the atmosphere (111). Copyright 2006, American Chemical Society. Reproduced with permission. (*b*) CCN activity plotted as a function of nuclei that are CCN active versus the total number of condensation nuclei as a function of percent supersaturation (112). Copyright 2006, American Geophysical Union. Reproduced with permission. Hygroscopic growth data comparing (*c*) 100-nm CaCO₃ (111, 112) and (*d*) 100-nm Ca(NO₃)₂ particles (111, 112). Panels *c* and *d* reproduced with permission. Copyright 2006, American Chemical Society; copyright 2006, American Geophysical Union. Abbreviations: APS, aerodynamic particle sizer; CPC, condensation particle counter; FTIR, Fourier transform infrared spectrometer; MCT, mercury cadmium telluride; RH, relative humidity; SMPS, scanning mobility particle sizer.

study and as a tool for understanding observations made during the study. ACE-Asia provided an opportunity to compare the model predictions with atmospheric measurements. **Figure 5** illustrates results from the three-dimensional global sulfur transport and deposition model used to predict the influence of heterogeneous chemistry on some important chemical species, including O_3, NO_2, SO_2, and HNO_3 (115). Each

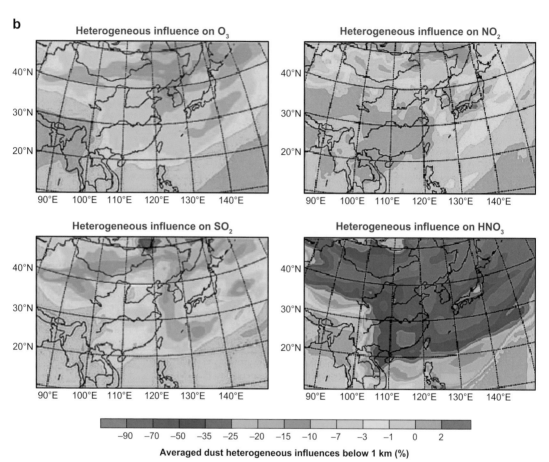

Figure 5

(*a*) The four heterogeneous reactions on mineral dust used by Tang et al. to model dust aerosol chemistry. (*b*) Contour plots showing averaged influences on O_3, NO_2, SO_2, and HNO_3 concentrations due to heterogeneous chemistry on mineral dust aerosol below 1 km (115). Panel *b* reproduced with the permission of the American Geophysical Union. Copyright 2004.

plot represents differences between the gas-phase concentrations of two model runs, one that includes four important heterogeneous reactions on mineral dust and another that includes only gas-phase reactions. The resulting contour plots show that heterogeneous reactions decrease the gas-phase composition of all these species. The greatest influence is on HNO_3 concentrations, which decrease by 90% in some regions.

Finally, new insights were acquired into the partitioning of chloride on atmospheric particles during ACE-Asia. Currently, chloride on filters is used to calculate the amount of unreacted sea salt present in the atmosphere. However, during ACE-Asia a new real-time observation showing the uptake of HCl released from sea salt by dust was measured using single-particle MS (116). This resulted in substantial amounts of Cl on dust rather than sea salt. Model predictions compared with the observations showed poor agreement. Thus, this heterogeneous process is now being studied in the laboratory to obtain kinetic information, which will be added to the model. This example demonstrates how investigators can use field observations, coupled with chemical models, to design laboratory studies to better understand such processes.

The modeling used during ACE-Asia represents a major effort to understand the role of heterogeneous reactions on mineral aerosol, but only four heterogeneous reactions were studied in the model. The incorporation of other species along with the effects of RH and dust composition will significantly improve these efforts in the future. In atmospheric chemistry models, heterogeneous chemistry needs to be treated in a more complex manner if the chemical composition of Earth's atmosphere is to be understood. Ideally, this understanding needs to be at a level that allows models to be used in a predictive fashion for various possible scenarios involving projected future anthropogenic activities and the consequences of these activities.

3. KEY ISSUES AND MEASUREMENT NEEDS

3.1. Marine Aerosols, Oceanic Productivity, and Clouds

As the ocean represents ~70% of Earth's surface, we need a better understanding of marine aerosol-cloud-climate interactions (117). A strong link exists between the chemical and biological processes in the ocean and the chemistry of the atmosphere above the ocean (118, 119). Thus, field studies focusing on marine environments are vital to furthering our understanding of marine aerosol production and cloud formation.

3.2. Organic Speciation

In the past few years, multidimensional gas chromatography has shown that tens of thousands of different organic species exist in aerosols collected in a single air mass (120). Thus, new methods emphasizing the speciation of organic aerosol are needed. Recent breakthroughs in the areas of near-edge X-ray absorption fine structure spectroscopy, Fourier transform infrared mapping, chemical ionization mass spectrometry, and high-resolution MS are shedding new light on the complexity of the organic fraction of aerosols (121, 122). After identifying a larger fraction of the organics in aerosols, the next step will involve quantifying the actual amounts of these species and better understanding their heterogeneity in particles using real-time measurements. Also, measurements are needed of the interactions between water and organic and mixed organic/inorganic/aqueous particles and the impact of phase changes, hygroscopic growth, cloud activation, and ice nucleation.

3.3. Understanding Cloud Formation and Reducing Uncertainties in the Indirect Effect

CCN formation is an area in which there is a significant disconnect between laboratory findings and field observations (123, 124). Making simplistic aerosols in the laboratory with one or two components has provided a better understanding of the fundamental properties that lead to more effective CCN. Direct measurements of the size-resolved single-particle mixing state of the actual cloud nuclei will improve our understanding of atmospheric CCN (125). For example, recent single-particle MS field measurements yielded new insight into how atmospheric aging affects the number of CCN formed in marine environments (126). Single-particle MS has also been employed to measure the chemistry of individual ice nuclei (95). Currently, it is believed that insoluble materials such as dust and soot play the largest role in ice nucleation (127–129). However, this has been difficult to test in field studies owing to the extremely low number concentrations of ice nuclei that are present. Future aircraft flights will focus on making further direct measurements of the chemistry of ice nuclei (130).

3.4. Linking Size-Resolved Mixing State with Optical Properties

Many field studies have investigated how chemical composition and size influence the radiative forcing in a particular region (10, 131–133). Most commonly, researchers measure particle size, optical properties, and chemistry using three separate techniques. Ideally, coupled size-shape-chemical-optical measurements would be made on the same individual particles. Steps have been taken in this direction in single-particle MS (92, 96). The ultimate measurement involves multiangle light-scattering measurements on the same individual particle, the power of which has been demonstrated for particles of known composition (134). Such direct measurements of the optical properties of a specific mixing state and size of real atmospheric aerosols are required as model inputs to reduce the uncertainties associated with regional and global radiative forcing on climate, as detailed in the recent Intergovernmental Panel on Climate Change report (135). Also, aerosols show strong vertical variability with multiple layers often observed; this vertical structure strongly impacts the overall amount of radiative forcing. To account for this, it is critical that we have a better understanding of how particle chemical mixing state varies as a function of altitude in different regions.

3.5. Nanoparticle Characterization

Particle formation via nucleation occurs in the smallest sized particles (<10 nm) on very rapid timescales. The processes involved in nucleation in urban environments are still under debate. Furthermore, with increased interest in the environmental health and safety, including toxicity, of engineered nanoparticles (136), investigators are developing methods that can measure the chemistry of single particles in the 10-nm size range (136, 137). The high sensitivity of MS allows one to detect the contents of an individual particle down to the sub-10-nm size range. However, particles this small cannot be detected using standard optical detection methods, which rely on the particle being larger than ~80 nm. One could condense water on the particles to grow them to a detectable size range and measure their chemistry via single-particle MS (138). However, charged aerosol detection has been developed as an alternative approach for detecting smaller particles than optical detection allows; it is being used as a universal detector in high-performance liquid chromatography analysis (139).

3.6. Miniaturization of Aerosol Chemistry Instruments

The size of the sampling platform has become very small, with the recent demonstration of using unmanned aerial vehicles to study aerosol-cloud interactions in the Indo-Asian region (140). This study did not obtain aerosol chemical information, but this may be possible in future missions. Time-of-flight mass spectrometers as small as 3 inches are now available (141). Optical spectroscopy techniques do not require pumping systems, and thus single-particle analysis using laser-induced breakdown spectroscopy may be able to meet smaller payload requirements. As is always the case, trade-offs exist between size and performance, but having some chemical information will be a major advance in studies of the spatial variability of aerosols on larger scales.

3.7. Determining the Factors Controlling Gas-Particle Partitioning of Real Atmospheric Aerosols

We need a better understanding of how the seed chemistry of a particle affects the partitioning and reactivity of aerosols (142). A number of recent studies have shown that far more species are associated with the particle phase than predicted from standard models (143). The acidity of the seed has been shown to affect the amount of species partitioned from the gas to particle phase and more recently to induce oligomerization in the particles (21, 143, 144). Because particles are highly concentrated and thus nonideal solutions, the behavior at the surface cannot be approximated, assuming no chemical interactions occur. Techniques that probe interfacial chemistry and interactions and rapid processes occurring at the molecular scale are required (145, 146). Also, we need a better understanding of how chemical mixing state influences the distribution of condensing SOA species.

3.8. Bioaerosols

There is a great deal of interest in the public health issues associated with bioaerosols, which include pollen, viruses, bacteria, and proteins. Issues range from understanding the spread of emerging infectious diseases to concerns over bioterrorism. Conventional analytical methods used for these aerosols include microscopy, protein staining, and microorganism cultivation. However, molecular-based studies have just recently begun and have the potential to provide mechanistic information on the immune response of these reactions (147). Recently, investigators used a single-particle mass spectrometer based on the ATOFMS design for the on-line analysis of single bioaerosols (148). Additionally, bioaerosols are known to undergo nitration when proteins are exposed to parts-per-billion levels of nitrogen dioxide (149). Such heterogeneous reactions may be important triggers for adverse immune reactions, such as allergies and asthma, in humans. However, little is known regarding the effects of the interaction of pollution and bioaerosols. Thus, this remains a largely unexplored area and should be considered in future field measurements and laboratory studies.

4. SUMMARY

As the field of atmospheric analysis moves forward, it is clear that atmospheric aerosols need to be better understood. Over the past decade, there has been a tremendous amount of growth in the number of analytical measurement techniques being developed to measure atmospheric aerosols. These advances are just beginning to unravel the complexity of atmospheric aerosols. Additional instruments that make rapid, real-time measurements of particle interfaces, as well as probe particle heterogeneity and phase, are needed to understand the processes contributing to aerosol formation in the atmosphere. Accurate information will ultimately allow for models to better understand and predict atmospheric aerosol composition and, in turn, to understand the impact these aerosols have on our climate on regional and global scales. In this review, we have highlighted the critical need for synergism between laboratory studies, field measurements, and modeling analysis to achieve the next level of understanding of atmospheric aerosols. Such a combined effort will put us in a much better position for developing sound environmental policies that protect our planet as we move into the future.

DISCLOSURE STATEMENT

The authors are not aware of any biases that might be perceived as affecting the objectivity of this review.

ACKNOWLEDGMENTS

Support for this work was provided by the National Science Foundation through 0503854, 0506679, and ATM-0625526. C.D.H. also received support from the University of Iowa Cardiovascular Center Institutional Research Fellowship. The authors would like to thank Gregory R. Carmichael, Elizabeth R. Gibson, Paula K. Hudson, Ryan C. Sullivan, and Sergio Guazzotti for their contributions to the studies presented in this work. Also, K.A.P. would like to thank Ryan Sullivan for helping edit the review.

LITERATURE CITED

1. Andreae MO, Crutzen PJ. 1997. Atmospheric aerosols: biogeochemical sources and role in atmospheric chemistry. *Science* 276:1052–58

2. Poschl U. 2005. Atmospheric aerosols: composition, transformation, climate and health effects. *Angew. Chem. Int. Ed. Engl.* 44:7520–40

3. Seinfeld JH. 2004. Air pollution: a half century of progress. *AIChE J.* 50:1096–108

4. Clemitshaw KC. 2004. A review of instrumentation and measurement techniques for ground-based and airborne field studies of gas-phase tropospheric chemistry. *Crit. Rev. Environ. Sci. Technol.* 34:1–108

5. Noble CA, Prather KA. 1996. Real-time measurement of correlated size and composition profiles of individual atmospheric aerosol particles. *Environ. Sci. Technol.* 30:2667–80

6. Sullivan RC, Guazzotti SA, Sodeman DA, Prather KA. 2007. Direct observations of the atmospheric processing of Asian mineral dust. *Atmos. Chem. Phys.* 7:1213–36

7. Gard EE, Kleeman MJ, Gross DS, Hughes LS, Allen JO, et al. 1998. Direct observation of heterogeneous chemistry in the atmosphere. *Science* 279:1184–87

8. Bauer SE, Balkanski Y, Schulz M, Hauglustaine DA, Dentener F. 2004. Global modeling of heterogeneous chemistry on mineral aerosol surfaces: influence on tropospheric ozone chemistry and comparison to observations. *J. Geophys. Res. Atmos.* 109:D02304. DOI:10.1029/2003JD003868

9. Dentener FJ, Carmichael GR, Zhang Y, Lelieveld J, Crutzen PJ. 1996. Role of mineral aerosol as a reactive surface in the global troposphere. *J. Geophys. Res. Atmos.* 101:22869–89

10. Quinn PK, Bates TS. 2003. North American, Asian, and Indian haze: similar regional impacts on climate? *Geophys. Res. Lett.* 30:1555. DOI:10.1029/2003GL016934

11. **Stier P, Seinfeld JH, Kinne S, Boucher O. 2007. Aerosol absorption and radiative forcing. *Atmos. Chem. Phys. Discuss.* 7:7171–233**

12. Cainey JM, Keywood M, Grose MR, Krummel P, Galbally IE, et al. 2007. Precursors to particles (P2P) at Cape Grim 2006: campaign overview. *Environ. Chem.* 4:143–50

13. Sullivan RC, Prather KA. 2005. Recent advances in our understanding of atmospheric chemistry and climate made possible by on-line aerosol analysis instrumentation. *Anal. Chem.* 77:3861–85

14. Ma Y, Weber RJ, Maxwell-Meier K, Orsini DA, Lee YN, et al. 2004. Intercomparisons of airborne measurements of aerosol ionic chemical composition during TRACE-P and ACE-Asia. *J. Geophys. Res. Atmos.* 109:D15S06. DOI:10.1029/2003JD003673

15. Middlebrook AM, Murphy DM, Lee SH, Thomson DS, Prather KA, et al. 2003. A comparison of particle mass spectrometers during the 1999 Atlanta Supersite project. *J. Geophys. Res. Atmos.* 108:8424. DOI:10.1029/2001JD000660

16. **McMurry PH. 2000. A review of atmospheric aerosol measurements. *Atmos. Environ.* 34:1959–99**

17. Llamas AM, Ojeda CB, Rojas FS. 2007. Process analytical chemistry—application of mass spectrometry in environmental analysis: an overview. *Appl. Spectrosc. Rev.* 42:345–67

11. An excellent overview of modeling the direct effect of aerosols and assumptions regarding mixing state.

16. Excellent review of analytical measurements of aerosol properties.

18. Fuzzi S, Andreae MO, Huebert BJ, Kulmala M, Bond TC, et al. 2006. Critical assessment of the current state of scientific knowledge, terminology, and research needs concerning the role of organic aerosols in the atmosphere, climate, and global change. *Atmos. Chem. Phys.* 6:2017–38

19. **Kanakidou M, Seinfeld JH, Pandis SN, Barnes I, Dentener FJ, et al. 2005. Organic aerosol and global climate modeling: a review. *Atmos. Chem. Phys.* 5:1053–123**

20. Claeys M, Graham B, Vas G, Wang W, Vermeylen R, et al. 2004. Formation of secondary organic aerosols through photooxidation of isoprene. *Science* 303:1173–76

21. Denkenberger K, Moffet RC, Holececk J, Rebotier TP, Prather KA. 2007. Real-time, single-particle measurements of oligomers in aged ambient aerosol particles. *Environ. Sci. Technol.* 41:5439–46

22. Grosse S, Letzel T. 2007. Liquid chromatography/atmospheric pressure ionization mass spectrometry with postcolumn liquid mixing for the efficient determination of partially oxidized polycyclic aromatic hydrocarbons. *J. Chromatogr. A* 1139:75–83

23. Muller C, Iinuma Y, Boge O, Herrmann H. 2007. Applications of CE-ESI-MS/MS analysis to structural elucidation of methylenecyclohexane ozonolysis products in the particle phase. *Electrophoresis* 28:1364–70

24. Poulain L, Monod A, Wortham H. 2007. Development of a new on-line mass spectrometer to study the reactivity of soluble organic compounds in the aqueous phase under tropospheric conditions: application to OH-oxidation of *N*-methylpyrrolidone. *J. Photochem. Photobiol. A Chem.* 187:10–23

25. Sullivan AP, Weber RJ, Clements AL, Turner JR, Bae MS, et al. 2004. A method for on-line measurement of water-soluble organic carbon in ambient aerosol particles: results from an urban site. *Geophys. Res. Lett.* 31:L13105. DOI:10.1029/2004GL019681

26. Ullah SMR, Takeuchi M, Dasgupta PK. 2006. Versatile gas/particle ion chromatograph. *Environ. Sci. Technol.* 40:962–68

27. Gross DS, Galli ME, Kalberer M, Prevot ASH, Dommen J, et al. 2006. Real-time measurement of oligomeric species in secondary organic aerosol with the aerosol time-of-flight mass spectrometer. *Anal. Chem.* 78:2130–37

28. Kalberer M, Paulsen D, Sax M, Steinbacher M, Dommen J, et al. 2004. Identification of polymers as major components of atmospheric organic aerosols. *Science* 303:1659–62

29. Baltensperger U, Kalberer M, Dommen J, Paulsen D, Alfarra MR, et al. 2005. Secondary organic aerosols from anthropogenic and biogenic precursors. *Faraday Discuss.* 130:265–78

30. Graber ER, Rudich Y. 2006. Atmospheric HULIS: How humic-like are they? A comprehensive and critical review. *Atmos. Chem. Phys.* 6:729–53

31. Hinz KP, Spengler B. 2007. Instrumentation, data evaluation and quantification in on-line aerosol mass spectrometry. *J. Mass Spectrom.* 42:843–60

32. Smith JN, Moore KF, Eisele FL, Voisin D, Ghimire AK, et al. 2005. Chemical composition of atmospheric nanoparticles during nucleation events in Atlanta. *J. Geophys. Res. Atmos.* 110:D22S03. DOI:10.1029/2005JD005912

19. Thorough review of organic aerosol with a focus on SOA and its importance for global climate modeling, including highlights of areas that need further study.

33. Hearn JD, Smith GD. 2006. Reactions and mass spectra of complex particles using aerosol CIMS. *Int. J. Mass Spectrom.* 258:95–103

34. DeCarlo PF, Kimmel JR, Trimborn A, Northway MJ, Jayne JT, et al. 2006. Field-deployable, high-resolution, time-of-flight aerosol mass spectrometer. *Anal. Chem.* 78:8281–89

35. Canagaratna MR, Jayne JT, Jimenez JL, Allan JD, Alfarra MR, et al. 2007. Chemical and microphysical characterization of ambient aerosols with the aerodyne aerosol mass spectrometer. *Mass Spectrom. Rev.* 26:185–222

36. Kondo Y, Miyazaki Y, Takegawa N, Miyakawa T, Weber RJ, et al. 2007. Oxygenated and water-soluble organic aerosols in Tokyo. *J. Geophys. Res. Atmos.* 112:D01203. DOI:10.1029/2006JD007056

37. Volkamer R, Jimenez JL, San Martini F, Dzepina K, Zhang Q, et al. 2006. Secondary organic aerosol formation from anthropogenic air pollution: rapid and higher than expected. *Geophys. Res. Lett.* 33:L17811. DOI:10.1029/2006GL026899

38. Williams BJ, Goldstein AH, Millet DB, Holzinger R, Kreisberg NM, et al. 2007. Chemical speciation of organic aerosol during the International Consortium for Atmospheric Research on Transport and Transformation 2004: results from in situ measurements. *J. Geophys. Res. Atmos.* 112:D10S26. DOI:10.1029/2006JD007601

39. Hopkins RJ, Tivanski AV, Marten BD, Gilles MK. 2007. Chemical bonding and structure of black carbon reference materials and individual carbonaceous atmospheric aerosols. *J. Aerosol Sci.* 38:573–91

40. Tivanski AV, Hopkins RJ, Tyliszczak T, Gilles MK. 2007. Oxygenated interface on biomass burn tar balls determined by single particle scanning transmission X-ray microscopy. *J. Phys. Chem. A* 111:5448–58

41. Kirchstetter TW, Novakov T. 2007. Controlled generation of black carbon particles from a diffusion flame and applications in evaluating black carbon measurement methods. *Atmos. Environ.* 41:1874–88

42. Chughtai AR, Kim JM, Smith DM. 2002. The effect of air/fuel ratio on properties and reactivity of combustion soots. *J. Atmos. Chem.* 43:21–43

43. Mochida M, Kawamura K. 2004. Hygroscopic properties of levoglucosan and related organic compounds characteristic to biomass burning aerosol particles. *J. Geophys. Res. Atmos.* 109:D21202. DOI:10.1029/2004JD004962

44. Hedberg E, Johansson C. 2006. Is levoglucosan a suitable quantitative tracer for wood burning? Comparison with receptor modeling on trace elements in Lycksele, Sweden. *J. Air Waste Manag. Assoc.* 56:1669–78

45. Havers N, Burba P, Lambert J, Klockow D. 1998. Spectroscopic characterization of humic-like substances in airborne particulate matter. *J. Atmos. Chem.* 29:45–54

46. Rudich Y, Donahue NM, Mentel TF. 2007. Aging of organic aerosol: bridging the gap between laboratory and field studies. *Annu. Rev. Phys. Chem.* 58:321–52

47. Roberts G, Mauger G, Hadley O, Ramanathan V. 2006. North American and Asian aerosols over the eastern Pacific ocean and their role in regulating cloud condensation nuclei. *J. Geophys. Res. Atmos.* 111:D13205. DOI:10.1029/2005JD006661

48. Petters MD, Kreidenweis SM. 2007. A single parameter representation of hygroscopic growth and cloud condensation nucleus activity. *Atmos. Chem. Phys.* 7:1961–71

49. Donahue NM, Hartz KEH, Chuong B, Presto AA, Stanier CO, et al. 2005. Critical factors determining the variation in SOA yields from terpene ozonolysis: a combined experimental and computational study. *Faraday Discuss.* 130:295–309

50. Zhang JY, Hartz KEH, Pandis SN, Donahue NM. 2006. Secondary organic aerosol formation from limonene ozonolysis: homogeneous and heterogeneous influences as a function of NOx. *J. Phys. Chem. A* 110:11053–63

51. Surratt JD, Murphy SM, Kroll JH, Ng NL, Hildebrandt L, et al. 2006. Chemical composition of secondary organic aerosol formed from the photooxidation of isoprene. *J. Phys. Chem. A* 110:9665–90

52. Weitkamp EA, Sage AM, Pierce JR, Donahue NM, Robinson AL. 2007. Organic aerosol formation from photochemical oxidation of diesel exhaust in a smog chamber. *Environ. Sci. Technol.* 41:6969–75

53. Prenni AJ, Petters MD, Kreidenweis SM, DeMott PJ, Ziemann PJ. 2007. Cloud droplet activation of secondary organic aerosol. *J. Geophys. Res. Atmos.* 112:D10223. DOI:10.1029/2006JD007963

54. Chung SH, Seinfeld JH. 2005. Climate response of direct radiative forcing of anthropogenic black carbon. *J. Geophys. Res. Atmos.* 110:D11102. DOI:10.1029/2004JD005441

55. Stier P, Seinfeld JH, Kinne S, Feichter J, Boucher O. 2006. Impact of nonabsorbing anthropogenic aerosols on clear-sky atmospheric absorption. *J. Geophys. Res. Atmos.* 111:D18201. DOI:10.1029/2006JD007147

56. Bond TC, Bergstrom RW. 2006. Light absorption by carbonaceous particles: an investigative review. *Aerosol Sci. Technol.* 40:27–67

57. Bond TC, Habib G, Bergstrom RW. 2006. Limitations in the enhancement of visible light absorption due to mixing state. *J. Geophys. Res. Atmos.* 111:D20211. DOI:10.1029/2006JD007315

58. Jacobson MZ. 2001. Strong radiative heating due to the mixing state of black carbon in atmospheric aerosols. *Nature* 409:695–97

59. Jacobson MZ. 2000. A physically-based treatment of elemental carbon optics: implications for global direct forcing of aerosols. *Geophys. Res. Lett.* 27:217–20. DOI:10.1029/1999GL010968

60. Chylek P, Videen G, Ngo D, Pinnick RG, Klett JD. 1995. Effect of black carbon on the optical properties and climate forcing of sulfate AEROSOLS. *J. Geophys. Res. Atmos.* 100:16325–32

61. Cheng YF, Eichler H, Wiedensohler A, Heintzenberg J, Zhang YH, et al. 2006. Mixing state of elemental carbon and non-light-absorbing aerosol components derived from in situ particle optical properties at Xinken in Pearl River Delta of China. *J. Geophys. Res. Atmos.* 111:D20204. DOI:10.1029/2005JD006929

62. Chandra S, Satheesh SK, Srinivasan J. 2004. Can the state of mixing of black carbon aerosols explain the mystery of 'excess' atmospheric absorption? *Geophys. Res. Lett.* 31:L19109. DOI:10.1029/2004GL020662

63. Lithgow GA, Robinson AL, Buckley SG. 2004. Ambient measurements of metal-containing PM2.5 in an urban environment using laser-induced breakdown spectroscopy. *Atmos. Environ.* 38:3319–28

64. Laskin A, Cowin JP, Iedema MJ. 2006. Analysis of individual environmental particles using modern methods of electron microscopy and X-ray microanalysis. *J. Electron Spectrosc. Relat. Phenom.* 150:260–74

65. Posfai M, Gelencser A, Simonics R, Arato K, Li J, et al. 2004. Atmospheric tar balls: particles from biomass and biofuel burning. *J. Geophys. Res. Atmos.* 109:D06213. DOI:10.1029/2003JD004169

66. Ma CJ, Choi KC. 2007. A combination of bulk and single particle analyses for Asian dust. *Water Air Soil Pollut.* 183:3–13

67. Semeniuk TA, Wise ME, Martin ST, Russell LM, Buseck PR. 2007. Hygroscopic behavior of aerosol particles from biomass fires using environmental transmission electron microscopy. *J. Atmos. Chem.* 56:259–73

68. Murphy DM. 2007. The design of single particle laser mass spectrometers. *Mass Spectrom. Rev.* 26:150–65

69. Sinha MP, Giffin CE, Norris DD, Estes TJ, Vilker VL, et al. 1982. Particle analysis by mass spectrometry. *J. Colloid Interface Sci.* 87:140–53

70. Gard E, Mayer JE, Morrical BD, Dienes T, Fergenson DP, et al. 1997. Real-time analysis of individual atmospheric aerosol particles: design and performance of a portable ATOFMS. *Anal. Chem.* 69:4083–91

71. Fergenson DP, Song XH, Ramadan Z, Allen JO, Hughes LS, et al. 2001. Quantification of ATOFMS data by multivariate methods. *Anal. Chem.* 73:3535–41

72. Zhao WX, Hopke PK, Qin XY, Prather KA. 2005. Predicting bulk ambient aerosol compositions from ATOFMS data with ART-2a and multivariate analysis. *Anal. Chim. Acta* 549:179–87

73. Qin XY, Bhave PV, Prather KA. 2006. Comparison of two methods for obtaining quantitative mass concentrations from aerosol time-of-flight mass spectrometry measurements. *Anal. Chem.* 78:6169–78

74. Bhave PV, Allen JO, Morrical BD, Fergenson DP, Cass GR, et al. 2002. A field-based approach for determining ATOFMS instrument sensitivities to ammonium and nitrate. *Environ. Sci. Technol.* 36:4868–79

75. Spencer MT, Prather KA. 2006. Using ATOFMS to determine OC/EC mass fractions in particles. *Aerosol Sci. Technol.* 40:585–94

76. Liu DY, Prather KA, Hering SV. 2000. Variations in the size and chemical composition of nitrate-containing particles in Riverside, CA. *Aerosol Sci. Technol.* 33:71–86

77. Wenzel RJ, Prather KA. 2004. Improvements in ion signal reproducibility obtained using a homogeneous laser beam for on-line laser desorption/ionization of single particles. *Rapid Commun. Mass Spectrom.* 18:1525–33

78. Murphy DM, Cziczo DJ, Hudson PK, Thomson DS. 2007. Carbonaceous material in aerosol particles in the lower stratosphere and tropopause region. *J. Geophys. Res. Atmos.* 112:D04203. DOI:10.1029/2006JD007297

79. Morrical BD, Fergenson DP, Prather KA. 1998. Coupling two-step laser desorption/ionization with aerosol time-of-flight mass spectrometry for the analysis of individual organic particles. *J. Am. Soc. Mass Spectrom.* 9:1068–73

80. Zachariah MR, Park K, Rai A, Kittelson DB, Miller A. 2005. Quantitative single particle mass spectrometry characterization of nanoaerosol composition and reactivity. *Abstr. Pap. Am. Chem. Soc.* 230:U1532

81. LaFranchi BW, Petrucci GA. 2006. A comprehensive characterization of photoelectron resonance capture ionization aerosol mass spectrometry for the quantitative and qualitative analysis of organic particulate matter. *Int. J. Mass Spectrom.* 258:120–33

82. Oktem B, Tolocka MP, Johnston MV. 2004. On-line analysis of organic components in fine and ultrafine particles by photoionization aerosol mass spectrometry. *Anal. Chem.* 76:253–61

83. Ellison GB, Tuck AF, Vaida V. 1999. Atmospheric processing of organic aerosols. *J. Geophys. Res. Atmos.* 104:11633–41

84. Semeniuk TA, Wise ME, Martin ST, Russell LM, Buseck PR. 2007. Water uptake characteristics of individual atmospheric particles having coatings. *Atmos. Environ.* 41:6225–35

85. Thornton JA, Abbatt JPD. 2005. N_2O_5 reaction on submicron sea salt aerosol: kinetics, products, and the effect of surface active organics. *J. Phys. Chem. A* 109:10004–12

86. Tervahattu H, Juhanoja J, Vaida V, Tuck AF, Niemi JV, et al. 2005. Fatty acids on continental sulfate aerosol particles. *J. Geophys. Res. Atmos.* 110:D06207. DOI:10.1029/2004JD005400

87. Woods E, Smith GD, Miller RE, Baer T. 2002. Depth profiling of heterogeneously mixed aerosol particles using single-particle mass spectrometry. *Anal. Chem.* 74:1642–49

88. Shields LG, Suess DT, Prather KA. 2007. Determination of single particle mass spectral signatures from heavy-duty diesel vehicle emissions for $PM_{2.5}$ source apportionment. *Atmos. Environ.* 41:3841–52

89. Bhave PV, Fergenson DP, Prather KA, Cass GR. 2001. Source apportionment of fine particulate matter by clustering single-particle data: tests of receptor model accuracy. *Environ. Sci. Technol.* 35:2060–72

90. Murphy DM, Cziczo DJ, Froyd KD, Hudson PK, Matthew BM, et al. 2006. Single-particle mass spectrometry of tropospheric aerosol particles. *J. Geophys. Res. Atmos.* 111:D23S32. DOI:10.1029/2006JD007340

91. Guazzotti SA, Suess DT, Coffee KR, Quinn PK, Bates TS, et al. 2003. Characterization of carbonaceous aerosols outflow from India and Arabia: Biomass/biofuel burning and fossil fuel combustion. *J. Geophys. Res. Atmos.* 108:4485. DOI:10.1029/2002JD003277

92. Moffet RC, Prather KA. 2005. Extending ATOFMS measurements to include refractive index and density. *Anal. Chem.* 77:6535–41

93. Spencer MT, Shields LG, Prather KA. 2007. Simultaneous measurement of the effective density and chemical composition of ambient aerosol particles. *Environ. Sci. Technol.* 41:1303–9

94. Zelenyuk A, Cai Y, Chieffo L, Imre D. 2005. High precision density measurements of single particles: the density of metastable phases. *Aerosol Sci. Technol.* 39:972–86

95. Cziczo DJ, Thomson DS, Thompson TL, DeMott PJ, Murphy DM. 2006. Particle analysis by laser mass spectrometry (PALMS) studies of ice nuclei and other low number density particles. *Int. J. Mass Spectrom.* 258:21–29

96. Murphy DM, Cziczo DJ, Hudson PK, Schein ME, Thomson DS. 2004. Particle density inferred from simultaneous optical and aerodynamic diameters sorted by composition. *J. Aerosol Sci.* 35:135–39

97. **Heard DE. 2006. Field measurements of atmospheric composition. In *Analytical Techniques for Atmospheric Measurement*. Oxford, UK: Blackwell, pp. 1–68**

98. Kahn R, Anderson J, Anderson TL, Bates T, Brechtel F, et al. 2004. Environmental snapshots from ACE-Asia. *J. Geophys. Res. Atmos.* 109:D19S14. DOI:10.1029/2003JD004339

99. Tegen I, Fung I. 1995. Contribution to the atmospheric mineral aerosol load from land-surface modification. *J. Geophys. Res. Atmos.* 100:18707–26

100. Bian H, Zender CS. 2003. Mineral dust and global tropospheric chemistry: Relative roles of photolysis and heterogeneous uptake. *J. Geophys. Res. Atmos.* 108:4672. DOI:10.1029/2002JD003143

101. Arimoto R, Kim YJ, Kim YP, Quinn PK, Bates TS, et al. 2006. Characterization of Asian dust during ACE-Asia. *Glob. Planet. Change* 52:23–56

102. Krueger BJ, Grassian VH, Cowin JP, Laskin A. 2004. Heterogeneous chemistry of individual mineral dust particles from different dust source regions: the importance of particle mineralogy. *Atmos. Environ.* 38:6253–61

103. Bates TS, Quinn PK, Coffman DJ, Covert DS, Miller TL, et al. 2004. Marine boundary layer dust and pollutant transport associated with the passage of a frontal system over eastern Asia. *J. Geophys. Res. Atmos.* 109:D19S19. DOI:10.1029/2003JD004094

104. Tang YH, Carmichael GR, Seinfeld JH, Dabdub D, Weber RJ, et al. 2004. Three-dimensional simulations of inorganic aerosol distributions in east Asia during spring 2001. *J. Geophys. Res. Atmos.* 109:D19S23. DOI:10.1029/2003JD004201

105. **Usher CR, Michel AE, Grassian VH. 2003. Reactions on mineral dust. *Chem. Rev.* 103:4883–939**

106. Cwiertny DM, Young MA, Grassian VH. 2007. Heterogeneous chemistry and photochemistry of mineral dust aerosol. *Annu. Rev. Phys. Chem.* 59:27–51

107. Vlasenko A, Sjogren S, Weingartner E, Gaggeler HW, Ammann M. 2005. Generation of submicron Arizona test dust aerosol: chemical and hygroscopic properties. *Aerosol Sci. Technol.* 39:452–60

108. Mashburn CD, Frinak EK, Tolbert MA. 2006. Heterogeneous uptake of nitric acid on Na-montmorillonite clay as a function of relative humidity. *J. Geophys. Res.* 111:D15213. DOI:101029/2005JD006525

97. An excellent book chapter discussing field measurement techniques, goals, and study locations.

105. A review of heterogeneous reactions on mineral dust aerosol.

109. Gibson ER, Cain JP, Wang H, Grassian VH, Laskin A. 2008. Kinetic study of heterogeneous reaction of $CaCO_3$ particles with gaseous HNO_3 using particle-on-substrate stagnation flow reactor approach. *J. Phys. Chem. A.* 112:1561–71

110. Santschi C, Rossi MJ. 2006. Uptake of CO_2, SO_2, HNO_3 and HCl on calcite ($CaCO_3$) at 300 K: mechanism and the role of adsorbed water. *J. Phys. Chem. A* 110:6789–802

111. Gibson ER, Hudson PK, Grassian VH. 2006. Physicochemical properties of nitrate aerosols: implications for the atmosphere. *J. Phys. Chem. A* 110:11785–99

112. Gibson ER, Hudson PK, Grassian VH. 2006. Aerosol chemistry and climate: laboratory studies of the carbonate component of mineral dust and its reaction products. *Geophys. Res. Lett.* 33:L13811. DOI:10.1029/2006GL026386

113. Al-Abadleh H, Grassian VH. 2003. Oxide surfaces as environmental interfaces. *Surf. Sci. Rep.* 52:63–161

114. Baltrusaitis J, Usher CR, Grassian VH. 2007. Reactions of sulfur dioxide on calcium carbonate single crystal and particle surfaces at the adsorbed water carbonate interface. *Phys. Chem. Chem. Phys.* 9:3011–24

115. Tang Y, Carmichael GR, Kurata G, Uno I, Weber RJ, et al. 2004. Impacts of dust on regional tropospheric chemistry during the ACE-Asia experiment: a model study with observations. *J. Geophys. Res.* 109:D19S21. DOI:10.1029/2003JD003806

116. Sullivan RC, Guazzotti SA, Sodeman DA, Tang Y, Carmichael GR, et al. 2007. Mineral dust is a sink for chlorine in the marine boundary layer. *Atmos. Environ.* 41: 7166–79

117. O'Dowd CD, De Leeuw G. 2007. Marine aerosol production: a review of the current knowledge. *Philos. Trans. R. Soc. A Math. Phys. Eng. Sci.* 365:1753–74

118. Mahowald NM, Baker AR, Bergametti G, Brooks N, Duce RA, et al. 2005. Atmospheric global dust cycle and iron inputs to the ocean. *Global Biogeochem. Cycles* 19:GB4025. DOI:10.1029/2004GB002402

119. An excellent, concise summary of the current understanding of iron-containing aerosols and their roles in global biogeochemical and climate cycles.

119. Jickells TD, An ZS, Andersen KK, Baker AR, Bergametti G, et al. 2005. Global iron connections between desert dust, ocean biogeochemistry, and climate. *Science* 308:67–71

120. Adahchour M, Beens J, Vreuls RJJ, Brinkman UAT. 2006. Recent developments in comprehensive two-dimensional gas chromatography (GC × GC) I. Introduction and instrumental set-up. *Trends Anal. Chem.* 25:438–54

121. Gilardoni S, Russell LM, Sorooshian A, Flagan RC, Seinfeld JH, et al. 2007. Regional variation of organic functional groups in aerosol particles on four US East Coast platforms during the International Consortium for Atmospheric Research on Transport and Transformation 2004 campaign. *J. Geophys. Res. Atmos.* 112:D10S27. DOI:10.1029/2006JD007737

122. Reinhardt A, Emmenegger C, Gerrits B, Panse C, Dommen J, et al. 2007. Ultra-high mass resolution and accurate mass measurements as a tool to characterize oligomers in secondary organic aerosols. *Anal. Chem.* 79:4074–82

123. Reviews the status of global models used for predicting the indirect effect and discusses how satellite measurements can be used for validation.

123. Lohmann U, Quaas J, Kinne S, Feichter J. 2007. Different approaches for constraining global climate models of the anthropogenic indirect aerosol effect. *Bull. Am. Meteorol. Soc.* 88:243–49

124. Lohmann U, Feichter J. 2005. Global indirect aerosol effects: a review. *Atmos. Chem. Phys.* 5:715–37

125. Medina J, Nenes A, Sotiropoulou REP, Cottrell LD, Ziemba LD, et al. 2007. Cloud condensation nuclei closure during the International Consortium for Atmospheric Research on Transport and Transformation 2004 campaign: effects of size-resolved composition. *J. Geophys. Res. Atmos.* 112:D10S31. DOI:10.1029/2006JD007588

126. Furutani H, Prather KA. 2007. Assessment of the relative importance of atmospheric aging on CCN activity derived from field observations. *Atmos. Environ.* In press

127. Sassen K, DeMott PJ, Prospero JM, Poellot MR. 2003. Saharan dust storms and indirect aerosol effects on clouds: CRYSTAL-FACE results. *Geophys. Res. Lett.* 30:1633. DOI:10.1029/2003GL017371

128. Richardson MS, DeMott PJ, Kreidenweis SM, Cziczo DJ, Dunlea EJ, et al. 2007. Measurements of heterogeneous ice nuclei in the western United States in springtime and their relation to aerosol characteristics. *J. Geophys. Res. Atmos.* 112:D02209. DOI:10.1029/2006JD007500

129. Cantrell W, Heymsfield A. 2005. Production of ice in tropospheric clouds: a review. *Bull. Am. Meteorol. Soc.* 86:795–807

130. Heymsfield A, Stith J, Rogers D, Field P, DeMott P, et al. 2005. *The Ice in Clouds Experiment: research plan. Scientific overview document.* http://www.eol.ucar.edu/~dcrogers/Ice-Init/ICE-SOD.pdf

131. Quinn PK, Bates TS. 2005. Regional aerosol properties: comparisons of boundary layer measurements from ACE 1, ACE 2, AEROSOLS99, INDOEX, ACE-Asia, TARFOX, and NEAQS. *J. Geophys. Res. Atmos.* 110:D14202. DOI:10.1029/2004JD004755

132. Bates TS, Anderson TL, Baynard T, Bond T, Boucher O, et al. 2006. Aerosol direct radiative effects over the northwest Atlantic, northwest Pacific and north Indian oceans: estimates based on in-situ chemical and optical measurements and chemical transport modeling. *Atmos. Chem. Phys.* 6:1657–732

133. Carrico CM, Kus P, Rood MJ, Quinn PK, Bates TS. 2003. Mixtures of pollution, dust, sea salt, and volcanic aerosol during ACE-Asia: radiative properties as a function of relative humidity. *J. Geophys. Res. Atmos.* 108:8650. DOI:10.1029/2003JD003405

134. Dick WD, Ziemann PJ, McMurry PH. 2007. Multiangle light-scattering measurements of refractive index of submicron atmospheric particles. *Aerosol Sci. Technol.* 41:549–69

135. Forster P, Ramaswamy V, Artaxo P, Berntsen T, Betts R, et al. 2007. Summary for policymakers. In *Climate Change 2007: The Physical Science Basis. Contribution of Working Group I to the Fourth Assessment Report of the Intergovernmental Panel on Climate Change*, ed. S Solomon, D Qin, M Manning, Z Chen, M Marquis, et al. Cambridge, UK: Cambridge Univ. Press, pp. 2–18

136. Smith JN, Moore KF, McMurry PH, Eisele FL. 2004. Atmospheric measurements of sub-20 nm diameter particle chemical composition by thermal desorption chemical ionization mass spectrometry. *Aerosol Sci. Technol.* 38:100–10

137. Wang SY, Zordan CA, Johnston MV. 2006. Chemical characterization of individual, airborne sub-10-nm particles and molecules. *Anal. Chem.* 78:1750–54

138. Hering SV, Stolzenburg MR. 2005. A method for particle size amplification by water condensation in a laminar, thermally diffusive flow. *Aerosol Sci. Technol.* 39:428–36

139. Dixon RW, Peterson DS. 2002. Development and testing of a detection method for liquid chromatography based on aerosol charging. *Anal. Chem.* 74:2930–37

140. Ramanathan V, Ramana MV, Roberts G, Kim D, Corrigan C, et al. 2007. Warming trends in Asia amplified by brown cloud solar absorption. *Nature* 448:575–78

141. Prieto MC, Kovtoun VV, Cotter RJ. 2002. Miniaturized linear time-of-flight mass spectrometer with pulsed extraction. *J. Mass Spectrom.* 37:1158–62

142. Seinfeld JH, Pankow JF. 2003. Organic atmospheric particulate material. *Annu. Rev. Phys. Chem.* 54:121–40

143. Lee S, Kamens RM, Jang MS. 2005. Gas and particle partitioning behavior of aldehyde in the presence of diesel soot and wood smoke aerosols. *J. Atmos. Chem.* 51:223–34

144. Jang MS, Carroll B, Chandramouli B, Kamens RM. 2003. Particle growth by acid-catalyzed heterogeneous reactions of organic carbonyls on preexisting aerosols. *Environ. Sci. Technol.* 37:3828–37

145. Butler JR, Mitchem L, Hanford KL, Truel L, Reid JP. 2007. In situ comparative measurements of the properties of aerosol droplets of different chemical composition. *Faraday Discuss.* 137:1–16

146. Gilman JB, Tervahattu H, Vaida V. 2006. Interfacial properties of mixed films of long-chain organics at the air-water interface. *Atmos. Environ.* 40:6606–14

147. Despres V, Nowoisky J, Klose M, Conrad R, Andreae MO, et al. 2007. Molecular genetics and diversity of primary biogenic aerosol particles in urban, rural and high-alpine air. *Biogeosci. Discuss.* 4:349–84

148. Tobias HJ, Pitesky ME, Fergenson DP, Steele PT, Horn J, et al. 2006. Following the biochemical and morphological changes of *Bacillus atrophaeus* cells during the sporulation process using bioaerosol mass spectrometry. *J. Microbiol. Methods* 67:56–63

149. Franze T, Weller MG, Niessner R, Poschl U. 2005. Protein nitration by polluted air. *Environ. Sci. Technol.* 39:1673–78

142. An overview of the key models and equations used to predict gas/particle partitioning of organic aerosols.

Multiplexed Spectroscopic Detections

Kyle D. Bake and David R. Walt

Department of Chemistry, Tufts University, Medford, Massachusetts 02155;
email: david.walt@tufts.edu

Annu. Rev. Anal. Chem. 2008. 1:515–47

First published online as a Review in Advance on March 4, 2008

The *Annual Review of Analytical Chemistry* is online at anchem.annualreviews.org

This article's doi:
10.1146/annurev.anchem.1.031207.112826

1936-1327/08/0719-0515$20.00

Key Words

arrays, fluorescence, surface plasmon resonance, surface-enhanced Raman spectroscopy, microparticles

Abstract

This review describes various platforms used for multiplexed spectroscopic analysis. We highlight the use of different types of spectroscopy for multiplexed detections, including Raman spectroscopy, surface-enhanced Raman spectroscopy, surface plasmon resonance, and fluorescence. This review also explores the use of cross-reactive sensors in combination with pattern-recognition algorithms to monitor multiple analytes in aqueous and vapor matrices. It also discusses applications of these techniques, paying special attention to their use in the detection of biologically relevant analytes.

1. INTRODUCTION

Multianalyte detection schemes have become an important analytical tool and have been applied in fields ranging from genetics (1, 2) to environmental monitoring (3, 4). In particular, many spectroscopic techniques have been developed that enable multi-analyte detection. Compared with other analytical methods (such as magnetic resonance, chromatography, or electrochemistry), spectroscopy lends itself to multiplexing because different colors of light can be easily created, separated, and detected simultaneously. Advances in photonics technologies have led to the development of handheld fluorimeters (5), handheld Raman spectrometers (6), and portable total reflection X-ray fluorescence spectrometers (7), and additional miniaturization and capability are expected. To empower these devices further, investigators are developing multiplexed detection schemes that enable essentially thousands of experiments to be conducted concurrently.

Multiplexed spectroscopic detection is performed in one of two modalities. First and simplest is the use of spectroscopy in which each analyte possesses a different spectroscopic signature. This method is a direct one as the intrinsic properties of the analytes give rise to different spectroscopic signatures. Second, different indicators can be used to bind analytes and report their presence. This approach is indirect in that binding or reaction between the analyte and the indicator provides the signal. In either scheme, the detection of multiple analytes simultaneously requires that light of different wavelengths be distinguishable from one another. This distinction can be made by resolving the color or other property (e.g., fluorescence lifetime) of the different analytes or indicators. In addition to these two schemes (direct and indirect), we can further categorize multiplexed spectroscopic analysis according to how the different analytes are resolved. In the first approach, resolution is accomplished spectrally. In this scheme, the analytes or the indicators are identified by their distinct spectroscopic signatures. Another approach is to use spatial resolution whereby different sensing regions are spatially separated. For example, different indicators can be placed at different regions of a substrate, and a color change at a particular position correlates with the presence and/or concentration of the analyte. Spatially resolved multiplexed substrates are called arrays.

This review covers recent advances in various spectroscopic methods used for multiplexed detection schemes. We discuss the platforms that make multiplexed detection possible, such as arrays and encoded microcarriers. Some applications of systems that use fluorescence, surface-enhanced Raman spectroscopy (SERS), and surface plasmon resonance (SPR) are discussed in terms of both commercial and research uses. Other reviews have covered the basic operating principles of the different types of spectroscopy used in these schemes, so we do not cover them here (8–13).

2. DIRECT MULTIPLEXED DETECTIONS

2.1. Spectral Separation

The direct measurement of multiple analytes by optical methods requires that the analytes of interest be distinguishable from the background as well as from each other. These methods typically involve collecting spectra and then applying deconvolution methods to separate the spectral signals arising from the different components in the sample. Fluorescence and absorption peaks tend to be broad, and the identification and quantification of multiple compounds are difficult in complex mixtures. For this reason, detection methods using Raman and X-ray fluorescence spectroscopy, whose peaks are narrow in comparison, have been developed for multiplexed detections. In addition, X-ray spectra are unique to specific elements.

Raman spectroscopy can directly detect many different analytes on the same platform, such as a variety of dyes used on ancient artifacts (14, 15), but it has not yet led to the development of a system capable of detecting multiple analytes simultaneously. Recently the Van Duyne group (16) proposed a method using SERS that may make it possible to simultaneously monitor both lactose and glucose at physiologically relevant concentrations. SERS has also been shown to be useful for the detection of bio- and chemical warfare agents (13, 17, 18). Although no multiplexing was performed, the technique was able to detect and identify *Bacillus anthracis*, *Yersinia pestis*, *Burkholderia mallei*, *Francisella tularensis*, *Brucella abortus*, and ricin on the same platform (19). The researchers first obtained Raman spectra for all live species and then employed pattern-recognition algorithms to identify all pathogens at the species level in blind trial evaluations (19). This Raman study shows great promise for the quick and accurate detection of multiple pathogens in real-world situations.

The detection of trace metals has become popular in the X-ray spectroscopic field (20–23). The detection of platinum, palladium, and rhodium, for example, is important in the recycling of complex materials, such as automotive catalysts. Van Meel et al. (23) used X-ray fluorescence to detect all three metals simultaneously with detection limits below 5 ppm for all metals. This technique was much faster than the typically used inductively coupled plasma optical emission spectrometry with a much easier sample preparation process (23). The simultaneous detection of all three elements is possible because each metal has unique emission peak locations.

These multiplexed detections of trace metals have also been applied to the study of various diseases, such as Parkinson's (24), amylotrophic lateral sclerosis (24), and prostate cancer (25). Through use of X-ray fluorescence, researchers were able to simultaneously detect potassium, calcium, iron, copper, zinc, and selenium in brain tissue (24). By using cluster and discriminant analysis, they classified samples as positive or negative for both Parkinson's and amylotrophic lateral sclerosis (24). Through similar data analysis (using X-ray fluorescence detection of iron, zinc, copper, and manganese), the investigators successfully classified prostate cancer in 97.7% of samples (25). The speed and accuracy with which these experiments can be performed may lead to novel diagnostic capabilities, as well as a better understanding of disease, especially if the concentrations of the metals can be correlated to biochemical changes within the cells (26).

3. INDIRECT MULTIPLEX DETECTIONS

3.1. Spectral Separation

The methods described above all employ an intrinsic spectroscopic property of the analyte of interest to carry out detection. In multiplexed detections, however, it is more common to use an indicator that interacts with the analyte of interest. These detection schemes are indirect because the signal does not come directly from the analyte of interest.

3.1.1. Dyes as indicators.

To detect multiple analytes at a single location, one must have a spectral separation of the indicators. An example of multicolor detection is the Sanger method for DNA sequencing (27). In this method, each of the four bases (ACGT) is labeled with a different dye, and the incorporation of each labeled base into the growing DNA strand results in a different color that can be easily distinguished to identify the base at that position. Similarly, most multianalyte sensors use different color indicators to make the analysis easier. One- and two-layer sensors make use of a single sensing region to detect multiple species. These layered sensors contain all the sensing chemistry in a single material but, to date, have only achieved duplexed detection. These sensors are made of a porous solid support, such as ethyl cellulose (28) or silica sol-gel (29), which houses fluorescent dyes that are sensitive to the analytes of interest. These materials are typically suspended in a polymer that does not interfere with the excitation or emission spectral properties of the sensing dyes. When the polymer is cast, it forms a layer that is a few micrometers thick. The dyes must undergo reversible reactions with the analyte of interest to express a change in a fluorescent property, such as lifetime (30), intensity (31), or emission wavelength.

Dual sensors detect two analytes by positioning the sensing chemistries for both targeted analytes at a single location. The first dual-layer configuration was reported in 1988 and consisted of two different sensing layers (31). One layer housed a dye sensitive to CO_2, whereas the other layer housed an O_2-sensitive dye. Signal differentiation was accomplished by using two fluorescent dyes with different emission spectra. Variations in the thickness of the sensing layers resulting from the production process led to errors in fluorescence intensity.

Since these initial results, researchers have made various efforts to remove this dependence on layer thickness. One method involved the use of a reference dye that does not participate in any analyte-dependent reaction and does not interfere with the sensing dye. This approach compares the fluorescence intensity from the sensing dye with that of a reference dye housed in the same layer (**Figure 1**). Carbon-dioxide sensing chemistry is based on dyes that exhibit changes in fluorescence intensity resulting from changes in pH due to the formation of carbonic acid within the sensing layer. Tetraoctylammonium and tetraoctylammonium hydroxide were used to monitor CO_2 at low concentrations (high pH), whereas 8-hydroxypyrene-1,3,6-trisulfonate was used to monitor high concentrations of CO_2 (low pH). All dyes were dissolved in ethyl cellulose and ground into microparticles. The reference dye, $Ir_2(C_{30})Cl_2$, was dissolved in the gas-impermeable

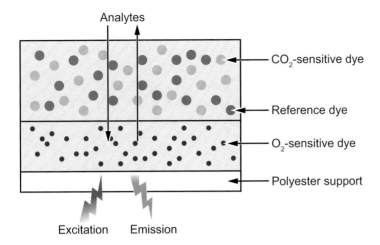

Analytes

— CO$_2$-sensitive dye

— Reference dye

— O$_2$-sensitive dye

— Polyester support

Excitation   **Emission**

Figure 1

Schematic of a two-layer dual sensor used to monitor CO$_2$ and O$_2$. Analytes are allowed to diffuse from top to bottom, whereas excitation and emission collection are performed from the bottom. Figure reproduced from Reference 32.

poly(acrylonitrile-coacrylic acid) and also ground into microparticles. Both the reference and sensing microparticles were held in polydimethylsiloxane to form the CO$_2$-sensing layer (**Figure 1**). The concentration of the analyte is calculated from the ratio of the two dyes and is therefore independent of the layer thickness (32). The two dyes have the same excitation and emission wavelengths but differ in their fluorescence lifetime. Dual-lifetime referencing is used to differentiate between the two signals. This technique works when the reference and sensing dyes have different (approximately three orders of magnitude) fluorescence lifetimes (33). For the oxygen-sensing layer, platinum(II)-5,10,15,20-tetrakis(2,3,4,5,6-pentafluorophenyl)porphyrin (Pt-TFPP), held in a polystyrene support, was used as the indicator (**Figure 1**) (32) as its fluorescence lifetime is dependent on oxygen concentration. The use of fluorescence lifetime enables the measurement to be independent of the sensing-layer thickness. Through the combination of dual-lifetime referencing and fluorescence-lifetime monitoring, researchers have produced sensors capable of monitoring bacterial growth by measuring both the consumption of O$_2$ and the production of CO$_2$ (32) and capable of monitoring both CO$_2$ and O$_2$ in aquatic systems (28).

Humidity (34) and temperature (35) changes result in large variations in the response of these indicators. Owing to the strong temperature dependence of many sensors, a method to monitor and compensate for temperature would be an important addition to the production of sensors capable of accurately monitoring multiple species in real-world situations (36). The dual-layer approach currently can monitor only two species simultaneously. For such sensors to find wider application in monitoring biological, clinical, or geological processes, they must be capable of monitoring more analytes simultaneously by using both spectral and temporal (dual-lifetime referencing) discrimination (37). The use of spectral separation alone will likely not be adequate for higher levels of multiplexing as fluorescence emission peaks tend to be broad and peak overlap is unavoidable.

3.1.2. Dyes as labels. Another approach to using spectral separation is to employ indicators as labels. This approach still separates the indicators spectrally, but they do not participate in a reaction with the analyte other than serving as labels. Multiplexing is accomplished by determining which indicators are bound, enabling a determination of the analytes present in the sample. Four different fluorescent dyes have been used for DNA sequencing as described above (38, 39); however, due to the broad peaks in fluorescence spectra, dyes with narrow bandwidths would be of significant utility. For example, multiplexed SERS detection can be performed effectively in this manner (40). SERS is well suited for multiplexing through the use of different labels because of the narrow (~1 nm, full width at half-maximum) bandwidths of Raman fingerprints (41). One can use most classes of chromophores as SERS labels; thus the number of possible Raman tags is extremely large (42). As shown by Irudayaraj and coworkers (43), the use of eight different nonfluorescent Raman tags could allow for a sensitive and selective simultaneous detection of eight different DNA sequences. These authors obtained SERS signals by attaching both the probe (a complementary DNA strand) and the label (a SERS-active tag) onto a gold nanoparticle (**Figure 2a**). The nanoparticle probes each contained a unique DNA strand that corresponded to a unique Raman-active molecule. As an initial test, they allowed small aliquots of mixtures containing two, four, and eight different probes to dry on a gold-covered glass slide and obtained Raman signals (**Figure 2b**). Many of the SERS tags had multiple overlapping peaks, thus reducing the number of unique identification peaks per label, and as the number of tags present increased, the peaks became less defined (**Figure 2c**). Probes such as these will likely lead to the ability to perform DNA detections using spectral separation. Coupling both probe and label to a common structure (**Figure 2a**) enables SERS to be utilized for in vitro or in vivo detections of various biological components. For example, breast cancer and floating leukemia cells were detected on the same platform through labeling receptors on the cell membranes with SERS tags (44).

The ability to employ Raman labels for multiplexed protein detections in human tissue was recently accomplished using composite organic-inorganic nanoparticles (COINs) (45). A COIN is formed when silver nanoparticles aggregate with the aid of Raman-active molecules (46). The Raman tags decrease the effective negative charge

---

Figure 2

(*a*) Illustration of a surface-enhanced Raman DNA probe, consisting of thiol-terminated DNA probes and Raman tags (*diamond shapes*) attached to a gold nanoparticle. (*b*) Normalized Raman intensity versus wavelength spectra for mixtures of tags 1 and 2 (*mixture 1*), tags 1–4 (*mixture 2*), and tags 1–8 (*mixture 3*). (*c*) Table of observed Raman peaks for each tag when examined alone and when multiplexed with the other seven tags. Figure reprinted with permission from Reference 43. Copyright 2007, American Chemical Society.

a

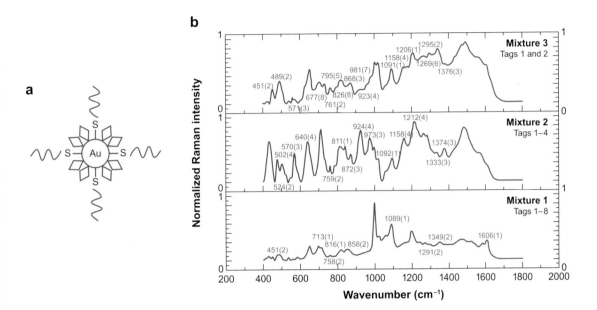

b

Normalized Raman intensity

Wavenumber (cm⁻¹)

Mixture 3
Tags 1 and 2

451(2) 489(2) 795(5) 981(7) 1158(4) 1206(1) 1295(2) 868(3) 1091(1) 1269(6) 1376(3)
677(8) 826(8) 923(4) 571(3) 761(2)

Mixture 2
Tags 1–4

502(4) 570(3) 640(4) 811(1) 924(4) 973(3) 1158(4) 1212(4) 1374(3) 1092(1) 1333(3) 524(2) 759(2) 872(3)

Mixture 1
Tags 1–8

451(2) 713(1) 816(1) 858(2) 1089(1) 1349(2) 1606(1) 758(2) 1291(2)

c

Probe	Observed Raman peaks (cm-1)	Observed peaks in eight-component mixture (cm⁻¹)
Tag-1	416, 494, 646, 712, 813, 998, 1089, 1204, 1465, 1606	1091, 1206
Tag-2	413, 450, 489, 524, 575, 593, 650, 691, 759, 821, 862, 1021, 1069, 1145, 1293, 1345, 1487	451, 489, 761, 1295
Tag-3	441, 484, 569, 726, 747, 870, 977, 1000, 1098, 1186, 1234, 1334, 1378, 1518, 1579	571, 868, 1376
Tag-4	438, 505, 565, 638, 710, 920, 1063, 1158, 1216, 1251, 1381, 1439, 1470, 1559, 1689	923, 1158
Tag-5	540, 660, 733, 794, 880, 968, 1016, 1061, 1198, 1242, 1335, 1543, 1607, 1636	795
Tag-6	483, 651, 728, 781, 1016, 1098, 1271, 1319, 1425, 1480, 1537	1269
Tag-7	495, 651, 687, 728, 848, 981, 1072, 1173, 1320, 1477	981
Tag-8	498, 540, 645, 675, 731, 784, 828, 853, 910, 1016, 1108, 1259, 1317, 1468, 1558	677, 826

on the surface of silver nanoparticles and reduce electrostatic repulsion, thus allowing aggregation to occur (**Figure 3a**). The COINs can be coated with a ~10-nm-thick layer of cross-linked bovine serum albumin and then functionalized with the desired probe molecule, such as an antibody (**Figure 3a**). The different COINs contain different Raman labels, enabling multiplexed detection. Each COIN's bovine serum albumin surface was functionalized with an antibody specific to a unique target protein through the use of pendant carboxylic acid groups (45). A two-COIN staining of a human prostate tissue sample was performed to detect the prostate-specific proteins PSA and CK18. The multiplexed spectrum in **Figure 3b** is an overlay of the signals from both COINs, in which individual signals are determined by a least-squares analysis. These studies demonstrate that it is possible to use SERS for multiplexed detection in cells and tissues, enabling their potential use in medical diagnostics and research.

Metallic nanoparticles are also capable of acting as optical labels. Metallic nanoparticles can exhibit strong SPR absorption peaks that are characteristic of their size and shape. When the environment surrounding such nanoparticles is altered, the spectral peak locations change. The development of simple methods for synthesizing gold nanostructures (47) with highly controllable surface plasmon wavelengths (48–51) has been crucial for enabling the use of SPR in multiplexed sensing platforms. Consequently, SPR is a convenient method for multiplexed detection because it is possible to produce gold nanorods that have well-defined and unique SPR absorption bands (47), which can be used for both encoding the particles and the analytical measurement step (52). The gold nanorods can be made into optical biosensors by attaching antibodies via thiol-gold binding. Obtaining an absorption spectrum of a solution of functionalized gold nanoparticles before and after sample introduction (**Figure 4**) enables the simultaneous detection of multiple analytes. Each absorption peak in the spectrum results from a different subpopulation of gold nanorods; the degree of red shift between the control and sample spectra is proportional to analyte binding and therefore to analyte concentration. The gold SPR spectral position is dependent on the solvent and the extent of equilibrium between free and bound analyte, and equilibrium must be reached prior to obtaining absorption spectra. Gold nanostructures can be made insensitive to solvent effects while still maintaining sensitivity to binding by coating them with a 1.5-nm layer of silica (53). This lab-in-a-tube concept may be useful for performing multiplexed in vitro and in vivo detections.

3.2. Spatial Separation

Researchers can perform multiplexed detection using spatial resolution by anchoring the sensing chemistry for different analytes at different locations on a solid support or by using many different solid supports, each unique for a single analyte. The solid support can be an encoded microparticle, containing a specific sensing molecule, or the different sensing molecules may share a support by being deposited or printed onto a substrate, such as plastic or glass. For the latter type of arrays, the position of a sensing element identifies its specificity.

Microparticle supports must be encoded to determine which molecules are attached to each microparticle. These encoding schemes all use light as a stimulus and

a

COIN encapsulation

(i)

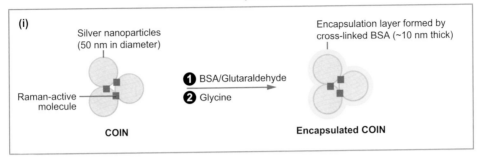

Silver nanoparticles
(50 nm in diameter)

Raman-active molecule

1 BSA/Glutaraldehyde
2 Glycine

Encapsulation layer formed by cross-linked BSA (~10 nm thick)

COIN

Encapsulated COIN

COIN functionalization

(ii)

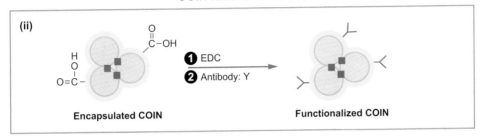

1 EDC
2 Antibody: Y

Encapsulated COIN

Functionalized COIN

b

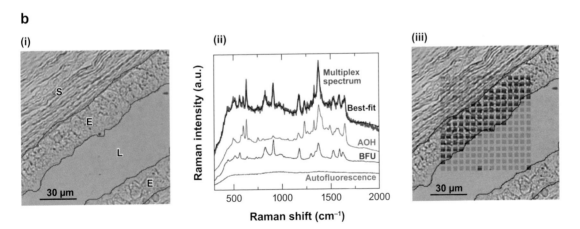

(i)

S
E
L
E

30 µm

(ii)

Raman intensity (a.u.)

Multiplex spectrum

Best-fit

AOH

BFU

Autofluorescence

500   1000   1500   2000

Raman shift (cm⁻¹)

(iii)

30 µm

Figure 3

(*a*) Diagram describing composite organic-inorganic nanoparticle (COIN) production (*i*) and surface functionalization with biological probes (*ii*). (*b, i*) Light microscopy images of human tissue samples showing stroma (S), epithelium (E), and lumen (L). (*ii*) The background (autofluorescence), raw Raman spectrum when both COINs are present, and the resulting deconvoluted spectra of the two probes, each with a unique Raman-active molecule, acridine orange (AOH) or basic fuchsin (BFU). (*iii*) Same image as in panel *b*, part *i*, with each individual square representing where a Raman spectrum was obtained. The red squares indicate the detection of AOH, whereas the blue squares represent the detection of BFU. Figure reprinted from Reference 45. Copyright 2007, American Chemical Society.

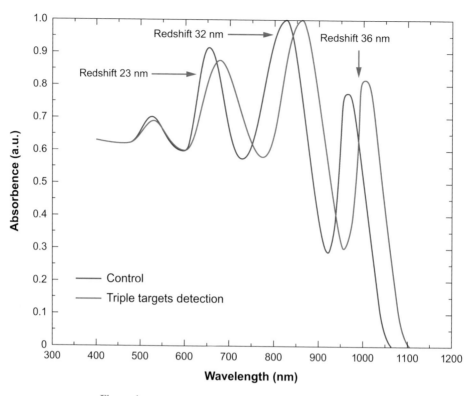

Figure 4

Absorption profile of a solution containing three gold nanorods before (control) and after (triplex targets detection) sample binding. Figure reprinted from Reference 52. Copyright 2007, American Chemical Society.

then read an optical response, such as fluorescence intensity, fluorescence lifetime, fluorescence emission location, SERS, or SPR. There are two types of optical responses from such encoded materials. First, the encoding signals provide information about which sensing chemistry is attached to each particle. Second, an orthogonal optical signal reports on the presence and concentration of the analyte that has bound to or reacted with the particle. Just like the encoding schemes, the analyte detection can be performed using the same set of spectroscopies. Although it is possible to use different types of spectroscopy for encoding and detecting, it is more typical to use the same spectroscopic technique for both functions.

3.3. Types of Array Platforms

Arrays are defined by the method used to distinguish the different sensing chemistries from one another. Suspension arrays consist of small particles that contain the sensing chemistry, whereas fixed arrays possess all the sensing chemistry on a common solid substrate.

3.3.1. Suspension arrays. Mixtures of various sensing chemistries attached to different small solid supports (typically in the nano- to micrometer size range) are known as suspension arrays. The readout process of a suspension array is performed through one of three methods: by using a flow cytometer (54), by passing the solution through a microfluidic platform such that only one particle is interrogated at a time (55, 56), or by transferring an aliquot of the suspension to a microscope slide and imaging the sample using a two-dimensional imager such as a charge-coupled device (57). Each particle type contains sensing chemistry for a single analyte of interest and is encoded through the use of some type of optical signal that does not interfere with the detection scheme. Each particle type contains the same sensing chemistry with identical encoding but differs from particles that contain sensing chemistry for a different analyte. Through the use of optical bar codes, investigators can distinguish particles from one another. Various types of optical bar codes have been used to encode particles for multiplex suspension detections. One of the simplest approaches is to use polymer microspheres containing entrapped dyes. Polymer microspheres can be swelled in organic solvents, allowing fluorescent dyes to be entrapped inside the spheres once the solvent has evaporated and the microspheres have shrunk back to their original size (54, 58). By using different dyes and/or different concentrations of a dye, researchers can produce unique small particles containing the sensing chemistry.

In 2001, the Natan group (59) developed optical bar codes that resemble the thick and thin stripes seen on department store price tags. By individually electrodepositing gold and silver in a preformed porous Al_2O_3 membrane, the researchers produced metallic bar codes by varying the thickness of the gold and silver stripes (**Figure 5a**). Silver reflects strongly at 405 nm compared with gold, which results in a reflection image showing dark and light areas on the rod that can be used for optical encoding. Thiol-terminated sensing molecules then can be easily attached to the metal surface to produce optical probes.

The shapes and composition of reflective solids have also been used to produce suspensions capable of multiplexed detections. Investigators used a two-step anodization process to produce an alumina template containing pores of various dimensions. They modified the pores in the template with silica and dissolved the alumina template to form silica nanotubes (SNTs) as seen in **Figure 5b** (60). The resulting size differences between the reflective SNTs allow them to be used for encoding. **Figure 5b**(ii) shows a dark-field image of SNTs produced with four segments synthesized in the presence of templates created using four anodization steps. The total length of each SNT is held constant at 6.30 μm, whereas the lengths of the four segments composing the different nanotubes are varied. As seen in the image, each segment has a different diameter and can be used as an encoding bit. For example, in **Figure 5b**, S4A comprises segments with lengths (in micrometers) of 0.8/2.4/1.6/1.6, whereas S4B has 1.6/1.6/1.6/1.6 and S4C has 2.4/0.8/1.6/1.6. From these ratios, it is possible to identify the different SNTs. Different sensing chemistries can be conjugated to the different SNTs by the use of well-established silica chemistry.

Both the gold and silver striped metal rods and SNTs provide an approach to optically encode particles containing sensing chemistry. Although time consuming, it is possible to produce large amounts of particles in these preparations. For example, in

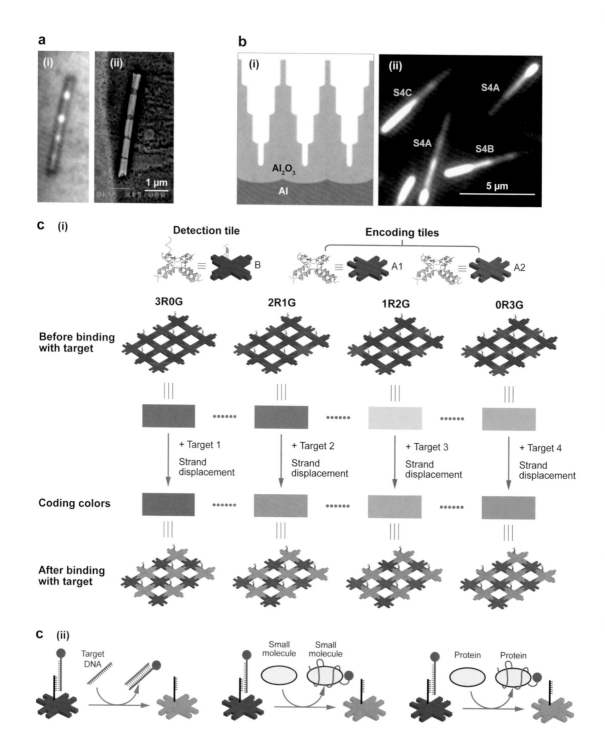

the case of the SNTs, each preparation contains enough particles to perform 100,000 immunoassays (60).

Lin et al. (61) recently created a fluorescently encoded suspension array by self-assembly using DNA as building tiles. Each building tile possessed its own unique fluorescent dye. By using two building tiles in different ratios, along with a third detection tile, they formed a fluorescent suspension array (**Figure 5c**). The detection tile housed a third fluorescent dye. This dye was attached to a DNA sequence that hybridized to a DNA strand on the detection tile. This hybridization was used in the detection scheme through a strand-displacement method. The binding of the dye-modified DNA to the target analyte, either DNA or a small molecule, is energetically favorable compared with its binding to the detection tile (61). Once the dye-labeled DNA is bound to the target, it is no longer associated with the detection tile (**Figure 5c**). A positive detection is measured as the decrease in fluorescence of the detection nanotile. Owing to the decrease in the fluorescent signal with a positive detection, these types of probes are known as turn-off probes.

The Doyle group (56) recently developed a microfluidic platform for the production of polymer microparticles that house both sensing chemistry and optical bar codes. The particles have two or three different regions all produced at the same time by photopolymerization inside a microfluidic channel (**Figure 6a**). The bar code region consists of orientation indicators (long stripes), as well as coding elements (squares) (**Figure 6b**). The analyte detection region is polymerized such that it has sensing chemistry attached to the polymer backbone. The researchers used a flow-through particle reader to read the individual particles after incubation with a fluorescently labeled DNA target. In principle, the dot-coding system (**Figure 6b**) can hold 2^{20} different bar codes. **Figure 6c** shows the results of a duplexed experiment using dye-labeled target DNA. This experiment demonstrated that the bar code region of the microparticles remains unchanged over a detection event, and only the region containing the complementary DNA strand increases in fluorescence when a positive detection event occurs (56).

3.3.2. Fixed arrays.
Other arrays have their sensing chemistries at specifically defined locations on a solid support; such arrays are known as fixed arrays. The different sensing chemistries can be physically placed at a desired spot on the solid support or

---

Figure 5

(*a*) Reflection (*i*) and field emission–scanning electron microscopy (*ii*) images of a gold-silver multistripe particle. The narrow, dark stripes in part *ii* are silver stripes of (*from top to bottom*) 240, 170, 110, and 60 nm. **Figure 5a** reprinted from Reference 59 with permission from AAAS. (*b*) Al_2O_3 template after third pore widening and fourth anodization (*i*) along with the resulting silica nanotubes under dark-field illumination (*ii*). **Figure 5b** reprinted from Reference 60, copyright 2007, American Chemical Society. (*c*) Schematic production of self-assembled encoded DNA nanotile suspension array components (*i*) and the corresponding detection events for DNA, small molecules, and proteins (*ii*, *left to right*) showing the turn-off probe style of detection. **Figure 5c** reprinted from Reference 61. Copyright 2007, American Chemical Society.

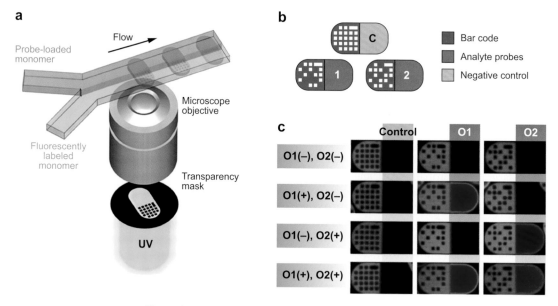

Figure 6

(*a*) Microfluidic platform for the production of encoded polymer microparticles that house both bar coding and detection regions synthesized by photopolymerization. (*b*) Schematic of control and sensing microparticles. (*c*) Fluorescent images of (*from top to bottom*) microparticles after hybridization with no target, the complement to 01, the complement to 02, and the complements to both 01 and 02. Figure taken from Reference 56. Reprinted with permission from AAAS.

by randomly distributing microparticles onto a preformed array of microwells. Arrays in which the researcher has control over the location of particular sensing molecules are known as directed arrays, whereas the arrays in which the location of sensing molecules must be determined after production of the array are known as randomly ordered arrays.

3.3.2.1. Directed arrays. Directed arrays were first developed in the mid-1990s by the Fodor (62–64) and Brown (65) groups. These early arrays were developed primarily for gene screening (63, 65). The Brown group, for example, was able to utilize a two-color fluorescence readout of 45 different genes simultaneously (65). A directed array is produced when the various sensing molecules are placed in a specific location on a solid support. This placement (66) can be done by contact printing (65, 67), noncontact printing (67, 68), or photolithography (64). Each position on directed arrays is specific for a particular analyte. These positions are fixed, and signals therefore can be attributed to the proper analyte by the location at which they occur. An important parameter for array technologies is the density of sensing spots per area as it is a major factor for controlling the number of analytes that can be measured in a given sample volume. The density is largely controlled by the feature size of the sensing spots. A spot size of 50 μm has been realized using commercially available pin printers, whereas photolithography (69) and photomasking techniques (70) have enabled the spot size to decrease to 10–18 μm. Recently, Bright and coworkers (71) demonstrated the ability to print array features of 9 μm by the use of a quartz pin printer, thus increasing the probe density possible for printed arrays.

3.3.2.2. Randomly ordered arrays. Randomly ordered arrays use self-assembly techniques to create arrays of microspheres (also called beads) in random locations in an ordered substrate. The substrate is prepared with micro- or nanosized wells that can be loaded with micro- or nanospheres matched to the well size and that contain the sensing chemistry for a particular analyte (72). One can produce the wells by etching a fiber-optic bundle in acidic solution (73) or by a breath-figure method (74). In the fiber-optic approach, a fiber-optic bundle is etched in acidic solution. The cladding of the optical fiber etches at a slower rate than the optical-fiber core. This differential etching produces wells at the end of the fiber bundle (73, 75, 76). The depth of the wells is dependent on the length of time and the concentration of the acid used. The breath-figure method forms an ordered array of pores by a three-step process (77). First, a solid support (such as a glass slide) is covered with a thin layer of a polymer dissolved in a hydrophobic solvent. This solvent layer is exposed to air with high humidity (~50%). As water droplets form on the surface of the hydrophobic solvent layer, the temperature of the surface and the water droplets equilibrates, and the water begins to sink into the hydrophobic solvent owing to its higher density. Ordered arrays are formed when both the water and hydrophobic solvent evaporate (74, 77).

To create a randomly ordered array, one needs to load microspheres or nanospheres into the well arrays. A typical procedure for loading microspheres into microwells involves a self-assembly method in which an aqueous solution containing microspheres is placed on the well array surface and allowed to evaporate. Capillary forces cause the beads to assemble into the wells, presumably to minimize surface free energy. Each bead contains sensing chemistry that is either specific to a particular analyte or reacts with many analytes. Because the array is assembled in a random fashion and contains beads with different specificities, it is necessary to determine the location of the different bead types. There are two general approaches for determining bead location—bead encoding and bead decoding. Encoding is accomplished using a number of approaches, including the incorporation of dyes in the beads by solvent swelling (72, 76), by attaching different ratios of different quantum dots (78), or by producing microspheres made from polymers containing Raman-active functional groups (known as bar coded resins) (79). In bead decoding, the different bead types are interrogated sequentially by exposing the beads to solutions containing molecules with attached labels that bind specifically to each bead type. For example, oligonucleotide arrays can be decoded by exposing them to pools of dye-labeled oligonucleotides (80) (**Figure 7**). All these encoding or decoding approaches result in the ability to distinguish the identity of each individual microsphere. Through this randomly ordered approach, investigators have realized array densities as high as 4.5×10^6 array elements mm^{-2} using 300-nm-diameter spheres with 500-nm separation (81).

3.3.3. Applications of arrays. Arrays have been extensively used as detection tools for biologically relevant species. Arrays have been produced that are capable of detecting cells, proteins, and nucleic acids. There has also been a considerable interest in the use of cross-reactive sensors for air and water monitoring.

3.3.3.1. Detection of biologically relevant analytes. The area that has benefited the most from advances in fluorescence-based multiplexed detections has been the biological sciences (8, 82). Detection of whole cells, proteins, RNA, and DNA can all

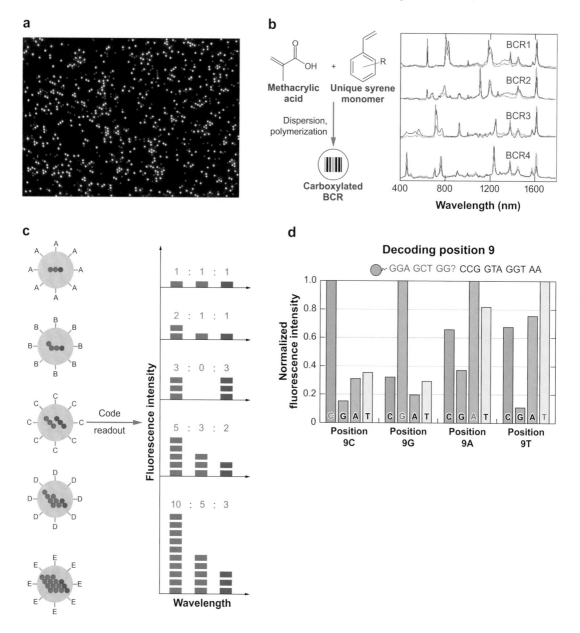

be accomplished using a microarray platform with fluorescence detection as the predominant signal transduction scheme, although other schemes have been employed. This review does not cover the large body of work enabled by commercially available microarrays used for genotyping and gene-expression analysis even though most such platforms are based on fluorescence detection of tens of thousands to millions of array features. Instead, we focus on a number of diverse and lower-density applications of multiplexed detections using spectroscopic techniques.

Investigators have detected cells (83–85) and oligonucleotides (86) using a sandwich assay format, whereby the target analyte is bound by an immobilized capture probe on the array surface and detected using a fluorescently labeled secondary probe. For increased sensitivity, polymerase chain reaction amplification has been used to amplify and label target oligonucleotides (87); the products have then been hybridized to microarrays (88–90).

Antibodies have been used with array platforms to detect proteins (55, 91, 92) and cells (55, 91, 93–95) in a multiplexed format. Recently, Fernandez-Calvo et al. (95) developed a competitive immunoassay for use in astrobiology for the detection and identification of analytes ranging from small compounds, such as naphthalene, to whole cells. This assay is performed by creating a competition for binding sites between antigens and the contents of the sample of interest. First, researchers allowed antigens used as competitors to bind to antibody tracers. They then allowed these complexes to bind to the array. Upon sample introduction, the antigens in the sample compete for the antibody tracers. After a rinse step, a fluorescently labeled protein binds to the tracers that remain on the array. The presence of the analyte of interest in the sample results in a lower fluorescent signal on the array (95).

The detection of whole cells on an antibody array has been used to detect biological threat agents (93, 94, 96). For the detection of *Salmonella typhimurium* (96) in irrigation water used for sprouting seeds, investigators used an optical-fiber-based technique to accomplish a normally labor-intensive procedure much faster (20 min) than typical microbiological methods that require a 48-h or longer culturing. The speed, accuracy, and simplicity of optical array-based techniques may accomplish the goal of "detect to protect" human populations from biological attacks (55, 94).

The ability to identify and quantify pathogens by their specific DNA or RNA sequences has also been exploited to detect food-borne pathogens (88, 97), biowarfare

---

Figure 7

(*a*) Fluorescent image of europium (fluorescent-dye) encoded beads of 0.1 M (*dim spheres*) and 0.5 M (*bright spheres*). **Figure 7a** reprinted with permission from Reference 92. Copyright 2006, Elsevier. (*b*) Production of bar coded resins (BCRs) and corresponding Raman spectra of four different BCRs. **Figure 7b** reprinted from Reference 79, copyright 2007, American Chemical Society. (*c*) Theoretical code readout of nanoparticle-encoded beads showing the ability to multiplex by varying the ratio of different quantum dots. **Figure 7c** reprinted by permission from Reference 78. Copyright 2001, Macmillan Publishers Ltd. (*d*) Normalized experimental results for decoding a position along the DNA sequence of many microspheres by a combinatorial approach using short oligonucleotides. **Figure 7d** reprinted from Reference 80. Copyright 2003, American Chemical Society.

agents (90, 98), and respiratory pathogens (61); to monitor harmful algae blooms in coastal waters (89); and to compare and contrast biological communities in natural samples (soil, river sediments, and marine sediments) (4). The use of a sandwich assay for DNA detection allows an extra level of selectivity, as both the capture probe and the labeled probe must bind to the target, thereby reducing the amount of nonspecific binding that occurs. Multiplexed detection schemes afford the capability to simultaneously detect many different DNA strands and to use multiple probes per organism. Through the use of two oligonucleotide probes for *Salmonella* spp., Ahn & Walt (88) employed a fiber-optic-bundle array to obtain a high level of discrimination between the target organisms and other related organisms with high sequence similarity that could have caused false-positive signals. These experiments were performed using a sandwich assay with a fluorescently labeled signal probe. Using the same fiber-optic-bundle platform but with a single capture probe sequence per organism, Ahn et al. (89) simultaneously detected three toxin-producing algae species with a high degree of specificity. These two studies demonstrate the ability to use multiplexed detection schemes both to rule out false positives by monitoring multiple probes per organism and to detect multiple species simultaneously.

Recently, Heath and coworkers (99) described how spotted DNA microarrays could be used as a platform to simultaneously detect DNA, proteins, and cell populations (**Figure 8**). They employed a variety of transduction schemes, but fluorescence detection could be used for all analytes. DNA detection on this platform was accomplished using a sandwich assay as described above. For proteins, antibodies were attached to DNA complementary to oligonucleotides on the microarray, thus forming a nucleic acid tether. This tether enabled the antibodies to be immobilized on the

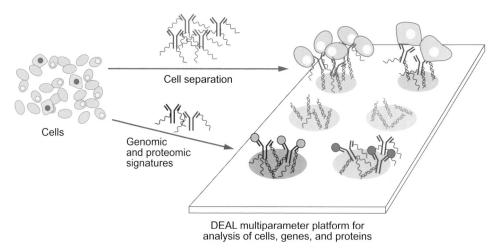

DEAL multiparameter platform for
analysis of cells, genes, and proteins

Figure 8

Illustration showing how a DNA microarray can be multiplexed to simultaneously detect cells, DNA, and proteins on the same platform. The immobilized oligonucleotides were used as both capture probes and tethers for antibodies. Figure reprinted from Reference 99. Copyright 2007, American Chemical Society.

DNA array, converting it to an antibody array. The antibodies are able to bind to the analytes of interest, such as proteins (99, 100) or cells (**Figure 8**) (99). In addition to the extremely high multiplexing capability, there are many advantages to producing arrays in this way. DNA microarrays are stable over much longer periods of time than protein and antibody arrays, and by keeping the antibodies in solution until use, one can maintain their binding activity. From a production perspective, the DNA surface can be used as a universal platform for any protein detection as it is only necessary to modify the proteins with the tethering oligonucleotide (100). This universal system capable of simultaneously detecting cells, DNA, RNA, and proteins is a major step forward toward using multiplexed analyte detection for diagnostics of cancer and cardiovascular disease (101).

Researchers have also found increased use for arrays in screening small molecules as inhibitors of RNA interactions. This screening is important in the search for new antibiotics and for probing the side effects of current ones, such as streptomycin. Typically the methods used for screening these interactions have been low throughput and have required large amounts of sample. Multiplexing technologies have been applied to simultaneously probe a large number of RNA sequences or small molecules, while taking advantage of the small sample requirement seen when using microarrays. Small-molecule microarrays (102) have been used to find new ligands (103) and protein inhibitors (104), as well as to determine the specificity of a protease (105) or to detect kinases (106). An important limitation of these arrays is the method used to immobilize the small molecules to the array as the attachment may alter their biochemical function. To remove this limitation, Liang et al. (107) designed an experiment to probe the ability of a single small molecule to inhibit RNA binding to many sequences on a DNA microarray. Although only one small molecule could be interrogated at a time, this method has the benefit of being able to screen multiple RNA hybridizations simultaneously (**Figure 9a**). A decrease in fluorescence signal (**Figure 9b**) of particular sequences on the array in the presence of neomycin or streptomycin when compared with a control experiment lacking these molecules indicated binding of the small molecule to the RNA strand, thereby inhibiting its binding to the DNA on the microarray. These array applications demonstrate the flexibility of DNA arrays because the signals from these arrays detect the inhibition of binding owing to small molecules as compared with the studies discussed above in which they were used to monitor the presence and concentration of DNA.

Although fluorescence-based arrays have dominated the literature, SPR (11) has also recently been utilized for multiplexed detections (108, 109). SPR arrays have been produced by the vapor deposition of thin layers of silver onto glass slides covered with a monolayer of silica nanospheres. Investigators followed silver deposition with sonication to remove the nanospheres, thus resulting in slides consisting of triangular silver nanoparticle arrays formed between the nanospheres prior to removal (**Figure 10**) (110). As proof of concept, the investigators printed a 2×1 carbohydrate-sensing array on this type of silver nanoparticle substrate to simultaneously monitor the binding of concanavalin A to mannose and galactose (110).

Endo et al. (111) accomplished higher levels of SPR multiplexing through the use of core-shell structured nanoparticle layers. As seen in **Figure 11**, they produced

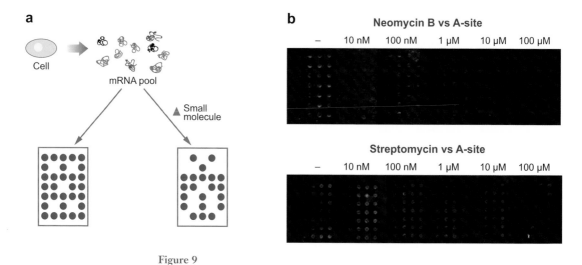

Figure 9

(*a*) Schematic of the method used for determining small-molecule interference of RNA binding and (*b*) experimental results from inhibition experiments performed with neomycin B and streptomycin ranging from 0 nM to 100 μM. Figure taken from Reference 107. Copyright 2006, National Academy of Sciences, U.S.A.

an array consisting of two layers of gold separated by silica nanoparticles, and the sensing chemistry was deposited by a nanoliter-dispensing instrument to form the printed array. White light was coupled down an optical fiber, and the reflected light was collected by the detection fiber in the same optical-fiber bundle. As shown in **Figure 11***a*, the absorbance of the array increased at ~550 nm as antigen binding occurred. Through this setup, the authors observed detection limits of 100 pg ml^{-1} for all targeted proteins with a range of detection spanning between 6 and 12 orders of magnitude (111). SPR does not require a secondary probe, but the signal can be amplified if one is used. For example, through use of an enzyme-labeled indicator, an RNA microarray demonstrated a picomolar detection limit (112).

3.3.3.2. Analyte detection using cross-reactive arrays. Another approach to performing multi-analyte sensing is to employ cross-reactive sensor arrays. In this approach, the different sensors in the array are not specific to a particular analyte but react with multiple analytes. The response pattern is unique to each analyte and provides the requisite specificity (113). This approach is based loosely on principles of the olfactory system and has been referred to as electronic or artificial noses for vapor sensing (75, 114) or electronic tongues for solution sensing (115). Both systems work by using cross-reactive sensors that respond differentially to various analytes (75). The sensor-array responses provide a combinatorial code that is analyzed with pattern-recognition algorithms, enabling the discrimination of more analytes than there are sensors. Both the electronic tongue and nose elucidate the identity of an unknown analyte or mixture by comparing the array responses to those from a library of responses obtained from known samples.

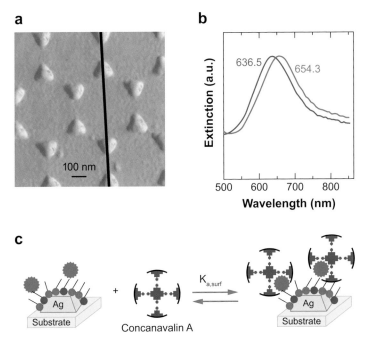

a

100 nm

b

636.5 654.3

Extinction (a.u.)

500 600 700 800

Wavelength (nm)

c

Ag
Substrate

+ Concanavalin A

$K_{a,surf}$

Ag
Substrate

Figure 10

(*a*) Tapping-mode atomic force microscopy image of a silver biosensor produced by the vapor deposition of silver onto a glass slide after the removal of the nanosphere monolayer. (*b*) Surface plasmon resonance spectra of silver nanoparticles after mannose modification ($\lambda_{max} = 636.5$ nm) and after binding of concanavalin A ($\lambda_{max} = 654.3$ nm) under nitrogen. (*c*) Scheme for concanavalin A detection. The green circles represent mannose, the pink circles represent alcohol-terminated self-assembled monolayers, and the blue circles represent self-assembled monolayers terminating in mannose. Figure reprinted from Reference 110. Copyright 2004, American Chemical Society.

There are two aspects of these types of sensors that directly affect their ability to provide accurate identification of analytes. The first is the reproducibility of a response from the sensors to a given analyte; the second lies in the interpretation of the obtained responses. Many types of pattern recognition techniques, such as principal component analysis (116, 117), K-nearest neighbor (118), and discriminant analysis (119, 120) are widely used for interpreting data. However, this list is far from exhaustive and there are ongoing efforts to create pattern recognition methods that will result in better classification (121). Although there are many different electronic-nose formats (10), only a few are based on multiplexed spectral detection.

The first optical nose was developed in our laboratory using solvatochromic dyes entrapped in different polymers on the ends of different optical fibers (75, 122). The more modern platform for these arrays is based on the randomly ordered microsphere arrays discussed above, with the microspheres containing different vapor-sensitive chemistries. Researchers can form arrays by using thousands of cross-reactive fluorescent microspheres deposited in wells (113, 123) or by printing dyes on silica

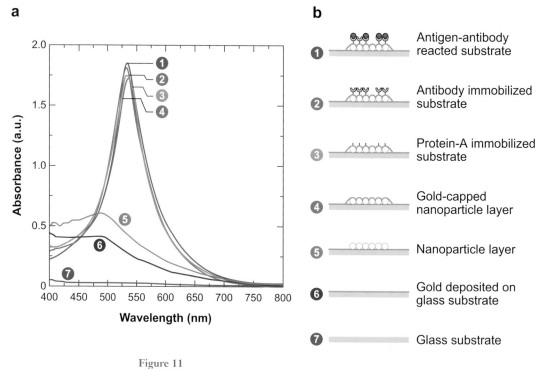

a

Absorbance (a.u.) vs Wavelength (nm)

b

① Antigen-antibody reacted substrate

② Antibody immobilized substrate

③ Protein-A immobilized substrate

④ Gold-capped nanoparticle layer

⑤ Nanoparticle layer

⑥ Gold deposited on glass substrate

⑦ Glass substrate

Figure 11

(*a*) Sequential absorbance spectra of the surface plasmon resonance biosensor as it is (*b*) formed from a glass substrate (*bottom panel*) to the binding of an analyte of interest (*top panel*). Figure reprinted from Reference 111. Copyright 2006, American Chemical Society.

thin-layer chromatography plates (124). Both these systems rely on the reproducible responses upon exposure to a particular vapor to identify analytes of interest. The printed arrays use a wide range of dyes, including metalloporphyrins, pH indicator dyes, and solvatochromic dyes (125). The diversity in the dye types enables these cross-reactive arrays to exhibit a high degree of sensitivity and selectivity to a wide range of analytes (125). The microsphere-based system monitors the change in fluorescence intensity over time and/or the emission wavelength maxima of each individual microsphere. These arrays have been able to identify complex vapor mixtures, such as various types of coffee-bean odors (126), as well as explosive vapors (127), with a high level of success. Perhaps the biggest advantage of these types of arrays is their ability to identify mixtures that they have been trained to detect (128). The level of confidence in a measurement is an important parameter for electronic noses as the number of false positives must be minimized if they are to be deployed for health and safety applications, such as for disease detection using breath samples or for explosives or chemical-agent detection in subway stations and airports. The flow environment, such as inside a model nasal cavity, can enhance vapor discrimination (129). Maintaining the long-term stability of electronic noses is also important if such systems are to find widespread applications for public safety. The longevity of a fluorescent sensor is largely limited by photobleaching of the indicator dyes. By limiting the power of the excitation source and only illuminating one subsection of the entire array at any given time, one could use a system for, theoretically, over 170,000 exposures. This increase in longevity represents an increase of three orders of magnitude compared with a traditional system (130).

Electronic tongues work in a similar manner to electronic noses, except they are designed for monitoring liquid environments (131–134). To demonstrate the ability of such arrays to differentiate between complex mixtures, researchers used an array consisting of eight colorimetric anion sensors immobilized in polyurethane distributed in drilled wells on a microscope slide to measure the concentrations of multiple anions. Several sensors were selective for fluoride and pyrophosphate, whereas others showed cross-reactivity for other anions. The array responses were able to distinguish between eight different types of toothpastes in water (135). Responses from all eight toothpastes were first obtained to determine the characteristic responses from the electronic tongue unique to each toothpaste brand. By comparing these known responses with the responses obtained from unknown samples using pattern-recognition techniques (such as principal-component analysis), the electronic tongue identified the particular brands of toothpaste.

Cross-reactive optical sensors in conjunction with pattern-recognition techniques have also been used to detect and identify amino acids (136, 137). The amino acids were identified by indicator displacement assays. The receptors in the assay are non-covalently bound to fluorescent indicators. The addition of different amino acids results in different degrees of displacement of the indicators. Unique patterns for each amino acid can be established and used to identify an unknown sample through the use of multiple receptors and indicators. A solution-based system has been reported for the detection of proteins (138). In this system, six different gold nanoparticle surfaces are noncovalently coated with fluorescent polymer conjugates. When bound to the nanoparticles, the polymer fluorescence is quenched. The different nanoparticle preparations are placed in different solutions. When proteins are added to these solutions, the proteins compete for the different surfaces and displace the fluorescent polymers to different extents, causing the solutions to recover their fluorescence. Linear discriminant analysis was used to process the response patterns for the array of solutions to determine which protein was present. These polymer-nanoparticle hybrids were able to identify and quantify seven different proteins individually and in mixtures using the fluorescent response patterns.

4. CONCLUSION

In this review, we discuss the present state-of-the-art techniques for producing platforms capable of multiplexed spectroscopic detection. We highlight the production and uses of directed, randomly ordered, and suspension arrays, especially for biologically relevant analytes. For suspension and random arrays, we discuss the various methods of particle encoding, such as fluorescent dyes, Raman-active microspheres, size and shape variations, and bar codes on microparticles. Finally, multiplexing techniques involving newer spectroscopies, such as SERS and SPR, are summarized.

The ability to continually monitor two analytes at the same location has been accomplished by dual-layer sensors, whereas the need to increase the number of analytes these systems can monitor has been made apparent. The use of cross-reactive sensors in an array format is a promising tool for monitoring air and water samples. Fluorescence-based DNA microarrays have been used for a wide range of applications, including detecting and identifying cell types by their DNA or RNA, producing antibody arrays, and probing small-molecule interactions with RNA. Multiplexed SERS and SPR detections, although not as prevalent as fluorescence systems, show promise toward becoming useful tools in medical diagnostics. As the development of new methods for producing multiplexed systems is realized and as more analytes are able to be detected, the use of multiplexed detection systems will become more useful both in the laboratory setting and in a variety of commercial applications.

SUMMARY POINTS

1. Multiplexed spectroscopic detection platforms have been used to measure thousands of analytes simultaneously.

2. Detections can be direct, in which spectroscopic signatures of the analytes are measured, or indirect, in which the measured spectral response results from the analyte reacting with an indicator molecule.

3. Dual-layer sensors continually monitor gases, both in solution and in air, using fluorescent dyes that undergo a reversible reaction with the analyte gases. These reactions change the fluorescence intensity, lifetime, or wavelength of the dye that can be correlated to analyte concentration.

4. Surface-enhanced Raman tags can be employed to simultaneously detect multiple analytes in the same location owing to the tags' narrow spectral bandwidths.

5. Multiple particles, such as microspheres carrying different sensing chemistries, must be encoded to identify the sensing chemistry on each particle.

6. SPR peaks of metallic nanoparticles can be used for both encoding and detecting lab-in-a-tube type multiplexed analysis.

7. Arrays have been used extensively for biologically relevant analytes, including DNA, RNA, proteins, and cells. The ability to detect all these analytes on the same platform simultaneously has recently been demonstrated.

8. Cross-reactive sensors can be employed along with pattern-recognition software to detect multiple analytes. These types of cross-reactive sensors are known as artificial noses and artificial tongues. These sensors must first be calibrated with known samples to obtain the characteristic pattern for each analyte of interest.

FUTURE ISSUES

1. There is a need to create higher-density arrays to enable smaller sample sizes and increase the number of analytes that can be detected.

2. There is a need to produce integrated systems for nonexperts, including arrays, sample and fluidic handling, and readout instruments that can perform analyses quickly and accurately.

3. The degree of multiplexing of dual-layer sensors must be increased to enable wider application of these systems.

4. The level of confidence in the identification of harmful vapors should be increased—as false positives and false negatives are highly disruptive and costly—before these systems can be used in airports, subway stations, or other public spaces where large numbers of people could be affected.

5. SPR and SERS probes should be further developed for in vivo and in vitro detections.

DISCLOSURE STATEMENT

David R. Walt is the Scientific Founder and a Director of Illumina, Inc., a manufacturer of microarrays using the random array technology discussed in this paper. All of the work reported in this paper is the result of research carried out in the investigator's laboratory at Tufts University.

ACKNOWLEDGEMENTS

The authors acknowledge support from the National Science Foundation (CHE-0518293).

LITERATURE CITED

1. van Dam RM, Quake SR. 2002. Gene expression analysis with universal n-mer arrays. *Genome Res.* 12:145–52

2. Hardenbol P, Yu FL, Belmont J, MacKenzie J, Bruckner C, et al. 2005. Highly multiplexed molecular inversion probe genotyping: over 10,000 targeted SNPs genotyped in a single tube assay. *Genome Res.* 15:269–75

3. Bartosiewicz M. 2001. Applications of gene assays in environmental toxicology: fingerprints of gene regulation associated with cadmium chloride, benzo(*a*)pyrene, and trichloroethylene. *Environ. Health Perspect.* 109:71–74

4. Wu LY, Thompson DK, Liu XD, Fields MW, Bagwell CE, et al. 2004. Development and evaluation of microarray-based whole-genome hybridization for detection of microorganisms within the context of environmental applications. *Environ. Sci. Technol.* 38:6775–82

5. Phillips TE, Bargeron CB, Benson RC, Carlson MA, Fraser AB, et al. 2002. Development of an automated handheld immunoaffinity fluorometric biosensor. *Proc. SPIE Int. Soc. Opt. Eng.* 3913:186–92

6. Cullum BM, Mobley J, Chi ZH, Stokes DL, Miller GH, Vo-Dinh T. 2000. Development of a compact, handheld Raman instrument with no moving parts for use in field analysis. *Rev. Sci. Instrum.* 71:1602–7

7. Kunimura S, Kawai J. 2007. Portable total reflection X-ray fluorescence spectrometer for nanogram Cr detection limit. *Anal. Chem.* 79:2593–95

8. Epstein JR, Biran I, Walt DR. 2002. Fluorescence-based nucleic acid detection and microarrays. *Anal. Chim. Acta* 469:3–36

9. Waggoner A. 2006. Fluorescent labels for proteomics and genomics. *Curr. Opin. Chem. Biol.* 10:62–66

10. Albert KJ, Lewis NS, Schauer CL, Sotzing GA, Stitzel SE, et al. 2000. Cross-reactive chemical sensor arrays. *Chem. Rev.* 100:2595–626

11. Willets KA, Van Duyne RP. 2007. Localized surface plasmon resonance spectroscopy and sensing. *Annu. Rev. Phys. Chem.* 58:267–97

12. Szaloki I, Osan J, Van Grieken RE. 2006. X-ray spectrometry. *Anal. Chem.* 78:4069–96

13. Haynes CL, Yonzon CR, Zhang XY, Van Duyne RP. 2005. Surface-enhanced Raman sensors: early history and the development of sensors for quantitative biowarfare agent and glucose detection. *J. Raman Spectrosc.* 36:471–84

14. Vandenabeele P, Moens L, Edwards HGM, Dams R. 2000. Raman spectroscopic database of azo pigments and application to modern art studies. *J. Raman Spectrosc.* 31:509–17

15. Whitney AV, Van Duyne RP, Casadio F. 2006. An innovative surface-enhanced Raman spectroscopy (SERS) method for the identification of six historical red lakes and dyestuffs. *J. Raman Spectrosc.* 37:993–1002

16. Shah NC, Lyandres O, Walsh JT, Glucksberg MR, Van Duyne RP. 2007. Lactate and sequential lactate-glucose sensing using surface-enhanced Raman spectroscopy. *Anal. Chem.* 79:6927–32

17. Stuart DA, Biggs KB, Van Duyne RP. 2006. Surface-enhanced Raman spectroscopy of half-mustard agent. *Analyst* 131:568–72

18. Zhang XY, Young MA, Lyandres O, Van Duyne RP. 2005. Rapid detection of an anthrax biomarker by surface-enhanced Raman spectroscopy. *J. Am. Chem. Soc.* 127:4484–89

19. Kalasinsky KS, Hadfield T, Shea AA, Kalasinsky VF, Nelson MP, et al. 2007. Raman chemical imaging spectroscopy reagentless detection and identification of pathogens: signature development and evaluation. *Anal. Chem.* 79:2658–73

20. Eba H, Numako C, Iihara J, Sakurai K. 2000. Trace chemical characterization using monochromatic X-ray undulator radiation. *Anal. Chem.* 72:2613–17

21. Sakurai K, Eba H, Inoue K, Yagi N. 2002. Wavelength-dispersive total-reflection X-ray fluorescence with an efficient Johansson spectrometer and an undulator X-ray source: detection of 10^{-16} g-level trace metals. *Anal. Chem.* 74:4532–35

22. Vincze L, Vekemans B, Brenker FE, Falkenberg G, Rickers K, et al. 2004. Three-dimensional trace element analysis by confocal X-ray microfluorescence imaging. *Anal. Chem.* 76:6786–91

23. Van Meel K, Smekens A, Behets M, Kazandjian P, Van Grieken R. 2007. Determination of platinum, palladium, and rhodium in automotive catalysts using high-energy secondary target X-ray fluorescence spectrometry. *Anal. Chem.* 79:6383–89

24. Chwiej J, Fik-Mazgaj K, Szczerbowska-Boruchowska M, Lankosz M, Ostachowicz J, et al. 2005. Classification of nerve cells from substantia nigra of patients with Parkinson's disease and amyotrophic lateral sclerosis with the use of X-ray fluorescence microscopy and multivariate methods. *Anal. Chem.* 77:2895–900

25. Banas K, Jasinski A, Banas AM, Gajda M, Dyduch G, et al. 2007. Application of linear discriminant analysis in prostate cancer research by synchrotron radiation-induced X-ray emission. *Anal. Chem.* 79:6670–74

26. Miller LM, Wang Q, Smith RJ, Zhong H, Elliott D, Warren J. 2007. A new sample substrate for imaging and correlating organic and trace metal composition in biological cells and tissues. *Anal. Bioanal. Chem.* 387:1705–15

27. Sanger F, Nicklen S, Coulson AR. 1977. DNA sequencing with chain-terminating inhibitors. *Proc. Natl. Acad. Sci.* 74:5463–67

28. Schroeder CR, Neurauter G, Klimant I. 2007. Luminescent dual sensor for time-resolved imaging of pCO_2 and pO_2 in aquatic systems. *Microchim. Acta* 158:205–18

29. Nivens DA, Schiza MV, Angel SM. 2002. Multilayer sol-gel membranes for optical sensing applications: single layer pH and dual layer CO_2 and NH_3 sensors. *Talanta* 58:543–50

30. Bambot S, Holavanahali R, Lakowicz JR, Carter GM, Rao G. 1994. Optical oxygen sensor using fluorescence lifetime measurement. *Adv. Exp. Med. Biol.* 361:197–205

31. Wolfbeis OS, Weis LJ, Leiner MJP, Ziegler WE. 1988. Fiber-optic fluorosensor for oxygen and carbon dioxide. *Anal. Chem.* 60:2028–30

32. Borisov SM, Krause C, Arain S, Wolfbeis OS. 2006. Composite material for simultaneous and contactless luminescent sensing and imaging of oxygen and carbon dioxide. *Adv. Mater.* 18:1511–16

33. Huber C, Klimant I, Krause C, Wolfbeis OS. 2001. Dual lifetime referencing as applied to a chloride optical sensor. *Anal. Chem.* 73:2097–103

34. Takato K, Gokan N, Kaneko M. 2005. Effect of humidity on photoluminescence from Ru(bpy)(3)(2+) incorporated into a polysaccharide solid film and its application to optical humidity sensor. *J. Photochem. Photobiol. A* 169:109–14

35. Schroder CR, Polerecky L, Klimant I. 2007. Time-resolved pH/pO_2 mapping with luminescent hybrid sensors. *Anal. Chem.* 79:60–70

36. Lee SM, Chung WY, Kim JK, Suh DH. 2004. A novel fluorescence temperature sensor based on a surfactant-free PVA/borax/2-naphthol hydrogel network system. *J. Appl. Polym. Sci.* 93:2114–18

37. Nagl S, Wolfbeis OS. 2007. Optical multiple chemical sensing: status and current challenges. *Analyst* 132:507–11

38. Lewis EK, Haaland WC, Nguyen F, Heller DA, Allen MJ, et al. 2005. Color-blind fluorescence detection for four-color DNA sequencing. *Proc. Natl. Acad. Sci.* 102:5346–51

39. Ju J, Kim DH, Bi L, Meng Q, Bai X, et al. 2006. Four-color DNA sequencing by synthesis using cleavable fluorescent nucleotide reversible terminators. *Proc. Natl. Acad. Sci.* 103:19635–40

40. Cao YWC, Jin RC, Mirkin CA. 2002. Nanoparticles with Raman spectroscopic fingerprints for DNA and RNA detection. *Science* 297:1536–40

41. Ni J, Lipert RJ, Dawson GB, Porter MD. 1999. Immunoassay readout method using extrinsic Raman labels adsorbed on immunogold colloids. *Anal. Chem.* 71:4903–8

42. Graham D, Mallinder BJ, Whitcombe D, Watson ND, Smith WE. 2002. Simple multiplex genotyping by surface-enhanced resonance Raman scattering. *Anal. Chem.* 74:1069–74

43. Sun L, Yu CX, Irudayaraj J. 2007. Surface-enhanced Raman scattering based nonfluorescent probe for multiplex DNA detection. *Anal. Chem.* 79:3981–88

44. Kim JH, Kim JS, Choi H, Lee SM, Jun BH, et al. 2006. Nanoparticle probes with surface enhanced Raman spectroscopic tags for cellular cancer targeting. *Anal. Chem.* 78:6967–73

45. Sun L, Sung KB, Dentinger C, Lutz B, Nguyen L, et al. 2007. Composite organic-inorganic nanoparticles as Raman labels for tissue analysis. *Nano Lett.* 7:351–56

46. Su X, Zhang J, Sun L, Koo TW, Chan S, et al. 2005. Composite organic-inorganic nanoparticles (COINs) with chemically encoded optical signatures. *Nano Lett.* 5:49–54

47. Perez-Juste J, Pastoriza-Santos I, Liz-Marzan LM, Mulvaney P. 2005. Gold nanorods: synthesis, characterization and applications. *Coord. Chem. Rev.* 249:1870–901

48. Liz-Marzan LM. 2006. Tailoring surface plasmons through the morphology and assembly of metal nanoparticles. *Langmuir* 22:32–41

49. Bukasov R, Shumaker-Parry JS. 2007. Highly tunable infrared extinction properties of gold nanocrescents. *Nano Lett.* 7:1113–18

50. El-Sayed MA. 2001. Some interesting properties of metals confined in time and nanometer space of different shapes. *Acc. Chem. Res.* 34:257–64

51. Oldenburg SJ, Averitt RD, Westcott SL, Halas NJ. 1998. Nanoengineering of optical resonances. *Chem. Phys. Lett.* 288:243–47

52. Yu CX, Irudayaraj J. 2007. Multiplex biosensor using gold nanorods. *Anal. Chem.* 79:572–79

53. Ruach-Nir I, Bendikov TA, Doron-Mor I, Barkay Z, Vaskevich A, Rubinstein I. 2007. Silica-stabilized gold island films for transmission localized surface plasmon sensing. *J. Am. Chem. Soc.* 129:84–92

54. Fulton RJ, McDade RL, Smith PL, Kienker LJ, Kettman JR. 1997. Advanced multiplexed analysis with the Flowmetrix™ system. *Clin. Chem.* 43:1749–56

55. McBride MT, Gammon S, Pitesky M, O'Brien TW, Smith T, et al. 2003. Multiplexed liquid arrays for simultaneous detection of simulants of biological warfare agents. *Anal. Chem.* 75:1924–30

56. Pregibon DC, Toner M, Doyle PS. 2007. Multifunctional encoded particles for high-throughput biomolecule analysis. *Science* 315:1393–96

57. Raschke G, Kowarik S, Franzl T, Sonnichsen C, Klar TA, et al. 2003. Biomolecular recognition based on single gold nanoparticle light scattering. *Nano Lett.* 3:935–38

58. Egner BJ, Sunil R, Smith H, Bouloc N, Frey JG, et al. 1997. Tagging in combinatorial chemistry: the use of coloured and fluorescent beads. *Chem. Commun.* 8:735–36

59. Nicewarner-Pena SR, Freeman RG, Reiss BD, He L, Pena DJ, et al. 2001. Submicrometer metallic barcodes. *Science* 294:137–41

60. He B, Son SJ, Lee SB. 2007. Suspension array with shape-coded silica nanotubes for multiplexed immunoassays. *Anal. Chem.* 79:5257–63

61. Lin CX, Liu Y, Yan H. 2007. Self-assembled combinatorial encoding nanoarrays for multiplexed biosensing. *Nano Lett.* 7:507–12

62. Fodor SPA, Rava RP, Huang XHC, Pease AC, Holmes CP, Adams CL. 1993. Multiplexed biochemical assays with biological chips. *Nature* 364:555–56

63. Lipshutz RJ, Morris D, Chee M, Hubbell E, Kozal MJ, et al. 1995. Using oligonucleotide probe arrays to access genetic diversity. *Biotechniques* 19:442–47

64. Pease AC, Solas D, Sullivan EJ, Cronin MT, Holmes CP, Fodor SPA. 1994. Light-generated oligonucleotide arrays for rapid DNA-sequence analysis. *Proc. Natl. Acad. Sci. USA* 91:5022–26

65. Schena M, Shalon D, Davis RW, Brown PO. 1995. Quantitative monitoring of gene-expression patterns with a complementary-DNA microarray. *Science* 270:467–70

66. Dufva M. 2005. Fabrication of high quality microarrays. *Biomol. Eng.* 22:173–84

67. Gutmann O, Niekrawietz R, Kuehlewein R, Steinert CP, Reinbold S, et al. 2004. Non-contact production of oligonucleotide microarrays using the highly integrated topspot nanoliter dispenser. *Analyst* 129:835–40

68. Okamoto T, Suzuki T, Yamamoto N. 2000. Microarray fabrication with covalent attachment of DNA using bubble jet technology. *Nat. Biotechnol.* 18:438–41

69. Hacia JG, Fan JB, Ryder O, Jin L, Edgemon K, et al. 1999. Determination of ancestral alleles for human single-nucleotide polymorphisms using high-density oligonucleotide arrays. *Nat. Genet.* 22:164–67

70. Nuwaysir EF, Huang W, Albert TJ, Singh J, Nuwaysir K, et al. 2002. Gene expression analysis using oligonucleotide arrays produced by maskless photolithography. *Genome Res.* 12:1749–55

71. Tehan EC, Higbee DJ, Wood TD, Bright FV. 2007. Tailored quartz pins for high-density microsensor array fabrication. *Anal. Chem.* 79:5429–34

72. Michael KL, Taylor LC, Schultz SL, Walt DR. 1998. Randomly ordered addressable high-density optical sensor arrays. *Anal. Chem.* 70:1242–48

73. Pantano P, Walt DR. 1996. Ordered nanowell arrays. *Chem. Mater.* 8:2832–35

74. Lu MH, Zhang Y. 2006. Microbead patterning on porous films with ordered arrays of pores. *Adv. Mater.* 18:3094–98

75. Dickinson TA, White J, Kauer JS, Walt DR. 1996. A chemical-detecting system based on a cross-reactive optical sensor array. *Nature* 382:697–700

76. Ferguson JA, Steemers FJ, Walt DR. 2000. High-density fiber-optic DNA random microsphere array. *Anal. Chem.* 72:5618–24

77. Karthaus O, Maruyama N, Cieren X, Shimomura M, Hasegawa H, Hashimoto T. 2000. Water-assisted formation of micrometer-size honeycomb patterns of polymers. *Langmuir* 16:6071–76

78. Han MY, Gao XH, Su JZ, Nie S. 2001. Quantum-dot-tagged microbeads for multiplexed optical coding of biomolecules. *Nat. Biotechnol.* 19:631–35

79. Raez J, Blais DR, Zhang Y, Alvarez-Puebla RA, Bravo-Vasquez JP, et al. 2007. Spectroscopically encoded microspheres for antigen biosensing. *Langmuir* 23:6482–85

80. Epstein JR, Ferguson JA, Lee KH, Walt DR. 2003. Combinatorial decoding: an approach for universal DNA array fabrication. *J. Am. Chem. Soc.* 125:13753–59

81. Tam JM, Song LN, Walt DR. 2005. Fabrication and optical characterization of imaging fiber-based nanoarrays. *Talanta* 67:498–502

82. Brogan KL, Walt DR. 2005. Optical fiber-based sensors: application to chemical biology. *Curr. Opin. Chem. Biol.* 9:494–500

83. Watts HJ, Lowe CR, Pollardknight DV. 1994. Optical biosensor for monitoring microbial cells. *Anal. Chem.* 66:2465–70

84. Rowe-Taitt CA, Golden JP, Feldstein MJ, Cras JJ, Hoffman KE, Ligler FS. 2000. Array biosensor for detection of biohazards. *Biosens. Bioelectron.* 14:785–94

85. Rowe-Taitt CA, Hazzard JW, Hoffman KE, Cras JJ, Golden JP, Ligler FS. 2000. Simultaneous detection of six biohazardous agents using a planar waveguide array biosensor. *Biosens. Bioelectron.* 15:579–89

86. Scholin C, Miller P, Buck K, Chavez F, Harris P, et al. 1997. Detection and quantification of *Pseudo-nitzschia australis* in cultured and natural populations using LSU rRNA-targeted probes. *Limnol. Oceanogr.* 42:1265–72

87. Bowers HA, Tengs T, Glasgow HB, Burkholder JM, Rublee PA, Oldach DW. 2000. Development of real-time PCR assays for rapid detection of *Pfiesteria piscicida* and related dinoflagellates. *Appl. Environ. Microb.* 66:4641–48

88. Ahn S, Walt DR. 2005. Detection of *Salmonella* spp. using microsphere-based, fiber-optic DNA microarrays. *Anal. Chem.* 77:5041–47

89. Ahn S, Kulis DM, Erdner DL, Anderson DM, Walt DR. 2006. Fiber-optic microarray for simultaneous detection of multiple harmful algal bloom species. *Appl. Environ. Microb.* 72:5742–49

90. Belgrader P, Benett W, Hadley D, Richards J, Stratton P, et al. 1999. Infectious disease: PCR detection of bacteria in seven minutes. *Science* 284:449–50

91. Delehanty JB, Ligler FS. 2002. A microarray immunoassay for simultaneous detection of proteins and bacteria. *Anal. Chem.* 74:5681–87

92. Rissin DM, Walt DR. 2006. Duplexed sandwich immunoassays on a fiber-optic microarray. *Anal. Chim. Acta* 564:34–39

93. King KD, Vanniere JM, Leblanc JL, Bullock KE, Anderson GP. 2000. Automated fiber optic biosensor for multiplexed immunoassays. *Environ. Sci. Technol.* 34:2845–50

94. Jung CC, Saaski EW, McCrae DA, Lingerfelt BM, Anderson GP. 2003. RAPTOR: a fluoroimmunoassay-based fiber optic sensor for detection of biological threats. *IEEE Sens. J.* 3:352–60

95. Fernandez-Calvo P, Nake C, Rivas LA, Garcia-Villadangos M, Gomez-Elvira J, Parro V. 2006. A multi-array competitive immunoassay for the detection of broad-range molecular size organic compounds relevant for astrobiology. *Planet. Space Sci.* 54:1612–21

96. Kramer MF, Lim DV. 2004. A rapid and automated fiber optic-based biosensor assay for the detection of *Salmonella* in spent irrigation water used in the sprouting of sprout seeds. *J. Food Protect.* 67:46–52

97. Kristensen R, Gauthier G, Berdal KG, Hamels S, Remacle J, Holst-Jensen A. 2007. DNA microarray to detect and identify trichothecene- and moniliformin-producing fusarium species. *J. Appl. Microbiol.* 102:1060–70

98. Song L, Ahn S, Walt DR. 2006. Fiber-optic microsphere-based arrays for multiplexed biological warfare agent detection. *Anal. Chem.* 78:1023–33

99. Bailey RC, Kwong GA, Radu CG, Witte ON, Heath JR. 2007. DNA-encoded antibody libraries: a unified platform for multiplexed cell sorting and detection of genes and proteins. *J. Am. Chem. Soc.* 129:1959–67

100. Boozer C, Ladd J, Chen SF, Yu Q, Homola J, Jiang SY. 2004. DNA directed protein immobilization on mixed ssDNA/oligo(ethylene glycol) self-assembled monolayers for sensitive biosensors. *Anal. Chem.* 76:6967–72

101. Tang L, Ren YJ, Hong B, Kang KA. 2006. Fluorophore-mediated, fiber-optic, multi-analyte, immunosensing system for rapid diagnosis and prognosis of cardiovascular diseases. *J. Biomed. Opt.* 11:021011

102. Uttamchandani M, Walsh DP, Yao SQ, Chang YT. 2005. Small molecule microarrays: recent advances and applications. *Curr. Opin. Chem. Biol.* 9:4–13

103. Barnes-Seeman D, Park SB, Koehler AN, Schreiber SL. 2003. Expanding the functional group compatibility of small-molecule microarrays: discovery of novel calmodulin ligands. *Angew. Chem. Int. Ed. Engl.* 42:2376–79

104. Kuruvilla FG, Shamji AF, Sternson SM, Hergenrother PJ, Schreiber SL. 2002. Dissecting glucose signalling with diversity-oriented synthesis and small-molecule microarrays. *Nature* 416:653–57

105. Salisbury CM, Maly DJ, Ellman JA. 2002. Peptide microarrays for the determination of protease substrate specificity. *J. Am. Chem. Soc.* 124:14868–70

106. Harris J, Mason DE, Li J, Burdick KW, Backes BJ, et al. 2004. Activity profile of dust mite allergen extract using substrate libraries and functional proteomic microarrays. *Chem. Biol.* 11:1361–72

107. Liang FS, Greenberg WA, Hammond JA, Hoffmann J, Head SR, Wong CH. 2006. Evaluation of RNA-binding specificity of aminoglycosides with DNA microarrays. *Proc. Natl. Acad. Sci. USA* 103:12311–16

108. Thiel AJ, Frutos AG, Jordan CE, Corn RM, Smith LM. 1997. In situ surface plasmon resonance imaging detection of DNA hybridization to oligonucleotide arrays on gold surfaces. *Anal. Chem.* 69:4948–56

109. Haes AJ, Stuart DA, Nie SM, Van Duyne RP. 2004. Using solution-phase nanoparticles, surface-confined nanoparticle arrays and single nanoparticles as biological sensing platforms. *J. Fluoresc.* 14:355–67

110. Yonzon CR, Jeoungf E, Zou SL, Schatz GC, Mrksich M, Van Duyne RP. 2004. A comparative analysis of localized and propagating surface plasmon resonance sensors: the binding of concanavalin a to a monosaccharide functionalized self-assembled monolayer. *J. Am. Chem. Soc.* 126:12669–76

111. Endo T, Kerman K, Nagatani N, Hiepa HM, Kim DK, et al. 2006. Multiple label-free detection of antigen-antibody reaction using localized surface plasmon resonance–based core-shell structured nanoparticle layer nanochip. *Anal. Chem.* 78:6465–75

112. Fang S, Lee HJ, Wark AW, Corn RM. 2006. Attomole microarray detection of microRNAs by nanoparticle-amplified SPR imaging measurements of surface polyadenylation reactions. *J. Am. Chem. Soc.* 128:14044–46

113. Stitzel SE, Cowen LJ, Albert KJ, Walt DR. 2001. Array-to-array transfer of an artificial nose classifier. *Anal. Chem.* 73:5266–71

114. Dickinson TA, White J, Kauer JS, Walt DR. 1998. Current trends in 'artificial-nose' technology. *Trends Biotechnol.* 16:250–58

115. Toko K. 1998. Electronic tongue. *Biosens. Bioelectron.* 13:701–9

116. Jin C, Kurzawski P, Hierlemann A, Zellers ET. 2007. Evaluation of multi-transducer arrays for the determination of organic vapor mixtures. *Anal. Chem.* 80:227–36

117. Park J, Groves WA, Zellers ET. 1999. Vapor recognition with small arrays of polymer-coated microsensors: a comprehensive analysis. *Anal. Chem.* 71:3877–86

118. Bencic-Nagale S, Walt DR. 2005. Extending the longevity of fluorescence-based sensor arrays using adaptive exposure. *Anal. Chem.* 77:6155–62

119. Doleman BJ, Lonergan MC, Severin EJ, Vaid TP, Lewis NS. 1998. Quantitative study of the resolving power of arrays of carbon black-polymer composites in various vapor-sensing tasks. *Anal. Chem.* 70:4177–90

120. Sisk BC, Lewis NS. 2006. Vapor sensing using polymer/carbon black composites in the percolative conduction regime. *Langmuir* 22:7928–35

121. Lavine B, Workman J. 2006. Chemometrics. *Anal. Chem.* 78:4137–45

122. White J, Kauer JS, Dickinson TA, Walt DR. 1996. Rapid analyte recognition in a device based on optical sensors and the olfactory system. *Anal. Chem.* 68:2191–202

123. Dickinson TA, Michael KL, Kauer JS, Walt DR. 1999. Convergent, self-encoded bead sensor arrays in the design of an artificial nose. *Anal. Chem.* 71:2192–98

124. Rakow NA, Suslick KS. 2000. A colorimetric sensor array for odour visualization. *Nature* 406:710–13

125. Janzen MC, Ponder JB, Bailey DP, Ingison CK, Suslick KS. 2006. Colorimetric sensor arrays for volatile organic compounds. *Anal. Chem.* 78:3591–600

126. Albert KJ, Walt DR, Gill DS, Pearce TC. 2001. Optical multibead arrays for simple and complex odor discrimination. *Anal. Chem.* 73:2501–8

127. Albert KJ, Myrick ML, Brown SB, James DL, Milanovich FP, Walt DR. 2001. Field-deployable sniffer for 2,4-dinitrotoluene detection. *Environ. Sci. Technol.* 35:3193–200

128. Walt DR. 2005. Electronic noses: Wake up and smell the coffee. *Anal. Chem.* 77:A45

129. Stitzel SE, Stein DR, Walt DR. 2003. Enhancing vapor sensor discrimination by mimicking a canine nasal cavity flow environment. *J. Am. Chem. Soc.* 125:3684–85

130. Bencic-Nagale S, Walt DR. 2005. Extending the longevity of fluorescence-based sensor arrays using adaptive exposure. *Anal. Chem.* 77:6155–62

131. Sohn YS, Goodey A, Anslyn EV, McDevitt JT, Shear JB, Neikirk DP. 2005. A microbead array chemical sensor using capillary-based sample introduction: toward the development of an "electronic tongue". *Biosens. Bioelectron.* 21:303–12

132. Goodey A, Lavigne JJ, Savoy SM, Rodriguez MD, Curey T, et al. 2001. Development of multianalyte sensor arrays composed of chemically derivatized polymeric microspheres localized in micromachined cavities. *J. Am. Chem. Soc.* 123:2559–70

133. Zhang C, Bailey DP, Suslick KS. 2006. Colorimetric sensor arrays for the analysis of beers: a feasibility study. *J. Agric. Food Chem.* 54:4925–31

134. Zhang C, Suslick KS. 2007. Colorimetric sensor array for soft drink analysis. *J. Agric. Food Chem.* 55:237–42

135. Palacios MA, Nishiyabu R, Marquez M, Anzenbacher P. 2007. Supramolecular chemistry approach to the design of a high-resolution sensor array for multianion detection in water. *J. Am. Chem. Soc.* 129:7538–44

136. Folmer-Andersen JF, Kitamura M, Anslyn EV. 2006. Pattern-based discrimination of enantiomeric and structurally similar amino acids: an optical mimic of the mammalian taste response. *J. Am. Chem. Soc.* 128:5652–53

137. Buryak A, Severin K. 2005. A chemosensor array for the colorimetric identification of 20 natural amino acids. *J. Am. Chem. Soc.* 127:3700–1

138. You CC, Miranda OR, Gider B, Ghosh PS, Kim IB, et al. 2007. Detection and identification of proteins using nanoparticle-fluorescent polymer 'chemical nose' sensors. *Nat. Nanotechnol.* 2:318–23

Terrestrial Analysis of the Organic Component of Comet Dust*

Scott A. Sandford

Astrophysics Branch, NASA Ames Research Center, Moffett Field, California 94035-1000; email: Scott.A.Sandford@nasa.gov

Annu. Rev. Anal. Chem. 2008. 1:549–78

The *Annual Review of Analytical Chemistry* is online at anchem.annualreviews.org

This article's doi:
10.1146/annurev.anchem.1.031207.113108

Key Words

comets, organics, astrochemistry, spectroscopy, ices, astrobiology, meteorites, interplanetary dust particles

Abstract

The nature of cometary organics is of great interest, both because these materials are thought to represent a reservoir of the original carbon-containing materials from which everything else in our solar system was made and because these materials may have played key roles in the origin of life on Earth. Because these organic materials are the products of a series of universal chemical processes expected to operate in the interstellar media and star-formation regions of all galaxies, the nature of cometary organics also provides information on the composition of organics in other planetary systems and, by extension, provides insights into the possible abundance of life elsewhere in the universe. Our current understanding of cometary organics represents a synthesis of information from telescopic and spacecraft observations of individual comets, the study of meteoritic materials, laboratory simulations, and, now, the study of samples collected directly from a comet, Comet P81/Wild 2.

1. INTRODUCTION

IDP: interplanetary dust particle

It is now understood that our Solar System formed from the collapse of a portion of a dense interstellar molecular cloud of gas, ice, and dust (1). The collapsing material formed a disk surrounding a central protostar. The material in this disk ultimately suffered one of several different fates. Much of the material was incorporated into the central protostar, which eventually became our Sun. A portion of the material was also ejected from the system by bipolar jets and gravitational interactions in the disk. Most of the remaining material was incorporated into small bodies, called planetesimals, which subsequently accreted to form the planets. However, some of these planetesimals escaped ejection from the system or incorporation into larger bodies and survived in the form of asteroids and comets (2, 3).

The importance of these small bodies far outweighs their minor contribution to the mass of our Solar System. All the material that ended up in the Sun and planets has been thoroughly reprocessed in the more than 4.5 billion years since these bodies formed. Other than clues about the bulk elemental abundances of the original materials from which they formed, these large bodies cannot provide many insights into the nature of the raw starting materials from which they were made.

The material in smaller bodies such as asteroids and comets has undergone considerably less parent-body processing and contains more pristine samples than planets. Fortunately, we can study cometary and asteroidal materials by a number of means. We receive samples of these objects on Earth in the form of meteorites and interplanetary dust particles (IDPs), and these samples generally demonstrate that comets and asteroids, indeed, comprise primitive materials (4, 5). In the case of meteorites, this is evidenced by their great ages (the clustering of meteorite ages around 4.56×10^9 years is taken to represent the formation time of our Solar System).

However, the study of meteorites has also demonstrated that they are not completely pristine samples of the original material from which everything else was made. Meteorites show evidence of varying degrees of thermal processing and aqueous alteration, most of which is thought to have occurred shortly after their parent asteroids formed (6). Also, most meteorites are from asteroids that formed in the region between Mars and Jupiter, a region warm enough that these bodies did not incorporate a full share of the more volatile components in the protosolar disk. Thus, meteorites provide insights into the nature of protosolar materials, but they have some limitations.

Comets are thought to have formed and been stored much further out in the Solar System. Comets probably contain a more representative portion of the volatile components of the original protoplanetary disk and have undergone less parent-body processing since formation (3). Thus, cometary materials may represent the best samples of pristine early solar system materials presently available for study and may provide powerful insights into the formation of the entire Solar System, not just comets.

The nature of cometary volatiles and organics is also of great astrobiological importance. The accretionary processes that made Earth are expected to have formed a molten body that destroyed any molecular complexity of the original starting materials and at least partially devolatilized the planet. Making Earth habitable may therefore have required the delivery of additional volatiles and organics to its surface after it had cooled. Although the collision of entire asteroids and comets should have delivered organics to the early Earth, this is likely to have been a relatively inefficient process. The large temperatures and pressures resulting from the hypervelocity impact of such large objects with the Earth probably destroyed much of the incoming material (7, 8). However, these same small bodies, particularly comets, also eject much of their mass into the interplanetary medium in the form of very small grains. These small grains can survive collision with the Earth's atmosphere while undergoing considerably less alteration (see Section 3.2). Thus, comets probably delivered organic materials to the early Earth both directly as large bodies and indirectly via small dust grains, with the latter probably supplying the main source of surviving organics.

Investigators have frequently discussed the idea that comets and asteroids delivered key volatiles and organics to the early cooling Earth that may have played a role in the origin and evolution of life (9–13), and it represents a focal point in the field of astrobiology. Thus, understanding the nature of cometary materials is important both because it provides information about the materials that formed the Solar System and because it may provide insights into the origin of life on Earth.

Because the composition of comets is thought to be the end result of a series of universal processes involving stellar, interstellar, and star-forming environments, it should be generally representative of the composition of these bodies in other stellar and planetary systems. Thus, insofar as comets have played a role in the formation of life on Earth, we expect them to be available to play a similar role in other planetary systems containing appropriate conditions for the origin of life (whatever those may be). Understanding comets in this context may therefore provide insights into the frequency of life elsewhere in the universe.

Our current understanding of comets and their constituent organics is the combined result of information gathered using a number of different techniques. These include the use of telescopic remote-sensing (largely spectroscopic) techniques, laboratory simulations, spacecraft flybys of individual comets, the study of meteoritic materials found on Earth, and the study of cometary samples returned to Earth from Comet 81P/Wild 2 by the Stardust spacecraft.

Here I concentrate on laboratory studies of actual samples, particularly on stratospheric IDPs thought to come from comets and samples from Comet 81P/Wild 2. I also restrict myself primarily to discussions of cometary organics and, to a lesser extent, cometary volatiles. Cometary minerals are only briefly mentioned, and then only when they bear on the discussion of associated organics. Those interested in pursuing the subject of cometary mineralogy can find a good introduction to the relevant literature elsewhere (14–17).

There are numerous advantages to sample return for the study and understanding of our Solar System. Having actual samples on hand in terrestrial laboratories allows for the use of state-of-the-art analytical techniques and equipment, providing for the ultimate current precision, sensitivity, resolution, and reliability. In contrast, spacecraft instruments making in situ measurements are, of necessity, not state of the art. Sample returns also avoid limitations associated with cost, power, mass, and reliability that are imposed on spacecraft instruments (some of the terrestrial analytical devices used to study Stardust samples were not only more massive than the Stardust spacecraft, they were more massive than Stardust's launch pad!). In addition, returned samples are a resource for current and future studies by a broad international community that can use many different analytical techniques in an iterative and fully adaptive fashion not limited by instrument designs or ideas current at the time of the spacecraft's launch. Also, the samples reside in terrestrial laboratories, so researchers can replicate and verify analyses using multiple, fully calibrated techniques and instruments. Finally, because the actual analyses are done on Earth, sample return takes advantage of a tremendous resource not fully available to nonreturn missions, namely the expertise of the world's analytical chemists, physicists, and meteoriticists.

The study of samples of this sort represents a significant analytical challenge. Both stratospheric IDPs and the Wild 2 comet samples consist of particles that are typically smaller than 25 μm in diameter and contain less than 1 ng of total material. To further complicate matters, these particles generally consist of highly heterogeneous aggregates of subgrains in the micrometer and submicrometer size range (see Sections 3.2 and 4.1). Despite the difficulty in measuring complex subnanogram samples, we have learned a surprising amount from these samples. Furthermore, if the past is any indication, there is every reason to expect that the challenge represented by these samples will be met by continued improvements in analytical techniques, leading to new revelations about the nature of our Solar System and its formation.

2. EVIDENCE FOR COMETARY ORGANICS FROM NONLABORATORY STUDIES

Although this review focuses on what we have learned about comets on the basis of laboratory studies of cometary samples, it is worth touching briefly on what we know about comets based on a number of other approaches. These are summarized in the subsections that follow.

2.1. Evidence for Cometary Volatiles and Organics from Telescopic Observations

Comets formed and have since spent most of their time in the outer Solar System, where they are difficult to reach by spacecraft or detect telescopically. As a result, much of what we know about comets comes from the small population of comets that have been perturbed into orbits that bring them close to the Sun. The telescopic imaging of comets was sufficient to suggest a model, first described by Whipple (18), of comets as "dirty snowballs" (i.e., they consist of a mixture of dust grains and volatile ices). These materials are inert when comets reside in the cold outer Solar System, but in the inner Solar System, the ices sublime and produce outgassing that forms the cometary dust and gas comae and tails associated with visible comets (19–21).

Telescopic studies of the composition of comets rely heavily on the use of spectroscopic detection of the molecules in cometary comae. The list of gas-phase species identified in cometary comae using telescopic ultraviolet (UV), infrared (IR), and submillimeter spectroscopy is dominated by simple, volatile molecules, the dominant species being H_2O followed by CO, CO_2, CH_3OH, CH_4, H_2CO, and NH_3. Their relative abundances are similar to the ices seen in dense interstellar clouds (22), consistent with the idea that comets contain relatively unprocessed nebular materials. However, molecules released from the comet are immediately exposed to solar radiation and are subject to photodissociation and fast ion-neutral reactions (19). Some species (e.g., H_2O) show spatial distributions that suggest they are primary ice components (i.e., they were released directly from the comet nucleus). Others, such as CN and H_2CO, show distributions suggesting injection in an extended volume around the nucleus, either through the delayed sublimation of ejected ice grains or through the sequential photolysis of more complex molecular species (23–25).

The presence of more complex organics is inferred from species showing extended coma distributions that suggest they are photofragments of more complex organic parent molecules (26). Also, the surfaces of comets are observed to have very low albedos, inconsistent with surfaces dominated by ice, but suggestive of the presence of complex organics (27–31). We can also infer the presence of complex organics from the identification of an IR emission feature centered near 2950 cm^{-1} thought to result from C-H stretching vibrations in organic dust grains being heated by the Sun following ejection from the nucleus (32–36). Finally, we see a variety of complex organics in various environments in the interstellar medium (ISM) (37–40), and one would expect some of these materials to survive incorporation into comets.

2.2. Evidence for Cometary Organics from Laboratory Simulations

As mentioned above, the ices in comets appear to have similar compositions to the ices seen in interstellar clouds. Extensive laboratory simulations of mixed-molecular ices having interstellar cloud compositions have demonstrated that UV and charged-particle irradiation of such materials results in the production of a host of more complex organic materials (41–47), many of which are of astrobiological interest (amino acids, amphiphiles, quinones, etc.). Thus, the radiation processing of cometary ices and their interstellar precursors is expected to have resulted in cometary volatiles containing a variety of more complex organics.

2.3. Evidence for Cometary Organics from Spacecraft Flybys

A small number of comets have been studied using spacecraft flybys. These include the Giotto and two Vega spacecraft flybys of Comet Halley (48, 49), the Deep Space 1 flyby of Comet Borrelly (50), the Stardust spacecraft flyby of Comet Wild 2 (28), and the Deep Impact spacecraft flyby of Comet Tempel 1 (31). Researchers have studied the organics in these comets using both IR spectroscopy and mass spectrometry. The IR spectra obtained by spacecraft are largely consistent with data obtained from more remote telescopes (32, 33).

Mass spectra taken by the Giotto and Vega spacecraft showed that the Comet Halley dust grains consisted of mineral grains, CHON particles (particles dominated

ISM: interstellar medium

CHON particle: particle rich in the elements C, H, O, and N

by C, H, O, and N), and mixtures of the two (51–55). The CHON particles appeared to be rich in O and N relative to C compared to normal meteoritic materials. The Stardust spacecraft was also equipped with a mass spectrometer (56), which obtained limited spectra that confirmed cometary particles contain organics and suggested they were rich in nitrogen (57). Further measurements of this type are expected from a high-resolution time-of-flight secondary ion mass spectrometer on the Rosetta spacecraft, currently on its way to a 2014 rendezvous with Comet 67 P/Churyumov-Gerasimenko (58).

3. EVIDENCE FOR COMETARY ORGANICS FROM ORPHANED SAMPLES: METEORITES AND STRATOSPHERIC DUST PARTICLES

3.1. Organics in Meteorites

Most meteorites probably come from asteroids, not comets. However, meteorites still represent materials that have undergone considerably less processing than planetary materials. Thus, organics found in the least processed meteorites might be expected to have similarities with cometary organics. The most appropriate meteorites for comparison are the carbonaceous chondrites (CCs), which are thought to be among the least processed meteorites (59).

CCs contain a complex population of organics (60, 61) that includes both soluble and insoluble phases. The vast majority of the carbon is present in an insoluble organic macromolecular material (IOM) (**Figure 1**). This insoluble material contains more than 90% of the carbon in CCs and can survive exposure to water and aqueous acids. IOM is highly aromatic and consists of disorganized aromatic, heteroaromatic, and hydroaromatic ring domains cross-linked by a wide variety of short methylene chains, biphenyl groups, ethers, sulfides, etc. (60, 62, 63). The fraction of C in aromatic domains within the CC Murchison IOM is between 0.61 and 0.66, and only ~30% of this C is bonded to H, indicating the aromatics are highly substituted (63). The abundance of N, O, and S in IOM varies between meteorites, but the IOM in Murchison ($C_{100}H_{71}N_{1.2}O_{12}S_2$) is representative. Oxygen is present in a wide range of functionalities, and nuclear magnetic resonance studies suggest that O-containing functional groups in the Murchison IOM are highly linked (63). The formation history of meteoritic IOM is currently poorly understood. Possibilities include formation in the ISM, the protosolar nebula, the meteoritic parent bodies, or some combination of these.

CCs also contain soluble organics. These have been more widely examined than the IOM because they are amenable to study by traditional wet-chemistry analytical techniques (60). Extractable organics that have been identified include amino acids, amines, and amides (61, 64, 65); aliphatic hydrocarbons (66, 67); polycyclic aromatic hydrocarbons (PAHs) (62, 68, 69); carboxylic, dicarboxylic, and hydrocarboxylic acids (64, 70, 71); amphiphiles (72, 73); quinones (70); nitrogen heterocycles (67); and sugars (74). Many of these species play important biochemical roles in life on Earth, particularly the amino acids, amphiphiles, and quinones (75, 76). This is one

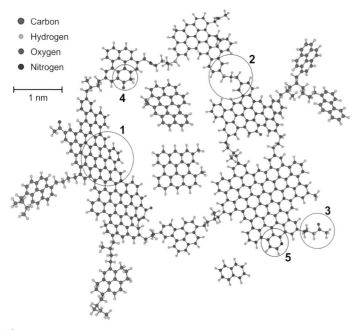

Figure 1

The carrier of the bulk of the solid carbon in the diffuse interstellar medium is thought to consist of a network of aromatic structures (such as that in circle 1) interlinked by short aliphatic bridges (circle 2) and containing aliphatic carbonyl groups (circle 3), aromatic carbonyl groups (circle 4), and aromatic nitrogen (circle 5), similar to the structures shown here. This structure is similar to that of the insoluble organic macromolecular material that composes the majority of the organics in primitive meteorites, although the meteoritic material appears to be considerably richer in oxygen. Figure adapted from Reference 40.

of the prime motivations for considering the possibility that accreted extraterrestrial materials may have played key roles in the origin of life.

One indication that meteoritic organics represent relatively unprocessed materials is the presence of nonterrestrial enrichments in D and ^{15}N in them (64, 71, 77). As discussed in more detail in Section 5, these enrichments suggest the presence of molecules produced in the ISM that have never fully equilibrated with normal solar system materials. Given that cometary materials have experienced even less parent-body processing than asteroidal materials (i.e., meteorites), one would anticipate that such isotopic anomalies might also be common in cometary samples.

3.2. Organics in Stratospheric Interplanetary Dust Particles

Dust particles ejected from comets and asteroids enter solar orbits that evolve under the influence of nongravitational forces (78–81), and some of these particles ultimately collide with Earth. Many particles vaporize as they enter Earth's atmosphere (shooting stars), but particles under a few hundreds of micrometers in diameter can survive deceleration high in Earth's atmosphere, where gas densities are low, without melting (82). These particles can be collected in the stratosphere using high-altitude aircraft (83–86). Typical collected stratospheric IDPs have diameters of <25 μm and masses on the order of 1 ng. This precludes the use of many analytical techniques. Nonetheless, over the past 25 years or so, investigators have successfully studied these particles using a variety of microanalytical techniques.

Studies have demonstrated that these particles are indeed extraterrestrial and can be separated into several different classes (85). Particles in one of these classes, referred to as anhydrous porous IDPs, are likely cometary dust grains (87–89). These IDPs consist of unequilibrated mixtures of anhydrous minerals (predominantly olivines, pyroxenes, and iron-rich sulfides) residing in porous aggregate structures (88–90) and are generally rich in C. Meteorites typically contain a few percent C by weight, but the average weight percentage of C in a suite of 19 IDPs was 12.5 ± 5.7%, and values as high as 24% were seen (88).

Organics in these particles have been studied using several analytical techniques. Two-step laser-desorption laser-ionization mass spectrometry (L^2MS) studies have shown that these particles contain variable and rich spectra of alkylated and unalkylated PAHs (91, 92). In some cases, the distribution of PAHs looks similar to CCs, but some IDPs show richer PAH populations that extend to higher masses (93, 94). Raman spectra of these PAHs confirm the presence of poorly ordered aromatic materials (95–97). Both aromatics and aliphatic $-CH_3$ and $-CH_2-$ groups have been detected in some particles based on the presence of their characteristic C-H stretching bands in the 3100–2800 cm^{-1} IR region (98, 99). These spectra generally show $-CH_2-/-CH_3$ band area ratios of >2, a considerably larger ratio than seen in IOM in primitive CCs (~1.1) and in the diffuse ISM (1.1–1.25) (38, 40). The higher $-CH_2-/-CH_3$ implies IDPs contain longer or less branched aliphatic chains.

Finally, carbon X-ray absorption near-edge spectroscopy (C-XANES) measurements of IDPs demonstrate that organics are present that have a variety of C-bonding states (98, 100, 101). They confirm the presence of abundant aromatic materials and show that IDP organics typically have higher O/C and N/C ratios than meteorites. In many cases, the organic matter appears to be the glue that holds subgrains in the IDP together. Overall, the IDP results indicate that most of the organics are inconsistent with formation by aqueous processing, but they already existed at the time that primitive, anhydrous dust was being assembled (98, 101). This implies the organics likely formed in either the solar nebula or in an interstellar environment.

In summary, the organics seen in stratospheric IDPs, many of which are thought to come from comets, appear to have many similarities to the materials seen in CCs. In particular, both appear to be dominated by aromatic materials. However, there are some distinct differences: The IDPs sometimes show more complex populations of PAHs, and the overall organic population in the anhydrous stratospheric IDPs has considerably higher O/C and N/C ratios than found in meteorites. The nature of the organics suggests the bulk of the material has interstellar or protosolar nebula origins rather than parent-body origins.

4. ORGANICS FROM COMET 81P/WILD 2: THE STARDUST COMET SAMPLE RETURN MISSION

Our understanding of cometary organics made a quantum leap forward with the January 15, 2006, return of samples from Comet 81P/Wild 2 by the Stardust spacecraft (102). One of the scientific goals of Stardust was to establish whether comets contained complex organic materials and to establish the abundance, chemical, and isotopic nature of any organics present (102–104). Such information provides insights into the formation and evolution of comets and places constraints on the environments and processes by which they were made. Characterization of these organics also allows for comparison of cometary materials with other extraterrestrial materials (meteorites, IDPs, and ISM organics).

4.1. The Collection of Material from Comet 81P/Wild 2

Stardust was the first mission in history to return solid samples from an astronomical body beyond the Earth-Moon system. The mission retrieved samples from Comet 81P/Wild 2, a comet currently in an orbit that approaches the orbits of both Jupiter and Mars. This ~4.5-km-diameter body was formed and spent most of the past 4.5 or so billion years in the Kuiper belt outside the orbit of Neptune. Only recently did it wander into the inner Solar System, where it had a close encounter with Jupiter on September 10, 1974, that placed it in its current orbit. The comet has an expected dynamical lifetime of ~10^4 years before it hits a larger object or is ejected from the Solar System (105). In its current orbit Comet 81P/Wild 2 approaches the Sun close enough that solar heating causes classic cometary activity near perihelion.

During the Stardust flyby on January 2, 2004, the comet was active, and images showed the presence of at least 20 dust jets coming from the nucleus (28, 106) (**Figure 2a**). Stardust approached to within 234 km of Wild 2's surface, and the encounter occurred at a solar distance of 1.86 AU. Particles ejected from the comet were exposed to space for only a few hours before collection, but solar heating probably vaporized most ices during transit from Wild 2 to Stardust. Particles were collected for return when they impacted at 6.12 km s^{-1} into silica aerogel, a porous glass comprised of nanometer-sized silica filaments with bulk density that varied from <0.01 g cm^{-3} at the impact surface to 0.05 g cm^{-3} at 3-cm depth. Stardust aerogel tiles collected over a thousand 5–300-μm (and many more smaller) comet particles. Onboard impact sensors indicate that most of the collected particles were associated with just a few specific dust jets (107).

Particle impacts into aerogel produced tracks whose shapes depended on the nature of the impacting particle (**Figure 2b**). Nonfragmenting particles produced carrot-shaped tracks with length/diameter ratios of >25. However, many tracks show bulbous upper regions and sometimes multiple roots. These tracks were produced by weakly bound aggregate particles that broke apart on impact with the aerogel (102, 108). The upper parts of tracks are lined with melted aerogel that contains dissolved projectile material; the mid-regions contain less melt and more preserved projectile material and compressed aerogel; and the track ends contain largely

a   **b**

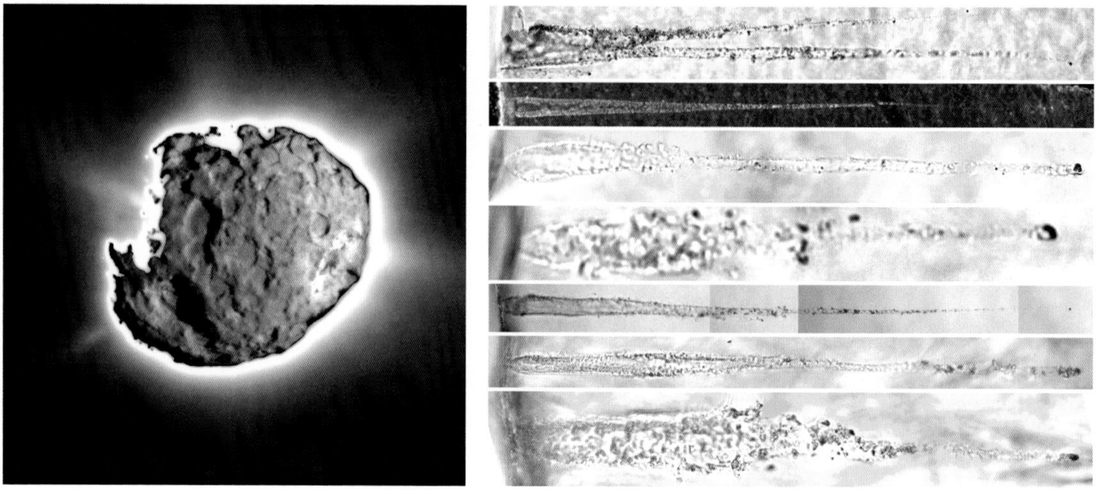

Figure 2

(*a*) A composite image of Comet 81P/Wild 2 consisting of a 10-ms exposure image (to capture the surface of the cometary nucleus) overlain on a 100-ms exposure image (to capture the large number of dust jets emerging from the nucleus). The nucleus is a very dark (albedo of ~0.03) ~5 km-diameter oblate spheroid covered with unique topographical structures. (*b*) Optical images of seven tracks produced by Wild 2 particles that impacted Stardust aerogel collector material. Solid particles left thin carrot tracks (see fifth track from top), but most impacting particles consisted of weakly bound aggregates of smaller grains that produced more bulbous tracks, often with multiple roots. Figure adapted from Reference 102.

unmelted materials (terminal particles). In most cases, the deepest penetrating particles are solid mineral grains or aggregates comprised of micrometer-size or larger grains. In addition to aerogel, approximately 15% of the Stardust collector surface was the aluminum frame and aluminum foils used to hold the aerogel. Impacts on the frame produced bowl-shaped craters lined with melted, and in some cases unmelted, projectile residues (108).

4.2. Measurement of Wild 2 Organics in the Returned Samples

Approximately 200 investigators around the world participated in the preliminary examination of the samples returned by Stardust, and their findings appeared in a special issue of *Science* (16, 17, 102, 108–111). The results presented in these studies were obtained using an enormous variety of analytical techniques, some of which did not exist when Stardust was launched.

As mentioned above, the analysis of the Wild 2 samples was a challenging prospect because the samples were complex, nanogram-sized aggregates that broke up into smaller particles and were distributed along the entire length of aerogel tracks. For organics analyses, investigators had to pay additional attention to the possibility of contaminants associated with the aerogel collector medium, flight of the spacecraft, and return and recovery of the sample return capsule. Fortunately, contaminants were found generally to be of low-enough abundance or were sufficiently well characterized that they can be distinguished from the cometary organics (111). Most problematic was the aerogel collector medium itself. Stardust aerogel consists primarily of amorphous SiO_2 but contains from one-quarter to a few weight percent C. Nuclear magnetic resonance studies indicate that this C is largely in the form of simple $Si–CH_3$ groups easily distinguishable from the cometary organics described below. It should be noted that not all the organics in the samples will be fully representative of the original cometary material because some may have been modified during aerogel impact. There is also evidence that at least some organic compounds were generated or altered by the impact heating of the aerogel itself (111–113) (see below). Despite these difficulties, we have learned a great deal about cometary organics and the origin of the Solar System from these samples.

Analytical techniques used during the preliminary examination of organics in the Wild 2 samples include L^2MS, liquid chromatography with UV fluorescence detection and time-of-flight mass spectrometry (LC-FD/TOF-MS), scanning transmission X-ray microscopy (STXM), XANES, IR and Raman spectroscopy, ion chromatography with conductivity detection (IC), secondary ion mass spectrometry (SIMS), and time-of-flight SIMS. A summary of the combined findings of these analyses is provided below. More complete discussions of the organics preliminary examination results can be found in Sandford et al. (111) and its associated supporting online material, and in a forthcoming special issue of *Meteoritics and Planetary Science*.

Multiple experimental techniques demonstrate that the samples contain PAHs. L^2MS mass spectra obtained from individual particles and on aerogel surfaces along impact tracks show PAHs and their alkylated derivatives, with two distinct types of PAH distributions distinguishable from low-aerogel backgrounds (111, 114, 115) (**Figure 3**). In some cases, PAH populations dominated by benzene and naphthalene (one- to two-ring PAHs), including alkylation out to several CH_3 additions, are observed in the absence of larger PAHs (**Figure 3a**). Such distributions resemble pyrolysis products of meteoritic IOM and are observed in high-laser-power L^2MS measurements of Stardust aerogel tiles, suggesting many of the lower-mass PAHs may originate from impact processing of C original to the aerogel (112, 113). The second type of PAH population shows complex distributions that strongly resemble those seen in some meteorites and IDPs (**Figure 3b**). The similarity to IDPs extends to masses beyond 300 a.m.u., although several track spectra show mass envelopes extending up to 800 a.m.u. with both odd and even mass peaks. Such high-mass envelopes in IDPs have been attributed to the polymerization of smaller aromatics within the samples by radiation processing during their extended exposure to interplanetary space or heating during atmospheric entry (91, 115). Similar polymerization of the original PAH population in the Stardust samples by impact heating may explain the higher-mass envelopes observed in them (111). The more complex comet mass spectra also include additional mass peaks not observed in meteorite mass spectra, but seen in some IDP spectra, that suggest the presence of O- and N-substituted aromatic species having heterofunctionality external to the aromatic structure.

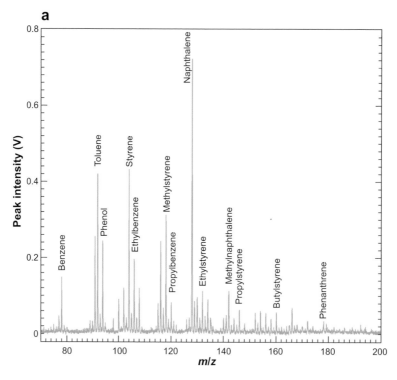

a

(Figure a: mass spectrum, Peak intensity (V) vs m/z, with labeled peaks: Benzene, Toluene, Phenol, Styrene, Ethylbenzene, Methylstyrene, Propylbenzene, Naphthalene, Ethylstyrene, Methylnaphthalene, Propylstyrene, Butylstyrene, Phenanthrene)

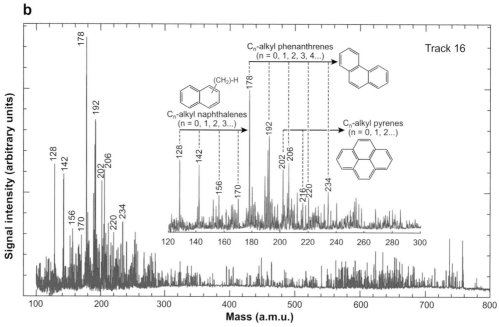

b

Track 16

C_n-alkyl phenanthrenes
(n = 0, 1, 2, 3, 4...)

C_n-alkyl naphthalenes
(n = 0, 1, 2, 3...)

C_n-alkyl pyrenes
(n = 0, 1, 2...)

(CH_2)-H

(Figure b: Signal intensity (arbitrary units) vs Mass (a.m.u.), with labeled peaks 128, 142, 156, 170, 178, 192, 202, 206, 216, 220, 234)

The presence of PAHs has also been confirmed by time-of-flight SIMS analyses of terminal particles extracted from aerogel tracks, a dissected aerogel keystone with a particle track split down the middle, and residues found in a crater in aluminum foil. The mass distribution shows a steep decrease in PAH abundances with an increasing number of C atoms (111, 116, 117). Raman spectra acquired from 12 Stardust particles also confirm the presence of aromatic materials (111, 118). All the Raman spectra are dominated by two broad bands centered at $\sim$1360 Δcm^{-1} and $\sim$1580 Δcm^{-1}. These D and G bands, respectively, are characteristic of graphite-like sp^2-bonded carbon in condensed carbon rings. The Raman spectra are similar to those of many IDPs and primitive meteorites, and the G band position and width parameters span the entire range observed in IDPs and meteorites. It is not clear whether this variation reflects heterogeneity in the cometary samples or variable processing during aerogel impact. However, the presence of G bands with low frequencies and large widths indicates that at least some organics were captured with relatively little alteration. Many Raman spectra of Stardust particles are characterized by very high backgrounds that increase with increasing Δcm^{-1} position, indicating that the samples may be rich in noncarbon heteroatoms such as N. In a few cases, aromatic materials were also identified by the detection of an aromatic CH stretching mode band using IR spectroscopy (111, 118, 119).

It was possible to extract many individual cometary particles from Stardust aerogel and microtome them into multiple slices thin enough to make them amenable to scanning transmission X-ray microscopy and C,N,O-XANES analyses. C-XANES spectra of many of these thin sections confirmed the presence of aromatic bonds. Full C,N,O-XANES spectra showed $1s$-π^* transitions consistent with variable abundances of aromatic, keto/aldehydic, and carboxyl moieties, as well as amides and nitriles (**Figure 4a**) (111, 120). Although confirming aromatics are present, the XANES data suggest that considerably less olefinic and aromatic material is present in Comet Wild 2 samples than seen in CCs and IDPs. Some particles contain abundant C, none of which is aromatic (particle 1 in **Figure 4a**).

XANES analyses also provided quantitative estimates of atomic O/C and N/C ratios present in the various functional groups identified (**Figure 4b**). Overall, the

---

Figure 3

(a) Polycyclic aromatic hydrocarbons (PAHs) found along the walls of tracks are dominated by simple distributions of one- and two-ring PAHs and their alkylated variants. These PAHs may result (all or in part) from the impact alteration of C in the original aerogel collector material. (b) The PAH populations found in extracted particles can show a much richer distribution of higher-mass species. Observed species include naphthalene ($C_{10}H_8$, two rings, 128 a.m.u.), phenanthrene ($C_{14}H_{10}$, three rings, 178 a.m.u.), and pyrene ($C_{16}H_{10}$, four rings, 202 a.m.u.), along with their alkylated homologs extending up to at least C_4-alkyl. Peaks at 101, 112, 155, and 167 a.m.u. (not shown), inconsistent with simple PAHs, were observed when an attenuated laser photoionization pulse was used to minimize photofragmentation. These peaks could result from O- and N-substituted aromatic species having heterofunctionality external to the aromatic structure (i.e., not N or O heterocyclics). Similar mass peaks have been observed in several interplanetary dust particles (9). Figure adapted from supporting online material associated with Reference 111.

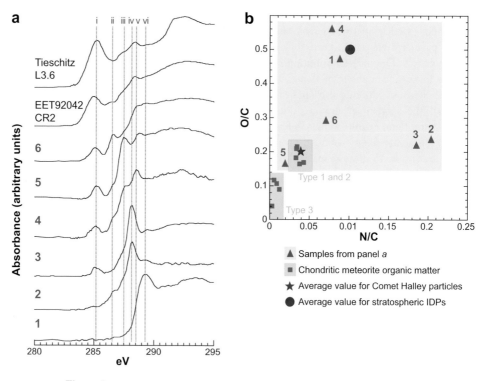

Figure 4

(*a*) C X-ray absorption near-edge spectroscopy spectra of six samples from Comet Wild 2 samples (*1 to 6*) compared with organic matter from two primitive meteorites. Positions associated with specific organic functional groups are marked with dashed vertical lines (i, C=C; ii, C=C–O; iii, C=O; iv, N–C=O; v, O–C=O; and vi, C–O). Sample chemistry clearly varies enormously between samples. (*b*) Atomic O/C and N/C for the six samples in panel *a* (*blue triangles*) compared with chondritic meteorite organic matter (*red squares*, in which the higher values correspond to petrologic type 1 and 2, and the lower values are type 3) and the average values for Comet Halley particles (*star*) and stratospheric interplanetary dust particles (IDPs) (*circle*). Figure adapted from Reference 111.

Wild 2 organics were found to be considerably richer in the heteroatoms O and N relative to both meteoritic organic matter and the average composition of Comet Halley particles measured by Giotto, but qualitatively similar to the average O/C and N/C reported for stratospheric IDPs (111, 120). Both the O and N exist in a wide variety of bonding states. Particles particularly rich in N exhibit abundant amide C in their XANES spectra.

The presence of N-containing compounds is further suggested by studies of collector aerogel using LC-FD/TOF-MS (111, 121). Stardust may have returned a diffuse sample of gas-phase molecules that struck the aerogel directly from the cometary coma or that diffused away from grains after impact. To test this possibility, investigators carried samples of flight aerogel through a hot-water extraction and acid hydrolysis procedure to determine if excess primary amine compounds were present in the aerogels. Only methylamine (MA), ethylamine (EA), and glycine were detected above background levels, and the absence of MA and EA in a flown aerogel witness coupon suggests that these amines are cometary in origin. The concentrations of MA and EA in aerogel adjacent to a track were similar to those present in aerogel not located near a track, suggesting that these amines, if cometary, originate from submicrometer particles or gas that directly impacted the collector. The presence of excess glycine may indicate a cometary origin for this amino acid as well. No MA, EA, or glycine was detected in non-acid-hydrolyzed aerogel extracts, suggesting that they are present in an acid-soluble bound form, rather than as a free primary amine. Compound specific isotope measurements will be necessary to constrain the origin of these amines.

IR spectra taken from tracks and individual extracted particles show the presence of both aromatic and nonaromatic chemical functional groups (16, 111, 118, 119). IR spectra of particles and tracks often contain absorption features at 3322 cm^{-1} (–OH), 3065 cm^{-1} (aromatic CH), 2968 cm^{-1} (–CH$_3$), 2923 cm^{-1} (–CH$_2$–), 2855 cm^{-1} (–CH$_3$ and –CH$_2$–), and 1706 cm^{-1} (C=O). One particle also showed a weak 2232 cm^{-1} band consistent with –C≡N stretching vibrations. Combined, the IR data indicate the presence of aromatic, aliphatic, carboxylic, and N-containing functional groups, consistent with the results of other analytical techniques. The observed –CH$_2$– (2923 cm^{-1})/ –CH$_3$ (2968 cm^{-1}) band depth ratios in the returned samples are typically ~2.5, corresponding to a functional group ratio of –CH$_2$–/–CH$_3$ ~ 3.7 in these cometary samples, assuming typical intrinsic band strengths for these features. This value is similar to that seen in the IR spectra of anhydrous IDPs, but considerably larger than seen in IOM in primitive CCs (~1.1) and the diffuse ISM (1.1–1.25), suggesting that the aliphatic moieties in particles from Comet 81P/Wild 2 are longer or less branched than those in meteorites and the diffuse ISM. The ratio of aromatic to aliphatic C–H is quite variable in the IR spectra, consistent with the variations implied by XANES data.

In several cases it was possible to make IR spectral maps of entire impact tracks and their surrounding aerogel. The organic components that produce the –OH, –CH$_3$, –CH$_2$–, and C=O IR absorption bands sometimes extend well beyond the visible track edge (**Figure 5**) (111, 119). This implies that the particles contained organics that volatilized and diffused into the surrounding aerogel during impact. Because similar length tracks are also seen in the same aerogels that show no IR-detectable organics beyond those seen in the original aerogel, this material is unlikely to result from impact-altered aerogel carbon.

Researchers using SIMS ion imaging have made elemental maps of Wild 2 particle sections (111). These maps demonstrated that N and S are associated with organic molecules and show that Wild 2 samples contain highly heterogeneous N distributions, with N/C ratios ranging from 0.005 to almost 1. Some particles exhibit the entire range of values, whereas others fall more uniformly at the high N/C end of the range. However, there are regions with high C content that are not rich in N. Sulfur is typically associated with C and N, but is also distributed in small hot spots presumably owing to sulfides, which are common in the samples (17).

SIMS was also used to make H, C, and N isotopic measurements of Wild 2 particles (110, 122). D/H enrichments up to approximately three times the terrestrial value are observed within approximately half the particles. D enrichments are seen

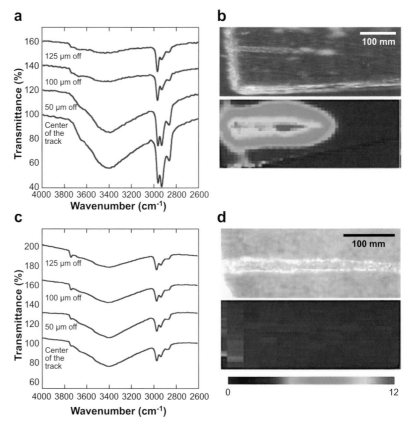

Figure 5

Infrared transmittance spectra and maps obtained from two particle tracks in Stardust aerogel. (*a*) Spectra taken along a line perpendicular to cometary impact track 59. In addition to aerogel features, the spectra across track 59 display peaks at 3322 (broad), 2968, 2923, 2855, and 1706 cm^{-1} (not shown), both inside the track and extending outward into the aerogel. (*b*) An optical image of track 59 (*top panel*) and a corresponding map showing the intensity distribution of the 2923 cm^{-1} peak due to excess cometary −CH$_2$− (*bottom panel*). (*c*) Spectra taken along a line perpendicular to cometary impact track 61. Unlike track 59, these spectra show no evidence for extended excess −CH$_2$−, but only show the features due to aerogel. (*d*) An optical image of track 61 (*top panel*) and a corresponding 2923 cm^{-1} peak map showing that no excess −CH$_2$− is present (*bottom panel*). The false-color image scale shown at the bottom is used in both maps (*b*, *d*). In both cases, the entrance of the cometary particle is from the left side. Figure adapted from Reference 111.

in meteorites and IDPs and are thought to result from materials with an interstellar/protostellar chemical heritage (123–126). The D enrichment in Wild 2 samples is heterogeneously distributed and associated with C, indicating the carrier is probably organic. The elevated D/H ratios are comparable with those of many IDP and meteorite samples, although none of the comet samples examined to date has shown ratios as extreme as the most anomalous IDPs and meteorites. Investigators also observed isotopic anomalies in N in the form of ^{15}N excesses (110). As with D, these anomalies are heterogeneously distributed and appear in the form of hot spots that differ from the surrounding material. When observed, the D and ^{15}N enhancements provide clear evidence of a cometary origin for the organics and suggest that cometary organics contain materials with an interstellar/protostellar chemical heritage.

4.3. An Overview of the Nature of Organics from Comet 81P/Wild 2

Overall, the organics in the Wild 2 cometary samples show many similarities with those seen in IDPs, and to a lesser extent with those in primitive CCs. However, there are some distinct differences; some of the cometary material appears to represent a new class of organics not previously seen in other extraterrestrial samples. These differences make the returned comet samples unique among currently available extraterrestrial samples.

Raman, XANES, and SIMS data all demonstrate that the distribution of organics (overall abundance, functionality, and relative elemental abundances of C, N, and O) is remarkably heterogeneous both within particles and between particles. The observed variations indicate that cometary organics represent a highly unequilibrated reservoir of materials. Similar to meteoritic organic matter, Wild 2 organics contain both aromatic and nonaromatic fractions. However, the Stardust samples exhibit a greater range of compositions (higher O and N concentrations) and include an abundant organic component that is poor in aromatics and a more labile fraction (possibly the same material). The nonaromatic fraction appears to be far more abundant relative to aromatics than in meteorites. The labile materials may be absent from stratospheric IDPs because they are lost and/or modified during atmospheric entry heating or by radiation during the IDPs' transit from their parent body to Earth. Given the O- and N-rich nature of the Stardust materials, these labile organics could represent a class of materials that have been suggested as parent molecules to explain the extended coma sources of some molecular fragments such as CN (23, 24). In general terms, the organics in Stardust samples appear to be even more primitive than those in meteorites and IDPs, at least in terms of being highly heterogeneous and unequilibrated.

The presence of high O and N contents and the high ratios of $-CH_2-/-CH_3$ seen in the IR data indicate that the Stardust organics are not identical to the organics seen in the diffuse ISM, which look more like the insoluble macromolecular material seen in primitive meteorites but with even lower O/C ratios (40). This implies that cometary organics are not the direct result of stellar ejecta or diffuse ISM processes, but rather the ultimate result of dense cloud and/or protosolar nebular processes. The overall composition is qualitatively consistent with what is expected from radiation processing of astrophysical ices in dense clouds and the polymerization of simple species such as HCO, H_2CO, and HCN (42, 127).

5. H, C, N, AND O ISOTOPES AND THE ORIGIN OF COMETARY ORGANICS

To understand the nature of cometary organics, we need to establish their isotopic systematics. This is important for both practical and scientific reasons. Practically speaking, the detection of nonterrestrial isotopic ratios in returned samples provides clear proof that the samples are extraterrestrial and not contamination. The D and ^{15}N enrichments seen in stratospheric IDPs and samples from Comet Wild 2 provide clear proof that they have an extraterrestrial origin. Scientifically, the identification of isotopic anomalies, and their carriers, in cometary organics provides important insights into the processes and environments responsible for their creation.

A number of interstellar chemical processes and environments can fractionate H, N, C, and O isotopes, and some of these environments are expected to leave characteristic isotopic fingerprints in their products. There are at least four different interstellar processes that can lead to D-enriched organics (124), largely at the expense of D loss from molecular H_2. These are gas-phase ion-molecule reactions (128, 129), gas-grain reactions (129–131), unimolecular photodissociation of species such as PAHs (37), and UV photolysis of D-enriched ices (132). The first three of these produce isotopic fractionations resulting from the zero-point energy differences of H and D bonds, whereas the fourth takes simple, previously enriched molecules and passes the enrichments on to more complex species. Each process involves different environments or compositional components of the ISM, and each leaves a distinct D signature in the resultant organic population (124). For example, in the case of PAHs, each process places excess D in different populations of PAHs and/or on different molecular sites on the enriched PAHs. Thus, the molecular identity of carriers of excess D, and the molecular positions of the excess D, can provide important clues about the carrier's natal environment.

Some of these interstellar processes can also drive isotopic fractionation involving C, N, and O. However, isotopic zero-point energy differences are significantly smaller for these elements, and their fractionations are expected to be smaller (129). Gas-phase ion-molecule reactions can enrich ^{15}N relative to ^{14}N in many molecules in cold interstellar clouds at the expense of N_2 and NH_3 (133, 134). Expectations for C and O are less clear, however. In the case of C, ion-molecule reactions and proton transfer reactions (135, 136) compete with isotope-selective photodissociation reactions (129) that produce the opposite $^{13}C/^{12}C$ fractionations. As a result, overall C isotope fractionation in the gas phase depends on the relative importance of different processes, an issue that is not fully understood. Oxygen in the gas phase resides primarily in CO, O, and O_2, and theoretical calculations suggest that little $^{16}O/^{18}O$ fractionation should occur in these species or their products via ion-molecule reactions (135). However, CO photolytic self-shielding could produce significant mass-independent fractionation of ^{16}O relative to both ^{17}O and ^{18}O in some environments (137, 138).

In summary, one would expect many organic molecules having interstellar origins to show enrichments in D and ^{15}N, and possibly ^{16}O. In the case of C, the existence of multiple reaction schemes that cause opposite fractionations suggests that C isotopes should not correlate with D fractionations in any simple way. Isotopic analyses of stratospheric IDPs and Wild 2 samples appear to be qualitatively consistent with these expectations. Significant enrichments are seen in both D and ^{15}N in both types of samples (92, 100, 110, 123, 139). In contrast, C isotopes show only minor variations (139), and O-isotopes are generally found to have solar abundances [although oxide minerals in a few particles have been seen to be enriched in ^{16}O (139, 140)].

The heterogeneous distribution of the observed D and ^{15}N anomalies in these samples suggests that materials having an interstellar heritage were mixed with solar nebular materials during the formation of the Solar System but that they were not exposed to significant processing after the component subgrains were assembled, either within the protosolar nebula or in the particle's parent body.

6. ANALYTICAL CHALLENGES FOR FUTURE STUDIES OF EXTRATERRESTRIAL SAMPLES

The successful return of samples from Comet 81P/Wild 2 is a fresh example of the power of sample return studies and reiterates the value of such samples as already demonstrated by NASA's lunar, cosmic dust, solar wind, and meteorite collections. The astrophysical community's appreciation of the value of such missions is evidenced by the large number of scientists who participated in the preliminary examination of these samples and the significant number of future sample return missions currently being considered by NASA. Given that currently available samples will continue to be studied for decades to come, and that new samples will almost certainly be coming from additional solar system objects, it is worth considering what analytical challenges are likely to be important for the study of extraterrestrial samples in the future.

6.1. Ever Smaller Samples

Certainly one challenge is the continued pursuit of the ability to study ever smaller samples. Typical Stardust particles had original diameters of <25 μm and total masses of ~1 ng, and had their component subgrains smeared out along millimeter-long tracks. Of this material, only a fraction was organic, and some of this material was further dispersed by being vaporized and redeposited in the surrounding aerogel. Clearly the ability to measure very small samples is a necessity. Many of the analytical techniques used on the Stardust samples did not exist or were immature when the spacecraft originally launched in 1999. Improving technologies allowed the samples returned in 2006 to be studied in a number of ways that were completely unanticipated by the mission's original co-investigators. Despite these successes, there are multiple future improvements in the study of small particles that are desirable. First, of course, there will undoubtedly be areas in which technological advances will allow for the use of whole new analytical techniques that have been previously restricted only to larger samples. Each of these will open a new window onto the study of these samples.

Continued improvements in sensitivity will also be helpful. However, for extraterrestrial samples, true sensitivity is not so much a measure of how little of an individual organic molecule can be detected, but whether that molecule can be detected and identified in the midst of a complex mixture of other organics. When pursuing possible analytical techniques that could be applied to IDPs and Stardust samples, I have more than once been disappointed to find that a technique advertised to be able to detect subnanogram samples actually needs the sample to be a pure compound or needs 1 g of material in a sample chamber to generate the nanogram that is measured, and so on. In the case of samples such as those from Comet 81P/Wild 2, techniques are needed that can study subnanogram samples that are compositionally complex and for which the nanogram of mixed material is all there is, which is indeed an analytical challenge.

6.2. Wedding Isotopic Analyses to Molecular Analyses

Understanding the H, N, C, and O isotopic ratios of extraterrestrial organics is important, both because the detection of anomalous ratios proves the materials are extraterrestrial and because the magnitude and molecular locations of the excesses provide important insights into the carrier's formation environment. Of particular interest are enrichments of D and ^{15}N.

To date, most studies of D and ^{15}N excesses in extraterrestrial materials have been restricted to identifying their presence, the magnitude of the anomaly, and, in some cases, mapping the distribution of the anomalous material within the overall sample. These anomalies are often associated with C, and the carriers are inferred to be organic in nature. In many cases, separate analytical studies of these same samples have shown that a variety of organic materials is indeed present. However, these independent observations neither prove the anomalous D and ^{15}N carriers are organics nor tell us about the molecular nature of the actual carriers. To realize the full potential of these materials for establishing the astrophysical environments in which these carriers were formed and evolved, we need the ability not only to detect D and ^{15}N enrichments, but also to establish (a) what specific molecular species are carrying the anomalies and (b) where the D and ^{15}N are sited on the carrier molecules. Wedding these two types of information from small samples that contain complex mixtures of organics is clearly an analytical challenge of the first order, but it is one that represents the frontier for our next leap forward for understanding how complex organics form and evolve in space.

7. CONCLUSIONS

Our understanding of cometary organics is based on information gathered from telescopic observations, comet flybys by spacecraft, the study of meteorites and IDPs, laboratory simulations, and the study of samples returned from Comet 81P/Wild 2 by Stardust. Our understanding of these materials is incomplete, although considerable progress has been made in recent years. Because one major advantage of sample return missions is that the samples continue to be available for further study, additional major progress can be expected as Wild 2 samples continue to be studied and analytical techniques improve.

When combined, the information gathered by these various approaches suggests that comets are rich in organic materials that come in a wide variety of forms, spanning a large range in volatility and molecular complexity. At one extreme are cometary ices that contain a variety of simple volatile molecules. The delivery of these volatiles (particularly H_2O) could have played a key role in making the early Earth habitable. These same volatiles may have played critical roles in the production of more complex organic materials via radiation processing. At the opposite extreme, comets also appear to contain organic materials that are qualitatively similar to the insoluble macromolecular material that dominates the organic fraction of primitive carbonaceous chondritic meteorites. Both materials are rich in interlinked aromatic domains. However, the cometary material differs from its meteoritic counterparts by being richer in both O and N, suggesting cometary materials have experienced less thermal processing. Comets also appear to contain an organic fraction of intermediate volatility that has no clear meteoritic counterpart. The nature of this material is currently poorly constrained, but its composition appears to be qualitatively consistent with the kinds of organics expected to be produced by radiation processing of mixed-molecular ices.

Because the processes that led to life on Earth are poorly understood, it is difficult to assess the importance cometary organics may have had for the origin of life. However, it is becoming increasingly clear that comets contain a complex population of organic materials. This population appears to include a wider range of organics than is found in primitive meteorites, which themselves contain many species of astrobiological interest (amino acids, quinones, amphiphiles, etc.). Clearly comets contain a unique population of components that have much to tell us about the nature of interstellar and protostellar chemistry, and insights already gathered from the recent studies of cometary samples returned by the Stardust spacecraft suggest exciting progress will continue in this field in the coming years.

SUMMARY POINTS

1. Comets appear to contain abundant organic materials.

2. Cometary organics represent a complex mixture of unequilibrated species.

3. Cometary organics must represent a variety of formation processes and environments.

4. Organics from Comet 81P/Wild 2 appear to have experienced little processing while residing in the comet parent body.

5. Cometary organics include both volatile and refractory species in a wide variety of bonding states.

6. Cometary organics are richer in O and N relative to C than is typical for meteorites.

7. The presence of D and ^{15}N enrichments in cometary materials indicates that at least some of the organics in these bodies have an interstellar chemical heritage.

FUTURE ISSUES

1. What are the molecular carriers and bonding sites of D and ^{15}N isotopic enrichments in primitive extraterrestrial organic samples?

2. What do these molecular carriers and bonding sites tell us about the processes and environments in which they were made?

DISCLOSURE STATEMENT

The author is not aware of any biases that might be perceived as affecting the objectivity of this review.

ACKNOWLEDGMENTS

This work was supported by NASA grants from the Origins of Solar System, Exobiology, Cosmochemistry, and Discovery Mission Programs. The author is grateful to the many, many people who made the Stardust Mission both a reality and a joy to work on. The author is also grateful for the helpful advice and review of early versions of this manuscript provided by Dr. R. N. Zare.

LITERATURE CITED

1. Mannings V, Boss AP, Russell SS, eds. 2000. *Protostars and Planets IV.* Tucson: Univ. Ariz. Press
2. Bottke WF Jr, Cellino A, Paolicchi P, Binel RP, eds. 2002. *Asteroids III.* Tucson: Univ. Ariz. Press
3. Festou MC, Keller HU, Weaver HA, eds. 2004. *Comets II.* Tucson: Univ. Ariz. Press
4. Kerridge JF, Matthews MS, eds. 1988. *Meteorites and the Early Solar System.* Tucson: Univ. Arizona Press
5. Lauretta DS, McSween HY Jr, eds. 2006. *Meteorites and the Early Solar System II.* Tucson: Univ. Ariz. Press
6. Krot AN, Hutcheon ID, Brearley AJ, Pravdivtseva OV, Petaev MI, Hohenberg CM. 2006. Timescales and settings for alteration of chondritic meteorites. See Ref. 5, pp. 525–53
7. Pierazzo E, Chyba CF. 1999. Amino acid survival in large cometary impacts. *Meteorit. Planet. Sci.* 34:909–18
8. Ross DS. 2006. Cometary impact and amino acid survival: chemical kinetics and thermochemistry. *J. Phys. Chem. A* 110:6633–37
9. Oro J, Holzer G, Lazcano-Araujo A. 1980. The contribution of cometary volatiles to the primitive earth. In *Life Sciences and Space Research*, Vol. 18: *Proc. Open Meet. Work. Group Space Biol., Bangalore, India, May 29–June 9, 1979*, ed. R Holmquist, pp. 67–82. Oxford: Pergamon
10. Chyba CF, Thomas PJ, Brookshaw L, Sagan C. 1990. Cometary delivery of organic molecules to the early earth. *Science* 249:366–73
11. Huebner WF, Boice DC. 1992. Comets as a possible source of prebiotic molecules. *Orig. Life Evol. Biosph.* 21:299–315
12. Chyba CF, McDonald GD. 1995. The origin of life in the solar system: current issues. *Annu. Rev. Earth Planet. Sci.* 23:215–49
13. Thomas PJ, Chyba CF, McKay CP, eds. 1996. *Comets and the Origin and Evolution of Life.* New York: Springer
14. Harker DE, Wooden DH, Woodward CE, Lisse CM. 2002. Grain properties of comet C/1995 O1 Hale-Bopp. *Astrophys. J.* 580:579–97

15. Hanner MS, Bradley JP. 2004. Composition and mineralogy of cometary dust. See Ref. 3, pp. 555–64.

16. Keller LP, Bajt S, Baratta GA, Borg J, Bradley JP, et al. 2006. Infrared spectroscopy of comet 81P/Wild 2 samples returned by stardust. *Science* 314:1728–31

17. Zolensky ME, Zega TJ, Yano H, Wirick S, Westphal AJ, et al. 2006. Mineralogy and petrology of comet 81P/Wild 2 nucleus samples. *Science* 314:1735–39

18. Whipple FL. 1950. A comet model I. The acceleration of comet Encke. *Astrophys. J.* 111:375–94

19. Combi MR, Harris WM, Smyth WH. 2004. Gas dynamics and kinetics in the cometary comae: theory and observations. See Ref. 3, pp. 523–52

20. Ip WH. 2004. Global solar wind interaction and ionospheric dynamics. See Ref. 3, pp. 605–29

21. Rodgers D, Charnley SB, Huebner WF, Boice DC. 2004. Physical processes and chemical reactions in cometary comae. See Ref. 3, pp. 505–22

22. Ehrenfreund P, Charnley SB, Wooden DH. 2004. From interstellar material to cometary particles and molecules. See Ref. 3, pp. 115–33

23. Mumma MJ, Weissman PR, Stern SA. 1993. Comets and the origin of the solar system: reading the Rosetta Stone. In *Protostars and Planets III*, ed. EH Levy, JI Lunine, MS Matthews, pp. 1171–252. Tucson: Univ. Ariz. Press

24. Strazzulla G. 1999. Ion irradiation and the origin of cometary materials. *Space Sci. Rev.* 90:269–74

25. Bocklee-Morvan D, Crovisier J, Mumma MJ, Weaver HA. 2004. The composition of cometary volatiles. See Ref. 3, pp. 391–423

26. Klavetter JJ, A'Hearn MF. 1994. An extended source for CN jets in comet P/Halley. *Icarus* 107:322–34

27. Keller HU, Arpigny C, Barbieri C, Bonnet RM, Cazes S, et al. 1986. First Halley multicolor camera imaging results from Giotto. *Nature* 321:320–26

28. Brownlee DE, Hörz F, Newburn RL, Zolensky M, Duxbury TC, et al. 2004. Surface of young Jupiter family comet 81 P/Wild 2: view from the Stardust Spacecraft. *Science* 304:764–69

29. Lamy PL, Toth I, Fernandez YR, Weaver HA. 2004. The sizes, shapes, albedos, and colors of cometary nuclei. See Ref. 3, pp. 223–64

30. Oberst J, Giese B, Howington-Kraus E, Kirk R, Soderblom L, et al. 2004. The nucleus of Comet Borrelly: a study of morphology and surface brightness. *Icarus* 167:70–79

31. A'Hearn MF, Belton MJS, Delamere WA, Kissel J, Klaasen KP, et al. 2005. Deep impact: excavating comet Tempel 1. *Science* 310:258–64

32. Combes M, Moroz VI, Crifo JF, Lamarre JM, Charra J, et al. 1986. Infrared sounding of comet Halley from Vega 1. *Nature* 321:266–68

33. Combes M, Moroz VI, Crovisier J, Encrenaz T, Bibring JP, et al. 1988. The 2.5–12 µm spectrum of comet Halley from the IKS-Vega experiment. *Icarus* 76:404–36

34. Chyba C, Sagan C. 1987. Infrared emission by organic grains in the coma of comet Halley. *Nature* 330:350–53

35. Danks AC, Encrenaz T, Bouchet P, Le Bertre T, Chalabaev A. 1987. The spectrum of comet p/Halley from 3.0 to 4.0 µm. *Astron. Astrophys.* 184:329–32

36. Lisse CM, VanCleve J, Adams AC, A'Hearn MF, Fernández YR, et al. 2006. Spitzer spectral observations of the deep impact ejecta. *Science* 313:635–40

37. Allamandola LJ, Tielens AGGM, Barker JR. 1989. Interstellar polycyclic aromatic hydrocarbons: the infrared emission bands, the excitation-emission mechanism and the astrophysical implications. *Ap. J. Suppl. Ser.* 71:733–55

38. Sandford SA, Allamandola LJ, Tielens AGGM, Sellgren K, Tapia M, Pendleton Y. 1991. The interstellar C-H stretching band near 3.4 μm: constraints on the composition of organic material in the diffuse interstellar medium. *Astrophys. J.* 371:607–20

39. Sandford SA, Pendleton YJ, Allamandola LJ. 1995. The galactic distribution of aliphatic hydrocarbons in the diffuse interstellar medium. *Astrophys. J.* 440:697–705

40. Pendleton YJ, Allamandola LJ. 2002. The organic refractory material in the diffuse interstellar medium: mid-infrared spectroscopic constraints. *Astrophys. J. Suppl. Ser.* 138:75–98

41. Dworkin JP, Deamer DW, Sandford SA, Allamandola LJ. 2001. Self-assembling amphiphilic molecules: synthesis in simulated interstellar/precometary ices. *Proc. Nat. Acad. Sci. USA* 98:815–19

42. Bernstein MP, Sandford SA, Allamandola LJ, Chang S, Scharberg MA. 1995. Organic compounds produced by photolysis of realistic interstellar and cometary ice analogs containing methanol. *Astrophys. J.* 454:327–44

43. Bernstein MP, Sandford SA, Allamandola LJ, Gillette JS, Clemett SJ, Zare RN. 1999. Ultraviolet irradiation of polycyclic aromatic hydrocarbons in ices: production of alcohols, quinones, and ethers. *Science* 283:1135–38

44. Bernstein MP, Dworkin JP, Sandford SA, Cooper GW, Allamandola LJ. 2002. The formation of racemic amino acids by ultraviolet photolysis of interstellar ice analogs. *Nature* 416:401–3

45. Bernstein MP, Moore MH, Elsila JE, Sandford SA, Allamandola LJ, Zare RN. 2003. Side group addition to the PAH coronene by proton irradiation in cosmic ice analogs. *Astrophys. J.* 582:L25–29

46. Cottin H, Szopa C, Moore MH. 2001. Production of hexamethylenetetramine in photolyzed and irradiated interstellar cometary ice analogs. *Astrophys. J.* 561:L139–42

47. Muñoz Caro GM, Meierhenrich UJ, Schutte WA, Barbier B, Arcones Segovia A, et al. 2002. Amino acids from UV irradiation of interstellar analogues. *Nature* 416:403–6

48. Reinhard R. 1986. The Giotto encounter with comet Halley. *Nature* 231:313–18

49. Sagdeev RZ, Blamont J, Galeev AA, Moroz VI, Shapiro VD, et al. 1986. Vega spacecraft encounters with comet Halley. *Nature* 231:259–62

50. Boice DC, Wegmann R. 2007. The Deep Space 1 encounter with comet 19P/Borrelly. *Adv. Space Res.* 39:407–12

51. Kissel J, Brownlee DE, Buchler K, Clark BC, Fechtig H, et al. 1986. Composition of comet Halley dust particles from Giotto observations. *Nature* 321:336–37

52. Huebner WF, Boice DC, Sharp CM. 1987. Polyoxymethylene in comet Halley. *Astrophys. J.* 320:L149–52

53. Kissel J, Krueger FR. 1987. The organic component in dust from comet Halley as measured by the PUMA mass spectrometer on board Vega 1. *Nature* 326:755–60

54. Jessberger EK, Christoforidis A, Kissel J. 1988. Aspects of the major element composition of Halley's dust. *Nature* 332:691–95

55. Fomenkova MN, Chang S, Mukhin LM. 1994. Carbonaceous components in the comet Halley dust. *Geochim. Cosmochim. Acta* 58:4503–12

56. Kissel J, Glasmachers A, Grün E, Henkel Höfner H, Haerendel G, et al. 2003. Cometary and interstellar dust analyzer for comet Wild 2. *J. Geophys. Res.* 108:8114

57. Kissel J, Krueger FR, Silen J. 2005. Analysis of cosmic dust by the 'cometary and interstellar dust analyser' (CIDA) onboard the Stardust Spacecraft. *Proc. Dust in Planetary Systems, September 26–28, 2005, Kaua'i, Hawaii*, p. 95 (Abstr.). LPI Contrib. No. 1280. Houston: Lunar Planet. Inst.

58. Kissel J, Altwegg K, Clark BC, Colangeli L, Cottin H, et al. 2007. COSIMA: high resolution time-of-flight secondary ion mass spectrometer for the analysis of cometary dust particles onboard Rosetta. *Space Sci. Rev.* 128:823–67

59. Dodd RT. 1981. Carbonaceous chondrites. In *Meteorites: A Petrologic–Chemical Synthesis*. Cambridge: Cambridge Univ. Press. pp. 29–76

60. Cronin JR, Pizzarello S, Cruikshank DP. 1988. Organic matter in carbonaceous chondrites, planetary satellites, asteroids, and comets. See Ref. 4, pp. 819–57

61. Pizzarello S, Cooper GW, Flynn GJ. 2006. The nature and distribution of the organic material in carbonaceous chondrites and interplanetary dust particles. See Ref. 5, pp. 625–51

62. Sephton MA, Gilmour I. 2001. Pyrolysis-gas chromatography-isotope ratio-mass spectrometry of macromolecular material in meteorites. *Planet. Space Sci.* 49:465–71

63. Cody GD, Alexander CMO, Tera F. 2002. Solid-state (1H and 13C) nuclear magnetic resonance spectroscopy of insoluble organic residue in the Murchison meteorite: a self-consistent quantitative analysis. *Geochim. Cosmochim. Acta* 66:1851–65

64. Epstein S, Krishnamurthy RV, Cronin JR, Pizzarello S, Yuen GU. 1987. Unusual stable isotope ratios in amino acid and carboxylic acid extracts from the Murchison meteorite. *Nature* 326:477–79

65. Pizzarello S, Krishnamurthy RV, Epstein S, Cronin JR. 1991. Isotopic analyses of amino acids from the Murchison meteorite. *Geochim. Cosmochim. Acta* 55:905–10

66. Yuen G, Blair N, Des Marais DJ, Chang S. 1984. Carbon isotopic composition of low molecular weight hydrocarbons and monocarboxylic acids from Murchison meteorite. *Nature* 307:252–54

67. Pizzarello S, Huang Y, Becker L, Poreda RJ, Nieman RA, et al. 2001. The organic content of the Tagish Lake meteorite. *Science* 293:2236–39

68. Hahn JH, Zenobi R, Bada JL, Zare RN. 1988. Application of two-step laser mass spectrometry to cosmochemistry: direct analysis of meteorites. *Science* 239:1523–25

69. Zenobi R, Philippoz JM, Buseck PR, Zare RN. 1989. Spatially resolved organic analysis of the Allende meteorite. *Science* 246:1026–29

70. Krishnamurthy R, Epstein S, Cronin J, Pizzarello S, Yuen G. 1992. Isotopic and molecular analyses of hydrocarbons and monocarboxylic acids of the Murchison meteorite. *Geochim. Cosmochim. Acta* 56:4045–58

71. Cronin JR, Pizzarello S, Epstein S, Krishnamurthy RV. 1993. Molecular and isotopic analyses of the hydroxy acids, dicarboxylic acids, and hydroxydicarboxylic acids of the Murchison meteorite. *Geochim. Cosmochim. Acta* 57:4745–52

72. Deamer DW. 1985. Boundary structures are formed by the organic components of the Murchison carbonaceous chondrite. *Nature* 317:792–94

73. Deamer DW, Pashley RM. 1989. Amphiphilic components of carbonaceous meteorites. *Orig. Life Evol. Biosph.* 19:21–33

74. Cooper G, Kimmich N, Belisle W, Sarinana J, Brabham K, Garrel L. 2001. Carbonaceous meteorites as a source of sugar-related organic compounds for the early Earth. *Nature* 414:879–83

75. Chyba CF, Sagan C. 1992. Endogenous production, exogenous delivery and impact-shock synthesis of organic molecules: an inventory for the origin of life. *Nature* 355:125–32

76. Deamer D, Dworkin JP, Sandford SA, Bernstein MP, Allamandola LJ. 2002. The first cell membranes. *Astrobiology* 2:371–81

77. Pizzarello S, Huang Y. 2004. The deuterium content of individual Murchison amino acids. *Lunar Planet. Sci.* 35:1212–13 (Abstr.)

78. Poynting JH. 1903. Radiation in the Solar System: its effect on temperature and its pressure on small bodies. *Philos. Trans. R. Soc. Lond. Ser. A* 202:525–52

79. Sandford SA. 1986. Solar flare track densities in interplanetary dust particles: the determination of an asteroidal versus cometary source of the zodiacal dust cloud. *Icarus* 68:377–94

80. Liou JC, Zook HA, Jackson AA. 1995. Radiation pressure, Poynting-Robertson drag, and solar wind drag in the restricted three-body problem. *Icarus* 116:186–201

81. Liou JC, Zook HA. 1996. Comets as a source of low eccentricity and low inclination interplanetary dust particles. *Icarus* 123:491–502

82. Fraundorf P. 1980. The distribution of temperature maxima for micrometeorites decelerated in the Earth's atmosphere without heating. *Geophys. Res. Lett.* 10:765–68

83. Brownlee DE. 1985. Cosmic dust: collection and research. *Annu. Rev. Earth Planet. Sci.* 13:147–73

84. Sandford SA. 1987. The collection and analysis of extraterrestrial dust particles. *Fundam. Cosmic Phys.* 12:1–73

85. Bradley JP, Sandford SA, Walker RM. 1988. Interplanetary dust particles. See Ref. 4, pp. 861–95

86. Brownlee DE, Sandford SA. 1992. Cosmic dust. In *Exobiology in Solar System Exploration*, ed. GC Carle, DE Schwartz, JL Huntington, pp. 145–57. Washington, DC: NASA

87. Sandford SA, Bradley JP. 1989. Interplanetary dust particles collected in the stratosphere: observations of atmospheric heating and constraints on their interrelationships and sources. *Icarus* 82:146–66

88. Thomas KL, Blanford GE, Keller LP, Klock W, McKay DS. 1993. Carbon abundance and silicate mineralogy of anhydrous interplanetary dust particles. *Geochim. Cosmochim. Acta* 57:1551–66

89. Bradley JP. 1994. Nanometer-scale mineralogy and petrology of fine-grained aggregates in anhydrous interplanetary dust particles. *Geochim. Cosmochim. Acta* 58:2123–34

90. Zolensky ME, Barrett R. 1994. Chondritic interplanetary dust particles: basing their sources on olivine and pyroxene compositions. *Meteoritics* 29:616–20

91. Clemett S, Maechling C, Zare R, Swan P, Walker R. 1993. Identification of complex aromatic molecules in individual interplanetary dust particles. *Science* 262:721–25

92. Messenger S, Clemett SJ, Keller LP, Thomas KL, Chillier XDF, Zare RN. 1995. Chemical and mineralogical studies of an extremely deuterium-rich IDP. *Meteoritics* 30:546–47

93. Clemett SJ, Zare RN. 1997. Microprobe two-step laser mass spectrometry as an analytical tool for meteoritic samples. In *Molecules in Astrophysics: Probes and Processes*, ed. EF van Dishoeck, pp. 305–20. Leiden, Neth.: IAU

94. Clemett SJ, Chillier XDF, Gillette S, Zare RN, Maurette M, et al. 1998. Observation of indigenous polycyclic aromatic hydrocarbons in 'giant' carbonaceous antarctic micrometeorites. *Orig. Life Evol. Biosph.* 28:425–48

95. Allamandola LJ, Sandford SA, Wopenka B. 1987. Interstellar polycyclic aromatic hydrocarbons and carbon in interplanetary dust particles and meteorites. *Science* 237:56–59

96. Wopenka B. 1988. Raman observations on individual interplanetary dust particles. *Earth Planet. Sci. Lett.* 88:221–31

97. Quirico E, Raynal PI, Borg J, d'Hendecourt L. 2005. A micro-Raman survey of 10 IDPs and 6 carbonaceous chondrites. *Planet. Space Sci.* 53:1443–48

98. Flynn GJ, Keller LP, Jacobsen C, Wirick S. 2004. An assessment of the amount and types of organic matter contributed to the Earth by interplanetary dust. *Adv. Space Res.* 33:57–66

99. Keller LP, Messenger S. 2004. On the origin of GEMS II: spectrum imaging of GEMS grains in interplanetary dust particles. *Meteorit. Planet. Sci.* 39:5186 (Abstr.)

100. Keller LP, Messenger S, Flynn GJ, Jacobsen C, Wirick S. 2000. Chemical and petrographic studies of molecular cloud materials preserved in interplanetary dust. *Meteorit. Planet. Sci.* 35:A86–87

101. Flynn GJ, Keller LP, Feser M, Wirick S, Jacobsen C. 2003. The origins of organic matter in the solar system: evidence from the interplanetary dust particles. *Geochim. Cosmochim. Acta* 67:4791–806

102. Brownlee D, Tsou P, Aléon J, Alexander CMO, Araki T, et al. 2006. Comet 81P/Wild 2 under a microscope. *Science* 314:1711–16

103. Brownlee DE, Tsou P, Anderson JD, Hanner MS, Newburn RL, et al. 2003. Stardust: comet and interstellar dust sample return mission. *J. Geophys. Res.* 108:8111

104. Tsou P, Brownlee DE, Sandford SA, Hörz F, Zolensky ME. 2003. Wild 2 and interstellar sample collection and Earth return. *J. Geophys. Res.* 108:8113

105. Levison HF, Duncan MJ. 1997. From the Kuiper belt to Jupiter-family comets: the spatial distribution of ecliptic comets. *Icarus* 127:13–32

106. Sekanina Z, Brownlee DE, Economou TE, Tuzzolino AJ, Green SF. 2004. Modeling the nucleus and jets of comet 81P/Wild 2 based on the Stardust encounter data. *Science* 304:769–74

107. Tuzzolino AJ, Economou TE, Clark BC, Tsou P, Brownlee DE, et al. 2004. Dust measurements in the coma of comet Wild 2 by the Dust Flux Monitor Instrument. *Science* 304:1776–80

108. Hörz F, Bastien R, Borg J, Bradley JP, Bridges JC, et al. 2006. Impact features on Stardust: implications for comet 81P/Wild 2 dust. *Science* 314:1716–19

109. Flynn GJ, Bleuet P, Borg J, Bradley JP, Brenker FE, et al. 2006. Elemental compositions of comet 81P/Wild 2 samples collected by Stardust. *Science* 314:1731–35

110. McKeegan KD, Aléon J, Bradley J, Brownlee D, Busemann H, et al. 2006. Isotopic compositions of cometary matter returned by Stardust. *Science* 314:1724–28

111. Sandford SA, Aléon J, Alexander CMO, Araki T, Bajt S, et al. 2006. Organics captured from comet 81P/Wild 2 by the Stardust spacecraft. *Science* 314:1720–24

112. Spencer MK, Zare RN. 2007. Comment on "Organics captured from comet 81P/Wild 2 by the Stardust spacecraft". *Science* 317:1680

113. Sandford SA, Brownlee DE. 2007. Response to Comment on "Organics captured from comet 81P/Wild 2 by the Stardust spacecraft". *Science* 317:1680

114. Spencer MK, Clemett SJ, Sandford SA, McKay DS, Zare RN. 2007. Organic compound detection along hypervelocity particle impact tracks in Stardust aerogel. *Meteorit. Planet. Sci.* In press

115. Clemett SJ, Spencer MK, Sandford SA, McKay DS, Zare RN. 2007. Complex aromatic hydrocarbons in STARDUST samples collected from the comet 81P/Wild-2. *Meteorit. Planet. Sci.* In press

116. Stephan T, Rost D, Vicenzi EP, Bullock ES, MacPherson GJ, et al. 2008. TOF-SIMS analysis of cometary matter in Stardust aerogel tracks. *Meteorit. Planet. Sci.* 38:2346

117. Stephan T, Flynn GJ, Sandford SA, Zolensky ME. 2008. TOF-SIMS analysis of cometary particles extracted from Stardust aerogel. *Meteorit. Planet. Sci.* 38:1126

118. Rotundi A, Baratta GA, Borg J, Brucato JR, Busemann H, et al. 2008. Combined micro-Raman, micro-IR and field emission scanning electron microscope analyses of comet 81P/Wild 2 particles collected by Stardust. *Meteorit. Planet. Sci.* In press

119. Bajt S, Sandford SA, Flynn GJ, Matrajt G, Snead CJ, et al. 2008. Infrared spectroscopy of Wild 2 particle hypervelocity tracks in Stardust aerogel: evidence for the presence of volatile organics in comet dust. *Meteorit. Planet. Sci.* In press

120. Cody GD, Ade H, Alexander CMO, Araki T, Butterworth A, et al. 2008. Quantitative organic and light element analysis of comet 81P/Wild 2 particles using C-, N-, and O-μ-XANES. *Meteorit. Planet. Sci.* In press

121. Glavin DP, Dworkin JP, Sandford SA. 2008. Detection of cometary amines in samples returned by Stardust. *Meteorit. Planet. Sci.* In press

122. Matrajt G, Ito M, Wirick S, Messenger S, Brownlee DE, et al. 2007. Carbon investigation of Stardust particles: a TEM, NanoSIMS and XANES study. *Meteorit. Planet. Sci.* 38:1338

123. Messenger S. 2000. Identification of molecular-cloud material in interplanetary dust particles. *Nature* 404:968–71

124. Sandford SA, Bernstein MP, Dworkin JP. 2001. Assessment of the interstellar processes leading to deuterium enrichment in meteoritic organics. *Meteorit. Planet. Sci.* 36:1117–33

125. Aléon J, Robert F. 2004. Interstellar chemistry recorded by nitrogen isotopes in Solar System organic matter. *Icarus* 167:424–30

126. Busemann H, Young AF, Alexander CMO, Hoppe P, Mukhopadhyay S, Nittler LR. 2006. Interstellar chemistry recorded in organic matter from primitive meteorites. *Science* 312:727–30

127. Schutte WA, Allamandola LJ, Sandford SA. 1993. Organic molecule production in cometary nuclei and interstellar ices by thermal formaldehyde reactions. *Icarus* 104:118–37

128. Dalgarno A, Lepp S. 1984. Deuterium fractionation mechanisms in interstellar clouds. *Astrophys. J.* 287:L47–50

129. Tielens AGGM. 1997. Deuterium and interstellar chemical processes. In *Astrophysical Implications of the Laboratory Study of Presolar Materials*, ed. TJ Bernatowicz, EK Zinner, pp. 523–44. Woodbury, NY: Am. Inst. Phys.

130. Tielens AGGM. 1983. Surface chemistry of deuterated molecules. *Astron. Astrophys.* 119:177–84

131. Tielens AGGM. 1992. The D/H ratio in molecular clouds. In *Astrochemistry of Cosmic Phenomena*, ed. PD Singh, pp. 91–95. Dordrecht: Kluwer

132. Sandford SA, Bernstein MP, Allamandola LJ, Gillette JS, Zare RN. 2000. Deuterium enrichment of PAHs by photochemically induced exchange with deuterium-rich cosmic ices. *Astrophys. J.* 538:691–97

133. Terzieva R, Herbst E. 2000. The possibility of nitrogen isotopic fractionation in interstellar clouds. *Mon. Not. R. Astron. Soc.* 317:563–68

134. Charnley SB, Rodgers SD. 2002. The end of interstellar chemistry as the origin of nitrogen in comets and meteorites. *Astrophys. J.* 569:L133–37

135. Langer WD, Graedel TE, Frerking MA, Armentrout PB. 1984. Carbon and oxygen isotope fractionation in dense interstellar clouds. *Astrophys. J.* 277:581–604

136. Langer WD, Graedel TE. 1989. Ion-molecule chemistry of dense interstellar clouds: nitrogen-, oxygen-, and carbon-bearing molecule abundances and isotopic ratios. *Astrophys. J. Suppl. Ser.* 69:241–69

137. Sheffer Y, Lambert DL, Federman SR. 2002. Ultraviolet detection of interstellar 12C17O and the CO isotopomeric ratios toward X Persei. *Astrophys. J.* 574:L171–74

138. Lyons JR, Young ED. 2005. CO self-shielding as the origin of oxygen isotope anomalies in the early solar nebula. *Nature* 435:317–20

139. McKeegan KD, Walker RM, Zinner E. 1985. Ion microprobe isotopic measurements of individual interplanetary dust particles. *Geochim. Cosmochim. Acta* 49:1971–87

140. McKeegan KD. 1987. Oxygen isotopes in refractory stratospheric dust particles: proof of extraterrestrial origin. *Science* 237:1468–71

RELATED RESOURCES

Stardust Mission website, **http://stardust.jpl.nasa.gov/home/index.html**
NASA Curatorial website, **http://www-curator.jsc.nasa.gov/**

High-Resolution Mass Spectrometers

Alan G. Marshall[1,2] and Christopher L. Hendrickson[1,2]

[1]National High Magnetic Field Laboratory, Florida State University, Tallahassee, Florida 32310; email: marshall@magnet.fsu.edu; hendrick@magnet.fsu.edu

[2]Department of Chemistry and Biochemistry, Florida State University, Tallahassee, Florida 32306

Annu. Rev. Anal. Chem. 2008. 1:579–99

First published online as a Review in Advance on March 14, 2008

The *Annual Review of Analytical Chemistry* is online at anchem.annualreviews.org

This article's doi: 10.1146/annurev.anchem.1.031207.112945

1936-1327/08/0719-0579$20.00

Key Words

orbitrap, FTMS, reflectron TOF, Fourier transform, ion cyclotron resonance, FT-ICR

Abstract

Over the past decade, mass spectrometry has been revolutionized by access to instruments of increasingly high mass-resolving power. For small molecules up to ~400 Da (e.g., drugs, metabolites, and various natural organic mixtures ranging from foods to petroleum), it is possible to determine elemental compositions ($C_c H_h N_n O_o S_s P_p \ldots$) of thousands of chemical components simultaneously from accurate mass measurements (the same can be done up to 1000 Da if additional information is included). At higher mass, it becomes possible to identify proteins (including posttranslational modifications) from proteolytic peptides, as well as lipids, glycoconjugates, and other biological components. At even higher mass (~100,000 Da or higher), it is possible to characterize posttranslational modifications of intact proteins and to map the binding surfaces of large biomolecule complexes. Here we review the principles and techniques of the highest-resolution analytical mass spectrometers (time-of-flight and Fourier transform ion cyclotron resonance and orbitrap mass analyzers) and describe some representative high-resolution applications.

1. INTRODUCTION

Mass-resolving power: $m/\Delta m_{50\%}$, in which $\Delta m_{50\%}$ is the mass spectral peak full-width at half-maximum peak height

TOF: time-of-flight

FT: Fourier transform

MALDI: matrix-assisted laser desorption/ionization

ESI: electrospray ionization

Mass resolution: the separation (in Daltons) between two mass spectral peaks (such that the valley between the sum of the peaks is equal to the height of the smaller individual peak)

The history of spectroscopy is the history of resolution. Mass spectrometers typically measure the mass-to-charge ratio (m/z) of an ion. As mass-resolving power ($m/\Delta m_{50\%}$, see below) increases, several new plateaus of chemical information become accessible (1):

1. At $m/\Delta m_{50\%} > 100$, there is a separation of different charge states for ions derived from the same neutral analyte [e.g., $(M+H)^+$ versus $(M+2H)^{2+}$, etc.];

2. at $m/\Delta m_{50\%} > 1000$, there is a separation of peaks of different nominal mass (e.g., 325 Da versus 326 Da);

3. at $m/\Delta m_{50\%} > 10,000$, the resolution of small (<2500 Da) peptides of the same nominal mass differ by one amino acid (except for isomeric leucine and isoleucine); and

4. at $m/\Delta m_{50\%} > 100,000$, there is a separation of peaks for nominally isobaric species (i.e., molecules of the same nominal mass differing in elemental composition, e.g., N_2 versus CO, both $\sim$28 Da).

As illustrated below, these capabilities have contributed to the creation of whole new analytical fields ranging from petroleomics to proteomics.

Mass analyzers discriminate among ions of different m/z by subjecting them to constant, pulsed, or periodically time-varying electric and/or magnetic fields (2). This review focuses on the highest-resolution mass analyzers (specifically those capable of routine broadband resolving power $>10,000$) for analytical chemical and biochemical applications, namely, reflectron time-of-flight (TOF) and Fourier transform (FT) (orbitrap and ion cyclotron resonance) instruments that are compatible with liquid sample introduction and associated ionization techniques [e.g., matrix-assisted laser desorption/ionization (MALDI), electrospray ionization (ESI), and other atmospheric pressure ionization methods]. Other analytical mass analyzers, such as electric/magnetic sectors, quadrupole mass filter, ion trap, triple quadrupole, single-pass TOF, and so on, have important uses but are not optimal for the highest-resolution applications.

1.1. Mass Resolution and Mass-Resolving Power

Mass resolution is defined as the minimum mass difference, $m_2 - m_1$, between two mass spectral peaks such that the valley between their sum is a specified fraction of the height of the smaller individual peak. For example, if two equal-magnitude peaks are separated by exactly the width at half-maximum height ($\Delta m_{50\%}$) of either peak (i.e., 50% valley definition; see **Figure 1a**), then the mass resolution $m_2 - m_1 = \Delta m_{50\%}$. Mass-resolving power may be defined either for a single peak of mass, m, as $m/\Delta m_{50\%}$, or for two equal-magnitude peaks as $m_2/(m_2 - m_1)$ (e.g., for 50% valley definition, $m_2/\Delta m_{50\%}$). (For multiply charged ions, m can be replaced by m/z in the above statements.) Mass-resolving power is useful for evaluating mass analyzer performance because it is a measure of precision over a wide range of m (or m/z), whereas mass resolution determines the ability to distinguish ions of different

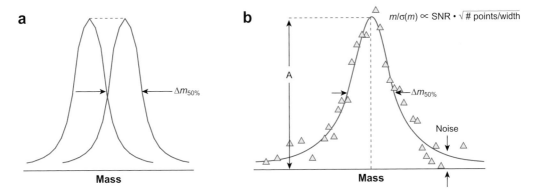

Figure 1

(*a*) Two equal-magnitude mass spectral peaks of equal width, separated by one peak width at half-maximum peak height, $\Delta m_{50\%}$. Mass resolution is typically defined as $\Delta m_{50\%}$, whereas mass-resolving power is typically defined as $m/\Delta m_{50\%}$. (*b*) Relation between the mass-measurement precision predicted for the average of many measurements, and the signal-to-noise ratio and number of data points per peak width for a single mass spectrum (see text for details).

elemental composition. For a given mass analyzer, it is important to specify the *m* (or *m/z*) value at which resolving power or resolution is reported.

1.2. Mass-Measurement Precision

Mass imprecision, $\sigma(m)$, may be defined as the root-mean-square deviation of many mass measurements. Conversely, mass precision is $1/\sigma(m)$. However, one would prefer not to have to make many measurements. For an analog (i.e., continuous) mass spectrum, mass precision is simply proportional to the spectral peak signal-to-noise ratio, S/N, for a single measurement. For a discrete mass spectrum (as in most current instruments), mass precision is given by

$$\text{Mass precision} = c(\text{S/N})\sqrt{\text{number of data points per peak width,}} \qquad (1)$$

in which *c* is a constant (of order unity) determined by the peak shape and spectral baseline noise is independent of signal (as in the FT instruments described below) (3). A similar expression applies to ion-counting mass analyzers, except that the peak height imprecision is the square root of the number of counts (4). Thus, it is possible to predict the precision that would be obtained from many measurements on the basis of the S/N and discrete sampling for a single measurement (see **Figure 1*b***).

Equation 1 shows that high mass-measurement precision requires the highest possible S/N and smallest digital point spacing. Thus, mass-measurement precision for low-magnitude peaks (as in many applications) is necessarily lower than for high-magnitude peaks (as typically advertised by mass spectrometer vendors). Equation 1 helps explain why TOF mass analyzer mass-measurement precision has improved greatly over the past decade, partly from increased resolving power but also from faster digitizers (to give more data points per peak width).

1.3. Mass Accuracy and Mass Calibration

Exact molecular mass is calculated from the combined atomic masses (usually based on the most abundant isotope for each atom) of the elemental composition (e.g., $C_cH_hN_nO_oS_sP_p$) of an ion or molecule, for example, $12.000000 + 4(1.007825) = 16.0313$ Da for CH_4. Note that exact mass of an element differs from the usual "chemical" definition of atomic mass, which is the abundance-weighted average overall isotopes of that element. Mass calibration consists of fitting the observed mass measurements to the accurate masses of two or more different ions. Calibration may be internal (i.e., the reference masses are for ions of known elemental composition in the same mass spectrum as the analyte) or external (i.e., reference masses from a mass spectrum of another analyte acquired under similar conditions). Internal calibration is typically at least twice as accurate as external calibration. The advantages of external and internal calibration may be combined if ions are alternately injected from analyte and reference samples but accumulated together, thereby allowing for real-time adjustment of the relative magnitudes of analyte and calibrant mass spectral peaks (5).

Finally, the above discussion has been limited to random noise. Systematic errors can affect the accuracy of measurements in separate spectra and even between different peaks in the same spectrum, for example, from space-charge effects due to Coulomb repulsion between ions at sufficiently high density. Thus, it is important to eliminate systematic errors, for example, by maintaining the same number of ions in each measurement (6).

2. TIME-OF-FLIGHT MASS ANALYZERS

TOF mass analysis is conceptually simple. If ions of the same initial position and velocity can be simultaneously accelerated (by a pulsed direct-current electric field) to a kinetic energy of zeV electron volts, and then allowed to fly freely (i.e., no external electric or magnetic fields) to a detector located d meters away, then ion TOF is related to m/z:

$$\text{TOF} = d\ \sqrt{1/2\ ze V}\sqrt{m/z}. \tag{2}$$

From Equation 1, it is easy to show that TOF mass-resolving power is related to time resolving power by

$$m/\Delta m = 1/2(t/\Delta t). \tag{3}$$

TOF mass analysis is inherently fast and sensitive because all masses are measured simultaneously (i.e., the multiplex advantage). Scanning instruments (for example, sectors and quadrupoles) sequentially focus only one ion mass on the detector while all others are lost.

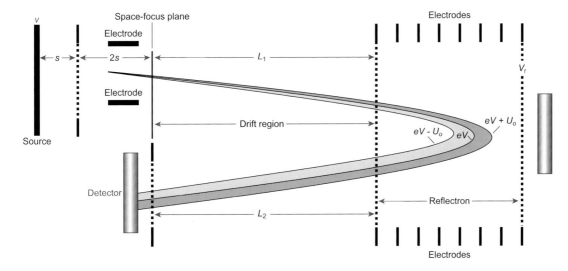

Figure 2

Schematic diagram of (*a*) a time-of-flight mass analyzer and (*b*) a reflectron time-of-flight mass analyzer. Figure adapted with permission from Reference 70.

2.1. Orthogonal Time-of-Flight Mass Spectrometry

Although it is not possible for different ions to be at the same position and velocity before acceleration, a good approximation is achieved by the orthogonal TOF design, in which a beam of ions (along, say, the x axis) is accelerated in a direction (say, the z axis) perpendicular to that of the ion beam (7). Thus, ions of different x-position and x-velocity have essentially the same initial z-position before acceleration. Collisional cooling in a radiofrequency multipole ion guide further improves the spatial and velocity distributions (8). TOF resolution can be further improved by tailoring the spatial and temporal accelerating voltage so as to focus ions of the same m/z but of different initial position and speed at a given postacceleration z-position (9).

2.2. Reflectron

Additional improvement in TOF resolution is based on an analogy to a mechanical pendulum. To a first approximation, the period of a pendulum is independent of its amplitude of oscillation. Similarly, if ions enter a region of spatially quadratic z-potential, then ions of a given m/z arriving at the same time but with different speed (and thus kinetic energy) will roll uphill until they slow to zero speed, then roll downhill and return to arrive at the detector simultaneously (see **Figure 2**). That reflectron principle (10) thus narrows the TOF distribution for each m/z and increases the TOF path length, thereby increasing mass-resolving power. Commercial orthogonal reflectron TOF mass analyzers can now routinely attain a mass-resolving power of >10,000 to yield mass accuracy of ~5–10 ppm if S/N is sufficiently high.

2.3. Multiple-Pass Time-of-Flight Mass Spectrometry

In principle, TOF reflectron mass-resolving power should increase by a factor of n, for an instrument configured to produce n reflections or energy-isochronous cycles (**Figure 3a**). In practice, a fraction of the ions is lost at each cycle, and the increase in resolving power with number of reflections is sublinear. Nevertheless, mass-resolving power of 350,000 at m/z 28 (N_2^+ versus CO^+) has been achieved for 501 cycles of a multiturn electric four-sector TOF instrument (total ion path, 644 m) (11). A second-generation toroidal design eliminates the need for numerous quadrupolar lenses (12).

If ions traverse the same path in each cycle, then the accessible mass spectral m/z range is necessarily reduced by a factor of n, so that the mass spectrum is inherently narrowband. Such devices are therefore especially suitable for targeted analysis as might be encountered in outer space and planetary exploration. Extension of the toroids in a third orthogonal direction allows for a spiral rather than planar ion trajectory (**Figure 3b**) without reduction in m/z range; an eight-turn spiral instrument has demonstrated resolving power of $m/\Delta m_{50\%} = 80,000$ at $m/z = 2564$ (the peptide ACTH18–39) (13).

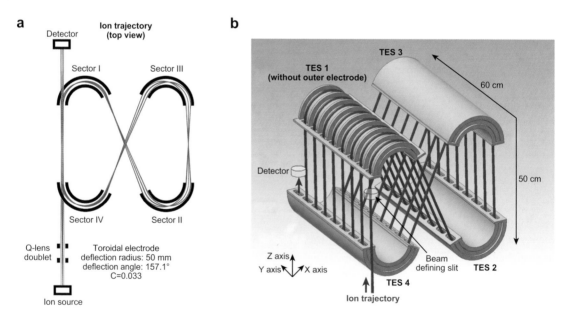

Figure 3

Two multiple-pass time-of-flight mass analyzer designs. (*a*) Multum II, with four toroidal sectors. Figure adapted with permission from Reference 71. (*b*) Spiral ion optical device. See text for discussion. Figure adapted with permission from Reference 13.

3. FOURIER TRANSFORM MASS ANALYZERS: COMMON FEATURES

3.1. Fourier Transform Spectroscopy

ICR: ion cyclotron resonance

Ion cyclotron resonance and orbitrap mass analyzers are based on measurements of rf rather than on ion deflection (electric/magnetic sectors), ion stability (quadrupole mass analyzer, quadrupole ion trap), or time of transit (TOF). To first order, ion cyclotron resonance (ICR) and orbitrap frequencies are independent of ion energy, and the induced signal is linearly proportional to ion motional amplitude. For such linear response systems, it is possible to obtain a frequency-domain spectrum by Fourier transformation of the time-domain signal for an initially spatially coherent ion packet. In ICR, initially spatially incoherent ions of a given m/z become spatially coherent following resonant rf electric field excitation; in the orbitrap, spatial coherence is achieved by the injection of ions of a given m/z in a time period that is short in comparison to one cycle of the axial trapping oscillation (see below).

Several common features of FT-ICR and orbitrap MS follow directly from FT data reduction (14). (*a*) Because signals from ions of a wide m/z range can be detected simultaneously, FT MS offers the multiplex advantage of yielding the entire mass spectrum at once, rather than requiring that each peak be scanned through separately. (*b*) Because the magnitude of the response is linearly proportional to the magnitude of the excitation, it is possible to achieve optimally selective and optimally flat-magnitude excitation by specifying its magnitude spectrum and phase-encoding it (usually a quadratic variation of phase with frequency), followed by inverse FT to generate a stored-waveform inverse Fourier transform (SWIFT) time-domain waveform (15, 16). For simple flat-magnitude excitation over a single m/z range, SWIFT reduces to linear frequency sweep excitation (17). Dipolar excitation/ejection in the orbitrap can also be done by SWIFT (18). (*c*) In the zero-collision limit, the time-domain signal is undamped, and FT-MS mass-resolving power varies directly with data acquisition period T (14). However, T is limited in two ways: On-line chromatographic sample introduction limits T to ~1 s, so as to be able to acquire data for several mass spectra during elution of a single chromatographic peak; and the time-domain signal typically decays exponentially with time constant τ. Thus, S/N decreases with time during acquisition, so that T should be no longer than 2–3 τ (19). (*d*) Digital spectral resolution may be improved by padding an N-point time-domain data set with another N zeroes (zero-filling) before discrete FT. However, further zero-fills distort the spectrum without adding new information (20). (*e*) Spectral peak shape may be smoothed by multiplying the time-domain data by any of several apodization waveforms that typically give more weight to the middle time-domain data than to the initial and later data (14). Note that apodization may make the spectra appear better, but actually reduces the information content—a good example of why one should not rely on a visual evaluation of FT spectral data. (For a truncated time-domain signal, apodization can aid in peak-picking, by reducing the Gibbs oscillation "wiggles" on either side of each peak.) In general, any weighting that effectively shortens the time-domain signal must correspondingly broaden the frequency-domain spectral peaks.

3.2. Duty Cycle: External Ion Accumulation

Both ICR and orbitrap mass analyzers typically require ~1 s for data acquisition, storage, and processing. However, many of the most useful ionization sources generate ions continuously. Therefore, to not lose ions arriving during mass analysis, it is useful to accumulate ions external to the mass analyzer and then inject them as soon as the preceding data acquisition/processing cycle is complete (in fact the same is true for TOF mass analysis). For ICR, ions are conveniently accumulated in a multipole electric ion trap (21), whereas the orbitrap employs a curved multipole electric ion trap termed the C-trap (22). The next problem is how to eject ions quickly from the external trap, to minimize spatial spreading of ions of a single *m/z*. For ICR, ion ejection may be synchronized by applying 10–30 V to tilted wires placed between the rods of a multipole electric ion trap (23), whereas for the orbitrap, ions are pulsed radially out of the C-trap by a direct-current voltage across two rods and by rapid switching of the rf potential to zero. An advantage of the orbitrap is that the external C-trap may be placed very close to the orbitrap to minimize TOF discrimination.

4. FOURIER TRANSFORM ION CYCLOTRON RESONANCE

Ions moving in a spatially uniform static magnetic field, B, rotate at a cyclotron frequency, ν_c (Hz), where

$$\nu_c = ezB/2\pi m, \qquad (4)$$

in which e is the elementary charge. At room temperature, typical ion cyclotron orbital radii are at the submillimeter level; moreover, ions of a given *m/z* rotate with random phase. Thus, to generate a detectable signal, it is necessary to resonantly excite the ions with an oscillating or rotating electric field, to yield a spatially coherent packet of ions of a given *m/z*. The motion of this ion packet gives rise to a time-domain signal consisting of the difference in current induced on a pair of opposed electrodes (see **Figure 4**). This signal is digitized and subjected to discrete fast Fourier transformation to yield a spectrum of ion cyclotron frequencies, which may then be converted to a spectrum of *m/z*.

4.1. Mass-Resolving Power

Introduced in 1974 (24), FT-ICR MS has evolved (25, 26) to become the highest-resolution broadband mass analysis technique (27). It is easy to see why. Mass-resolving power in a magnetic sector or TOF mass analyzer depends upon ion path length during the experiment. The highest-resolution double-focusing electric/magnetic sector instrument had an ion path length of ~7 m; a typical TOF instrument has a path length of ~ 1 m, and the multiturn TOF instruments cited above have path lengths ranging from ~20 to 600 m. By comparison, at 9.4 T (a magnetic field equivalent to 400 MHz for proton nuclear magnetic resonance), ions with an *m/z* of 1000 have a cyclotron frequency of 144,346 Hz and travel $2\pi \times 0.01 \times 144{,}346 = 9070$ meters in one second. Stated another way, FT-ICR resolving power (as defined above) is simply the number of cyclotron orbits during the data acquisition

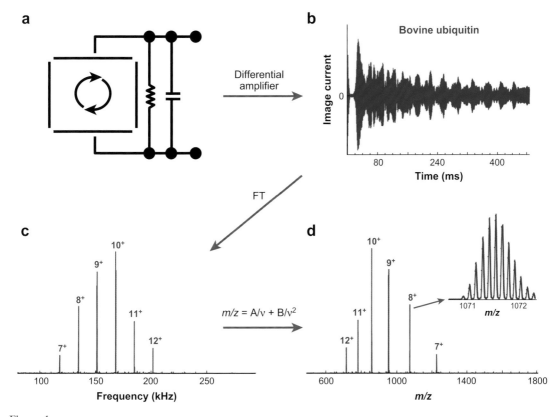

Figure 4

(*a*) Schematic representation of excited ion cyclotron rotation, (*b*) time-domain image-current signal from opposed detection electrodes, (*c*) frequency-domain spectrum obtained by fast Fourier transform of the digitized time-domain signal, and (*d*) Fourier transform–ion cyclotron resistance *m/z* spectrum obtained by calibrated frequency-to-*m/z* conversion. A full-range mass spectrum (including computation) is typically generated in ~1 s.

period (28): Thus, observation of the same 1000 *m/z* ions for 3 s yields a potential mass-resolving power of up to $m/\Delta m_{50\%} \approx 400,000$. [Even without FT, ion cyclotron frequency detected at liquid helium temperature has achieved single-ion detection and mass accuracy to within better than 10^{-9} Da (29), but for low-mass ions and very narrow bandwidth.]

From Equation 4, it is readily shown that ICR mass-resolving power varies as $1/(m/z)$. Moreover, broadband rf excitation and detection electronics are typically flat over approximately a factor of 10 in frequency, and the ion optics that accumulate (21) and transmit ions from the ionization source to the ICR cell in the bore of the magnet also limit the *m/z* range. Thus, the instrument response is optimally flat over a particular mass range ($200 < m/z < 1000$; $400 < m/z < 2000$, etc.) according to the choice of excitation amplifier, detection preamplifier, and voltage amplitude/frequency applied to ion collection and transmission multipoles.

4.2. Nonideal Electric and Magnetic Field Profiles

FT-ICR MS provides much higher resolution and mass accuracy (sub-parts-per-million) than one might expect, given that (*a*) the spatial inhomogeneity of the magnet is 5–20 ppm over the volume of the ICR trap, (*b*) the electrostatic trapping potential is not perfectly quadrupolar, and (*c*) the rf excitation electric field is not spatially uniform. First, the cyclotron rotation of the ions averages out *xy* magnet inhomogeneity, much as for spinning the sample in FT-NMR spectroscopy, and the axial oscillation of ions along the magnetic field averages out magnet axial inhomogeneity. Second, the trapping potential in an ICR cell of virtually any shape (cubic, cylindrical, rectilinear, etc.) approaches quadrupolar near the center of the cell (30); thus, limiting ICR orbital radius to ∼1/2 the cell radius reduces nonquadrupolarity. Third, the excitation electric field may be made uniform either by capacitively coupling the end cap and excitation electrodes (31) or by segmenting the end caps (infinity cell) (32).

4.3. Magnetic Mirror Effect

The high-field superconducting magnet of an FT-ICR instrument presents a special problem, namely, how to inject externally formed ions through the magnetic field gradient on their way to the ICR cell. The magnetic mirror effect can slow and even reflect off-axis ions of low kinetic energy (33). The solution is to keep the ions focused near-axis by enveloping them in multipole [e.g., quadrupole (34) or octopole (35)] electric ion guides or focusing them along the magnetic field lines with electrostatic lenses (36). The advantage of rf multipoles is that they continuously constrain ions along the magnet axis so that small misalignment between the multipole axis and magnet axis is negligible, whereas electrostatic lenses must be more accurately aligned with the magnetic field. However, multipoles limit the *m/z* range of transmitted ions.

5. ORBITRAP

The orbitrap confines ions in an electrostatic quadrologarithmic potential well (37) created between carefully shaped coaxial central and outer electrodes [see **Figure 5** (38)]. Ions are pulsed into the device so that they rotate around the central electrode and oscillate along it with axial frequencies of ∼50–150 kHz for *m/z* of 200–2000. The outer electrode is split into two halves to allow differential image-current detection (39). Unlike ICR, ion excitation is not necessary to induce large amplitude, coherent oscillations, but advantages of and uses for dipolar excitation are being explored and high-resolution ion isolation has been achieved (40, 41). Instead, ions are injected with the requisite coherent motion [by use of the C-trap (42, 43)], and detection occurs immediately after all ions have been injected into the trap and after voltage on the central electrode has stabilized (44). As for the orbitrap, it is possible to detect coherent axial oscillation in a Penning (ICR) ion trap and to FT the signal to obtain a mass spectrum (45); however, because the ion cyclotron rotational frequency is typically much greater than the axial oscillation frequency in an ICR trap, cyclotron

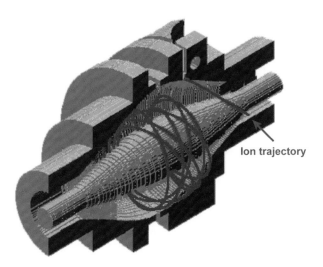

 Ion trajectory

Figure 5

Schematic diagram of an orbitrap, showing ion injection (*upper right*) and subsequent ion trajectory. Detection is based on splitting the outer ring electrode at the trap midplane and detecting the difference in charge induced on the two halves. Figure adapted with permission from Reference 38.

rotation is preferred for detection owing to its greater frequency dispersion (and thus higher mass-resolving power).

5.1. Mass-Resolving Power

As for ICR frequency, the orbitrap axial oscillation frequency is independent of ion energy and amplitude and is thus ideal for m/z analysis. As shown in Equation 5, the axial frequency depends only upon the ion m/z and the field curvature k. From Equation 5, it can be shown that mass-resolving power is one-half the frequency resolving power and is proportional to $1/(m/z)^{1/2}$ (37). In this respect, the orbitrap is unlike ICR (and similar to TOF), with the consequence that mass-resolving power decreases more slowly with m/z (or, conversely, improves less at lower m/z):

$$\nu_z = (ezk/m)^{1/2}/2\pi. \tag{5}$$

The orbitrap typically achieves a mass-resolving power of $\sim$60,000 at m/z of 400 in a 750-ms detection period. Resolving power is limited by the observation period duration, collisions with residual gas molecules, imperfections in the electric field (caused by, for example, the ion injection slot and/or inaccuracy of machining), and instability of the high-voltage power supply.

5.2. Mass Accuracy

Orbitrap mass inaccuracy is typically <5 ppm for externally calibrated mass spectra and <2 ppm for internally calibrated mass spectra (43). Because variability of external calibration is dominated by instability of the inner electrode voltage, the orbitrap and its high voltage supply are thermally regulated. Internal calibration is limited by space-charge effects. However, accurate mass measurement (<5 ppm) is maintained for dynamic ranges greater than 5000 (46), which is similar to that for ICR measurements and at least an order of magnitude higher than that for TOF (47).

5.3. Mass Range

The mass range for a single mass spectrum may be extended by the principle of electrodynamic squeezing (37, 38). Ions are injected with several keV of kinetic energy while the voltage on the central electrode is ramped monotonically to higher amplitude so that the ion motional amplitudes shrink by a few percent during an axial oscillation period, preventing ion loss due to collisions with the outer electrode and allowing ion injection to proceed over many oscillation periods (e.g., tens of microseconds for a typical mass range of $m/z = 200$–2000). A further delay of some tens of milliseconds is required for voltage stabilization before image-current detection proceeds.

6. SELECTED APPLICATIONS

6.1. High Mass Accuracy: What It Can and Cannot Provide

Every isotope of every chemical element has a different mass defect (i.e., difference between exact mass and the nearest integer). Thus, sufficiently accurate mass measurement (to within $\sim$0.0001 Da) can uniquely identify elemental composition ($C_c H_h N_n S_s O_o \ldots$). However, mass differences between isomers are too small (a difference of $\sim$1 eV in the heat of formation corresponds to a difference of $\sim$$10^{-9}$ Da) to measure in a broadband mass spectrum. The number of chemically distinct primary structures of molecules less than $\sim$1000 Da in mass has been estimated to be greater than 10^{60} (48). Even if the building blocks of a molecule were known (e.g., a peptide consisting only of amino acids), more than half the pairs of peptides differing by up to three amino acids would be positional isomers (e.g., ABC versus ACB versus BCA) or substitution isomers (e.g., LeuAsn versus ValGln) (49). Thus, mass measurement cannot always yield a unique amino acid composition for a given peptide, and mass-based protein identification typically requires MS/MS (see below) to yield a sequence tag of 4–5 consecutive amino acids.

6.2. Need for High Mass Resolution

High mass accuracy necessarily requires resolution to yield a single mass spectral peak. For example, **Figure 6** shows MS/MS product ion spectra from a marine toxin. If only a single species were present, it would be possible to locate the center of a TOF mass spectral peak to within (say) 1/100 of the peak width, to yield a mass-measurement accuracy of a few parts per million. However, the TOF measurement in this case can never identify the analyte because the signal consists of five unresolved components.

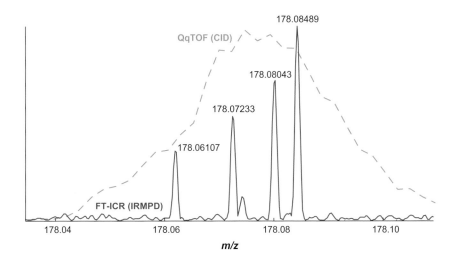

Figure 6

Isobaric species at *m/z* of 178 in the infrared multiphoton dissociation product ion spectrum of neosaxitoxin. The ion structures as well as their calculated exact masses are shown below the spectrum. For comparison, the quadrupole mass filters (Qq; see text) time-of-flight collision-induced dissociation (CID) spectrum is shown as a dashed line. This figure demonstrates the need for high resolution prior to mass measurement. Figure adapted with permission from Reference 72. Abbreviation: FT-ICR, Fourier transform–ion cyclotron resonance.

Thus, for any initially unknown sample, it is essential to perform high-resolution mass analysis first, to determine the number of species present.

6.3. Choice of Mass Analyzer

TOF mass analyzers have (in principle) no upper *m/z* limit and are thus particularly advantageous for the detection of singly charged ions greater than 5000 Da in mass (as generated by MALDI) and/or for applications requiring acquisition of more than 1 mass spectrum per second. TOF, ICR, and orbitrap mass analyzers each require a pulsed ion source: either inherently pulsed (as for MALDI) or by sudden ejection of ions accumulated continuously (as for ESI) in an external multipole ion trap. FT mass analyzers (ICR and orbitrap) are optimal for ions of *m/z* < 5000 (e.g., singly charged drugs, metabolites, and other organics and multiply charged biomacromolecules).

Relative to FT-ICR MS, the orbitrap exhibits lower mass resolution and mass accuracy, but higher sensitivity and *m/z* range when ions are injected from an external source.

6.4. Tandem Mass Spectrometry

For FT-ICR, ions may be dissociated either in the ICR cell (infrared multiphoton dissociation, electron capture dissociation) or in an external electric multipole ion trap (collision-induced dissociation, CID). (CID can be conducted in the ICR cell, but subsequent high-resolution detection requires lengthy pump-down to remove the collision gas.) Advantages of in-cell dissociation include the absence of ion loss due to transmission, and TOF discrimination between the external ion trap and the ICR cell. Moreover, electron capture dissociation (see below) is optimally conducted with simultaneous or consecutive infrared heating (50). With the orbitrap, dissociation is conducted external to the orbitrap because the product ions form over a long period compared to that of an axial oscillation, and therefore are not spatially coherent for subsequent detection. Typical MS/MS modes for the orbitrap are CID and electron transfer dissociation. Orthogonal TOF mass spectrometers typically employ quadrupole mass filters (Qq) before the TOF for CID.

7. SELECTED APPLICATIONS

7.1. Proteomics

Improvements in MS instrumentation have led to tremendous growth in the field of proteomics. High resolution (to resolve overlapping isotopic distributions and identify charge state) and accurate mass measurement [to improve identification confidence and limit search space for faster data processing (51)] have led to new proteomic methodologies. Top-down proteomics analyzes intact proteins (instead of enzymatically digested peptides) to better characterize the protein state (52, 53), and has recently been coupled to on-line liquid chromatography (54). Use of accurate mass tags for digested proteins minimizes the need for time-consuming MS/MS-based peptide identification and can, in principle, improve throughput (55). Protein quantitation is also improved by accurate mass analysis (56, 57).

7.2. Protein Modifications

Posttranslational modifications (PTMs) play a critical role in many protein functions. High-resolution, accurate mass, and sophisticated tandem MS have combined to allow identification, localization, and characterization of a variety of biologically relevant protein modifications (58). Electron capture dissociation (59, 60) and electron transfer dissociation (61) extensively cleave peptide backbone bonds while retaining labile PTMs such as phosphorylation and glycosylation, thereby enabling PTM localization. Complementary collisional or infrared multiphoton dissociation results in extensive characterization of glycosylation (50, 62). Each MS/MS technique benefits from high resolution and accurate mass, which are required to determine ion charge state, resolve closely spaced mass doublets, and confidently assign isobaric PTMs.

7.3. Metabolomics

A metabolome is the complete set of low-molecular-weight ($<\sim$1500 Da) metabolites in an organism and varies with genotype, cell cycle stage, and environment. A recently published database of the human metabolome includes $\sim$2600 endogenous metabolites (and an additional 3300 drugs and food additives) (63) that vary over many orders of magnitude in concentration and require multiple ionization techniques and both mass spectrometer polarities to observe (64). Accurate mass analysis is a critical but insufficient component of metabolite identification, owing to the presence of structural isomers and the lack of elemental constraints. Chromatographic separation reduces mass spectral complexity, improves dynamic range, and can separate isomeric structures (65). Furthermore, isotopic relative abundances and probabilistic elemental constraints can reduce the number of possible elemental compositions

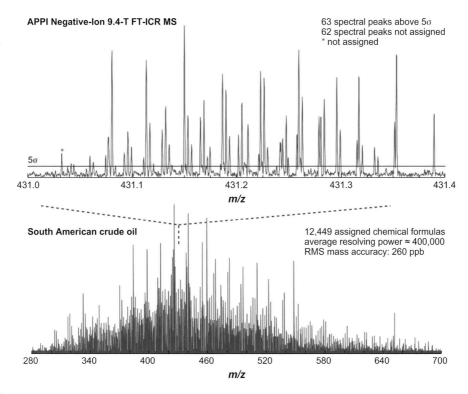

Figure 7

Atmospheric pressure photoionization negative ion 9.4-T Fourier transform–ion cyclotron resonance (FT-ICR) mass spectrum of a South American crude oil, showing the largest total number (and largest number spanning one Dalton) of assigned elemental compositions published to date. Figure adapted with permission from Reference 73.

(66, 67). For example, 3 ppm mass-measurement error coupled with 2% relative abundance accuracy reduces the number of possible elemental compositions below that of a mass spectrometer capable of 0.1 ppm mass error (FT-ICR MS is approaching that value).

7.4. Petroleomics

The most complex organic mixtures (i.e., largest number of elemental compositions within a concentration dynamic range of $\sim10^4$:1) are derived from petroleum and its products. Ultrahigh-resolution (average $m/\Delta m_{50\%} > 300,000$ for $200 < m/z < 1000$) MS can resolve and identify more than 10,000 different elemental compositions in a single mass spectrum (see **Figure 7**). Such capability has spawned the new field of petroleomics (68, 69), namely, the correlation and, ultimately, prediction of the properties and behavior of petroleum and its products on the basis of knowledge of its detailed chemical composition.

FUTURE ISSUES

1. Much of the development of high-resolution mass analyzers to date has been guided by simulated single-ion trajectories in electric and magnetic fields. Future improvements will rely on simulations that include ion-ion Coulomb interactions as well as ion image charges on nearby electrodes.

2. Up to another twofold increase in mass-resolving power can be realized by implementation of absorption-mode rather than magnitude-mode FT spectral display (14).

3. Additional potential improvement in determination of spectral peak positions and amplitudes may become available from nonuniform time-domain sampling (the FT requires equally spaced time-domain data points) and from non-FT data reduction (14).

4. Because mass-measurement accuracy is directly proportional to mass spectral peak height-to-noise ratio (3), any improvement in ionization efficiency should translate into improved mass accuracy.

DISCLOSURE STATEMENT

The authors are not aware of any biases that might be perceived as affecting the objectivity of this review.

ACKNOWLEDGMENTS

This work was supported by National Science Foundation grant DMR-06-54118, Florida State University, and the National High Magnetic Field Laboratory in Tallahassee, Florida.

LITERATURE CITED

1. Marshall AG, Hendrickson CL, Shi SD-H. 2002. Scaling MS plateaus with high-resolution FTICR MS. *Anal. Chem.* 74:252A–59A

2. Gross JH. 2004. *Mass Spectrometry: A Textbook*. Berlin: Springer. 541 pp.

3. Chen L, Cottrell CE, Marshall AG. 1986. Effect of signal-to-noise ratio and number of data points upon precision in measurement of peak amplitude, position, and width in Fourier transform spectrometry. *Chemom. Intell. Lab. Syst.* 1:51–58

4. Lee H-N, Marshall AG. 2000. Theoretical maximal precision for mass-to-charge ratio, amplitude, and width measurement in ion-counting mass analyzers. *Anal. Chem.* 72:2256–60

5. Flora JW, Harris JC, Muddiman DC. 2001. High-mass accuracy of product ions produced by SORI-CID using a dual electrospray ionization source coupled with FTICR mass spectrometry. *Anal. Chem.* 73:1247–51

6. Syka JEP, Marto JA, Bai DL, Horning S, Senko MW, et al. 2004. Novel linear quadrupole ion trap/FT mass spectrometer: performance characterization and use in the comparative analysis of histone H3 post-translational modifications. *J. Proteome Res.* 3:621–26

7. Verentchikov AN, Ens W, Standing KG. 1994. Reflecting time-of-flight mass spectrometer with an electrospray ion source and orthogonal extraction. *Anal. Chem.* 66:126–33

8. Krutchinsky AN, Chernushevich IV, Spicer VL, Ens W, Standing KG. 1998. Collisional damping interface for an electrospray ionization time-of-flight mass spectrometer. *J. Am. Soc. Mass Spectrom.* 9:569–79

9. Cotter RJ. 1997. *Time-of-Flight Mass Spectrometry: Instrumentation and Applications in Biological Research*. Washington, DC: Amer. Chem. Soc. 326 pp.

10. Mamyrin BA, Karataev VI, Shmikk DV, Zaulin VA. 1973. The mass-reflectron, a new nonmagnetic time-of-flight mass spectrometer with high resolution. *Sov. Phys. JETP* 37:45–48

11. Toyoda M, Ishihara M, Okumura D, Yamaguchi S, Katakuse I. 2001. Investigation of a multi-turn time-of-flight mass spectrometer MULTUM linear plus. *Adv. Mass Spectrom.* 15:437–39

12. Toyoda M, Okumura D, Ishihara M, Katakuse I. 2003. Multi-turn time-of-flight mass spectrometers with electrostatic sectors. *J. Mass Spectrom.* 38:1125–42

13. Satoh T, Sato T, Tamura J. 2007. Development of a high-performance MALDI-TOF mass spectrometer utilizing a spiral ion trajectory. *J. Am. Soc. Mass Spectrom.* 18:1318–23

14. Marshall AG, Verdun FR. 1990. *Fourier Transforms in NMR, Optical, and Mass Spectrometry: A User's Handbook*. Amsterdam: Elsevier. 460 pp.

15. Marshall AG, Wang T-CL, Ricca TL. 1985. Tailored excitation for Fourier transform ion cyclotron resonance mass spectrometry. *J. Amer. Chem. Soc.* 107:7893–97

16. Guan S, Marshall AG. 1993. Stored waveform inverse Fourier transform (SWIFT) axial excitation/ejection for quadrupole ion trap mass spectrometry. *Anal. Chem.* 65:1288–94

17. Comisarow MB, Marshall AG. 1974. Frequency-sweep Fourier transform ion cyclotron resonance spectroscopy. *Chem. Phys. Lett.* 26:489–90

18. Julian RK Jr, Cooks RG. 1993. Broad band excitation in the quadrupole ion trap mass spectrometer using shaped pulses with the inverse Fourier transform. *Anal. Chem.* 65:1827–33

19. Marshall AG, Comisarow MB, Parisod G. 1979. Relaxation and spectral line shape in Fourier transform ion cyclotron resonance spectroscopy. *J. Chem. Phys.* 71:4434–44

20. Bartholdi E, Ernst RR. 1973. Zero-filling analysis for FT/NMR. *J. Magn. Reson.* 11:9–19

21. Senko MW, Hendrickson CL, Emmett MR, Shi SD-H, Marshall AG. 1997. External accumulation of ions for enhanced electrospray ionization Fourier transform ion cyclotron resonance mass spectrometry. *J. Am. Soc. Mass Spectrom.* 8:970–76

22. Olsen JV, de Godoy LMF, Li G, Macek B, Mortensen P, et al. 2005. Parts per million mass accuracy on an orbitrap mass spectrometer via lock mass injection into a C-trap. *Mol. Cell. Proteomics* 4:2010–21

23. Wilcox BE, Hendrickson CL, Marshall AG. 2002. Improved ion extraction from a linear octopole ion trap: SIMION analysis and experimental demonstration. *J. Am. Soc. Mass Spectrom.* 13:1304–12

24. Comisarow MB, Marshall AG. 1974. Fourier transform ion cyclotron resonance spectroscopy. *Chem. Phys. Lett.* 25:282–83

25. Marshall AG. 2000. Milestones in Fourier transform ion cyclotron resonance mass spectrometry technique development. *Int. J. Mass Spectrom.* 200:331–36

26. Marshall AG, Hendrickson CL. 2002. Fourier transform ion cyclotron resonance detection: principles and experimental configurations. *Int. J. Mass Spectrom.* 215:59–75

27. Marshall AG, Hendrickson CL, Jackson GS. 1998. Fourier transform ion cyclotron resonance mass spectrometry: a primer. *Mass Spectrom. Rev.* 17:1–35

28. Shi SD-H, Drader JJ, Freitas MA, Hendrickson CL, Marshall AG. 2000. Comparison and interconversion of the two most common frequency-to-mass calibration functions for Fourier transform ion cyclotron resonance mass spectrometry. *Int. J. Mass Spectrom.* 195/196:591–98

29. Rainville S, Thompson JK, Pritchard DE. 2004. An ion balance for ultra-high-precision atomic mass measurements. *Science* 303:334–38

30. Guan S, Marshall AG. 1995. Ion traps for FT-ICR/MS: Principles and design of geometric and electric configurations. *Int. J. Mass Spectrom. Ion Process.* 146/147:261–96

31. Beu SC, Laude DA Jr. 1992. Open trapped ion cell geometries for FT/ICR/MS. *Int. J. Mass Spectrom. Ion Process.* 112:215–30

32. Caravatti P, Allemann M. 1991. RF shim by trap segmentation. *Org. Mass Spectrom.* 26:514–18

33. McIver RT Jr. 1990. Trajectory calculations for axial injection of ions into a magnetic field: overcoming the magnetic mirror effect with an RF quadrupole lens. *Int. J. Mass Spectrom. Ion Process.* 98:35–50

34. McIver RT Jr, Hunter RL, Bowers WD. 1985. Coupling a quadrupole mass spectrometer and a Fourier transform mass spectrometer. *Int. J. Mass Spectrom. Ion Process.* 64:67–77

35. Senko MW, Hendrickson CL, Pasa-Tolic L, Marto JA, White FM, et al. 1996. Electrospray ionization FT-ICR mass spectrometry at 9.4 tesla. *Rapid Commun. Mass Spectrom.* 10:1824–28

36. Alford JM, Williams PE, Trevor DJ, Smalley RE. 1986. Metal cluster ICR: combining supersonic metal cluster beam technology with FT-ICR. *Int. J. Mass Spectrom. Ion Process.* 72:33–51

37. Makarov A. 2000. Electrostatic axially harmonic orbital trapping: a high-performance technique of mass analysis. *Anal. Chem.* 72:1156–62

38. Hu Q, Noll RJ, Li H, Makarov A, Hardman M, Cooks RG. 2005. The orbitrap: a new mass spectrometer. *J. Mass Spectrom.* 40:430–43

39. Comisarow MB. 1978. Signal modeling for ion cyclotron resonance. *J. Chem. Phys.* 69:4097–104

40. Hu Q, Cooks RG, Noll RJ. 2007. Phase-enhanced selective ion ejection in an orbitrap mass spectrometer. *J. Am. Soc. Mass Spectrom.* 18:980–83

41. Hu Q, Makarov A, Cooks RG, Noll RJ. 2006. Resonant ac dipolar excitation for ion motion control in the orbitrap mass analyzer. *J. Phys. Chem. A* 110:2682–89

42. Kholomeev A, Makarov A, Denisov E, Lange O, Balschun W, Horning S. 2006. *Squeezing a camel through the eye of a needle: a curved linear trap for pulsed ion injection into an orbitrap analyzer.* Presented at Amer. Soc. Mass Spectrom. Conf. Mass Spectrom. & Allied Top., 54th, Seattle, Washington

43. Makarov A, Denisov E, Kholomeev A, Balschun W, Lange O, et al. 2006. Performance evaluation of a hybrid linear ion trap/orbitrap mass spectrometer. *Anal. Chem.* 78:2113–20

44. Makarov A. 2003. Interfacing the orbitrap mass analyzer to an electrospray ion source. *Anal. Chem.* 75:1699–705

45. Schweikhard L, Blundschling M, Jertz R, Kluge H-J. 1989. Fourier transform mass spectrometry without ion cyclotron resonance: direct observation of the trapping frequency of trapped ions. *Int. J. Mass Spectrom. Ion Process.* 89:R7–R12

46. Makarov A, Denisov E, Lange O, Horning S. 2006. Dynamic range of mass accuracy in LTQ orbitrap hybrid mass spectrometer. *J. Am. Soc. Mass Spectrom.* 17:977–82

47. Blom KF. 2001. Estimating the precision of exact mass measurements in an orthogonal time-of-flight mass spectrometer. *Anal. Chem.* 73:715–19

48. Hertkorn N, Ruecker C, Meringer M, Gugisch R, Frommberger M, et al. 2007. High-precision frequency measurements: indispensable tools at the core of the molecular-level analysis of complex systems. *Anal. Bioanal. Chem.* 389:1311–27

49. He F, Emmett MR, Håkansson K, Hendrickson CL, Marshall AG. 2004. Theoretical and experimental prospects for protein identification based solely on accurate mass measurement. *J. Proteome Res.* 3:61–67

50. Håkansson K, Chalmers MJ, Quinn JP, McFarland MA, Hendrickson CL, Marshall AG. 2003. Combined electron capture and infrared multiphoton dissociation for multistage MS/MS in an FT-ICR mass spectrometer. *Anal. Chem.* 75:3256–62

51. Zubarev R, Mann M. 2007. On the proper use of mass accuracy in proteomics. *Mol. Cell. Proteomics* 6:377–81

52. Kelleher NL. 2004. Top down proteomics. *Anal. Chem.* 76:196A–203A

53. Waanders MB, Olsen JV, Mann M. 2006. Top-down protein sequencing and MS3 on a hybrid linear quadrupole ion trap–orbitrap mass spectrometer. *Mol. Cell. Proteomics* 5:949–58

54. Parks BA, Jiang L, Thomas PM, Wenger CD, Roth MJ, et al. 2007. Top-down proteomics on a chromatographic time scale using linear ion trap Fourier transform hybrid mass spectrometers. *Anal. Chem.* 79:7984–91

55. Bogdanov B, Smith RD. 2005. Proteomics by FT-ICR mass spectrometry: top down and bottom up. *Mass Spectrom. Rev.* 24:168–200

56. Gruhler A, Olsen JV, Mohammed S, Mortensen P, Faergeman NJ, et al. 2005. Quantitative phosphoproteomics applied to the yeast signaling pathway. *Mol. Cell. Proteomics* 4:310–27

57. Venable JD, Wholschlegel J, McClatchey DB, Park SK, Yates JR. 2007. Relative quantification of stable isotope labeled peptides using a linear ion trap–orbitrap hybrid mass spectrometer. *Anal. Chem.* 79:3056–64

58. Meng F, Forbes AJ, Miller LM, Kelleher NL. 2005. Detection and localization of protein modifications by high-resolution tandem mass spectrometry. *Mass Spectrom. Rev.* 24:126–34

59. Cooper HJ, Håkansson K, Marshall AG. 2005. The role of electron capture dissociation in biomolecular analysis. *Mass Spectrom. Rev.* 24:201–22

60. Zubarev RA, Kelleher NL, McLafferty FW. 1998. Electron capture dissociation of multiply charged protein cations: a nonergodic process. *J. Am. Chem. Soc.* 120:3265–66

61. Syka JEP, Coon JJ, Schroeder MJ, Shabanowitz J, Hunt DF. 2004. Peptide and protein sequence analysis by electron transfer dissociation mass spectrometry. *Proc. Natl. Acad. Sci. USA* 101:9528–33

62. Håkansson K, Cooper HJ, Emmett MR, Costello CE, Marshall AG, Nilsson CL. 2001. Electron capture dissociation and infrared multiphoton dissociation MS/MS of an N-glycosylated tryptic peptide to yield complementary sequence information. *Anal. Chem.* 73:4530–36

63. Wishart DS. 2007. Proteomics and the human metabolome project. *Expert Rev. Proteomics* 4:333–35

64. Aharoni A, Ric de Vos CH, Verhoeven HA, Maliepaard CA, Kruppa G, et al. 2002. Nontargeted metabolome analysis by use of Fourier transform ion cyclotron resonance mass spectrometry. *Omics* 6:217–34

65. Ding J, Sorensen CM, Zhang Q, Jiang H, Jaitly N, et al. 2007. Capillary LC coupled with high-mass measurement accuracy mass spectrometry for metabolic profiling. *Anal. Chem.* 79:6081–93

66. Kind T, Fiehn O. 2006. Metabolomic database annotations via query of elemental compositions: mass accuracy is insufficient even at less than 1 ppm. *BMC Bioinformatics* 7:234

67. Kind T, Fiehn O. 2007. Seven golden rules for heuristic filtering of molecular formulas obtained by accurate mass spectromtry. *BMC Bioinformatics* 8:105

68. Marshall AG, Rodgers RP. 2004. Petroleomics: the next grand challenge for chemical analysis. *Acc. Chem. Res.* 37:53–59

69. Rodgers RP, Marshall AG. 2006. Chapter 3. Petroleomics: advanced characterization of petroleum-derived materials by Fourier transform ion cyclotron resonance mass spectrometry (FT-ICR MS). In *Asphaltenes, Heavy Oils, and Petroleomics*, ed. OC Mullins, EY Sheu, A Hammami, AG Marshall. New York: Springer. pp. 63–93

70. Cotter RJ. 1999. The new time-of-flight mass spectrometry. *Anal. Chem.* 71:445A–51A

71. Okamura D, Toyoda M, Ishihara M, Katakuse I. 2004. A compact sector-type multi-turn time-of-flight mass spectrometer 'MULTUM II.' *Nucl. Instrum. Methods Phys. Res. A* 519:331–37

72. Sleno L, Volmer DA, Marshall AG. 2005. Assigning product ions from complex MS/MS spectra: the importance of mass uncertainty and resolving power. *J. Am. Soc. Mass Spectrom.* 16:183–98

73. Purcell JM, Hendrickson CL, Rodgers RP, Marshall AG. 2006. Atmospheric pressure photoionization Fourier transform ion cyclotron resonance mass spectrometry for complex mixture analysis. *Anal. Chem.* 78:5906–12

Surface-Enhanced Raman Spectroscopy

Paul L. Stiles, Jon A. Dieringer, Nilam C. Shah, and Richard P. Van Duyne

Department of Chemistry, Northwestern University, Evanston, Illinois 60208;
email: vanduyne@chem.northwestern.edu

Annu. Rev. Anal. Chem. 2008. 1:601–26

First published online as a Review in Advance on
March 18, 2008

The *Annual Review of Analytical Chemistry* is online
at anchem.annualreviews.org

This article's doi:
10.1146/annurev.anchem.1.031207.112814

Key Words

nanoparticles, sensing, plasmonics, electromagnetic enhancement, nanofabrication, vibrational spectroscopy

Abstract

The ability to control the size, shape, and material of a surface has reinvigorated the field of surface-enhanced Raman spectroscopy (SERS). Because excitation of the localized surface plasmon resonance of a nanostructured surface or nanoparticle lies at the heart of SERS, the ability to reliably control the surface characteristics has taken SERS from an interesting surface phenomenon to a rapidly developing analytical tool. This article first explains many fundamental features of SERS and then describes the use of nanosphere lithography for the fabrication of highly reproducible and robust SERS substrates. In particular, we review metal film over nanosphere surfaces as excellent candidates for several experiments that were once impossible with more primitive SERS substrates (e.g., metal island films). The article also describes progress in applying SERS to the detection of chemical warfare agents and several biological molecules.

1. INTRODUCTION

SERS: surface-enhanced Raman spectroscopy

Raman scattering: inelastic scattering of a photon from a molecule in which the frequency change precisely matches the difference in vibrational energy levels

LSPR: localized surface plasmon resonance

It has been 30 years since it was recognized that the Raman spectra of submonolayer coverages of molecules could be acquired on electrochemically roughened coinage metal surfaces (1, 2). Since then, the field of surface-enhanced Raman spectroscopy (SERS) has grown dramatically, demonstrating its power as an analytical tool for the sensitive and selective detection of molecules adsorbed on noble metal nanostructures. Over 5000 research articles, 100 review articles, and several books on SERS have appeared in the literature. Such a large research database provides ample testimony of the impact of SERS on both fundamental and applied studies in fields as diverse as chemistry, physics, materials science, surface science, nanoscience, and the life sciences (3–16).

Understanding the mechanisms of SERS has been a struggle since the early days of its inception, when the primary goal was merely explaining the 10^6-fold intensity enhancement of the normal Raman scattering cross-section. At the time, the enhancement factor, $EF_{SERS} = 10^6$, could be understood as the product of two contributions: (*a*) an electromagnetic enhancement mechanism and (*b*) a chemical enhancement mechanism. These two mechanisms arise because the intensity of Raman scattering is directly proportional to the square of the induced dipole moment, μ_{ind}, which, in turn, is the product of the Raman polarizability, α, and the magnitude of the incident electromagnetic field, E. As a consequence of exciting the localized surface plasmon resonance (LSPR) of a nanostructured or nanoparticle metal surface, the local electromagnetic field is enhanced by a factor of 10, for example. Because Raman scattering approximately scales as E^4, the electromagnetic enhancement factor is of order 10^4. Researchers viewed the chemical enhancement factor of 10^2 as arising from the excitation of adsorbate localized electronic resonances or metal-to-adsorbate charge-transfer resonances (e.g., resonance Raman scattering). It is also worthwhile to note that surface-enhanced resonance Raman scattering with combined SERS and resonance Raman scattering enhancement factors in the 10^9–10^{10} range was possible at the time.

Despite this enhancement, which made SERS orders of magnitude more sensitive than normal Raman spectroscopy, the full power of SERS unfortunately was not utilized until this past decade. Early systems suffered immensely from irreproducibility owing to ill-defined substrates fabricated by electrochemical roughening. Recent advances in nanofabrication and the 1997 discovery of single-molecule SERS (31, 32), the ultimate limit of detection, have caused an explosion of new research and the extension of SERS from an interesting physical phenomenon to a robust and effective analytical technique. In addition, surface preparation and modification techniques (26–28, 77) have also allowed for analyte selectivity. As such, investigators have applied SERS to many analytical systems in recent years, including anthrax detection (29, 77), chemical warfare–stimulant detection (17), in vitro (6, 30, 79) and in vivo (78) glucose sensing, environmental monitoring (18), the monitoring of heterogeneous catalytic reactions (19), and explosive-agent detection (20). The need for new SERS-based sensing, detecting, and monitoring platforms has also driven new instrumental techniques such as the integration of SERS active substrates into fiber-optic assemblies (18).

This review first describes the SERS electromagnetic mechanism (EM) from a practical standpoint, not dwelling on lengthy equations and derivations, by taking a simple spherical model to extract the analytical features of the EM such as wavelength dependence, distance dependence, and overall enhancement factor. Then, we give an overview of modern nanofabrication techniques for SERS active surfaces. Finally, we conclude with examples of the analytical applications of SERS performed by our research group.

2. FUNDAMENTALS OF THE SURFACE-ENHANCED RAMAN SPECTROSCOPY ELECTROMAGNETIC MECHANISM

When an electromagnetic wave interacts with a metal surface, the fields at the surface are different than those observed in the far field. If the surface is rough, the wave may excite localized surface plasmons on the surface, resulting in amplification of the electromagnetic fields near the surface. If one assumes that there is enhancement of the intensity of the incident and scattered fields (albeit at different wavelengths), then the possibility of a large enhancement of Raman scattering intensity arises. This notion is commonly known as the EM of SERS and has served as an important model in the understanding of SERS since its discovery in the 1970s (1, 2, 21). Although this mechanism has been the topic of multiple reviews (5, 22–25) and is reasonably well understood, the reinvigorated interest in SERS due to better-defined substrates (26–28), its application to sensing (29, 30), and the observation of single-molecule surface-enhanced resonance Raman scattering (31, 32) has driven a new wave of experiments designed to lay a stronger foundation and provide deeper insight into the EM. These experiments include testing the wavelength dependence of SERS (33, 34) and the distance dependence of SERS (35, 36) and making theoretical calculations to link the small, arbitrarily shaped noble-metal nanoparticle electromagnetic enhancement to observed SERS enhancement factors (37, 38). We begin our treatment of the EM with the origin of electromagnetic enhancement on the nanoscale, the LSPR.

2.1. The Localized Surface Plasmon Resonance

The LSPR occurs when the collective oscillation of valence electrons in a coinage metal nanoparticle is in resonance with the frequency of incident light (**Figure 1**). Understanding features in the characteristic extinction spectrum (absorption plus scattering) can be complex owing to the effects of the dielectric environment, size, and shape on the full width at half-maximum and spectral location. As such, more detailed explanations are available elsewhere, and we concentrate only on features of the LSPR that are directly applicable to the SERS EM (39). Here we consider a quasistatic approach using a spherical nanoparticle of radius a, irradiated by z polarized light of wavelength λ, in the long wavelength limit ($a/\lambda < 0.1$). This leads to the assumption that the electric field around the nanoparticle is uniform (**Figure 1**), which allows Maxwell's equations to be replaced by the Laplace equation of electrostatics (25). Although this approach is useful for predicting the properties of the SERS EM,

Electromagnetic mechanism (EM): theory describing the magnitude of the SERS enhancement factor arising from field enhancement of both the incident and scattered fields

Localized surface plasmon: electromagnetic field–driven coherent oscillation of the surface conduction electrons in a material with negative real and near-zero imaginary dielectric constants

Extinction spectrum: absorption plus elastic scattering spectrum

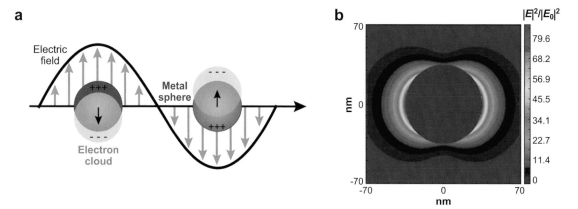

Figure 1

(*a*) Illustration of the localized surface plasmon resonance effect. (*b*) Extinction efficiency (ratio of cross section to effective area) of a spherical silver nanoparticle of 35-nm radius in vacuum $|E|^2$ contours for a wavelength corresponding to the plasmon extinction maximum. Peak $|E|^2 = 85$.

it should not be used as a quantitative treatment as the LSPR extinction of small spherical particles lies in the ultraviolet and is damped by interband transitions and other excitations (12, 40). The resulting analytical solution for the magnitude of the electromagnetic field outside the particle, E_{out}, is given by

$$E_{out}(x, y, z) = E_0\hat{\mathbf{z}} - \alpha E_0 \left[\frac{\hat{\mathbf{z}}}{r^3} - \frac{3z}{r^5} \left(x\hat{\mathbf{x}} + y\hat{\mathbf{y}} + z\hat{\mathbf{z}} \right) \right], \tag{1}$$

where x, y, and z are the usual Cartesian coordinates; r is the radial distance; $\hat{\mathbf{x}}$, $\hat{\mathbf{y}}$, and $\hat{\mathbf{z}}$ are the Cartesian unit vectors; and α is the metal polarizability expressed as

$$\alpha = ga^3, \tag{2}$$

where a is the radius of the sphere and g is defined as

$$g = \frac{\varepsilon_{in} - \varepsilon_{out}}{(\varepsilon_{in} + 2\varepsilon_{out})}. \tag{3}$$

Here ε_{in} is the dielectric constant of the metal nanoparticle, and ε_{out} is the dielectric constant of the external environment. The magnitude of the enhancement is wavelength dependent, owing to the strong wavelength dependence of the real portion of the metal nanoparticle dielectric constant. As such, the maximum enhancement occurs when the denominator of g approaches zero ($\varepsilon_{in} \approx -2\varepsilon_{out}$). Examining Equation 1, we also can see that the field enhancement decays with r^{-3}, implying the existence of a finite sensing volume around the nanoparticle. (We discuss these implications in later sections.)

According to Mie theory, the extinction spectrum, $E(\lambda)$, of an arbitrarily shaped nanoparticle is given by

$$E(\lambda) = \frac{24\pi^2 N a^3 \varepsilon_{out}^{3/2}}{\lambda \ln(10)} \left[\frac{\varepsilon_i(\lambda)}{(\varepsilon_r(\lambda) + \chi \varepsilon_{out})^2 + \varepsilon_i(\lambda)^2} \right], \tag{4}$$

where ε_r and ε_i are the real and imaginary components of the metal dielectric function ε_{in}, respectively (41, 42). Again, the real portion of the metal dielectric constant is wavelength dependent. This equation is also generalized from the small sphere solution by replacing the factor of 2 that appears with ε_{out} in Equation 3 with χ, where χ is the shape factor that accounts for deviation from spherical particle geometries into higher aspect ratio structures. This assumption is made because Mie theory cannot be solved analytically for structures other than spheroids. Because χ amplifies ε_{out}, the shape factor is generalized as the sensitivity of the LSPR extinction spectrum to the dielectric environment. The value of χ is 2 for the case of a sphere, but it can take on values as large as 20 (43). The dielectric resonance condition ($\varepsilon_r \approx -\chi \varepsilon_{out}$) is met in the visible region of the spectrum for nanoparticles of high aspect ratio (large χ) comprising coinage metals such as silver and gold. Therefore, it is trivial to perform SERS on pre-existing Raman spectroscopic instrumentation without modification when using these substrates.

Because one can only solve the extinction spectrum analytically for spheres and spheroids, it must be approximated for all other geometries (28, 43). Researchers have developed numerical methods that represent the nanoparticle target with N finite polarizable elements commonly referred to as dipoles. These coupled dipoles interact with the applied field, and by applying the Claussius-Mossotti polarizabilities with some radiative reaction correction (44, 45), one can calculate the extinction and scattering spectra of the target. These methods include the discrete dipole approximation and the finite-difference time domain methods (37, 45, 46), and calculated results typically match well with experiment.

2.2. E^4 Enhancement

In Raman spectroscopy, the scattered intensity is linear with the incident field intensity, E_0^2 (47). Because this field intensity is magnified at the nanoparticle surface, the Raman intensity is therefore related to the absolute square of E_{out} evaluated at the surface of the nanoparticle ($r = a$). Manipulating Equations 1–3, this is given for a small metal sphere by

$$|\mathbf{E}_{out}|^2 = E_0^2[|1 - g|^2 + 3\cos^2\theta(2\mathrm{Re}(g) + |g|^2)], \tag{5}$$

where θ is the angle between the incident field vector and the vector to the position of the molecule on the surface. The peak enhancement occurs when θ is $0°$ or $180°$, corresponding to the molecule being on the axis of light propagation, and in cases in which g is large, the maximum enhancement approaches $|\mathbf{E}_{out}|^2 = 4E_0^2 |g|^2$. **Figure 2** shows the field intensity as a function of position for a spherical particle. The ratio of

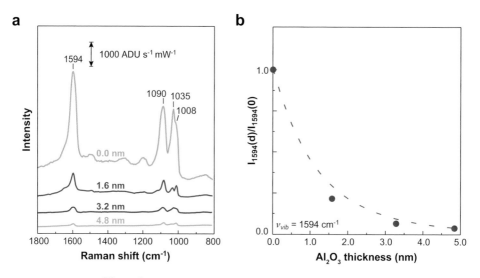

a

1594

1000 ADU s⁻¹ mW⁻¹

1090 1035
            1008

0.0 nm

1.6 nm

3.2 nm

4.8 nm

Intensity

1800 1600 1400 1200 1000 800

Raman shift (cm⁻¹)

b

1.0

$I_{1594}(d)/I_{1594}(0)$

ν_{vib} = 1594 cm⁻¹

0.0

0.0   1.0   2.0   3.0   4.0   5.0

Al₂O₃ thickness (nm)

Figure 2

(*a*) Surface-enhanced Raman spectra of pyridine adsorbed to silver film over nanosphere samples treated with various thicknesses of alumina (0.0 nm, 1.6 nm, 3.2 nm, and 4.8 nm). ex = 532 nm, P = 1.0 mW, and acquisition time = 300 s. (*b*) Plot of surface-enhanced Raman spectroscopy intensity as a function of alumina thickness for the 1594 cm⁻¹ band (circles). The dashed curved line is a fit of this data to Equation 8 (35). Reproduced with permission from the Royal Society of Chemistry.

the maximum to minimum intensity on a metal sphere is 4, and the radially averaged intensity is $|\overline{\mathbf{E}}_{out}|^2 = 2E_0^2|g|^2$.

In Raman scattering, the applied field induces an oscillating dipole in the molecule on the surface. This dipole then radiates, and there is a small probability that the radiated light is Stokes shifted by the vibrational frequency of the molecule. Although Equation 5 gives a general expression for the enhancement of the incident field, the emission of radiation from the dipole may be enhanced. The treatment for the enhanced emission intensity is more complex and has received proper treatment by Kerker and colleagues (48, 49), but a first-order approximation is to use an expression similar to Equation 5, except evaluated at the Raman Stokes-shifted frequency. Therefore, in this approximation, we can write the following expression:

$$EF = \frac{|\mathbf{E}_{out}|^2|\mathbf{E}'_{out}|^2}{|\mathbf{E}_0|^4} = 4|g|^2|g'|^2, \tag{6}$$

Enhancement factor (EF): magnitude of increase in Raman scattering cross section when the molecule is adsorbed to a SERS-active substrate

where the primed symbols refer to the field evaluated at the scattered frequency. We define this expression as the theoretical SERS EM enhancement factor (EF), and if the Stokes shift is small, both g and g' are at approximately the same wavelength, and the EF scales as g^4. In the literature, this is commonly referred to as E⁴ enhancement or the fourth power of field enhancement at the nanoparticle surface. Although this expression was derived using certain approximations, it is identical to that of Kerker's more rigorous approach. Assuming $|g|$ is approximately 10 for a small sphere in this model, then the magnitude of the SERS EM enhancement is 10^4 to 10^5, and $|g|$ can be much larger in higher-order silver nanostructures in which the EM EFs can approach 10^8 (34).

Although this theoretical treatment of SERS is useful in determining the origin of the often-stated E^4 enhancement approximation, in practical use, it is often simpler to experimentally measure the EF analytically than to predict it theoretically. The EF for a SERS system can be described by

$$EF = \frac{[I_{SERS}/N_{surf}]}{[I_{NRS}/N_{vol}]}, \tag{7}$$

which, evaluated at a single excitation wavelength, describes the average Raman enhancement and accounts for the enhancement of both the incident excitation and the resulting Stokes-shifted Raman fields, where I_{SERS} is the surface-enhanced Raman intensity, N_{surf} is the number of molecules bound to the enhancing metallic substrate, I_{NRS} is the normal Raman intensity, and N_{vol} is the number of molecules in the excitation volume (34, 50). Practically, for a given molecule, one must measure I_{SERS} and I_{NRS} independently and be careful when evaluating spot size and probe volume to determine the EF analytically. McFarland et al. (34) give a detailed approach for measuring SERS EFs.

2.3. Distance Dependence

The SERS distance dependence is critical both mechanistically and practically. The EM predicts that SERS does not require the adsorbate to be in direct contact with the surface but within a certain sensing volume. From a practical perspective, there are certain experiments, such as those involving surface-immobilized biological molecules (51), in which direct contact between the adsorbate of interest and the surface is not possible because the surface is modified with a capture layer for specificity or biocompatibility. Because the field enhancement around a small metal sphere decays with r^{-3}, using the E^4 approximation, the overall distance dependence should scale with r^{-12}. Taking into account the increased surface area scaling with r^2 as one considers shells of molecules at an increased distance from the nanoparticle, one should experimentally observe the r^{-10} distance dependence:

$$I_{SERS} = \left(\frac{a+r}{a}\right)^{-10}, \tag{8}$$

where I_{SERS} is the intensity of the Raman mode, a is the average size of the field-enhancing features on the surface, and r is the distance from the surface to the adsorbate (36).

The ideal distance-dependence experiment is one in which the thickness of the spacer layer could be easily varied in thickness from a few angstroms to tens of nanometers. Furthermore, the spacers would be conformal to handle roughened and nanostructured surfaces, pinhole free, and chemically uniform. Atomic layer

Atomic layer deposition (ALD): self-limiting growth process performed by alternating dosing and saturation of precursors

FON: film over nanosphere

deposition (ALD) is such a spacer fabrication method that produces highly uniform and controlled thin films (52). Precursor gases are alternately pulsed through a reactor and purged away, resulting in a self-limiting growth process that constructs one film layer at a time. A specific number of layers of Al_2O_3 can be deposited with subnanometer-thickness resolution.

We have deposited Al_2O_3 multilayers onto Ag film over nanosphere (AgFON) surfaces to probe the distance dependence of SERS. **Figure 3a** shows the SER spectra for pyridine adsorbed on AgFON surfaces coated with four different thicknesses

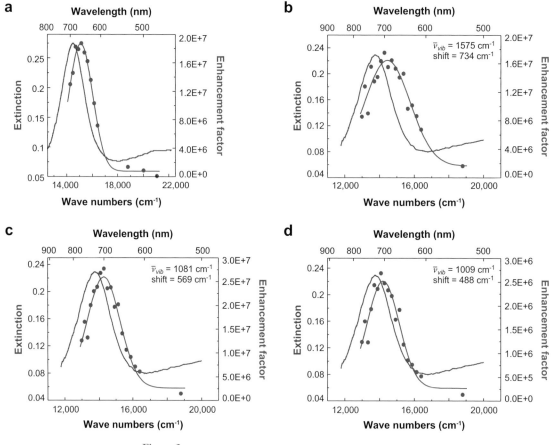

Figure 3

(a) Surface-enhanced Raman excitation spectra with Si as intensity standard. Localized surface plasmon resonance (LSPR) $\lambda_{max} = 690$ nm, profile fit maximum at $\lambda_{ex,max} = 662$ nm. (b–d) Effect of Stokes Raman shift. (b) Profile of the 1575 cm^{-1} vibrational mode of benzenethiol. Distance between LSPR λ_{max} and excitation profile fit line $\lambda_{ex,max} = 734$ cm^{-1}. Enhancement factor (EF) $= 1.8 \times 10^7$. (c) 1081 cm^{-1} vibrational mode, shift $= 569$ cm^{-1}, and EF $= 2.8 \times 10^7$. (d) 1009 cm^{-1} vibrational mode, shift $= 488$ cm^{-1}, and EF $= 2.7 \times 10^6$. Reproduced with permission from Reference 34. Copyright 2005, American Chemical Society.

of ALD-deposited Al_2O_3. **Figure 3b** shows a plot of the relative intensity of the 1594 cm^{-1} band as a function of Al_2O_3 thickness. Fitting the experimental data to Equation 8 leads to the average size of the enhancing particle, $a = 12.0$ nm. The term d_{10} defines the surface-to-molecule distance required to decrease the SERS intensity by a factor of ten. The data presented in this work clearly show that SERS is a long-range effect with a d_{10} value for this particular surface nanostructure of 2.8 nm. This value was derived assuming that a complete monolayer was formed with each ALD cycle. Quartz crystal microbalance measurements have shown that the average thickness of the ALD sequence is 1.1 Å and consequently a complete layer of alumina is not formed with each cycle when repeated layers are formed. The mechanism for the formation of alumina on silver is still debated, but assuming that the quartz crystal microbalance measurements accurately model the first few layers of ALD deposition on silver, the measured value of d_{10} could be as low as 1 nm.

2.4. Surface-Enhanced Raman Spectroscopy Wavelength Dependence

The E^4 enhancement approximation predicts that the best spectral location of the LSPR for maximum EM enhancement is coincident with the laser excitation wavelength. This would lead to maximum enhancement of the incident field intensity at the nanoparticle surface. In practice, this is not the case. According to the theoretical treatment presented in Section 2.2, it is necessary to achieve electromagnetic enhancement of both the incident field and the radiated field, which, for Raman scattering, is at a different wavelength. When the vibrational energy spacing corresponds to the fingerprint region (500–1500 cm^{-1}), the small Stokes-shift approximation breaks down and, consequently, so does the E^4 approximation. Although experiments were proposed at the inception of SERS to measure this effect, studies performed were inconclusive owing to limitations in instrumentation and poor definition of the SERS substrates (53–55). The ideal experiment is to have a continuously tunable excitation and detection scheme over the bandwidth of a well-defined LSPR.

A recently completed systematic study (34) investigated the optimum excitation wavelength as a function of the spectral position of the maximum LSPR extinction. This was made possible by recent advances in nanofabrication and better tunability in both the SERS excitation and detection. **Figure 3a** shows a characteristic wavelength scanned excitation profile of the 1081 cm^{-1} peak from benzenethiol. The excitation profile shows the highest SERS EF occurs when the excitation wavelength is higher in energy than the spectral maximum of the LSPR extinction spectrum. A Gaussian fit was made to the profile and distance in energy of the maximum of the excitation profile, and the LSPR extinction spectrum is 613 cm^{-1}, which is on the order of half the vibrational energy, 1081 cm^{-1}. To generalize the magnitude of the profile shift, the experiment was repeated for three vibrational modes in benzenethiol: 1009 cm^{-1}, 1081 cm^{-1}, and 1575 cm^{-1} (**Figure 3b–d**). As the vibrational energy increases, so does the difference in energy of the maximum of the excitation profile and the maximum of the LSPR extinction spectrum. For all three modes, the magnitude of the separation was approximately half, which agrees with the EM as it predicts that the enhancement of both the incident and scattered fields is important to maximize the SERS EF. The enhancement of both fields occurs optimally when the frequencies of the incident and scattered fields straddle the LSPR extinction spectrum.

The experiment was also performed on multiple silver nanoparticle sizes and shapes with extinction spectra at various locations in the visible region of the electromagnetic spectrum. The data measured in the experiment are available elsewhere (34), but in all cases, the magnitude of the energy separation between the excitation profile maximum and the LSPR extinction maximum is roughly half. Consequently, to achieve maximum EM enhancement, one should either prepare the sample such that the LSPR extinction is in the proper location for fixed laser frequency apparatuses or set the wavelength to a higher frequency than the LSPR extinction in tunable systems. In the wavelength-scanned experiment, the relevant LSPR extinction spectrum is the one observed after adsorption of the analyte as the LSPR extinction spectrum is sensitive to the dielectric environment. These experiments illustrate the importance of optimizing the plasmon and excitation wavelengths to achieve maximum SERS enhancements.

3. EXPERIMENTAL DETAILS

3.1. Instrumentation

Figure 4 gives a schematic representation of two instrumental approaches to the measurement of SER spectra. The first approach (**Figure 4a**) is that of the macro-Raman configuration. In this setup, a laser is focused on the SERS substrate at a glancing angle, while the Raman light is collected by a large collection lens. The light is then focused through the entrance slit of a spectrometer and detected using a liquid-nitrogen-cooled charged-coupled device camera. The instrumental approach shown in **Figure 4a** is typically used for experiments requiring low spatial resolution and high raw SERS intensity.

For experiments that require higher spatial resolution, the micro-Raman configuration is used (**Figure 4b**). Laser light is both focused and collected through the same high numerical–aperture objective, after which the scattered light is passed through a notch filter for the removal of Rayleigh-scattered light. Finally the light is focused and directed to a spectrometer and detector. In experiments scanning SER spectra over many different wavelengths, the schematic shown in **Figure 4c** is used. This setup employs the use of a series of flipper mirrors for the quick switching between lasers to access nearly the entire visible and near-infrared range of the spectrum.

3.2. Nanofabrication

In the early stages of SERS, most substrates were made using randomly deposited metal films or electrochemically roughened electrodes (5). For example, a thin layer of gold or silver, typically between 5 and 10 nm, was vapor deposited onto a slide, resulting in a collection of thin island films capable of supporting surface plasmons. Although a small amount of tunability is possible through the modification of film

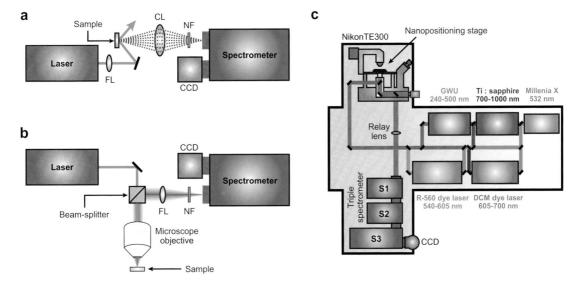

Figure 4

Schematic diagram of typical instrumentation used for surface-enhanced Raman spectroscopy experiments. (*a*) Macro-Raman configuration used when 100-μm-to-millimeter spatial resolution is sufficient. (*b*) Micro-Raman setup, which uses a microscope objective for illumination and collection, increasing the spatial resolution of the instrument. (*c*) Schematic diagram of the experimental setup as laid out on the laser table. The setup allows for easy tunability over the visible wavelengths, using a combination of lasers while maintaining alignment on the sample. The triple spectrograph analyzes Raman scattering excited at wavelengths throughout the visible spectrum. Abbreviations: CCD, charge-coupled device; CL, collection lens; FL, focusing lens; NF, notch filter.

thickness and solvent annealing, variations and a lack of information regarding the SERS active sites prevented a quantitative application of SERS (56, 57).

The past decade has seen significant advances in both the understanding and applications of SERS. One contributing factor in this reinvigoration has been the advances in nanofabrication. The ability to control the shape and orientation of nanoparticles on a surface has reduced many of the complex variables related to SERS and has greatly enhanced our understanding of this phenomenon. A particular nanofabrication method used to great effect in our laboratory is nanosphere lithography (NSL) (58). This method is both cost effective and relatively easy to implement. The process involves drop coating polymer nanospheres on a substrate and then allowing the nanospheres to self-assemble into a close-packed hexagonal array. This array is then used as a mask for the creation of several different SERS active substrates (**Figure 5**). To fabricate periodic nanotriangle arrays, one deposits metal (15–100 nm) directly through the gaps in the nanosphere mask, and upon removal of the mask, a periodic array of nanoscale metal is left. A relatively thick layer of a plasmonic metal (~200 nm) can be deposited directly onto the nanosphere mask. Such metal FON surfaces have an extremely high density of SERS active sites. These FON surfaces are particularly

Nanosphere lithography (NSL): deposition mask formed by self-assembly of hexagonally close-packed nanospheres

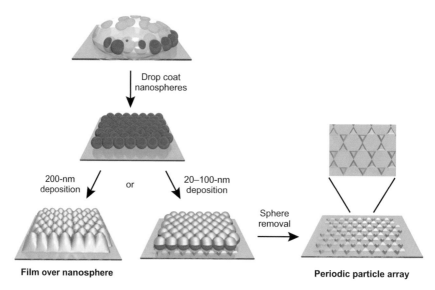

Drop coat
nanospheres

200-nm
deposition

or

20–100-nm
deposition

Sphere
removal

Film over nanosphere

Periodic particle array

Figure 5

Schematic representation of the nanosphere lithography process for fabricating metal film over nanosphere and periodic particle arrays of metal nanotriangles.

robust and can withstand large environmental perturbations, as discussed below. Finally, the nanosphere layer can be used as an etch mask. Reactive ion etching through the mask creates small wells, into which metal can be deposited (59).

3.3. Optimized Surface-Enhanced Raman Spectroscopy Surfaces

The fundamental requirement for SERS is a substrate that supports a surface plasmon resonance. The search for the optimal SERS substrate is an ongoing endeavor, with many such surfaces being reported in the literature (60–65). A full review of all the current SERS substrates would be impractical, but we briefly mention that nanopatterning techniques have greatly expanded the list of possible surfaces and have uncovered many of the characteristics required for maximum signal enhancement.

With such a large volume of SERS substrates reported, a systematic method for evaluating the surface's SERS characteristics is needed. In essence, the goal is to have the largest amount of SERS active sites combined with the greatest EM EF for each site. To illustrate this simple, but many times forgotten, requirement for SERS surfaces, we compare the EFs and density of SERS active sites on three surfaces.

For a proper comparison to be carried out, we must define an expression that accounts for all the many different variables on a SERS substrate. The following equation contains all the factors that dictate the intensity of a SERS signal:

$$I(\omega_S) = N A \Omega \frac{d\sigma(\omega_S)}{d\Omega} P_L(\omega_L)\varepsilon(\omega_L)^{-1} Q(\omega_S) T_m T_0 EF, \qquad (9)$$

where N is the molecule surface density, A is the excitation area, $I(\omega_S)$ is the SERS intensity at Stokes frequency ω_S, Ω is the solid angle of photon collection, $\frac{d\sigma(\omega_S)}{d\Omega}$ is the Raman scattering cross section, $P_L(\omega_L)$ is the radiant flux at excitation frequency, $\varepsilon(\omega_L)$ is the energy of the incident photon, $Q(\omega_S)$ is the quantum efficiency of the detector, T_m is the transmission efficiency of the spectrometer, T_0 is the transmission efficiency of the collection optics, and EF is the enhancement factor.

We can further simplify the above multiplicative expression by conveniently combining several of the known and approximated constants into a single constant, leaving the following:

$$I(\omega_S) = 5.4 \times 10^9 (\text{cm}^{-2} \times \text{torr}^{-1} \times \text{s}^{-2}) \times P_i \times t_{\exp} \times A \times S_i \times a_i \times EF, \quad (10)$$

where it has been assumed that $\Omega = 1$ sr; $\frac{d\sigma(\omega_S)}{d\Omega} = 10^{-30} \frac{\text{cm}^2}{\text{sr}\times\text{molecule}}$; $P_L(\omega_L) = 20$ W cm^{-2}; $Q(\omega_S) = 0.93$; $T_m = 0.75$; $T_0 = 0.86$; and finally a sticking probability (S_i), an estimate of fraction of SERS active sites (a_i), and exposure time ($t_{\exp}$) at partial pressure (P_i) have been added to approximate the number of molecules in the probe area (A).

With a simple expression that includes all the relevant factors that dictate SERS intensity, it is now possible to compare the theoretical limits of detection for SERS active surfaces. In the case of the AgFON, the EF has been measured to be 10^7 (66), whereas the fraction of molecules residing in SERS active sites is essentially unity. Using Equation 9, assuming a 1-s exposure time and an illuminated area of 10^{-4} cm^2, the calculated SERS intensity is 4×10^5 counts s^{-1} ppb^{-1}. With regard to the triangular nanoparticle array fabricated with NSL (**Figure 5**), larger EFs have been reported and are typically on the order of 10^8 (25). Contrary to the FON, there are much fewer active sites on the NSL fabricated surface, and if the fraction of molecules residing on SERS active sites within the illuminated area is assumed to be 0.07, then the SERS intensity is calculated to be 3×10^5 counts s^{-1} ppb^{-1}, slightly smaller than the AgFON surface. To further illustrate the importance of EFs as well as SERS active sites, we examine the theoretical SERS intensity of a substrate drop coated with colloidal silver aggregate. These randomly prepared aggregates provide extremely large EFs of 10^{14} and have been found to be responsible for single-molecule SERS (31, 32). Given that the SERS enhancement relies on the uncontrolled orientation of the silver colloids and the random binding of the analyte molecules to one of the SERS hot spots, the fraction of molecules within the illuminated area is quite small. If the estimated fraction is 10^{-10}, the SERS intensity turns out to be 4×10^2 counts s^{-1} ppb^{-1}, much less than either the FON or NSL fabricated triangle array. The above comparison illustrates the importance of including the density of SERS active sites on a substrate as well as the EF when considering the optimal substrate for a given application.

4. FUNCTIONALIZED SURFACES AND APPLICATIONS

4.1. Surface-Enhanced Raman Spectroscopy Sensing

SERS has great potential for chemical and biological sensing applications because it is selective and sensitive and gives little interference from water. Owing to the

SAM: self-assembled
monolayer

distance dependence of SERS, an important criterion for these sensors is that the analyte of interest must be within a few nanometers of the nanostructured surface. This can be achieved by either drop coating the analyte directly on bare-metal FONs or using various surface functionalization techniques to bring the analyte closer to the noble-metal structure.

Although some substances can be detected on bare FONs, these substrates are not stable owing to oxidation of the metal surface (67, 68). In addition, some important analytes, such as glucose, have low affinity toward the bare metal surface. Furthermore, bare FON substrates do not have any mechanism for isolating the compound of interest from interferents. To overcome these limitations, investigators have used several methods to functionalize SERS substrates. SERS detection is usually facilitated by the use of various coatings ranging from simple alkanethiolates to complex macrocyclic molecules (69). These molecules are anchored to the noble-metal surface by a thiolate group and form self-assembled monolayers (SAMs). The SAM can separate the analyte of interest from interfering analytes and bring it closer to the nanostructured surface, analogous to the use of a stationary phase in high-performance liquid chromatography. Although SAMs have been useful for many applications, thermal desorption (70, 71) and photooxidation (72–74) can result in defects in the coatings and thermodynamically unstable substrates (75).

ALD offers an alternative technique for forming functionalized SERS substrates that can overcome many of the limitations of SAMs (35). Highly controlled thin films are produced by a self-limiting growth process. ALD broadens the scope of SERS sensors by offering new functionalities to the SERS substrates. One such substrate is alumina-functionalized AgFON substrates. Alumina is commonly used as a stationary phase in chromatography and has a predictable affinity based on polarity. Quartz crystal microbalance measurements have demonstrated a uniform growth rate of $\sim$1 Å per deposition cycle (52). Because the SERS signal is highly distance dependent, this subnanometer thickness is highly advantageous for preserving the sensitivity of SERS. In addition, alumina is extremely stable against oxidation and high temperatures and can significantly increase the lifetime of SERS substrates (76, 77). Research is also underway to coat substrates with other ALD materials, such as titania, introducing even more functionalities for SERS sensing.

An example of a SERS-based sensor functionalized with a SAM is the in vivo glucose sensor. AgFON substrates are modified with a mixed SAM consisting of decanethiol (DT) and mercaptohexanol (MH) (30). DT/MH has dual hydrophobic and hydrophilic properties, making it ideal for in vivo glucose detection. To successfully monitor glucose fluctuations throughout the day, an in vivo glucose sensor must be reversible and stable, have a quick temporal response, and be able to detect physiologically relevant glucose concentrations accurately. We demonstrated the reversibility of the DT/MH AgFON sensor by alternatively exposing the sensor to 0- and 100-mM aqueous glucose solutions (pH $\sim$ 7) (30). **Figure 6a** shows the SERS spectra of a DT/MH-functionalized AgFON in phosphate-buffered saline. **Figure 6b,c** shows the difference spectra demonstrating reversible partitioning and departitioning. In addition, we demonstrated 10-day stability and a temporal response of <30 s (30). Quantitative detection has also been demonstrated in vitro as well as in vivo using

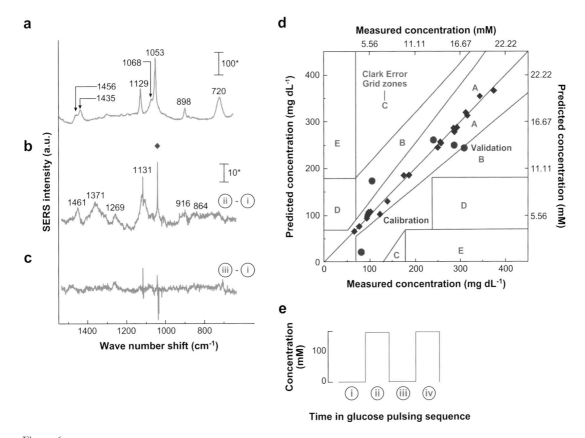

Figure 6

Glucose reversibility and partial least-squares leave-one-out analysis. Glucose pulsing sequence on the self-assembled monolayer–modified Ag film over nanosphere (AgFON) surface for reversibility experiments. (*a*) Surface-enhanced Raman spectra of the decanethiol/ mercaptohexanal (DT/MH)-functionalized AgFON in 0-mM phosphate buffered saline solution. $\lambda_{ex} = 532$ nm, $P_{laser} = 10$ mW, $t = 20$ min, and pH ~ 7. (*b*) Difference spectra showing partitioning of glucose. (*c*) Difference spectra showing departitioning of glucose. The red diamond marks the imperfect subtraction of the narrow band at 1053 cm^{-1} due to nitrate, resulting in a sharp peak in the difference spectra. Asterisks denote analog-to-digital units (mW^{-1} min^{-1}). (*d*) Calibration (*blue diamonds*) and validation (*red circles*) plot using a single substrate and a single spot on a DT/MH-functionalized AgFON. Calibration plot was constructed using 21 data points, and validation plot was constructed using five data points taken over a range of glucose concentrations (10–450 mg dl^{-1}) in vivo (rat). RMSEC = 7.46 mg dl^{-1} (0.41 mM) and RMSEP = 53.42 mg dl^{-1} (2.97 mM) with four loading vectors. $\lambda_{ex} = 785$ nm, $P_{laser} = 50$ mW, and $t = 2$ min. Figure reproduced with permission from Reference 78. Copyright 2006, American Chemical Society.

partial least-squares chemometric analysis (30, 78). **Figure 6*d*** depicts the in vivo calibration and validation models of a DT/MH-functionalized AgFON substrate implanted in a rat. Finally, we have extended the scope of the DT/MH-functionalized AgFON sensor for multi-analyte detection. The experiment was conducted by

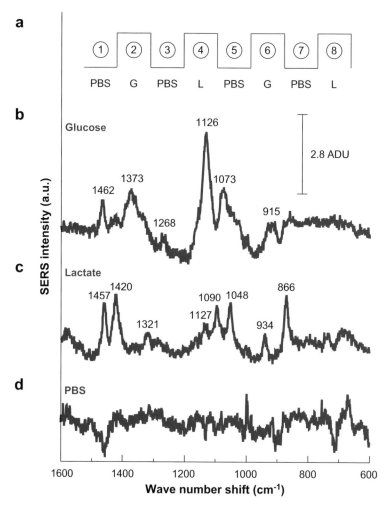

Figure 7

Sequential glucose and lactate pulsing. (*a*) Step changes of glucose (G) and lactate (L) concentrations (0–100 mM) introduced into the sensor. (*b*) Mean difference spectrum (average of difference between steps 2 and 1, and 6 and 5) demonstrating partitioning of glucose. (*c*) Mean difference spectrum (average of difference between steps 4 and 3, and 8 and 7) demonstrating partitioning of lactate. (*d*) Representative difference spectrum of consecutive rinsing steps (3 and 1) demonstrating departitioning of glucose. PBS, phosphate buffered saline; a.d.u., analog-to-digital units. $\lambda_{ex} = 532$ nm, $P = 13$ mW, and $t_{acq} = 10$ min. Figure reproduced with permission from Reference 79. Copyright 2007, American Chemical Society.

alternately injecting 100-mM glucose and 100-mM lactate solutions into the flow cell and rinsing the surface with phosphate-buffered saline between each step. The results indicate that both analytes partition and departition successfully from the DT/MH-functionalized SAM (**Figure 7**) (79).

A SERS-based sensor using alumina-functionalized AgFON substrates has also been fabricated to quantitatively detect an anthrax biomarker, calcium dipicolinate (CaDPA). In this case, CaDPA was initially sensed on bare AgFON substrates and then optimized using alumina. CaDPA (a carboxylic acid) was extracted from *Bacillus subtilis*, a harmless analog of *B. anthracis*, by sonicating in dilute nitric acid and drop coating on bare AgFON substrates. The average intensity of the 1020 cm^{-1} peak was used to construct the adsorption isotherm and determine the limit of detection of CaDPA. Researchers found a limit of detection of 2600 spores and a temporal stability of 3–10 days for the bare AgFON substrates (29). The use of an alumina capture layer produced by ALD significantly increased the sensitivity and stability of the anthrax sensor. Alumina imparts a new functionality to AgFON substrates because of its high affinity for carboxylic acids. Researchers have demonstrated that two cycles of alumina increase the limit of detection to1400 spores (77). The alumina layer also improves the temporal stability of the substrate to at least 9 months by preventing oxidation of the AgFON surface (77).

4.2. Transition-Metal Surface-Enhanced Raman Spectroscopy

Although investigators have demonstrated SERS using several different transition metals (80–82), the vast majority of work has relied on the coinage metals: Cu, Ag, and Au. Such metals support plasmons in the visible region and are resistant to corrosion. Although resistance to degradation might be a desirable characteristic in the simple identification of molecular adsorbates to the surface, it is a definite drawback when trying to apply SERS to the monitoring of heterogeneous catalytic processes.

In an effort to apply SERS to more diverse systems and to possibly use the technique to track a catalytic reaction, several groups have turned to thin overlayers of various transition metals on top of plasmonic metal surfaces or nanoparticles (9, 82, 83). The idea is to retain the surface plasmon of the coinage metal while changing the reactivity of the surface. This technique has been dubbed borrowed SERS, as the catalytically active metal film is thin enough to borrow the EM enhancement of the underlying plasmonic metal. The technique faces many challenges. For one, to avoid the dampening of the plasmon of the underlying metal, one must ensure that the catalytically active overlayer is sufficiently thin. Conversely, the layer must be thick enough to ensure full coverage and minimize any pinholes. Such pinhole-free overlayers will minimize any SERS signal arising from the analyte binding directly to the SERS active metal, instead of the catalytically active metal. Recently, reports on SERS using pure transition metals, such as palladium and platinum, have appeared in the literature (9, 80, 82). These surfaces generally have relatively weak EFs, on the order of 1000, and owe much of that enhancement to the lightning-rod effect (80). With the application of many recent nanofabrication techniques, a refinement in the understanding of transition-metal SERS and borrowed SERS is likely to arise.

4.3. Metal Film over Nanosphere: Universal Surface-Enhanced Raman Spectroscopy Substrate

As it is now well known, the EF of a particular SERS active surface results primarily from the nanostructures on it. A SERS substrate must possess the structures necessary to support either SPRs or LSPRs. Although it is now possible to engineer surfaces of many different structures, many early SERS substrates (vapor-deposited metal island films and electrochemically roughened surfaces) were plagued with irreproducible EFs. The most common limitation of these early substrates was the so-called irreversible loss phenomenon. Specifically, the nanostructures on the surfaces were only metastable, and once a large-enough perturbation was applied to the surface, the nanostructuring was lost, along with the SERS activity. As a result, SERS has been systematically ignored as a detection solution for many applications.

In our laboratory we have demonstrated that, with the proper substrate, SERS can indeed be used in many applications that were once thought impossible. One example application is the coupling of SERS to temperature-programmed desorption in ultrahigh-vacuum environments (84). It was once thought that SERS would be practically useless for a temperature-programmed-desorption substrate because once the temperature was ramped to high-enough levels, the nanostructures would anneal and SERS activity would be irreversibly lost. However, a relatively recent publication demonstrated that an Ag film deposited over silica nanospheres (AgFON) could be used to create high EF SERS substrates (see **Figure 5**) that were ultrahigh-vacuum compatible. These AgFON surfaces were reproducible, stable, and predictable over large temperature ranges. For a demonstration of the utility of AgFON surfaces in ultrahigh-vacuum, temperature-programmed-desorption studies, benzene, pyridine, and C_{60} were deposited and then thermally desorbed for several cycles while monitoring their SERS intensity. **Figure 8** presents the SERS intensity of the 992 cm^{-1} band of benzene as a representative spectrum. As seen in **Figure 8a**, after an initial

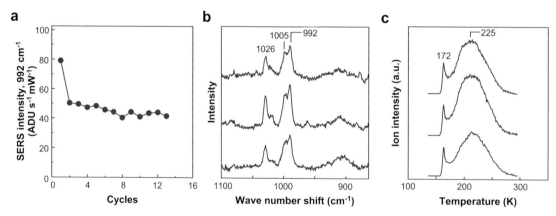

Figure 8

Stability of Ag film over nanosphere (AgFON) toward replicate adsorption-desorption cycles of benzene and pyridine. (*a*) Surface-enhanced Raman spectroscopy (SERS) intensity of the 992 cm^{-1} peak of benzene. Each cycle consists of adsorption of benzene at $T_{surf} = 110$ K and thermal desorption at 300 K with $\beta = 4$ Ks^{-1}. SERS spectra were taken with 18 mW of $\lambda_{ex} = 514.5$ nm for 60 s. (*b*) Replicate SERS spectra in the 900–1100 cm^{-1} region of pyridine on the AgFON surface collected using 8 mW of $\lambda_{ex} = 632.8$ nm for 60 s. (*c*) The corresponding temperature-programmed-desorption spectra of pyridine desorbing from the AgFON at $\beta = 4$ Ks^{-1}.

decline, the SERS intensity remains virtually constant over 13 cycles ranging from 110 K to 300 K. **Figure 8b** shows the corresponding SER spectra after a typical cycle. The experiments clearly demonstrate the possibility of using AgFON-based SERS substrates in experiments requiring large temperature ranges, as the surface displayed both temporal stability and reproducibility.

The AgFON was not only resistant to large temperature perturbations (as discussed above), but it resisted electrochemical perturbations as well (26, 51, 85). Using typical electrochemically roughened SERS electrodes for electrochemically based SERS experiments has proven to be difficult, due mainly to irreproducibility problems and an irreversible loss of the SERS activity of the surface as potentials and electrolyte concentrations reach extreme levels. Similar to the large temperature changes discussed above, these large potential and electrolyte changes presumably perturb the metastable nanostructure of the SERS electrode, causing an irreversible loss of the nanostructures that support the surface plasmons. As a result, most electrochemical SERS experiments, using traditional metal oxidation-reduction cycle (MORC) roughened electrodes, were limited to relatively modest potential ranges. SERS electrodes based on the metal FON do not show the same signal losses at large potentials measured using MORC electrodes (26). **Figure 9** shows a representation of the difference between the metal FON and the MORC-based electrodes. The AgFON surface (**Figure 9a**) clearly shows that SERS intensity is retained as the potential ranges from extremely negative (–1200 mV) back to more positive, whereas the AgORC-based electrode has its SERS intensity decrease dramatically at negative potentials, never to return.

With the difficulties using MORC and metal island films as SERS substrates, it might be tempting to turn to metal colloids as an alternative. This would allow the investigation of many biomolecular systems that denature on bare Ag or Au surfaces (86, 87). Although metal colloids support surface plasmons and therefore should be viable SERS substrates, they still have many difficulties with regard to reproducibility and stability to changing environmental conditions. For example, when using silver or gold colloids, changes in the solution ionic strength, pH, electrolyte, composition, and target molecule concentration can cause uncontrollable aggregation of the colloid. Again, owing to its stability over wide temperature, potential, and concentration ranges, the AgFON can be used for the SERS investigation of many biologically important molecules. Given the tendency of large biomolecules to denature over bare Ag or Au surfaces, the functionalization of the AgFON surface with an alkanethiol-based SAM is often required. The SAM effectively separates the bare metal from the biomolecule of interest, thus preventing its breakdown. This method has been demonstrated recently by functionalizing an AgFON with carboxylic acid–terminated alkanethiolates and investigating the distance and orientation dependence of electron transfer in cytochrome c (51). The important practical significance of the work demonstrated the possibility of SERS spectroelectrochemistry experiments over a wide range of ionic strength, pH, electrolyte, and redox potentials. The stability of the functionalized AgFON enables the SERS studies of adsorbed biomolecules as a function of these important biophysical variables, which have not been previously accessible.

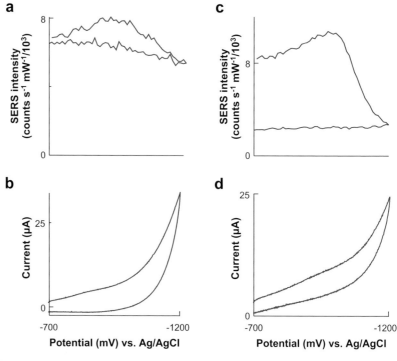

Figure 9

Surface-enhanced Raman spectroscopy (SERS) detected cyclic voltammograms. SERS intensity of the 1008 cm^{-1} band of pyridine (ring breathing mode) versus potential. (*a*) Ag film over nanosphere (AgFON) electrode and (*c*) in situ Ag oxidation-reduction cycle (AgORC) electrode (25 mC cm^{-2}). Laser excitation was 632.8 nm at 4 mW, 0.1-s dwell time, 50-mM pyridine in 0.1-mM KCl. The onset of H$_2$ evolution from (*b*) AgFON and (*d*) AgORC electrodes is indicated by the increase in current between –1.1 and –1.2 V.

The development of these AgFON-based SERS experiments is important for one practical reason: They demonstrate that the irreversible loss phenomenon, previously thought to be universal to all SERS active substrates, is in fact particular to the SERS substrate under investigation. The primary cause for the irreducible loss of SERS activity of metal island films and MORC-type surfaces can be traced to the alteration of the metastable nanostructures responsible for the surface plasmons. The AgFON SERS substrate, conversely, is reproducible, stable, cost effective, and robust enough to remain SERS active over large temperature, potential, and concentration ranges. Given these results, we believe that many of the applications thought to be impractical for SERS-based detection can now be considered.

DISCLOSURE STATEMENT

The authors receive research funding under the following grants: NSF grants DMR-0520513, EEC-0647560, CHE-0414554, and BES-0507036; AFOSR/MURI grant

F49620-02-1-0381; DTRA JSTO Program (grant FA9550-06-1-0558); DOE grant DE-FG02-03ER15457; NIH grant 4 R33 DK066990-02; and grant U54CA119341 from the NIH National Cancer Institute.

ACKNOWLEDGMENTS

The authors gratefully acknowledge Professor George C. Schatz for his assistance with theoretical calculations and many helpful discussions. This work was supported by NSF grants DMR-0520513, EEC-0647560, CHE-0414554, and BES-0507036; AFOSR/MURI grant F49620-02-1-0381; DTRA JSTO Program (Grant FA9550-06-1-0558); DOE grant DE-FG02-03ER15457; NIH grant 4 R33 DK066990-02; and grant number U54CA119341 from the NIH National Cancer Institute. The contents of the review are solely the responsibility of the authors and do not necessarily represent the official views of the NSF, AFOSR, DOE, or the NIH.

LITERATURE CITED

1. Jeanmarie DL, Van Duyne RP. 1977. Surface Raman spectroelectrochemistry, part 1: heterocyclic, aromatic, and aliphatic amines adsorbed on the anodized silver electrode. *J. Electroanal. Chem.* 84:120

2. Albrecht MG, Creighton JA. 1977. Anomalously intense Raman spectra of pyridine at a silver electrode. *J. Am. Chem. Soc.* 99:5215–17

3. Campion A, Kambhampati P. 1998. Surface-enhanced Raman scattering. *Chem. Soc. Rev.* 27:241–50

4. Kneipp K, Kneipp H, Itzkan I, Dasari RR, Feld MS. 1999. Surface-enhanced Raman scattering: a new tool for biomedical spectroscopy. *Curr. Sci.* 77:915–24

5. Moskovits M. 1985. Surface-enhanced spectroscopy. *Rev. Mod. Phys.* 57:783–826

6. Stuart DA, Yonzon CR, Zhang XY, Lyandres O, Shah NC, et al. 2005. Glucose sensing using near-infrared surface-enhanced Raman spectroscopy: gold surfaces, 10-day stability, and improved accuracy. *Anal. Chem.* 77:4013–19

7. Braun G, Lee SJ, Dante M, Nguyen TQ, Moskovits M, Reich N. 2007. Surface-enhanced Raman spectroscopy for DNA detection by nanoparticle assembly onto smooth metal films. *J. Am. Chem. Soc.* 129:6378–79

8. Braun G, Pavel I, Morrill AR, Seferos DS, Bazan GC, et al. 2007. Chemically patterned microspheres for controlled nanoparticle assembly in the construction of SERS hot spots. *J. Am. Chem. Soc.* 129:7760–61

9. Ren B, Liu GK, Lian XB, Yang ZL, Tian ZQ. 2007. Raman spectroscopy on transition metals. *Anal. Bioanal. Chem.* 388:29–45

10. Chen LC, Ueda T, Sagisaka M, Hori H, Hiraoka K. 2007. Visible laser desorption/ionization mass spectrometry using gold nanorods. *J. Phys. Chem. C* 111:2409–15

11. Willets KA, Van Duyne RP. 2007. Localized surface plasmon resonance spectroscopy and sensing. *Annu. Rev. Phys. Chem.* 58:267–97

12. Schatz GC, Young MA, Van Duyne RP. 2006. Electromagnetic mechanism of SERS. See Ref. 88, pp. 19–45

13. Kneipp K, Kneipp H, Bohr HG. 2006. Single-molecule SERS spectroscopy. See Ref. 88, pp. 261–77

14. Kneipp J. 2006. Nanosensors based on SERS for applications in living cells. See Ref. 88, pp. 335–49

15. Yonzon CR, Lyandres O, Shah NC, Dieringer JA, Van Duyne RP. 2006. Glucose sensing with surface-enhanced Raman spectroscopy. See Ref. 88, pp. 367–79

16. Vo-Dinh T, Yan F, Wabuyele MB. 2006. Surface-enhanced Raman scattering for biomedical diagnostics and molecular imaging. See Ref. 88, pp. 409–26

17. Stuart DA, Biggs KB, Van Duyne RP. 2006. Surface-enhanced Raman spectroscopy of half-mustard agent. *Analyst* 131:568–72

18. Stokes DL, Alarie JP, Ananthanarayanan V, Vo-Dinh T. 1999. Fiber optic SERS sensor for environmental monitoring. *Proc. SPIE.* 3534:647–54

19. Kundu S, Mandal M, Ghosh SK, Pal T. 2004. Photochemical deposition of SERS active silver nanoparticles on silica gel and their application as catalysts for the reduction of aromatic nitro compounds. *Journal of Colloid and Interface Science.* 272:134–44

20. Sylvia JM, Janni JA, Klein JD, Spencer KM. 2000. Surface-enhanced Raman detection of 2,4-dinitrotoluene impurity vapor as a marker to locate landmines. *Anal. Chem.* 72:5834–40

21. Fleischmann M, Hendra PJ, McQuillan AJ. 1974. Raman spectra of pyridine adsorbed at a silver electrode. *Chem. Phys. Lett.* 26:163–66

22. Kerker M. 1984. Electromagnetic model for surface-enhanced Raman-scattering (SERS) on metal colloids. *Acc. Chem. Res.* 17:271–77

23. Metiu H, Das P. 1984. The electromagnetic theory of surface enhanced spectroscopy. *Annu. Rev. Phys. Chem.* 35:507–36

24. Moskovits M. 2005. Surface-enhanced Raman spectroscopy: a brief retrospective. *J. Raman Spectrosc.* 36:485–96

25. Schatz GC, Van Duyne RP. 2002. Electromagnetic mechanism of surface-enhanced spectroscopy. In *Handbook of Vibrational Spectroscopy*, ed. JM Chalmers, PR Griffiths, pp. 759–74. New York: Wiley

26. Dick LA, McFarland AD, Haynes CL, Van Duyne RP. 2002. Metal film over nanosphere (MFON) electrodes for surface-enhanced Raman spectroscopy (SERS): improvements in surface nanostructure stability and suppression of irreversible loss. *J. Phys. Chem. B* 106:853–60

27. Jensen TR, Malinsky MD, Haynes CL, Van Duyne RP. 2000. Nanosphere lithography: tunable localized surface plasmon resonance spectra of silver nanoparticles. *J. Phys. Chem. B* 104:10549–56

28. Malinsky MD, Kelly KL, Schatz GC, Van Duyne RP. 2001. Nanosphere lithography: effect of substrate on the localized surface plasmon resonance spectrum of silver nanoparticles. *J. Phys. Chem. B* 105:2343–50

29. Zhang X, Young MA, Lyandres O, Van Duyne RP. 2005. Rapid detection of an anthrax biomarker by surface-enhanced Raman spectroscopy. *J. Am. Chem. Soc.* 127:4484–89

30. Lyandres O, Shah NC, Yonzon CR, Walsh JT, Glucksberg MR, Van Duyne RP. 2005. Real-time glucose sensing by surface-enhanced Raman spectroscopy

in bovine plasma facilitated by a mixed decanethiol/mercaptohexanol partition layer. *Anal. Chem.* 77:6134–39

31. Nie SM, Emery SR. 1997. Probing single molecules and single nanoparticles by surface-enhanced Raman scattering. *Science* 275:1102–6

32. Kneipp K, Wang Y, Kneipp H, Perelman LT, Itzkan I, et al. 1997. Single molecule detection using surface-enhanced Raman scattering (SERS). *Phys. Rev. Lett.* 78:1667–70

33. Haynes CL, Van Duyne RP. 2003. Plasmon-sampled surface-enhanced Raman excitation spectroscopy. *J. Phys. Chem. B* 107:7426–33

34. McFarland AD, Young MA, Dieringer JA, Van Duyne RP. 2005. Wavelength-scanned surface-enhanced Raman excitation spectroscopy. *J. Phys. Chem. B* 109:11279–85

35. Dieringer JA, McFarland AD, Shah NC, Stuart DA, Whitney AV, et al. 2005. Surface enhanced Raman spectroscopy: new materials, concepts, characterization tools, and applications. *Faraday Discuss.* 132:9–26

36. Kennedy BJ, Spaeth S, Dickey M, Carron KT. 1999. Determination of the distance dependence and experimental effects for modified SERS substrates based on self-assembled monolayers formed using alkanethiols. *J. Phys. Chem. B* 103:3640–46

37. Yang W-H, Schatz GC, Van Duyne RP. 1995. Discrete dipole approximation for calculating extinction and Raman intensities for small particles with arbitrary shapes. *J. Chem. Phys.* 103:869–75

38. Zeman EJ, Schatz GC. 1987. An accurate electromagnetic theory study of surface enhancement factors for silver, gold, copper, lithium, sodium, aluminum, gallium, indium, zinc, and cadmium. *J. Phys. Chem.* 91:634–43

39. Kelly KL, Coronado E, Zhao L, Schatz GC. 2003. The optical properties of metal nanoparticles: the influence of size, shape, and dielectric environment. *J. Phys. Chem. B* 107:668–77

40. Kreibig U, Vollmer M. 1995. *Optical Properties of Metal Clusters.* Berlin: Springer

41. Bohren CR, Huffman DF. 1983. *Absorption and Scattering of Light by Small Particles.* New York: Wiley

42. Mie G. 1908. Contributions to the optics of turbid media, especially colloidal metal solutions. *Ann. Phys.* 25:377–445

43. Link S, El-Sayed MA. 1999. Spectral properties and relaxation dynamics of surface plasmon electronic oscillations in gold and silver nano-dots and nano-rods. *J. Phys. Chem. B* 103:8410–26

44. Purcell EM, Pennypacker CR. 1973. Scattering and absorption of light by non-spherical dielectric grains. *Astrophys. J.* 186:705–14

45. Draine BT, Flatau PJ. 1994. Discrete-dipole approximation for scattering calculations. *J. Opt. Soc. Am. A* 11:1491–99

46. Taflove A. 1995. *Computational Electrodynamics: The Finite-Difference Time Domain Method.* Boston: Artech House

47. McCreery RL. 2000. *Raman Spectroscopy for Chemical Analysis.* New York: Wiley Intersci. 420 pp.

48. Kerker M, Wang DS, Chew H. 1980. Surface enhanced Raman-scattering (SERS) by molecules adsorbed at spherical particles. *Appl. Opt.* 19:4159–74

49. Wang DS, Kerker M. 1981. Enhanced Raman scattering by molecules adsorbed at the surface of colloidal spheroids. *Phys. Rev. B* 24:1777–90

50. Van Duyne RP. 1979. Laser excitation of Raman scattering from adsorbed molecules on electrode surfaces. *Chem. Biochem. Appl. Lasers* 4:101–85

51. Dick LA, Haes AJ, Van Duyne RP. 2000. Distance and orientation dependence of heterogeneous electron transfer: a surface-enhanced resonance Raman scattering study of cytochrome *c* bound to carboxylic acid terminated alkanethiols adsorbed on silver electrodes. *J. Phys. Chem. B* 104:11752–62

52. Whitney AV, Elam JW, Zou SL, Zinovev AV, Stair PC, et al. 2005. Localized surface plasmon resonance nanosensor: a high-resolution distance-dependence study using atomic layer deposition. *J. Phys. Chem. B* 109:20522–28

53. Blatchford CG, Campbell JR, Creighton JA. 1982. Plasma resonance enhanced Raman scattering by adsorbates on gold colloids: the effects of aggregation. *Surf. Sci.* 120:435–55

54. Felidj N, Aubard J, Levi G, Krenn JR, Hohenau A, et al. 2003. Optimized surface-enhanced Raman scattering on gold nanoparticle arrays. *Appl. Phys. Lett.* 82:3095–97

55. Vlckova B, Gu XJ, Moskovits M. 1997. SERS excitation profiles of phthalazine adsorbed on single colloidal silver aggregates as a function of cluster size. *J. Phys. Chem. B* 101:1588–93

56. Doron-Mor I, Barkay Z, Filip-Granit N, Vaskevich A, Rubinstein I. 2004. Ultrathin gold island films on silanized glass: morphology and optical properties. *Chem. Mater.* 16:3476–83

57. Pieczonka NPW, Aroca RF. 2005. Inherent complexities of trace detection by surface-enhanced Raman scattering. *Chemphyschem* 6:2473–84

58. Haynes CL, Van Duyne RP. 2001. Nanosphere lithography: a versatile nanofabrication tool for studies of size-dependent nanoparticle optics. *J. Phys. Chem. B* 105:5599–611

59. Hicks EM, Zhang XY, Zou SL, Lyandres O, Spears KG, et al. 2005. Plasmonic properties of film over nanowell surfaces fabricated by nanosphere lithography. *J. Phys. Chem. B* 109:22351–58

60. Baker GA, Moore DS. 2005. Progress in plasmonic engineering of surface-enhanced Raman-scattering substrates toward ultra-trace analysis. *Anal. Bioanal. Chem.* 382:1751–70

61. Aroca RF, Alvarez-Puebla RA, Pieczonka N, Sanchez-Cortez S, Garcia-Ramos JV. 2005. Surface-enhanced Raman scattering on colloidal nanostructures. *Adv. Colloid Interface Sci.* 116:45–61

62. Kneipp K, Kneipp H, Kneipp J. 2006. Surface-enhanced Raman scattering in local optical fields of silver and gold nanoaggregates: from single-molecule Raman spectroscopy to ultrasensitive probing in live cells. *Acc. Chem. Res.* 39:443–50

63. He L, Musick MD, Nicewarner SR, Salinas FG, Benkovic SJ, et al. 2000. Colloidal Au-enhanced surface plasmon resonance for ultrasensitive detection of DNA hybridization. *J. Am. Chem. Soc.* 122:9071–77

64. Felidj N, Truong SL, Aubard J, Levi G, Krenn JR, et al. 2004. Gold particle interaction in regular arrays probed by surface enhanced Raman scattering. *J. Chem. Phys.* 120:7141–46

65. Gunnarsson L, Bjerneld EJ, Xu H, Petronis S, Kasemo B, Kall M. 2001. Interparticle coupling effects in nanofabricated substrates for surface-enhanced Raman scattering. *Appl. Phys. Lett.* 78:802–4

66. Haes AJ, Haynes CL, McFarland AD, Schatz GC, Van Duyne RR, Zou SL. 2005. Plasmonic materials for surface-enhanced sensing and spectroscopy. *MRS Bull.* 30:368–75

67. Von Raben KU, Chang RK, Laube BL, Barber PW. 1984. Wavelength dependence of surface-enhanced Raman scattering from Ag colloids with adsorbed CN^- complexes, SO_3^{2-}, and pyridine. *J. Phys. Chem. B* 88:5290–96

68. Fornasiero D, Grieser F. 1987. Analysis of the visible absorption and SERS excitation spectra of silver sols. *J. Chem. Phys.* 87:3213–17

69. Carron K, Peitersen L, Lewis M. 1992. Octadecylthiol-modified surface-enhanced Raman spectroscopy substrates: a new method for the detection of aromatic compounds. *Environ. Sci. Technol.* 26:1950–54

70. Ishida T, Hara M, Kojima I, Tsuneda S, Nishida N, et al. 1998. High resolution X-ray photoelectron spectroscopy measurements of octadecanethiol self-assembled monolayers on Au(111). *Langmuir* 14:2092–96

71. Zhang ZS, Wilson OM, Efremov MY, Olson EA, Bruan PV, et al. 2004. Heat capacity measurements of two-dimensional self-assembled hexadecanethiol monolayers on polycrystalline gold. *Appl. Phys. Lett.* 84:5198–200

72. Lewis M, Tarlov M, Carron K. 1995. Study of the photooxidation process of self-assembled alkanethiol monolayers. *J. Am. Chem. Soc.* 117:9574–75

73. Schoenfisch MH, Pemberton JE. 1998. Air stability of alkanethiol self-assembled monolayers on silver and gold surfaces. *J. Am. Chem. Soc.* 120:4502–13

74. Zhang Y, Terrill RH, Tanzer TA, Bohn PW. 1998. Ozonolysis is the primary cause of UV photooxidation of alkanethiolate monolayers at low irradiance. *J. Am. Chem. Soc.* 120:2654–55

75. Love JC, Estroff LA, Kriebel JK, Nuzzo RG, Whitesides GM. 2005. Self-assembled monolayers of thiolates on metals as a form of nanotechnology. *Chem. Rev.* 105:1103–69

76. King F. 1987. *Aluminum and Its Alloys.* New York: Ellis Harwood. 313 pp.

77. Zhang X, Zhao J, Whitney A, Elam J, Van Duyne RP. 2006. Ultrastable substrates for surface-enhanced Raman spectroscopy fabricated by atomic layer deposition: improved anthrax biomarker detection. *J. Am. Chem. Soc.* 128:10304–9

78. Stuart DA, Yuen JM, Shah NC, Lyandres O, Yonzon CR, et al. 2006. In vivo glucose measurement by surface-enhanced Raman spectroscopy. *Anal. Chem.* 78:7211–15

79. Shah NC, Lyandres O, Walsh JT, Glucksberg MR, Van Duyne RP. 2007. Lactate and sequential lactate-glucose sensing using surface-enhanced Raman spectroscopy. *Anal. Chem.* 79:6927–32

80. Abdelsalam ME, Mahajan S, Bartlett PN, Baumberg JJ, Russell AE. 2007. SERS at structured palladium and platinum surfaces. *J. Am. Chem. Soc.* 129:7399–406

81. Mrozek MF, Xie Y, Weaver MJ. 2001. Surface-enhanced Raman scattering on uniform platinum-group overlayers: preparation by redox replacement of underpotential-deposited metals on gold. *Anal. Chem.* 73:5953–60

82. Tian ZQ, Yang ZL, Ren B, Li JF, Zhang Y, et al. 2006. Surface-enhanced Raman scattering from transition metals with special surface morphology and nanoparticle shape. *Faraday Discuss.* 132:159–70

83. Park S, Yang PX, Corredor P, Weaver MJ. 2002. Transition metal–coated nanoparticle films: vibrational characterization with surface-enhanced Raman scattering. *J. Am. Chem. Soc.* 124:2428–29

84. Litorja M, Haynes CL, Haes AJ, Jensen TR, Van Duyne RP. 2001. Surface-enhanced Raman scattering detected temperature programmed desorption: optical properties, nanostructure, and stability of silver film over SiO_2 nanosphere surfaces. *J. Phys. Chem. B* 105:6907–15

85. Hulteen JC, Young MA, Van Duyne RP. 2006. Surface-enhanced hyper-Raman scattering (SEHRS) on Ag film over nanosphere (FON) electrodes: surface symmetry of centrosymmetric adsorbates. *Langmuir* 22:10354–64

86. Cotton TM, Schultz SG, Van Duyne RP. 1980. Surface-enhanced resonance Raman-scattering from cytochrome *c* and myoglobin adsorbed on a silver electrode. *J. Am. Chem. Soc.* 102:7960–62

87. Copeland RA, Fodor SPA, Spiro TG. 1984. Surface-enhanced Raman spectra of an active flavo enzyme: glucose oxidase and riboflavin binding protein on silver particles. *J. Am. Chem. Soc.* 106:3872–74

88. Kneipp K, Moskovits M, Kneipp H, eds. 2006. *Surface-Enhanced Raman Scattering: Physics and Applications*. Berlin: Springer

Time-Resolved Microdialysis for In Vivo Neurochemical Measurements and Other Applications

Kristin N. Schultz and Robert T. Kennedy

Department of Chemistry, University of Michigan, Ann Arbor, Michigan 48109;
email: rtkenn@umich.edu

Annu. Rev. Anal. Chem. 2008. 1:627–61

First published online as a Review in Advance on March 18, 2008

The *Annual Review of Analytical Chemistry* is online at anchem.annualreviews.org

This article's doi:
10.1146/annurev.anchem.1.031207.113047

Key Words

capillary electrophoresis, liquid chromatography, enzyme assays, amino acids, dopamine, neuropeptides

Abstract

Monitoring changes in chemical concentrations over time in complex environments is typically performed using sensors and spectroscopic techniques. Another approach is to couple sampling methods, such as microdialysis, with chromatographic, electrophoretic, or enzymatic assays. Recent advances of such coupling have enabled improvements in temporal resolution, multianalyte capability, and automation. In a sampling and analysis method, the temporal resolution is set by the mass sensitivity of the analytical method, analysis time, and zone dispersion during sampling. Coupling methods with high speed and mass sensitivity to microdialysis sampling help to reduce some of these contributions to yield methods with temporal resolution of seconds. These advances have been primarily used in monitoring neurotransmitters in vivo. This review covers the problems associated with chemical monitoring in the brain, recent advances in using microdialysis for time-resolved in vivo measurements, sample applications, and other potential applications of the technology such as determining reaction kinetics and process monitoring.

1. INTRODUCTION

Temporal resolution: the frequency by which data are collected and analyzed

Extracellular space: the fluid compartment that surrounds and bathes neurons and glial cells in the brain

Understanding normal and abnormal functioning of the brain unquestionably represents one of the greatest challenges to scientists. Behavioral control, cognition, and emotions are ultimately encoded by neurotransmission, or chemical communication between neurons, which involves the movement of ions across membranes and chemicals across the synapse between two neurons (**Figure 1**). During neurotransmission, chemicals (neurotransmitters) released from a neuron diffuse across the synaptic gap to interact with receptors found on neighboring neurons. These interactions can either promote or inhibit action potential generation in the receiving neuron, thus affecting its chemical release (for a review of neurotransmission, see Reference 1). In addition to the acute effects on action potential generation, released chemicals can alter other cellular functions such as gene expression, growth, connectivity, metabolism, and responsiveness to further stimuli (2, 3). Neurotransmitter release from other parts of neurons (such as the dendrites and cell soma) and glial cells is also important in signaling (1).

We have gained many insights into neurotransmission from ex vivo experiments on brain tissue slices and cultured cells, but ultimate understanding of the complex interconnections of the brain requires the study of living, intact subjects. Performing in vivo measurements makes it possible to discern the regulation of neurotransmission in the presence of complete neuronal circuits (e.g., inputs from different brain regions) and cellular milieu created by the activity of all cells in a brain region. Furthermore, in vivo measurements allow the correlation of neurochemistry to behavior and emotional states.

Among the factors to consider when developing and evaluating in vivo methods are chemical heterogeneity, spatial resolution, and temporal resolution. The brain extracellular space comprises a complex chemical soup that includes neurotransmitters, growth factors, and metabolites. Neurotransmitters run the gamut from small gaseous molecules (e.g., nitrous oxide), to small organic molecules (e.g., glutamate), to oligopeptides (e.g., neurotensin) with concentrations from 1 pM to 1 mM. This chemical variety places great demands on the selectivity, dynamic range, sensitivity, and versatility of the analytical methods used for in vivo chemical measurement. The structural heterogeneity of the brain places significant demands on the spatial resolution of a chemical measurement. Distinct brain regions <0.5 mm^3 can be critical for controlling a particular behavior, for example. To accurately study neurotransmission, the established method must ensure measurements are taken solely from the region of interest. Finally, temporal resolution is critically important for in vivo brain chemical measurements. Individual neurons release neurotransmitters by exocytosis, and the neurotransmitters are rapidly eliminated from the synapse by reuptake or enzymatic degradations. Fluctuations in neurotransmitter levels can occur on the millisecond time scale by this process, although sustained changes in neuronal firing or reuptake (such as what may occur during a behavioral activation) can result in longer-lived changes in neurotransmitter levels. Cellular responses to neurotransmission and the effects of pharmacological agents also occur over a longer time scale (see **Figure 2**). Thus, the temporal resolution that can be achieved in measuring neurochemicals

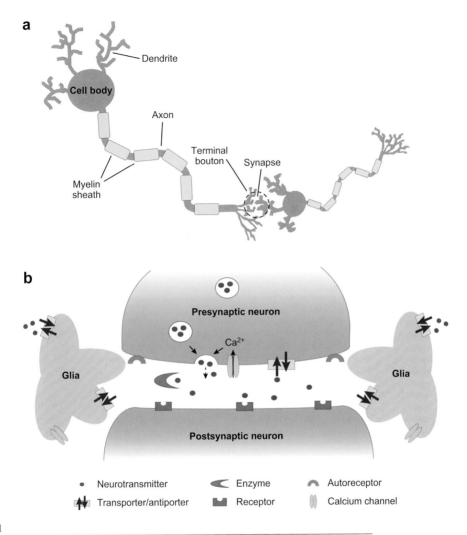

Figure 1

Neuroanatomy and neurotransmission. (*a*) Depiction of basic neuronal structure and synaptic features. Neurons consist of dendrites, cell bodies, axons, and terminal boutons. Dendrites collect chemical input from neighboring neurons, which influences the membrane potential of the neuron. Once the threshold membrane potential is reached, an action potential is generated and travels from the cell body down the axon toward the terminal boutons. Myelin ensheathment of neurons increases the speed of action potential propagation. Once the action potential reaches the presynaptic terminal, it triggers the opening of voltage-sensitive Ca^{2+} channels. The influx of calcium causes synaptic vesicles to migrate toward the synaptic cleft and release their cargo into the synapse. (*b*) Depiction of neurotransmission at synapses. Neurotransmitters diffuse across the synapse to interact with and bind to receptors on the postsynaptic neuron. These receptors can be on dendrites, cell bodies, or on another axon. If a neurotransmitter binds to an autoreceptor on the presynaptic neuron, a negative feedback loop is initiated to inhibit further release. Signaling is terminated via enzyme degradation or reuptake into the presynaptic terminal by transporters. Other supportive neuronal cells such as glia can also release and take up neurotransmitters (e.g., glutamate).

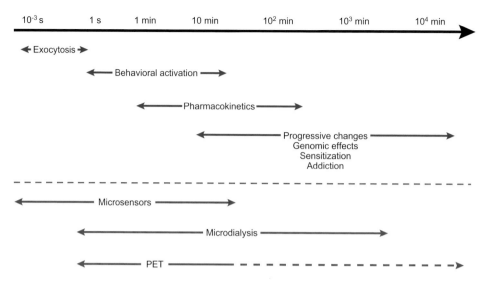

Figure 2

Temporal resolution of neurochemical events and analytical methods. The time scale of neurochemical events or their effects ranges from less than 1 s to days. The temporal resolutions of relevant in vivo chemical measurement methods are listed below the dashed line. Because it is noninvasive, positron emission tomography (PET) allows multiple measurements over many days in the same subject so that longer-term changes can be detected.

limits the types of questions that a particular method can address. In this review, we cover recent technological advances in the use of microdialysis sampling coupled with chemical measurement techniques to measure rapid chemical fluctuations.

2. IN VIVO DETECTION OF NEUROTRANSMITTERS

Researchers have developed a variety of techniques to measure neurotransmission in vivo. Prominent among these methods are positron emission tomography (PET), microsensors, and microdialysis sampling (for reviews, see References 4, 5, 6). **Figure 2** illustrates the temporal resolution in these methods.

Besides temporal resolution, other factors are also important when comparing these methods. PET and microsensors offer the advantages of noninvasive imaging and high spatial resolution, respectively. PET is expensive and is presently limited to just a few neurotransmitter systems. In addition, spatial resolution is limited when working on small laboratory animals such as rats and mice. Microsensors have excellent spatial and temporal resolution. The weaknesses of microsensors include an unproven ability to determine basal concentrations, long-term changes, and multiple analytes.

Microdialysis can provide complementary information to microsensors and PET. As a general sampling method, it can be used for most transmitters, offering unsurpassed versatility and multianalyte capability when coupled with sensitive and selective

PET: positron emission tomography

analytical techniques such as high-performance liquid chromatography (HPLC). It provides basal concentrations as well as dynamic changes. The spatial resolution is worse than sensors because the probes are generally larger. The temporal resolution has typically been in the 10–30-min range; however, recent work has pushed this to the seconds range (see **Table 1**). These properties, along with its ease of use, have made microdialysis the workhorse for in vivo neurochemical measurements, with over 11,000 published studies using this method. In addition, the recent advances in temporal resolution have potentially opened the possibility for many new applications.

HPLC: high-performance liquid chromatography

Concentration detection limit: the minimum concentration that can be detected at known confidence

Mass detection limit: the minimum mass that can be detected at known confidence

2.1. Microdialysis Overview

Microdialysis is but one of several sampling methods used in the brain. Methods that preceded microdialysis include the cortical cup (7, 8) and push-pull perfusion (9). These other methods have become less popular owing to limited spatial resolution and tissue damage associated with the collection, respectively. Microdialysis emerged as a modification of push-pull perfusion probes (**Figure 3**). In this approach, a membrane is placed around the outside of the push-pull arms to serve as a barrier between the perfusion fluid (normally pumped at 0.3 to 2 μL min^{-1}) and the surrounding tissue, thereby decreasing the tissue damage induced by perfusion (10–12). As fluid is perfused through the interior lumen of the dialysis probe, it washes out the material that diffused into the probe, thus creating a concentration gradient from the outside of the probe to the inside. Therefore, microdialysis samples chemicals through diffusional gradients rather than physically removing fluid from the extracellular space. Molecules with molecular weight below the cutoff of the membrane are sampled.

When considering microdialysis for temporally resolved measurements, analyte recovery by the probe is an important issue. Recovery can be defined as absolute or relative. Absolute recovery refers to the mass of analyte collected over a period, whereas relative recovery refers to the concentration of analyte in the dialysate divided by the concentration in the sampled media. Perfusion flow rate strongly controls recovery such that increasing flow rates through the dialysis probe increase the absolute recovery while decreasing the relative recovery. Absolute recovery increases because higher flow rates create a steep concentration gradient between the probe and the extracellular fluid, thus enhancing the flux of molecules to the probe. Relative recovery decreases as the flow rate is increased because there is less time for equilibrium to be reached between the solution flowing through the probe and the extracellular space. Therefore, one must choose a flow rate that allows both adequate relative recovery to meet the instrument's concentration detection limit and adequate absolute recovery to meet the instrument's mass detection limit. Besides flow rate, other factors also affect recovery. Recovery increases with increasing molecular-weight cutoff of the membrane (up to a limit), temperature, and active membrane length. In general, higher recovery (especially absolute recovery) aids in achieving high temporal resolution, as discussed below.

Table 1 High–temporal resolution dialysis methods

Analyte[a]	Experiment type	Method[b]	Temporal resolution(s)	Off-line	On-line	Reference(s)
Glu	Method development	CE-LIF	1	x		64
Glu	Method development	CE-LIF	6	x		63
D- and L-Asp	Method development	CE-LIF	3		x	95
Glu	Ascorbate effects on electrically stimulated release	CE-LIF	3		x	96
Neuroactive amines and AAs	Method development	cLC-EC	10	x		32
Glu, Asp, and DA	Apomorphine, PDC, NMDA, and nomifensine effects	CE-LIF	10	x		97
Glu and Asp	Electrical stimulation	CE-LIF	12		x	69
Neuroactive amines and AAs	Alcohol and estrogen effects	CE-LIF	15		x	98–100
Neuroactive amines and AAs	Behavioral	CE-LIF	15		x	87, 101, 102
Neuroactive amines and AAs	Method development	CE-LIF	20		x	67, 74
NA, DA, Glu, Asp, and GABA	Method development	CE-LIF	20	x		62
Neuroactive amines and AAs	Method development	Microfluidic CE-LIF	30		x	73
Neuroactive amines and AAs	Method development	MEKC-LIF	30		x	70
Glu	Behavioral	CE-LIF	30	x		103
Glu and GABA	Behavioral	CE-LIF	30	x		59
Gabapentin	Pharmacokinetics	CE-LIF	30	x		104
Glu	Haloperidol effects on flash-evoked release	CE-LIF	30	x		105
Arg, Glu, and Asp	Pain response to formalin test	CE-LIF	30	x		106
NA, DA, Glu, and Asp	Method development	CE-LIF	30	x		60
Glu	Electrical stimulation	CE-LIF	30	x		107
Glucose and lactate	Neuronal metabolism	Enzyme assay–EC	30		x	79, 83, 108
GABA, Glu, and Asp	Pain response in human spinal cord	CE-LIF	60	x		109
Glu, GABA, Arg, and Asp	Pain response in rat spinal cord	CE-LIF	60	x		110
Glu	Behavioral	CE-LIF	60	x		58
Glu and Asp	NMDA effects	CE-LIF	60	x		111

(Continued)

Table 1 (*Continued*)

Analyte[a]	Experiment type	Method[b]	Temporal resolution(s)	Off-line	On-line	Reference(s)
Isoproterenel	Pharmacokinetics	CE-EC	60	x		112
Lactate	Glutamate uptake effects	Enzyme assay–EC	60		x	113
Glucose	Method development	Enzyme assay–EC	60		x	114
Glutamate	Hypoxia response	Enzyme assay–FLU	60		x	115
Glucose	Behavioral	Enzyme assay–FLU	60		x	85
Lactate	Behavioral	Enzyme assay–FLU	60		x	116
Ethanol	Pharmacokinetics	GC	60	x		117
5-HT and 5-HIAA	Method development	cLC-EC	60–120	x		118
Acetaminophen and caffeine	Pharmacokinetics	HPLC-UV	60		x	119
Glu and Asp	Method development	HPLC-FLU	60	x		120
Glu and GABA	Hypoxia response	HPLC-EC	60	x		21
5-HT	Hypoxia response	HPLC-EC	90	x		21
DA	Behavior	HPLC-EC	60	x		30
DA	Cocaine effects	HPLC-EC	60	x		31
Adenosine	Hypoxia response	HPLC-UV	90	x		21
DA	Method development	MEKC-LIF	90		x	71
SR 4233 and SR 4317	Pharmacokinetics	MEKC-LIF	90		x	121
DA, NA, Glu, and Asp	Behavioral	CE-LIF	120	x		57
Glu and Asp	Circadian rhythm effects	CE-LIF	120	x		122
PEA, Glu, and Asp	Method development	CE-LIF	120	x		72
NA and Glu	Method development	CE-LIF	120	x		61
Glucose	Method development	Enzyme assay–EC	120		x	123
DA	Behavioral	HPLC-EC	120	x		124, 125
DA and 5-HT	Behavioral	HPLC-EC	120	x		22

[a]Amino acids are represented by their three-letter code. 5-HT, serotonin; 5-HIAA, 5-hydroxyindoleacetic acid; AA, amino acid; DA, dopamine; NA, noradrenaline; PEA, phosphoethanolamine.

[b]CE: capillary electrophoresis; cLC, capillary liquid chromatography; EC, electrochemical detection; FLU, fluorescence detection; GC, gas chromatography;

HPLC, high-performance liquid chromatography; LIF, laser-induced fluorescence; MEKC, micellar electrokinetic chromatography; UV, ultraviolet-visible absorbance detection.

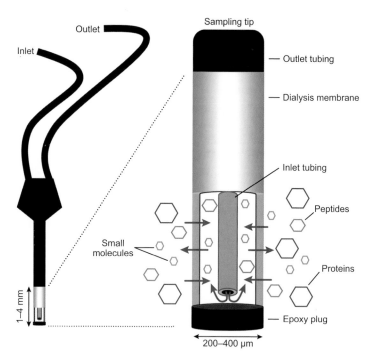

Figure 3

A depiction of a concentric microdialysis probe along with a magnified image of the sampling tip. To perform microdialysis sampling, one pushes a perfusion buffer into the inlet that mimics the ionic content of the extracellular space. When this buffer flows past the active area of the membrane, this creates a concentration gradient between the tissue and the lumen of the probe. Analytes that have a mass below the molecular-weight cutoff of the membrane diffuse into the probe and can be collected for further analysis. Dialysate exits the probe through the outlet to be collected off-line into a vial or to be sent directly toward the instrument's detector. Local pharmacological manipulations can also be performed simultaneously by adding the agent to the perfusion buffer.

2.2. Temporal Resolution of Microdialysis

The ability of microdialysis measurements to monitor fast changes (second-to-minute time scale) is greatly affected by the mass detection limits of the analytical method used for assaying dialysate. To increase temporal resolution, one collects smaller fractions, and the moles of analyte per sample decrease (**Figure 4**). For example, if a sampled compound has a concentration of 1 μM, and a flow rate of 1 μL min^{-1} is used, then 10-min temporal resolution requires the detection of 10.0 pmol of analyte in 10-μL samples, whereas 10-s temporal resolution requires the detection of 0.2 pmol of analyte in 167-nL samples. Thus, high temporal resolution requires high mass sensitivity and the analysis of small volumes. Increasing absolute recovery, especially by increasing flow rate, may help improve temporal resolution; however, this comes at the cost of more dilute samples (due to lower relative recovery).

Mass sensitivity: the smallest increment of mass that can be measured

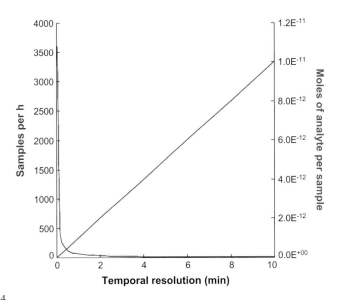

Figure 4

Effects of temporal resolution on the mass of analyte/sample and the number of samples generated. A plot of the number of samples generated and moles per sample as a function of temporal resolution for a microdialysis measurement. This plot assumes 1.0 μM of analyte in the dialysate stream, and the calculations were done using a dialysis flow rate of 1.0 μL min^{-1}. As temporal resolution increases, moles available to detect decrease and the number of samples dramatically increases.

In addition to mass detection limits, one needs to consider the method's throughput because raising the temporal resolution also rapidly increases the number of samples that need to be analyzed (**Figure 4**). For methods with 5-min temporal resolution, 12 samples are collected per hour. An increase in temporal resolution to 10 s makes the sample load 360 per hour! These two problems, diminishing analyte mass and increasing sample load, create substantial demands on the analytical system when attempting to improve the temporal resolution of microdialysis methods.

As discussed below, the use of high-throughput methods with excellent mass sensitivity has pushed the temporal resolution of microdialysis methods into the realm of seconds. With the advent of these methods, other factors can begin to limit the temporal resolution of the method. For example, adsorption to the membrane or other fluidic components can cause slow dynamics. A fundamental limit is the dispersion of a concentration plug owing to diffusion and parabolic flow (i.e., Taylor dispersion) as it is transported from the sampling probe to the analytical system or fraction collector. Previous studies have indicated that this type of broadening would limit temporal resolution to approximately 16 s if the flow rate is 1 μL min^{-1} and 85 s if the flow rate is 0.2 μL min^{-1} (13). (These limits of temporal resolution were achieved with short distances between the probe and instrument. For freely moving animals, such short distances are often not feasible, so for practical purposes, the limits of temporal resolution may be higher.) Because lower flow rates give worse temporal resolution, a trade-off in relative recovery and temporal resolution is necessary. Although we can envision a variety of modifications to reduce this limit of temporal resolution, most work to date has focused on improving mass detection limits because this is usually the limiting factor.

Assays of dialysate samples may be performed either on-line or off-line. Off-line methods have the advantage of higher overall throughput because multiple sampling experiments can be performed simultaneously. This is realized because sampling is a separate step from detection. Off-line analysis also allows flexibility in assays. For example, assays that are slow, perhaps because of a long derivatization step or slow separation, may be used with off-line fraction collection. In addition, the problem of dynamic range can be reduced by assaying the same sample multiple times under different gain settings. The main difficulty of off-line analysis is sample handling because hundreds of submicroliter samples can be easily generated in a day.

On-line methods offer the advantage of minimal manual sample handling because the dialysate is continuously pumped toward the detector. In addition, with an on-line method, it is possible to monitor the output during the experiment to ensure proper sample collection. Such on-line monitoring can be useful to catch difficulties as they arise during an experiment. On-line monitoring may also be useful for providing rapid feedback in certain situations such as chemical monitoring during surgery. On-line analysis, however, places stringent requirements on the speed of the analytical method. Thus, assay time can limit the temporal resolution rather than the mass detection limit or Taylor dispersion.

3. LIQUID CHROMATOGRAPHY METHODS FOR ANALYZING DIALYSATE

Liquid chromatography methods are the most popular methods to analyze dialysate. These methods are used not only for measuring the release of neurotransmitters, but also for studying pharmacodynamics and metabolic changes. Microdialysis works well for coupling to HPLC because the dialysate is protein free, which limits biofouling and makes further purification unnecessary prior to analysis. In addition, for pharmacokinetic studies, microdialysis only samples the unbound drug, which is a measurement of the therapeutically active component of the drug. Although conventional-sized HPLC columns remain the most popular approach for analysis, advances in column miniaturization such as microbore HPLC and capillary liquid chromatography (cLC) have allowed temporal resolution to be improved from 5–30 min to 10–180 s by virtue of the improved mass sensitivity associated with these small columns.

3.1. High-Performance Liquid Chromatography Methods

The mass limit of detection (LOD) of HPLC dictates that samples are usually collected in 5–30-min fractions, thus limiting temporal resolution (14–20). The mass sensitivity strongly depends on the detector used, with fluorescence and electrochemical detectors offering substantial improvements over ultraviolet absorbance when applicable. In some cases, high-sensitivity detection can allow temporal resolutions of 1–2 min. Glutamate, GABA, serotonin, adenosine (21), and dopamine (22) have been monitored at this temporal resolution, for example.

One advance that increased mass sensitivity was the development of microbore HPLC columns. Whereas traditional HPLC columns have inner diameters of 4–5 mm, microbore columns have inner diameters of 0.3–1.0 mm. With smaller column diameters, a given mass of material is diluted less before reaching the detector, resulting in improved detection limits. Studies have shown improved temporal resolution when switching from an HPLC column to a microbore column (23, 24). Most microbore LC methods offer temporal resolutions of 3–5 min (25–29); however, some microbore LC assays have been developed for 1–2-min monitoring of dopamine (30, 31).

GABA: γ-aminobutyric acid

MS: mass spectrometry

3.2. Capillary Liquid Chromatography

Within the past few years, cLC has emerged as a potential method for assaying dialysate samples. cLC columns have inner diameters of 25–150 μm and usually are prepared in the laboratory using them, although commercial systems are starting to become available. The greatly reduced column bore magnifies the effect on mass sensitivity such that attomole detection limits are possible with cLC columns. The use of cLC also improves compatibility with mass spectrometry (MS) and reduces mobile phase consumption.

Investigators have applied cLC with electrochemical detection to determine neuroactive amines and amino acids in dialysate. This method had LODs of 20–80 amol and as a result could achieve 10-s temporal resolution with off-line fraction collection (32). This method resolved 16 amino acids including glutamate, aspartate, and GABA from a 200-nL sample (diluted to 2 μL). Presently, the lack of automation and long analysis time have prevented this method from gaining popularity.

Most of the work using cLC for dialysate analysis has utilized reversed-phase LC columns; however, more exotic stationary phases have also been reported. One interesting method utilized an antiadenosine aptamer as a stationary phase to selectively preconcentrate and retain this purine in dialysate samples. The resulting method allowed 5-min temporal resolution for monitoring this neurotransmitter (33). In principle, by altering the aptamer, one could then apply this approach to a variety of neurotransmitters or compounds.

3.3. Capillary Liquid Chromatography–Mass Spectrometry

The low flow rates of cLC columns make them particularly amendable to coupling with mass spectrometry, allowing mass detection limits in the low attomole range. Such low mass detection limits are attainable because the background signal can be greatly reduced through selective monitoring of a specific mass-to-charge ratio. In addition, if tandem MS is available, then the sensitivity can be increased further because the chemical noise decreases faster than the signal with higher stages of MS

(34). We have used cLC-MS to achieve 2.4-min temporal resolution for acetylcholine measurements in dialysate (35). Notably, this method did not require the use of acetyl-cholinesterase inhibitors to artificially increase the acetylcholine level in dialysate to achieve this resolution. In principle, the resolution could be 30 s. By comparison, other LC-MS methods achieved 20-min temporal resolution for acetylcholine (36).

The high sensitivity possible with cLC-MS is especially useful for measuring neuropeptide release in vivo. Neuropeptides have picomolar basal concentrations in the brain. As a result, in dialysis samples, just a few attomoles of neuropeptide may be available. This makes studying neuropeptides an analytical challenge. Radioimmunoassays have been the main method for measuring neuropeptides in dialysate (for a review, see Reference 37). However, with detection limits of ~100 amol, one could only use these methods for ~30-min temporal resolution and measure only one neuropeptide at once. Advances in cLC and LC with electrochemical detection have allowed some neuropeptides to be detected in dialysate (38–40); however, the low concentration makes such detection difficult. Furthermore, the method is limited to peptides that either are naturally electroactive or can react to form an electroactive product. In contrast, cLC-MS is adaptable to detecting multiple peptides in one assay with sequence specificity and, in principle, is applicable to all neuropeptides without derivatization.

Caprioli's group (41, 42) first used capillary LC-MS to detect endogenous neuropeptides. This method improved the sensitivity of traditional electrospray ionization sources by creating a microelectrospray source that accommodated nanoliter flow rates (which increases the ionization efficiency of the sample). Their source consisted of a 50-μm silica capillary containing a silica frit and a 2-cm bed of reversed-phase particles. Dialysate samples were preconcentrated and desalted on this bed before being injected into the mass spectrometer. The investigators reported a detection limit for neurotensin of less than 30 amol. Moreover, miniaturization and changes in operation have further improved this method. In the cLC-MS of neuropeptides, a practical difficulty is the time required to preconcentrate microliter samples onto columns that have total volumes of just a few nanoliters. A method for met- and leu-enkephalin utilized an automated two-pressure LC system to reduce analysis time (43). One pump had a high flow rate to hasten sample loading and column rinsing, and the other pump had a low flow rate to perform the separation. Because both pumps were pressure equalized, there was no time wasted waiting for pump pressure to stabilize when switching between loading and separation. In addition, smaller column diameter and emitter dimensions, along with a lower electrospray flow rate, helped to increase separation efficiency and mass sensitivity compared with previous methods. The mass sensitivity of this method was such that a temporal resolution of 3.3 min was possible; however, as this was an on-line method, temporal resolution was limited by the analysis time, which included time required to load the sample, separate analytes, and then re-equilibrate the column (30 min). The sampling interval was cut to approximately 16.5 min by changing the gradient and other parameters (34). cLC-MS has also been utilized to study other neuropeptides, including angiotensin (44) and neurotensin (42).

3.4. Current Status of Capillary Liquid Chromatography for Microdialysis and Future Directions

Although cLC offers higher mass sensitivity and therefore higher temporal resolution than HPLC, several drawbacks limit the applicability of this method. The stability of capillary columns is far less than commercial HPLC columns because their smaller size makes it easier to clog or break them. Typically a new capillary column is required for each set of dialysate samples collected. The cLC methods also require specialized LC equipment to load the sample onto the column and perform the analysis. The continued commercialization of this technology, especially with MS interfaces, will likely make it more feasible for dialysate analysis in the future.

The enhanced temporal resolution for microdialysis monitoring by cLC is undermined by the relatively long separation times, which can create a bottleneck in routinely performing high–temporal resolution methods, either off-line or on-line. Recent advances, however, may alleviate this problem. For example, the development of ultra-high-pressure LC (for review, see Reference 45) has made faster separations possible. The use of elevated temperatures along with more stable stationary phases can also improve the separation speed (46, 47). Although these innovations have not been applied to microdialysis samples, they likely could be used to make high–temporal resolution monitoring more routine.

4. CAPILLARY ELECTROPHORESIS METHODS FOR ANALYZING DIALYSATE

Capillary electrophoresis (CE) methods are also popular methods for coupling to dialysis. CE is well suited for achieving high temporal resolution because it has both high mass sensitivity and the potential for high throughput. Mass detection limits by CE with laser-induced fluorescence (LIF) detection can be in the yoctomole range (48), suggesting the potential for subsecond temporal resolution. Such impressive detection limits are generally achieved only with LIF, but electrochemical detection can achieve attomole LODs, and ultraviolet detection can achieve femtomole LODs. The throughput of CE can enable rapid analysis, off-line or on-line, of the many samples collected by a microdialysis experiment. High throughput can be achieved through rapid separation. As recently reviewed (49), separation speed in CE can be in the seconds range compared with the minutes typically required for equivalent separations by HPLC. CE is also amenable to parallel operation (50), an innovation that is an outgrowth of the Human Genome Project, which, in principle, could improve the throughput of the off-line analysis of dialysate; however, this has not been implemented. Besides the high-throughput capability, CE is also amenable to automation, which further facilitates its use for dialysate analysis. For these reasons, CE has been highly useful for temporally resolved microdialysis measurements.

4.1. Detection and Derivatization Methods

With CE methods, LIF detection is the most popular detection method for dialysate. Some work couples CE to electrochemical detection (51–53), but this method is not

CE: capillary electrophoresis

LIF: laser-induced fluorescence

as popular because of the lack of commercial availability. For good chromophores, or compounds present at sufficiently high concentration, ultraviolet absorbance detection has also been used (54).

Although LIF provides impressive detection limits, it can only detect fluorescent compounds. Because native fluorescence of most neurotransmitters is inconsequential, a derivatization step is needed. As many transmitters contain primary amines, amine-reactive reagents have been used extensively. The relatively rapid reaction kinetics of orthophthaldialdehyde/β-mercaptoethanol (10–30 s) and naphthalene-2,3-dicarboxaldehyde/cyanide (180 s) make them good derivatization agents for on-line CE methods. These methods are also advantageous because they are fluorogenic, meaning that they have low fluorescence until they react. Fluorescent tags such as fluorescein isothiocyanate offer high quantum yields, but their reaction kinetics are extremely slow, and their native fluorescence tends to create many background peaks. The long reaction time prevents these tags from being used for on-line measurements. Fluorescent derivatization agents are not limited to being amine-reactive agents. Thiol-reactive agents such as methanolic monobromobimane are available and have been used in microdialysis/CE-LIF methods (55).

4.2. Off-Line Capillary Electrophoresis Methods for Studying Neurotransmission

The first methods developed to couple microdialysis to CE were off-line methods. Early work involved collecting dialysate into a vial, adding derivatization solution, and then injecting the derivatized sample onto the capillary (56–59). Although these approaches used CE, they required up to 7 μL of sample for injection (even though only a few nanoliters are injected) and therefore did not take advantage of the potential for high temporal resolution. The main advantage over HPLC in these instances was the analysis time.

Off-line methods have achieved temporal resolutions of 1 to 30 s by using various strategies for derivatizing and manipulating smaller samples. In continuous-flow derivatization, the dialysate stream is mixed on-line with reagents and then collected into small fractions for reaction and storage. The on-line addition of reagents is highly reproducible for small-volume fractions and creates a larger volume, at the expense of some dilution, which facilitates off-line injection and sample manipulation. Strategies for continuous-flow derivatization have evolved over time and are summarized in **Figure 5**.

On-line mixing chambers were originally a series of three small pieces of polyethylene tubing glued between pieces of capillary and connected to the outlet of the microdialysis probe (60, 61). Syringes containing internal standards and fluorescent derivatization agents were then connected to these reaction chambers via capillary tubing. As dialysate flowed from the probe toward the collection vial, the sample was continuously derivatized. This work took advantage of the off-line collection and dilution to utilize a discontinuous buffer system and achieve excellent concentration detection limits.

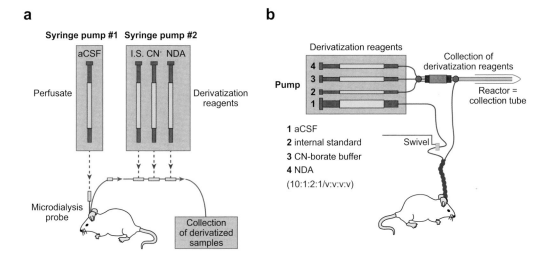

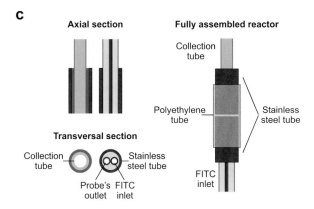

a

Syringe pump #1 Syringe pump #2

aCSF   I.S. CN⁻ NDA

Perfusate   Derivatization
                             reagents

Microdialysis
probe

Collection
of derivatized
samples

b

Derivatization reagents

Pump

4
3
2
1

Collection of
derivatization reagents

Reactor =
collection tube

1 aCSF
2 internal standard
3 CN-borate buffer
4 NDA

(10:1:2:1/v:v:v:v)

Swivel

c

Axial section   **Fully assembled reactor**

Collection
tube

Transversal section

Polyethylene
tube

Stainless
steel tube

Collection
tube

Stainless
steel tube

FITC
inlet

Probe's FITC
outlet   inlet

Figure 5

Evolution of the continuous-flow derivatization reactor. (*a*) The first continuous-flow
derivatization agent reactor was composed of a series of three polyethylene tubes connected to
the outlet of the microdialysis probe. The capillaries from the syringes were inserted into the
polyethylene tubing. The derivatization reaction (naphthalene-2,3-dicarboxaldehyde/cyanide)
occurs within the polyethylene tubing. **Figure 5*a*** reprinted from Reference 61, used with
permission from Elsevier. (*b*) In the second-generation reactor, the dialysate is derivatized
inside the collection tube rather than inside the polyethylene tubing. With this new reactor,
the outlet of the probe is glued to the side of a larger bore capillary that delivers the fluorescent
derivatization agents to the vial. **Figure 5*b*** reprinted from Reference 62, used with permission
from Elsevier. Abbreviations: aCSF, artificial cerebral spinal fluid; CN, cyanide; I.S., internal
standard; NDA, naphthalene-2,3-dicarboxaldehyde. (*c*) The Hernandez group reduced the
volume of the reactor to 26 nL. The reactor is a small gap between two pieces of stainless steel
tubing. In one side of the reactor, the dialysate and the fluorescent derivatization agent (FITC)
are introduced through separate capillaries. On the other side of the reactor is a collection
capillary. At the end of the experiment, the collection capillary is cut into 4-mm pieces
representing 1-s dialysis samples. The derivatized dialysate is then transferred to a vial via
centrifugation. **Figure 5*c*** reprinted from Reference 64, used with permission from Elsevier.

To further improve temporal resolution, researchers decreased the dead volume of the reaction chamber. In this modified method, dialysate collected from a freely moving animal is derivatized in the collection vial instead of inside the polyethylene tubing (**Figure 5b**) (62). The outlet capillary from the microdialysis probe is glued to the outside of a larger bore capillary that contains the fluorescent derivatization agents as well as the internal standards. Samples were collected for 20 s corresponding to 940 nL of total volume.

The Hernandez group (63) modified the concept of continuous-flow derivatization to include three separate chambers. Their precolumn reactor resembled a three-barreled micropipette with one barrel connected to the outlet of the probe, the second barrel connected to a syringe filled with buffer, and the third barrel connected to a syringe of fluorescent derivatization solution. The solutions were mixed at the tip of the reactor, at which a capillary then delivered the samples into a collection vial for 6 s per fraction. This system used fluorescein isothiocyanate for derivatization requiring a 16-h reaction time. After this time, samples were diluted and injected onto a CE instrument.

A modification of this precolumn reactor allowed an increase in the temporal resolution to 1 s. (This temporal resolution was based on the mass sensitivity and fraction size achieved and discounts the Taylor dispersion discussed above.) The modified system reduced the volume of the reactor to 26 nL (64). In addition, samples were collected continuously into a capillary instead of as individual fractions (see **Figure 5c**). After sample collection, the capillary was cut into short pieces (4 mm), each containing a sample fraction corresponding to 1 s of dialysate collection time. The sample was transferred to a collection vial via centrifugation. After 18 h of reaction time with fluorescein isothiocyanate, the sample was diluted and then injected onto a CE instrument.

4.3. On-Line Capillary Electrophoresis Methods for Studying Dialysate Samples

A significant research effort has also been devoted to developing on-line coupling of microdialysis and CE. In this approach, sample collection, derivatization, and separation all occur continuously and automatically, resulting in no manual sample handling. For this approach to be successful, on-line derivatization has to be rapid; therefore, reagents such as fluorescein isothiocyanate are not acceptable. Furthermore, separation time has to be short.

The Lunte group was the first to utilize on-line sample derivatization (using naphthalene-2,3-dicarboxaldehyde/cyanide) and detection. They used this approach to measure glutamate and aspartate with 70-s temporal resolution (65). This method uses two mixing crosses, one as a premixer for derivatization agents and the other as reactor that mixes the dialysate with the derivatization agents. The derivatized sample then travels to a microinjection valve, which alternatively sends the sample or running buffer to the injection interface.

The Kennedy group has developed a system in which dialysate is derivatized on-line and then periodically injected onto a CE system with a flow-gate

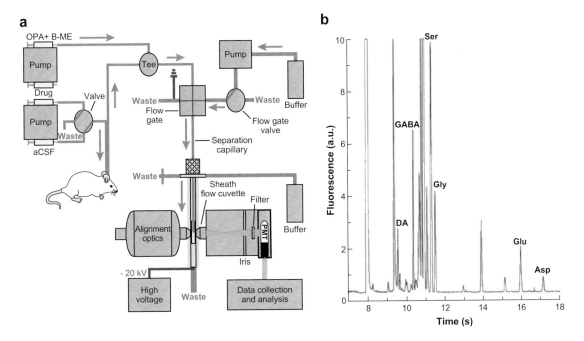

Figure 6

On-line capillary electrophoresis–laser-induced fluorescence instrument with flow gating.
(*a*) Block diagram of a complete on-line capillary electrophoresis instrument. The outlet of the
microdialysis probe is connected to a T-shaped junction, at which dialysate is mixed with the
fluorescent derivatization agents. The derivatized dialysate is then electrokinetically injected
onto a separation capillary using a flow-gate interface. The separation capillary (10 μm inner
diameter) is threaded inside of a sheath flow cuvette, allowing for off-column laser-induced
fluorescence detection. The fluorescent signal is collected by a photomultiplier tube (PMT),
and LabView software is used to collect and analyze the data. Abbreviations: aCSF, artificial
cerebral spinal fluid; B-ME, β-mercaptoethanol; OPA, orthophthaldialdehyde. (*b*) A
representative in vivo electropherogram collected during a potassium stimulation. Five
neuroactive amines (glutamate, aspartate, GABA, taurine, and glutamine) are separated in
under 20 s. **Figure 6*b*** reprinted with permission from Reference 67. Copyright Wiley-VCH
Verlag GmbH & Co. KGaA. Abbreviations: DA, dopamine; Ser, serine; Gly, glycine; Glu,
glutamate; Asp, aspartate.

interface developed by the Jorgenson lab (66). The use of the flow-gate interface plus narrow-bore capillaries (10 μm) has enabled the obtainment of separations of 3–100 s. **Figure 6** presents an overview of the instrument. In this
method, the dialysate sample and fluorescent derivatization agents are mixed at
a T-shaped junction or in coaxial capillaries. The sample is allowed to react inside of a larger bore reaction capillary, which enters one side of the flow gate. By
incorporating a flow-gate interface (for a description of operation, see Reference
67), both the temporal resolution and separation efficiencies were significantly improved over previous interfaces. High efficiencies were possible even at low dialysis flow rates (75 or 155 μL min^{-1}) (68). At such low flow rates, nearly 100%

relative recovery was achieved, allowing quantitative monitoring with good temporal resolution.

This method has evolved over time. Shorter columns and high voltages allowed 3-s temporal resolution for glutamate and aspartate analysis (69). Improvements in detector sensitivity and injection protocols allowed extremely high efficiencies (300,000 to 500,000 theoretical plates) and the resolution of five neuroactive amino acids in approximately 20 s (67). Alterations in buffer conditions (such as use of micellar electrokinetic chromatography, different derivatization reagents, and different detectors) have allowed for the detection of different compounds. Of particular note is the resolution of 17 amino acids in 30 s (70), the resolution of dopamine with 62 other amines (71), the detection of glutathione and cysteine using methanolic-monobromobimane derivatization (55), and the detection of ascorbate and lactate in dialysate (54).

The on-line methods that have been developed utilize custom-built instrumentation. This instrumentation creates a barrier to the wide use of this approach. Investigators have worked to make these on-line methods more accessible to laboratories that do not specialize in CE instrumentation. Off-line continuous-flow derivatization coupled on-line to a commercial CE instrument has been developed to meet this end (72). Instead of three separate reaction chambers, this method uses only one piece of polyethylene tubing to serve as the reaction/mixing chamber. Making this method fully on-line, the derivatized sample is sent directly to the sample loop (630 nL) of an automatic injection valve instead of a vial. An injection interface was used to electrokinetically inject the sample onto the separation capillary to simultaneously measure glutamate, aspartate, and phosphoethanolamine with 2-min temporal resolution. This is the first method reported to couple microdialysis on-line to a commercial instrument for routine sample analysis with minimal sample handling. A flow-gate system has also been assembled using almost all commercial components, which facilitates its use (70).

Another approach that may allow more widespread use of on-line electrophoretic analysis is the use of microfluidic chips. Both glass and polydimethylsulfoxane chips have been coupled to microdialysis sampling for in vivo (73–77) and in vitro experiments (78). In principle, using these chips should be fairly straightforward as all the fluidic connections can be premade. A National Institutes of Health–funded National Resource Center, the Center for Neural Communication Technology (**http://www.cnct.engin.umich.edu/**), is presently investigating the feasibility of providing such chips for user laboratories.

5. ENZYME ASSAY METHODS FOR ANALYZING DIALYSATE SAMPLES

In addition to separations-based methods, enzyme assays have also been frequently used for coupling to microdialysis. Such methods offer dedicated analysis to a small number of analytes and can be rapid and sensitive enough for high–temporal resolution measurement. In general, enzyme assays can be miniaturized with little sensitivity loss. As a result, their mass detection limits can be extremely high. Also, it is straightforward to automate such assays as it generally only requires continuous-flow mixing.

Researchers have developed several enzyme assay methods to study glucose and lactate metabolism in the brain to understand stroke and other neuronal injuries. The sampling frequency ranges from 15 s to 2 min, depending on the analyte and the method. One common approach involves the formation of a bed of particles that contain immobilized enzymes (79–82). For the assays, horseradish peroxidase and an oxidase such as glucose oxidase are immobilized on the particles. Dialysate is injected onto the packed bed along with the electrochemical mediator ferrocene monocarboxylic acid. The oxidase enzyme creates hydrogen peroxide, which is then used by the horseradish peroxidase to reduce the ferrocene mediator to ferricinium, which is measured by a downstream glassy electrode. This change in current as the electrode reduces the ferricinium ion back to ferrocene is proportional to the amount of analyte in the sample. Modifications to this method allow multiple assays to be performed at once through the use of an intricate injection valve that can alternatively send dialysate to one specific enzyme assay (83).

Rather than immobilizing the enzymes on particles to create a packed bed, Rhemrev-Boom et al. (84) have immobilized enzymes in poly(1,3-phenylenediamine) to create a permselective membrane. This immobilization creates a low-volume (20-nL) biosensor. The fabrication of this biosensor involves inserting two platinum wires (one auxiliary and one working electrode) and an Ag/AgCl reference electrode into a piece of Tygon tubing. After washing steps, a solution of poly(1,3 phenylenediamine) plus the oxidoreductase enzyme is pushed into the Tygon tubing. After electrochemical polymerization, the permselective membrane is created. The oxidoreductase enzyme forms hydrogen peroxide when the analyte of interest flows through the biosensor, which is measured by the platinum working electrode. This biosensor can be connected to the outlet of the dialysis probe to allow real-time measurement. The authors chose poly(1,3-phenylenediamine) as the immobilization agent because it does not hinder the sensor from reducing the hydrogen peroxide.

Fluorescence enzyme assays have also been developed to selectively measure the compound of interest. In these methods, dialysate is mixed with a variety of compounds to produce NADPH, which is naturally fluorescent (85). For glucose these compounds include ATP, hexokinase, glucose-6-phosphate dehydrogenase, Mg, and NADP. When these compounds react, two reactions take place: Hexokinase converts glucose to glucose-6-phosphate, which is then used in the second enzyme reaction with glucose-6-phosphate dehydrogenase to reduce NADP to NADPH. The production of NADPH is then proportional to the glucose present. This method also can detect lactate in the so-called lactography method (86). Unfortunately, with this method, glucose levels below basal levels (25 μM) are not detectable owing to either reaction efficiency or absorption. However, this method makes up for this in its simplicity and low cost compared with its electrochemical counterparts. In principle, this approach would be amenable to almost any enzyme reaction that uses NADP as a cofactor.

NADP: nicotinamide adenine dinucleotide phosphate

6. APPLICATIONS OF HIGH–TEMPORAL RESOLUTION MICRODIALYSIS

The evolution of analytical methods with sufficient sensitivity and throughput to allow microdialysis methods good temporal resolution has resulted in new opportunities for studying brain neurochemistry that are just starting to be exploited. **Table 1** summarizes some of the analytes, methods, and applications of methods with temporal resolution better than 2 min. We discuss some illustrative examples below.

An important target of microdialysis measurements has been dopamine, a transmitter known to be involved in reward, reinforcement, and addiction behavior. The development of microbore HPLC methods has improved the temporal resolution of dopamine and its metabolite 3,4-dihydroxyphenylacetic acid (DOPAC) from 10–20 min to 1 min, enabling the correlation of rapidly changing behaviors with dopamine release and metabolism (30). In one example, dopamine and DOPAC were monitored at 1-min temporal resolution as rats self-administered cocaine (30). Dopamine levels stayed above basal levels throughout the self-administration, and animals administered another dose of cocaine to maintain these elevated levels. The duration of this elevated dopamine response was proportional to the dose of cocaine administered, with higher doses having longer latencies between shots. These results are consistent with the hypothesis that dropping dopamine levels (and not dopamine depletion) trigger subsequent cocaine-seeking behavior. A subsequent finding in this experiment was that DOPAC levels were lower during the cocaine administration than prior to drug administration, suggesting that dopamine metabolism is reduced.

The even higher temporal resolution possible with microdialysis coupled on-line to CE-LIF detection has enabled several experiments correlating neurochemistry with a variety of behaviors (see **Table 1**). One experiment examined changes in behavior and neurochemical release at 15-s intervals during the presentation of predator odor known to evoke strong behavioral activation in rats (87). One subset of animals (high responders) immediately started to dig and burrow upon presentation of the odor and showed a large biphasic response in glutamate and GABA release that lasted just 180 s but was slightly delayed from the onset of behavior. Another subset of animals (low responders) showed no behavioral response. These animals also had no neurochemical response to the odor. Indeed, an extremely high correlation between the amount of amino acid release and behavioral activation was observed. Thus, individual differences in behavior and neurochemical release were detectable, suggesting a relationship between chemical release and behavior. Because the glutamate release occurred after the behavior started, it was hypothesized that the release resulted from a response to a stress hormone. Such observations were made possible by the temporal resolution.

Interestingly, previous experiments had performed similar tests while monitoring glutamate at 10-min intervals by microdialysis (88). In those experiments, the maximal increase was 150%, maintained for 30 min after odor presentation. By contrast, at high temporal resolution, the peak increase was 500%—slightly delayed compared to odor presentation (and behavior)—biphasic, and lasted for just 3 min. When the data for the 15-s temporal resolution measurements were binned into 10-min fractions,

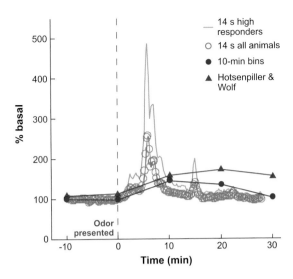

Figure 7

Capillary electrophoresis–laser-induced fluorescence method for measuring transient amino acid changes in response to predator odor. Dialysate measurements are presented in 14-s and 10-min bins to show how glutamate levels change when the predator odor (2,5-dihydro-2,4,5-trimethylthiazoline, from fox) is present. Microdialysis was performed in the nucleus accumbens. The dashed line represents the time at which the odor was presented to the rat. Rats with a high behavioral response to the odor showed an increase in glutamate that lasted only 3 min and peaked at 500%. This response is reduced if the data from both the high and low responders are grouped. Averaging the data into 10-min bins gives a response similar to that seen when using high-performance liquid chromatography to analyze fractions collected over 10-min intervals [data replotted with permission from Hotsenpiller & Wolf (88)]. Decreasing the temporal resolution from 20 s to 10 min dilutes the magnitude of the glutamate change and masks the transient biphasic nature of the response. Figure reprinted from Reference 87, used with permission from Wiley-Blackwell.

they closely matched the previous results (see **Figure 7**), proving that the differences observed resulted from temporal resolution and not the experiment. These differences dramatically highlight the significance of improved temporal resolution.

Continuous-flow derivatization with off-line CE-LIF has also been used for behavioral studies. One study measured GABA and glutamate release in the lateral hypothalamus during feeding behaviors (59). This experiment collected 500-nL dialysate samples corresponding to 30-s bins. Rats were deprived of food for 16 h before the microdialysis experiments began. Samples were taken premeal, during the meal, and after the meal. Glutamate levels spiked once the animal began to eat and then declined, whereas GABA levels slowly rose as the animal reached satiation (**Figure 8**). These findings suggest that glutamate release in the lateral hypothalamus may drive an animal to eat, whereas GABA release in this brain region opposes the drive to eat.

Off-line CE-LIF has also been used to measure amino acid and catecholamine release across the sleep-wake cycle in rats (57). Microdialysis probes were implanted in both the medial prefrontal cortex and the nucleus accumbens, and dialysate was collected off-line for 2 min. Each sample was then sorted as to whether it was collected

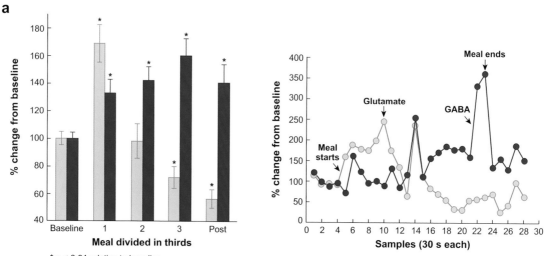

* p < 0.01 relative to baseline

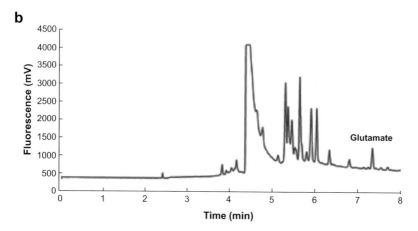

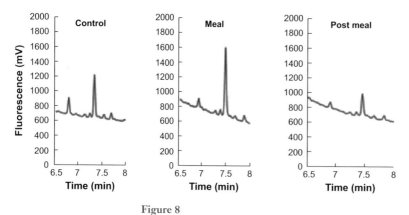

Figure 8

(*Continued*)

during waking, slow-wave sleep, or REM sleep according to electrophysiological readings. Two different separation conditions were used on each sample, one optimized for amino acids and the other for amines. The excitatory amino acids glutamate and aspartate only changed in the nucleus accumbens, in which both amino acids decreased as the animal went into slow-wave sleep and REM sleep. This change was mirrored by norepinephrine in both structures. However, dopamine release was higher in both structures during waking and REM sleep compared with slow-wave sleep. These changes in dopamine levels were hypothesized to be a result of the active cognitive processes.

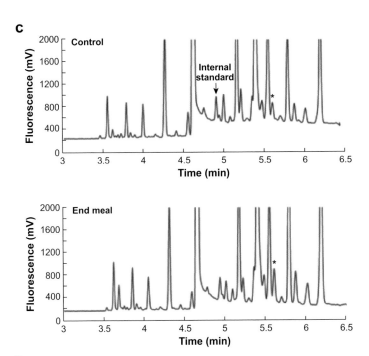

Figure 8

Off-line capillary electrophoresis–laser-induced fluorescence (CE-LIF) method for measuring changes in GABA and glutamate during eating. (*a*) Both the average data (*left panel*) and representative trace from a single animal (*right panel*) demonstrate how extracellular GABA (*dark blue*) and glutamate (*light blue*) levels change throughout a meal. Samples were collected every 30 s as the animal ate and were analyzed using an off-line CE-LIF method. Microdialysis was performed in the lateral hypothalamus. Both panels show that glutamate levels rose for the first third of the meal and then decreased, whereas GABA levels rose at the end of the meal (*$p < 0.01$ relative to baseline). (*b*) A representative in vivo electropherogram identifying the glutamate peak. Each dialysate sample had a separation time of 10 min. Electropherograms showing changes in the glutamate peak height across the meal are depicted as well. (*c*) Sample electropherograms showing the change in GABA levels across the meal. The GABA peak is marked with an asterisk. Figure adapted with permission from Reference 59. Copyright 2003 by the American Psychological Association. The use of APA information does not imply endorsement by APA.

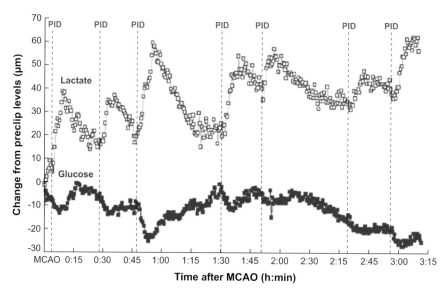

Figure 9

Microdialysis coupled to a dual enzyme assay for lactate and glucose. The occurrence of peri-infarct depolarizations (PIDs) was compared to both glucose and lactate levels after occlusion of the medial cerebral artery. Using a custom-made valve, dialysate was sent every 30 s to either a glucose or lactate assay. The occurrence of PIDs correlates well with an increase in lactate concentration and a decrease in glucose concentration. Figure reprinted with permission from Macmillan Publishers Ltd: *J. Cereb. Blood Flow Metab.* (Reference 79), copyright 2005. Abbreviation: MCAO, middle cerebral artery occlusion.

Enzyme bioassays can measure both lactate and glucose with 15–30-s temporal resolution (79). This method is applicable for clinical measurements because fast metabolite measurements can give insight into the physical state of a patient. In this experiment, the researchers occluded the middle cerebral artery in anesthetized cats and monitored the resulting changes in glucose and lactate every 30 s. When blood flow is stopped, spontaneous depolarizations occur in the surrounding tissue. These depolarizations may cause the size of the infarct to increase. Interestingly, the researchers were able to correlate changes in both glucose and lactate to the occurrence of these depolarizations. When the depolarization took place, lactate increased, while glucose decreased (**Figure 9**). These characteristic changes in glucose and lactate could then be used in the clinical setting to diagnose these spontaneous depolarizations.

As these examples show, high–temporal resolution measurements have been used in a wide variety of applications, including stress responses, learning, feeding, sleep, addiction, and stroke pathophysiology. These initial studies, however, represent a tiny fraction of all microdialysis work, suggesting that we are just starting to see the impact of high–temporal resolution dialysis methods. Especially promising is the ability to correlate behaviors, which are typically rated on 10–30-s intervals, with the neurochemical changes occurring on the same time scale. Other examples, such as rapid pharmacological responses and short-lived processes, may also be studied. Ultimately, it may be possible to also improve the spatial resolution of such measurements. New sampling methods such as low-flow push-pull perfusion (89) and direct sampling (90) have made smaller brain regions accessible to monitoring; however, at present the temporal resolution of such measurements is limited by the flow dynamics of these systems. That is, they require exceedingly low flow rates, which results in high levels of Taylor dispersion during sampling.

Although this review focuses on using microdialysis to perform in vivo central-nervous-system measurements, this technique is applicable to other in vivo environments, in vitro systems, and even industrial processes. Microdialysis has also been used in in vitro experiments specifically investigating enzyme kinetics (91–93) and drug metabolism (94). Enzyme kinetics methods traditionally use spectrophotometry to monitor the appearance and disappearance of products and reactants. The reaction is initiated by adding the substrate, the sample is incubated, and the reaction is quenched by removing the protein or adding an enzyme inhibitor. This procedure can be arduous because, to make accurate time points, all the enzyme must be removed or the reaction will continue. Moreover, some of the analytes of interest could be lost in the separation step when the protein is precipitated out. Microdialysis can circumvent this separation process because it samples only the smaller reaction products and reactants. The membrane keeps the enzymes out, thus quenching the reaction. The sample is protein free, so it can then be injected onto an HPLC column to be separated and analyzed. With this chromatography step, one can separate analytes with similar absorbance spectra that could not be separated using spectrophotometric methods. The continuous sampling provided by microdialysis also makes these kinetic measurements more automatic because the reaction does not need to be manually quenched. These microdialysis enzyme methods have been coupled to both LC and MS with temporal resolutions of 1–10 min. The temporal resolution could possibly be improved by utilizing some of the methods outlined above.

SUMMARY POINTS

1. Microdialysis is a versatile sampling method that can be coupled on-line or off-line to a variety of methods such as LC, CE, MS, and enzyme bioassays for monitoring in vivo neurochemical events.

2. Temporal resolution is important in neurochemical studies because neurotransmitter dynamics cover a wide range, from milliseconds to days.

3. The challenge in creating high–temporal resolution dialysis methods is that as the temporal resolution increases, the mass of analyte available in the sample decreases, and the number of samples to be analyzed increases.

4. Achieving temporal resolution as high as 10 s in microdialysis is possible, but requires methods with attomolar detection limits and high throughput.

5. The temporal resolution possible with microdialysis is ultimately limited by the broadening of concentration plugs owing to flow and diffusion, as they are transported to the analytical system or fraction collector.

6. Capillary-based separation methods (cLC and CE) and enzymatic assays have high mass sensitivities, which facilitate the creation of high–temporal resolution dialysis methods.

7. Recent improvements in temporal resolution have enabled new experiments correlating neurochemical changes with rapidly changing behavior and detecting transient neurochemical events.

8. Microdialysis sampling can also be useful for in vitro experiments such as enzyme kinetics assays to make methods more simple and automated.

FUTURE ISSUES

1. Is it possible to create instruments and methods that allow high temporal resolution that are easy to use and affordable for most neuroscience laboratories?

2. Will advances in commercial HPLC equipment and the development of ultrahigh-pressure LC instrumentation improve the separation time enough to have temporal resolutions comparable with CE methods?

3. Can band broadening associated with sampling be reduced or eliminated to achieve temporal resolution better than 1 s by sampling methods?

4. Most assay development has focused on amino acids and catecholamines. What other molecules should be targeted, and can assays be developed for them?

5. Is there a way to reduce the tissue response to microdialysis probes so that measurements can be taken over several days to weeks?

6. Will improvements in sampling methods such as low-flow push-pull perfusion create high temporal methods for analyzing brain regions that are too small for dialysis probes?

7. What new neuroscience questions will be addressed by the high–temporal resolution methods currently available?

8. Will the new technology make microdialysis of sufficient interest for widespread use in clinical and surgical applications?

DISCLOSURE STATEMENT

The authors are not aware of any biases that might be perceived as affecting the objectivity of this review.

LITERATURE CITED

1. Zoli M, Torri C, Ferrari R, Jansson A, Zini I, et al. 1998. The emergence of the volume transmission concept. *Brain Res. Rev.* 26:136–47

2. Stenberg D. 2007. Neuroanatomy and neurochemistry of sleep. *Cell. Mol. Life Sci.* 64:1187–204

3. Falluel-Morel A, Chafai M, Vaudry D, Basille M, Cazillis M, et al. 2007. The neuropeptide pituitary adenylate cyclase-activating polypeptide exerts antiapoptotic and differentiating effects during neurogenesis: Focus on cerebellar granule neurones and embryonic stem cells. *J. Neuroendocrinol.* 19:321–27

4. Sossi V, Ruth TJ. 2005. Micropet imaging: in vivo biochemistry in small animals. *J. Neural Transm.* 112:319–30

5. Wightman RM. 2006. Probing cellular chemistry in biological systems with microelectrodes. *Science* 311:1570–74

6. Watson CJ, Venton BJ, Kennedy RT. 2006. In vivo measurements of neurotransmitters by microdialysis sampling. *Anal. Chem.* 78:1391–99

7. Moroni F, Pepeu G. 1984. The cortical cup technique. In *Measurement of Neurotransmitter Release In Vivo*, ed. CA Marsden, pp. 63–79. Chichester: Wiley

8. Mulas A, Mulas ML, Pepeu G. 1974. Effect of limbic system lesions on acetylcholine release from cerebral cortex of rat. *Psychopharmacologia* 39:223–30

9. Myers RD, Adell A, Lankford MF. 1998. Simultaneous comparison of cerebral dialysis and push-pull perfusion in the brain of rats: a critical review. *Neurosci. Biobehav. Rev.* 22:371–87

10. Ungerstedt U, Herreramarschitz M, Zetterstrom T. 1982. Dopamine neurotransmission and brain function. *Prog. Brain Res.* 55:41–49

11. Zetterstrom T, Vernet L, Ungerstedt U, Tossman U, Jonzon B, Fredholm BB. 1982. Purine levels in the intact rat brain: studies with an implanted perfused hollow fiber. *Neurosci. Lett.* 29:111–15

12. Tossman U, Ungerstedt U. 1986. Microdialysis in the study of extracellular levels of amino acids in the rat brain. *Acta Physiol. Scand.* 128:9–14

13. Lada MW, Vickroy TW, Kennedy RT. 1997. High temporal resolution monitoring of glutamate and aspartate in vivo using microdialysis on-line with capillary electrophoresis with laser-induced fluorescence detection. *Anal. Chem.* 69:4560–65

14. Crick EW, Osorio I, Bhavaraju NC, Linz TH, Lunte CE. 2007. An investigation into the pharmacokinetics of 3-mercaptopropionic acid and development of a steady-state chemical seizure model using in vivo microdialysis and electrophysiological monitoring. *Epilepsy Res.* 74:116–25

15. Huang YJ, Liao JF, Tsai TH. 2005. Concurrent determination of thalidomide in rat blood, brain and bile using multiple microdialysis coupled to liquid chromatography. *Biomed. Chromatogr.* 19:488–93

16. Shi ZQ, Zhang QH, Jiang XG. 2005. Pharmacokinetic behavior in plasma, cerebrospinal fluid and cerebral cortex after intranasal administration of hydrochloride meptazinol. *Life Sci.* 77:2574–83

17. Wu YT, Tsai TR, Lin LC, Tsai TH. 2007. Liquid chromatographic method with amperometric detection to determine acteoside in rat blood and brain microdialysates and its application to pharmacokinetic study. *J. Chromatogr. B Anal. Technol. Biomed. Life Sci.* 853:281–86

18. Chang YL, Tsai PL, Chou YC, Tien JH, Tsai TH. 2005. Simultaneous determination of nicotine and its metabolite, cotinine, in rat blood and brain tissue using microdialysis coupled with liquid chromatography: pharmacokinetic application. *J. Chromatogr. A* 1088:152–57

19. Lin LC, Chen YF, Tsai TR, Tsai TH. 2007. Analysis of brain distribution and biliary excretion of a nutrient supplement, gastrodin, in rat. *Anal. Chim. Acta* 590:173–79

20. Gilinsky MA, Faibushevish AA, Lunte CE. 2001. Determination of myocardial norepinephrine in freely moving rats using in vivo microdialysis sampling and liquid chromatography with dual-electrode amperometric detection. *J. Pharm. Biomed. Anal.* 24:929–35

21. Richter DW, Schmidt-Garcon P, Pierrefiche O, Bischoff AM, Lalley PM. 1999. Neurotransmitters and neuromodulators controlling the hypoxic respiratory response in anaesthetized cats. *J. Physiol. Lond.* 514:567–78

22. Bradberry CW, Rubino SR. 2004. Phasic alterations in dopamine and serotonin release in striatum and prefrontal cortex in response to cocaine predictive cues in behaving rhesus macaques. *Neuropsychopharmacology* 29:676–85

23. McLaughlin KJ, Faibushevich AA, Lunte CE. 1999. Microdialysis sampling with on-line microbore HPLC for the determination of tirapazamine and its reduced metabolites in rats. *Analyst* 125:105–10

24. Sharp T, Carlsson A, Zetterstrom T, Lundstrom K, Ungerstedt U. 1986. Rapid measurement of dopamine release using brain dialysis combined with microbore HPLC. *Ann. N. Y. Acad. Sci.* 473:512–15

25. Pych JC, Chang Q, Colon-Rivera C, Haag R, Gold PE. 2005. Acetylcholine release in the hippocampus and striatum during place and response training. *Learn. Mem.* 12:564–72

26. Steele KM, Lunte CE. 1995. Microdialysis sampling coupled to online microbore liquid chromatography for pharmacokinetic studies. *J. Pharm. Biomed. Anal.* 13:149–54

27. Yan QS. 1999. Extracellular dopamine and serotonin after ethanol monitored with 5-minute microdialysis. *Alcohol* 19:1–7

28. Liang XZ, Palsmeier RK, Lunte CE. 1995. Dual-electrode amperometric detection for the determination of SR4233 and its metabolites with microbore liquid chromatography. *J. Pharm. Biomed. Anal.* 14:113–19

29. Lorrain DS, Riolo JV, Matuszewich L, Hull EM. 1999. Lateral hypothalamic serotonin inhibits nucleus accumbens dopamine: implications for sexual satiety. *J. Neurosci.* 19:7648–52

30. Wise RA, Newton P, Leeb K, Burnette B, Pocock D, Justice JB. 1995. Fluctuations in nucleus-accumbens dopamine concentration during intravenous cocaine self-administration in rats. *Psychopharmacology* 120:10–20

31. Newton AP, Justice JB. 1994. Temporal response of microdialysis probes to local perfusion of dopamine and cocaine followed with one-minute sampling. *Anal. Chem.* 66:1468–72

32. Boyd BW, Witowski SR, Kennedy RT. 2000. Trace-level amino acid analysis by capillary liquid chromatography and application to in vivo microdialysis sampling with 10-s temporal resolution. *Anal. Chem.* 72:865–71

33. Deng Q, Watson CJ, Kennedy RT. 2003. Aptamer affinity chromatography for rapid assay of adenosine in microdialysis samples collected in vivo. *J. Chromatogr. A* 1005:123–30

34. Baseski HM, Watson CJ, Cellar NA, Shackman JG, Kennedy RT. 2005. Capillary liquid chromatography with MS3 for the determination of enkephalins in microdialysis samples from the striatum of anesthetized and freely-moving rats. *J. Mass Spectrom.* 40:146–53

35. Shackman HM, Shou M, Cellar NA, Watson CJ, Kennedy RT. 2007. Microdialysis coupled on-line to capillary liquid chromatography with tandem mass spectrometry for monitoring acetylcholine in vivo. *J. Neurosci. Methods* 159:86–92

36. Hows MEP, Organ AJ, Murray S, Dawson LA, Foxton R, et al. 2002. High-performance liquid chromatography/tandem mass spectrometry assay for the rapid high sensitivity measurement of basal acetylcholine from microdialysates. *J. Neurosci. Methods* 121:33–39

37. Nyberg F, Nylander I, Terenius L. 1987. Measurements of opioid peptides in human body fluids by radioimmunoassay. In *Handbook of Experimental Pharmacology.* ed. C Patrono, BA Peskar, pp. 227–53. Berlin: Springer-Verlag

38. Jung MC, Shi GY, Borland L, Michael AC, Weber SG. 2006. Simultaneous determination of biogenic monoamines in rat brain dialysates using capillary high-performance liquid chromatography with photoluminescence following electron transfer. *Anal. Chem.* 78:1755–60

39. Shen H, Lada MW, Kennedy RT. 1997. Monitoring of met-enkephalin in vivo with 5-min temporal resolution using microdialysis sampling and capillary liquid chromatography with electrochemical detection. *J. Chromatogr. B Anal. Technol. Biomed. Life Sci.* 704:43–52

40. Shen H, Witowski SR, Boyd BW, Kennedy RT. 1999. Detection of peptides by precolumn derivatization with biuret reagent and preconcentration on capillary liquid chromatography columns with electrochemical detection. *Anal. Chem.* 71:987–94

41. Emmett MR, Caprioli RM. 1994. Micro-electrospray mass spectrometry: ultrahigh-sensitivity analysis of peptides and proteins. *J. Am. Soc. Mass Spectrom.* 5:605–13

42. Andren PE, Caprioli RM. 1999. Determination of extracellular release of neurotensin in discrete rat brain regions utilizing in vivo microdialysis/electrospray mass spectrometry. *Brain Res.* 845:123–29

43. Haskins WE, Wang ZQ, Watson CJ, Rostand RR, Witowski SR, et al. 2001. Capillary LC-MS2 at the attomole level for monitoring and discovering endogenous peptides in microdialysis samples collected in vivo. *Anal. Chem.* 73:5005–14

44. Lanckmans K, Sarre S, Smolders I, Michotte Y. 2007. Use of a structural analogue versus a stable isotope labeled internal standard for the quantification of angiotensin IV in rat brain dialysates using nano-liquid chromatography/tandem mass spectrometry. *Rapid Commun. Mass Spectrom.* 21:1187–95

45. Thompson JW, Mellors JS, Eschelbach JW, Jorgenson JW. 2006. Recent advances in ultrahigh-pressure liquid chromatography. *LC GC North Am.* 24:16–20

46. Yang XQ, Ma LJ, Carr PW. 2005. High temperature fast chromatography of proteins using a silica-based stationary phase with greatly enhanced low pH stability. *J. Chromatogr. A* 1079:213–20

47. Wang XL, Barber WE, Carr PW. 2006. A practical approach to maximizing peak capacity by using long columns packed with pellicular stationary phases for proteomic research. *J. Chromatogr. A* 1107:139–51

48. Whitmore C, Olsson U, Larsson E, Hindsgaul O, Palcic M, Dovichi N. 2007. Yoctomole analysis of ganglioside metabolism in PC12 cellular homogenates. *Electrophoresis* 28:3100–4

49. Kennedy RT, Watson CJ, Haskins WE, Powell DH, Strecker RE. 2002. In vivo neurochemical monitoring by microdialysis and capillary separations. *Curr. Opin. Chem. Biol.* 6:659–65

50. Sinville R, Soper SA. 2007. High resolution DNA separations using microchip electrophoresis. *J. Sep. Sci.* 30:1714–28

51. Rose MJ, Lunte SM, Carlson RG, Stobaugh JF. 2003. Amino acid and peptide analysis using derivatization with *p*-nitrophenol-2,5-dihydroxyphenylacetate *bis*-tetrahydropyranyl ether and capillary electrophoresis with electrochemical detection. *J. Pharm. Biomed. Anal.* 30:1851–59

52. Vandaveer WR, Pasas-Farmer SA, Fischer DJ, Frankenfeld CN, Lunte SM. 2004. Recent developments in electrochemical detection for microchip capillary electrophoresis. *Electrophoresis* 25:3528–49

53. Arnett SD, Osbourn DM, Moore KD, Vandaveer SS, Lunte CE. 2005. Determination of 8-oxoguanine and 8-hydroxy-2′-deoxyguanosine in the rat cerebral cortex using microdialysis sampling and capillary electrophoresis with electrochemical detection. *J. Chromatogr. B Anal. Technol. Biomed. Life Sci.* 827:16–25

54. Lada MW, Kennedy RT. 1995. Quantitative in vivo measurements using microdialysis on-line with capillary zone electrophoresis. *J. Neurosci. Methods* 63:147–52

55. Lada MW, Kennedy RT. 1997. In vivo monitoring of glutathione and cysteine in rat caudate nucleus using microdialysis on-line with capillary zone electrophoresis-laser induced fluorescence detection. *J. Neurosci. Methods* 72:153–59

56. Sauvinet V, Parrot S, Benturquia N, Bravo-Moraton E, Renaud B, Denoroy L. 2003. In vivo simultaneous monitoring of γ-aminobutyric acid, glutamate, and L-aspartate using brain microdialysis and capillary electrophoresis with laser-induced fluorescence detection: analytical developments and in vitro/in vivo validations. *Electrophoresis* 24:3187–96

57. Lena I, Parrot S, Deschaux O, Muffat-Joly S, Sauvinet V, et al. 2005. Variations in extracellular levels of dopamine, noradrenaline, glutamate, and aspartate across the sleep-wake cycle in the medial prefrontal cortex and nucleus accumbens of freely moving rats. *J. Neurosci. Res.* 81:891–99

58. Rada P, Moreno SA, Tucci S, Gonzalez LE, Harrison T, et al. 2003. Glutamate release in the nucleus accumbens is involved in behavioral depression during the Porsolt swim test. *Neuroscience* 119:557–65

59. Rada P, Mendialdua A, Hernandez L, Hoebel BG. 2003. Extracellular glutamate increases in the lateral hypothalamus during meal initiation, and GABA peaks during satiation: microdialysis measurements every 30 s. *Behav. Neurosci.* 117:222–27

60. Bert L, Robert F, Denoroy L, Stoppini L, Renaud B. 1996. Enhanced temporal resolution for the microdialysis monitoring of catecholamines and excitatory amino acids using capillary electrophoresis with laser-induced fluorescence detection: analytical developments and in vitro validations. *J. Chromatogr. A* 755:99–111

61. Robert F, Bert L, Lambas-Senas L, Denoroy L, Renaud B. 1996. In vivo monitoring of extracellular noradrenaline and glutamate from rat brain cortex with 2-min microdialysis sampling using capillary electrophoresis with laser-induced fluorescence detection. *J. Neurosci. Methods* 70:153–62

62. Parrot S, Sauvinet V, Riban V, Depaulis A, Renaud B, Denoroy L. 2004. High temporal resolution for in vivo monitoring of neurotransmitters in awake epileptic rats using brain microdialysis and capillary electrophoresis with laser-induced fluorescence detection. *J. Neurosci. Methods* 140:29–38

63. Tucci S, Rada P, Sepulveda MJ, Hernandez L. 1997. Glutamate measured by 6-s resolution brain microdialysis: capillary electrophoretic and laser-induced fluorescence detection application. *J. Chromatogr. B* 694:343–49

64. Rossell S, Gonzalez LE, Hernandez L. 2003. One-second time resolution brain microdialysis in fully awake rats: protocol for the collection, separation and sorting of nanoliter dialysate volumes. *J. Chromatogr. B Anal. Technol. Biomed. Life Sci.* 784:385–93

65. Zhou SY, Zuo H, Stobaugh JF, Lunte CE, Lunte SM. 1995. Continuous in vivo monitoring of amino acid neurotransmitters by microdialysis sampling with online derivatization and capillary electrophoresis separation. *Anal. Chem.* 67:594–99

66. Hooker TF, Jorgenson JW. 1997. A transparent flow gating interface for the coupling of microcolumn LC with CZE in a comprehensive two-dimensional system. *Anal. Chem.* 69:4134–42

67. Bowser MT, Kennedy RT. 2001. In vivo monitoring of amine neurotransmitters using microdialysis with on-line capillary electrophoresis. *Electrophoresis* 22:3668–76

68. Lada MW, Kennedy RT. 1996. Quantitative in vivo monitoring of primary amines in rat caudate nucleus using microdialysis coupled by a flow-gated interface to capillary electrophoresis with laser-induced fluorescence detection. *Anal. Chem.* 68:2790–97

69. Lada MW, Vickroy TW, Kennedy RT. 1998. Evidence for neuronal origin and metabotropic receptor-mediated regulation of extracellular glutamate and aspartate in rat striatum in vivo following electrical stimulation of the prefrontal cortex. *J. Neurochem.* 70:617–25

70. Shou M, Smith AD, Shackman JG, Peris J, Kennedy RT. 2004. In vivo monitoring of amino acids by microdialysis sampling with on-line derivatization by naphthalene-2,3-dicarboxyaldehyde and rapid micellar electrokinetic capillary chromatography. *J. Neurosci. Methods* 138:189–97

71. Shou MS, Ferrario CR, Schultz KN, Robinson TE, Kennedy RT. 2006. Monitoring dopamine in vivo by microdialysis sampling and on-line CE-laser-induced fluorescence. *Anal. Chem.* 78:6717–25

72. Robert F, Bert L, Parrot S, Denoroy L, Stoppini L, Renaud B. 1998. Coupling on-line brain microdialysis, precolumn derivatization and capillary electrophoresis for routine minute sampling of O-phosphoethanolamine and excitatory amino acids. *J. Chromatogr. A* 817:195–203

73. Cellar NA, Kennedy RT. 2006. A capillary-PDMS hybrid chip for separations-based sensing of neurotransmitters in vivo. *Lab Chip* 6:1205–12

74. Cellar NA, Burns ST, Meiners JC, Chen H, Kennedy RT. 2005. Microfluidic chip for low-flow push-pull perfusion sampling in vivo with on-line analysis of amino acids. *Anal. Chem.* 77:7067–73

75. Li MW, Huynh BH, Hulvey MK, Lunte SM, Martin RS. 2006. Design and characterization of poly(dimethylsiloxane)-based valves for interfacing continuous-flow sampling to microchip electrophoresis. *Anal. Chem.* 78:1042–51

76. Sandlin ZD, Shou MS, Shackman JG, Kennedy RT. 2005. Microfluidic electrophoresis chip coupled to microdialysis for in vivo monitoring of amino acid neurotransmitters. *Anal. Chem.* 77:7702–8

77. Huynh BH, Fogarty BA, Nandi P, Lunte SA. 2006. A microchip electrophoresis device with on-line microdialysis sampling and on-chip sample derivatization by naphthalene 2,3-dicarboxaldehyde/2-mercaptoethanol for amino acid and peptide analysis. *J. Pharm. Biomed. Anal.* 42:529–34

78. Huynh BH, Fogarty BA, Martin RS, Lunte SM. 2004. On-line coupling of microdialysis sampling with microchip-based capillary electrophoresis. *Anal. Chem.* 76:6440–47

79. Hopwood SE, Parkin MC, Bezzina EL, Boutelle MG, Strong AJ. 2005. Transient changes in cortical glucose and lactate levels associated with peri-infarct depolarisations, studied with rapid-sampling microdialysis. *J. Cereb. Blood Flow Metab.* 25:391–401

80. Jones DA, Ros J, Landolt H, Fillenz M, Boutelle MG. 2000. Dynamic changes in glucose and lactate in the cortex of the freely moving rat monitored using microdialysis. *J. Neurochem.* 75:1703–8

81. Miele M, Boutelle MG, Fillenz M. 1996. The source of physiologically stimulated glutamate efflux from the striatum of conscious rats. *J. Physiol. Lond.* 497:745–51

82. Boutelle MG, Fellows LK, Cook C. 1992. Enzyme packed-bed system for the online measurement of glucose, glutamate, and lactate in brain microdialysate. *Anal. Chem.* 64:1790–94

83. Parkin MC, Hopwood SE, Jones DA, Hashemi P, Landolt H, et al. 2005. Dynamic changes in brain glucose and lactate in pericontusional areas of the human cerebral cortex, monitored with rapid sampling on-line microdialysis: relationship with depolarisation-like events. *J. Cereb. Blood Flow Metab.* 25:402–13

84. Rhemrev-Boom MM, Jonker MA, Venema K, Jobst G, Tiessen R, Korf J. 2001. On-line continuous monitoring of glucose or lactate by ultraslow microdialysis combined with a flow-through nanoliter biosensor based on poly(*m*-phenylenediamine) ultrathin polymer membrane as enzyme electrode. *Analyst* 126:1073–79

85. van der Kuil JHF, Korf J. 1991. On-line monitoring of extracellular brain glucose using microdialysis and a NADPH-linked enzymatic assay. *J. Neurochem.* 57:648–54

86. Elekes O, Venema K, Postema F, Dringen R, Hamprecht B, Korf J. 1996. Evidence that stress activates glial lactate formation in vivo assessed with rat hippocampus lactography. *Neurosci. Lett.* 208:69–72

87. Venton BJ, Robinson TE, Kennedy RT. 2006. Transient changes in nucleus accumbens amino acid concentrations correlate with individual responsivity to the predator fox odor 2,5-dihydro-2,4,5-trimethylthiazoline. *J. Neurochem.* 96:236–46

88. Hotsenpiller G, Wolf ME. 2003. Baclofen attenuates conditioned locomotion to cues associated with cocaine 123 administration and stabilizes extracellular glutamate levels in rat nucleus accumbens. *Neuroscience* 118:123–34

89. Kottegoda S, Shaik I, Shippy SA. 2002. Demonstration of low flow push-pull perfusion. *J. Neurosci. Methods* 121:93–101

90. Kennedy RT, Thompson JE, Vickroy TW. 2002. In vivo monitoring of amino acids by direct sampling of brain extracellular fluid at ultralow flow rates and capillary electrophoresis. *J. Neurosci. Methods* 114:39–49

91. Modi SJ, LaCourse WR. 2006. Monitoring carbohydrate enzymatic reactions by quantitative in vitro microdialysis. *J. Chromatogr. A* 1118:125–33

92. Zhou J, Shearer EC, Hong J, Riley CM, Schowen RL. 1996. Automated analytical systems for drug development studies. 5. A system for enzyme kinetic studies. *J. Pharm. Biomed. Anal.* 14:1691–98

93. Kerns EH, Volk KJ, Klohr SE, Lee MS. 1999. Monitoring in vitro experiments using microdialysis sampling on-line with mass spectrometry. *J. Pharm. Biomed. Anal.* 20:115–28

94. Gunaratna C, Kissinger PT. 1997. Application of microdialysis to study the in vitro metabolism of drugs in liver microsomes. *J. Pharm. Biomed. Anal.* 16:239–48

95. Thompson JE, Vickroy TW, Kennedy RT. 1999. Rapid determination of aspartate enantiomers in tissue samples by microdialysis coupled on-line with capillary electrophoresis. *Anal. Chem.* 71:2379–84

96. Rebec GV, Witowski SR, Sandstrom MI, Rostand RD, Kennedy RT. 2005. Extracellular ascorbate modulates cortically evoked glutamate dynamics in rat striatum. *Neurosci. Lett.* 378:166–70

97. Bert L, Parrot S, Robert F, Desvignes C, Denoroy L, et al. 2002. In vivo temporal sequence of rat striatal glutamate, aspartate and dopamine efflux during apomorphine, nomifensine, NMDA and PDC in situ administration. *Neuropharmacology* 43:825–35

98. Smith A, Watson CJ, Frantz KJ, Eppler B, Kennedy RT, Peris J. 2004. Differential increase in taurine levels by low-dose ethanol in the dorsal and ventral striatum revealed by microdialysis with on-line capillary electrophoresis. *Alcohol. Clin. Exp. Res.* 28:1028–38

99. Smith A, Watson CJ, Kennedy RT, Peris J. 2003. Ethanol-induced taurine efflux: low dose effects and high temporal resolution. In *Taurine 5: Beginning the 21st Century*, ed. JB Lombardini, SW Schaffer, J Azuma, pp. 485–92. New York: Kluwer Academic

100. Hu M, Watson CJ, Kennedy RT, Becker JB. 2006. Estradiol attenuates the K^+ induced increase in extracellular GABA in rat striatum. *Synapse* 59:122–24

101. Venton BJ, Robinson TE, Kennedy RT, Maren S. 2006. Dynamic amino acid increases in the basolateral amygdala during acquisition and expression of conditioned fear. *Eur. J. Neurosci.* 23:3391–98

102. Presti MF, Watson CJ, Kennedy RT, Yang M, Lewis MH. 2004. Behavior-related alterations of striatal neurochemistry in a mouse model of stereotyped movement disorder. *Pharmacol. Biochem. Behav.* 77:501–7

103. Tucci S, Rada P, Hernandez L. 1998. Role of glutamate in the amygdala and lateral hypothalamus in conditioned taste aversion. *Brain Res.* 813:44–49

104. Rada P, Tucci S, Perez J, Teneud L, Chuecos S, Hernandez L. 1998. In vivo monitoring of gabapentin in rats: a microdialysis study coupled to capillary electrophoresis and laser-induced fluorescence detection. *Electrophoresis* 19:2976–80

105. Reyes E, Rossell S, Paredes D, Rada P, Tucci S, et al. 2002. Haloperidol abolished glutamate release evoked by photic stimulation of the visual cortex in rats. *Neurosci. Lett.* 327:149–52

106. Silva E, Hernandez L, Quinonez B, Gonzalez LE, Colasante C. 2004. Selective amino acids changes in the medial and lateral preoptic area in the formalin test in rats. *Neuroscience* 124:395–404

107. Robert F, Parisi L, Bert L, Renaud B, Stoppini L. 1997. Microdialysis monitoring of extracellular glutamate combined with the simultaneous recording of evoked field potentials in hippocampal organotypic slice cultures. *J. Neurosci. Methods* 74:65–76

108. Bhatia R, Hashemi P, Razzaq A, Parkin MC, Hopwood SE, et al. 2006. Application of rapid-sampling, online microdialysis to the monitoring of brain metabolism during aneurysm surgery. *Neurosurgery* 58:313–20

109. Parrot S, Sauvinet V, Xavier JM, Chavagnac D, Mouly-Badina L, et al. 2004. Capillary electrophoresis combined with microdialysis in the human spinal cord: a new tool for monitoring rapid peroperative changes in amino acid neurotransmitters within the dorsal horn. *Electrophoresis* 25:1511–17

110. Dmitrieva N, Rodriguez-Malaver AJ, Hernandez L. 2004. Differential release of neurotransmitters from superficial and deep layers of the dorsal horn in response to acute noxious stimulation and inflammation of the rat paw. *Eur. J. Pain* 8:245–52

111. Parrot S, Bert L, Renaud B, Denoroy L. 2003. Glutamate and aspartate do not exhibit the same changes in their extracellular concentrations in the rat striatum after N-methyl-D-aspartate local administration. *J. Neurosci. Res.* 71:445–54

112. Hadwiger ME, Torchia SR, Park S, Biggin ME, Lunte CE. 1996. Optimization of the separation and detection of the enantiomers of isoproterenol in microdialysis samples by cyclodextrin-modified capillary electrophoresis using electrochemical detection. *J. Chromatogr. B Biomed. Appl.* 681:241–49

113. Demestre M, Boutelle M, Fillenz M. 1997. Stimulated release of lactate in freely moving rats is dependent on the uptake of glutamate. *J. Physiol. Lond.* 499:825–32

114. Kaptein WA, Zwaagstra JJ, Venema K, Korf J. 1998. Continuous ultraslow microdialysis and ultrafiltration for subcutaneous sampling as demonstrated by glucose and lactate measurements in rats. *Anal. Chem.* 70:4696–700

115. Dijk SN, Kropvangastel W, Obrenovitch TP, Korf J. 1994. Food deprivation protects the rat striatum against hypoxia ischemia despite high extracellular glutamate. *J. Neurochem.* 62:1847–51

116. Krugers HJ, Jaarsma D, Korf J. 1992. Rat hippocampal lactate efflux during electroconvulsive shock or stress is differently dependent on entorhinal cortex and adrenal integrity. *J. Neurochem.* 58:826–30

117. Nurmi M, Kiianmaa K, Sinclair JD. 1994. Brain ethanol in AA, ANA, and Wistar rats monitored with one-minute microdialysis. *Alcohol* 11:315–21

118. Parrot S, Lambas-Senas L, Sentenac S, Denoroy L, Renaud B. 2007. Highly sensitive assay for the measurement of serotonin in microdialysates using capillary high-performance liquid chromatography with electrochemical detection. *J. Chromatogr. B Anal. Technol. Biomed. Life Sci.* 850:303–9

119. Chen AQ, Lunte CE. 1995. Microdialysis sampling coupled online to fast microbore liquid chromatography. *J. Chromatogr. A* 691:29–35

120. Kehr J. 1998. Determination of glutamate and aspartate in microdialysis samples by reversed-phase column liquid chromatography with fluorescence and electrochemical detection. *J. Chromatogr. B* 708:27–38

121. Hogan BL, Lunte SM, Stobaugh JF, Lunte CE. 1994. Online coupling of in vivo microdialysis sampling with capillary electrophoresis. *Anal. Chem.* 66:596–602

122. Parrot S, Bert L, Renaud B, Denoroy L. 2001. Large interexperiment variations in microdialysate aspartate and glutamate in rat striatum may reflect a circannual rhythm. *Synapse* 39:267–69

123. Rhemrev-Boom RM, Tiessen RG, Jonker AA, Venema K, Vadgama P, Korf J. 2002. A lightweight measuring device for the continuous in vivo monitoring of glucose by means of ultraslow microdialysis in combination with a miniaturised flow-through biosensor. *Clin. Chim. Acta* 316:1–10

124. Bradberry CW, Rubino SR. 2006. Dopaminergic responses to self-administered cocaine in rhesus monkeys do not sensitize following high cumulative intake. *Eur. J. Neurosci.* 23:2773–78

125. Bradberry CW. 2000. Acute and chronic dopamine dynamics in a nonhuman primate model of recreational cocaine use. *J. Neurosci.* 20:7109–15

Applications of Ultrafast Lasers for Optical Measurements in Combusting Flows

James R. Gord,[1] Terrence R. Meyer,[2] and Sukesh Roy[1]

[1] Air Force Research Laboratory, Propulsion Directorate, Wright-Patterson Air Force Base, Ohio 45433; email: james.gord@wpafb.af.mil, sroy@woh.rr.com

[2] Mechanical Engineering Department, Iowa State University, Ames, Iowa 50011; email: trm@iastate.edu

Annu. Rev. Anal. Chem. 2008. 1:663–87

First published online as a Review in Advance on March 18, 2008

The *Annual Review of Analytical Chemistry* is online at anchem.annualreviews.org

This article's doi: 10.1146/annurev.anchem.1.031207.112957

Key Words

combustion diagnostics, laser-induced fluorescence, pump/probe, ballistic imaging, coherent anti-Stokes Raman scattering, wave mixing

Abstract

Optical measurement techniques are powerful tools for the detailed study of combustion chemistry and physics. Although traditional combustion diagnostics based on continuous-wave and nanosecond-pulsed lasers continue to dominate fundamental combustion studies and applications in reacting flows, revolutionary advances in the science and engineering of ultrafast (picosecond- and femtosecond-pulsed) lasers are driving the enhancement of existing diagnostic techniques and enabling the development of new measurement approaches. The ultrashort pulses afforded by these new laser systems provide unprecedented temporal resolution for studies of chemical kinetics and dynamics, freedom from collisional-quenching effects, and tremendous peak powers for broad spectral coverage and non-linear signal generation. The high pulse-repetition rates of ultrafast oscillators and amplifiers allow previously unachievable data-acquisition bandwidths for the study of turbulence and combustion instabilities. We review applications of ultrafast lasers for optical measurements in combusting flows and sprays, emphasizing recent achievements and future opportunities.

1. INTRODUCTION

CARS: coherent
anti-Stokes Raman
scattering

RFWM: resonant
four-wave mixing

Propulsion systems represent a substantial fraction of the cost, weight, and complexity of aircraft and spacecraft. The vast majority of these propulsion systems are powered through fuel combustion; therefore, the detailed study of fundamental combustion phenomena has emerged as a highly relevant and important field of endeavor. Today's combustion scientists and engineers devote much of their work to improving propulsion-system performance while simultaneously reducing pollutant emissions. Increasing the affordability, maintainability, and reliability of these critical propulsion systems is a major driver of activity as well.

Although our efforts in the Combustion and Laser Diagnostics Research Complex at Wright-Patterson Air Force Base are focused primarily on combustion phenomena associated with air-breathing and rocket propulsion, a host of other applications drive advances in combustion science as well. These include internal-combustion engines, land- and sea-based power generation, industrial processing, combustion-based synthesis, waste incineration, and fire safety, for example. Clean, efficient combustion technologies and alternative fuels (e.g., Fischer-Tropsch fuels and biofuels) are critical for meeting current and future energy demands while reducing our dependence on fossil fuels and minimizing such environmental impacts as smog, particulates, acid rain, greenhouse gases, and global warming.

Advanced measurement techniques that exploit lasers and optics have become well-established tools for characterizing combusting flows (1–4). Such noninvasive measurement approaches are often ideally suited for visualizing complex reacting flows and quantifying key chemical-species concentrations, temperature, and fluid-dynamic parameters. The fundamental information these techniques provide is essential for achieving a detailed understanding of the chemistry and physics of combustion processes.

Many successful optical measurements achieved to date in combusting flows have been based on the use of conventional continuous-wave and nanosecond-pulsed laser systems, including Q-switched Nd:YAG lasers, excimer lasers, and associated YAG- and excimer-pumped dye lasers. These systems have been the workhorses in most experiments involving such optical measurement techniques as planar laser-induced fluorescence, particle-image velocimetry, laser-induced incandescence, coherent anti-Stokes Raman scattering (CARS) spectroscopy, and resonant four-wave mixing (RFWM). These laser systems afford high pulse energies required for sheet lighting in planar techniques and for nonlinear interactions such as those in CARS and RFWM. They also provide the relatively narrow spectral bandwidths required for spectroscopic studies of key gas-phase combustion species (e.g., OH, CH, NO, and CO).

Although the impact of continuous-wave and nanosecond-pulsed lasers systems on the modern science of combustion measurements is undeniable, continuing revolutionary advances in the science and engineering of ultrafast lasers (i.e., picosecond- and femtosecond-pulsed lasers) (5–8) have enhanced the capabilities and utility of existing combustion-diagnostic techniques while enabling the development and application of new measurement methodologies previously unachievable. Early practitioners pioneered ultrafast combustion measurements based on modelocked argon-ion and Nd:YAG lasers with synchronously pumped dye lasers, but the advent of modelocked titanium:sapphire (Ti:sapphire) oscillators and amplifier systems based on regenerative and multipass configurations and chirped-pulse amplification has changed the landscape dramatically, accelerating the development and application of new ultrafast laser–based combustion diagnostics.

PS: polarization spectroscopy

Regardless of the architecture, two key features of ultrafast lasers are responsible for the tremendous utility they afford for combustion measurements: ultrashort pulses and high pulse-repetition rates. Myriad beneficial characteristics stem from the ultrashort picosecond and femtosecond pulses delivered by modern ultrafast laser systems. These advantages of ultrashort pulses are realized in terms of the time resolution achievable and the temporal duration of the optical combustion measurements. In addition, ultrashort pulses enable tremendous instantaneous power from laser systems of moderate to low average power.

The time resolution afforded by ultrafast laser systems has been exploited to study the kinetics and dynamics of combustion chemistry and energy-transfer processes. Investigators have used separation in time of various pump and probe pulses to discriminate against nonresonant background signals, enhancing measurement sensitivity and selectivity and enabling the determination of key minor-species concentrations. Because of the ultrashort duration of some of these measurements, signals can be acquired that are largely free of collisional and pressure effects. This key feature of ultrafast combustion measurements addresses one of the major limitations of conventional nanosecond-pulsed diagnostics, and it enables quantitative measurements of parameters such as number densities and temperature in high-pressure, turbulent flames characteristic of most practical combustion devices. In these systems the collisional-quenching environment is typically highly inhomogeneous and rapidly changing in both space and time.

The instantaneous power from these ultrafast laser systems has been exploited to drive many desirable nonlinear phenomena. Researchers have utilized such nonlinearities to expand the spectral coverage available from these laser systems and to achieve various novel higher-order signals based on multiple-wave mixing. Amplified Ti:sapphire-based systems can deliver usable radiation throughout an impressive spectral region across the ultraviolet (UV), visible, and infrared. Through continuing advances involving laser-matter interactions and higher-order harmonic generation, extreme UV radiation and X-rays can be produced for diagnostic applications. At the opposite end of the spectrum, terahertz radiation has been generated using ultrafast lasers and applied to combustion measurements. Nonlinear signal generation in combusting flows has been explored with ultrafast lasers and wave-mixing techniques that include polarization spectroscopy (PS), CARS, and RFWM (described in detail below).

Many ultrafast laser systems deliver very high pulse-repetition rates. Modelocked oscillators feature repetition rates of the order ~100 GHz, and commercially available amplifiers deliver pulses at rates up to 1–300 kHz. Fluctuation timescales of the order 1–100 μs characterize the high-pressure, turbulent combustion environments found in most practical devices. Although conventional 10-Hz, nanosecond-pulsed laser systems can be used to study these turbulent combustion environments, they provide only probability density functions for combustion parameters and cannot capture time correlations describing fluctuating turbulent combustion. Investigators have used measurements based on high-repetition-rate, ultrafast laser systems, conversely, to capture probability density functions, as well as time series, time correlations, and power spectral densities (PSDs) describing the frequency content of turbulent phenomena of interest.

In the sections below, we review the advantages of ultrafast laser systems for optical combustion measurements with special emphasis on applications involving ultrafast laser-induced fluorescence (LIF), linear pump/probe techniques, time-gated ballistic imaging, ultrafast CARS spectroscopy, and resonant and Raman PS and wave mixing. Advances stemming from the ultrashort pulses and high pulse-repetition rates provided by ultrafast lasers are evident throughout these applications.

2. ULTRAFAST LASER-INDUCED FLUORESCENCE

LIF has been used extensively for measurements of minor-species concentrations and gas temperature in reacting and nonreacting flows (1–4). Measurements of minor species are important for understanding flame chemistry and validating models of ignition, heat release, flame propagation, pollutant formation, and flame extinction. The availability of picosecond lasers for time-resolved LIF has played a critical role in the study of energy-transfer processes, in extending LIF measurements to the deep UV, and in improving the quantitative nature of LIF in unknown quenching environments.

The quantitative interpretation of LIF signals requires detailed knowledge of molecular energy-transfer processes such as electronic quenching, rotational energy transfer (RET), and vibrational energy transfer. One can easily investigate these processes, which compete with the radiative decay of the excited state, at low pressures using nanosecond lasers (9). At atmospheric pressure, however, transfer rates must be measured with picosecond resolution given gas collision rates that yield fluorescence lifetimes of the order 1 ns.

A typical setup for OH LIF comprises a chirped-pulse amplified Ti:sapphire laser system (1.5-ps pulse width, 0.5 mJ per pulse) that is frequency tripled to the UV near 284 nm (1-ps pulse width, 20 μJ per pulse) (10). Other investigators have used a Raman-excimer laser in combination with stimulated Brillouin scattering (11), passive and active modelocking in combination with stimulated Brillouin scattering (12), or regenerative amplification with or without a distributed-feedback dye laser (13, 14) to achieve ~100-ps pulses with sufficient spectral resolution (of the order 0.5 cm^{-1} at atmospheric pressure) to isolate individual rotational transitions. Another advantage of picosecond lasers is the ability to access species such as H atoms (12), O atoms (15, 16), CO (17), and NO (18) through two-photon excitation. Typical detection schemes include fast photomultiplier tubes, streak cameras, and fast-gated optical imagers for time-resolved studies of collisional quenching, RET, and vibrational energy transfer (9–18).

In addition to fundamental studies of energy-transfer processes for LIF, ultrafast lasers have also been applied for quantitative measurements in flames of practical interest. In particular, a number of researchers have used fluorescence lifetimes for quantitative measurements of minor-species concentrations in flames in which the temperature and colliding-species concentrations may be unknown. These lifetimes, which are largely dominated by collisional quenching for atmospheric-pressure flames, are typically of the order ~1–3 ns and require the use of a picosecond laser for sufficient time resolution. Application of this approach is particularly important in unsteady, turbulent flames. Studies have been performed for vortex/flame interactions (19) and two-dimensional imaging (20) using picosecond LIF with streak cameras. In turbulent flames, it is also desirable to compute the PSD to analyze the frequency content of number-density fluctuations. To obtain longer time series for PSD measurements of minor species, one can employ a picosecond Ti:sapphire oscillator along with a multichannel photon-counting system for on-the-fly quenching corrections based on fluorescence-lifetime measurements (21, 22). This eliminates uncertainties due to the variation of temperature and colliding-species concentrations in turbulent flames. Time-series measurements have been made for CH (23) and OH (24), as well as OH and temperature (25). **Figure 1** shows a schematic of the optics layout and time-gating approach for the latter, and **Figure 2** shows typical time-resolved OH and temperature measurements in a vortex/flame burner.

Picosecond lasers have been instrumental in improving our understanding of molecular energy-transfer processes in LIF. In addition, picosecond LIF has proven to be useful in turbulent flames in which LIF signals are influenced by local variations in the rate of collisional quenching.

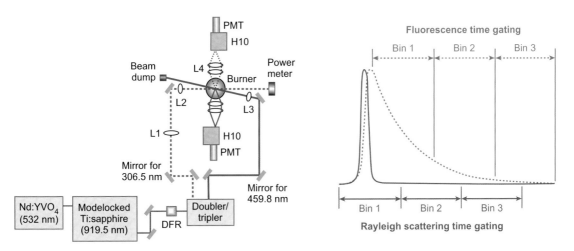

Figure 1

Optical layout (*left panel*) and time-gating diagram (*right panel*) for simultaneous Rayleigh scattering and lifetime-corrected OH laser-induced fluorescence. Bins 1–3 integrate over 3.5-ns bins to detect Rayleigh scattering and resolve fluorescence lifetime assuming single exponential decay. Abbreviations used: DFR, double Fresnel rhomb; H10, 0.1-m monochromator; L1, 1-m lens; L2, 20-cm lens; L3, 20-cm lens; L4, 15-cm lens; PMT, photomultiplier tube. Figure reprinted with permission from Reference 25. Copyright 2007, Optical Society of America.

Figure 2

Simultaneous
measurements of
temperature and lifetime-
corrected OH laser-
induced fluorescence
during vortex/flame
interaction.

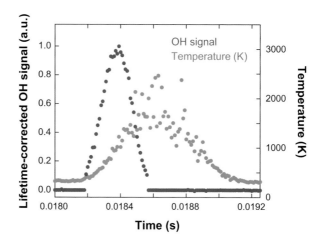

3. LINEAR PUMP/PROBE TECHNIQUES

A number of groups have developed and applied linear pump/probe techniques to measure absolute number densities of key species as well as temperature in combusting flows. These techniques are linear in the sense that the combustion analyte under study interacts linearly with the pump beam (i.e., a one-photon pump-analyte interaction) and linearly with the probe beam (i.e., a one-photon probe-analyte interaction). Typically in these experiments, the pump beam interacts with the analyte in a linear, one-photon absorption process, transferring some population from the ground electronic state to an excited electronic state, thereby creating a transient hole in the ground-state population and a transient excited-state population. The probe beam subsequently interacts with this transient population distribution through one or more of three processes, based in large part on the spectral characteristics of the probe beam. A probe beam spectrally resonant with ground-state absorption features of the analyte experiences a transient reduction in absorption upon interaction with the pump-modified population distribution. This is manifested as a transient gain in the transmitted intensity of the probe beam. A probe beam spectrally resonant with emission features of the analyte experiences a transient enhancement through stimulated emission upon interaction with the pump-modified population distribution. This is also manifested as a transient gain in the transmitted intensity of the probe beam. A probe beam spectrally resonant with excited-state absorption features of the analyte experiences a transient increase in absorption upon interaction with the pump-modified population distribution. This is manifested as a transient bleach in the transmitted intensity of the probe beam. As the redistribution of population induced by the pump beam relaxes in time, so too do the transient phenomena associated with the probe-beam transmission.

In practice investigators often observe these transient phenomena experimentally by modulating the pump-beam intensity. This modulation is transferred from the pump beam to the analyte population distribution and thereby onto the probe beam. Lock-in detection of the transmitted probe beam reveals the extent of modulation transfer from the pump beam to the probe beam, and the lock-in signal scales linearly with the analyte concentration. The time delay between the pump and probe beams can be adjusted to explore the temporal characteristics of the population-relaxation dynamics.

ASOPS: asynchronous optical sampling

Lytle and coworkers (26, 27) developed and applied an ingenious scheme for scanning the pump/probe delay and exploring the temporal evolution of the population relaxation using a technique they termed asynchronous optical sampling (ASOPS). In ASOPS, two separate laser oscillators are used—one for the pump beam and one for the probe beam. The optical cavities of the two oscillators are adjusted to slightly different lengths such that the two oscillators operate at slightly different pulse-repetition rates. This difference in pulse-repetition rates is manifested as a repetitive phase walkout between the pump and probe pulses that repeats at Δf, the difference between the two pulse-repetition rates. A no-moving-parts pump/probe delay is achieved without the need for an optomechanical delay line (see **Figure 3**).

The ASOPS technique complements the LIF techniques described in Section 2 above and similar to those techniques can be used to measure absolute number densities and explore the detailed, time-evolving collisional-quenching environments in turbulent combustion through the determination of the population lifetime decay. Fiechtner and coworkers (28–31) applied ASOPS to the measurement of atomic sodium and OH in various laboratory flames. They developed rate-equation models to extract quantitative number densities and collisional-quenching rates from these ASOPS experiments (30, 32).

Fiechtner & Linne (33) pursued a simplified pump/probe arrangement with a fixed pump/probe delay rather than a scanning delay. Although this approach does not reveal the temporal evolution of population relaxation, it can be configured to yield measurements of absolute number densities of key chemical species free from the effects of collisional quenching, provided the fixed pump/probe delay is set such that the probe beam interacts with the pump-modified analyte on a timescale that is short with respect to the collisional timescale. In this fashion, Settersten and coworkers

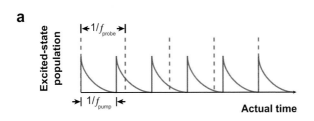

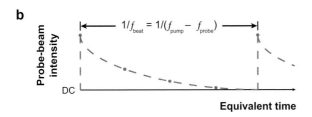

Figure 3

Conceptual asynchronous optical sampling timing diagram depicting (*a*) excited-state population and (*b*) probe-beam intensity upon stimulated emission. Pump pulses are indicated in blue, and probe pulses are indicated in red.

THz-TDS: terahertz
time-domain spectroscopy

(34, 35) achieved measurements of potassium and other species in various laboratory flames and modeled the characteristics of the pump/probe signal using rate equations and density-matrix equations (34–37). By forming the pump beam into a laser light sheet and crossing that pump-beam sheet with an upcollimated probe beam in a premixed methane-air flame maintained on a Meeker-type burner, Linne et al. (38) achieved ultrafast linear pump/probe two-dimensional imaging of potassium seeded into the flame.

Advances in ultrafast laser systems have driven tremendous innovations in sensing with terahertz radiation (39). We can also consider terahertz time-domain spectroscopy (THz-TDS) as a linear pump/probe technique, although the pump and probe interactions differ from those discussed above. In the pump/probe experiments described above, the pump and probe beams interact directly with the analyte; however, the pump and probe beams serve different purposes in THz-TDS. In these experiments, the pump beam from a femtosecond-pulsed laser interacts with a target material to generate broadband terahertz radiation, either through photoconduction in a biased semiconductor (e.g., low-temperature-grown GaAs) or through optical rectification (a process described in detail in Reference 39) in a nonlinear medium. When the pulsed terahertz radiation generated in this fashion is transmitted through an absorbing analyte, the spectral and temporal characteristics of the terahertz pulse are modified through linear absorption processes. The probe beam is utilized for time-gated detection of the transmitted terahertz pulse, either through photoconductive or electro-optic sampling. By scanning the pump/probe delay, one can capture the temporal characteristics of the transmitted terahertz pulse, and the Fourier transformation of that pulse yields its frequency-resolved transmission spectrum from which terahertz absorption features of the analyte are determined.

Cheville & Grishchkowsky (40, 41) measured species concentrations and temperature in premixed propane-air flames with THz-TDS techniques based on a traditional optomechanical scanning delay line for variation of the pump/probe delay. Brown et al. (42) adopted the ASOPS scheme described above for H_2O-vapor measurements with a no-moving-parts pump/probe delay. The potential for future applications of THz-TDS in combustion is promising. The terahertz spectral region is rich in absorption features of interest for the quantification of key combustion species (especially H_2O) and temperature. Furthermore, hydrocarbon fuels and carbonaceous soot exhibit little or no absorption in this spectral region, suggesting that THz-TDS should be an ideal technique for measurements in liquid-hydrocarbon-fueled, highly sooting combustion environments such as those characteristic of most practical devices.

4. TIME-GATED BALLISTIC IMAGING

As light propagates through a turbid medium, its direction, polarization, and phase are altered owing to gradients in the index of refraction. Diffuse photons pass through the sample volume with significant multiple scattering events and emerge with a shift in location and direction. This is shown schematically in **Figure 4a** and leads to the blurring of internal features within the medium. Snake photons are altered to a

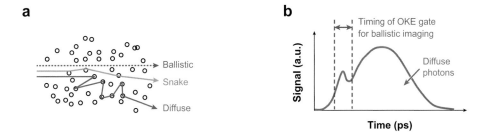

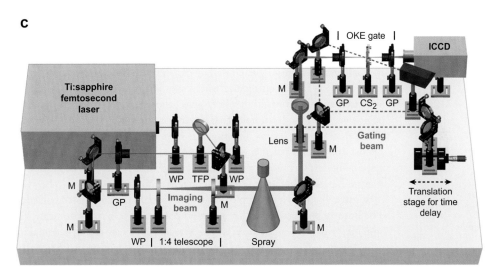

Figure 4

Conceptual picture of (*a*) ballistic, snake, and diffuse photons; (*b*) time trace of signal on detector; and (*c*) optical layout for time-gated ballistic-photon imaging. Abbreviations: GP, Glan polarizer; ICCD, intensified charge-coupled device camera; M, mirror; OKE, optical Kerr effect; TFP, thin film polarizer; WP, half-wave plate.

lesser degree as they propagate with relatively few scattering events, whereas ballistic photons pass through without deviation and maintain their direction of propagation, polarization, and phase.

Interest in utilizing ballistic photons for imaging applications is driven by the possibility that one could use visible and infrared light sources to image turbid media without the need for ionizing radiation or synchrotron sources. The general approach is to separate ballistic or snake photons from diffuse photons using differences in direction, polarization, coherence, or temporal properties of the light passing through the medium. It is possible, for example, to employ a spatial filter to reduce photons that have lost collimation or confocal properties after exiting the sample volume (43, 44). Polarization gating involves the use of an analyzer rather than a spatial filter (45, 46), whereas coherence gating can be accomplished using nonlinear mixing processes (e.g., second-harmonic generation or holography) to discriminate against scattered photons (47–50).

In many cases, spatial filtering and polarization gating may not provide sufficient discrimination against diffuse photons, whereas coherence gating may be too restrictive in that it can eliminate all but the ballistic photons. For unsteady multiphase flows, it is often necessary to use both snake and ballistic photons to have sufficient signal levels for instantaneous two-dimensional imaging of transient mixing processes. One approach, enabled by the availability of ultrafast lasers, is to distinguish photons based on passage time through the turbid medium. A femtosecond laser beam that undergoes significant scattering exits with longer pulse width (order picosecond) because diffuse photons travel a longer path through the sample volume. This longer path is shown schematically in **Figure 4a** along with a conceptual view in **Figure 4b** of short time gating to discriminate against diffuse photons.

Early work on time-domain photon discrimination demonstrated that one could use picosecond lasers to achieve gate widths on the order of 7–10 ps using an optical Kerr-effect (OKE) time gate consisting of a CS_2 cell placed between a pair of crossed polarizers (51–53). Birefringence with a short relaxation time is induced in the CS_2 upon excitation with an ultrafast laser pulse, allowing light to pass through the crossed polarizers through adjustment of a translation stage to vary the time delay between the imaging and gating beams. Investigators have used this approach for discriminating against diffuse photons in a number of applications (54, 55). With the availability of amplified femtosecond laser systems, it has become possible to achieve time gates as short as 2 ps with high transmission efficiency (56–60).

Figure 4c shows an example of the optical setup for time-gated ballistic imaging. The linearly polarized 1 mJ per pulse output of a 1 kHz repetition rate Ti:sapphire amplifier with 80-fs pulse width is split 90% and 10% into gating and imaging beams, respectively, using a wave plate–polarizer combination. The imaging path includes a Glan polarizer, a half-wave plate, and a beam-expanding telescope. After passing through a dense spray, the imaging beam is weakly focused through the OKE gate, then spatially filtered, and relay imaged directly into an intensified charge-coupled device camera. The gating beam passes through a half-wave plate and a mechanical time-delay stage before arriving at the 1-cm-thick, 2.5-cm-diameter CS_2 cell. The induced birefringence across the imaging beam is kept fairly constant within the CS_2 cell because the imaging beam is relatively small (<500-μm diameter) and passes through the centroid of the relatively large gating beam ($\sim$6-mm diameter). The transmission efficiency of the OKE gate is $\sim$30% when activated by the 80-fs laser pulse, which is sufficient for signal-to-leakage ratios of $\sim$20:1. This arrangement allows instantaneous two-dimensional imaging of turbid media, in this case a liquid spray. Laser sources with repetition rates as high as 10 kHz with 1 mJ per pulse are now available and will enable ballistic imaging at unprecedented data rates.

This approach has been used for measurements of liquid breakup phenomena in diesel sprays (58), liquid jets in gaseous crossflow (59), and coaxial rocket injectors. **Figure 5** shows an example of coaxial rocket injectors, comparing a rocket spray shadowgram with and without time gating. Internal structures that were previously not visible due to diffuse scattering are revealed with the use of ultrafast time gating. This has implications for the study of liquid jet breakup and gas-liquid mixing processes in multiphase reacting flows.

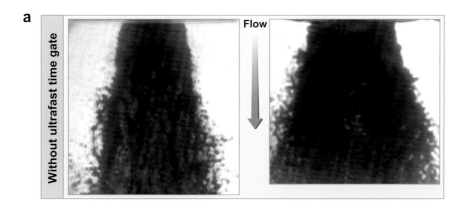

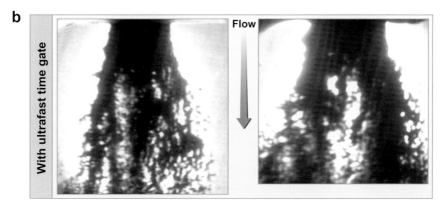

Figure 5

Sample images of rocket spray (*a*) without ultrafast time gate and (*b*) with ultrafast time gate. Flow is from top to bottom.

Advances in ultrafast laser technology have led to significant growth in the implementation of time-gated ballistic-imaging systems for visualizing hidden structures in optically dense media. A number of innovations such as two-color ballistic imaging (59) and dual-pulse ballistic imaging for velocimetry (60) have also been demonstrated recently. Future work could involve the use of three or more consecutive pulses to acquire the acceleration of interfacial regions as well as velocity within dense media. Finally, efforts are underway to use Monte Carlo simulations to predict the properties of ultrafast laser–light propagation through scattering media (61), including implementation for cases with inhomogeneously distributed scatterers.

5. ULTRAFAST COHERENT ANTI-STOKES RAMAN SCATTERING SPECTROSCOPY

CARS spectroscopy is widely used for temperature and major-species-concentration measurements in reacting flows and plasmas (1–3). Because of the phase-matching

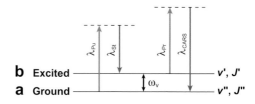

Figure 6

Energy-level diagram for coherent anti-Stokes Raman scattering signal-generation process, where *a* and *b* denote ground and excited levels of molecule, respectively, and ω_v corresponds to vibrational frequency of molecule. Here v' and J' refer to excited-state vibrational and rotational quantum number, respectively, and v'' and J'' refer to ground-state vibrational and rotational quantum number, respectively.

requirement and laser-like nature of the signal, CARS is ideally suited for reacting flows with significant background emission because one can easily isolate the CARS signal spectrally, spatially, and temporally from the flame emission. The technique also provides spatially and temporally resolved information with high accuracy.

An energy-level diagram for the CARS signal-generation process is shown in **Figure 6**. In CARS the wavelengths of the pump and Stokes beams are chosen to excite either the vibrational or rotational transitions of the molecule. We can also describe this excitation process as creating coherence in the medium with a pump-Stokes pair after which the coherence evolves according to the interaction of the molecules with the surrounding medium. When a probe beam interacts with the excited molecules, it is scattered at an anti-Stokes-shifted frequency to yield the CARS signal.

Until recently most of the CARS work in reacting flows was performed using nanosecond lasers to determine gas temperature and the concentrations of major species such as N_2, O_2, CO_2, CO, H_2, and H_2O (1, 62–65). Electronic-resonance-enhanced CARS using nanosecond lasers has also been demonstrated for determining the concentration of flame radicals such as OH, NO, and C_2H_2 (66–68). Traditional nanosecond CARS uses a narrowband ($\sim$0.001 cm^{-1}) transform-limited pump laser and a broadband ($\sim$150 cm^{-1}) Stokes laser to excite the entire rovibrational manifold of the molecule (**Figure 7a**). For example, a narrowband laser at 532 nm and a broadband laser at $\sim$607 nm excite the rovibrational energy levels of N_2, which is typically targeted for temperature measurements because of its abundance in air-fed reacting flows. The band head of the $v'= 1 \rightarrow v''= 0$ transition in the ground electronic state falls at $\sim$2330 cm^{-1}. As shown in **Figure 7a**, only one pump-Stokes pair contributes to the excitation of the coherence for a particular transition. However, nanosecond CARS has several disadvantages that challenge its application in high-pressure, turbulent reacting flows: (*a*) interference of the nonresonant background signal with the resonant signal, which affects the accuracy and sensitivity of the measurements, especially for hydrocarbon-fueled combustion (1, 69); (*b*) the low repetition rates of the lasers used, which complicate efforts to study the temporal characteristics of turbulent flames and explore combustion instabilities; and (*c*) the need to understand the collisional environment and associated

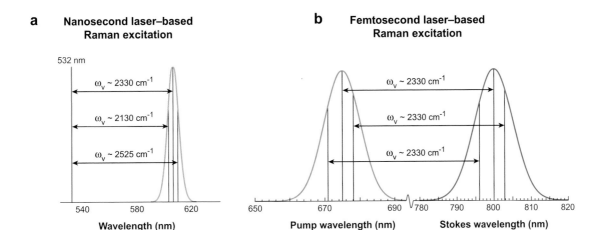

Figure 7

Raman excitation scheme for gas-phase N_2 using (*a*) nanosecond laser–based multiplex coherent anti-Stokes Raman scattering and (*b*) femtosecond pump and Stokes lasers.

dephasing and relaxation processes for quantitative interpretation of the CARS signal.

The use of ultrafast lasers to address these issues in reacting flows has been the subject of continuing research activities in the Air Force Research Laboratory's Combustion and Laser Diagnostics Research Complex at Wright-Patterson Air Force Base. In picosecond or femtosecond CARS, nonresonant background signals are observed only when all three laser beams are temporally and spatially coincident. These nonresonant interferences can be suppressed by delaying the probe beam with respect to the Raman-excitation beams. Roy et al. (69) performed picosecond multiplex CARS experiments with the probe beam delayed by ∼150 ps with respect to the Raman-excitation beams. In this arrangement, the nonresonant background signal is reduced by more than three orders of magnitude, whereas the resonant signal is reduced by only a factor of three to yield a tremendous improvement in signal-to-noise ratio. However, in picosecond CARS, the quantitative interpretation of the signal still requires some knowledge of the collisional physics, and the repetition rate of the lasers used in these particular experiments is only of the order 10–20 Hz.

Femtosecond CARS spectroscopy has the potential to overcome the problems associated with nanosecond CARS for combustion applications, as evidenced by recent studies of femtosecond CARS in noncombusting environments (70–75). When two femtosecond laser pulses are used to create coherence in the medium (as in the case of CARS), the ground and excited states are coupled efficiently because of the availability of a large number of pump-Stokes pairs within the bandwidth of the laser pulses contributing to the excitation of the same coherence (**Figure 7***b*) (75). **Figure 7** shows pump-Stokes pairs for the excitation of N_2 in the ground electronic state, as mentioned above. This specific feature of femtosecond laser–based Raman excitation, along with the suppression of the nonresonant background using a delayed probe beam, holds the potential for making the femtosecond CARS technique suitable for detecting minor species in reacting flows. Moreover, femtosecond CARS allows measurements to be made at rates of 1 kHz or greater and time-resolved CARS signals to be acquired over a time period that is short with respect to the collisional timescale, thereby eliminating the need to understand collisional broadening, line narrowing, and other dephasing and relaxation processes due to collisions.

Dantus et al. (76) eloquently describe the excitation of vibrational and rotational coherences by femtosecond lasers and the subsequent decay of these coherences either through frequency-spread dephasing or loss of alignment. Initially it was anticipated that the large bandwidths characteristic of femtosecond lasers would be problematic for molecular spectroscopy in reacting flows because of the associated lack of selectivity (broadband excitation of many transitions of one or more molecules) and relatively inefficient coupling of these spectrally broad pulses to individual transitions as compared with coupling of narrowband nanosecond pulses more closely matched to the line width of these transitions. However, the excitation process depicted in **Figure 7b** and recent research activities in this field have shown that the bandwidth of femtosecond lasers is actually an advantage rather than a hindrance.

In 1987 femtosecond CARS was first used to study molecular beat phenomena in liquid-phase benzene, cyclohexane, and pyridine (77); subsequently, Hayden & Chandler (78) demonstrated its application to the investigation of gas-phase molecular dynamics. Lang et al. (71) focused their work with femtosecond CARS on determining the molecular parameters and gas-phase temperature from the time-resolved oscillatory pattern of the Raman coherence following pump-Stokes excitation of H_2. They determined those parameters from the width and relative heights of the coherence recurrence peaks. The measurement of temperature from these peaks at $\sim$320 ps, as described by Lang et al. (70), requires a detailed understanding of the collisional dephasing and relaxation physics of the probe molecule within its surrounding environment. Researchers have also used femtosecond CARS to make measurements in dense media to investigate RET processes (72), to determine the concentrations of *ortho*- and *para*-deuterium (79), and to measure single-shot temperature by probing H_2 using a chirped probe pulse (80). The technique has also been used for microscopy (81), the selective control of molecular structure (82), detection of bacterial spores (83), and investigation of the ground- and excited-state dynamics of molecules (84).

The focus of our efforts is the application of time-resolved femtosecond CARS for temperature measurements in high-temperature flames, based on the frequency-spread dephasing rate after the initial impulsive excitation of the Raman coherence in N_2 by femtosecond pump and Stokes beams. After the initial excitation, all in-phase Raman coherences excited by the nearly transform-limited laser pulses begin to oscillate out of phase with respect to each other as a result of slight differences in their frequencies. Because of the frequency differences between the neighboring transitions, the resulting coherence begins to dephase; the dephasing rate depends on temperature only and is completely insensitive to collisions (73). **Figure 8** shows time-resolved femtosecond CARS signals during the first few picoseconds after the initial impulsive excitation as a function of temperature. The coherence dephases at a faster rate with increasing temperature as a result of the contribution

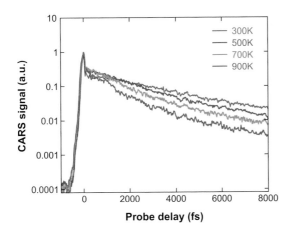

Figure 8

Temperature-dependent
time-resolved femtosecond
coherent anti-Stokes Raman
scattering signal of N_2 (73).

from many energy levels; higher rotational and vibrational levels are populated with increasing temperature according to the Boltzmann distribution. Roy and colleagues (73–74) reported detailed theoretical and experimental results related to this concept and discussed the application of this technique in reacting flows (75). They found the estimated absolute accuracy and precision of the measurement technique to be ± 40 K and ± 50 K, respectively, over the temperature range 1500–2500 K (75). **Figure 9** illustrates the temporal evolution of time-resolved femtosecond CARS signals during the first few picoseconds after the initial excitation as a function of pressure. It is evident from **Figure 9** that the coherence dephasing rate during the first few picoseconds is insensitive to collisions. However, collisions begin to influence the dephasing rates when the pressure is increased beyond 20 bar (72). We have designed current research activities in our laboratory to address these key issues: (*a*) single-shot temperature measurements at rates of 1 kHz or greater using a spectrally chirped probe pulse, (*b*) the influence of other molecules excited by the broadband femtosecond laser pulses on measurements of temperature and species

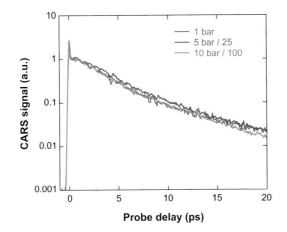

Figure 9

Pressure dependence of
time-resolved N_2
femtosecond coherent
anti-Stokes Raman
scattering signal during first
few picoseconds after
impulsive excitation of the
coherence.

concentrations, and (*c*) concentration measurements of minor species such as C_2H_2, C_6H_6, and other flame molecules and radicals using shaped laser pulses.

6. RESONANT AND RAMAN-INDUCED POLARIZATION SPECTROSCOPY AND WAVE MIXING

PS has emerged as a valuable spectroscopic tool for measuring the concentrations of minor species such as OH, CH, and NH in reacting flows (1–3, 85–88) and plasmas (89). PS is a nonlinear pump/probe technique in which the probe beam is linearly polarized prior to interacting with the medium of interest. PS uses either a circularly or a linearly polarized pump beam for selective pumping of the population from the ground to the excited states; in the latter case, the pump-beam polarization is rotated $45°$ with respect to the probe-beam polarization. Because of the anisotropy induced by the pump beam, the probe-beam polarization becomes slightly elliptical or slightly rotated while passing through the medium. As a consequence, some of the probe beam leaks through a polarization analyzer whose transmission axis is orthogonal to the original probe-beam polarization; this leakage is the PS signal. To illustrate the introduction of anisotropy by selective pumping, **Figure 10** shows an energy-level diagram of the $P_1(2)$ transition of OH. A linearly polarized pump beam couples the $\Delta M = 0$ transitions, and a right or left circularly polarized pump beam couples either the $\Delta M = +1$ or the $\Delta M = -1$ transitions, respectively.

Two distinct advantages of using ultrafast lasers for PS are (*a*) the reduction in collisional dependence and (*b*) the determination of the state-specific rotational, orientation, and alignment relaxation rates from time-resolved measurements. Roy and colleagues (86, 91) showed that when using an ultrafast laser (laser pulse width $\tau_L < \tau_C$ characteristic collision time), the collision-rate dependence of the PS signal is significantly decreased as compared with that in the long-pulse laser case

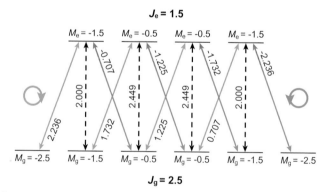

Figure 10

Energy-level diagram for Zeeman state structure of $P_1(2)$ transition. Allowed $\Delta M = 0$ transitions are indicated by dashed arrows; $\Delta M = \pm 1$ transitions are indicated by solid arrows. J and M represent rotational and projection quantum numbers, respectively. Strengths and phases of transitions are indicated by numerical value of x or z components of geometry-dependent part of the dipole matrix element (90). Figure redrawn from Reference 91.

$(\tau_L > \tau_C)$ for a nonsaturating pump beam. For a saturating pump beam, the picosecond PS signal is nearly independent of collisions (91).

The polarization-dependent selective pumping shown in **Figure 10** creates various types of anisotropies such as orientation and alignment in the medium with relaxation rates due to elastic (M_J changing) and inelastic RET collisions that can significantly affect the resulting PS signal and signals associated with other laser techniques that exploit laser-induced anisotropy. Orientation, which describes the net helicity or spin of the system, and alignment, which describes the spatial distribution of angular momentum, are proportional to the dipole and quadrupole moments of the angular momentum distributions, respectively (92). To illustrate the anisotropies created by the pump laser, **Figure 11** shows the population distribution in the excited Zeeman states for the $P_1(8)$ transition of OH. **Figure 11a** shows the oriented distribution of the excited-state population when pumped by a right circularly polarized beam, whereas **Figure 11b** shows the aligned distribution when pumped by a linearly polarized beam. The population distribution is shown for a time at which the 100-ps (full width at half-maximum) pump laser reaches the peak intensity of $5 \times 10^9 \, \mathrm{W \, m^{-2}}$ and was calculated using the density-matrix numerical code described by Roy et al. (91).

The use of picosecond lasers enables experimental investigation of the rates at which these anisotropies are destroyed in collisional environments. Dreizler and colleagues (93, 94) used PS and RFWM to determine the population, orientation, and alignment relaxation rates of OH in reacting flows. The RFWM technique, in which the two pump photons originate from two different pump beams, is similar to PS, in which both pump photons originate from the same pump beam. Unlike PS, the RFWM technique allows measurements of the population, orientation, and alignment relaxation rates independently through the control of polarization settings for

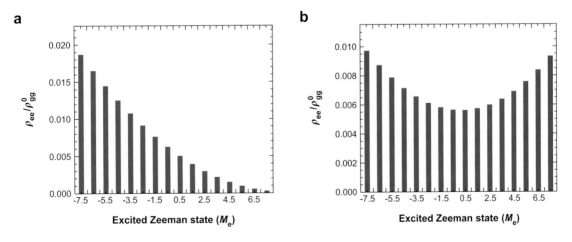

Figure 11

Population distribution in excited Zeeman states for $P_1(8)$ transition pumped by (*a*) right circularly polarized beam and (*b*) linearly polarized beam. ρ_{gg}^0 is ground-state population at $t = 0$ s.

each laser beam. In an experiment in which both pump and probe beams are resonant with the same transitions, the measured relaxation rates have a contribution from both the ground and the excited states as reported by Dreizler and colleagues (93, 94). To measure the ground-state orientation and alignment relaxation rates of OH, Chen and coworkers (95) and Costen & McKendrick (96) employed two-color PS in which the pump and probe beams are coupled through an intermediate level in the ground electronic state. Chen & Settersten (97) also demonstrated two-color RFWM for determining the ground-state population, orientation, and alignment relaxation rates by varying the polarization of two pump beams and the probe beam independently. Recently, picosecond laser–based, two-color, two-photon PS was used for the detection of atomic hydrogen in an atmospheric-pressure H_2-air flame; because of its high reactivity and diffusivity, the hydrogen atom plays an important role in combustion chemical kinetics (98).

Femtosecond laser–based Raman-induced PS and degenerate four-wave mixing were used to study the ground-level RET in N_2, O_2, and CO_2 (99, 100). The broadband femtosecond lasers create a rotational wave packet by simultaneously exciting multiple rotational Raman transitions. The delayed probe beam then probes the misalignment and realignment of this rotational coherence due to elastic and inelastic collisions. These studies will have a significant impact in determining the RET rates for species relevant to reacting flows. Raman-induced degenerate four-wave mixing of H_2, in which the rotational transitions of H_2 were excited using a sub-20-fs laser pulse, has also been used to determine temperature from the relative intensities of the beat frequencies between the Stokes and anti-Stokes transitions (101).

7. CONCLUSION

Revolutionary ultrafast laser technologies are redefining the combustion-diagnostics arena. Unprecedented new measurement capabilities are emerging as researchers exploit the ultrashort pulses and high pulse-repetition rates delivered by these systems. This review highlights recent advances in ultrafast LIF, linear pump/probe techniques, time-gated ballistic imaging, ultrafast CARS, and resonant and Raman-induced PS and wave mixing. Continuing developments of ultrafast laser systems promise to provide further reductions in cost, size, and complexity with increased robustness and stability and improved user-friendliness. All these enhancements will allow the transition of ultrafast laser–based combustion diagnostics from the laboratory to the field for applications that include test-cell and shop-floor measurements, nondestructive evaluation/inspection, and onboard sensing and control.

SUMMARY POINTS

1. Detailed studies of combustion chemistry and physics are critical for the continued advancement of numerous combustion-related applications, including propulsion, power generation, industrial processing, waste incineration, and fire safety.

2. Optical combustion diagnostics are ideal tools for noninvasive characterization of reacting and nonreacting flows.

3. Although continuous-wave and nanosecond-pulsed lasers are the current sources of choice for most combustion diagnostics, emerging ultrafast (picosecond- and femtosecond-pulsed) laser technologies are driving new developments in combustion measurements.

4. Ultrashort pulses enable time-resolved measurements with unprecedented temporal resolution and freedom from collisional effects that plague measurements with nanosecond-pulsed lasers. Peak powers achievable with these pulses allow nonlinear frequency conversion, expanding the spectral coverage of these systems to new heights. Similarly, these peak powers enable nonlinear signal generation in combusting systems with lasers of moderate to low average power.

5. High pulse-repetition rates afforded by ultrafast oscillators and amplifiers provide the data-acquisition bandwidth necessary to study the temporal evolution of turbulent fluctuations and combustion instabilities. Time series and PSDs can be constructed to characterize these flame phenomena.

6. Ultrafast combustion measurements achieved to date include LIF, linear pump/probe measurements, time-gated ballistic imaging, CARS, PS, and wave mixing. In many cases the ultrafast variants described in this review enjoy numerous and significant advantages over their continuous-wave and nanosecond-pulsed analogs.

7. Continuing and future developments of ultrafast lasers hold tremendous promise for even greater spectral coverage, single-shot high-speed measurements, and application to complex real-world systems for test-cell and shop-floor measurements, nondestructive evaluation/inspection, and on-board sensing and control.

DISCLOSURE STATEMENT

The authors are not aware of any biases that might be perceived as affecting the objectivity of this review.

ACKNOWLEDGMENTS

The authors gratefully acknowledge editorial assistance provided by Ms. Marian Whitaker and continuing long-term financial support and encouragement from Dr. Julian Tishkoff, Air Force Office of Scientific Research. We also gratefully recognize a host of outstanding colleagues and collaborators with whom we have interacted over the course of the past many years. Their tremendous expertise and enthusiasm have contributed immeasurably to many of the efforts described in this review. Of particular note are the contributions of Dr. G.J. Fiechtner, Prof. G.B. King, Dr. M.A. Linne, Prof. R.P. Lucht, and Dr. T.B. Settersten.

LITERATURE CITED

1. Eckbreth AC. 1996. *Laser Diagnostics for Combustion Temperature and Species*. Amsterdam: Gordon & Breach. 596 pp. 2nd ed.
2. Kohse-Höinghaus K, Jeffries JB. 2002. *Applied Combustion Diagnostics*. New York: Taylor & Francis. 705 pp.
3. Linne MA. 2002. *Spectroscopic Measurement: An Introduction to the Fundamentals*. New York: Academic. 414 pp.
4. Laurendeau NM. 2005. *Statistical Thermodynamics: Fundamentals and Applications*. New York: Cambridge Univ. Press. 448 pp.
5. Diels J-C, Rudolph W. 2006. *Ultrashort Laser Pulse Phenomena*. New York: Academic. 680 pp. 2nd ed.
6. Rullière C, ed. 2005. *Femtosecond Laser Pulses: Principles and Experiments*. New York: Springer. 426 pp. 2nd ed.
7. Fermann ME, Galvanauskas A, Sucha G. 2003. *Ultrafast Lasers: Technology and Applications*. New York: Marcel Dekker. 784 pp.
8. Hobbs PCD. 2000. *Building Electro-Optical Systems: Making It All Work*. New York: Wiley & Sons. 727 pp.
9. Hartlieb AT, Markus D, Kreutner W, Kohse-Höinghaus K. 1997. Measurement of vibrational energy transfer of OH ($A^2\Sigma^+, v' = 1 \to 0$) in low pressure flames. *Appl. Phys. B* 65:81–91
10. Beaud P, Radi PP, Franzke D, Frey H-M, Mischler B, et al. 1998. Picosecond investigation of the collisional deactivation of OH $A^2\Sigma^+$ ($v' = 1, N' = 4, 12$) in an atmospheric-pressure flame. *Appl. Opt.* 37:3354–67
11. Nielsen T, Bormann F, Burrows M, Andresen P. 1997. Picosecond laser-induced fluorescence measurement of rotational energy transfer of OH $A^2\Sigma^+$ ($v' = 2$) in atmospheric pressure flames. *Appl. Opt.* 36:7960–69
12. Agrup S, Ossler F, Aldén M. 1995. Measurements of collisional quenching of hydrogen atoms in an atmospheric-pressure hydrogen oxygen flame by picosecond laser-induced fluorescence. *Appl. Phys. B* 61:479–87
13. Settersten TB, Patterson BD, Kronemayer H, Sick V, Schulz C, Daily JW. 2006. Branching ratios for quenching of nitric oxide $A^2\Sigma^+$ ($v' = 0$) to $X^2\Pi$ ($v'' = 0$). *Phys. Chem. Chem. Phys.* 8:5328–38
14. Brockhinke A, Kreutner W, Rahmann U, Kohse-Höinghaus K, Settersten TB, Linne MA. 1999. Time-, wavelength-, and polarization-resolved measurements of OH $A^2\Sigma^+$ picosecond laser-induced fluorescence in atmospheric-pressure flames. *Appl. Phys. B* 69:477–85
15. Agrup S, Aldén M. 1994. Two-photon laser-induced fluorescence and stimulated emission measurements from oxygen atoms in a hydrogen/oxygen flame with picosecond resolution. *Opt. Commun.* 113:315–23
16. Frank JH, Chen X, Patterson BD, Settersten TB. 2004. Comparison of nanosecond and picosecond excitation for two-photon laser-induced fluorescence imaging of atomic oxygen in flames. *Appl. Opt.* 43:2588–97
17. Settersten TB, Dreizler A, Farrow RL. 2002. Temperature- and species-dependent quenching of CO B probed by two-photon laser-induced fluorescence using a picosecond laser. *J. Chem. Phys.* 117:3173–79

18. Settersten TB, Patterson BD, Gray JA. 2006. Temperature- and species-dependent quenching of NO $A^2\Sigma^+$ ($v' = 0$) probed by two-photon laser-induced fluorescence using a picosecond laser. *J. Chem. Phys.* 124:234308

19. Brockhinke A, Bülter A, Rolon JC, Kohse-Hoinghaus K. 2001. ps-LIF measurements of minor species concentration in a counter diffusion flame interactive with a vortex. *Appl. Phys. B* 72:491–96

20. Ossler F, Metz T, Martinsson L, Aldén M. 1998. Two-dimensional visualization of fluorescence lifetimes by use of a picosecond laser and a streak camera. *Appl. Opt.* 37:2303–14

21. Klassen MS, Thompson BD, Reichardt TA, King GB, Laurendeau NM. 1994. Flame concentration measurements using picosecond time-resolved laser-induced fluorescence. *Combust. Sci. Technol.* 97:391–403

22. Pack SD, Renfro MW, King GB, Laurendeau NM. 1999. Laser-induced fluorescence triple-integration method applied to hydroxyl concentration and fluorescence lifetime measurements. *Combust. Sci. Technol.* 140:405–25

23. Renfro MW, King GB, Laurendeau NM. 2000. Scalar time-series measurements in turbulent $CH_4/H_2/N_2$ nonpremixed flames: CH. *Combust. Flame* 122:139–50

24. Renfro MW, King GB, Laurendeau NM. 1999. Quantitative hydroxyl concentration time-series measurements in turbulent nonpremixed flames. *Appl. Opt.* 38:4596–608

25. Meyer TR, King GB, Gluesenkamp M, Gord JR. 2007. Simultaneous high-speed measurement of temperature and lifetime-corrected OH laser-induced fluorescence in unsteady flames. *Opt. Lett.* 32:2221–23

26. Elzinga PA, Lytle FE, Jiang Y, King GB, Laurendeau NM. 1987. Pump/probe spectroscopy by asynchronous optical sampling. *Appl. Spectrosc.* 41:2–4

27. Elzinga PA, Kneisler RJ, Lytle FE, Jiang Y, King GB, Laurendeau NM. 1987. Pump/probe method for fast analysis of visible spectral signatures utilizing asynchronous optical sampling. *Appl. Opt.* 26:4303–9

28. Kneisler RJ, Lytle FE, Fiechtner GJ, Jiang Y, King GB, Laurendeau NM. 1989. Asynchronous optical sampling: a new combustion diagnostic for potential use in turbulent, high-pressure flames. *Opt. Lett.* 14:260–62

29. Fiechtner GJ, King GB, Laurendeau NM, Lytle FE. 1992. Measurements of atomic sodium in flames by asynchronous optical sampling: theory and experiment. *Appl. Opt.* 31:2849–64

30. Fiechtner GJ. 1992. *Quantitative concentration measurements in atmospheric-pressure flames by picosecond pump/probe absorption spectroscopy*. PhD thesis. Purdue Univ.

31. Fiechtner GJ, King GB, Laurendeau NM. 1995. Quantitative concentration measurements of atomic sodium in an atmospheric hydrocarbon flame with asynchronous optical sampling. *Appl. Opt.* 34:1117–26

32. Fiechtner GJ, King GB, Laurendeau NM. 1995. Rate-equation model for quantitative concentration measurements in flames by picosecond pump-probe absorption spectroscopy. *Appl. Opt.* 34:1108–16

33. Fiechtner GJ, Linne MA. 1994. Absolute concentrations of potassium by picosecond pump/probe absorption in fluctuating, atmospheric pressure flame. *Combust. Sci. Technol.* 100:11–27

34. Settersten TB. 1999. *Picosecond pump/probe diagnostics for combustion.* PhD thesis. Colo. School of Mines

35. Settersten TB, Linne MA. 2002. Picosecond pump-probe absorption spectroscopy in gases: models and experimental validation. *Appl. Opt.* 41:2869–78

36. Settersten T, Linne M, Gord J, Fiechtner G. 1999. Density matrix and rate equation analyses for picosecond pump/probe combustion diagnostics. *AIAA J.* 37:723–31

37. Settersten TB, Linne MA. 2002. Modeling pulsed excitation for gas-phase laser diagnostics. *J. Opt. Soc. Am. B* 19:954–64

38. Linne MA, Morse DC, Skilowitz JL, Fiechtner GJ, Gord JR. 1995. Two-dimensional pump-probe imaging in reacting flows. *Opt. Lett.* 20:2414–16

39. Mittleman D, ed. 2003. *Sensing with Terahertz Radiation.* New York: Springer. 337 pp.

40. Cheville RA, Grischkowsky D. 1995. Far-infrared, terahertz time-domain spectroscopy of flames. *Opt. Lett.* 20:1646–48

41. Cheville RA, Grischkowsky D. 1998. Observation of pure rotational absorption spectra in the v_2 band of hot H_2O in flames. *Opt. Lett.* 23:531–33

42. Brown MS, Fiechtner GJ, Rudd JV, Zimdars DA, Warmuth M, Gord JR. 2006. Water-vapor detection using asynchronous THz sampling. *Appl. Spectrosc.* 60:261–65

43. Schmitt JM, Knüttel A, Yadlowsky M. 1994. Confocal microscopy in turbid media. *J. Opt. Soc. Am. A* 11:2226–35

44. Kempe M, Genack AZ, Rudolph W, Dorn P. 1997. Ballistic and diffuse light detection in confocal and heterodyne imaging systems. *J. Opt. Soc. Am. A* 14:216–23

45. Demos SG, Alfano RR. 1996. Temporal gating in highly scattering media by the degree of optical polarization. *Opt. Lett.* 21:161–63

46. Mujumdar S, Ramachandran H. 2004. Imaging through turbid media using polarization modulation: dependence on scattering anisotropy. *Opt. Commun.* 241:1–9

47. Fujimoto JG, De Silvestri S, Ippen EP, Puliafito CA, Margolis R, Oseroff A. 1986. Femtosecond optical ranging in biological systems. *Opt. Lett.* 11:150–52

48. Yoo KM, Xing Q, Alfano RR. 1991. Imaging objects hidden in highly scattering media using femtosecond second-harmonic-generation cross-correlation time gating. *Opt. Lett.* 16:1019–21

49. Bordenave E, Abraham E, Jonusauskas G, Oberlé J, Rullière C. 2002. Longitudinal imaging in biological tissues with a single laser shot correlation system. *Opt. Exp.* 10:35–40

50. Kippelen B, Marder SR, Hendrickx E, Maldonado JL, Guillemet G, et al. 1998. Infrared photorefractive polymers and their applications for imaging. *Science* 279:54–57

51. Duguay MA, Mattick AT. 1971. Ultrahigh speed photography of picosecond light pulses and echoes. *Appl. Opt.* 10:2162–70

52. Sala K, Richardson MC. 1975. Optical Kerr effect induced by ultrashort pulses. *Phys. Rev. A* 12:1036–47

53. Ho PP, Alfano RR. 1979. Optical Kerr effect in liquids. *Phys. Rev. A* 20:2170–87

54. Wang L, Ho PP, Liu C, Zhang G, Alfano RR. 1991. Ballistic 2-D imaging through scattering walls using an ultrafast optical Kerr gate. *Science* 253:769–71

55. Galland PA, Liang X, Wang L, Breisacher K, Liou L, et al. 1995. Time-resolved optical imaging of jet sprays and droplets in highly scattering medium. *Proc. Am. Soc. Mech. Eng.* HTD-321:585–88

56. Paciaroni M, Linne M. 2004. Single-shot, two-dimensional ballistic imaging through scattering media. *Appl. Opt.* 43:5100–9

57. Paciaroni M, Linne M, Hall T, Delplanque J-P, Parker T. 2006. Single-shot two-dimensional ballistic imaging of the liquid core in an atomizing spray. *At. Sprays* 16:51–70

58. Linne M, Paciaroni M, Hall T, Parker T. 2006. Ballistic imaging of the near field in a diesel spray. *Exp. Fluids* 40:836–46

59. Linne MA, Paciaroni M, Gord JR, Meyer TR. 2005. Ballistic imaging of the liquid core for a steady jet in crossflow. *Appl. Opt.* 44:6627–34

60. Sedarsky DL, Paciaroni ME, Linne MA, Gord JR, Meyer TR. 2006. Velocity imaging for the liquid-gas interface in the near field of an atomizing spray: proof of concept. *Opt. Lett.* 31:906–8

61. Wang X, Wang LV, Sun C-W, Yang C-C. 2003. Polarized light propagation through scattering media: time-resolved Monte Carlo simulations and experiments. *J. Biomed. Opt.* 8:608–17

62. Roy S, Meyer TR, Brown MS, Velur VN, Lucht RP, Gord JR. 2003. Triple-pump coherent anti-Stokes Raman scattering (CARS): temperature and multiple-species concentration measurements in reacting flows. *Opt. Commun.* 224:131–37

63. Meyer TR, Roy S, Lucht RP, Gord JR. 2005. Dual-pump dual-broadband CARS for exhaust-gas temperature and CO_2-O_2-N_2 mole-fraction measurements in model gas-turbine combustors. *Combust. Flame* 142:52–61

64. Hall RJ, Shirley JA. 1983. Coherent anti-Stokes Raman spectroscopy of water vapor for combustion diagnostics. *Appl. Spectrosc.* 37:196–202

65. Rahn LA, Zych LJ, Mattern PL. 1979. Background-free CARS studies of carbon monoxide in a flame. *Opt. Commun.* 30:249–52

66. Attal-Trétout B, Schmidt SC, Crété E, Dumas P, Taran JP. 1990. Resonance CARS of OH in high-pressure flames. *J. Quant. Spectrosc. Radiat. Transf.* 43:351–64

67. Roy S, Kulatilaka WD, Naik SV, Laurendeau NM, Lucht RP, Gord JR. 2006. Effects of quenching on electronic-resonance-enhanced coherent anti-Stokes Raman scattering of nitric oxide. *Appl. Phys. Lett.* 89:104105

68. Chai N, Naik SV, Kulatilaka WD, Laurendeau NM, Lucht RP, et al. 2007. Detection of acetylene by electronic resonance-enhanced coherent anti-Stokes Raman scattering. *Appl. Phys. B* 87:731–37

69. Roy S, Meyer TR, Gord JR. 2005. Time-resolved dynamics of resonant and non-resonant broadband picosecond coherent anti-Stokes Raman scattering signals. *Appl. Phys. Lett.* 87:264103

70. Lang T, Motzkus M, Frey HM, Beaud P. 2001. High resolution femtosecond coherent anti-Stokes Raman scattering: determination of rotational constants, molecular anharmonicity, collisional line shifts, and temperature. *J. Chem. Phys.* 115:5418–26

71. Lang T, Kompa KL, Motzkus M. 1999. Femtosecond CARS on H_2. *Chem. Phys. Lett.* 310:65–72

72. Knopp G, Beaud P, Radi P, Tulej M, Bougie B, et al. 2002. Pressure-dependent N_2 Q-branch fs-CARS measurements. *J. Raman Spectrosc.* 33:861–65

73. Lucht RP, Roy S, Meyer TR, Gord JR. 2006. Femtosecond coherent anti-Stokes Raman scattering measurement of gas temperatures from frequency-spread dephasing of the Raman coherence. *Appl. Phys. Lett.* 89:251112

74. Lucht RP, Kinnius PJ, Roy S, Gord JR. 2007. Theory of femtosecond coherent anti-Stokes Raman scattering spectroscopy of gas-phase resonant transitions. *J. Chem. Phys.* 127:044316

75. Roy S, Kinnius PJ, Lucht RP, Gord JR. 2008. Temperature measurements in reacting flows by time-resolved femtosecond coherent anti-Stokes Raman scattering (fs-CARS) spectroscopy. *Opt. Commun.* 281:319–25

76. Dantus M, Bowman RM, Zewail AH. 1990. Femtosecond laser observations of molecular vibration and rotation. *Nature* 343:737–39

77. Leonhardt R, Holzapfel W, Zinth W, Kaiser W. 1987. Terahertz quantum beats in molecular liquids. *Chem. Phys. Lett.* 133:373–77

78. Hayden CC, Chandler DW. 1995. Femtosecond time-resolved studies of coherent vibrational Raman scattering in large gas-phase molecules. *J. Chem. Phys.* 103:10465–72

79. Knopp G, Kirch K, Beaud P, Mishima K, Spitzer H, et al. 2003. Determination of the ortho-/para deuterium concentration ratio with femtosecond CARS. *J. Raman Spectrosc.* 34:989–93

80. Lang T, Motzkus M. 2002. Single-shot femtosecond coherent anti-Stokes Raman-scattering thermometry. *J. Opt. Soc. Am. B* 19:340–44

81. Cheng JX, Xie XS. 2004. Coherent anti-Stokes Raman scattering microscopy: theory, instrumentation, and applications. *J. Phys. Chem. B* 108:827–40

82. Weiner AM, Leaird DE, Wiederrecht GP, Nelson KA. 1990. Femtosecond pulse sequences used for optical manipulation of molecular motion. *Science* 247:1317–19

83. Scully MO, Kattawar GW, Lucht RP, Opatrný T, Pilloff H, et al. 2002. FAST CARS: engineering a laser spectroscopic technique for rapid identification of bacterial spores. *Proc. Nat. Acad. Sci. USA* 99:10994–1001

84. Schmitt M, Knopp G, Materny A, Kiefer W. 1998. The application of femtosecond time-resolved coherent anti-Stokes Raman scattering for the investigation of ground and excited state molecular dynamics of molecules in the gas phase. *J. Phys. Chem. A* 102:4059–65

85. Demtröder W. 2002. *Laser Spectroscopy.* New York: Springer. 987 pp. 3rd ed.

86. Reichardt TA, Teodoro FD, Farrow RL, Roy S, Lucht RP. 2000. Collisional dependence of polarization spectroscopy with a picosecond laser. *J. Chem. Phys.* 113:2263–69

87. Keifer J, Li Z, Zetterberg J, Linvin M, Aldén M. 2007. Simultaneous laser-induced fluorescence and sub-Doppler polarization spectroscopy of the CH radical. *Opt. Commun.* 270:347–52

88. Tobai J, Dreier T. 1999. Measurement of relaxation times of NH in atmospheric pressure flames using picosecond pump-probe degenerate four-wave mixing. *J. Mol. Struct.* 480–481:307–10

89. Danzmann K, Grützmacher K, Wende B. 1986. Doppler-free two-photon polarization-spectroscopic measurement of the Stark-broadened profile of the hydrogen La line in a dense plasma. *Phys. Rev. Lett.* 57:2151–53

90. Sargent M III, Scully MO, Lamb WE Jr. 1974. *Laser Physics*. Boulder, CO: Westview Press. 464 pp.

91. Roy S, Lucht RP, Reichardt TA. 2002. Polarization spectroscopy using short-pulse lasers: theoretical analysis. *J. Chem. Phys.* 116:571–80

92. Zare RN. 1988. *Angular Momentum: Understanding Spatial Aspects in Chemistry and Physics*. New York: Wiley & Sons. 368 pp.

93. Dreizler A, Taddy R, Suvernev AA, Himmelhaus M, Dreier T, Foggi P. 1995. Measurement of orientational relaxation times of OH in a flame using picosecond time-resolved polarization spectroscopy. *Chem. Phys. Lett.* 240:315–23

94. Tadday R, Dreizler A, Suvernev AA, Dreier T. 1997. Measurement of orientational relaxation times of OH ($A^2\Sigma$–$X^2\Pi$) transitions in atmospheric pressure flames using picosecond time-resolved nonlinear spectroscopy. *J. Mol. Struct.* 410–411:85–88

95. Chen X, Patterson BD, Settersten TB. 2004. Time-domain investigation of OH ground-state energy transfer using picosecond two-color polarization spectroscopy. *Chem. Phys. Lett.* 388:358–62

96. Costen ML, McKendrick KG. 2005. Orientation and alignment moments in two-color polarization spectroscopy. *J. Chem. Phys.* 122:164309

97. Chen X, Settersten TB. 2007. Investigation of OH $X^2\Pi$ collisional kinetics in a flame using picosecond two-color resonant four-wave-mixing spectroscopy. *Appl. Opt.* 46:3911–20

98. Kulatilaka WD, Lucht RP, Roy S, Gord JR, Settersten TB. 2007. Detection of atomic hydrogen in flames using picosecond two-color two-photon-resonant six-wave-mixing spectroscopy. *Appl. Opt.* 46:3921–27

99. Morgan M, Price W, Hunziker L, Ludowise P, Blackwell M, Chen Y. 1993. Femtosecond Raman-induced polarization spectroscopy studies of rotational coherence in O_2, N_2 and CO_2. *Chem. Phys. Lett.* 209:1–9

100. Frey HM, Beaud P, Gerber T, Mischler B, Radi PP, Tzannis AP. 1999. Femtosecond nonresonant degenerate four-wave mixing at atmospheric pressure and in a free jet. *Appl. Phys. B* 68:735–39

101. Hornung T, Skenderovic H, Kompa KL, Motzkus M. 2004. Prospect of temperature determination using degenerate four-wave mixing with sub-20 fs pulses. *J. Raman Spectrosc.* 35:934–38

Matrix-Assisted Laser Desorption/Ionization Imaging Mass Spectrometry for the Investigation of Proteins and Peptides

Kristin E. Burnum,* Sara L. Frappier,*
and Richard M. Caprioli

Mass Spectrometry Research Center, Departments of Chemistry and Biochemistry,
Vanderbilt University, Nashville, Tennessee 37221;
email: Kristin.E.Burnum@vanderbilt.edu, Sara.L.Frappier@vanderbilt.edu,
R.Caprioli@vanderbilt.edu

Annu. Rev. Anal. Chem. 2008. 1:689–705

First published online as a Review in Advance on
March 18, 2008

The *Annual Review of Analytical Chemistry* is online
at anchem.annualreviews.org

This article's doi:
10.1146/annurev.anchem.1.031207.112841

1936-1327/08/0719-0689$20.00

*These authors contributed equally to this work.

Key Words

imaging, MALDI, mass spectrometry, proteins, peptides

Abstract

Mass spectrometry (MS) is an excellent technology for molecular imaging because of its high data dimensionality. MS can monitor thousands of individual molecular data channels measured as mass-to-charge (m/z). We describe the use of matrix-assisted laser desorption/ionization (MALDI) MS for the image analysis of proteins, peptides, lipids, drugs, and metabolites in tissues. We discuss the basic instrumentation and sample preparation methods needed to produce high-resolution images and high image reproducibility. Matrix-addition protocols are briefly discussed along with normal operating procedures, and selected biological and medical applications of MALDI imaging MS are described. We give examples of both two- and three-dimensional imaging, including normal mouse embryo implantation, sperm maturation in mouse epididymis, protein distributions in brain sections, protein alterations as a result of drug administration, and protein changes in brain due to neurodegeneration and tumor formation. Advantages of this technology and future challenges for its improvement are discussed.

1. INTRODUCTION

MS: mass spectrometry

MALDI: matrix-assisted laser desorption/ionization

IMS: imaging mass spectrometry

Over the past decade, proteomics has become a vital complement to genetic analysis in the investigation of nearly all aspects of the life sciences. These include the elucidation of cellular processes in both health and disease (1–3) and the discovery and evaluation of pharmaceutical compounds (4–9). Mass spectrometry (MS) has emerged as an essential analytical tool for the investigation of these molecular processes. Indeed, new advances in MS now provide the opportunity for investigative studies of molecular interactions in intact tissue. Unlike studies conducted with intact tissue decades ago, studies performed today can take advantage of the exquisite molecular specificity offered by MS. In particular, matrix-assisted laser desorption/ionization (MALDI) imaging mass spectrometry (IMS) allows investigators to analyze the spatial distribution of proteins directly in tissue specimens.

IMS can be used to localize specific molecules such as drugs, lipids, peptides, and proteins directly from fresh-frozen tissue sections with lateral image resolution of 30–50 μm. Thin frozen sections (10–15 μm thick) are cut and thaw-mounted on target plates; subsequently, an energy-absorbing matrix is applied. Areas with a typical target spot size of about 50 μm in diameter are ablated with a UV laser, thereby giving rise to ionic molecular species that are recorded according to their mass-to-charge (m/z) values. Thus, a single mass spectrum is acquired from each ablated spot in the array. Signal intensities at specific m/z values can be exported from this array to give a two-dimensional ion-density map, or image, constructed from the specific coordinate location of that signal and its corresponding relative abundance. For high-resolution images, matrix is deposited in a homogeneous manner to the surface of the tissue in such a way as to minimize the lateral dispersion of the molecules of interest. This can be achieved either by automatically printing arrays of small droplets or by robotically spraying a continuous coating. Each micro spot or pixel coordinate is then automatically analyzed by MALDI MS (**Figure 1**). From the analysis of a single section, images at virtually any molecular weight may be obtained, provided that there is sufficient signal intensity to record.

One of the most compelling aspects of IMS is that it provides the ability to simultaneously visualize the spatial arrangement of hundreds of analytes directly from tissue without any prior knowledge or need for target specific reagents such as antibodies. IMS enables the visualization of posttranslational modifications and proteolytic processing while retaining spatial localization. Other MS techniques, such as secondary ionization mass spectrometry (SIMS), have also been used for a variety of imaging applications. One of the major advantages of SIMS is that it is capable of high-resolution imaging (50–100 nm) for elements and small molecules (m/z <1000 Da). However, thus far it has not been shown to be effective for the analysis of proteins and large peptides.

This review focuses on MALDI IMS analysis of proteins and peptides in terms of basic instrumentation, sample preparation, and recent applications. MALDI MS is an effective technology for both qualitative and quantitative analysis of normal and diseased tissue and for assessing temporal changes in biological systems. In addition, this technology has been applied to the generation of three-dimensional protein

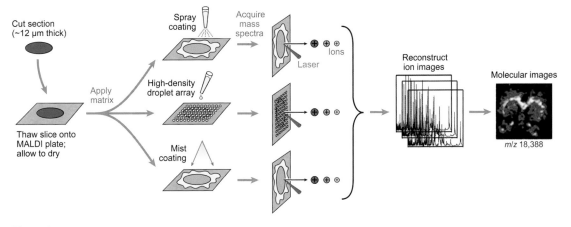

Figure 1

Scheme of a typical direct tissue imaging experiment. Three types of matrix application are shown: spray coating, high-density droplet array, and mist coating. Subsequent image acquisition is shown for each technique. Abbreviation: MALDI, matrix-assisted laser desorption/ionization.

images of brain, whole-body reconstructions, and the measurement of protein changes in specific tissues after systemic drug administration.

2. THE TECHNOLOGY

The MS instrumentation best suited for the analysis of peptides and proteins directly from tissues is MALDI time-of-flight (TOF) technology. The ablation process directed by the focused laser beam, together with the high frequency of the laser pulse, has rendered TOF the most widely used ionization method for imaging. The duty cycle of a modern TOF analyzer is an ideal match for the pulsed laser process and also has the advantages of a theoretically unlimited mass range, high ion-transmission efficiency, multiplex detection capability, and simplicity in instrument design and maintenance (10).

A brief description of a MALDI TOF MS experiment follows for those who are not currently practitioners and who seek introductory information; more detailed treatments can be found in other works (11, 13). A typical analysis of proteins directly from tissue is described for illustrative purposes. There are two main experimental approaches: profiling and imaging. Profiling involves analyzing discrete areas of the tissue sections and subjecting the resulting protein profiles to computational analysis. Typically, this uses 5–20 spots, each measuring approximately 0.2–1 mm in diameter. These experiments are designed to make comparisons between representative areas on pieces of tissue, such as a normal healthy area versus a diseased area, or between two different specimens. Thus, in the profiling mode, fine spatial resolution is not required. A sufficient number of areas must be sampled to gain statistical confidence in the results; the number varies depending on the specific experiment. On the other hand, high-resolution imaging of a tissue requires that the entire tissue section be

TOF: time of flight

analyzed through an ordered array of spots, or raster, in which spectra are acquired at intervals that define the image resolution (e.g., every 50 μm in both the x and the y directions). Two-dimensional ion-intensity maps, or images, can then be created by plotting the intensity of any signal obtained as a function of its xy coordinates. The resulting images allow rapid assessment of protein localization differences between and among samples.

Tissues used for analysis should be frozen in liquid nitrogen immediately after resection to preserve morphology and minimize protein degradation through prote-olysis. The tissue is usually sectioned in a cryostat to give 10- to 12-μm-thick sections and is then thaw-mounted onto an electrically conductive sample plate (11). Sample plates include gold-coated or stainless steel metal plates and glass slides that have a conductive coating. The tissue may be gently rinsed with ethanol acting as a fixative and a wash to remove lipids and salts. Alternatively, IMS-compatible tissue-staining protocols can be used in conjunction with the optically transparent glass slides, allowing correlation of IMS data with histological features of the same section by optical microscopy (14).

MALDI IMS requires the application of energy-absorbing matrix. The matrix is typically a small organic molecule that cocrystallizes with the analytes of interest on the tissue surface. The matrix is capable of absorbing laser energy, thereby causing the analyte to desorb from the sample surface and ionize. The most commonly used matrices include 3,5-dimethoxy-4-hydroxy-cinnamic acid (SA), 2,5-dihydroxybenzoic acid (DHB), and α-cyano-4-hydroxy-cinnamic acid (HCCA). Solvent combinations and the matrix used vary depending on the type of molecule and tissue being investigated (11). A 50:50 (v/v) acetonitrile/water or ethanol/water mixture is generally recommended for use in tissue analysis (12). SA is an excellent matrix for the analysis of protein (11), and DHB and HCCA are primarily used for the analysis of peptides and lower-molecular-weight analytes.

For high-resolution imaging, the matrix solution should be homogenously deposited across the tissue section in such a manner as to avoid significant lateral migration of analytes. Currently, this is achieved by applying matrix solution to the tissue in either a spotted array or a homogenous spray coating (11). A continuous and homogenous spray coating allows the highest spatial resolution, but densely spotted arrays yield higher reproducibility and generally better spectra quality. Heterogenous matrix coating gives rise to hot and cold spots for ablation due to random crystal formation, therefore producing poor, highly pixilated images. Numerous robotic spotting devices are commercially available and utilize acoustic (15), piezoelectric (16), ink-jet printer (17), and capillary deposition techniques (18). Several robotic spray coating devices are also commercially available and utilize a mist-nebulizing method (19) or a thermally assisted spray method (20).

Protein analysis is usually performed on a linear TOF instrument to achieve the highest possible sensitivity. Ions formed and desorbed during the laser pulse are extracted and accelerated into the field-free region of the TOF analyzer. Ions are usually detected by a multichannel plate detector and the TOF of the various ions is inversely proportional to their m/z values. This time measurement is then converted to m/z through appropriate calibration procedures. For the analysis of low-molecular-weight species, an ion mirror or reflectron can be used in the ion flight path to compensate for the initial velocity/energy distribution and can thus improve resolution (21).

Other analyzer combinations that have been used with MALDI IMS include TOF-TOF (16), orthogonal TOF and orthogonal quadrupole-TOF (22, 23), ion mobility (24), Fourier transform ion cyclotron resonance (25, 26) and ion trap technologies (27, 28). These additional tools have provided capabilities for protein identification, high mass resolution acquisition, and detection of small molecules such as drugs and metabolites. Although it is beyond the scope of this article to describe these techniques in more detail, the reader is referred to other works for more thorough descriptions (16, 22–25, 27).

3. QUANTITATION

To assess changes in protein concentrations in a given tissue, pixel-to-pixel reproducibility must be high; that is, two pixels close together having the same protein concentrations should give the same spectra within acceptable standards. Although these standards vary from experiment to experiment, typically variations of $\pm 15\%$ (or less) are deemed acceptable. Factors bearing directly on this aspect of the analysis include ionization efficiency of a given molecule, ion-suppression effects, extraction efficiency of the matrix-deposition process, and the robustness and effectiveness of postacquisition processing. High pixel-to-pixel reproducibility can be achieved if careful attention is paid to sample preparation and matrix application. In addition, instrument parameters such as voltage settings and laser power must be kept constant within a given experiment. A robotic matrix application device is invaluable in removing operator-to-operator variations. In a recent study of protein distribution in brain, relative standard errors of 3.7–9.6% were obtained between striatal regions in multiple animals (22). However, one must keep in mind that because ionization processes can be affected by the physical and chemical processes of molecules, it is difficult to estimate relative concentrations of two completely different proteins by comparing their peak heights alone.

4. DATA PROCESSING AND ANALYSIS

It is important to assess the reproducibility of the mass spectra so that variations in peak intensities can be correlated to biological endpoints. To do so, one must perform the two general types of data analysis: preprocessing and statistical analysis. The preprocessing step reduces the experimental variance between spectra through the removal of background, normalization of the peak intensity to the total ion current, and peak alignment algorithms. Normalization of the spectra minimizes variation arising from day-to-day instrument fluctuations, differences in matrix crystallization across tissue sections, and changes in sample preparation and the chemical properties of the underlying tissue (29). Various algorithms are employed for all of the spectra processing steps: baseline subtraction, peak alignment, normalization and peak picking. Ion images are generated directly from these processed data sets.

The next phase of data processing involves downstream events such as calculation of average spectra for each specimen type or targeted area in the tissue (as determined by the investigator) and subsequent statistical analysis procedures. Principal components analysis can be performed to reduce the dimensionality of a given data set using processes to extract, display, and rank the variance within the data set. By identifying patterns in data, the investigator can more easily determine similarities and differences (30). Significance analysis of microarrays (SAM) can be performed to assign a score to each feature on the basis of change in its spectral intensity relative to the standard deviation of repeated measurements. For features with a score greater than a predetermined threshold, SAM uses permutations of the repeated measurements to estimate the percentage of features identified by chance (the false discovery rate) (31). This algorithm is used to elucidate features that change significantly between two groups of specimens. A hierarchical clustering analysis (HCA) may also be applied to the data set, enabling samples to be grouped blindly according to their expression profiles. HCA functions by calculating the dissimilarity between the individual analyses (32).

5. MOLECULAR IMAGES OF DISEASE

To date, considerable effort has been focused on finding molecular markers that represent early indicators of disease. MALDI IMS provides a means to visualize molecularly specific information while maintaining spatial integrity. For example, cancer progression depends on essential characteristics such as the presence of growth factors, insensitivity to growth-inhibiting signals, evasion of apoptosis, high replicate potential, sustained angiogenesis, and tissue invasion and metastasis (33). Alterations in protein expression, proteolytic processing, and posttranslational modifications all contribute to cellular transformation. IMS analyses of tissue sections reflect the overall status of the tissue; therefore, analyses of tissues in various states can reveal differences in the expression of proteins that otherwise cannot be predicted. IMS has been used to image protein distributions in multiple types of cancer. Imaging analysis has been used to probe proteome changes in mouse breast and brain tumor (34, 35), glioblastoma biopsies (36), and human lung metastasis to the brain (12) and prostate (34, 37). Identifying features that reveal differential expression patterns between cancerous and normal tissue can provide valuable insight into the molecular mechanisms of cancer, provide molecular diagnostic and prognostic signatures, and identify possible drug targets in implicated pathways.

Ion images obtained from a mouse brain with a developed glioblastoma tumor in the right hemisphere after injection with GL26 glioma cells (S.L. Frappier, unpublished data) are shown in **Figure 2a**. Histone H4 localizes in the tumor tissue in contrast to guanine nucleotide-binding protein γ7 and cytochrome c oxidase polypeptide VIIc, which localize in the surrounding normal brain tissue. Hundreds of ion images were produced from a single data-acquisition process.

Three-dimensional MALDI IMS was first reported in 2005 for the depiction of myelin basic protein isoform 8 in the corpus callosum of a mouse brain (38); more recently, it was used to explore a three-dimensional protein volume in the

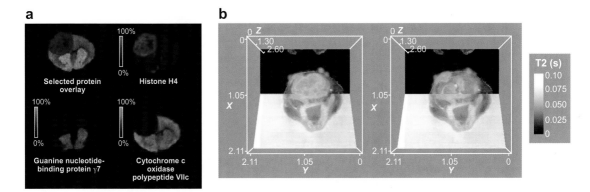

a

Selected protein overlay · Histone H4

Guanine nucleotide-binding protein γ7 · Cytochrome c oxidase polypeptide VIIc

b

T2 (s)

Figure 2

Imaging mass spectrometry (IMS) analysis of a glioma tumor mouse brain. (*a*) Ion images from a two-dimensional IMS experiment of a mouse brain with a glioma tumor located in the right hemisphere. (*b*) Spatially resolved three-dimensional IMS volumes coregistered to in vivo magnetic resonance imaging data and the optical volume of a glioma tumor mouse brain.

substantia nigra in a mouse model of Parkinson's disease (39). The acquisition of three-dimensional molecular images also raises the exciting possibility of combining these IMS images with magnetic resonance imaging (MRI) and perhaps positron emission tomography (PET) imaging, so that the molecularly specific MS data can be coregistered with the MRI data. The latter technique relies on the relaxation properties of excited hydrogen nuclei in water and lipids after exposure to a powerful, uniform magnetic field (40).

Figure 2*b* shows the results of experiments wherein MS and MRI were combined to analyze a mouse brain injected with glioma cells and to visualize the data in three dimensions (41). The entire mouse head was sectioned from the olfactory bulb to the beginning of the spinal cord. Coronal sections of the mouse brain were collected at 20-μm increments as a high-resolution camera acquired blockface images every 40 μm. Tissue sections for IMS analysis were collected every 160 μm, coated with SA, and imaged at 150-μm lateral resolution. In the left panel of **Figure 2*b***, histone H2B is associated with the tumor as it presents in the contralateral ventricle. Conversely, the right panel of **Figure 2*b***, showing myelin basic protein isoform 4, demonstrates that corpus callosum can be accurately spatially correlated in a three-dimensional sample-specific context. A good correlation of the tumor between MALDI ion intensities, contrast variation in the magnetic resonance parameters, and the optical volume was observed.

IMS has also been utilized to study neurodegenerative diseases such as Alzheimer's and Parkinson's. Neurodegeneration is widespread in Alzheimer's disease, affecting several brain regions. β-Amyloid plaques accumulate causing plaque formation and subsequent neurodegeneration (42, 43). A mutation of the amyloid precursor protein gene causes higher production of β-amyloid and β-amyloid peptides in the cortical and hippocampal structures of the brain. MALDI images, prepared by coating tissue with SA and imaging at a resolution of 100 μm, show that the β-amyloid peptides

MRI: magnetic resonance imaging

PET: positron emission tomography

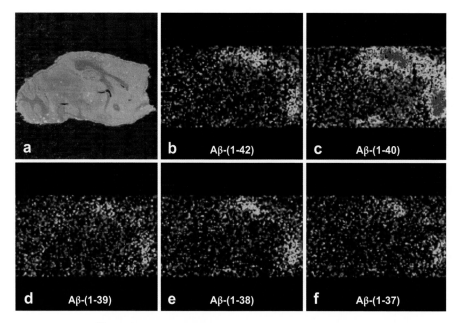

Figure 3

Imaging mass spectrometry images from a diseased brain section from an Alzheimer's patient. (*a*) Optical image of the sagittal diseased brain section. (*b*) Aβ-(1-42) molecular image (*m/z* 4515.1). (*c*) Aβ-(1-40) molecular image (*m/z* 4330.9). (*d*) Aβ-(1-39) molecular image (*m/z* 4231.7). (*e*) Aβ-(1-38) molecular image (*m/z* 4132.6). (*f*) Aβ-(1-37) molecular image (*m/z* 4075.5). Reprinted with permission from Reference 44.

(I-40) and (I-42) are the most abundant amyloid peptides in those brain regions in an Alzheimer's disease mouse model (44) (**Figure 3**). The extensive deposits of the Aβ (I-40) are seen localized in the cortex of the sagittal section.

From previous studies of Parkinsonism, a condition that affects regions of the brain that control motor function (such as the substantia nigra and the striatum), it was found that dopamine neurons in the nigrostriatal pathway are gradually decreased in the striatum. IMS technology has been applied to the 1-methyl 4-phenyl 1,2,3,6-tetrahydropyridine (MPTP) animal model of Parkinsonism (45); the investigators found that the protein PEP-19, a neuronal calmodulin-binding protein, was significantly reduced (>30%) in the striatum compared to control regions in normal brain. Further study is needed to determine PEP-19's specific involvement in neurodegeneration.

6. NEW INSIGHTS INTO BIOLOGY

MALDI IMS of tissue sections can provide information on the spatial and temporal action of proteins involved in processes that take place in organs or tissue substructures containing heterogeneous cell types. A major advantage of MALDI IMS is that it allows the investigator to visualize molecular events while retaining spatial

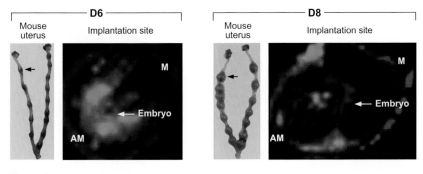

Figure 4

Optical images of a day 6 (D6) pregnant mouse uterus and day 8 (D8) pregnant mouse uterus along with the corresponding ion images (arrows indicate the implantation site). D6 implantation site and D8 implantation site ion-density maps obtained for: m/z 5923 (*purple*), m/z 4039 (*green*), m/z 6717 (*blue*), m/z 8193 (*red*). See text for details. Abbreviations: AM, antimesometrial; M, mesometrial.

information that would be lost using techniques that require tissue homogenization and extraction.

We have used IMS to study the mammalian reproductive process involving implantation and early development of the embryo. The heterogeneous cell types of the uterus, each of which has a unique function, pose a challenge for studying events during early pregnancy. In rodents and humans, one limiting factor is the availability of adequate amounts of tissue for biochemical and molecular biology experiments. Studying the uterine molecular landscape using conventional approaches presents an arduous task as some proteins are localized to particular cell types on a specific day of pregnancy and do not reappear later in the process. We studied implantation sites from mice on day 6 (D6) and day 8 (D8) of pregnancy and obtained protein distributions that were unique to embryo, surrounding muscle, and uterine stroma within an area measuring 2 mm² (K. Burnum, unpublished data). The ion-intensity maps presented in **Figure 4** depict how protein location and expression change during early pregnancy. The ions depicted in red and green are expressed in the presence of the embryo only on D6 through D8 of pregnancy. The ion shown in red is expressed at the site of the embryo and increases by D8. The uterus has two poles; the mesometrial pole (M), where the blood supply enters the uterus, and the antimesometrial pole (AM), where the embryo implants. The ion depicted in purple remains at the AM pole in these images, whereas the ion depicted in green shifts solely to the AM pole by D8. The ion shown in blue represents the muscle surrounding the uterus. Disregulation of events prior to, during, or immediately after implantation contributes to poor pregnancy rates. Therefore, understanding protein localization during implantation is essential to providing potential targets for treating infertility and developing novel contraceptive approaches.

In another project involving mammalian reproduction, researchers studied the spatial distribution of proteins in the sexually mature mouse epididymis (46), a structure that contains a coiled tube attached to the back of the testis. Immature sperm

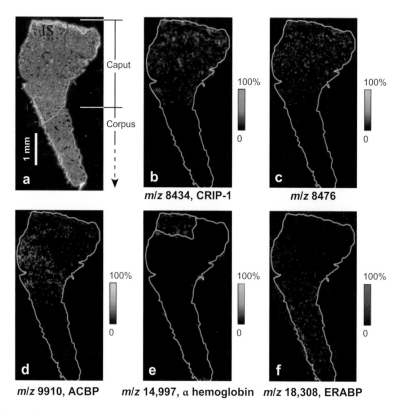

Figure 5

High-resolution imaging mass spectrometry of caput epididymis. (*a*) Optical image of a 12-μm section mounted on a sample plate prior to matrix application. (*b–f*) Ion-density maps obtained for *m/z* 8434 (CRIP-1), *m/z* 8476, *m/z* 9910 (ACBP), *m/z* 14,997 (α hemoglobin), and *m/z* 18,308 (ERABP). Reprinted with permission from Reference 46.

becomes competent for fertilization as it travels through the epididymis and interacts with proteins located along this tubule. The epididymis consists of a head (caput), body (corpus), and tail (cauda). Molecular images were generated for the whole epididymis and specific proteins were mapped over this structure. Evidence for the regional localization of some of the major epididymal proteins in the mouse as observed by IMS are shown in **Figure 5**. From these and additional images, 30 protein signals that showed regionalized patterns along the epididymis were targeted and identified.

7. PROTEOME RESPONSE TO DRUG ADMINISTRATION

In addition to assisting in disease differentiation and diagnosis, the proteome signature of a tissue can also be used to determine the effects of drug/small-molecule administration to an animal model or patient. Over the past decade, proteins have become principal targets for drug discovery, and proteomics-oriented drug research has come to the forefront of activity in this area. Proteomics play a major role in drug development, specifically in target identification, target validation, drug design, lead optimization, and preclinical and clinical development (47).

Pharmaceutical drug discovery is a laborious and tedious process: It can take 8–10 years, at great expense, to bring a drug candidate into the market (4). Obviously, it is essential to find ways to expedite this process. With early assessment of the distribution of a drug candidate in targeted tissues, IMS can greatly assist in the discovery and validation of processes related to drug administration, distribution, metabolism, and excretion (48, 49). For example, IMS can detect the presence and location of an individual drug and its metabolites in a label-free protocol, a significant advantage over other small-molecule imaging techniques that typically require the addition of a radioactive tag to the molecule of interest, such as in autoradiography. Another advantage of IMS is that it is capable of providing information on both the pharmacological and biological effects of a drug in that it can detect molecular features that may be markers of drug efficacy or toxicity. Other imaging techniques provide little information on the molecular identity of these biological endpoints. Thus, IMS can monitor the analyte of interest and also the corresponding proteome response. For example, investigators have discovered transthyretin as a marker for gentamicin-induced nephrotoxicity in rat (50). Gentamicin-induced nephrotoxicity is seldom fatal and is usually reversible but often results in long hospital stays. Thus, there is great interest in finding potential markers of early toxicity and also in helping to elucidate the molecular mechanism. Investigators utilized MALDI IMS to determine differential protein expression within the rat kidney (cortex, medulla, and papilla), identified features of interest between dosed and control rat, and then applied downstream protein identification procedures. Transthyretin was significantly increased in the treated mouse kidney over control; these findings were validated with Western blot and immunohistochemistry.

To study the relationship between drug distribution in tumors and the resulting protein alterations, mass spectral images were obtained from MMTV/HER2 tumors excised from mice (35). Investigators were able to identify markers that indicate a response of the tumor to administration of erbB receptor inhibitors OSI-774 and Herceptin®. Inhibition of tumor cell proliferation and induction of apoptosis and tumor reduction were predicted by a >80% reduction in thymosin β4 and ubiquitin levels that were detectable after 16 h of a single drug dose before any evidence of in situ cellular activity.

The same procedures can also be applied to whole-body animal tissue sections for a system-wide analysis in a single experiment (22). For whole-body sagittal tissue sections, using the same sample preparation and analysis conditions described for IMS experiments of tissues, signals unique to individual organs were detected and used to produce a two-dimensional protein map of a control rat (**Figure 6**). By expanding the capabilities of MALDI IMS to investigate multiple tissue types simultaneously across a whole-body tissue section, distinct protein patterns can be identified and used to monitor whole-body system dynamics.

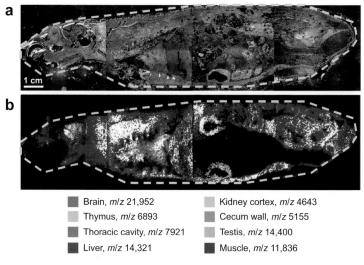

a

1 cm

b

■ Brain, *m/z* 21,952   ■ Kidney cortex, *m/z* 4643
■ Thymus, *m/z* 6893   ■ Cecum wall, *m/z* 5155
■ Thoracic cavity, *m/z* 7921   ■ Testis, *m/z* 14,400
■ Liver, *m/z* 14,321   ■ Muscle, *m/z* 11,836

Figure 6

Whole-body protein analysis of rat sagittal tissue section by imaging mass spectrometry.
(*a*) Optical image of rat sagittal tissue section across four gold matrix-assisted laser
desorption/ionization target plates. (*b*) Ion image overlay of unique organ signals. Reprinted
with permission from Reference 22.

8. IN SITU PROTEOMICS

Because the simple linear MALDI IMS analysis only provides a measurement of the
molecular weight of a given protein, off-line techniques involving extraction, LC separation, digestion, and LC/MS/MS analyses with database searches are performed to
unequivocally identify a protein. These approaches are effective, but are very time
consuming and costly. Direct in situ protein identification would provide a significant advantage in such cases. We have developed a protocol that allows digestion of
proteins and peptides directly on thin tissue sections using well-defined microspotted
arrays of trypsin that can subsequently be spotted with matrix for IMS and MALDI
MS/MS analyses (16). In one such study (**Figure 7**), we applied trypsin in 150-μm
diameters in an array with spots located 200 μm apart on a section of mouse brain.
For two specific proteins, PEP19 and neurogranin, four tryptic peptides from each
were identified and sequenced by MALDI MS/MS directly from the tissue digest
spots. The image of each peptide was then constructed (**Figure 7**).

The application of IMS to the analysis of formalin-fixed, paraffin-embedded
(FFPE) tissue sections has also been undertaken. FFPE is the most convenient method
to preserve samples in hospital tissue banks; this method is of great interest as millions
of tissue samples are stored worldwide, many of which are linked to detailed patient
histories and outcomes. However, analysis of the proteins in formalin-fixed tissues
is a major challenge: The fixation process alters the structures of the proteins by
cross-linking, thereby changing the molecular weight, often in unpredictable ways.
Recent studies have focused on protocols to analyze such samples and thus far have

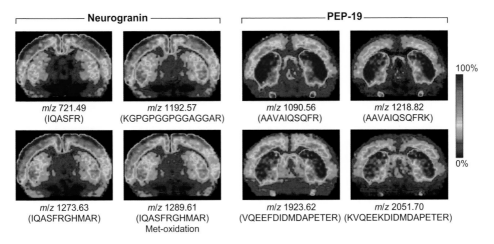

Neurogranin

PEP-19

100%

0%

m/z 721.49
(IQASFR)

m/z 1192.57
(KGPGPGGPGGAGGAR)

m/z 1090.56
(AAVAIQSQFR)

m/z 1218.82
(AAVAIQSQFRK)

m/z 1273.63
(IQASFRGHMAR)

m/z 1289.61
(IQASFRGHMAR)
Met-oxidation

m/z 1923.62
(VQEEFDIDMDAPETER)

m/z 2051.70
(KVQEEKDIDMDAPETER)

Figure 7

Examples of on-tissue tryptic digestion and subsequent matrix-assisted laser desorption/ ionization imaging mass spectrometry analysis. *Left*: Tryptic peptides generated from the digestion of the 7.5-kDa protein, neurogranin. *Right*: Tryptic peptides generated from the digestion of the 6.7-kDa protein, PEP-19. Reprinted with permission from Reference 16.

shown promise. For example, IMS was used to analyze two-year-old archived FFPE rat brain tissues and the results showed the same localization of various proteins as that of the proteins within frozen tissue sections (51).

9. PERSPECTIVES AND CONCLUSIONS

The molecular complexity of biological systems obviates the need for molecularly specific tools to elucidate proteomic events in both spatial and temporal distributions. One of the most effective ways to present such information is in the form of images, as demonstrated through MRI, PET, and confocal microscopy. Similarly, images of the molecular composition of living systems will allow us to gain a more comprehensive view of biological processes and reveal complex molecular interactions. This approach will be vital in elucidating molecular aspects of disease and also drug effectiveness by providing spatial and relevant time-based information.

MALDI IMS provides unique information on peptides, proteins, lipids, metabolites, xenobiotics, and other endogenous compounds directly from complex tissue samples, bringing powerful new capabilities to biological research. Sample preparation is relatively easy and is amenable to robotics for high-throughput analysis. Most importantly, it is an excellent discovery tool, as the identity of the analytes do not need to be known prior to analysis. Many hundreds of signals can be recorded from a single experiment with high mass measurement accuracy. Because the spatial integrity of the tissue is maintained, molecular images and patterns can be correlated with histopathological analyses and other molecular imaging modalities.

Given the current rapid advances in both physics and chemistry, the prospects for MALDI IMS are bright. Such advances are sure to bring higher spatial resolution, increased dynamic range of detection, and improved visualization of the molecular complexities within living systems. This emerging technology, and enabling variations of it, will continue to be an essential way to present complex data. Moreover, this technology appeals to the human intellect and to our ability to understand complexity in a visual format.

DISCLOSURE STATEMENT

The authors are not aware of any biases that might be perceived as affecting the objectivity of this review.

ACKNOWLEDGMENTS

This work was supported by RMC grants NIH/NIGMS 5R01 GM58008-09 and DOD W81XWH-05-1-0179. K.B. is supported by NICHD training grant no. HD007043-31A2. The authors would like to thank Dr. Pierre Chaurand and Dr. Shannon Cornett for providing insights throughout this work.

LITERATURE CITED

1. Marko-Varga G, Fehniger TE. 2004. Proteomics and disease: the challenges for technology and discovery. *J. Proteome Res.* 3:167–78
2. Oehr P. 2006. 'Omics'-based imaging in cancer detection and therapy. *Pers. Med.* 3:19–32
3. Lescuyer P, Hochstrasser D, Rabilloud T. 2007. How shall we use the proteomics toolbox for biomarker discovery? *J. Proteome Res.* 6:3371–76
4. Lewin DA, Weiner MP. 2004. Molecular biomarkers in drug development. *Drug Discov. Today* 9:976–83
5. Bieck PR, Potter WZ. 2005. Biomarkers in psychotropic drug development: integration of data across multiple domains. *Annu. Rev. Pharmacol. Toxicol.* 45:227–46
6. Strand KJ, Khalak H, Strovel JW, Ebner R, Augustus M. 2006. Expression biomarkers for clinical efficacy and outcome prediction in cancer. *Pharmacogenomics* 7:105–15
7. Pusch W, Flocco MT, Leung SM, Thiele H, Kostrzewa M. 2003. Mass spectrometry–based clinical proteomics. *Pharmacogenomics* 4:463–76
8. Scaros O, Fisler R. 2005. Biomarker technology roundup: From discovery to clinical applications, a broad set of tools is required to translate from the lab to the clinic. *Biotechniques* 38:S30–32
9. Kamb A, Wee S, Lengauer C. 2007. Opinion: Why is cancer drug discovery so difficult? *Nat. Rev. Drug Discov.* 6:115–20
10. Dass C. 2001. *Principles and Practice of Biological Mass Spectrometry*. New York: Wiley
11. Schwartz SA, Reyzer ML, Caprioli RM. 2003. Direct tissue analysis using matrix-assisted laser desorption/ionization mass spectrometry: practical aspects of sample preparation. *J. Mass Spectrom.* 38:699–708

12. Johnson MD, Floyd JL, Caprioli RM. 2006. Proteomics in diagnostic neuropathology. *J. Neuropathol. Exp. Neurol.* 65:837–45

13. Chaurand P, Schwartz SA, Caprioli RM. 2004. Profiling and imaging proteins in tissue sections by MS. *Anal. Chem.* 76:86A–93A

14. Chaurand P, Schwartz SA, Billheimer D, Xu BGJ, Crecelius A, Caprioli RM. 2004. Integrating histology and imaging mass spectrometry. *Anal. Chem.* 76:1145–55

15. Aerni HR, Cornett DS, Caprioli RM. 2006. Automated acoustic matrix deposition for MALDI sample preparation. *Anal. Chem.* 78:827–34

16. Groseclose MR, Andersson M, Hardesty WM, Caprioli RM. 2007. Identification of proteins directly from tissue: in situ tryptic digestions coupled with imaging mass spectrometry. *J. Mass Spectrom.* 42:254–62

17. Baluya DL, Garrett TJ, Yost RA. 2007. Automated MALDI matrix deposition method with inkjet printing for imaging mass spectrometry. *Anal. Chem.* 79:6862–67

18. Guo J. 2005. *Discovery of biomarkers for endometrial carcinoma by direct profiling of proteins in tissue sections.* Presented at Annu. Am. Soc. Mass Spectrom., 53rd, San Antonio, TX

19. Schuerenberg M, Luebbert C, Deininger S-O, Ketterlinus R, Detlev, Suckau D. 2007. MALDI tissue imaging: mass spectrometric localization of biomarkers in tissue slices. *Nat. Methods* 4:iii

20. Nye GJ, Norris JL, Nickerson S. 2006. *Optimization of the performance of automated matrix spraying for MALDI imaging mass spectrometry.* Presented at Annu. Am. Soc. Mass Spectrom., 54th, Seattle, WA

21. Mamyrin BA. 1994. Laser-assisted reflectron time-of-flight mass spectrometry. *Int. J. Mass Spectrom. Ion Process.* 131:1–19

22. Khatib-Shahidi S, Andersson M, Herman JL, Gillespie TA, Caprioli RM. 2006. Direct molecular analysis of whole-body animal tissue sections by imaging MALDI mass spectrometry. *Anal. Chem.* 78:6448–56

23. Li Y, Shrestha B, Vertes A. 2007. Atmospheric pressure molecular imaging by infrared MALDI mass spectrometry. *Anal. Chem.* 79:523–32

24. McLean JA, Ridenour WB, Caprioli RM. 2007. Profiling and imaging of tissues by imaging ion mobility-mass spectrometry. *J. Mass Spectrom.* 42:1099–105

25. Taban IM, Altelaar AFM, Van der Burgt YEM, McDonnell LA, Heeren RMA, et al. 2007. Imaging of peptides in the rat brain using MALDI-FTICR mass spectrometry. *J. Am. Soc. Mass Spectrom.* 18:145–51

26. Altelaar AFM, Taban IM, McDonnell LA, Verhaert P, de Lange RPJ, et al. 2007. High-resolution MALDI imaging mass spectrometry allows localization of peptide distributions at cellular length scales in pituitary tissue sections. *Int. J. Mass Spectrom.* 260:203–11

27. Verhaert PD, Conaway MCP, Pekar TM, Miller K. 2007. Neuropeptide imaging on an LTQ with vMALDI source: the complete 'all-in-one' peptidome analysis. *Int. J. Mass Spectrom.* 260:177–84

28. DeKeyser SS, Kutz-Naber KK, Schmidt JJ, Barrett-Wilt GA, Li LJ. 2007. Imaging mass spectrometry of neuropeptides in decapod crustacean neuronal tissues. *J. Proteome Res.* 6:1782–91

29. Norris JL, Cornett DS, Mobley JA, Andersson M, Seeley EH, et al. 2007. Processing MALDI mass spectra to improve mass spectral direct tissue analysis. *Int. J. Mass Spectrom.* 260:212–21

30. Jolliffe IT. 2002. *Principal Component Analysis.* New York: Springer-Verlag

31. Tusher VG, Tibshirani R, Chu G. 2001. Significance analysis of microarrays applied to the ionizing radiation response. *Proc. Natl. Acad. Sci.* 98:5116–21

32. Meunier B, Dumas E, Piec I, Bechet D, Hebraud M, Hocquette J-F. 2006. Assessment of hierarchical clustering methodologies for proteomic data mining. *J. Proteome Res.* 6:358–66

33. Hanahan D, Weinberg RA. 2000. The hallmarks of cancer. *Cell* 100:57–70

34. Stoeckli M, Chaurand P, Hallahan DE, Caprioli RM. 2001. Imaging mass spectrometry: a new technology for the analysis of protein expression in mammalian tissues. *Nat. Med.* 7:493–96

35. Reyzer ML, Caldwell RL, Dugger TC, Forbes JT, Ritter CA, et al. 2004. Early changes in protein expression detected by mass spectrometry predict tumor response to molecular therapeutics. *Cancer Res.* 64:9093–100

36. Chaurand P, Sanders ME, Jensen RA, Caprioli RM. 2004. Proteomics in diagnostic pathology: profiling and imaging proteins directly in tissue sections. *Am. J. Pathol.* 165:1057–68

37. Schwamborn K, Krieg RC, Reska M, Jakse G, Knuechel R, Wellmann A. 2007. Identifying prostate carcinoma by MALDI-imaging. *Int. J. Mol. Med.* 20:155–59

38. Crecelius AC, Cornett DS, Caprioli RM, Williams B, Dawant BM, Bodenheimer B. 2005. Three-dimensional visualization of protein expression in mouse brain structures using imaging mass spectrometry. *J. Am. Soc. Mass Spectrom.* 16:1093–99

39. Andersson M, Groseclose MR, Deutch AY, Caprioli RM. 2008. Imaging mass spectrometry of proteins and peptides: 3D volume reconstruction. *Nature Methods* 5:101–8

40. Haacke EM, Brown RW, Thompson ML, Venkatesan R. 1999. *Magnetic Resonance Imaging: Physical Principles and Sequence Design.* New York: Wiley

41. Sinha TK, Khatib-Shahidi S, Yankeelov TE, Mapara K, Ehtesham M, et al. 2008. Integrating spatially resolved three-dimensional MALDI IMS with in vivo magnetic resonance imaging. *Nature Methods* 5:57–59

42. Ingelsson M, Fukumoto H, Newell KL, Growdon JH, Hedley-Whyte ET, et al. 2004. Early Aβ accumulation and progressive synaptic loss, gliosis, and tangle formation in AD brain. *Neurology* 62:925–31

43. Schonheit B, Zarski R, Ohm TG. 2004. Spatial and temporal relationships between plaques and tangles in Alzheimer pathology. *Neurobiol. Aging* 25:697–711

44. Rohner TC, Staab D, Stoeckli M. 2005. MALDI mass spectrometric imaging of biological tissue sections. *Mech. Ageing Dev.* 126:177–85

45. Skold K, Svensson M, Nilsson A, Zhang XQ, Nydahl K, et al. 2006. Decreased striatal levels of PEP-19 following MPTP lesion in the mouse. *J. Proteome Res.* 5:262–69

46. Chaurand P, Fouchécourt S, DaGue BB, Xu BJ, Reyzer ML, et al. 2003. Profiling and imaging proteins in the mouse epididymis by imaging mass spectrometry. *Proteomics* 3:2221–39

47. Walgren JL, Thompson DC. 2004. Application of proteomic technologies in the drug development process. *Toxicol. Lett.* 149:377–85

48. Rudin M, Rausch M, Stoeckli M. 2005. Molecular imaging in drug discovery and development: potential and limitations of nonnuclear methods. *Mol. Imag. Biol.* 7:5–13

49. Hsieh Y, Chen J, Korfmacher WA. 2007. Mapping pharmaceuticals in tissues using MALDI imaging mass spectrometry. *J. Pharmacol. Toxicol. Methods* 55:193–200

50. Meistermann H, Norris JL, Aerni H-R, Cornett DS, Friedlein A, et al. 2006. Biomarker discovery by imaging mass spectrometry: Transthyretin is a biomarker for gentamicin-induced nephrotoxicity in rat. *Mol. Cell Proteomics* M500399–200

51. Lemaire R, Desmons A, Tabet JC, Day R, Salzet M, Fournier I. 2007. Direct analysis and MALDI imaging of formalin-fixed, paraffin-embedded tissue sections. *J. Proteome Res.* 6:1295–305

Formation and Characterization of Organic Monolayers on Semiconductor Surfaces

Robert J. Hamers

Department of Chemistry, University of Wisconsin at Madison, Madison, Wisconsin 53706; email: rjhamers@wisc.edu

Annu. Rev. Anal. Chem. 2008. 1:707–36

First published online as a Review in Advance on March 21, 2008

The *Annual Review of Analytical Chemistry* is online at anchem.annualreviews.org

This article's doi: 10.1146/annurev.anchem.1.031207.112916

1936-1327/08/0719-0707$20.00

Key Words

silicon, biosensors, interfaces, microelectronics, infrared spectroscopy, scanning tunneling microscopy (STM)

Abstract

Organic-semiconductor interfaces are playing increasingly important roles in fields ranging from electronics to nanotechnology to biosensing. The continuing decrease in microelectronic device feature sizes is raising an especially great interest in understanding how to integrate molecular systems with conventional, inorganic microelectronic materials, particularly silicon. The explosion of interest in the biological sciences has provided further impetus for learning how to integrate biological molecules and systems with microelectronics to form true bioelectronic systems. Organic monolayers present an excellent opportunity for surmounting many of the practical barriers that have hindered the full integration of microelectronics technology with organic and biological systems. Of all the semiconductor materials, silicon and diamond stand out as unique. This review focuses upon the preparation and characterization of organic and biomolecular layers on semiconductor surfaces, with special emphasis on monolayers formed on silicon and diamond.

1. INTRODUCTION

Organic-semiconductor interfaces are playing increasingly important roles in fields ranging from electronics to nanotechnology to biosensing (1–5). The continuing decrease in microelectronic device feature sizes is raising especially great interest in understanding how to integrate molecular systems with conventional, inorganic microelectronic materials, particularly silicon. The explosion of interest in the biological sciences has provided further impetus for learning how to integrate biological molecules and systems with microelectronics to form true bioelectronic systems. Yet, practical barriers to such an integration remain due to the inherent instability of silicon and most other semiconductors in aqueous environments.

Organic monolayers present an excellent opportunity for surmounting many of the practical barriers that have hindered the full integration of microelectronics technology with organic and biological systems. Monolayers comprised simply of alkyl chains can act as nearly impervious barriers to aqueous media due to their hydrophobic nature. Thin monolayers can act either as electrically resistive layers or as conductive layers, depending on the nature of the bonding within the layers.

Of all the semiconductor materials, silicon and diamond stand out as unique; monolayers on these materials are consequently the most widely studied and therefore constitute the primary emphasis of this review. Silicon is of paramount importance because it is the foundation of the microelectronics field and is likely to remain so for the foreseeable future. Diamond is an especially appealing material because in addition to being the hardest substance known, it is also remarkably stable in harsh chemical and electrochemical environments; this makes it a nearly ideal material for applications involving exposure to wet environments, especially those involving biological biomolecules. Here, I present a selective review of the formation and characterization of organic monolayers, with an emphasis on monolayers formed on silicon and diamond.

2. FORMATION OF ORGANIC MONOLAYERS ON SEMICONDUCTOR SURFACES

The formation of well-defined organic layers on silicon and other semiconductors typically follows one of two routes. Modification by adsorption in vacuum most commonly begins with "clean" surfaces that are often prepared by annealing under ultrahigh vacuum conditions. The formation of a free surface leaves the surface atoms with low coordination numbers and typically leads to unsaturated "dangling bonds" at the interface. The semiconductor surfaces undergo atomic rearrangements (also called reconstructions) to reduce the number of exposed dangling bonds, but in most cases the resulting surfaces remain highly reactive. Grafting of molecular layers under ambient conditions typically involves first passivating the surfaces via either wet-chemical or plasma methods, which produces surfaces in which the unsaturated dangling bonds are terminated with hydrogen, chlorine, or another substituent that reduces the reactivity to facilitate handling in ambient conditions. This process must then be followed by reaction with the organic molecule of interest.

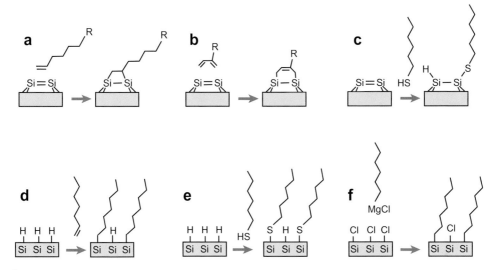

Figure 1

Commonly used methods for forming organic monolayers on semiconductor surfaces. Panels *a–c* depict methods used in ultrahigh vacuum studies, where the starting surfaces are clean. Panels *d–f* depict methods used in ambient conditions on hydrogen-passivated or halogen-passivated surfaces.

Figure 1 presents some of the most commonly used methods for forming organic monolayers. **Figure 1a–c** depicts methods used in ultrahigh vacuum studies, where the starting surfaces are clean. The precise arrangements of atoms depend on the semiconductor and the specific crystal face; those shown in **Figure 1a–c** are appropriate for the (001) surfaces of silicon, diamond, and germanium, which share a common bonding motif. **Figure 1d–f** depicts methods used in ambient conditions on hydrogen-passivated or halogen-passivated surfaces; most of these studies have been performed using silicon, although recent studies have extended these methods to diamond and other semiconductors.

2.1. Formation of Semiconductor-Organic Monolayers in Vacuo

In ultrahigh vacuum, the most widely studied surface has been the (001) crystal face of silicon, as it represents the starting point for virtually all silicon-based microelectronic devices. The (001) surfaces of silicon, germanium, and the diamond allotrope of carbon all undergo surface reconstructions in which adjacent atoms move toward one another, forming dimers (6–8). Because the bulk-truncated surfaces have two broken bonds per atom, the formation of the dimers is, to a first approximation, a rehybridization of these two dangling bonds to allow formation of a Si=Si, C=C, or Ge=Ge double bond at the surface. Although this description is vastly oversimplified, it highlights the fact that there are important conceptual connections between these surfaces and the more common C=C, Si=Si, and Ge=Ge bonds of organic and organometallic compounds (9, 10). Of particular interest has been the desire to use bond-forming reactions, such as the well-known cycloaddition reactions of organic chemistry, in order to facilitate a wider array of surface modifications (11–39).

The reaction between organic alkenes and the (001) face of silicon, germanium, and diamond appears similar to a [2+2] cycloaddition reaction, in which two electrons from the C=C π bonds of the reactant molecules and two electrons from the π bond of the surface dimers link together in a pericyclic manner, thereby linking each molecule to the surface with two bonds and forming a four-member ring at the interface (**Figure 1***a*) (15). A second type of reaction, technically equivalent to a [4+2] Diels-Alder reaction, can also take place if there are conjugated double bonds present in the reactant molecule. For example, four electrons of the molecule and two electrons of the substrate interact to form a six-member ring at the interface (shown in **Figure 1***b*) (21–24, 30, 40).

The [2+2], [4+2], and other cycloaddition reactions represent important pathways for organic monolayer formation on silicon and other semiconductors. An important practical feature of this approach is the fact that these are pure addition reactions, so there is no cleavage of C–H or other bonds nor any byproducts other than the simple coupling. This feature makes these reactions ideally suited to formation of well-defined monolayers at room temperature and below. Because the (001) surfaces of Si, Ge, and C (diamond) consist of dimers, the formation of two covalent bonds between each molecule and the underlying surface also implies that when the molecules are linked to the surface, the orientation of the underlying dimers is translated into the orientation of the upper molecular layer. By choosing small cyclic molecules that are highly symmetric and relatively rigid, it is possible to fabricate molecular layers that are essentially epitaxial with the underlying silicon substrate (19, 41).

One interesting aspect of these cycloaddition reactions has been gaining an understanding of the reaction mechanisms, as well as identifying to what extent the well-known Woodward-Hoffman symmetry rules can apply at a surface (42). For typical alkenes, the reaction between two alkenes is forbidden in the highest-symmetry geometry, in which one molecular approaches another from the top. However, the C=C, Si=Si, and Ge=Ge dimers that form on silicon are distorted away from planar configurations. These distorted configurations lead to increased radical character and also facilitate low-symmetry reaction pathways that are not subject to Woodward-Hoffman symmetry rules (14, 21, 30). The [2+2] reactions of simple alkenes with Si=Si and Ge=Ge dimers occur with a reaction probability close to unity (15, 18–20, 39, 43). The reactions on diamond are less favorable, requiring significantly greater exposure (17). Nevertheless, reactions do occur on all three surfaces.

Another class of reaction, the reaction of a 1,3-diene with these surfaces, has also been investigated. This reaction is formally equivalent to a concerted [4+2] (Diels-Alder) reaction. Although normal alkenes react easily with 1,3-dienes but not with other simple alkenes (due to the aforementioned symmetry constraints), it was originally expected that the Si, Ge, and diamond surfaces would exhibit a similar preference (23, 40). Because the surface distortions render the [2+2] reaction (involving a single C=C bond reacting with the surface) unexpectedly facile, in practice the reaction of 1,3-dienes with these surfaces leads to multiple products that include both [2+2] reaction products (i.e., reaction via a single C=C bonds) and reaction via the [4+2] route (21, 30, 44). Consequently, it is now recognized that the formation of well-organized monolayers is generally more successful when using simple alkenes, except perhaps in the case of diamond (25, 45, 46, 99). Whereas most studies have focused on linking simple hydrocarbons to the surface, the ideas above have been extended to include other types of unsaturated bonds, including carbonyl groups (47).

The formation of molecular layers incorporating conjugated π–electron structures has also been of great interest, in part because electron transfer through π-conjugated systems is expected to be more facile than through saturated organic systems (3). Benzene itself bonds to the surface via a number of different configurations (48–51). Simple substitution to form toluene and/or xylenes leads to further disruption of the structure (11). However, if there is a secondary reactive group present, in most cases the organic monolayers use these groups to bond to the surface, leaving the aromatic character intact. The most notable example is styrene, which reacts almost completely via the C=C group external to the aromatic ring, leaving the π conjugation within the ring intact. Other variants, such as phenyl isothiocyanate (12) and benzonitrile (52–54) also preferentially link through the external substituent.

On the basis of these studies, a number of generalizations can be made. The first is that the reactivity of clean silicon and germanium surfaces in vacuum is sufficiently high that almost any molecule within a reactive group, including simple C=C bonds, will react with the surfaces. Thus, the key question in forming well-defined mono-layers becomes one of controlling the overall selectivity of reaction. Because they have only one reactive group, simple alkenes provide the best-defined layers. Dienes and other molecules bearing more than one unsaturated group generally leave poorly defined layers, due to the competition between [4+2] and [2+2] reactions. How-ever, molecular-bearing aromatic groups have a much higher propensity to adopt a well-defined structure that preserves the aromaticity.

A second class of reactions in vacuo involves simple functional groups, such as the thio, amino, and hydroxy groups, which have labile hydrogen atoms that can be easily removed, allowing the S, N, or O atom to link to the semiconductor. Following extensive studies on self-assembled monolayers on metals, many experiments have focused on thiol groups interacting with silicon (55) and other semiconductors such as GaN (56), InP (57, 58), and GaAs (59–62). In each case, bonding to the surface necessitates cleavage of an S–H, N–H, or O–H bond, with the hydrogen remaining bound to the surface. This approach can also be used to link aromatic systems to semiconductor surfaces using reactants such as benzenethiol (55), phenol (63), and aniline (64, 65).

Although most investigations of monolayer formation in vacuo started with highly reactive clean surfaces, a number of studies have investigated reactions with hydrogen-terminated surfaces, in which the undercoordinated surface bonds are bonded to H atoms. These H-passivated surfaces are much less reactive than the clean surfaces; consequently, most molecules will react selectively where the H-passivation has been eliminated. Local desorption of hydrogen can be achieved by low-energy electrons, such as those emitted from the tip of a scanning tunneling microscope. The initiation of a single reaction leads to a propagation of the reaction: As the reaction proceeds via a radical intermediate in which a molecular radical state is able to abstract an H atom from the adjacent dimer, its reactivity toward subsequent molecules is thereby greatly enhanced. This suggests the possibility of making a variety of extended molecular structures (66).

Studies of monolayer formation on Si(111)-(7 × 7) have also been conducted. On Si(111), the large amount of charge transfer that occurs between different atoms within the unit cells allows reactions to proceed via a diradical pathway (67).

2.2. Formation of Semiconductor-Organic Monolayers under Ambient Conditions

Organic monolayers on semiconductor surfaces can also be made under ambient conditions via a wide range of methods (shown in **Figure 1d–f**). In order for the monolayers to form, it is usually necessary to remove oxide layers and passivate the surface against additional reaction. This is typically achieved using simple monatomic reagents to coordinately saturate all surface bonds. The most common passivating agent is hydrogen (68), but other halogens including chlorine (69, 70) and iodine (71) have been used. The goal of the passivation step is to provide all surface atoms with a nearly ideal coordination, thereby reducing their reactivity toward the ambient atmosphere. Hydrogen is a common passivating agent because H-terminated surfaces of silicon exhibit very low densities of mid-gap surface electronic states (68) and because silicon samples, when immersed in a solution of HF or NH_4F, become H-terminated. HF is excellent at removing surface oxides but does not etch the underlying silicon very quickly. NH_4F etches the silicon as well, but is highly anisotropic. Typically, dilute (~2%–10%) HF is used to passivate the Si(001) surface and ~40% NH_4F is used on the (111) face (72). Most ambient-condition studies using silicon use the (111) face because the anisotropic nature of the NH_4F leads to a spontaneous smoothing of the Si(111) surface, but leads to roughening of the (001) surface (73–75).

Once the starting surface is obtained, reaction with alkenes and other unsaturated compounds can occur as depicted in **Figure 1d–f**; however, reactions require some form of activation. The reactions of Si–H bonds with alkenes are often analogous to classical hydrosilylation reactions, in which Lewis acids such as $EtAlCl_2$ are able to enhance the reactions of H-terminated silicon with organic alkenes and alkynes (76, 77).

The reaction of H-terminated surfaces with simple alkenes can be initiated by diacyl peroxides, which decompose when heated to form radicals that can abstract hydrogen atoms from the surface (78, 79), thereby exposing coordinatively unsaturated "dangling bonds." Similarly, Bowden and colleagues used 4-(decanoate)-2,2,6,6-tetramethylpiperidinooxy (TEMPO) as a radical initiator to link unprotected carboxylic acids to the Si(111) surfaces (80). As an alternative to chemical activation, thermal activation has been successfully used, typically by immersing the H-terminated Si surfaces in the alkenes of interest and heating them for extended periods (hours to days) at ~200°C (72, 78, 79, 81). Recently, there has been great interest in the use of photochemical activation, initiated by UV light at wavelengths of ~254 nm (82–86). The use of light is an attractive option because it provides a way to pattern the surfaces with different molecular groups to form molecular arrays. Electrografting can also be used to link alkenes and alkynes to H-terminated Si surfaces, although it appears that there are important mechanistic differences between these (87, 88). These studies have demonstrated that there are multiple ways to achieve the reaction of alkenes with silicon and diamond surfaces.

In addition to alkenes, a number of other functional groups including thiols, alcohols, and aldehydes react directly with the H-terminated surface with suitable activation, forming Si–S–C Si–O–C linkages and/or siloxane esters (89–91). Although most early studies of the functionalization of silicon focused on forming surfaces terminated with alkyl chains, the practical utility of monolayers on silicon often requires the ability to form a molecular layer containing well-defined functional groups at the distal end. In most cases involving two reactive groups, such as a terminal alkene with an alcohol or carboxylic acid group at the distal end, the resulting layers are typically disordered and consist of a mixture of both groups. Consequently, to achieve well-defined functional monolayers it is necessary to identify chemical groups that are unreactive toward the surface during the grafting process, but which can be deprotected after grafting to expose synthetically useful reactive groups (72). The choice of protecting group and deprotection conditions often can be based on known wet-chemical methods (92).

Sieval et al. were the first to successfully demonstrate this approach. They functionalized silicon with olefins bearing ester groups using thermal activation, and showed that they could be reduced to form alcohol-modified surfaces or oxidized to carboxylic acids (72). Boukherroub et al. showed that photochemical reactions with ethyl undecylenate with Si(111) led to ester-modified surfaces that could subsequently be converted to produce carboxylic acid–modified surfaces (93). Strother and colleagues used a similar process to link esters to the surface (83); they showed that by deprotecting the samples and using standard biochemical methods, it was possible to prepare well-defined DNA-modified surfaces exhibiting good selectivity toward binding of complementary versus noncomplementary sequences in solution, as well as good stability.

Whereas H-terminated silicon surfaces are relatively unreactive, Cl-terminated surfaces are more reactive and therefore more amenable to additional methods of modification. Bansal and Lewis (69, 70) formed organic monolayers on Cl-terminated silicon surfaces followed by reactions with Grignard reagents and with organolithium compounds (shown in **Figure 1f**). Use of this method has yielded layers exhibiting excellent passivation properties (69).

3. CHARACTERIZATION OF ORGANIC MONOLAYERS ON SEMICONDUCTORS

The characterization of organic monolayers typically involves a combination of multiple techniques. For chemical analysis, the most commonly used methods are Fourier transform infrared spectroscopy (FTIR) and X-ray photoelectron spectroscopy (XPS). These techniques complement one another in terms of the types of information they provide. Although the frequencies and intensities of the vibrational modes observed in FTIR provide detailed information about the nature of the chemical functional groups, FTIR spectra are difficult to relate to absolute numbers of molecules and functional groups because of the absence of reliable standards for surface measurements and because the electromagnetic boundary conditions, quantum-mechanical selection rules, and coupling between molecules at semiconductor interfaces are complicated. XPS provides excellent information about the elements and their oxidation states and is generally more useful than FTIR for quantitative analysis because the selection rules for excitation of core levels are rather straightforward and because standards are readily available.

For spatial information, scanned-probe methods such as scanning tunneling microscopy (STM) provide state-of-the-art spatial resolution, but work well only on very flat surfaces. STM in particular can be applied only to organic layers that are sufficiently conductive for electrons to tunnel through. Recent studies show that scanning electron microscopy (SEM) can provide useful information about molecular layers on submicron length scales; such data are interesting in part because the SEM imaging process involves transmission of low-energy electrons through the layers.

3.1. Scanning Tunneling Microscopy

STM remains one of the primary tools for obtaining direct, atomic-level information about bonding at organic-semiconductor interfaces. Using STM as a tool for surface analysis, several types of information can be obtained, such as the location of molecules on the surface, the number and atomic geometry of different bonding configurations, and the electronic structure of individual adsorbed molecules. However, because STM images represent a convolution of the electronic density of states and the local barrier to electron tunneling, it is often difficult to extract sufficiently quantitative information. Nevertheless, STM is unparalleled in its ability to observe and characterize the structure and electronic properties of individual molecules and molecular layers.

STM has been widely applied to the investigation of molecular layers (15, 16, 18–21, 34, 39, 41, 94). In cases where the molecules are clearly asymmetric in shape, the STM images directly reflect the molecular shape, thereby providing useful information about the bonding locations on the surface. For example, **Figure 2b,c** shows high-resolution STM images of the molecule 1,5-cyclooctadiene (COD) bonded to the Si(001) surface via a [2+2] cycloaddition reaction (19), yielding an ordered, epitaxial layer of COD molecules. Infrared measurements show that only one C=C bond is involved in bonding, leaving a structure similar to that shown in profile in **Figure 2a**. In the high-resolution STM images, the ordering of the individual molecules into rows (commensurate with the underlying lattice of Si=Si dimers) can be clearly seen. At the highest resolution (**Figure 2c**), each molecule is shown to consist of two clear lobes of higher intensity; thus, the STM images can even provide direct information about the intramolecular structure.

At low coverage, the STM is able to provide a measure of the apparent height of individual molecules. The apparent height involves a convolution of geometry and electronic structure, but is of interest for understanding charge transport

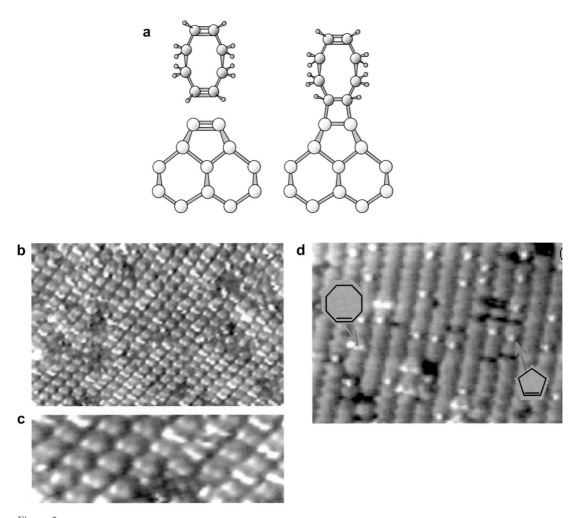

Figure 2

(*a*) Side-view representation of the bonding of 1,5-cyclooctadiene with the Si(001) surface.
(*b, c*) High-resolution scanning tunneling microscopy (STM) images of the resulting molecular
layers. (*d*) High-resolution STM images showing two different molecules, cyclopentene
and cyclooctene, bonded to the Si(001) surface. The STM images reveal the locations
of bonding with respect to the underlying dimer rows, and also show that the cyclopentene
and cyclooctene molecules have different shapes. Reproduced with permission from
Reference 41.

through molecules. Images of surfaces containing two different molecules provide
information about the relative conductivity of the different molecules. For example,
Figure 2*d* shows a Si(001) surface that was exposed to a mixture of cyclopentene and
cyclooctene (41). The STM image clearly shows molecules with two distinct shapes:
Some molecules appear as individual bright spots centered along the rows of underlying Si=Si dimers (which are barely resolved in this image), whereas the others appear

as closely spaced pairs of lobes. By varying the amount of each of the two molecules in the dosing gas, it is easy to identify each molecule in order to confirm their identities. Thus, STM clearly has the ability to distinguish between different molecular species on a surface.

The ability to identify molecules on an a priori basis is also possible in principle. However, the images from the STM reflect a somewhat complex convolution of information about the electronic density of states of the sample and tip, combined with an understanding of the electronic barrier between them (48). Continued interest in understanding charge transport through individual molecules and the fabrication of single-molecule devices (95) has led to further efforts to develop practical theoretical formalisms for predicting the apparent shapes, heights, and other features of molecules as viewed via STM. Much of the original work in the field applied the Tersoff-Hamann theory (96, 97) which is a derivative of the broader Bardeen theory of electron tunneling. In the Tersoff-Hamann theory, the tunneling current is controlled by the density of states and by the probability of electron tunneling, which in turn is controlled primarily by the sample-tip separation and by the local barrier height (work function). However, there are a number of approximations (nearly planar surface, spherical tip) that render this theory inappropriate for understanding images of single molecules, especially on semiconducting surfaces and at higher bias. Ratner and colleagues have been particularly successful using scattering methods incorporating Green's functions (98, 99).

Measurements of the current as a function of the applied voltage, commonly referred to as tunneling spectroscopy (100, 101), can be used to probe the electronic properties of individual molecules (101, 102). This approach has been widely used to characterize the current-voltage characteristics of molecules and molecular layers. Of particular interest has been the observation of negative differential resistance over certain types of molecular structures (103, 104). Tunneling spectroscopy measurements (measurements of current as a function of voltage at specific locations) are complementary to voltage-dependent STM images (measurements of the variation in apparent height at different applied voltages); both provide information on the electrical characteristics as a function of space and applied voltage.

3.2. X-Ray Photoelectron Spectroscopy

XPS is one of the principal methods used to obtain information about the chemical composition of surfaces and monolayers (105). A significant advantage of XPS over almost all other surface analytical methods is that XPS can be performed in a quantitative manner. Detailed analysis of chemical shifts can provide additional insights into the effective oxidation states of surface species. More importantly, the X-ray absorption and photoelectron emission cross sections of core-level transitions are not very sensitive to the local chemical environment; consequently, it is possible to tabulate the relative sensitivity factors for various elements and to use these data to relate the observed intensities to actual concentrations in a relatively straightforward manner.

XPS at low-energy resolution provides an elemental analysis of the surface species. At higher resolution, additional chemical information is obtained through the

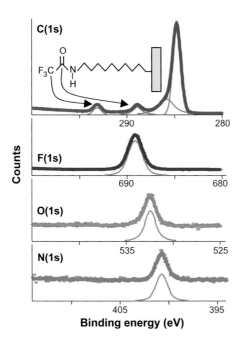

Figure 3

X-ray photoelectron spectroscopy (XPS) spectra of a monolayer of trifluoroacetamide-protected aminodecene covalently bonded to a diamond surface. The different oxidation states of carbon in the $-CF_3$ and carbonyl group can be easily resolved. The sharpness of the peaks for the other elements confirms the chemical homogeneity of the monolayer.

core-level shift. Surface species undergo shifts in their core-level binding energies due to the presence of electron-withdrawing or electron-donating groups. The bonding of electron-withdrawing groups increases the XPS binding energies, typically by $\sim$1–2 eV per unit change in the formal oxidation state. **Figure 3**, for example, shows an XPS spectrum of a monolayer of trifluoroacetamide-protected aminodecene. In this case, bonding of the molecule to the surface leads to at least four distinct types of carbon. The strong electron-withdrawing nature of the F atoms in the CF_3 group increases the binding energy from $\sim$285 eV (characteristic of the alkyl chain and the underlying diamond substrate) to 293 eV, as the formal oxidation state of the C atoms changes from 0 to +3.

The use of quantitative XPS measurements to characterize monolayers on semiconductor surfaces is relatively straightforward. By measuring the ratio of a core-level characteristic of the organic monolayer [such as the C(1s) layer] and another core-level characteristic of the underlying bulk [such as Si(2p)], the number of molecules per unit area can be obtained. The electron escape depths in XPS are typically $\sim$2 nm; this thickness is comparable to that of most molecular monolayers such that the inelastic scattering of the electrons must be accounted for in order to perform quantitative analysis. For a uniform thin film, the electrons from the bulk are attenuated exponentially with distance, so that the integrated intensity of photoemitted electrons from the bulk is given by:

$$A_B = \int_t^\infty \sigma_B \rho_B \exp^{\frac{-(z-t)}{\lambda_{B,B}\sin\theta}} \exp^{\frac{-t}{\lambda_{B,L}\sin\theta}} dz = \sigma_B \rho_B \lambda_{B,B} \sin\theta \exp^{\frac{-t_L}{\lambda_{B,L}\sin\theta}}.$$

The corresponding intensity from the molecular layer is:

$$A_L = \int_0^t \sigma_L \rho_L \exp^{\frac{-t}{\lambda_{L,L}\sin\theta}} dz = \sigma_L \lambda_{L,L} \rho_L \sin\theta \left(1 - \exp^{\frac{-t_L}{\lambda_{L,L}\sin\theta}}\right).$$

The observed ratio of intensities is then given by:

$$\frac{A_L}{A_B} = \frac{\rho_L}{\rho_B} \frac{\lambda_{L,L}}{\lambda_{B,B}} \frac{\sigma_L}{\sigma_B} \frac{\left(1 - \exp^{\frac{-t_L}{\lambda_{L,L}\sin\theta}}\right)}{\exp^{\frac{-t_L}{\lambda_{B,L}\sin\theta}}}.$$

In these equations, $\lambda_{B,B}$, $\lambda_{B,L}$, and $\lambda_{L,B}$ are the inelastic mean free paths of electrons of the bulk phase passing through the bulk [i.e., Si(2p) photoelectrons in silicon], electrons of the bulk phase passing through the layer [i.e., Si(2p) photoelectrons in the organic layer], and electrons of the monolayer phase passing through the monolayer [i.e., C(1s) electrons in the organic film], respectively. σ_B and σ_L are the photoelectron cross sections (often represented as sensitivity factors), ρ_B and ρ_L are the volume densities of the bulk and surface, and A_B and A_L are the measured peak areas. Constant factors such as the incident excitation intensity are omitted.

Perhaps the biggest challenge to quantitative analysis via this method is the uncertainty in the effective scattering probability within the organic film. In some cases, the scattering probability has been measured directly by making two samples and directly comparing the intensity of emission of the bulk peaks on the "bare" and functionalized samples (105).

3.3. Valence-Band Photoemission and Inverse Photoemission Electron Spectroscopy

Although XPS provides information about the elemental composition of the sample and effective oxidation state within molecular layers, measurements of valence electronic states are more typically performed using ultraviolet photoemission spectroscopy (UPS), also often referred to as valence-band photoemission. Functionally, this method involves replacing the X-ray source of XPS with a UV excitation source, typically a He-discharge lamp or, for higher resolution and count rates, a synchrotron source. UPS has higher energy resolution than XPS and is better suited to analysis of the valence bands. UPS can very easily determine the position of the valence-band edge with respect to the Fermi energy and can measure the work function. Its complementary technique, inverse photoemission electron spectroscopy (IPES), involves sending an electron of known energy onto the sample and measuring the intensity of emitted photons. IPES yields information about the empty electronic states, such as the semiconductor conduction band and, more importantly, the unfilled states associated with the surface electronic states. **Figure 4a** depicts the energy schemes for both UPS and IPES measurements.

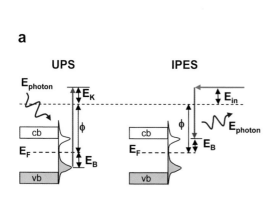

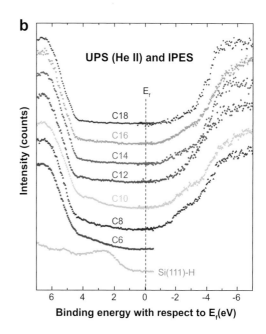

Figure 4

(*a*) Energy level schematic for ultraviolet photoemission spectroscopy (UPS) and inverse photoemission electron spectroscopy (IPES). (*b*) UPS and IPES data for alkyl chains of different lengths bonded to Si(001) surface.

Recent experiments by Segev et al. investigated both the filled and empty valence orbitals and have compared them using theoretical calculations (106). An important consequence of these studies is the first identification of electronic state within the bandgap induced by the interface arising from hybridization of the electronic states of the silicon and the organic layer. **Figure 4b** shows UPS and IPES data for monolayers of different lengths; the significant intensity within the bulk bandgap of the semiconductor reveals the presence of new electronic states associated with the Si-C interface. Such states are important in controlling charge transport in molecular systems.

3.4. Infrared Spectroscopy of Organic Monolayers

Infrared spectroscopy has been widely used to characterize the properties of organic monolayers at semiconductor surfaces. Measurements can be performed in a number of experimental geometries, including single-bounce external reflection and multiple internal reflection (MIR) geometries (107). Many aspects of the theory of infrared spectroscopy at surfaces have been reviewed by Chabal (107). One important aspect of infrared spectroscopy at semiconductor surfaces is that the selection rules are quite different from metal surfaces. The boundary conditions are different for s-polarized light (which has the electric field perpendicular to the plane formed by the incident and outgoing beams, and therefore also has the electric field parallel to the surface) and p-polarized light (which has the electric field in the plane formed by the incident and outgoing beams). For metal surfaces, the electromagnetic boundary conditions cause a cancellation effect for s-polarized light and an enhancement for p-polarized light. Hence, techniques such as polarization-modulation infrared spectroscopy can be used to enhance the sensitivity at metal surfaces and to provide a well-defined background for the infrared spectroscopy measurements. At semiconductor surfaces, however, the dielectric nature of the materials leads to quite distinct selection rules. Because both s- and p-polarized light can interact with molecular species, measurements using polarized light can provide useful information about the orientation of molecules, both along the surface normal and within the sample plane.

Information may be obtained from the frequencies, intensities, and polarization dependences of the spectra. At submonolayer coverages, FTIR is particularly useful for helping to identify molecular bonding configurations. For example, numerous studies have investigated the interaction of organic alkenes with silicon surfaces. Because the C–H stretching modes of alkene are higher than those of saturated alkanes, the presence or absence of the alkene stretch can be used to determine whether or not the molecules bind to the surface via the alkene group (15).

FTIR can also be used to identify molecular orientations on the surface. To gain insight into the azimuthal orientation of molecular functional groups, a substrate with a uniform orientation must be used. The (001) surface of silicon consists of Si=Si dimer units whose orientation rotates by 90° across each atomic step; consequently, most surfaces consist of an equal mixture of dimers in each of two configurations. However, by using a sample that is intentionally miscut by ~4° from the (001) plane toward the <110> direction, it is possible to prepare a single-domain surface on which the Si=Si dimers are aligned with their axes all in the same direction (108). Consequently, any orientation dependence in the interaction between the surface and the organic layer can be templated into the orientation of the molecules; in essence, the surface becomes a way of holding molecules in precise orientations, allowing their spectroscopic properties to be probed by controlling the polarization of light (15). Whereas this technique was first used to investigate the interaction of cyclopentene with the Si(001) surface (15), a particularly dramatic example was obtained through the interaction of acetonitrile (CH_3CN) bonded to the Si(001) surface. In this case the interaction with the Si(001) surface produced a C=N bond oriented parallel to the Si=Si dimers. **Figure 5b** shows FTIR spectra measured using s-polarized and p-polarized light; very high strong absorbance for the s-polarized light, but not for p-polarized light, demonstrates that the C=N bond is oriented parallel to the Si=Si dimers. It is interesting to note that in this example, the most intense absorption occurred using s-polarized light, which on metallic substrates undergoes a cancellation at the interface due to the image field induced within the metal. This unexpected result illustrates how different the selection rules at semiconductor surfaces are from those at metal surfaces due to the differences in dielectric response of the substrates.

For longer-chain molecules, information on the tilt angle can be obtained from measurements of the reflectivity using s-polarized and p-polarized light. The angle of the infrared transition dipole α can be obtained from the absorbances of s- and

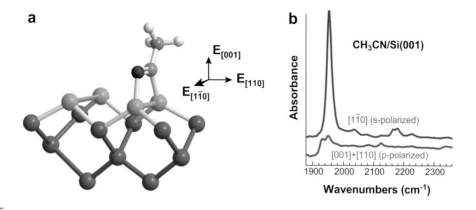

Figure 5

(*a*) Schematic drawing showing the bonding configuration produced by a [2+2] cycloaddition reaction of acetonitrile (CH$_3$CN) onto the Si(001) surface. (*b*) Polarized Fourier transform infrared spectroscopy (FTIR) data showing a strong absorbance using s-polarized light with the electric field oriented parallel to the C=N bond on the surface.

p-polarized light (A$_s$ and A$_p$, respectively) via $\alpha_{TD} = \tan^{-1}\sqrt{\left(\dfrac{2\frac{A_s}{A_p}E_z^2}{E_y^2 - \frac{A_s}{A_p}E_x^2}\right)}$, where E$_x$, E$_y$,

and E$_z$ represent the magnitude of the electric fields along the x, y, and z directions, respectively (72). These values depend on the index of refraction and the angle of incidence and can be readily calculated using the well-known Fresnel laws. From these measurements, the orientation of the transition dipole for the symmetric (α_a) and antisymmetric (α_{as}) methylene vibrational modes is measured. Once the orientation of these transition dipoles is known, the tilt of the molecules within the layer can be calculated by simple geometry:

$$\phi_{tilt} = 1 - \cos^2\alpha_s - \cos^2\alpha_{as}.$$

A third piece of information that can be gleaned from infrared spectra is the degree to which the molecules adopt an all-trans configuration. In pure solid alkenes, all of the molecules in the hydrocarbon chains are in an all-trans configuration such that the carbon backbone of each molecule lies in a single plane. However, in liquid form there is substantial twisting about the individual bonds; these out-of-plane twists, or "gauche defects," alter the frequencies of the methylene (−CH$_2$−) vibrational modes. The infrared vibrational properties have been extensively studied for a variety of alkanethiols on gold (109–111). On gold surfaces, the alkyl chains of self-assembled monolayers give rise to asymmetric and symmetric C–H stretching modes near 2920 cm^{-1} and 2850 cm^{-1}. However, local structural information is contained in subtle shifts from these values, with the −CH$_2$− asymmetric stretching mode shifting from 2924 cm^{-1} to 2918 cm^{-1} and the symmetric stretching mode shifting from 2855 cm^{-1} to 2851 cm^{-1} when going from liquid-like to solid state (111). Similar ideas have been applied to monolayers grafted onto semiconductor surfaces. Experimental data from photochemical grafting of 1-octene to silicon yielded C–H modes with frequencies as low as 2920 cm^{-1}, and 1-octadecene yielded 2917 cm^{-1} (85), which is identical (within experimental error) to the frequency of a pure solid alkane. Perring and colleagues showed that undecylenic acid yields 2923 cm^{-1} and 2853 cm^{-1} (80); according to Wang et al., photochemically activated grafting of 1-dodecene to polycrystalline diamond yielded frequencies of 2922 cm^{-1} and 2850 cm^{-1} (112).

3.5. Scanning Electron Microscopy of Monolayers

Although SEM is commonly used as a structural analysis tool, it can also be used to directly image molecular layers on semiconductor surfaces (80, 112–114). This ability stems from the fact that molecular layers alter the potential barriers that commonly exist at surfaces and therefore can significantly impact the yield of secondary electrons. The use of SEM to image the spatial distribution of molecules can also provide important insights into the mechanisms of reactions.

One important question revolves around the nature of contrast in SEM images of molecular monolayers. Early studies of monolayers on gold (115, 116) showed clear contrast between different layers, but in those studies the possible influence of electron beam damage and the presence of surface contamination layers made it difficult to identify the origins of contrast. Studies of thicker layers produced by proteins (117, 118) showed that the contrast was dominated by the scattering of electrons within the molecular layers and was correlated with the mass coverage. On gold, the secondary electrons are essentially all generated within the gold, and the organic layers basically act as an inelastic scattering layer. SEM imaging of patterned molecular layers on semiconductor surfaces has been demonstrated using silicon (80, 119), diamond (112, 120, 121), and gallium nitride (122).

With regard to semiconductors, Saito et al. (113) investigated the contrast among alkoxysilanes with different terminal groups after bonding to oxidized silicon surfaces and found that the contrast among different molecules was determined primarily by the electron affinity of the molecules involved, as calculated via computational chemistry methods. Based on this, the authors proposed that the emission of secondary electrons occurred by excitation from bulk states to normally unoccupied energy levels, followed by excitation into vacuum; consequently, the lowest unoccupied molecular orbital (LUMO) levels closer to the vacuum level (higher in energy) were more effective at enhancing secondary electron yield compared with lower-lying levels.

Recent studies of molecular layers on diamond surfaces have shown a similar trend (112). One particularly interesting aspect of the work on diamond is that contrast is also developed between the H-terminated diamond surface and monolayers of simple hydrocarbons. **Figure 6** shows an SEM image of a diamond sample that was produced using photochemical grafting with a mask in order to produce a central region functionalized with 1-dodecene, whereas the surrounding region consisted of H-terminated diamond. The 1-dodecene monolayer significantly decreased the secondary electron emission. H-terminated diamond has negative electron affinity (i.e., the conduction band is higher in energy than the vacuum level) and is therefore a facile electron emitter; however, pure hydrocarbons also have negative electron affinity. Why, then, does 1-dodecene reduce the electron yield from H-terminated

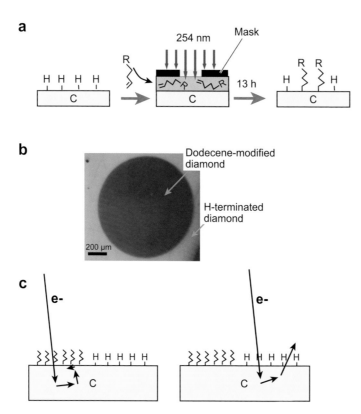

Figure 6

(*a*) Photopatterning of a molecular layer using a simple shadow mask. (*b*) Scanning electron microscopy (SEM) image demonstrating its ability to discriminate functionalized from nonfunctionalized regions. (*c*) Likely mechanism of image contrast, based on electron scattering at the diamond-molecule interface.

diamond? It is likely that the answer lies in treating the electron as a wave. Studies of binary monolayers on gold surface have shown that disorganized layers formed by mixing two different molecules on the surface give rise to lower electron emission than well-ordered layers of either molecule in pure form (123, 124). On diamond and other semiconductor surfaces, most molecular layers are disordered near the semiconductor-molecule boundary as most molecules cannot pack closely enough together to bind to each surface site. The resulting disorder at the semiconductor-organic interface leads to increased scattering electrons at the interface, reducing the secondary electron yield.

3.6. X-Ray Reflectivity and Grazing Incidence X-Ray Diffraction

X-ray methods can be used to probe the variations in electron density perpendicular to and parallel to the interface. By measuring reflectivity as a function of the momentum transfer vector, it is possible to quantify properties such as the density and thickness of the monolayer, the molecular tilt, and interfacial roughness (125). Tidswell et al. (125) developed much of the methodology for analyzing the properties of organic thin films, having obtained their initial data from alkoxysilane monolayers on oxidized silicon surfaces.

More recent experiments have characterized a variety of organic monolayers on semiconductor surfaces. In specular reflectivity measurements, the reflectivity is measured as a function of the momentum transfer vector $Q(\theta) = \frac{4\pi}{\lambda} \sin(\theta)$. For very small values of momentum transfer, total reflection is observed. At high values of momentum transfer, the reflectivity follows the Fresnel laws corresponding to reflection from a single sharp interface. At intermediate values, however, the reflectivity depends on the profile of electron density perpendicular to the surface. By fitting the data, information about the thickness of the film, roughness of the semiconductor-organic interface, and electron density profile can be obtained. The minima in the X-ray reflectivity measurements correspond to $Q = \frac{(2n+1)\pi}{L}$, such that the thickness corresponds to half-integral multiples of the wavelength; consequently, measuring the position and depth of these minima provides a rapid means of identifying the thickness and average density of the molecular layers. Sieval and colleagues examined molecular layers produced by thermal grafting of monolayers to H-terminated Si(100) and Si(111) surfaces and succeeded in determining the thickness of the layers (72, 81). Knowledge of the molecular structure and bond lengths of the alkyl chain was sufficient to determine that the organic layers were tilted with respect to the surface normal by 27°, which is in remarkable agreement with measurements based on the dichroic ratio in infrared spectroscopy. Hersam and colleagues similarly characterized the thickness of 4-bromostyrene monolayers formed on Si(111) surfaces (126).

Whereas X-ray reflectivity measurements typically provide information about the electron density perpendicular to the surface, there is a critical angle below which X-rays undergo total reflection. At these very low angles (on the order of 0.2° or less), the reflectivity provides information about the in-plane ordering. Early studies of monolayers produced by reaction of alkoxysilanes on oxidized silicon showed clear diffraction peaks due to local ordering of the molecular chains. Studies of alkenes ranging from dodecene (C10) to octadecene (C18) showed that the molecular layers are amorphous regardless of length (127). It is noteworthy that although FTIR frequencies and grazing incidence X-ray diffraction (GIXD) data both probe the structural order within a film, the FTIR measurements infer disorder from the presence of gauche defects, whereas GIXD directly probes the spatial periodicity of the surface electron density. Thus, a layer that appears crystalline by FTIR may not necessarily appear ordered by X-ray reflectivity.

3.7. Biomolecular Recognition at Bio-Functionalized Semiconductor Surfaces

Although most studies have focused on single monolayers, more recent studies have addressed many issues associated with the fabrication of organic films for biological applications, especially the use of bio-modified semiconductors for applications such as electronic sensing systems. Early work in this field was targeted toward chemical modification of the silicon oxide surface (5, 128, 129). Wagner et al. conducted early studies of biologically active monolayers linked to monolayers on silicon (130). Strother et al. used photochemical grafting of protected alkenes to single-crystal

Si(001) and Si(111) surfaces via two routes. One route consisted of grafting of alkenyl esters followed by deprotection and conversion to an amine-terminated surface (83), and the second consisted of an alkene with a protected amine group that was subsequently deprotected and linked to DNA. Subsequent deprotection exposed the reactive carboxylic acid and amine groups that were then covalently linked to DNA oligonucleotides (83). More recent studies have shown that other semiconductors, including GaN(123) and diamond (131–134), can be modified via the same route. For biologically modified semiconductors, the primary measures of interest are typically the selectivity of binding (i.e., the degree to which a bio-modified surface will recognize and bind to a complementary target molecule, while rejecting others), the stability of the interfaces, and the electronic properties of the resulting interface. Early studies on silicon usually modified the silicon oxide (5), but the inherent instability of SiO_2 in aqueous solutions and the presence of charged OH^- groups at the surface led to poor sensitivity and poor stability. Organic monolayers bonded directly to Si improve the stability because the organic monolayers act as a hydrophobic barrier layer that prevents water, OH−, and other reactive species from penetrating to the silicon-organic interface. However, even with the protective layers degradation still occurs, which continues to fuel interest in alternative materials such as diamond. Studies by Yang et al. showed that the photochemical functionalization of diamond with organic monolayers, followed by subsequent chemical modification steps, yielded DNA-modified surfaces exhibiting stability superior to that of functionalized silicon, gold, and SiO_2 (134). Stability was also demonstrated at temperatures of 60°C (135), above the DNA-denaturing temperature. A primary motivating factor for biological functionalization of semiconductors is the opportunity to take advantage of the electronic properties of semiconductors for hybrid bio-electronic sensing devices. The case of diamond, however, is even more interesting because the extreme hardness of diamond makes it useful in applications such as biomedical implants, where mechanical properties as well as biological response may be important to achieve long-term functionality.

Using organic monolayers as interfaces between semiconductors and biomolecules, several groups have demonstrated the ability to achieve direct detection of biomolecules such as DNA (4, 136, 137) and antibodies (131, 133). In such studies, the binding of DNA is typically measured in two ways. The first is to link one strand of DNA to the surface and to measure the intensity of fluorescence from the surface after exposure to molecules having a complementary sequence with a fluorescent tag. The second is to investigate the changes in electrical response. The ability to directly detect biological binding processes arises from the fact that binding of charged biomolecules near the semiconductor surface changes the electrostatic potential, which in turn leads to a change in the band-bending and the associated conductivity near the surface of the semiconductor. This change in conductivity can be measured in two ways. One way is to measure the electrical impedance between the modified semiconductor surface and a counter-electrode in solution, usually using a three-electrode potentiostat. Studies have been performed on silicon (4, 136, 137) and diamond (132, 133) surfaces. These studies showed that at frequencies between ~1 kHz and ~1 MHz, the impedance of the system was dominated by the impedance of the semiconductor space-charge layer; consequently, binding of molecules to the surface altered the resistance and capacitance of the space-charge layer, producing a significant change in the overall impedance of the system. This impedance measurement has been used to detect antibody-antigen binding (133). The second method of characterizing the electrical response is to use a field-effect transistor geometry, in which one deposits two metallic contacts (the source and drain of the transistor) onto the surface and then measures the change in current parallel to the surface (131).

Fundamentally, electrochemical impedance spectroscopy (EIS) and the field effect transistor (FET) both measure the semiconductor space-charge layers but differ as to whether the impedance is measured perpendicular to (EIS) or parallel to (FET) the interface. The EIS geometry does not require any special fabrication techniques, but in order to be sensitive to the semiconductor space-charge region it is necessary to operate at the correct frequency. The FET geometry requires special fabrication techniques but can be performed at any frequency, including dc.

For many biological applications, the ability to control the selectivity of binding, especially to eliminate the nonspecific binding of proteins and other biomolecules, is critical. Early studies of self-assembled monolayers on gold showed that ethylene glycol oligomers were very effective at restricting the nonspecific binding of many proteins. Lasseter and colleagues (139) grafted alkenes bearing ethylene glycol groups to surfaces of silicon, diamond, and gold; a quantitative comparison showed that functionalized diamond and gold surfaces were comparable to or possibly even better than those reported previously on modified gold surfaces (138, 139). Lasseter et al. also demonstrated that by making mixed monolayers consisting of ethylene glycol groups with interspersed attachment points covalently linked to biotin, it was possible to detect the complementary protein avidin in undiluted serum with only minimal interference (139).

3.8. Electronic Properties of Molecules at Semiconductor Surfaces

Understanding the electrical properties of molecules at surfaces is crucial for a number of emerging applications, including molecular electronics and the use of semiconductors as a basis for chemical and biological sensing. A variety of classical and nonclassical electrochemical measurements have been used to characterize the properties of organic monolayers (69, 70, 140–146). Impedance measurements, in which the interfacial impedance is measured as a function of frequency, are especially useful as a means to determine the resistance and capacitance of organic monolayers and their interfaces. The most common method, EIS, uses a small ac potential to characterize the small-signal response as a function of frequency and dc potential, typically performed in electrolyte solution using a three-electrode potentiostat (147). EIS measurements are useful because the electronic properties of semiconductor-organic interfaces typically contain contributions from the resistance and capacitance of the semiconductor space-charge region, the capacitance and electron-transfer resistance of the monolayer, and the capacitance of the electrochemical double layer.

By measuring the response over a wide range of frequencies and fitting it to circuit models, it is possible to extract the values of the individual components (137). Two parameters of particular interest have been the capacitance and resistance of the organic monolayers. Bansal et al. found that alkyl monolayers on silicon have excellent passivation properties against electrochemical oxidation (69, 70). The possibility of using organic monolayers as a dielectric has led to interest in characterizing the capacitance of individual monolayers and organosilane monolayers electrochemically grafted to silicon surfaces (147, 148). Yu et al. investigated organic monolayers and found that they formed molecular capacitors on Si with a dielectric constant of approximately 3.3. By adding a known quantity of a redox agent such as dimethylferrocene, it is possible to quantitatively evaluate the extent to which an organic monolayer blocks electron-transfer reactions (69).

4. SUMMARY

The use of organic monolayers provides a unique opportunity to achieve improved integration of semiconductors with organic and biological molecules. Continued advances in the synthesis and characterization of these layers will further expand the range of possible applications of molecular monolayers as interfaces between semiconductors and a wider range of organic and/or biological materials.

DISCLOSURE STATEMENT

The author is not aware of any biases that might be perceived as affecting the objectivity of this review.

ACKNOWLEDGMENTS

This work is based in part on research funded by the National Science Foundation, grant nos. CHE-0613010 and DMR-0425880.

LITERATURE CITED

1. Allara DL. 1995. Critical issues in applications of self-assembled monolayers. *Biosens. Bioelectron.* 10:771–83
2. Mirkin CA, Ratner MA. 1992. Molecular electronics. *Annu. Rev. Phys. Chem.* 43:719–54
3. Aviram A, Ratner MA. 1974. Molecular rectifiers. *Chem. Phys. Lett.* 29:277–83
4. Fritz J, Cooper EB, Gaudet S, Sorger PK, Manalis SR. 2002. Electronic detection of DNA by its intrinsic molecular charge. *Proc. Nat. Acad. Sci. USA* 99:14142–46
5. Bergveld P. 1996. The future of biosensors. *Sens. Actuators A.* 56:65–73
6. Appelbaum JA, Baraff GA, Hamann DR. 1976. The Si(100) surface. III. Surface reconstruction. *Phys. Rev. B* 14:588–601
7. Appelbaum JA, Hamann DR. 1976. Electronic structure of solid surfaces. *Rev. Mod. Phys.* 48:479–96

8. Hamers RJ, Tromp RM, Demuth JE. 1986. Scanning tunneling microscopy of Si(001). *Phys. Rev. B* 34:5343–57

9. Buriak JM. 2001. Diamond surfaces: just big organic molecules? *Angew. Chem. Int. Ed.* 40:532–34

10. Liu H, Hamers RJ. 1997. Stereoselectivity in molecule-surface reactions: adsorption of ethylene on silicon(001) surfaces. *J. Am. Chem. Soc.* 119:7593–94

11. Coulter SK, Hovis JS, Ellison MD, Hamers RJ. 2000. Reactions of substituted aromatic hydrocarbons with the Si(001) surface. *J. Vac. Sci. Tech. A* 18:1965–70

12. Ellison MD, Hamers RJ. 1999. Adsorption of phenyl isothiocyanate on Si(001): a 1,2-dipolar surface addition reaction. *J. Phys. Chem. B* 103:6243–51

13. Hamers RJ, Butler JE, Lasseter T, Nichols BM, Russell JN, et al. 2005. Molecular and biomolecular monolayers on diamond as an interface to biology. *Diamond Relat. Mater.* 14:661–68

14. Hamers RJ, Coulter SK, Ellison MD, Hovis JS, Padowitz DF, et al. 2000. Cycloaddition chemistry of organic molecules with semiconductor surfaces. *Acct. Chem. Res.* 33:617–24

15. Hamers RJ, Hovis J, Lee S, Liu H, Shan J. 1997. Formation of ordered, anisotropic organic monolayers on the Si(001) surface. *J. Phys. Chem.* 101:1489–92

16. Hovis J, Lee S, Liu H, Hamers RJ. 1997. Controlled formation of organic layers on semiconductor surfaces. *J. Vac. Sci. Technol. B* 15(4):1153–58

17. Hovis JS, Coulter SK, Hamers RJ, D'Evelyn MP, Russell JN, Butler JE. 2000. Cycloaddition chemistry at surfaces: reaction of alkenes with the diamond(001)–2 × 1 surface. *J. Am. Chem. Soc.* 122:732–33

18. Hovis JS, Hamers RJ. 1998. Structure and bonding of ordered organic monolayers of 1,3,5,7-cyclooctatetraene on the Si(001) surface: surface cycloaddition chemistry of an antiaromatic molecule. *J. Phys. Chem. B* 102:687–92

19. Hovis JS, Hamers RJ. 1997. Structure and bonding of ordered organic monolayers of 1,5-cyclooctadiene on the silicon(001) surface. *J. Phys. Chem. B* 101:9581–85

20. Hovis JS, Liu H, Hamers RJ. 1998. Cycloaddition chemistry and formation of ordered organic monolayers on silicon(001) surfaces. *Surf. Sci.* 402–404:1–7

21. Hovis JS, Liu HB, Hamers RJ. 1998. Cycloaddition chemistry of 1,3-dienes on the silicon(001) surface: competition between [4+2] and [2+2] reactions. *J. Phys. Chem. B* 102:6873–79

22. Konecny R, Doren DJ. 1998. Cycloaddition reactions of unsaturated hydrocarbons on the Si(100)–2 × 1 surface: theoretical predictions. *Surf. Sci.* 417:169–88

23. Teplyakov AV, Kong MJ, Bent SF. 1997. Vibrational spectroscopic studies of Diels-Alder reactions with the Si(100)–2 × 1 surface as a dienophile. *J. Am. Chem. Soc.* 119:11100–1

24. Teplyakov AV, Kong MJ, Bent SF. 1998. Diels-Alder reactions of butadienes with the Si(100)–2 × 1 surface as a dienophile: vibrational spectroscopy, thermal desorption and near edge X-ray adsorption fine structure studies. *J. Chem. Phys.* 108:4599–606

25. Wang GT, Bent SF, Russell JN, Butler JE, D'Evelyn MP. 2000. Functionalization of diamond(100) by Diels-Alder chemistry. *J. Am. Chem. Soc.* 122:744–45

26. Bent SF. 2002. Organic functionalization of group IV semiconductor surfaces: principles, examples, applications, and prospects. *Surf. Sci.* 500:879–903

27. Russell JN, Butler JE, Wang GT, Bent SF, Hovis JS, et al. 2001. Pi-bond vs radical character of the diamond(100)–2 × 1 Surface. *Mat. Chem. Phys.* 72:147–51

28. Filler MA, Bent SF. 2003. The surface as molecular reagent: organic chemistry at the semiconductor interface. *Prog. Surf. Sci.* 73:1–56

29. Filler MA, Mui C, Musgrave CB, Bent SF. 2003. Competition and selectivity in the reaction of nitriles on Ge(100)–2 × 1. *J. Am. Chem. Soc.* 125:4928–36

30. Choi CH, Gordon MS. 1999. Cycloaddition reactions of 1,3-cyclohexadiene on the silicon(001) surface. *J. Am. Chem. Soc.* 121:11311–17

31. Schwartz MP, Hamers RJ. 2007. Reaction of acetonitrile with the silicon(001) surface: a combined XPS and FTIR study. *Surf. Sci.* 601:945–53

32. Schwartz MP, Hamers RJ. 2002. The role of pi-conjugation in attachment of organic molecules to the silicon(001) surface. *Surf. Sci.* 515:75–86

33. Schwartz MP, Halter RJ, McMahon RJ, Hamers RJ. 2003. Formation of an atomically abrupt interface between a polycyclic aromatic molecule and the silicon(001) surface via direct Si-C linkage. *J. Phys. Chem. B* 107:224–28

34. Schwartz MP, Ellison MD, Coulter SK, Hovis JS, Hamers RJ. 2000. Interaction of pi-conjugated organic molecules with pi-bonded semiconductor surfaces: structure, selectivity, and mechanistic implications. *J. Am. Chem. Soc.* 122:8529–38

35. Schwartz MP, Barlow DE, Weidkamp KP, Russell JN, Butler J, et al. 2006. 1,3-H-transfer for CHD on Si(001). *J. Am. Chem. Soc.* 128:11054–61

36. Schwartz MP, Barlow DE, Russell JN, Weidkamp KP, Butler JE, et al. 2006. Semiconductor surface-induced 1,3-hydrogen shift: the role of covalent vs Zwitterionic character. *J. Am. Chem. Soc.* 128:11054–61

37. Schwartz MP, Barlow DE, Russell JN, Butler JE, D'Evelyn MP, Hamers RJ. 2005. Adsorption of acrylonitrile on diamond and silicon(001)–(2 × 1) surfaces: effects of dimer structure on reaction pathways and product distributions. *J. Am. Chem. Soc.* 127:8348–54

38. Fang LA, Liu JM, Coulter S, Cao XP, Schwartz MP, et al. 2002. Formation of pi-conjugated molecular arrays on silicon(001) surfaces by heteroatomic Diels-Alder chemistry. *Surf. Sci.* 514:362–75

39. Lee SW, Hovis JS, Coulter SK, Hamers RJ, Greenlief CM. 2000. Cycloaddition chemistry on germanium(001) surfaces: the adsorption and reaction of cyclopentene and cyclohexene. *Surf. Sci.* 462:6–18

40. Konecny R, Doren DJ. 1997. Theoretical prediction of a facile Diels-Alder reaction on the Si(100)–2 × 1 surface. *J. Am. Chem. Soc.* 119:11098–99

41. Padowitz DF, Hamers RJ. 1998. Voltage-dependent STM images of covalently bound molecules on Si(100). *J. Phys. Chem. B* 102:8541–45

42. Woodward RB, Hoffmann R. 1970. *The Conservation of Orbital Symmetry*. New York: Academia

43. Lee SW, Nelen LN, Ihm H, Scoggins T, Greenlief CM. 1998. Reactions of 1,3-cyclohexadiene with the Ge(100) surface. *Surf. Sci.* 410:L773–78

44. Choi CH, Gordon MS. 2002. Cycloaddition reactions of acrylonitrile on the Si(001) Surface. *J. Am. Chem. Soc.* 124:6162–67

45. Fitzgerald DR, Doren DJ. 2000. Functionalization of diamond(100) by cycloaddition of butadiene: first-principles theory. *J. Am. Chem. Soc.* 122:12334–39

46. Hossain MZ, Aruga T, Takagi N, Tsuno T, Fujimori N, et al. 1999. Diels-Alder reaction on the clean diamond(100)–2 × 1 surface. *J. Appl. Phys. Pt. 2 Lett.* 38:L1496–98

47. Barriocanal JA, Doren DJ. 2001. Cycloaddition of carbonyl compounds on Si(100): new mechanisms and approaches to selectivity for surface cycloaddition reactions. *J. Am. Chem. Soc.* 123:7340–46

48. Hofer WA, Fisher AJ, Lopinski GP, Wolkow RA. 2001. Adsorption of benzene on Si(100)–(2 × 1): adsorption energies and STM image analysis by ab initio methods. *Phys. Rev. B* 63:085314

49. Jung YS, Gordon MS. 2005. Cycloaddition of benzene on Si(100) and its surface conversions. *J. Am. Chem. Soc.* 127:3131–39

50. Lee JY, Cho JH. 2005. Conversion between two binding states of benzene on Si(001). *Phys. Rev. B* 72:235317

51. Shimomura M, Munakata M, Honma K, Widstrand SM, Johansson L, et al. 2003. Structural study of benzene adsorbed on Si(001) surface by photoelectron diffraction. *Surf. Rev. Lett.* 10:499–503

52. Qu YQ, Han KL. 2004. Theoretical studies of benzonitrile at the Si(100)–2 × 1 surface. *J. Phys. Chem. B* 108:8305–10

53. Tao F, Wang ZH, Chen XF, Xu GQ. 2002. Selective attachment of benzonitrile on Si(111)–7 × 7: configuration, selectivity, and mechanism. *Phys. Rev. B* 65:115311

54. Takeuchi N, Selloni A. 2005. Density functional theory study of one-dimensional growth of styrene on the hydrogen-terminated Si(001)–(3 × 1) surface. *J. Phys. Chem. B* 109:11967–72

55. Coulter SK, Schwartz MP, Hamers RJ. 2001. Sulfur atoms as tethers for selective attachment of aromatic molecules to silicon(001) surfaces. *J. Phys. Chem. B* 105:3079–87

56. Bermudez VM. 2002. Functionalizing the GaN(0001)–(1 × 1) surface I. The chemisorption of aniline. *Surf. Sci.* 499:109–23

57. Yamamoto H, Butera RA, Gu Y, Waldeck DH. 1999. Characterization of the surface to thiol bonding in self-assembled monolayer films of $C_{12}H_{25}SH$ on InP(100) by angle-resolved X-ray photoelectron spectroscopy. *Langmuir* 15:8640–44

58. Lim H, Carraro C, Maboudian R, Pruessner MW, Ghodssi R. 2004. Chemical and thermal stability of alkanethiol and sulfur passivated InP(100). *Langmuir* 20:743–47

59. Jun Y, Zhu XY, Hsu JWP. 2006. Formation of alkanethiol and alkanedithiol monolayers on GaAs(001). *Langmuir* 22:3627–32

60. Lunt SR, Santangelo PG, Lewis NS. 1991. Passivation of GaAs surface recombination with organic thiols. *J. Vac. Sci. Tech. B* 9:2333–36

61. Hou T, Greenlief M, Keller SW, Nelen L, Kauffman JF. 1997. Passivation of GaAs(100) with an adhesion promoting self-assembled monolayer. *Chem. Mater.* 9:3181–86

62. Gu Y, Waldeck DH. 1998. Electron tunneling at the semiconductor-insulator-electrolyte interface: photocurrent studies of the n-InP-alkanethiol-ferrocyanide system. *J. Phys. Chem. B* 102:9015–28

63. Casaletto MP, Carbone M, Piancastelli MN, Horn K, Weiss K, Zanoni R. 2005. A high resolution photoemission study of phenol adsorption on Si(100)–2 × 1. *Surf. Sci.* 582:42–48

64. Rummel RM, Ziegler C. 1998. Room temperature adsorption of aniline (C6H5NH2) on Si(100)(2 × 1) observed with scanning tunneling microscopy. *Surf. Sci.* 418:303–13

65. Cao XP, Coulter SK, Ellison MD, Liu HB, Liu JM, Hamers RJ. 2001. Bonding of nitrogen-containing organic molecules to the silicon(001) surface: the role of aromaticity. *J. Phys. Chem. B* 105:3759–68

66. Lopinski GP, Wayner DDM, Wolkow RA. 2000. Self-directed growth of molecular nanostructures on silicon. *Nature* 406:48–51

67. Lu X, Wang XL, Yuan QH, Zhang Q. 2003. Diradical mechanisms for the cycloaddition reactions of 1,3-butadiene, benzene, thiophene, ethylene, and acetylene on a Si(111)–7 × 7 surface. *J. Am. Chem. Soc.* 125:7923–29

68. Yablonovitch E, Allara DL, Chang CC, Gmitter T, Bright TB. 1986. Unusually low surface-recombination velocity on silicon and germanium surfaces. *Phys. Rev. Lett.* 57:249–52

69. Bansal A, Lewis NS. 1998. Electrochemical properties of (111)-oriented n-Si surfaces derivatized with covalently attached alkyl chains. *J. Phys. Chem. B* 102:1067–70

70. Bansal A, Li XL, Lauermann I, Lewis NS, Yi SI, Weinberg WH. 1996. Alkylation of Si surfaces using a two-step halogenation Grignard route. *J. Am. Chem. Soc.* 118:7225–26

71. Cai W, Lin Z, Strother T, Smith LM, Hamers RJ. 2002. Chemical modification and patterning of iodine-terminated silicon surfaces using visible light. *J. Phys. Chem. B* 106:2656–64

72. Sieval AB, Demirel AL, Nissink JWM, Linford MR, van der Maas JH, et al. 1998. Highly stable Si-C linked functionalized monolayers on the silicon (100) surface. *Langmuir* 14:1759–68

73. Higashi GS, Becker RS, Chabal YJ, Becker AJ. 1991. Comparison of Si(111) surfaces prepared using aqueous solutions of NH4F versus HF. *Appl. Phys. Lett.* 58:1656–58

74. Dumas P, Chabal YJ, Gunther R, Ibrahimi AT, Petroff Y. 1995. Vibrational characterization and electronic properties of long range–ordered, ideally hydrogen-terminated Si(111). *Prog. Surf. Sci.* 48:313–24

75. Dumas P, Chabal YJ. 1991. Electron–energy loss characterization of the H-terminated Si(111) and Si(100) surfaces obtained by etching in NH4F. *Chem. Phys. Lett.* 181:537–43

76. Buriak JM, Allen MJ. 1998. Lewis acid mediated functionalization of porous silicon with substituted alkenes and alkynes. *J. Am. Chem. Soc.* 120:1339–40

77. Buriak JM, Stewart MP, Geders TW, Allen MJ, Choi HC, et al. 1999. Lewis acid mediated hydrosilylation on porous silicon surfaces. *J. Am. Chem. Soc.* 121:11491–502

78. Linford MR, Chidsey CED. 1993. Alkyl monolayers covalently bonded to silicon surfaces. *J. Am. Chem. Soc.* 115:12631–32

79. Linford MR, Fenter P, Eisenberger PM, Chidsey CED. 1995. Alkyl monolayers on silicon prepared from 1-alkenes and hydrogen-terminated silicon. *J. Am. Chem. Soc.* 117:3145–55

80. Perring M, Dutta S, Arafat S, Mitchell M, Kenis PJA, Bowden NB. 2005. Simple methods for the direct assembly, functionalization, and patterning of acid-terminated monolayers on Si(111). *Langmuir* 21:10537–44

81. Sieval AB, Linke R, Zuilhof H, Sudholter EJR. 2000. High-quality alkyl monolayers on silicon surfaces. *Adv. Mater.* 12:1457–60

82. Effenberger F, Gotz G, Bidlingmaier B, Wezstein M. 1998. Photoactivated preparation and patterning of self-assembled monolayers with 1-alkenes and aldehydes on silicon hydride surfaces. *Angew. Chem. Int. Ed.* 37:2462–64

83. Strother T, Cai W, Zhao XS, Hamers RJ, Smith LM. 2000. Synthesis and characterization of DNA-modified silicon (111) surfaces. *J. Am. Chem. Soc.* 122:1205–09

84. Strother T, Hamers RJ, Smith LM. 2000. Covalent attachment of oligodeoxyribonucleotides to amine-modified Si(001) surfaces. *Nucleic Acids Res.* 28:3535–41

85. Cicero RL, Linford MR, Chidsey CED. 2000. Photoreactivity of unsaturated compounds with hydrogen-terminated silicon(111). *Langmuir* 16:5688–95

86. Stewart MP, Buriak JM. 2001. Exciton-mediated hydrosilylation on photoluminescent nanocrystalline silicon. *J. Am. Chem. Soc.* 123:7821–30

87. Robins EG, Stewart MP, Buriak JM. 1999. Anodic and cathodic electrografting of alkynes on porous silicon. *Chem. Comm.* 2479–80

88. Wang D, Buriak JM. 2005. Electrochemically driven organic monolayer formation on silicon surfaces using alkylammonium and alkylphosphonium reagents. *Surf. Sci.* 590:154–61

89. Cleland G, Horrocks BR, Houlton A. 1995. Direct functionalization of silicon via the self-assembly of alcohols. *J. Chem. Soc. Faraday Trans.* 91:4001–3

90. Boukherroub R, Morin S, Bensebaa F, Wayner DDM. 1999. New synthetic routes to alkyl monolayers on the Si(111) surface. *Langmuir* 15:3831–35

91. Boukherroub R, Morin S, Sharpe P, Wayner DDM, Allongue P. 2000. Insights into the formation mechanisms of Si-OR monolayers from the thermal reactions of alcohols and aldehydes with Si(111)-H. *Langmuir* 16:7429–34

92. Wuts PGM, Greene TW, eds. 2006. *Protective Groups in Organic Synthesis*. New York: Wiley

93. Boukherroub R, Wayner DDM. 1999. Controlled functionalization and multistep chemical manipulation of covalently modified Si(111) surfaces. *J. Am. Chem. Soc.* 121:11513–15

94. Hamers RJ, Hovis JS, Greenlief CM, Padowitz DF. 1999. Scanning tunneling microscopy of organic molecules and monolayers on silicon and germanium (001) surfaces. *Jpn. J. Appl. Phys.* 38:3879–87

95. McCreery RL. 2004. Molecular electronic junctions. *Chem. Mater.* 16:4477–96
96. Tersoff J, Hamann DR. 1983. Theory and application for the scanning tunneling microscope. *Phys. Rev. Lett.* 50:1998–2001
97. Tersoff J, Hamann DR. 1985. Theory of the scanning tunneling microscope. *Phys. Rev. B* 31:805–13
98. Mujica V, Kemp M, Ratner MA. 1994. Electron conduction in molecular wires. 2. Application to scanning tunneling microscopy. *J. Chem. Phys.* 101:6856–64
99. Mujica V, Kemp M, Ratner MA. 1994. Electron conduction in molecular wires.1. A scattering formalism. *J. Chem. Phys.* 101:6849–55
100. Hamers RJ, Tromp RM, Demuth JE. 1986. Surface electronic structure of Si(111)–(7 × 7) resolved in real space. *Phys. Rev. Lett.* 56:1972–75
101. Hamers RJ. 1989. Atomic-resolution surface spectroscopy with the scanning tunneling microscope. *Ann. Rev. Phys. Chem.* 40:531–59
102. Wolkow RA. 1999. Controlled molecular adsorption on silicon: laying a foundation for molecular devices. *Ann. Rev. Phys. Chem.* 50:413–41
103. Rakshit T, Liang GC, Ghosh AW, Hersam MC, Datta S. 2005. Molecules on silicon: self-consistent first-principles theory and calibration to experiments. *Phys. Rev. B* 72:125305
104. Guisinger NP, Basu R, Baluch AS, Hersam MC. 2003. Molecular electronics on silicon: an ultrahigh vacuum scanning tunneling microscopy study. In *Molecular Electronics III*, ed. J Reimers, C Picconatto, J Ellebogen, R Shashidar, pp. 227–34. New York: N.Y. Acad. Sci.
105. Terry J, Linford MR, Wigren C, Cao RY, Pianetta P, Chidsey CED. 1999. Alkyl-terminated Si(111) surfaces: a high-resolution, core level photoelectron spectroscopy study. *J. Appl. Phys.* 85:213–21
106. Segev L, Salomon A, Natan A, Cahen D, Kronik L, et al. 2006. Electronic structure of Si(111)-bound alkyl monolayers: theory and experiment. *Phys. Rev. B* 74:165323
107. Chabal YJ. 1988. Surface infrared spectroscopy. *Surf. Sci. Rep.* 8:211–357
108. Alerhand OL, Berker AN, Joannopoulos JD, Vanderbilt D, Hamers RJ, Demuth JE. 1990. Finite temperature phase diagram of vicinal Si(100) surfaces. *Phys. Rev. Lett.* 64:2406–9
109. Nuzzo RG, Dubois LH, Allara DL. 1990. Fundamental studies of microscopic wetting on organic-surfaces. 1. Formation and structural characterization of a self-consistent series of polyfunctional organic monolayers. *J. Am. Chem. Soc.* 112:558–69
110. Hostetler MJ, Stokes JJ, Murray RW. 1996. Infrared spectroscopy of three-dimensional self-assembled monolayers: N-alkanethiolate monolayers on gold cluster compounds. *Langmuir* 12:3604–12
111. Porter MD, Bright TB, Allara DL, Chidsey CED. 1987. Spontaneously organized molecular assemblies. 4. Structural characterization of normal alkyl thiol monolayers on gold by optical ellipsometry, infrared spectroscopy, and electrochemistry. *J. Am. Chem. Soc.* 109:3559–68
112. Wang X, Colavita PE, Metz KM, Butler JE, Hamers RJ. 2007. Direct photopatterning and SEM imaging of molecular monolayers on diamond surfaces:

mechanistic insights into UV-initiated molecular grafting. *Langmuir* 23:11623–30

113. Saito N, Wu Y, Hayashi K, Sugimura H, Takai O. 2003. Principle in imaging contrast in scanning electron microscopy for binary microstructures composed of organosilane self-assembled monolayers. *J. Phys. Chem. B* 107:664–67

114. Wu YY, Hayashi K, Saito N, Sugimura H, Takai O. 2003. Imaging micropatterned organosilane self-assembled monolayers on silicon by means of scanning electron microscopy and Kelvin probe force microscopy. *Surf. Interface Anal.* 35:94–98

115. Lopez GP, Biebuyck HA, Whitesides GM. 1993. Scanning electron microscopy can form images of patterns in a self-assembled monolayer. *Langmuir* 9:1513–16

116. Wollman EW, Frisbie CD, Wrighton MS. 1993. Scanning electron-microscopy for imaging photopatterned self-assembled monolayers on gold. *Langmuir* 9:1517–20

117. Lopez GP, Biebuyck HA, Harter R, Kumar A, Whitesides GM. 1993. Fabrication and imaging of 2-dimensional patterns of proteins adsorbed on self-assembled monolayers by scanning electron microscopy. *J. Am. Chem. Soc.* 115:10774–81

118. Mack NH, Dong R, Nuzzo RG. 2006. Quantitative imaging of protein adsorption on patterned organic thin-film arrays using secondary electron emission. *J. Am. Chem. Soc.* 128:7871–81

119. Streifer JA, Colavita P, Hamers RJ. 2008. Evidence for two distinct mechanisms in photochemical grafting of molecular layers to silicon surfaces. Manuscript in preparation

120. Whelan CS, Lercel MJ, Craighead HG, Seshadri K, Allara DL. 1996. Improved electron-beam patterning of Si with self-assembled monolayers. *Appl. Phys. Lett.* 69:4245–47

121. Smith RK, Lewis PA, Weiss PS. 2004. Patterning self-assembled monolayers. *Prog. Surf. Sci.* 75:1–68

122. Kim H, Colavita PE, Metz KM, Nichols BM, Sun B, et al. 2006. Photochemical functionalization of gallium nitride thin films with molecular and biomolecular layers. *Langmuir* 22:8121–26

123. Kadyshevitch A, Naaman R. 1996. The interactions of electrons with organized organic films studied by photoelectron transmission. *Thin Solid Films* 288:139–46

124. Kadyshevitch A, Ananthavel SP, Naaman R. 1997. The role of three dimensional structure in electron transmission through thin organic layers. *J. Chem. Phys.* 107:1288–90

125. Tidswell IM, Ocko BM, Pershan PS, Wasserman SR, Whitesides GM, Axe JD. 1990. X-ray specular reflection studies of silicon coated by organic monolayers (alkylsiloxanes). *Phys. Rev. B* 41:1111–28

126. Basu R, Lin JC, Kim CY, Schmitz MJ, Yoder NL, et al. 2007. Structural characterization of 4-bromostyrene self-assembled monolayers on Si(111). *Langmuir* 23:1905–11

127. Ishizaki T, Saito N, SunHyung L, Ishida K, Takai O. 2006. Study of alkyl organic monolayers with different molecular chain lengths directly attached to silicon. *Langmuir* 22:9962–66

128. Souteyrand E, Martin JR, Martelet C. 1994. Direct detection of biomolecules by electrochemical impedance measurements. *Sens. Actuators B* 20:63–69

129. Wei F, Sun B, Guo Y, Zhao XS. 2003. Monitoring DNA hybridization on alkyl modified silicon surface through capacitance measurement. *Biosens. Bioelectron.* 18:1157–63

130. Wagner P, Nock S, Spudich JA, Volkmuth WD, Chu S, et al. 1997. Bioreactive self-assembled monolayers on hydrogen-passivated Si(111) as a new class of atomically flat substrates for biological scanning probe microscopy. *J. Struc. Biol.* 119:189–201

131. Yang WS, Hamers RJ. 2004. Fabrication and characterization of a biologically sensitive field effect transistor using a nanocrystalline diamond thin film. *Appl. Phys. Lett.* 85:3626–28

132. Yang WS, Butler JE, Russell JN, Hamers RJ. 2004. Interfacial electrical properties of DNA-modified diamond thin films: intrinsic response and hybridization-induced field effects. *Langmuir* 20:6778–87

133. Yang WS, Butler JE, Russell JN, Hamers RJ. 2007. Direct electrical detection of antigen-antibody binding on diamond and silicon substrates using electrical impedance spectroscopy. *Analyst* 132:296–306

134. Yang WS, Auciello O, Butler JE, Cai W, Carlisle JA, et al. 2002. DNA-modified nanocrystalline diamond thin-films as stable, biologically active substrates. *Nat. Mater.* 1:253–57

135. Lu MC, Knickerbocker T, Cai W, Yang WS, Hamers RJ, Smith LM. 2004. Invasive cleavage reactions on DNA-modified diamond surfaces. *Biopolymers* 73:606–13

136. Souteyrand E, Cloarec JP, Martin JR, Wilson C, Lawrence I, et al. 1997. Direct detection of the hybridization of synthetic homo-oligomer DNA sequences by field effect. *J. Phys. Chem. B* 101:2980–85

137. Cai W, Peck JR, van der Weide DW, Hamers RJ. 2004. Direct electrical detection of hybridization at DNA-modified silicon surfaces. *Biosens. Bioelectron.* 19:1013–19

138. Clare TL, Clare BH, Nichols BM, Abbott NL, Hamers RJ. 2005. Functional monolayers for improved resistance to protein adsorption: oligo(ethylene glycol)–modified silicon and diamond surfaces. *Langmuir* 21:6344–55

139. Lasseter TL, Clare BH, Abbott NL, Hamers RJ. 2004. Covalently modified silicon and diamond surfaces: resistance to nonspecific protein adsorption and optimization for biosensing. *J. Am. Chem. Soc.* 126:10220–21

140. Forbes MDE, Lewis NS. 1990. Real-time measurements of interfacial charge-transfer rates at silicon liquid junctions. *J. Am. Chem. Soc.* 112:3682–83

141. Lewis NS. 2005. Chemical control of charge transfer and recombination at semiconductor photoelectrode surfaces. *Inorg. Chem.* 44:6900–11

142. Prokopuk N, Lewis NS. 2004. Energetics and kinetics of interfacial electron-transfer processes at chemically modified InP/liquid junctions. *J. Phys. Chem. B* 108:4449–56

143. Royea WJ, Juang A, Lewis NS. 2000. Preparation of air-stable, low recombination velocity Si(111) surfaces through alkyl termination. *Appl. Phys. Lett.* 77:1988–90

144. Fajardo AM, Lewis NS. 1997. Free-energy dependence of electron-transfer rate constants at Si/liquid interfaces. *J. Phys. Chem. B* 101:11136–51

145. Fajardo AM, Lewis NS. 1996. Rate constants for charge transfer across semiconductor-liquid interfaces. *Science* 274:969–72

146. Tse KY, Nichols BM, Yang WS, Butler JE, Russell JN, Hamers RJ. 2005. Electrical properties of diamond surfaces functionalized with molecular monolayers. *J. Phys. Chem. B* 109:8523–32

147. Koiry SP, Aswal DK, Saxena V, Padma N, Chauhan AK, et al. 2007. Electrochemical grafting of octyltrichlorosilane monolayer on Si. *Appl. Phys. Lett.* 90:113118

148. Yu HZ, Morin S, Wayner DDM, Allongue P, de Villeneuve CH. 2000. Molecularly tunable "organic capacitors" at silicon/aqueous electrolyte interfaces. *J. Phys. Chem. B* 104:11157–61

Nanoscopic Porous Sensors

John J. Kasianowicz,[1] Joseph W.F. Robertson,[1]
Elaine R. Chan,[1] Joseph E. Reiner,[1]
and Vincent M. Stanford[2]

[1] National Institute of Standards and Technology, Semiconductor Electronics
Division, Electronics and Electrical Engineering Laboratory, Gaithersburg,
Maryland 20899-8120; email: john.kasianowicz@nist.gov

[2] National Institute of Standards and Technology, Information Access Division,
Information Technology Laboratory, Gaithersburg, Maryland 20899-8940

Annu. Rev. Anal. Chem. 2008. 1:737–66

The *Annual Review of Analytical Chemistry* is online
at anchem.annualreviews.org

This article's doi:
10.1146/annurev.anchem.1.031207.112818

1936-1327/08/0719-0737$20.00

Key Words

analyte detection, Coulter counter, DNA sequencing, ion channel,
nanopore-based sensor, resistive-pulse detection

Abstract

There are thousands of different nanometer-scale pores in biology,
many of which act as sensors for specific chemical agents. Recent
work suggests that protein and solid-state nanopores have many potential uses in a wide variety of analytical applications. In this review
we survey this field of research and discuss the prospects for advances
that could be made in the near future.

1. INTRODUCTION

Nanopore chemical sensors are miniaturized descendents of the Coulter counter (1), a device that measures resistive pulses to detect microscopic particles, such as red blood cells, in a narrow capillary. The classical Coulter counter was able to detect ∼10-μm-size particles in ∼100-μm-diameter capillaries. These techniques were first applied to the nanoscale (∼100 nm) with nuclear-track etched pores in the early 1970s by DeBlois and Bean (2). Other mesofluidic structures with diameters of less than 1 μm also have the potential for use in the analysis of macromolecules, colloids, and bioparticles measuring >100 nm (3–5).

Nanopore-based sensors are fundamentally chemical in nature because the interaction time of the analyte with the pore, when governed by physics alone, is too short to be accurately measured using electronics. Resistive-pulse techniques require an analyte to enter into and reside within a capillary for a period of time long enough to be detected with ionic current measurements. As resistive-pulse sensors are miniaturized to the molecular scale (1 to 10 nm), the characteristic diffusion time for a molecule becomes quite short (∼50 to 500 ns), and only ∼50 to 500 ions pass the molecule in a nanopore with a 1-nS conductance. For an analyte molecule to reside within the pore long enough for detection, there must be either an appreciable binding (or adsorption) of the analyte to the interior of the pore or a physical means to inhibit the partitioning of the analyte out of the pore. Therefore, optimizing the interfacial chemistry between an analyte and the nanopore interior must factor significantly into any successful detection scheme.

The first truly molecular-scale nanopores to be used experimentally were protein ion channels (**Figure 1a**), the study of which formed the foundation of biophysics. Channels can selectively transport particular species of ions across cell membranes and alter their conductance state by changing the transmembrane electrostatic potential. By virtue of these two properties, channels form the molecular basis of many processes, including the propagation of neural impulses (6), muscle activity (7), and protein translocation across cell membranes (8, 9). More recent developments demonstrated that these nanodevices have the potential for use as chemical sensors.

The chemical affinity between protein ion channels embedded in planar lipid bilayer membranes and a variety of analytes has permitted the detection and quantification of H^+ and D^+ ions (10, 11), divalent cations (12, 13), single-stranded RNA and DNA molecules (14–19), small organic molecules (20), specific sugar molecules (21), poly(ethylene glycol) (PEG) (22–29), and anthrax toxins (30).

Whereas biological nanopores offer precisely controlled structures and interfacial chemistry, solid-state nanopores in silicon nitride have also been developed (**Figure 1b**) (31–37) to take advantage of the potentially improved stability offered by semiconductor materials. Solid-state nanopores were initially used to detect individual double-stranded DNA (dsDNA) molecules (32), which are too large to transport through many ion channels (14). Other nanopores fabricated from carbon nanotubes (3) or by heavy ion bombardment combined with chemical etching (38, 39) have also made inroads into nanoscale pore sensor elements.

a

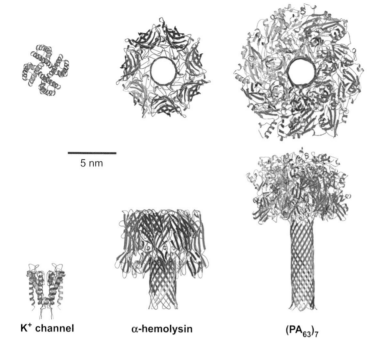

Figure 1

(*a*) Biological and
(*b*) solid-state nanopores.
(*a*) *From left to right*: crystal
structures of a K⁺-selective
ion channel (118), and the
channel formed by
Staphylococcus aureus
α-hemolysin (61) and a
molecular model for the
channel formed by *Bacillus
anthracis* protective antigen
63 (119). (*b*) *From left to
right*: synthetic silicon
nitride nanopore (36),
carbon multiwall nanotube
in an epoxy matrix (3), and
nanopores formed in
track-etched polyimide,
polycarbonate or
poly(ethylene terephthalate)
membranes (38, 39).

5 nm

K⁺ channel   **α-hemolysin**   **(PA₆₃)₇**

b

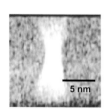

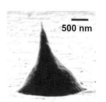

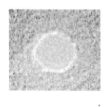

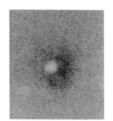

5 nm

100 nm

500 nm

E-beam   **Multiwall carbon
nanotube**   **Electroless
AU**

The experiments and theoretical methods described herein suggest that both protein and solid-state nanopores will continue to act as excellent platforms for single-molecule analytical measurements.

2. ELECTRONIC DETECTION OF MOLECULES USING SINGLE NANOMETER-SCALE PORES: SIGNAL PROCESSING

One advantage of the nanopore-based analyte detection method is that the resistive-pulse measurements are electronic. Thus, they have the potential to be performed at relatively low cost, and the resulting data are readily amenable to a wide variety of statistical signal processing algorithms.

Consider a model system consisting of a single nanopore in a high-impedance matrix. In the absence of analytes, the ionic current that flows through the pore in response to a fixed value of the applied potential is stable. Analytes that interact with the nanopore can alter the latter's conductance via electrostatic or steric interactions. Below, we identify three typical signals that are obtained in single-nanopore experiments and discuss the methods used to analyze the data.

2.1. Steady-State and Kinetic Analysis: Estimation of Event Amplitudes and Dwell Times

In the simplest case, the reversible interaction between an analyte and a single nanopore causes the pore conductance to fluctuate between two well-defined mean values. The analyte concentration can be determined by estimating the time spent by the nanopore in each of the two states, for instance via a conventional dwell-time approach (40–46). Because reaction kinetics are characteristic of the chemical interactions between analytes and a binding site, the dynamics of the current fluctuations also reveal information about the type of analyte.

At the extremes of analyte concentration (i.e., $[A] \ll K$ and $[A] \gg K$, where $[A]$ is the analyte concentration, the binding constant of the reaction in mol/L is defined by $K = k_{off}/k_{on}$, and k_{on} and k_{off} are the rate constants for the association with and dissociation of analyte from the nanopore, respectively), the current will be virtually always in one state or the other (**Figure 2a**, *left*). When the analyte concentration $[A] = K$, the channel spends, on average, half the time in each of the two conductance states. In a manner similar to the Henderson-Hasselbalch relation for the concentration of aqueous protons in a buffer solution, the time-averaged nanopore ionic current will vary monotonically from one extreme of the current to the other as a function of analyte concentration (**Figure 2a**, *center*). Thus, calibration of a particular nanopore with known concentrations of an analyte enables the determination of the analyte concentration in a test solution (11).

The kinetics of the current fluctuations provide additional information to help determine the identity of the analyte. **Figure 2a** (*right*) illustrates two possible dwell-time distributions determined from two different hypothetical ionic current time series. If the reaction between the analyte and a nanopore can be described by a simple reversible chemical reaction, the lifetime distribution for events in the bound state would be described by a single exponential. If the analyte concentration were low (i.e., [A] $\ll K$), the mean lifetime derived from the distribution would be $\sim 1/k_{off}$, which would be characteristic of the analyte type. For other types of interactions between the analyte and the nanopore (e.g., transport of analyte through the nanopore), the lifetime distribution may be better described by another function, such as a Gaussian.

2.2. Spectral Analysis

Fourier analysis has provided keen insight into the mechanism by which the neural impulse propagates across a synapse (47, 48), and is particularly useful for the analytical applications described herein for two reasons. First, it can be used to analyze current fluctuations that are not completely resolved due to bandwidth limitations (**Figure 2b**, *left*). Second, it provides a direct measurement of the frequency content in the time series and hence of the characteristic timescales of the analyte-nanopore interactions.

For a random telegraph two-state system, the power spectral density (PSD) of the current noise is nearly white (i.e., frequency independent) at low frequencies; at higher frequencies, the PSD decreases as $1/f^2$ (49) (**Figure 2b**, *center, blue trace*). The transition between those two regimes is characterized by the corner frequency, f_c, which is the frequency at which the PSD decreases twofold. In general, f_c provides information about the two timescales of the reaction between the analyte and the nanopore: (1) the mean time the analyte is bound to the pore (obtained when [A] $\ll K$), and (2) the mean time the analyte takes to find and react with the nanopore (which depends on [A]) (50). For a given experiment, fitting a simple theoretical expression to an experimental PSD data set provides estimates for the low frequency noise, $S(0)$, and f_c (**Figure 2b**, *center, black trace*).

As shown in **Figure 2b** (*right, blue trace*), at the extreme values of the analyte concentration, the value of $S(0)$ is minimal. For [A] $\approx K$, $S(0)$ approaches its maximum value. This makes intuitive sense: When there is a shortage or an excess of analyte, the current fluctuations must be minimal and the maximum of the current fluctuations must occur when the rates of analyte association with and dissociation from the pore are equal (11). The kinetic information in the reaction is determined from the characteristic relaxation time for the interaction between the analyte and the nanopore, defined as $\tau = 1/2\pi f_c$. In the limit of [A] ≈ 0, $\tau \sim 1/k_{off}$. For increasing values of [A], the second characteristic time of the reaction (related to k_{on} [A]) starts to dominate.

A least-squares fit of simple equations to the distributions of $S(0)$ and τ, derived from the PSD data as a function of analyte concentration, provides estimates for several reaction parameters, including the number of binding sites for analyte in the nanopore, the thermodynamic information about the reaction (i.e., the pK), and the kinetic information in terms of k_{on} and k_{off} (11, 24, 50). This method discriminates particularly well among the different analytes that bind to the nanopore because it makes use of both the thermodynamic (pK) and kinetic (k_{on} and k_{off}) information.

Finally, the distribution of $S(0)$ over a wide range of [A] could also help determine whether there is more than one characteristic binding constant. This is important because even if the binding sites are chemically identical, the narrow confines of a nanoscopic pore could cause the binding of analyte to one site to depend on the occupancy state of nearby sites. That is, the binding of more than one analyte of the same type to the nanopore may be either a cooperative or an anticooperative process.

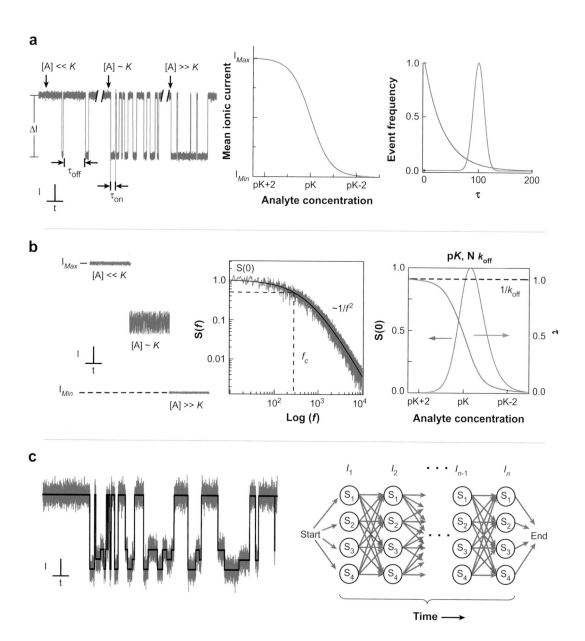

2.3. Hidden Markov Models and Viterbi Decoding Analysis

As discussed below, the interaction of polymers with a nanopore can generate ionic current time series that are much more complicated than those depicted in **Figure 2a** (*left*) and **2b** (*left*). For example, consider the noisy, multistate current recording in **Figure 2c** (*left, blue trace*). One might ask what produces the variety of conductance states and the sequence of conductance value transitions within each event.

For analytical purposes, there are two questions to be addressed. First, can persistent conductance states and patterns in the data be identified in order to reduce the events in the time series to a small number of parameters? If so, data reduction and interpretation can be greatly simplified. Second, what can be learned about the sample that caused the current blockades, given the observed data? Specifically, can the ionic current time series help determine whether the number of distinctly different types of analytes in a sample, the concentration(s) of the analyte(s), the physical or chemical information encoded in the analyte(s), and the properties of the analyte(s) change over time?

One method of analyzing this class of time series is to use the architecture of the Hidden Markov Model (HMM) (51, 52). An HMM is a statistical model of a Markov source that generates a time series of observations that are probabilistically related to the states. The observable time series is hypothesized to be made up of random variables drawn from a set of state-transition-dependent probability distributions (i.e., output distributions). In this case, a Markov source is a matrix of transition probabilities between possible states of the analyte-induced nanopore current levels. The problem is that the state of the Markov source is not directly observable (i.e., it is hidden). Instead, we can only observe a variable that is statistically related to the hidden states, namely the ionic current.

In the examples described in this review, the conductance time series is related to a particular analyte interacting with the nanopore expressed through its characteristic blockade level. The advantage of using HMM methodologies rather than more simplistic techniques becomes evident as the signals become more complex, for instance,

---

Figure 2

Classes of signals from nanopore-based detection and characterization of analytes. (*a*) Ionic current time series for a single nanopore in the presence of an analyte at three different concentrations (*left*); dependence of the mean current on analyte concentration (*center*); two possible lifetime distributions estimated from blockade event dwell-time analysis (*right*). (*b*) Ionic current time series with indistinct current fluctuations caused by an analyte (*left*); power spectral density of the current fluctuations (*center*); dependence of the kinetic, thermodynamic, and structural parameters derived from spectral density analysis (*right*). (*c*) Characterizing complex current time series with Hidden Markov Model (HMM)–Gaussian Mixture Model statistical analysis. *Left*: Time series data (*blue*) with a corresponding decoded state sequence (*black*). *Right*: Viterbi graph search algorithm that finds the most probable state sequence given the HMM and the data. The values of $I_1, \ldots, I_n$ represent the conductance time series from the pore that is interrogated by analyte(s), and the nodes labeled $S_1, \ldots, S_4$ represent the states in the conductance distribution. The gray arrows represent the probabilities for making transitions from one state to another in one time step. The red arrows represent one solution for a given HMM and data set.

current blockades that are characterized by steps through multiple conductance states or events that are heavily buried in noise and cannot be delineated by simple threshold methods.

The hypothetical data in **Figure 2c** (*left, blue trace*) illustrate analyte-induced ionic current blockades with overlapping states. The use of thresholding mechanisms to describe the system (i.e., assigning representative conductance values for the time series based on whether the conductance has changed by an arbitrary number of standard deviations from arbitrary mean values) results in many physically unrealistic state changes (*not shown*). In contrast, the HMM method determines the state transition probability matrix and incorporates it into a state-decoding algorithm (53), resulting in a superior description of the data set in **Figure 2c** (*left, black trace*). The Viterbi decoding algorithm is particularly useful because it makes decisions based on the whole sequence of data and is not overly influenced by outlier events in the time series or by a degree of overlap in the output distributions.

The classical Baum-Welsh training procedure, often used for HMM decoding, requires supervised training, i.e., human-labeled training sets, to optimize the parameters of the HMM. Furthermore, it requires the assumption that the noise of the system is white and independent of the blockade level, conditions that are not necessarily met for nanopore-based conductance measurements. Various research groups have proposed variations on the Baum-Welsh training procedure or have eliminated it altogether (54). These variations allow more accurate parameter estimation in the presence of correlated input data (i.e., state value plus noise) (55, 56). These methods were extensively developed for other time-correlated fields such as speech pattern recognition (57, 58) as well as for other complex time-series problems, such as economic forecasting (59).

Figure 2c (*right*) illustrates an HMM graph of a first-order Markov model. In this type of model, a particular state depends only on the immediate prior states, and this assumption is reasonable for reversible chemical reactions. One solution of the HMM is shown by the pathway defined by the red arrows in the graph. In the GMM-based HMM, the best state sequence of the HMM given the data is estimated by a Maximum A Posteriori (MAP) estimation procedure (60).

3. DETECTION AND CHARACTERIZATION OF NUCLEIC ACIDS WITH SINGLE NANOPORES

Figure 1a shows only three of the many protein ion channels found in nature. One of these, the *Staphylococcus aureus* α-hemolysin ion channel, has become an excellent model system for analytical applications for two reasons. First, unlike many channels, studies have shown that this nanopore can remain fully open for periods of up to several hours because its voltage-dependent gating behavior (i.e., the spontaneous switching between different conductance states) can be completely suppressed (11, 54). Therefore, any conductance fluctuations observed in the presence of analyte can be attributed to that agent and not to the pore itself. Second, some polymers can reside ~500-fold longer inside this pore (~100 μs) than would be expected from a simple one-dimensional diffusion calculation (24). The latter result is significant in that it would otherwise be impossible to directly observe polymer-induced ionic current blockades. Specifically, without such analyte-binding interactions, there would be a statistically insignificant number of ions (~100 or fewer) that would flow past the molecule while it is inside the nanopore.

A nanopore that does not gate and that interacts with analytes that enter it is ideally suited to the task of interrogating polymers (24). For example, the α-hemolysin nanopore can also be used to detect and characterize single-stranded RNA and DNA polynucleotides. **Figure 3a** illustrates ionic current blockades caused by individual poly[U] RNA molecules that were added to the aqueous phase, bathing one side of the nanopore. The polymers were driven into the pore by an applied electric field (14, 16). The blockades were well defined in both amplitude and lifetime. For a poly[U] sample that was relatively monodisperse in length, the distribution of blockade lifetimes (**Figure 3b**, *inset*) was described well by a three-component Gaussian Mixture Model (GMM). Similar experiments performed with different monodisperse lengths of poly[U] demonstrated that the two longest characteristic lifetimes were proportional to the number of bases in the polymer (**Figure 3b**) (14). Because the contour lengths of these polymers were >52 nm, and the α-hemolysin channel had to be at least as long as a lipid bilayer is wide (~4 nm), it was assumed that the polymers were threading completely through the α-hemolysin channel.

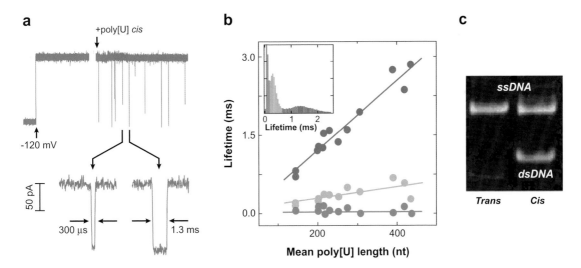

Figure 3

Detection and characterization of single-stranded polynucleotides using a single protein nanopore. (*a*) Transient ionic current blockades caused by single-stranded RNA of one length. (*b*) Residence times for different-length RNAs in the nanopore. The colors correspond to the three mean residence times estimated from blockade-event lifetime histograms for each length poly[U]. *Inset*: Residence-time histogram for 210-nucleotide-long poly[U]. (*c*) Polymerase chain reaction demonstrates that single-stranded DNA (ssDNA), but not double-stranded DNA (dsDNA), is transported through the α-hemolysin channel from the *cis* to the *trans* side. Adapted from Reference 14 with permission.

Similar results were obtained with homopolymers of single-stranded DNA (ssDNA). However, in this case the conductance blockade lifetime distributions were described by only two Gaussians. The lifetime of the longer-lived events was again proportional to the polynucleotide length. Blunt-ended double-stranded DNA (dsDNA) only caused short-lived blockades and were assumed not to thread through the pore, as single-stranded RNA and DNA polynucleotides were thought to do. More compelling evidence for this interpretation was obtained using the polymerase chain reaction (PCR) technique (**Figure 3c**). By adding both ssDNA and dsDNA to the solution bathing one side of the nanopore, PCR was used to verify that ssDNA but not dsDNA was transported through the α-hemolysin channel (14). These results are consistent with the diameter of the α-hemolysin pore estimated from the channel's crystal structure (61), which was unknown at the time of the DNA transport measurements. dsDNA has been shown to thread through larger ion channels in a similar manner (62, 63).

If poly[U] RNA threads completely through the pore, why should there be three characteristic lifetimes for each polymer length? The events with the shortest lifetimes, which are independent of the polynucleotide length, are most likely caused by polymers entering and leaving the same pore entrance. It had been suggested that the two different longer lifetimes, which are proportional to the polymer contour length, might be due to different rates of transport for poly[U] molecules that enter the nanopore via either their 3′- or 5′-ends (14). More recent experiments and molecular dynamics simulations demonstrated that this is indeed the case (18).

4. SEQUENCING DNA WITH SINGLE NANOPORES

Because single-stranded polynucleotides can thread through a single α-hemolysin channel at ~2 μs/base (**Figure 3b**) essentially as straight rods, it was suggested that a single nanopore might prove useful for rapidly sequencing DNA if each of the four bases in a polynucleotide decrease the nanopore conductance by an amount that correlates with base type (i.e., A, T, G, and C for DNA). In the simplest possible scheme, the sequence would be read directly from the ionic current time series (14). If that were possible, then it would take only ~2 ms to sequence a kilobase-long piece of DNA and only ~6000 s to sequence an entire human genome! However, the limitations of that scheme have been noted (14). These constraints include the low signal-to-noise ratio (e.g., only ~500 ions flow past any one base inside a nanopore) and that the rate of polymer transport through the nanopore is probably not uniform.

Recently, homopolymers containing cytosine or adenosine were shown to cause distinctly different blockade patterns in the single α-hemolysin channel current (15). Specifically, poly[C]-induced current blockades were greater in magnitude and shorter in duration than those caused by poly[A]. Moreover, a diblock copolymer of poly[C]:poly[A] caused predominantly two-step blockades that were characteristic of the poly[C] signature followed by that of the poly[A] segment (15). Some researchers suggested that this represented evidence for sequencing individual bases of DNA with a single nanopore. However, it is possible that different solution structures of poly[C] and poly[A] (64) are responsible for the varying degrees of current blockade and residence times for the polymers driven through the nanopore.

Other schemes for sequencing DNA with nanopores were subsequently proposed. According to these hypotheses, the DNA sequence could be determined from either the transverse tunneling current flowing through a single base and two electrodes at opposite sides of a pore entrance (**Figure 4**, center) (65), the change in the voltage resulting from a single base moving across a dielectric barrier (**Figure 4**, right) (66), or the flow of current through a single-electron transistor (67). Further details on each of these DNA sequencing schemes can be found in a recent review (68).

The reliability of nanopore-based DNA sequencing using conductance measurements is limited in part by the low signal-to-noise ratio caused by too few ions flowing past each base in the pore. Conceivably, averaging the signal by using an oscillating electric field to repeatedly floss the ssDNA molecule back and forth through the pore may provide a solution to this problem (17, 69, 70). Oscillating electric fields have also been suggested to provide a means for precisely controlling the position of ssDNA

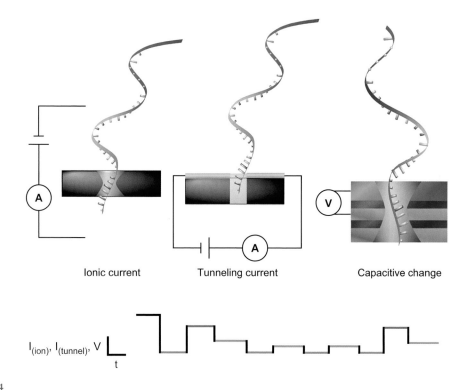

Ionic current Tunneling current Capacitive change

$I_{(ion)}$, $I_{(tunnel)}$, V

t

Figure 4

Proposed nanopore-based DNA sequencing schemes. *Top, from left to right*: direct DNA sequence readout via measurements of ionic current (14), tunneling current (65), and voltage differences (107). *Bottom*: highly simplified cartoon illustration of how each of the four different bases of DNA might produce characteristic time series recordings in each of the above schemes.

within a nanoconstriction (71). These methodologies have not yet been reduced to practice.

5. ANALYTE-INDUCED SIGNAL COMPLEXITY: FRIEND OR FOE?

Figure 5a illustrates conductance blockades caused by identical-length homopolymers of single-stranded poly[T], poly[C], and poly[A] as they are driven into a single α-hemolysin nanopore (17). The three polymer types can be distinguished easily from one another by visual inspection of their respective signals. Like poly[U]-induced blockades (**Figure 3a**), the amplitude distributions for events caused by poly[C] and poly[A] can be described by single Gaussians. In contrast, the blockade distributions and patterns caused by poly[T] are much more complex. Also, the signal patterns for poly[T] depend on the direction in which the polymer is driven through the nanopore (**Figure 5b**).

An HMM analysis (see Section 2.3, above) demonstrated that poly[T]-induced blockades, with event lifetimes as long as 2 s, were described by a GMM with ~38 components for a polymer driven through the pore in one direction and ~18 components when it was transported in the opposite direction (54). There is some solace in the ansatz that the complex blockade signatures are most likely related to the structure of the α-hemolysin nanopore interior (54). However, the question remains whether this degree of signal complexity is cause for celebration or grief.

Figure 5c illustrates a series of poly[T]-induced current blockade amplitude histograms over a wide range of blockade lifetime intervals. When presented in this manner, the relatively large number of Gaussian components needed to describe the data does not appear to be so daunting. Rather, the data take on the appearance of a spectral fingerprint (60).

It is possible that the complexity of a polymer's signals (e.g., the distribution of blockade amplitudes, as shown in **Figure 5c**) and the HMM matrix that describes the probabilities of making a transition between a given conductance state and each of the many others may provide the means of identifying, with a high degree of certainty, a particular analyte type (60). If this is indeed the case, then it may become possible to determine how cells "think" in real time by observing changes in the cytoplasmic mRNA content in response to different stimuli (72). Of course, the nanopore would first have to be calibrated or trained with the kinds of mRNA that are thought to be produced by the cell in question.

6. ANALYTE DETECTION AND QUANTITATION USING MOLECULAR ADAPTERS

The sensor functionality of single protein nanopores can be altered with genetic engineering. For example, the native α-hemolysin channel can be rendered relatively insensitive to heavy metal divalent cations (12, 13, 73). By placing novel amino acid side chains at one entrance to the nanopore, the channel is sensitized to relatively low concentrations of divalent cations. However, this method limits the range of analytes that

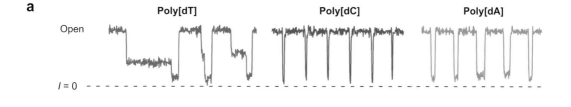

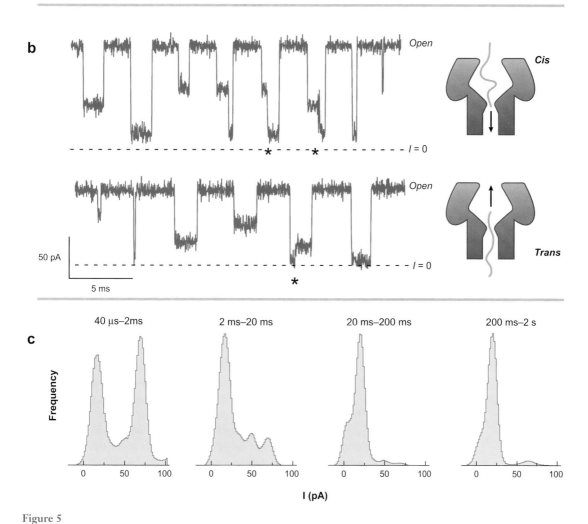

Figure 5

Complex signals caused by polynucleotides driven into a single protein nanopore. (*a*) Ionic current blockades caused by identical-length single-stranded DNA homopolymers of poly[dT], poly[dC], and poly[dA]. (*b*) Poly[dT] current blockade patterns depend on the direction in which the polynucleotide enters the pore. Note the reverse patterns for blockades are denoted by (*). (*c*) Conductance-state histograms for poly[dT] in the nanopore over a range of residence times. Adapted from References 16 and 17 with permission.

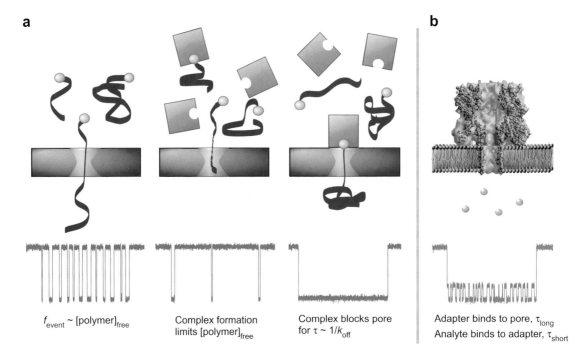

a

$f_{\text{event}} \sim [\text{polymer}]_{\text{free}}$

Complex formation
limits $[\text{polymer}]_{\text{free}}$

Complex blocks pore
for $\tau \sim 1/k_{\text{off}}$

b

Adapter binds to pore, τ_{long}
Analyte binds to adapter, τ_{short}

Figure 6

Detection schemes with analyte binding sites on pore-permeant macromolecules. The binding site (*green sphere*) is attached to either (*a*) a polymer whose ability to enter (or transport through) the pore depends on the presence of analyte (*orange*), or (*b*) a molecular adapter that binds to both the pore interior and an analyte of choice. Adapted from References 17 and 20 with permission.

can be detected and quantitated with a nanopore, and it is difficult to rationally design stereospecific analyte binding sites located at the pore entrance or inside the pore (13).

One way to avoid these issues is to position the analyte binding site on something other than the nanopore. Two approaches to this solution are illustrated in **Figure 6**. In one method, the binding site is placed on a polymer (**Figure 6a**). In the absence of analyte, the polymer is driven into and through the nanopore (14), and the frequency of blockades increases with the concentration of free polymer in solution (16, 17). The presence of analyte changes the manner in which the polymer interacts with the nanopore. If the polymer is sufficiently short, the analyte inhibits the entry of polymer into the pore, decreasing the number of polymer-induced blockades per unit time in a stoichiometric fashion. Thus, the reduction in time-averaged blockade frequency is directly related to the analyte concentration (17). If the polymer is long enough to enter the pore even when bound with analyte, then the mean time that the complex is bound can be determined from the distribution of polymer–analyte-induced current blockades (17). Because different polymers give rise to different classes of nanopore current blockades, this method can be used for the simultaneous detection of multiple analytes.

The binding site can also be located within cyclic molecules (e.g., β-cyclodextrins) that can react with several types of small molecules (20). In such a case the cyclic molecule, which by chance binds to the pore interior (see Reference 24), acts as a molecular adapter. When the adapter binds to the nanopore, the conductance decreases. The subsequent binding of analytes to the adapter, when it is inside the pore, further modulates the nanopore ionic current blockade (**Figure 6b**).

The use of polymers or molecular adapters that bind analytes confers an additional advantage over positioning the binding site on the nanopore itself: The analyte to be detected can be changed merely by replacing the polymer or adapter in solution (17).

7. SINGLE-MOLECULE FORCE SPECTROSCOPY VIA SINGLE NANOPORES

Placing a large macromolecule on one end of a polynucleotide inhibits (16, 17) or retards (17) the translocation of a polynucleotide through the nanopore (illustrated schematically in **Figure 6a**, *right*). It has been shown that if a polymer-macromolecule complex is driven into the pore with a relatively high magnitude of the applied potential and if the voltage difference is subsequently decreased, the polymer can take a surprisingly long time (i.e., many seconds) to back-diffuse out of the nanopore (16). This finding suggested a new method to probe the interactions between the polymer and the nanopore and/or intrapolymer interactions (**Figure 7**). In fact, single nanopores provide a possible improvement for single-molecule force spectroscopy (18), because they can be used to perform many more experiments per unit time than can be done with the test molecule tethered to a solid support. In general, the ability of a molecule to enter and/or translocate through a single nanopore can be modulated by a suite of forces (e.g., electrical, chemical, optical tweezer). By modulating any of these forces, the residence time of a polymer in the pore can be altered.

A simplified version of this method (17) was used to estimate the strength of DNA hairpins using single α-hemolysin nanopores (19). The free end of the polymer could enter the pore. However, the hairpin, which is a dynamic entity, did not immediately follow the free end past the α-hemolysin channel vestibule. This study demonstrated that the lifetime of the DNA hairpin correlated with the free energy of the hairpin formation and was substantially altered by a single base mismatch. A variation on this experiment was used to estimate the time it takes to unzip duplex DNA that has an ssDNA overhang (74).

Figure 7a illustrates how the forces of an applied potential and an optical tweezer acting on a DNA molecule–polystyrene bead complex can be balanced when the DNA is inside the nanopore (75). This technique should eventually prove useful for controlling the rate at which individual polymers thread through a single nanopore. In addition, the ability to measure the force on the polymer inside the pore (using optical tweezers) provides another analytical tool to probe inter- and intramolecular interactions.

Figure 7b illustrates how nanopore-based force spectroscopy can be used to study the interactions between DNA binding proteins and polynucleotides. Once the DNA end of the DNA-protein complex is inside the pore, the applied potential is rapidly

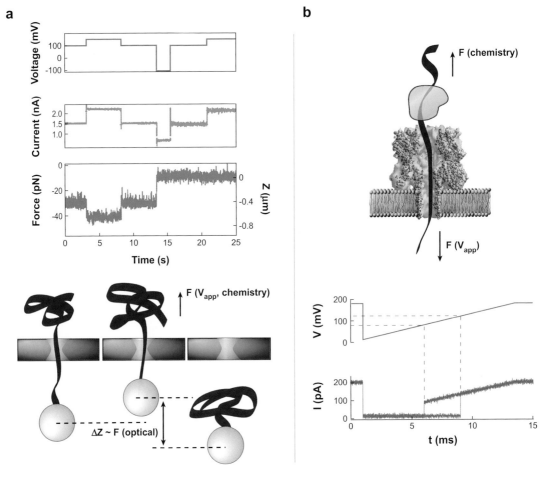

Figure 7

Two nanopore-based single-molecule force spectroscopy techniques. (*a*) The ability to apply forces (e.g., via a transmembrane electrical potential difference or an optical tweezer) to a polymer in the nanopore will enable quantitative analysis of intramolecular interactions within the polymer and of intermolecular interactions between a polymer and other molecules. (*b*) Force spectroscopy on DNA-protein complexes. *Top:* DNA-binding protein-mediated force delays the voltage driven transport of the polynucleotide through the nanopore. *Center:* Time-dependent voltage applied to an individual protein-DNA complex. *Bottom:* Single channel current from two independent voltage ramp experiments. Dissociation of the complex causes the abrupt increase in the current (*dashed lines*) to the open pore conductance. Adapted from References 16, 75, and 76 with permission.

decreased and then slowly increased. The time it takes the complex to dissociate, which leads to the complete transport of the polynucleotide through the nanopore (see, e.g., Reference 17) (**Figure 6***a*), is determined (76).

Similar experiments (77) were performed to determine the rate at which polynucleotides unzip as a function of the applied potential force loading rate. Researchers have suggested that this approach may prove useful in estimating DNA sequences by measuring the time it takes to unzip complementary versus noncomplementary DNA strands (78).

8. DETECTION AND CHARACTERIZATION OF PROTEINS AND ANTIBODIES

Single nanopores have also been used to detect proteins and determine whether they are in a native or unfolded state. For example, by placing a binding site for a protein or antibody on one end of a polynucleotide (**Figure 6a**), investigators showed that a single nanopore can be used to detect protein and antibodies with a protein ion channel that normally has no affinity for either analyte (17). In contrast, single α-hemolysin channels were used to directly detect short, charged polypeptides that have no known physiological relationship to the nanopore (79, 80).

It is also possible to detect full-length proteins with high specificity by using ion channels that normally interact with them. *Bacillus anthracis*, the bacteria that causes anthrax-related cell death, secretes three toxins: protective antigen (PA83, ~83 kg/mol), lethal factor (LF, ~89 kg/mol), and edema factor (EF, ~90 kg/mol). *In vivo*, PA83 binds to cells and is cleaved into two fragments. The 63-kg/mol fragment, PA63, remains associated with cell membranes and forms a transmembrane ion channel. Either LF or EF binds to the PA63 channel (**Figure 1a**) to form lethal or edema toxins, respectively; these complexes are subsequently endocytosed. The pH of the endosome is decreased and LF or EF is transported into the cytosol, causing cell death (81). The binding of either LF or EF to a PA63 ion channel in planar bilayer membranes converts the nanopore's current-voltage relationship from slightly nonlinear to strongly rectifying. This effect was graded with either the LF or the EF concentration, and the apparent binding constants were ~50 pM (30). Also, an antibody against PA63, which has no effect on the nanopore's *I-V* relationship, completely inhibits the ability of LF to alter the nanopore's conductance at any voltage (30). Thus, the anthrax PA63 ion channel can be used as a sensitive detector for LF or EF and as a high-throughput screening device for potential therapeutic agents against anthrax infection.

In addition to detecting protein fragments and full-length proteins, single nanopores can be used to determine whether individual soluble proteins are in their folded (native) conformation, or completely unfolded. This task is usually addressed using circular dichroism on an ensemble of the protein. As is illustrated in **Figure 8a**, in the absence of guanidinium chloride [Gdm-HCl], the single α-hemolysin channel current was quiescent, even in the presence of a maltose binding protein from *E. coli*. The latter protein, which contains no stabilizing disulfide bonds, comprises 370 amino acid side chains (~40 kg/mol molecular mass) and is negatively charged at physiological pH. Interestingly, for concentrations of Gdm-HCl greater than 0.8 M, the protein caused well-defined current blockades (82). Control experiments demonstrated that 0.8 M of Gdm-HCl denatures the maltose binding protein but does not significantly affect the α-hemolysin channel conductance. Thus, the denatured protein appears to be driven into the α-hemolysin nanopore

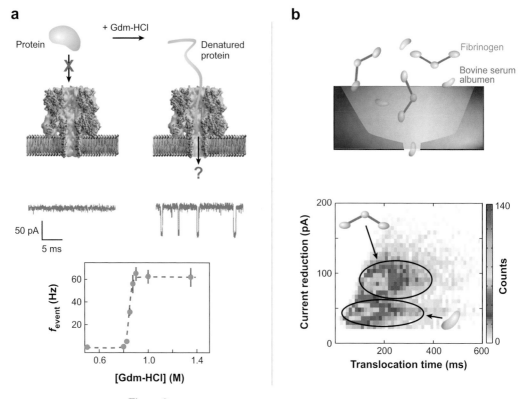

Figure 8

Direct detection of proteins with single nanopores. (*a*) A protein in its native conformation is too large to enter a single α-hemolysin nanopore (*top left*). Addition of guanidinium chloride greater than 0.8 M denatures the protein. The denatured protein can then enter the nanopore and cause transient current blockades (*top right, middle*). The frequency of blockades increases with increasing denaturation concentration (*bottom*). (*b*) The ability of two different proteins (fibrinogen and bovine serum albumin) to partition into a solid-state nanopore was determined from single channel recordings. The results are plotted in terms of the mean current blockade amplitudes versus the residence times of the proteins in the nanopore. Adapted from References 82 and 86 with permission.

in a manner similar to polynucleotides (14, 16, 84) and dextran sulfate molecules (83).

Solid-state nanopores have also proven to be effective for detecting proteins. A single solid-state elliptical nanopore with an orifice ∼58 nm by 50 nm and 20 nm thick was used to detect bovine serum albumin (BSA) at low concentration (85). Smaller single solid-state nanopores (∼16 nm in diameter, ∼10 nm long) were used to detect two different proteins: BSA (∼66 kg/mol) and fibrinogen (340 kg/mol) (86). Distinctly different patterns in a plot of the protein-induced current blockade amplitude versus the protein residence times in the pore (**Figure 8b**) were observed for fibrinogen and BSA, and thereby suggest that the nanopore was able to discriminate between the two proteins. Based on the influence of BSA on the pore conductance as a function of pH, the charge on that protein was estimated for each pH value. The transport of the protein through the nanopore, as judged by conductance measurements, was confirmed using chemiluminescence with fluorescently labeled versions of the polypeptide.

9. SINGLE-MOLECULE MASS SPECTROMETRY USING A NANOPORE

Nonelectrolyte polymers have been used to estimate the size of ion channels by comparing the relative conductivity of an ion channel with the addition of polymers having a range of molecular weights or sizes (22, 24, 26, 27). By approaching the problem from the reverse (i.e. probing the size of the polymer molecules with a nanopore of known geometry), then a single nanopore can provide the basis for accurate measurements for the sizes of individual molecules in solution.

The principle of sizing nanopores with polymers is simple. Some polymers (e.g., PEG) decrease the bulk electrolyte conductivity and those molecules that are small enough to enter the pore will decrease the single-channel conductance. The size of the pore is then estimated by determining the largest polymer that can enter the pore and knowing the hydrodynamic radii of the polymers.

In principle, the polymer-induced decrease in pore conductance should scale with the size of the polymer in the pore: Larger polymers should decrease the conductance more than smaller ones. The question, then, is how small of a difference between the polymer molecular masses can one determine using a nanopore.

A representation of the experiment is illustrated in **Figure 9a**. PEG added to the solution bathing one side of the membrane caused well-defined polymer-induced fluctuations in the ionic current (**Figure 9b**). The average residence time of the polymers in the pore was on the order of 1 ms (24, 28). Representative data for polydisperse (MW_{avg} = 1500 g/mol) and monodisperse (MW = 1294 g/mol) PEG were collected until a large number ($>10^5$) of individual polymer-pore interactions were observed to ensure a statistically significant sampling of the data (29). There were clearly discernible differences in the depths of the current blockades caused by the polydisperse PEG sample, whereas the monodisperse PEG caused blockades that were virtually all of the same mean conductance value. An all-points histogram of the entire current time series does not permit accurate decoding of the blockades caused by each of the differently sized PEGs in the dispersion (*not shown*).

To resolve the individual components within the mixture, each blockade event is represented by its mean current value. A histogram made from the mean current blockade amplitudes clearly resolves $\sim$24 PEG n-mers ranging from n = 25 to n = 49 ($HO(CH_2CH_2O)_nH$) (**Figure 9c**, *red*) and correlated 1:1 with a matrix-assisted laser desorption/ionization–time-of-flight mass spectrum of the sample. Calibration of the mass of the sample was achieved by an identical analysis of PEG-1294 (**Figure 9c**, *blue*). Experimental data were reduced with an HMM/GMM Viterbi decoding analysis (**Figure 9c**, *black*), which allowed access to the residence time for each polymer

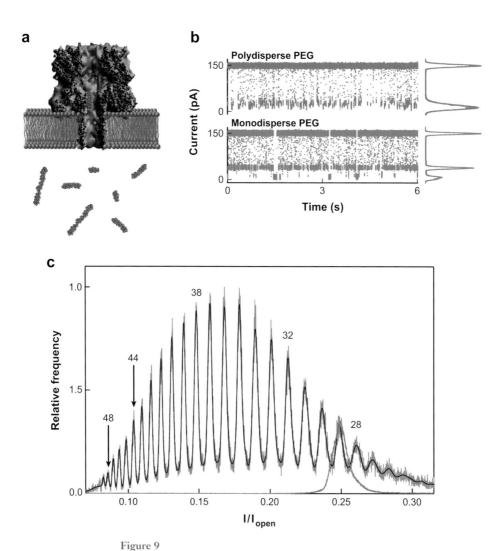

Figure 9

(*a*) Representation of single-molecule mass spectrometry of polymers using a single α-hemolysin nanopore. (*b*) The dispersion of conductance blockade amplitudes for polydisperse poly(ethylene glycol) (PEG) (*red*) is much greater than that for a monodisperse sample of the polymer (*blue*). (*c*) Conductance-based mass distributions for polydisperse (*red*) and monodisperse (*blue*) PEG. The numbers correspond to the degree of polymerization of the PEGs. Adapted from Reference 29 with permission.

in the mixture, thereby permitting two dimensions of discrimination for a single molecular component (blockade depth and time). This technique can be easily extended to any polymer analytes [such as poly(styrene sulfonate); see Reference 87] that interact with the interior of the nanopore.

10. THEORETICAL ADVANCES IN MODELING POLYMER TRANSPORT THROUGH SINGLE NANOPORES

The development of analytical theories and significant advances in theoretical chemistry and molecular simulation techniques have provided valuable insights into the mechanisms of polymer transport through nanopores. Initially, several analytical treatments describing the problem of polymer translocation through a narrow pore were undertaken. Some of those studies focused on predicting the dependence of polynucleotide residence time in a nanopore as a function of the polymer length (88, 89). Three different models (**Figure 10b**) suggest that (*a*) the dynamics of the polymer chain in the bulk gives rise to an entropic barrier (90, 91), (*b*) the polymer chain strongly interacts with the nanopore (i.e., the theory ignores the part of the chain that is free in solution) (92), and (*c*) the motion of the flexible polymer over the free-energy barrier is treated as a one-dimensional piece or kink within the chain (the kink provides a time-dependent solution to the Kramers problem) (93–95). Interestingly, several of these theories predict that the residence time of a polymer in the pore increases in proportion to the polymer contour length in the presence of an applied electrostatic potential, as was determined experimentally (14). Clearly, there is a need for new experiments to better distinguish among these different models.

Recently, a theory describing nanopore-based single-molecule force spectroscopy experiments was developed (96) and applied to the unzipping of DNA hairpins through a single biological nanopore (97). The authors applied either a linear cubic or cusp free energy surface to the Kramers model to find an analytical expression for the rate of activation as a function of the applied force. This expression leads to the probability distribution of the rupture forces that can be compared to experimental results. A purely theoretical treatment was also used to determine the error associated with the unzipping of DNA (91, 92, 98) and the estimation of the DNA sequence (90, 99–102). The error in this approach decreases exponentially with the number of unzipping attempts (103).

In addition to the analytical developments summarized above, microscopically detailed molecular simulations of macromolecular transport through biological (104, 105) and solid-state (106–108) nanopores have provided further insight into these processes that have remained elusive when studied with analytical methods alone. The simulations enable a visualization, at the atomic level of detail, of how ions and macromolecules stochastically electrodiffuse through very small structures (**Figure 10a**). In one study (104), the average ionic occupancy of the transmembrane pore, osmotic permeability, current-voltage relationships, and selectivity for charged ions can be estimated. These powerful computational methods are also being used to aid the rational design of complex solid-state nanopores for DNA sequencing efforts (107, 109) as well as to study the effects of single-nucleotide polymorphisms and the interactions between DNA and DNA-binding proteins (110).

In contrast to brute-force, atomistically detailed simulations, the Poisson-Nernst-Planck theory (PNP) is a simple, coarse-grained approach for modeling ion transport in single nanopores. Significant advances in the development of computational algorithms for carrying out PNP calculations in biological nanopores have occurred over

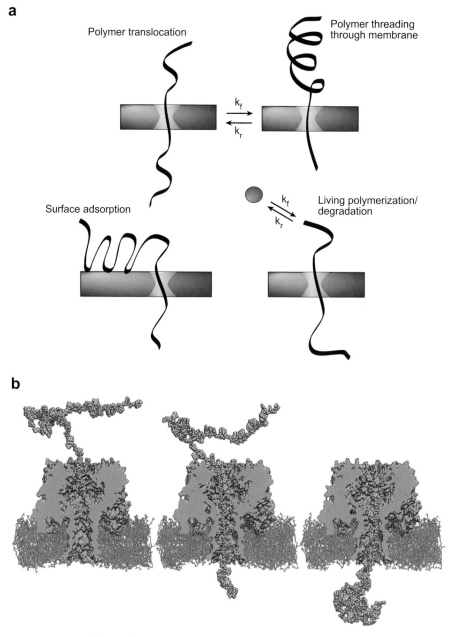

Figure 10

Theoretical methods for polymer-nanopore studies. (*a*) Analytical approaches for the study of polymer translocation through single nanopores also provide keen insight into the process. (*b*) Molecular dynamics simulation of a polynucleotide being driven through a single α-hemolysin nanopore. Courtesy of Aleksei Aksimientiev (University of Illinois at Urbana-Champaign). Adapted from References 90, 91, and 120 with permission.

the past two decades (111–115), and the theory has recently been shown to be an effective component of hybrid and multiscale modeling frameworks (105, 116, 117). These results will undoubtedly enhance our ability to understand the fundamental physics of macromolecular transport through nanopores. The properties predicted from the calculations will also aid in the design of nanopores for many analytical purposes.

11. SUMMARY

This review describes some of the experimental and theoretical efforts applied to the development of single protein and solid-state nanopores for a wide variety of applications, including the detection and quantitation of ions, proteins, polypeptides and polynucleotides; single-molecule force spectroscopy; high-throughput screening against biological warfare agents; and a conductance-based mass spectrometry that should prove to be complementary to existing mass spectrometry techniques. Single nanopores also have the potential to provide linchpin technologies useful for rapid DNA sequencing and proteomic analysis.

The nanoporous sensor community is currently divided into two camps largely based on the origin of the nanopores used. Although both biological and solid-state pores have been successfully applied to chemical analysis, biological nanopores have a ~3.8-billion-year head start on development, and have perfected precision chemical modification of the interior of the pore wall. Solid-state nanopores offer the promise of improved long-term stability, yet precise physical and chemical modification remains an elusive goal. Indeed, if the α-hemolysin protein ion channel could speak, it might be justified in quoting Mark Twain: "The reports of my death have been greatly exaggerated."

As with many other analytical techniques, the choice of nanopore will be determined by the application. It is also possible that a combination of biological and solid-state nanopores in a hybrid platform will become a formidable partnership for precision analytical measurements.

Nature demonstrates that a good strategy for the unambiguous identification of a wide range of analytes is to employ many different ion channels, each with a high degree of selectivity for a particular analyte. This approach seems likely to be true if nanopore-based sensors can be made practical in robust matrices (121).

ACKNOWLEDGMENTS

Supported in part by the Office of Law Enforcement Standards at the National Institute of Standards and Technology.

DISCLOSURE STATEMENT

The authors have filed a patent application for the technology described in Section 9 (polymer analysis using nanopores).

LITERATURE CITED

1. Coulter WH. 1953. *U.S. Patent No. 2,656,508*
2. DeBlois RW, Bean CP. 1970. Counting and sizing of submicron particles by resistive pulse technique. *Rev. Sci. Instrum.* 41:909–16
3. Ito T, Sun L, Crooks RM. 2003. Simultaneous determination of the size and surface charge of individual nanoparticles using a carbon nanotube–based Coulter counter. *Anal. Chem.* 75:2399–406
4. Saleh OA, Sohn LL. 2003. Direct detection of antibody-antigen binding using an on-chip artificial pore. *Proc. Natl. Acad. Sci. USA* 100:820–24
5. Uram JD, Ke K, Hunt AJ, Mayer M. 2006. Label-free affinity assays by rapid detection of immune complexes in submicrometer pores. *Angew. Chem. Intl. Ed.* 45:2281–85
6. Hodgkin AL, Huxley AF. 1952. Currents carried by sodium and potassium ions through the membrane of the giant axon of *Loligo*. *J. Physiol. Lond.* 116:449–72
7. Katz B. 1966. *Nerve, Muscle, Synapse*. New York: McGraw-Hill
8. Henry JP, Chich JF, Goldschmidt D, Thieffry M. 1989. Blockade of a mitochondrial cationic channel by an addressing peptide: an electrophysiological study. *J. Membr. Bio.* 112:139–47
9. Simon SM, Blobel G. 1991. A protein-conducting channel in the endoplasmic reticulum. *Cell* 65:371–80
10. Bezrukov SM, Kasianowicz JJ. 1993. Current noise reveals protonation kinetics and number of ionizable sites in an open protein ion channel. *Phys. Rev. Lett.* 70:2352–55
11. Kasianowicz JJ, Bezrukov SM. 1995. Protonation dynamics of the α-toxin ion channel from spectral analysis of pH-dependent current fluctuations. *Biophys. J.* 69:94–105
12. Braha O, Walker B, Cheley S, Kasianowicz JJ, Song LZ, et al. 1997. Designed protein pores as components for biosensors. *Chem. Biol.* 4:497–505
13. Kasianowicz JJ, Burden DL, Han LC, Cheley S, Bayley H. 1999. Genetically engineered metal ion binding sites on the outside of a channel's transmembrane β-barrel. *Biophys. J.* 76:837–45
14. Kasianowicz JJ, Brandin E, Branton D, Deamer DW. 1996. Characterization of individual polynucleotide molecules using a membrane channel. *Proc. Natl. Acad. Sci. USA* 93:13770–73
15. Akeson M, Branton D, Kasianowicz JJ, Brandin E, Deamer DW. 1999. Microsecond time-scale discrimination among polycytidylic acid, polyadenylic acid, and polyuridylic acid as homopolymers or as segments within single RNA molecules. *Biophys. J.* 77:3227–33
16. Henrickson SE, Misakian M, Robertson B, Kasianowicz JJ. 2000. Driven DNA transport into an asymmetric nanometer-scale pore. *Phys. Rev. Lett.* 85:3057–60
17. Kasianowicz JJ, Henrickson SE, Weetall HH, Robertson B. 2001. Simultaneous multianalyte detection with a nanometer-scale pore. *Anal. Chem.* 73:2268–72
18. Mathe J, Aksimentiev A, Nelson DR, Schulten K, Meller A. 2005. Orientation discrimination of single-stranded DNA inside the α-hemolysin membrane channel. *Proc. Natl. Acad. Sci. USA* 102:12377–82

19. Vercoutere W, Winters-Hilt S, Olsen H, Deamer D, Haussler D, Akeson M. 2001. Rapid discrimination among individual DNA hairpin molecules at single-nucleotide resolution using an ion channel. *Nat. Biotech.* 19:248–52

20. Gu LQ, Braha O, Conlan S, Cheley S, Bayley H. 1999. Stochastic sensing of organic analytes by a pore-forming protein containing a molecular adapter. *Nature* 398:686–90

21. Kullman L, Winterhalter M, Bezrukov SM. 2002. Transport of maltodextrins through maltoporin: a single-channel study. *Biophys. J.* 82:803–12

22. Krasilnikov OV, Sabirov RZ, Ternovsky VI, Merzliak PG, Muratkhodjaev JN. 1992. A simple method for the determination of the pore radius of ion channels in planar lipid bilayer membranes. *FEMS Microbiol. Immun.* 105:93–100

23. Bezrukov SM, Vodyanoy I, Parsegian VA. 1994. Counting polymers moving through a single-ion channel. *Nature* 370:279–81

24. Bezrukov SM, Vodyanoy I, Brutyan RA, Kasianowicz JJ. 1996. Dynamics and free energy of polymers partitioning into a nanoscale pore. *Macromolecules* 29:8517–22

25. Movileanu L, Bayley H. 2001. Partitioning of a polymer into a nanoscopic protein pore obeys a simple scaling law. *Proc. Natl. Acad. Sci. USA* 98:10137–41

26. Bezrukov SM, Kasianowicz JJ. 2002. Dynamic partitioning of neutral polymers into a single ion channel. In *Structure and Dynamics of Confined Polymers*, ed. JJ Kasianowicz, MSZ Kellermayer, DW Deamer, pp. 117–30. Dordrecht, Netherlands: Kluwer

27. Krasilnikov OV. 2002. Sizing channels with neutral polymers. In *Structure and Dynamics of Confined Polymers*, ed. JJ Kasianowicz, MSZ Kellermayer, DW Deamer, pp. 97–116. Dordrecht, Netherlands: Kluwer

28. Krasilnikov OV, Rodrigues CG, Bezrukov SM. 2006. Single polymer molecules in a protein nanopore in the limit of a strong polymer-pore attraction. *Phys. Rev. Lett.* 97:018301

29. Robertson JWF, Rodrigues CG, Stanford VM, Rubinson KA, Krasilnikov OV, Kasianowicz JJ. 2007. Single-molecule mass spectrometry in solution using a solitary nanopore. *Proc. Natl. Acad. Sci. USA* 104:8207–11

30. Halverson KM, Panchal RG, Nguyen TL, Gussio R, Little SF, et al. 2005. Anthrax biosensor: protective antigen ion channel asymmetric blockade. *J. Biol. Chem.* 280:34056–62

31. Li J, Stein D, McMullan C, Branton D, Aziz MJ, Golovchenko JA. 2001. Ion-beam sculpting at nanometre length scales. *Nature* 412:166–69

32. Li JL, Gershow M, Stein D, Brandin E, Golovchenko JA. 2003. DNA molecules and configurations in a solid-state nanopore microscope. *Nat. Mater.* 2:611–15

33. Storm AJ, Chen JH, Ling XS, Zandbergen HW, Dekker C. 2003. Fabrication of solid-state nanopores with single-nanometre precision. *Nat. Mater.* 2:537–40

34. Heng JB, Ho C, Kim T, Timp R, Aksimentiev A, et al. 2004. Sizing DNA using a nanometer-diameter pore. *Biophys. J.* 87:2905–11

35. Fologea D, Gershow M, Ledden B, McNabb DS, Golovchenko JA, Li JL. 2005. Detecting single-stranded DNA with a solid-state nanopore. *Nano Lett.* 5:1905–9

36. Kim MJ, Wanunu M, Bell DC, Meller A. 2006. Rapid fabrication of uniformly sized nanopores and nanopore arrays for parallel DNA analysis. *Adv. Mater.* 18:3149–53
37. Kim MJ, McNally B, Murata K, Meller A. 2007. Characteristics of solid-state nanometre pores fabricated using a transmission electron microscope. *Nanotechnology* 18:205302
38. Li NC, Yu SF, Harrell CC, Martin CR. 2004. Conical nanopore membranes: preparation and transport properties. *Anal. Chem.* 76:2025–30
39. Siwy Z, Dobrev D, Neumann R, Trautmann C, Voss K. 2003. Electro-responsive asymmetric nanopores in polyimide with stable ion-current signal. *Appl. Phys. A* 76:781–85
40. Horn R, Lange K. 1983. Estimating kinetic constants from single channel data. *Biophys. J.* 43:207–23
41. Chay TR, Kang HS, Chay SC. 1988. Analyzing stochastic events in multi-channel patch clamp data. *Biol. Cyber.* 58:19–33
42. Ball FG, Sansom MSP. 1989. Ion-channel gating mechanisms: model identification and parameter-estimation from single channel recordings. *Proc. Royal Soc. Lond. B* 236:385–416
43. Magleby KL, Weiss DS. 1990. Estimating kinetic parameters for single channels with simulation. *Biophys. J.* 58:1411–26
44. Colquhoun D, Hawkes AG. 1981. On the stochastic properties of single ion channels. *Proc. Royal Soc. Lond. B* 211:205–35
45. Qin F, Auerbach A, Sachs F. 1996. Estimating single-channel kinetic parameters from idealized patch-clamp data containing missed events. *Biophys. J.* 70:264–80
46. Qin F, Auerbach A, Sachs F. 1997. Maximum likelihood estimation of aggregated Markov processes. *Biophys. J.* 264:375–83
47. Stevens CF. 1972. Inferences about membrane properties from electrical noise measurements. *Biophys. J.* 12:1028–47
48. DeFelice LJ. 1981. *Introduction to Membrane Noise*. New York: Plenum
49. Machlup S. 1954. Noise in semiconductors: spectrum of a 2-parameter random signal. *J. Appl. Phys.* 25:341–43
50. Kasianowicz JJ, Henrickson SE, Misakian M, Lerman JC, Panchal RG, et al. 2007. The detection and characterization of ions, DNA, and proteins using nanometer-scale pores. In *Handbook of Biosensors and Biochips*, ed. RS Marks, DC Cullen, I Karube, CR Lowe, HH Weetall. New York: Wiley
51. Baum L. 1972. An inequality and associated maximization technique in statistical estimation of probabilistic functions of a Markov process. *Inequalities* 3:1–8
52. Redner RA, Walker HF. 1984. Mixture densities, maximum likelihood and the EM algorithm. *SIAM Rev.* 26:195–237
53. Vitterbi AJ. 1967. Error bounds for convolutional codes and an asymmetrically optimum decoding algorithm. *IEEE Trans. Inf. Theory* IT-13:260–67
54. Kasianowicz JJ, Henrickson SE, Misakian M, Weetall HH, Robertson B, Stanford V. 2002. Physics of DNA threading through a nanometer pore and applications to simultaneous multianalyte sensing. In *Structure and Dynamics of Confined Polymers*, ed. JJ Kasianowicz, MSZ Kellermayer, DW Deamer, pp. 141–64. Dordrecht, Netherlands: Kluwer

55. Qin F, Auerbach A, Sachs F. 2000. Hidden Markov modeling for single channel kinetics and correlated noise. *Biophys J.* 79:1928–44

56. Venkataramanan L, Walsh JL, Kuc R, Sigworth FJ. 1998. Identification of Hidden Markov Models for ion channel currents. Part I: Colored background noise. *IEEE Trans. Sig. Proc.* 46:1901–15

57. Rabiner LR. 1989. A tutorial on Hidden Markov Models and selected applications in speech recognition. *Proc. IEEE* 77:257–86

58. Rabiner LR, Juang BH. 1986. An introduction to Hidden Markov Models. *IEEE ASSP Mag.* 3:4–16

59. Fraser AM, Dimitriadis A. 1993. Forecasting probability densities by using Hidden Markov Models with mixed states. In *Time Series Prediction*, ed. AS Weigend, NA Gershenfeld. Santa Fe, NM: Addison-Wesley

60. Stanford VM, Kasianowicz JJ. 2004. *Transport of DNA through a single nanometer-scale pore: evolution of signal structure*. Presented at IEEE Workshop Genomic Signal Process. Stat., Baltimore, MD

61. Song LZ, Hobaugh MR, Shustak C, Cheley S, Bayley H, Gouaux JE. 1996. Structure of staphylococcal α-hemolysin, a heptameric transmembrane pore. *Science* 274:1859–66

62. Szabo I, Bathori G, Tombola F, Brini M, Coppola A, Zoratti M. 1997. DNA translocation across planar bilayers containing *Bacillus subtilis* ion channels. *J. Biol. Chem.* 272:25275–82

63. Szabo I, Bathori G, Tombola F, Coppola A, Schmehl I, et al. 1998. Double-stranded DNA can be translocated across a planar membrane containing purified mitochondrial porin. *FASEB J.* 12:495–502

64. Saenger W. 1984. *Principles of Nucleic Acid Structure*. New York: Springer-Verlag

65. Zwolak M, Di Ventra M. 2005. Electronic signature of DNA nucleotides via transverse transport. *Nano Lett.* 5:421–24

66. Gracheva ME, Aksimentiev A, Leburton JP. 2006. Electrical signatures of single-stranded DNA with single base mutations in a nanopore capacitor. *Nanotechnology* 17:3160–65

67. Mali P, Lal RK. 2004. The dnaSET: A novel device for single-molecule DNA sequencing. *IEEE Trans. Electron Dev.* 51:2004–12

68. Zwolak M, Di Ventra M. 2007. Physical approaches to DNA sequencing and detection. *Rev. Mod. Phys.* 80:141–65

69. Kasianowicz JJ. 2004. Nanopores. Flossing with DNA. *Nature Materials* 3:355–65

70. Sigalov G, Comer J, Timp G, Aksimentiev A. 2008. Detection of DNA sequences using an alternating electric field in a nanopore capacitor. *Nano Lett.* In press

71. Polonsky S, Rossnagel S, Stolovitzky G. 2007. Nanopore in metal-dielectric sandwich for DNA position control. *Appl. Phys. Lett.* 91:153103

72. Kasianowicz JJ. 2002. Nanometer-scale pores: potential applications for analyte detection and DNA characterization. *Dis. Markers* 18:185–91

73. Menestrina G, Mackman N, Holland IB, Bhakdi S. 1987. *Escherichia coli* hemolysin forms voltage-dependent ion channels in lipid membranes. *Biochim. Biophys. Acta* 905:109–17

74. Sauer-Budge AF, Nyamwanda JA, Lubensky DK, Branton D. 2003. Unzipping kinetics of double-stranded DNA in a nanopore. *Phys. Rev. Lett.* 90:238101

75. Keyser UF, Koeleman BN, Van Dorp S, Krapf D, Smeets RMM, et al. 2006. Direct force measurements on DNA in a solid-state nanopore. *Nat. Phys.* 2:473–77

76. Hornblower B, Coombs A, Whitaker RD, Kolomeisky A, Picone SJ, et al. 2007. Single-molecule analysis of DNA-protein complexes using nanopores. *Nat. Methods* 4:315–17

77. Bates M, Burns M, Meller A. 2003. Dynamics of DNA molecules in a membrane channel probed by active control techniques. *Biophys. J.* 84:2366–72

78. Baldazzi V, Bradde S, Cocco S, Marinari E, Monasson R. 2007. Inferring DNA sequences from mechanical unzipping data: the large-bandwidth case. *Phys. Rev. E* 75:011904

79. Movileanu L, Schmittschmitt JP, Scholtz JM, Bayley H. 2005. Interactions of peptides with a protein pore. *Biophys. J.* 89:1030–45

80. Stefureac R, Long YT, Kraatz HB, Howard P, Lee JS. 2006. Transport of α-helical peptides through α-hemolysin and aerolysin pores. *Biochemistry* 45:9172–79

81. Ascenzi P, Visca P, Ippolito G, Spallarossa A, Bolognesi M, Montecucco C. 2002. Anthrax toxin: a tripartite lethal combination. *FEBS Lett.* 531:384–88

82. Oukhaled G, Mathe J, Biance AL, Bacri L, Betton JM, et al. 2007. Unfolding of proteins and long transient conformations detected by single nanopore recording. *Phys. Rev. Lett.* 98:158101

83. Oukhaled AG, Bacri L, Mathe J, Pelta J, Auvray L. 2008. Effect of the screening on the transport of polyelectrolytes through nanopores. *Europhys. Lett.* In press

84. Ambjornsson T, Apell SP, Konkoli Z, Di Marzio EA, Kasianowicz JJ. 2002. Charged polymer membrane translocation. *J. Chem. Phys.* 117:4063–73

85. Han AP, Schurmann G, Mondin G, Bitterli RA, Hegelbach NG, et al. 2006. Sensing protein molecules using nanofabricated pores. *Appl. Phys. Lett.* 88:093901

86. Fologea D, Ledden B, McNabb DS, Li J. 2007. Electrical characterization of protein molecules by a solid-state nanopore. *Appl. Phys. Lett.* 91:053901

87. Murphy RJ, Muthukumar M. 2007. Threading synthetic polyelectrolytes through protein pores. *J. Chem. Phys.* 126:051101

88. de Gennes PG. 1999. Problems of DNA entry into a cell. *Physica A* 274:1–7

89. de Gennes PG. 1999. Passive entry of a DNA molecule into a small pore. *Proc. Natl. Acad. Sci. USA* 96:7262–64

90. Muthukumar M. 1999. Polymer translocation through a hole. *J. Chem. Phys.* 111:10371–74

91. Sung W, Park PJ. 1996. Polymer translocation through a pore in a membrane. *Phys. Rev. Lett.* 77:783–86

92. Lubensky DK, Nelson DR. 1999. Driven polymer translocation through a narrow pore. *Biophys. J.* 77:1824–38

93. Sebastian KL, Paul AKR. 2000. Kramers problem for a polymer in a double well. *Phys. Rev. E* 62:927–39

94. Sebastian KL. 2000. Kink motion in the barrier crossing of a chain molecule. *Phys. Rev. E* 61:3245–48

95. Lee S, Sung WY. 2001. Coil-to-stretch transistion, kink formation and efficient barrier crossing of a flexible chain. *Phys. Rev. E* 63:021115

96. Dudko OK, Hummer G, Szabo A. 2006. Intrinsic rates and activation free energies from single-molecule pulling experiments. *Phys. Rev. Lett.* 96:108101

97. Dudko OK, Mathe J, Szabo A, Meller A, Hummer G. 2007. Extracting kinetics from single-molecule force spectroscopy: nanopore unzipping of DNA hairpins. *Biophys. J.* 92:4188–95

98. Kotsev S, Kolomeisky AB. 2006. Effect of orientation in translocation of polymers through nanopores. *J. Chem. Phys.* 125:084906

99. Kong CY, Muthukumar M. 2002. Modeling of polynucleotide translocation through protein pores and nanotubes. *Electrophoresis* 23:2697–703

100. Kong CY, Muthukumar M. 2004. Polymer translocation through a nanopore. II. Excluded volume effect. *J. Chem. Phys.* 120:3460–66

101. Muthukumar M. 2001. Translocation of a confined polymer through a hole. *Phys. Rev. Lett.* 86:3188–91

102. Muthukumar M. 2002. Theory of sequence effects on DNA translocation through proteins and nanopores. *Electrophoresis* 23:1417–20

103. Baldazzi V, Cocco S, Marinari E, Monasson R. 2006. Interference of DNA sequences from mechanical unzipping: an ideal case study. *Phys. Rev. Lett.* 96:128102

104. Aksimentiev A, Schulten K. 2005. Imaging α-hemolysin with molecular dynamics: ionic conductance, osmotic permeability, and the electrostatic potential map. *Biophys. J.* 88:3745–61

105. Muthukumar M, Kong CY. 2006. Simulation of polymer translocation through protein channels. *Proc. Natl. Acad. Sci. USA* 103:5273–78

106. Aksimentiev A, Heng JB, Timp G, Schulten K. 2004. Microscopic kinetics of DNA translocation through synthetic nanopores. *Biophys. J.* 87:2086–97

107. Gracheva ME, Xiong AL, Aksimentiev A, Schulten K, Timp G, Leburton JP. 2006. Simulation of the electric response of DNA translocation through a semiconductor nanopore-capacitor. *Nanotechnology* 17:622–33

108. Heng JB, Aksimentiev A, Ho C, Marks P, Grinkova YV, et al. 2006. The electromechanics of DNA in a synthetic nanopore. *Biophys. J.* 90:1098–106

109. Heng JB, Aksimentiev A, Ho C, Dimitrov V, Sorsch TW, et al. 2005. Beyond the gene chip. *Bell Labs Tech. J.* 10:5–22

110. Zhao Q, Sigalov G, Dimitrov V, Dorvel B, Mirsaidov U, et al. 2007. Detecting SNPs using a synthetic nanopore. *Nano Lett.* 7:1680–85

111. Barcilon V. 1992. Ion flow through narrow membrane channels. 1. *SIAM J. Appl. Math.* 52:1391–404

112. Barcilon V, Chen DP, Eisenberg RS. 1992. Ion flow through narrow membrane channels. 2. *SIAM J. Appl. Math.* 52:1405–25

113. Eisenberg RS. 1996. Computing the field in proteins and channels. *J. Membr. Biol.* 150:1–25

114. Graf P, Kurnikova MG, Coalson RD, Nitzan A. 2004. Comparison of dynamic lattice Monte Carlo simulations and the dielectric self-energy Poisson-Nernst-Planck continuum theory for model ion channels. *J. Phys. Chem. B* 108:2006–15

115. Coalson RD, Kurnikova MG. 2005. Poisson-Nernst-Planck theory approach to the calculation of current through biological ion channels. *IEEE Trans. Nanobiosci.* 4:81–93

116. Mamonov AB, Coalson RD, Nitzan A, Kurnikova MG. 2003. The role of the dielectric barrier in narrow biological channels: a novel composite approach to modeling single-channel currents. *Biophys. J.* 84:3646–61

117. Shilov IY, Kurnikova MG. 2003. Energetics and dynamics of a cyclic oligosaccharide molecule in a confined protein pore environment. a molecular dynamics study. *J. Phys. Chem. B* 107:7189–201

118. Doyle DA, Cabral JM, Pfützner RA, Kuo AL, Gulbis JM, et al. 1998. The structure of the potassium channel: molecular basis of K^+ conduction and selectivity. *Science* 280:69–77

119. Nguyen TL. 2004. Three-dimensional model of the pore form of anthrax protective antigen: structure and biological implications. *J. Biomol. Struct. Dyn.* 22:253–65

120. Di Marzio EA, Kasianowicz JJ. 2003. Phase transitions within the isolated polymer molecule: coupling of the polymer threading a membrane transition to the helix-random coil, the collapse, the adsorption, and the equilibrium polymerization transitions. *J. Chem. Phys.* 119:6378–87

121. Shenoy DK, Barger W, Singh A, Panchal RG, Misakian M, et al. 2005. Functional reconstitution of ion channels in polymerizable lipid membranes. *Nano Lett.* 5:1181–85

Combining Self-Assembled Monolayers and Mass Spectrometry for Applications in Biochips

Zachary A. Gurard-Levin and Milan Mrksich

Department of Chemistry and Howard Hughes Medical Institute, University of Chicago, Chicago, Illinois 60637; email: mmrksich@uchicago.edu

Annu. Rev. Anal. Chem. 2008. 1:767–800

First published online as a Review in Advance on March 28, 2008

The *Annual Review of Analytical Chemistry* is online at anchem.annualreviews.org

This article's doi:
10.1146/annurev.anchem.1.031207.112903

Key Words

high-throughput assays, protein arrays, SAMDI, surface chemistry, systems biology

Abstract

Biochip arrays have enabled the massively parallel analysis of genomic DNA and hold great promise for application to the analysis of proteins, carbohydrates, and small molecules. Surface chemistry plays an intrinsic role in the preparation and analysis of biochips by providing functional groups for immobilization of ligands, providing an environment that maintains activity of the immobilized molecules, controlling nonspecific interactions of analytes with the surface, and enabling detection methods. This review describes recent advances in surface chemistry that enable quantitative assays of a broad range of biochemical activities. The discussion emphasizes the use of self-assembled monolayers of alkanethiolates on gold as a structurally well-defined and synthetically flexible platform for controlling the immobilization and activity of molecules in an array. The review also surveys recent methods of performing label-free assays, and emphasizes the use of matrix-assisted laser desorption/ionization mass spectrometry to directly observe molecules attached to the self-assembled monolayers.

1. PERSPECTIVES AND OVERVIEW

Assays of biochemical activities are fundamental to biological research, drug discovery, clinical diagnostics, food and environmental safety, biological warfare and other areas (1–5). The development and application of bioanalytical methods therefore continue to be dominant themes in analytical chemistry. Current research is motivated by several goals, including the analysis of small sample volumes (such as lysates from individual cells and other complex samples), massively parallel assays of large families of activities, and label-free detection formats. The development of biochips—or patterned arrays of immobilized molecules—addresses these themes and represents a significant achievement in genomic analysis. Such development also offers promising opportunities in proteomics, glycomics, and related topics.

This review provides a chemical perspective on the status of biochip arrays. We begin with an overview of the preparation and use of biochips, discuss current work with self-assembled monolayers (SAMs) to tailor the interfacial layer, and emphasize current efforts to apply label-free approaches to the biochip arrays.

1.1. Enter Biochip Microarrays: Oligonucleotides

The first high-density oligonucleotide microarray was reported by Fodor and colleagues in 1991 (6). The oligonucleotides were directly synthesized on a glass slide and lithographic masks were used to photochemically activate designated spots on the substrate for coupling each nucleotide. The first arrays had densities of 100,000 spots/cm^2 (6) and enabled the broad-scale identification of RNA transcripts in cellular samples. The sample DNA was fluorescently tagged, and hybridization of DNA to the immobilized oligonucleotides was detected with a fluorescence scanner and quantitated to reveal the pattern of gene expression associated with a cellular activity (7). Within ten years, these DNA arrays were a common tool in life sciences laboratories, providing unprecedented information on the global patterns of activities in cells (8–10). The technology is now mature, as arrays that comprise several million oligonucleotides (11) are now commercially available.

1.2. Extension to Protein and Small-Molecule Arrays

The rapid impact of DNA chips on the field of biology provided a strong motivation to develop arrays based on other classes of molecules, including peptides, proteins, carbohydrates, and small molecules. Yet, the development of these seemingly analogous arrays has proven much more difficult and still awaits broader commercialization. These difficulties with development begin with the tendency of proteins to adsorb nonspecifically to essentially all man-made materials (12–15). Unwanted adsorption is the source of background signals in many assays of biological samples (16). It is not an issue with most gene chip experiments because the isolation and amplification of DNA in samples exclude significant levels of protein. For protein arrays, this nonspecific adsorption is often accompanied by denaturation and therefore contributes to a loss of activity of the immobilized proteins (17). The immobilization of proteins or small molecules also presents challenges that DNA arrays do not. With regard to the latter, the uniform structure of oligonucleotides allows the development of straightforward immobilization strategies that apply equally well to all members of the array. The varied structures of proteins—which differ in molecular weight, charge, stability, and aggregation state—substantially complicate the development of universal strategies for immobilization. The methods that are now used, and which are discussed in this review, are limited in that they do not provide for a uniform activity of the proteins in the array. Solutions for these challenges in developing methods to prepare and apply biochip arrays outside the genomics arena are found in surface engineering, and are a central topic of this review.

1.3. Applications of Protein and Small-Molecule Arrays

Notwithstanding the challenges identified above, early examples of protein and small-molecule arrays presage the value that these tools will bring to the life sciences. Snyder and colleagues prepared the first protein arrays for proteome-wide surveys of biochemical interactions (**Figure 1**) (18). This group cloned a collection of 5800 yeast open-reading frames into a yeast high-copy expression vector to express each protein as its glutathione-S-transferase (GST) fusion. These proteins were spotted on glass slides presenting glutathione ligands and were then assayed to identify those sets of proteins that were binding partners for calmodulin and phosphoinositide. The arrays identified several activities, both known and previously unknown. However, these early examples were encumbered by significant levels of false negative and false positive findings (19). Because the proteins were not individually purified and characterized, several of the intended proteins were not present in the array and others

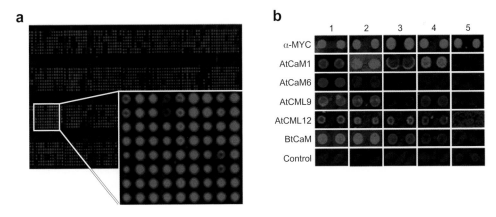

Figure 1

(*a*) Example of a high-density protein microarray presenting 1133 recombinant proteins on a nitrocellulose-coated glass slide. The array has been stained with a fluorescently labeled antibody that recognizes the MYC affinity tag. (*b*) The array was probed with fluorescently labeled calmodulin (CaM) and calmodulin-like proteins (CML) to identify the binding partners of these proteins.

were likely present in altered forms due to denaturation, proteolysis, or association with other partners, all of which can compromise specific interactions.

MacBeath and colleagues reported arrays having approximately 100 proteins, which represented the set of SH2 adaptor proteins from the human proteome (20), and addressed the ambiguities associated with protein expression by individually cloning, sequencing, and purifying each protein. These proteins were immobilized to aldehyde-coated glass slides and probed with a set of fluorescently labeled peptides to map out the consensus sequences and specificities of each adaptor domain (21). The significance of this work is that it permitted determination of the relative binding affinities of the peptides for each protein target and revealed specificity profiles of the receptors that may be involved in cancer. By probing the arrays with short peptide ligands, the authors avoided the complications that arise with nonspecific adsorption of macromolecules. Another common approach, based on antibody arrays, has been demonstrated for profiling cytokine levels in breast cancer cells (22) and blood (23) as well as for profiling human prostate cancer sera (24). Ligler and colleagues, for example, have developed a multianalyte-array biosensor that detects and identifies multiple analytes simultaneously from real-world samples (25).

The development of small-molecule arrays has also been slow due to the difficulty in arraying thousands of molecules with different functionalities and controlling non-specific interactions with the surface (5). Shair et al. created reaction microarrays in which they determined the enantiomeric excess of tens of thousands of compounds simultaneously by comparing the relative affinities of two chiral fluorescent tags for each compound (26). Schreiber and colleagues prepared a library of 3780 molecules using the "one bead–one stock solution" approach, then created a small-molecule array by immobilizing this collection to a glass slide using a quill-pin contact printing robot (27). The arrays were treated with a fluorescently labeled protein with the aim of identifying those spots to which the protein localized, and therefore the small molecules with selective affinity for the target protein. To control the nonspecific adsorption of the protein, the array was first treated with the protein bovine serum albumin (BSA)—a common strategy used to passivate sites for nonspecific adsorption (28)—yet this experiment was still plagued by substantial levels of nonspecific adsorption.

2. BIOCHIP ARRAYS: COMPONENTS

In this section, we present a primer on the component technologies that are necessary for preparing and applying biochips. For each area, we describe the methods that are currently the most commonly used and identify the benefits and limitations specific to each. This section is followed by a detailed survey of SAMs and their combination with mass spectrometry (MS) detection methods, the topic that is the focus of this review.

2.1. Surface Chemistries

The surface of the biochip is not a passive element in the assay; rather, it optimally fulfills several roles. These roles include binding of the molecules that make up the array; providing an environment that optimizes the activities of immobilized molecules; reducing nonspecific interactions of molecules in the sample with the chip; and providing substrates that are compatible with specific detection methods. A broad range of surface chemistries have been used and can be categorized as follows: (1) surface chemistries that employ polymeric materials, (2) those that use hydrogels, and (3) those that use SAMs as supports for the array. The first group includes the use of polystyrene, polycarbonate, and poly(ethylenimine) to immobilize molecules (29). The properties of these surfaces can be modified through reactive processes, including aqueous oxidation, plasma treatment, and mechanical roughening. In general, however, the resulting surfaces are complex in that they comprise several chemical functional groups and therefore require an empirical procedure for optimizing their use. Hydrogels are lightly cross-linked polymers that undergo enormous swelling in water—typically with more than 100-fold increases in volume—and therefore provide a three-dimensional environment that more closely approximates that present in solution (30). These materials, which include polyacrylamide, dextran, and agarose, are typically grafted onto a polymeric substrate and offer the benefits of being relatively effective at preventing nonspecific protein adsorption and of providing a higher loading capacity of immobilized species. The use of a hydrogel in assays, however, can be complicated by mass transport effects of soluble proteins through the gel and therefore necessitates caution when used in quantitative assays of biochemical activities. SAMs allow the most stringent control over surface structure and therefore provide a well-defined environment around immobilized molecules. These substrates are prepared by the self-assembly of precursor molecules, either alkanethiols or alkylsilanes, onto gold or hydroxylated substrates, respectively, and are discussed further below (31–34).

2.2. Immobilization Strategies

An important concern with chip-based assays is that immobilization alters the activity of the molecule, sometimes resulting in an almost complete loss of activity. This consequence for activity is of most concern with proteins, although it applies to small molecules as well. There are several reasons for this. First, on immobilization, part of a protein is oriented towards the substrate and is not available for interactions with soluble partners. Second, for surfaces that promote nonspecific adsorption—as is the case for the overwhelming majority of all materials—immobilized proteins will undergo denaturation. The presence of the surface can also inhibit large-scale conformational changes of proteins that can be required for their biochemical activity. We discuss below the four classes of immobilization strategies that are used, and group them according to whether the immobilization relies on specific or nonspecific interactions and whether it results in a covalent or noncovalent linkage of the molecule to the substrate.

2.2.1. Nonspecific and noncovalent immobilization. The oldest and experimentally most straightforward methods are those that rely on physical adsorption of proteins or molecules to surfaces. As a general rule, essentially all proteins will adsorb to essentially all surfaces. This is true for surfaces that are hydrophobic or are positively or negatively charged; those that present hydrogen bond donors or acceptors; or even those that are modified with Teflon-like chemistries (12–15). Adsorption is usually fast—sometimes near the diffusion limit—and often yields a single layer of protein (13). The resulting layer is typically heterogeneous in structure, both in the orientation and denaturation state of the presented protein, and is dependent upon the conditions employed in the adsorption (35). For example, the use of solutions having higher concentrations of protein results in less denaturation because the proteins have less time to unfold (with an accompanying larger footprint) before the neighboring sites become occupied with adjacent proteins (36). Hence, these methods require substantial tuning to optimize the activities of protein. Whereas these methods can be valuable for assays of single proteins, the demand to optimize many immobilization events makes them impractical for the preparation of arrays. A final concern with proteins that are immobilized noncovalently is that they can exchange with soluble proteins during an assay. This concern, known as the Vroman effect, has been characterized extensively for materials that contact blood and remains a concern in bioassays (37).

2.2.2. Nonspecific and covalent immobilization.

Surfaces presenting functional groups that can be used to covalently link the molecule to the chip avoid the exchange of immobilized proteins, and can be applied to low-molecular-weight molecules (including peptides) that would otherwise have poor affinity for the surface. Indeed, glass slides modified with N-hydroxysuccinimide (NHS) esters or with aldehydes are available commercially and are frequently used in the preparation of DNA and protein arrays, respectively (38–41). When used to immobilize arrays of proteins, these slides result in a heterogeneous display of the immobilized species, as each protein on average has several side chains that carry reactive functional groups, and require that the molecules be purified prior to arraying (42).

2.2.3. Specific and noncovalent immobilization.

The use of surfaces that present molecular groups that can selectively interact with a tag on the protein provides for more control over the immobilization process and eases the requirement to purify the molecules prior to arraying. Attachment of biotin-tagged ligands to substrates that are modified with a layer of streptavidin is the most common example of a specific but noncovalent strategy for immobilizing molecules (38). The many reagents available to biotinylate ligands together with the specificity and high affinity of the ligand-protein interaction make this technique effective for biochip applications (43). For proteins, it is now possible to express recombinant forms with a tag that can be biotinylated with the enzyme BirA from *Escherichia coli* (44). The expression of proteins that have a hexahistidine tag allows direct immobilization to surfaces presenting a chelated Ni(II) ion (45, 46). The his-Ni(II) interaction has high affinity (45, 47) and is reasonably stable on the time scale of most biochip assays, making it the most important strategy used in protein arrays (48). The binding of GST fusions to glutathione-modified substrates has also been a common approach to protein immobilization, but is limited by the weak affinity of this ligand-protein complex. With a dissociation rate constant of about 0.1 s^{-1} (49), the immobilized proteins would be expected to dissociate from the substrate during an experiment. The apparent stability of chips prepared using GST fusions is likely dependent on nonspecific interactions between the protein and surface and on the possible dimerization of GST with an increased affinity for the substrate. Jiang and coworkers have described a strategy to use proteins tagged with an oligonucleotide for immobilization to an oligonucleotide array (50). The specificity inherent in hybridization of oligonucleotides allows the array to be prepared from a mixture of all protein-DNA conjugates.

2.2.4. Specific and covalent immobilization. The above-described strategies that make use of immobilization domains provide the best control in positioning proteins at surfaces and reduce the demand for purification by allowing proteins to be selectively immobilized from mixtures, although they may not always provide sufficient stability for biochemical assays, particularly when stringent wash conditions are employed. To address this last limitation, we investigated a strategy wherein the immobilization domain of a fusion protein was made to interact with an irreversible inhibitor, leading to a selective and covalent attachment of the protein to the substrate. We demonstrated this method with the selective binding of the serine esterase cutinase to a class of phosphonate ligands to give a covalent adduct between the ligand and an active site serine residue (**Figure 2**) (51). The ability to prepare protein reagents using recombinant methods and to selectively immobilize target proteins without rigorous purification also prevents the loss of activity that often accompanies the manipulations used in purifying proteins. Johnsson and coworkers reported a similar approach using the human DNA repair protein O^6-alkylguanine–DNA alkyltransferase as the immobilization domain (52, 53). Another study used the substrate binding domain of poly(hydroxyalkanoate) depolymerase (54). Finally, Camarero and coworkers developed a method based on a *trans*-slicing process. The protein of interest was fused to an N-intein that was complementary to a C-intein on the monolayer. Upon association of the two domains, a splicing reaction resulted in release of the intein and a covalent attachment of the protein to the monolayer (55).

2.3. Content

A challenge in assembling biochip arrays that is often skirted by investigators concerns the preparation of the library of reagents to be arrayed. Whereas oligonucleotides are commercially available in high volume and are inexpensive (approximately $2 for 1 μg of a 20-mer), peptides, proteins, carbohydrates, and other small molecules are not generally available commercially. When they are, however, the cost for large libraries can be prohibitive. Solid-phase synthetic methods are important for generating large pools of small molecules and have been applied successfully to peptides and, more recently, to certain classes of oligosaccharides and small molecules (56, 57). Proteins, with typical masses ranging from 10 to 100 kD but occasionally as high as 1000 kD, must be prepared using biological methods (58). The preparation of large pools of proteins using parallel automated methods are complicated by the relatively poor yields inherent to cloning and expression (even in the *E. coli* host with good expression vectors, approximately one-third of the expected proteins are produced), the

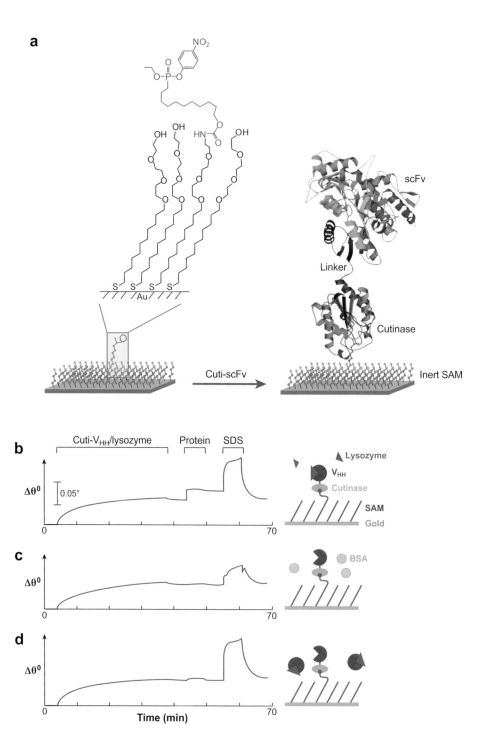

improper folding of proteins and resulting insoluble particles, and the purification of proteins. Further, larger proteins and those requiring posttranslational modifications often require mammalian host expression systems (58). Early studies of protein arrays used massively redundant arrays (40), and the sizes of arrays are still almost always limited by the available molecular content. Clearly, the generation of large numbers of functional reagents remains a bottleneck in the biochip field.

2.4. Arraying

Commercial oligonucleotide arrays are now available with 1-μm feature sizes, whereas the still-limited content available for other classes of molecules eases the requirement to spot sizes of 50–100 μm. The current methods can be categorized as those that apply solutions of reagents to discrete regions of a slide, those that use light to activate regions of a slide for immobilization of reagents, and those that rely on self-assembly of the array. The first group of methods has benefited from the development of a variety of mechanical devices for delivering nanoliter-scale reagent droplets to defined regions of a surface (59); these methods include pin arrayers and ink-jet printing tools (60, 61). These robotic arrayers can reliably deliver molecules to arrays with feature sizes approaching 100 μm, but these arrays often have poor uniformity with regard to size and composition of the spots. Spots often have a bright ring of immobilized molecules around the perimeter, likely derived from the rapid evaporation of the droplet after it is delivered to the substrate, and can compromise quantitative comparison of spots across the array (62). Microfluidic devices prevent the evaporation of drops and therefore represent an exciting alternative method for the controlled delivery of biomolecules to surfaces (63–65). Delamarche and coworkers reported a microfluidic capillary system that autonomously transports submicroliter volumes capable of fabricating cellular microarrays that can measure cellular secretions after exposure to stimuli using the authors' "micromosaic immunoassay" (66, 67). Finally, contact methods that rely on applying a stamp or other applicator to a substrate can be effective. Crooks and coworkers described an interesting example wherein a stamp was used to pick up an array of oligonucleotides—which was initially hybridized to an array of oligonucleotides—and then to deliver the molecules to a target plate, where they were immobilized in the geometry of the array (68).

The combination of photolithography, which can irradiate a substrate with a pattern of light at micrometer-scale resolution, with photochemical protecting groups

---

Figure 2

(*a*) Illustration of a scheme to use self-assembled monolayers (SAMs) presenting an irreversible phosphonate inhibitor of the serine esterase cutinase to immobilize a single-chain antibody. Surface plasmon resonance spectroscopy shows the immobilization of the fusion protein and subsequent association of antigen (*b*). Control experiments show that the surfaces prevent the nonspecific adsorption of bovine serum albumin (BSA) (*c*) and that soluble inhibitors can block specific interactions at the surface (*d*). In each case, treatment of the monolayers with detergent (SDS) resulted in the removal of the antigen, but not the covalently immobilized fusion protein.

can be applied to the synthesis of an array of molecules directly on the biochip surface. This strategy was used to prepare the first DNA chips and is still important for preparation of the highest density DNA chips. Gulari and coworkers prepared peptide arrays by using photogenerated acids to spatially direct the deprotection of amino groups during peptide synthesis (69). For smaller-sized arrays, a series of droplets can sequentially be applied and removed from regions of the surface. Frank developed this SPOT method for preparing arrays of peptides on nitrocellulose resins (70), and Chang and coworkers have used this method to prepare arrays of small molecules (71). These methods are exciting because of the efficiency with which complex arrays can be prepared, but they are challenging to develop. Because the final molecules are attached to the substrate and cannot be purified, the syntheses must proceed in high yield. The small number of molecules present on the surface, however, makes it extremely challenging to assess the quality of the synthesis and to optimize reaction yields. Below, we describe the development of MS methods that address this limitation.

Monk and Walt have described a clever approach to preparing arrays by self-assembly, wherein they allowed oligonucleotide-modified beads to attach to the ends of a bundled optical fiber (72). Using split-pool synthesis, a large library of beads was created such that each bead presented a unique sequence of oligonucleotides. Assembly of the beads on the fiberoptic bundle yielded an array, but with the curious problem that the address of each oligonucleotide was unknown. A training procedure, whereby known mixtures of fluorescently tagged oligonucleotides were hybridized with the array, was then used to map the locations of oligonucleotides within the array. This application is commercially viable and shows that the expense associated with the training procedure can be lower than that associated with the deterministic synthesis or placement of high-density arrays.

2.5. Detection Methods

Biochip arrays are used to identify binding interactions between soluble molecules and their partners on the array and to identify those molecules on the array that are substrates for an enzyme. Most experiments rely on detection of a fluorescent signal to identify these interactions. In protein binding experiments, for example, the target protein is labeled and then applied to the biochip array, after which analysis with a flatbed scanner identifies those spots that have retained the soluble protein (40, 73). Similarly, the identification of molecules in an array that are substrates for an enzyme can be performed using fluorescently labeled antibodies that bind the product (74, 75). Fluorescence detection methods offer the benefits of being fast and sensitive and of using commercially available scanners. The use of a fluorescent chromophore, however, can alter or block the function of the labeled protein, can lead to increased nonspecific interactions of the protein with the substrate, and precludes the simultaneous analysis of several target proteins. A larger concern with this and other label-dependent methods is that unanticipated biochemical activities cannot be identified because the choice of labeling strategy requires knowledge of the activity to be assayed. We note that radioisotopes can be employed in similar ways, but because of safety concerns this approach is less common (48, 76).

The limitations inherent to the use of labels have motivated the development of label-free detection methods that can monitor interactions of unmodified analytes. The most common approaches are based on optical strategies to monitor changes in the refractive index of the medium adjacent to the biosensor, which increases as proteins bind to the surface and displace solvent (77–80). Surface plasmon resonance spectroscopy (SPR) now represents a standard method for monitoring the interactions of a soluble and an immobilized binding partner and has the benefit of providing real-time, and therefore kinetic, measurements on interactions (81–83). Corn et al. have developed imaging SPR instruments to monitor the hybridization of oligonucleotides (84). To improve sensitivity, the authors harnessed an enzymatic amplification technique (85) to achieve a 10^6-fold improvement over nonamplified formats (86). Georgiadis and coworkers have applied imaging SPR to obtain kinetic and thermodynamic measurements on the specific binding of drugs to DNA arrays and were able to discriminate between different binding sites on the same DNA biosensor (87).

Below, we discuss the development of mass spectrometric methods for label-free detection of biochips. In addition, we note that several methods are now in early stages of development and may offer alternative label-free strategies for analyzing biochips. These approaches are based on electrochemical detection (88, 89), piezo resonance sensors (90), and calorimetric methods (91, 92).

3. SELF-ASSEMBLED MONOLAYERS FOR BIOCHIPS

We and others have made extensive use of SAMs in biochip applications because of the unmatched control over surface structure and properties that these films provide (93). In this section, we describe the characteristics that make SAMs well suited for use in preparing biochips, including the chemical flexibility to tailor the interface and the availability of inert surface chemistries. We review examples that demonstrate monolayers in quantitative assays of biochemical activities. The subsequent section continues this discussion with an overview of MS as a label-free method for analyzing monolayers and with a review of applications to biochip assays.

SAMs of alkanethiolates of gold were first described nearly 25 years ago and have remained the most important strategy for preparing structurally well-defined and complex organic surfaces (94, 95). The monolayers assemble from a solution of terminally substituted long-chain alkanethiols, and the assembly completes in several hours. Extensive work has shown that the alkanethiolates anchor to the (1,1,1) surface of gold in a hexagonal lattice to give a well-packed array of alkyl chains in the *trans*-extended conformation (31, 96–98). This arrangement positions the terminal functional group of the alkanethiolates at the surface and therefore provides a straightforward strategy for engineering the chemistry of a surface. Moreover, monolayers may be prepared from a mixture of two or more alkanethiols to introduce multiple functional groups onto the surface and to control the density of the active functional group (99–101). The monolayers are stable under the conditions employed in biochip applications, but they desorb at temperatures of 80°C or with irradiation of UV light (102, 103).

3.1. Inert Surfaces

The most significant finding concerning the monolayers that are well suited to bio-analytical applications were the reports by Prime and Whitesides that monolayers presenting short oligomers of the ethylene glycol group prevented the nonspecific adsorption of protein (104, 105). The mechanisms that underlie this property are not fully elucidated—it is not yet possible, for example, to design a new inert surface—but they are believed to be related to the conformational entropy of the glycol oligomers and the structure of solvent in proximity to the surface (106, 107). Empirical methods have been used to identify additional examples of functional groups that render monolayers inert (108). These include the mannitol group, which can maintain inertness for several weeks (109); however, the glycol groups remain the most important for biochip applications.

3.2. Immobilization Chemistries

Several chemistries have been developed for immobilizing ligands to monolayers (**Figure 3**). Surfaces that present the immobilization group at 1–2% density among the oligo(ethylene glycol) chains are generally effective in optimizing the amount of immobilized molecule (and therefore signal in an assay), providing for a uniform environment of molecule (by avoiding significant crowding), and reducing nonspecific interactions with the surface (110). The use of monolayers that present a maleimide group against a background of tri(ethylene glycol) groups is particularly convenient for the immobilization of peptides (111). An important benefit with this approach is that the density of ligand is determined by the density of the reactive group on the monolayer—not by the kinetics of the coupling reaction, which depend on the concentration of the reagent—and therefore arrays that present a multitude of molecules do so at uniform density. This property allows direct comparison of activities across the array. We have also used the Diels-Alder reaction for immobilization of diene-conjugated peptides to monolayers that present the benzoquinone group (76). Other immobilization methods have used the cycloaddition of azide and terminal alkyne groups (112) and the reaction of amino-substituted ligands with monolayers presenting anhydride groups (113).

Numerous methods have been described for immobilizing proteins, including the use of monolayers presenting a NTA-Ni(II) ligand for the immobilization of His-tagged proteins (46). Another approach has used proteins engineered to present a single cysteine residue on their surface that can mediate immobilization to monolayers that present maleimide groups (114). Abbott and coworkers prepared a variant of RNase A wherein a cysteine residue was activated as a mixed disulfide with 2-amino-5-thiobenzoic acid and which then reacted with a thiol on the monolayer to immobilize the protein with a disulfide tether (115).

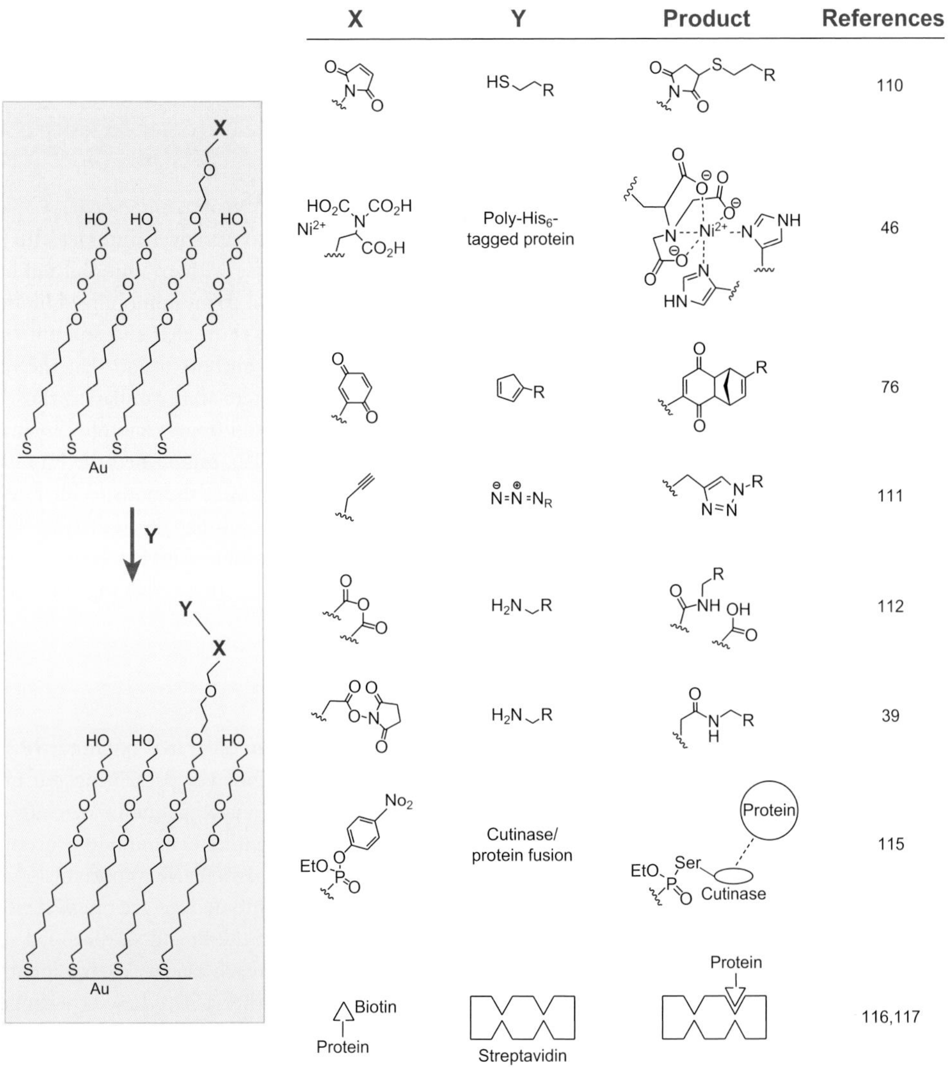

X	Y	Product	References
	HS~R		110
Ni²⁺ HO₂C N CO₂H CO₂H	Poly-His₆-tagged protein		46
			76
	⊖N=N⁺=N_R		111
	H₂N~R		112
	H₂N~R		39
EtO O P O NO₂	Cutinase/protein fusion		115
Biotin Protein	Streptavidin	Protein	116,117

Figure 3

A chart summarizing several chemistries used to immobilize ligands to self-assembled monolayers. In each case, monolayers presenting the group X are used to immobilize ligands conjugated to group Y.

We have applied the cutinase fusion protein strategy described earlier to prepare a small antibody array (116). By uniformly orienting the antibodies on the surface, each antibody had a specific activity greater than 90% and the density could be controlled to optimize the activities of the antibodies. For example, we found that the fraction of antibodies that bound antigen was constant for antibodies that bound small antigens, but this fraction decreased with increasing densities of an antibody that bound large antigens. This example reflects the lateral crowding at the surface when the antigen has a larger footprint than the immobilized antibody. Taken together, these methods provide controlled chemistries for optimizing the activities of immobilized proteins.

3.3. Molecular Recognition on Monolayers

Early examples of monolayers that were designed to selectively interact with proteins demonstrated the association of carbonic anhydrase with an immobilized benzene-sulfonamide ligand and the binding of streptavidin to an immobilized biotin group (117, 118). For example, one study used SPR to show that the amount of bound protein increased with the density of ligand; the authors found that the rate constants for association and dissociation of the protein were similar to those for the homogeneous phase interaction. Indeed, several additional examples, including the binding of proteins to clustered carbohydrates (119), established the importance of immobilizing ligands to inert surfaces so as to minimize the nonspecific interactions that contribute to background signal, as well as to optimize the activity of the immobilized molecules. These examples served as the starting point for the development of monolayers for the preparation of biochip arrays and for a wide range of biochemical assays.

3.4. Quantitative Assays with Monolayers

In the first report of a carbohydrate array, we demonstrated quantitative assays of protein binding and enzyme activity using a monolayer presenting ten monosaccharides. The carbohydrates were prepared with a cyclopentadiene group that was used to immobilize the sugars to a monolayer presenting benzoquinone groups (120). The uniform density that this strategy ensures is particularly important for assays of carbohydrate-binding proteins, as many of these interactions are oligovalent and the association constants can vary with the density of the ligand. We probed the array with a panel of fluorescently labeled lectins and in each case we clearly identified the binding specificity of each protein (**Figure 4**). Further, when these experiments were performed in the presence of a soluble carbohydrate, we measured a dose-dependent inhibition of the binding to immobilized ligands; these data permitted a quantitative analysis of binding affinities. We also used the arrays to profile the specificity of a galactosyltransferase enzyme. By profiling the array with a panel of lectins both before and after the enzyme reaction, we inferred the carbohydrates that were modified by the enzyme. As part of this work, we demonstrated that the yield for the reaction was constant for densities of ligand up to 70%, but that the yield decreased at higher densities (121). This finding again reflects the crowding of ligands that renders some of them inaccessible to the enzyme.

In another experiment, we demonstrated an assay of the src kinase (76). We again used the Diels-Alder reaction to immobilize a peptide substrate for the kinase and monitored the phosphorylation reaction using a ^{32}P-labeled phosphate group. Application of a sample containing src and ATP resulted in phosphorylation of the peptide,

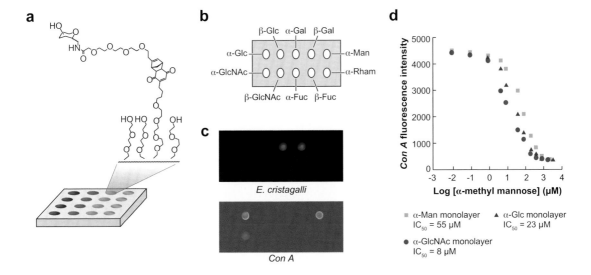

Figure 4

(*a*) Carbohydrate array prepared by immobilizing diene-conjugated carbohydrates to monolayers presenting a benzoquinone group. (*b*) The resulting arrays were treated with *Erythrina cristagalli* and *Concanavlin A* (*Con A*), two fluorescently labeled lectin proteins, to reveal (*c*) the binding specificities of the proteins. The addition of soluble carbohydrates gives a dose-dependent inhibition of the binding of lectin to the monolayer and (*d*) provides quantitative information on the binding affinities.

whereas a control peptide whose active tyrosine was substituted with a phenylalanine residue was inactive. Significantly, we observed levels of phosphorylation that were 75-fold greater than the background count, which again reflects the effectiveness of the glycol groups in preventing nonspecific interactions. By applying an array of microliter droplets, each of which contained the kinase and varying concentrations of a known inhibitor, we obtained titration curves that revealed the dissociation constant of the inhibitor (76).

These examples demonstrate the high performance of monolayers that combine immobilized ligands with oligo(ethylene glycol) layers in biochemical assays of protein binding and enzyme activity. The control over presentation of the ligand, which includes the density and orientation of the immobilized molecules, and the effectiveness of the ethylene glycol layer in preventing unwanted interactions at the surface and in maintaining the activities of immobilized molecules are very well suited to the construction of biochip arrays for a broad range of applications and represent a significant advance over many of the current—and commercially available—surface chemistries now in use. The preceding examples have all used label-dependent methods to determine biochemical activities on the chip and therefore carry the same limitations discussed earlier. In the next section we describe the development of a mass spectrometric technique to characterize monolayers and describe several classes of assays that can be performed using this label-free method.

4. ASSAYS WITH SELF-ASSEMBLED MONOLAYERS FOR MATRIX-ASSISTED LASER DESORPTION/IONIZATION TIME-OF-FLIGHT MASS SPECTROMETRY

Monolayers serve as efficient substrates for matrix-assisted laser desorption/ionization (MALDI) MS. Upon irradiation with a laser, the alkanethiolates are released from the gold surface (through a loss of the sulfur-gold bond) and give rise to peaks whose masses correspond to the terminally substituted alkanethiolates. Early work by the Wilkins and Hanley groups investigated laser desorption of alkanethiolates and showed that both the monomers and dimers of the alkanethiolates were observed, along with fragments of these species (122, 123). We find that the use of matrix substantially reduces fragmentation of the intact alkanethiolates, thereby producing clean and easily interpretable spectra. Most significantly, the use of monolayers that are functionalized with appropriate chemical and biological functionality permits a broad range of assays. This section gives several examples of these assays using the technique termed self-assembled monolayers for matrix-assisted laser desorption/ionization (SAMDI) MS.

4.1. Enzyme Activity Assays

Most enzymes act on substrates and generate products whose masses are distinct from the substrate. For example, kinases add phosphate groups to hydroxyl-bearing side chains of proteins, proteases cleave proteins by hydrolyzing an amide bond, and acetylases convert the amino groups of lysine residues to the corresponding acetamides (124). SAMDI offers the ability to perform label-free assays of enzyme activities, starting with the immobilization of the relevant enzyme substrate to monolayers that are otherwise inert. Treatment of the monolayers with the enzyme and any required cofactors leads to the product, which can then be observed in the mass spectrum. **Figure 5** shows several examples of enzyme activity assays.

In a kinase activity assay, we immobilized a cysteine-terminated peptide to a monolayer presenting the maleimide group (125). A SAMDI spectrum showed clear peaks corresponding to the immobilized peptide and to the background tri(ethylene glycol)–terminated alkanethiolates, and also showed a lack of peaks for the initial maleimide-terminated alkanethiolate; this demonstrated that the immobilization reaction was complete. After the surface was treated with src kinase, a SAMDI spectrum revealed

---

Figure 5

Examples of enzyme activity assays performed with self-assembled monolayers for matrix-assisted laser desorption/ionization mass spectrometry (SAMDI MS), wherein monolayers presenting carbohydrate or peptide ligands were treated with an enzyme. In each case, mass spectra revealed peaks corresponding to the masses of the substituted alkanethiolates before and after modification of the enzyme. Examples are shown for the galactosylation of an immobilized carbohydrate by (*a*) β(1,4)-galactosyl transferase and (*b*) proteolysis of a peptide by caspase-3, as well as (*c*) phosphorylation of a peptide by src kinase.

a

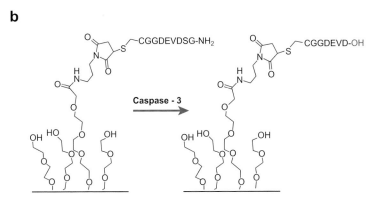

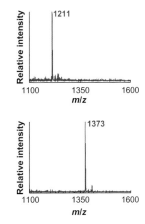

b

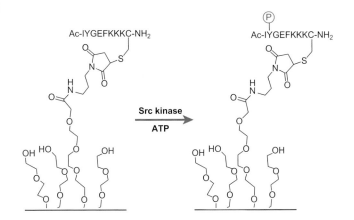

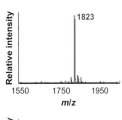

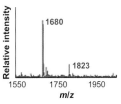

c

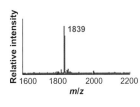

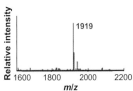

that the peaks representing peptide-alkanethiol conjugates shifted by 80 Da, as expected for the increase in mass following phosphorylation. In this and most other examples, multiple peaks were observed for the anticipated analyte that represented the alkanethiolates in their monomeric and disulfide forms. We do not know whether the disulfides formed upon desorption of the monolayer from the gold surface, which is well precedented in thermal reactions (126), or whether the disulfides formed in the ion cloud following desorption. We also observed adducts of the alkanethiols with multiple counter ions, including proton, sodium, and potassium. In any event, we did not see significant fragmentation of the functionalized alkanethiolates and consequently obtain spectra that are straightforward to interpret. Other examples have characterized the methylation of arginine by protein arginine methyltransferase 1 (127), the protease activity of caspase-3 (128), and the modification of carbohydrates by glycosyltransferase (120).

These examples illustrate several benefits inherent to the SAMDI method. First, the use of MS provides a label-free method for assaying a broad range of biochemical activities. The lack of labels eliminates several steps in the assay and avoids the need for developing antibodies to label an intended analyte, thereby significantly reducing development time and decreasing the risk that introduction of the label will interfere with the biological activity being assayed. MS also permits simultaneous assay of distinct enzyme activities—which would normally be incompatible because of different labeling strategies—to be performed on the same chip and thus with a single sample.

MS also provides instructive information about the analyte. Whereas SPR and related optical methods provide information on the amount of protein that interacts with an immobilized ligand, they do not depend on the composition of those proteins. The ability of SAMDI to identify each species at the surface according to its molecular weight allows a straightforward discrimination between signals due to specific analytes and to background. This ability also allows observation of both the products and substrates of an enzyme and therefore can verify that the immobilized molecule was indeed on the chip (thereby identifying possible false negatives); further, SAMDI can provide a better assessment of the yield of the enzymatic reaction. Moreover, the mass resolution of this method also permits multianalyte assays to be performed. In one example, we immobilized a mixture of four peptides (**Figure 6**), each of which was clearly resolved in the SAMDI spectrum. Treatment of the monolayer with mixtures of kinases resulted in selective phosphorylation reactions, which could be analyzed by SAMDI to determine which kinases were present in the sample (125).

4.2. Solution-Phase Assays

A common concern with chip-based assays is that immobilization of a molecule may compromise its biochemical activity. For example, immobilization of a protein in an improper orientation can prevent its association with a binding partner, and immobilization of a substrate using a short tether can prevent its access to a buried active site of an enzyme. We performed an assay of a methyl transferase enzyme using a monolayer that presented a peptide substrate; we found that the peptide was fully inactive (127). We then repeated the reaction in solution using a cysteine-terminated

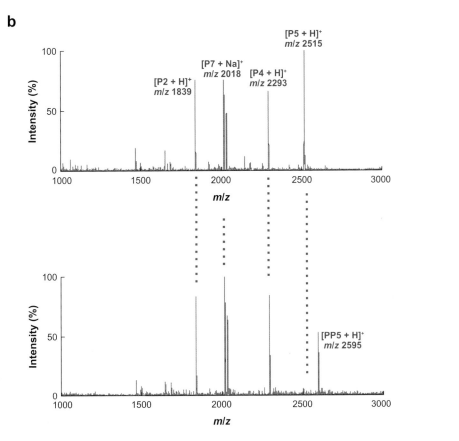

a

Kinase	Substrate peptide	Molecular weight	Molecular weight on surface (H$^+$)
PKA	P2: Ac-LRRASLGC-NH$_2$	915.5	1839
CaMKII	P4: Ac-KRQQSFDLFBC-NH$_2$	1368.7	2293
CKI	P5: Ac-CKRRALpSVASLPGL-NH$_2$	1590.8	2515
Erk	P7: Ac-TGPLSPGPFGC-NH$_2$	1072.5	1996

b

(Top spectrum, Intensity (%) vs m/z)

[P2 + H]$^+$ m/z 1839
[P7 + Na]$^+$ m/z 2018
[P4 + H]$^+$ m/z 2293
[P5 + H]$^+$ m/z 2515

(Bottom spectrum, Intensity (%) vs m/z)

[PP5 + H]$^+$ m/z 2595

Figure 6

An example of multianalyte assays enabled by self-assembled monolayers for matrix-assisted laser desorption/ionization mass spectrometry (SAMDI MS). A mixture comprising four cysteine-terminated peptides, each of which was a selective substrate for one of four kinases (*a*) was immobilized to a monolayer presenting maleimide groups. The four peptides had distinct masses and each gave a distinct peak in the mass spectrum. Treatment of the monolayer with CKI kinase (*b*) resulted in a monolayer where only P5 had shifted by 80 Da, revealing the selective phosphorylation of this peptide.

peptide. At the end of the reaction, we applied the reaction mixture directly to the monolayer. Because there were no other thiols in the assay mixture, the peptide selectively immobilized to a monolayer presenting a maleimide group, after which the monolayer was rinsed and the peptide was analyzed by SAMDI.

Indeed, in this format the peptide was efficiently methylated and a time course of the reaction—obtained by spotting microliter aliquots from a single reaction mixture to a monolayer at different times—revealed a kinetic profile that agreed with that obtained using conventional assays based on high-pressure liquid chromatography. This approach should be applicable to other assays provided that the substrate contains an immobilization tag that permits selective reaction with a functional group on the monolayer. Significantly, this method combines the advantages offered by homogeneous phase reactions for maintaining biochemical activity and liquid handling in microtiter plates with those offered by immobilized format assays for simplifying sample preparation and analysis.

4.3. High-Throughput Screening

MS methods have not been applied to the discovery of inhibitors of enzymes. MS has significant benefits, but they are limited by the need to enrich the sample with the intended analyte and remove salt. This makes sample preparation impractical for screening applications that require many tens of thousands of independent experiments.

Because the SAMDI method uses substrates wherein the intended analyte is covalently attached, sample preparation requires only that the substrate be rinsed and then spotted with matrix, both of which are compatible with high-throughput manipulations. We demonstrated this approach by screening a library of 10,000 compounds to identify selective inhibitors against the anthrax lethal factor toxin, a protease that cleaves MAP kinase proteins in the host cell (**Figure 7**) (129). A peptide substrate for lethal factor was immobilized to a monolayer modified with a 10×10 array of circular islands. Droplets containing the enzyme and a pool of eight compounds from the library were applied to each island, allowed to stand for 1 h, and then rinsed. Mass spectra acquired from islands present on a dozen plates showed complete cleavage of the immobilized peptide on the majority of spots, whereas 1% of the spots showed partial or absent cleavage, revealing that those pools contained an inhibitor of the protease. Compounds from the active pools were then assayed individually, and a

---

Figure 7

Self-assembled monolayers for matrix-assisted laser desorption/ionization (SAMDI) used to perform a screen of 10,000 small molecules to identify inhibitors of the anthrax lethal factor (LF) protease. (*a*) A peptide substrate for LF was immobilized to a monolayer presenting maleimide groups. (*b*) Treatment of the monolayer with recombinant protease resulted in cleavage of the peptide, which was then analyzed by SAMDI mass spectrometry (MS). (*c*) Chemical screens were performed by arraying 100 droplets containing the protease and eight compounds from the library, followed by analysis of the spots with MS. (*d*) The analysis clearly identified the spots that had an inhibitor in the reaction mixture.

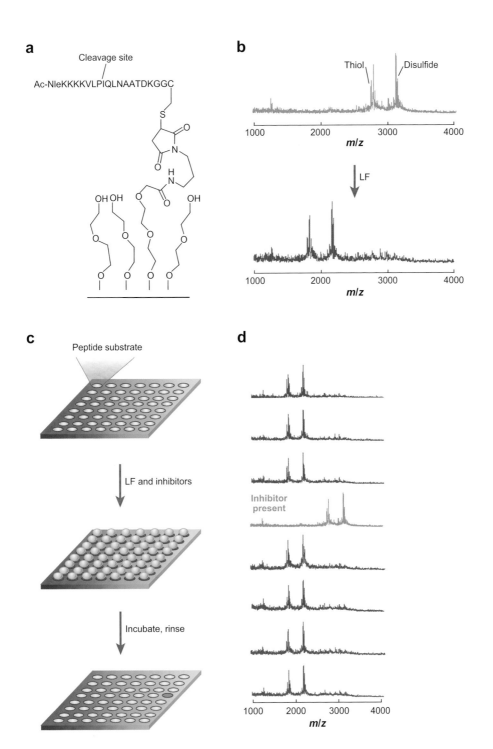

a

Cleavage site

Ac-NleKKKKVLPIQLNAATDKGGC

b

Thiol   Disulfide

LF

c

Peptide substrate

LF and inhibitors

Incubate, rinse

d

Inhibitor
present

single candidate with 1 μM dissociation constant was identified and found to be active in cell culture assays.

This example highlights the benefits of performing high-throughput screens with mass spectrometric methods. Most importantly, the approach avoids the high rate of false positive results that are common to fluorescence methods (because compounds in the chemical library are fluorescent at the wavelengths used in the assay) and minimizes the time required to develop and format a new assay. The SAMDI method, however, has a significantly lower throughput than fluorescent methods, requiring approximately 1 h to analyze a single plate. We expect advances in the sensitivity and automation of instruments to accelerate the throughput, but not to the levels now common in fluorescent approaches.

4.4. Chemical Reaction Screening

The task of reaction discovery in synthetic chemistry is also rooted (in part) in numerous trial-and-error experiments to identify reagents and conditions that promote a desired reaction. Yet, the same chip-based tools that are under intense development for biological applications have not, with one notable exception, been applied to reaction discovery.

Liu and coworkers reported a strategy that uses oligonucleotides that are functionalized with common functional groups (130). They used a panel of reagents that permitted each pair of functional groups to be brought into proximity via the hybridization of complementary oligonucleotides. When treated with reagents that promoted a reaction resulting in the joining of the functional groups, the resulting cross-linked oligonucleotides could be isolated and applied to an oligonucleotide array to identify the sequences of DNA (and, therefore, the identity of the functional groups that had reacted). This work is significant because it can identify unanticipated reactions. Most current methods in chemistry instead begin with a known transformation and a product that can be detected, then screen for reagents that efficiently promote the reaction (131, 132).

The SAMDI method complements that reported by Liu et al. in that both methods are able to identify unanticipated reactions; however, the MS method can identify those reactions (1) that do not result in coupling of the two reagents, (2) that require a stoichiometry of the reagents other than 1:1, and (3) that use reagents and solvents that are otherwise incompatible with DNA templates. Indeed, a recent report showed the utility of SAMDI for developing reactions of immobilized molecules with soluble reagents (133). Whereas current methods require a combination of several analytical methods to characterize the products of interfacial reactions (with a substantial effort required for each reaction), SAMDI rapidly provides information for the products formed as well as the approximate yields. We recently reported 15 new reactions that were developed with the aid of the SAMDI method, including palladium-mediated cross-coupling of aromatic halides. In each case, we were able to rapidly identify conditions that gave a high-yielding conversion. For example, treatment of a monolayer presenting terminal alkyne groups with sodium methoxide in deuterated water promoted the exchange of the terminal hydrogen with a deuterium atom. The mass spectrum clearly showed the mass increase associated with the addition of a single neutron to the alkyne. To investigate the application of this method to the identification of unanticipated reactions, we prepared monolayers that presented a single functional group at low density against a nonreactive background, and treated the monolayers with an array of common reagents. The spots were then analyzed by SAMDI to identify those that gave a high-yielding conversion to a new product whose structure was not readily deduced from the mass. This screen identified a novel reaction, wherein a primary amine reacted with three equivalents of an aldehyde under mild conditions to provide an N-alkylpyridinium product (133).

4.5. Assays of Cellular Activities

Many applications seek to measure enzyme or protein binding activities in complex samples, including extracts prepared from cell culture and bodily fluids taken for clinical analysis, and are often complicated by significant levels of nonspecific signal. In these cases, the MS approaches are especially valuable in that specific and nonspecific analytes can be distinguished based on their masses. In one instance, we used a SAMDI assay to monitor the activities of caspase proteases that initiate the apoptotic pathways in cells (several caspase enzymes undergo sequential activation when the cell death machinery is activated). The assay now widely used in biology relies on the collection of cell lysate and the addition of fluorescently tagged tetrapeptides that are substrates for the caspases. The substrates incorporate the fluorogenic reporter at the amide bond that undergoes hydrolysis and therefore forces a large nonnatural residue into the enzyme active site. In consequence, the peptide substrates have poor specificity for the enzymes, leading to cleavage by several of the caspase family members. Using the SAMDI assay, we immobilized longer peptide sequences that spanned both sides of the cleavage site (128). Treatment of the monolayers with lysates from cells that had been stimulated to activate the apoptotic pathway resulted in cleavage of the peptides and quantitation of caspase activity in the cells. A direct comparison of the SAMDI and fluorogenic assays revealed that the use of longer peptides in SAMDI gave improved enzyme-type specificity and also showed that the SAMDI assay can measure endogenous enzyme activities in complex cell lysates. Another study reported on the measurement of kinase activities in cell lysates (134).

4.6. Clinical Immunoassays

MS has been an important method in clinical diagnostic laboratories. Significant early work performed by Nelson and colleagues used MS to perform immunoassays on SAMs (135–139). The development of surface-enhanced laser desorption/ionization (SELDI) MS (140), which is based on the partially selective enrichment of proteins to substrates with chemistries that are electrostatic or hydrophobic to varying degrees, has prompted many efforts to apply MS for identifying and analyzing biomarkers (141). However, the presence of mixtures on the SELDI plates still complicates analysis of the spectra. Furthermore, these techniques are limited by reliance on either physical adsorption or the covalent attachment of random amino acids, which can

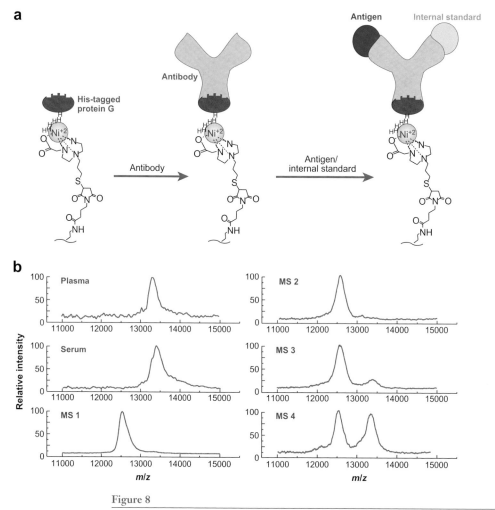

Figure 8

A mass spectrometry (MS) assay used to perform an immunoassay of cystatin C in patient clinical samples. (*a*) A His-tagged protein G is immobilized to a monolayer and used to immobilize an antibody against cystatin. (*b*) Treatment of the immunosensor with plasma, serum, or cerebral spinal fluid from patients with multiple sclerosis reveals the proteolytic form of the antigen associated with multiple sclerosis.

lead to nonspecific adsorption and a lack of control over orientation and density. The use of the surface chemistries described above can repair these limitations and may accelerate the use of MS methods in clinical immunoassays. We applied this approach to the analysis of cystatin (an inhibitor of cysteine proteases found in several tissues and a candidate biomarker for multiple sclerosis) in cerebral spinal fluids taken from healthy patients and from patients diagnosed with multiple sclerosis (142). The protein was present in a truncated form in the latter population, and is therefore a good candidate for analysis by MS (**Figure 8**). We prepared a monolayer that presented an antibody with equal affinity for both the full and truncated forms of cystatin. Application of 1 μL of sample to the monolayer resulted in capture of the protein antigen, and analysis by SAMDI clearly discriminated between the two forms of protein and identified the samples taken from the patients with multiple sclerosis.

5. FUTURE PERSPECTIVES

In this review, we have addressed the status of biochip arrays from a surface chemistry perspective, emphasized the use of SAMs to control the immobilization of molecules and their activities, and discussed the ability of MS to perform a broad range of biochemical assays. The use of well-defined surface chemistries allows the activities of immobilized molecules to be optimized and used for quantitative measurement of binding affinities and enzyme activity. Analyzing the biochips with MS substantially simplifies the formatting of new assays and permits distinct biochemical activities to be profiled on the same chip. These benefits will be particularly important for applications in higher-density biochips, where current methods still suffer from high rates of false positive and negative information and where applications are often constrained by the strategies available for labeling biochemical activities.

Current efforts emphasize the preparation and analysis of monolayers that have hundreds to thousands of molecules in the array and thus approach the common densities of current protein and small-molecule arrays. Monolayers are compatible with many of the arraying tools—with the exception of those wherein the pin contacts the substrate and can damage the monolayer—that are important for preparation of biochips on glass slides. A difficulty with the monolayers stems from the lower throughput inherent to mass spectrometric methods as compared to fluorescence-imaging methods, which use flatbed scanners. Entry-level instruments can acquire data from the monolayers at a rate of approximately 250 spots/h. Modern instruments with higher sensitivity are expected to accelerate this rate by a factor of five; the incorporation of multiple assays in each spot may lead to a further tenfold throughput, allowing most protein and small-molecule arrays to be analyzed in one day. As with current MALDI MS, better methods for uniformly depositing matrix on the substrates will be important.

The early work in biochip arrays for applications outside of nucleic acids has been exciting and points to the many important applications that these tools will enable. Advances in surface chemistry will be vital to realizing these opportunities and to expanding the applications that may be targeted. This review provides an early status report on the surface chemistries that are most relevant to biochips and which will be important to furthering this important technology in bioanalytical chemistry.

DISCLOSURE STATEMENT

The authors are not aware of any biases that might be perceived as affecting the objectivity of this review.

LITERATURE CITED

1. Liotta LA, Espina V, Mehta AI, Calvert V, Rosenblatt K, et al. 2003. Protein microarrays: meeting analytical challenges for clinical applications. *Cancer Cell* 3:317–25

2. Lueking A, Cahill DJ, Mullner S. 2005. Protein biochips: a new and versatile platform technology for molecular medicine. *Drug Discov. Today* 10:789–94

3. Petrik J. 2006. Diagnostic applications of microarrays. *Transfus. Med.* 16:233–47

4. Robinson WH. 2006. Antigen arrays for antibody profiling. *Curr. Opin. Chem. Biol.* 10:67–72

5. Ma HC, Horiuchi KY. 2006. Chemical microarray: a new tool for drug screening and discovery. *Drug Discov. Today* 11:661–68

6. Fodor SPA, Read JL, Pirrung MC, Stryer L, Lu AT, Solas D. 1991. Light-directed, spatially addressable parallel chemical synthesis. *Science* 251:767–73

7. Schena M, Shalon D, Davis RW, Brown PO. 1995. Quantitative monitoring of gene-expression patterns with a complementary-DNA microarray. *Science* 270:467–70

8. Brown PO, Botstein D. 1999. Exploring the new world of the genome with DNA microarrays. *Nat. Genet.* 21:33–37

9. Chittur SV. 2004. DNA microarrays: tools for the 21st century. *Comb. Chem. High T. Scr.* 7:531–37

10. Lockhart DJ, Winzeler EA. 2000. Genomics, gene expression and DNA arrays. *Nature* 405:827–36

11. Haushalter RC. 2004. Method for in situ, on-chip chemical synthesis. *U.S. Patent No. 20040175710*

12. Mrksich M, Sigal GB, Whitesides GM. 1995. Surface-plasmon resonance permits in-situ measurement of protein adsorption on self-assembled monolayers of alkanethiolates on gold. *Langmuir* 11:4383–85

13. Tengvall P, Lundstrom I, Liedberg B. 1998. Protein adsorption studies on model organic surfaces: an ellipsometric and infrared spectroscopic approach. *Biomaterials* 19:407–22

14. Elwing H. 1998. Protein absorption and ellipsometry in biomaterial research. *Biomaterials* 19:397–406

15. Margel S, Vogler EA, Firment L, Watt T, Haynie S, Sogah DY. 1993. Peptide, protein, and cellular interactions with self-assembled monolayer model surfaces. *J. Biomed. Mater. Res.* 27:1463–76

16. Haab BB, Dunham MJ, Brown PO. 2001. Protein microarrays for highly parallel detection and quantification of specific proteins and antibodies in complex solutions. *Gen. Biol.* 2:0004.1–04.13

17. Kusnezow W, Hoheisel JD. 2003. Solid supports for microarray immunoassays. *J. Mol. Recognit.* 16:165–76

18. Zhu H, Bilgin M, Bangham R, Hall D, Casamayor A, et al. 2001. Global analysis of protein activities using proteome chips. *Science* 293:2101–5

19. Michaud GA, Salcius M, Zhou F, Bangham R, Bonin J, et al. 2003. Analyzing antibody specificity with whole proteome microarrays. *Nat. Biotechnol.* 21:1509–12

20. Machida K, Thompson CM, Dierck K, Jablonowski K, Karkkainen S, et al. 2007. High-throughput phosphotyrosine profiling using SH2 domains. *Mol. Cell* 26:899–915

21. Jones RB, Gordus A, Krall JA, MacBeath G. 2006. A quantitative protein interaction network for the ErbB receptors using protein microarrays. *Nature* 439:168–74

22. Lin Y, Huang R, Chen L, Li S, Shi Q, et al. 2004. Identification of interleukin-8 as estrogen receptor regulated factor involved in breast cancer invasion and angiogenesis by protein arrays. *Int. J. Cancer* 109:570–15

23. Brooks RA, Wimhurst JA, Rushton N. 2002. Endotoxin contamination of particles producing misleading inflammatory cytokine responses from macrophages in vitro. *J. Bone Joint Surg.* 84:295–99

24. Miller JC, Zhou HP, Kwekel J, Cavallo R, Burke J, et al. 2003. Antibody microarray profiling of human prostate cancer sera: antibody screening and identification of potential biomarkers. *Proteomics* 3:56–63

25. Taitt CR, Golden JP, Shubin YS, Shriver-Lake LC, Sapsford KE, et al. 2004. A portable array biosensor for detecting multiple analytes in complex samples. *Microbial Ecol.* 47:175–85

26. Korbel GA, Lalic G, Shair MD. 2001. Reaction microarrays: a method for rapidly determining the enantiomeric excess of thousands of samples. *J. Am. Chem. Soc.* 123:361–62

27. Kuruvilla FG, Shamji AF, Sternson SM, Hergenrother PJ, Schreiber SL. 2002. Dissecting glucose signalling with diversity-oriented synthesis and small-molecule microarrays. *Nature* 416:653–57

28. Peterfi Z, Kocsis B. 2000. Comparison of blocking agents for an ELISA for LPS. *J. Immunoassay* 21:341–54

29. Oh SJ, Hong BJ, Choi KY, Park JW. 2006. Surface modification for DNA and protein microarrays. *OMICS: J. Interg. Biol.* 10:327–43

30. Rubina AY, Dementieva EI, Stomakhin AA, Darii EL, Pan'kov SV, et al. 2003. Hydrogel-based protein microchips: manufacturing, properties, and applications. *Biotechniques* 34:1008–22

31. Vericat C, Vela ME, Salvarezza RC. 2005. Self-assembled monolayers of alkanethiols on Au(111): surface structures, defects and dynamics. *Phys. Chem. Chem. Phys.* 7:3258–68

32. Schaeferling M, Schiller S, Paul H, Kruschina M, Pavlickova P, et al. 2002. Application of self-assembly techniques in the design of biocompatible protein microarray surfaces. *Electrophoresis* 23:3097–105

33. Wink T, van Zuilen SJ, Bult A, van Bennekom WP. 1997. Self-assembled monolayers for biosensors. *Analyst* 122:R43–R50

34. Love JC, Estroff LA, Kriebel JK, Nuzzo RG, Whitesides GM. 2005. Self-assembled monolayers of thiolates on metals as a form of nanotechnology. *Chem. Rev.* 105:1103–69

35. Johnson RD, Arnold FH. 1995. The Temkin isotherm describes heterogeneous protein absorption. *Biochim. Biophys. Acta* 1247:293–97

36. Seigel RR, Harder P, Dahint R, Grunze M, Josse F, et al. 1997. On-line detection of nonspecific protein adsorption at artificial surfaces. *Anal. Chem.* 69:3321–28

37. Turbill P, Beugeling T, Poot AA. 1996. Proteins involved in the Vroman effect during exposure of human blood plasma to glass and polyethylene. *Biomaterials* 17:1279–87

38. Yam CM, Deluge M, Tang D, Kumar A, Cai CZ. 2006. Preparation, characterization, resistance to protein adsorption, and specific avidin-biotin binding of poly (amidoamine) dendrimers functionalized with oligo(ethylene glycol) on gold. *J. Colloid. Interf. Sci.* 296:118–30

39. Patel N, Davies MC, Hartshorne M, Heaton RJ, Roberts CJ, et al. 1997. Immobilization of protein molecules onto homogeneous and mixed carboxylate-terminated self-assembled monolayers. *Langmuir* 13:6485–90

40. MacBeath G, Schreiber SL. 2000. Printing proteins as microarrays for high-throughput function determination. *Science* 289:1760–63

41. Tessier PM, Lindquist S. 2007. Prion recognition elements govern nucleation, strain specificity and species barriers. *Nature* 447:556–61

42. Rao SV, Anderson KW, Bachas LG. 1998. Oriented immobilization of proteins. *Mikrochim. Acta* 128:127–43

43. Smith CL, Milea JS, Nguyen GH. 2006. Immoblization of nucleic acids using biotin-strept(avidin) systems. *Top. Curr. Chem.* 261:63–90

44. Viens A, Mechold U, Lehrmann H, Harel-Bellan A, Ogryzko V. 2004. Use of protein biotinylation in vivo for chromatin immunoprecipitation. *Anal. Biochem.* 325:68–76

45. Nieba L, Nieba-Axmann SE, Persson A, Hamalainen M, Edebratt F, et al. 1997. BIACORE analysis of histidine-tagged proteins using a chelating NTA sensor chip. *Anal. Biochem.* 252:217–28

46. Sigal GB, Bamdad C, Barberis A, Strominger J, Whitesides GM. 1996. A self-assembled monolayer for the binding and study of histidine tagged proteins by surface plasmon resonance. *Anal. Chem.* 68:490–97

47. Lata S, Reichel A, Brock R, Tampe R, Piehler J. 2005. High-affinity adaptors for switchable recognition of histidine-tagged proteins. *J. Am. Chem. Soc.* 127:10205–15

48. Zhu H, Klemic JF, Chang S, Bertone P, Casamayor A, et al. 2000. Analysis of yeast protein kinases using protein chips. *Nat. Genet.* 26:283–89

49. Armstrong RN. 1991. Glutathione S-transferases: reaction-mechanism, structure, and function. *Chem. Res. Toxicol.* 4:131–40

50. Boozer C, Ladd J, Chen S, Yu QM, Homola J, Jiang S. 2004. DNA direction protein immobilization of mixed ssDNA/oligo(ethylene glycol) self-assembled monolayers for sensitive biosensors. *Anal. Chem.* 76:6967–72

51. Hodneland CD, Lee YS, Min DH, Mrksich M. 2002. Selective immobilization of proteins to self-assembled monolayers presenting active site-directed capture ligands. *Proc. Natl. Acad. Sci. USA* 99:5048–52

52. Kindermann M, George N, Johnsson N, Johnsson K. 2003. Covalent and selective immobilization of fusion proteins. *J. Am. Chem. Soc.* 125:7810–11

53. Sielaff I, Arnold A, Godin G, Tugulu S, Klok HA, Johnsson K. 2006. Protein function microarrays based on self-immobilizing and self-labeling fusion proteins. *ChemBioChem* 7:194–202

54. Lee SJ, Park JP, Park TJ, Lee SY, Lee S, Park JK. 2005. Selective immobilization of fusion proteins on poly(hydroxyalkanoate) microbeads. *Anal. Chem.* 77:5755–59

55. Kwon Y, Coleman MA, Camarero JA. 2006. Selective immobilization of proteins onto solid supports through split-intein-mediated protein trans-splicing. *Angew. Chem. Int. Edit.* 45:1726–29

56. de Paz JL, Noti C, Seeberger PH. 2006. Microarrays of synthetic heparin oligosaccharides. *J. Am. Chem. Soc.* 128:2766–67

57. Duffner JL, Clemons PA, Koehler AN. 2007. A pipeline for ligand discovery using small-molecule microarrays. *Curr. Opin. Chem. Biol.* 11:74–82

58. Hunt I. 2005. From gene to protein: a review of new and enabling technologies for multi-parallel protein expression. *Protein Expres. Purif.* 40:1–22

59. Haab BB. 2001. Advances in protein microarray technology for protein expression and interaction profiling. *Curr. Opin. Drug Disc.* 4:116–23

60. Roda A, Guardigli M, Russo C, Pasini P, Baraldini M. 2000. Protein microdeposition using a conventional ink-jet printer. *Biotechniques* 28:492–96

61. Barbulovic-Nad I, Lucente M, Sun Y, Zhang MJ, Wheeler AR, Bussmann M. 2006. Bio-microarray fabrication techniques: a review. *Crit. Rev. Biotechnol.* 26:237–59

62. Deng Y, Zhu XY, Kienlen T, Guo A. 2006. Transport at the air/water interface is the reason for rings in protein microarrays. *J. Am. Chem. Soc.* 128:2768–69

63. Wang PC, DeVoe DL, Lee CS. 2001. Integration of polymeric membranes with microfluidic networks for bioanalytical applications. *Electrophoresis* 22:3857–67

64. Hansen C, Quake SR. 2003. Microfluidics in structural biology: smaller, faster... better. *Curr. Opin. Struc. Biol.* 13:538–44

65. Weibel DB, Whitesides GM. 2006. Applications of microfluidics in chemical biology. *Curr. Opin. Chem. Biol.* 10:584–91

66. Lovchik R, von Arx C, Viviani A, Delamarche E. 2007. Cellular microarrays for use with capillary-driven microfluidics. *Anal. Bioanal. Chem.* 390(3):801–8

67. Bernard A, Michel B, Delamarche E. 2001. Micromosaic immunoassays. *Anal. Chem.* 73:8–12

68. Kim J, Crooks RM. 2007. Replication of DNA microarrays prepared by in situ oligonucleotide polymerization and mechanical transfer. *Anal. Chem.* 79:7267–74

69. Gao XL, Zhou XC, Gulari E. 2003. Light directed massively parallel on-chip synthesis of peptide arrays with t-Boc chemistry. *Proteomics* 3:2135–41

70. Frank R. 1992. Spot-synthesis: an easy technique for the positionally addressable, parallel chemical synthesis on a membrane support. *Tetrahedron* 48:9217–32

71. Walsh DP, Chang YT. 2004. Recent advances in small molecule microarrays: applications and technology. *Comb. Chem. High T. Scr.* 7:557–64

72. Monk DJ, Walt DR. 2004. Optical fiber-based biosensors. *Anal. Bioanal. Chem.* 379:931–45

73. Lynch JL, deSilva CJS, Peeva VK, Swanson NR. 2006. Comparison of commercial probe labeling kits for microarray: towards quality assurance and consistency of reactions. *Anal. Biochem.* 355:224–31

74. Sapsford KE, Charles PT, Patterson CH, Ligler FS. 2002. Demonstration of four immunoassay formats using the array biosensor. *Anal. Chem.* 74:1061–68

75. Arenkov P, Kukhtin A, Gemmell A, Voloshchuk S, Chupeeva V, Mirzabekov A. 2000. Protein microchips: use for immunoassay and enzymatic reactions. *Anal. Biochem.* 278:123–31

76. Houseman BT, Huh JH, Kron SJ, Mrksich M. 2002. Peptide chips for the quantitative evaluation of protein kinase activity. *Nat. Biotechnol.* 20:270–74

77. Cooper MA. 2003. Label-free scanning of bio-molecular interactions. *Anal. Bioanal. Chem.* 377:834–42

78. Haake HM, Schutz A, Gauglitz G. 2000. Label-free detection of biomolecular interaction by optical sensors. *Fresen. J. Anal. Chem.* 366:576–85

79. Schaferling M, Nagl S. 2006. Optical technologies for the read out and quality control of DNA and protein microarrays. *Anal. Bioanal. Chem.* 385:500–17

80. Fang Y. 2006. Label-free cell-based assays with optical biosensors in drug discovery. *Assay Drug. Dev. Techn.* 4:583–95

81. Homola J, Vaisocherova H, Dostalek J, Piliarik M. 2005. Multi-analyte surface plasmon resonance biosensing. *Methods* 37:26–36

82. Fujiwara K, Watarai H, Itoh H, Nakahama E, Ogawa N. 2006. Measurement of antibody binding to protein immobilized on gold nanoparticles by localized surface plasmon spectroscopy. *Anal. Bioanal. Chem.* 386:639–44

83. Boozer C, Kim G, Cong SX, Guan HW, Londergan T. 2006. Looking towards label-free biomolecular interaction analysis in a high-throughput format: a review of new surface plasmon resonance technologies. *Curr. Opin. Biotech.* 17:400–5

84. Goodrich TT, Lee HJ, Corn RM. 2004. Direct detection of genomic DNA by enzymatically amplified SPR imaging measurements of RNA microarrays. *J. Am. Chem. Soc.* 126:4086–87

85. Goodrich TT, Lee HJ, Corn RM. 2004. Enzymatically amplified surface plasmon resonance imaging method using RNase H and RNA microarrays for the ultrasensitive detection of nucleic acids. *Anal. Chem.* 76:6173–78

86. Lee HJ, Li Y, Wark AW, Corn RM. 2005. Enzymatically amplified surface plasmon resonance imaging detection of DNA by exonuclease III digestion of DNA microarrays. *Anal. Chem.* 77:5096–100

87. Wolf LK, Gao Y, Georgiadis RM. 2007. Kinetic discrimination of sequence-specific DNA-drug binding measured by surface plasmon resonance imaging and comparison to solution phase measurements. *J. Am. Chem. Soc.* 129:10503–11

88. Yang LJ, Li YB, Erf GF. 2004. Interdigitated array microelectrode-based electrochemical impedance immunosensor for detection of *Escherichia coli* O157: H7. *Anal. Chem.* 76:1107–13

89. Xu DK, Xu DW, Yu XB, Liu ZH, He W, Ma ZQ. 2005. Label-free electrochemical detection for aptamer-based array electrodes. *Anal. Chem.* 77:5107–13

90. Rabe U, Kopycinska M, Hirsekorn S, Arnold W. 2002. Evaluation of the contact resonance frequencies in atomic force microscopy as a method for surface characterisation (invited). *Ultrasonics* 40:49–54

91. Jelesarov I, Bosshard HR. 1999. Isothermal titration calorimetry and differential scanning calorimetry as complementary tools to investigate the energetics of biomolecular recognition. *J. Mol. Recognit.* 12:3–18

92. Clas SD, Dalton CR, Hancock BC. 1999. Differential scanning calorimetry: applications in drug development. *Pharm. Sci. Technol.* 2:311–20

93. Whitesides GM, Kriebel JK, Love JC. 2005. Molecular engineering of surfaces using self-assembled monolayers. *Sci. Prog.* 88:17–48

94. Nuzzo RG, Allara DL. 1983. Adsorption of bifunctional organic disulfides on gold surfaces. *J. Am. Chem. Soc.* 105:4481–83

95. Finklea HO, Avery S, Lynch M, Furtsch T. 1987. Blocking oriented monolayers of alkyl mercaptans on gold electrodes. *Langmuir* 3:409–13

96. Porter MD, Bright TB, Allara DL, Chidsey CED. 1987. Spontaneously organized molecular assemblies. 4. Structural characterization of normal-alkyl thiol monolayers on gold by optical ellipsometry, infrared spectroscopy, and electrochemistry. *J. Am. Chem. Soc.* 109:3559–68

97. Strong L, Whitesides GM. 1988. Structures of self-assembled monolayer films of organosulfur compounds adsorbed on gold single crystals: electron-diffraction studies. *Langmuir* 4:546–58

98. Poirier GE, Pylant ED. 1996. The self-assembly mechanism of alkanethiols on Au(111). *Science* 272:1145–48

99. Bain CD, Whitesides GM. 1988. Formation of 2-component surfaces by the spontaneous assembly of monolayers on gold from solutions containing mixtures of organic thiols. *J. Am. Chem. Soc.* 110:6560–61

100. Biebuyck HA, Bian CD, Whitesides GM. 1994. Comparison of organic monolayers on polycrystalline gold spontaneously assembled from solutions containing dialkyl disulfides or alkenethiols. *Langmuir* 10:1825–31

101. Yan L, Marzolin C, Terfort A, Whitesides GM. 1997. Formation and reaction of interchain carboxylic anhydride groups on self-assembled monolayers on gold. *Langmuir* 13:6704–12

102. Ryan D, Parviz BA, Linder V, Semetey V, Sia SK, et al. 2004. Patterning multiple aligned self-assembled monolayers using light. *Langmuir* 20:9080–88

103. Bensebaa F, Ellis TH, Badia A, Lennox RB. 1998. Thermal treatment of n-alkanethiolate monolayers on gold, as observed by infrared spectroscopy. *Langmuir* 14:2361–67

104. Prime KL, Whitesides GM. 1991. Self-assembled organic monolayers: model systems for studying adsorption of proteins at surfaces. *Science* 252:1164–67

105. Prime KL, Whitesides GM. 1993. Adsorption of proteins onto surfaces containing end-attached oligo(ethylene oxide): a model system using self-assembled monolayers. *J. Am. Chem. Soc.* 115:10714–21

106. Herrwerth S, Eck W, Reinhardt S, Grunze M. 2003. Factors that determine the protein resistance of oligoether self-assembled monolayers: internal hydrophilicity, terminal hydrophilicity, and lateral packing density. *J. Am. Chem. Soc.* 125:9359–66

107. Welle A, Grunze M, Tur D. 1998. Plasma protein adsorption and platelet adhesion on poly[bis(trifluoroethoxy)phosphazene] and reference material surfaces. *J. Colloid. Interf. Sci.* 197:263–74

108. Ostuni E, Chapman RG, Holmlin RE, Takayama S, Whitesides GM. 2001. A survey of structure-property relationships of surfaces that resist the adsorption of protein. *Langmuir* 17:5605–20

109. Luk YY, Kato M, Mrksich M. 2000. Self-assembled monolayers of alkanethiolates presenting mannitol groups are inert to protein adsorption and cell attachment. *Langmuir* 16:9604–08

110. Houseman BT, Mrksich M. 2001. The microenvironment of immobilized Arg-Gly-Asp peptides is an important determinant of cell adhesion. *Biomaterials* 22:943–55

111. Houseman BT, Gawalt ES, Mrksich M. 2003. Maleimide-functionalized self-assembled monolayers for the preparation of peptide and carbohydrate biochips. *Langmuir* 19:1522–31

112. Devaraj NK, Miller GP, Ebina W, Kakaradov B, Collman JP, et al. 2005. Chemoselective covalent coupling of oligonucleotide probes to self-assembled monolayers. *J. Am. Chem. Soc.* 127:8600–01

113. Chapman RG, Ostuni E, Yan L, Whitesides GM. 2000. Preparation of mixed self-assembled monolayers (SAMs) that resist adsorption of proteins using the reaction of amines with a SAM that presents interchain carboxylic anhydride groups. *Langmuir* 16:6927–36

114. Hong HG, Jiang M, Sligar SG, Bohn PW. 1994. Cysteine-specific surface tethering of genetically engineered cytochromes for fabrication of metalloprotein nanostructures. *Langmuir* 10:153–58

115. Luk YY, Tingey ML, Dickson KA, Raines RT, Abbott NL. 2004. Imaging the binding ability of proteins immobilized on surfaces with different orientations by using liquid crystals. *J. Am. Chem. Soc.* 126:9024–32

116. Kwon Y, Han ZZ, Karatan E, Mrksich M, Kay BK. 2004. Antibody arrays prepared by cutinase-mediated immobilization on self-assembled monolayers. *Anal. Chem.* 76:5713–20

117. Mrksich M, Grunwell JR, Whitesides GM. 1995. Biospecific adsorption of carbonic-anhydrase to self-assembled monolayers of alkanethiolates that present benzenesulfonamide groups on gold. *J. Am. Chem. Soc.* 117:12009–10

118. Haussling L, Ringsdorf H, Schmitt FJ, Knoll W. 1991. Biotin-functionalized self-assembled monolayers on gold: surface-plasmon optical studies of specific recognition reactions. *Langmuir* 7:1837–40

119. Horan N, Yan L, Isobe H, Whitesides GM, Kahne D. 1999. Nonstatistical binding of a protein to clustered carbohydrates. *Proc. Natl. Acad. Sci. USA* 96:11782–86

120. Houseman BT, Mrksich M. 2002. Carbohydrate arrays for the evaluation of protein binding and enzymatic modification. *Chem. Biol.* 9:443–54

121. Houseman BT, Mrksich M. 1999. The role of ligand density in the enzymatic glycosylation of carbohydrates presented on self-assembled monolayers of alkanethiolates on gold. *Angew. Chem. Int. Edit.* 38:782–85

122. Trevor JL, Lykke KR, Pellin MJ, Hanley L. 1998. Two-laser mass spectrometry of thiolate, disulfide, and sulfide self-assembled monolayers. *Langmuir* 14:1664–73

123. Gong WH, Elitzin VI, Janardhanam S, Wilkins CL, Fritsch I. 2001. Effect of laser fluence on laser desorption mass spectra of organothiol self-assembled monolayers on gold. *J. Am. Chem. Soc.* 123:769–70

124. Walsh CT. 2006. *Posttranslational Modifications of Proteins: Expanding Nature's Inventory.* Greenwood Village, Colo.: Roberts

125. Min DH, Su J, Mrksich M. 2004. Profiling kinase activities by using a peptide chip and mass spectrometry. *Angew. Chem. Int. Edit.* 43:5973–77

126. Di Valentin C, Scagnelli A, Pacchioni G. 2005. Theory of nanoscale atomic lithography: an ab initio study of the interaction of "cold" Cs atoms with organthiols self-assembled monolayers on Au(111). *J. Phys. Chem. B* 109:1815–21

127. Min DH, Yeo WS, Mrksich M. 2004. A method for connecting solution-phase enzyme activity assays with immobilized format analysis by mass spectrometry. *Anal. Chem.* 76:3923–29

128. Su J, Rajapaksha TW, Peter ME, Mrksich M. 2006. Assays of endogenous caspase activities: a comparison of mass spectrometry and fluorescence formats. *Anal. Chem.* 78:4945–51

129. Min DH, Tang WJ, Mrksich M. 2004. Chemical screening by mass spectrometry to identify inhibitors of anthrax lethal factor. *Nat. Biotechnol.* 22:717–23

130. Kanan MW, Rozenman MM, Sakurai K, Snyder TM, Liu DR. 2004. Reaction discovery enabled by DNA-templated synthesis and in vitro selection. *Nature* 431:545–49

131. Copeland GT, Miller SJ. 2001. Selection of enantioselective acyl transfer catalysts from a pooled peptide library through a fluorescence-based activity assay: an approach to kinetic resolution of secondary alcohols of broad structural scope. *J. Am. Chem. Soc.* 123:6496–502

132. Petra DGI, Reek JNH, Kamer PCJ, Schoemaker HE, van Leeuwen PWNM. 2000. IR spectroscopy as a high-throughput screening technique for enantioselective hydrogen-transfer catalysts. *Chem. Commun.* 683–84

133. Li J, Thiara PS, Mrksich M. 2007. Rapid evaluation and screening of interfacial reactions on self-assembled monolayers. *Langmuir* 23:11826–35

134. Su J, Bringer MR, Ismagilov RF, Mrksich M. 2005. Combining microfluidic networks and peptide arrays for multi-enzyme assays. *J. Am. Chem. Soc.* 127:7280–81

135. Nedelkov D. 2006. Mass spectrometry-based immunoassays for the next phase of clinical applications. *Exp. Rev. Prot.* 3:631–40

136. Nedelkov D, Nelson RW. 2001. Analysis of native proteins from biological fluids by biomolecular interaction analysis mass spectrometry (BIA/MS): exploring the limit of detection, identification of nonspecific binding and detection of multi-protein complexes. *Biosens. Bioelectron.* 16:1071–78

137. Nelson RW, Krone JR, Bieber AL, Williams P. 1995. Mass-spectrometric immunoassay. *Anal. Chem.* 67:1153–58

138. Nedelkov D, Kiernan UA, Niederkofler EE, Tubbs KA, Nelson RW. 2006. Population proteomics: the concept, attributes, and potential for cancer biomarker research. *Mol. Cell Proteomics* 5:1811–18

139. Nedelkov D, Kiernan UA, Niederkofler EE, Tubbs KA, Nelson RW. 2005. Investigating diversity in human plasma proteins. *Proc. Natl. Acad. Sci. USA* 102:10852–57

140. Engwegen JYMN, Gast MCW, Schellens JHM, Beijnen JH. 2006. Clinical proteomics: searching for better tumour markers with SELDI-TOF mass spectrometry. *Trends Pharmacol. Sci.* 27:251–59

141. Albrethsen J, Bogebo R, Olsen J, Raskov H, Gammeltoft S. 2006. Preanalytical and analytical variation of surface-enhanced laser desorption-ionization time-of-flight mass spectrometry of human serum. *Clin. Chem. Lab. Med.* 44:1243–52

142. Patrie SM, Mrksich M. 2007. Self-assembled monolayers for MALDI-TOF mass spectrometry for immunoassays of human protein antigens. *Anal. Chem.* 79:5878–87

Liposomes: Technologies and Analytical Applications

Aldo Jesorka and Owe Orwar

Department of Chemical and Biological Engineering, Chalmers University of Technology, SE-41296 Göteborg, Sweden; email: aldo.jesorka@chalmers.se, owe.orwar@chalmers.se

Annu. Rev. Anal. Chem. 2008. 1:801–32

First published online as a Review in Advance on April 1, 2008

The *Annual Review of Analytical Chemistry* is online at anchem.annualreviews.org

This article's doi:
10.1146/annurev.anchem.1.031207.112747

Key Words

liposomes, liposome analytics, liposome technologies, phospholipid vesicles, nanotube-vesicle networks

Abstract

Liposomes are structurally and functionally some of the most versatile supramolecular assemblies in existence. Since the beginning of active research on lipid vesicles in 1965, the field has progressed enormously and applications are well established in several areas, such as drug and gene delivery. In the analytical sciences, liposomes serve a dual purpose: Either they are analytes, typically in quality-assessment procedures of liposome preparations, or they are functional components in a variety of new analytical systems. Liposome immunoassays, for example, benefit greatly from the amplification provided by encapsulated markers, and nanotube-interconnected liposome networks have emerged as ultrasmall-scale analytical devices. This review provides information about new developments in some of the most actively researched liposome-related topics.

1. INTRODUCTION

Since the first observation of phospholipid vesicles (liposomes) in 1965 by Bangham and colleagues, who effectively determined the character of lipid membrane–enclosed volumes and their connection to biological cells (1), liposome-related topics have become commonplace in the literature (see, for example, References 2–4).

The nature of lipid vesicles as well as their relevance in biological environments have been extensively studied (5–7), and applications have extended into many fields. Today, liposome-derived technologies are established as one of the cornerstones of bionanotechnology (8). The unique versatility of lipid vesicles with respect to composition, size variety, and capacity for embedding and encapsulating materials has led to applications in chemical and biochemical analytics and even to industrial-scale applications in drug delivery, cosmetics, food technology, and proteomics (9). Liposomes are commonly utilized as precursors in the fabrication of suspended bilayers (10, 11), and recently, nanotubes-conjugated liposome networks have emerged as novel biomimetic chemical reactor systems with capabilities for single-molecule analysis (12, 13). In this review, we give an overview of the fields of liposome research, technologies, and applications, particularly in analytical chemistry, covering fundamentals and advances in preparation and characterization and new techniques in areas such as chromatography and biosensors.

2. LIPOSOMES

2.1. Phospholipid Membrane Vesicle (Liposome) Characteristics and Properties

Liposomes are spherical soft-matter particles consisting of one or more bilayer membrane(s), and are most commonly composed of phospholipids encapsulating a volume of aqueous medium. The aqueous medium is typically the same as that in which the liposomes are suspended, but each can often be individually exchanged, for instance by microinjection or dialysis. Liposomes are readily prepared in the laboratory (5, 14).

In liposome formation, dissolved lipid molecules, consisting of a hydrophilic headgroup and a hydrophobic tail, self-assemble into bimolecular lipid leaflets upon decreasing their solubility in the surrounding medium. Whereas ordinary amphiphiles have critical micelle concentrations (CMCs) of 10^{-2}–10^{-4} M, the CMC of bilayer forming lipids is four to five orders of magnitude smaller, meaning that the water solubility of these materials is extremely low. The lipid's headgroups are exposed to the aqueous phase and the hydrophilic hydrocarbon moieties are forced to face each other in the bilayer. This free energy–driven process is recognized as one of the most powerful mechanisms in bottom-up engineering (15).

Liposomes as analogs of natural membranes are generally assembled by spontaneous self-organization from pure lipids or lipid mixtures. **Figure 1** schematically depicts the fundamental assembly process, in this case leading to a unilamellar vesicle.

The low water solubility of lipid molecules greatly affects the dynamics of lipid exchange between the bilayer and the surrounding medium. Residence times for

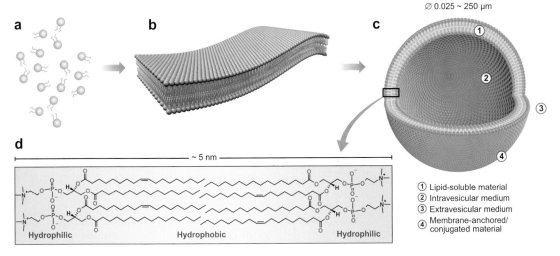

a

b

c

⌀ 0.025 ~ 250 μm

①
②
③
④

d

— ~ 5 nm —

Hydrophilic · · · · · · · Hydrophobic · · · · · · · Hydrophilic

① Lipid-soluble material
② Intravesicular medium
③ Extravesicular medium
④ Membrane-anchored/conjugated material

Figure 1

Schematic illustration of the fundamental self-assembly process from individual phospholipid molecules (*a*) to bilayer membrane leaflets (*b*), followed by transformation into liposomes (*c*). A single bilayer is typically ~5 nm thick and consists of neatly arranged individual lipid molecules with their hydrophobic tails facing each other and their hydrophilic headgroups facing toward the internal and external aqueous medium (*d*). Apart from the structural characteristics of the lipid molecules themselves, the properties and functionality of liposomes are largely defined by their size and the composition of the four distinct regions highlighted in panel *c*. Various natural and synthetic lipid molecules are available for the preparation of bilayer membranes and liposomes (140).

phospholipids are on the order of 10^4 s, compared to 10^{-4} s of typical micelle-forming amphiphiles (16).

The most commonly utilized lipids are phospholipids, in particular the charge-neutral phosphatidylcholine and the negatively charged phosphatidic acid, phosphatidylglycerol, phosphatidylserine, and phosphatidylethanolamine, each of which has a different combination of fatty acid chains in the hydrophobic region of the molecule. Stearylamine can be employed when cationic liposomes are needed. Apart from charge, the nature of the fatty acid residues in each lipid molecule, particularly the number of double bonds in the chain, is responsible for fundamental bilayer properties such as phase behavior and elasticity.

To obtain liposome preparations, crude or purified plant or animal lipid extracts are often employed. Plant and cyanobacteria lipid extracts contain mostly phosphoglycerides and glycosylglycerides, whereas extracts from erythrocytes or liver cells are mainly composed of phosphoglycerides, sphingolipids, and sterols. Unpurified extracts contain other hydrophobic molecules, such as membrane protein material as contaminants, and are therefore of limited use (17, 18).

A highly specific group of phospholipids are synthetic lipids with designed functionalities. To this group belong lipids that are headgroup modified for coupling reactions (19), chelating metal ions (20), and even multifunctional lipids (21).

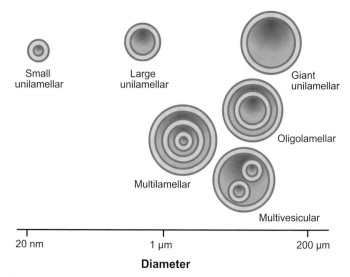

Figure 2

Schematic representation of the commonly applied classification scheme for liposomes. Small unilamellar vesicles (∼0.02 μm to ∼0.2 μm), large unilamellar vesicles (∼0.2 μm to ∼1 μm), and giant unilamellar vesicles (>1 μm) are the three most important groups for analytical applications. Multilamellar vesicles are frequently used in pharmaceutical and cosmetic applications (56). Multivesicular vesicles are giant vesicles encapsulating smaller liposomes and have been used in nanoreactor assemblies (141) and as drug delivery tools (vesosomes) (142). The drawings are not to scale.

Embedding functional regions such as reactive double bonds into the hydrophobic tail section has led to the creation of classes of polymerizeable lipids that function as cross-linking elements and, under UV or radical polymerization conditions, lead to membranes and vesicles of enhanced stability (22–24).

Liposomes are, from a morphological perspective, most frequently classified by their size and number of membrane bilayers (lamellae) (**Figure 2**).

Unilamellar vesicles are of special interest in research, mostly due to their well-characterized membrane properties and facile preparation. They are divided into three size types: small, large, and giant. Controlled processes for formation of oligo-lamellar vesicles are rare, although an example of a polymer-induced transformation of unilamellar to bilamellar vesicles has been reported (25). Multilamellar vesicles often show physical properties and behavior that are very different from the unilamellar species, and they are commonly used for industrial applications such as drug delivery. They are also employed as lipid reservoirs in the process of nanotube-vesicle network (NVN) fabrication (26).

Liposomes are generally not considered to be at thermodynamic equilibrium because curvature energy is being confined in the vesicles as they are produced. The curvature free energy, also known as the bending energy or curvature elastic energy of the liposome, is determined by the bending rigidity and curvature of the membrane and is responsible for the large variety of shapes that liposomes can take. Three

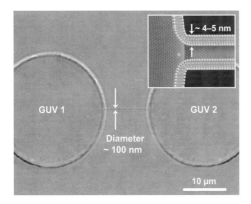

Figure 3

Differential interference contrast micrograph of a phosphatidylcholine lipotube suspended between two giant unilamellar vesicles (GUVs). The tube is only stable in the suspended state and will retract and merge with the connected GUV when cut off and released on one end. *Inset*: A detailed schematic view of the high-curvature region where the nanotube is connected to the vesicle.

models have been developed to describe these dependencies: the spontaneous curvature model (27), the bilayer couple model (28), and the area-difference elasticity model (29). Many other physical aspects of liposome structure and dynamics of formation were described in a recent overview (30).

2.2. Lipotubes (Lipid-Membrane Nanotubes)

An important feature of the liposome is the fluidity of its membrane, which allows for incidents of membrane deformation without disrupting bilayer integrity. A practical consequence of this property is the formation of lipotubes (lipid membrane nanotubes) when a point load is applied to a vesicle (31). Lipotube formation involves first-order shape transitions, wherein the obtained shape represents a minimum in surface free energy, i.e., the surface-to-volume ratio of the tubular structure is optimized.

Lipotubes are flexible membrane conduits with a diameter of ∼100 nm and a variable length of up to several hundred micrometers. The diameter of the structure depends largely on the lipid composition, which determines membrane tension and curvature. Transport of biopolymer material and detection of single molecules in lipotubes have been demonstrated in the context of NVN technology, as discussed below. **Figure 3** shows a lipotube tethered from a phosphatidyl choline liposome and suspended between two giant unilamellar vesicles (GUVs), as used for single-molecule transport experiments (32).

2.3. Vesicular Gels

Vesicular phospholipid gels (VPGs) are semisolid, aqueous phospholipid dispersions wherein the lipid phase consists of liposomes (33). They can be viewed as a hydrogel variant, but, unlike other hydrogels, tightly packed vesicles rather than macromolecular constituents cause the gel-like rheological behavior. In VPGs, the vesicles are so densely packed that the aqueous volume between the vesicles is reduced to a minimum, giving rise to steric interactions between neighboring vesicles and thus to a semisolid consistency. VPGs are distinctly different from liposome gels, wherein a hydrophilic polymer forms an embedding matrix.

VPGs are capable of storing and releasing drugs in a controlled manner over extended periods of time and may serve as depot implants for controlled release. When VPGs are exposed to excess aqueous medium, they are readily transformed into small unilamellar vesicle (SUV) dispersions, where the SUVs display high encapsulation efficiency, irrespective of the molecular weight, charge, or membrane permeability of the loaded material (34).

3. LIPOSOME PREPARATION AND ANALYTICAL CHARACTERIZATION

The beneficial properties of liposomes as drug carriers were first recognized in the 1970s and 1980s, and the number of their biopharmaceutical applications has increased rapidly (35). The need for liposomal drug carrier formulations in particular has stimulated research on liposome-preparation procedures. At least four major and several minor methodologies are in use today, including methods based on dry lipid films or emulsions and methods involving the use of micelle-forming detergents or the principles of solvent injection. Some commonly used procedures (**Figure 4**) are outlined in the following section.

The specificity, homogeneity, and laboratory-scale availability of liposomes containing natural or synthetic lipids has had an immediate impact on studies of diverse cellular and biophysical phenomena. A multitude of methods exists for the

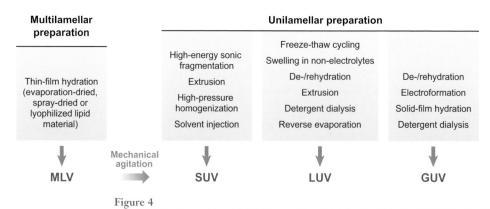

Figure 4

Common preparation techniques for different types of liposomes, categorized by lamellarity and size range. Multilamellar vesicles (MLVs) can be transformed into unilamellar vesicles by various means of mechanical treatment (5). Abbreviations: GUV, giant unilamellar vesicle; LUV, large unilamellar vesicle; SUV, small unilamellar vesicle.

preparation of liposomes of different size and lamellarity. Reviews and books have been published describing the protocols in great detail (7, 9, 14, 36).

Several fabrication methods start from a dry film that is subsequently hydrated with an aqueous solution containing the substance to be encapsulated (37); others are based on the reverse-evaporation technique (38). Using the film-hydration method, (mostly) multilamellar vesicles (MLVs) are formed with high encapsulation efficiency, but also with great variations in vesicle size, size distribution, and lamellarity. Some approaches replace a water-insoluble organic solvent that contains the dissolved lipids with an aqueous solution containing the substance to be encapsulated (39). This rapid solvent exchange is a novel strategy featuring direct transfer of the lipid mixture from the organic solvent to an aqueous buffer. The method is very fast and provides considerable entrapment efficiency. The solvent-injection method (40) and variants (41) involve the very slow injection of, typically, an ethanolic solution of lipids into an aqueous phase, leading to the formation of unilamellar vesicles. Liposomes obtained in this manner display a well-defined size distribution and a high encapsulation efficiency; residual ethanol is then removed by repeated dialysis steps against distilled water. Another methodology of liposome formation is founded on addition of detergent (42) or chaotropic ions (43) to the aqueous/organic mixture of lipids, and the subsequent removal of these reagents by gel filtration or dialysis.

New procedures have been reported, including a coacervation (phase-separation) technique based on water mixing with lipid solutions in different soluble alcohols (44) and microfluid-directed liposome formation (45).

3.1. Multilamellar Vesicle Preparation

The preparation of MLVs involves a simple and robust film-hydration technique (46). Lipid solution is initially dried, either by using an evaporator or by spray drying/ lyophilization (47) for larger-scale preparations. Alternatively, this step can be performed at very low pressure for a few hours in the presence of a neutral dessicant. For subsequent MLV formation, the sample is mechanically agitated in the presence of a hydration medium such as buffer. MLVs in the micrometer-size range are spontaneously formed when the film is exposed to an excess volume of aqueous buffer.

For transformation into unilamellar vesicles, MLVs can be dispersed by various mechanical methods, providing means for easy scale-up, large capacity, and fast processing (up to hundreds of liters per hour). Disadvantages are possible sample degradation and residual lipid particles of different size ranges.

3.2. Extrusion

Extrusion is a technique suitable for generating unilamellar liposomes of well-defined size (48). When MLVs are forced through narrow-pore membrane filters under pressure, membrane rupture and resealing occur and encapsulated content leaks out. Therefore, extrusion is performed in the presence of medium containing the final load concentration, and external solute is removed only after formation is complete (49).

The hydrated vesicles are initially subjected to several cycles of a freeze-thaw procedure and are subsequently forced through double-stacked polycarbonate membranes of decreasing pore size (e.g., sequentially 200, 100, and 50 nm) at elevated temperatures. The vesicles are commonly extruded five to ten times through each double-stacked membrane until the desired size is reached.

On-chip extrusion of micrometer-sized vesicles and extraordinarily long lipotubes through fabricated surface apertures has recently been reported (50). This method is a novel fabrication approach based on a combination of top-down and bottom-up processes.

3.3. Sonication

Sonication is a simple method for reducing the size of liposomes (51). The common laboratory method involves treating hydrated vesicles for several minutes with a titanium-tipped probe sonicator in a temperature-controlled environment. The vesicle preparations are subsequently passed through a 0.45-μm membrane filter to remove titanium particles that have detached from the tip surface during the procedure.

3.4. Freeze Drying of Monophase Solutions

Stability-influencing effects such as aggregation, fusion, and phospholipid hydrolysis are mainly due to the aqueous environment, as liposome preparations obtained by traditional methods are buffered suspensions. To improve this situation, a strategy for preparing proliposomes was developed (52). This method involves the coating of dry carrier powders with phospholipids that, upon hydration, form liposomes of a broad size distribution. A similar novel procedure for the preparation of large unilamellar vesicles (LUVs) of narrow size distribution is based on the initial formation of a uniphase solution of lipids and water-soluble support materials, such as sucrose in a t-butanol/H$_2$O cosolvent system, which is then lyophilized to remove the solvents (53). Upon the addition of water, the lyophilized product rapidly forms homogenous submicrometer-sized vesicles.

3.5. Hydration, Dehydration, and Swelling

Hydration/dehydration methods are based on the swelling of dried lipid films upon exposure to an aqueous medium. Because of the presence of buffer salts within the dehydrated film, an osmotic pressure gradient forces water between the individual bilayers, and the lamellae separate to form liposomes. This procedure is frequently used to obtain mixtures of multilamellar and giant unilamellar liposomes (54, 55). A critical factor that affects the yield of formation is the degree of bilayer separation, which is influenced by temperature, lipid composition, and the ionic composition of the surrounding medium. Inclusion of negatively charged lipids is known to enhance separation of the lamellae. A relatively low-ionic-strength buffer (10–50 mM) and the absence of multivalent ions that interlink charged lipids are essential. The process is comparatively fast (minutes compared to hours or days with other protocols), and unilamellar vesicles are produced in high yield.

The most commonly proposed mechanism for liposome formation from a dehydrated film is the detachment (56) of membrane sheets, which eventually separate and close upon themselves, remaining connected to the film via membrane tethers. Mechanical agitation is needed to detach the liposomes from the film.

3.6. Electroformation

There is no available method to exclusively produce GUVs, although the electroformation method yields high numbers of unilamellar vesicles (57, 58). Some lipid mixtures are very difficult to swell into vesicles. Electroformation, wherein an alternating electric field is applied while swelling the lipid film (57), is an efficient means of overcoming this problem. The field is believed to cause fluctuations in the bilayers, thus inducing detachment of lamellae and formation of vesicles. Lipid-film thickness, peak-to-peak voltage, duration of treatment, and frequency influence the process. The method is limited to low ionic-strength buffers, but a method to increase the ionic strength while preserving osmolality has also been reported (59).

Electroformation in combination with spincoating is a recent addition to the methodology (60). Very uniform lipid films are generated over a large substrate area, and film thickness can be controlled by spin speed, lipid concentration, and solvent viscosity. Films are typically 25–50 μm thick, which is reported to be the optimum thickness for electroformation. GUVs obtained by this method are considerably (two to five times) larger than those formed with standard techniques, and even lipids that do not readily form liposomes can be successfully assembled.

3.7. Ink-Jet Injection into an Aqueous Phase

By utilizing ink-jet technology, it is possible to prepare and suspend SUVs directly in one step. In this method, an ink-jet cartridge injects droplets of lipid solution in a water-miscible solvent with high reproducibility into an aqueous volume. Vesicle formation proceeds via a nucleation (micellization) and growth mechanism. The well-controlled distribution of monodisperse droplets leads to a stable level of supersaturation, which determines the number of nuclei and thus the final vesicle size. Routinely, monodisperse vesicle dispersions with high reproducibility of the mean particle diameter can be prepared for a given ink-jet cartridge type and amphiphile concentration. Volumes of $1-1.5 \text{ mL/min}^{-1}$ are possible to obtain (61).

3.8. Chemical Analysis and Characterization of Liposome Preparations

Five important aspects of liposome analysis have been pointed out in a recent review (62): lamellarity determination, size determination, quantitative lipid analysis, encapsulant determination, and characterization of liposomes with respect to manufacturing (quality assurance). A detailed description of today's most commonly used methods and of novel techniques for the treatment of these aspects, including advantages and limitations, is provided therein.

Particle mean size and size distribution, measured with flow field–flow fractionation coupled to multiangle light scattering (63) or other detection methods (64, 65), and polydispersity are the most important parameters to consider when describing a liposome or any other colloidal dispersion. For liposomes, size distribution is closely related to the characteristics of the lipid bilayer, including spontaneous curvature, the Helfrich elastic moduli, or even molecular parameters such as the extended length of a surfactant chain or the area per headgroup (66, 67). Even though size distribution is normally measured by dynamic light scattering, accurate values for liposome suspensions cannot be reliably obtained from light or other scattering methods. Transmission electron microscopy at cryogenic temperature (cryo-TEM), freeze-fracture TEM, and quasi-elastic light scattering are complementary methods to determine mean radii, polydispersity, and size distribution (68).

A powerful method is gel exclusion chromatography, in which the hydrodynamic radius can be accurately determined. Commercial columns can separate liposomes in the size range of 30–300 nm. Other columns can separate SUVs from micellar contaminants. However, these columns with their colloidal polymer particles are prone to clogging; there is a risk of electrostatic interactions of positively charged packing material with the (sometimes) slightly negatively charged medium. In addition, high salt content can cause sample precipitation.

The lamellarity of liposomes is determined by electron microscopy or by spectroscopic techniques. Nuclear magnetic resonance (P-NMR or F-NMR) spectroscopy is applied with or without the addition of a paramagnetic agent that shifts or bleaches the signal of the observed nuclei.

Encapsulation efficiency is commonly measured by encapsulating a hydrophilic marker (i.e., radioactive sugar, ion, fluorescent dye), sometimes using single-molecule detection (69). Electron spin resonance methods allow for the measurement of the internal volume of preformed vesicles. The surface potential is accessible via the zeta potential, and osmolality can be determined by vapor pressure osmometry. Phase transition and phase separations are measured by fluorescence-based pH indicators, NMR, fluorescence methods, Raman spectroscopy, or electron spin resonance (6).

Limiting factors for some applications are physical (70), chemical, and biological liposome stability (71). Especially in clinical applications of liposome preparations, shelf-life stability is of importance, and the lifetime of chromatography and sensor/assay components is also influenced by vesicle instability. Physical stability includes size stability and the ratio of lipid to encapsulated or membrane-bound agent; it is often improved by low temperature storage.

Chemical instability arises primarily from hydrolysis and oxidation of lipid molecules, but also from digestion by degrading enzymes. Hydrolysis removes the fatty acid residues, and oxidation is an influential factor in the presence of unsaturated lipids.

Biological stability of liposomes is one of the most problematic issues and is generally rather limited. For example, depending on the type of lipids present in a formulation, leakage, aggregation, and binding interactions with other solution constituents are common. Fluorescent markers such as carboxyfluorescein or pyrene have been used to investigate membrane fluidity and permeability in terms of solute leakage from liposomes by optical methods (72).

4. LIPOSOME APPLICATIONS IN ANALYTICAL SCIENCES

The beneficial structural and functional characteristics of liposomes have led to intense and widespread use within the analytical sciences. These characteristics include biocompatibility, a flexible membrane that has affinity to accommodate lipophilic or polymeric molecules, a self-assembly mechanism of membrane formation that is flexible enough to allow for embedding of material within the membrane, and the ability to conjugate a large variety of functional species directly to single lipid molecules (19). Probably most importantly, the secluded internal volume can hold encapsulated material, such as water-soluble marker molecules, which may be released under well-defined conditions.

Several very recent reviews have systematically surveyed the area of analytical chemistry with respect to the role of liposomes in analytical applications (73–75). In the following section, we briefly discuss important applications. Liposomes can either function as analytes, i.e., the target of the analytical process, or can be used as a tool, component, or device in analytical applications. A recent comprehensive review (62) is entirely dedicated to the analytical options available to investigate liposomes (see Sections 3.8 and 5.5).

Interest in liposome technologies and applications for analytical purposes is constantly growing, and four major areas of activity can be identified: liquid chromatography (LC), capillary electrophoresis (CE), immunoassays, and biosensors. Moreover, applications of liposomes in single-molecule spectroscopy, imaging, and cell biology are emerging, and liposome-nanotube networks have proven to be flexible, reconfigurable nanobioreactor systems (26).

4.1. Liposome-Affinity Chromatography and Capillary Electrophoresis

LC and electrophoresis applications are some of the most important methods in the analytical sciences and are central to product analytics in the pharmaceutical and biotechnology industries. To study the interactions between certain analytes and phospholipid membranes, liposomes have been utilized in liquid chromatographic and capillary electromigration techniques. The close structural resemblance of liposomes to natural cell membranes makes them suitable tools to exploit specific interactions. Several reviews summarize the latest progress in these separation sciences (74–76).

There are various well-known ways to immobilize phospholipids and liposomes onto LC columns. Immobilized artificial membranes are composed of propylaminosilica particles onto which phospholipids are covalently linked. Stationary phases containing immobilized liposomes can be prepared sterically or dynamically or by using hydrophobic ligands, covalent binding, or the avidin-biotin technique. A growing area of interest is the development of column gels with larger pore sizes to allow the immobilization of larger liposomes, as an increase in external liposome surface area leads to a better approximation of natural membranes. Only liposomes containing relatively simple phospholipids have been used for immobilization so far; an extension to more complex compositions is one of the goals of further development.

Vesicles have also been employed as coated stationary phases and pseudostationary phases in various electrokinetic chromatography (EKC) applications (75). Most applications so far have focused on the development of analytical techniques and on the characterization of vesicle phases. In addition, the use of vesicles in EKC has the potential to greatly improve separation; to determine the lipophilic or hydrophobic character of drugs and other analytes; to monitor liposome preparations in the pharmaceutical industry; and to aid the modeling of interactions between biological membranes and membraneophilic material, such as hormones and proteins. Of particular interest is the use of liposomes as models for biological membranes and lipoprotein particles and the potential to modify the vesicle composition to mimic natural biological nanoparticles. Many methods in affinity CE may be applied to the analysis of these interactions for quantitative determination of binding constants and binding sites.

In contrast to applications where liposomes are used as an analytical tool in electrophoresis, which is still uncommon, CE has already become a powerful and established technique for the characterization of vesicle preparations, yielding information about properties such as size, polydispersity, surface charge, permeability, homogeneity, rigidity, and composition (76). However, the mechanism of vesicle electromigration is not yet fully understood and better models are needed.

In summary, there are as yet no firmly established separation techniques utilizing phospholipids and liposomes. Until now, researchers have focused upon developing techniques, characterizing the vesicular phases employed, and evaluating the separation capabilities towards specific types of analytes. Future studies in this challenging field, presumably aided by advances in nanotechnology as well as micro- and nanofluidics, can be expected to successfully identify areas of opportunity for separations on immobilized lipid phases (phospholipids, liposomes, and proteoliposomes).

4.2. Liposome-Based Immunoassays and Biosensors

Two other areas in analytics where liposomes have been very successfully applied are immunoassays and biosensors (77). Liposomes can enclose a large number of fluorescence-marker molecules and amplify the fluorescence signal. One of the success factors in the sensor applications area is the absence of pronounced stability problems. Liposomes remain stable for weeks or months when handled under controlled conditions and stored under nitrogen at low temperature. Moreover, due to the large available membrane surface area, they can accommodate a great number of receptor molecules.

Recent reviews give an overview of applications of liposomes in homogeneous and heterogeneous immunoassays, including methods to conjugate liposomes to antigens or antibodies, and discuss applications of flow-injection liposome immunoassays as well as liposome immunosensors (73, 78). Several typical examples are described in detail therein.

The beneficial amplification properties of liposomes have become apparent in a number of biosensor applications, for example for the determination of pH and oxygen or the detection of bacteria, alkaloids, and other agents (79), and they employ a broad range of detection systems such as microgravimetry and optical, electrochemical, and densitometric methods.

4.3. Liposomes for In Vivo Imaging Applications

Scintigraphic techniques based on various radiolabels, in particular ^{99m}Tc, are useful tools for the noninvasive analysis of the in vivo behavior of liposomes. Using these techniques, quantitative information regarding the in vivo movement, distribution, and fate of the liposomes becomes readily available.

Several techniques for labeling liposomes have been developed, the most promising of which are the afterloading methods. One successful afterloading method for the labeling of liposomes with ^{99m}Tc is based on preformed liposomes loaded with a reducing agent. A lipophilic chelator, hexamethylpropyleneamine oxime (HMPAO), is applied to the label, which is then mixed with the liposomes. After entering the membrane, the lipophilic complex is converted into a hydrophilic form, causing the label to be released into the internal volume. A second afterloading method, which also yields liposomes with high radiochemical purity and good stability, is based on the fixation of the radiolabel by a lipid-chelator conjugate in the lipid bilayer.

Another promising bioimaging concept that does not rely on radionuclides is the use of magnetoliposomes for in vivo magnetic resonance imaging. This method was recently employed to obtain direct evidence of the stealthiness of poly(ethylene) glycol (PEG)-ylated magnetic fluid–loaded liposomes (80).

4.4. Liposome Reactors and Networks

Surfactant nanotube-vesicle networks represent some of the smallest and most structurally flexible devices known for performing controlled chemistry (12). Our group has developed GUV-lipotube networks as a means of transporting reactive material between containers, for initiation and control of chemical reactions in ultrasmall volumes, and as analytical devices with a resolution down to the single-molecule level. Fabrication, functionalization, and analytical capabilities of these systems have been reviewed recently (26); here, we briefly discuss aspects of their application in the area of ultrasmall-scale chemical analysis.

4.4.1. Network formation and modification. A self-organization methodology to control geometry, dimensionality, topology, and functionality in surfactant

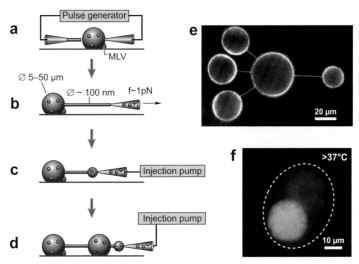

Figure 5

Schematic drawings showing micropipette-assisted formation and internal functionalization of networks of vesicles and nanotubes. (*a*) A micropipette is inserted into a unilamellar vesicle by means of electroinjection. (*b*) The pipette is then pulled away, forming a lipid nanotube. (*c*) A new vesicle is formed by injecting buffer into the nanotube orifice at the tip of the pipette. Repeating this procedure forms networks of vesicles. (*d*) By exchanging the solution in the pipette during the network formation, the interior contents of the vesicles can be differentiated. (*e*) Micrograph of a five-vesicle network with a central container and four daughter containers connected by single lipotubes. The membrane is stained with a fluorescent dye to enhance visibility, and the multilamellar reservoir was removed from the structure. (*f*) Micrograph of a giant unilamellar vesicle with two individually internalized hydrogel compartments as a means of interior functionalization. Each hydrogel compartment encloses fluorescent nanoparticles.

membranes for nanoscale soft-matter device fabrication is the foundation of the network concept (81) (**Figure 5**). Unconventional yet highly effective fabrication routes yield three-dimensional functionalized liposomal nanodevices. In the last few years, our group, as well as other groups, has extended the scope and contributed to the development of novel functionalization, subcompartmentalization, and content control methods, for example by using macromolecular hydrogel–forming (82) or phase-separation systems (83).

4.4.2. Transport phenomena in container networks. In order to take full advantage of lipid NVNs as systems for performing chemical operations, controlled means of material transport through the nanotubes are of central importance. Three fundamental mechanisms have been established so far. The first, Marangoni transport, is based on membrane tension gradients and utilizes the dynamic and fluid character of the bilayer membrane. The second is electrophoretic transport, and the third is based on diffusion, a very effective means of transport over short distances. In **Figure 6a**, a two-container network for diffusive transport is shown. **Figure 6b-c** displays a network with a small particle entrapped in the interconnecting

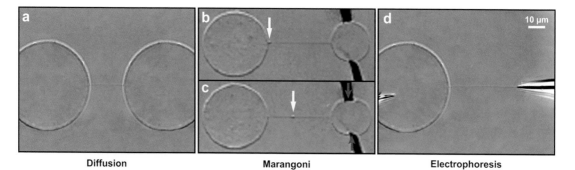

a	b	c	d
			10 μm

Diffusion   **Marangoni**   **Electrophoresis**

Figure 6

Photomicrographs of giant unilamellar vesicles (GUVs) connected by a lipid nanotube. Three
different approaches to performing nanotube-mediated transport are shown. (*a*) Diffusion-
driven transport. Using diffusion as a way of transport means that there is no flow of fluid in
the system; instead, chemical potentials strive to eliminate concentration gradients across the
nanotube. (*b*, *c*) Tension-driven lipid flow induces solvent Marangoni flow inside the nanotube
and concomitant intratubular fluid transport. Carbon fibers are used to increase the tension in
the vesicle to the right (*red arrows*), thereby establishing a tension gradient across the
nanotube. This leads to a lipid flow and a concomitant fluid flow inside the nanotube
(illustrated by the particle indicated by the white arrow). (*d*) Electrophoretic transport. A
GUV is connected to a pipette via a lipid nanotube. By applying an electric field across the
lipid nanotube, charged species such as DNA can be made to move through the nanotube.

tube, utilized to demonstrate the membrane-coupled transport upon the mechanical
creation of a tension gradient. The particle, indicated by a white arrow, travels along
the tube in the direction of the gradient, as the membrane is moved to reduce tension
at its source. **Figure 6d** shows the basic setup of an electrophoresis experiment, with
two electrode-pipettes penetrating, on one side, a vesicle and, on the other side, the
opening of a suspended lipid nanotube.

4.4.3. Liposome-nanotube networks mimicking exocytosis.

Small, lipid
nanotube–connected vesicles contained within larger GUVs have been used to mimic
and analyze the membrane mechanics in a cell during exocytosis, a fundamental cellular process (84, 85). In particular, the dynamics of pore opening was successfully
studied using this system. **Figure 7** outlines the experimental setup. A microinjection pipette is electro-inserted into the interior of a unilamellar vesicle, then pushed
out through the opposing wall and finally pulled back into the interior. Spontaneous
formation of a lipid nanotube and of an artificial exocytosis vesicle occurs at the tip
of the micropipette. Fluid injection at a constant flow rate results in growth of the
newly formed vesicle with a simultaneous shortening of the nanotube until vesicle
opening, the final stage of exocytosis, takes place. After releasing the contents of a
vesicle in this system, a new vesicle is formed at the pipette tip through the attached
nanotube, and the experiment can be repeated in several consecutive cycles. Highly
quantitative and temporal measurements of released catechol have been carried out
by amperometry using a carbon fiber electrode pair, as shown in **Figure 7**. These

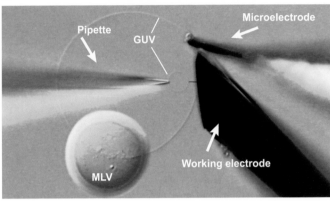

b

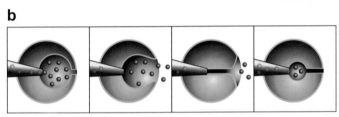

Figure 7

Vesicle-nanotube network used for modeling exocytosis. (*a*) A microphotograph showing a small pipette-attached giant unilamellar vesicle (GUV) inside a larger surface-attached GUV. A small dark line represents the lipid nanotube, which is difficult to observe under a light microscope. The vesicles are linked via a lipid nanotube and a multilamellar vesicle (MLV) is attached as a membrane reservoir. Also shown is a carbon fiber microelectrode used for electroinjection of the pipette into the vesicle and a carbon fiber working electrode used for measuring electrochemically active substances released from the exocytosis event. (*b*) Schematics illustrating the event of nanotube-mediated merging of the two vesicles and subsequent release of material from the small pipette-attached vesicle to the outside of the large vesicle. To repeat the process, a new vesicle is generated at the pipette tip. Reprinted with permission (18).

measurements have provided evidence that membrane mechanics can account for the temporal aspects of the final stage of vesicle opening in exocytosis, and that lipid flow is important in regulating release via the fusion pore.

4.4.4. Chemical reactions in complex nanotube-vesicle networks. For certain enzyme-catalyzed reactions, strong confinement leads to unusual phenomena such as oscillatory behavior in product formation or enhanced enzyme stability (86). Thus, size and dimensionality of a reactor are important properties that can shape the dynamic properties of a chemical reaction. NVNs can be used to conveniently study the dynamics of closely confined chemical reactions (87).

Concentration of reactants, mixing, and relevant time scales are controlled by injection protocols and geometric factors of the networks (e.g., length, diameter of the

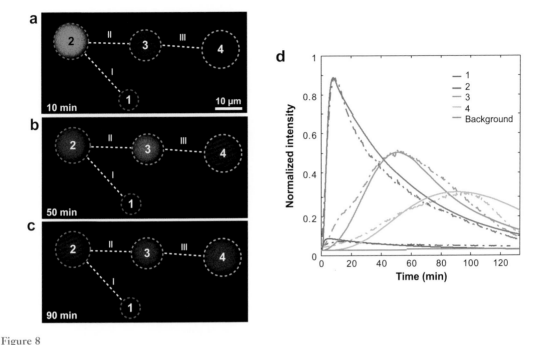

Figure 8

(*a–c*) Fluorescence microscopy images showing product formation from a 3,6,–fluorescein diphosphate–alkaline phosphatase reaction in a linear network. (*d*) The normalized fluorescence intensities of the corresponding measurements for each vesicle in the network versus time. Initially, the enzyme-filled vesicle is vesicle 1; the other vesicles are filled with a nonfluorescent substrate that becomes fluorescent when converted to product. The dashed lines in panel *d* represent the theoretical fit to the experimentally measured product formation. Fluorescence images were digitally edited to improve quality.

nanotubes, diameter of the vesicles, size of nanotube junctions). The system also offers smooth transitions from three-dimensional space (vesicles and large inner-diameter nanotubes) to one-dimensional space (small inner-diameter nanotubes). Container size and geometry, connectivity, and topology can be controlled in the networks, and polymers can be included to create crowded environments (88).

Experiments have been performed to explore enzyme-catalyzed reactions in solitary vesicles as well as in GUVs, and enzyme-catalyzed reaction-diffusion systems were studied theoretically and experimentally in different network geometries (13, 89). It could be shown that the transition from a compact geometry (sphere) to a structured geometry (several spheres connected by nanoconduits) induces an ordinary enzyme-catalyzed reaction to display wave-like or front-propagating properties (**Figure 8**). In the network depicted, an enzyme is introduced into one node and diffuses down its concentration gradient into neighboring nodes through nanotubes. The chemical potential and the directionality of enzyme diffusion can be exactly controlled. The temporal pattern of front propagation as well as the rate of reaction depends upon the network geometry.

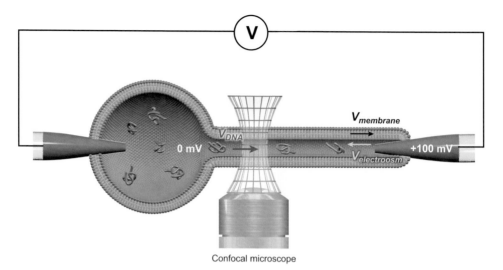

Confocal microscope

Figure 9

Schematic depicting the geometry of the nanotube-vesicle network system and confocal fluorescence detection used in the single-molecule transport and detection experiments. The nanotube is aligned across the confocal excitation/detection spot of the confocal microscopy setup and an electric potential is applied between the electrodes, with the nanotube-coupled electrode having positive potential. The fluorescence signal from the DNA molecules is detected as the molecules pass the excitation/detection spot. V_{DNA} is the velocity of the DNA molecules, $V_{membrane}$ is the velocity of the membrane, and $V_{electroosm}$ is the velocity of the electroosmotic flow. Because the latter two cancel each other out, no net liquid flow occurs in the nanotube.

4.4.5. Single-molecule detection in nanotube-vesicle networks. The recent considerable progress in single-molecule detection techniques has opened up a new era of biological and biophysical research. In single-molecule studies, results are not obscured by the ensemble-averaged measurements inherent to classical biochemical investigations (90). The electrophoretic transport technique, in conjunction with ultrasmall electrodes operating in the millivolt range, confocal microscopy, and single-molecule fluorescence detection, has been used by Tokarz et al. to transport and detect large (5.4–166 kbp) double-stranded DNA molecules with a detection efficiency close to unity (32). The size dependence of the DNA conformation inside the nanotube can be elucidated from the fluorescence bursts originating from the individual DNA molecules as they are detected during intratubular transport (**Figure 9**).

Three parameters were extracted from the electropherograms: the maximum fluorescence signal detected from each DNA molecule (peak intensity I_p), the duration of the peak (transit time Δt), and the integrated fluorescence intensity (peak area A_I). The average peak area increases linearly with increasing DNA size, which is consistent with a strong confinement inside the nanotube because it ensures that the entire molecule passes through the detection volume. Therefore, the system is capable of single-file transport and counting of DNA molecules. The average peak intensity is observed to increase linearly with DNA size as well. The transit time exhibits a linear dependence on DNA size, too, but in this case with an intercept with the ordinate axis at $t_0 = 20.0$ ms. This corresponds to the transit time for a small DNA molecule across the finite detection volume (0.8 μm in diameter), yielding a velocity of 40 μm/s towards the positive electrode. From these parameters it is possible to characterize the electrophoretic system and to interrogate the conformation of the DNA molecules during transport.

A theoretical description of semiflexible polymers, such as DNA, in soft tubes has been proposed (91). It predicts that small polymers will be squeezed by the tube into an elongated shape that can be described by the blob model (92), but if a polymer is large enough it will deform the tube and retain a globule-like conformation. The coil is spherical in shape, but compacted compared with the unconfined polymer coil because of forces arising from the wall of the expanded tube.

5. LIPOSOME TECHNOLOGIES

The progress arising from basic liposome research, namely the ability to (*a*) prepare liposomes of controlled size distribution; (*b*) control liposome permeability; (*c*) achieve high biocompatibility by polymer-modified "stealth" liposomes; (*d*) control temperature and pH sensitivity by an accordingly defined lipid composition; and (*e*) control liposome aggregation and fusion has almost immediately led to technological solutions for industrial applications of liposomes, most notably in a medical and pharmaceutical context (93). Steric stabilization by surface-grafted hydrophilic molecules has beneficially influenced the pharmacokinetics and bioavailability of liposomes and their contents (94). Remote-loading methods have allowed stable liposome encapsulation at high drug-to-lipid ratios with considerable practical impact. Numerous technologies and methodologies are presently known and have been extensively reviewed in a new series of books (9). Novel applications are constantly being developed, following closely each new step in progress in basic lipid research and in analytical and physicochemical sciences of liposomes. In this section, we discuss liposome drug and gene delivery, transdermal delivery, and GUV proteomics applications.

5.1. Liposome-Based Drug and Macromolecule Delivery

Liposomes and other types of vesicles such as virosomes, niosomes, transfersomes, proteosomes, archaeosomes, and phospholipid-alkylresorcinol liposomes (PLARosomes) are commonly applied as substance-delivery tools (95). The liposomal shell can enclose or bind many different classes of substances; liposomes are therefore successfully utilized as therapeutic agents for the delivery of antibacterial, antiviral, and anticancer drugs, as well as hormones, enzymes, and nucleotides (96–98).

The objective of using liposomes as drug carriers is to achieve a high localization of active compounds at disease sites such as tumors or inflammations. In addition, the liposome-encapsulated or -associated drug should become available to the target cells. In this respect, the liposome differs from many other controlled-release strategies, in which drug release occurs either in plasma or directly at the site of administration. Liposomes have many superior features, particularly biocompatibility, biodegradability, low toxicity, and structural variability. However, certain problems limit manufacture and development (11): Apart from factors that complicate all liposome applications, such as stability and undesirable size distribution, there are several other critical factors of importance in industrial applications, such as low entrapment efficiency, irreproducibility of preparation conditions, sterility in the context of clinical use, upscaling difficulties, and short circulation half-lives. Conventional liposomes in particular, which are prepared from neutral phospholipids, are comparatively unstable structures. They also have low load capabilities and tend to leak the internalized substances through their boundary. Cholesterol is often added to the formulations with the aim to stabilize these structures (99, 100). Stealth liposomes, also known as sterically stabilized or PEG-ylated liposomes, are phospholipid vesicles modified by incorporation of hydrophilic polymers, such as PEG, in the bilayer (80, 96, 101). The resulting coated liposome surface provides stabilization enhancement and additional protection of the encapsulated substances. The circulation half-life in vivo as compared to conventional substance delivery liposomes is enhanced.

Active liposomes are vesicles containing substances that provide a selective, controlled release of the liposome content. When one of two main controlled-release mechanisms is activated in vivo, a change in the liposome structure or in the liposome reactivity promotes release of entrapped material from its interior. One mechanism of targeted release is based on the development of an affinity reaction. Another, the triggered-release mechanism, involves incorporating an element into the liposome; this promotes structural changes in the bilayer membrane as a consequence of an external stimulus, for example pH (102), temperature variation (103), or light irradiation (104). These triggered responses involve loss of membrane integrity and in situ liberation of the entrapped compound. Use of stealth liposomes is essential in both release principles.

Lipoplexes, used in gene delivery, are liposomes containing charged phospholipids. These liposomes provide covalent interactions with oppositely charged macromolecules such as DNA, RNA, and proteins, and allow for internalization of high-molecular-weight substances.

In addition, liposomes currently have widespread applications in the cosmetics industry (for transdermal substance delivery) and as food ingredients (105–107).

5.2. Lipoplexes (Gene Delivery)

Progress in identification of functional genes has stimulated the application of new and modern medical procedures such as gene vaccination and gene therapy (108, 109). Although gene therapy as a treatment for certain severe medical conditions continues to hold great promise, progress in developing effective clinical protocols has been rather slow. Problems such as upscaling, batch reproducibility, and other industrial parameters remain to be addressed (110). Nevertheless, in the last ten years progress has been considerable (111, 112), and a large step forward has been made from the initial cell culture studies performed in the 1990s.

The ideal liposomes for therapeutic gene delivery will encapsulate plasmid DNA with high efficiencies, shield the DNA from enzymatic degradation, and feature a narrow size distribution of ~100 nm or less in diameter to be able to access extravascular regions. Only very recently have there been significant advances in the formulation of plasmid DNA into relatively small, stable plasmid DNA–containing lipidic particles or liposomes that protect plasmid DNA from enzymatic degradation. For example, the addition of polycationic polymers such as polylysine to plasmid DNA, prior to or during the addition of cationic liposomes, results in the formation of sub-100-nm particles (113).

The interaction of cationic liposomes with anionic DNA is based on the electrostatic attraction between the positive charges of the vesicle and the negative charges of the phosphate backbone of the DNA. Upon mixing of the two species, neutralization of the negative charge of the DNA results in condensation of the lipid with the DNA. Still, the structure of the lipid/DNA complex depends on several factors, mainly the chemical composition of both lipid and buffer and the lipid-to-plasmid ratio. Additionally, the formation of lipoplexes depends not only on the method of liposome preparation (114), but also on the procedure used to obtain lipid/DNA complexes. Audouy and Hoekstra (115) and Segura and Shea (116) have reviewed aspects of the successful in vivo use of cationic lipid–based gene delivery systems to enhance our understanding of fundamental and structural parameters that govern transfection efficiency. This includes lipid/DNA complex formation (117), bioavailability and -stability, complex target-membrane interactions and translocation, and gene integration into the nucleus (118).

5.3. Transdermal Delivery Applications

In order to make more pharmaceutics and cosmetics bioavailable via the transdermal route, novel substance-delivery systems are actively being developed. Vesicular systems are steadily gaining importance (119, 120), complementing other dermatological methods such as iontophoresis and microneedles.

Liposomes improve penetration of agents into the skin, but not permeation into the lower epidermal layers. Interestingly, most experts working in the field of active dispersions seem to agree that liposomes do not penetrate as intact vesicles into or permeate through the skin. Liposomes are believed to be deformed and transformed into fragments (121), and are merely considered to be a convenient means of working with phospholipids. Therefore, size, shape, and lamellarity of liposomes are not critically relevant for dermatological and cosmetic formulations.

5.4. Giant Liposome–Based Proteomics

A range of methods has been developed for reconstitution of membrane proteins in unilamellar vesicles (122–124) as well as in GUVs that can be used to build liposome networks (125). Reconstitution in GUVs includes fusion of proteoliposomes generated by detergent-mediated reconstitution or insertion of proteins into preformed GUVs via peptide-induced fusion (125–128). This approach has certain limitations. Proteins have to be removed from their native environment, usually by application of detergents, where it is often not possible to ensure that all the proteins are reconstituted uniformly in their functional orientation. Furthermore, as the isolation procedures can be quite harsh, protein activity may be reduced or completely lost. The most beneficial approach with respect to these problems is the formation of NVNs directly from a native cell membrane. Simplified access to the cell membrane can be obtained by exploiting the natural reactions of biological cells to particular means of stress. Upon chemical or mechanical stress, cells can form unilamellar micrometer-sized protrusions (plasma membrane–derived vesicles). If there is a local defect in membrane-cytoskeleton attachment, a protrusion is extruded through inflation of the detached membrane by intracellular fluid flow while the overall cell volume stays nearly constant. These cell-derived membrane structures are compatible with the micromanipulation procedures and tools used for manipulation and modification of GUVs, and serve as precursors for NVN formation.

In order to utilize plasma membrane–derived vesicles for hybrid structures combining the cellular plasma membrane with vesicle networks, they must be induced to form. Formation can be triggered in various ways, for instance chemically by applying a combination of dithiothreitol and formaldehyde. Membrane-derived vesicles can be used to build surface-adhered networks of plasma membrane vesicles, with typical vesicle diameters of 5–10 μm and tube lengths of several tens of micrometers, using the electroinjection technique developed for NVN fabrication. Native membrane composition, orientation, and function are fully preserved.

Figure 10a is a schematic representation of an adherent cell with a protruding cell-derived vesicle, connected via a nanotube to a daughter vesicle that is still attached to the micropipette. The cell-derived vesicle is used as a membrane source for formation of a daughter vesicle. Organelles and cytoskeletal structures remain largely in the adherent cell, whereas the membrane protrusion most likely encloses low-molecular-weight cytosolic components. The initial protrusion and the subsequently formed daughter vesicle have membrane proteins embedded in the membrane that are properly oriented and fully functional. Using this method, networks of complex connectivity entirely derived from plasma membrane material can be constructed. However, larger structures such as organelles, if present in the mother protrusion, cannot always pass through the nanotubes because of size constraints. The presence of membrane glycoproteins in the network boundary has recently been verified, in this case by a selective dye (WGA-Alexa 488) with affinity to sialic acid and N-acetylglucosaminyl residues (129).

5.5. Analytical Procedures for Quality Control in Technological Applications

The use of liposome-based drug delivery systems, just like small-scale lipid preparations for research purposes, requires determination of relevant chemical and physical characteristics, such a size distribution and composition, encapsulation efficiency, and chemical stability (73, 130).

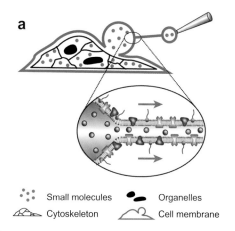
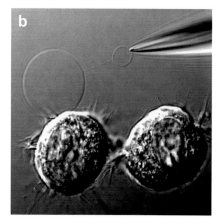

Figure 10

(*a*) Schematic showing the principle of cell–nanotube-vesicle networks. (*b*) Simple
two-container network constructed using lipid directly from a cultured NG 108-15 cell.

Today, photon correlation spectroscopy is widely employed to obtain mean particle size and polydispersity of liposomal delivery formulations. Electron paramagnetic
resonance imaging methods (131) are commonly employed to investigate the influence of liposome size on transport kinetics and bioavailability of active substances.
The most commonly applied method to determine the encapsulation efficiency of
loaded liposomes involves removal of the nonencapsulated drug. This is achieved
by size exclusion chromatography (132), dialysis, filtration, or centrifugation. Proton NMR spectroscopy is an alternative technique that allows the determination of
liposomal content without the need for physical separation of entrapped and nonen-
trapped material (133).

An important stability-related measurement is the determination of the pH of
aqueous liposomal preparations, as nonphysiological pH-values promote lipid hydrolysis, in some cases catalyzed by the loaded drug material. Lipid hydrolysis products can be quantitatively analyzed by high-performance thin-layer chromatography
with densitometric detection. The optimal pH for the encapsulated material should
ideally be compatible with that for the phospholipids, which is in practice often difficult to achieve. Another important stability parameter, namely oxidation damage to
the polyunsaturated fatty acid residues in lipids molecules, can be judged from the
presence of conjugated dienes and hydroperoxides, alkyl peroxides, aldehydes, and
ketones, which can be detected and quantified by UV/visible spectroscopy, either directly or after derivatization (134, 135). Another method is to determine the change
in concentration of unsaturated fatty acids by differential scanning calorimetry (136).

Analytical methods applied to the study of lipoplexes include electron microscopy,
calorimetry, and scattering methods to characterize the liposomes. The interaction
between DNA and cationic liposomes is determined using the PicoGreen® dye exclusion assay (137). Lipoplexes in liquid formulation are known not to have a very long

shelf-life; aggregation and loss of transfection efficiency occur rather rapidly (138, 139).

6. SUMMARY AND OUTLOOK

Liposomes are currently being successfully applied in many areas of chemistry, medicine, and biotechnology, often in larger batch sizes, as in drug delivery or transdermal drug application at low toxicity levels.

Analytical applications of liposomes are predominant in the areas of liposome-affinity chromatography, sensors, and immunoassays, which largely benefit from signal enhancement due to the liposomes' capacity to hold comparatively large amounts of internalized marker compounds.

Liposomal delivery systems are established as effective vehicles to incorporate active compounds into compartmentalized structures, such as living cells and foods. This technological area has a variety of analytical methods for quality control associated to monitor their features and to determine their active components and ingredients. However, analytical procedures commonly measure only the total drug concentration and do not distinguish between free and encapsulated material. Vesicular phospholipid gels are, in some areas, a potential remedy to the problem, as the amount of noninternalized material in the gel is exceptionally low.

Progress is also apparent in single-molecule spectroscopy and in liposome-based imaging. Emerging technologies are micro- and nanoreactor concepts based on liposome-encapsulated, lipotube-interconnected volumes. NVNs allow for controlled transport of material down to the single-molecule level by diffusion or field-driven transport. Hybrid structures of biological cells and giant liposomes directly derived from cells open pathways to nondestructive mobilization and analysis of membrane proteins.

DISCLOSURE STATEMENT

The authors are not aware of any biases that might be perceived as affecting the objectivity of this review.

ACKNOWLEDGMENTS

The Royal Swedish Academy of Sciences, the Swedish Science Research Council (VR), and the Swedish Foundation for Strategic Research (SSF) supported this work.

LITERATURE CITED

1. Bangham AD, Standish MM, Watkins JC. 1965. Diffusion of univalent ions across lamellae of swollen phospholipids. *J. Mol. Biol.* 13:238–52
2. Baumgart T, Hess ST, Webb WW. 2003. Imaging coexisting fluid domains in biomembrane models coupling curvature and line tension. *Nature* 425:821–24

3. Veatch SL, Keller SL. 2005. Miscibility phase diagrams of giant vesicles containing sphingomyelin. *Phys. Rev. Lett.* 94:148101

4. Lasic DD, Joannic R, Keller BC, Frederik PM, Auvray L. 2001. Spontaneous vesiculation. *Adv. Colloid Interface Sci.* 89:337–49

5. Torchilin V, Weissig V, eds. 2003. *Liposomes: A Practical Approach*. Oxford/New York: Oxford Univ. Press

6. Basu SC, Basu ME. 2002. *Liposome Methods and Protocols*. Totowa, NJ: Humana

7. Luisi PL, Walde P. 1999. *Giant Vesicles*. New York: Wiley

8. Leduc PR, Wong MS, Ferreira PM, Groff RE, Haslinger K, et al. 2007. Towards an in vivo biologically inspired nanofactory. *Nat. Nanotechnol.* 2:3–7

9. Gregoriadis G. 2006. *Liposome Technology*. New York: Taylor & Francis/Informa

10. Tampé R, Dietrich C, Gritsch S, Elender G, Schmitt L. 1996. Biofunctionalized membranes on solid surfaces. In *Nanofabrication and Biosystems: Integrating Materials Science, Engineering, and Biology*, ed. LWJHC Hoch, HG Craighead, pp. 201–21. Cambridge, UK: Cambridge Univ. Press

11. Reimhult E, Hook F, Kasemo B. 2003. Intact vesicle adsorption and supported biomembrane formation from vesicles in solution: influence of surface chemistry, vesicle size, temperature, and osmotic pressure. *Langmuir* 19:1681–91

12. Karlsson R, Karlsson A, Ewing A, Dommersnes P, Joanny JF, et al. 2006. Chemical analysis in nanoscale surfactant networks. *Anal. Chem.* 78:5960–68

13. Sott K, Lobovkina T, Lizana L, Tokarz M, Bauer B, et al. 2006. Controlling enzymatic reactions by geometry in a biomimetic nanoscale network. *Nano Lett.* 6:209–14

14. Chatterjee S, Banerjee DK. 2002. Preparation, isolation, and characterization of liposomes containing natural and synthetic lipids. In *Liposome Methods and Protocols*, ed. SC Basu, M Basu, pp. 3–16. Totowa, NJ: Humana

15. Shimomura M, Sawadaishi T. 2001. Bottom-up strategy of materials fabrication: a new trend in nanotechnology of soft materials. *Curr. Opin. Colloid Interface Sci.* 6:11–16

16. Israelachvili JN. 1992. *Intermolecular and Surface Forces*. London, New York, San Diego: Academic

17. Gurr MI, Frayn KN, Harwood J. 2002. *Lipid Biochemistry*. Oxford: Blackwell Sci.

18. Gunstone FD, Harwood JL, Dijkstra AJ. 2007. *The Lipid Handbook*. Boca Raton, London, New York: CRC

19. Hassane FS, Frisch B, Schuber F. 2006. Targeted liposomes: convenient coupling of ligands to preformed vesicles using "click chemistry." *Bioconjug. Chem.* 17:849–54

20. Chikh GG, Li WM, Schutze-Redelmeier MP, Meunier JC, Bally MB. 2002. Attaching histidine-tagged peptides and proteins to lipid-based carriers through use of metal-ion-chelating lipids. *Biochim. Biophys. Acta: Biomembranes* 1567:204–12

21. Roy BC, Peterson R, Mallik S, Campiglia AD. 2000. Synthesis and fluorescence properties of new fluorescent, polymerizable, metal-chelating lipids. *J. Org. Chem.* 65:3644–51

22. Yan F, Texter J. 2006. Polymerization of and in mesophases. *Adv. Colloid Interface Sci.* 128:27–35

23. Lasic DD, Bolotin E, Brey RN. 2000. Polymerized liposomes: from biophysics to applications. Part I. *Chim. Oggi-Chem. Today* 18:48–51

24. Lasic DD, Bolotin E, Brey RN. 2001. Polymerized liposomes: from biophysics to applications. Part II. *Chim. Oggi-Chem. Today* 19:45–48

25. Lee JH, Agarwal V, Bose A, Payne GF, Raghavan SR. 2006. Transition from unilamellar to bilamellar vesicles induced by an amphiphilic biopolymer. *Phys. Rev. Lett.* 96:048102

26. Karlsson M, Davidson M, Karlsson R, Karlsson A, Bergenholtz J, et al. 2004. Biomimetic nanoscale reactors and networks. *Annu. Rev. Phys. Chem.* 55:613–49

27. Deuling HJ, Helfrich W. 1976. Curvature elasticity of fluid membranes: catalog of vesicle shapes. *J. Phys.* 37:1335–45

28. Seifert U, Berndl K, Lipowsky R. 1991. Shape transformations of vesicles: phase-diagram for spontaneous-curvature and bialyer-coupling models. *Phys. Rev. A* 44:1182–202

29. Miao L, Seifert U, Wortis M, Dobereiner HG. 1994. Budding transitions of fluid-bilayer vesicles: the effect of area-difference Elasticity. *Phys. Rev. E* 49:5389–407

30. Gradzielski M. 2003. Vesicles and vesicle gels: structure and dynamics of formation. *J. Phys.-Condens. Matter* 15:R655–97

31. Heinrich V, Bozic B, Svetina S, Zeks B. 1999. Vesicle deformation by an axial load: from elongated shapes to tethered vesicles. *Biophys. J.* 76:2056–71

32. Tokarz M, Akerman B, Olofsson J, Joanny JF, Dommersnes P, Orwar O. 2005. Single-file electrophoretic transport and counting of individual DNA molecules in surfactant nanotubes. *Proc. Natl. Acad. Sci. USA* 102:9127–32

33. Brandl M. 2007. Vesicular phospholipid gels: a technology platform. *J. Liposome Res.* 17:15–26

34. Brandl M, Massing U. 2006. Vesicular phospholipid gels. In *Liposome Technology*, ed. Gregoriadis G, pp. 241–57. New York: Taylor & Francis/Informa

35. Maurer N, Fenske DB, Cullis PR. 2001. Developments in liposomal drug delivery systems. *Expert Opin. Biol. Ther.* 1:923–47

36. Lasch JWV, Brandl M. 2003. Preparation of liposomes. In *Liposomes*, ed. Torchilin V, Weissig V, pp. 3–30. New York: Oxford Univ. Press

37. Gregoriadis G, Dasilva H, Florence AT. 1990. A procedure for the efficient entrapment of drugs in dehydration-rehydration liposomes (DRVs). *Int. J. Pharm.* 65:235–42

38. Szoka F, Papahadjopoulos D. 1978. Procedure for preparation of liposomes with large internal aqueous space and high capture by reverse-phase evaporation. *Proc. Natl. Acad. Sci. USA* 75:4194–98

39. Buboltz JT, Feigenson GW. 1999. A novel strategy for the preparation of liposomes: rapid solvent exchange. *Biochim. Biophys. Acta-Biomembr.* 1417:232–45

40. Pons M, Foradada M, Estelrich J. 1993. Liposomes obtained by the ethanol injection method. *Int. J. Pharm.* 95:51–56

41. Wagner A, Vorauer-Uhl K, Kreismayr G, Katinger H. 2002. The crossflow injection technique: an improvement of the ethanol injection method. *J. Liposome Res.* 12:259–70

42. Lasch J. 1995. Interaction of detergents with lipid vesicles. *Biochim. Biophys. Acta—Rev. Biomembr.* 1241:269–92

43. Tatulian SA, Ermakov YA, Gordeliy VI, Sokolova AE, Syrykh AG. 1992. Influence of Calcium and Some Chaotropic Anions on the Interactions of Dipalmitoyl Phosphatidylcholine Membranes. *Biol. Membr.* 9:741–55

44. Ishii F, Takamura A, Ishigami Y. 1995. Procedure for preparation of lipid vesicles (liposomes) using the coacervation (phase-separation) technique. *Langmuir* 11:483–86

45. Jahn A, Vreeland WN, DeVoe DL, Locascio LE, Gaitan M. 2007. Microfluidic directed formation of liposomes of controlled size. *Langmuir* 23:6289–93

46. Kirby C, Gregoriadis G. 1984. Dehydration-rehydration vesicles: a simple method for high-yield drug entrapment in liposomes. *Bio-Technol.* 2:979–84

47. Kikuchi H, Yamauchi H, Hirota S. 1991. A spray-drying method for mass-production of liposomes. *Chem. Pharm. Bull.* 39:1522–27

48. Berger N, Sachse A, Bender J, Schubert R, Brandl M. 2001. Filter extrusion of liposomes using different devices: comparison of liposome size, encapsulation efficiency, and process characteristics. *Int. J. Pharm.* 223:55–68

49. Mui B, Chow L, Hope MJ. 2003. Extrusion technique to generate liposomes of defined size. *Methods Enzym.* 367:3–14

50. Dittrich PS, Heule M, Renaud P, Manz A. 2006. On-chip extrusion of lipid vesicles and tubes through microsized apertures. *Lab Chip* 6:488–93

51. Woodbury DJ, Richardson ES, Grigg AW, Welling RD, Knudson BH. 2006. Reducing liposome size with ultrasound: bimodal size distributions. *J. Liposome Res.* 16:57–80

52. Payne NI, Timmins P, Ambrose CV, Ward MD, Ridgway F. 1986. Proliposomes: a novel solution to an old problem. *J. Pharm. Sci.* 75:325–29

53. Li CL, Deng YJ. 2004. A novel method for the preparation of liposomes: freeze drying of monophase solutions. *J. Pharm. Sci.* 93:1403–14

54. Criado M, Keller BU. 1987. A membrane fusion strategy for single-channel recordings of membranes usually non-accessible to patch-clamp pipette electrodes. *FEBS Lett.* 224:172–76

55. Karlsson M, Nolkrantz K, Davidson MJ, Stromberg A, Ryttsen F, et al. 2000. Electroinjection of colloid particles and biopolymers into single unilamellar liposomes and cells for bioanalytical applications. *Anal. Chem.* 72:5857–62

56. Lasic D. 1993. *Liposomes: From Physics to Applications.* Amsterdam: Elsevier

57. Angelova MI, Dimitrov DS. 1986. Liposome electroformation. *Faraday Discuss. Chem. Soc.* 81: 303–11

58. Kuribayashi K, Tresset G, Coquet P, Fujita H, Takeuchi S. 2006. Electroformation of giant liposomes in microfluidic channels. *Meas. Sci. Technol.* 17:3121–26

59. Estes DJ, Mayer M. 2005. Giant liposomes in physiological buffer using electroformation in a flow chamber. *Biochim. Biophys. Acta-Biomembr.* 1712:152–60

60. Estes DJ, Mayer M. 2005. Electroformation of giant liposomes from spin-coated films of lipids. *Colloids Surf. B-Biointerfaces* 42:115–23

61. Hauschild S, Lipprandt U, Rumplecker A, Borchert U, Rank A, et al. 2005. Direct preparation and loading of lipid and polymer vesicles using inkjets. *Small* 1:1177–80

62. Edwards KA, Baeumner AJ. 2006. Analysis of liposomes. *Talanta* 68:1432–41

63. Korgel BA, van Zanten JH, Monbouquette HG. 1998. Vesicle size distributions measured by flow field–flow fractionation coupled with multiangle light scattering. *Biophys. J.* 74:3264–72

64. Williams SKR, Lee D. 2006. Field-flow fractionation of proteins, polysaccharides, synthetic polymers, and supramolecular assemblies. *J. Sep. Sci.* 29:1720–32

65. Yohannes G, Pystynen KH, Riekkola ML, Wiedmer SK. 2006. Stability of phospholipid vesicles studied by asymmetrical flow field–flow fractionation and capillary electrophoresis. *Anal. Chim. Acta* 560:50–56

66. Israelachvili JN, Marcelja S, Horn RG. 1980. Physical principles of membrane organization. *Q. Rev. Biophys.* 13:121–200

67. Bergstrom M. 2001. Molecular interpretation of the mean bending constant for a thermodynamically open vesicle bilayer. *Langmuir* 17:7675–86

68. Coldren B, van Zanten R, Mackel MJ, Zasadzinski JA, Jung HT. 2003. From vesicle size distributions to bilayer elasticity via cryo-transmission and freeze-fracture electron microscopy. *Langmuir* 19:5632–39

69. Sun BY, Chiu DT. 2005. Determination of the encapsulation efficiency of individual vesicles using single-vesicle photolysis and confocal single–molecule detection. *Anal. Chem.* 77:2770–76

70. Armengol X, Estelrich J. 1995. Physical stability of different liposome compositions obtained by extrusion method. *J. Microencapsul.* 12:525–35

71. Sulkowski WW, Pentak D, Nowak K, Sulkowska A. 2005. The influence of temperature, cholesterol content and pH on liposome stability. *J. Mol. Struct.* 744:737–47

72. Gliozzi A, Relini A, Chong PLG. 2002. Structure and permeability properties of biomimetic membranes of bolaform archaeal tetraether lipids. *J. Membr. Sci.* 206:131–47

73. Gomez-Hens A, Fernandez-Romero JM. 2005. The role of liposomes in analytical processes. *Trac-Trends Anal. Chem.* 24:9–19

74. Owen RL, Strasters JK, Breyer ED. 2005. Lipid vesicles in capillary electrophoretic techniques: characterization of structural properties and associated membrane-molecule interactions. *Electrophoresis* 26:735–51

75. Wiedmer SK, Jussila MS, Riekkola ML. 2004. Phospholipids and liposomes in liquid chromatographic and capillary electromigration techniques. *Trac–Trends Anal. Chem.* 23:562–82

76. Bilek G, Kremser L, Blaas D, Kenndler E. 2006. Analysis of liposomes by capillary electrophoresis and their use as carrier in electrokinetic chromatography. *J. Chromatogr. B-Anal. Technol. Biomed. Life Sci.* 841:38–51

77. Diaz–Gonzalez M, Gonzalez-Garcia MB, Costa-Garcia A. 2005. Recent advances in electrochemical enzyme immunoassays. *Electroanalysis* 17:1901–18

78. Rongen HAH, Bult A, van Bennekom WP. 1997. Liposomes and immunoassays. *J. Immunol. Methods* 204:105–33

79. Chen JR, Miao YQ, He NY, Wu XH, Li SJ. 2004. Nanotechnology and biosensors. *Biotechnol. Adv.* 22:505–18

80. Martina MS, Fortin JP, Menager C, Clement O, Barratt G, et al. 2005. Generation of superparamagnetic liposomes revealed as highly efficient MRI contrast agents for in vivo imaging. *J. Am. Chem. Soc.* 127:10676–85

81. Karlsson M, Sott K, Cans AS, Karlsson A, Karlsson R, Orwar O. 2001. Micropipet-assisted formation of microscopic networks of unilamellar lipid bilayer nanotubes and containers. *Langmuir* 17:6754–58

82. Markstrom M, Gunnarsson A, Orwar O, Jesorka A. 2007. Dynamic microcompartmentalization of giant unilamellar vesicles by sol gel transition and temperature induced shrinking/swelling of poly(N-isopropyl acrylamide). *Soft Matter* 3:587–95

83. Long MS, Jones CD, Helfrich MR, Mangeney-Slavin LK, Keating CD. 2005. Dynamic microcompartmentation in synthetic cells. *Proc. Natl. Acad. Sci. USA* 102:5920–25

84. Cans AS, Wittenberg N, Karlsson R, Sombers L, Karlsson M, et al. 2003. Artificial cells: unique insights into exocytosis using liposomes and lipid nanotubes. *Proc. Natl. Acad. Sci. USA* 100:400–04

85. Cans AS, Wittenberg N, Eves D, Karlsson R, Karlsson A, et al. 2003. Amperometric detection of exocytosis in an artificial synapse. *Anal. Chem.* 75:4168–75

86. Wang G, Yau ST. 2007. Spatial confinement induced enzyme stability for bioelectronic applications. *J. Phys. Chem. C* 111:11921–26

87. Karlsson A, Sott K, Markstrom M, Davidson M, Konkoli Z, Orwar O. 2005. Controlled initiation of enzymatic reactions in micrometer-sized biomimetic compartments. *J. Phys. Chem. B* 109:1609–17

88. Jesorka A, Markstrom M, Karlsson M, Orwar O. 2005. Controlled hydrogel formation in the internal compartment of giant unilamellar vesicles. *J. Phys. Chem. B* 109:14759–63

89. Konkoli Z, Karlsson A, Orwar O. 2003. The pair approach applied to kinetics in restricted geometries: Strengths and weaknesses of the method. *J. Phys. Chem. B* 107:14077–86

90. Ishijima A, Yanagida T. 2001. Single molecule nanobioscience. *Trends Biochem. Sci.* 26:438–44

91. Brochard–Wyart F, Tanaka T, Borghi N, de Gennes PG. 2005. Semiflexible polymers confined in soft tubes. *Langmuir* 21:4144–48

92. Degennes PG. 1979. Brownian motions of flexible polymer-chains. *Nature* 282:367–70

93. Janoff AS. 1998. *Liposomes: Rational Design*. Boca Raton: CRC

94. Dos Santos N, Allen C, Doppen AM, Anantha M, Cox KAK, et al. 2007. Influence of poly(ethylene glycol) grafting density and polymer length on liposomes: relating plasma circulation lifetimes to protein binding. *Biochim. Biophys. Acta-Biomembr.* 1768:1367–77

95. Ranade VV, Hollinger MA. 2004. Site-specific drug delivery using liposomes as carriers. In *Drug Delivery Systems*. Boca Raton: CRC

96. Lasic DD, Vallner JJ, Working PK. 1999. Sterically stabilized liposomes in cancer therapy and gene delivery. *Curr. Opin. Mol. Ther.* 1:177–85

97. Eckstein F. 2007. The versatility of oligonucleotides as potential therapeutics. *Expert Opin. Biol. Ther.* 7:1021–34

98. Weissig V, Boddapati SV, Cheng SM, D'Souza GGM. 2006. Liposomes and liposome-like vesicles for drug and DNA delivery to mitochondria. *J. Liposome Res.* 16:249–64

99. Huang JY, Buboltz JT, Feigenson GW. 1999. Maximum solubility of cholesterol in phosphatidylcholine and phosphatidylethanolamine bilayers. *Biochim. Biophys. Acta-Biomembr.* 1417:89–100

100. Veatch SL, Keller SL. 2002. Organization in lipid membranes containing cholesterol. *Phys. Rev. Lett.* 89:268101

101. Hashizaki K, Itoh C, Sakai H, Yokoyama S, Taguchi H, et al. 2000. Freeze-fracture electron microscopic and calorimetric studies on microscopic states of surface-modified liposomes with poly(ethylene glycol) chains. *Colloids Surf. B-Biointerfaces* 17:275–82

102. Hafez IM, Ansell S, Cullis PR. 2000. Tunable pH-sensitive liposomes composed of mixtures of cationic and anionic lipids. *Biophys. J.* 79:1438–46

103. Anyarambhatla GR, Needham D. 1999. Enhancement of the phase transition permeability of DPPC liposomes by incorporation of MPPC: a new temperature-sensitive liposome for use with mild hyperthermia. *J. Liposome Res.* 9:491–506

104. Benkoski JJ, Jesorka A, Edvardsson M, Hook F. 2006. Light-regulated release of liposomes from phospholipid membranes via photoresponsive polymer-DNA conjugates. *Soft Matter* 2:710–15

105. Barenholz Y, Lasic DD. 1996. *Handbook of Nonmedical Applications of Liposomes*, Vol. 1–4. Boca Raton: CRC

106. Roy CC, Bouthillier L, Seidman E, Levy E. 2004. New lipids in enteral feeding. *Curr. Opin. Clin. Nutr. Metab. Care* 7:117–22

107. Honeywell–Nguyen PL, Groenink HWW, Bouwstra JA. 2006. Elastic vesicles as a tool for dermal and transdermal delivery. *J. Liposome Res.* 16:273–80

108. Templeton NS, Lasic DD. 1999. New directions in liposome gene delivery. *Mol. Biotechnol.* 11:175–80

109. de Lima MCP, Neves S, Filipe A, Duzgunes N, Simoes S. 2003. Cationic liposomes for gene delivery: from biophysics to biological applications. *Curr. Med. Chem.* 10:1221–31

110. Garidel P, Peschka-Suss R. 2006. Lipoplexes in gene therapy under the considerations of scaling up, stability issues, and pharmaceutical requirements. In *Liposome Technology, Vol. 1: Liposome Preparation and Related Techniques*, ed. G Gregoriadis, pp. 97–138. New York: Taylor & Francis/Informa

111. Patil SD, Rhodes DG, Burgess DJ. 2005. DNA-based therapeutics and DNA delivery systems: a comprehensive review. *Aaps J.* 7:E61–E77

112. Zhdanov RI, Podobed OV, Vlassov VV. 2002. Cationic lipid-DNA complexes—lipoplexes—for gene transfer and therapy. *Bioelectrochemistry* 58:53–64

113. Gao X, Huang L. 1996. Potentiation of cationic liposome-mediated gene delivery by polycations. *Biochemistry* 35:1027–36

114. Goncalves E, Debs RJ, Heath TD. 2004. The effect of liposome size on the final lipid/DNA ratio of cationic lipoplexes. *Biophys. J.* 86:1554–63

115. Audouy S, Hoekstra D. 2001. Cationic lipid-mediated transfection in vitro and in vivo. *Mol. Membr. Biol.* 18:129–43

116. Segura T, Shea LD. 2001. Materials for non-viral gene delivery. *Annu. Rev. Mater. Res.* 31:25–46

117. Chesnoy S, Huang L. 2000. Structure and function of lipid-DNA complexes for gene delivery. *Annu. Rev. Biophys. Biomol. Struct.* 29:27–47

118. Duzgunes N, de Ilarduya CT, Simoes S, Zhdanov RI, Konopka K, de Lima MCP. 2003. Cationic liposomes for gene delivery: novel cationic lipids and enhancement by proteins and peptides. *Curr. Med. Chem.* 10:1213–20

119. Langer R. 2004. Transdermal drug delivery: past progress, current status, and future prospects. *Adv. Drug Delivery Rev.* 56:557–58

120. Choi MJ, Maibach HI. 2005. Liposomes and niosomes as topical drug delivery systems. *Skin Pharmacol. Physiol.* 18:209–19

121. Lautenschlaeger H. 2001. Liposomes. In *Handbook of Cosmetic Science and Technology*, ed. AO Barel, M Paye, HI Maibach, pp. 201–09. New York, Basel: Marcel Dekker

122. Seddon AA, Curnow P, Booth PJ. 2004. Membrane proteins, lipids and detergents: not just a soap opera. *Biochim. Biophys. Acta-Biomembr.* 1666:105–17

123. Rigaud JL, Pitard B, Levy D. 1995. Reconstitution of membrane-proteins into liposomes: application to energy-transducing membrane-proteins. *Biochim. Biophys. Acta-Bioenerg.* 1231:223–46

124. de Planque MRR, Mendes GP, Zagnoni M, Sandison ME, Fisher KH, et al. 2006. Controlled delivery of membrane proteins to artificial lipid bilayers by nystatin-ergosterol modulated vesicle fusion. *IEE Proc. Nanobiotechnol.* 153:21–30

125. Davidson M, Karlsson M, Sinclair J, Sott K, Orwar O. 2003. Nanotube-vesicle networks with functionalized membranes and interiors. *J. Am. Chem. Soc.* 125:374–78

126. Doeven MK, Folgering JHA, Krasnikov V, Geertsma ER, Van Den Bogaart G, Poolman B. 2005. Distribution, lateral mobility and function of membrane proteins incorporated into giant unilamellar vesicles. *Biophys. J.* 88:1134–42

127. Girard P, Pecreaux J, Lenoir G, Falson P, Rigaud JL, Bassereau P. 2004. A new method for the reconstitution of membrane proteins into giant unilamellar vesicles. *Biophys. J.* 87:419–29

128. Kahya N, Pecheur EI, de Boeij WP, Wiersma DA, Hoekstra D. 2001. Reconstitution of membrane proteins into giant unilamellar vesicles via peptide-induced fusion. *Biophys. J.* 81:1464–74

129. Bauer B, Davidson M, Orwar O. 2006. Direct reconstitution of plasma membrane lipids and proteins in nanotube-vesicle networks. *Langmuir* 22:9329–32

130. Gomez-Hens A, Fernandez-Romero JM. 2006. Analytical methods for the control of liposomal delivery systems. *Trac. Trends Anal. Chem.* 25:167–78

131. Sentjurc M, Vrhovnik K, Kristl J. 1999. Liposomes as a topical delivery system: the role of size on transport studied by the EPR imaging method. *J. Controll. Release* 59:87–97

132. Grabielle-Madelmont C, Lesieur S, Ollivon M. 2003. Characterization of loaded liposomes by size exclusion chromatography. *J. Biochem. Biophys. Methods* 56:189–217

133. Zhang XM, Patel AB, de Graaf RA, Behar KL. 2004. Determination of liposomal encapsulation efficiency using proton NMR spectroscopy. *Chem. Phys. Lipids* 127:113–20

134. Lang JK, Vigopelfrey C. 1993. Quality control of liposomal lipids with special emphasis on peroxidation of phospholipids and cholesterol. *Chem. Phys. Lipids* 64:19–29

135. Hamilton RJ, Kalu C, McNeill GP, Padley FB, Pierce JH. 1998. Effects of tocopherols, ascorbyl palmitate, and lecithin on autoxidation of fish oil. *J. Am. Oil Chem. Soc.* 75:813–22

136. Ulkowski M, Musialik M, Litwinienko G. 2005. Use of differential scanning calorimetry to study lipid oxidation. 1. Oxidative stability of lecithin and linolenic acid. *J. Agric. Food Chem.* 53:9073–77

137. Lee H, Williams SKR, Allison SD, Anchordoquy TJ. 2001. Analysis of self-assembled cationic lipid-DNA gene carrier complexes using flow field–flow fractionation and light scattering. *Anal. Chem.* 73:837–43

138. Lai E, van Zanten JH. 2002. Real time monitoring of lipoplex molar mass, size and density. *J. Controll. Release* 82:149–58

139. Lai E, van Zanten JH. 2002. Evidence of lipoplex dissociation in liquid formulations. *J. Pharm. Sci.* 91:1225–32

140. Fahy E, Subramaniam S, Brown HA, Glass CK, Merrill AH, et al. 2005. A comprehensive classification system for lipids. *J. Lipid Res.* 46:839–61

141. Bolinger PY, Stamou D, Vogel H. 2004. Integrated nanoreactor systems: triggering the release and mixing of compounds inside single vesicles. *J. Am. Chem. Soc.* 126:8594–95

142. Kisak ET, Coldren B, Evans CA, Boyer C, Zasadzinski JA. 2004. The vesosome: a multicompartment drug delivery vehicle. *Curr. Med. Chem.* 11:199–219

Fundamentals of Protein Separations: 50 Years of Nanotechnology, and Growing

David A. Egas and Mary J. Wirth

Department of Chemistry, University of Arizona, Tucson, Arizona 85721;
email: egas@email.arizona.edu, mwirth@email.arizona.edu

Annu. Rev. Anal. Chem. 2008. 1:833–55

First published online as a Review in Advance on April 10, 2008

The *Annual Review of Analytical Chemistry* is online at anchem.annualreviews.org

This article's doi:
10.1146/annurev.anchem.1.031207.112912

Key Words

proteomics, microarrays, electrophoresis, isoelectric focusing, chromatography, biomarkers

Abstract

The separation of proteins in biology samples has long been recognized as an important and daunting endeavor that continues to have enormous impact on human health. Today's technology for protein separations has its origins in the early nanotechnology of the 1950s and 1960s, and the methods include immunoassays and other affinity extractions, electrophoresis, and chromatography. What is different today is the need to resolve and identify many low-abundance proteins within complex biological matrices. Multidimensional separations are the rule, high speed is needed, and the separations must be able to work with mass spectrometry for protein identification. Hybrid approaches that combine disparate separation tools (including recognition, electrophoresis, and chromatography) take advantage of the fact that no single class of separation can resolve the proteins in a biological matrix. Protein separations represent a developing area technologically, and understanding the principles of protein separations from a molecular and nanoscale viewpoint will enable today's researchers to invent tomorrow's technology.

1. INTRODUCTION

The sequencing of the human genome has greatly accelerated the field of protein analysis. Whereas the genome is the blueprint for the organism, the proteins carry out the functions of biological cells. The analysis of proteins has emerged as a vital area of medical research and clinical diagnostics. Every disease can potentially be described by its unique set of proteins and their concentrations. The magnitude of the analytical task is daunting: The human genome codes for over 20,000 proteins, and there are scores of physiologically relevant post-translational modifications. Consequently, protein analysis represents an enormous challenge to the separation scientist. We begin this review with illustrations of why protein analysis is such a vital and active field today, and we show the principles of the separations used in these analyses. We demonstrate that nanotechnology pervades protein separations, and the design of new separation technology rests on a fundamental understanding of basic principles.

2. EXTRACTIONS OF SINGLE PROTEINS

Disease is associated with a change in protein levels. This means that the diagnosis of disease, in principle, can be made through the detection of unusual protein levels. As a common example, diseases of the thyroid and pituitary glands are diagnosed in part by the analysis of the protein thyrotropin, which is a hormone secreted by the pituitary gland, stimulating the thyroid gland (1). Clinical labs use a commercial kit that tests for this protein in blood samples by immunoassay, which is the most common technique for clinically sensing protein levels in body fluids. A protein, or any other species, whose concentration is indicative of a disease is called a biomarker. In this case, multiple biomarkers are used to distinguish between thyroid and pituitary diseases. Other common clinical immunoassays include biomarkers for heart attack, prostate cancer, infectious disease, and pregnancy. Immunoassays are among the most powerful separations today, selectively extracting one type of protein at nanomolar concentration levels out of a mixture of tens of thousands of proteins. Immunoassays are not normally discussed in the analytical chemistry literature as separation tools, but they are extractions, and the fundamentals are common to all extractions: Selectivity and fast mass transport are the primary technological issues.

Yalow & Berson (2) originated the concept of the immunoassay in 1959, earning Yalow the Nobel Prize in 1977 for this invention, which revolutionized clinical analysis. The immunoassay is based on nature's nanotechnology, in which an antibody specifically binds to the protein through a spatial arrangement of functional groups that fit like a lock and key with the functional groups of the protein or other analyte of interest (3). **Figure 1** illustrates the commonly used sandwich immunoassay, in which two different antibodies recognize different parts of a protein, which is the analyte. The capture antibody is bound to the surface, and the protein of interest is selectively removed from the complex mixture, such as blood serum, by virtue of the high specificity and strength of the binding. The immunoassay is essentially a highly selective solid-phase extraction. The orientation of the capture antibody cannot be controlled so nicely in practice, and the analysis relies on a sufficient number of capture antibodies that have their recognition sites accessible to the protein of interest.

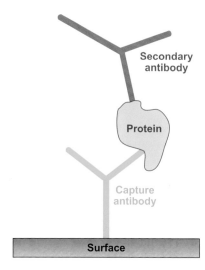

Figure 1

Depiction of antibodies in an immunoassay of a protein.

The binding constant, K_b, is defined in Equation 1 for the antibody, antigen, and complex concentrations ([Ab], [Ag] and [AbAg], respectively):

$$K_b = \frac{[AbAg]}{[Ab][Ag]},$$ (1)

where K_b ranges from 10^5 to 10^9. Binding equilibria allow the analysis of concentrations approximately two orders of magnitude lower than $1/K_b$, which is the nanomolar to picomolar scale for immunoassays. High binding constants correlate with the slow kinetics of dissociation of the complex because K_b is the ratio of rate constants for binding, k_b, and dissociation, k_d:

$$K_b = \frac{k_b}{k_d}.$$ (2)

Because the rate constant for dissociation is so small, unwanted species can be rinsed from the substrate with minimal loss of the targeted protein. To detect the protein that has been extracted, one uses a secondary antibody. This antibody is typically labeled to use fluorescence or chemiluminescence for signal transduction. In developing new immunoassays, the selectivity (or its opposite, the cross-reactivity) must be characterized because proteins sometimes have similar sequences or even a subunit in common with another protein.

The speed of immunoassays can be undesirably slow. The full expression for the rate of formation of the antibody-antigen complex, AbAg, has three terms, which result from the rate of diffusion of antigen to the antibody, the rate of binding of antigen to antibody, and the rate of dissociation of the complex (4):

$$\frac{d[AbAg]}{dt} = D \cdot \nabla^2 [Ag] + k_b \Gamma_{Ab} [Ag] - k_d [AbAg],$$ (3)

where D is the diffusion coefficient of the antigen, and Γ is the surface concentration of the accessible antibody. As an illustration, a study of the binding kinetics of the HIV-1 capsid protein p24 to a monoclonal antibody revealed that $k_b = 2.25 \times 10^5\ M^{-1}\ s^{-1}$ and $k_d = 0.0062\ s^{-1}$ (5). One can see from this very low rate constant for dissociation that rinsing removes the loosely bound species without significantly reducing the selectively bound species. One can also see that low antigen concentrations give low rates of binding despite the high rate constant. For example, if the concentration of antigen were 1 nM, then the time constant, $(k_b[Ag])^{-1}$, would be 1300 s. The antibody can also deplete the surrounding solution of antigen, in which case diffusion is required to supply antigen, and this can be the slow step. Modeling Equation 3 using these parameters, or even a 100-fold lower value of k_b (4), shows that diffusion is typically the slow step. Therefore, researchers who want to decrease analysis times assist the mass transport with flow, electromigration, or the use of beads as substrates, which can be shaken in the solution of analyte.

Unmet needs for immunoassays include higher sensitivity, smaller quantities of reagents (particularly the antibodies), faster binding times, less nonspecific adsorption, and the development of inexpensive synthetic antibodies. Aptamers are an attempt to address this last need (6).

By the time a protein analysis has made its way into the clinical lab as an immunoassay, the research to discover the protein biomarkers, and the research to understand why the disease is correlated with each biomarker, has already been done. Protein separations that can analyze large numbers of proteins are needed for this research, and the first such separation we discuss below is the protein microarray.

3. PROTEIN MICROARRAYS

Investigators can study large numbers of proteins using microarrays of capture antibodies, which are now commercially available from a variety of sources. A protein microarray is an organized high-density assortment of microscopic spots of proteins immobilized onto a support. This can be a solid planar support (7–10), microspherical beads (11), and polymers or gels (12). Ink-jet printing is commonly used for the impregnation of proteins onto the supports (13). Other procedures include microcontact printing (8) and self-assembling protein microarrays (14). All methods of printing give pronounced deformities in spot shapes, often giving rings rather than filled spots (10), which can be a challenge in quantitative interpretation. Additives prevent the protein from denaturing when the microarray is dried after printing (15). In the use of protein microarrays (such as arrays of antibodies), the microscopic spots of printed protein bind the analytes from solution, and the same considerations discussed above for immunoassays (such as enhancing mass transport and rinsing weakly bound species) also apply for microarrays (16). Protein microarrays represent a rapidly developing technology.

As an example of one type of protein microarray, cytokine arrays are promising in medicine (17). Cytokines are small proteins involved in the inflammatory response, which is activated by disease. Because cytokine biology is complex, typically 12 cytokines need to be measured simultaneously to characterize the inflammatory process, hence the attractiveness of microarrays. One interesting application is in tailoring HIV therapy to the patient. HIV destroys a certain type of white blood cell called a CD4 T cell, and current treatment to kill the virus, if effective, restores the levels of these white blood cells. Some HIV patients continue to have low CD4 levels despite the therapy lowering the viral load. Sachdeva et al. (18) used a cytokine array to study the differences between the two populations of patients having low viral loads but differing levels of CD4 cells. The array recognized 12 different cytokines. The authors found that two cytokines had significantly different levels in the two groups, and the identities of these two proteins point to greater immune damage in the case of the patients whose CD4 cell levels had not recovered. With further research, insight into the biology of HIV could lead to improved therapy for the group whose CD4 cells did not recover. Most applications of protein arrays are in the research stage because it is a new technology, and as with DNA microarrays, the technology will likely result in new clinical tests and drugs tailored to the patient.

Unlike DNA, proteins have no amplification reaction that is analogous to polymerase chain reaction (PCR); therefore, sensitivity is a problem that has slowed the development of protein microarrays. Consequently, the flat glass substrates widely used for DNA arrays are used less for proteins arrays. The substrates for protein arrays are usually thin, hydrophilic polymer films, which allow for more moles per area to be printed to increase sensitivity (12). The polymeric substrates for protein arrays are compatible with the same microarray printers used for the well-established technology of DNA microarrays. The printed spots spread more on the polymers, but the density requirements for protein arrays are not as great as they are for DNA arrays. Polymer substrates bind antibodies and other proteins well, and they leave the proteins in their native conformation to retain binding specificity. The binding is not understood: Printed proteins appear to physisorb (15). As with DNA arrays, researchers must use a blocking solution after printing to deactivate the surface so that it does not bind the analytes. New substrate materials for protein microarrays would be valuable because polymers suffer from low optical transmission and fluorescence background.

Silica colloidal crystals, which are porous crystals of silica nanoparticles, have recently been introduced as a substrate for protein arrays (19). They offer a high binding capacity, optical transparency, no fluorescence, and fast mass transport. Silica colloidal crystals occur in nature as opals, which makes them another example of nature's nanotechnology. **Figure 2** compares a natural opal of gem quality with the synthetic silica colloidal crystal shown in **Figure 2b**. Scanning electron micrographs, for both the natural and synthetic opals, show that the silica nanoparticles are arranged in a face-centered cubic lattice structure. Because the sizes of the silica particles are on the order of the wavelength of light, their crystalline order gives Bragg diffraction at visible wavelengths (20). Scientists can make a crystal of monodisperse nanoparticles that has just one color, whereas nature creates a patient mixing of crystallites of different sized nanoparticles and orientations. For protein analysis, the size of the nanoparticle determines the pore size, and simple geometry reveals the limiting pore diameter to be 15% of the particle diameter. One would probably want a pore diameter of at least 40 nm to accommodate the sandwiches of three proteins for an antibody

Figure 2

(*a*) Gem-quality opal (*left panel*) (courtesy of George R. Rossman, California Institute of Technology) and scanning electron micrograph (SEM) image (*right panel*) (courtesy of Chi Ma, California Institute of Technology).
(*b*) Synthetic opal (*left panel*) and SEM image of the synthetic opal (*right panel*).

array. This would require nanoparticles on the order of 250 nm in diameter, which are readily synthesized and crystallized. Researchers compared a silica colloidal crystal and a flat surface for the capture of the protein streptavidin, using biotin bound to the surfaces of both types of substrates. The signal from labeled streptavidin was increased by a factor of 80 relative to that of the flat surface for a colloidal crystal 11 μm in thickness, made with nanoparticles 250 nm in diameter (19). This high enhancement agreed with theory, and the increase in fluorescence background was negligible. These materials are being evaluated currently as substrates for antibody arrays.

For both immunoassays and antibody arrays, it is necessary to know the identities of the proteins before the measurements can be made. Determining which proteins are involved in a disease, and discovering new biomarkers for disease, requires the separation and identification of unknown proteins from the complex sample. These separations must resolve enormous numbers of proteins, including the post-translationally modified forms of proteins. In addition, the separation has to be compatible with mass spectrometry to enable identification of the proteins. There is no one type of separation with adequate power to separate protein mixtures as complex as human serum or cell lysates. To address the challenge, researchers use combinations of powerful separations, each based on a complementary property of the protein. There is no one approach that works for all applications, and the field continues to develop. The next sections discuss the types of separations used.

4. GEL ELECTROPHORESIS: INTRODUCTION

Complex protein analyses use at least two orthogonal separations, combined with mass spectrometry. A given separation can be described by its peak capacity, which is the number of peaks that can be fit into the separation window. A high peak capacity is 100. In two-dimensional (2D) separations, if one dimension has a peak capacity of n proteins and the other dimension has a peak capacity of m proteins, then the total peak capacity is the product, $n \times m$. This assumes that the mechanisms of the separations are independent of one another (i.e., they are orthogonal). If the mechanisms are perfectly correlated, the peak capacity remains on the order of n or m. Aside from this condition of orthogonality, the random overlap of peaks along one dimension reduces n and m each to 18% of their ideal values, based on statistics alone (21). The limited peak capacities, relative to the large number of proteins in serum samples, make multidimensional separations a requirement. Subsequent identification by mass spectrometry provides some additional resolution of overlapped peaks.

The peak capacity in one dimension is limited by the similarity of the properties of the components (22). For example, if the components are similar in size, a separation based on size inherently gives a small peak capacity. For multidimensional separations, Giddings (23) defined the term sample dimensionality as the number of independent types of properties that can be used to achieve a separation. For example, a polydisperse sample of polymer, such as polystyrene, has one dimension because only the molecular weight of polystyrene varies, not its composition. Proteins have multiple properties that vary because of the variety of amino acids. The most notable variables are molecular weight, charge, and hydrophobicity. A 3D separation of proteins is thus expected to be possible for complex protein samples.

In a typical 2D gel electrophoresis separation, one axis is the isoelectric point (i.e., charge) and the other axis is the molecular weight. These are quite orthogonal to one another. The third dimension of hydrophobicity is not used in conjunction with 2D gels because the proteins are denatured by sodium dodecyl sulfate (SDS) in the size separation, which masks the hydrophobicity of the protein. After separation by the 2D gel, the separated spots of denatured proteins are individually removed and digested by trypsin into peptides. These peptides are then separated by reversed-phase liquid chromatography (RPLC). The unique combination of peptides is used to identify the protein. Some overlap of protein spots in the gel can be tolerated because the mass spectrometer can resolve multiple peptides within overlapping peaks, and the algorithms for protein identification consider multiple proteins contributing to the collection of peptides from a given gel spot. One problem is that a high-abundance protein can hide a low-abundance protein. Nonetheless, owing to the orthogonality of its two dimensions and the power of mass spectrometers, the 2D polyacrylamide gel in combination with mass spectrometry is the workhorse of proteomics.

Investigators have used 2D gel separations in the search for biomarkers. Using a 2D gel technique called differential in-gel electrophoresis (DIGE), one can search for potential biomarkers by labeling proteins from healthy and diseased cells with different color dyes (24–26). The chief difficulty is that there are 12 high-abundance proteins in serum that compose 85% of the total serum protein, and these proteins are approximately six orders of magnitude more abundant than the known biomarkers (27). These give broad bands in the gel owing to precipitation, and they swamp out the signals of lower-abundance proteins. To alleviate this problem, researchers remove the high-abundance proteins by passing the serum through an affinity column containing

Figure 3

Differential in-gel electrophoresis image before and after the immunodepletion of abundant serum proteins. Figure taken with permission from Yu et al. (29).

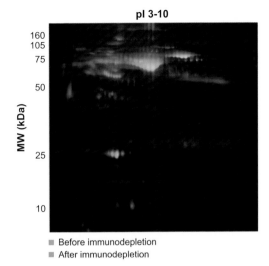

pI 3-10

MW (kDa)

160
105
75

50

25

10

■ Before immunodepletion
■ After immunodepletion

antibodies that bind the high-abundance proteins (28). **Figure 3** illustrates an example of a separation by 2D DIGE, comparing the total serum protein with serum that has been depleted of the 12 most abundant proteins (29). One high-abundance protein, human serum albumin, is responsible for the bright peaks at approximately 66 kDa. The loss of the high-abundance proteins allows one to scale the intensity to reveal many proteins that were not visible in the whole serum sample.

In one study of biomarkers in pancreatic cancer, DIGE revealed apolipoprotein E as a potential biomarker because it is 7.7-fold higher in cancer patients than in healthy patients (29). This protein, however, was not validated as a biomarker after the study of a larger number of serum samples. Aside from the low number of serum samples that is reasonable for studies with DIGE, there may be a more fundamental reason why this method has not fulfilled its promise in biomarker discovery. Clinically successful biomarkers are very-low-abundance proteins (e.g., nanomolar concentration levels), and even with serum depletion, DIGE is not sensitive to such low levels (27). The potential biomarker mentioned above, apolipoprotein E, is 100-fold higher in abundance than prostate specific antigen, which is a widely used clinical biomarker whose concentration is on the nanomolar scale when indicative of cancer. More separation power to achieve a higher dynamic range is needed. In addition, much higher speeds are needed because it takes 1 day just to separate with DIGE. Below we examine the principles of electrophoresis to understand the factors that limit its resolution and speed.

5. GEL ELECTROPHORESIS: MECHANISMS

Isoelectric focusing takes advantage of the fact that the velocity, v, of the protein depends on the charge, z, of the protein:

$$v = zFE/f = \mu_e E, \qquad (4)$$

where E is the applied electric field, f is the friction coefficient for the protein, and F is the Faraday constant. The terms z, F, and f are usually lumped together as the electrophoretic mobility, μ_e. The charge of the protein is dependent on pH, and every protein has a pH at which its charge is zero (i.e., its isoelectric pH). Isoelectric focusing uses a medium that has a pH gradient spatially imposed, and the protein electromigrates until it reaches the region of its isoelectric point, and then it stops. As it diffuses in either direction, it picks up a charge that causes it to electromigrate back to the position of its isoelectric pH, hence the term focusing. Resolution depends on the steepness of the pH gradient (dpH/dx), the change in protein charge with pH (dz/dpH), and the temperature (30):

$$ R_s = \frac{\Delta p I}{4} \left(\frac{FE}{RT} \frac{-dz/d\mathrm{pH}}{d\mathrm{pH}/dx} \right)^{1/2}. \tag{5} $$

Extraordinary resolution is possible: $\Delta\mathrm{pH} = 0.001$ can be resolved for a shallow pH gradient and a large change in charge with pH.

Isoelectric focusing is typically performed in gel strips for proteomics, and this separation is allowed to run overnight. In principle, one could apply an enormous electric field and separate the proteins quickly. High fields are used for isoelectric focusing in capillaries, giving focusing in 7 min (31) and more recently in 30 s (32). The high initial currents are a problem for the larger cross-sectional areas of gels because the Joule heat cannot be dissipated quickly. Capillaries are thinner, giving less current for a given voltage and faster heat transfer. Any approach to speeding up 2D electrophoresis will likely involve miniaturizing the volume to minimize the effects of Joule heating.

The second dimension of the gel electrophoresis separation is sieving. This separates proteins based on their sizes, using the porous nature of the gel. Referring to Equation 4, the friction coefficient of the protein is related by Stokes' law to its radius, r:

$$ f = 6\pi\eta r, \tag{6} $$

where η is the viscosity of the medium. Because the migration is electrically driven, the charges of the proteins must be the same for the separation to depend only on size. This is accomplished by denaturing the protein with the surfactant SDS, which binds to the protein in a ratio of 1.4 g of SDS per 1 g of protein (33). Because the charge per gram is constant, the electrophoretic mobility is constant. Separating the SDS-denatured proteins based on size now requires a sieving medium.

Investigators use nanotechnology to sieve proteins because of their nanoscale sizes. To achieve sieving, the pores through which the protein passes must be on the order of the size of the proteins. Researchers developed sieving media well before the use of SDS, beginning in the 1950s with starch gels (34). We can consider these starch gels to be the advent of nanotechnology in separations. A few years later, in 1959, polyacrylamide gels were introduced for protein separations (35), and agarose gels were introduced in 1966 (36). Investigators demonstrated the use of cross-linked polyacrylamide gel to sieve SDS-denatured proteins as early as 1967 (37). The polyacrylamide gels and SDS denaturation of the proteins remain the dominant technology in proteomics.

The principles of sieving electrophoresis explain why polyacrylamide gels have stood the test of time, and how a new material might be developed to improve the speed. Giddings (30) described the partition coefficient, K, of a protein, as it distributes between nanoscale pores and free solution by relating K to the fraction of accessible volume in the pore:

$$K = \frac{\text{accessible volume}}{\text{true volume}} = \frac{\pi \left(R^2 - r^2 \right) \cdot L}{\pi R^2 \cdot L}, \tag{7}$$

where r and R are the radii of the protein and pore, respectively; and L is the pore length. This partition coefficient describes the mechanism of size-exclusion chromatography (SEC). Rodbard & Chrambach (38) showed that the electrophoretic mobility, μ_e, relative to that in free solution, μ_e^0, is equal to the same K as in SEC (i.e., both are described by the ratio of accessible to true volume of the pore):

$$K = \frac{\text{accessible volume}}{\text{true volume}} = \frac{\mu_e}{\mu_e^0}. \tag{8}$$

The ability of protein to electromigrate in a gel is thus dictated by the accessible volume in the gel pores. An intuitive way to understand this is that the velocity of the protein has to be zero when it is contact with a wall, and the accessible volume is proportional to the probability that the protein is not in contact with a wall. Entropy thus controls sieving separations.

The pores of polyacrylamide gels are not monodisperse. Ogston (39) first modeled the random structure of the gel fibers. Using this structural description, we arrive at the Ogston-Morris-Rodbard-Chrambach model for gel electrophoresis, which describes the electrophoretic mobility relative to that in free solution:

$$\frac{\mu_e}{\mu_e^0} = \exp \left[-\frac{\pi}{4} \left(\frac{d+r}{R} \right)^2 \right], \tag{9}$$

where d is the radius of the gel fibers; and r and R are again defined as the protein and pore radii, respectively. This model agrees well with experiment (40). Other models give comparable agreement with experiment (40), so one should not take Equation 9 as the law of the land. Even using Equation 7 (which would ignore the distribution of pores), one can obtain a similar dependence of mobility on protein size.

The essential issue for any sieving model is that the radius of the pore must be on the same order as the radius of the protein to give a sieving separation. Experimentally, the pore size of the gel is controlled by the percentage of both monomer and cross-linker in the gel (41). To illustrate the role of pore size, **Figure 4** shows a series of plots of computed mobility versus protein radius (using Equation 9) for three different pore radii. In each case, the fiber radius was assumed to be equal to the pore radius for simplicity. The 10-nm pore radius is shown to be well suited to the molecular weight range of 10 to 200 kDa that is typically studied in proteomics. **Figure 4** also shows the same relation for the monodisperse pores, which require a somewhat larger pore radius to span this range of molecular weights. The graph illustrates how similar the behaviors are for the gel and the monodisperse pores. The flatter slope indicates that the monodisperse pores give somewhat greater selectivity than the distribution

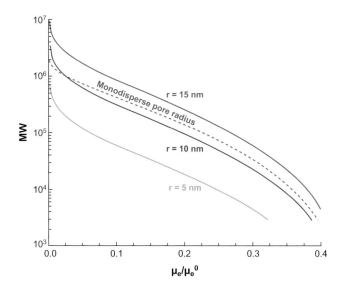

Figure 4

Calculated dependence of
electrophoretic mobility on
molecular weight (MW) for
gel sieving with three
different pore radii (*solid
lines*). The dotted line is the
same curve, but for
monodisperse pore radius.

of pores. Another difference is the abrupt cutoff at $r = R$ for monodisperse pores, which is smoothed in the case of the gel because the distribution of pores provides a path for proteins larger than the average pore size. Overall, the differences between gels and monodisperse pores are rather small, and both illustrate that nanoscale pores enable a size-based separation of proteins.

Researchers have used atomic force microscopy (AFM) to image the nanoscale structure of gels used in electrophoresis, using tapping mode, which imparts minimal force to the soft material (42). **Figure 5a** shows an AFM image of a 3% agarose gel, revealing a web structure. This image gives us a visual way of thinking about gels. Image analysis revealed the distribution of pore diameters in **Figure 5b**, which agrees with the Gaussian distribution in Ogston's (39) structural model. The AFM results further showed that the pore-size distribution narrowed with higher agarose concentration, decreasing by a factor of two for a 5% agarose solution. The pore diameters in all cases are much larger than the protein diameter. AFM images of polyacrylamide gels show the expected small pore sizes (43). The wide pores of the agarose introduce an important question: Why can such a wide-pore medium sieve small proteins?

The answer is interesting, and it has consequences for speeding up sieving separations. Even lower concentrations of agarose give size-based separations of small proteins (44). In general, large hydrophilic polymers in homogeneous solution impart a size-based separation without forming permanent pores. The explanation is that agarose is a soluble polymer, and at sufficiently high concentrations, the polymer chains become entangled to form transient pores (45, 46). The larger pores obvious in **Figure 5a** are not the same pores responsible for the sieving of much smaller proteins. Many hydrophilic polymers of high molecular weight are now available for sieving of proteins and DNA fragments, including polyethylene glycol, cellulose derivatives and linear polyacrylamide, and copolymers (47). These polymer solutions have been called replaceable gel media, non-cross-linked polymers, and entangled polymer

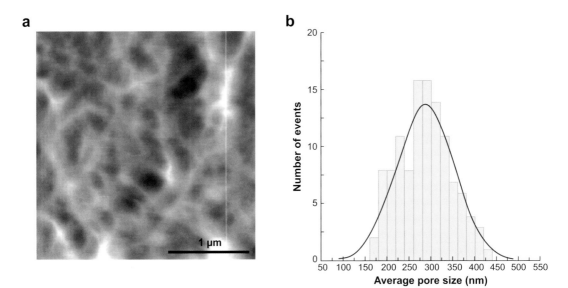

Figure 5

(*a*) Atomic force microscopy (AFM) image of agarose gel. (*b*) Pore size distribution determined from image analysis of the AFM image. Figure taken with permission from Pernodet et al. (42).

solutions. These media have enabled the routine use of gels in capillaries for fast and automated DNA sequencing (45, 48–50). This advance led to the completion of the human genome project ahead of schedule (51). Entangled polymer solutions are now used commercially for fast microchip-based protein sieving (52). In the latter case, a set of protein molecular weight standards was easily resolved in 40 s using an electric field of 340 V cm^{-1}. This is a tenfold higher electric field than a slab gel can tolerate. The microchip can accommodate the higher field strength because it is only 13 μm deep, giving lower current and faster heat dissipation. The ability to avoid preparing a gel, which must be thick to be reasonably uniform, is thus possible by using entangled polymer solutions in thin capillaries and thin channels of microchips. The drawback is that these media cannot be cast readily in a 2D format to allow 2D electrophoresis. The polymer solution is also difficult to couple with other separations or with mass spectrometry.

Silica colloidal crystals have been introduced as a thin separation medium that can be cast in a 2D format (53, 54). These materials are described above in the context of proteins arrays. The particle size can be tailored to give optimal pore sizes for sieving. Investigators demonstrated the separation of proteins by sieving through the pores of colloidal crystals in a microchip: The channel was filled with a colloidal crystal of 160-nm silica particles, and proteins from 20 to 205 kDa were separated using a field strength of 31 V cm^{-1} (54). Much higher fields can be used, and the authors did not observe any deleterious heating effects in electrochromatography up to 1000 V cm^{-1} (55). The surfaces of silica nanoparticles can be made to be as high in quality as the best chromatographic silica gel (56); in fact, silica nanoparticles are constituents of the highest-quality silica gel particles (57). The use of these materials in separations will advance as better and faster means are being developed for making these crystals, such as calcining to enable crack-free material crystals (58) and spin coating to enable the deposition of crystals in less than 1 min (59). These materials have the potential for 2D protein separations because they are inherently 2D in format, and they have the potential for interfacing with mass spectrometry, either through matrix-assisted laser desorption/ionization or protein extraction. This could increase the sensitivity of DIGE because the colloidal crystals are intrinsically low in fluorescence (19).

There are many unmet needs for protein electrophoresis and 2D gel electrophoresis. Obviously the speed and resolution of the 2D separation need to be improved. Another unmet need is low cost. DIGE analyses in particular are expensive, with one 2D gel separation typically costing approximately $2000 without the mass spectrometry. Devices for fast 2D separations on the benchtop in researchers' labs that allow detection of lower-abundance proteins would be valuable.

6. CHROMATOGRAPHY OF PROTEINS

Gel electrophoresis utilizes size and charge as the two properties on which to base a protein separation. In high-performance liquid chromatography (HPLC), it is common to use the third distinguishing property of proteins, which is hydrophobicity. This is achieved by RPLC. These separations use a hydrophobic stationary phase, typically a monolayer of short hydrocarbon chains covalently bonded to the silica surface through a siloxane linkage. This monolayer is on the order of 1 nm in thickness, which allows fast mass transport to the mobile phase. These monolayer bonded phases represent another example of the early use of nanotechnology in separations (60). Separations of proteins using stationary phases of hydrophobic monolayers are best achieved by gradient elution RPLC, for which the gradient begins with a mobile-phase composition that is mostly water and ends with a composition that is mostly organic, typically acetonitrile. For peptides, the elution occurs in order of increasing hydrophobicity (61). This necessarily would correlate with size if two peptides or proteins had similar amino acid compositions, but overall the correlation with size is not high. For example, for bovine serum, albumin (66 kDa) elutes earlier than the much smaller myoglobin (16.7 kDa) (62). RPLC thus contributes in a complementary way to selectivity in protein separations by separating on the basis of hydrophobicity.

Chromatography can also separate on the basis of size. SEC is based on the same fundamental principles as sieving electrophoresis, as detailed above. SEC offers one potential advantage over sieving electrophoresis: There is no need for denaturing with SDS. The need for SDS denaturation in electrophoresis arises because the native proteins have different charges, and so much SDS is adsorbed that it masks the differences in protein charges. Because there is no electric field in SEC, the charges of the proteins do not contribute to the retention times; therefore, SDS is not needed in SEC. The problem with using SDS in electrophoresis is that it prevents the use of mass spectrometry on whole proteins because there is no effective way of removing SDS from the gel spots. Size exclusion, by avoiding this denaturing, would allow the proteins to be directly analyzed by mass spectrometry.

There are two reasons why size exclusion is not used nearly as extensively as sieving for protein separations. First, chromatography columns do not lend themselves readily to 2D separations. Second, the peak capacity of SEC is low. We can overcome the first issue to some extent by using sequential columns to achieve 2D separations, in which one column is eluted slowly and the other column analyzes each peak quickly (63). Coupling two 1D columns to achieve a 2D separation requires more planning and design than separations that have a spatially 2D format, such as a slab gel or a thin layer chromatography plate. As one illustration, Moore & Jorgenson (64) performed a 3D separation of peptides from a tryptic digest of ovalbumin using SEC as the first dimension, followed by reversed-phase HPLC, and finally by capillary electrophoresis. These are the three orthogonal properties pointed out above for proteins, and the same idea applies to peptides. **Figure 6a** shows the SEC separation of the peptides, illustrating that this technique does not separate the rather limited number of peptides well. (We discuss the reasons below.) **Figure 6b** shows the final

Figure 6

(*a*) Size-exclusion chromatogram of peptides. (*b*) Display of three-dimensional separation using size-exclusion chromatography (SEC), reversed-phase liquid chromatography (RPLC), and capillary zone electrophoresis (CZE) for the three dimensions. Figure taken with permission from Moore & Jorgenson (64).

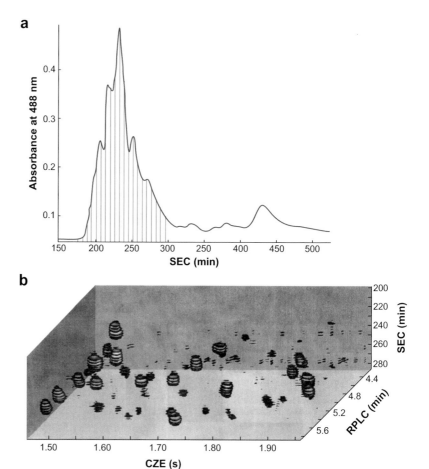

3D separation. The vertical discs result from the fact that the broad, slowly eluting SEC peaks are sampled discretely by the reversed-phase column. The peptides are well resolved in three dimensions. **Figure 6b** illustrates that the three dimensions are orthogonal to one another. For any of the three axes correlated with one another, the effect would be that the data points lie within a plane rather than in 3D space. The figure shows that the points fill 3D space rather uniformly, demonstrating that these three separation modes are orthogonal. **Figure 6b** also illustrates that the peak capacity is smallest for the SEC dimension.

Coupling three 1D chromatographic separations together is an engineering feat that must be carefully planned. For the 3D peptide separation, each successive dimension has a progressively faster timescale: The SEC scale spans 80 min, the reversed-phase axis spans 1.5 min, and the capillary zone electrophoresis axis spans 27 s. This contrasts with the 2D gel, in which sampling and timing are not issues. Also, the mobile phases must be compatible in successive dimensions. To our knowledge, a 3D separation has not been used for proteins. A nice example of coupling two 1D separations of proteins uses two capillaries; in this study, Hu et al. (65) coupled micellar electrokinetic chromatography (based on hydrophobicity) and capillary zone electrophoresis (based on charge) for fast 2D separations in proteomics studies of single cells. In this example, the authors achieved very high sensitivity and low loss to enable single cells to be analyzed. New and well-grounded ideas, such as the ones presented here, to achieve multidimensional protein separations will ultimately enable the resolution of low-abundance proteins for mechanistic studies of diseases and biomarker discovery.

The limited peak capacity of SEC is an interesting problem because the same physical phenomenon gives rise to sieving separations in gels, yet sieving gels have peak capacities on the order of 50, whereas SEC can give a peak capacity as high as 10. The fundamentals of SEC explain why this is so. First, SEC has a smaller window in which the peaks must fit, and second, the peaks are broad compared to the width of this window. This can be explained from first principles. Recalling our above discussion of sieving, we give the equation for the partition coefficient in size exclusion here for the retention volume, V_R; the mobile-phase volume, V_0; and the total volume (mobile phase plus pores), V_T:

$$K = \frac{\text{accessible volume}}{\text{true volume}} = \frac{\pi \left(R^2 - r^2 \right) \cdot L}{\pi R^2 \cdot L} = \frac{V_R - V_0}{V_T - V_0}. \tag{10}$$

All the peaks in the chromatogram fit within the window bounded by V_0 and V_T. The largest proteins, which do not fit into the pores, elute at V_0, and the lowest–molecular weight species enter these pores without any size exclusion and elute at V_T. To estimate how many peaks fit between these two extremes, we consider a typical set of column parameters. If a column is randomly packed with spheres as the stationary phase, $V_0 = 40\% \ V_{sphere}$ (66). The maximum possible pore volume equals the sphere volume, which gives $V_T = V_0 + V_{sphere}$. In this limiting case, the range of retention volumes is V_0 to $V_0 \times 2.5$. More typically, the range is V_0 to $V_0 \times 1.5$ (67). Retention volumes larger than the theoretical maximum can be observed if adsorption occurs, which is irrelevant to this retention mechanism. The peak capacity, n, neglecting statistical overlap, is the ratio of this range to the average peak width, 4σ. Converting retention volume to retention time makes for a more convenient expression:

$$n = \frac{1.5 \times V_0 - V_0}{4\sigma_V} = \frac{0.5 \times t_0}{4\sigma_t}. \qquad (11)$$

For pure size exclusion, let us suppose that $L = 10$ cm and $V_0 = 0.3$ mL. If the plate height is estimated to be 10 μm (which is about as good as one can do with SEC columns), we can calculate the peak width through the relation $H = \sigma^2/L$, where the width is 4σ. These conditions give a peak capacity of 5. This illustrates the problem with SEC: The window is so narrow that even for this high efficiency of $H = 10$ μm, there are few resolved peaks. One way to increase the number of resolved peaks would be to increase the column length to 25 cm, but the number of resolved peaks would go up only as $L^{1/2}$, which is a factor of 1.6. Another way would be to use smaller particles, 1.8 μm in diameter, if these were available, which would give a factor of 3 at the most by reducing H to an impressive 1 μm. There are not any anticipated factors at this time that would increase the peak capacity enough to catch up with sieving. Fundamentally, because the window is narrow relative to the peak width, the peak capacities of SEC are not likely to approach those of sieving electrophoresis or those of other HPLC techniques.

7. SUMMARY OF PROTEIN SEPARATION TECHNIQUES

The various techniques described above, among others, are shown in **Table 1**. The properties used by each technique are also shown. The additional techniques

Table 1   Protein separation techniques and their properties

Technique	Property			
	Size	**Charge**	**Hydrophobicity**	**Affinity**
PAGE	✓			
IEF		✓		
2D-PAGE[a]	✓	✓		
CZE	✓[b]	✓[b]		
RPLC			✓	
SCX		✓		
MudPIT[c]		✓	✓	
SEC	✓			
IMAC				✓
Microarrays				✓

[a]Sequential combination of IEF and PAGE in a two-dimensional gel.
[b]Separation based on ratio of charge to size.
[c]Sequential combination of SCX and RPLC.
Abbreviations: 2D, two-dimensional; CZE, capillary zone electrophoresis; IEF, isoelectric focusing; IMAC, immobilized metal affinity chromatography; MudPIT, multidimensional protein identification technology; PAGE, polyacrylamide gel electrophoresis; RPLC; reversed-phase liquid chromatography; SCX; strong cation exchange; SEC, size exclusion chromatography.

presented include immobilized metal ion affinity chromatography (IMAC), strong cation exchange (SCX), and multidimensional protein identification technology (MudPIT). In IMAC the specific binding of histidine residues to metal ions is utilized as the separation principle, and proteins are typically genetically engineered to have a hexahistine tag for purification by IMAC. In SCX the proteins adsorb to cationic sites in a resin and are differentially eluted with step increases in ionic strength of the mobile phase.

8. COMBINATION OF SEPARATIONS

In proteomics, investigators often use 2D electrophoresis or, alternatively, 2D HPLC. Combining the strengths of diverse separation tools is a promising approach in proteomics. One illustration is the proteomics of glycoproteins. These species are interesting because altered glycosylation is the hallmark of cancer (68). Glycosylation is a common type of post-translational modification that attaches a carbohydrate to a protein. A novel approach for biomarker discovery uses this idea to analyze only the glycoproteins of serum samples, in which the salient concept is that the biomarker does not need to be a protein whose level has changed owing to the disease; instead, it is the structure of the carbohydrate attached to the protein that has changed (69). To search for proteins of altered carbohydrate structure, Zhao et al. (69) applied a combination of affinity extraction, RPLC, and protein microarrays to serum samples of patients with pancreatic cancer. They first used affinity extractions, removing the high-abundance serum proteins with a multiple antibody column and then used a lectin column to extract the glycoproteins. Lectins are proteins that selectively bind glycoproteins, and a mixture of lectins in the affinity column ensured that a variety of glycoproteins were extracted. The study of only the protein bound to the lectin column further reduces the amount of high-abundance proteins and simplifies the mixture to the relevant proteins.

Subsequent gradient elution RPLC of the lectin-extracted protein sample gave protein peaks that were rather well resolved. The authors used a column of nonporous 1.8-μm silica particles for this separation. The smaller particle diameter is needed to achieve a reasonable surface area for separation. This is because in chromatography the resolution, R_s, is proportional to surface area, A, in the limit of low surface area:

$$R_s = \frac{\sqrt{L/H}}{4}(\alpha - 1)\frac{K \cdot A/V_m}{1 + K \cdot A/V_m}. \tag{12}$$

This equation strictly holds for isocratic elutions, but we can think of the gradient elution as having a progressively changing K for each protein as mobile-phase composition is varied. The nonporous particles provide a lower plate height by eliminating broadening due to intraparticle diffusion. The other terms are the selectivity, α, and the volume of the mobile phase, which are comparable to conventional gradient elution RPLC. Their minimizing H while maintaining sufficient surface area enabled approximately 24 protein peaks to be observed in RPLC.

The researchers collected fractions from the RPLC separation to isolate the glycoproteins, which were then spotted onto a substrate to make a glycoprotein array. It

Figure 7

Protein array using five different lectins (SNA, CONA, PNA, MAL, and AAL) to detect glycoproteins in reversed-phase liquid chromatography (RPLC) fractions of the same retention time for 24 different patient serum samples. AAL detects the most glycosylation, and the samples from cancer patients are the most extensively glycosylated. Figure taken with permission from Zhao et al. (69).

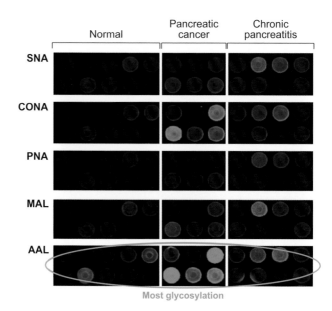

is called a reverse protein array when the analytes, rather than the capture agents, are spotted onto the substrate. The protein array was probed by labeled lectins, with each type of lectin chosen to recognize a different type of carbohydrate. **Figure 7** shows a section of the protein array corresponding to one protein fraction from the serum samples of 24 patients, probed by five different lectins. One can see that the six samples from cancer patients typically had more glycosylation than either the ten samples from healthy patients or the eight samples from patients with pancreatitis. For this protein fraction, the most glycosylation was observed for the lectin AAL (*Aleuria aurentia*), which selects for fucosylation. The microarray revealed multiple proteins whose fucosylation was altered. Mass spectrometry of the corresponding RPLC protein fractions enabled identification of the proteins, showing that biomarkers originate from the liver rather than the pancreas, apparently owing to the inflammatory reaction associated with pancreatic cancer. Higher-sensitivity detection might enable the identification of pancreatic proteins. This study illustrates the use of a strategic combination of a wider variety of separations tools, and this approach could have a significant impact on biomarker discovery and mechanistic studies of disease if even higher sensitivity can be achieved.

9. CONCLUSIONS

The era of proteomics has placed large demands on the analysis of highly complex protein samples, requiring multidimensional separations coupled with mass spectrometry. The modern evolution of separations toward both higher speed and higher resolution through miniaturization has improved proteomics mainly in the case of chromatographic separations, in which particles smaller than 2 μm provide sharper peaks and faster elution. Immunodepletion columns and protein microarrays are new manifestations of antibody-antigen interactions, and these provide valuable new tools for proteomics. Gel electrophoresis has evolved little, and this is one area in which breakthrough might have the most impact. Higher resolution in all types of protein separations is a general need for proteomics.

DISCLOSURE STATEMENT

The authors are not aware of any biases that might be perceived as affecting the objectivity of this review.

ACKNOWLEDGMENTS

This work was supported by NIH R01 GM65980. We thank Professor George R. Rossman of the California Institute of Technology for kindly providing the images for **Figure 2**.

LITERATURE CITED

1. Lepage R, Albert C. 2006. Fifty years of development in the endocrinology laboratory. *Clin. Biochem.* 39:542–57

2. Yalow RS, Berson SA. 1959. Assay of plasma insulin in human subjects by immunological methods. *Nature* 184:1648–49

3. Woof JM, Burton DR. 2004. Human antibody–Fc receptor interactions illuminated by crystal structures. *Nat. Rev. Immunol.* 4:89–99

4. Hu GQ, Gao YL, Li DQ. 2007. Modeling micropatterned antigen-antibody binding kinetics in a microfluidic chip. *Biosens. Bioelectron.* 22:1403–9

5. Glaser RW, Hausdorf G. 1996. Binding kinetics of an antibody against HIV p24 core protein measured with real-time biomolecular interaction analysis suggest a slow conformational change in antigen p24. *J. Immunol. Methods* 189:1–14

6. Tombelli S, Minunni M, Mascini M. 2007. Aptamers-based assays for diagnostics, environmental and food analysis. *Biomol. Eng.* 24:191–200

7. Zhu H, Snyder M. 2001. Protein arrays and microarrays. *Curr. Opin. Chem. Biol.* 5:40–45

8. MacBeath G, Schreiber SL. 2000. Printing proteins as microarrays for high-throughput function determination. *Science* 289:1760–63

9. Zhou FX, Bonin J, Predki PF. 2004. Development of functional protein microarrays for drug discovery: progress and challenges. *Comb. Chem. High Throughput Screen.* 7:539–46

10. Liu YS, Li CM, Yu L, Chen P. 2007. Optimization of printing buffer for protein microarrays based on aldehyde-modified glass slides. *Front. Biosci.* 12:3768–73

11. Melton L. 2004. Protein arrays: proteomics in multiplex. *Nature* 429:101–7

12. Angenendt P, Glokler J, Murphy D, Lehrach H, Cahill DJ. 2002. Toward optimized antibody microarrays: a comparison of current microarray support materials. *Anal. Biochem.* 309:253–60

13. Lee BH, Nagamune T. 2004. Protein microarrays and their applications. *Biotechnol. Bioprocess Eng.* 9:69–75

14. Ramachandran N, Hainsworth E, Bhullar B, Eisenstein S, Rosen B, et al. 2004. Self-assembling protein microarrays. *Science* 305:86–90

15. Wu P, Grainger DW. 2006. Comparison of hydroxylated print additives on antibody microarray performance. *J. Proteome Res.* 5:2956–65

16. Kusnezow W, Syagailo YV, Ruffer S, Baudenstiel N, Gauer C, et al. 2006. Optimal design of microarray immunoassays to compensate for kinetic limitations: theory and experiment. *Mol. Cell. Proteomics* 5:1681–96

17. Huang RP. 2007. An array of possibilities in cancer research using cytokine antibody arrays. *Expert Rev. Proteomics* 4:299–308

18. Sachdeva N, Yoon HS, Oshima K, Garcia D, Goodkin K, Asthana D. 2007. Biochip array-based analysis of plasma cytokines in HIV patients with immunological and virological discordance. *Scand. J. Immunol.* 65:549–54

19. Zheng SP, Zhang H, Ross E, Van Le T, Wirth MJ. 2007. Silica colloidal crystals for enhanced fluorescence detection in microarrays. *Anal. Chem.* 79:3867–72

20. Jiang P, Bertone JF, Hwang KS, Colvin VL. 1999. Single-crystal colloidal multilayers of controlled thickness. *Chem. Mater.* 11:2132–40

21. Davis JM, Giddings JC. 1983. Statistical theory of component overlap in multicomponent chromatograms. *Anal. Chem.* 55:418–24

22. Davis JM, Giddings JC. 1985. Statistical method for estimation of number of components from single complex chromatograms: application to experimental chromatograms. *Anal. Chem.* 57:2178–82

23. Giddings JC. 1995. Sample dimensionality: a predictor of order-disorder in component peak distribution in multidimensional separation. *J. Chromatogr. A* 703:3–15

24. Morita A, Miyagi E, Yasumitsu H, Kawasaki H, Hirano H, Hirahara F. 2006. Proteomic search for potential diagnostic markers and therapeutic targets for ovarian clear cell adenocarcinoma. *Proteomics* 6:5880–90

25. Queloz PA, Crettaz D, Thadikkaran L, Sapin V, Tissot JD. 2006. Two-dimensional gel electrophoresis based technologies for potential biomarkers identification in amniotic fluid: a simple model. *Protein Pept. Lett.* 13:959–63

26. Huang HL, Stasyk T, Morandell S, Dieplinger H, Falkensammer G, et al. 2006. Biomarker discovery in breast cancer serum using 2-D differential gel electrophoresis/MALDI-TOF/TOF and data validation by routine clinical assays. *Electrophoresis* 27:1641–50

27. Hoffman SA, Joo WA, Echan LA, Speicher DW. 2007. Higher dimensional (Hi-D) separation strategies dramatically improve the potential for cancer biomarker detection in serum and plasma. *J. Chromatogr. B Anal. Technol. Biomed. Life Sci.* 849:43–52

28. Huang L, Harvie G, Feitelson JS, Gramatikoff K, Herold DA, et al. 2005. Immunoaffinity separation of plasma proteins by IgY microbeads: meeting the needs of proteomic sample preparation and analysis. *Proteomics* 5:3314–28

29. Yu KH, Rustgi AK, Blair IA. 2005. Characterization of proteins in human pancreatic cancer serum using differential gel electrophoresis and tandem mass spectrometry. *J. Proteome Res.* 4:1742–51

30. Giddings JJ. 1991. *Unified Separation Science*. New York: Wiley & Sons

31. Liu Z, Lemma T, Pawliszyn J. 2006. Capillary isoelectric focusing coupled with dynamic imaging detection: a one-dimensional separation for two-dimensional protein characterization. *J. Proteome Res.* 5:1246–51

32. Yao B, Yang HH, Liang QL, Luo G, Wang LD, et al. 2006. High-speed, whole-column fluorescence imaging detection for isoelectric focusing on a microchip using an organic light emitting diode as light source. *Anal. Chem.* 78:5845–50

33. Reynolds JA, Tanford C. 1970. Binding of dodecyl sulfate to proteins at high binding ratios: possible implications for state of proteins in biological membranes. *Proc. Natl. Acad. Sci. USA* 66:1002–7

34. Smithies O. 1955. Grouped variations in the occurrence of new protein components in normal human serum. *Nature* 175:307–8

35. Raymond S, Weintraub L. 1959. Acrylamide gel as a supporting medium for zone electrophoresis. *Science* 130:711–11

36. Laurell CB. 1966. Quantitative estimation of proteins by electrophoresis in agarose gel containing antibodies. *Anal. Biochem.* 15:45–52

37. Shapiro AL, Vinuela E, Maizel JV. 1967. Molecular weight estimation of polypeptide chains by electrophoresis in SDS-polyacrylamide gels. *Biochem. Biophys. Res. Commun.* 28:815–20

38. Rodbard D, Chrambach A. 1970. Unified theory for gel electrophoresis and gel filtration. *Proc. Natl. Acad. Sci. USA* 65:970–77

39. Ogston AG. 1958. The spaces in a uniform random suspension of fibres. *Trans. Faraday Soc.* 54:1754–57

40. Kopecka K, Drouin G, Slater GW. 2004. Capillary electrophoresis sequencing of small ssDNA molecules versus the Ogston regime: fitting data and interpreting parameters. *Electrophoresis* 25:2177–85

41. Holmes DL, Stellwagen NC. 1991. Estimation of polyacrylamide-gel pore size from Ferguson plots of linear DNA fragments. 2. Comparison of gels with different cross-linker concentrations, added agarose and added linear polyacrylamide. *Electrophoresis* 12:612–19

42. Pernodet N, Maaloum M, Tinland B. 1997. Pore size of agarose gels by atomic force microscopy. *Electrophoresis* 18:55–58

43. Suzuki A, Yamazaki M, Kobiki Y. 1996. Direct observation of polymer gel surfaces by atomic force microscopy. *J. Chem. Phys.* 104:1751–57

44. Acevedo F, Marin V, Wasserman M. 1995. Electrophoretic size separation of proteins treated with sodium dodecyl sulfate in 1% agarose gels. *Electrophoresis* 16:1394–400

45. Tietz D, Gottlieb MH, Fawcett JS, Chrambach A. 1986. Electrophoresis on uncrosslinked polyacrylamide: molecular sieving and its potential applications. *Electrophoresis* 7:217–20

46. Zhou D, Wang YM. 2006. Mechanisms of DNA separation by capillary electrophoresis in nongel sieving matrices. *Prog. Chem.* 18:987–94

47. Chu BJ, Liang DH. 2002. Copolymer solutions as separation media for DNA capillary electrophoresis. *J. Chromatogr. A* 966:1–13

48. Best N, Arriaga E, Chen DY, Dovichi NJ. 1994. Separation of fragments up to 570 bases in length by use of 6′5 T-non-cross-linked polyacrylamide for DNA sequencing in capillary electrophoresis. *Anal. Chem.* 66:4063–67

49. Ganzler K, Greve KS, Cohen AS, Karger BL, Guttman A, Cooke NC. 1992. High-performance capillary electrophoresis of SDS protein complexes using UV-transparent polymer networks. *Anal. Chem.* 64:2665–71

50. Wu D, Regnier FE. 1992. Sodium dodecyl sulfate capillary gel electrophoresis of proteins using non-cross-linked polyacrylamide. *J. Chromatogr.* 608:349–56

51. Dovichi NJ, Zhang JZ. 2000. How capillary electrophoresis sequenced the human genome. *Angew. Chem. Int. Ed. Engl.* 39:4463–68

52. Bousse L, Mouradian S, Minalla A, Yee H, Williams K, Dubrow R. 2001. Protein sizing on a microchip. *Anal. Chem.* 73:1207–12

53. Zhang H, Wirth MJ. 2005. Electromigration of single molecules of DNA in a crystalline array of 300-nm silica colloids. *Anal. Chem.* 77:1237–42

54. Zeng Y, Harrison DJ. 2007. Self-assembled colloidal arrays as three-dimensional nanofluidic sieves for separation of biomolecules on microchips. *Anal. Chem.* 79:2289–95

55. Zheng SP, Ross E, Legg MA, Wirth MJ. 2006. High-speed electroseparations inside silica colloidal crystals. *J. Am. Chem. Soc.* 128:9016–17

56. Le TV, Ross EE, Velarde TRC, Legg MA, Wirth MJ. 2007. Sintered silica colloidal crystals with fully hydroxylated surfaces. *Langmuir* 23:8554–59

57. Kirkland JJ, Kohler J. 1989. *U.S. Patent No. 4,874,518*

58. Chabanov AA, Jun Y, Norris DJ. 2004. Avoiding cracks in self-assembled photonic band-gap crystals. *Appl. Phys. Lett.* 84:3573–75

59. Mihi A, Ocana M, Miguez H. 2006. Oriented colloidal-crystal thin films by spin-coating microspheres dispersed in volatile media. *Adv. Mater.* 18:2244–49

60. Kirkland JJ. 1971. High speed liquid-partition chromatography with chemically bonded organic stationary phases. *J. Chromatogr. Sci.* 9:206–14

61. Krokhin OV, Craig R, Spicer V, Ens W, Standing KG, et al. 2004. An improved model for prediction of retention times of tryptic peptides in ion pair reversed-phase HPLC: its application to protein peptide mapping by off-line HPLC-MALDI MS. *Mol. Cell. Proteomics* 3:908–19

62. Koyama J, Nomura J, Shiojima Y, Ohtsu Y, Horii I. 1992. Effect of column length and elution mechanism on the separation of proteins by reversed-phase high-performance liquid chromatography. *J. Chromatogr.* 625:217–22

63. Bushey MM, Jorgenson JW. 1990. Automated instrumentation for comprehensive two-dimensional high-performance liquid-chromatography capillary zone electrophoresis. *Anal. Chem.* 62:978–84

64. Moore AW, Jorgenson JW. 1995. Comprehensive three-dimensional separation of peptides using size-exclusion chromatography reversed-phase liquid-chromatography optically gated capillary zone electrophoresis. *Anal. Chem.* 67:3456–63

65. Hu S, Michels DA, Fazal MA, Ratisoontorn C, Cunningham ML, Dovichi NJ. 2004. Capillary sieving electrophoresis/micellar electrokinetic capillary chromatography for two-dimensional protein fingerprinting of single mammalian cells. *Anal. Chem.* 76:4044–49

66. Schure MR, Maier RS. 2006. How does column packing microstructure affect column efficiency in liquid chromatography? *J. Chromatogr. A* 1126:58–69

67. Yau WW, Ginnard CR, Kirkland JJ. 1978. Broad-range linear calibration in high-performance size-exclusion chromatography using column packings with bimodal pores. *J. Chromatogr.* 149:465–87

68. Gerber-Lemaire S, Juillerat-Jeanneret L. 2006. Glycosylation pathways as drug targets for cancer: glycosidase inhibitors. *Mini-Rev. Med. Chem.* 6:1043–52

69. Zhao J, Patwa TH, Qiu WL, Shedden K, Hinderer R, et al. 2007. Glycoprotein microarrays with multi-lectin detection: unique lectin binding patterns as a tool for classifying normal, chronic pancreatitis and pancreatic cancer sera. *J. Proteome Res.* 6:1864–74

Functional and Spectroscopic Measurements with Scanning Tunneling Microscopy

Amanda M. Moore and Paul S. Weiss

Departments of Chemistry and Physics, Pennsylvania State University, University Park, Pennsylvania 16802; email: stm@psu.edu

Annu. Rev. Anal. Chem. 2008. 1:857–82

The *Annual Review of Analytical Chemistry* is online at anchem.annualreviews.org

This article's doi:
10.1146/annurev.anchem.1.031207.112932

Key Words

STM manipulation, scanning tunneling spectroscopy, substrate-mediated interactions

Abstract

Invented as a surface analytical technique capable of imaging individual atoms and molecules in real space, scanning tunneling microscopy (STM) has developed and advanced into a technique able to measure a variety of structural, functional, and spectroscopic properties and relationships at the single-molecule level. Here, we review basic STM operation and image interpretation, techniques developed to manipulate single atoms and molecules with the STM to measure functional properties of surfaces, local spectroscopies used to characterize atoms and molecules at the single-molecule level, and surface perturbations affecting surface coverage and surface reactions. Each section focuses on determining the identity and function of chemical species so as to elucidate information beyond topography with STM.

1. INTRODUCTION

STM: scanning tunneling microscopy

The drive of analytical chemistry to measure compositions, morphologies, and concentrations of molecules from the bulk to the atomic scale has led to the development of techniques to characterize molecular and atomic arrangements on surfaces. Scanning tunneling microscopy (STM), invented by Binnig & Rohrer in 1981 (1) (and for which they won the Nobel prize in 1986), is able to determine atomic arrangements by rastering an atomically sharp tip across a flat conducting or semiconducting surface, thus rendering real-space images. This ability has made STM a widely used tool for surface characterization. Although many books (2–4) and reviews (5–7) have been written on STM development, theory, and use, the goal of this review is to focus on STM measurements beyond its most simple use as a real-space molecular and atomic probe. These measurements include the interpretation of STM images and the ability to manipulate single atoms and molecules, to obtain local spectroscopic data, and to characterize substrate-adsorbate interactions.

2. SCANNING TUNNELING MICROSCOPY OPERATION

An overview of STM operation is detailed schematically in **Figure 1**, which presents a one-dimensional metal-vacuum-metal STM tunnel junction. An atomically sharp conducting metal tip, typically tungsten or a platinum/iridium alloy, is brought within

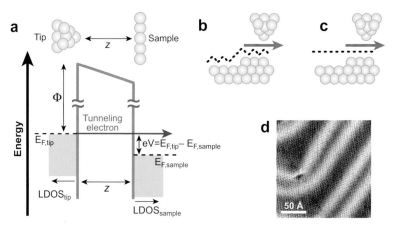

Figure 1

(*a*) An energy level diagram for a one-dimensional electron-tunneling junction. The Fermi energy levels (E_F) of the tip and sample are offset by the applied bias voltage (V) times the electron charge (e). The resultant current is exponentially dependent on the distance between the sample and the tip (z). LDOS, local density of states; Φ, work function of the metal. (*b*) Schematic of a scanning tunneling microscopy (STM) tip rastering across a metal surface in constant-current mode. The tip is extended and retracted, maintaining a constant tunneling current between tip and sample. (*c*) Schematic of a STM tip rastering across a metal surface in constant-height mode, maintaining a constant tip-sample separation, measuring current. (*d*) Atomically resolved topographic STM image operating in constant current mode of a Au{111} surface with herringbone reconstruction (95 Å × 95 Å V_{sample} = −0.05 V; I_{tunnel} = 200 pA).

a short distance (~3–10 Å) of a conducting or semiconducting sample using piezo-electric ceramic materials for probe tip placement and motion. An applied bias voltage (V) offsets the Fermi levels (E_F) of the tip and sample; the polarity of the bias voltage determines the direction of the electron flow. For instance, if $E_{F,tip}$ is greater than $E_{F,sample}$ (as shown in **Figure 1a**), this shift allows electrons to tunnel from the occupied states of the tip into unoccupied states of the sample. By changing the bias polarity, electron flow occurs in the opposite direction (from sample to tip).

In classical mechanics, an electron traveling between the tip and the sample would require an energy greater than the work function of the tip/sample material to overcome the barrier between them. However, quantum-mechanically, electrons are able to tunnel across the barrier. The state of an electron in a one-dimensional junction with a rectangular barrier is

$$\psi(z) = \psi(0)e^{-\kappa z},$$

where

$$\kappa = \frac{\sqrt{2m(V - E)}}{\hbar},$$

ψ is the electronic wavefunction, z is the tip-sample separation, m is the electron mass, V is the potential in the barrier, E is the energy of the tunneling electron, and $\hbar$ is Planck's constant. Through the Born interpretation of the wavefunction, the square of the wavefunction is proportional to the probability distribution of the electron, and thus, the probability of tunneling yielding a tunneling current I:

$$|\psi(z)|^2 \propto e^{-2\kappa z} \propto I(z).$$

For small biases, the quantity (V-E) can be approximated as the work function (Φ) of the metal (~5 eV) (8); thus, the tunneling current decreases by about an order of magnitude for every 1-Å change in z. The molecular and atomic resolution of the STM is enabled through this exponential current decay; therefore, the tunneling current is extremely localized, and the STM is sensitive to both lateral and vertical changes in topography (often tenths of angstroms or less) (9, 10).

The STM tip is rastered across a surface in one of two modes: constant current (**Figure 1b**) or constant height (**Figure 1c**). In constant-current (the most common) mode, the feedback loop (FBL) maintains a set current by adjusting the tip-sample separation. The recorded topography is dependent on both the geometric structure and the local density of states (LDOS) of the tip and sample (9, 11, 12). Constant-current mode is able to measure surface features with higher precision (atomic resolution and better) as described above, but is generally slower due to the need for mechanical manipulation of the scanner in the z direction, controlled through the FBL. **Figure 1d** presents an atomic resolution image of a Au{111} herringbone–reconstructed substrate obtained in constant-current mode. In constant-height mode, the tip is maintained at a set distance from the sample during scanning, recording the current and allowing for much faster measurements; constant-height mode is most useful for relatively smooth surfaces.

Fermi level: the highest occupied electron energy state, where at 0 K no electrons will have enough energy to rise above this level

FBL: feedback loop

LDOS: local density of states

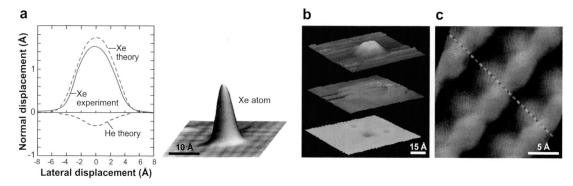

a — **Normal displacement (Å)** vs **Lateral displacement (Å)**: Xe theory, Xe experiment, He theory

b — Xe atom, 10 Å

c — 15 Å, 5 Å

Figure 2

(*a*) (*Left*) Theoretically calculated apparent heights for Xe and He and experimentally measured apparent height for Xe. (*Right*) Scanning tunneling microscopy (STM) image of a Xe atom on Ni{110} (40 Å × 40 Å; V_{sample} = −0.020 V; I_{tunnel} = 1 nA). Adapted and reprinted with permission from References 11 and 13. Copyright 1991 and 1986, the American Physical Society. (*b*) STM images of Ni_3 clusters on MoS_2 (60 Å × 60 Å; V_{sample} from top to bottom: +2.0 V, +1.4 V, −2.0 V; I_{tunnel} from top to bottom: 100 pA, 100 pA, 200 pA). Reprinted with permission from Reference 16. Copyright 1998, the American Chemical Society. (*c*) Two superimposed STM images of GaAs{110}. Shown are the unoccupied states centered around the Ga atoms (false-colored *green*) and the occupied states centered around the As atoms (false-colored *red*). (V_{sample} = +1.9 and −1.9 V, recorded simultaneously.) Reprinted with permission from Reference 18. Copyright 1987, the American Physical Society.

Figure 2 illustrates systems wherein interpretation of the STM images depend upon the molecules present and the scanning conditions used to acquire the images, as STM images are a convolution of topography and electronic structure (LDOS). In the case of Xe adsorbed on Ni{110} (**Figure 2*a***), the most significant contribution to imaging at energies near the Fermi level is from the *6s*-state of the adsorbate because the spatial extent is much larger for *6s* than for *5p*; the Lorentzian tail of each orbital crosses the Fermi level, although each is centered ∼4 eV away (11, 13, 14). Calculations of the LDOS (**Figure 2*a***, *left*) showed that for adsorbed He atoms, although they physically protrude from a surface, their apparent height in STM would be negative because He has a closed valence shell lower in energy and thus produces a decrease in the LDOS (13); this was confirmed experimentally (P.S. Weiss & D.M. Eigler, unpublished observations).

The location of an adsorbate on the surface and the STM tunneling conditions also influence the apparent height and shape of atoms at the surface at different adsorption sites (15, 16). As shown in **Figure 2*b***, a Ni_3 cluster was adsorbed on a MoS_2 surface and imaged at three sample bias voltages: +2.0 V, +1.4 V, and −2.0 V. At +2.0 V, the cluster appeared as a three-lobed protrusion, indicating that it enhanced the LDOS at this energy. When the sample bias was +1.4 V, the Ni_3 cluster did not appear as a protrusion from the MoS_2, but rather there appeared a diffuse ring resulting from a perturbation of the MoS_2 by the surface electronic structure of the cluster (17). At −2.0 V, the cluster appeared as a depression due to the depletion of the local density of filled states (16).

One of the most illuminating early examples of the convolution of topography and local electronic structure was found for images of GaAs(110). Here, Feenstra et al. showed that, depending on the bias voltage polarity, images of stoichiometric GaAs{110} surfaces display either the Ga atoms or the As atoms (18). **Figure 2c** is an overlay of two images acquired simultaneously at sample bias voltages of +1.9 V and −1.9 V. Calculations and electronegativity indicate that the unoccupied local state density is centered on the Ga atoms, whereas the occupied local state density is centered on the As atoms, consistent with their relative electronegativities (9, 18, 19).

3. ATOM AND MOLECULE MANIPULATION

Manipulating atoms and molecules can lead to insights into surface properties and binding sites (20–24). Several methods have been developed to manipulate atoms and molecules using STM, including lateral, vertical, and inelastic tunneling–induced manipulation. Typically, manipulations are performed at low temperature and in vacuum (20–22, 25), although there are recent examples of large-molecule manipulation at room temperature in vacuum (26–29) and in liquid (30). Manipulation is a slow, serial process for building up structures, yet new and interesting phenomena have been discovered through manipulation.

3.1. Manipulation Types

The first example of STM manipulation was lateral manipulation (LM) used to move Xe atoms on metal surfaces (20, 21, 24); this technique has since been performed on many systems (23, 29, 31–40). The LM procedure consists of three steps. First, the probe tip is brought towards the surface, thereby increasing the tip-atom (tip-molecule) interaction. Second, the tip is moved to the desired location across the surface, thus moving the atom (molecule) under its influence. Finally, the tip is retracted, leaving the manipulated atom (molecule) at its final location on the surface (**Figure 3a**, *top*) (31). Rieder and colleagues determined different modes of LM, measuring the tunneling current and feedback error signal during manipulation. These traces corresponded to: (1) pulling, where the STM tip is placed in front of the adatom, and the atom follows the tip due to an attractive tip-atom interaction; (2) pushing, where the atom or molecule is repelled and moves in front of the tip due to the interaction with the tip; and (3) sliding, where the atom or molecule is bound or trapped by the tip and moves along smoothly (shown schematically in **Figure 3a**, *bottom*) (31).

A second manipulation technique transfers the atom or molecule vertically between the STM tip and the surface. Vertical manipulation (VM) (**Figure 3b**, *top*) is performed by applying an electric field between the tip and sample when the tip is over the atom (molecule) to be moved, thus transferring the atom (molecule) to the STM tip. The tip is then moved to the new location and the electric field bias polarity is reversed, returning the atom (molecule) to the surface (21, 41–43). It is also possible to perform VM through mechanical contact, thereby picking up the atom or molecule with the tip (44, 45). **Figure 3b**, *bottom* shows the current versus time plot for moving a Xe atom between the STM tip and a Ni{110} surface (41).

Inelastic tunneling: tunneling between two states with different energies where energy is conserved through transfer, causing excitation (or deexcitation)

LM: lateral manipulation

VM: vertical manipulation

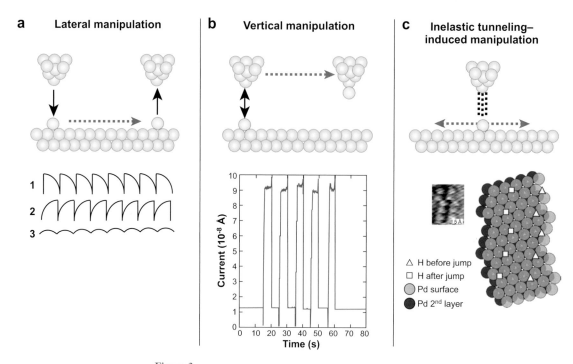

a Lateral manipulation

b Vertical manipulation

c Inelastic tunneling–induced manipulation

1

2

3

Current (10^{-8} A)

10
9
8
7
6
5
4
3
2
1
0

0 10 20 30 40 50 60 70 80
Time (s)

5 Å

△ H before jump
☐ H after jump
◯ Pd surface
● Pd 2nd layer

Figure 3

Uses of scanning tunneling microscopy (STM) to manipulate atoms and molecules. (*a*) Lateral manipulation, wherein the tip (1) pushes, (2) pulls, or (3) slides the manipulated species across the surface due to tip/sample interactions. (*b*) *Top*: Vertical manipulation, wherein the atom or molecule is vertically displaced by the STM tip. *Bottom*: Current versus time plot for picking up and replacing a Xe atom on a Ni{110} surface. Reprinted with the permission of Macmillan Publishers Ltd: *Nature* (41). Copyright 1991. (*c*) *Top*: Inelastic tunneling–induced manipulation, wherein tunneling electrons excite rotational, vibrational, or electronic transitions resulting in motion. *Bottom*: Hydrogen atoms manipulated through inelastic tunneling on a Pd surface. Reprinted with permission from Reference 49. Copyright 2006, American Chemical Society.

Inelastic electron tunneling (IET)–induced manipulation is a process wherein tunneling electrons or holes (depending on bias polarity) are injected from the tip positioned above the adsorbed atom (molecule), thus moving it (**Figure 3c**, *top*). The electron energy is transferred through an excited state, leading to excitation (rotational, vibrational, or electronic), the rate of which is controlled by the applied bias and the tunneling current (33, 46–48). **Figure 3c**, *bottom*, presents a schematic of H-atom manipulation on Pd{111} using IET (49, 50).

3.2. Functional Measurements Using Manipulation

IET: inelastic electron tunneling

Manipulation techniques have been used to study otherwise inaccessible systems by simply adsorbing molecules to surfaces. **Figure 4a** details how trajectories in LM are guided by the surface potential energy landscape on the atomic scale (40). In this example, single Ag adatoms were manipulated on a Ag{111} surface at different angles (θ) relative to the close-packed [110] surface direction.

Moving atoms at a rate of 10 Å/s with a tunneling resistance of 15 kΩ for all motion, the small steps in the tip trace shown in **Figure 4a** indicate motion of the atom between face-centered cubic (fcc) and hexagonally close-packed (hcp) sites (*yellow arrows*), whereas larger steps (*white arrows*) indicate jumps from fcc to fcc (or hcp to hcp) sites due to tip motion away from the close-packed direction. When the angle from the close-packed direction is increased (θ = 20°), less time was spent traveling in the close-packed direction and the adatoms made larger jumps (*red arrows*) due to the surface corrugation. Prior to these larger jumps, the lateral force (seen in the tip traces) increased in order to give the atom enough energy to overcome the barriers to motion over the surface. Once the adatom hopped, the probe tip quickly retracted in response to the FBL. For θ = 30° manipulation, the adatom did not travel along the close-packed direction; rather, theory predicted that the minimum energy path for this motion is to slide between threefold hollow sites around the surface atoms (*dotted white arrows*) (40, 51).

The LM for a single adatom described above involved sliding and slip-stick motions. Kelly, Tour, and colleagues have manipulated larger molecules known as nanocars, which contain fullerene (C_{60}) "wheels," and found that they roll across Au{111} surfaces (29). The chassis and axles of these nanocars were built from oligo(phenylene-ethynylene) (OPE) linkers and were attached to four fullerene wheels. The spacings between the fullerenes in the nanocars (each fullerene appeared as a protrusion) indicated the orientation of the car; the direction of travel was determined from sequences of images. **Figure 4b,c** displays STM images in which a single nanocar was manipulated by the STM tip. Only "pulling" motions (tip in front of the car) perpendicular to the axle direction moved the nanocar. Pushing the nanocar or pulling it in a direction parallel to the axles did not result in motion. By clever molecular design of three-fullerene molecules, Kelly, Tour, and colleagues determined that fullerene rotation (rather than sliding) was responsible for the motion of the nanocars (29).

In addition to building an understanding of motion on surfaces, VM has been used to change the LDOS of the tip, thus enhancing observation of the electronic properties. As the LDOS of the tip and sample are both important for STM imaging, VM has been used to attach molecules to an STM tip; using this assembly, surfaces can then be imaged, resolving surface structures not obtained from bare probe tips (45). Kelly et al. created fullerene-terminated STM tips by vacuum-depositing a fullerene film onto highly oriented pyrolytic graphite (HOPG), then bringing an STM tip (Pt:Rh) into contact with the film to lift a C_{60} molecule onto the tip. Using these functionalized tips, HOPG surfaces containing defects created by low-energy Ar^+ bombardment were imaged (**Figure 4d**), revealing theoretically predicted (52) threefold scattering patterns and a $\sqrt{3} \times \sqrt{3}$ superlattice arising from the electrons scattering along hexagonal crystallographic directions of the two-dimensional graphite layers. At room temperature, these images were only observed using a fullerene-functionalized tip, effectively as a result of sharpening of the electronic structure of the tip, an effect that was reproduced with an unfunctionalized tip at 77 K (45, 53).

Electric field–induced manipulation has been useful not only for moving molecules from the surface to the tip or from one location to another, but also for changing the orientation of a molecule bound to a surface. We have studied functionalized OPE molecules inserted into host *n*-alkanethiolate and 3-mercapto-*N*-nonyl-propionamide (1ATC9) matrices to understand how they function as conductance switches (54–60). In an *n*-alkanethiolate matrix, OPE molecules exhibit stochastic conductance switching between two conductance states (defined as ON and OFF), depending on the packing of the matrix around the inserted OPE molecules, and limited control was possible via the bias (54, 55). Nitro-functionalized OPE molecules were inserted into a 1ATC9 host containing buried hydrogen-bonding amide functionality, thus stabilizing the conductance states and enabling observations of switching

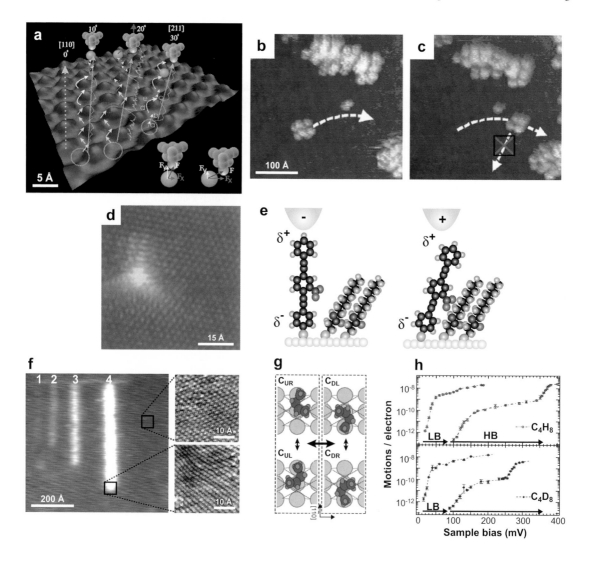

driven by the applied electric field between the STM tip and substrate. At positive sample bias, the nitro-functionalized OPE molecules were driven to the ON conductance state, and at negative sample bias, they were switched to the OFF conductance state (**Figure 4e**). If the OPE molecules were not functionalized (i.e., did not contain the nitro-group on the middle phenylene ring), the bias polarity could not drive switching. We determined that the combination of the applied field interaction with the molecular dipole and the ability of the nitro-functionality to hydrogen-bond with the host matrix mediated the observed conductance switching. This was tested by designing molecules with a range of molecular dipole magnitudes and signs (56).

Depositing H_2 onto Pd{111} resulted in H atoms both adsorbed to the surface and absorbed into the bulk Pd (49, 50). **Figure 3c** illustrates how surface-adsorbed H can be manipulated using IET. Interestingly, the H atoms absorbed into the bulk Pd can be manipulated to just below the top layer of Pd using IET (**Figure 4f**). The occupation of these sites was characterized using differential conductance imaging (described below). Differential conductance images recorded over bare Pd showed no distortion; however, when measured over the area patterned with subsurface H atoms, the hexagonal arrays of the surface Pd atoms were distorted (49, 50).

Action spectra can also be acquired using IET manipulation, shown experimentally using *cis*-2-butene (C_4H_8) (61). Here, Kawai and colleagues induced vibrationally

---

Figure 4

Examples of scanning tunneling microscopy (STM) lateral manipulation (LM) (*a-c*), vertical manipulation (VM) (*d-e*), and inelastic tunneling manipulation (*f*). (*a*) Superposition of an Ag{111} surface with tip-path manipulation signals. Lateral force components are shown (*bottom, right*). Also shown are the original positions of the adatoms (*dotted yellow circles*), the path of the tip motion (*dotted yellow lines*), the experimental tip path traces (*solid yellow lines*), and the hopping movements (*arrows*). See text for details. Reprinted with permission from Reference 40. Copyright 2003, the American Physical Society. (*b-c*) STM manipulation of a single nanocar (380 Å × 380 Å; $V_{sample} = 0.1$ V; $I_{tunnel} = 30$ nA; $I_{manipulation} = 3.5$ nA). The nanocar was pulled as indicated by the white arrow (*b*) to its location in panel *c*. The nanocar failed to move when the tip was moved 90° to the previous motion. Reprinted with permission from Reference 29. Copyright 2005, the American Chemical Society.
(*d*) STM image of a defect on highly oriented pyrolytic graphite (HOPG) imaged with a fullerene-functionalized tip (56 Å × 56 Å; $V_{sample} = -100$ mV; $I_{tunnel} = 1$ nA). Reprinted with permission from Reference 45. Copyright 1996, AAAS.
(*e*) Schematic of bias-controlled switching of nitro-functionalized oligo(phenylene-ethynylene) (OPE) molecules inserted into a 3-mercapto-*N*-nonyl-propionamide (1ATC9) host matrix using the STM electric field. (*f*) Manipulation of subsurface H atoms from the bulk Pd using STM at 4 K. (700 Å × 700 Å; $V_{sample} = 0.025$ V; $I_{tunnel} = 50$ pA.) The four lines were created by moving the tip over the surface at sample bias 0.7 V with currents of 1, 10, 50, and 150 pA for lines 1–4, respectively. The insets represent differential conductance images showing atomic resolution of the Pd{111} surface (30 Å × 30 Å; $V_{sample} = 0.018$ V; $I_{tunnel} = 200$ pA). *Top inset*: The hexagonal array of the Pd atoms over the surface was not distorted. *Bottom inset*: The hexagonal array is distorted over the patterned regions. Reprinted with permission from Reference 50. Copyright 2005, National Academy of Sciences, USA. (*g*) Proposed structural orientations of *cis*-2-butene on Pd{110}. (*h*) Action spectra for C_4H_8 (*top*) and C_4D_8 (*bottom*) ($I_{tunnel} = 3$ nA for C_4H_8 and 2 nA for C_4D_8). Reprinted with permission from Reference 61. Copyright 2005, the American Physical Society.

STS: scanning tunneling
spectroscopy

mediated motion, moving the C_4H_8 between four equivalent orientations on Pd{110}. As shown in **Figure 4g** (*left*), motions between the pairs C_{UR} and C_{UL} or C_{DL} and C_{DR} were found to be low-barrier (LB) motions characterized by small changes in the tunneling current (<1 nA), whereas motions from one pair to the other pair were found to be high-barrier (HB) motions, with measured tunneling current changes of ~3 nA. Action spectra (**Figure 4g**, *right*) with the motion yield (molecular motions per injected electron) as a function of sample bias were recorded for normal and deuterated (C_4D_8) *cis*-2-butene. Changes in the slopes of the action spectra indicated vibrational modes of the C_4H_8 molecules. The energies at the changes in the slope of the action spectra were compared with high-resolution electron energy loss spectroscopy (HREELS) data (62). From these data, the motion yields were assigned for the LB action spectra as the metal-carbon stretch for the C_4H_8 (C_4D_8) at ~37 mV (~31 mV) and for the C-H bending mode at ~115 mV (~95 mV) (with the expected isotopic shifts for the deuterated molecules). For the HB action spectra, the ~115 mV (~110 mV) was assigned as the C-C stretching mode, weakly influenced by the isotope effect, and the ~360 mV (~270 mV) was assigned as the C-H stretching mode. The carbon double-bond stretching mode was not visible in the action spectra.

4. SCANNING TUNNELING SPECTROSCOPY

Scanning tunneling spectroscopy (STS) has been used to characterize the electronic and vibrational properties of surfaces and adsorbed molecules. The advantage of STS lies in its ability to probe single atoms and molecules; however, it can be difficult to obtain STS spectra that are unaffected by outside influences. In many cases, STS measurements are taken in ultrastable instruments in ultrahigh vacuum (UHV) so as to isolate the sample from acoustic and vibrational noise; this is done at cryogenic temperatures to reduce the thermal spread of electron energies. To perform spectroscopy using the STM tunneling junction, the tunneling gap is typically held at a constant set point by interrupting the FBL while a voltage ramp is applied and the tunneling current is measured as a function of the applied bias voltage. The first and second derivatives of the conductance (I/V) can be measured using a lock-in amplifier (LIA); this results in differential conductance (dI/dV), which can be normalized to approximate the LDOS and inelastic scanning tunneling spectroscopy (d^2I/dV^2). The resulting peaks can sometimes be assigned as vibrational modes of adsorbed molecules, or for atomic-scale magnetic structures, they can be assigned as spin excitation spectra.

Early I/V measurements were performed on semiconducting surfaces (63–68). Conductance measurements were shown to depend on tip-sample separation (63, 64). Avouris and Lyo used this distance dependence to tune the interactions between tip and substrate, thereby gaining an understanding of reaction pathways for the oxidation of Si{111} 7 × 7 and manipulating adsorbates (63, 65). Further I/V measurements recorded over doped silicon surfaces displayed negative differential resistance over locations containing dopant molecules (66, 67).

Small changes in the slope of I/V spectra (**Figure 5a**) arise from changes in the LDOS of the tip and sample and can be measured through the differential

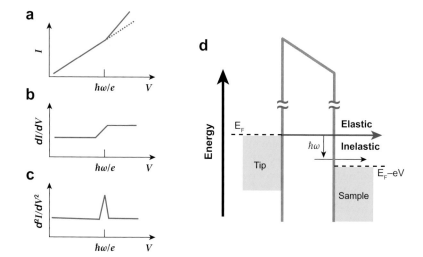

Figure 5

(*a*) Scanning tunneling
spectroscopy (STS) for
current versus voltage (*I/V*)
measuring conductance.
(*b*) Differential conductance
(*dI/dV*) measuring the local
density of states. (*c*) Inelastic
STS (*d²I/dV²*) measuring
molecular vibrations.
(*d*) Energy level diagram for
inelastic STS. When the
electron energy rises above
$\hbar\omega$, inelastic tunneling
channels are accessible in
the molecule, leading to
molecular vibrations.

conductance (**Figure 5*b***). The current dependence on *z* may cause these features
to be obscured, but they can be normalized using the ratio of the differential conductance to the (integral) conductance (*dI/dV/I/V*) (64, 69). Eigler and colleagues
showed that for Fe adsorbed onto Pt{111} using three different tips, the normalized
differential conductance spectra could be used to identify single Fe adatoms. The
authors demonstrated this by acquiring spectra over both the Fe adatoms and the
bare Pt{111}; peaks displayed in the Fe spectra and not the Pt{111} spectra were
attributed to the Fe adatoms and were consistent with the expected state density
(70).

Inelastic electron tunneling spectroscopy (IETS), measured as d^2I/dV^2 versus
V spectra, produces vibrational spectra for molecules on surfaces. IETS spectra
(**Figure 5*c***) are measured using phase-sensitive detection of the tunneling current
at the second harmonic of the modulated bias voltage using a LIA (71–73). Such
IETS spectra have been difficult to obtain due to the need for extreme mechanical
stability to isolate the system acoustically and vibrationally (74), and for cryogenic
temperatures to reduce the thermal spread of electronic energies (71–73). As shown
schematically in **Figure 5*d***, the tunneling electrons lose energy ($\hbar\omega$) to vibrational
modes of the molecules within the tunneling junction, giving rise to an inelastic
tunneling channel. Examples include acetylene on Cu{100} (71), CO in constructed
Fe(CO) species (72), and substrate-adsorbate vibrations for benzene on Ag{100} (73).
Although IETS measurements for vibrational spectra have been performed, the selection rules and phase differences between elastic and inelastic channels are not yet
understood (71–77). Furthermore, IETS measurements have been recorded for spin
excitation spectra using a low-temperature UHV STM coupled with a high magnetic
field. Tunneling electrons in this system lose energy to spin-flip excitations of single (or chains of) magnetic atoms within the tunneling junction with respect to the
magnetic field applied to the sample (78–80).

IETS: inelastic electron
tunneling spectroscopy

4.1. Scanning Tunneling Spectroscopy of Local Electronic Structure

With micro- and nanoelectronics reaching ever smaller scales, and with the ultimate limit of single molecules acting as electronic components, it has become important to measure and to compare the conductivities within and among families of single molecules. Highly conjugated molecules such as OPE molecules (described above) and oligo(phenylene-vinylene) (OPV) molecules have been of interest due to their potential use as molecular wires and switches (54, 55, 60, 81). Although the conductivities of alkanethiolates, OPEs, and OPVs can be measured as full monolayers (82–84), it is interesting to compare these measurements to the conductances of single molecules of the same species (81). Blum et al. used STS to measure the conductances of isolated α,ω-dithiols of dodecane, OPE, and OPV molecules inserted into host undecanethiolate monolayers. After insertion, the pendant thiols of the inserted molecules were exposed to solutions of gold nanoparticles, creating a second contact for each inserted molecule. Conductance measurements were obtained, showing that OPVs were the most conductive, followed by OPE molecules, with alkanethiolates the least conductive, thus scaling with bulk measurements over hundreds of molecules (81), consistent with electrochemical conductance measurements (85–87).

Ligand-stabilized metal nanoparticle systems represent another area in which charge transport dynamics are of interest (88–92). These systems, similar to the conductance measurements described above, contain a dithiol linker to tether a nanoparticle to a substrate. Isolated or tethered nanoparticles are able to exhibit single-electron transfer, and the energy required to add an extra electron to the particle is greater than $k_B T$, where k_B is Boltzmann's constant and T is absolute temperature (88, 93, 94). The energy required to charge a metallic nanoparticle with a single electron is inversely proportional to the particle's capacitance:

$$E_{charge} = \frac{e^2}{2C},$$

where $e = 1.602 \times 10^{-19}$ C. The particle's capacitance can be calculated as

$$C = 4\pi\varepsilon\varepsilon_0 r \left(1 + r/2L\right),$$

where ε is the dielectric constant of the ligand shell, ε_0 is the dielectric constant of vacuum, r is the radius of the particle, and L is the thickness of the tunneling junction (93–95).

A single nanoparticle addressed by STM can exhibit Coulomb blockade, which is the increase of the differential resistance near zero bias, and Coulomb staircase, which is quantized electron charging of the particle, both of which can be observed in the I/V spectra (89, 93–95). The potential around the Coulomb blockade region where current begins to flow is known as the threshold voltage. Coulomb blockade occurs in tunneling junctions and is described as resistor-capacitor circuits in series modeling two tunneling barriers controlling the electron transport through this junction; the first barrier is between the probe tip and the cluster, and the second barrier is between the cluster and the substrate. We have analyzed precise structures of isolated, ligand-stabilized undecagold clusters (Au$_{11}$) stabilized by triphenylphosphine ligands

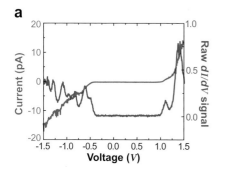
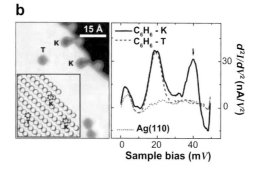

Figure 6

(*a*) Current versus voltage (*I/V*) exhibiting Coulomb blockade and differential conductance (*dI/dV*) displaying single-electron charging plots for a Au_{11} cluster bound to a Au{111} substrate via a decanedithiolate tether. Significant spectral diffusion is observed between spectra for these clusters. (*b*) *Left:* Scanning tunneling microscopy image of benzene adsorbed on Ag{110} (70 Å × 54 Å; V_{sample} = 0.1 V; I_{tunnel} = 1.6 nA). *Inset:* Model of the proposed location of the labeled molecules. *Right:* Inelastic scanning tunneling spectroscopy acquired over benzene molecules at K sites (*solid black line*), at T sites (*dotted red line*), and over the Ag{110} (*dotted green line*) (V_{sample} = 0.1 V; I_{tunnel} = 2 nA; V_{AC} = 4 mV rms at a frequency of 641 Hz). Reprinted with permission from Reference 73. Copyright 2001, the American Physical Society.

immobilized on octanethiolate self-assembled monolayers via α,ω-decanedithiolate tethers (89). Using ultrastable UHV STM operating at cryogenic temperatures, we were able to record high-resolution images as well as current-voltage and differential conductance spectra over the nanoparticle assemblies (**Figure 6a**). Interestingly, for a single tethered Au_{11} cluster, as well as across many clusters, significant spectral diffusion was observed, including through the Coulomb blockade region. This study demonstrates that even in a well-defined, carefully controlled system, the electronic structure is still complex (89).

4.2. Scanning Tunneling Spectroscopy as a Local Vibrational Spectroscopy

Local vibrational spectroscopy was first demonstrated by Ho and colleagues, who measured isolated acetylene (C_2H_2) and deuterated acetylene (C_2D_2) on Cu{100} surfaces (71). They found a peak in the d^2I/dV^2 spectrum at 358 mV corresponding to the C-H stretching mode. Furthermore, taking vibrational spectra over the C_2D_2 resulted in the expected isotopic shift of the peak for the C-D stretching mode to 266 mV (71). The authors were also able to show hindered rotation of CO on Cu{100} and Cu{110} (76). They constructed mono- and dicarbonyl species of Fe on Ag{110} and measured the concomitant shifts in the C-O stretch for these species (72).

Rust and colleagues used IETS to study benzene molecules adsorbed to Ag{110} not only to identify the molecule on the surface, but to elucidate each molecule's chemical state and local environment. Using VM, benzene molecules were picked up with the STM tip and imaging was performed with the tip-molecule system. Benzene molecules were found to be adsorbed to the Ag{110} surface at atop sites. Vibrational IETS spectra were recorded for isolated chemisorbed benzene molecules on terraces and at step edges, as well as for benzene adsorbed as a dense hexagonally packed ordered monolayer, which hindered lateral movement of the benzene molecules. For isolated benzene on a terrace, IETS peaks appeared at 4 and 19 mV due to excitation of adsorbate-substrate vibrational modes. When the benzene molecules were hexagonally packed, these peaks shifted to 7 and 44 mV, showing that IETS is sensitive to the chemical environment of the adsorbate. Furthermore, when comparing IETS spectra of molecules adsorbed on a terrace (T) to spectra of molecules adsorbed near a substrate step edge (K), the latter exhibited an additional peak at 39 mV, a vibration not visible at other sites. In the STM images (**Figure 6b**), molecules adsorbed at the K sites had an asymmetric shape due to their interactions with nearby kink sites, which was believed to contribute to the additional vibrational peak (73).

4.3. Scanning Tunneling Spectroscopy of Local Magnetic Structure

To record spin-excitation spectra, Mn atoms were adsorbed to NiAl(110) surfaces containing Al_2O_3 islands (78). Differential conductance spectra were recorded over the individual Mn atoms. After applying a magnetic field, the Mn adsorbed on the oxide islands exhibited a dip in the differential conductance spectra centered at zero bias. The Al_2O_3 insulation layer was necessary to isolate the Mn atoms from the conducting surface for the IETS spectra to be observed. The IETS spin-flip signature was observed as a step in the *dI/dV* spectra corresponding to the spin-flip transistion. The observed conductance steps shifted to higher energy with increased magnetic field strength (78).

Further experiments used STM manipulation to create linear chains of 1 to 10 Mn atoms over which IETS spin-excitation spectra were recorded (79). In this study, the Mn atoms were deposited on a Cu{100} surface containing insulating islands of copper nitride. The *dI/dV* spectra recorded for the chains displayed dependence on the parity of the chains (i.e., odd-length chains exhibited a dip around zero bias, whereas this dip was absent from the even-length chain spectra). A spin-flip excitation required a total spin greater than zero, which was available in the odd-length chains. In the even-length chains, however, the ground state spin was zero. Both odd- and even-length chains displayed steps in the *dI/dV* spectra at larger energies. These steps corresponded to the spin-changing transitions from the ground state (singlet) to an excited state (triplet) (79).

In addition to the Mn chains, individual Fe and Mn atoms were deposited on CuN islands on Cu{100} to determine the magnetic anisotropy of single atomic spins (80). For Fe atoms at zero-field, a strong excitation was observed, indicating strong magnetic anisotropy in the system—stronger than that observed for the Mn atoms. Using STM manipulation, the Fe and Mn atoms were positioned so that the magnetic field was oriented along two different spatial directions. The excitation energies for the spectral steps changed depending on the binding site location, and the Fe and Mn atoms could be manipulated between the two binding sites, thereby switching between the excitation spectra. The dependence of the spin excitations on the field direction represented further evidence of strong magnetic anisotropy (80).

5. SUBSTRATE-MEDIATED INTERACTIONS

An important and beautiful aspect of STM is that it can be used to image specific empty and filled orbitals in much the way that a mobile adsorbate might "see" them. This enables one to probe the local electronic structures that drive the chemistry, dynamics, and other surface processes. Substrate-mediated interactions have been shown theoretically and experimentally to influence adsorbate bonding and dynamics at ranges far beyond chemical bonding distances (96–98). Surface-state electrons of close-packed surfaces of noble metals are known to behave as two-dimensional electron gases (99–101). These surface-state electrons are scattered by substrate steps, defects, and adsorbates, creating standing waves that can mediate interactions between adsorbates (17, 102–110). Here, we describe how the STM is used to map these electronic perturbations and how they are used to understand reactions and motion at the single-molecule level.

5.1. Surface Electronic Structure Perturbation

Figure 7 illustrates three ways in which the electronic structure of a surface may be perturbed. The strongest of these (**Figure 7*a***), known as the Smoluchowski effect, occurs at substrate step edges, where the electron distribution at the step is smoothed by a charge transfer from the top to the bottom of the step edge (111). This effect is illustrated in **Figure 7*a*** (*bottom*), wherein benzene molecules are shown to have adsorbed in two rows, one at the top and one at the bottom of a Cu{111} step at 77 K (102, 105). A second, intermediate effect (**Figure 7*b***) is the adsorption of atoms on the surface perturbing the local surface electronic structure (15, 102, 104, 105). The STM image (**Figure 7*b***, *bottom*) shows an isolated benzene molecule on Pt{111}. The LDOS surrounding this molecule is perturbed several substrate lattice sites away from the molecule. Note that this effect is also seen in **Figure 7*a*** (*bottom*), where these local perturbations constructively and destructively interfere. The third and weakest effect (**Figure 7*c***, *top*) on the electronic structure is the scattering of surface-state electrons from features such as step edges, surface defects, and adsorbates (**Figure 7*c***, *bottom*) with scattering from Br islands on Cu{111} (99, 102, 112–114).

Electronic structures are imaged with STM at biases close to the Fermi level or through differential conductance imaging. At low biases, the STM images approximate the LDOS at the Fermi level (9, 99). Using *dI/dV* imaging, the Avouris and Eigler groups found that the wavelength of the oscillation in the surface LDOS changed as a function of energy; they subsequently mapped the (previously known) dispersion curves (99, 113).

Lau and Kohn theoretically predicted that long-range interactions were mediated by partially filled surface-state bands leading to oscillatory interaction potentials, known as Friedel oscillations, with periods of half the Fermi wavelength ($\lambda_F/2$) of the substrate (115). These results were initially confirmed with field ion microscopy (97).

Friedel oscillations: the distribution of electrons around a surface impurity

Fermi wavelength: the wavelength of the carriers that dominate electrical transport

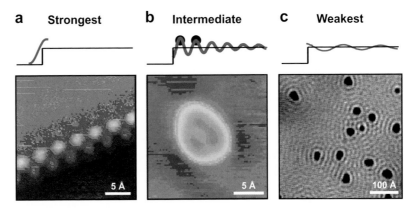

a Strongest   **b** Intermediate   **c** Weakest

5 Å   5 Å   100 Å

Figure 7

Substrate-mediated interactions on surfaces. (*a*) The strongest interactions occur at step edges where the electron distribution is smoothed by a charge transfer from the top to the bottom of the step edge. The scanning tunneling microscopy (STM) image shows benzene on Cu{111} (30 Å × 30 Å; V_{sample} = 100 mV; I_{tunnel} = 100 pA). Reprinted with permission from Reference 17. Copyright 2003, the American Chemical Society. (*b*) An intermediate substrate-mediated interaction occurs when atoms adsorb onto a surface, perturbing the local surface electronic structure. The STM image shows a single benzene atom as a three-lobed feature on Pt{111} with local electronic perturbations, imaged as depressions along the direction of the lobes of the benzene molecules (20 Å × 20 Å; V_{sample} = 0.05 V; I_{tunnel} = 100 pA). Reprinted with permission from Reference 15. Copyright 1993, the American Physical Society. (*c*) The weakest of the substrate-mediated interactions is the scattering of the electrons in the surface state from surface features such as step edges. The STM image shows Br adatom islands (*depressions*) on Cu{111} with local electronic perturbations imaged close to the Fermi level (420 Å × 420 Å; V_{sample} = −50 mV; I_{tunnel} = 20 pA). Reprinted with permission from Reference 104. Copyright 2007, the American Physical Society.

Later, with STM, single adatoms of Cu and Co were studied for their contributions to Friedel oscillations; it was found that the substrate surface-state electrons were most important to long-range interactions, as opposed to the identity of the adsorbate (109, 110). The coherence length (λ_C) for the surface-state electron standing wave pattern is inversely proportional to the temperature (T) of the surface and can be calculated using the following equation:

$$\lambda_C \cong \frac{\hbar^2 k_F}{3.5m^* k_B T},$$

where k_F is the Fermi level wave vector of the surface state, k_B is Boltzmann's constant, and m^* is the effective mass of a surface-state electron (116, 117). This yields an estimate of the distance at which molecules can communicate across a surface.

Coherence length: the propagation distance over which a wave maintains its ability to exhibit interference

5.2. Substrate-Mediated Interactions between Adsorbed Molecules

The high symmetry, relatively simple electronic structure, and high STM contrast resolution of benzene have made it a useful probe molecule for studying substrate-adsorbate interactions (73, 102, 105, 118, 119). At low temperatures, the rate of diffusion for benzene molecules can be lowered sufficiently for individual molecules to be observed. Isolated benzene adsorbed on Pt{111} displays three distinct molecular shapes, which are attributed to benzene lying flat at three different surface sites and are due to changes in the interactions between the adsorbed molecules and the surface electronic states (15, 120). As shown in **Figure 7a**, when benzene is adsorbed on Cu{111} it decorates both the tops and bottoms of monatomic step edges, perturbing the surface electronic structure and thereby setting up adsorption sites for subsequent rows of molecules (102, 105). When a higher coverage of benzene is deposited on the Cu{111} surface and the bias voltage is increased, the STM tip sweeps benzene from the terrace to the step, resulting in the third and fourth "phantom" rows observed in the STM images (**Figure 8a**). These rows appear noisy in STM images, indicating that benzene molecules diffuse into and out of, or between, binding sites created by the adsorption of the first row of molecules. This diffusion occurs faster than the time scale of the STM images (17, 105). Benzene adsorbed on Au{111} at low coverage (0.1 monolayer) shows adsorption only at surface defect sites and at the top of Au{111} step edges, indicating that benzene seeks areas of high empty-state density, consistent with its nucleophilicity (121). Finally, if benzene is adsorbed on Ag{110} at low coverage, molecules only adsorb at the [1$\bar{1}$0] step edges (**Figure 8b**) (73, 118, 122) due to the necessity of available free charge and free sites for the charge to occupy. For Ag{110}, anisotropy exists in the bulk density of states leading to reduction in the charge transfer for [001] steps, thereby reducing the ability of the step sites to accept charge from the benzene (123). Only the [1$\bar{1}$0] steps, which are the ends of silver closed-packed atomic rows, allow adsorption (122).

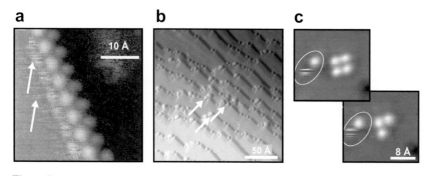

Figure 8

(*a*) Scanning tunneling microscopy (STM) image of benzene adsorbed at a Cu{111} step edge. The STM tip has moved additional benzene molecules to the second and third rows above the step (*white arrows*) ($V_{\text{sample}} = 100$ mV; $I_{\text{tunnel}} = 100$ pA). Reprinted with permission from Reference 105. Copyright 1995, Elsevier. (*b*) STM image of benzene deposited on Ag{110} at 66 K. Benzene molecules adsorb only on the [1$\bar{1}$0] step edges (*white arrows*) ($V_{\text{sample}} = 0.05$ V; $I_{\text{tunnel}} = 4$ nA). Reprinted with permission from Reference 17. Copyright 2003, the American Chemical Society. (*c*) STM images showing clusters of phenyl intermediates on Cu{111} at 77 K ($V_{\text{sample}} = 200$ mV; $I_{\text{tunnel}} = 80$ pA). Reprinted with permission from Reference 17. Copyright 2003, the American Chemical Society.

The surface-catalyzed formation of biphenyl from halobenzene molecules adsorbed on Cu{111} proceeds through a series of steps. First, the halobenzene (e.g., C_6H_5Br, C_6H_5I) adsorbs on the copper surface, dissociating above 180 K to form surface-bound halogen atoms and phenyl intermediates (124). Between 180 K and 300 K, the phenyl groups diffuse on the surface and above 300 K they react, forming biphenyl. Rieder and colleagues used manipulation techniques to control the dissociation, position, orientation, and coupling of iodobenzene molecules adsorbed adjacent to Cu{111} steps in each step of the reaction (125). We exposed a Cu{111} crystal to iodobenzene at 293 K, then cooled the crystal to 77 K for STM measurements (119). At this temperature, phenyl binds to both step edges and at terrace sites with moderate diffusion on open terraces. Phenyl pairs are observed to form and to break, as shown in **Figure 8c**. This indicates relatively weak intermolecular interactions and suggests that the intermediates have not coupled to form covalent bonds. The geometry of the dimer indicates that the phenyl unsatisfied bonds are facing either towards or away from one another. The role of the substrate-mediated interaction is twofold: The surface is needed to create the phenyl intermediates and the phenyl species on the surface experience intermolecular forces due to loosely bound complexes that position the phenyl intermediates in close proximity and prealign them for the reaction to form biphenyl.

The atomic halide intermediates in this reaction were also studied (104). In this experiment, bromobenzene was adsorbed to a Cu{111} surface at 293 K, and Br adatom islands were formed by ramping the substrate temperature to 600 K. At 4 K, the standing wave interference patterns of the surface state scattering from the Br adatom islands were imaged using STM, and the long-range interactions were quantified by determining the interisland distances. The interaction potential (E_r) between the islands was determined as

$$E_r = -k_B T \ln[(g(r))],$$

where k_B is Boltzmann's constant, T is the temperature at which the islands were formed, and $g(r)$ is the pair correlation function, which was extracted by dividing the experimental pair distribution function by the theoretical function for similar noninteracting species. The coherence length of the Cu was calculated to be $\sim$22 Å, which is consistent with the elevated temperature at which the Br atoms are mobile and form islands. This analysis showed the oscillations in the interisland distance interaction potential to have a periodicity of ($\lambda_F/2$) of Cu{111} up to a distance of $\sim$56 Å (104).

6. CONCLUSIONS AND FUTURE PROSPECTS

STM has been developed into an analytical technique not only capable of imaging compositions and structures at surfaces, but also useful for understanding binding sites and reaction processes. Here, we have presented an overview of STM image interpretation, STM atom and molecule manipulation, STS, and substrate-mediated interactions. Future work will soon make STM an even more powerful analytical technique by simultaneously coupling secondary signals, including photons and microwaves, into and out of the STM tunneling junction (126–130). Early results in these areas bode well for future applications of STM.

SUMMARY POINTS

1. STM images are a convolution of topography and electronic structure, which must be understood in order to interpret images.

2. Atomic and molecular manipulations can be used to elucidate the properties of a surface and adsorbates including the electronic structure and the vibrational modes of molecules at the surface.

3. STS gives insights into the conductances of molecules, the LDOS of the surface and adsorbates, and the vibrational modes of molecules.

4. Substrate-mediated interactions modulate adsorbate potentials, determine adsorption sites, and guide adsorbates and reaction pathways for catalytic reactions.

FUTURE ISSUES

1. Selection rules for IETS are still being determined.

2. Coupling secondary signals into the scanning tunneling microscope tunneling junction will lead to unprecedented abilities to characterize systems at the single-molecule level.

DISCLOSURE STATEMENT

The authors are not aware of any biases that might be perceived as affecting the objectivity of this review.

ACKNOWLEDGMENTS

The authors gratefully acknowledge support from the Air Force Office of Scientific Research, the Army Research Office, the Defense Advanced Research Projects Agency, the National Institutes for Standards and Technology, the National Science Foundation, and the Office of Naval Research.

LITERATURE CITED

1. Binnig G, Rohrer H, Gerber C, Weibel E. 1982. Surface studies by scanning tunneling microscopy. *Phys. Rev. Lett.* 49:57–61

2. Bonnell DA, ed. 2001. *Scanning Probe Microscopy and Spectroscopy: Theory, Techniques, and Applications.* New York: Wiley

3. Chen CJ. 1993. *Introduction to Scanning Tunneling Microscopy*. New York: Oxford Univ. Press

4. Stroscio JA, Kaiser WJ. 1993. *Scanning Tunneling Microscopy*. San Diego: Academic

5. Binnig G, Rohrer H. 1986. Scanning tunneling microscopy. *IBM J. Res. Develop.* 30:355–69

6. Binnig G, Rohrer H. 1987. Scanning tunneling microscopy: from birth to adolescence. *Rev. Mod. Phys.* 59:615–25

7. Griffith JE, Kochanski GP. 1990. Scanning tunneling microscopy. *Annu. Rev. Mater. Sci.* 20:219–44

8. Coombs JH, Welland ME, Pethica JB. 1988. Experimental barrier heights and the image potential in scanning tunneling microscopy. *Surf. Sci.* 198:L353–58

9. Tersoff J, Hamann DR. 1985. Theory of the scanning tunneling microscope. *Phys. Rev. B* 31:805–13

10. Garcia N. 1986. Theory of scanning tunneling microscopy and spectroscopy: resolution, image and field states, and thin oxide layers. *IBM J. Res. Develop.* 30:533–42

11. Eigler DM, Weiss PS, Schweizer EK. 1991. Imaging Xe with a low-temperature scanning tunneling microscope. *Phys. Rev. Lett.* 66:1189–91

12. Tersoff J, Hamann DR. 1983. Theory and application for the scanning tunneling microscope. *Phys. Rev. Lett.* 50:1998–2001

13. Lang ND. 1986. Theory of single-atom imaging in the scanning tunneling microscope. *Phys. Rev. Lett.* 56:1164–67

14. Lang ND. 1987. Apparent size of an atom in the scanning tunneling microscope as a function of bias. *Phys. Rev. Lett.* 58:45–48

15. Weiss PS, Eigler DM. 1993. Site dependence of the apparent shape of a molecule in scanning tunneling microscope images: benzene on Pt{111}. *Phys. Rev. Lett.* 71:3139–42

16. Kushmerick JG, Weiss PS. 1998. Mobile promoters on anisotropic catalysts: nickel on MoS_2. *J. Phys. Chem. B* 102:10094–97

17. Sykes ECH, Han P, Kandel SA, McCarty GS, Weiss PS. 2003. Substrate-mediated interactions and intermolecular forces between molecules adsorbed on surfaces. *Acc. Chem. Res.* 36:945–53

18. Feenstra RM, Stroscio JA, Tersoff J, Fein AP. 1987. Atom-selective imaging of the GaAs(110) surface. *Phys. Rev. Lett.* 58:1192–95

19. Lubinsky AR, Duke CB, Lee BW, Mark P. 1976. Semiconductor surface reconstruction: the rippled geometry of GaAs(110). *Phys. Rev. Lett.* 36:1058–61

20. Weiss PS, Eigler DM. 1992. Adsorption and accommodation of Xe on Pt{111}. *Phys. Rev. Lett.* 69:2240–43

21. Weiss PS, Eigler DM. 1993. What is underneath? Moving atoms and molecules to find out. In *NATO ASI Series E: Applied Sciences*, ed. T Binh, N Garcia, K Dransfeld, 213–17. Amsterdam: Kluwer Academic

22. Hla SW. 2005. Scanning tunneling microscopy single atom/molecule manipulation and its application to nanoscience and technology. *J. Vac. Sci. Technol. B* 23:1351–60

23. Manoharan HC, Lutz CP, Eigler DM. 2000. Quantum mirages formed by coherent projection of electronic structure. *Nature* 403:512–15

24. Eigler DM, Schweizer EK. 1990. Positioning single atoms with a scanning tunneling microscope. *Nature* 344:524–26

25. Jensen SC, Baber AE, Tierney HL, Sykes ECH. 2007. Adsorption, interaction, and manipulation of dibutyl sulfide on Cu{111}. *ACS Nano* 1:22–29

26. Cuberes MT, Schlittler RR, Gimzewski JK. 1996. Room-temperature repositioning of individual C_{60} molecules at Cu steps: operation of a molecular counting device. *Appl. Phys. Lett.* 69:3016–18

27. Jung TA, Schlittler RR, Gimzewski JK, Tang H, Joachim C. 1996. Controlled room-temperature positioning of individual molecules: molecular flexure and motion. *Science* 271:181–84

28. Sloan PA, Palmer RE. 2005. Two-electron dissociation of single molecules by atomic manipulation at room temperature. *Nature* 434:367–71

29. Shirai Y, Osgood AJ, Zhao Y, Kelly KF, Tour JM. 2005. Directional control in thermally driven nanocars. *Nano Lett.* 5:2330–34

30. Griessl SJH, Lackinger M, Jamitzky F, Markert T, Hietschold M, et al. 2004. Room-temperature scanning tunneling microscopy manipulation of single C-60 molecules at the liquid-solid interface: playing nanosoccer. *J. Phys. Chem. B* 108:11556–60

31. Bartels L, Meyer G, Rieder KH. 1997. Basic steps of lateral manipulation of single atoms and diatomic clusters with a scanning tunneling microscope tip. *Phys. Rev. Lett.* 79:697–700

32. Gross L, Moresco F, Alemani M, Tang H, Gourdon A, et al. 2003. Lander on Cu(211)-selective adsorption and surface restructuring by a molecular wire. *Chem. Phys. Lett.* 371:750–56

33. Hla SW, Braun KF, Wassermann B, Rieder KH. 2004. Controlled low-temperature molecular manipulation of sexiphenyl molecules on Ag(111) using scanning tunneling microscopy. *Phys. Rev. Lett.* 93:208302

34. Joachim C, Tang H, Moresco F, Rapenne G, Meyer G. 2002. The design of a nanoscale molecular barrow. *Nanotechnology* 13:330–5

35. Lagoute J, Kanisawa K, Fölsch S. 2004. Manipulation and adsorption-site mapping of single pentacene molecules on Cu(111). *Phys. Rev. B* 70:245415

36. Meyer G, Zöphel S, Rieder KH. 1996. Scanning tunneling microscopy manipulation of native substrate atoms: a new way to obtain registry information on foreign adsorbates. *Phys. Rev. Lett.* 77:2113–16

37. Moresco F, Meyer G, Rieder KH, Tang H, Gourdon A, et al. 2001. Recording intramolecular mechanics during the manipulation of a large molecule. *Phys. Rev. Lett.* 87:088302

38. Stroscio JA, Celotta RJ. 2004. Controlling the dynamics of a single atom in lateral atom manipulation. *Science* 306:242–47

39. Stroscio JA, Eigler DM. 1991. Atomic and molecular manipulation with the scanning tunneling microscope. *Science* 254:1319–26

40. Hla SW, Braun KF, Rieder KH. 2003. Single-atom manipulation mechanics during a quantum corral construction. *Phys. Rev. B* 67:201402

41. Eigler DM, Lutz CP, Rudge WE. 1991. An atomic switch realized with the scanning tunneling microscope. *Nature* 352:600–3

42. Bartels L, Meyer G, Rieder KH. 1997. Controlled vertical manipulation of single CO molecules with the scanning tunneling microscope: a route to chemical contrast. *Appl. Phys. Lett.* 71:213–15

43. Bartels L, Meyer G, Rieder KH. 1998. Dynamics of electron-induced manipulation of individual CO molecules on Cu(111). *Phys. Rev. Lett.* 80:2004–7

44. Dujardin G, Mayne A, Robert O, Rose F, Joachim C, et al. 1998. Vertical manipulation of individual atoms by a direct STM tip-surface contact on Ge(111). *Phys. Rev. Lett.* 80:3085–88

45. Kelly KF, Sarkar D, Hale GD, Oldenburg SJ, Halas NJ. 1996. Threefold electron scattering on graphite observed with C_{60}-adsorbed STM tips. *Science* 273:1371–73

46. Komeda T, Kim Y, Kawai M, Persson BNJ, Ueba H. 2002. Lateral hopping of molecules induced by excitation of internal vibration mode. *Science* 295:2055–58

47. Lauhon LJ, Ho W. 2000. Control and characterization of a multistep unimolecular reaction. *Phys. Rev. Lett.* 84:1527–30

48. Stipe BC, Rezaei MA, Ho W. 1997. Single-molecule dissociation by tunneling electrons. *Phys. Rev. Lett.* 78:4410–13

49. Fernández-Torres LC, Sykes ECH, Nanayakkara SU, Weiss PS. 2006. Dynamics and spectroscopy of hydrogen atoms on Pd{111}. *J. Phys. Chem. B* 110:7380–84

50. Sykes ECH, Fernández-Torres LC, Nanayakkara SU, Mantooth BA, Nevin RM, et al. 2005. Observation and manipulation of subsurface hydride in Pd{111} and its effect on surface chemical, physical and electronic properties. *Proc. Natl. Acad. Sci.* 102:17907–11

51. Kühnle A, Meyer G, Hla SW, Rieder KH. 2002. Understanding atom movement during lateral manipulation with the STM tip using a simple simulation method. *Surf. Sci.* 499:15–23

52. Mizes HA, Foster JS. 1989. Long-range electronic perturbations caused by defects using scanning tunneling microscopy. *Science* 244:559–62

53. Kushmerick JG, Kelly KF, Rust H-P, Halas NJ, Weiss PS. 1999. Observations of anisotropic electron scattering on graphite with a low-temperature scanning tunneling microscope. *J. Phys. Chem. B* 103:1619–22

54. Donhauser ZJ, Mantooth BA, Kelly KF, Bumm LA, Monnell JD, et al. 2001. Conductance switching in single molecules through conformational changes. *Science* 292:2303–07

55. Donhauser ZJ, Mantooth BA, Pearl TP, Kelly KF, Nanayakkara SU, et al. 2002. Matrix-mediated control of stochastic single molecule conductance switching. *Jpn. J. Appl. Phys.* 41:4871–77

56. Lewis PA, Inman CE, Maya F, Tour JM, Hutchison JE, et al. 2005. Molecular engineering of the polarity and interactions of molecular electronic switches. *J. Am. Chem. Soc.* 127:17421–28

57. Lewis PA, Inman CE, Yao YX, Tour JM, Hutchison JE, et al. 2004. Mediating stochastic switching of single molecules using chemical functionality. *J. Am. Chem. Soc.* 126:12214–15

58. Moore AM, Dameron AA, Mantooth BA, Smith RK, Fuchs DJ, et al. 2006. Molecular engineering and measurements to test hypothesized mechanisms in single molecule conductance switching. *J. Am. Chem. Soc.* 128:1959–67

59. Dameron AA, Ciszek JW, Tour JM, Weiss PS. 2004. Effect of hindered internal rotation on packing and conductance of self-assembled monolayers. *J. Phys. Chem. B* 108:16761–67

60. Bumm LA, Arnold JJ, Cygan MT, Dunbar TD, Burgin TP, et al. 1996. Are single molecular wires conducting? *Science* 271:1705–7

61. Sainoo Y, Kim Y, Okawa T, Komeda T, Shigekawa H, et al. 2005. Excitation of molecular vibrational modes with inelastic scanning tunneling microscopy processes: examination through action spectra of *cis*-2-butene on Pd(110). *Phys. Rev. Lett.* 95:246102

62. Kawai M, Komeda T, Kim Y, Sainoo Y, Katano S. 2004. Single-molecule reactions and spectroscopy via vibrational excitation. *Phil. Trans. R. Soc. Lond. A* 362:1163–71

63. Avouris P, Lyo IW. 1991. Probing and inducing surface chemistry on the atomic scale using the STM. *Am. Inst. Phys. Conf. Proc.* 241:283–97

64. Stroscio JA, Feenstra RM, Fein AP. 1986. Electronic structure of the Si(111)-2 × 1 surface by scanning tunneling microscopy. *Phys. Rev. Lett.* 57:2579–82

65. Avouris P, Lyo IW. 1991. Probing and inducing surface chemistry with the STM: the reactions of Si(111)-7 × 7 with H_2O and O_2. *Surf. Sci.* 242:1–11

66. Lyo IW, Avouris P. 1989. Negative differential resistance on the atomic scale: implications for atomic scale devices. *Science* 245:1369–71

67. Bedrossian P, Chen DM, Mortensen K, Golovchenko J. 1989. Demonstration of the tunnel-diode effect on an atomic scale. *Nature* 342:258–60

68. Feenstra RM. 1989. Electronic states of metal atoms on the GaAs(110) surface studied by scanning tunneling microscopy. *Phys. Rev. Lett.* 63:1412–15

69. Lang ND. 1986. Spectroscopy of single atoms in the scanning tunneling microscope. *Phys. Rev. B* 34:5947–50

70. Crommie MF, Lutz CP, Eigler DM. 1993. Spectroscopy of a single adsorbed atom. *Phys. Rev. B* 48:2851–54

71. Stipe BC, Rezaei MA, Ho W. 1998. Single-molecule vibrational spectroscopy and microscopy. *Science* 280:1732–35

72. Lee HJ, Ho W. 2000. Structural determination by single-molecule vibrational spectroscopy and microscopy: contrast between copper and iron carbonyls. *Phys. Rev. B* 61:R16347–50

73. Pascual JI, Jackiw JJ, Song Z, Weiss PS, Conrad H, et al. 2001. Adsorbate-substrate vibrational modes of benzene on Ag(110) resolved with scanning tunneling spectroscopy. *Phys. Rev. Lett.* 86:1050–53

74. Ferris JH, Kushmerick JG, Johnson JA, Yoshikawa Youngquist MG, Kessinger RB, et al. 1998. Design, operation, and housing of an ultrastable, low-temperature, ultrahigh vacuum scanning tunneling microscope. *Rev. Sci. Instrum.* 69:2691–95

75. Stipe BC, Rezaei MA, Ho W. 1998. Coupling of vibrational excitation to the rotational motion of a single adsorbed molecule. *Phys. Rev. Lett.* 81:1263–66

76. Lauhon LJ, Ho W. 1999. Single-molecule vibrational spectroscopy and microscopy: CO on Cu(001) and Cu(110). *Phys. Rev. B* 60:R8525–28

77. Moresco F, Meyer G, Rieder KH. 1999. Vibrational spectroscopy of CO/Cu(211) with a CO terminated tip. *Mod. Phys. Lett. B* 13:709–15

78. Heinrich AJ, Gupta JA, Lutz CP, Eigler DM. 2004. Single-atom spin-flip spectroscopy. *Science* 306:466–69

79. Hirjibehedin CF, Lutz CP, Heinrich AJ. 2006. Spin coupling in engineered atomic structures. *Science* 312:1021–24

80. Hirjibehedin CF, Lin C-Y, Otte AF, Ternes M, Lutz CP, et al. 2007. Large magnetic anisotropy of a single atomic spin embedded in a surface molecular network. *Science* 317:1199–203

81. Blum AS, Yang JC, Shashidhar R, Ratna BR. 2003. Comparing the conductivity of molecular wires with the scanning tunneling microscope. *Appl. Phys. Lett.* 82:3322–24

82. Kushmerick JG, Holt DB, Pollack SK, Ratner MA, Yang JC, et al. 2002. Effect of bond-length alternation in molecular wires. *J. Am. Chem. Soc.* 124:10654–55

83. Kushmerick JG, Naciri J, Yang JC, Shashidhar R. 2003. Conductance scaling of molecular wires in parallel. *Nano Lett.* 3:897–900

84. Kushmerick JG, Allara DL, Mallouk TE, Mayer TS. 2004. Electrical and spectroscopic characterization of molecular junctions. *MRS Bull.* 29:396–402

85. Sikes HD, Smalley JF, Dudek SP, Cook AR, Newton MD, et al. 2001. Rapid electron tunneling through oligophenylenevinylene bridges. *Science* 291:1519–23

86. Dudek SP, Sikes HD, Chidsey CED. 2001. Synthesis of ferrocenethiols containing oligo(phenylenevinylene) bridges and their characterization on gold electrodes. *J. Am. Chem. Soc.* 123:8033–38

87. Sachs SB, Dudek SP, Hsung RP, Sita LR, Smalley JF, et al. 1997. Rates of interfacial electron transfer through π-conjugated spacers. *J. Am. Chem. Soc.* 119:10563–64

88. Simon U. 1998. Charge transport in nanoparticle arrangements. *Adv. Mater.* 10:1487–92

89. Smith RK, Nanayakkara SU, Woehrle GH, Pearl TP, Blake MM, et al. 2006. Spectral diffusion in the tunneling spectra of ligand-stabilized undecagold clusters. *J. Am. Chem. Soc.* 128:9266–67

90. Wang B, Wang H, Li H, Zeng C, Hou JG, et al. 2001. Tunable single-electron tunneling behavior of ligand-stabilized gold particles on self-assembled monolayers. *Phys. Rev. B* 63:035403

91. Xue YQ, Ratner MA. 2003. Microscopic theory of single-electron tunneling through molecular-assembled metallic nanoparticles. *Phys. Rev. B* 68:235410

92. Zhang HJ, Schmid G, Hartmann U. 2003. Reduced metallic properties of ligand-stabilized small metal clusters. *Nano Lett.* 3:305–7

93. Andres RP, Bein T, Dorogi M, Feng S, Henderson JI, et al. 1996. 'Coulomb staircase' at room temperature in a self-assembled molecular nanostructure. *Science* 272:1323–25

94. Dorogi M, Gomez J, Osifchin R, Andres RP, Reifenberger R. 1995. Room-temperature Coulomb blockade from a self-assembled molecular nanostructure. *Phys. Rev. B* 52:9071–77

95. Andres RP, Datta S, Dorogi M, Gomez J, Henderson JI, et al. 1996. Room temperature Coulomb blockade and Coulomb staircase from self-assembled nanostructures. *J. Vac. Sci. Technol. A* 14:1178–83

96. Einstein TL, Schrieffer JR. 1973. Indirect interaction between adatoms on a tight-binding solid. *Phys. Rev. B* 7:3629–48

97. Tsong TT. 1973. Field-ion microscope observations of indirect interaction between adatoms on metal surfaces. *Phys. Rev. Lett.* 31:1207–11

98. Watanabe F, Ehrlich G. 1989. Direct mapping of adatom-adatom interactions. *Phys. Rev. Lett.* 62:1146–49

99. Crommie MF, Lutz CP, Eigler DM. 1993. Imaging standing waves in a two-dimensional electron gas. *Nature* 363:524–27

100. Gartland PO, Slagsvold BJ. 1975. Transitions conserving parallel momentum in photoemission from the (111) face of copper. *Phys. Rev. B* 12:4047–58

101. Memmel N. 1998. Monitoring and modifying properties of metal surfaces by electronic surface states. *Surf. Sci. Rep.* 32:91–163

102. Kamna MM, Stranick SJ, Weiss PS. 1996. Imaging substrate-mediated interactions. *Science* 274:118–19

103. Merrick ML, Luo WW, Fichthorn KA. 2003. Substrate-mediated interactions on solid surfaces: theory, experiment, and consequences for thin-film morphology. *Prog. Surf. Sci.* 72:117–34

104. Nanayakkara SU, Sykes ECH, Fernández-Torres LC, Blake MM, Weiss PS. 2007. Long-range electronic interactions at high temperature: bromine adatom islands on Cu(111). *Phys. Rev. Lett.* 98:206108

105. Stranick SJ, Kamna MM, Weiss PS. 1995. Interactions and dynamics of benzene on Cu{111} at low temperature. *Surf. Sci.* 338:41–59

106. Silly F, Pivetta M, Ternes M, Patthey F, Pelz JP, et al. 2004. Creation of an atomic superlattice by immersing metallic adatoms in a two-dimensional electron sea. *Phys. Rev. Lett.* 92:01610

107. Wahlström E, Ekvall I, Olin H, Wallden L. 1998. Long-range interaction between adatoms at the Cu(111) surface imaged by scanning tunnelling microscopy. *Appl. Phys. A* 66:S1107–10

108. Kulawik M, Rust H-P, Heyde M, Mantooth BA, Weiss PS. 2005. Interaction of CO molecules with surface state electrons on Ag{111}. *Surf. Sci.* 590:L253–58

109. Repp J, Moresco F, Meyer G, Rieder KH, Hyldgaard P, et al. 2000. Substrate mediated long-range oscillatory interaction between adatoms: Cu/Cu(111). *Phys. Rev. Lett.* 85:2981–84

110. Knorr N, Brune H, Epple M, Hirstein A, Schneider MA, et al. 2002. Long-range adsorbate interactions mediated by a two-dimensional electron gas. *Phys. Rev. B* 65:115420–21

111. Smoluchowski R. 1941. Anisotrophy of the electronic work function of metals. *Phys. Rev.* 60:661–74

112. Crommie MF, Lutz CP, Eigler DM. 1993. Confinement of electrons to quantum corrals on a metal surface. *Science* 262:218–20

113. Hasegawa Y, Avouris P. 1993. Direct observation of standing-wave formation at surface steps using scanning tunneling spectroscopy. *Phys. Rev. Lett.* 71:1071–74

114. Avouris P, Lyo IW, Molinás-Mata P. 1995. STM studies of the interaction of surface-state electrons on metals with steps and adsorbates. *Chem. Phys. Lett.* 240:423–28

115. Lau KH, Kohn W. 1978. Indirect long-range oscillatory interaction between adsorbed atoms. *Surf. Sci.* 75:69–85

116. Fujita D, Amemiya K, Yakabe T, Nejoh H, Sato T, et al. 1997. Anisotropic standing-wave formation on an Au(111)-(23 × √3) reconstructed surface. *Phys. Rev. Lett.* 78:3904–7

117. Jeandupeux O, Burgi L, Hirstein A, Brune H, Kern K. 1999. Thermal damping of quantum interference patterns of surface-state electrons. *Phys. Rev. B* 59:15926–34

118. Pascual JI, Jackiw JJ, Kelly KF, Conrad H, Rust H-P, et al. 2000. Local electronic structural effects and measurements on the adsorption of benzene on Ag(110). *Phys. Rev. B* 62:12632–35

119. Weiss PS, Kamna MM, Graham TM, Stranick SJ. 1998. Imaging benzene molecules and phenyl radicals on Cu{111}. *Langmuir* 14:1284–89

120. Sautet P, Bocquet ML. 1994. A theoretical analysis of the site dependence of the shape of a molecule in STM images. *Surf. Sci.* 304:L445–L50

121. Sykes ECH, Mantooth BA, Han P, Donhauser ZJ, Weiss PS. 2005. Substrate-mediated intermolecular interactions: a quantitative single molecule analysis. *J. Am. Chem. Soc.* 127:7255–60

122. Pascual JI, Jackiw JJ, Song Z, Weiss PS, Conrad H, et al. 2002. Adsorption and growth of benzene on Ag(110). *Surf. Sci.* 502:1–6

123. Urbach LE, Percival KL, Hicks JM, Plummer EW, Dai HL. 1992. Resonant surface 2nd harmonic generation: surface-states on Ag(110). *Phys. Rev. B* 45:3769–72

124. Ullmann F, Bielecki J. 1901. Über Synthesen in der Biphenylreihe. *Ber. Dtsch. Chem. Ges.* 34:2174–85

125. Hla SW, Bartels L, Meyer G, Rieder K-H. 2000. Inducing all steps of a chemical reaction with the scanning tunneling microscope tip: towards single molecule engineering. *Phys. Rev. Lett.* 85:2777–80

126. Grafström S. 2002. Photoassisted scanning tunneling microscopy. *Appl. Phys. Rev.* 91:1717–53

127. Stranick SJ, Weiss PS, Parikh AN, Allara DL. 1993. Alternating current scanning tunneling spectroscopy of self-assembled monolayers on gold. *J. Vac. Sci. Technol. A* 11:739–41

128. Stranick SJ, Weiss PS. 1994. Alternating current scanning tunneling microscopy and nonlinear spectroscopy. *J. Phys. Chem.* 98:1762–64

129. Stranick SJ, Weiss PS. 1994. A tunable microwave frequency alternating-current scanning tunneling microscope. *Rev. Sci. Instrum.* 65:918–21

130. Bumm LA, Weiss PS. 1995. Small cavity nonresonant tunable microwave frequency alternating current scanning tunneling microscope. *Rev. Sci. Instrum.* 66:4140–45

Coherent Anti-Stokes Raman Scattering Microscopy: Chemical Imaging for Biology and Medicine

Conor L. Evans and X. Sunney Xie

Department of Chemistry and Chemical Biology, Harvard University, Cambridge, Massachusetts 02138; email: xie@chemistry.harvard.edu

Annu. Rev. Anal. Chem. 2008. 1:883–909

The *Annual Review of Analytical Chemistry* is online at anchem.annualreviews.org

This article's doi:
10.1146/annurev.anchem.1.031207.112754

1936-1327/08/0719-0883$20.00

Key Words

Raman, CARS, molecular imaging, biomedical imaging, metabolic imaging, skin imaging

Abstract

Coherent anti-Stokes Raman scattering (CARS) microscopy is a label-free imaging technique that is capable of real-time, nonperturbative examination of living cells and organisms based on molecular vibrational spectroscopy. Recent advances in detection schemes, understanding of contrast mechanisms, and developments of laser sources have enabled superb sensitivity and high time resolution. Emerging applications, such as metabolite and drug imaging and tumor identification, raise many exciting new possibilities for biology and medicine.

1. INTRODUCTION

Advances in optical imaging techniques have revolutionized our ability to study the microscopic world. Simple microscopy techniques, such as bright field and differential interference contrast microscopy, have played a large role in cellular and molecular biology experiments but do not provide chemical specificity. Imaging modalities capable of identifying specific molecules have significantly improved our understanding of biological processes on the microscopic scale. Many of these techniques, however, require the use of exogenous labels that often perturb the system of interest. Intrinsic imaging techniques such as native fluorescence imaging (1) offer molecular specificity, but the number of endogenous fluorophores are limited.

Vibrational microscopy techniques offer intrinsic chemical selectivity, as different molecules have specific vibrational frequencies. Infrared microscopy (2) has seen rapid development, but it is limited by a number of difficulties including low sensitivity due to non-background-free detection, low spatial resolution associated with the long infrared wavelengths, and water absorption of the infrared light. Raman microscopy has been extensively explored and has found biomedical applications in glucose detection (3), tumor diagnostics (4, 5), DNA detection (6), and microendoscopy (7), among others (**Figure 1**).

Raman microscopy does, however, have a major limitation. The Raman effect is extremely weak (typical photon conversion efficiencies for Raman are lower than 1 in

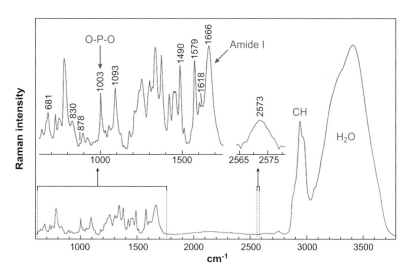

Figure 1

The Raman spectrum of the P22 virus showing characteristic vibrational frequencies observed in biological samples. Several vibrational modes of particular interest in vibrational microscopy are labeled (*blue*). The O-P-O stretching vibration arises from the vibration of the DNA backbone. The amide-I band is characteristic of proteins and can be used to map out protein density. The CH-stretching band is typically used to image lipids in biological samples. The H_2O-stretching vibrations of water are important for following water flow and density. Adapted with permission from Reference 79.

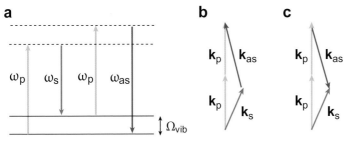

Figure 2

(*a*) Diagram of the coherent anti-Stokes Raman scattering (CARS) process. When the difference between the pump and Stokes frequencies ($\omega_p - \omega_s$) matches the molecular vibrational frequency, Ω_{vib}, the anti-Stokes signal is generated at a frequency $\omega_{as} = 2\omega_p - \omega_s$. (*b*) Phase-matching condition for forward-generated CARS. (*c*) Phase-matching condition for backward- (epi-)generated CARS. *k* is known as the wavevector, and is given by $k = 2\pi / \lambda$. Here, k_p, k_s, and k_{as} represent the pump, Stokes, and anti-Stokes wavevectors, respectively.

10^{18}); therefore, data acquisition times are long. Raman microscopy images require high laser powers and long integration times of 100 ms to 1 s per pixel (8). These factors severely limit the application of Raman microscopy to the study of living systems.

Far stronger vibrational signals can be obtained with coherent anti-Stokes Raman scattering (CARS), which was first reported by Maker and Terhune at the Ford Motor Company in 1965 (9). Ironically, it was not named CARS until almost ten years later (10). In the CARS process, a pump beam at frequency ω_p and a Stokes beam at frequency ω_s interact with a sample via a wave-mixing process. When the beat frequency $\omega_p - \omega_s$ matches the frequency of a Raman active molecular vibration, the resonant oscillators are coherently driven by the excitation fields, thereby generating a strong anti-Stokes signal at $\omega_{as} = 2\omega_p - \omega_s$ (**Figure 2**).

The Reintjes group at the Naval Research Laboratory was the first to use CARS as a contrast mechanism for microscopy (11). Due to technical difficulties there were no further developments until 1999, when CARS microscopy was revived at Pacific Northwest National Laboratory (12) with a new method. Since then, CARS microscopy has been used to visualize living cells with contrast of different vibrational modes, including the phosphate stretch vibration (DNA), amide I vibration (protein) (13), OH stretching vibration (water) (14), and the CH group of stretching vibrations (lipids) (15, 16). Among these modes, the signal from lipids is so high that single phospholipid bilayers can be visualized (17, 18). Meanwhile, CARS has proved to be a powerful imaging modality for studying tissues in vivo (19, 20).

The advantages of CARS are summarized as follows:

1. It provides contrast based on the intrinsic molecular vibrations of a specimen, circumventing the need for extrinsic labels.
2. It is orders of magnitude more sensitive than spontaneous Raman microscopy (21–23), permitting video-rate vibrational imaging at moderate excitation powers.

CARS: coherent anti-Stokes Raman scattering

3. The nonlinear nature of the CARS process automatically grants it the capability of three-dimensional sectioning (12), which is essential for imaging thick tissues or cellular structures.
4. The anti-Stokes signal is blue-shifted from the pump and Stokes frequencies, and is therefore easily detected in the presence of one-photon fluorescence.
5. When using near-infrared excitation wavelengths, CARS microscopy can penetrate to depths of nearly 0.4 mm, allowing imaging in thick tissues.
6. As the CARS process occurs on the ground electronic state, sample photodamage is minimized (19), especially when picosecond pulses are used to reduce multiphoton effects.

In this article we summarize some of the recent advances in CARS microscopy, with an emphasis on new applications in the biomedical sciences. In Section 2 we review the fundamentals of CARS microscopy, and in Section 3 we discuss laser sources. Sections 4 and 5 focus on recent innovations and applications of CARS to cellular and biological imaging.

2. CARS SPECTROSCOPY

In CARS, the pump and Stokes fields coherently drive all resonant oscillators in the excitation volume at $\omega_p - \omega_s$ with a well-defined phase. The coherent superposition of the microscopic induced dipoles generates a macroscopic third-order polarization $P^{(3)}$ at the anti-Stokes frequency. The anti-Stokes field arises from the nonlinear interactions of the pump and Stokes fields, E_p and E_s, respectively, and is given by $P^{(3)}(\omega_{as}) \propto \chi^{(3)} E_p^2 E_s^*$, where the complex proportionality constant $\chi^{(3)}$ is known as the third-order susceptibility. By solving the wave equation, assuming plane pump and Stokes waves, one obtains the anti-Stokes signal intensity,

$$I_{AS} \propto \left| \chi^{(3)} \right|^2 I_p^2 I_s \left(\frac{\sin(\Delta k z / 2)}{\Delta k / 2} \right)^2 \tag{1}$$

where z is the sample thickness, $k_i = 2\pi / \lambda_i$ is the wavevector, and Δk, the wavevector mismatch, is defined as $\Delta k = k_{as} - (2k_p - k_s)$ and gives the velocity difference of the three frequencies. The sinc function is maximized when $\Delta k z$ is close to zero, which is known as the phase-matching condition. **Figure 2b,c** shows the phase-matching conditions for forward- and backward (epi)-detected CARS signal. Although $\chi^{(3)}$ is linearly dependent on the number of oscillators, the CARS signal depends upon $|\chi^{(3)}|^2$ and is therefore proportional to the square of the number of vibrational oscillators. This makes CARS different from Raman, which is linearly dependent on the number of vibrational oscillators (24).

Even when $\omega_p - \omega_s$ is tuned far from vibrational resonances, the pump and Stokes fields can induce a macroscopic polarization at the anti-Stokes frequency due to the electronic response of the material. When ω_p and ω_s are far from electronic resonance, this polarization leads to a vibrationally nonresonant contribution to the CARS signal. When $\omega_p - \omega_s$ is tuned to a particular vibrational frequency, the anti-Stokes signal is enhanced. Therefore, $\chi^{(3)}$ has two terms, one resonant ($\chi_R^{(3)}$) and one nonresonant ($\chi_{NR}^{(3)}$):

$$\chi^{(3)} = \chi_{NR}^{(3)} + \frac{\chi_R^{(3)}}{\Delta - i\Gamma}, \qquad (2)$$

where Δ is the detuning $\Delta = \omega_p - \omega_s - \Omega_R$ (Raman shift), and where Ω_R is the center frequency of a homogenously broadened Raman line with bandwidth Γ.

2.1. Contributions to the CARS Response

As the CARS intensity is proportional to $|\chi^{(3)}|^2$, the intensity of the anti-Stokes signal can be written as:

$$I_{CARS}(\Delta) \propto \left|\chi_{NR}^{(3)}\right|^2 + \left|\chi_R^{(3)}(\Delta)\right|^2 + 2\chi_{NR}^{(3)} \, \mathrm{Re}\, \chi_R^{(3)}(\Delta), \qquad (3)$$

where $\mathrm{Re}\, \chi_R^{(3)}$ is the real part of resonant term of $\chi^{(3)}$. The first term is independent of the Raman shift, and is known as the nonresonant background. The second term contains only resonant information, and is the dominant contribution when probing strong and/or concentrated resonant scatterers. Mixing between the nonresonant and resonant contributions creates the third term, which contains the real part of the vibrational response. Plotted in **Figure 3a** is the spectral response of each term, showing its individual contribution. As the shape of the third term is dispersive, the addition of the three terms creates a redshift of the maximum of the CARS spectral peak and causes a negative dip at the blue end (25) (**Figure 3b**). The redshift in peak position, which depends upon the relative intensity of the resonant and nonresonant contributions, makes it difficult to apply the wealth of information in Raman literature to assigning CARS spectra (26).

The nonresonant contribution also introduces an offset that gives CARS microscopy images a background (**Figure 3c,d**). The blue-end dip is not desirable as it gives negative contrast (**Figure 3e**). Spectral interference between two or more resonances can result in distorted line shapes, and precludes an immediate quantitative interpretation of the spectrum, as the neighboring peaks influence one another's intensity. In congested spectral regions, this leads to nearly uninterpretable CARS spectra. It is possible to glean Raman spectra from the CARS signal by extracting the $\chi^{(3)}$ through interferometry (21, 22), although such methods can complicate a CARS imaging system.

2.2. Imaging versus Spectroscopy

Although it would be ideal to collect a complete spectrum for every object in a CARS microscopy image, in practice these are difficult to obtain. In recent CARS microspectroscopy experiments, a broadband femtosecond laser source is used in conjunction with a monochromator to collect pixel-by-pixel spectroscopic data (27, 28). This results in integrations times of milliseconds to seconds per pixel for many samples, causing significant photodamage. In addition, current detectors used in spectroscopy experiments have long readout times that limit their acquisition speed. These limits currently cap CARS microspectroscopy imaging experiments to frame rates of several minutes for 256×256 pixel image (29, 30). Such acquisition rates are too slow for

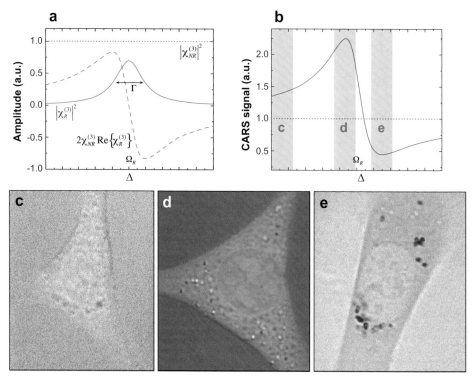

Figure 3

(*a*) Three components of the coherent anti-Stokes Raman scattering (CARS) signal plotted as a function of detuning. Shown here are the purely resonant term (*solid line*), the nonresonant background term (*dotted line*), and the mixing term (with a dispersive shape) (*dashed line*). The plotted curves were calculated for $\chi_{NR}^{(3)} = 1.2\ \chi_{R}^{(3)}(\Delta = 0)$. (*b*) The total CARS signal. The solid line represents the sum of the contributions from panel *a*, while the dotted line represents the nonresonant background. (*c–d*) Forward-propagating CARS images of 3T3-L1 cells that display contrast corresponding to the highlighted regions of Raman shifts in panel *b*. Panel *c* presents a cell imaged off resonance; only nonresonant contrast is observed. Panel *d* shows a cell imaged at 2845 cm^{-1}, the CH_2 symmetric stretching vibration. Numerous lipid bodies, including lipid droplets, can be seen. (*e*) Cell imaged at the blue dip of the CH-stretching band at 2950 cm^{-1}. Resonant features appear dark against the nonresonant background.

most biomedical experiments and are prohibitive for studying dynamics in biological systems. Even when a spectrum can be collected, it cannot immediately be translated to quantitative information without significant off-line processing (31).

We believe that it is best to utilize a narrowband imaging approach for high-resolution CARS microscopy. Instead of collecting a spectrum at each point in the image, one can collect an image for selected points in the spectrum. Doing so capitalizes on the strength of the CARS response, allowing for vibrationally selective imaging with high time resolution. By tuning in the laser system to only one resonance, CARS microscopy can be carried out at video-rate speeds (19) with diffraction-limited resolution. This narrowband approach has far greater applicability as a biomedical imaging technique than broadband excitation schemes (29–34). Many specimens, such as tissues, are highly scattering and easily distort the spectral phase and polarization properties needed for broadband interferometric techniques. Narrowband CARS microscopy has been widely applied, is not limited to thin or weakly scattering samples, and has the capability for future use in clinical settings as an in situ imaging technique.

NA: numerical aperture

2.3. Tight Focusing Relaxes the Phase-Matching Condition

The first CARS microscope was built using a nonlinear geometry (11). This cross-beam arrangement, however, suffered from low spatial resolution and sensitivity, and its complexity was not ideal for a microscopic imaging modality. It was not until 1999 that a discovery cleared the way for modern CARS microscopy (12). The effective axial point spread function for near-infrared (NIR) light used with a high numerical aperture (NA) lens (>0.8) is about 1 μm long. As the NA becomes larger than 0.2, the $sinc^2$ function in Equation 1 asymptotically approaches a maximum, rendering non-collinear phase-matching geometries, such as BOXCARS unnecessary. Therefore, it is possible to use a collinear pump and Stokes beams to generate anti-Stokes signal at the tight focus. These collinear beams are raster-scanned over a sample using commercial confocal scanning microscopes (13) (**Figure 4a**).

2.4. Forward- and Epi-CARS Generation Mechanisms

Due to the coherent nature of the signal buildup in CARS, the far-field radiation pattern is more complicated than those observed in incoherent imaging techniques such as fluorescence and Raman microscopy, where the signal is emitted in all directions. The radiation pattern in CARS microcopy is highly dependent on a number of parameters, including the size and shape of the scattering objects, the nonlinear susceptibilities of the object, and the local environment (35). Each vibrationally oscillator can be considered as a Hertzian dipole at the anti-Stokes frequency (**Figure 5a**). For extremely thin slabs of oscillators, the radiation pattern becomes more directional, and propagates equally in the forward and epi directions (**Figure 5b**). As the sample thickness, z, increases, constructive interference occurs in the forward direction (36) (**Figure 5c**), as Δk is small (**Figure 2b**), resulting in forward-propagating CARS (F-CARS). At the same time, destructive interference occurs in the backward direction as Δk is large (**Figure 2c**) (35, 37), leading to no epi-CARS signal for bulk objects (**Figure 5d**).

It would seem, then, that the CARS signal would be generated in the forward direction only. This is not the case, however, as epi-directed signal (38) has been observed in CARS microscopy arising from three different mechanisms. In the first mechanism, epi-CARS signal is generated from objects whose size is small enough ($\lambda_p/3$) for incomplete destructive interference to occur in the backward direction. If the size of scatterer is larger, the epi-directed fields from dipoles across the object run out of phase with each other, leading to destructive interference of the epi-CARS signal (**Figure 5e**). We note that the phase-matching condition still holds for this mechanism; while Δk is large, the path length, z, is small enough to maximize the sinc function of Equation 1. In the second mechanism for epi-CARS generation,

www.annualreviews.org • CARS Microscopy   *889*

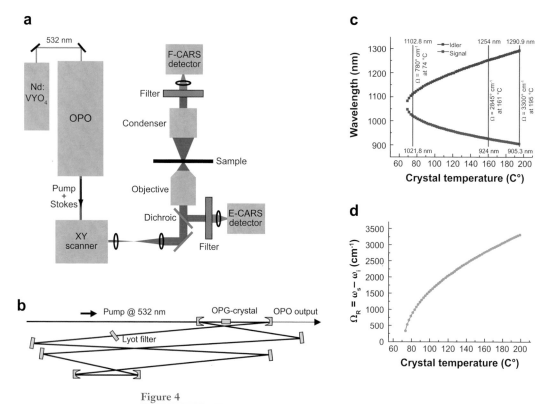

Figure 4

(*a*) Diagram of a collinear beam-scanning coherent anti-Stokes Raman scattering (CARS) microscope. An Nd:Vanadate laser pumps a dual-wavelength optical parametric oscillator (OPO). The combined pump and Stokes beam from the OPO is raster-scanned over the sample by an *XY* scanner, typically a set of galvanometric mirrors. The CARS signal generated in the forward direction is collimated by a condenser, separated from the excitation beams by a filter, and focused onto a detector for collection. The epi-CARS signal is detected by a large area detector placed after a filter, or by a descanned detector in a manner similar to epi-fluorescence detection (not shown). (*b*) Cavity design for the dual-wavelength OPO. Both noncritically phase-matched LBO (LiB$_3$O$_5$) and PP-KTP (periodically poled KTiOPO$_4$) crystals have been successfully used for parametric downconversion. When LBO is used, the output wavelengths can be simultaneously tuned from 670 nm to 980 nm for the signal and 1100 nm to 1350 nm for the idler. Fine-wavelength tuning is accomplished using a Lyot filter set at Brewster's angle. (*c*) Signal and idler tuning curves as a function of temperature for a PP-KTP based OPO. (*d*) Energy difference between the signal and idler frequencies as a function of crystal temperature for PP-KTP. Abbreviation: F-CARS, forward-propagating coherent anti-Stokes Raman scattering.

backward-propagating anti-Stokes signal is generated at sharp discontinuities in $\chi^{(3)}$. Edges or discontinuities give rise to epi-CARS signal, as they are essentially infinitely small objects that break the symmetry of the focal volume (35) (**Figure 5b,f**). There is a third mechanism for generating epi-directed photons, although it is not related to coherent signal buildup. In this mechanism, a sample that contains many local changes in the index of refraction can redirect forward-propagating photons in the

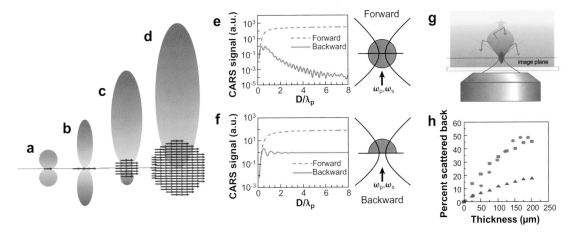

Figure 5

The effects of coherent signal buildup in the focal volume. (*a*) The radiation pattern of a single oscillating dipole sends equal intensities in both the forward and backward (epi) directions. (*b*) The fields from a plane of dipoles coherently add to generate equal signal levels in the forward and epi directions. (*c*) A few induced dipoles together generate a forward-propagating component as well as a weak epi-directed signal. (*d*) Many dipoles in the focal spot, such as in bulk material, coherently interfere to generate only a strong forward signal. (*e*) As the size of the scattering object increases, the epi signal generated by Mechanism I drops dramatically while the forward signal continues to build to an asymptotic level defined by the focal volume. Adapted from Reference 35 with permission. (*f*) At an interface, Mechanism II generates an epi signal that is constant in intensity once the object size exceeds the focal volume. (*g*) Mechanism III: Forward-propagating coherent anti-Stokes Raman scattering (F-CARS) photons can be backscattered by a turbid sample and redirected in the epi direction. (*b*) Collection efficiency of epi detection by Mechanism III as a function of sample thickness in a tissue phantom (intralipid suspension) and mouse skin tissue (19). The focus was placed 1 μm deep for both theory and experiment. Shown are the percentage of forward signals redirected backward as the thickness of the intralipid layer was increased (*gray dots*); the results of a Monte Carlo simulation of the experiment (*red squares*); and a curve calculated by the same Monte Carlo simulation, this time using the tissue parameters of mouse skin ($\mu_s = 150$ cm^{-1}, $\mu_a = 0.1$ cm^{-1}, and $g = 0.85$) (*blue triangles*). Nearly 15% of the F-CARS signal generated in mouse skin was estimated to scatter backward.

epi direction. In turbid samples such as skin, initially forward-propagating photons undergo multiple linear scattering events, which redirect a large number of photons back toward the objective (**Figure 5g**). This mechanism was found to be the primary contributor to epi-CARS signal from tissues (19) (**Figure 5b**). We note that the backward signal collected by the objective, which is often out of focus, is only generated at the laser excitation volume.

3. LASER SOURCES FOR CARS

Recent advances in CARS microscopy have been facilitated by the development of new light sources. We believe that the ideal sources for CARS microscopy are picosecond tunable pulsed laser systems operating in the NIR.

3.1. Laser Parameters

An important consideration for laser wavelength choice is the nonresonant background. Nonresonant CARS signal can be generated when the pump wavelength is near the peak of a two-photon resonance. Sources in the NIR minimize these two-photon interactions and therefore provide images with better signal-to-noise ratios. Another advantage of using NIR laser sources is low multiphoton absorption–induced photodamage. Multiphoton (typically two-photon) absorption is the major cause of sample damage in most samples (16, 39). CARS sources employing pump beams shorter than 800 nm can generate a large amount of multiphoton absorption by ultraviolet electronic resonances. Researchers have found that most multiphoton damage significantly decreases when the pump wavelength is raised above 800 nm (19, 39).

Near-infrared excitation has another advantage. Tissue samples are typically highly scattering, limiting the penetration depth of CARS microscopy. As the pump and Stokes beams converge in a specimen, scattering in the tissue leads to a loss of laser intensity accompanied by an increase in the focal spot size (40). This results in a deteriorated CARS signal due to its nonlinear intensity dependence. The reduced scattering experienced by NIR pump and Stokes wavelengths maximizes the sample penetration depth, allowing for deep CARS imaging even in turbid samples.

Pulsed lasers systems are needed as CARS signal scales cubically with the intensity of the incident laser light. As such, the choice of pulse width is worthy of careful consideration. Vibrational linewidths are typically on the order of $10-20$ cm^{-1}, whereas $\sim$100-fs-duration pulses are about 150 cm^{-1} in bandwidth in the near IR. A femtosecond pulsed system centered on a resonance will, therefore, use only a small part of its spectral components to pump the narrow Raman line, but will generate a large nonresonant background signal that often obscures chemically specific contrast. Moreover, because the nonresonant signal coherently mixes with the resonant signal of interest, it cannot simply be subtracted from an image. A good compromise between signal strength and spectral resolution is to use pulses $\sim$3 ps in duration (38). **Table 1** lists our recommendations for optimum laser parameters for CARS microscopy.

Table 1 Parameters of laser light sources for CARS microscopy

Parameter	Optimal Range
Pump wavelength range	780–980 nm
Stokes wavelength range	1000–1300 nm
Pulse duration	2–7 ps
Spectral bandwidth	3–5 cm^{-1}
Pulse energy	0.1–1 nJ
Pulse repetition rate	50–100 MHz

3.2. Optimum Laser Sources for CARS Microscopy

Our latest CARS source was introduced by Ganikhanov et al. (41) and uses the signal and idler outputs of an optical parametric oscillator (OPO) as the pump and Stokes frequencies, respectively, for CARS microscopy. The most recent design (**Figure 4b**) is a broadly tunable picosecond OPO (commercialized as the Levante Emerald by APE-Berlin), based on a noncritically phase-matched crystal for parametric down-conversion, and is synchronously pumped by the second-harmonic (532 nm) output of a mode-locked Nd:YVO$_4$ laser (HighQ Laser, Austria). The signal and idler frequencies are continuously temperature tunable about the degeneracy point (signal and idler wavelengths at 1064 nm) to 670 nm for its signal beam and to 2300 nm for its idler beam, making it possible to cover the entire chemically important vibrational frequency range of 200–3600 cm^{-1} without cavity adjustment (**Figure 4c,d**). Unlike previous light sources such as synchronized Ti:Sapphire lasers (13, 42, 43), the pump (signal) and Stokes (idler) pulse trains are inherently optically synchronized and collinearly overlapped at the output of the OPO. This considerably reduces the complexity of a CARS setup, as the combined output beam can be immediately coupled into a microscope (**Figure 4a**).

OPO: optical parametric oscillator

A drawback of the NIR source is the low spatial resolution of ~350 nm. Additionally, anti-Stokes signal can be generated at wavelengths well into the NIR for many vibrational modes that are not accessible with standard detectors. Fortunately, the wide tuning range of this source enables it to run at two different wavelength regimes. In order to achieve higher-resolution imaging or to reach vibrational bands in the fingerprint region, the signal output of the OPO can be combined with the 1064-nm output of the Vanadate laser. This optically synchronized wavelength pair can achieve a microscopic resolution of 300 nm, and the Raman shift can be tuned as low as 800 cm^{-1}. As the OPO essentially delivers the wavelength tunability of two sources, its extreme flexibility allows the study of samples with a range of resolutions and depth penetrations.

Looking into the future, fiber lasers offer great potential as they are simple, reliable, and low in cost compared with free-space systems. Fiber lasers could find use as pump lasers for OPOs, or as sources of pump and Stokes beams directly.

4. INCREASING SENSITIVITY

The first three-dimensional CARS microscope in 1999 (12) required nearly 30 minutes to acquire a single image. As of 2008, CARS microscopy systems run at acquisition rates of 30 frames per second, an improvement in sensitivity of over four orders of magnitude. Although many improvements to CARS microscopy have been made over the past few years, it is still limited by the nonresonant background whose fluctuation due to laser noise can often obscure weak resonant signals. Numerous methods have been developed to decrease this nonresonant contribution. Polarization-sensitive detection (44) was developed to completely suppress the nonresonant contribution through polarization control, but the technique significantly

FM-CARS: frequency modulation coherent anti-Stokes Raman scattering

attenuates the resonant signal of interest. Epi-CARS microscopy (45) eliminates the nonresonant background from the bulk material, but not the nonresonant contribution from small scatters. Time-resolved CARS (46) has also been used for nonresonant background rejection, but it is cumbersome to implement. Interferometric CARS approaches, although successful in eliminating the nonresonant background by separating the real and imaginary components of the third-order susceptibility $\chi^{(3)}$ (21, 22, 47), are difficult to use with heterogeneous samples due to index-of-refraction changes. All of the above approaches are limited in sensitivity as polarization, temporal phase, spatial phase, and spectral phase can all be influenced by sample heterogeneity and therefore are not always reliable.

We have developed a more sensitive approach called frequency modulation CARS (FM-CARS). The nonresonant background, $|\chi_{NR}^{(3)}|^2$, the first term in Equation 1, is constant over all Raman shifts and dominates the CARS signal when the resonant signal is weak, i.e. when the second term in Equation 1 is negligible. Under this condition, we exploit the dispersive nature of the mixing term by modulating the Raman shift (Δ) around the center frequency Ω_R. This results in an amplitude modulation of the CARS signal, which can be detected using a lock-in amplifier. By toggling the Raman shift, a resonant spectral feature becomes a frequency modulation (FM) to amplitude modulation (AM) converter (**Figure 6a**). The spectrally flat nonresonant background does not contribute to the detected modulated signal, and therefore is suppressed. This frequency modulation CARS technique is general enough to be applied to either forward or backward CARS in any sample. Additionally, if the modulation is carried out at rates exceeding 500 kHz, the modulated signal can be separated from laser, mechanical, and most electrical noise sources. Using this approach, FM-CARS

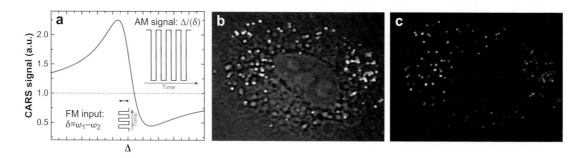

Figure 6

(*a*) Schematic of the frequency modulation coherent anti-Stokes Raman scattering (FM-CARS) process. Represented here are the sum of the contributions from **Figure 2a** (*solid line*) and the nonresonant background (*dotted line*). The resonance acts as an FM-to-AM converter, resulting in an amplitude-modulated signal that can be detected by a lock-in amplifier. (*b*) Forward-CARS image of a fixed A549 human lung cancer cell cultured with deuterium-labeled oleic acids taken at a Raman shift of 2100 cm^{-1}. (*c*) FM-CARS image obtained when toggling between 2060 cm^{-1} and 2100 cm^{-1}. Nonresonant background components have been significantly reduced by the FM-CARS method. Abbreviations: FM, frequency modulation; AM, amplitude modulation.

systems have been shown to provide sensitivities nearly three orders of magnitude greater than conventional CARS imaging (23).

As FM-CARS can be easily incorporated into any detection geometry, it can be immediately applied to biomedical imaging. To demonstrate the capabilities of FM-CARS, deuterated lipids were isolated from the nonresonant contribution in human lung cancer cells (23). An F-CARS image of these cells at the CD_2 stretching frequency shows a number of cellular features, complicating identification of the cellular components that contain the deuterated oleic acid (OA) (**Figure 6b**). When FM-CARS is used, however, the nonresonant signal is suppressed and the deuterated components can be easily identified (**Figure 6c**). Continued progress in FM-CARS promises to improve the sensitivity limit, allowing for future applications such as drug and metabolite imaging that will have significant benefits for cell biology and biomedicine.

OA: oleic acid

5. APPLICATIONS OF COHERENT ANTI-STOKES RAMAN SCATTERING MICROSCOPY TO BIOLOGY AND MEDICINE

Developments over the past several years have enabled the application of CARS microscopy to the chemical, materials, biological, and medical fields. Chemical applications include many studies on lipid vesicles (31, 48, 49), lipid layers (50), and the ordering of lipid domains (18, 39). In the materials field, CARS has been used to examine the dynamics of water in organic environments (51, 52) and has been applied to the processing of photoresists (51) and the ordering of liquid crystals (53). The most exciting recent applications of CARS have been in the fields of biological and medical imaging, and are the focus of this section.

5.1. Imaging Cells with Chemical Selectivity

CARS provides new views into cellular structures. A recent example is the imaging of plant cells. Plant cell walls are primarily composed of polysaccharides (such as cellulose), lignin, and glycoproteins. Lignin is largely responsible for the resistance to chemical/enzymatic degradation of cellulose into short chain sugar molecules in the process of biomass conversion to biofuel. However, it is difficult to image lignin using conventional imaging methods. In order to improve the conversion efficiency, one requires an imaging technique with contrast based on chemical composition for real-time monitoring. The structure of lignin (**Figure 7a**) gives rise to a Raman spectrum (**Figure 7b**) with a band at 1600 cm^{-1} due to the aryl symmetric ring stretching vibration, which can serve as a sensitive probe for lignin. **Figure 7c** shows a CARS image of corn stover tuned into the 1600 cm^{-1} stretch, which reveals the distribution of lignin within the walls of individual cells.

Far more work has been done on mammalian cells, where the cellular organelles can be imaged by CARS. The strongest observed signal comes from the CH bonds, which are abundant in lipids. This strong signal allows the observation of dynamic processes with high time resolution. For example, rapid intracellular transport in

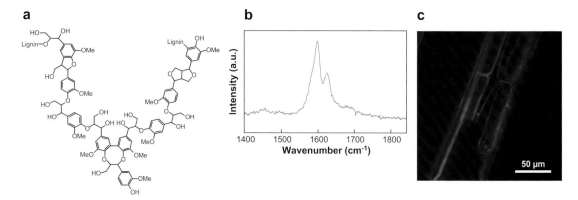

a **b** **c**

Figure 7

(*a*) Chemical structure of the lignin polymer. (*b*) Raman spectrum of lignin, with prominent bands near 1600 cm^{-1} arising from the aryl ring stretching vibrations. (*c*) Coherent anti-Stokes Raman scattering microscope image at 1600 cm^{-1} showing the distribution of lignin in the cell walls surrounding the plant cells in corn stover.

mammalian cells is driven by molecular motors that bind to and move select cellular organelles along the cellular cytoskeleton (54, 55). Organelle trafficking is an essential process, and even small changes to the cellular environment can have large effects on intracellular transport. In particular, the transport of lipid droplets (LDs) (56–58) can be difficult to selectively visualize, as fluorescent LD labels may have perturbative effects (59). CARS microscopy, with its exquisite sensitivity to lipids, has proven to be an excellent tool for LD tracking. In a recent study by Nan et al. (16), CARS microscopy was used to monitor the transport of LDs in steroidogenic Y-1 mouse adrenal cortical cells by tuning into the lipid CH$_2$ stretching frequency (2845 cm^{-1}). LDs were observed to undergo caged, subdiffusive motion due to trapping in the cytoskeleton along with the rapid, superdiffusive motion generated by a molecular motor. In particular, a correlation between cell shape and LD transport activity was observed, which is thought to increase the collision rate of LDs with mitochondria for steroidogenesis. It is evident that CARS is complementary to existing microscopy techniques, such as fluorescence microscopy, for interrogating cellular structure and dynamics.

5.2. Metabolic Imaging

Metabolites are small molecules that are difficult to image using conventional microscopy techniques. Labels such as fluorophores are typically comparable in size to most metabolites, and will thus significantly alter their behavior in vivo. Knowledge of the transport, delivery, and localization of such molecules is critical for the study of disease and for the development of effective treatments. The vibrational selectivity of CARS microscopy makes it a powerful tool for such studies. For example, CARS microscopy's sensitivity to long-chain hydrocarbons makes it an ideal method for studying the metabolism of lipids. In an early experiment, CARS

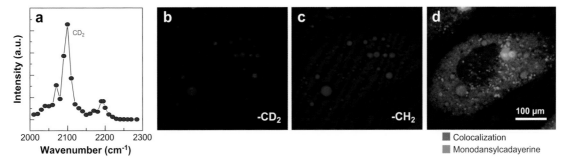

Figure 8

Lipid metabolism studied with coherent anti-Stokes Raman scattering (CARS) microscopy. (*a*) The CARS spectrum of deuterated oleic acid. (*b-d*) Rat hepatocyte cells were incubated with both eicosapentaenoic acid and deuterated oleic acid. (*b*) CARS image taken at the CD_2 symmetric stretching vibration at 2090 cm^{-1}. (*c*) CARS image taken at the CH_2 symmetric stretching vibration at 2845 cm^{-1}. (*d*) Composite image of panels *b* and *c*, showing colocalization along with two-photon fluorescence from monodansylcadayerine, a label for degradative lysozymes.

was used to follow the differentiation of cells into adipocytes, following the growth of LDs in cells over the course of several days (15). Two recent studies have used CARS microscopy to investigate the changes in lipid metabolism caused by the hepatitis C virus (60, 61). Recent work by Hellerer et al. (62) demonstrates the ability of CARS microscopy to image lipid metabolism in a living organism by monitoring lipid storage in different strains of *Caenorhabditis elegans*.

The above experiments used CH contrast. For many studies, however, a method is needed for imaging a specific small molecule inside cells. Isotope labeling via deuterium substitution has recently been demonstrated as a powerful method (63). The deuterium in a CD bond is heavier than hydrogen, placing CD vibrational stretching frequencies into an otherwise "silent" region of the biological Raman spectra near 2100 cm^{-1} (**Figure 8*a***). By substituting nonreactive hydrogens with deuterium to avoid perturbation and tuning into these unique CD stretching frequencies, it is possible to selectively image a specific molecule using CARS contrast.

This approach was used to selectively image the metabolism of different lipids in a study investigating the effects of omega-3 fatty acids. Fish oil is rich in polyunsaturated fats, such as omega-3 fatty acids, and has been found to affect lipid biochemistry in humans by lowering blood triglyceride levels. To study the effect of omega-3 fatty acids on cellular lipid metabolism, rat hepatoma cells were treated with a combination of normal (OA) and omega-3 (eicosapentaenoic acid, EPA) fatty acids. Because the two molecules have similar CARS spectra, deuterium substitution in OA was used to specifically image the monounsaturated fatty acid. Cells grown with only oleic acid in the culture media were found to sequester OA into small LDs. When cells were cultured only with EPA, the EPA was incorporated into lysosomes (acidic organelles that digest cellular components). Interestingly, when cells were grown with OA and

EPA: eicosapentaenoic acid

EPA together, both fatty acids were found colocalized in lysosomes (**Figure 8*b,c,d***), indicating that omega-3 fatty acids such as EPA change the way cells process normal fatty acids (63). This CARS metabolic study, using deuterium substitution for specific nonperturbative labeling, is important for understanding the health benefits of polyunsaturated fatty acids such as omega-3s.

5.3. Biomedical Imaging

Over the past several years, many applications of CARS microscopy to biomedicine have emerged. CARS imaging is especially useful for in vivo and in situ investigations, wherein the use of selective labels might be impossible or prohibitive. Compared to techniques such as magnetic resonance imaging, CARS does not have a large penetration depth; instead, it offers subcellular spatial resolution and high time resolution.

CARS imaging in vivo was first demonstrated on the skin of a mouse, and utilized a real-time video-rate CARS imaging system (19). By tuning into the CH_2 vibrational stretching frequency, CARS microscopy was able to visualize the abundant lipid structures throughout the 120-μm depth of mouse ear skin. At the skin surface, the bright polygonal stratum corneum was visible due to the presence of the intracellular "mortar" that holds the many surface corneocytes together. This intracellular material is rich in lipids, ceramides, and cholesterol and gave rise to a strong CARS signal (**Figure 9*a***). Multicellular sebaceous glands appeared 20 μm below the surface of the skin (**Figure 9*b***). These glands are packed with micrometer-sized granules of sebum, a compound rich in triglycerides and wax esters (**Figure 9*e***). At depths of 60 μm, large adipocytes were clearly visible, many aligned along blood vessels (**Figure 9*c***). At the bottom of the dermis, small adipocytes forming the subcutaneous fat layer could be seen (**Figure 9*d***). The use of a video-rate CARS imaging system allowed for rapid, three-dimensional reconstruction of the entire tissue depth (**Figure 9*f***). This study was also able to track, in real time, chemical diffusion into skin by following the application of baby oil.

Recent studies have investigated the transdermal delivery of retinol, a drug that stimulates collagen growth in skin. The conjugated polyene structure of the drug (**Figure 10*a***) gives rise to a strong vibrational band (**Figure 10*b***) that can be used for specific imaging with CARS. **Figure 10*c*** shows the distribution of a 10% retinol solution applied to mouse ear skin. The drug is seen to localize in the intercellular space between the corneocytes of the stratum corneum, which is a pathway for entry into the dermis (64).

As demonstrated by the brightness of the adipocytes in CARS images (**Figure 9**), adipose tissue yields strong CARS signal. White adipose tissue of a mouse omentum majus, for example, generates intense CARS signals from the large (>50 μm) adipocytes (**Figure 11*a***).

CARS microscopy has also been used to visualize the microstructure of excised mouse lungs (**Figure 11*b***). Lung tissue is primarily composed of small air sacs called alveoli that are coated with a lipid-rich surfactant. CARS images of lung tissue, tuned into the symmetric CH_2 stretching vibration, show these alveoli along with numerous

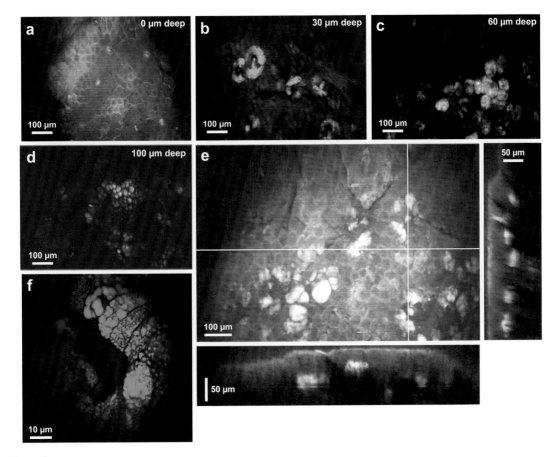

Figure 9

Coherent anti-Stokes Raman scattering images of mouse skin at the lipid band (2845 cm^{-1}) in vivo. (*a*) The surface of hairless mouse skin imaged at the lipid band. The outlines of the corneocytes are clearly visible due to the lipid-rich intracellular "mortar" of the stratum corneum. (*b*) Sebaceous glands imaged at ~30 μm deep. (*c*) Adipocytes at approximately ~60 μm deep, in the dermis. (*d*) Subcutaneous fat composed of many small adipocytes, nearly 100 μm deep. (*e*) Two-dimensional projection of 60 images from a depth stack taken every 2 μm. *YZ* and *XZ* cross sections (*right* and *bottom* panels, respectively) were constructed from the depth stack along the white lines. The cross sections are presented in inverse contrast to show better detail. (*f*) Three-dimensional rendering of a mouse sebaceous gland. The crescent-shaped sebaceous gland surrounding a hair shaft is composed of multiple cells, each filled with numerous micrometer-sized CH$_2$-rich sebum granules.

lipid-rich cells, most likely surfactant cells (type II pneumoctyes), Clara cells, and macrophages (66, 67).

Tissues of the kidney give excellent contrast when imaged with CARS microscopy. Adipose tissue, visualized on the surface of the kidney, stands out prominently in CARS images taken at the lipid band (**Figure 11c**). Beneath the kidney surface, at depths of approximately 40 μm, proximal and distal renal tubules are clearly visible

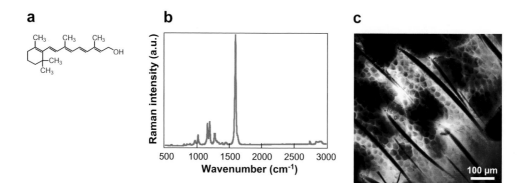

a **b** **c**

Figure 10

(*a*) Structure of retinol. (*b*) Raman spectrum of retinol, showing the strong characteristic peak at 1594 cm^{-1} arising from the conjugated polyene structure. (*c*) 620 µm × 620 µm image of mouse ear skin treated with a 10% retinol in myritol 318 solution. Myritol 318 is a compound typically found in skin creams (65).

(**Figure 11***d*). Close examination of the tubule walls shows the many rounded nuclei of the tubules' epithelial cells, which appear dark due to their low lipid content.

The retina is composed of multiple layers of lipid-rich neurons, each with a different function and microscopic structure that can be readily identified using CARS microscopy (**Figure 11***e*). The photoreceptor, inner and outer nuclear, and inner and outer plexiform layers are easily seen in cross-sectional images. En face CARS depth stacks (**Supplemental Movie 1**; follow the **Supplemental Material link** from the Annual Reviews home page at **http://www.annualreviews.org**) allow full three-dimensional reconstruction of retinal tissue, in which it is possible to visualize the nerve fiber layer and ganglion cells. Capillaries crossing the retina surface, many containing red blood cells, are easily seen with lipid contrast (**Figure 11***f*).

Numerous CARS microscopy studies have focused on nerve bundle structure and function. For example, excised spinal cords have been visualized using the CH$_2$ stretching vibration (68), and the sciatic nerve of living mice has been imaged using minimal surgical techniques (69). Recent studies have even used CARS contrast to study the breakdown of nerve structures in demyelination disorders (70).

A new and exciting biomedical application of CARS microscopy is the imaging of brain tissue (20). Brain tissue is lipid dense as it is composed of billions of neurons and support cells. Using CH$_2$ stretching contrast, CARS microscopy has been used to visualize a number of brain structures. A coronal section of mouse brain, taken 2.8 mm from the bregma, shows a number of brain structures when imaged with CARS microscopy. In order to maintain cellular resolution and image the full organ, the brain mosaics shown were built from individual 700 µm × 700 µm CARS images (**Figure 12***a*). White matter tracts, such as the association fiber bundle in the centrum semiovale, the corpus callosum, and corticospinal tracts, are rich in myelin and give rise to intense lipid band CARS signals. The white matter regions in the diencephalon

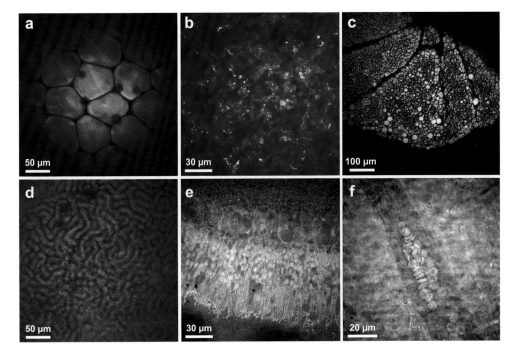

Figure 11

Coherent anti-Stokes Raman scattering (CARS) imaging of various tissues ex vivo with CH_2 contrast. (*a*) Epi-CARS image of white adipose tissue of mouse omentum majus. These large adipose cells are packed with fatty acids and give rise to strong CARS signals. (*b*) Epi-CARS microscopy of mouse lung tissue, showing the individual alveoli. CARS signals are thought to arise from the lipid-rich surfactant cells, Clara cells, and macrophages. (*c*) Epi-CARS image of the surface of the adipocyte-covered mouse kidney. (*d*) Epi-CARS image of mouse kidney taken at a depth of 40 μm reveals many renal tubules. (*e*) Forward-propagating CARS image of a fixed bovine retina in cross section. The first several layers of the retina can be identified. (*f*) En face epi-CARS image of a fixed human retina taken at the retinal surface. A capillary just above the nerve fiber layer contains numerous erythrocytes.

and deep brain nuclei can also be identified by their CARS signal strength. In order to compare the contrast from lipid band CARS imaging to the gold standard in biological imaging, hematoxylin and eosin (H&E) histological preparations were carried out on CARS-visualized brain samples. **Figure 12***b* shows a 700 μm × 700 μm image of the corpus callosum and surrounding structures, which are compared with the corresponding H&E-stained section (**Figure 12***c*), demonstrating the gray-white matter contrast and revealing the microstructural anatomical information available from CARS microscopy.

This study also demonstrated the ability of CARS to distinguish healthy and diseased brain tissue (20). A large astrocytoma is readily seen in a lipid band CARS image due to the lipid-poor nature of the tumor (**Figure 12***d*). Close examination of the tumor margins (**Figure 12***e*) reveals the highly invasive nature of the astrocytoma as it infiltrates the surrounding healthy white matter. Such studies open

H&E: hematoxylin and eosin

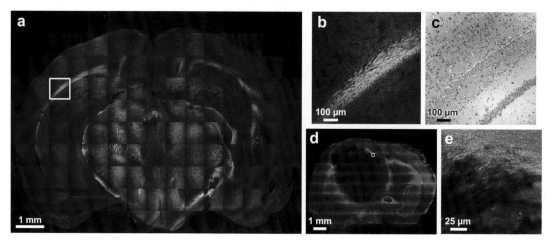

Figure 12

Epi-CARS microscopy applied to brain tissue imaging. (*a*) Mosaic image of a mouse brain coronal section taken at the lipid band showing numerous brain structures. (*b*) A single zoomed-in image corresponding to the white box in panel *a*. (*c*) Hematoxylin and eosin (H&E) image of the same region of the same mouse brain. The structures visible in both images, from upper left to bottom right, are the cortex, corpus callosum, oriens layer, and pyramidal layer. The corpus callosum, a myelinated brain structure, gives rise to a strong CARS signal. (*d*) A mosaic CARS image of astrocytoma in a mouse sacrificed four weeks after inoculation with tumor cells. (*e*) A zoomed-in image corresponding to the white box in panel *d* shows tumor infiltration at the margin.

the door to a number of potential clinical applications, wherein CARS microscopy could one day replace traditional histopathology in brain imaging. In particular, a CARS microendoscope (71) offers the ability to probe deep into brain tissue for diagnostic imaging and could reduce the need for brain tissue resectioning in the future.

6. THE FUTURE OF COHERENT ANTI-STOKES RAMAN SCATTERING MICROSCOPY

Advances over the past several years have made CARS microscopy a state-of-the-art technique. Its ability to perform label-free chemical imaging is beginning to make contributions to biology and medicine. Current FM-CARS systems are capable of probing several important metabolites. CARS microscopy, specifically the FM-CARS technique, also has the potential to play a large role in drug research as it enables the study of drug localization and uptake at the subcellular level.

Another important application of CARS will be in cancer diagnostics. Tumors often show distinct chemical and morphological differences from their host tissue (72, 73). CARS microscopy is capable of providing chemically specific information along with tissue morphology to identify and study cancerous lesions. CARS can also be combined with other intrinsic imaging techniques such as multiphoton fluorescence microscopy (74), second harmonic generation microscopy (75), and third harmonic generation microscopy (76) to provide a multimodal platform for tissue diagnostics.

CARS endoscopy will also be used as an in situ surgical probe for tumor margins. A CARS microendoscope, as part of a fiber probe or needle biopsy, could provide a means of sampling a surgical region prior to resection. Such an approach might be applied to the diagnosis of breast tumors, which are characterized by changes in lipid content and morphology (77). The sensitivity of CARS to lipids could also be used to detect and analyze atherosclerotic plaques in situ (78).

In 1999, the use of CARS microscopy to study living systems was just beginning. There has been a vision to move CARS imaging from the laser lab into the hospital, where it could provide direct benefits to patients. At this point, CARS microscopy is well on its way to reaching that goal. Given the creativity and innovation of the ever-growing community of CARS enthusiasts, CARS microscopy will become a valuable tool for biomedicine for many years to come.

SUMMARY POINTS

1. CARS microscopy provides chemically selective information by tuning into characteristic vibrational resonances in samples without the use of labels or the complication of photobleaching.

2. The stimulated coherent excitation of many vibrational oscillators gives rise to a much stronger signal than that of conventional Raman microscopy, allowing for real-time imaging of living cells or organisms at tolerable laser powers.

3. The CARS signal is only generated at the focal spot, allowing 3D sectioning of thick tissues.

4. Epi-directed CARS signal is generated via three different mechanisms: incomplete destructive interference by objects smaller than the wavelength of light, discontinuity of the third-order nonlinear susceptibility $\chi^{(3)}$ at the interface of two media, and backscattering of initially forward-propagating photons in turbid specimens.

5. The optimum light source for CARS imaging is a picosecond pulsed laser that operates in the NIR range, preferably above 800 nm to avoid multiphoton damage of specimens and to allow for deep penetration in thick samples.

6. The sensitivity of CARS microscopy has been improved by frequency modulation CARS detection.

7. Isotope substitution by deuterium offers a well-isolated CD stretching frequency for mapping the distribution of metabolites or drugs.

8. As a medical imaging technique, CARS is capable of whole-organ imaging at subcellular resolutions in real time.

FUTURE ISSUES

1. What is the ultimate sensitivity of CARS detection?

2. Can fiber delivery and fiber-based laser sources reduce the cost and complexity of CARS microscopy systems?

3. Can the penetration depth of CARS microscopy be extended through adaptive optics?

4. Is it possible to apply the capabilities of CARS to intraoperative endoscopy?

DISCLOSURE STATEMENT

Patents and patent applications held by X.S.X. have been licensed to multiple microscope manufacturers.

ACKNOWLEDGMENTS

We would like to extend a special thanks to our collaborators who have made much of this work possible. Prof. Charles Lin and his colleagues Prof. Daniel Côté and Dr. Mehron Pouris'haag at Massachusetts General Hospital were instrumental in developing video-rate CARS microscopy for skin imaging. Dr. Robert Farris of the National Institutes of Health Eye Institute provided retina samples. Dr. Geoffrey Young, Prof. Stephen Wong, Dr. Xiaoyin Xu, and Dr. Santosh Kesari of Harvard Medical School were collaborators in the brain imaging work. Prof. Zemin Yao at the University of Ottawa was our key collaborator in the lipid metabolism study. We worked closely with Prof. John Pezaski's group at the National Research Council in Ottawa on the hepatitis C projects. Dr. Yiwei Jia of Olympus was of critical assistance in building our first beam-scanning CARS microscope, and has been a very helpful colleague. Dr. Christa Ackermann of Pfizer has worked closely with us on the development of the next-generation FM-CARS system with funding from the Pfizer corporation. Dr. Shuliang Zhang of Unilever has collaborated on several imaging projects with funding from the Unilever Corporation.

We must also acknowledge many Xie group members, past and present, whose hard work and dedication made the many advances presented here possible: Prof. Andreas Zumbush, Dr. Gary Holtom, Dr. Andreas Volkmer, Dr. Lewis Book, Prof. Ji-Xin Cheng, Prof. Eric O. Potma, Prof. François Légaré, Prof. Feruz Ganikhanov, Dr. Silvia Carrasco, Dr. Xiaolin Nan, Dr. Wei Yuan Yang, Brian G. Saar, and Christian Freudiger. Funding for the development of CARS microscopy has come from the National Institutes of Health (R01 GM62536–02), the National Science Foundation (DBI-0649892, DBI-0138028) and a National Institutes of Health Director's Pioneer Award awarded to X.S.X. C.L.E. wishes to acknowledge the National Science Foundation for a Graduate Research Fellowship.

LITERATURE CITED

1. Zipfel WR, Williams RM, Christie R, Nikitin AY, Hyman BT, Webb WW. 2003. Live tissue intrinsic emission microscopy using multiphoton-excited native

fluorescence and second harmonic generation. *Proc. Natl. Acad. Sci.* 100:7075–80

2. Miller L, Smith G, Carr G. 2003. Synchrotron-based biological microspectroscopy: from the mid-infrared through the far-infrared regimes. *J. Biol. Phys.* 29:219–30

3. Berger A, Itzkan I, Feld M. 1997. Feasibility of measuring blood glucose concentration by near-infrared Raman spectroscopy. *Spectrochim. Acta A Mol. Biomol. Spectrosc.* 53:287–92

4. Huang Z, McWilliams A, Lui H, McLean DI, Lam S, Zeng H. 2003. Near-infrared Raman spectroscopy for optical diagnosis of lung cancer. *Int. J. Cancer* 107:1047–52

5. Nijssen A, Bakker Schut TC, Heule F, Caspers PJ, Hayes DP, et al. 2002. Discriminating basal cell carcinoma from its surrounding tissue by Raman spectroscopy. *J. Invest. Dermatol.* 119:64–69

6. Cao Y, Jin R, Mirkin C. 2002. Nanoparticles with Raman spectroscopic fingerprints for DNA and RNA detection. *Science* 297:1536–40

7. Shim MG, Song LMWK, Marcon NE, Wilson BC. 2000. In vivo near-infrared Raman spectroscopy: demonstration of feasibility during clinical gastrointestinal endoscopy. *Photochemistry Photobiol.* 72:146–50

8. van Manen HJ, Kraan YM, Roos D, Otto C. 2005. Single-cell Raman and fluorescence microscopy reveal the association of lipid bodies with phagosomes in leukocytes. *Proc. Natl. Acad. Sci.* 102:10159–64

9. Maker PD, Terhune RW. 1965. Study of optical effects due to an induced polarization third order in the electric field strength. *Phys. Rev.* 137:A801–18

10. Begley RF, Harvey AB, Byer RL. 1974. Coherent anti-Stokes Raman scattering. *Appl. Phys. Lett.* 25:387–90

11. Duncan MD, Reintjes J, Manuccia TJ. 1982. Scanning coherent anti-Stokes Raman microscope. *Opt. Lett.* 7:350–52

12. Zumbusch A, Holtom GR, Xie XS. 1999. Three-dimensional vibrational imaging by coherent anti-Stokes Raman scattering. *Phys. Rev. Lett.* 82:4142–45

13. Cheng JX, Jia YK, Zheng G, Xie XS. 2002. Laser-scanning coherent anti-Stokes Raman scattering microscopy and applications to cell biology. *Biophys. J.* 83:502–9

14. Dufresne ER, Corwin EI, Greenblatt NS, Ashmore J, Wang DY, et al. 2003. Flow and fracture in drying nanoparticle suspensions. *Phys. Rev. Lett.* 91:224501

15. Nan X, Yang WY, Xie XS. 2004. CARS microscopy: lights up lipids in living cells. *Biophotonics Int.* 11:44

16. Nan X, Potma EO, Xie XS. 2006. Nonperturbative chemical imaging of organelle transport in living cells with coherent anti-Stokes Raman scattering microscopy. *Biophys. J.* 91:728–35

17. Potma EO, Xie XS. 2003. Detection of single lipid bilayers in coherent anti-Stokes Raman scattering (CARS) microscopy. *J. Raman Spectrosc.* 34:642–50

18. Potma EO, Xie XS. 2005. Direct visualization of lipid phase segregation in single lipid bilayers with coherent anti-Stokes Raman scattering microscopy. *Chem. Phys. Chem.* 6:77–79

19. Evans CL, Potma EO, Puoris'haag M, Cote D, Lin CP, Xie XS. 2005. Chemical imaging of tissue in vivo with video-rate coherent anti-Stokes Raman scattering microscopy. *Proc. Natl. Acad. Sci. USA* 102:16807–12

20. Evans CL, Xu X, Kesari S, Xie XS, Wong STC, Young GS. 2007. Chemically-selective imaging of brain structures with CARS microscopy. *Opt. Expr.* 15:12076–87

21. Evans CL, Potma EO, Xie XS. 2004. Coherent anti-Stokes Raman scattering spectral interferometry: determination of the real and imaginary components of nonlinear susceptibility $\chi^{(3)}$ for vibrational microscopy. *Opt. Lett.* 29:2923–25

22. Potma EO, Evans CL, Xie XS. 2006. Heterodyne coherent anti-Stokes Raman scattering (CARS) imaging. *Opt. Lett.* 31:241–43

23. Ganikhanov F, Evans CL, Saar BG, Xie XS. 2006. High sensitivity vibrational imaging with frequency modulation coherent anti-Stokes Raman scattering (FM-CARS) microscopy. *Opt. Lett.* 31:1872–74

24. Hellwarth RW. 1977. Third-order optical susceptibilities of liquids and solids. *Prog. Quantum Electron.* 5:1–68

25. Maeda S, Kamisuki T, Adachi Y. 1988. Condensed phase CARS. In *Advances in Non-Linear Spectroscopy*, ed. RJH Clark, RE Hester, pp. 253–97. New York: Wiley

26. Lin-Vien D, Colthup NB, Fateley WG, Grasselli JG. 1991. *The Handbook of Infrared and Raman Characteristic Frequencies of Organic Molecules*. San Diego: Academic

27. Wurpel GWH, Schins JW, Müller M. 2002. Chemical specificity in 3D imaging with multiplex CARS microscopy. *Opt. Lett.* 27:1093–95

28. Cheng JX, Volkmer A, Book LD, Xie XS. 2002. Multiplex coherent anti-Stokes Raman scattering microspectroscopy and study of lipid vesicles. *J. Phys. Chem. B.* 106:8493–98

29. Kee TW, Cicerone MT. 2004. Simple approach to one-laser, broadband coherent anti-Stokes Raman scattering microscopy. *Opt. Lett.* 29:2701–3

30. Lim SH, Caster AG, Leone SR. 2005. Single-pulse phase-control interferometric coherent anti-Stokes Raman scattering spectroscopy. *Phys. Rev. A* 72:41803

31. Vartiainen EM, Rinia HA, Müller M, Bonn M. 2006. Direct extraction of Raman line-shapes from congested CARS spectra. *Opt. Expr.* 14:3622–30

32. Oron D, Dudovich N, Silberberg Y. 2003. Femtosecond phase-and-polarization control for background-free coherent anti-Stokes Raman spectroscopy. *Phys. Rev. Lett.* 90:213902

33. Frumker E, Oron D, Mandelik D, Silberberg Y. 2004. Femtosecond pulse-shape modulation at kilohertz rates. *Opt. Lett.* 29:890–92

34. Frumker E, Tal E, Silberberg Y, Majer D. 2005. Femtosecond pulse-shape modulation at nanosecond rates. *Opt. Lett.* 30:2796–98

35. Cheng JX, Volkmer A, Xie XS. 2002. Theoretical and experimental characterization of coherent anti-Stokes Raman scattering microscopy. *J. Opt. Soc. Am. B* 19:1363–75

36. Boyd RW. 2003. *Nonlinear Optics*. London: Academic

37. Potma EO, Boeij WPD, Wiersma DA. 2000. Nonlinear coherent four-wave mixing in optical microscopy. *J. Opt. Soc. Am. B* 17:1678–84

38. Cheng J, Volkmer A, Book LD, Xie XS. 2001. An epi-detected coherent anti-Stokes Raman scattering (E-CARS) microscope with high spectral resolution and high sensitivity. *J. Phys. Chem. B* 105:1277–80

39. Fu Y, Wang H, Shi R, Cheng JX. 2006. Characterization of photodamage in coherent anti-Stokes Raman scattering microscopy. *Opt. Expr.* 14:3942–51

40. Beaurepaire E, Oheim M, Mertz J. 2001. Ultra-deep two-photon fluorescence excitation in turbid media. *Opt. Commun.* 188:25–29

41. Ganikhanov F, Carrasco S, Xie XS, Katz M, Seitz W, Kopf D. 2006. Broadly tunable dual-wavelength light source for coherent anti-Stokes Raman scattering microscopy. *Opt. Lett.* 31:1292–94

42. Potma EO, Jones DJ, Cheng JX, Xie XS, Ye J. 2002. High-sensitivity coherent anti-Stokes Raman microscopy with two tightly synchronized picosecond lasers. *Opt. Lett.* 27:1168–70

43. Jones DJ, Potma EO, Cheng JX, Burfeindt B, Pang Y, et al. 2002. Synchronization of two passively mode-locked, ps lasers within 20 fs for coherent anti-Stokes Raman scattering microscopy. *Rev. Sci. Instrum.* 73:2843–48

44. Cheng JX, Book LD, Xie XS. 2001. Polarization coherent anti-Stokes Raman scattering microscopy. *Opt. Lett.* 26:1341–43

45. Volkmer A, Cheng J, Xie XS. 2001. Vibrational imaging with high sensitivity via epi-detected coherent anti-Stokes Raman scattering microscopy. *Phys. Rev. Lett.* 87:23901

46. Volkmer A, Book LD, Xie XS. 2002. Time-resolved coherent anti-Stokes Raman scattering microscopy: imaging based on Raman free induction decay. *Appl. Phys. Lett.* 80:1505–7

47. Andresen ER, Keiding SR, Potma EO. 2006. Picosecond anti-Stokes generation in a photonic-crystal fiber for interferometric CARS microscopy. *Opt. Expr.* 14:7246–51

48. Potma EO, Xie XS. 2003. Detection of single lipid bilayers with coherent anti-Stokes Raman scattering (CARS) microscopy. *J. Raman Spectrosc.* 34:642–50

49. Wurpel GWH, Rinia HA, Muller M. 2005. Imaging orientational order and lipid density in multilamellar vesicles with multiplex CARS microscopy. *J. Microsc.* 218:37–45

50. Wurpel GWH, Schins JM, Müller M. 2004. Direct measurement of chain order in single phospholipid mono- and bilayers with multiplex CARS. *J. Phys. Chem. B* 108:3400–3

51. Potma EO, Xie XS, Muntean L, Preusser J, Jones D, et al. 2004. Chemical imaging of photoresists with coherent anti-Stokes Raman scattering (CARS) microscopy. *J. Phys. Chem. B* 108:1296–1301

52. Dufresne ER, Corwin EI, Greenblatt NA, Ashmore J, Wang DY, et al. 2003. Flow and fracture in drying nanoparticle suspensions. *Phys. Rev. Lett.* 91:224501

53. Saar BG, Park H-S, Xie XS, Lavrentovich OD. 2007. Three-dimensional imaging of chemical bond orientation in liquid crystals by coherent anti-Stokes Raman scattering microscopy. *Opt. Expr.* 15(21):13585–96

54. Schliwa M, Woehlke G. 2001. Molecular motors: switching on kinesin. *Nature* 411:424–25

55. Vale RD. 2003. The molecular motor toolbox for intracellular transport. *Cell* 112:467–80

56. Murphy DJ. 2001. The biogenesis and functions of lipid bodies in animals, plants and microorganisms. *Prog. Lipid Res.* 40:325–438

57. Liu P, Ying Y, Zhao Y, Mundy DI, Zhu M, Anderson RGW. 2004. Chinese hamster ovary K2 cell lipid droplets appear to be metabolic organelles involved in membrane traffic. *J. Biol. Chem.* 279:3787–92

58. Fujimoto Y, Itabe H, Sakai J, Makita M, Noda J, et al. 2004. Identification of major proteins in the lipid droplet–enriched fraction isolated from the human hepatocyte cell line HuH7. *Biochim. Biophys. Acta Mol. Cell Res.* 1644:47–59

59. Fukumoto S, Fujimoto T. 2002. Deformation of lipid droplets in fixed samples. *Histochem. Cell Biol.* 118:423–28

60. Rakic B, Sagan SM, Noestheden M, Belanger S, Nan X, et al. 2006. Peroxisome proliferator–activated receptor α antagonism inhibits hepatitis C virus replication. *Chem. Biol.* 13:23–30

61. Nan X, Tonary AM, Stolow A, Xie XS, Pezacki JP. 2006. Intracellular imaging of HCV RNA and cellular lipids by using simultaneous two-photon fluorescence and coherent anti-Stokes Raman scattering microscopies. *ChemBioChem* 7:1895–97

62. Hellerer T, Axang C, Brackmann C, Hillertz P, Pilon M, Enejder A. 2007. Monitoring of lipid storage in *Caenorhabditis elegans* using coherent anti-Stokes Raman scattering (CARS) microscopy. *Proc. Natl. Acad. Sci.* 104:14658–63

63. Xie XS, Yu J, Yang WY. 2006. Living cells as test tubes. *Science* 312:228–30

64. Pudney PDA, Melot M, Caspers PJ, Van Der Pol A, Puppels GJ. 2007. An in vivo confocal Raman study of the delivery of trans-retinol to the skin. *Appl. Spectrosc.* 61:804–11

65. Barry BW. 1991. Lipid-protein-partitioning theory of skin penetration enhancement. *J. Control. Release* 15:237–48

66. Young B, Heath JW. 2000. *Wheater's Functional Histology: A Text and Colour Atlas.* London: Churchill Livingstone

67. Heinrich C, Bernet S, Ritsch-Marte M. 2007. Wide-field coherent anti-Stokes Raman scattering microscopy with non-phase-matching illumination. *Opt. Lett.* 32:3468–69

68. Wang H, Fu Y, Zickmund P, Shi R, Cheng JX. 2005. Coherent anti-Stokes Raman scattering imaging of axonal myelin in live spinal tissues. *Biophys. J.* 89:581–91

69. Huff TB, Cheng JX. 2007. In vivo coherent anti-Stokes Raman scattering imaging of sciatic nerve tissue. *J. Microsc.* 225:175–82

70. Fu Y, Wang H, Huff TB, Shi R, Cheng JX. 2007. Coherent anti-Stokes Raman scattering imaging of myelin degradation reveals a calcium-dependent pathway in lyso-PtdCho-induced demyelination. *J. Neurosci. Res.* 85:2870–81

71. Légaré F, Evans CL, Ganikhanov F, Xie XS. 2006. Towards CARS endoscopy. *Opt. Expr.* 14:4427–32

72. Haka AS, Shafer-Peltier KE, Fitzmaurice M, Crowe J, Dasari RR, Feld MS. 2005. Diagnosing breast cancer by using Raman spectroscopy. *Proc. Natl. Acad. Sci.* 102:12371–76

73. Gniadecka M, Philipsen PA, Sigurdsson S, Wessel S, Nielsen OF, et al. 2004. Melanoma diagnosis by Raman spectroscopy and neural networks: structure alterations in proteins and lipids in intact cancer tissue. *J. Invest. Dermatol.* 122: 443–49

74. Zipfel WR, Williams RM, Webb WW. 2003. Nonlinear magic: multiphoton microscopy in the biosciences. *Nat. Biotechnol.* 21:1369–77

75. Campagnola PJ, Millard AC, Terasaki M, Hoppe PE, Malone CJ, Mohler WA. 2002. Three-dimensional high-resolution second-harmonic generation imaging of endogenous structural proteins in biological tissues. *Biophys. J.* 81:493–508

76. Squier JA, Muller M, Brakenhoff GJ, Wilson KR. 1998. Third harmonic generation microscopy. *Opt. Expr.* 3:315–24

77. Shafer-Peltier KE, Haka AS, Fitzmaurice M, Crowe J, Myles J, et al. 2002. Raman microspectroscopic model of human breast tissue: implications for breast cancer diagnosis in vivo. *J. Raman Spectrosc.* 33:552–63

78. Le TT, Langohr IM, Locker MJ, Sturek M, Cheng JX. 2007. Label-free molecular imaging of atherosclerotic lesions using multimodal nonlinear optical microscopy. *J. Biomed. Opt.* 12:054007

79. Thomas GJ Jr. 1999. Raman spectroscopy of protein and nucleic acid assemblies. *Ann. Rev. Biophys. Biomol. Struct.* 28:1–27

ANNUAL REVIEWS
Intelligent Synthesis of the Scientific Literature

Annual Reviews – Your Starting Point for Research Online
http://arjournals.annualreviews.org

- Over 1150 Annual Reviews volumes—more than 26,000 critical, authoritative review articles in 35 disciplines spanning the Biomedical, Physical, and Social sciences— available online, including all Annual Reviews back volumes, dating to 1932

- Current individual subscriptions include seamless online access to full-text articles, PDFs, Reviews in Advance (as much as 6 months ahead of print publication), bibliographies, and other supplementary material in the current volume and the prior 4 years' volumes

- All articles are fully supplemented, searchable, and downloadable — see http://anchem.annualreviews.org

- Access links to the reviewed references (when available online)

- Site features include customized alerting services, citation tracking, and saved searches

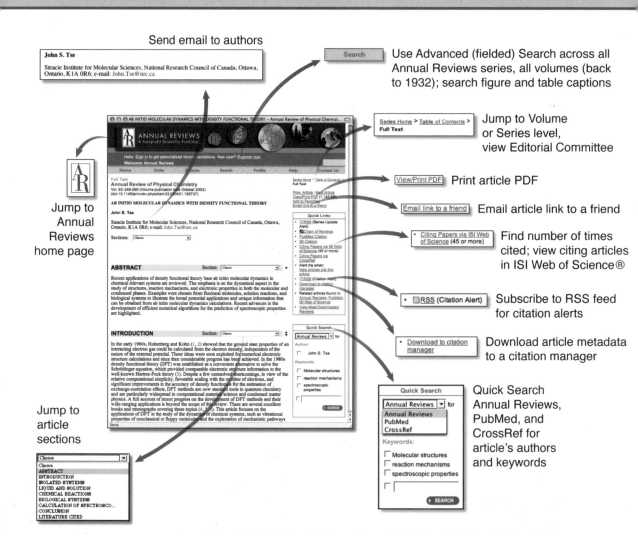

Send email to authors

Use Advanced (fielded) Search across all Annual Reviews series, all volumes (back to 1932); search figure and table captions

Jump to Volume or Series level, view Editorial Committee

Print article PDF

Email article link to a friend

Find number of times cited; view citing articles in ISI Web of Science®

Subscribe to RSS feed for citation alerts

Download article metadata to a citation manager

Quick Search Annual Reviews, PubMed, and CrossRef for article's authors and keywords

Jump to Annual Reviews home page

Jump to article sections